HYPERSPECTRAL REMOTE SENSING OF VEGETATION

HYPERSPECTRAL REMOTE SENSING OF VEGETATION

Edited by
Prasad S. Thenkabail
John G. Lyon
Alfredo Huete

CRC Press
Taylor & Francis Group
Boca Raton   London   New York

CRC Press is an imprint of the
Taylor & Francis Group, an **informa** business

CRC Press
Taylor & Francis Group
6000 Broken Sound Parkway NW, Suite 300
Boca Raton, FL 33487-2742

© 2012 by Taylor & Francis Group, LLC
CRC Press is an imprint of Taylor & Francis Group, an Informa business

No claim to original U.S. Government works

Printed in the United States of America on acid-free paper
Version Date: 2011908

International Standard Book Number: 978-1-4398-4537-0 (Hardback)

Visit the Taylor & Francis Web site at
http://www.taylorandfrancis.com

and the CRC Press Web site at
http://www.crcpress.com

Contents

PART I   Introduction and Overview

PART II   Hyperspectral Sensor Systems

PART III   Data Mining, Algorithms, Indices

v

PART IV   Leaf and Plant Biophysical and Biochemical Properties

PART V   Vegetation Biophysical Properties

PART VI   Vegetation Processes and Function (ET, Water Use, GPP, LUE, Phenology)

PART VII   Species Identification

PART VIII   Land Cover Applications

PART IX   Detecting Crop Management, Plant Stress, and Disease

PART X   *Hyperspectral Data in Global Change Studies*

PART XI   *Hyperspectral Remote Sensing of Outer Planets*

PART XII   *Conclusions and Way Forward*

Foreword

The publication of this book, *Hyperspectral Remote Sensing of Vegetation*, marks a milestone in the application of imaging spectrometry to studies of 70% of the Earth's landmass that is vegetated. This book shows not only the breadth of international involvement in the use of hyperspectral data, but also the breadth of innovative application of mathematical techniques to extract information from the image data.

Imaging spectrometry evolved from a combination of insights from the vast heterogeneity of reflectance signatures from the Earth's surface seen in the ERTS-1 (Landsat-1) four-band images and the field spectra that were acquired to help more fully understand the causes of the signatures. It was not until 1979 when the first hybrid area array detectors, mercury-cadmium-telluride on silicon CCDs, became available that it was possible to build an imaging spectrometer capable of operating at wavelengths beyond $1.0\,\mu m$. The AIS (airborne imaging spectrometer), developed at NASA/JPL, had only 32 cross-track pixels but that was enough for geologists to clamor for its development to see *between* the bushes to determine the mineralogy of the substrate. In those early years, vegetation cover was just a nuisance!

In the early 1980s, spectroscopic analysis was driven by the interest to identify mineralogical composition by exploiting absorptions found in the SWIR region from overtone and combination bands of fundamental vibrations found in the mid-IR region beyond $3\,\mu m$ and the electronic transitions in transition elements appearing, primarily, short of $1.0\,\mu m$. The interests of the geologists had been incorporated in the Landsat TM sensor in the form of the add-on band 7 in the $2.2\,\mu m$ region based on field spectroscopic measurements. However, one band, even in combination with the other six, did not meet the needs for mineral identification. A summary of mineralogical analyses is presented by Vaughan et al. in this volume. A summary of the historical development of hyperspectral imaging can be found in Goetz (2009).

At the time of the first major publication of the AIS results (Goetz et al., 1985), very little work on vegetation analysis using imaging spectroscopy had been undertaken. The primary interest was in identifying the relationship of the chlorophyll absorption red-edge to stress and substrate composition that had been seen in airborne profiling and in field spectral reflectance measurements. Most of the published literature concerned analyzing NDVI, which only required two spectral bands.

In the time leading up to the 1985 publication, we had only an inkling of the potential information content in the hundreds of contiguous spectral bands that would be available to us with the advent of AVIRIS (airborne visible and infrared imaging spectrometer). One of the authors, Jerry Solomon, presciently added the term "hyperspectral" to the text of the paper to describe the "…multidimensional character of the spectral data set," or, in other words, the mathematically, overdetermined nature of hyperspectral data sets. The term hyperspectral as opposed to multispectral data moved into the remote sensing vernacular and was additionally popularized by the military and intelligence community.

In the early 1990s, as higher quality AVIRIS data became available, the first analyses of vegetation using statistical techniques borrowed from chemometrics, also known as NIRS analysis used in the food and grain industry, were undertaken by John Aber and Mary Martin of the University of New Hampshire. Here nitrogen contents of tree canopies were predicted from reflectance spectra by regression techniques using reference measurements from laboratory wet chemical analyses of needle and leaf samples acquired by shooting down branches. At the same time, the remote sensing community began to recognize the value of "too many" spectral bands and the concomitant wealth of spatial information that was amenable to information extraction by statistical techniques. One of them was Eyal Ben-Dor who pioneered soil analyses using hyperspectral imaging and who is one of the contributors to this volume.

As the quality of AVIRIS data grew, manifested in increasing SNR, an ever-increasing amount of information could be extracted from the data. This quality was reflected in the increasing number of nearly noiseless principal components that could be obtained from the data or, in other words, its dimensionality. The explosive advances in desktop computing made possible the application of image processing and statistical analyses that revolutionized the uses of hyperspectral imaging. Joe Boardman and others at the University of Colorado developed what has become the ENVI software package to make possible the routine analysis of hyperspectral image data using "unmixing techniques" to derive the relative abundance of surface materials on a pixel by pixel basis.

Many of the analysis techniques discussed in this volume, such as band selection and various indices, are rooted in principal components analysis. The eigenvector loadings or factors indicate which spectral bands are the most heavily weighted allowing others to be discarded to reduce the noise contribution. As sensors become better, more information will be extractable and fewer bands will be discarded. This is the beauty of hyperspectral imaging, allowing the choice of the number of eigenvectors to be used for a particular problem. Computing power has reached such a high level that it is no longer necessary to choose a subset of bands just to minimize the computational time.

As regression techniques such as PLS (partial least squares) become increasingly adopted to relate a particular vegetation parameter to reflectance spectra, it must be remembered that the quality of the calibration model is a function of both the spectra and the reference measurement. With spectral measurements of organic and inorganic compounds under laboratory conditions, we have found that a poor model with a low coefficient of determination (r^2) is most often associated with inaccurate reference measurements, leading to the previously intuitive conclusion that "spectra don't lie."

Up to this point, AVIRIS has provided the bulk of high-quality hyperspectral image data but on an infrequent basis. Although Hyperion has provided some time series data, there is no hyperspectral imager yet in orbit that is capable of providing routine, high-quality images of the whole Earth on a consistent basis. The hope is that in the next decade HyspIRI will be providing VNIR and SWIR hyperspectral images every three weeks and multispectral thermal data every week. This resource will revolutionize the field of vegetation remote sensing since so much of the useful information is bound up in the seasonal growth cycle. The combination of the spectral, spatial, and temporal dimensions will be ripe for the application of statistical techniques and the results will be extraordinary.

Dr. Alexander F. H. Goetz, PhD
Chairman and Chief Scientist, ASD Inc.
Boulder, Colorado

REFERENCES

Goetz, A.F.H. 2009. Three decades of hyperspectral imaging of the Earth: A personal view, *Remote Sensing of Environment*, **113**, S5–S16.

Goetz, A.F.H., G. Vane, J. Solomon, and B.N. Rock. 1985. Imaging spectrometry for Earth remote sensing, *Science*, **228**, 1147–1153.

Preface

Over the years, I have seen a real need for a book on hyperspectral remote sensing (imaging spectroscopy) of vegetation, especially given the recent rapid advances made in imaging spectroscopy and opportunities for unique applications hitherto thought to be infeasible using broadband remote sensing. The need for a book to catalogue these knowledge advances in the study of terrestrial vegetation using hyperspectral narrowband data is of critical importance to a wide spectrum of scientific community, students, and professional application practitioners. Given this need, the goal of this book was to provide a comprehensive set of chapters documenting knowledge advances made in applying hyperspectral remote sensing technology in the study of terrestrial vegetation. This is a very practical offering about a complex subject that is rapidly advancing its knowledge-base and practical utility in wide array of applications. In a very practical way, the book demonstrates the experience, utility, methods, and models used in studying vegetation using hyperspectral data. Written by leading experts in the global arena, each chapter (a) focuses on specific applications, (b) reviews "state-of-the-art" knowledge, (c) highlights the advances made, and (d) provides guidance for appropriate use of hyperspectral (or imaging spectroscopy) data in the study of vegetation such as crop yield modeling, crop biophysical and biochemical property characterization, crop moisture assessment, species identification, spectrally separating vegetation types, and modeling biophysical and biochemical quantities.

This book focuses specifically on hyperspectral remote sensing as applied to terrestrial vegetation applications. This is a big market area, and the chapters discuss in detail a wide array of applications such as agricultural croplands; study of crop moisture and forests; and numerous other applications such as droughts, crop stress, crop productivity, and water productivity. To the best of our knowledge, there is no comparable book, source, and/or organization that has brought this body of knowledge together in one place, making this a "must buy" for professionals. This is clearly a unique contribution whose time has come. The highlights of the book include:

1. Best global expertise on hyperspectral remote sensing of vegetation, agricultural crops, crop water use, plant species detection, crop productivity, wetland studies, forest type and species characterization, carbon flux assessments from vegetation, and water productivity mapping and modeling.
2. Clear articulation of methods to conduct the work; very practical.
3. Comprehensive review of the existing technology and clear guidance on how best to use hyperspectral data for various applications.
4. Case studies from a variety of continents with their own subtle requirements.
5. Complete solutions from methods to applications inventory, modeling, and mapping.

Hyperspectral narrow-band spectral data, as discussed in various chapters of this book, are already fast emerging as practical solutions in modeling and mapping vegetation. Recent research has demonstrated the effectiveness of hyperspectral data, as discussed in the 28 chapters of this book, in (a) quantifying agricultural crops with regard to their biophysical and harvest yield characteristics, (b) modeling forest canopy biochemical properties, (c) establishing plant and soil moisture conditions, (d) detecting crop stress and disease, (e) mapping leaf chlorophyll content as it influences crop production, (f) identifying plants affected by contaminants such as arsenic, (g) demonstrating sensitivity to plant nitrogen content, and (h) invasive species mapping. The ability to significantly better quantify, model, and map plant chemical, physical, and water properties using hyperspectral narrowband data is well established and has great utility. The chapters discuss in-depth approaches and methods of modeling and mapping vegetation using optimal hyperspectral narrowbands (by dropping redundant bands), hyperspectral vegetation indices, and whole spectral analysis.

Even though these accomplishments and capabilities have been reported in various places, the need for a collective "knowledge bank" that links these various advances in one place was missing. Further, most scientific papers address specific aspects of research, neither providing a comprehensive assessment of advances that have been made nor providing any information as to how the professional can implement those advances in their work. For example, even though scientific journals report practical applications of hyperspectral narrow-bands, one has still to canvass the literature broadly to obtain the pertinent facts. Since several papers report this, there is a need to synthesize these findings so that the reader can get the correct picture of issues such as selecting the best wavebands for their practical applications, best approaches and methods for efficient classification of hyperspectral data to study vegetation, and best models for plant biophysical and biochemical quantities. Approaches and strategies to tackle these issues are discussed in details in various chapters. Studies may differ when describing the best methods for detecting parameters such as crop moisture variability, chlorophyll content, and stress levels. Thereby, a professional will need the sort of synthesis and detail provided in each chapter of this book in order to adopt best practices for their own work.

This book can be used for advanced graduate courses as well as by professionals, policy makers, governments, and research organizations when in need of proven methods to address a wide array of issues pertaining to study of the Planet Earth using hyperspectral (imaging spectroscopy) narrowband data.

Dr. Prasad S. Thenkabail
Editor in Chief
Hyperspectral Remote Sensing of Vegetation

Acknowledgments

The book was made possible by sterling contributions from leading professionals from around the world in the area of hyperspectral remote sensing (or imaging spectroscopy) of vegetation and agricultural crops. As can be seen from the list of authors and coauthors, these are basically the "who is who" in hyperspectral remote sensing of vegetation. All the contributors have written insightful chapters, which is an outcome of years of dedicated research, to make the book appealing to a broad section of readers dealing with remote sensing. My gratitude goes to (mentioned in no particular order) the following authors (names of lead authors of the chapters appear in bold): **Drs. Fred Ortenberg** (Technion, Israel Institute of Technology, Israel), **Jiaguo Qi** (Michigan State University, United States), **Sreekala Bajawa** (University of Arkansas, United States), **Antonio Plaza** (University of Extremadura, Spain), **Anatoly Gitelson** (University of Nebraska-Lincoln, United States), **Yongqin Zhang** (University of Toronto, Canada), **Yan Zhu** (Nanjing Agricultural University, China), **Izaya Numata** (South Dakota State University, United States), **Roberto Colombo** (University of Milan-Bicocca, Italy), **Daniela Stroppiana** (Institute for Electromagnetic Sensing of the Environment, Italy), **Elizabeth Middleton** (NASA Goddard Space Flight Center, United States), **Victor Alchanatis** (Agricultural Research Organization, Volcani Center, Israel), **Dar Roberts** (University of California at Santa Barbara, United States), **Pamela Nagler** (U.S. Geological Survey [USGS], United States), **Lênio Soares Galvão** (Instituto Nacional de Pesquisas Espaciais [INPE]—Brazil), **Matthew L. Clark** (Sonoma State University, United States), **Ruiliang Pu** (University of South Florida, United States), **Valerie Thomas** (Virginia Tech., United States), **Elijah W. Ramsey III** (USGS, United States), **Eyan Ben Dor** (Tel Aviv University, Israel), **Terry Slonecker** (USGS, United States), **Jianlong Li** (Nanjing University, China), **Haibo Yao** (Mississippi State University, United States), **Tomoaki Miura** (University of Hawaii, United States), **R. Greg Vaughan** (USGS, United States), Yoshio Inoue (National Institute for Agro-Environmental Sciences, Japan), Dr. Narumon Wiangwang (Royal Thai Government, Thailand), Subodh Kulkarni (University of Arkansas, United States), Javier Plaza (University of Extremadura, Spain), Gabriel Martin (University of Extremadura, Spain), Segio Sánchez (University of Extremadura, Spain), Wei Wang (Nanjing Agricultural University, China), Xia Yao (Nanjing Agricultural University, China), Busetto Lorenzo (Università Milano-Bicocca, Italy), Meroni Michele (Università Milano-Bicocca, Italy), Rossini Micol (Università Milano-Bicocca, Italy), Panigada Cinzia (Università Milano-Bicocca, Italy), Fava, F. (Università degli Studi di Sassari, Italy), Boschetti, M. (Institute for Electromagnetic Sensing of the Environment, Italy), Brivio, P. A. (Institute for Electromagnetic Sensing of the Environment, Italy), K. Fred Huemmrich (University of Maryland, Baltimore County, United States), Yen-Ben Cheng (Earth Resources Technology, Inc., United States), Hank A. Margolis (Centre d'Études de la Forêt, Canada), Yafit Cohen (Agricultural Research Organization, Volcani Center, Israel), Kelly Roth (University of California at Santa Barbara, United States), Ryan L. Perroy (University of Wisconsin-La Crosse, United States), Wei Wang (Nanjing Agricultural University, China), Dr. Xia Yao (Nanjing Agricultural University, China), Keely L. Roth (University of California, Santa Barbara, United States), B. B. Marithi Sridhar (Bowling Green University, United States), Aaryan Dyami Olsson (Northern Arizona University, United States), Willem Van Leeuwen (University of Arizona, United States), Edward Glenn (University of Arizona, United States), **José Carlos Neves Epiphanio** (INPE, Brazil), **Fábio Marcelo Breunig** (INPE, Brazil), **Antônio Roberto Formaggio** (INPE, Brazil), Amina Rangoonwala (IAP World Services, Lafayette, Los Angeles), Cheryl Li (Nanjing University, China), Deghua Zhao (Nanjing University,

China), Cengcheng Gang (Nanjing University, China), Lie Tang (Mississippi State University, United States), Lei Tian (Mississippi State University, United States), Robert Brown (Mississippi State University, United States), Deepak Bhatnagar (Mississippi State University, United States), Thomas Cleveland (Mississippi State University, United States), Hiroki Yoshioka (Aichi Prefectural University, Japan), T. N. Titus (USGS, United States), J. R. Johnson (USGS, United States), J. J. Hagerty (USGS, United States), L. Gaddis (USGS, United States), L. A. Soderblom (USGS, United States), and P. Geissler (USGS, United States).

My two coeditors, Professor John G. Lyon and Professor Alfredo Huete, have done an outstanding job. Their enormous knowledge of hyperspectral remote sensing, combined with the vastness and depth of their understanding of remote sensing, in general, and hyperspectral remote sensing, in particular, made my job that much easier. I have learned a lot from them and continue to do so. Both of them edited each of the 28 chapters of the book and also helped structure the chapters for a flawless reading. I am indebted to their insights, guidance, support, motivation, and encouragement throughout the project. Each chapter of the book had two or three peer reviewers. So, for all the 28 chapters there were about 70 peer reviewers. I am grateful to these anonymous reviewers who contributed their valuable time and insights to the peer-review process.

My coeditors and I are grateful to Dr. Alexander F. H. Goetz for writing the foreword for the book. Dr. Goetz is one of the pioneers of hyperspectral remote sensing and certainly needs no introduction. He started his career working on spectroscopic reflectance and emittance studies of Moon and Mars. He was a principal investigator of Apollo-8 and Apollo-12 multispectral photography studies. He later turned his attention to remote sensing of the Earth working in collaboration with Dr. Gene Shoemaker to map geology of Coconino County (Arizona) using Landsat-1 data. He then became an investigator in further Landsat, Skylab, Shuttle, and EO-1 missions. At NASA/JPL, he pioneered field spectral measurements and initiated the development of hyperspectral imaging. He spent 21 years on the faculty of the University of Colorado, Boulder, and retired in 2006 as an emeritus professor of geological sciences and an emeritus director of the Center for the Study of Earth from Space. Since then he has been chairman and chief scientist of ASD Inc., a company that has provided more than 850 research laboratories in over 60 countries with field spectrometers. His foreword is a must read for anyone studying this book.

I am blessed to have had the support and encouragement (professional and or personal) of my U.S. Geological Survey (USGS) colleagues. In particular, I would like to mention Mr. Edwin Pfeifer, Dr. Susan Benjamin, Dr. Dennis Dye, Mr. Miguel Velasco, Dr. Chandra Giri, and Dr. Thomas Loveland. There are many other colleagues who made my job at USGS that much easier.

My wife (Sharmila Prasad) and daughter (Spandana Thenkabail) are the two great pillars of my life. I am always indebted to their patience, support, and love.

Finally, kindly bear with me for sharing a personal turmoil. The year 2010 was a difficult year for me personally and for my family in particular. Just when I started working on the book project, I learned that I had colon cancer. I had to undergo a major surgery (relatively easy part) and six months of chemotherapy (difficult part). It is quite satisfying that on my last day of chemotherapy, we delivered the book by FedEx to Taylor & Francis Group! Luckily, we (editors) never had to postpone this project. I am very grateful to some extraordinary people who helped me through these difficult times: Dr. Parvasthu Ramanujam (surgeon); Dr. Paramjeet K. Bangar (oncologist); three great nurses (Irene, Becky, and Maryam) at Banner Boswell Hospital (Sun City, AZ, United States); the courage, love, patience, and prayers from my wife, daughter, and several family members, friends, and colleagues; and support from numerous others that I have not named here. During this phase, I learned a lot about cancer, and it gave me an enlightened perspective of life. I would certainly not have completed editing this book without being a survivor. Yes, my prayers were answered. I learned a great deal about life—good and bad. I pray for all those cancer and other patients with dire illnesses, and I hope that we will soon find a simple cure.

Finally I would like to thank a number of people at Taylor & Francis/CRC Press. They include Irma Shagla; Arunkumar Aranganathan, project manager, SPi Global, and his team of editors; and Jennifer Ahringer. Shagla got the approval for the book project and played a pivotal role in supporting it; Ahringer coordinated book materials; and Aranganathan and his team did an outstanding job in editing the book. Their highly professional efforts are deeply appreciated.

Dr. Prasad S. Thenkabail, PhD
Editor in Chief
Hyperspectral Remote Sensing of Vegetation

Editors

Dr. Prasad S. Thenkabail has more than 25 years experience working as a well recognized international expert in remote sensing and geographic information systems (RS\GIS) and its applications to agriculture, natural resource management, water resources, sustainable development, and environmental studies. His work experience spans over 25 countries spread across West and Central Africa (Republic of Benin, Burkina Faso, Cameroon, Central African Republic, Côte d'Ivoire, Gambia, Ghana, Mali, Nigeria, Senegal, and Togo), Southern Africa (Mozambique, South Africa), South Asia (Bangladesh, India, Myanmar, Nepal, and Sri Lanka), Southeast Asia (Cambodia), Middle East (Israel, Syria), East Asia (China), Central Asia (Uzbekistan), North America (United States), South America (Brazil), and Pacific (Japan).

Dr. Thenkabail has a wealth of work experience in premier global institutes, holding key lead research positions. Currently, he is a research geographer at the U.S. Geological Survey (USGS). His roles include being a lead researcher for a number of projects (e.g., irrigated cropland water productivity in California, global croplands and their water use, imaging spectroscopy of vegetation), coordinator (2010–present) of the Committee for Earth Observation Systems (CEOS) Agriculture Societal Beneficial Area (SBA), and a science advisor (2010–present) to the Land Surface Imaging Constellation for CEOS, GEO, GEOSS. He co-leads an IEEE "Water for the World" project. He is also an adjunct professor, Department of Soil, Water, and Environmental Science (SWES), University of Arizona (UoA).

Prior to USGS, Dr. Thenkabail worked as a principal researcher in the Global Research Division and as head of the Remote Sensing and Geographic Information Systems (RS\GIS) at the International Water Management Institute (IWMI) headquartered in Sri Lanka with a network of offices in Asia and Africa. During this period that he led projects such as the global irrigated area mapping (GIAM); Global Map of Rainfed Cropland Areas (GMRCA) (http://www.iwmigiam.org); water productivity mapping in Central Asia; wetland mapping in Africa; drought monitoring system for Afghanistan, Pakistan, and parts of India; and the IWMI data storehouse pathway (http://www.iwmidsp.org). All of these projects were pioneering efforts using advanced remote sensing data, methods, and approaches conceptualized, led, and managed by Dr. Thenkabail.

Dr. Thenkabail did pioneering work on hyperspectral remote sensing of vegetation and agricultural crops, especially when he worked as associate research scientist (a research faculty position) at the Yale Center for Earth Observation (YCEO) at the Yale University, New Haven, Connecticut. In this position, he worked in NASA-funded research in Africa and Asia. He was the principal investigator for the NASA-funded project called "Characterization of Eco-regions in Africa (CERA). He also worked on hyperspectral remote sensing and carbon stock estimations from remote sensing in African rain forests and savannas. This work resulted in Dr. Thenkabail's appointment to the scientific advisory board of Rapideye, a private German Earth Resources Satellite Company. He played a pivotal role in recommending the design of wave bands in the Rapideye sensor onboard a constellation of five satellites [TACHYS (Rapid), MATI (Eye), CHOMA (Earth), CHOROS (Space), TROCHIA (Orbit)] launched recently by Rapideye. His research played a key role in the selection of "red-edge" band. Prior to this, for nearly five years, he led the remote sensing programs at the International Institute of Tropical Agriculture (IITA), working mostly in West and Central African countries based in Nigeria and the International Center for Integrated Mountain Development (ICIMOD) working in the Hindu-Kush Himalayan countries based in Katmandu, Nepal. During this period, he led the remote sensing component of the inland valley wetland characterization and mapping using Landsat TM and SPOT HRV data for the West and Central African Nations.

Dr. Thenkabail is the main editor of the book entitled *Remote Sensing of Global Croplands for Food Security* (Publisher: Taylor & Francis Group, 2009). He coedited a special issue for the *Journal of Remote Sensing* on the subject of "global croplands" that has 21 excellent papers on the topic of global cropland and their water use (http://www.mdpi.com/journal/remotesensing/special_issues/croplands/). The USGS and NASA selected him to be on the Landsat Science Team for a period of 5 years starting 2006 (http://ldcm.usgs.gov/intro.php). He is also one of the editors of *Remote Sensing of Environment*. In June 2007, Dr. Thenkabail's team was recognized by the Environmental System Research Institute (ESRI) for "special achievement in GIS" (SAG award) for their tsunami-related work and for their innovative spatial data portals (http://www.iwmidsp.org) and science applications (http://www.iwmigiam.org). In 2008, he and his coauthors were the second place recipients of the 2008 John I. Davidson ASPRS President's Award for practical papers (for their paper on spectral matching techniques used in mapping global irrigated areas). He won the 1994 Autometric Award of the American Society of Photogrammetric Engineering and Remote Sensing (ASPRS) for superior publication in remote sensing. Dr. Thenkabail's publications were selected as one of the best five papers consecutively for three years (2004–2006) in the IWMI's annual research meeting (ARM). His team was also awarded the "best team" at IWMI during ARM 2006.

Early in his career, Dr. Thenkabail worked as a scientist with the National Remote Sensing Agency (NRSA), Department of Space, and Government of India. He began his professional career as a lecturer in hydrology, water resources, hydraulics, hydraulics laboratory, and open channel flow in the colleges affiliated to Bangalore and Mysore University in India. He has more than 80+ publications, mostly peer-reviewed and published in major international remote sensing journals.

Dr. John G. Lyon's research has involved advanced remote sensing and GIS applications to water and wetland resources, agriculture, natural resources, and engineering applications. He is the author of books on wetland landscape characterization, wetland and environmental applications of GIS, and accuracy assessment of GIS and remote sensing technologies. Lyon was educated at Reed College in Portland, OR, and the University of Michigan, Ann Arbor, and has previously served as a professor of civil engineering and natural resources at Ohio State University (1981–1999). For approximately eight years, he was the director (SES) of the U.S. Environmental Protection Agency Office of Research and Development's (ORD) Environmental Sciences Division, which conducts research on remote sensing and GIS technologies as applied to environmental issues, including landscape characterization and ecology, and hazardous wastes. Lyon currently serves as a senior scientist (ST) in the EPA Office of the Science Advisor in Washington, District of Columbia, and is co-lead for work on the Group on Earth Observations and the Global Earth Observation System of Systems, and research on geospatial issues in the agency.

Dr. Alfredo Huete is currently a professor in the Faculty of Science, Plant Functional Biology and Climate Change Cluster, at the University of Technology Sydney, Australia. Prior to this appointment, he was professor of soils and remote sensing in the Department of Soil, Water, and Environmental Science at the University of Arizona, Tucson, Arizona.

Dr. Huete received his professional degrees at the University of California at Berkeley (MSc) and University of Arizona (PhD). He serves on the editorial review boards of *Remote Sensing of Environment*, Revisa de Teledetección de la Asociación Española de Teledetección (AET), and the online journal *Remote Sensing*. He is a member of NASA-EOS MODIS science team and has led the development and implementation of the MODIS vegetation index products. He is also part of the Japanese JAXA GCOM-SGLI science team and European PROBA-V IUC user expert group. He was also a member of the EO-1 Hyperion science team and NASA advisory team to evaluate NPOESS and NPP vegetation index products for land monitoring, environmental data records (EDRs) and long-term climate data records (CDRs). Professor Huete is also an active member of the International Society of Photogrammetry and Remote Sensing (ISPRS) Commission on Remote Sensing Applications and Policies, where he currently serves as chair of Working Group VIII on Land.

Dr. Huete's research interests focus on understanding large-scale soil–vegetation–climate interactions, processes, and changes with remotely sensed measurements from satellites. He is also involved with field-based and tower optical instrumentation in support of remote sensing studies coupling satellite observations with eddy covariance tower flux measurements. His research areas of interest also include phenology measures and shifts in seasonality in response to climate forcings, land use activities, and their coupling to carbon and water models. He has done extensive research in the phenology of tropical rain forests and savannas in the Amazon and Southeast Asia and has over 100 research publications in peer-reviewed journals, a book, and more than 20 chapter contributions.

List of Acronyms and Abbreviations

ν	stretching vibration
δ	bending vibration
1DL_DGVI	first-order derivative green vegetation index derived using local baseline
1DZ_DGVI	first-order derivative green vegetation index derived using zero baseline
3S	HEAD
5S	simulation of satellite signal in the solar spectrum
6S	second simulation of the satellite signal in the solar spectrum
ACI	anthocyanin content index
ACORN	atmospheric CORrection now program
ADEOS	advanced earth observing satellite
AERONET	aerosol robotic network
AET	actual evapotranspiration
AISA	airborne imaging spectroradiometer for application
ALI	advanced land imager
AMEE	automatic morphological endmember extraction
ANC	abundance non-negativity constraint
ANN	artificial neural networks
ANOVA	one-way analysis of variance
AOT	aerosol optical thickness
AOTF	acousto-optic tunable filter
APAR	absorbed photosynthetically active radiation
AR HTBVI	atmospherically resistant hyperspectral two-band vegetation indices
ARI	anthocyanin reflectance index
ARVI	atmospherically resistant vegetation index
ASC	abundance sum-to-one constraint
ASD	analytical spectral devices
ASI	Agenzia Spaziale Italiana
ASTER	advanced space-borne thermal emission and reflection radiometer
ATCOR	ATmospheric CORrection program
ATREM	ATmospheric REMoval program
ATSAVI	adjusted transformed soil-adjusted vegetation index
AVHRR/NOAA-17	advanced very high resolution radiometer/national oceanic and atmospheric administration-17
AVHRR	advanced very high resolution radiometer
AVIRIS	airborne visible/infrared imaging spectrometer
BB-PAC	biophysical and biochemical properties of agricultural crops
BBVI	broadband vegetation index models
BD-RDP	beamlet-decorated recursive dyadic partitioning
BDRF	bidirectional reflectance function
BE	blue edge
BmND	derivative-based modified normalized difference
BmSR	derivative-based modified simple ratio
BRDF	bidirectional reflectance distribution function
BRDI	bromus distachyon
CAI	cellulose absorption index

CAI	cloud aerosol imager
CAO	Carnegie airborne observatory
CAPY	carduus pychnocephalus
CARI	chlorophyll absorption reflectance index
CART	classification and regression tree
CASI	compact airborne spectrographic imager
CBERS-2	China–Brazil earth resources satellite
CCA	Convex cone analysis
CC	chlorophyll content
CCCI	canopy chlorophyll content index
CCD/CBERS-2	charge-coupled device/China–Brazil earth resources satellite-2
CCD	charge-coupled device
CCSM	cross correlogram spectral matching
CDA	canonical discriminant analysis
Cd	cadmium
Chl_{green}	chlorophyll index using green reflectance
$Chl_{red-edge}$	chlorophyll index using red-edge reflectance
CHRIS/PROBA	compact high-resolution imaging spectrometer/project for on board autonomy
CHRIS	compact high-resolution imaging spectrometer
CI_{green} and $CI_{red\ edge}$	green- and red-edge chlorophyll indices, respectively
CIR	color infrared
$CI_{red\ edge}$	chlorophyll red-edge index
CMF	color matching functions
CMG	climate modeling grid
CNES	Centre National d'Etudes Spatiales
CNPq	Conselho nacional de desenvolvimento científico e tecnológico
CP	crude protein (%)
CRDR	continuum removal derivative reflectance
CRI1 and 2	carotenoid reflectance index
CRI	carotenoid reflectance index
Cu	copper
D	absorption band depth
DAIS	digital airborne imaging spectrometer
DD	double difference
DEM	digital elevation model
DLR	German aerospace agency
DN	digital number
DOAS	differential optical absorption spectroscopy
DoD	Department of Defense
DT	decision tree
DVI	difference vegetation index
DWAB	dry weight of aboveground biomass
DWSI	disease water stress index
DWT	discrete wavelet transform
E	radiance
ECHO	extraction and classification of homogenous objects`
ED	Euclidean distance
EGU	European Geoscience Union
EMS	electromagnetic spectrum
EnMAP	environmental mapping and analysis program

ENVI	environment for visualizing images
ENVISAT	environmental satellite
EO-1	earth observing-1
EO-1	earth observing-1 satellite
EOS	earth observing system
EPPD	effective photon penetration depth
ERDAS	earth resource data analysis system
ERS	earth remote sensing
ESA	European Space Agency
ET	evapotranspiration
ETM+/Landsat-7	enhanced thematic mapper plus/Landsat-7
ETM+	Landsat-7 enhanced thematic mapper plus
EUFAR	EUropean Facility for Airborne Research
EVI2	two-band enhanced vegetation index
EVI	enhanced vegetation index
FAO	Food and Agriculture Administration
FAPAR	fraction of absorbed photosynthetically active radiation
FAPESP	Fundação de amparo a pesquisa do estado de são paulo
FDR	first derivative reflectance
FEDM	frequent domain electromagnetic
FLAASH	ENVI's fast line-of-sight atmospheric analysis of spectral hypercubes
FLAASH	fast line-of-sight atmospheric analysis of spectral hypercubes
FNIR	far-near infrared (1100–1300 nm)
FORMOSAT	Taiwanese satellite operated by Taiwanese National Space Organization (NSPO). Data marketed by SPOT
FOV	field of view
FPAR	fraction of photosynthetically active radiation
FPGAs	field programmable gate arrays
FR	full resolution
FS	Fort Sherman, Panama
FSI	full spectral imaging
FTHSI	Fourier transform hyperspectral imager
FTIR	Fourier transform infrared
FTS	Fourier transform spectrometers
GAC	global area coverage
GA	genetic algorithm
GCOM-C	global change observation mission-climate
GEOEYE-1 and 2	providing data in 0.25 to 1.65 m resolution
GERIS	geophysical and environmental research imaging spectrometer
GI	green index
GIS	geographic information system
GLI	global imager
GO-RT	geometrical optical and radiative transfer
GO	geometrical optical
GOME	Global Ozone Monitoring Experiment
GOSAT	greenhouse gases observing satellite
GP	green peak
GPP	gross primary production
GPR	ground-penetrating radar
GPS	global positioning system
GPS	ground positioning system

GPUs	graphics processing units
GV	green vegetation
HATCH	high-accuracy ATmosphere correction for hyperspectral data
HDGVI	hyperspectral derivative greenness vegetation indices
HD	hard disk
HHVI	hyperspectral hybrid vegetation indices
HICO	hyperspectral imager for the coastal ocean
HIS	hyperspectral Imagers
HMBM	hyperspectral multiple-band models
HPLC	high-performance liquid chromatography
HRG/SPOT-5	high geometric resolution instrument/systeme pour l'observation de la terre-5
HRS	hyperspectral remote sensing
HS	hyperspectral
HSR	hyper spectral remote sensing
HSS	hyperspectral sensor
HTBVI	hyperspectral two-band vegetation index
HTV	H-2 Transfer Vehicle
HVI	hyperspectral vegetation index
HVIST	hyperspectral vegetation indices of SWIR and TIR bands
HYDICE	hyperspectral digital imagery collection experiment
HyMap	airborne hyperspectral scanners
HyMAP	hyperpectral MAPpping sensor
HYMAP	hyperspectral mapper
HYPER-I-NET	Hyperspectral imaging network
Hyperion	first spaceborne hyperspectral sensor onboard earth observing-1(EO-1)
HypspIRI	hyperspectral imaging spectrometer and infrared imager
HySI	hyperspectral imager
HyspIRI	hyperspectral infrared imager
ICA	independent component analysis
ICAMM	independent component analysis-based mixed model
ICARE	International Conference on Airborne Research for the Environment
IEA	iterative error analysis
IFOV	instantaneous field of view
IG	inverted Gaussian
IKONOS	Greek word for "image"
IKONOS	high-resolution satellite operated by GeoEye
IMZ	intensive measurement zones
INS	inertial navigation system
IPS	invasive plant species
IR	infrared
IRS-1C/D-LISS	Indian remote sensing satellite/linear imaging self-scanner
IRS-P6-AWiFS	Indian remote sensing satellite/advanced wide field sensor
IS	imaging spectroscopy
ISS	International Space Station
ITC	individual tree crown
JAXA	Japan Aerospace Exploration Agency
JD	Julian day
JPSS	joint polar satellite system
K-T	Kaufman-Tanré aerosol retrieval
KFD	Kernel Fisher Discriminant

KOMFOSAT	Korean multipurpose satellite. Data marketed by SPOT image
L	irradiance
LAD	leaf angle distribution
LAI	leaf area index
LAI	leaf area index (m^2/m^2)
Landsat-1, 2, 3 MSS	multispectral scanner
Landsat-4, 5 TM	thematic mapper
Landsat-7 ETM+	enhanced thematic mapper plus
LANDSAT-TM	land satellite thematic mapper sensor
Landsat	land remote sensing satellite program
LANDSAT	MSS-land satellite multi spectral sensor
LCI	leaf chlorophyll index
LDA	linear discriminant analysis
LEO	low earth orbit
LFM	live fuel moisture
LICOR	An instrument-used to measure leaf are index
LiDAR	light detection and ranging
LI	Lepidium Index
LNA	leaf nitrogen accumulation
LNC	leaf nitrogen concentration
LOWTRAN	LOW resolution model for prediction atmosphere TRANsition
LSBS	La selva biological station, Costa Rica
LUE	light use efficiency
LUT	lookup table
LWVI-2	leaf water vegetation index-2
MACI	modified anthocyanin content index
MAE	mean absolute error
MARI	modified anthocyanin reflectance index
MaxAE	maximum absolute error
MCARI	modified chlorophyll absorption ratio index
MDA	multiple discriminant analysis
MERIS'	medium resolution imaging spectrometer
MESMA	multiple-endmember spectral mixture analysis
MF	Matched Filter
MIA	mutual information analysis
MIC	mutual information criterion
MightySat	Mighty Satellite
MLC	maximum likelihood classification
MLP	multilayer perceptron
MLR	multiple linear regression
MLR	multivariate linear regression
MMI	minimal mutual information
mND680	modified normalized difference
mND705	modified normalized difference
mND	modified normalized difference
MNDVI	modified normalized differential vegetation index
MNF	minimum noise fraction
MODIS	moderate imaging spectral radiometer
MODIS	moderate resolution imaging spectrometer
MODIS	moderate resolution imaging spectroradiometer
MODTRAN	MODerate resolution atmospheric TRANsmittance and radiance

MRF	Markov random field
MSAVI2	modified second soil-adjusted vegetation index
MSAVI	improved soil-adjusted vegetation index
MSI	moisture stress index
MSMISat	multi-sensor micro-satellite imager satellite
MS	multispectral
mSR705	modified simple ratio
mSR	modified simple ratio
MTCI	MERIS terrestrial chlorophyll index
MTMF	mixture tuned matched filtering
MWIR	medium-wave infrared
N	nitrogen
NASA	National Aeronautics and Space Administration
NDII	normalized difference infrared index
NDI	normalized difference index
NDLI	normalized difference lignin index
NDNI	normalized difference nitrogen index
ND	normalized difference
NDRE	normalized difference red edge
NDVI	normalized difference vegetation index
NDWI	normalized difference water index
NEE	net ecosystem carbon dioxide exchange
NE	noise equivalent
NIR	near-infrared reflectance
NIRS	near-infrared spectroscopy
nm	nanometer
NMP	NASA's new millennium program
NN	neural network
NOAA	National Oceanic and Atmospheric Administration
NPCI	normalized pigment chlorophyll ratio index
NPOESS	National Polar-Orbiting Operational Environmental Satellite System
NPP	NPOESS preparatory project
NPVAI	non-photosynthetic vegetation area index
NPV	non-photosynthetic vegetation
NRL	Naval Research Laboratory
NSA	normalized spectral area
NSMI	normalized soil moisture index
OLS	ordinary least squares
OMI	ozone monitoring instrument
OM	organic matter
ORASIS	optical real-time adaptive spectral identification system
OSAVI	optimized soil-adjusted vegetation indices
OSP	orthogonal subspace projection
PAR	photosynthetically active radiation
PBI	plant biochemical index
Pb	lead
PCA	principal component analysis
PC	principal component
PCR	principal components regression
PET	potential evapotranspiration
PHI	azimuth angle

PI2	pigment index 2
PIMA	field portable infrared spectrometer
PLNTHT	plant height (mm)
PLS	partial least squares
PLSR	partial least square regression
PNM	Parque Natural Metropolitano, Panama
POS	penetrating optical sensor
PPI	pixel purity index
PP	projection pursuit
PPR	plant pigment ratio
PRI	photochemical/physiological reflectance index
PRI	photochemical reflectance index
PRI	photosynthetic reflectance index
PRI	physiological reflectance index
PRISMA	hyperspectral precursor and application mission
PRISMA	PRecursore IperSpettrale della Missione Applicativa
PROBA	project for on-board autonomy
PSF	point spread function
PSND	pigment-specific normalized difference
PSRI	plant senescence reflectance index
PSSR	pigment-specific spectral ratio
PV	photosynthetic vegetation
QUICKBIRD	satellite from DigitalGlobe, a private company in the United States
R	reflectance
R1	reproductive stage 1 (beginning bloom)
R2	coefficient of determination
R3	reproductive stage 3 (beginning pod)
RAPID EYE – A/E	satellite constellation from Rapideye, a German company
RARS	ratio analysis of reflectance spectra
RBF:	radial basis function
RDP	recursive dyadic partitioning
RE	red edge
RENDVI	red-edge normalized difference vegetation index
REP	red-edge position
RESOURSESAT	satellite launched by India
RGB	red green blue
RGRI	red/green ratio index
RGR	red:green ratio
RMSE	root mean square error
RMS	root mean square
ROI	region of interest
ROSIS	reflective optics spectrographic imaging system
RPD	ratio of prediction to deviation
RRMSE	relative root mean square
RT	radiative transfer
RTM	radiative transfer models
RVI	ratio vegetation index
RVIhyp	hyperspectral ratio VI
RVSI	red-edge vegetation stress index
RWC	relative water content
SA HTBVI	soil-adjusted hyperspectral two-band vegetation indices

SAM	spectral angle mapper
SAVI2	second soil-adjusted vegetation index
SAVI	soil-adjusted vegetating index
SBFS	sequential backward floating selection
SBS	sequential backward selection
SBUV	solar backscatter ultraviolet
SCIAMACHY	scanning imaging absorption spectrometer for atmospheric cartography
SCM	spectral correlation measure
SCR	spatially coherent regions
SeaWiFS	sea-viewing wide field-of-view sensor
SFFS	sequential forward floating selection
SFS	sequential forward selection
SGI	sum green index
SGLI	second generation global imager
SGR	summed green reflectance
SIPI	structurally insensitive pigment index
SLA	specific leaf area
SLR	stepwise linear regression
SMA	spectral mixture analysis
SMGM	soil moisture Gaussian model
SNR	signal-to-noise ratio
SPECIM	SPECtral Imaging
SPOT	satellites pour l'observation de la terre or earth-observing satellites
SPP	spatial preprocessing
SPSS	statistical product and service solutions
SR	simple ratio
SSEE	spatial spectral endmember extraction
SVD	singular value decomposition
SVM	support vector machine
SWIR	shortwave infrared
SWIR	shortwave infrared (1300–2500 nm)
SZA	solar zenith angle
TAU	Tel Aviv University
TCARI/OSVAI	transformed chlorophyll absorption in reflectance index/optimized soil-adjusted vegetation index
TEM	total entropy measure
TES	tropospheric emission spectrometer
TF	tropical forest
TIR	thermal infrared
TML	total metal level
TOA	top of atmosphere
TOC	top of canopy
TOMS	total ozone mapping spectrometer
TSAVI	transformed soil-adjusted vegetation index
TSVMs	transductive SVMs
TVI	triangular vegetation index
U.S.	United States
UAV	unmanned aerial vehicle
UHF	ultrahigh frequency
UMV	unmanned vehicle
UNFCCC	United Nations Framework Convention on Climate Change

USAD	United States Department of Agriculture
USGS	United States Geological Survey
UV	ultraviolet
UV	ultra wave
VARI	vegetation atmospherically resistant index
VARI	visible atmospherically resistant index
VCA	vertex component analysis
VD	virtual dimensionality
VF	vegetation fraction
VI	(spectral) vegetation index
VIGreen	vegetation index green
VIg	visible green index
VIIRS	visible infrared imager radiometer suite
VIP	variable importance for projection
VIS	visible
VMC	volumetric moisture content
VNIR	visible near-infrared sensor
VNIR	visual and near infrared
VOG-1	Vogelmann red-edge index-1
VPD	vapor pressure deficit
VSWIR	visible and short wave infrared
VZA	view zenith angle
W	water
WAA	water absorption area
WAD	water absorption depth
WBI	water band index
WBM	wet biomass (kg/m2)
WDRVI	wide dynamic range vegetation index
WFIS	wide field-of-view imaging spectrometer
WI	water index
WORLDVIEW	DigitalGlobe's earth imaging satellite
WSC	World Soil Congress
WT	wavelet transform
YE	yellow edge
YI	yellowness index
Zn	zinc

Contributors

Victor Alchanatis
Institute of Agricultural Engineering
Agricultural Research Organization
Bet Dagan, Israel

Sreekala G. Bajwa
Division of Agriculture
University of Arkansas
Fayetteville, Arkansas

E. Ben-Dor
Department of Geography
Tel-Aviv University
Tel-Aviv, Israel

Deepak Bhatnagar
U.S. Department of Agriculture
Agricultural Research Service
Southern Regional Research Center
New Orleans, Louisiana

M. Boschetti
National Research Council
Institute for Electromagnetic Sensing
    of the Environment
Milan, Italy

Fábio Marcelo Breunig
Instituto Nacional de Pesquisas Espaciais
São José dos Campos, São Paulo, Brazil

P.A. Brivio
National Research Council
Institute for Electromagnetic Sensing
    of the Environment
Milan, Italy

Robert L. Brown
U.S. Department of Agriculture
Agricultural Research Service
Southern Regional Research Center
New Orleans, Louisiana

Yen-Ben Cheng
Earth Resources Technology, Inc.
Laurel, Maryland

Panigada Cinzia
Remote Sensing of Environmental Dynamics
    Laboratory
Dipartimento di Scienze dell'Ambiente e del
    Territorio
Università Milano-Bicocca
Milano, Italy

Matthew L. Clark
Center for Interdisciplinary Geospatial
    Analysis
Department of Geography and Global Studies
Sonoma State University
Rohnert Park, California

Thomas E. Cleveland
U.S. Department of Agriculture
Agricultural Research Service
Southern Regional Research Center
New Orleans, Louisiana

Yafit Cohen
Institute of Agricultural Engineering
Agricultural Research Organization
Bet Dagan, Israel

José Carlos Neves Epiphanio
Instituto Nacional de Pesquisas Espaciais
São José dos Campos, São Paulo, Brazil

F. Fava
Desertification Research Group
Università degli Studi di Sassari
Sassari, Italy

Antônio Roberto Formaggio
Instituto Nacional de Pesquisas Espaciais
São José dos Campos, São Paulo, Brazil

Lisa R. Gaddis
Astrogeology Science Center
U.S. Geological Survey
Flagstaff, Arizona

Lênio Soares Galvão
Instituto Nacional de Pesquisas Espaciais
São José dos Campos, São Paulo, Brazil

Chengcheng Gang
College of Life Science
Nanjing University
Nanjing, People's Republic of China

Paul E. Geissler
Astrogeology Science Center
U.S. Geological Survey
Flagstaff, Arizona

Anatoly A. Gitelson
University of Nebraska-Lincoln
Lincoln, Nebraska

Edward P. Glenn
Department of Soil, Water, and Environmental
    Science
Environmental Research Laboratory
The University of Arizona
Tucson, Arizona

Justin J. Hagerty
Astrogeology Science Center
U.S. Geological Survey
Flagstaff, Arizona

K. Fred Huemmrich
University of Maryland, Baltimore County
College Park, Maryland

and

Joint Center for Earth Systems Technology
Baltimore, Maryland

Alfredo Huete
School of Environmental Sciences
University of Technology Sydney
Sydney, New South Wales, Australia

and

Department of Soil, Water, and Environmental
    Sciences
The University of Arizona
Tucson, Arizona

Yoshio Inoue
Agro-Ecosystem Informatics Research
National Institute for Agro-Environmental
    Sciences
Tsukuba, Ibaraki, Japan

Jeffery R. Johnson
Astrogeology Science Center
U.S. Geological Survey
Flagstaff, Arizona

Subodh S. Kulkarni
Division of Agriculture
University of Arkansas
Fayetteville, Arkansas

Willem J.D. van Leeuwen
School of Geography and Development
and
School of Natural Resources
    and the Environment
Office of Arid Lands Studies
Arizona Remote Sensing Center
The University of Arizona
Tucson, Arizona

Cherry Li
College of Life Science
Nanjing University
Nanjing, People's Republic of China

Jianlong Li
College of Life Science
Nanjing University
Nanjing, People's Republic of China

Busetto Lorenzo
Remote Sensing of Environmental Dynamics
    Laboratory
Dipartimento di Scienze dell'Ambiente e del
    Territorio
Università Milano-Bicocca
Milano, Italy

John G. Lyon
Las Vegas Laboratory
United States Environmental Protection
    Agency
Las Vegas, Nevada

Hank A. Margolis
Faculté de Foresterie
Centre d'Études de la Forêt
de Géographie et de Géomatique
Université Laval
Laval, Quebec, Canada

Gabriel Martín
Hyperspectral Computing Laboratory
Department of Technology of Computers
    and Communications
Escuela Politécnica de Cáceres
University of Extremadura
Cáceres, Spain

Meroni Michele
Remote Sensing of Environmental Dynamics
    Laboratory
Dipartimento di Scienze dell'Ambiente e del
    Territorio
Università Milano-Bicocca
Milano, Italy

Rossini Micol
Remote Sensing of Environmental Dynamics
    Laboratory
Dipartimento di Scienze dell'Ambiente e del
    Territorio
Università Milano-Bicocca
Milano, Italy

Elizabeth M. Middleton
Goddard Space Flight Center
National Aeronautics and Space
    Administration
Greenbelt, Maryland

Tomoaki Miura
Department of Natural Resources
    and Environmental Management
University of Hawaii at Manoa
Honolulu, Hawaii

Pamela Lynn Nagler
U.S. Geological Survey
Southwest Biological Science Center
Sonoran Desert Research Station
Tucson, Arizona

Izaya Numata
South Dakota State University
Brookings, South Dakota

Aaryn Dyami Olsson
Laboratory of Landscape Ecology
    and Conservation Biology
College of Engineering, Forestry & Natural
    Sciences
Northern Arizona University
Flagstaff, Arizona

Fred Ortenberg
Technion
Israel Institute of Technology
Haifa, Israel

Ryan L. Perroy
Department of Geography and Earth Science
University of Wisconsin-La Crosse
La Crosse, Wisconsin

Antonio Plaza
Hyperspectral Computing Laboratory
Department of Technology of Computers
    and Communications
Escuela Politécnica de Cáceres
University of Extremadura
Cáceres, Spain

Javier Plaza
Hyperspectral Computing Laboratory
Department of Technology of Computers
    and Communications
Escuela Politécnica de Cáceres
University of Extremadura
Cáceres, Spain

Ruiliang Pu
Department of Geography
University of South Florida
Tampa, Florida

Jiaguo Qi
Department of Geography
and
Center for Global Change & Earth
    Observations
Michigan State University
East Lansing, Michigan

Elijah Ramsey III
U.S. Geological Survey
National Wetland Research Center
Lafayette, Louisiana

Amina Rangoonwala
Five Rivers Services, LLC
U.S. Geological Survey
National Wetland Research Center
Lafayette, Louisiana

Colombo Roberto
Remote Sensing of Environmental Dynamics
    Laboratory
Dipartimento di Scienze dell'Ambiente e del
    Territorio
Università Milano-Bicocca
Milano, Italy

Dar A. Roberts
Department of Geography
University of California
Santa Barbara, California

Keely L. Roth
Department of Geography
University of California
Santa Barbara, California

Sergio Sánchez
Hyperspectral Computing Laboratory
Department of Technology of Computers
    and Communications
Escuela Politécnica de Cáceres
University of Extremadura
Cáceres, Spain

E. Terrence Slonecker
Eastern Geographic Science Center
U.S. Geological Survey
Reston, Virginia

Laurence A. Soderblom
Astrogeology Science Center
U.S. Geological Survey
Flagstaff, Arizona

B.B. Maruthi Sridhar
Department of Geology
Bowling Green State University
Bowling Green, Ohio

Daniela Stroppiana
National Research Council
Institute for Electromagnetic Sensing
    of the Environment
Milan, Italy

Lie Tang
Department of Agricultural and Biosystems
    Engineering
Iowa State University
Ames, Iowa

Prasad S. Thenkabail
Western Geographic Science Center
U.S. Geological Survey
Flagstaff, Arizona

Valerie Thomas
Department of Forest Resources
    and Environmental Conservation
Virginia Tech
Blacksburg, Virginia

Lei Tian
Department of Biological and Agricultural
    Engineering
University of Illinois at Urbana Champaign
Champaign, Illinois

Timothy N. Titus
Astrogeology Science Center
U.S. Geological Survey
Flagstaff, Arizona

R. Greg Vaughan
Astrogeology Science Center
U.S. Geological Survey
Flagstaff, Arizona

Wei Wang
Jiangsu Key Laboratory for Information
    Agriculture
National Engineering and Technology Center
    for Information Agriculture
College of Agriculture
Nanjing Agricultural University
Nanjing, Jiangsu, People's Republic of China

Narumon Wiangwang
Department of Fisheries
Information Technology Center
Royal Thai Government
Bangkok, Thailand

Haibo Yao
Stennis Space Center
Geosystems Research Institute
Mississippi State University
Starkville, Mississippi

Xia Yao
Jiangsu Key Laboratory for Information
    Agriculture
National Engineering and Technology Center
    for Information Agriculture
College of Agriculture
Nanjing Agricultural University
Nanjing, Jiangsu, People's Republic of China

Hiroki Yoshioka
Department of Information Science
    and Technology
Aichi Prefectural University
Aichi, Japan

Yongqin Zhang
Division of Biological and Physical Sciences
Delta State University
Cleveland, Mississippi

Dehua Zhao
College of Life Science
Nanjing University
Nanjing, People's Republic of China

Yan Zhu
Jiangsu Key Laboratory for Information
    Agriculture
National Engineering and Technology Center
    for Information Agriculture
College of Agriculture
Nanjing Agricultural University
Nanjing, Jiangsu, People's Republic of China

Part I

Introduction and Overview

1 Advances in Hyperspectral Remote Sensing of Vegetation and Agricultural Croplands

Prasad S. Thenkabail, John G. Lyon, and Alfredo Huete

CONTENTS

1.1 INTRODUCTION AND RATIONALE

Recent advances in hyperspectral remote sensing (or imaging spectroscopy) demonstrate a great utility for a variety of land monitoring applications. It is now possible to be diagnostic in sensing species and plant communities using remotely sensed data and to do so in a direct and informed manner using modern tools and analyses. Hyperspectral data analyses are superior to traditional broadband analyses in spectral information. Many investigations explore and document remote sensing of vegetation and agricultural croplands. Some examples include (a) detecting plant stress [1], (b) measuring chlorophyll content of plants [2], (c) identifying small differences in percent of green vegetation cover [3], (d) extracting biochemical variables such as nitrogen and lignin [2,4–6], (e) discriminating land-cover types [7], (f) detecting crop moisture variations [8], (g) sensing subtle variations in leaf pigment concentrations [2,9,10], (h) modeling biophysical and yield characteristics of agricultural crops [6,11,12], (i) improving the detection of changes in sparse vegetation [13], and (j) assessing absolute water content in plant leaves [14]. This is a fairly detailed list but not exhaustive, meant to provide the reader with a measure of the current, proven experimental capabilities, and operational applications, and stimulate investigations of new, ambitious applications.

The spectral properties of vegetation are strongly determined by their biophysical and biochemical attributes, such as leaf area index (LAI), the amount of live biomass and senesced biomass, moisture content, pigments (e.g., chlorophyll), and spatial arrangement of structures [15,16]. We are capable of measuring those phenomenon and processes to test hypotheses and valuable applications on a variety of ecosystems. For example, assessment of biophysical and biochemical properties of vegetation such as rangelands [17,18], agricultural crops [1,7,11,12,19], and weeds [7] are essential for evaluating productivity, providing information needed for local farmers and institutions, and assessing grazing potential for livestock. Even though remote sensing has been recognized as a reliable method for estimating these biophysical and biochemical vegetation variables, existing broadband sensors have proven inadequate or supplied limited information for the purpose [1,7,20–23]. Clearly, broadbands have limitations in providing adequate information on properties such as crop growth stage identification, crop type differentiation, generation of agricultural crop statistics, forest type and species identification, characterizing complex forest versus nonforest interactions, and detail mapping of land cover classes of interest to diverse scientific and other user communities (1,11). The limitations of broadband analyses are illustrated by vegetation indices (VIs), which saturate beyond a certain level of biomass and LAI [12]. For example, VIs typically increase over an LAI range from 0 to between 3 and 5 before an asymptote is reached. While extremely useful over the years, the upper limit of this sensitivity apparently differs among vegetation types and can only be driven so far to a solution for a given application. Saturation is more pronounced for planophile canopies [11,12]. However, compared with erectophile canopies of the same LAI, planophile canopies are less influenced by soil brightness variations [19]. In contrast, hyperspectral datasets allow identification of features, direct measurement of canopy variables, such as biochemical content (e.g., chlorophyll, nitrogen, lignin), forest species, chemistry distribution, timber volumes, water, etc. [2], and biophysical (e.g., LAI, biomass) and yield characteristics [7,10,12,24].

Hyperspectral sensors gather near-continuous spectra from imaging spectrometers such as the National Aeronautics and Space Administration's (NASA)-designed Airborne Visible-infrared Imaging Spectrometer (AVIRIS) and Compact Airborne Spectrographic Imager (CASI). This new generation of sensors offers tremendous improvements in spatial, spectral, radiometric, and temporal resolutions as well as improvements in optics and mechanics when compared with older generation of sensors (Table 1.1). The promise and potential of hyperspectral narrowband sensors for a wide array of Earth resource applications has motivated design and also the launch of spaceborne sensors such as Hyperion on board the Earth Observing-1 (EO-1) [25,26], and the upcoming Hyperspectral Imaging Spectrometer and Infrared Imager (HyspIRI). These sensors gather data in 210–220 narrowbands from 380 to 2500 nm at 60 m resolution or better (Table 1.1). The HyspIRI's Thermal

TABLE 1.1
Broadband and Narrowband Satellite Sensor Spatial, Spectral, Radiometric, Waveband, and Other Data Characteristics[a]

Sensor	Spatial (m)	Spectral (#)	Radiometric (Bit)	Band Range (µm)	Band Widths (µm)	Irradiance (W m⁻² sr⁻¹ µm⁻¹)	Data Points (# per ha)	Frequency of Revisit (Days)
A. Coarse resolution broadband sensors								
1. AVHRR	1,000	4	11	0.58–0.68	0.10	1,390	0.01	daily
				0.725–1.1	0.375	1,410		
				3.55–3.93	0.38	1,510		
				10.30–10.95	0.65	0		
				10.95–11.65	0.7	0		
B. Coarse resolution narrowband sensors								
2. MODIS	250, 500, 1,000	36/7	12	0.62–0.67	0.05	1,528.2	0.16, 0.04, 0.01	daily
				0.84–0.876	0.036	974.3	0.16, 0.04, 0.01	
				0.459–0.479	0.02	2,053		
				0.545–0.565	0.02	1,719.8		
				1.23–1.25	0.02	447.4		
				1.63–1.65	0.02	227.4		
				2.11–2.16	0.05	86.7		
C. Multispectral broadband sensors								
3. Landsat-1, 2, 3 MSS	56 × 79	4	6	0.5–0.6	0.1	1,970	2.26	16
				0.6–0.7	0.1	1,843		
				0.7–0.8	0.1	1,555		
				0.8–1.1	0.3	1,047		
4. Landsat-4, 5 TM	30	7	8	0.45–0.52	0.07	1,970	11.1	16
				0.52–0.60	0.80	1,843		
				0.63–0.69	0.60	1,555		
				0.76–0.90	0.14	1,047		
				1.55–1.74	0.19	227.1		
				10.4–12.5	2.10	0		
				2.08–2.35	0.25	80.53		

(continued)

TABLE 1.1 (continued)
Broadband and Narrowband Satellite Sensor Spatial, Spectral, Radiometric, Waveband, and Other Data Characteristics[a]

Sensor	Spatial (m)	Spectral (#)	Radiometric (Bit)	Band Range (μm)	Band Widths (μm)	Irradiance (W m^{-2} sr^{-1} μm^{-1})	Data Points (# per ha)	Frequency of Revisit (Days)
5. Landsat-7 ETM+	30	8	8	0.45–0.52	0.65	1,970	44.4, 11.1	16
				0.52–0.60	0.80	1,843		
				0.63–0.69	0.60	1,555		
				0.50–0.75	0.150	1,047		
				0.75–0.90	0.200	227.1		
				10.0–12.5	2.5	0		
				1.75–1.55	0.2	1,368		
				0.52–0.90 (p)	0.38	1,352.71		
6. ASTER	15, 30, 90	15	8	0.52–0.63	0.11	1,846.9	44.4, 11.1, 1.23	16
				0.63–0.69	0.06	1,546.0		
				0.76–0.86	0.1	1,117.6		
				0.76–0.86	0.1	1,117.6		
				1.60–1.70	0.1	232.5		
				2.145–2.185	0.04	80.32		
				2.185–2.225	0.04	74.96		
				2.235–2.285	0.05	69.20		
				2.295–2.365	0.07	59.82		
				2.360–2.430	0.07	57.32		
			12	8.125–8.475	0.35	0		
				8.475–8.825	0.35	0		
				8.925–9.275	0.35	0		
				10.25–10.95	0.7	0		
				10.95–11.65	0.7	0		

7. ALI	30	10	12	0.048–0.69 (p)	0.64	1,747.8600	11.1	16
				0.433–0.453	0.20	1,849.5		
				0.450–0.515	0.65	1,985.0714		
				0.425–0.605	0.80	1,732.1765		
				0.633–0.690	0.57	1,485.2308		
				0.775–0.805	0.30	1,134.2857		
				0.845–0.890	0.45	948.36364		
				1.200–1.300	1.00	439.61905		
				1.550–1.750	2.00	223.39024		
				2.080–2.350	2.70	78.072727		
8. SPOT-1	2.5–20	15	16	0.50–0.59	0.09	1,858	1,600, 25	3–5
-2				0.61–0.68	0.07	1,575		
-3				0.79–0.89	0.1	1,047		
-4				1.5–1.75	0.25	234		
				0.51–0.73 (p)	0.22	1,773		
9. IRS-1C	23.5	15	8	0.52–0.59	0.07	1,851.1	18.1	16
				0.62–0.68	0.06	1,583.8		
				0.77–0.86	0.09	1,102.5		
				1.55–1.70	0.15	240.4		
				0.5–0.75 (P)	0.25	1,627.1		
10. IRS-1	23.5	15	8	0.52–0.59	0.07	1,852.1	18.1	16
				0.62–0.68	0.06	1,577.38		
				0.77–0.86	0.09	1,096.7		
				1.55–1.70	0.15	240.4		
				0.5–0.75 (P)	0.25	1,603.9		
11. IRS-P6-AWiFS	56	4	10	0.52–0.59	0.07	1,857.7	3.19	16
				0.62–0.68	0.06	1,556.4		
				0.77–0.86	0.09	1,082.4		
				1.55–1.70	0.15	239.84		

(continued)

TABLE 1.1 (continued)
Broadband and Narrowband Satellite Sensor Spatial, Spectral, Radiometric, Waveband, and Other Data Characteristics[a]

Sensor	Spatial (m)	Spectral (#)	Radiometric (Bit)	Band Range (µm)	Band Widths (µm)	Irradiance (W m^{-2} sr^{-1} µm^{-1})	Data Points (# per ha)	Frequency of Revisit (Days)
12. CBERS-2 -3B	20 m pan / 20 m MS		11	0.51–0.73	0.22	1,934.03	25, 25	
-3 -4	5 m pan / 20 m MS			0.45–0.52	0.07	1,787.10	400, 25	
				0.52–0.59	0.07	1,587.97		
				0.63–0.69	0.06	1,069.21		
				0.77–0.89	0.12	1,664.3		
D. Hyperspectral narrowband sensors								
13. Hyperion	30	220 (196b)	16	196 effective Calibrated bands VNIR (band 8–57 427.55–925.85 nm SWIR (band 79–224) 932.72–2,395.53 nm	10 nm wide (approx.) for all 196 bands	See data in Neckel and Labs (1984). Plot it and obtain values for Hyperion bands	11.1	16
14. ASD spectroradiometer	1,134 cm² at 1.2 m Nadir view 18° Field of view	2,100 bands 1 nm width between 400–2,500 nm	16	2,100 effective bands	1 nm wide (approx.) in 400–2500 nm	See data in Neckel and Labs (1984). Plot it and obtain values for Hyperion bands	88,183	5–16
15. HyspIRI VSWIR	60	210	16	210 bands in 380–2,500 nm	10 nm wide (approx.) for all 210 bands	See data in Neckel and Labs (1984). Plot it	2.77	19

Sensor	Spatial res.	No. of bands	No. of bands	7 bands in 7,500–12,000 nm and 1 band in 3,000–5,000 nm (3,980 nm center)	7 bands in 7,500–12,000 nm	Center wavelength		Revisit
16. HyspIRI TIR	60	8	16	7 bands in 7,500–12,000 nm and 1 band in 3,000–5,000 nm (3,980 nm center)	0.71; 0.89; 0.66; 0.96	See data in Neckel and Labs (1984). Plot it	2.77	5
E. Hyperspatial broadband sensors								
17. IKONOS	1–4	4	11	0.445–0.516; 0.506–0.595; 0.632–0.698; 0.757–0.853	0.07; 0.08; 0.06; 0.13	1,930.9; 1,854.8; 1,156.5; 1,156.9	10,000, 625	5
18. QUICKBIRD	0.61–2.44	4	11	0.45–0.52; 0.52–0.60; 0.63–0.69; 0.76–0.89		1,381.79; 1,924.59; 1,843.08; 1,574.77	14,872, 625	5
19. RESOURSESAT	5.8	3	10	0.52–0.59; 0.62–0.68; 0.77–0.86	0.07; 0.06; 0.09	1,853.6; 1,581.6; 1,114.3	33.64	24
20. RAPID EYE – A – E	6.5	5	12	0.44–0.51; 0.52–0.59; 0.63–0.68; 0.69–0.73; 0.77–0.89	0.07; 0.07; 0.05; 0.04; 0.12	1,979.33; 1,752.33; 1,499.18; 1,343.67; 1,039.88	236.7	1–2
21. WORLDVIEW	0.55	1	11	0.45–0.51	0.06	1,996.77	40,000	1.7–5.9
22. FORMOSAT-2	2–8	5	11	0.45–0.52; 0.52–0.60; 0.63–0.69; 0.76–0.90; 0.45–0.90 (p)	0.07; 0.08; 0.06; 0.14; 0.45	1,974.93; 1,743.12; 1,485.23; 1,041.28; 1,450	2,500, 156.25	daily

(continued)

TABLE 1.1 (continued)
Broadband and Narrowband Satellite Sensor Spatial, Spectral, Radiometric, Waveband, and Other Data Characteristics[a]

Sensor	Spatial (m)	Spectral (#)	Radiometric (Bit)	Band Range (μm)	Band Widths (μm)	Irradiance (W m⁻² sr⁻¹ μm⁻¹)	Data Points (# per ha)	Frequency of Revisit (Days)
23. KOMPSAT-2	1–4	5	10	0.5–0.9	0.4	1,379.46	10,000, 625	3–28
				0.45–0.52	0.07	1,974.93		
				0.52–0.6	0.08	1,743.12		
				0.63–0.59	0.04	1,485.23		
				0.76–0.90	0.14	1,041.28		

Source: Edited and adapted from Thenkabail, P.S. et al., Global croplands and their water use remote sensing and non-remote sensing perspectives. Book Chapter. Chapter 16. In: Weng, Q. (ed.), *Advances in Environmental Remote Sensing: Sensors, Algorithms, and Applications*, Taylor & Francis, Boca Raton, FL, 2010; Melesse, A.M. et al., *Sens. J.* 7, 3209, 2007. http://www.mdpi.org/sensors/papers/s7123209.pdf

ASD, Analytical Spectral Devices Inc. Provider of hand-held spectroradiometers; ASTER, Advanced Spaceborne Thermal Emission and Reflection Radiometer; AVHRR, Advanced Very High Resolution Radiometer; CBERS-2, China-Brazil Earth Resources Satellite; CP, Crude Protein (%); FNIR, Far near-infrared (1100–1300 nm); FORMOSAT, Taiwanese Satellite Operated by Taiwanese National Space Organization NSPO. Data Marketed by SPOT; GEOEYE-1 and 2, Providing Data in 0.25–1.65 m resolution; Hyperion, First Spaceborne Hyperspectral Sensor Onboard Earth Observing-1(EO-1); HyspIRI, Hyperspectral Infrared Imager; IKONOS, High-Resolution Satellite Operated by GeoEye; IRS-1C/D-LISS, Indian Remote Sensing Satellite/Linear Imaging Self Scanner; IRS-P6-AWiFS, Indian Remote Sensing Satellite/Advanced Wide Field Sensor; KOMFOSAT, Korean Multipurpose Satellite. Data Marketed by SPOT Image; LAI, Leaf area index (m²/m²); Landsat-1, 2, 3 MSS, Multi Spectral Scanner; Landsat-4, 5 TM, Thematic Mapper; Landsat-7 ETM+, Enhanced Thematic Mapper Plus; LiDAR, Light Detection and Ranging; LNA, Leaf Nitrogen Accumulation; MODIS, Moderate Imaging Spectral Radio Meter; NIR, Near-infrared (740–1100 nm); N, Nitrogen (%); PLNTHT, Plant Height (mm); QUICKBIRD, Satellite from DigitalGlobe, a private company in the USA; RAPID EYE – A/E, Satellite Constellation from Rapideye, a German company; RESOURSESAT, Satellite Launched by the India; SPOT, Satellites Pour l'Observation de la Terre or Earth-Observing Satellites; SWIR, Short Wave Infrared (1300–2500 nm); TIR, Thermal Infrared; VNIR, Visible Near-Infrared Sensor; VSWIR, Visible and Short Wave Infrared; WBM, Wet Biomass (kgm²); WORLDVIEW, DigitalGlobe's Earth Imaging Satellite.

[a] Of the 242 bands, 196 are unique and calibrated. These are (a) band 8 (427.55 nm) to band 57 (925.85 nm) that are acquired by visible and near-infrared (VNIR) sensor; and (b) band 79 (932.72 nm) to band 224 (2395.53 nm) that are acquired by the short wave infrared (SWIR) sensor.

Infrared (TIR) has 7 bands in 7,500–12,000 nm that saturate at 400 k and 1 band in 3,000–5,000 nm (centered at 3,980 nm) that saturates at 1400 k and acquires data with 60 m spatial resolution.

However, it must be noted that using hyperspectral data is much more complex than multispectral data. Hyperspectral systems collect large volumes of data in a short time leading to a number of issues that need to be addressed. For example, Hyperion, the first spaceborne hyperspectral sensor, onboard EO-1 launched by the NASA's New Millennium Program (NMP) gathers near-continuous data in 220 discrete narrowbands along the 400–2500 nm spectral range at 30 m spatial resolution and in 12 bits. Each image is 7.5 km in swath by 100 km along track. The volume of data collected using Hyperion for an area equivalent to Landsat TM image area is roughly 37 times the data volume of the TM scene.

Increases in data volume pose great challenges in data handling. The issues include data storage volume, data storage rate, or transmission bandwidth, real-time analog to digital bandwidth and resolution, computing bottle necks in data analysis, and the need for new algorithms for data utilization (e.g., atmospheric correction is more complicated) [1,11]. These issues make it imperative that methods and techniques be advanced and developed to handle higher-dimensional datasets.

Future generations of satellites may carry specialized optimal sensors designed to gather data for targeted applications. Or they may carry a narrow-waveband hyperspectral sensor like Hyperion and HyspIRI from which users with different application needs can extract appropriate optimal wavebands. However, having continuous spectral coverage with many narrowbands does not necessarily mean more information. Indeed, most of these bands, and especially the ones that are close to one another, provide redundant information. This redundancy can require users to devote substantial time in data mining, complex processing to identify and remove redundant bands, and puts a heavy burden on computing-processing-storage resources.

A far better option is to focus on the design of an optimal sensor for a given application, such as for vegetation studies, and by excluding redundant bands. Even when the data are acquired in full range of hundreds or thousands of hyperspectral narrowbands, a priori knowledge of optimal bands for a particular application helps. Investigators can quickly select these bands and more efficiently spend time and expertise resources in using these for the required application. Optimal hyperspectral sensors will help reduce data volumes, eliminate the problems of high-dimensionality of Hyperspectral datasets, and make it feasible to apply traditional classification methods on a few selected bands (optimal bands) that capture most of the information of crop characteristics [4,7,9,11,27]. Thereby, knowledge of application specific "optimal bands" for high-dimensional datasets, such as Hyperion and HyspIRI is crucial to reduce costs in data analysis and computer resources.

Table 1.1 compares the spectral and spatial resolution of narrowband and broadband data that are currently in use and those soon to be used. A number of recent studies have indicated the advantages of using discrete narrowband data from a specific portion of the spectrum when compared with broadband data to arrive at optimal quantitative or qualitative information on crop or vegetation characteristics [2,4,7,9,11,12,27]. Hence, this approach or direction has caught the attention of investigators and the remote sensing community is moving in this direction.

The overarching goal of this chapter is to explore and determine the optimal hyperspectral narrowbands for use in the study of vegetation and agricultural crops and to enumerate methods and approaches. This is partially to overcome the "curse" of high dimensionality or "Hughes phenomenon," where the ratio of the number of pixels with known class identity (i.e., training pixels) and the number of bands must be maintained at or above minimum value to achieve statistical confidence and functionality. In hyperspectral data, with hundreds or even thousands of wavebands, the number of training pixels needed grows exponentially (Hughes phenomenon), making it very difficult to address this spectral diversity.

Our first obstacle is to identify hyperspectral narrowbands that are best suited for studying natural vegetation and agricultural croplands. In the process, we have detected and eliminated redundant bands or examples that supply little knowledge to the application. We then highlight optimal hyperspectral wavebands, in 400–2500 nm range, best suited to study vegetation and agricultural crops.

There are a number of studies [1,4,5,7,20,27–29] that indicate that the narrow wavebands located in specific portions of the spectrum have the ability to provide required optimal information sought for a given application. However, there is a clear need for synthesis of the studies conducted in different parts of the world to find a general consensus of optimal wavebands that applies to the varying vegetation types and agricultural practices. Optimal Hyperion wavebands determined for vegetation studies established in this study will help reduce data volumes, eliminate the problems of high-dimensionality of Hyperion datasets, and make it feasible to apply traditional classification methods on a few selected bands (optimal bands) that capture most of the information of vegetation and agricultural croplands.

Secondly, we will present and discuss methods and approaches of hyperspectral data analysis. We advocate the quick elimination of redundant bands and identify most useful bands. For example, when a user receives hyperspectral data for a vegetation and/or agricultural study, selected wavebands can be analyzed and all other wavebands can be ignored. This process establishes categorization approaches to achieve highest accuracies, develop indices and wavebands that best model biophysical and biochemical quantities, and identify and eliminate redundant bands. This process involves an exhaustive review of the performance of hyperspectral vegetation indices (HVIs) that will help establish the wavebands associated with these indices in the study of vegetation and agricultural crops.

1.2 HYPERSPECTRAL REMOTE SENSING OF VEGETATION AND AGRICULTURAL CROPS

Agricultural crops are significantly better characterized, classified, modeled, and mapped using hyperspectral data. There are many studies supporting this, conducted on a wide array of crops and their biophysical and biochemical variables, such as yield [32], chlorophyll a and b [33,34], total chlorophyll [35], nitrogen content [36], carotenoid pigments [29], plant stress [10], plant moisture [37], above ground biomass [1,7,11,12], and biophysical variables [17–19,38,39].

Hyperspectral remote sensing is well suited for early detection of Nitrogen (N) as an important determinant of crop productivity and its quality. Knowing nitrogen levels early in the growing season is critical for effective farming. Leaf nitrogen is a key element in monitoring crop growth stage, its fertilization status, and assessing productivity. Leaf N accumulation (LNA), as a product of leaf N content and leaf weight, reflect not only information on leaf N status, but also vegetation coverage during crop growth [20]. For example, nitrogen deficiency can be evaluated using ratios such as R_{743}/R_{1316} [40].

Thenkabail et al. [11,12] studied numerous crops (e.g., Figure 1.1a) that included cotton (*Gossypium*), potato (*Solanum erianthum*), soybeans (*Glycine max*), corn (*Zea mays*), sunflower (*Helianthus* spp.), barley (*Hordeum vulgare* L.), wheat (*Triticum aestivum* L. or *Triticum durum* Desf.), lentil (*Lens esculenta* (Moench)). Or *Lens orientale* (Boiss.), *Schmalh. Or Lens culinaris* (Medikus), cumin (*Cuminum cyminum* L.), chickpea (*Cicer arietinum* L.), and vetch (*Vicia narbonensis* L.). The parameters studied and found useful included LAI: the leaf area index (m²/m²), WBM: wet biomass (kg/m²), PLNTHT: aboveground plant height (meter), CP: plant crude protein (%), N: nitrogen (%), and CC: mean canopy cover (%). Thousands of spectral measurements were made. These comprehensive studies lead to identification of 12 optimal narrowbands in 400–1050 nm [11,12].

Forest and savanna biotic variables (e.g., height, basal area, biomass, and LAI) are modeled and mapped with significantly greater accuracies using hyperspectral data [41–45]; e.g., Figure 1.1b through d) when compared to broadband data. The ability to distinguish species type, assess age, structure (e.g., leaf angle), and phenology (e.g., leaf pigmentation, flowering), and map shade trees and understory crops is becoming increasingly possible as a result of hyperspectral narrowband data [46,47].

When there are significant topographic effects, such as shadow ridges, LiDAR data combined with hyperspectral data is extremely useful [48]. Thenkabail et al. [1,7] conducted a hyperspectral study of vegetation in the African savannas and rainforests using Hyperion and Analytical Spectral Devices (ASD) spectroradiometers, both collecting data in 400–2500 nm (Table 1.1). They collected spectra from thousands of locations (e.g., few examples in Figure 1.1) that included an extensive and comprehensive series of vegetation types such as (a) weeds (*Imperata cylindrical* the most prominent weed of African savannas and *Chromolenea odorata* the most prominent weed in African rainforests), (b) shrubs (e.g., *Pteleopsis habeensis, Piliostigma thonningii, Isoberlinia tomentosa*; Figure 1.1b), (c) grasses (e.g., *Brachiaria jubata, Brachiaria stigmatisata, Digitaria* sp.), (d) agricultural crops (corn, rice, cowpea, groundnut, soybean, cassava), (e) other weeds (e.g., *Ageraturm conyzoids, Aspilia africana, Tephrosia bracteolate, Cassia obtusifolia*), (f) agricultural fallows (e.g., of different age years), (g) primary forests (e.g., pristine, degraded), (h) secondary forests (e.g., young, mature, mixed), (i) regrowth forests (e.g., <3, 3–5, 5–8, >8 years), (j) slash and burn agriculture, and (k) forest tree species (e.g., *Piptadenia africana, Discoglypremna coloneura, Antrocaryon klaineanum, Pycnantus angolensis, Rauwolfia macrophylla, Alstonia congensis, Cissus* spp., *Lacospermas secundii, Haumania delkelmaniana, Alchornea floribunda, Lacospermas secundii, Alchornea*

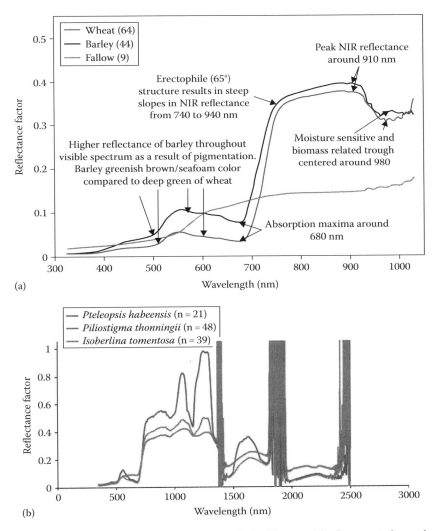

FIGURE 1.1 (**See color insert.**) Hyperspectral characteristics illustrated for few vegetation and agricultural crops. Hyperspectral narrowband data obtained from ASD spectroradiometer illustrated for (a) agricultural crops, (b) shrub species, and

(*continued*)

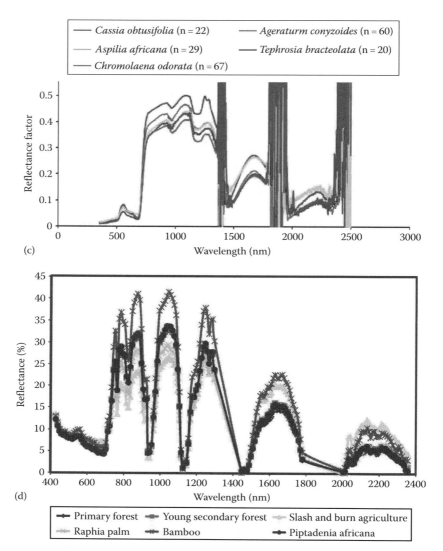

FIGURE 1.1 (continued) **(See color insert.)** (c) weed species. Hyperspectral narrowband data obtained from the Hyperion sensor onboard Earth Observing-1 (EO-1) satellite illustrated for (d) tropical rainforest vegetation. (From Thenkabail, P.S. et al., *Remote Sens. Environ.*, 91, 354, 2004; Thenkabail, P.S. et al., *Remote Sens. Environ.*, 90, 23, 2004; Thenkabail, P.S. et al., *Remote Sens. Environ.*, 71, 158, 2000. With permission.)

cordifolia, Elaeis guineensis, Pteredium aquilinium, Megaphrenium spp., *Pteredium aquilinium, Megaphrenium* spp., *Epatorium adoratum, Aframomum giganteum, Pteredium acquilinium*).

The opportunities to distinguish various species using hyperspectral data are many. For example, the shrub species *Isoberlinia tomentosa* has high reflectivity in the red and NIR bands when compared with *Piliostigma thonningii*, but in the SWIR bands it is just the opposite (Figure 1.1b). The shrub species *Pteleopsis habeensis* has dramatically high reflectivity relative to other two shrub species (*Piliostigma thonningii, Isoberlinia tomentosa*) but beyond 2000 nm. *Isoberlinia tomentosa* has higher reflectivity than other two species (Figure 1.1b). Similarly, the weed species *Aspilia africana* has higher reflectivity relative to other weed species (*Ageraturm conyzoids, Tephrosia bracteolate, Cassia obtusifolia*) only beyond 2000 nm (Figure 1.1c). These results clearly imply the value of using data from distinct parts of the electromagnetic spectrum. These studies resulted in recommending 22–23 optimal bands to study vegetation in 400–2500 nm [1,7].

The relationship between vegetation characteristics and spectral indices can vary considerably based on vegetation types, crop types, and species types [47].

Biomass and NDVI relationships within and between agricultural crops and various vegetation types are very distinct and are useful discriminators. These differences occur as a result of distinctive features, such as [12] (a) plant structure (e.g., erectophile and planophile; e.g., Figure 1.1a), (b) plant composition (e.g., nitrogen, lignin, chlorophyll a and b), and (c) quantitative characteristics (e.g., biomass per unit area, LAI, and plant height). Thereby, a more specific and accurate attempt at biomass estimations using remotely sensed data of various characteristics will need to take into consideration specific vegetation, land cover, crop type, and species types [11].

Radiative transfer models are used to derive quantities, such as LAI and biomass of forests [47,49]. These models require surface reflectance and land cover products as inputs, for example, PROSPECT model [87]. Accuracy of modeled quantities will depend on a number of land covers that can be mapped. Recent studies have also established the possibilities of direct biomass measurements using LiDAR that will provide within-canopy structure, ground topography, and tree height, thus enabling the indirect estimation of biomass and LAI [48].

From the literature and hard work of investigators, it has been established that hyperspectral narrowbands provide significant additional information when compared with similar information obtained from broadband sensors in estimating biophysical characteristics [32,42], biochemical properties [29,50], forest classification [51], and in other vegetation studies involving forests, grasses, shrubs, and weeds [1,7]. This is due to many limitations of broadband data.

For example, most of the indices derived using the older generation of sensors tend to saturate [12] resulting in significant limitations in distinguishing species or modeling biophysical and biochemical quantities. Nondestructive measurements of vegetation characteristics, such as canopy chemical content (e.g., chlorophyll, nitrogen), forest species, chemistry distribution, timber volumes, and plant moisture [24,35,51,52], and biophysical and yield characteristics [1,7,11,12,29] are best measured using hyperspectral data.

Hyperspectral narrowband data have also the potential to separate vegetation into taxonomic levels [53] and forest ecotopes [4]. But, this aspect still needs considerably more study.

These efforts cover a wide array of agricultural crops and vegetation classes from forests, rangelands, savannas, and other ecosystems. They also cover a wide range of agro-ecologies and ecosystems spread across the world. As a result, a more holistic view of hyperspectral data performance in the study of agricultural crops across agro-ecological systems as well as forests and other vegetation across varied ecosystems [54,55] is presented. During this process, we also took into consideration that the high dimensionality of hyperspectral data can, at times, lead to over-fitting of statistical models [12] to an over-optimistic view of their power [56]. This can be unfortunate, but once the issue is understood, approaches can be elucidated and employed to avoid difficulties.

Taking all these factors into consideration, we set forth two key objectives in writing this chapter: (a) establish optimal hyperspectral narrowbands best suited to study vegetation and agricultural crops by pooling collective findings from comprehensive set of investigations, and (b) identify and discuss a suite of methods and approaches that helps to rapidly identify hyperspectral narrowbands best suited to study through data mining of hundreds or thousands of wavebands.

1.3 HYPERSPECTRAL DATA COMPOSITION FOR STUDY OF VEGETATION AND AGRICULTURAL CROPS

Hyperspectral narrowbands typically contain 100–1500 wavebands and collect data in near-continuous spectrum from several regions of the electromagnetic spectrum (ultraviolet, visible, near-, mid-, and far-infrared) (Figure 1.1) that offers many opportunities to study specific vegetation variables, such as LAI [57], biomass [19,56], vegetation fraction [58], canopy biochemistry [43], forest structure and characteristics (e.g., diameter at breast height, tree height, crown size, tree density) [44,59], and plant species [46,60]. These data in hundreds or even thousands of bands are composed into single hyperspectral

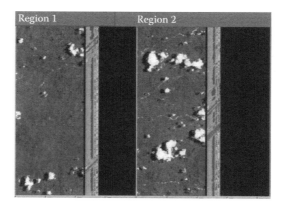

FIGURE 1.2   (See color insert.) Hyperspectral data cube (HDC). The 242 band Hyperion data composed as HDC for two areas of African rainforests. Spectral signatures derived for few classes from this figure are illustrated in Figure 1.1d.

data cube (HDC; Figure 1.2). For example, a click on any pixel of Hyperion HDC (e.g., Figure 1.2) will provide continuous spectrum of various vegetation categories (e.g., Figure 1.1d).

The principles of hyperspectral data composition and analysis can also be applied to time-series multi-spectral data. For example, a MODIS NDVI monthly maximum value composite for 1 year is composed as a single mega-file data cube (MFDC) (e.g., Figure 1.3). The time-series spectra of a

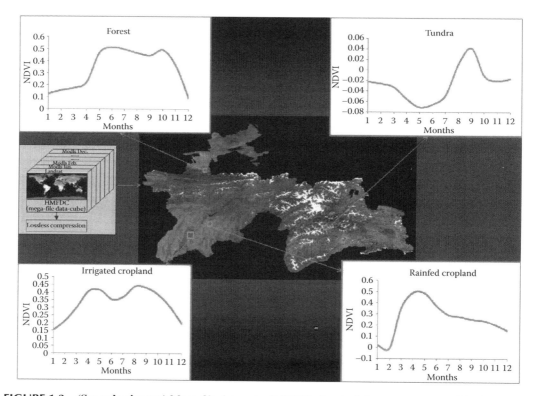

FIGURE 1.3   (See color insert.) Mega-file data cube (MFDC). Spectral signatures extracted for few classes from the 12 band MODIS monthly NDVI maximum value composite time-series MFDC. Akin to hyperspectral data (see Figures 1.1 and 1.2), this multispectral time-series is composed as hyperspectral data cube. All hyperspectral data analysis techniques can be applied here. Note: Illustrated for Tajikistan.

few classes extracted from this simple 12-band MFDC is illustrated in Figure 1.3, akin to the hyperspectral data (Figure 1.1d) extracted from HDC (Figure 1.2).

1.4 METHODS AND APPROACHES OF HYPERSPECTRAL DATA ANALYSIS FOR VEGETATION AND AGRICULTURAL CROPS

The goal of dimension reduction is to reduce the number of features without sacrificing significant information. Care must be taken to retain all key wavebands and avoid reducing the discrimination power that results in models that lead to lower accuracies, and lead to models that fail to explain maximum variability.

1.4.1 LAMBDA (λ_1) BY LAMBDA (λ_2) PLOTS

Hyperspectral narrowband data has high interband correlation that results in multiple measurements of same quantity [61]. Thereby, one of the key steps in hyperspectral data analysis is to identify and remove redundant bands. Using all the bands will only add to computing resources and does not add to additional information content.

For example, wavebands that are close to one another (e.g., 680 vs. 690 nm), typically, are highly correlated (e.g., R^2 value of 0.99) and provide similar information. Thereby, using one of the two bands will suffice. An innovative Lambda (λ_1 = 400–2500 nm) by Lambda (λ_2 = 400–2500 nm) plot of R^2 values is used to determine (Figure 1.4): (a) redundant bands and (b) unique bands. Figure 1.4 shows interband correlation of rainforest vegetation gathered using 242 Hyperion bands (in 400–2500 nm with each band having 10 nm bandwidth) plotted as lambda 1 (λ_1 = 400–2500 nm) versus lambda 2 (λ_2 = 400–2500 nm).

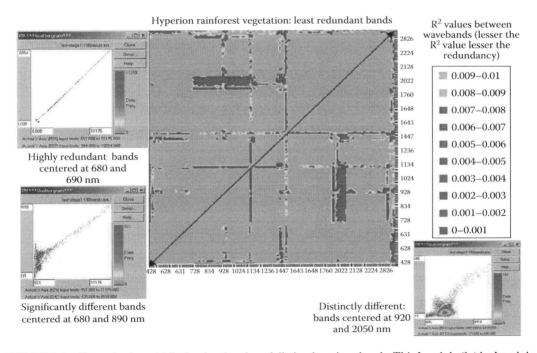

FIGURE 1.4   (See color insert.) Redundant bands and distinctly unique bands. This Lambda (λ_1) by Lambda (λ_2) plot of Hyperion bands show redundant bands (higher the correlation higher the redundancy) and distinctly unique bands (lower the correlation between the bands greater the uniqueness). For example, an R-square value of 1 indicates that the two wavebands are perfectly correlated and provide the same information. Thus, only one of the two bands should be used. An R-square value of 0 indicates that the two wavebands provide unique information, and thus it is relevant to use both wavebands.

In Figure 1.4, the least redundant bands (R^2 values of <0.004) are shown in magenta. For example, band 680 nm and band 690 nm were highly correlated, and hence, redundant (meaning we only need to use one of the 2 bands). However, 680 nm is significantly different than 890 nm, and 920 nm and 2050 nm are distinctly different bands. This means, it will be valuable to have bands 890, 920, 2050 nm, and either 680 or 690 nm. Therefore, it will suffice to select the least correlated bands (or most distinct bands) (e.g., magenta, green, and yellow in Figure 1.4) for hyperspectral analysis mining of all redundant bands.

1.4.2 PRINCIPAL COMPONENT ANALYSIS

Principal component analysis (PCA) plays two key roles in hyperspectral data analysis for vegetation and agricultural crops. It helps in

1. Selecting best wavebands to model biophysical and biochemical quantities
2. Eliminating redundant bands (by highlighting key wavebands)

Through this process, we are left with best bands, able to eliminate redundant bands, and help reduce data volumes.

PCA is a method in which original data is transformed into a new coordinate system, which acts to condense the information found in the original intercorrelated variables into a few uncorrelated variables, called principal components (PCs) [62]. Typically, first few PCs explain overwhelming proportion of variability (explained by eigenvalue) in data. Often, adjacent hyperspectral wavebands contain redundant information [11]. The original high-dimensional data is thus transformed to a few bands that contain most of the information in the original bands.

The importance of the hyperspectral wavebands in each PC are determined based on the magnitude of eigenvectors or factor loadings (the higher the eigenvector, the higher the importance of the band) for every vegetation and crop biophysical and biochemical variables. It is important to take these PCs and look at eigenvectors of the wavebands. Again, for each PC, the first few bands possess high eigenvectors. The importance of hyperspectral wavebands is established based on high factor loadings (or eigenvectors) associated with these wavebands [1,7,63]. The importance of the wavebands will vary depending on the biophysical and biochemical quantities. Thus, PCA helps in determining (a) wavebands that have the greatest influence in PCA1, PCA2, and so on (based on factor loadings or eigenvectors of each PC); and (b) percent variability explained by each PC (eigenvalues).

1.4.3 OTHER HYPERSPECTRAL DATA MINING ALGORITHMS

There are numerous other methods of hyperspectral data analysis to reduce the large number of wavebands to a manageable number of wavebands while still retaining optimal information. They are discussed in other chapters and will not be discussed here. But it is worthwhile mentioning the other key data reduction approaches that still retain optimal information for a particular application. These are

1. *Uniform Feature Design* [64]: The main goal of the uniform feature design (UFD) is to reduce the dimensionality of the data set while still retaining as much spectral shape information as possible in hyperspectral data. However, UMD averages band is not an optimal feature extraction method, and loses some of the physical meaning (e.g., spectral shape).
2. *Wavelet Transforms* [65–67]: The basic idea behind wavelet transform is to analyze data at different scales or resolutions. Wavelet analysis is specially suited for phenology studies,

such as in (a) identifying the start of growing season, duration of crop growth, and the time of harvest; and (b) removing the high frequency noise caused by the frequent cloud-cover in a highly automated way, especially in the tropics. Wavelet analysis is optimal in terms of detecting transient events in data and adapts well to conditions where responses change significantly in amplitude during experiments. However, some of the wavelet methods are based on the best approximation for data representation and hence a limitation in determining the actual condition.

3. *Artificial Neural Networks* [68–70]: The artificial neural network (ANN) is one of the first methods used in data mining and consists of a system of simple, interconnected neurons or nodes. It is a model representing a nonlinear mapping between an input vector and an output vector. The biggest difficulty in ANN is "training the nodes" based on substantial and accurate knowledge base. Since hyperspectral data consist of hundreds or thousands of bands, training large number of nodes can be painstaking.

All of these methods have advantages and disadvantages. In the study of the biophysical and biochemical properties of vegetation, we advocate the most useful band reduction approach that still provides optimal information by retaining key bands and removing redundant bands is the lambda (λ_1) by lambda (λ_2) plots (Section 1.4.1). The main advantage is in retaining the identity of original bands that can then be linked clearly to biophysical and biochemical attributes and the physical basis of sensitivity of these variables to wavebands clearly explained.

1.5 OPTIMAL HYPERSPECTRAL NARROWBANDS: HYPERSPECTRAL VEGETATION INDICES TO STUDY VEGETATION AND CROP BIOPHYSICAL AND BIOCHEMICAL PROPERTIES

Spectral VIs are computed by using broadband data as well as narrowband data. Even though VIs are computed using the same algebraic manipulations, their calculated values are different, thus affecting their stability in predicting agronomic variables, such as total green LAI [10]. Previous research has demonstrated their usefulness and potential for agricultural applications, such as estimating and forecasting crop yields, monitoring crop conditions, classifying and mapping crop types, and assisting precision-farming activities [27].

However, there are two significant limitations of broadband (e.g., Landsat, SPOT, IRS) derived VIs. First, the VIs saturate at high vegetation levels [71]. So, beyond a certain LAI or biomass or canopy cover, they are asymptotic [12,71]. Secondly, an overwhelming proportion of the broadband VIs is constructed with red and NIR spectral measurements or even otherwise offer only a few VIs. The broadband VIs are significantly correlated with crop agronomic variables, such as LAI, above ground biomass, and chlorophyll content. However, a large proportion of variability in modeling biophysical and biochemical quantities is not explained by broadband VIs.

The HVIs overcome both these limitations to a significant degree [1,4,7,11,12,17,50]. The HVIs have greater dynamic range. For example, an NDVI computed using a targeted narrowband (<5 nm range), such as 682 and 910 nm, has greater dynamic range of NDVI than a broadband NDVI, such as that from Landsat. This will help to better model the biophysical and biochemical properties of crops and to explain a significantly higher proportion of their variability. Further, a large number of HVIs offer greater opportunity in finding a right index for studying a vegetation variable (e.g., moisture sensitivity using an index involving 970 nm, stress index involving red-edge band around 720 nm, and nitrogen deficiency using a band that involves 1316 nm).

Given the aforementioned facts, four distinct types of HVIs are recommended to study any vegetation or agricultural crop biophysical and biochemical quantities. These indices are discussed in the following paragraphs.

1.5.1 HYPERSPECTRAL TWO BAND VEGETATION INDEX

The hyperspectral two band vegetation index (HTBVI) for narrowbands i and j will be [11,12]:

$$\text{Narrowband HTBVI}_{ij} = \frac{(R_j - R_i)}{(R_j + R_i)} \tag{1.1}$$

where i, j = 1, N, with N = number of narrowbands. For Hyperion 220 bands (N), for example, the total number of possible HTBVIs will be 48,400 (N × N). Since the indices above and below the diagonal in an N × N matrix "mirror" each other, it will suffice to consider indices that are either below or above the diagonal of a matrix. The number of unique indices will be 24,090 [(48,400–220)/2]. In comparison, the six nonthermal Landsat bands will have a meager 36 VIs (N × N), of which only 15 are unique. However, whereas such a large number of indices (24,090) provides many opportunities to study the biophysical and biochemical properties of vegetation and agricultural crop, the best information is contained in only a few selected bands or indices with the rest becoming redundant.

For example, 490 narrowbands each of 1.43 nm wide spread across 350–1050 nm were available for computing HTBVIs (Figure 1.5). These 490 bands were reduced to 49 bands (each band 10 nm wide) that resulted in 1176 unique HTBVIs. Each of these 1176 unique indices were correlated with wet biomass (WMB) of soybean and cotton crop. The resulting R^2 values are plotted in a contour plot (Figure 1.5) for soybean crop (above diagonal) and cotton (below diagonal). Such a contour plot of lambda 1 (λ_1 = 350–1050 nm) versus lambda 2 (λ_2 = 350–1050 nm) depicts "the bulls-eyes" of waveband centers and waveband widths that provide highest R^2-values (Figure 1.5). This will

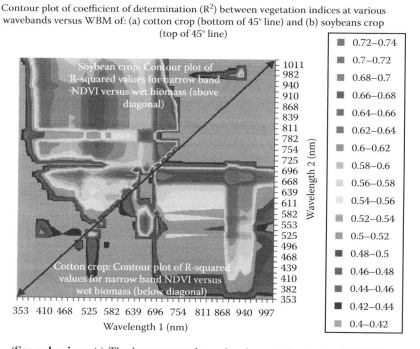

FIGURE 1.5 (See color insert.) The hyperspectral two-band vegetation index (HTBVI) versus the crop biophysical variable depicting areas of rich information content. The HTBVIs are correlated with crop wet biomass (WBM) and contour plots of R^2 values depicted for the soybean crop (above diagonal) and the cotton crop (below diagonal). The "bulls-eye" features help us determine waveband center and waveband width with highest R^2 values. These are the best bands to model the biophysical and biochemical quantities of crops. (From Thenkabail, P.S. et al., *Remote Sens. Environ.*, 71, 158, 2000. With permission.)

help us identify hyperspectral narrowband centers and widths that are best suited to model the biophysical and biochemical variables of crops.

1.5.2 Hyperspectral Multiple-Band Models

The hyperspectral multiple-band models (HMBMs) are computed as follows [11,12]:

$$HMBM_i = \sum_{J=1}^{N} a_{ij}R_j \qquad (1.2)$$

where
  HMBMs is the crop variable i
  R is the reflectance in bands j (j = 1 to N with N = 220 for Hyperion)
  a is the coefficient for reflectance in band j for ith variable

The process involves running stepwise linear regression models (e.g., using MAXR algorithm in Statistical Analysis System (SAS) with any one biophysical or biochemical variable as dependant variable and the numerous hyperspectral narrowbands as independent variables [72,73]. The MAXR method begins by finding a narrow waveband variable (R_j) that produces the highest coefficient of determination (R^2) [74]. Then, another narrowband variable, the one that yields the greatest increase in R^2 value, is added. Once the two narrowband model is obtained, each of the narrowband variables in the model are compared to each narrowband variable not in the model.

For each comparison, MAXR determines if removing one narrowband variable and replacing it with the other narrowband variable increases R^2. Comparisons begin again, and the process continues until MAXR finds that no replacement could increase R^2. The two narrowband model, thus achieved, is considered the best two narrowband model. Another variable is then added to the model, and the comparing-and-switching process is repeated to find the best three narrowband model, and so forth until the best n narrowband model is determined [73,74].

There is chance of over-fitting when using HMBMs. This is overcome when additional narrowbands are considered only when they significantly increase R^2 value from a previous model [12]. For example, if the R^2 value of a two narrowband model was 0.75, we consider a three narrowband model only if the R^2 value goes equal to or beyond 0.80 (an R^2 value increase of 0.05). In a study involving many biophysical quantities of several crops, Thenkabail et al. [1] demonstrated the following distinct advantages of the using HMBVIs. These were

1. The variability of the biophysical variables of several crop were explained at 95% or above using HMBMs.
2. HMBMs explained up to 27% greater variability in the biophysical variables of crop than similar broadband models.
3. HMBMs explained up to 11% greater variability in the biophysical variables of crop than HTBVI models.

1.5.3 Hyperspectral Derivative Greenness Vegetation Indices

When a difference index is formulated using two closely spaced band centers, it is indicative of the slope of the reflectivity with respect to wavelength and is often referred to as a hyperspectral derivative vegetation greenness index [HDGVIs; 75,76]. First derivatives (HDGVI1) are approximated by dividing the difference in reflectance value between spectrally adjacent bands by the corresponding difference in band central wavelength [77]. Second derivatives (HDGVI2) are determined likewise from d1 spectra [77]. The performance of two indices was shown to be practically identical

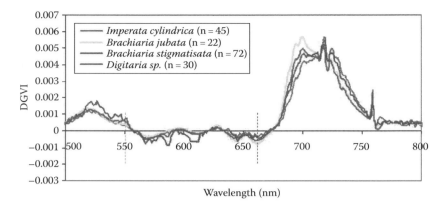

FIGURE 1.6 **(See color insert.)** First-order hyperspectral derivative greenness vegetation index (HDGVI1) computed along the 500–800 nm range for certain weed species.

[78,11,12], and, thus, it will suffice to calculate only one of these. HDGVI1 measure the amplitude of the chlorophyll and are computed by taking near-continuous spectra, such as 626–795 nm [75] or along the chlorophyll red-edge (0.700–0.740 μm). The chlorophyll red-edge portion is considered to have the maximum sensitivity to changes in green vegetation per unit change in wavelength in the electromagnetic spectrum [11]. The first-order (HDBVI1) are computed as follows:

$$\text{HDGVI} = \sum \frac{\lambda_n(\rho'(\lambda_i) - (\rho'(\lambda j))}{\lambda_1 \Delta \lambda_I} \tag{1.3}$$

where
  i and j are band numbers
  λ is the center of wavelength, lambda 1 (e.g., λ_1 = 626 nm) versus lambda 2 (e.g., λ_2 = 795 nm)
  ρ' is the first derivative reflectance

One can integrate HDGVIs using different waveband ranges (e.g., λ_1 = 700 nm, λ_2 = 740 nm or λ_1 = 940 nm, λ_2 = 980 nm). HDGVI1 computed for the range of 500–800 nm is illustrated for certain weed species in Figure 1.6.

Indications are that derivative indices are of particular importance in monitoring plant stress [11], complex vegetation conditions (mixture of green and brown; [76]), and grassland or weed canopies [75,78].

1.5.4 Hyperspectral Hybrid Vegetation Indices

1.5.4.1 Soil-Adjusted Hyperspectral Two Band Vegetation Indices

Soil-adjusted vegetation indices (SAVIs) were first proposed by Huete [79,80] and were developed to account for changes of the optical properties of the background in an attempt to align the VI isolines with the greenness isolines (usually expressed in terms of LAI) over the entire dynamic range of the greenness measure [76]. SAVI is defined as [79]

$$\text{SAVI} = \frac{\text{NIR} - \text{Red}}{\text{NIR} + \text{Red} + \text{L}}(1 + \text{L})$$

where
  NIR is the near infrared band reflectance
  Red is the red band reflectance
  L is the soil adjustment factor that varies according to the vegetation cover or LAI

L = 0 for a full canopy cover (here, SAVI= NDVI) and L = 1 for a zero canopy cover. However, an L = 0.5 is often taken as a default value.

For example, it is possible to compute SA HTBVIs for each of the 24,090 HTBVIs computed in Section 1.5.1. To reduce the computing time and optimize resources, SA HTBVIs can be computed only for the best HTBVIs (Figure 1.5). The value of computing SA HTBVIs is to enhance the sensitivity of HTBVIs by normalizing soil background effects, thus helping to explain greater variability in crop biophysical and biochemical variables, such as biomass, LAI, nitrogen, and chlorophyll a.

1.5.4.2 Atmospherically Resistant Hyperspectral Two Band Vegetation Indices

An atmospherically resistant vegetation index (ARVI) was first proposed by Kaufmann and Tanre [81] and to account for aerosol effects by using the difference in blue and red reflectance to derive the surface red reflectance. This index is especially useful to study leaf pigment. Galvão et al. [24] define ARVI for Hyperion data as

$$\frac{(\lambda_{864} - (2 * \lambda_{671} - \lambda_{467}))}{(\lambda_{864} + (2 * \lambda_{671} - \lambda_{467}))}$$

Another approach to minimize atmospheric effects on NDVI is to use the middle-infrared wavelength region (1300–2500 nm) as a substitute for the red band since longer wavelengths are much less sensitive to smoke and aerosols [82].

Again, it is possible to compute AR HTBVIs for each of the 24,090 HTBVIs computed in Section 1.5.1. To reduce the computing time and optimize resources, AR HTBVIs can be computed only for the best HTBVIs (Figure 1.5). The value of computing AR HTBVIs is to enhance the sensitivity of HTBVIs by normalizing atmospheric aerosol and moisture effects, thus helping to explain greater variability in crop biophysical and biochemical variables, such as biomass, LAI, nitrogen, and chlorophyll a.

1.5.4.3 Hyperspectral Vegetation Indices of SWIR and TIR Bands

Obtaining narrowband data beyond 1100 nm, from shortwave infrared (SWIR) (FNIR and SWIR; 1100–2500 nm) and thermal infrared (TIR; 3000–14000 nm), are important unique and/or complementing and/or supplementing information in addition to information on vegetation gained in visible and near infrared (VNIR) discussed in previous paragraphs. The nonphotosynthetic vegetation (NPV), such as litter, senesced leaves, and other dry vegetation are best differentiated based on ligno–cellulose bands in the SWIR [83,84] but are not spectrally separable from soil in the visible and near-infrared wavelength region [85]. Separating forest types and cropland classes, categorizing forest age classes or croplands at various stages of growth, modeling forest and cropland biotic factors, such as canopy height, basal area, biomass, and LAI are often best predicted through a combination of visible and SWIR [50,59]. Moisture and plant stress properties are best quantified by including TIR bands along with SWIR and VNIR [24,44].

1.6 OTHER METHODS OF HYPERSPECTRAL DATA ANALYSIS

There are numerous other methods of hyperspectral data analysis for studying the biophysical and biochemical properties of vegetation and agricultural crop. These include

1. *Independent Component Analysis* is an unsupervised temporal unmixing methodology that helps decompose spectra. For example, it can recover both the time profile and area distribution of different crop types [86].
2. *Wavelet Transform* can detect automatically the local energy variation of a spectrum in different spectral bands at each scale and provide useful information for hyperspectral image classification [66].

3. *Radiative Transfer Models* such as PROSPECT [87] and LIBERTY [88] are used to simulate wide range of hyperspectral signatures and their impact on vegetation and crop variables.

4. *Minimum Noise Fraction (MNF) Transformation* separates noise from the data by using only the coherent portions, thus improving spectral processing results.

5. *Spectral Unmixing Analysis* [SMA; 89] uses reference spectra (referred to as "end members") in order to "unmix" characteristics of spectra within each pixel. The biotic and abiotic characteristics of the constituents within a pixel will be gathered through ground truth data. The spectra of the individual constituents within the pixel can be obtained using spectroradiometer and/or from spectral libraries. SMA classifications are known to map land cover more accurately than maximum likelihood classification [89] and have the advantage of assessing the within pixel composition. Once all the materials in the image are identified, then it is possible to use linear spectral unmixing to find out how much of each material is in each pixel [89].

1.7 BROADBAND VEGETATION INDEX MODELS

It was determined by Thenkabail et al. [12] that the VIs computed for a wide range of existing broadband sensors, such as Landsat MSS, Landsat TM, SPOT HRV, NOAA AVHRR, and IRS-1C, using simulated data for crops from a spectroradiometer, were highly correlated (R^2 value = 0.95 or higher). Hence, computing VIs for any one of these sensors will provide nearly the same information as similar indices computed for other sensors. The most common categories of broadband indices are

1. NIR and Red-based indices [90]
2. Soil-adjusted indices [79,80]
3. Atmospheric-resistant indices [91,92]
4. Mid infrared-based indices [38,39]

1.8 SEPARATING VEGETATION CLASSES AND AGRICULTURAL CROPS USING HYPERSPECTRAL NARROWBAND DATA

1.8.1 CLASS SEPARABILITY USING UNIQUE HYPERSPECTRAL NARROWBANDS

There are fewer opportunities in separating vegetation classes using broadbands from sensors, such as Landsat TM (e.g., Figure 1.7a) when compared to numerous opportunities offered by narrowbands from sensors, such as EO-1 Hyperion (e.g., Figure 1.7b). For example, the Landsat TM

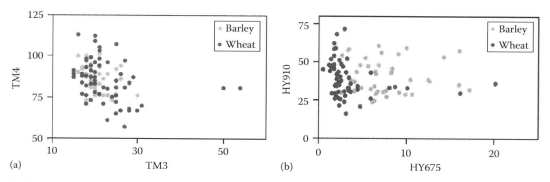

FIGURE 1.7 (See color insert.) Separating two agricultural crops using broadbands versus narrowbands. The two broadbands fail to separate wheat from barley (Figure 1.7a) whereas two distinct narrowbands separate wheat from barley (Figure 1.7b).

broadbands of near infrared (TM4) versus red (TM3) could not separate wheat crop from barley crop (Figure 1.7a). However, with two narrowbands involving near infrared (centered at 910 nm with narrowband-width of 10 nm) versus red (centered at 675 nm with narrowband-width of 10 nm), an overwhelming proportion of the wheat crop fields were separated from barley crop fields (Figure 1.7b). Since hyperspectral sensors have hundreds of wavebands, the likelihood of finding wavebands that can separate vegetation/crop types (e.g., Figure 1.7b) or various groups of vegetation/crop biophysical and biochemical quantities increases drastically.

1.8.2 Class Separability Using Statistical Methods

Stepwise discriminant analysis (SDA) is a powerful statistical tool for separating vegetation categories and agricultural crop types based on their quantitative and qualitative characteristics, such as biophysical quantities, biochemical compositions, structural properties, and species type. The independent hyperspectral waveband variables are chosen to enter or leave the model using (a) significance level of F-test analysis of covariance, where the variables already chosen act as covariates and the variable under consideration is the dependent variable, or (b) the squared partial correlation for predicting the variable under consideration from the CLASS variable, controlling the effects of the variables selected for the model [74]. Stepwise selection begins with no variable in the model. At each step, if a variable already in the model fails to meet the criterion to stay, the worst such variable is removed. Otherwise, the variable that contributes most to the discriminatory power of the model is entered. Finally, when all the variables that meet the criterion stay and all the variables that do not meet the criterion are eliminated, the stepwise selection process stops [74]. The stepwise discriminate analysis ([91,92], Chapter 6) will be performed using PROC STEPDISC algorithm of SAS [74] through a stepwise selection [93].

Class separability in SDA can be expressed, most powerfully and lucidly using (a) Wilks' lambda and (b) Pillai trace. Wilks' lambda and Pillai trace are based on the eigenvalues Γ of $A*W^{-1}$ where A is the among SS and cross-products matrix, and W the pooled SS and cross-products matrix:

$$\text{Wilks' } \Lambda = \frac{\Pi 1}{(1 + \lambda_i)}$$

$$\text{Pillai's trace} = \sum \frac{\lambda_i}{(1 + \lambda_i)}$$

Determinants (variance) of the S matrices are found. Wilks' Lambda is the test statistic preferred for multivariate analysis of variance (MANOVA), and is found through a ratio of the determinants.

$$\Lambda = \frac{|S_{error}|}{|S_{effect} + S_{error}|}$$

where S is a matrix which is also known as "sum-of-squares (SS) and cross-products," "cross-products," or "sum-of-products" matrices.

Wilks' Lambda is the most commonly available and reported; however, Pillai's criterion is more robust, and therefore, more appropriate when there are small or unequal sample sizes. When separating two classes

1. The lower the value of Wilks' lambda, the greater the separability between two classes
2. The higher the value of Pillai's trace, the greater the separability between two classes.

Class separability using hyperspectral data is illustrated for few vegetation types in Figure 1.8. In each case, with increase in number of wavebands, the separability also increases (Figure 1.8; note: the lower

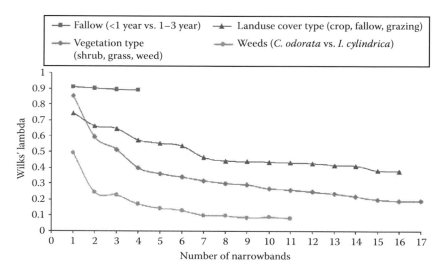

FIGURE 1.8   (See color insert.) Class separability using hyperspectral narrowbands determined based on Wilks' lambda. Lower the Wilks' lambda, the greater is the separability. So, with the addition of wavebands, the separability increases, reaching an optimal point beyond which the addition of wavebands does not make a difference. (From Thenkabail, P.S. et al., *Remote Sens. Environ.*, 91, 354, 2004. With permission.)

the Wilks' lambda, the greater the separability). For example, vegetation types (shrubs vs. grasses vs. weeds) are best separated using around 17 hyperspectral narrowbands (Figure 1.8), whereas the two weeds (*Chromolenea odorata* vs. *Imperata cylindrica*) are separated best using 11 hyperspectral narrowbands. It is possible to have small incremental increase in spectral separability beyond 11 bands for weeds and beyond 17 bands for vegetation types, but those increases are very small and statistically insignificant. Figure 1.8 also shows that the fallows (1 year fallow vs. 1–3 year fallow) are not well separated by even hyperspectral narrowbands since Wilks' lambda remains high.

In each case, it is possible to determine the exact wavebands that help increase separability (see [1,7,11,12]).

1.8.3   Accuracy Assessments of Vegetation and Crop Classification Using Hyperspectral Narrowbands

Classification accuracies of vegetation and agricultural crops attained using hyperspectral narrowband data are substantially higher than broadbands [1,47,49,94]. For example, above 90% classification accuracies have been obtained in classifying five agricultural crops using about 20 selected hyperspectral narrowband data, relative to just about 60% accuracies obtained using six nonthermal broadbands (e.g., Figure 1.9). Similarly, Dalponte et al. [5] studied a forest area in Italy characterized by 23 different classes reaching accuracies of about 90% with hyperspectral data acquired at a spectral resolution of 4.6 nm in 126 bands. Clark et al. [95] studied seven deciduous tree species with the HYDICE sensor, using three different classifiers, reaching accuracies to the order of 90%.

Also, sensitivity to aboveground biomass and ability to map forest successional stages are achieved using hyperspectral narrowbands [43,44]. Earlier, we have established that selected hyperspectral narrowbands can play a crucial role in separating two vegetation categories (e.g., Figure 1.7b) when two broadbands fail to do so (e.g., Figure 1.7a). Also, the hyperspectral narrowbands explain about 10%–30% greater variability in modeling quantitative biophysical quantities (e.g., LAI, biomass) relative to broadband data [4,11]. The specific narrowbands, relative to broadbands, that

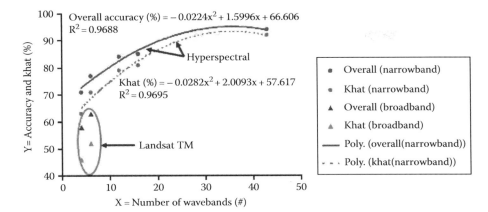

FIGURE 1.9   (See color insert.) Classification accuracies using hyperspectral narrowbands versus Landsat broadbands. The accuracies increased by 25%–30% when about 25 hyperspectral narrowbands were used to classify five agricultural crops when compared to six broad-Landsat bands. Classification accuracies reach about 95% with 30 bands, beyond which accuracies do not increase. (From Thenkabail, P.S. et al., *Remote Sens. Environ.*, 91, 354, 2004. With permission.)

improve crop classification accuracies, explain greater variability in crop biophysical models, and help better separate crop types are identified in Table 1.2.

1.9   OPTIMAL HYPERSPECTRAL NARROWBANDS IN STUDY OF VEGETATION AND AGRICULTURAL CROPS

Optimal hyperspectral narrowbands in the study of vegetation and agricultural croplands (Table 1.2) are determined based on an exhaustive review of literature that includes (a) identifying redundant bands (e.g., Figure 1.4); (b) modeling by linking crop biophysical and biochemical variables with hyperspectral indices and wavebands (e.g., Figure 1.5); (c) establishing wavebands that best help separate vegetation and crop types (e.g., Figure 1.7); (d) identifying wavebands, through statistical and other approaches, that best separate vegetation and crop types and characteristics (e.g., Figure 1.8); and (e) establishing classification accuracies of vegetation and crop classes and identifying wavebands that best help enhance these accuracies (e.g., Figure 1.9). The 28 wavebands listed in Table 1.2 are based on their frequency of occurrences in numerous studies discussed in this paper. These prominent wavebands include (i) two red bands: centered at 675 nm (chlorophyll absorption maxima) and 682 nm (most sensitive to biophysical quantities and yield); (ii) three NIR bands: 845 nm (mid-point of NIR "shoulder" that is most sensitive to biophysical quantities and yield), 915 nm (peak spectrum in NIR, useful for computing crop moisture sensitivity index with 975 nm), and 975 nm (moisture and biomass sensitive band). The NIR portion of the spectrum is highly sensitive to changes in biophysical quantities and plant structure.

For example, the planophile (30°) structure plant leaf contributes significantly to greater reflectance in NIR and greater absorption in red when compared with erectophile (65°) structure [11]. The erectophile structure leads to significant slope changes in spectra in the region of 740–940 nm; (iii) five green bands: 515, 520, 525, 550, and 575 nm. These bands are overwhelmingly sensitive to biochemical properties. $\lambda = 550$ nm is strongly correlated with total chlorophyll. The green band λ centered at 520 nm provides the most rapid positive change in reflectance per unit change in wavelength anywhere in the visible portion of the spectrum. The green band λ centered at 575 nm provides the most rapid negative change in reflectance per unit change in wavelength anywhere in the visible portion of the spectrum. Overall, the green bands are very sensitive to plant/leaf nitrogen

TABLE 1.2

Optimal Hyperspectral Narrowbands Recommended in the Study of Vegetation and Agricultural Crops

Waveband Number	Waveband Center (nm)	Importance in Vegetation and Agricultural Cropland Studies	References
A. Blue Bands			
1	375	*fPAR, leaf water:* fraction of photosynthetically active radiation (fPAR), leaf water content	[1,7,11,12]
2	466	*Chlorophyll:* chlorophyll a and b in vegetation	[1,7,11,12]
3	490	*Senescing and loss of chlorophyll/browning, ripening, crop yield:* sensitive to loss of chlorophyll, browning, ripening, senescing, and soil background effects	[1,7]
B. Green Bands			
4	515	*Nitrogen:* leaf nitrogen, wetland vegetation studies	[1,7]
5	520	*Pigment, biomass changes:* positive change in reflectance per unit change in wavelength of this visible spectrum is maximum around this green waveband. Sensitive to pigment	[1,29]
6	525	*Vegetation vigor, pigment, nitrogen:* positive change in reflectance per unit change in wavelength in maximum as a result of vegetation vigor, pigment, and nitrogen	
7	550	*Chlorophyll and biomass:* total chlorophyll; Chlorophyll/carotenoid ratio, vegetation nutritional and fertility level; vegetation discrimination; vegetation classification. Biophysical (e.g., biomass) quantity modeling	[4,11,27]
8	575	*Vegetation vigor, pigment, and nitrogen:* negative change in reflectance per unit change in wavelength is maximum as a result of sensitivity to vegetation vigor, pigment, and N	
C. Red Bands			
9	675	*Chlorophyll absorption maxima:* greatest crop-soil contrast is around this band for most crops in most growing conditions. Strong correlations with chlorophyll a and b	[1,4,7]
10	682	*Biophysical quantities and yield and chlorophyll absorption:* leaf area index, wet and dry biomass, plant height, grain yield, crop type, crop discrimination. Strong correlations with chlorophyll a and b	[1,7]
D. Red-Edge Bands			
11	700	*Stress and chlorophyll:* nitrogen stress, crop stress, crop growth stage studies	[1,4,7]
12	720	*Stress and chlorophyll:* nitrogen stress, crop stress, crop growth stage studies. Red shift for healthy vegetation, blue shift for stressed vegetation	[11,21,37]
13	740	*Nitrogen accumulation:* leaf nitrogen accumulation. Red shift for healthy vegetation, blue shift for stressed vegetation	[1,4,7]

	Band	Wavelength	Description	Reference
E. Near-infrared (NIR) Bands				
	14	845	*Biophysical quantities and yield*: leaf area index, wet and dry biomass, plant height, grain yield, crop type, and crop discrimination. Often used in an index with 682 nm	[21]
	15	915	*Biophysical quantities and yield*: peak NIR reflectance. Useful for computing crop biophysical quantities along with a band such as 682 nm	[11,37]
	16	975	*Moisture and biomass*: center of moisture sensitive "trough"; water band index, leaf water, biomass	[20]
F. Far NIR (FNIR) Bands				
	17	1100	*Biophysical quantities*: sensitive to biomass and leaf area index. A point of the most rapid rise in spectra with unit change in wavelength in far near infrared (FNIR)	[40]
	18	1215	*Moisture and biomass*: a point of the most rapid fall in spectra with unit change in wavelength in FNIR. Sensitive to plant moisture	[20]
	19	1245	*Water sensitivity*: water band index, leaf water, biomass. Reflectance peak in 1050–1300 nm	[11,75]
G. Short-Wave Infrared (SWIR) Bands				
	20	1316	*Nitrogen*: leaf nitrogen content of crops	[11]
	21	1445	*Vegetation classification and discrimination*: ecotype classification; plant moisture sensitivity. Moisture absorption trough in early short wave infrared (ESWIR)	[10,1,7]
	22	1518	*Moisture and biomass*: a point of the most rapid rise in spectra with unit change in wavelength in SWIR. Sensitive to plant moisture	[11,20,21]
	23	1725	*Lignin, biomass, starch, moisture*: sensitive to lignin, biomass, starch. Discriminating crops and vegetation.	[7]
	24	2035	*Moisture and biomass*: moisture absorption trough in far short-wave infrared (FSWIR)	[1,4,7]
	25	2173	*Protein, nitrogen*	[11,75]
	26	2260	*Moisture and biomass*: moisture absorption trough in far short-wave infrared (FSWIR). A point of most rapid change in slope of spectra based on vegetation vigor, biomass	[4,12,20]
	27	2295	*Stress*: sensitive to soil background and plant stress	[1,7,37]
	28	2359	*Cellulose, protein, nitrogen*: sensitive to crop stress, lignin, and starch	

and pigment. There are dramatic shifts in crop-soil spectral behavior, around 568 and 520 nm, when there is change in pigment content and chloroplast for different crop types, growth stages, and growing conditions; (iv) three red-edge bands: 700 and 720 nm (sensitive to vegetation stress), and 740 nm (sensitive to vegetation nitrogen content). The red-edge bands are especially sensitive to crop stress and changes in total chlorophyll and have potential to form a useful drought index. They are sensitive to senescing rates, chlorophyll changes, browning, ripening, carotenoid, and soil background effects; (v) three FNIR bands: 1100 nm (sensitive to biophysical quantities), 1215 nm (sensitive to moisture), and 1245 nm (sensitive to water); (vi) nine SWIR bands (1316, 1445, 1518, 1725, 2035, 2173, 2260, 2295, and 2359 nm). These wavebands are overwhelmingly sensitive to moisture and biochemical properties; and (vii) three blue bands (375, 466, and 490 nm) that are specially sensitive to fPAR, and senescing. $\lambda = 490$ nm is especially sensitive to carotenoid, leaf chlorophyll, and senescing conditions.

A nominal bandwidth ($\Delta\lambda$) of 5–10 nm can be used for all wavebands. Too narrow waveband widths seem to lead to lower signal-to-noise ratio. Keeping a band-width ($\Delta\lambda$) of 5–10 nm will ensure that optimal information on a particular feature is captured rather than average conditions captured in broadbands. Bandwidth as narrow as 3 nm are proven to be more optimal, but often have significant noise issues.

Overall, there is overwhelming evidence that the hyperspectral wavebands and indices involving specific narrow wavebands provide significantly better estimates on vegetation and crop biophysical quantities (e.g., LAI, biomass, plant height, canopy cover), biogeochemistry (e.g., lignin, chlorophyll a, chlorophyll b, photosynthesis, respiration, nutrient use (e.g., through LAI, and fAPAR), canopy water content, and improved discrimination of species and land cover.

1.10 CONCLUSIONS

An overwhelming proportion of the hyperspectral narrowband (imaging spectroscopy) data in vegetation and agricultural cropland studies can be redundant. So, it is extremely important to identify and remove the redundant bands from further analysis to ensure the most effective and efficient use of hyperspectral data in vegetation, agricultural cropland, and other resource applications by, for example, overcoming *the curse* of high dimensionality or the Hughes phenomenon (*needing a large volume of training samples for classification, yet with no guarantee to higher classification accuracies as a result of a large number of redundant bands*). Through an series of studies reported in this chapter, we have identified 28 optimal hyperspectral narrowbands (Table 1.2) best suited in the study of agricultural crops. The wavebands were identified based on their ability to (a) best model biophysical and biochemical properties; (b) distinctly separate vegetation and crops based on their species type, structure, and composition; and (c) accurately classify crop types, crop dominance, and crop species. These optimal narrowbands (Table 1.2) are best applicable to agricultural croplands. We refer the reader to Table 28.1 of Chapter 28 where we recommend optimal hyperspectral narrowbands in study of terrestrial vegetation including croplands. Table 28.1 was composed based on broader consensus from hundreds of researchers studying varied vegetation categories. We want to emphasize that the optimal hyperspectral narrowbands recommended in Table 1.2 of this chapter should be applied only for agricultural cropland studies. Ideally, narrowband widths should be 3 nm but no more than 5–10 nm wide to achieve the best results in quantifying, modeling, and mapping agricultural crops.

We highlighted computation of unique HVIs and have also identified the most valuable wavebands for studying agricultural crops (Table 1.2). The specific importance of each of the wavebands to the biophysical and biochemical properties of vegetation and agricultural crops are also highlighted in Table 1.2. We recognize that there is more than one waveband of importance for each crop variable. For example, biophysical quantities (e.g., LAI, Biomass) and grain yield are best modeled using narrow wavebands centered at 682 nm and 910 nm (alternatively 845 nm or 1100 nm). Similarly, we identify that moisture sensitivity in leaf/plant is studied by several wavebands centered at 975,

1215, 1245, 1518, 2035, and 2260 nm. Specific narrowband centers to study other crop quantities and conditions such as pigments, chlorophyll, nitrogen, stress, biomass, yield, starch, and cellulose are listed in Table 1.2. The 28 wavebands (Table 1.2) will also allow us to compute 378 unique hyperspectral two-band vegetation indices (HTBVIs). Some of the biophysical and biochemical indices that can be computed from the narrow wavebands listed in Table 1.2 are defined in various chapters of this book (e.g., Tables 8.7, 10.1, 14.1, 17.2, 19.2, 23.2, and 23.3). Further, combination of few of these narrowbands, often best classify crop types, crop dominance, or crop species. Accuracies of such classifications are substantially higher than same classification accuracies obtained using broadband data. Thereby, there is clear evidence that the 28 identified narrow wavebands (Table 1.2) and/or various HVIs computed from them will best characterize, classify, model, and map a wide array of agricultural crop types and their biophysical and biochemical properties.

We explain the need to take caution in eliminating redundant bands, since some of the redundant bands in one application maybe useful in some other. Taking this into consideration, methods and approaches of data mining to quickly identify key wavebands important for a given application (and as a result help eliminate redundant bands for that particular application) are highlighted and discussed.

ACKNOWLEDGMENTS

The authors would like to thank Zhouting Wu of Northern Arizona University (NAU) for preparing Figure 1.3. Comments from reviewers are much appreciated. Editing by Dr. Susan Benjamin of the U.S. Geological Survey is acknowledged with gratitude.

REFERENCES

1. Thenkabail, P.S., Enclona, E.A., Ashton, M.S., and Van Der Meer, V. 2004. Accuracy assessments of hyperspectral waveband performance for vegetation analysis applications. *Remote Sensing of Environment*. 91(2–3): 354–376.
2. Blackburn, A.G. and Ferwerda, J.G. 2008. Retrieval of chlorophyll concentration from leaf reflectance spectra using wavelet analysis. *Remote Sensing of Environment*. 112(4): 1614–1632.
3. Chen, J., Wang, R., and Wang, C. 2008. A multiresolution spectral angle-based hyperspectral classification method. *International Journal of Remote Sensing*. 29(11): 3159–3169.
4. Chan, J.C. and Paelinckx, D. 2008. Evaluation of Random Forest and Adaboost tree-based ensemble classification and spectral band selection for ecotope mapping using airborne hyperspectral imagery. *Remote Sensing of Environment*. 112(6): 2999–3011.
5. Dalponte, M., Bruzzone, L., Vescovo, L., and Gianelle, D. 2009. The role of spectral resolution and classifier complexity in the analysis of hyperspectral images of forest areas. *Remote Sensing of Environment*. 113(11): 2345–2355.
6. Houborg, R. and Boegh, E. 2008. Mapping leaf chlorophyll and leaf area index using inverse and forward canopy reflectance modelling and SPOT reflectance data. *Remote Sensing of Environment*. 112: 186–202.
7. Thenkabail, P.S., Enclona, E.A., Ashton, M.S., Legg, C., and Jean De Dieu, M. 2004. Hyperion, IKONOS, ALI, and ETM+ sensors in the study of African rainforests. *Remote Sensing of Environment*. 90: 23–43.
8. Colombo, R., Meroni, M., Marchesi, A., Busetto, L., Rossini, M., Giardino, C., and Panigada, C. 2008. Estimation of leaf and canopy water content in poplar plantations by means of hyperspectral indices and inverse modeling. *Remote Sensing of Environment*. 112(4): 1820–1834.
9. le Maire, G., François, C., Soudani, K., Berveiller, D., Pontailler, J.Y., Bréda, N., Genet, H., Davi, H., and Dufrêne, E. 2008. Calibration and validation of hyperspectral indices for the estimation of broadleaved forest leaf chlorophyll content, leaf mass per area, leaf area index and leaf canopy biomass. *Remote Sensing of Environment*. 112: 3846–3864.
10. Zhao, D., Huang, L., Li, J., and Qi, J. 2007. A comparative analysis of broadband and narrowband derived vegetation indices in predicting LAI and CCD of a cotton canopy. *ISPRS Journal of Photogrammetry and Remote Sensing*. 62(1): 25–33.
11. Thenkabail P.S., Smith, R.B., and De-Pauw, E. 2002. Evaluation of narrowband and broadband vegetation indices for determining optimal hyperspectral wavebands for agricultural crop characterization. *Photogrammetric Engineering and Remote Sensing*. 68(6): 607–621.

12. Thenkabail P.S., Smith, R.B., and De-Pauw, E. 2000. Hyperspectral vegetation indices for determining agricultural crop characteristics. *Remote Sensing of Environment.* 71: 158–182.

13. Lyon, J.G., Yuan, D., Lunetta, R.S., and Elvidge, C.D. 1998. A change detection experiment using vegetation indices. *Photogrammetric Engineering and Remote Sensing.* 64: 143–150.

14. Jollineau, M.Y. and Howarth, P.J. 2008. Mapping an inland wetland complex using hyperspectral imagery. *International Journal of Remote Sensing.* 29(12): 3609–3671.

15. Hill, M.J. 2004. Grazing agriculture: Managed pasture, grassland, and rangeland. In: Ustin, S.L. (ed.), *Manual of Remote Sensing. Volume 4 Remote Sensing for Natural Resource Management and Environmental Monitoring*, John Wiley & Sons, Hoboken, NJ, pp. 449–530.

16. Asner, G.P. 1998. Biophysical and biochemical sources of variability in canopy reflectance. *Remote Sensing of Environment.* 64: 234–253.

17. Darvishzadeh, R., Skidmore, A., Schlerf, M., Atzberger, C., Corsi, F., and Cho, M. 2008. LAI and chlorophyll estimation for a heterogeneous grassland using hyperspectral measurements. *ISPRS Journal of Photogrammetry and Remote Sensing.* 63: 409–426.

18. Darvishzadeh, R., Skidmore, A., Schlerf, M., and Atzberger, C. 2008. Inversion of a radiative transfer model for estimating vegetation LAI and chlorophyll in a heterogeneous grassland. *Remote Sensing of Environment.* 112: 2592–2604.

19. Darvishzadeh, R., Skidmore, A., Atzberger, C., and van Wieren, S. 2008. Estimation of vegetation LAI from hyperspectral reflectance data: Effects of soil type and plant architecture. *International Journal of Applied Earth Observation and Geoinformation.* 10: 358–373.

20. Yao, X., Zhu, Y., Tian, Y.C., Feng, W., and Cao, W.X. 2010. Exploring hyperspectral bands and estimation indices for leaf nitrogen accumulation in wheat. *International Journal of Applied Earth Observation and Geoinformation.* 12(2): 89–100.

21. Maire, G.L., François, C., Soudani, K., Berveiller, D., Pontailler, J.V., Bréda, N., Genet, H., Davi, H., and Dufrêne, E. 2008. Calibration and validation of hyperspectral indices for the estimation of broadleaved forest leaf chlorophyll content, leaf mass per area, leaf area index and leaf canopy biomass. *Remote Sensing of Environment.* 112(10): 3846–3864.

22. Vaiphasa, C.K., Skidmore, K.A., de Boer, W.F., and Vaiphasa, T. 2007. A hyperspectral band selector for plant species discrimination. *ISPRS Journal of Photogrammetry and Remote Sensing.* 62(3): 225–235.

23. Thenkabail, P.S. 2003. Biophysical and yield information for precision farming from near-real time and historical Landsat TM images. *International Journal of Remote Sensing.* 24(14): 2879–2904.

24. Galvão, L.S., Roberts, D.A., Formaggio, A.R., Numata, I., and Breunig, F.M. 2009. View angle effects on the discrimination of soybean varieties and on the relationships between vegetation indices and yield using off-nadir Hyperion data. *Remote Sensing of Environment.* 113(4): 846–856.

25. Pearlman, J.S., Barry, P.S., Segal, C.C., Shepanski, J., Beiso, D., and Carman, S.L. 2003. Hyperion, a space-based imaging spectrometer. *IEEE Transactions on Geoscience and Remote Sensing.* 41: 1160–1173.

26. Ungar, S.G., Pearlman, J.S., Mendenhall, J.A., and Reuter, D. 2003. Overview of the Earth Observing One (EO-1) mission. *IEEE Transactions on Geoscience and Remote Sensing.* 41: 1149–1159.

27. Yang, F., Li, J., Gan, X., Qian, Y., Wu, X., and Yang, Q. 2009. Assessing nutritional status of *Festuca arundinacea* by monitoring photosynthetic pigments from hyperspectral data. *Computers Electronics in Agriculture.* 70(1): 52–59.

28. Nolin, A.W. and Dozier, J. 2000. A hyperspectral method for remotely sensing the grain size of snow. *Remote Sensing of Environment.* 74(2): 207–216.

29. Blackburn, G.A. 1998. Quantifying chlorophylls and carotenoids at leaf and canopy scales: An evaluation of some hyperspectral approaches. *Remote Sensing of Environment.* 66: 273–285.

30. Thenkabail, P.S., Hanjra, M.A., Dheeravath, V., and Gumma, M. 2010. Global croplands and their water use remote sensing and non-remote sensing perspectives. Book Chapter. Chapter 16. In: Weng, Q. (ed.), *Advances in Environmental Remote Sensing: Sensors, Algorithms, and Applications*, Taylor & Francis, Boca Raton, FL.

31. Melesse, A.M., Weng, Q., Thenkabail, P., and Senay, G. 2007. Remote sensing sensors and applications in environmental resources mapping and modelling. Special Issue of Remote Sensing of Natural Resources and the Environment. *Sensors Journal.* 7: 3209–3241. http://www.mdpi.org/sensors/papers/s7123209.pdf

32. Wang F.M., Huang J.F., and Wang X.Z. 2008. Identification of optimal hyperspectral bands for estimation of rice biophysical parameters. *Journal of Integrative Plant Biology.* 50(3): 291–299.

33. Delegido, J., Alonso, L., González, G., and Moreno, J. 2010. Estimating chlorophyll content of crops from hyperspectral data using a normalized area over reflectance curve (NAOC). *International Journal of Applied Earth Observation and Geoinformation.* 12(3): 165–174.

34. Zhu, Y., Zhou, D., Yao, X., Tian, Y., and Cao, W. 2007. Quantitative relationships of leaf nitrogen status to canopy spectral reflectance in rice. *Australian Journal of Agricultural Research.* 58(11): 1077–1085. doi:10.1071/AR06413

35. Haboudane, D., Miller, J.R., Pattey, E., Zarco-Tejada, P.J., and Strachan, I.B. 2004. Hyperspectral vegetation indices and novel algorithms for predicting green LAI of crop canopies: Modeling and validation in the context of precision agriculture. *Remote Sensing of Environment.* 90: 337–352.

36. Rao, N.R., Garg, P.K., and Ghosh, S.K. 2007. Estimation of plant chlorophyll and nitrogen concentration of agricultural crops using EO-1 Hyperion hyperspectral imagery. *Journal of Agricultural Science.* 146: 1–11.

37. Penuelas, J., Filella, I., Lloret, P., Munoz, F., and Vilajeliu, M. 1995. Reflectance assessment of mite effects on apple trees. *International Journal of Remote Sensing.* 16: 2727–2733.

38. Thenkabail, S.P., Ward, A.D., and Lyon, J.G. 1994. LANDSAT-5 Thematic Mapper models of soybean and corn crop characteristics. *International Journal of Remote Sensing.* 15(1): 49–61.

39. Thenkabail, S.P., Ward, A.D., Lyon, J.G., and Merry, C.J. 1994. Thematic Mapper vegetation indices for determining soybean and corn crop growth parameters. *Photogrammetric Engineering and Remote Sensing.* 60(4): 437–442.

40. Abdel-Rahman, E.M., Ahmed, F.B., and Van den Berg, M. 2010. Estimation of sugarcane leaf nitrogen concentration using in situ spectroscopy. *International Journal of Applied Earth Observation and Geoinformation.* 12(1): S52–S57.

41. Stagakis, S., Markos, N., Sykioti, O., and Kyparissis, A. 2010. Monitoring canopy biophysical and biochemical parameters in ecosystem scale using satellite hyperspectral imagery: An application on a *Phlomis fruticosa* Mediterranean ecosystem using multiangular CHRIS/PROBA observations. *Remote Sensing of Environment.* 114(5): 977–994.

42. Asner, G.P. and Martin, R.E. 2009. Airborne spectranomics: Mapping canopy chemical and taxonomic diversity in tropical forests. *Frontiers in Ecology and the Environment.* 7: 269–276.

43. Asner, G.P., Martin, R.E., Ford, A.J., Metcalfe, D.J., and Liddell, M.J. 2009. Leaf chemical and spectral diversity in Australian tropical forests. *Ecological Applications.* 19: 236–253.

44. Kalacska, M., Sanchez-Azofeifa, G.A., Rivard, B., Caeilli, T., White, H.P., and Calvo-Alvarado, J.C. 2007. Ecological fingerprinting of ecosystem succession: Estimating secondary tropical dry forest structure and diversity using imaging spectroscopy. *Remote Sensing of Environment.* 108: 82–96.

45. Schlerf, M., Atzberger, C., and Hill, J. 2005. Remote sensing of forest biophysical variables using HyMap imaging spectrometer data. *Remote Sensing of Environment.* 95(2): 177–194.

46. Oldeland, J., Wesuls, D., Rocchini, D., Schmidt, M., and Jürgens, N. 2010. Does using species abundance data improve estimates of species diversity from remotely sensed spectral heterogeneity? *Ecological Indicators.* 10: 390–396.

47. Papes, M., Tupayachi, R., Martínez, P., Peterson, A.T., and Powell, G.V.N. 2010. Using hyperspectral satellite imagery for regional inventories: A test with tropical emergent trees in the Amazon Basin. *Journal of Vegetation Science.* 21(2): 342–354.

48. Anderson, J.E., Plourde, L.C., Martin, M.E., Braswell, B.H., Smith, M.-L., Dubayah, R.O. et al., 2008. Integrating waveform LiDAR with hyperspectral imagery for inventory of a northern temperate forest. *Remote Sensing of Environment.* 112: 1856–1870.

49. Schull, M.A., Knyazikhin, Y., Xu, L., Samanta, A., Carmona, P.L., Lepine, L., Jenkins, J.P., Ganguly, S., and Myneni, R.B. 2010. Canopy spectral invariants, Part 2: Application to classification of forest types from hyperspectral data. *Journal of Quantitative Spectroscopy and Radiative Transfer.* doi:10.1016/j.jqsrt.2010.06.004

50. Cho, M.A. 2007. Hyperspectral remote sensing of biochemical and biophysical parameters: The derivative red-edge "double peak feature," a nuisance or an opportunity? PhD dissertation. International Institute for Geo-information science and earth observation (ITC), Enschede, the Netherlands, pp. 206.

51. Zarco-Tejada, P.J., Berjón, A., López-Lozano, R., Miller, J.R., Martín, P., Cachorro, V., González, M.R., and de Frutos, A. 2005. Assessing vineyard condition with hyperspectral indices: Leaf and canopy reflectance simulation in a row-structured discontinuous canopy. *Remote Sensing of Environment.* 99: 271–287.

52. Haboudane, D., Miller, J.R., Tremblay, N., Zarco-Tejada, P.J., and Dextraze, L. 2002. Integrated narrow-band vegetation indices for prediction of crop chlorophyll content for application to precision agriculture. *Remote Sensing of Environment.* 81: 416–426.

53. Ustin, S.L., Roberts, D.A., Gamon, J.A., Asner, G.P., and Green, R.O. 2004. Using imaging spectroscopy to study ecosystem processes and properties. *BioScience.* 54: 523–534.

54. Hamilton, A.J. 2005. Species diversity or biodiversity? *Journal of Environmental Management*. 75: 89–92.

55. Gillespie, T.W., Foody, G.M., Rocchini, D., Giorgi, A.P., and Saatchi, S. 2008. Measuring and modelling biodiversity from space. *Progress in Physical Geography*. 32: 203–221.

56. Lee, K.S., Cohen, W.B., Kennedy, R.E., Maiersperger, T.K., and Gower, S.T. 2004. Hyperspectral versus multispectral data for estimating leaf area index in four different biomes. *Remote Sensing of Environment*. 91(3–4): 508–520.

57. Koger, C.J., Bruce, L.M., Shaw, D.R., and Reddy, K.N. 2003. Wavelet analysis of hyperspectral reflectance data for detecting pitted morning glory (*Ipomoea lacunosa*) in soybean (*Glycine max*). *Remote Sensing of Environment*. 86(1): 108–119.

58. Asner, G.P. and Heidebrecht, K.B. 2002. Spectral unmixing of vegetation, soil and dry carbon cover in arid regions: Comparing multispectral and hyperspectral observations. *International Journal of Remote Sensing*. 23: 3939–3958.

59. White, J.C., Gómez, C., Wulder, M.A., and Coops, N.C. 2010. Characterizing temperate forest structural and spectral diversity with Hyperion EO-1 data. *Remote Sensing of Environment*. 114(7): 1576–1589.

60. Martin, M., Newman, S., Aber, J., and Congalton, R. 1998. Determining forest species composition using high resolution remote sensing data. *Remote Sensing of Environment*. 65: 249–254.

61. Melendez-Pastor, I., Navarro-Pedreño, J., Koch, M., and Gómez, I. 2010. Applying imaging spectroscopy techniques to map saline soils with ASTER images. *Geoderma*. 158(1–2): 55–65.

62. Zhao, G. and Maclean, L. 2000. A comparison of canonical discriminant analysis and principal component analysis for spectral transformation. *Photogrammetric Engineering and Remote Sensing*. 66: 841–847.

63. Ferwerda, J.G., Skidmore, A.K., and Mutanga, O. 2005. Nitrogen detection with hyperspectral normalized ratio indices across multiple plant species. *International Journal of Remote Sensing*. 26(18): 4083–4095

64. Filippi, A.M. and Jensen, J.R. 2006. Fuzzy learning vector quantization for hyperspectral coastal vegetation classification. *Remote Sensing of Environment*. 100(4): 512–530.

65. Sakamoto, T., Yokozawa, M., Toritani, H., Shibayama, M., Ishitsuka, N., and Ohno, H. 2005. A crop phenology detection method using time-series MODIS data, *Remote Sensing of Environment*. 96: 366–374.

66. Hsu, P.-H. 2007. Feature extraction of hyperspectral images using wavelet and matching pursuit. *ISPRS Journal of Photogrammetry and Remote Sensing*. 62(2): 78–92.

67. Martínez, B. and Gilabert, M.A. 2009. Vegetation dynamics from NDVI time series analysis using the wavelet transform. *Remote Sensing of Environment*. 113(9): 1823–1842.

68. Ingram, J.C., Dawson, T.P., and Whittaker, R.J. 2005. Mapping tropical forest structure in southeastern Madagascar using remote sensing and artificial neural networks. *Remote Sensing of Environment*. 94(4): 491–507.

69. Trombetti, M., Riaño, D., Rubio, M.A., Cheng, Y.B., and S.L. Ustin. 2008. Multi-temporal vegetation canopy water content retrieval and interpretation using artificial neural networks for the continental USA. *Remote Sensing of Environment*. 112(1): 203–215.

70. Liu, Z.-Y., Wu, H.-F., and Huang, J.-F. 2010. Application of neural networks to discriminate fungal infection levels in rice panicles using hyperspectral reflectance and principal components analysis. *Computers and Electronics in Agriculture*. 72(2): 99–106.

71. Jiang, Z. and Huete, A.R. 2010. Linearization of NDVI based on its relationship with vegetation fraction. *Photogrammetric Engineering and Remote Sensing*. 76(8): 965–975.

72. Fava, F., Colombo, R., Bocchi, S., Meroni, M., Sitzia, M., Fois, N., and Zucca, C. 2009. Identification of hyperspectral vegetation indices for Mediterranean pasture characterization. *International Journal of Applied Earth Observation and Geoinformation*. 11(4): 233–243.

73. Guisan, A., Edwards, T.C. Jr., and Hastie, T. 2002. Generalized linear and generalized additive models in studies of species distributions: Setting the scene. *Ecological Modelling*. 157(2–3): 89–100.

74. SAS Institute Inc., 2010, *SAS/STAT User's Guide*, Version 6. 4th edn., Vol. 1, Cary, NC.

75. Elvidge, C.D. and Chen, Z. 1995. Comparison of broadband and narrowband red and near-infrared vegetation indices. *Remote Sensing of Environment*. 54: 38–48.

76. Broge, N.H. and Leblanc, E. 2001. Comparing prediction power and stability of broadband and hyperspectral vegetation indices for estimation of green leaf area index and canopy chlorophyll density. *Remote Sensing of Environment*. 76(2): 156–172.

77. Lucas, K.L. and Carter, G.A. 2008. The use of hyperspectral remote sensing to assess vascular plant species richness on Horn Island, Mississippi. *Remote Sensing of Environment*. 112(10): 3908–3915.

78. Curran, P.J., Foody, G.M., Lucas, R.M., Honzak, M., and Grace, J. 1997. The carbon balance of tropical forests: From the local to the regional scale. In: van Gardingen, P.R., Foody, G.M., and Curran, P.J. (eds.), *Scaling-Up from Cell to Landscape*, Cambridge University Press, Cambridge, U.K., pp. 201–227.
79. Huete, A.R. 1988. A soil-adjusted vegetation index (SAVI). *Remote Sensing of the Environment*. 25: 295–309.
80. Huete, A.R. 1989. Soil influences in remotely sensed vegetation-canopy spectra. In: Asrar, G. (ed.), *Theory and Applications of Optical Remote Sensing*, Wiley, New York, pp. 107–141.
81. Kaufman, Y.J. and Tanre, D. 1996. Strategy for direct and indirect methods for correcting the aerosol effect on remote sensing from AVHRR to EOS-MODIS, *Remote Sensing of Environment*. 55: 65–79.
82. Jiang, J., Huete, A. R., Didan, K., and Miura, T. 2008. Development of a two-band enhanced vegetation index without a blue band. *Remote Sensing of Environment*. 112(10): 3833–3845.
83. Numata, I., Roberts, D.A., Chadwick, O.A., Schimel, J.P., Galvão, L.S., and Soares, J.V. 2008. Evaluation of hyperspectral data for pasture estimate in the Brazilian Amazon using field and imaging spectrometers. *Remote Sensing of Environment*. 112(4): 1569–1683.
84. Numata, I., Roberts, D.A., Chadwick, O.A., Schimel, J., Sampaio, F.R., Leonidas, F.C., and Soares, J.V. 2007. Characterization of pasture biophysical properties and the impact of grazing intensity using remotely sensed data. *Remote Sensing of Environment*. 109: 314–327.
85. Asner, G.P. and Lobell, D.B. 2000. A biogeophysical approach for automated SWIR unmixing of soils and vegetation. *Remote Sensing of Environment*. 74: 99–112.
86. Ozdogan, M. 2010. The spatial distribution of crop types from MODIS data: Temporal unmixing using independent component analysis. *Remote Sensing of Environment*. 114(6): 1190–1204.
87. le Maire, G., Francois, C., and Dufrene, E. 2004. Towards universal broad leaf chlorophyll indices using PROSPECT simulated database and hyperspectral reflectance measurements. *Remote Sensing of Environment*. 89: 1–28.
88. Coops, N.C. and Stone, C. 2005. A comparison of field-based and modelled reflectance spectra from damaged *Pinus radiata* foliage. *Australian Journal of Botany*. 53: 417–429.
89. Pacheco, A. and McNairn, H. 2010. Evaluating multispectral remote sensing and spectral unmixing analysis for crop residue mapping. *Remote Sensing of Environment*. 114(10): 2219–2228.
90. Tucker, C.J. 1977. Spectral estimation of grass canopy variables. *Remote Sensing of Environment*. 6: 11–26.
91. Draper, N.R. and Smith, H. 1981. *Applied Regression Analysis*, Wiley, New York.
92. Kaufman, Y.J. and Tanré, D. 1992. Atmospherically resistant vegetation index (ARVI) for EOS-MODIS. *IEEE Transactions on Geoscience and Remote Sensing*. 30(2): 261–270.
93. Klecka, W.R. 1980. *Discriminant Analysis*. Quantitative Applications in Social Sciences Series, No. 19. Sage Publications, Thousand Oaks, CA.
94. Bork, E.W., West, N.E., and Price, K.P. 1999. Calibration of broad- and narrowband spectral variables for rangeland covers component quantification. *International Journal of Remote Sensing*. 20: 3641–3662.
95. Clark, M.L., Roberts, D.A., and Clark, D.B. 2005. Hyperspectral discrimination of tropical rain forest tree species at leaf to crown scales. *Remote Sensing of Environment*. 96: 375–398.

Part II

Hyperspectral Sensor Systems

2 Hyperspectral Sensor Characteristics: Airborne, Spaceborne, Hand-Held, and Truck-Mounted; Integration of Hyperspectral Data with LIDAR

Fred Ortenberg

CONTENTS

2.1 INTRODUCTION

This chapter sets forth the fundamental concepts of hyperspectral (HS) imaging—a powerful tool to collect the precise remote sensing data involved in detection, recognition, and examination of the different objects of scientific, economic, and military character. The development of the relevant equipment is also discussed as well as current and future applications of HS technology.

Hyperspectral sensor (HSS) is one of the devices in use in the remote sensing activity, aimed at observing the Earth's surface with ground-based, airborne, or spaceborne imaging gear [1–7]. In outline, remote sensing consists in photo- or electronic recording of the spatial and spectral distributions of electromagnetic radiation, as emitted by any object under monitoring either on the Earth's surface or underground. Present-day technologies enable such a recording throughout the entire spectrum, from the radio waveband to the x-ray and gamma regions. The respective receivers differ widely, for each of them is based on specific physical principles.

Basically, three distinct spectral regions are used in remote sensing, namely, visible light with very-near infrared (IR), thermal, and radio. In each of them, an image of any object is unique; it is defined by the object's specifics as to emission, reflection, absorption, and scattering of the electromagnetic waves. In the visible and very-near IR regions, it is the object's capacity to reflect the solar energy in the first place, determined by the chemical composition of its surface.

The thermal IR (TIR) region is characterized by radiated energy, the latter being directly dependent on the object temperature. In the ultra high frequency (UHF) radio region—that of the radars—the reflection is determined by the surface smoothness and texture.

The sensor, which operates in visible waveband, can either cover the entire region in a so-called panchromatic imaging or be confined to some of the spectrum zones (e.g., green or red). In the first case, the basic sensor's characteristic is its resolution. At present, the sensors are developed of a very-high-resolution (VHR) class. A camera of such a class onboard Low Earth Orbit (LEO) satellite allows Earth's surface imaging with resolution of less than 1 m. In the second case, a multizonal image yields a set of distinct zone-specific images, which can be directly synthesized into color variants to bring out the details of interest, such as roads, installations, water surfaces, and vegetation.

The development of such equipment does not consist any more in the enhancement of the images' spatial resolution, but in the increase of the number of the images taken in different spectral zones, that is, of the number of spectral channels, at a reasonable spatial resolution of each image. The more zones, the more useful information! With many narrow spectral zones, the sensor becomes HS.

It should be borne in mind, however, that the region accessible to the human eye is but a minor source of such information. The ability of bees to single out melliferous herbs in a multihued summer meadow relates to the fact that their vision can extend into the ultra violet (UV) region. With such a "super-vision" directed at the Earth's surface, it is possible to tell healthy vegetation from degraded one and man-made objects from natural.

TIR radiation also carries particular information unattainable with the human eye. The natural radiation of any object indicates its surface temperature: the higher the latter, the lighter the shade of gray in the image. The eye being more sensitive to variations in hue than to those in brightness, the image contrasts are often deepened by means of a "scale" of the colors commonly associated with temperatures—from violet or blue ("cold") to red or brown ("hot"). Thermal photography is a widely used tool for detection of the heat leaks in both industrial installations and in built-up residential areas. They are also useful in reconnaissance: an underground military plant can be invisible in the ordinary photograph, but since its functioning entails heat generation, it would shine brightly in the thermal region. Finally, such a picture of a person can provide valuable information on his/her state of health.

Regarding the very-near IR region, it is possible to take any of its zones and substitute it for one of the primary colors in an ordinary color image. Such an operation, known as color synthesis, is performed by the "false-color" films used in remote sensing. The inexperienced eye may find it hard to recognize familiar objects in synthesized images, but on the other hand, they help overcome the limitations of human vision.

The most promising method of remote sensing is HS imaging, which produces tens or even hundreds of images in narrow zones. The absorption spectra of substances and materials being specific, this technique makes it possible to identify vegetation, minerals, geological formations, soils, structural materials in buildings and pavements via the physical/chemical compositions of the objects. Thanks to the extrahigh resolution (of the order of that of a lab spectrometer) the volume of obtainable information can be increased by a four-figure factor.

Unlike visible- and IR-region imaging devices, the radars belong to the active-sensor class. Whereas the former just passively capture either reflected and scattered solar radiation or that of the Earth's surface, the latter emit their own electromagnetic waves and record their echo off the object. Radar imaging has features absent in other remote sensing techniques: it cannot be hampered by blanket cloud cover, its images have their particular geometric distortions, its coherence makes it possible to obtain detailed relief patterns accurate to tens of centimeters, and it can operate both in day time and at night.

HS imaging is an evolutionary product of multispectral (MS) systems, whereby—by means of new technologies—the number of information collection channels can be increased from 3–10 to 100–1000 with high spectral resolution of 1–10 nm. The result is a multidimensional spatial–spectral image in which each element (pixel) is characterized by its individual spectrum. Such an image is called an information cube, with two of its dimensions representing the projection of the imaged area on the plane, and the third—the frequency of the received radiation. A HS imaging device is an optico–electronic multichannel system designed for simultaneous independent generation of an image and corresponding video signal in a discrete or continuous sequence of spectrum intervals, supplemented by radiometric, spectral, and spatial image parameters pertaining to those ranges.

In other words, present-day MS and HS scanning devices are radiometrically calibrated multichannel video spectrometers. The values of brightness, registered by the imaging system for some object in the different spectral zones, together with their graphical mappings as spectral curves, allows to clearly discern the object and mark it off on the image.

The term "hyperspectral" has been repeatedly criticized, for such a collocation seems to be incongruous, without any physical sense. More justified would be using the prefix "hyper-" with a quantitative characterization to imply "too many," as in the word "hypersonic," that is, highly above the speed of sound. But in our case "spectrum" has a physico–mathematical, rather than a quantitative meaning. Similarly, this is why the words like "hyperoptical," "hyperoily," "hypermechanical," "hyperaerial" etc., are devoid of sense. Consequently, the word "hyperspectral" might represent rather a metaphor to express the excitement of the researchers with regard to new technological means able to obtain a multitude of the images of the same object, each of them in a very narrow spectral band. Spectroradiometer imaging technology emerged at the time when the panchromatic imagery was sufficiently advanced, and the picture quality was defined by the achieved spatial resolution. The images taken from the satellite orbit were said to be either of "high resolution" (1 m) or of "very high resolution" ($\ll$1 m). By the term "resolution," the spatial one was meant by default. Since the image sets have been provided by the spectroradiometer, the technique of obtaining these data cubes began to be referred to as the "hyper-resolution imaging." Until now, some researches keep using this term, even without specifying what resolution they mean—spatial or spectral. The expression "spectral hyper-resolution imaging" could have the meaning of visual spectroradiometer data, but the specialists have chosen instead this somewhat simplistic and inappropriate derivation—"HS." In any case, the unsuitable term has taken root and cannot be canceled, nor is there a real need for that. Still, it should be made clear that what matters is not the discovery of a new phenomenon, but the development of the advanced video spectrometers with high spectral resolution.

The interest in such optico–electronic devices is caused not only by their spectacular applications, but also by the fact that these technical means, in many respects, imitate the elements of the visual apparatus of higher animals and humans. It is known that all living beings acquire 80%–90% of the information about their environment through vision. Therefore, the paths of developing automatic systems of the technical vision for detection, recognition, and classification of the different objects are under close attention.

HS imaging, being able to extract more precise and detailed information as compared with other techniques, is one of the most efficient and fast-developing directions of Earth remote sensing (ERS). The data on the energy reflected from the ground objects provide extensive material for detailed analysis. Still, it is worth mentioning that the development of the needed HS equipment proved to be a very complex task and hard to implement. Designing HSS, in particular for spaceborne applications, was of purely theoretical interest for a long time, for the following reasons:

- The quality of the equipment precluded realization of HSS with high spatial and spectral resolution simultaneously.
- Reception of the vast volumes of HS data, their transmission through the communication channels, and their ground-based processing presented serious difficulties.
- The complexity of the data-flow containing the HS measurements required specialized and highly sophisticated processing software.

Nowadays, these difficulties are largely surmountable. At the time of writing, the way is already opened for the development of HS equipment in the optical and IR regions with high informational and operational indices, which in turn would mean a higher solution level for the problems to be solved.

During the last three decades, the hi-tech industries in many countries were considerably involved in HS subjects. The dedicated element base was created for manufacturing HS equipment. Various HSS, initially for ground and airborne applications, and later on for the spaceborne ones, were designed, tested, and put to use. The modern video spectrometer concept is based on the latest optical-system solutions involving large-format CCD matrices. Currently available are imaging systems operating in hundreds of spectral bands, with signal-to-noise ratio providing information in 12 bits per count. The advances in HS equipment and data processing software in the recent years has significantly extended the range of possible applications to be provided by airborne and spaceborne imaging.

The specifics just mentioned of putting HSS into space resulted in smaller availability of HS imagery as compared with other ERS data. One of the reasons for that is the fewer number of space-crafts with HS imaging systems, like Hyperion onboard the EO-1 spacecraft (NASA, Washington, DC), or CHRIS onboard PROBA (ESA). The increased HS-connected activity in space can be seen, as of late, to overcome the backlog. Spaceborne HSSs are expected to find a use as general purpose instruments, providing data for a broad range of end users (agriculture, mineralogy, etc.).

The present-day airborne and spaceborne HSS instrumentation presents devices operating in the optical and IR spectral band and combining high spatial, spectral, and radiometric resolutions, indispensable in recording distinctive features of the Earth's surface, such as vegetation cover, land-scape status, and anthropogenic impacts. Its broad functional potential derives from the fact that the object's spectral parameters as well as their surface distributions are the most prolific sources of information on the object's state and the changes it is undergoing in the course of exploitation, or its vital activity. The primary contribution of HS imaging techniques, which are evolving and get-ting introduced into new fields, will be in the exploration and development of the new applications via the selection of optimal spectral band parameters (bands position and widths). Practical opera-tional considerations (sensor cost, data volume, data processing costs, etc.) of most of both current and future applications are favorable for the use of HS systems, taking into account its economical efficiency. Aerial and space HSS provide an economical alternative to traditional ground-based observation techniques, permitting the data to be acquired for large areas in a minimum time frame, to enable faster application turnaround.

By virtue of its ability to provide information beyond the capacity of traditional panchromatic and MS photography, HSS is highly advantageous in solving economic and military problems. Experts estimate [8] that HSS is capable of solving up to 70% of all Earth-observation problems, while visual information with high spatial resolution will only be 30%.

As will be shown later in this chapter, additional information about the reflecting surfaces can be delivered by the return signals measured by the Light Detecting and Ranging systems (LIDAR). This is why the new HS applications are based on fusion of two imaging data types, as obtained by joint airborne or spaceborne HS and LIDAR systems.

2.2 HSS CONCEPT

HSS combines the following photonic technologies: conventional imaging, spectroscopy, and radiometry—to produce images together with spectral signatures associated with any spatial resolution element (pixel). The position of spectral imaging relative to related technologies is shown in Figure 2.1.

Data produced by a spectral imager create a cube, with position along two axes and wavelength along the third, as depicted in Figure 2.2. By proper calibration, the recorded values of the data cube can be converted to radiometric quantities that are related to the scene phenomenology

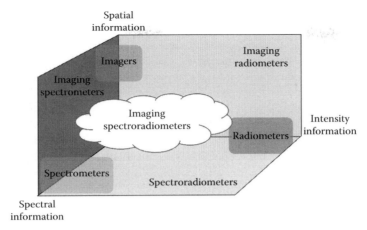

FIGURE 2.1 Relationship among radiometric, spectrometric, and imaging techniques.

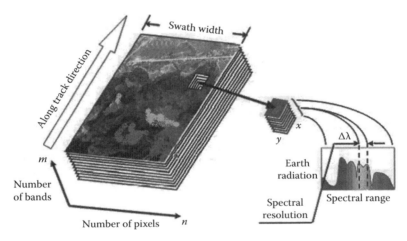

FIGURE 2.2 **(See color insert.)** Imaging spectroradiometer concept—data cube composed of individual images recorded over "m" spectral bands.

(e.g., radiance, reflectance, emissivity, etc.). The latter provides a link to spatial and spectral analytical models, spectral libraries, etc., to support various applications. The motion of the spacecraft is used to move the image line across the surface of the Earth, and the CCD is read out continuously to provide a complete HS dataset per line. The imagery for the whole area can be represented in the form of data cubes or a set of images, each image containing information from one wavelength.

To illustrate the strength of this technique, the combined spectral/spatial analysis allows detection in an image of optically unresolved (subpixel size) objects. Its applications range from the ERS to early cancer detection. It is obvious why the combination of imaging and spectroscopy is so attractive: conjoining of the two analyses, which can be considered as a data fusion, makes for enhanced image perception. Conventional color imagery touches upon the idea, except that it is based on the very broad bands, so the colors are often achieved at the expense of spatial resolution (e.g., by use of Red, Green, Blue (RGB) color-striped CCDs).

HS imaging systems have the advantage of providing data at high resolution over a large number of bands and, hence, are suitable for a wide variety of applications. Their MS counterparts, operating over less than 15 bands, are less expensive, yield smaller datasets and have higher signal-to-noise ratio. They are often tailored to a specific application, so that for other purposes, their bands may be suboptimal or even totally unsuitable. However, once the optimal bands have been identified for the task at hand, such a system can provide an overall superior solution.

In HS imaging, the decisive factor is not the number of bands (channels), but the band width—the narrower, the better—and the sequence of the takes. In other words, a 15-channel system would count as HS if it covers the band from 500 to 700 nm, with each zone not exceeding 10 nm, while a similar 15-channel system covering the short-wave and visible range and the three IR ranges would count as MS.

The HS imaging system breaks down the incoming optical radiation into hundreds of bands and searches for specific signatures unique to a scene. Those are processed in a way that permits reliable identification of the objects standing out against their backgrounds, which generate a visual representation of the "hidden" source in the same manner as IR displays. By this means, it is possible to "penetrate" dense foliage and locate underground installations, tunnels, pipelines, etc. Incidentally, this has led to the term "hyperspectral" being interpreted as hypersensitive to buried or underground objects, which is obviously incorrect.

According to the terms of reference approved by the UN, monitoring of the environment consists in regular observation of its constituents in space and time, for specific purposes and within the framework of programs worked out in advance. In the context of the vegetation cover, this means observation of its status, control of its current dynamics, forecast for the future, advance warning of catastrophic upheavals, recommendations on mitigation, or prevention of damage. In the same context, HS measurements can serve the purposes as control of meteorological and climatic factors, pollution control of the atmosphere, water bodies and soil, operational control of extraordinary man-made and natural situations, and information on land use and reforms.

HS data have a high commercial potential. Through them, significant savings are expected in the development and operation of monitoring systems, which is a task both complex and expensive, demanding a collaborative effort by the scientific and manufacturing sectors at the national levels. A promising possibility in this respect is recourse to the already-developed or planned dual-purpose space complexes. A survey of the monitoring services market makes it evident that there is little difference between the civil and military sectors regarding their requirements as to the quality and usability of the provided information. This, in turn, points to an integrated HS technology covering the two sectors and supported by multipurpose observation systems accommodating the broadest spectrum of problems. Economic considerations indicate that the dual-purpose version is the answer.

At the same time, it should be noted that the terms of reference of such HS systems give rise to conflicts between

- The required coverage and the particularity of the image
- The required spatial and spectral resolutions
- The required and real quantization orders of the spectral channel signals
- The video data volume and the limited traffic capacity of the current satellite communication channels

Specific features inherent in HS airborne and spaceborne applications are as follows:

- Through them, the detection and recognition steps yield accompanying spectral features. These admit lower spatial-resolution levels, which in turn allows recursing to smaller and lighter sensors. The latter can be accommodated on mini-vehicles, which, operating in clusters, permit reduction of the observation periodicity for a tract to some hours or less. Another result of the reduced spatial resolution level is a larger grid unit for the same photo-receiver format.
- HS data are readily amenable to computerized processing with involvement of geo-informational technologies. This, in turn, contributes to time-saving and facilitates the employment of less-qualified decoding personnel. Advanced software has been developed, incorporating geo-informational technologies.

2.3 HSS PHYSICS, PRINCIPLE, AND DESIGN

Imaging spectrometers typically use a 2D matrix array (e.g., a CCD), and produce progressive data cubes through successive recording—either of full spatial images, each at a different wavelength, or of narrow image swaths (1 pixel wide, multiple pixels long) with the corresponding spectral signature for each pixel in the swath. Remote imagers are designed to focus and measure the light reflected from contiguous areas on the Earth's surface. In many digital imagers, sequential measurements of small areas are taken in a consistent geometric pattern as the sensor platform moves, and subsequent processing is required to assemble them into a single image.

An optical dispersing element in the spectrometer splits light into narrow, adjoining wavelength bands, the energy in each band being measured by a separate detector. Using hundreds, or even thousands of detectors, spectrometers can accommodate bands as narrow as 1 nm over a wide wavelength range, typically at least 400–2400 nm (from visible to middle-IR). HS sensors cover bands narrower than their MS counterparts. Image data from several hundred bands are recorded concurrently, offering much higher spectral resolution than that provided by sensors covering broader bands. Sensors under development, designed to cover thousands of bands with even narrower bandwidth than HS, have a special name—ultraspectral.

The most common modes of image acquisition by means of HS sensors are known as "pushbroom scanning" (electronical) and "whiskbroom scanning" (electro-mechanical).

- Pushbroom scanners (Figure 2.3) use a line of detectors over a 2D scene. The number of pixels equals that of ground cells for a given swath. The motion of the carrier aircraft or satellite realizes the scan in the along-track direction, thus the inverse of the line frequency equals the pixel dwell time. In a 2D detector, one dimension can represent the swath width (spatial dimension, y) and the other the spectral range. Pushbroom scanners are lighter, smaller, and less complex than their whiskbroom counterparts because of fewer moving parts. They also have better radiometric and spatial resolution. Their major disadvantage is a large number of detectors amenable to calibration. These imaging spectrometers can be subdivided into Wide Field Imagers (MERIS, ROSIS) and Narrow Field Imagers (HSI, PRISM). These imager-types conduce to high or frequent global coverage when HSSs are installed on the board of a LEO satellite.

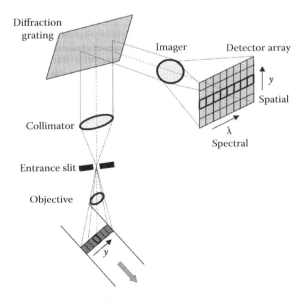

FIGURE 2.3 Principle of pushbroom scanning.

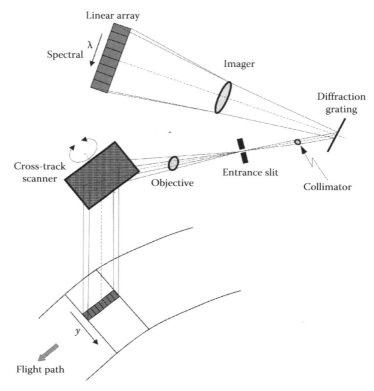

FIGURE 2.4   Principle of whiskbroom scanning.

- Whiskbroom scanners (Figure 2.4). On-axis optics or telescopes with scan mirrors sweep from one edge of the swath to the other. The field-of-view (FOV) of the scanner can be covered by a single detector or a single-line detector, the carrier's motion implementing the sweep. This means that the dwell time for each ground cell must be very short at a given instantaneous field of view (IFOV), because each scan line consists of multiple ground cells to be covered. Whiskbroom scanners tend to be large and complex. The moving mirrors bring in spatial distortions to be corrected by preprocessing before delivering the information to the user. An advantage of the whiskbroom scanners is that they have fewer detectors subject to calibration, as compared to other types of sensors. Well-known examples are AVHRR, Landsat, and SeaWiFS.

Two approaches to HS sensors were launched and operational for many years in the new millennium: the dispersion approach (e.g., EO-1) and the Fourier Transform approach (e.g., MightySAT).

- *Dispersion*: By this technique, the spectral images are collected by means of a grating or a prism. The incoming electromagnetic radiation is separated under different angles. The spectrum of each ground pixel is dispersed and focused at different locations of the 1D detector array. This technique is used for both image acquisition modes (pushbroom and whiskbroom scanners). HS imagers use mainly gratings as the dispersive element (HSI, SPIM).
- *Fourier Transform Spectrometer* (*FTS*): Spatial Domain FTSs, like the well-known sensors SMIFTS or FTHSI on MightySat II spacecraft, use the principle of the monolithic Sagnac interferometer (Figure 2.5). Unlike conventional FTSs, the Earth observation spectrometers in LEO operate with fixed mirrors. The optical scheme distributes the interferogram (spectrum) along one dimension of the detector, the other dimension representing the swath width.

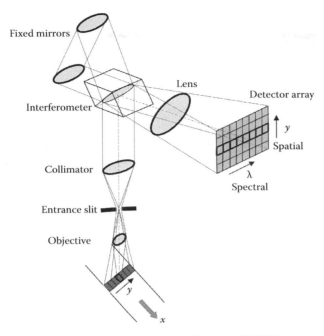

Fixed mirrors

Lens

Detector array

Interferometer

y

Spatial

Collimator

λ

Spectral

Entrance slit

Objective

y

x

FIGURE 2.5 Principle of Fourier Transform Hyperspectral Imager (FTHSI).

Classical HS imaging is performed by one of two basic principles: scanning by a pushbroom approach or by a tuneable filter approach, for example, using an Acousto-Optic Tuneable Filter (AOTF). In a pushbroom scanner, a line camera is used where the light is diffracted into different color/wavelength components by means of a spectrograph onto a focal plane array. In the tuneable filter approach, instead of scanning the image stepwise in one spatial dimension, the entire spatial images are taken, as the camera scans through the different wavelengths. This is achieved by placing the AOTF between the image and the camera [9]. AOTF is a solid-state device comprising a crystal filter whose transmission wavelength is controlled by attached piezoelectric transducers. The transducers create pressure waves in the crystal that changes its refractive index. These transducers are controlled by careful manipulation of the frequency of an applied electric field to vary the transmission wavelength of the filter in a stepwise manner. An image captured at each wavelength step is read into memory to form a slab of the hypercube. The data is then processed in order to produce the desired output image. The spectrograph in a pushbroom system provides better spectral resolution than the tuneable filter system, but at a cost of speed, while the latter device, with no moving parts, builds up the hypercube more swiftly.

For detailed studies of the Earth's surface in the reflection spectral band of 400–2400 nm, the onboard HS equipment should simultaneously provide high spatial resolution of 1–5 m, high spectral resolution of 5–10 nm in tens and even hundreds of channels, a swath of up to 100 km, and a signal-to-noise ratio ≥100. At present, neither HS device meets all these requirements. One remedy consists in replacing the fixed-parameter sensors by adaptive systems. A major consideration in developing the latter systems is that they would operate under widely varying conditions, including parameters of the atmosphere, characteristics and types of the Earth's surface, and the space platform motion and attitude. These variations are stochastic, which tends to impair both the efficacy of the observation process and the quality of the obtained images. Under the adaptive regimes, the information parameters of the system are selected and adjusted in real time for the problem in question. Among the adjustable information parameters are the spatial and spectral resolutions as well as the signal quantization width. The performance level of the observation missions can be raised to a qualitatively new stage by implementing the advanced operation modes for the onboard equipment.

One more prerequisite to this end is to reduce the volume of the transmitted imagery, combined with prevention of the information losses. Redundancy should be curtailed by a three- or even four-figure factor, which is unfeasible by the traditional means alone. The proposed solution consists in real-time quality assessment of the observation data and selection of the informative onboard channels to screen out the data with low informational content. This approach not only enables redundancy reduction, but also enhances the information output.

At present, the spectral intervals, their number, and their layout are chosen at the design stage and remain practically unchanged afterward. New methods of interval selection, currently under development, are based on the analysis of the spectral reflection characteristics. Unlike the conventional procedure, the intervals and their limits are selected through a compromise between the required spectral and spatial resolutions. This is implemented, for example, in Full Spectral Imaging (FSI)—a new method for acquiring, preprocessing, transmitting, and extracting information from full-spectrum remote-sensing data by the "spectral curve" approach, instead of the current "bytes-per-band" one [10]. By virtue of the bandwidths' optimization, FSI admits simplified instrument characteristics and calibration as well as reduced data transmission and storage requirements. These improvements may be accomplished without loss of remote sensing information. FSI is neither "hyperspectral," nor "superspectral" or "ultraspectral" imaging; it is an end-to-end system that involves the whole technological sequence, from the observation technique to the data processing. FSI belongs to the class of HS-type imagers based on a dispersive element and a pushbroom image acquisition system. Instrument throughput, spatial coverage, and spectral response can be enhanced using multiple focal planes. The multiple entrance slits would be optically connected to several spectrometer-and-detector combinations. Each spectrometer would have identical optical elements except for the dispersion characteristics of the grating and the detector. The operation of such systems, with user's demands taken into account, will be described in detail in the sequel.

An important task is screening of the received HS data to obtain more trustworthy representation of the Earth's surface. The information involves random, systematic, as well as system-defined distortions caused by the atmosphere, curvature of the Earth, displacement of the HS equipment relative the Earth's surface at the moments of imaging, and physical characteristics of the sensors and channels. To remove these multiple distortions, a number of corrections should be applied, based on radiation, radiometric, geometrical, and calibration techniques. Radiation correction is related to adjustment of the EM energy amount as received by each sensor, thereby compensating for diverse atmosphere transparency to the different frequency bands. Radiometric correction represents the elimination of the imaging system-defined distortions inserted both by the sensors and the transceivers. Geometrical correction, or image transformation, is intended to remove the distortions due to the Earth's curvature and rotation, inclination of the satellite's orbit with respect to the equatorial plane, and for high-resolution images, the terrain relief. This correction can be done automatically with the HSS position known. More precise transformation and image linking to a specific coordinate frame is usually achieved by interactively setting the reference points. In the transformation process, the pixel coordinates conversion to a new raster frame may cause some changes in the objects' shapes. In editing and joint processing of the different imaging data types as well as the images of the same terrain taken at different times, it is internationally accepted to use, as an exchange standard, a so-called *orthoplan* projection. Calibration consists in converting the dimensionless data, as obtained by the sensors of various spectral bands, into true normalized values of the reflected or emitted energy.

2.4 HSS OPERATIONAL MODES

The recent decades have been marked by a rapid advance in HS techniques. The development of the technical means for HS imaging went all the way from the simple lab devices with limited capabilities to sophisticated space complexes that enable excellent imaging of the Earth's surface

FIGURE 2.6 **(See color insert.)** Landscape fantastic HS control.

simultaneously in hundreds of the narrow spectral bands and successful downloading of this information to the Earth. Today, there are three basic methods of acquiring the HS images of the Earth: ground-based, airborne, and spaceborne imaging. Each method has peculiarities, advantages, shortcomings, and the range of problems of its own. Imaging method depends on the sensors' structural features imposed by their operating conditions—on the ground, in the atmosphere, or in space. The maximum efficiency can be attained when there are all the tools available for working in all three spheres as well as the opportunity of choosing the optimum way to achieve the goal. None of the cited methods ducked away from the progress, and their simultaneous improvement brought an impressive rise in the HS technique (Figure 2.6).

Some general descriptions and results obtained by the HS techniques in the different handheld, truck-mounted, airborne, and spaceborne applications are presented in the following.

2.4.1 Ground-Based HS Imaging

HS imagery is generated by the instruments called imaging spectroradiometers. The development of these sophisticated sensors involves the convergence and synergy of two related but distinct technologies: spectroscopy and remote imaging of the Earth. Spectroscopy is a study of emission, absorption, and reflection of light by the matter as a function of radiation wavelength. As applied to the field of optical remote sensing, spectroscopy deals with the spectrum of sunlight that is diffusely reflected (scattered) by the different materials on the Earth's surface. Ground-based or lab measurements of the light reflected from the test material became the first successful application of the spectrometers or spectroradiometers.

On-ground HS imagers are ideal tools for the identification of obscured or resolution-limited targets, defeating camouflage, and recognition of both solid and gaseous chemicals. There are manufacturing processes that can make use of MS- or HS data, which include inspection of color and paint quality, detection of rust, or the flaw detection of thin film coatings, etc. Another interesting sphere of application for HS techniques is related to functional mapping of the brain. Some of such exotic HS imaging applications would be presented in the next chapters of this book. However, to create these compact, robust, low-power devices, innovative approaches have been to be worked out for the traditional technology of spatial or spectral imaging instruments, initially bulky and costly. In this chapter, only ordinary applications of HS-imaging, such as terrain-scanning, will be considered, which contribute to the refinement of the HS sensors and expand their employment to the atmosphere and space.

Initially, during the introduction of the airborne and spaceborne HS sensors into practice, an important part of their primary calibration was the lab and ground field measurements of the optical characteristics of the agricultural materials. HS measurements made it possible to derive a continuous spectrum for each image cell. Once the adjustments for the sensor, atmospheric, and terrain effects applied, these image spectra can be compared with field or lab reflectance spectra, in order to recognize and map surface materials, for example, specific types of vegetation. Collections of spectral curves for both the materials and the objects, routinely measured under lab or field conditions, are stored in spectral libraries. These curves can be used for classification or automatic identification of objects and materials. Until recently, the imagers' resolution capability was restricted due to the limitations imposed both by the detector design and the requirements for data storage, transmission, and processing. Recent advances in these areas have contributed to the imagers' design, so their spectral ranges and resolution are now comparable with those of the ground-based spectrometers.

Surface Optics Corporation (SOC), which specializes in the characterization and exploitation of the optical properties of surfaces, has developed a wide range of commercial products including HS video imagers and processors. Its SOC-700 family of HS imaging systems is devised to address the real-time processing requirements of various applications, such as machine vision, biological screening, or target detection. The HS imager (Figure 2.7) is a high-quality, portable, and easy-to-use spectral imaging instrument, radiometrically calibrated, with a software intended for both the analysis and viewing. The imager captures a 640×640 pixel image that has 120 spectral bands deep in as fast as 4 s. HSS can acquire and process up to 100 lines of 12 bit data for all 120 bands and return an answer to be used in controlling the process, screening the cell samples, or telling apart friend from foe. The HS analysis package provides the tools for serious spectral work, including complete radiometric calibration, reflectance calculations, matched filtering, and atmospheric correction. The sensor is set up within minutes and can withstand rugged outdoor environments. Since its inception, it has been used under rainforest and desert conditions and has been drop-tested from 4 ft height.

Design and tests of handheld HS devices improved the real-time detection and quantification of chemical and biological agents. Such HS scanners were developed for wide spectrum of applications, from analytical purposes to medical. The staring HS imager, produced by Bodkin Design, can capture both the spectral and the spatial information of the scene instantly, without scanning [11]. It is designed for high-speed HS imaging of transient events and observation from moving platforms. The device has no moving components, making the system immune to mechanical failure. Its unique design simplifies the detector readout and the optical system, and provides a fully registered, HS data cube on every video frame. It is ideally suited for airborne, handheld, and vehicle-mounted systems.

The number of HSS applications in various areas of science and technology constantly extends. Thus, Terrax Inc. yields the Theia—a low-power, lightweight HS imaging system with automated

FIGURE 2.7 HS target detection system on a field trial.

image analysis and data optimization for real-time target detection and HS data comparison. It consists of cameras and a HS real-time image-capture and decoding subsystem. The Theia is designed to fit within the size, weight, and power envelope of a portable environment as well as fixed locations. It is robust and includes several onboard, real-time target detection and data optimization algorithms based on a calibration library of 100 substance cubes.

The system, called foveal HS imaging, was presented in report [12], which employs a HS fovea with a panchromatic periphery. This approach results in a low-cost and compact handheld HS imager suitable for applications, such as surveillance and biological imaging. In these applications, the panchromatic peripheral image may be used for situational awareness and screening of the potential threats of the HS fovea to that region of interest for recognition or detailed characterization. As was demonstrated in the applications, the real-time capability of the foveal HS imaging shows particular promises including surveillance, retinal imaging, and medicine.

The HS imager is a compact asymmetric anamorphic imaging spectrometer covering broad spectral regions available from a specialty manufacturer. The complete imaging system is contained in a variety of packages suitable for deployment in a mobile vehicle environment or with the imaging head remote from the processing stack, as in building perimeter security systems. The detection capability and corresponding display varies with the model and the database of cubes representing calibration for specific chemical or material signatures. Detection and determination can reliably work over distances from a few feet to 300 ft.

The Norwegians have built interesting technology demonstrator for HS target detection [13]. Developed was also the ground-to-ground demonstrator system, which is currently being used in tests. The package contains a HySpex VNIR HS camera, a line scanner monochrome camera, and a motorized turntable mounted on a tripod. The rotation axis and the linear FOV are vertical, so that the cameras scan around the horizon. The HS system that was successfully tested on a field trial in northern Norway is shown in Figure 2.7. The system is designed for detection of the small targets as well as for search and rescue operations. Concerning the last option, it should be noted that in many cases there is no need to identify a specific target—what is needed is to detect the presence of a target with unknown *a priori* characteristics. Search and rescue operations provide good examples of such an approach. For instance, consider a hiker lost in the mountains or a boat on the high seas to be promptly detected. Rather than conducting an extensive analysis of HS data cubes, which is quite a time-consuming task, a cursory technique based on "anomaly detection" can be applied.

Some researches with the ground remote sensing systems have been carried out using sensors attached to long hydraulic booms hoisted above the crop canopy from the ground (the University of Illinois). The images collected from such a close distance have resolutions much higher than those taken from aircraft or satellites. Other ground-based systems use vehicle-mounted sensors that control variable-rate applicators in real time. For example, the remote sensors distinguishing the weeds from the crop are mounted on sprayers that change the application rate of herbicides, to be applied on the go. This form of remote sensing technology, called machine vision, is now in wide use.

Most of the HS imagers operate in the visible to shortwave IR (SWIR) bands. HS imaging in IR spectrum range have multiple applications, and there is an urgent need to develop relevant instrumentation. Many industrial firms develop and turn such a production out. One such product, which can operate both on the ground and on board a plane, is presented in the following. The Aerospace Corporation has designed and built a state-of-the-art narrowband HS Fourier Transform IR (FTIR) scanner—Spatially Enhanced Broadband Array Spectrograph System (SEBASS), which offers some unique capabilities within the HS remote sensing arena. SEBASS can collect data either from the transportable ground-based sensor or from a low-flying aircraft, depending on the application at hand. The SEBASS capability in the mid-wave and long-wave IR (LWIR) is intended to remotely identify materials in the 3–13 μm "chemical fingerprint" spectral region. HS sensor also provides high-resolution temperature data. SEBASS captures 128 spectral bands in IR spectral range with its IFOV of 1 mrad, and FOV of 128 mrad (7.3°).

Its typical remote IR products include spectrally and radiometrically calibrated, atmospherically corrected HS data cubes. SEBASS has three operation modes:

- *Vehicle Mounted Measurements*—four mobile 2D scanning LWIR FTIR vehicle-mounted sensors. The vehicles can be driven to the site of interest to collect close-range HS data from 8 to 12 μ. The scanning FTIR sensors have been successfully deployed on over 100 ground-based data collections.
- *Handheld Measurements*—provide in situ ground spectral measurements using handheld instruments to verify material identifications derived from airborne measurements.
- *Laboratory Measurements*—to assist in material characterization of samples taken from regions of interest, establish ground truth, and improve interpretation of remote sensing data.

Incidentally, as was noticed by many researchers, the best results are those obtained by sharing the data as obtained by various HSSs at different positions with respect to the target [14]. In this way, for example, sugar beet disease was detected by combining multitemporal HS remote sensing data, as provided by the different airborne-, tractor-, and handheld spectroradiometers.

Imaging spectrometry data are well established for detailed mineral mapping from airborne and spaceborne systems. Overhead data, however, have substantial additional potential when used together with ground-based HS measurements. U.S. researchers used the HS scanner system to acquire airborne data, outcrop scans, and to image boxed drill core and rock chips at approximately 6 nm nominal spectral resolution in 360 channels from 400 to 2450 nm. Analysis results using standardized HS methodologies demonstrate rapid extraction of representative mineral spectra and mapping of mineral distributions and abundances. A case history highlights the capabilities of these integrated datasets for developing an improved understanding of relations between geology, alteration, and spectral signatures in 3D.

2.4.2 AIRBORNE HSS

Till 2000, a fleet of over 40 airborne sensor systems gave the final polished form of future data acquisition opportunities. During most of the last decade of the past millennium, HS imaging was an area of active research and development, so the images were available only to researchers. With the recent appearance of commercial airborne systems and with applications in resource management, agriculture, mineral exploration, and environmental monitoring, the HS imagery is poised to enter the mainstream of remote sensing [15–17]. But its effective use requires an understanding of the nature and limitations of both this data type and various strategies involved in its processing and interpretation.

Like the ground equipment, the airborne HS scanners are also designed to measure the intensity of the radiation scattered by the Earth's surface in the given spectral band—any pixel of the acquired image simultaneously contains both spatial and spectral information on the inspected terrain. The input radiation flux is split into components, according to a wavelength. For each waveband, certain lines of the matrix are reserved. These data are digitized and written on the hard disc of the unified control, data storage, and power supply module.

For spatial linking of the collected imagery, the inertial navigation system (INS) is involved in the post-flight processing of its own data together with those of the imaging scanner, to attribute the adequate geographic coordinates to each pixel. More advanced survey complexes are supplied with navigation systems that have the relevant software to control the airborne imagery process. Thus, along with imagery planning and recording, the system also controls both the aircraft and the imaging sensors during the flight. The estimation of the navigation and imagery precision is carried out in real time. To the credit of such imagery control systems and their high quality and reliability is the fact, that today, there is about 150 of them operational throughout the world.

Companies–producers of such equipment in many countries have mastered manufacturing of the guide-beams for the HS scanners, including those to operate in visual and near-IR (VNIR), SWIR, medium-wave IR (MWIR), and TIR bands. The airborne scanners, which are currently becoming operational, are notable for their perfection and variety, the number of their different types supposedly exceeding one hundred. The scanning HS sensors of the leading companies have reliable, stable, and functional components; a software handy for operation; high capacity of digitizing and readout; high-precision and efficient optics; stable and reproducible parameters of the sensor calibration, etc. The customers are often supplied by the companies not just with the imaging hardware, but with the complex imaging technologies. Such integrated digital airborne systems represent the off-the-shelf solutions for obtaining the final product.

Generally, each HS scanner consists of a sensor (opticoelectronic unit), power supply, data accumulation and control unit, vibration damping platform, operator display, GPS receiver, inertial measurement unit, and data processing device. A matter of priority is also to maintain functional completeness of the HS scanners' spectrum, that is, a possibility to use them in all the applications somehow connected with the spectral analysis. Sure enough, a mandatory requirement for the HS scanners under development by the present-day companies is their metrological provision.

It is only in this case that the quantitative assessments can be made of the spectral characteristics pertaining both to the underlying surface and specific ground objects, which, in turn, is extremely important for obtaining trustworthy results at the stage of applied analysis. This is why the companies are heavily investing in the development of their own optico–electronic tract, including CCD-receivers with high signal-to-noise ratio, stable radiometric calibration, etc. The equipment design and the operational software make it mostly possible to measure the solar radiation intensity at the moment of imaging and to account for it later on.

The airborne HS imaging ensures the recognition of the objects by their physicochemical makeup and enablesconducting identification of the plant species and the vegetation status, keeps under control the natural habitat of different plants, including the weeds and drug plants, spots the vegetation and soil disturbances, finds wetlands and salted soils, etc.

Aircraft, drone, helicopter, and zeppelin—each of them has some useful advantages as a platform for remote sensing systems. Aircrafts can fly at relatively low altitudes, thus admitting a submeter sensor spatial resolution. The airborne sensors can be used more flexibly because of the variation of height (flight level) and the forward scan velocity. Therefore, spatial resolution and swath width can be easily adapted to the task requirements in the course of flight.

Aircraft can easily change their imagery schedule to avoid weather problems, such as clouds, which may block a passive sensor's FOV, and to allow for sun illumination. Sensor maintenance, repair, and configuration changes are easily applicable to aircraft platforms. There are no bounds for an aircraft but the state borders. At the same time, for a low-flying aircraft with a narrow sensor's FOV, a number of flybys are needed to cover a large ground area. The turnaround time it takes to get the data to the user is delayed due to necessity of landing the aircraft before transferring the raw imagery to the data processing facility. Besides, the HS data can be easily and efficiently integrated with those obtained by the airborne LIDAR system or the sensors operating in other spectral bands.

The HS scanners are the most advanced and sophisticated airborne optical devices of ERS. Here, we name but a few of state-of-the-art HS airborne imagers with excellent characteristics, such as AVIRIS, HYDICE, AISA, HyMAP, ARES, CASI 1500, and AisaEAGLET and give basic specifications of the some interesting among them.

The Avionic Multisensors HS System, manufactured by SELEX Galileo company, includes four HS/MS optical heads (e.g., VNIR, SWIR, MWIR, and TIR), which provide, in different configurations, a spectral coverage from the visible (400 nm) to the TIR (12,000 nm) band; the inertial platform with integrated GPS (INS/GPS) to take records of the aircraft position and attitude, and the Instrument Control Unit/Preprocessing Computer to control the optical head and store data in internal HD memory. The "modular" approach allows, with just a change of mechanical interface, a flexible arrangement of instrument accommodation, making it suitable for use on different platforms, including unmanned aerial vehicle (UAV) and ultralight aircrafts.

HS System can be applied both in airborne and ground operations, although its basic configuration is designed for airborne platforms, with pushbroom scanning to build up the spectral data cubes. To derive georegistered images, the flight data coming from a dedicated GPS/INS unit are continuously logged in sync with the HS data. In the ground system, the same optical head is used as in the airborne system, as well as scanning platform synchronized with the image acquisition. In this mode, the instruments can be used as "static" cameras for applications, where the linear platform movement needed for pushbroom mode is not available. The development follows a modularity principle aimed at creation of a flexible system to be utilized on manned or unmanned platforms for different applications, such as UAV, light aircraft, medium-range aircraft for maritime and coastal surveillance and patrol, and multi-mission maritime aircraft for high-altitude surveillance.

SPECIM is one of the world's leading companies for HS imaging instruments, from UV through VNIR and SWIR up to TIR spectral bands. The company provides imaging spectrographs, spectral cameras, and HS imaging technology to a rapidly increasing number of industrial customers and large scientific clientele. SPECIM airborne HS sensors provide market leading solutions for remote sensing, from small UAV systems to full-featured commercial, research, and military tools. SPECIM brought to the market the AisaEAGLET HSS, particularly designed to meet space and weight limitations of UAVs and small piloted aircrafts. This HSS is the most compact and complete airborne HS imaging system to acquire full, contiguous VNIR HS data with a high spatial resolution of 1600 pixels and excellent sensitivity even under low-light conditions, with the imager's spectral range of 400–1000 nm and spectral resolution of 3.3 nm (up to 200 spectral bands). Its FOV is ~30°, IFOV—0.02°; resolution on the Earth's surface ~0.35 m from 1000 m altitude. The system has a total mass of 10 kg, including an HS head (3.5 kg), compact data acquisition computer, and GPS/INS unit. Sensor head (Figure 2.8) contains a wavelength and radiometric calibration file. System output has 12 bits digital, SNR in the range of 130:1–300:1 (depending on the band configuration), and a power consumption <100 W, DC 10–30 V. The throughput of the high-efficiency imaging spectrograph is practically independent of polarization. The operator can create HS and MS mode, apply specific band configurations, and quickly change from one mode or configuration to another in flight operation. The system's modular design facilitates its integration with different payloads. Company provides support for the implementation of the sensor system control from ground through a telemetry link to UAV. In addition to its high performance and compactness, its cost effectiveness makes it an exceptional and versatile remote sensing tool

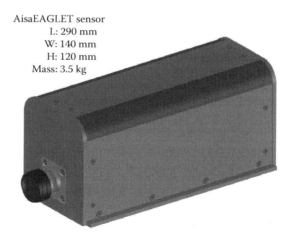

AisaEAGLET sensor
L: 290 mm
W: 140 mm
H: 120 mm
Mass: 3.5 kg

FIGURE 2.8   Optical head of AisaEAGLET HSS destined for operation on small aircrafts.

for environmental, forestry, agricultural, security, and defense applications. The AisaEAGLET HSS can be easily used in ground-based applications thanks to company proprietary scanning devices, which allow scanning of the target from a static platform, like a tripod or security monitoring facility.

SpecTIR Company presents the pushbroom-imaging spectral instrument ProSpecTIR-VS for remote sensing. The rugged high-performance instruments have superior spectral imaging capability and are built with components for maximum performance and utility—great performing dispersive optics, high dynamic range imaging devices, integrated GPS/INS sensor, and durable housing, all integrated with flight operations and recording hardware. The HS instrument has dual sensors individually covering VNIR wavelengths of 400–1000 nm and SWIR in the 1000–2500 nm wavelength range. The dual sensors are co-boresighted and include all hardware, acquisition and processing software for flight operations and spectral mapping. The HSS described can be installed in almost any light aircraft with aerial camera capability. The imagery is navigated with the integrated DGPS/IMU and after processing provides georeferenced, radiance, and reflectance files readily imported into spectral analysis programs. Instrument's main characteristics are spectral resolution of 2.9 nm, 250 spectral bands, 320 spatial pixels, FOV of 24°, IFOV of 1.3 mrad, SNR of 350–500 VNIR, and 800 SWIR.

An airborne HS sensor HyMap, developed in Australia and deployed for commercial operations around the world, provides 128 bands across the reflective solar wavelength region of 0.45–2.5 nm with contiguous spectral coverage (except for the atmospheric water vapor bands) and bandwidths between 15 and 20 nm. The sensor operates in a 3-axis gyro stabilized platform to minimize image distortions due to aircraft motion. The system, to be transported between international survey sites by air freight, can be rapidly adapted into any aircraft with a standard aerial camera port. The HyMap provides a signal-to-noise ratio (>500:1) and image quality setting the industry standard. Lab calibration and operational system monitoring ensures that the imagery is calibrated as required for demanding spectral mapping tasks. Geolocation and image geocoding are based on differential GPS and INS. Sensor's IFOV is 2.5 mrad along track and 2.0 mrad cross track; FOV is 61.3° (512 pixels); and ground IFOV is 3–10 m for typical operational range.

New AISA HS imager in dual mode is capable of acquiring 492 bands of 2–6 nm bandwidths across the spectral range of 395–2503 nm, at pixel sizes between 75 cm and 4 m. It is mounted on a twin-engine Navajo along with a LIDAR system with 10 cm vertical accuracy and a digital camera. A known HYDICE sensor collects data of 210 bands over the range of 400–2500 nm with FOV of 320 pixels wide at an IFOV, projection of which on the ground (pixel size) is of 1–4 m, depending on the aircraft altitude and ground speed.

Let us give some more examples of the current-day HSSs that contributed in raising the technological level in the area. Successfully tested in the United States was the HSS onboard the reconnaissance UAV designed for spotting the tanks, missile launchers, and other camouflaged military equipment located on the colorful background. The system developed by the TRW Company provides imaging in 384 working spectral bands. Another example is the ERS lab, with high spatial and spectral resolution, operated worldwide and lately, for example, successfully commissioned in Kazakhstan. Its equipment consists of a widescreen 136-Mpixel UltraCAM X camera—one of the most powerful airborne devices of its kind in the world market—and high-resolution CASI 1500 HS scanner, with 288 programmable channels in the visible and IR bands. By its technological level, the equipment meets standards of the satellite-borne ERS, the only difference being the carrier involved. The advantage of placing the equipment onboard the aircraft lies in the high resolution on the terrain (3–50 cm) and in highly accurate 5 cm linking, both in plane and in height, as ensured by the onboard navigation system data. The coverage resolution achieved by the airborne HS imaging is within 25–400 cm. At the lab maximum flight altitude, 8 km swath can be covered, as with the high-resolution OrbView-3 satellite. To cope with huge data streams, the onboard equipment includes also the 13-processor supercomputer with 4 Tb memory.

2.4.3 Spaceborne Imaging

Big hopes are pinned on the imaging from space. For a long time, the MS systems have been performing onboard the spacecraft, generating several images in separate wide spectral bands, from visual to IR. Till now, MS imaging from the new generation spacecraft (GeoEye: —one to four spectral zones, RapidEye: five zones, and WorldView: —two to eight zones) is of a great practical interest. Nowadays, there are opportunities of developing spaceborne HS equipment for observations in optical range, with high informational and operational characteristics, which would enhance the solution level of space monitoring tasks [18–22].

Satellite platforms enable placement of various sensors for observation and provide coverage of the Earth's surface and periodic monitoring of the areas of interest during the satellite revisits. The image resolution is defined by the satellite orbital parameters; its altitude in the first place. For satellites, there are no political borders, so they can sweep any part of the globe, regardless of the attitude of the states the onboard imaging equipment flies over. The basic weakness of the spaceborne imagery is its multihundred million dollars cost to develop the system as well as ground support facilities, while keeping in mind the spacecraft's relatively short operating life of about 5 years or less. Decrease of the satellite mass can result in substantial cost saving, primarily due to less powerful and low-cost launchers to be involved in putting the satellite into orbit. This is why the concept of small satellite-based HS imaging was developed and implemented by means of mini-satellites (100–500 kg).

The basic characteristics of the spacebased HS instruments are presented in Table 2.1. The list includes both flight-proven devices and those under development for future operations in space [23–35]. Most of them are devised for LEO small satellite missions, though HSS placing possibilities onboard the large satellites and satellites on geostationary orbit also is considered [36]. The first 14 rows in Table 2.1 pertain to HS satellite with a swath on the Earth's surface of 5–20 km, the number of spectral bands ~200, spatial and spectral resolution of 10–30 m and 10–20 nm, respectively. The SUPERSPEC and VENμS sensors represent MS imagers of the next generation; it was put in the table for comparison with previous HS instruments. The focal plane of MS cameras consists of a multiline array, one for each spectral band with relative narrow band spectral filters. Along with the narrow FOV HS instruments, some specimens of the wide FOV HS imaging systems would also be operational. Examples of such HS missions are shown in the last two rows of Table 2.1. It is obvious that space applications of HS remote sensors are under constant development.

Indicative example of remote sensor advancement is aforementioned joint Israeli/French scientific and technological project Vegetation and Environment New Micro Satellite (VENμS), with scheduled launch time in the coming years. It is designed for high-definition photography of agricultural tracts as a means of ecological control. VENμS launch weight is 260 kg, with its expected lifetime in space of more than 4 years. It has superior parameters in terms of the number and width of its bands, spatial resolution, imaged strip width, mass energy and information capacities. The 90 W camera weighs 45 kg and discerns the objects down to 5.3 m in size from 720 km altitude, scanning across 27.5 km swaths. Photographing will be carried out in 12 narrow spectral channels in the 415–910 nm range, with each band varying from 16 to 40 nm. In this mission, most satellite and sensor design parameters are optimized.

Let us briefly review the worldwide state of the art in the field. At the time of writing, there is a small number of experimental operational spacecrafts equipped with HSS, viz., EO-1 (under the NASA program) and MightySat II.1, aka Sindri (under U.S. Air Force program); two other military projects—ORBView-4/Warfighter and NEMO—are at various stages of readiness. The information capacity for different spectral bands was assessed by processing the measurement data as obtained in simultaneous imaging of the test objects by HSS and traditional optical technique (EO-1 with Landsat 7 and Terra; MightySat with the Keyhole visual reconnaissance satellite).

TABLE 2.1

Space-Based Hyperspectral Mission Overview: Operational and Planned Instruments

Instrument (Satellite)	Altitude, km	Pixel Size, m	Number Bands	Spectral Range, nm	Spectral Resolution, nm	IFOV, μrad	Swath, km
HIS (SIMSA)	523	25	220	430–2400	20	47.8	7.7
FTHSI (MightySatII)	565	30	256	450–1050	10–50	50	13
Hyperion (EO-1)	705	30	220	400–2500	10	42.5	7.5
CHRIS (PROBA)	580	25	19	400–1050	1.25–11.0	43.1	17.5
COIS (NEMO)	605	30	210	400–2500	10	49.5	30
ARIES-1 (ARIES-1)	500	30	32	400–1100	22	60	15
			32	2000–2500	16		
			32	1000–2000	31		
UKON-B	400	20	256	400–800	4–8	50	15
Warfighter-1 (OrbView-4)	470	8	200	450–2500	11	20	5
			80	3000–5000			
EnMAP	675	30	92	420–1030	5–10	30	30
			108	950–2450	10–20		
HypSEO (MITA)	620	20	~210	400–2500	10	40	20
MSMI (SUNSAT)	660	15	~200	400–2350	10	22	15
PRISMA	695	30	250	400–2500	<10	40	30
ARTEMIS (TacSat-3)	425	4	400	400–2500	5	70	~10
HyspIRI	~700	60	>200	380–2500	10	80	145
SUPERSPEC (MYRIADE)	720	20	8	430–910	20	30	120
VENμS	720	5.3	12	415–910	16–40	8	27.5
Global Imager (ADEOS-2)	802	250–1000	36	380–1195	10–1000	310–1250	1600
WFIS (like MODIS)	705	1400	630	400–1000	1–5	2000	2400

Investing in commercial satellites can be a risky business. As known, TRW's Lewis satellite with HS sensors was lost shortly after launch in 1997. In the same year, EarthWatch lost its EarlyBird satellite 4 days after launch. In spite of the risk, every firm feels obligated to incorporate such systems in future spacecraft meant for ERS.

For example, in Russia there is a metrological basis for the subject-related processing of spectrometer data. Specific prototypes of the HS equipment were developed by the AFAR scientific production enterprise and the Reagent scientific-technical center [37]. The operation of the Yukon-B video-spectrometric system, with mass of 1.5 kg, is based on the optoacoustic effect in the visible band (500–800 nm). Overall number of zones is 256, from which 10–100, with width of 4–8 nm, are program-selected. Spatial resolution at 400 km altitude will be 20 m, and the frame size of the surface area − 15 × 12 km, with one zonal image size amounting to 4–5 Mbit. Yukon-B is a component of the projected Yukon-UVIT complex designed to cover the UV, visible, and thermal spectral bands. As a platform for Yukon-B, the service module of the ISS Russian segment can be used as well as small D33 spacecraft of the Monitor-E model manufactured by the Khrunichev Plant, and Kondor-E of the Scientific Production Enterprise of the Machine Building.

Another Russian project is the Astrogon mini-satellite, under development at the Research Institute of the Electric Machinery at Istra, near Moscow, as a part of the Gazprom Trust aerospace monitoring program. The Reagent (see above) HS video-camera has the 700-channel capacity (with margin for increase up to 1000), with a spatial resolution of 5 m and adaptive control along both the spectral and spatial coordinates. The programmed control system permits the selection of resolution level from 0.5–100 nm in the spectral bands of 200–900 and 1000–1400 nm as well as one of two possible signal polarization modes. For the signal space-frequency transformation, both re-arrangeable optoacoustic filters and micro-channel photon detectors are used. The total mass of the imaging equipment is 4 kg, with power consumption of 15 W. The designed precision of the geometric matching of the images is 5 m. The camera control system makes it possible to carry out both instrumental and virtual stereo-imaging.

Two more developments, pertaining to the HS field, are at different stages of implementation in the Russian space industry, viz.:

- The Astrogon project, involving mini-satellite formation for space monitoring of wildfires, volcano eruptions, and other calamities, and tested in the experiments onboard planes and helicopters, is under consideration in a joint development with Germany. Involved in the project is IR technology applied by the German party on the small BIRD satellite. The monitoring would make it possible to discern and locate the sources of the excess energy flux of both natural and anthropogenic origins.
- The Vavilov State Optical Institute (St. Petersburg, Russia) has developed a compact onboard HS camera for the small spacecraft, to receive simultaneously in ~100 spectral channels of the VNIR bands (200–1000 nm) a 2D distribution of the brightness field with angular resolution ~20″, which corresponds to 100 m spatial resolution on the Earth's surface for 1000 km orbit. The instrument provides detailed spectral information on the small gaseous constituents of the atmosphere, the status of inland waters, and vegetation and soil specifics.

A remarkable achievement in the HS-assisted space exploration was the successful launch in May 2009 of the multipurpose TacSat-3 US Air Force satellite from the NASA Wallops Flight Facility on the Virginia coast [38]. The satellite (Figure 2.9), orbiting at 425 km altitude, is required to deliver HS images within 10 min after a shot. The main component of its specialized equipment, the Advanced Responsive Tactically-Effective Military Imaging Spectrometer (ARTEMIS), designed to operate in 400 spectral intervals in the 400–2500 nm band (the entire visible and very-near IR), is an advanced version of the Hyperion scanning spectrometer (NASA, 2000), which performs in 220 intervals with spatial resolution of 30 m/pixel. Space photos with such a resolution form the basis for scanning coverage of the popular geointerfaces, like Google Earth. Besides the spectrometer, the ARTEMIS system includes a telescope and a signal processor with 16 Gb onboard memory. The secondary mirror of the telescope is integrated into the focal plane light-receiving matrix, common for all the bands, unlike its separate counterparts in the prior generation of HS detectors. This improvement enabled a

FIGURE 2.9 **(See color insert.)** Artist's impression of TacSat-3 satellite in orbit.

simpler design of the telescope optical setup, easier data processing, and a cost saving. Duration of the TacSat-3 satellite demonstration mission was 1 year, 5 months, and 8 days.

Let us return to HS missions and see into some problems connected with the development of the HS satellites and their HS sensors. Energy resolution, defined as a number of resolved levels of the object's brightness, or of its image illumination, is chosen in accordance with required signal-to-noise ratio. It is well known that detection and recognition probability rises together with the increase in ratio. Besides, the interconnection between spatial, spectral, and energy resolutions, which takes place in live HS sensors, should be taken into consideration. For example, if higher spatial resolution is achieved by reducing the size of the image element, then the element would get lesser share of the energy, needed for splitting by operating spectral ranges and obtaining the required signal-to-noise ratio in each of them. This is why various combinations of the basic characteristics can be found in the real devices, depending on the task at hand.

The true parameters—bandwidths and instantaneous fields of view (IFOV)—of existing HS imagery systems in the visible spectral range are presented in Figure 2.10. The interrelation between the spectral and spatial resolutions, as calculated for an HS instrument with a 20 cm aperture, suitable for operation on small satellites, is drawn by a solid line. As seen, there is a good conformity between the theory and practice.

The major obstacle in developing HS sensors is well-known difficulty with designing the relatively cheap high sensitive photo-detectors, operating in a wide spectral range, with high spatial, spectral, and time resolution. This is also relevant with regard to constructing efficient, durable, and compact cooling systems for the radiation receivers as well as to some more scientific and technical problems. Not all the attempts made thus far to design HSS as a Fourier-spectrometer were successful, since the requirement of real-time performance failed to be met.

Another field of application for the small satellites may be the installation of a compact coarse-resolution imager to correct imagery, obtained by other sensors, for atmospheric variability caused mostly by water vapor and aerosols. Such atmospheric corrector (LEISA AC) was used in the EO-1

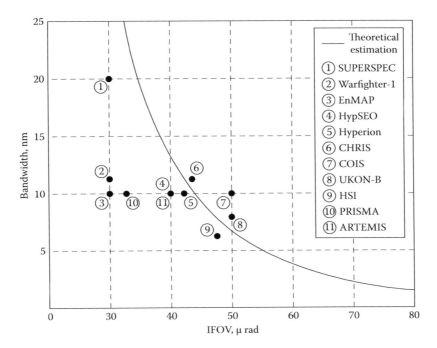

FIGURE 2.10 Bandwidth/IFOV relation for the visible range.

land imaging mission. Using wedged filter technology, the AC provided the spectral coverage of 890–1580 nm with moderate resolutions (spectral ∼5 nm and spatial ∼250 m at nadir).

Conceiving the onboard communication system with high-rate download of as many simultaneous images as there are channels, powered by additional solar panels, it should be taken into consideration, that enhancing spatial resolution would imply the weight increase. Likewise, adding more channels would also result in more weight as well as in more power consumption. Moreover, with the orbit altitude increase (e.g., from 400 to 1200 km), or with wavelength increase (from visible to IR), the imager aperture diameter should also be enlarged. For higher ground resolution, the satellite dimensions are defined not only by dimensions of the communication and power units, but also by those of the imager. Therefore, the requirements to the transmitting antenna, optical system, total power consumption, etc., would impose operational constraints on the mission design, which should be eventually translated into restrictions on the dimensions and mass of the spacecraft.

Given the existing physical and technological constraints, the resulting ambiguity as to the satellite dimensions urged us to determine the estimation criteria for the satellite's mass versus spatial resolution and channel array. The feasible combinations of the HS imagers with LEO observation satellites of various sizes and weights, depending on the devices' spatial and spectral resolutions, are depicted in Figure 2.11. Three different domains represent the conventional satellite classes, viz., micro (up to 100 kg), mini (100–500 kg), and large (over 500 kg). The completed analysis clearly shows that it is impossible to carry out all conceivable LEO HS missions with microsatellites alone. Meanwhile, most scientific, commercial, and military HS tasks can be implemented just with mini-satellites on the basis of conventional technology. According to one of the recent assessments, the manufacturing cost of such a mini-satellite is US$ 30 K/kg. However, a real total cost of a mini-satellite suitable for HS imaging payload is considerable as discussed above. As an example, the cost of the relatively inexpensive small satellite TacSat-3 with HSS ARTEMIS, described earlier, is reportedly US$ 90M.

Still, all spatial, spectral, and radiation requirements growing higher, the mass value would exceed the upper limit for mini-satellites, therefore ensuring 200–300 spectral bands with spatial resolution of some meters (especially in the IR region at high-LEO altitudes) would be feasible only with large satellites. Microsatellites would match well only in the visible region, with large limitations on the resolution (the ground resolutions of 30–60 m and the number of spectral bands 30–100, respectively). The spatial and spectral resolutions of the well-known operated HS satellites EO-1, MightySat II, Proba-1, and TacSat-3 are also drawn on Figure 2.11

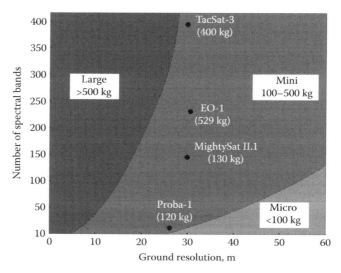

FIGURE 2.11 HSS accommodation on satellites of different subclasses as function of imager's ground resolution and number of bands.

with bullet points, their respective weights being put in brackets. As can be seen, the point's positions match well with the estimations.

However, used onboard the small satellites were special-purpose instruments with high spectral resolution in a limited number of wavelengths. As an example is the atmospheric ozone mapping by the small satellites, in a short spectral range of ozone bands in UV using a simple filter wheel photometer. Similar experiments with poor spatial resolution are available in small satellites [39].

Instrumentation mass and size, affordable by the present technology level, having such an impact on the future missions feasibility, big efforts are made to reduce the power consumption, mass, volume, and the unit costs of the HS sensors under development, as well as to lower demands to the data downlink. Only time will tell whether these efforts will lead to the emergence of miniature HS instruments suitable for the small satellites' missions.

2.5 LIDAR AND HS DATA INTEGRATION

Virtually all remote sensing techniques, including those presented in the previous paragraphs, rely upon passive sensing of the amount of solar radiation as reflected toward the sensor by clouds, ocean, or solid land, and IR-radiation emitted from the natural or artificial thermal sources on the Earth's surface. Unlike them, the Light Detection And Ranging (LIDAR) system is an active sensor. Since LIDAR carries a source of radiation of its own, it can determine where and when, in the daytime or at night, to take the measurements. LIDAR is similar to radar in the sense that it also can track any target of interest, from airplanes in flight to thunderstorms. But unlike the radar, LIDAR uses short pulses of coherent laser light, with very short wavelengths and high instant capacity, so its laser beam does not disperse while traveling away from the source, as the ordinary light does. Besides, lasers offer great advantages over conventional light sources in terms of peak power and narrow spectral bandwidth.

A primary functional part of LIDAR is the laser, its other components being the scanner, collimator, lens, receiver, amplifiers, samplers, and other optical and electronic elements. The main function of the laser is to generate emission, which, having been reflected by the Earth's surface or the ground objects, returns to the source, to be detected by highly sensitive receiver. The response time being directly proportional to the target range, it is equivalent to measuring the distance between the emitting source and the reflecting object. The wavelength choice depends on the laser function as well as on the safety and security requirements. The most commonly used are Nd:YAG lasers, which operate at wavelengths, such as 1550 nm, 1064 nm—near-IR, 532 nm—green light, and 355 nm—near-UV.

While ground-based LIDAR instruments profile a single viewing site, the airborne LIDAR systems offer one of the most accurate, expedient, and cost-effective ways to capture upward- or downward-looking data over a wider area. Vertical accuracies of less than 5 cm are possible at 500 m above ground, amounting to less than 20 cm at 3 km altitude. This enables LIDAR to be involved in various applications, such as surveys of the corridors like pipeline routes, roads, power lines, highways and railways, urban environments and mapping, flood plain surveys, forestry (mapping of tree canopies), archaeology, seismic exploration, coastal zone surveys, and oil and gas exploration.

In the airborne LIDAR systems, the laser range-finder is mounted over an opening in the aircraft floor and scans beneath the aircraft, producing a wide swath, over which the distance to the ground is measured. The scanning device controls the direction of the laser pulses' propagation, to provide coverage of some predetermined scan swath. In most cases, the cross-track scanning is implemented by using the oscillating mirror, while the along-track scanning—by the motion of the carrier aircraft itself along its operational path.

All the aircraft movements are recorded by its INS, to be used in the data postprocessing. The LIDAR can also be installed on a stabilized platform. By merging laser ranging, GPS positioning, and inertial attitude technologies, LIDAR can directly measure the shape of the Earth's surface beneath the aircraft's flight path. Data collection rates, accuracy, and other characteristics of some existing airborne LIDARs correspond to user's requirements and allow carrying out various researches.

A fundamental limitation of airborne laser scanners and imagery consists in their generally straight-down perspective, which makes impossible accurate and detailed mapping of vertical structures such as cliff faces, coastal ridges, or any vertical side of a natural or manmade structure. To reveal the vertical faces of structures from the air, imaging from different perspectives should be introduced in the LIDAR data acquisition process by means of some specialized mounting.

The airborne applications being limited to comparatively small regions, it is the spaceborne LIDAR systems that can provide continuous geospatial data and offer a truly global view of the Earth's surface. Furthermore, unlike airborne laser, a spaceborne one poses no hazard to the general public, even when viewing with the naked eye, binoculars, or small telescopes.

The first LIDAR system in space was the Lidar In-Space Technology Experiment (LITE) instrument, designed and built by NASA. From 260 km space shuttle orbit, its pencil-wide laser beam spreads approximately 300 m wide at the surface—about the size of three football fields. The data obtained by LITE were used in Earth atmosphere studies, to measure vegetative cover and to distinguish various types of surfaces. LIDAR is a very reliable tool for active remote sensing from space. A future spaceborne high-performance imaging LIDAR for global coverage is now under development.

While early LIDAR scanning systems were capable of registering a single return pulse for each transmitted pulse, most systems in operation today are able to record multiple discrete returns, which occurs when emitted LIDAR beam encounters in its path an object like a forest canopy. The data may be classified as first, second, third, or last returns, and bare-Earth. A return might be generated from a reflection caused by the top of the tree, but a sufficient amount of laser light energy is able to reflect from lower portions of the tree, or finally, from the ground. These multiple returns can be analyzed and classified to produce information about the objects as well as the bare ground surface. Buildings, trees, and power lines are individually discernible features and may be classified separately. Most often used are the first and last returns, the first one showing the highest features, such as the tree canopy, buildings, etc., and the last generally assumed to refer to the ground level. Being digital, it can be directly processed into detailed bare-Earth digital elevation model at vertical accuracies within 0.15–1 m. Derived products include contour maps, slope aspect, 3D topographic images, virtual reality visualizations, and more.

The LIDAR instruments have an option to measure the intensity of the returned light. This captured information can be displayed as an orthorectified image—a 256-gray level mapping of the light pulse reflectance amplitude. Such an image, comparable with a coarse photograph of the area under survey, can be painted according to the elevation levels.

The amplitudes of the laser scan return signals, as measured by LIDAR, do not allow proper reconstruction of the ground reflectance mainly because of spreading and absorption terms, which depend on the laser-target distance. This prevents not only accurate imaging of the ground, but also the integration of the intensity information with the segmentation/classification process. This difficulty can be overcome by a suitable calibration of the laser scan return values. A number of signal calibration procedures are known, insufficiently precise as they are. Anyway, the radiometric information of the laser scan is relatively poor in terms of information content, being the laser light monochromatic.

Some researchers utilize the return intensities to extract more information about the reflecting surfaces, than can be done by using LIDAR solely as a ranging tool. Using intensity measurements combined with height data for classification of vegetation types, it was found out, for example, that the reflectance values, as shown by pure broadleaf forests, are significantly higher than those for pure conifer forests.

The exploratory data analysis was carried out to assess the potential of the laser return type and intensity as variables for classification of individual trees or forest stands according to species. The evaluation of the irregular behavior of some major ground indices, fulfilled on the basis of integrating the LIDAR and HS data, made it possible to refine the technique of discovering new archeological sites. In land-cover classification, it proved effective to process LIDAR intensity data clouds together with high quality, large-coverage images provided by the airborne cameras.

Besides the intensity, the primary products derived from the raw LIDAR data include also the waveform returns. New full-waveform laser scanning technique offers advanced data analysis options not available in traditional LIDAR sensors. This technique, implemented in the latest generation of scanners, will promote evolution in such areas as continuous multilayer vegetation modeling, waveform-based feature classification, single-pulse slope determination, etc.

The LIDAR images, including orthophotos, can be seamlessly integrated with MS, HS, and panchromatic data sets. Combined with Geographic Information System (GIS) data and other surveying information, LIDAR imagery enables generating complex geomorphic-structure mapping products, building renderings, advanced 3D Earth modeling, and many more high-quality mapping products. Galileo is one of a few companies providing highly accurate fused image data products. Using high spatial-resolution LIDAR and HS imaging data, the maps were improved by adding a 3D perspective.

As was discovered, the airborne LIDAR can be used for monitoring biofluorescence when coupled with a HS tunable filter [40]. The promising technique of HS fluorescence imaging, as induced by a 632 nm laser, was applied for assessing soluble solids content of fruit. Laser-induced HS biofluorescence imaging of different objects is now under study in lab experiments, airborne applications, and fieldwork.

The efficiency of the advanced geospatial technology, based on the fusion of HS imagery and LIDAR data, was assessed lately in Canada and United States, as applied to the mapping of the location and condition of different tree species, for both the effective management of the forested ecosystems and counteraction to invasive tree pests. The HS imaging was obtained by Airborne Imaging Spectrometer for Applications (AISA) dual system with Eagle (395–970 nm) and Hawk (970–2503 nm) sensors, enabling simultaneous acquisition in 492 narrow spectral channels. LIDAR data were collected using a TRSI Mark II two-return sensor. A collection of ground spectral signatures were also used in data processing.

German and Israeli scientists are implementing the project on fusion of HS images and LIDAR data for engineering structure monitoring [41]. They confirm the assessed degradation of urban materials in artificial structures by exploring possible chemical and physical changes using spectral information across the VIS-NIR-SWIR spectral region (400–2500 nm). This technique provides the ability for easy, rapid, and accurate in situ assessment of many materials on a spatial domain under near real-time condition and high temporal resolution. LIDAR technology, on the other hand, offers precise information about the geometrical properties of the surfaces within the areas under study and can reflect different shapes and formations of the complex urban environment. Generating a monitoring system based on the integrative fusion between HSS and LIDAR data may enlarge the application envelop of each technology separately and contribute valuable information on urban runoff and planning.

The LIDAR and HS data were also investigated from the viewpoint of their potential to predict the canopy chlorophyll and carotenoid concentrations in a spatially complex boreal mixed wood. Using canopy scale application of HS reflectance and derivative indices, LIDAR data analysis was conducted to identify structural metrics related to chlorophyll concentration. Then the LIDAR metrics and HS indices were combined to determine whether concentration estimates could be further improved. Integrating mean LIDAR first-return heights for the 25th percentile with the HS derivative chlorophyll index enabled further strengthening of the relationship to canopy chlorophyll concentrations. Maps of the total chlorophyll concentration for the study site revealed distinct spatial patterns, indicative of the spatial distribution of species at the site.

The project [42] applied advanced geospatial technology, including high-resolution airborne AISA HSS and LIDAR data collection, in conjunction with analytical applications of GIS, to develop new tools needed for improved species mapping, risk assessment, forest health monitoring, rapid early detection, and management of invasive species. Similar technologies play an increasingly important role in offering accurate, timely, and cost-effective solutions to such problems.

Airborne LIDAR has become a fully operational tool for hydrographic surveying. Present-day airborne laser bathymetry systems can simultaneously measure both water depth and adjacent surface topography. HS imagery from aforementioned Compact Airborne Spectrographic Imager (CASI) (Section 2.4.2) has proven to be a valuable instrument for coastal measurements and analysis. CASI spectral resolution of 288 bands for each spatial pixel allows for extraction of a vast amount of information, such as water clarity and temperature, bottom type, bathymetry, as well as water quality (chlorophyll, dissolved organic carbon, and suspended minerals), soil types, and plant species. In order to achieve a comprehensive hydrographic capability, the LIDAR and HSS CASI sensors were integrated, to provide the different agencies with the relevant information. Naval Oceanographic Office, for example, uses the airborne laser bathymetry system to collect hydrographic information about the littoral zone for the warfighter and, by adding the HS capability, enhances its efficiency in rendering a quick and more adequate environmental picture.

An interesting research pertains to the integration of LIDAR and HS data to improve the discovery of new archeological sites [43]. Under investigation was the possibility of using such an integrated dataset in evaluation of the irregular behavior of some major ground indices. While HS data allow identifying specific humidity, vegetation, and thermal conditions in the target area, LIDAR data provide accurate geometric information. In fact, accurate filtering of the laser scanning data allows the computation of Digital Terrain Model, while Lambertian-based calibration of LIDAR intensity enhances the automatic data segmentation and thus the detection of possible sites of interest. In order to fully extract HS information from MIVIS and AISA sensor data, some specific procedures have been implemented together with adequate computer analysis. Recently, such data processing has been applied to both already discovered archeological sites of the ancient city of Aquileia, and the new areas in its northern part. When integrated, the resulting datasets revealed with sensible accuracy, the presence of a surface/below surface archaeological heritage. As was confirmed by the archaeologists, currently available airborne imaging, namely, LIDAR-assisted and the HS, is the most technologically advanced approach. In some cases, these instruments were put together as an integrated unique sensor mounted on a helicopter.

New remote sensing technologies, on the basis of LIDAR data integration with optical HS and MS imaging data, were developed for natural hazard detection and decision support systems. The innovative imagery fusion refines identification and mapping of past geologic events, such as landslides and faults, while providing also quicker and simpler processes for forecasting and mitigating future environmental hazards.

The U.S. researchers used remote sensing data to improve mapping and characterizing of the mechanical properties of rocks. In early 2007, simultaneous HS and LIDAR imagery was obtained from the plane over Cuprite, Nevada. The imaging instrumentation included the Optech LIDAR, with operational wavelength of $1.064\,\mu m$, next-generation HSS—Mapping Reflected-energy Spectrometer, working in the 0.4–$2.5\,\mu m$ spectral range, and the Nikon D2X digital camera to capture high-spatial resolution true-color images. Results of the data analysis suggested, for some surfaces, a correlation between mineral content and surface roughness, although the LIDAR resolution ($\sim 1\,m$ ground sampling distance) turned out to be too coarse to extract surface texture properties of clay minerals in some of the alluvial fans captured in the imagery. Such experiments may provide valuable information about the mechanical properties of the surface cover in addition to generating another variable of use for material characterization, image classification, and scene segmentation.

Short description of the successful researches, based on joint HSS and LIDAR data, does not pretend to be complete; it just demonstrates high efficiency of this imaging technique in various applications. Future mission planning should include consideration of determining optimal ground sampling to be used by LIDAR and HS systems [44–45]. The fusion of LIDAR elevation data and MS and HS classification results is a valuable tool for imagery analysis and should be explored more extensively.

2.6 SUMMARY AND OUTLOOK

HS imaging sensors are designed to provide a detailed analysis of the entire light spectrum from visible light up to far IR. Present-day sensors can break this wavelength range into hundreds or even thousands of sections for individual analysis and examine such extremely narrow sections, which makes it possible to look for specific chemical compositions that reflect light only in the chosen wavebands. HSSs simultaneously capture both spectral and spatial information of the scene and provide a fully registered spatial and spectral data cube for every video frame. HSSs are equally well suited for ground, airborne, and space operations. Due to their unique capabilities, they can collect data from hand-held and vehicle-mounted ground systems, low- and high-altitude aircraft, as well as LEO or even GEO satellites, according to the application in question. One shortcoming of last flight operations with HSS, also with aerial- and space-based LIDAR, is the need for a relatively cloud-free condition over the target area.

To verify the identifications derived from the airborne HS imagery, ground spectral measurements are to be carried out at the sites of interest by hand-held or truck-mounted HSS instruments, providing the ground truth, that is, the material characterization of the natural land samples in UV-VNIR-SWIR-MWIR-LWIR spectral bands.

Many airborne HSSs have been built and have become operational worldwide since 1987 virgin flight of AVIRIS—Airborne Visible/IR Imaging Spectrometer. Nowadays, HS images can be taken from aircrafts, UAVs, helicopters, or zeppelins. During the flights, the systems successfully fulfill their task of obtaining the HS imagery, although on a very short tracks available to every aircraft. Usually, the spaceborne HSS have their airborne counterparts involved in the development of the application data products before the satellite launch, and in calibration and data processing procedures.

Spaceborne HS imaging has huge potential with respect to duration of the observations, Earth's surface coverage, and abundance of the valuable information. Still, in its 10 year history there were, unfortunately, only four successful HS space missions. Described in this chapter were civilian, dual-use, and military HSSs that have been put into orbit as well as the sensors currently under development or design and some systems planned but never implemented, such as NEMO and Warfighter. Active promotion of the HS initiatives into space is constrained by prohibitive costs of such missions. For example, the estimated cost of the HyspIRI HS IR Imager, as proposed by NASA-JPL for 2013–2016, amounts to as much as US$ 300M. This is why many HS projects were left unfinished, and a number of interesting HSS missions are still waiting for further funding.

An expertise in the HSS architecture, engineering, design, development, acquisition, operations, and program management was set forth in this chapter. There are several types of HSSs, including pushbroom, whiskbroom, Fourier transform, and tunable filter approach sensors. Most of the airborne and spaceborne HS imagers operate in the Visible to SWIR bands. The Mid-Wave IR (MWIR) and LWIR capabilities are aimed at identification of materials in 3000–13,000 nm "chemical fingerprint" spectral region and high-resolution temperature data acquisition.

At present, most of the HS servicing firms are ready to assist civil and commercial clients in their remote sensing data demands. They can support the entire end-to-end data collection effort, starting with site identification, observation planning, imaging operations, and data analysis. Typical remote HS products include spectrally and radiometrically calibrated HS data cubes, georeference files for all data cubes, atmospherically corrected HS data cubes, field spectra from hand-held spectrometers, laboratory spectra of samples, and apparent emissivity data cubes for IR sensors.

Without doubt, the HS data from operating sensors provide an important contribution for the research community [4,46–47]. The current state-of-the-art HS data are far ahead of the first tentative steps taken about 30 years ago. New HS technology, including instruments performance, processing schemes and calibration, improved significantly. Well-established spheres of HS application include agriculture, forestry, water resources, atmosphere, geology, mineralogy, wetlands, environment, management of coastal waters, military and security, and urban areas. Here are some examples of successful HS commercial applications:

- Drug enforcement-like vehicle-mounted monitoring of illicit substance cultivation
- Marine biosphere monitoring in a hand-held or vehicle-mounted underwater configuration
- Forensic science and crime scene investigation and detection of counterfeit notes
- Environmental and toxicological monitoring of airborne or waterborne pollutants
- Mining and petroleum exploration
- Mapping of forest fires
- Medical applications for retinal imaging, colonoscopy, and skin cancer detection

HS mission and sensor performance parameters are derived from the requirements for specific applications. For spaceborne HS imaging, along with cost considerations, this leads to the choice of the orbit, relook capability, SNR, spatial resolution, data volume per time period, spectral resolution, spectral sampling interval, swath width, radiometric accuracy and stability, spectral range of interest, required time for the data delivery to the users, and processing steps.

Individual users' HS imagery and related data requirements, which are essential to the event under examination, differ from each other. The most important for all users proved to be not HSS and imagery qualities, but better service and reduced equipment costs.

Let us consider some customer overviews that reflect the users' requirements for the modern and future HSS characteristics depending on the imagery application. The required number of the spectral bands ranges 200–3000 over the entire spectral region of VNIR-TIR. Whereas agriculture, limnology, land use, and vegetation users are content with 200–300 bands, the geological, atmospheric, and some vegetation applications obviously will need more. Atmospheric application requirements heavily depend on the observation target (reaching from some spectral bands for aerosol studies to a huge number of bands for trace gas retrieval) and purpose (spatially coarse resolution for global observations and high for the urban ones).

The required spectral resolution in the VNIR ranges 0.05–30 nm, in the SWIR 0.2–40 nm, and in the MIR/TIR 4–400 nm. The high spectral resolution in the geology applications is significantly different as compared to the other applications, especially in the MIR/TIR waveband. The spatial resolution is expected to be between 4–20 m in the VNIR and SWIR and between 10–30 m in the MIR and TIR for atmospheric applications. The requirements for vegetation and geology applications are very similar, with SNR of 400–500 in the VNIR, in contrast to limnology and atmospheric applications, which would need SNR of 700 and 1500, respectively. HS images delivered from LEO satellite generally have coarser spatial resolution and wider swath, and vice versa. Aside from military application, in some civil applications, such as agriculture, of urgent importance is the in-time delivery of imagery products to the end-users.

Increasing advances in HS imagers has resulted more civil and military satellites planned to be put into orbit in the coming decade. These may include German EnMAP, Indian TWSat with HY SI-T coarse HS imager, South African MSMI HSS, satellites with sensors like Hyperon or CHRIS, wile-area synoptic sensor similar to MODIS and MERIS with finer spectral resolution, NASA HyspIRI and ESA FLEX, in case of funding. Future Landsat and NPOESS satellites might have the HS capability as well as U.K. Disaster Monitoring and German RapidEye constellations. Italy, China, Israel, Canada, and other countries have also announced their HS space projects. Along with this, the development of the airborne HS imaging for civilian and public-good applications is also anticipated.

Measuring the returned light intensities by LIDAR provides additional information about the reflecting surfaces, which can be utilized either by itself or together with HS data. The new HSS applications are focused on the airborne or spaceborne multiple sensor technologies, including LIDAR data, which enables combining information from different sensors in one solution.

REFERENCES

1. Glackin D.L., Peltzer G.R., *Civil, Commercial, and International Remote Sensing Systems and Geoprocessing*, The Aerospace Corporation, AIAA, Reston, VA, p. 89, 1999.
2. Kramer H.J., *Observation of the Earth and Its Environment, Survey of Missions and Sensors*, 4th edn., Springer, Berlin, Germany, 2002.
3. Nieke J., Schwarzer H., Neumann A., Zimmermann G., Imaging Spaceborne and Airborne Sensor Systems in the Beginning of the Next Century, *Conference on Sensors, Systems and Next Generation Satellites III*, 3221–71, SPIE, London, UK, 1997.
4. Puschell J.J., Hyperspectral imagers for current and future missions, *Proceedings of SPIE*, San Diego, CA, Vol. 4041, pp. 121–132, 2000.
5. Lillesand T.M., Kiefer R.W., Chipman J.W., *Remote Sensing and Image Interpretation*, 5th edn., Wiley, New York, 763 pp., 2004.
6. Chang C.-I., *Hyperspectral Imaging. Techniques for Spectral Detection and Classification*, Kluwer Academic/Plenum Publishers, New York, 370 pp., 2003.
7. Borengaser M., *Hyperspectral Remote Sensing*, Lewis, Boca Raton, FL, 2004.
8. А.Кучейко, Российские перспективы в гиперспектре, Новости космонавтики, Т.11, N7 (222), с. 24, 2001 (Kuchako A., Russian hyperspectral perspective, *Journal of Novosti Kosmonavtiki*, Vol. 11, N7, 222, 24, 2001).
9. Alsberg B., Hyperspectral chemical and property imaging, Patent 20090015686, USPC Class 3482221, January 15, 2009.
10. Bolton J.F., Full spectral imaging: A revisited approach to remote sensing, *SPIE Conference on Remote Sensing*, Barcelona, Spain, September 2003.
11. Bodkin A., Development of a miniature, hyperspectral imaging digital camera, Navy SBIR FY2005.1, Proposal N051-971-0772, 2005.
12. Fletcher-Holmes D.W., Harvey A.R., Real-time imaging with a hyperspectral fovea, *Journal of Optics A: Pure and Applied Optics*, 7, 298–302, 2005.
13. Skauli T., Kåsen I., Haavardsholm T., Kavara A., Tarabalka Y., Farsund O., Status of the Norwegian hyperspectral technology demonstrator, NATO OTAN, RTO-MP-SET-130, Norwegian Defence Research Establishment.
14. Yang C., Everitt J.H., Davis M.R., Mao, C., A CCD camera-based hyperspectral imaging system for stationary and airborne applications, *Geocarto International Journal*, 18(2), 71–80, 2003.
15. Anger C.D., Airborne hyperspectral remote sensing in the future, *Proceedings of the 4th International Airborne Remote Sensing Conference and Exhibition/21st Canadian Symposium on Remote Sensing*, Vol. 1, ERIM International Inc., Ann Arbor, MI, pp. 1–5, 1999.
16. Buckingham R., Staenz K., Hollinger A., A review of Canadian airborne and space activities in hyperspectral remote sensing, *Canadian Aeronautics and Space Journal*, 48(1), 115–121, 2002.
17. Jianyu W., Rong S., Yongqi X., The development of Chinese hyperspectral remote sensing technology, *SPIE*, 5640, 358, 2005.
18. Briottet X., et al., Military applications of hyperspectral imagery, *Proceedings of SPIE Defense & Security Symposium*, Orlando, FL, Paper No. 62390B, 2004.
19. Jason S., Cutter M., Meerman M., Curiel A.S., Low Cost Hyperspectral Imaging from a Microsatellite, *15th Annual AIAA/USU Conference on Small Satellites*, SSC01-II-1, Ogden, UT, 2001.
20. Curiel A.S., Cawthorne A., Sweeting M., Progress in small satellite Technology for Earth Observation missions, *5th IAA Symposium on Small Satellites for Earth Observation*, IAA-B5–0301, Berlin, Germany, 2005.
21. Cutter M., Review of a Small Satellite Hyper-Spectral Mission, *19th Annual AIAA/USU Conference on Small Satellites*, SSC05-IV-2, Logan, Utah, 2005.
22. Ortenberg F., Guelman M., Small satellite's role in future hyperspectral Earth Observation missions, *Acta Astronautica*, 64(11–12), 1251–1262, 2009.
23. Pearlman J., Barry P., Segal C., Shepanski J., Beiso D., Carman S., Hyperon, a space-based imaging spectrometer, *IEEE Transactions on Geoscience and Remote Sensing*, 41(6), 1160–1173, 2003.
24. Vidi R., Chiaratini L., Bini A., Hyperresolution: An hyperspectral and high resolution imager for Earth observation. *Proceedings of the 5th International Conference on Space Optics (ICSO 2994)*, Toulouse, France, pp. 105–111, 2004.
25. Davis C. et al., Ocean PHILLS hyperspectral imager: Design, characterization and calibration, *Optics Express*, 10(4), 210–221, 2002.
26. Hollinder A. et al., Resent developments in hyperspectral environment and resource observer (HERO) mission. *Proceedings of the International Geoscience and Remote Sensing Symposium and 27th Canadian Symposium on Remote Sensing*, Denver, CO, July, pp. 1620–1623, 2006.

27. Preti G., Cisbani A., De Cosmo V., Galeazzi C., Labate D., Melozzi M., Hyperspectral Instruments for Earth Observation, *International Conference on Space Optics (ICSO)*, Toulouse, France, 2008.

28. Cutter M., Sweeting M., The CHRIS Hyperspectral Mission—Five Years In-Orbit Experience, IAC-07-B1.3.02, *58th International Astronautical Congress*, Hyderabad, India, pp. 24—28, September 2007.

29. Stuffer T. et al., The EnMAP Hyperspectral Imager—An Advanced Optical Payload for Future Applications in Earth Observation Programmes, *Acta Astronautica*, 61, 115–120, 2007.

30. Sigernes F. et al., Proposal for a new hyper spectral imaging micro satellite: SVALBIRD, *5th IAA Symposium on Small Satellites for Earth Observation*, IAA-B5–0503, Berlin, Germany, 2005.

31. Morea G.D., Sabatini P., Perspectives and advanced projects for small satellite missions at Carlo Gavazzi Space, *Proceedings of the 4S Symposium Small Satellites Systems and Services*, La Rochelle, France, 2004.

32. Poinsignon V., Duthil P., Poilve H., A superspectral micro satellite system for GMES land cover applications, *Proceedings of the 4S Symposium Small Satellites Systems and Services*, La Rochelle, France, 2004.

33. Haring R.E., Pollock R., Cross R.M., Greenlee T., Wide-field-of-view imaging spectrometer (WFIS): From a laboratory demonstration to a fully functional engineering model, *Proceedings of SPIE*, Denver, CO, Vol. 4486, pp. 403–410, 2002.

34. Schoonwinkel A., Burger H., Mostert S., Integrate hyperspectral, multispectral and video imager for microsatellites, *19th Annual AIAA/USU Conference on Small Satellites*, SSC05-IX-6, Logan, Utah, 2005.

35. Cooley T., Lockwood R., Gardner J., Nadile R., Payton A., ARTEMIS: A rapid response hyperspectral imaging payload, *Paper No. RS4–2006–5002, 4th Responsive Space Conference*, Los Angeles, CA, April 24–27, 2006.

36. Li J., Sun F., Schmit T., Venzel W., Gurka J., Study of the hyperspectral environmental suit (HES) on GOES-R, *Proceedings of the 20th International Conference on Interactive Information and Processing System for Meteorology, Oceanography, and Hydrology*, January, Seattle, WA, Paper p2.21.6, 6 pp, 2004.

37. Astapenko V.M., Ivanov V.I., Khorolsky P.P., Review of current status and prospects of hyperspectral satellite imaging, *Space Science and Technology*, 8(4), 73, 2002.

38. Davis T., Straight S., Development of the Tactical Satellite 3 for responsive space missions, *Proceedings of the 4th Responsive Space Conference*, Los Angeles, CA, 10 pp., April 24–27, 2006.

39. Guelman M., Ortenberg F., Shiryaev A., Waler R., Gurwin-Techsat: Still alive and operational after nine years in orbit, *Acta Astronautica*, 65(1–2), 157–164, 2009.

40. Liu M., Zhang L., Guo E., Hyperspectral Laser-induced Fluorescence Imaging for Nondestructive Assessing Soluble Solids Content of Orange, *Computer and Computing Technologies in Agriculture*, 1, 51–59, 2008.

41. Brook A., Ben-Dor E., Richter R., Fusion of hyperspectral images and LIDAR data for civil engineering structure monitoring, *Commission VI, WG VI/4, Proceedings of ISPRS XXXVIII-1–4–7, W5, Paper 127*, Hannover, Germany, 2009.

42. Souci J., Hanou I., Puchalski D., High-resolution remote sensing image analysis for early detection and response planning for emerald ash borer, *Photogrammetric Engineering and Remote Sensing*, August, 905–907, 2009.

43. Johnson J.K., Sensor and data fusion technologies in archaeology, *ISPRS Workshop on Remote Sensing Methods for Change Detection and Process Modeling*, Cologne, Germany, 2010.

44. Fujii T., Fukuchi T., *Laser Remote Sensing*, Taylor & Francis, Boca Raton, FL, pp. 888, 2005.

45. Zhang J., Multi-source remote sensing data fusion: Status and trends, *International Journal of Image and Data Fusion*, 1(1), 5–24, March 2010.

46. Nieke J., Seiler B., Itten K., Ils Reusen I., Adriaensen S., Evaluation of user-oriented attractiveness of imaging spectroscopy data using the value-benefit analysis (VBA), *5th EARSeL Workshop on Imaging Spectroscopy*, Bruges, Belgium, April 23–25, 2007.

47. Buckingham R., Staenz K., Review of current and planned civilian space hyperspectral sensors for EO, *Canadian Journal of Remote Sensing*, 34(1), S187–S197, 2008.

3 Hyperspectral Remote Sensing in Global Change Studies

Jiaguo Qi, Yoshio Inoue, and Narumon Wiangwang

CONTENTS

3.1 INTRODUCTION

Few decades ago hyperspectral imagery data and processing software were available to only spectral remote sensing experts. Nowadays, the research and development of hyperspectral sensors for data acquisition and the associated software for data analysis is one of the fastest growing technologies in the field of remote sensing [1]. Unlike multispectral imaging systems (e.g., Landsat, SPOT, IKONOS type) that capture reflected or emitted incoming from the Earth's surface in a few broad wavelength bands across the electromagnetic spectrum, hyperspectral imagers (HSI) measure reflected radiation at numerous narrow, contiguous wavelength channels. The substantially finer spectral resolution data from hyperspectral sensors enhance the capability to characterize the Earth's surface more effectively than do the broadband multispectral data [2].

The distinction between hyper- and multispectral is based on the narrowness and contiguous nature of the measurements, not the "number of bands" [3]. For example, a sensor that measures only 20 spectral bands can be considered hyperspectral if those bands are narrow (e.g., 10 nm) and contiguous. On the other hand, if a sensor measures 20 wider spectral bands (e.g., 100 nm), or is separated by nonmeasured wavelength ranges, the sensor is no longer considered hyperspectral [1]. The detailed contiguous range of spectral bands of a hyperspectral sensor provides an ability to produce contiguous "spectrum," which is one of the characteristics that distinguishes it from multispectral sensors.

Radiances measured by multispectral sensors are generally adequate for rough discrimination of surface cover into categories; however, they are rather limited in the amount of quantitative information that can be inferred from the spectral content of the data. The spectra, or *spectral reflectance curves,* from hyperspectral remote sensors provide much more detailed information about absorption regions of the surface of interest very much like the spectra that would be measured in a spectroscopy laboratory. This unique characteristic of hyperspectral data is useful for a wide range of applications, such as mining, geology, forestry, agriculture, and environmental assessment.

This chapter is to focus on existing hyperspectral remote sensing systems, global change requirements, application examples, and challenges ahead.

3.2 HYPERSPECTRAL SENSORS AND CHARACTERISTICS

In the 1970s, space-based multispectral remote sensors were launched and produced images of the Earth's surface. Even only a few broad wavelength bands, the images greatly improved our understanding of our planet surface. The idea of developing hyperspectral, imaging sensors, also known as *imaging spectroscopy*, emerged in the early 1980s to improve our ability to better characterize the Earth's surface. The Airborne Visible/Infrared Imaging Spectrometer (AVIRIS) developed at the NASA Jet Propulsion Laboratory in California was the first spectrometer used on moving platforms such as aircrafts.

In the early imaging spectroscopy era, most of the hyperspectral sensors were mounted on aircrafts (e.g., AVIRIS). After decades of research and development, hyperspectral technology was expanded to space-based remote sensing systems and several satellite hyperspectral sensors were proposed and subsequently launched. The very first spaceborne hyperspectral sensors were NASA's Hyperion sensor on Earth Observing 1 (EO-1) satellite, and the U.S. Air Force Research Lab's Fourier Transform Hyperspectral Imager (FTHSI) on the MightySat II satellite. With more satellite-based sensors being planned, more hyperspectral imagery will be available to provide near global coverage at regular repeated cycles [1] suitable for global change studies.

3.2.1 SPACEBORNE SYSTEMS

Spaceborne hyperspectral sensors aboard satellites may provide continuous acquisition of Earth surface images at low cost. However, wide spatial coverage by a hyperspectral sensor is often compromised with its spatial resolution or ground sampling interval and other challenges. Consequently, repeated hyperspectral images are not widely available. Table 3.1 lists currently available spaceborne hyperspectral instruments, with a wide range of a number of spectral bands, spectral range, and swath width.

TABLE 3.1

Operational and Planned Satellite Hyperspectral Instruments

Instrument (Satellite)	Altitude (km)	Pixel Size (m)	Number of Bands	Spectral Range (nm)	Spectral Resolution (nm)	IFOV (μrad)	Swath (km)
Hyperion (EO-1)	705	30	220	400–2500	10	42.5	7.5
FTHSI (MightySatII)	575	30	256	450–1050	10–50	50	13
CHRIS (PROBA)	580	25	19	400–1050	1.25–11.0	43.1	17.5
OMI (AURA)	705	13000	780	270–500	0.45–1.0	115	2600
HICO	~390	92	102	380–900	5.7	<20	42 × 192
COIS (NEMO)	605	30	210	400–2500	10	49.5	30
HIS (SIMSA)	523	25	220	430–2400	20	47.8	7.7
Warfighter-1 (OrbView-4)	470	8	200	450–2400	11	20	5
	470	8	80	3000–5000	25	20	5
EnMAP (Scheduled 2014)	650	30	94	420–1000	5–10	30	30
	650	30	155	900–2450	10–20	30	30
HypSEO (MITA)	450	20	210	400–2500	10	40	20
MSMI (SUNSAT)	660	15	200	400–2350	10	22	15
PRISMA	695	30	250	400–2500	10	43	30
Global Imager (ADEOS-2)	803	250–1000	36	380–1195	10–1000	310–1250	1600
WFIS	705	1400	630	360–1000	1–5	2000	2400

The FTHSI on MightySat II, the first mission of the U.S. Air Force program initiated in 1995, was successfully launched in July 2000. FTHSI is the only Department of Defense (DoD) space-based HSI to discern spectrally unique objects with the Fourier transform technique. The width of the image footprint is 13 km with a spatial resolution 30 m, covering a spectral range from 450–1050 nm by 256 spectral bands (Table 3.1).

Later in November 2000, NASA launched Hyperion sensor onboard EO-1 satellite as part of a 1 year technology validation/demonstration mission. The Hyperion imaging spectrometer has a 30 m spatial resolution, 7.7 km swath width, and 10 nm contiguous spectral resolution (Table 3.1). With its high radiometric accuracy of 220 spectral bands, complex landscapes of the Earth's surface can be imaged and spectrally characterized.

Compact High Resolution Imaging Spectrometer (CHRIS) was launched in October 2001 onboard Project for Onboard Autonomy (PROBA) platform (Table 3.1). The sensor was developed by the United Kingdom Company with support from the British National Space Centre. The sensor was designed for the study of atmospheric aerosols, land surfaces, coastal, and inland waters with its 62 spectral bands ranging from 400 to 1050 nm and a spatial resolution of 17 m. Despite the fact that the mission was designed for a 1 year life, the sensor has been in operation for almost 10 years by 2010 [4].

The Global Imager (GLI) was part of the Advanced Earth Observing Satellite II (ADEOS II) mission, an international satellite mission led by the Japan Aerospace Exploration Agency (JAXA) with participations from United States (NASA) and France Centre Nationale d'Etudes Spatiales (CNES) (Table 3.1). Its spectral range was from 250 to 1000 nm and its image size is very large (1600 km). GLI mission is to collect data that aid better understanding of water, energy, and carbon circulations in order to contribute to global environmental change studies. ADEOS II was launched on December 14, 2002, but unfortunately the mission ended 10 months later, due to a failure of the solar panel on October 24, 2003.

Ozone Monitoring Instrument (OMI) equipped on the Earth Observing System (EOS) Aura spacecraft was designed to measure atmospheric composition (Table 3.1). The sensor was launched on July 15, 2004 into an ascending node 705 km sun-synchronous polar orbit. OMI is a nadir push-broom hyperspectral imaging sensor that observes solar backscatter radiation in the UV and visible wavelengths (264–504 nm). It has 780 spectral bands with a swath large enough to provide global coverage in one day (14 orbits) at a spatial resolution of 13 × 24 km at nadir. The key air quality components include NO_2, SO_2, and aerosol characteristics, as well as ozone (O_3) profiles.

Hyperspectral Imager for the Coastal Ocean (HICO), the Navy's "Sea Strike" was designed and built by the Naval Research Laboratory (NRL) to be the first spaceborne imaging spectrometer optimized for scientific investigation of the coastal ocean and nearby land regions with high signal-to-noise ratios (SNRs) in the blue spectral region and full coverage of water-penetrating wavelengths (Table 3.1). Due to the fact that water absorbs most of the light in the electromagnetic spectrum, visible light is the only part of the spectrum that sufficiently penetrates the water column to sense the water and the bottom surface properties. The sensor was launched on the H-2 Transfer Vehicle (HTV) and was rendezvoused with the International Space Station (ISS) in September 2009. The sensor has been serving as a spaceborne hyperspectral method to detect submerged objects, to provide environmental data products to naval forces, and to develop the coupled physical and bio-optical models of coastal ocean regions globally.

A few more sensors have been proposed for the near future launches, including PRecursore IperSpettrale della Missione Applicativa (PRISMA) under development by the Italian Space Agency (ASI). Listed in Table 3.1, PRISMA is a push-broom hyperspectral sensor and is devoted to derive information about land cover, soil moisture and agricultural land uses, quality of inland waters, status of coastal zones, pollution, and the carbon cycle [5]. The hyperspectral instrument is to acquire images in 250 spectral bands at 30 m spatial resolution. When combined with panchromatic camera, a higher spatial resolution (5 m) can be produced.

Environmental Mapping and Analysis Program (EnMAP) is a German push-broom hyperspectral satellite (Table 3.1) with a pointing feature for fast target revisit (4 days), providing high quality hyperspectral image data on a timely and frequent basis. Aboard EnMAP are HSI sensors that are designed to derive surface physical parameters on a global scale, with accuracy not achievable by currently available spaceborne sensors. Data from the 249 spectral channels, with a 30 m pixel size, are to be assimilated in physically based ecosystem models, and ultimately to provide information products reflecting the status of various terrestrial ecosystems [6].

Wide field-of-view imaging spectrometer (WFIS) is a proposed sensor that represents another aspect of imaging spectrometers (Table 3.1), by providing a possibility for an entire global surface study with frequent revisit like the Moderate Resolution Imaging Spectroradiometer (MODIS) aboard Terra and Aqua satellites. WFIS is a push-broom sensor that was designed to operate in the visible and near infrared spectral region (360–1000 nm) with an approximately 1 nm [7] sampling interval.

3.2.2 AIRBORNE SYSTEMS

Airborne hyperspectral imagery is becoming more easily accessible due to the increasing number of companies operating hyperspectral spectrometers. Similar to the scanning mechanism of a spaceborne hyperspectral sensor, an airborne sensor generates hundreds (often image columns) of individual pixels along the scan line direction (often perpendicular to flight direction). As the aircraft moves forward new array of pixels is generated along flight direction. As such, geometric quality of airborne images can be affected by environmental conditions such as wind and/or by flight operations such as aircraft speed and alignment. Therefore, airborne hyperspectral image processing/calibration may be complex and occasionally bring in errors to the analysis [8].

Airborne hyperspectral sensors are more flexible than satellite-based sensors in terms of acquisition schedules adjustable to weather conditions, spectral and spatial resolution requirements, and flight line arrangements. However, airborne hyperspectral images can be costly for large area coverage, due to limited swath width and slow speed of the carrier.

In comparison with spaceborne hyperspectral sensors, the configuration of airborne systems varies widely in terms of spectral range, number of spectral bands, manufactures, and spatial and temporal coverage. A survey of existing airborne hyperspectral sensors and their spectral characteristics is presented in Table 3.2. This is not an exhaustive list but it represents what is available in airborne systems.

TABLE 3.2

Current Airborne Hyperspectral Sensors and Data Providers

Airborne Sensors	Manufacturer	Number of Bands	Spectral Range (μm)
AISA EAGLE (Airborne Imaging Spectrometer)	Spectral Imaging	Up to 488	0.40–0.97
AISA EAGLET (Airborne Imaging Spectrometer)	Spectral Imaging	Up to 410	0.40–1.00
AISA HAWK (Airborne Imaging Spectrometer)	Spectral Imaging	254	0.97–2.50
AISA DUAL (Airborne Imaging Spectrometer)	Spectral Imaging	Up to 500	0.40–2.50
AISA OWL (Airborne Imaging Spectrometer)	Spectral Imaging	Up to 84	8.00–12.00
AVIRIS (Airborne Visible/Infrared Imaging Spectrometer)	NASA Jet Propulsion Lab	224	0.40–2.50
CASI-550 (Compact Airborne Spectrographic Imager)	ITRES Research	288	0.40–1.00
CASI-1500 Wide-Array (Compact Airborne Spectrographic Imager)	ITRES Research	288	0.38–1.05
SASI-600 (Compact Airborne Spectrographic Imager)	ITRES Research	100	0.95–2.45
MASI-600 (Compact Airborne Spectrographic Imager)	ITRES Research	64	3.00–5.00
TASI-600 (Compact Airborne Spectrographic Imager)	ITRES Research	32	8.00–11.5
DAIS 7915 (Digital Airborne Imaging Spectrometer)	GER Corporation	32	0.43–1.05
		8	1.50–1.80
		32	2.00–2.50
		1	3.00–5.00
		6	8.70–12.3
DAIS 21115 (Digital Airborne Imaging Spectrometer)	GER Corporation	76	0.40–1.00
		64	1.00–1.80
		64	2.00–2.50
		1	3.00–5.00
		6	8.00–12.0
EPS-H (Environmental Protection System)	GER Corporation	76	0.43–1.05
		32	1.50–1.80
		32	2.00–2.50
		12	8.00–12.5
HYDICE (Hyperspectral Digital Imagery Collection Experiment)	Naval Research Lab	210	0.40–2.50
HyMap	Analytical Imaging and Geophysics	32	0.45–0.89
		32	0.89–1.35
		32	1.40–1.80
		32	1.95–2.48
HySpex	Norsk Elektro Optikk	128 (VIS/NIR1)	0.40–1.00
		160 (VIS/NIR2)	0.40–1.00
		160 (SWIR1)	0.90–1.70
		256 (SWIR2)	1.30–2.50
PROBE-1	Earth Search Sciences Inc.	128	0.40–2.50

TABLE 3.3
Current Handheld Hyperspectral Sensors and Data Providers

Handheld Sensors	Manufacturer	Spectral Resolution	Spectral Range (µm)
FieldSpec 3 Hi-Res Portable Spectroradiometer	ASD Inc. (Analytical Spectral Devices)	3 nm at 700 nm 8.5 nm at 1400 nm 6.5 nm at 2100 nm	0.35–2.50
FieldSpec 3 Max Portable Spectroradiometer	ASD Inc. (Analytical Spectral Devices)	3 nm at 700 nm 10 nm at 1400/2100 nm	0.35–2.50
HandHeld 2 Portable Spectroradiometer	ASD Inc. (Analytical Spectral Devices)	<3 nm at 700 nm	0.325–1.075
UV-VIS Spectrometers (USB4000-UV-VIS)	Ocean Optics Inc.	~1.5	0.20–0.85
VIS-NIR Spectrometers (USB4000-VIS-NIR)	Ocean Optics Inc.	~1.5	0.35–1.00
UV-NIR Spectrometers (HR4000CG)	Ocean Optics Inc.	0.75	0.20–1.10
UV-VIS Hyperspectral USB Spectrometer	Edmund Optics Inc.	1.5	0.20–0.72
VIS-NIR Hyperspectral USB Spectrometer	Edmund Optics Inc.	2	0.35–1.05

3.2.3 GROUND-BASED SYSTEMS

Ground-based hyperspectral systems are available from a few commercial companies and their spectral characteristics are listed in Table 3.3. These systems can be mounted on low-elevation platforms such as trucks or be held by hand, due to their lightweight and small size. In general, the spectral range of these systems is from ultraviolet (UV) to middle infrared (200–2500 nm) with varying bandwidth or spectral resolution. The ground sampling interval or the footprint of these sensors varies depending on the height of the sensor and the total field-of-view (FOV) of the front optics. One of the advantages of the ground-based sensing systems is that they are flexible in deployment and can be used for both field-based and in-lab spectral measurements. Specifications on the operating system requirements, software support, as well as data storage, also vary greatly from sensor to sensor and manufacturer to manufacturer.

3.3 HYPERSPECTRAL REMOTE SENSING METHODS

Many of the methods developed for multispectral imagery analysis and processing can be adopted for hyperspectral images. However, the following is more specifically developed for hyperspectral analysis.

3.3.1 SUPPORT VECTOR MACHINES

The Support Vector Machines (SVMs) are new methods that have been successfully used for hyperspectral data classification [9–13]. The SVMs were designed to cope with the general problems of hyperspectral image classification. These approaches can efficiently work with large input, handle noise-attached data in a robust way, and produce sparse solutions [14,15]. SVMs are based on the *kernel methods* that map data from the original input feature space to a kernel feature space of higher dimensionality and then solve for a linear problem in that space [15]. These methods allow an interpretation of learning algorithms geometrically in the kernel space (which is nonlinearly related to the input space), thus combining statistics and geometry in an effective way [15] to take advantages of hyperspectral imagery.

3.3.2 KERNEL FISHER DISCRIMINANT ANALYSIS

The Kernel Fisher Discriminant (KFD) analysis is another new and effective method for hyperspectral data classification [16]. The KFD method adopts the same concept of kernel used in SVMs to obtain nonlinear solutions; however, KFD minimizes a different function than do the SVMs, and, thus, the solution is expressed in a different way for more accurate classification.

3.3.3 MATCHED FILTERING

Some hyperspectral applications are only focused on searching for the existence or fractional abundance of one or a few single target materials. Matched filtering (MF) is a type of unmixing procedure that identifies only targets of interest [17]. It is sometimes called a *Partial Unmixing* because spectra of all endmembers in an image are not required. MF was originally developed to compute abundances of targets of interest that are relatively rare in the image [1].

MF algorithms perform a mathematical transformation to maximize the contribution of the target spectrum while minimizing the background [1,18]. Therefore, the approaches perform best when target material is rare and does not contribute significantly to the background signature [1]. A modified version of MF uses derivatives of the spectra to enhance the differentiation ability [18]. The output image presents the fraction of the pixel that contains the target material [1].

3.3.4 LIBRARIES MATCHING TECHNIQUES

Spectral libraries are collections of reflectance spectra measured from materials of known composition, usually in the field or laboratory [1] that are highly desirable for hyperspectral image analysis. Laboratory-derived spectra may be found at the ASTER Spectral Library (2004), which contains the NASA Jet Propulsion Laboratory Spectral Library and the U.S. Geological Survey spectral library. Other publicly accessible reference spectral libraries are also available, such as those in digital image processing software (e.g., ENVI, ERDAS, PCI Geomatica) and other sources [19–25].

Absorption features of specific materials within a given instantaneous field of view (IFOV) are typically present in the spectra. These absorption features provide the information needed for the Earth's surface characterization based on the surface spectral absorption locations, relative depths, and widths [26,27]. Characterization and automatic detection of such absorption features on the basis of the spectral similarity between the pixel and target spectra is the fundamental principle of Library Matching techniques [26]. A measured spectrum may be divided into several spectral regions before absorption features in each of the regions are detected and matched with those from the spectral library. The algorithm assigns the pixel to the class that its spectrum most closely resembles [26]. Libraries matching techniques perform best when the scene includes extensive areas of pure materials that have corresponding reflectance spectra in the reference library.

Comparing the spectral properties in a hyperspectral image with the one stored in libraries can take significant amount of time and computing resources. Therefore, coding techniques (e.g., *Binary Spectral Encoding*) are developed to represent a pixel spectrum, which has a high degree of redundancy, in a simple and effective manner [27]. Library matching techniques may not perform well when mixtures of targets present in one pixel, or some materials have very similar spectral characteristics. Sample mixed spectra can be included in the library to improve the accuracy; however, it is not likely that all possible mixtures (and all mixture proportions) can be included in the reference library [18].

3.3.5 Derivative Spectroscopy

Among the techniques developed in remote sensing analysis, derivative spectroscopy is particularly promising for use with hyperspectral image data. Differentiation of a spectral curve estimates the slope at each wavelength over the entire spectral range. A first-order derivative is the rate of change of the absorbance with respect to wavelength. Although differentiation of the spectra does not provide more information than the original spectra, it can emphasize the target features while suppressing other unwanted information [26]. Second order derivative spectra, which are insensitive to substrate reflectance, have been used to mitigate soil background effect in vegetation studies [28,29].

Second or higher derivatives are relatively insensitive to variations in illumination (due to cloud cover), solar angle variance, or topographic effects [26]. Although some of the researches have used high order derivatives, first and second order derivatives have been the most common [30,31]. Talsky [32] suggested that the SNR decreases as derivative order increases. Spectral derivatives have successfully been used in remote sensing applications for decades [28,33,34]. In addition, several studies used this method directly toward specific applications, such as water quality assessment [35–37].

3.3.6 Narrowband Spectral Indices

Hyperspectral indices have been developed for quantification of biophysical parameters, based on specific absorption features that best describe the biophysical indicators. Examples of such indices are provided in the following [26].

3.3.6.1 Normalized Difference Vegetation Index (NDVI)

The traditional NDVI has been modified or computed with narrow spectral bands such as the following one to emphasize the sensitivity to green vegetation density.

$$\text{NDVI}_{\text{narrowband}} = \frac{\rho(860\,\text{nm}) - \rho(660\,\text{nm})}{\rho(860\,\text{nm}) - \rho(660\,\text{nm})}$$

3.3.6.2 Yellowness Index

The yellowness index (YI) is sensitive to leaf chlorosis and, therefore, is an indicator of stresses in plant leaves. The YI measures the change in shape of reflectance spectra between the 550 nm (maximum green reflectance band) and the 650 nm (maximum red absorption band). The YI uses only wavelengths in the visible spectrum, a region that is relatively insensitive to change in leaf water content and structure [26,33,38].

3.3.6.3 Normalized Difference Water Index

Normalized Difference Water Index (NDWI) is used to determine vegetation liquid water content, and can be derived from narrow spectral bands that are sensitive to water content.

$$\text{NDWI}_{\text{narrowband}} = \frac{\rho(860\,\text{nm}) - \rho(1240\,\text{nm})}{\rho(860\,\text{nm}) - \rho(1240\,\text{nm})}$$

3.3.6.4 Red Edge Position Determination

The Red Edge Position Determination (REP) is defined as the point of maximum slope on a vegetation reflectance spectrum between the red and near-infrared wavelengths. The REP is strongly correlated with foliar chlorophyll content and, therefore, can be a sensitive indicator of vegetation stress [26]. A linear method, based on narrow spectral bands features, was proposed by Clevers [39] to highlight the red edge changes in a given spectrum:

$$REP = 700 + 40 \left[\frac{\rho(860\,nm) - \rho(1240\,nm)}{\rho(860\,nm) + \rho(1240\,nm)} \right]$$

where

$$\rho_{(red\,edge)} = \frac{\rho(670\,nm) + \rho(780\,nm)}{2}$$

3.3.6.5 Crop Chlorophyll Content Prediction

This narrowband vegetation index, developed by Haboudane et al. [40], integrates the advantage of indices that minimize soil background effects and indices that are sensitive to chlorophyll concentration. The commonly used indices include the Transformed Chlorophyll Absorption in Reflectance Index (TCARI) [41] and the Optimized Soil-Adjust Vegetation Index (OSAVI) [42].

$$Crop\ Chlorophyll\ Content = \frac{TCARI}{OSAVI}$$

$$TCARI = 3 \left[(\rho_{700} - \rho_{670}) - 0.2 (\rho_{700} - \rho_{550}) \left(\frac{\rho_{700}}{\rho_{670}} \right) \right]$$

$$OSAVI = \frac{(1 + 0.16)(\rho_{800} - \rho_{670})}{(\rho_{800} + \rho_{670} + 0.16)}$$

3.3.7 Neural Network

The Neural Network (NN) is one of the promising feature selection methods. NNs are mathematical models that simulate brain dynamics [43] that are supposed to be quite powerful in remote sensing imagery analysis, especially in image classification, due to their nonlinear properties. It should be noted that NN is highly sensitive to the Hughes phenomenon (the *curse of dimensionality*), which is particularly a problem for hyperspectral images, and may not work effectively when dealing with a high number of spectral bands [15]. Moreover, the use of NN for hyperspectral image classification has been limited primarily because of the lengthy computational time required for the training process. Nonetheless, several researches have successfully used NN algorithms to estimate vegetation types and biophysical parameters, such as in coastal and ocean waters [44–46].

3.4 GLOBAL CHANGE REQUIREMENTS AND APPLICATIONS

Hyperspectral imagery has been used to assess, analyze, detect, and monitor several key global environmental change variables. For example, hyperspectral data have been used to estimate sediments, chlorophyll *a*, and algal type information in oceans and inland waters [37,47–50], identify vegetation species [51], study plant canopy chemistry [52–56], detect vegetation stress [57,58], monitor biogeochemical and greenhouse gases cycles [59–63], improve land cover classification accuracy and more details in plant species recognition [64]. Geologists also use imaging spectroscopy for soil organic matter estimation, salinity and moisture content detection, and mineral mapping [51,65].

3.4.1 Global Change Requirements

Because the Earth is a dynamic system, sufficient understanding of the complex interactions among physical and ecological processes is needed for global change studies. To achieve this goal, both long- and short-term observations are required to quantify, analyze, and subsequently understand the spatial and temporal variability, trend, and magnitudes of changes in ecosystems dynamics. Multispectral remote sensing systems such as NASA's Landsat and National Oceanic and Atmospheric Administration (NOAA)'s Advanced Very High Resolution Radiometer (AVHRR) sensors have been instrumental and inspirational to provide global coverage for systematic analysis of the Earth's dynamics [61,63]. However, the spectral characteristics and the sensor design of these systems limit their applications in the areas that require specific and more accurate assessment of, for example, nutrient deficiencies in plants, algal information of lakes and streams, invasive species identification and detection [66], soil composition, and specific atmospheric gas concentration. For example, reflectance spectra of agro-ecosystems and the seasonal changes of the spectra in a paddy rice field, shown in Figure 3.1, are much better spectrally characterized by hyperspectral signatures than by that of multispectral data [67]. Broadband spectral signatures would not be able to detect such subtle changes, but hyperspectral measurements enable the detection and quantification of plant, soil, and ecosystem variables due to their high spectral resolution and continuity. These requirements lead to the exploration of hyperspectral sensing systems where spectral information is much richer for enhanced and new applications in global change studies than multispectral data.

3.4.2 Global Change Applications

Hyperspectral data have been used in numerous applications to specifically address global environmental issues. The following are not meant to be an exhausted application list; rather they are examples demonstrating the type of issues one can address with hyperspectral data.

3.4.2.1 Water Quantity and Quality

Imaging spectrometry is a cost-effective technique for water quality studies over large areas of aquatic systems, such as lakes, coastal areas, bays, estuaries, and even oceans. For example, spaceborne and airborne hyperspectral images have been used to assess the trophic status of lakes and to map the spatial distribution of water quality parameters, such as temperature, chlorophyll a, turbidity, and total suspended solids, over large areas [37,47,68–70]. These studies are possible because of narrow, unique spectral absorption features of water bodies that are only detectable by hyperspectral data [37]. Owing to the unique spectral signatures of algal pigments, compositions of algal populations in aquatic systems can be detected by analyzing absorption properties in the region between 400 and 700 nm [49]. Various methods are effective in mapping water quality, including pigment-specific absorption algorithm, spectral angle mapping algorithm, spectral libraries comparison, principle component analysis, derivative spectroscopy, regression techniques, and other spectral indicators like band ratios [30,31,37,71,72].

3.4.2.2 Carbon Sequestration and Fluxes

Carbon sequestration is a process of removing carbon from the atmosphere and depositing it in a reservoir or carbon sink. The main natural sinks are photosynthesis by terrestrial plants and physicochemical and biological absorption of carbon dioxide (CO_2) by the oceans. The oceans are

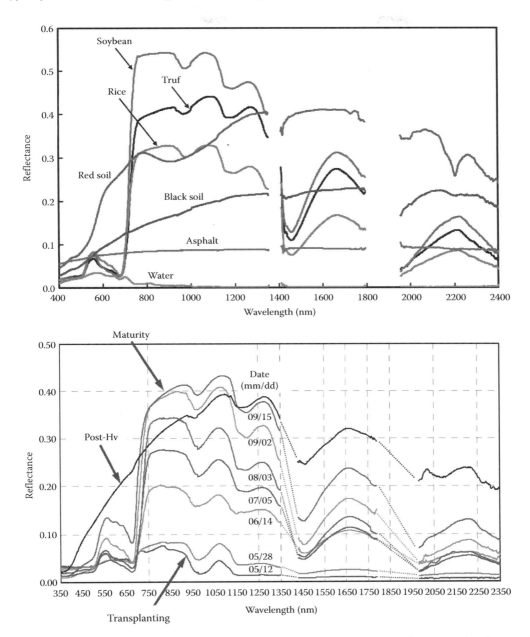

FIGURE 3.1 **(See color insert.)** Typical reflectance spectra in agro-ecosystem surfaces (upper) and seasonal change of spectra in a paddy rice field (lower).

the largest active carbon sinks on Earth, absorbing more than a quarter of the CO_2 released from human activities [4].

Hyperspectral data proved to be useful to estimate and map primary production in the oceans and other open surface waters [48–50] while multispectral data are equally suitable for terrestrial primary production estimation. The use of hyperspectral remote sensing to measure chlorophyll *a* from space has been a highly successful technique for mapping phytoplankton distribution on a global basis, which could be used to estimate the amount of carbon sequestration in the oceans.

3.4.2.3 Greenhouse Gas Emissions

Greenhouse gases, by definition of United Nations Framework Convention on Climate Change (UNFCCC), are "the atmospheric gases responsible for causing global warming and climate change. The major greenhouse gases are carbon dioxide (CO_2), methane (CH_4), and nitrous oxide (N_2O). Less prevalent—but very powerful—greenhouse gases are hydrofluorocarbons (HFCs), perfluorocarbons (PFCs) and sulphur hexafluoride (SF_6)." Hyperspectral remote sensing has shown to be successful in detecting CH_4 gas concentrations [59,60,62], despite some technical challenges. Although CH_4 is transparent in the visible part of the electromagnetic spectrum, it does contain a number of spectral features at longer wavelengths that can be detected by only hyperspectral sensors. Three significant absorption features, at 3.31, 3.28, and 3.21 μm, were detected in the CH_4 spectrum [59]. Some of the spectral features may appear to be obscured by water vapor absorption wavelength regions; however, the obscured spectral features are still detectable at high CH_4 concentrations [73]. The CH_4 absorption features at 0.88 and 7.7 μm have also been proposed for atmospheric studies with significant water vapor content [73]. Once the spectral bands associated with absorption features are determined, band ratios techniques could be used to develop indicators of CH_4 concentration.

3.4.2.4 Atmospheric Chemistry

Although multispectral remote sensing has been used for atmospheric monitoring, hyperspectral imagery proved to be more suitable due to its fine spectral resolution and sampling intervals. Space measurements of the O_3 column have been operated since the 1970s with the series of Solar Backscatter Ultraviolet (SBUV) and Total Ozone Mapping Spectrometer (TOMS) sensors [74]. These spectrometers have been used to monitor the O_3 layer, measure stratospheric dynamics, and detect tropospheric O_3 pollution on a regional scale [75]. The SBUV and TOMS, with 1 nm spectral resolution, measure backscattered radiance at two wavelengths and estimate the O_3 column [76,77]. In 1995, Global Ozone Monitoring Experiment (GOME) was launched and operated as the first space instruments that measure the UV and visible part of the spectrum with a high spectral resolution [75]. In 2002, the Scanning Imaging Absorption Spectrometer for Atmospheric Cartography (SCIAMACHY) was launched onboard Environmental Satellite (ENVISAT) to facilitate the measurement of atmospheric absorptions from the UV to the near-infrared spectral range (240–2380 nm), providing knowledge about the composition, dynamics, and radiation balance of the atmosphere.

The OMI hyperspectral sensor aboard the EOS Aura satellite was successfully launched in 2004, as a continuation of the long-term global total O_3 monitoring from satellite measurements that began in 1970 with SBUV and TOMS. Data from OMI were used to derive O_3 columns using Differential Optical Absorption Spectroscopy (DOAS) [75,78–80]. In comparison, TOMS estimates the spatial distribution of the total O_3 by observing changes of solar radiation in several near UV spectral bands. DOAS derives the O_3 column by fitting a reference O_3 absorption cross section to the measured sun-normalized radiance. The main advantages of DOAS compared to the original SBUV/TOMS techniques are that DOAS is less sensitive to the radiometric calibration of the instrument and less sensitive to disturbing factors like absorbing aerosols [75].

OMI can distinguish between aerosol types, such as smoke, dust, and sulfates, and can measure cloud pressure and coverage. Other instruments on Aura satellite, such as Tropospheric Emission Spectrometer (TES) may provide global measurements of tropospheric O_3 and its photochemical precursors. With the entire spectrum from 3.2 to 15.4 μm at high spectral resolution, many other gases such as CO, ammonia, and organics can be retrieved [81]. Aura satellite is providing the next level of atmospheric measurements in the stratospheric and tropospheric layers in order to understand the recovering process of the stratospheric O_3 layer, the composition of chemistry in the troposphere, and the roles of upper tropospheric aerosols, water vapor, and O_3 in climate change. The four instruments on Aura provide valuable data to global change studies as well as continuing important atmospheric composition monitoring that began earlier with other satellites such as TOMS [81,82].

3.4.2.5 Vegetation Ecology

Vegetation is a key attribute of land use and land cover change dynamics not only for its role as the food production but also for its role in land–atmosphere interactions. The exchange of biochemical between terrestrial ecosystems and the atmosphere could be determined by the properties of vegetation. Therefore, an accurate assessment of vegetation properties and temporal dynamics is very important in the Earth's system science [83]. Due to the extended spectral dimension, hyperspectral remote sensing enabled improved modeling, monitoring, and understanding of vegetation canopies [52,67] and enhanced quantitative mapping of key vegetation properties [54–56]. For example, data from hyperspectral sensors have been used to derive plant species composition, biological and biochemical properties of the forests [54], such as chlorophyll and nitrogen concentrations. These studies suggest that hyperspectral remote sensing brings new capabilities to estimate vegetation properties that otherwise would not be possible with traditional multispectral sensors. Still, interest has increased in using hyperspectral remote sensing for biodiversity monitoring [84,85] and invasive species mapping [66].

3.4.2.6 Vegetation Biochemical Properties

Foliar chemistry of plant canopies allows a better understanding of ecosystem function and service, since many biochemical processes, such as photosynthesis, respiration, and litter decomposition, are related to the foliar chemistry of plants. Important chemical components of vegetation foliage that could be used as bioindicators are the concentration of nitrogen and carbon, and the content of water [86]. For example, hyperspectral remote sensing combined with canopy radiative transfer models provided consistent and accurate information of these chemical compositions [54]. Forest biomass and aboveground carbon stocks have been estimated using hyperspectral imagery from AVIRIS sensor together with the partial least squares (PLS) regression method [87].

Plant pigments, such as chlorophyll and carotenoids, have specific absorption spectra, which play essential roles in the photochemical cycle in plant leaves. These specific absorption features are only detectable with hyperspectral sensors, as demonstrated by various studies. For example, Inada [88] showed that the narrow spectral band ratio of 800/550 nm is the most effective index for estimating the leaf chlorophyll content of rice. The reflectance at 675 and 550 nm has been used to determine chlorophyll content as well as for nitrogen stress detection.

The total nitrogen content of a canopy can be estimated using narrow bandwidth measurements in visible and near-infrared wavelengths such as 480, 620, and 840 nm [89], and the estimate accuracy can be further improved by using sharp absorption features in the shortwave infrared wavelengths, 1650 and 2200 nm [90].

3.4.2.7 Invasive Plant Species Detection

Species invasion is recognized as a significant threat to global biodiversity and ecosystem health. In some cases, invasive species could irreversibly change the structure and functioning of entire ecosystems and result in biological diversity loss [66,91]. Ecosystem research suggests that invasive and aggressive plant species may be the result of general ecosystem stress related to changes in the frequency of landscape disturbance, such as road construction, deforestation, land use conversion to agricultural or urban development, or other hydrologic alterations [91,92]. Identification of the extent of landscape being stressed by invasive plant species using spectral signatures can help target vulnerable area in need of restoration or protection [91]. Contrary to multispectral remote sensing, which can only detect invasions when the effects has spread out in a wide area, hyperspectral imagery offers a unique potential for analyzing the signals of ecological changes at an early stage and, therefore, an indication of biological invasion and biogeochemical change [66,92]. For example, nonnative species ceplant, jubata grass, fennel, and giant reed from a range of habitats in California were mapped with AVIRIS imagery at relatively high accuracy [93] and pure pixels of Brazilian pepper was detected with hyperspectral imagery [94]. It was also demonstrated that an early detection of invasive weeds (spotted knapweed and baby'sbreath) was also possible [95] using hyperspectral sensors.

3.4.2.8 Vegetation Health

Information about vegetation health, such as disease, fire disturbance, and insect attack, is crucial in ecosystem protection and management and hyperspectral remote sensing can provide diagnostic indicators for early detection. For example, Lawrence and Labus [96] successfully used high spatial resolution imagery to identify different levels of tree stress resulting from Douglas-fir beetle attack. Koetz et al. [97] also mapped spatially distributed fuel moisture content and fuel properties with inversion of radiative transfer models to serve as input for forest fire spread and mitigation models. Diagnostic analysis of specific disease, however, has been a challenging issue with hyperspectral remote sensing as sensors only "see" plant symptoms rather than the causes. Discrimination of diseases may be possible with knowledge of the physiological effect of the disease on leaf and canopy elements. For example, necrotic diseases can cause a darkening of leaves in the visible spectrum and a cell collapse that would decrease near-infrared reflectance. Chlorosis-induced diseases (mildews and some virus), for example, cause marked changes in the visible reflectance (similar to N deficiency). Other diseases, however, may be detected by their effects on canopy geometry.

3.5 HYPERSPECTRAL REMOTE SENSING CHALLENGES

There are numerous challenges facing hyperspectral remote sensing, ranging from system design, data processing, to methodological developments. Because little is available from literature, the following reflects general issues that can be addressed for broader and improved applications of hyperspectral remote sensing imagery.

3.5.1 SYSTEM DESIGN CHALLENGES

From global change perspective, the design of a hyperspectral sensor entails the configuration of the following key parameters and requirements: spectral regions, number of spectral bands, spectral bandwidth, spatial resolution, swath width, revisit cycle, SNRs, onboard storage, and data downlinks to ground stations. These system parameters will determine the appropriateness of a specific global change application.

There is a challenge in balancing the spectral bandwidth and SNR. Existing hyperspectral sensors operate in the spectral region from approximately 200–2500 nm. The spectral bandwidth widens from UV to shortwave infrared, in order to maintain an acceptable SNR in the longer wavelength region. This widening presents a challenge in methodological development of hyperspectral remote sensing as significant sharp absorption features will likely to be not detectable with wider spectral bandwidth.

SNR needs to be balanced with spatial resolution requirements as well. As global change studies increasingly require quantitative information about earth surface properties, there is a need to acquire hyperspectral images at high spatial resolution and over large coverage areas. High spatial resolution requires small IFOV of the sensor, but smaller IFOV results in lower SNR and compromises the sensor's ability to have large geographic area coverage.

There is also a conflict in a number of spectral bands, spatial resolution, swath width, and onboard data storage. The requirements for large number of contiguous spectral bands, high spatial resolution of large geographic coverage undoubtedly increase the data volume, which presents challenges on both onboard storage capacity and the time required for downloading to ground stations. For global change studies, the priority ought to be on the geographic extent of coverage as this allows a broader and diverse ecosystem analysis and at the same time increases global access to hyperspectral imagery.

3.5.2 PROCESSING AND VISUALIZATION CHALLENGES

For a given geographical area imaged, the data can be viewed as a two-dimensional image that represents spatial location and spectral information. Displaying hyperspectral data is more challenging than is for multispectral data. Hyperspectral images contain far more spectral bands than can be displayed with a standard red, green, and blue (RGB) display. A convenient visualization approach is to reduce the dimensionality of the image (from tens to hundreds dimensions) to three dimensions at the expense of information losses [98,99]. The optimal hyperspectral display methods for quantitative and qualitative analysis of the data should enhance natural colors, preserve natural edges or contours of the features, highlight target features of interest, and enable simple and quick computational processing [99].

Several different techniques have been proposed and implemented for useful dimensionality reduction of hyperspectral images. *Color matching functions* (*CMF*) is one of these methods that specify how much of each of three primary colors must be mixed to create the color sensation of a monochromatic light at a particular wavelength [98]. The technique linearly projected hyperspectral data in visible range onto the CMF to determine the amount of the three primary colors that would create the same color sensation as viewing the original spectrum. It creates consistent images where hue, brightness, and saturation have interpretable and relevant meaning [98]. A disadvantage of the CMF is that there might be a decrease in sensitivity of human vision at the edges of the visible spectrum [98].

Principal component analysis (*PCA*) is also used to reduce hyperspectral data dimensionality by assigning the first three principal components to RGB [100,101]. Recent work found that the use of *wavelets* to de-noise the spectra before applying PCA could improve visualization [102]. Disadvantages of PCA include the difficulty to interpret the displayed image because the displayed colors represent principal components that do not typically represent natural colors of the features. The colors change drastically, depending on the data, and they do not correlate strongly with data variation. The standard saturation used in PCA display leads to simultaneous contrast problems and the computational complexity is high [98].

A number of *Linear Methods* were used to optimize the hyperspectral imagery display [72,99,101,103–105]. Jacobson and Gupta [105] used fixed linear spectral weighting envelops to create natural looking palette while other information can still easily be added using highlight colors. The method maximizes usefulness for human analysis while maintaining natural look of the imagery [98]. Another data dimensionality reduction method is the *artificial neural networks* (*ANN*). After the NN training, images can be processed very quickly, making it reasonable to use for real-time analysis. However, a disadvantage of ANN is that it is unclear how the neurons handle new spectral inputs that were not in the training dataset.

3.5.3 Data Volumes and Redundancy

Hyperspectral images are composed of a large number of spectral bands in order to generate fine enough spectral resolution needed to characterize the spectral properties of surface materials. As a result, the volume of data in a single scene can be overwhelming. Although hyperspectral imagery provides the potential for more accurate and detailed information extraction than possible with other types of remotely sensed data, it can be spectrally overdetermined. Tremendous amount of the data in a scene are redundant and much of the additional data do not add to the inherent information content for a particular application [27]. Spectral *redundancy* means that the information content of one spectral band can be fully or partly predicted from other bands within the scene [27]. The adjacent spectral bands are often found to be highly correlated to one another and their reflectance values, therefore, appear nearly identical. One way to identify spectral redundancy is by computing the correlation matrix for the image where high correlation values between bands indicate high degrees of redundancy or dimensionality [27].

The greater the number of bands in an image, the more storage and processing time is required for the analysis. Therefore, developing effective tools and approaches to reduce the dimensionality of hyperspectral data, while retaining the information content in the imagery, remains a challenge. When analyzing a hyperspectral image, the focus has been on extracting spectral information within individual pixels, rather than spatial variations within each band. The traditional statistical classification methods that have been developed and used for multispectral image analysis may not be suitable for hyperspectral images unless they are modified to account for the high dimensionality nature of the hyperspectral data.

3.5.4 RADIOMETRIC CALIBRATION

One of the most critical steps in hyperspectral data analysis is to convert the measured radiance data to surface reflectance so that individual spectra can be compared directly with laboratory or field data for appropriate interpretation [18,26,106]. A comprehensive conversion method must account for the solar irradiance spectrum, lighting effects due to solar angle and topography, atmospheric transmission, sensor gain and offset, and path radiance due to atmospheric scattering.

Because hyperspectral sensors acquire data at near continuous wavelength, atmospheric correction should take into account the atmospheric absorption properties as shown in Figure 3.2. These absorption regions are dominated by water vapor (1.4 and 1.9 μm) with smaller contributions from CO_2, O_3, and other gases [18,107]. For example, narrow atmospheric water absorption bands in the visible and near-infrared spectrum at 0.69, 0.72, and 0.76 μm, an oxygen (O_2) absorption band at 0.76 μm, and CO_2 absorption bands in the shortwave infrared region at 2.005 and 2.055 μm have been used in atmospheric correction algorithms [106,108].

3.5.5 METHODOLOGICAL CHALLENGES

Hyperspectral images provide rich information about earth surfaces and therefore are desirable for global change studies. However, several issues should be considered in analysis and interpretation of such data. The large number of spectral bands in hyperspectral imagery and the small number of known target spectra in most image scenes create the problem known as the *curse of dimensionality*. The use of traditional image classification methods developed for multispectral analysis, such as Maximum Likelihood Classifier (MLC) and Multiple Linear Regression (MLR or ordinary least squares OLS), without a modification to account for the high dimensionality of the hyperspectral data usually result in low efficiency and accuracy of the classification process. The MLR method assumes no intercorrelation between the independent variables, and the number of samples (endmembers) should be larger than the number of independent variables (spectral bands). Therefore, if independent variables (spectral bands) have significant correlations among each other, which are common for hyperspectral data, MLR technique will be subjected to multicollinearity issue [109].

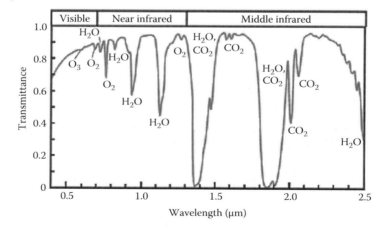

FIGURE 3.2   (See color insert.) Plot of atmospheric transmittance versus wavelength for typical atmospheric conditions.

For hyperspectral data that normally have tremendous number of bands, it would be very difficult to have adequate number of endmembers to make MLR work effectively. A solution is to engage in feature selection to reduce the dimensionality of the data set and remove redundant spectral bands and arrive at a data set with enough bands to address the application but not overwhelm the system with redundancy as discussed in Section 5.3.

3.6 DISCUSSION AND FUTURE DIRECTIONS

Accurate and timely information about land cover dynamics is essential for global change studies. This chapter shows that the hyperspectral data, either from ground-based, airborne, or spaceborne systems, provide much greater potential for detailed information extraction pertaining to global change than that can be achieved from multispectral imagery. Significant progress has been made in the development of new hyperspectral sensors, new technologies for data processing, new methods for analysis, and new models for enhanced information extraction for global change studies using hyperspectral data over the past decade. However, challenges exist in hyperspectral data access, data storage, data visualization, and analytical methodologies. These challenges are to be addressed by continued research efforts and new technological inventions in such areas as sensor design, data compression, data storage, and data compression.

Immediate emphasis will be in methodological developments with focus on information extraction algorithms from hyperspectral images. One would also see an increase in hyperspectral image availability for research and application development, as more and more agencies plan to launch hyperspectral sensors in coming years. Continued progress in sensor design, data availability, and new analytical methods will further promote broader hyperspectral applications in global change studies.

REFERENCES

1. Shippert, P., Introduction to hyperspectral image analysis, *Online Journal of Space Communication, Issue No. 3: Remote Sensing of Earth via Satellite*, 2003.
2. Birk, R.J. and McCord, T.B., Airborne hyperspectral sensor systems, *Aerospace and Electronic Systems Magazine*, 9(10):26–33, 1994.
3. Goetz, A.F.H. et al., Imaging spectrometry for earth remote sensing, *Science*, 228:1147–1153, 1985.
4. Earth Institute News, Oceans' uptake of manmade carbon may be slowing, *Earth Institute News*, The Earth Institute, Columbia University, New York, 2009, http://www.earth.columbia.edu/articles/view/2586
5. Labate, D. et al., The PRISMA payload optomechanical design, a high performance instrument for a new hyperspectral mission, *Acta Astronautica*, 65(9–10):1429–1436, 2009.
6. Stuffler, T. et al., The EnMAP hyperspectral imager—An advanced optical payload for future applications in Earth observation programmes, *Acta Astronautica*, 61(1–6):115–120, 2007.
7. Haring, R.E. et al., WFIS: A wide field-of-view imaging spectrometer, *Proceeding SPIE*, Vol. 3759, 1999. doi:10.1117/12.372678.
8. Cetin, H., Comparison of spaceborne and airborne hyperspectral imaging systems for environmental mapping, *Proceeding of ISPRS Congress Istanbul*, Istanbul, Turkey, 2004.
9. Gualtieri, J.A. and Cromp, R.F., Support vector machines for hyperspectral remote sensing classification, *Proceedings SPIE of 27th AIPR Workshop Advance in Computer Assisted Recognition*, Vol. 3584, pp. 221–232, 1998. ISBN: 9780819430540.
10. Gualtieri, J.A. et al., Support vector machine classifiers as applied to AVIRIS data, *Summaries of the Eighth JPL Airborne Earth Science Workshop*, JPL Publication, Pasadena, CA, 99-17:217–227, 1999.
11. Huang, C., Davis, L.S., and Townshend, J.R.G., An assessment of support vector machines for land cover classification, *International Journal of Remote Sensing*, 23(4):725–749, 2002.
12. Camps-Valls, G. et al., Robust support vector method for hyperspectral data classification and knowledge discovery, *IEEE Transactions on Geoscience and Remote Sensing*, 42(7):1530–1542, 2004.
13. Melgani, F. and Bruzzone, L., Classification of hyperspectral remote sensing images with support vector machines, *IEEE Transactions on Geoscience and Remote Sensing*, 42(8):1778–1790, 2004.

14. Cristianini, N. and Shawe-Taylor, J., *An Introduction to Support Vector Machines*, Cambridge University Press, Cambridge, U.K., 2000.
15. Camps-Valls, G., Kernel-based methods for hyperspectral image classification. *IEEE Transactions on Geoscience and Remote Sensing*, 43(6):1351–1362, 2005.
16. Mika, S. et al., Invariant feature extraction and classification in kernel spaces, *Advances in Neural Information Processing Systems*, Vol. 12, MIT Press, Cambridge, MA, 1999.
17. Boardman, J.W., Kruse, F.A., and Green, R.O., Mapping target signatures via partial unmixing of AVIRIS data, *Summaries of the Fifth JPL Airborne Earth Science Workshop*, 95(1):23–26, 1995.
18. Smith, R.B., *Introduction to Hyperspectral Imaging*, MicroImages, Inc., Lincoln, Nebraska, pp. 1–24, 2006.
19. Elvidge, C.D., Visible and infrared reflectance characteristics of dry plant materials, *International Journal of Remote Sensing*, 11(10):1775–1795, 1990.
20. Salisbury, J.W., D'Aria, D.M., and Jarosevich, E., Midinfrared (2.5–13.5 micrometers) reflectance spectra of powdered stony meteorites, *International Journal of Solar System Studies (Icarus)*, 92:280–297, 1991.
21. Salisbury, J.W. et al., *Infrared (2.1–25 Micrometers) Spectra of Minerals*, John Hopkins University Mineral Library, Johns Hopkins University Press, Baltimore, MD, 294 p., 1991.
22. Grove, C.I., Hook, S.J., and Paylor, E.D., *Laboratory Reflectance Spectra for 160 Minerals 0.4–2.5 Micrometers*, JPL Publication, Pasadena, CA, 1992.
23. Clark, R.N. et al., The U. S. Geological Survey, digital spectral library, version 1: 0.2 to 3.0 microns, *U.S. Geological Survey Open File Report 93–592*, 1340 p., 1993.
24. Salisbury, J.W., Wald, A., and D'Aria, D.M., Thermal-infrared remote sensing and Kirchhoff's law 1. Laboratory measurements, *Journal of Geophysical Research*, 99:11,897–11,911, 1994.
25. Korb, A.R. et al., Portable FTIR spectrometer for field measurements of radiance and emissivity, *Applied Optics*, 35:1679–1692, 1996.
26. Jensen, J.R., *Introductory Digital Image Processing: A Remote Sensing Perspective*, Pearson Prentice Hall, Upper Saddle River, NJ, 526 p., 2005.
27. Richards, J.A. and Jia, X., *Remote Sensing Digital Image Analysis: An Introduction*, 4th edn., Springer, Berlin, Germany, 439 p., 2006.
28. Demetriades-Shah, T.H., Steven, M.D., and Clark, J.A., High resolution derivative spectra in remote sensing, *Remote Sensing of Environment*, 33:55–64, 1990.
29. Li, Y. et al., Use of second derivatives of canopy reflectance for monitoring prairie vegetation over different soil backgrounds, *Remote Sensing of Environment*, 44:81–87, 1993.
30. Butler, W.L. and Hopkins, D.W., Higher derivative analysis of complex absorption spectra, *Photochemistry and Photobiology*, 12:439–450, 1970.
31. Fell, A.F. and Smith, G., Higher derivative methods in ultraviolet, visible and infrared spectrophotometry, *The Analytical Proceedings*, 19:28–32, 1982.
32. Talsky, G., *Derivative Spectrophotometry: Low and Higher Order*, VCH Publishers, New York, 228 p., 1994.
33. Philpot, W.D., The derivative ratio algorithm: Avoiding atmospheric effects in remote sensing, *IEEE Transactions on Geoscience and Remote Sensing*, 29(3):350–357, 1991.
34. Penuelas, J. et al., Reflectance indices associated with physiological changes in nitrogen- and water-limited sunflower leaves, *Remote Sensing of Environment*, 48:135–146, 1994.
35. Dick, K. and Miller J.R., Derivative analysis applied to high resolution optical spectra of freshwater lakes, *Proceedings of 14th Canadian Symposium on Remote Sensing*, Calgary, Alberta, CA, 1991.
36. Chen, Z., Curran, P.J., and Hansom, J.D., Derivative reflectance spectroscopy to estimate suspended sediment concentration, *Remote Sensing of Environment*, 40:67–77, 1992.
37. Wiangwang, N., Hyperspectral data modeling for water quality studies in Michigan's inland lakes, PhD Dissertation, Department of Geography, Michigan State University, East Lansing, MI, 243 p., 2006.
38. Adams, M.L., Philpot, W.D., and Norvell, W.A., Yellowness index: An application of the spectral second derivative to estimate chlorosis of leaves in stresses vegetation, *International Journal of Remote Sensing*, 20(18):3663–3675, 1999.
39. Clevers, J.G., Imaging spectrometry in agriculture: Plant vitality and yield indicators. *Imaging Spectrometry: A Tool for Environmental Observations*, Kluwer Academic, Alphen aan den Rijn, the Netherlands, pp. 193–219, 1994.
40. Haboudane, D. et al., Integrated narrow-band vegetation indices for prediction of crop chlorophyll content for application to precision agriculture, *Remote Sensing of Environment*, 81(2–3):416–426, 2002.
41. Daughtry, C.S.T. et al., Estimating corn leaf chlorophyll concentration from leaf and canopy reflectance, *Remote Sensing of Environment*, 74(2):229–239, 2000.

42. Rondeaux, G., Steven, M., and Baret, F., Optimization of soil-adjusted vegetation indices, *Remote Sensing of Environment*, 55:95–107, 1996.

43. Demouth, H. and Beale, M., *Neural Network Toolbox User's Guide Version 4*, The Math Works Inc., Natick, MA, 840 p., 2003.

44. Cipollini, P. et al., Retrieval of sea water optically active parameters from hyperspectral data by means of generalized radial basis function neural networks, *IEEE Transactions on Geoscience and Remote Sensing*, 39(7):1508–1524, 2001.

45. Zhang, Y. et al., Application of an empirical neural network to surface water quality estimation in the Gulf of Finland using combined optical data and microwave data, *Remote Sensing of Environment*, 81(2–3):327–336, 2002.

46. Zhang, Y. et al., Application of empirical neural networks to chlorophyll-*a* estimation in coastal waters using remote optosensors, *IEEE Sensors Journal*, 3(4):376–382, 2003.

47. Malthus, T.J. et al., An evaluation of the airborne thematic mapper sensor for monitoring inland water quality, *Proceeding of the 22nd Annual Conference of the Remote Sensing Society*, University of Durham, Durham, NC, 317–342, 1996.

48. Mumby, P.J. et al., A bird's-eye view of the health of coral reefs, *Nature*, 413:36–37, 2001.

49. Richardson, L.L., Hyperspectral imaging sensors and the marine coastal zone, *Hyperspectral Remote Sensing of the Ocean (Proceedings Volume)*, 4154:115–123, 2001.

50. Davis, C. et al., Ocean PHILLS hyperspectral imager: Design, characterization, and calibration, *Optics Express*, 10(4):210–221, 2002.

51. Clark, R.N. and Swayze, G.A., Mapping minerals, amorphous materials, environmental materials, vegetation, water, ice, and snow, and other materials: The USGS tricorder algorithm, *Summaries of the Fifth Annual JPL Airborne Earth Science Workshop*, 95-1(1):39–40, 1995.

52. Curran, P.J., Remote sensing of foliar chemistry, *Remote Sensing of Environment*, 30:271–278, 1989.

53. Ustin, S., *Remote Sensing for Natural Resource Management and Environmental Monitoring (Manual of Remote Sensing—Third Edition) Volume 4*, John Wiley & Sons, Chichester, U.K., pp. 679–729, 2004.

54. Goodenough, D.G., Li, J.Y., and Dyk, A., Combining hyperspectral remote sensing and physical modeling for applications in land ecosystems, *IEEE International Conference on Geoscience and Remote Sensing Symposium*, IGARSS'06, Denver, CO, pp. 2000–2004, 2006.

55. Koetz, B., Estimating biophysical and biochemical properties over heterogeneous vegetation canopies—radiative transfer modeling in forest canopies based on imaging spectrometry and lidar, PhD Thesis, Remote Sensing Laboratories, Department of Geography, University of Zurich, Zurich, Switzerland, 2006.

56. Huber, S. et al., The potential of spectrodirectional CHRIS/PROBA data for biochemistry estimation, *Envisat Symposium 2007*, Montreux, Switzerland, 6 p., 2007.

57. Merton, R.N., Multi-temporal analysis of community scale vegetation stress with imaging spectroscopy, PhD Thesis, Geography Department, University of Auckland, Auckland, New Zealand, 492 p, 1999.

58. Apan, A. et al., Detecting sugarcane 'orange rust' disease using EO-1 Hyperion hyperspectral imagery, *International Journal of Remote Sensing*, 25:489–498, 2004.

59. Krier, A. and Sherstnev, V.V., Powerful interface light emitting diodes for methane gas detection, *Journal of Physics D: Applied Physics*, 33(2):101–106, 2000.

60. Wei, H., The seasonal variation of column abundance of atmospheric CH_4 and precipitable water derived from ground-based IR solar spectra, *Infrared Physics and Technology*, 41:313–319, 2000.

61. Roy, J., Saugier, B., and Mooney, H.A., *Terrestrial Global Productivity*, Academic Press, San Diego, CA, 573 p., 2001.

62. Barnhouse, W.D., Methane plume detection using passive hyper-spectral remote sensing, MS Thesis, Bowling Green State University, Bowling Green, OH, 141 p., 2005.

63. Grace, J., Role of forest biomes in the global carbon balance, *The Carbon Balance of Forest Biomes*, Taylor & Francis Group, Boca Raton, FL, pp. 19–46, 2005.

64. Goodenough, D.G. et al., Processing Hyperion and ALI for forest classification, *IEEE Transactions on Geoscience and Remote Sensing*, 41:1321–1331, 2003.

65. Clark, R.N., Swayze, G.A., and Gallagher, A., Mapping the mineralogy and lithology of Canyonlands, Utah with imaging spectrometer data and the multiple spectral feature mapping algorithm, *Summaries of the Third Annual JPL Airborne Geoscience Workshop*, JPL publications, Pasadena, California, 92–14(1):11–13, 1992.

66. Asner, G.P. and Vitousek, P.M., Remote analysis of biological invasion and biogeochemical change, *Proceedings of the National Academy of Sciences of the United States of America*, 102:4383–4386, 2005.

67. Ustin, S.L. et al., (Eds.), Remote sensing of environment: State of the science and new directions, in *Remote Sensing of Natural Resources Management and Environmental Monitoring*, John Wiley & Sons, New York, pp. 679–729, 2004.

68. Jupp, D.L.B., Kirk, J.T.O., and Harris, G.P., Detection, identification and mapping of cyanobacteria—Using remote sensing to measure the optical quality of turbid inland waters, *Australian Journal of Marine and Freshwater Research*, 45:801–828, 1994.

69. Roelfsema, C. et al., Remote sensing of a cyanobacterial bloom (*lyngbya majuscule*) in Moreton Bay, Australia, *IEEE International Transactions on Geoscience and Remote Sensing Symposium, IGARSS'01 Proceedings*, pp. 613–615, 2001.

70. Wiangwang, N., Water clarity/trophic condition monitoring using satellite remote sensing data, Master's thesis, Department of Geography, Michigan State University, East Lansing, MI, 152 p., 2003.

71. Tsai, F. and Philpot, W., Derivative analysis of hyperspectral data, *Remote Sensing of Environment*, 66:41–51, 1998.

72. Richards, J.A. and Jia, X., *Remote Sensing Digital Image Analysis: An Introduction*, 3rd edn., Springer, New York, 363 p., 1999.

73. Des Marais, D.J. et al., Remote sensing of planetary properties and biosignatures on extrasolar terrestrial planets, *Astrobiology*, 2(2):153–181, 2002.

74. Heath, D.F. and Park, H., The solar backscatter ultraviolet (SBUV) and Total Ozone Mapping Spectrometer (TOMS) experiment, *The Nimbus-7 Users Guide*, NASA Goddard Space Flight Center, Greenbelt, MD, pp. 175–211, 1978.

75. Veefkind, J.P. et al., Total ozone from the Ozone Monitoring Instrument (OMI) using the DOAS technique, *IEEE Transactions on Geoscience and Remote Sensing*, 44(5):1239–1244, 2006.

76. Bhartia, P.K., OMI ozone product, *NASA Goddard Space Flight Center, OMI Algorithm Theoretical Basis Document Volume II*, 91 p., 2002.

77. Bhartia, P.K. et al., Highlights of the version 8 SBUV and TOMS datasets released at this symposium, *Proceedings of the XX Quadrennial Ozone Symposium*, Athens, Greece, p. 294, 2004.

78. Burrows, J.P. et al., The global monitoring experiment (GOME): Mission concept and first scientific results, *Journal of Atmospheric Sciences*, 56(2):151–175, 1999.

79. Piters, A.J.M. et al., GOME ozone fast delivery and value-added products, version 3.0, *KNMI Report, GOFAP-KNMI-ASD-01*, De Bilt, the Netherlands, 2000.

80. Spurr, R., Thomas, W., and Loyola, D., GOME level 1–2 algorithms description, *DLR Technical Note ER-TN-DLR-GO- 0025*, Oberpfaffenhofen, Germany, 2002.

81. Hilsenrath, E. et al., Early data from Aura and continuity from UARS and TOMS, *Space Science Reviews*, 125(1–4):417–430, 2006.

82. Levelt, P.F. et al., The ozone monitoring instrument, *IEEE Transactions on Geoscience and Remote Sensing*, 44(5):1093–1101, 2006.

83. Itten, K.I. et al., APEX—The hyperspectral ESA airborne prism experiment, *Sensors*, 8(10):6235–6259, 2008.

84. Turner, W. and Spector, S., Remote sensing for biodiversity science and conservation trends, *Ecology and Evolution*, 18:306–314, 2003.

85. Schaepman, M. and Malenovsky, Z., Bridging scaling gaps for the assessment of biodiversity from space, *The Full Picture*, Group on Earth Observations (GEO), Geneva, Switzerland, pp. 258–261, 2007.

86. Huber, S., Estimating foliar biochemistry from hyperspectral data in mixed forest canopy, *Forest Ecology and Management*, 256:491–501, 2008.

87. Goodenough, D.G. et al., Mapping forest biomass with AVIRIS and evaluating SNR impact on biomass prediction, *Natural Resources Canada*, internal report (presented at *NASA JPL AVIRIS Workshop*), Ottawa, Canada, 2005.

88. Inada, K., Spectral ratio of reflectance for estimating chlorophyll content of leaf, *Japanese Journal of Crop Science*, 154:261–265, 1985.

89. Shibayama, M. and Akiyama, T.A., Spectroradiometer for field use. VII. Radiometric estimation of nitrogen levels in filed rice canopies, *Japanese Journal of Crop Science*, 55:433–438, 1986.

90. Inoue, Y., Moran, M.S., and Horie, T., Analysis of spectral measurements in rice paddies for predicting rice growth and yield based on a simple crop simulation model, *Plant Production Science*, 1:269–279, 1998.

91. Lopez, R.D., *An Ecological Assessment of Invasive and Aggressive Plant Species in Coastal Wetlands of the Laurentian Great Lakes: A Combined Field-based and Remote Sensing Approach*, The United States Environmental Protection Agency, Environmental Sciences Division, Oak Ridge, TN, 2001.

92. Lopez, R.D. et al., *Using Landscape Metrics to Develop Indicators of Great Lakes Coastal Wetland Condition*. The United States Environmental Protection Agency, EPA/600/X-06/002, Washington, DC, 31 p., 2006.

93. Ustin, S.L. et al., Hyperspectral remote sensing for invasive species detection and mapping, *Geoscience and Remote Sensing Symposium*, 3:1658–1660, 2002.

94. Lass, L.W. and Prather, T.S., Detecting the Locations of Brazilian Pepper Trees in the Everglades with a Hyperspectral Sensor, *Weed Technology* 18(2):437–442, 2004.

95. Lass L.W. et al., A review of remote sensing of invasive weeds and example of the early detection of spotted knapweed (*Centaurea maculosa*) and babysbreath (*Gypsophila paniculata*) with a hyperspectral sensor, *Weed Science* 53(2):242–251, 2005.

96. Lawrence, R. and Labus, M., Early detection of Douglas-fir beetle infestation with subcanopy resolution hyperspectral imagery, *Western Journal of Applied Forestry*, 18:202–206, 2003.

97. Koetz, B. et al., Radiative transfer modeling within a heterogeneous canopy for estimation of forest fire fuel properties, *Remote Sensing of Environment*, 92:332–344, 2004.

98. Wyszecki, G. and Stiles, W.S., *Color Science: Concepts and Methods, Quantitative Data and Formulae*, 2nd edn., John Wiley & Sons, New York, 968 p. 2000.

99. Jacobson, N.P., Gupta, M.R., and Code, J.B., Linear fusion of image sets for display, *IEEE Transactions on Geoscience and Remote Sensing*, 45(10):3277–3288, 2007.

100. Ready, P.J. and Wintz, P.A., Information extraction, SNR improvement, and data compression in multi-spectral imagery, *IEEE Transactions on Communications*, 21(10):1123–1131, 1973.

101. Tyo, J.S. et al., Principal components-based display strategy for spectral imagery, *IEEE Transactions on Geoscience and Remote Sensing*, 41(3):708–718, 2003.

102. Kaewpijit, S., Moigne, J.L., and El-Ghazawi, T., Automatic reduction of hyperspectral imagery using wavelet spectral analysis, *IEEE Transactions on Geoscience and Remote Sensing*, 41:863–871, 2003.

103. Harsanyi, J.C. and Chang, C.I., Hyperspectral image classification and dimensionality reduction: An orthogonal subspace projection approach, *IEEE Transactions on Geoscience and Remote Sensing*, 32:779–785, 1994.

104. Tyo, J.S. and Olsen, R.C., Principal-components-based display strategy for spectral imagery, *IEEE Workshop on Advances in Techniques for Analysis of Remotely Sensed Data*, NASA Goddard Space Flight Center, Greenbelt, Maryland, 276–281, 2003.

105. Jacobson, N.P. and Gupta, M.R., Design goals and solutions for display of hyperspectral images, *IEEE Transactions on Geoscience and Remote Sensing*, 43(11):2684–2692, 2005.

106. Gao, B.C. et al., Atmospheric correction algorithms for hyperspectral remote sensing data of land and ocean, *Remote Sensing of Environment*, 113(S1):S17–S24, 2009.

107. Gao, B.C. and Goetz, A.F.H., Column atmospheric water vapor and vegetation liquid water retrievals from airborne imaging spectrometer data, *Journal of Geophysical Research*, 95(D4):3549–3564, 1990.

108. Kruse, F.A., Imaging spectrometer data analysis—A tutorial, *Proceedings of the International Symposium on Spectral Sensing Research (ISSSR)*, San Diego, CA, pp. 44–54, 1994.

109. Centner, V. et al., Comparison of multivariate calibration techniques applied to experimental NIR data sets, *Applied Spectroscopy*, 54:608–623, 2000.

Part III

Data Mining, Algorithms, Indices

4 Hyperspectral Data Mining

Sreekala G. Bajwa and Subodh S. Kulkarni

CONTENTS

4.1 INTRODUCTION

Why data mining? In hyperspectral remote sensing, data are collected in numerous (hundreds to thousands) narrow wavebands in one or more regions of the electromagnetic spectrum, and large numbers of data are collected. While offering tremendous potential, there are a variety of issues that must be addressed to "mine or extract the information of interest from hyperspectral data."

Each hyperspectral data set can take up hundreds of megabytes of storage space. For example, see a relatively small scene from an AVIRIS image in Figure 4.1 (92AV3C from https://engineering. purdue.edu/~biehl/MultiSpec/hyperspectral.html) with 145 × 145 pixels and 220 bands in the spectral region of 400–2500 nm. The image can be viewed as a cube with two spatial dimensions of length and width of the scene, with the third dimension being the spectral dimension. The spectral dimension has a large number of levels, 220 in this case, which is typical of hyperspectral data. In addition to the high dimensionality, hyperspectral data are also multicollinear or redundant. It means that all the information in those hyperspectral bands is not unique. Both the high dimensionality and data redundancy could pose major difficulties for the users of such data. For example, for a user interested in mapping a certain species of vegetation (or other similar applications), it may be difficult to know what bands to analyze, or how to extract relevant information from such a large data set. Some of the problems associated with the use of hyperspectral remote sensing are listed as follows.

1. *Data handling issues*: The users of hyperspectral should have the capability to store and handle large size data sets. They would require high performance computer with large storage capacity. Although data storage is becoming less of a problem with the decreasing cost of storage media, processing of such large data sets is still a big problem.
2. *Data redundancy problems*: It often refers to the fact that the information contained in each band of the hyperspectral image is not unique. On the contrary, many bands are very similar or redundant. Hyperspectral data redundancy can be visualized through covariance or correlation between bands. The band correlation surface shown in Figure 4.2 reveals high positive correlations between adjacent bands. It also shows negative correlation between bands in different spectral regions. For example,

FIGURE 4.1   (See color insert.) A hyperspectral image cube of AVIRIS scene from June 12, 1992 (https:// engineering.purdue.edu/~biehl/MultiSpec/hyperspectral.html) displayed with bands 40, 25, and 15 as R, G, and B for the top layer.

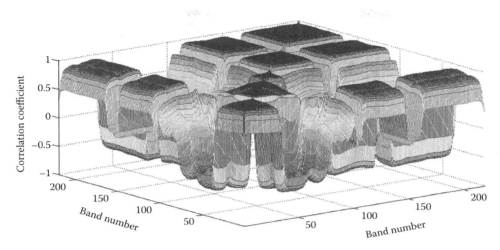

FIGURE 4.2 **(See color insert.)** Correlation between the bands of hyperspectral image shown in Figure 4.1. Only the alternate bands are used to compute the correlation.

the correlation between visible and near-infrared (NIR) bands is strongly negative ($r < -0.8$). Such high data redundancy indicates that data size could be significantly reduced by removing the redundant information.

3. *The curse of dimensionality*: As the number of bands in an image increases, the number of observations required to train a classifier increases exponentially to maintain the classification accuracies. This is called Hughes phenomenon [1], which refers to the loss of classifiability of an image with the same fixed number of training samples when the dimensionality of data increases [2]. Often, it is not possible to increase the size of training data due to time and budgetary constraints. In such cases, the data dimensionality must be reduced using an appropriate feature selection method.

What is data mining? Data mining refers to extracting and preserving previously unknown, potentially useful and reliable information or patterns contained in a hyperspectral image that are relevant to a specific application, without significant redundancy.

4.2 DATA MINING METHODS

Most hyperspectral image data mining procedures include a process of feature selection, followed by a process of information extraction [3]. The word feature is used throughout this chapter since band selection implies only selection of a subset of bands from the original image bands, and do not explicitly include other features that are developed by transforming two or more bands.

Both feature selection and information extraction methods could be either supervised or unsupervised (Figure 4.3). The unsupervised methods do not necessarily require ground data or prior knowledge of the target characteristics or phenomenon of interest to the user. Most unsupervised methods for feature selection are based on one or more feature characteristics such as variance, entropy, correlation, covariance, similarity measures, and natural grouping within the data [4–5].

Supervised methods use prior knowledge or training data on target characteristics or a phenomenon of interest to identify a group of best features and to train a classifier. Some of the supervised feature selection methods include divergence measures, correlation between features and ground data, support vector machines (SVM), etc. The output from both supervised and unsupervised methods of information extraction is usually a map indicating spatial patterns of interest or a table.

Data mining methods could also be either parametric or nonparametric. Generally speaking, parametric methods assume a data distribution, usually a normal distribution. A majority of the

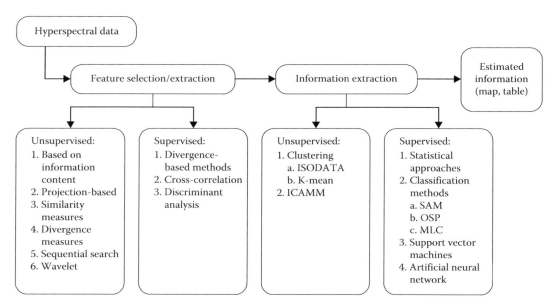

FIGURE 4.3   A summary of hyperspectral data mining methods.

methods used for data mining have an underlying assumption that the data are normally distributed. On the other hand, nonparametric methods make no such assumptions, and may be more appropriate for some applications.

4.3   FEATURE SELECTION/EXTRACTION METHODS

A feature refers to an individual band or the result of transforming one or more bands. For example, vegetation indices (VI) and principal components (PCs) are features. Hyperspectral data can potentially have hundreds to thousands of bands. Many thousands of features can be derived from these bands. Therefore, the feature space of a hyperspectral image could potentially have tens of thousands of features.

Interestingly, all those features may not be sensitive to the target variable or phenomenon of interest. Also, inclusion of all of these features in an information extraction process such as classification can make the process extremely slow, less accurate (explained by Hughes phenomenon), and computationally expensive. Therefore, feature selection/extraction is an important step in hyperspectral image data mining.

There are many methods available for feature selection or extraction. A feature selection method results in a subset of the original features, whereas a feature extraction method would provide a combination of new and/or reduced set of features that may include original features as well as newly extracted ones. A feature selection/extraction method should be able to identify the optimal set of features that would extract the information of interest with the highest possible accuracy and reliability in the least amount of time and computational effort and cost.

Feature extraction typically involves evaluation of a superset of features such as all bands and their transformations using an index of performance. Then, a subset of these features is selected based on this index of performance. The index of performance is a measure of the capability of the band to indicate the information of interest.

There are many methods available for feature selection from hyperspectral data. The choice of a method for a specific application may depend on availability of ground data as well as the scale of the problem. If the feature space has up to 19 features in it, it is called a small scale problem, whereas a medium scale has 20–49 features and large scale has 50 to infinite number of features in the feature space [6].

4.3.1 Feature Selection Based on Information Content

This group of methods uses various measures of information content of individual features to select a subset most appropriate for information extraction. All methods listed under this group of feature selection are unsupervised methods, meaning they do not require any training data. Although there are numerous methods scientists have experimented with, only the most commonly used methods are listed in the following.

4.3.1.1 Selection Based on Theoretical Knowledge

Feature extraction and feature selection can be made on theoretical knowledge of how the phenomenon or target characteristics of interest interact with radiation. For example, in optical remote sensing, a biochemical molecule that is indicative of the target characteristic of interest may have a fundamental band, one or more overtone bands, and combination bands. If the fundamental band for excitation of the molecule is known, it is easy to calculate the overtones and combination bands.

Information on the location of such bands could be used for extracting features from hyperspectral images in optical remote sensing. For example, the fundamental wavelength for three different vibrational modes of excitation for water vapor are 6270, 2738, and 2662 nm, with overtones at 3173 and 906 nm, and combination bands at 1876, 1135, 942, 906, 823, 796, and 652 nm [7]. Similarly, overtone or combination bands of other compounds include 1940 nm for liquid water, 2270 nm for lignin, 2336 nm for cellulose, etc. Many primary and secondary minerals and biochemical constituents of vegetation exhibit electronic (in visible wavelengths) and vibrational spectral features in infrared (IR) and NIR bands.

Several features such as VI have been derived based on the knowledge of the overtones or combination bands of chemical constituents of interest. In vegetation, chlorophyll-a and b, carotenoids, phytochromes, cellulose, nitrogen, canopy water content, etc. could be used as indicator of vegetation condition including growth stage, diseases, water and nutrient stresses, etc.

For example, a majority of the VIs are developed based on the knowledge of optical interaction of the target characteristics of interest. Almeida and De Souza Filho [8] have developed several VIs using the knowledge of overtone and combination bands of specific molecules of interest. These VI include R_{461}/R_{422} and R_{807}/R_{638} for chlorophyll-a, R_{520}/R_{470} and R_{807}/R_{648} for chlorophyll-b, R_{520}/R_{442} for α-carotene, R_{539}/R_{490} and R_{807}/R_{490} for carotenoids, R_{510}/R_{530} for anthocyanin, R_{845}/R_{730} for phytochrome P730, R_{778}/R_{658} for phytochrome P660, R_{1028}/R_{2101} for lignin, R_{2211}/R_{2400} for cellulose, R_{1731}/R_{1691} for nitrogen, and R_{1066}/R_{1452} for leaf water content.

4.3.1.2 Band Variance

In hyperspectral data, the variance of digital numbers (*DN*) within an image can be considered as a simplest measure of its information content. Therefore, feature variance can be used as a basis for selecting hyperspectral features [9]. The variance of *DN* within an image band can be calculated with Equation 4.1.

$$\text{Variance of a sample population, } s^2 = \frac{\sum_{i=1}^{N} (DN_i - \overline{DN})^2}{N-1} \tag{4.1}$$

where
DN_i is the digital number of ith pixel in the image band
$\overline{DN}$ is the average value of *DN*
N is the number of pixels in the image band

The same equation can be applied to any feature, with the *DN* replaced by the pixel values of the feature.

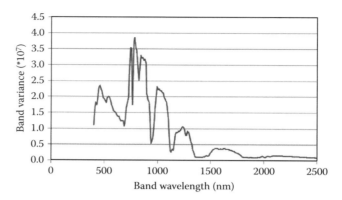

FIGURE 4.4   Band variance of the AVIRIS image shown in Figure 4.1.

The purpose of remote sensing is to understand a spatially dynamic phenomenon or variable. A feature sensitive to the spatial variability in a target should represent that information through a corresponding variability in its values. Therefore, selecting a set of features or bands with the most variability is a simple method of band selection. For example, the band variance of the hyperspectral image shown in Figure 4.1 indicates that the bands in the far red and NIR region have the highest variance (Figure 4.4).

A high information content in red and NIR is expected for an image scene consisting of soil and vegetation primarily. However, these adjacent bands can be very similar to each other since they are highly correlated (Figure 4.2). This brings up a major drawback of this method—it does not exclude redundant bands. Therefore, the information carried by the features selected with this method could be highly redundant. Also, if the variable or phenomenon of interest is not the cause of the dominating variability in the image scene but causes rather subtle variability, this method may not capture that information or bands carrying such information.

4.3.1.3 Information Entropy

Information entropy is a simple measure of information content of an image band [4,9–11]. It is calculated based on the probability of occurrence of each distinct *DN*. Entropy of an image band can be calculated using Equation 4.2.

$$\text{Information entropy, } H(X) = -\sum_{i=1}^{N} p_i \ln(p_i) \tag{4.2}$$

where
  p_i is the probability of occurrence of the *i*th *DN* in the hyperspectral image band *X*
  N is the number of distinct *DN*s in this band

If the entropy of an image band is high, then that band is considered to have high information content. For example, the entropy of a hyperspectral image of an agricultural field [9] indicates that the red bands have the highest entropy (Figure 4.5).

Entropy measure is very similar to variance measure in both benefits and drawbacks. Both methods are simple and easy to use, but do not consider redundancy of information in hyperspectral image bands.

Entropy measure can be expanded or combined with other measures to account for information redundancy between bands. Examples of such modification of entropy measure include total entropy measure (TEM) that combines band entropy with spectral derivative [11] and mutual information criterion (MIC). Wang and Angelopoulou [11] indicated that feature

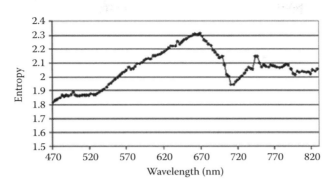

FIGURE 4.5 Entropy of a hyperspectral image of a corn field with 120 bands in the visible and NIR region. The image was acquired early in the season while the seeds were still germinating. (From Bajwa, S.G. et al. *Trans. ASABE*, 47, 895, 2004. With permission.)

selection with TEM resulted in better or comparable accuracies to feature selection with MIC, when both were compared under a classification scheme.

4.3.2 Projection-Based Methods

This group of methods is different from methods based on information content of the feature in that it transforms or projects the high-dimensional data into a low-dimensional space based on some constraint. A majority of the projection-based methods use projected bands that are linear combinations of the original bands.

4.3.2.1 Projection Pursuit

High-dimensional data space is mostly empty and therefore parametric methods are not very effective in image classification. High-dimensional data can be projected in numerous ways to lower dimensions. In case of projections of structured data into one or two dimensions, normality is not a stringent requirement because the data in the lower dimension represent a shadow of the high-dimensional data with a certain distribution. However, a parametric projection pursuit (PP) is usually used for projecting hyperspectral data to a reasonable lower dimension with higher dimensionality than two features for subsequent classification.

PP is a common method of obtaining lower dimensional and interesting projections of high-dimensional data. It is an unsupervised method that can be used for dimensionality reduction [12] and also to detect anomalies or patterns in hyperspectral imagery [13]. Both PCA and discriminant analysis could be considered as special cases of PP.

The PP uses linear combinations of original features to maximize some index of performance. This is equivalent to rotating the feature space. Scientists have used different types of PP index for different purposes for which the data were projected.

For classifying an image, the PP index should measure the ability of the projected data to form meaningful clusters. On the other hand, the PP index should help in identifying the outliers if the purpose of PP is to detect anomalies in a hyperspectral image. The PP index reported include a relative entropy image [14], Shannon entropy [15], Legendre's index [13], ratios of moments of a standard normal distribution [16], and the Bhattacharyya distance [12].

4.3.2.2 Principal Component Analysis

The principal component analysis (PCA) is a multivariate method commonly used for reducing data redundancy and dimensionality. PCs are linear combinations of the original group of bands or features such that the PCs are orthogonal to each other and the information content is maximized in each PC. The PCs are sorted based on their variance such that the first PC has the highest variance

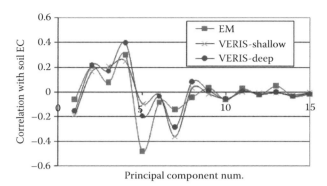

FIGURE 4.6   (See color insert.) Correlation between principal components of a hyperspectral image of an agricultural field and apparent soil electrical conductivity at two different depths. (From Bajwa, S.G. et al., *Trans. ASABE*, 47, 895, 2004. With permission.)

and it diminishes in successive PCs. Thus, a majority of the information contained in hundreds of bands of the hyperspectral image is captured in a few PCs, thus, achieving data dimensionality reduction. The decorrelated PCs also ensure that there is no data redundancy, which is otherwise a problem in remote sensing data. PCA can be applied to all or a subset of original bands or transformed bands such as VI [8–9] in order to understand a scene characteristic [17,18].

PCA is an unsupervised method. If the user is interested in a phenomenon or variable that causes subtle differences in target reflectance, then, PCA is not the best method for feature selection. For example, in an agricultural field with bare soil or relatively low vegetation coverage, soil electrical conductivity can cause subtle differences in reflectance. When a hyperspectral image of such a field with 120 bands in the visible and NIR region was subjected to PCA, the first three PC represented 99.6% of variability in the scene.

However, soil electrical conductivity showed the highest cross-correlation with PC5 and PC4 (Figure 4.6) [9]. The PC5 and PC4 represented only 0.2% and 0.1% of variability in the image, respectively. PCA is also not ideal for detecting small classes that are represented by relatively fewer pixels. In both these cases, the variable of interest generated relatively small variability in band *DN* compared to the dominant variability in the image. In such cases, PCA may not be able to correctly preserve the information of interest.

4.3.2.3 Independent Component Analysis

Independent component analysis (ICA) is a method superficially similar to PCA in that both methods reduce data redundancy and dimensionality. The ICA is different from PCA in that the PCA decorrelates the image bands, whereas the ICA uses independence of components to estimate them.

The ICA reveals hidden factors that underlie a set of measurements or signals [19]. The ICA assumes that each band is a linear mixture of independent hidden components and proceeds to recover the original factors or independent features through a linear unmixing operation [20]. The ICA can be performed on hyperspectral data transformed with PCA or on the original data under minimal mutual information (MMI) framework [21]. If the target classes of interest are not a major source of variability in the data, ICA on the original bands is recommended.

The limitation of ICA is that it can only separate linearly mixed models. It also assumes that the independent components are non-Gaussian. This would mean that the bands, which are linear combinations of non-Gaussian independent components could be Gaussian distributed according to the Central Limit Theorem.

Another limitation of ICA is that neither the variances (energies) nor the order of the independent components can be determined [22]. The advantage of ICA is that even when the components are not entirely independent, the ICA finds a transformation that maximizes the degree of independence between the components [22].

4.3.3 DIVERGENCE MEASURES

A majority of the divergence measures are supervised methods that require training data. Even in multispectral remote sensing, the supervised classification methods start with a separability analysis on training data, which are based on divergence measures. Measures of divergence are mostly distance-based.

4.3.3.1 Distance-Based Measures

Although distance measures are commonly used for supervised feature selection, they can also be used for unsupervised cluster analysis. Distance-based measures usually use training data to select a subset of features that have the best discrimination based on the distance between the mean class vectors [23].

There are many measures of distance between pixel vectors or between a pixel vector and class mean or between class means. The most commonly used distance measures include Euclidean, city block, Angular, Mahalanobis, divergence, Bhattacharyya, Kolmogorov variational distance, and Jeffries-Matusita distance (Table 4.1). Among these distance measures, the city block, Euclidean, and Angular measurements ignore the covariance C of the classes and do not make assumptions

---

TABLE 4.1

Distance Measures between Two Class Distributions a and b, and the Equations to Calculate These Distance Measures

Distance Measure	Equation
City block distance	$L_{CB} = \|\mu_a - \mu_b\| = \sum_{i=1}^{k} \|m_{ai} - m_{bi}\|$
Euclidean distance	$L_E = \|\mu_a - \mu_b\| = \left[(\mu_a - \mu_b)^T (\mu_a - \mu_b) \right]^{1/2}$
Angular distance	$\theta = a\cos\left[\dfrac{\mu_a^T \mu_b}{\|\mu_a\| \|\mu_b\|} \right]$
Normalized city block distance	$L_{NCB} = \sum_{i=1}^{k} \dfrac{\|m_{ai} - m_{bi}\|}{\left(\sqrt{c_{ai}} + \sqrt{c_{bi}}\right)/2}$
Mahalanobis distance	$L_M = \left[(\mu_a - \mu_b)^T \left(\dfrac{C_a + C_b}{2} \right)^{-1} (\mu_a - \mu_b) \right]^{1/2}$
Divergence	$D = \dfrac{1}{2} tr\left[(C_a - C_b)(C_b^{-1} - C_a^{-1}) \right] + tr\left[(C_a^{-1} + C_b^{-1})(\mu_a - \mu_b)(\mu_a - \mu_b)^T \right]$
Transformed divergence	$D^t = 2(1 - e^{-D/8})$
Kolmogorov variational distance	$K_a = \int \|P_a(\omega) - P_b(\omega)\| d\omega$, where $P_a(\omega)$ and $P_b(\omega)$ are joint conditional cumulative distribution functions of class a and b
Bhattacharyya distance	$L_B = \dfrac{L_M}{8} + \dfrac{1}{2} \ln\left[\dfrac{\|(C_a + C_b)/2\|}{\left(\|C_a\| \|C_b\|\right)^{1/2}} \right]$
Jeffries-Mausita	$L_{JM} = [2(1 - e^{-L_B})]^{1/2}$

The two class means are $\mu_a = (m_{a1}, \ldots, m_{ak})$ and $\mu_b = (m_{b1}, \ldots, m_{bk})$, where k is the number of features, and C is the covariance matrix

about the distribution of the classes. The remaining methods are considered more robust since they rely either on probability distribution or covariance.

In addition to the aforementioned measures, there are scatter matrix-based measures. Consider that S_w, S_b, and S_t are the within-class, between class, and total scatter matrices that are computed as shown as follows:

$$S_w = \sum_{i=1}^{L} P(\omega_i)C_i \tag{4.3}$$

$$S_b = \sum_{i=1}^{L} P(\omega_i)(\mu_i - \mu_0)(\mu_i - \mu_0)^T \tag{4.4}$$

$$S_t = S_w + S_b \tag{4.5}$$

where
  μ_i is the mean of ith class
  μ_0 is the mean of all classes together
  C_i is the class covariance
  L is the number of classes
  $P(\omega_i)$ is the class probability

Two divergence measures based on scatter matrix include $tr(S_w^{-1}S_b)$ and $det(S_w^{-1}S_b)$. These measures are also used for feature selection.

4.3.4 SIMILARITY MEASURES

Similarity measures use an index of performance that measures the degree of similarity between pairs of bands or features. Measures of similarity include correlation coefficient, mutual information entropy, and spectral derivative analysis.

4.3.4.1 Correlation Coefficient

If the correlation between two bands is high, they are considered redundant and hence one band may be adequate to represent the information [24]. The correlation coefficient is a measure of linear dependency between two variables x and y, and is widely used as the statistical measure of similarity between the two spectral bands x and y. It is defined as

$$\rho(x, y) = \frac{C(x, y)}{\sqrt{\sigma(x)\sigma(y)}} \tag{4.6}$$

where
  ρ is the correlation coefficient between x and y
  $C(x, y)$ is the covariance between x and y
  σ is the variance

Correlation between hyperspectral features can be used as a measure of their common information content or redundancy. The correlation coefficient can vary from −1 to +1. A correlation coefficient of 0 indicates no linear dependency whereas a +1 or −1 indicates a 100% dependency. For a pair of features that are highly correlated, one can be eliminated without losing any information. Spatial autocorrelation of a band can also be used as an indicator of information content of the band and hence as a criterion for band selection [23]. The bands with the highest spatial auto-correlation are regarded as the ones with the highest information content.

In supervised feature selection applications, the cross-correlation between the target character-istics and the superset of features can be used as an index for selection of the best set of features for target characterization [25]. If a feature shows high positive or negative correlation with the target characteristic, it should be included in the final set of features for estimating those characteristics. This method has been adopted for estimating many vegetation biophysical variables and for crop classification [24–26].

The drawbacks of correlation coefficient include its sensitivity to rotation of the scatter plot dia-gram in the (X, Y) plane. It is also invariant to scaling and translation of the variables. Because of these two properties, the correlation coefficient is somewhat unsuitable for feature selection in some applications. In spite of these drawbacks, cross-correlation is a widely adopted method for selecting features for developing simple or multivariate regression models to estimate target characteristics from hyperspectral data.

4.3.4.2 Mutual Information Analysis

The mutual information analysis (MIA) is a modification of the entropy-based method, and it con-siders the mutual information in any combination of two bands in a hyperspectral image [27]. If band X has M levels of DN and band Y has N levels of DN, then the mutual information entropy can be calculated as

$$H(X,Y) = -\sum_{i=1}^{M}\sum_{j=1}^{N} p_{i,j}\ln(p_{i,j}) \tag{4.7}$$

$$I(X,Y) = N\ln(N) + -\sum_{i=1}^{M}\sum_{j=1}^{N} F_{i,j}\ln(F_{i,j}) - \sum_{i=1}^{M} F_{i,+}\ln(F_{i,+}) - \sum_{j=1}^{N} F_{+,j}\ln(F_{+,j}) \tag{4.8}$$

$$MI = \frac{I(X,Y)}{I(Y)} \tag{4.9}$$

where
 $p_{i,j}$ is the probability of occurrence of pixels with a DN of i in band X and j in band Y
 $F_{i,+}$ is marginal summary of level i
 $F_{+,j}$ is marginal summary of level j
 $I(X,Y)$ is the mutual information between two bands X and Y
 MI is the percentage of mutual information expressed by band X with respect to band Y

If X and Y are probabilistically independent, $I(X,Y)$ would be zero. Conversely, if X and Y have perfect association, then $I(X,Y) = I(X) = I(Y)$. Using this method, bands with highest entropy and MMI can be selected. This method was shown to work well in identifying satellite image channels and outperformed entropy-based and correlation-based methods [27,28].

4.3.4.3 Spectral Derivative Analysis

The bandwidth of each band can be a variable in hyperspectral sensor design. The spectral deriva-tive method explores the bandwidth variable as a function of added information. It is apparent that if two adjacent bands do not differ much, then the underlying geospatial phenomenon can be char-acterized with only one band [9–10].

Although higher order derivatives can be calculated, the first and second order derivatives are commonly used in identifying spectral features. The mathematical descriptions of the first and second spectral derivatives are illustrated in Equations 4.10 and 4.11.

$$\text{First derivative, } D_{1\lambda} = \frac{\partial(DN(x,\lambda))}{\partial\lambda} \tag{4.10}$$

$$\text{Second derivative, } D_{2\lambda} = \frac{\partial^2(DN(x,\lambda))}{\partial\lambda^2} \tag{4.11}$$

where
  DN represents the digital number of a pixel in the hyperspectral image
  x is the spatial location of the pixel
  λ is the band characteristic or central wavelength

If D_1 is equal to zero, then one of the bands is redundant. In general, adjacent bands that differ a lot should be preserved for characterization, while adjacent bands similar to a specific band can be eliminated [10]. The second derivative identifies bands that can be represented by a linear combination of adjacent bands. Thus, if two adjacent bands can linearly interpolate the third band, then the third band is redundant. The larger the deviation from a linear model, the higher the information value of the band.

The drawback of the derivative analysis is that it usually compares only adjacent bands although higher order derivatives can be implemented to include larger number of bands.

Spectral derivatives can be combined with other feature selection methods such as divergence measures to identify useful derivative features. Tsai and Philpot [29] reported improved accuracies for land use land cover classification by adding spectral derivatives selected based on Jeffries-Matusita distance between training classes. Derivative analysis has been successfully utilized to identify optimal spectral bands for differentiating coastal wetland vegetation [30] and soil properties [10].

4.3.5 SEQUENTIAL SEARCH METHODS

Selection of features with most discriminatory power from a superset of features may include a search method and a criterion for selection. An important index of performance that is commonly used in multispectral remote sensing is the error of estimation of the target characteristics, or classification accuracy or clustering ability, depending on the purpose of feature extraction and application of interest.

Estimation of the classification accuracy (or error of estimation) for all combinations of features in hyperspectral remote sensing is a prohibitive task because of the tens of thousands of potential features. If there are L features in a data set, there will be $(2^L - 1)$ a combination of features that will need to be evaluated. For example, a feature space with a mere 100 features, there could be 1.3×10^{30} feature combinations. An appropriate search methodology can reduce some of the computational needs. Sequential search algorithms explore the search methodology rather than the index of performance.

Most popular sequential search methodologies include sequential forward selection (SFS) and sequential backward selection (SBS) [31]. In SFS method, the search starts with an empty feature space. Each feature is added one by one until a desired cardinality is obtained.

In SBS method, the search starts with the full feature space and features are removed one at a time until a desired cardinality is achieved. Both of these methods are suboptimal with the serious drawback that they cannot remove a feature already selected in a previous step (in case of SFS) or reconsider a feature that has been removed from the feature subset (in case of SBS). They could also be highly time consuming and computationally inefficient. Mao [32] proposed a modified selection algorithm called orthogonal forward selection and backward elimination that incorporated Gram-Schmidt and Givens orthogonal transformation into the SFS and SBS procedure. The advantage of this method was that it decorrelated the data in the transformed space.

Pudil [33] proposed a modification on the sequential forward/backward selection method by dynamically changing the number of features included or removed in a step. These methods are referred to as sequential forward floating selection (SFFS) and sequential floating backward selection (SFBS). These methods are more computationally efficient than SFS and SBS, and allow adding a feature that has been removed previously in the backward selection method, or removing a feature that has been previously added in the forward selection method. In addition to these floating selection methods, other selection methods such as genetic algorithms (GA) for moderately large dimensionality [34–36], simulated annealing [37,38], and steepest ascent [39] were also proposed for feature selection from high-dimensional data sets. Kudo and Sklansky [6] indicated that sequential search methods are appropriate for small and medium-scale problems, whereas GA is appropriate for large-scale problems.

4.3.6 OTHER METHODS

There are many other less frequently used methods for feature selection/extraction and dimensionality reduction. One such method is increasing the bandwidth by combining adjacent bands of a hyperspectral image that were selected based on one of the band selection methods. Another such method is applying spectral filters on the image data to remove noise and less useful features. Wavelet decomposition is another important method for identifying important features from a hyperspectral image.

4.3.6.1 Wavelet Decomposition Method

The wavelet decomposition is a kernel-based method that works with the frequency components of hyperspectral signals. Therefore, it is an ideal approach to feature extraction where a multiresolution approach is desirable [40]. The inherent ability of the wavelet function to vary the width of the operator allows it to separate fine scale and large scale information in a hyperspectral dataset. The fundamental operator used in wavelet transform (WT) is referred to as a *mother wavelet*. Any function $\psi(\lambda)$ can be used as a mother wavelet provided it satisfies the following admissibility condition:

$$\int_{-\infty}^{+\infty} \frac{|\Im(\psi(\lambda))|^2}{|\omega|} d\omega < \infty \tag{4.12}$$

where
$\Im$ indicates the Fourier transform of function $\psi(\lambda)$
ω is the Fourier domain variable

In other words, the mother wavelet function must oscillate with an average value of zero, while exhibiting an exponential decay and compact support. There are many mother wavelets available for one to choose. Some of the more common ones include biorthogonal spline, Haar, Daubechies, Gaussian, Symlet, Meyer and Coiflet, to name a few.

After choosing a mother wavelet, the wavelet transformation can be applied to a one-dimensional or two-dimensional signal using a discrete or continuous WT where the width and scale can be systematically changed. Therefore, WT is a useful tool to separate and identify features that are associated with a phenomenon or variable that may present in different scales.

For example, in a hyperspectral image scene, the land covers (crops, forests, grass, soil, water, impervious, etc.) may cause the large-scale variation. However, if the user is interested in tree species or evapotranspiration, the variation may be occurring at a finer scale than that caused by land cover types.

WT offers an efficient method to perform derivative analysis on hyperspectral data [40]. It can also be combined with statistical classifiers such as linear discriminant analysis (LDA) for supervised classification that can provide better classification accuracy than PCA-based classification [41]. A discrete wavelet transform (DWT) followed by LDA and maximum likelihood classification was able to successfully discriminate between soil, crop (soybean), and pitted morning-glory, a weed [41]. Also, a comparison of WT with PCA and band selection indicated that WT provided the best features for accurate mapping of forest crown closure and leaf area index (LAI) [42].

4.4 INFORMATION EXTRACTION METHODS

Important information such as biophysical variables related to vegetation can be obtained using two major groups of information extraction methods. The first groups include deterministic or stochastic radiative transfer models (RTM). These are process-based models that utilize our knowledge about the optical interaction of the target material to inference on characteristics we are interested in mapping. Typically, a model is trained with known data on the optical properties of the target, and then, it is inversed to develop an inverse model to estimate the target property of interest [43–45].

The second group of methods for information extraction is based on statistical or heuristic methods. These methods typically utilize a subset of features to develop a quantitative model or a classification protocol. Two major groups of statistical/heuristic information extraction methods include the following.

1. Modeling: If the target characteristic to quantify is numerical and the user prefers quantitative output, a common method used is development of a decision support system using a mathematical relationship. Examples include estimation of variables such as LAI or biomass, canopy nitrogen content, or evapo-transpiration from hyperspectral data.
2. Classification: If the target characteristic of interest is categorical, then a classification approach is most appropriate for information extraction. If the variable of interest is distribution of different vegetation species, identification of a specific invasive species, or land cover types, then classification is an appropriate strategy.

4.4.1 STATISTICAL METHODS

4.4.1.1 Multivariate and Partial Least Square Regression

Both multivariate and partial least square regression (PLSR) methods can be used for developing empirical models of a variable of interest based on hyperspectral image data. Multivariate regression can be only used when the number of observations in the ground data or training data is considerably larger than the number of features used as independent variables. Multivariate regression uses the method of least square to estimate the parameters (intercept and regression coefficient) of regression in a model of the form

$$y = b_0 + b_1 x_1 + b_2 x_2 + b_3 x_3 + \cdots + b_k x_k \qquad (4.13)$$

where
  y is the variable you are interested in estimating (biomass, evapo-transpiration, canopy nitrogen content, etc.)
  k is the number of features
  x_i is the ith feature
  b_i is the ith coefficient of regression with b_0 being the intercept

One problem with multivariate regression on image features is the high dependency or correlation between the features, which tends to exaggerate the goodness of fit. Since the PCs are uncorrelated, they can be used as features that eliminate this problem. Simple and multivariate regressions are extensively used for estimating biophysical variables such as LAI, biomass, foliar biochemical concentration, etc. [46–48].

PLSR is a popular method of modeling optically active chemical constituents of a target material with spectral data in spectroscopy. It is also ideal for high-dimensional data such as hyperspectral image data. When the number of observations in the training data is comparable or fewer than the number of features used as independent variables, then PLSR is a good method to develop a predictive model.

In PLSR, the extracted features are called partial least square (PLS) factors. The original bands are transformed to PLS factors such that the covariance between the dependent variable and the PLS factors in the training data set is maximized. This transformation is similar to PCA; the only difference is that the covariance is maximized here instead of the variance. The PLS factors are uncorrelated to each other. After transforming the data, a least square regression model is developed between the dependent variable and the PLS factors. Another advantage of this method is that more than one y variable could be modeled in one step. The PLSR has been applied on both spectroscopic and hyperspectral image data to estimate biophysical variables relating to vegetation [25,49–51].

Both the regression methods described here assume normal distribution of the data, and can be implemented only for modeling numerical variables. It also assumes that the error term in regression is homoscedastic or has constant variance for all X since they come from the same population. However, spatially dynamic phenomena (that are often represented in an image) are heteroscedastic, with error variance dependent on the spatial lag (distance) between observations. Therefore, another regression approach for modeling spatially dynamic phenomena is the spatial regression. Spatial regression approaches are explained in detail by Cressie et al. [52–65] and applied in remote sensing of vegetation by Bajwa et al. [48,56–57].

4.4.1.2 Discriminant Analysis

It works fast with high-dimensional hyperspectral data, and finds features that maximize the separation of the classes of interest [58]. It is widely used in pattern recognition applications [59–60]. The most commonly used discriminant analysis method is Fisher's LDA, which employs Fisher's ratio, which is the ratio of between-class matrix to within-class scatter matrix (refer to Equations 4.3 and 4.5).

To group hyperspectral data into L classes, the discriminant analysis identifies $L - 1$ eigen vectors as discriminant features. If there are k bands in the hyperspectral data, each pixel can be represented as a vector X with a $k \times 1$ dimension. If A is defined as a $k \times L - 1$ matrix with its columns representing the discriminant features, the original data set X can be projected onto the discriminant feature as $Y = A^T X$. The columns of A or the discriminant features are the eigen vectors of $(S_b^{-1} S_w)$ with nonzero eigen values, where S_b is the between-class scatter matrix and S_w within-class scatter matrix, which are calculated as shown in Equations 4.3 through 4.5.

The major problem with discriminant analysis is that it will identify only $L - 1$ features. In other words, the number of features cannot equal or exceed the number of classes. In hyperspectral images, the number of bands or features often exceeds the number of classes of interest. Also, discriminant analysis does not work well when distinguishing classes that have subtle differences, or very similar mean vectors. In many applications such as classifying similar species of vegetation, or growth stages of vegetation, or discriminating different types of stresses in the vegetation, the differences in reflectance are often very subtle. Such differences may not be easily distinguished with discriminant analysis since the class means may be very similar and discriminant analysis can use only relatively fewer ($L - 1$) features.

Fisher LDA can be combined with constrained energy minimization [59] methods to develop linear constrained distance-based discriminant analysis [58]. Although discrimnant analysis is mostly implemented as a supervised information extraction method, it can also be implemented for unsupervised information extraction using an automatic target detection method called target generation process [59].

4.4.2   Unsupervised Classification Methods

Unsupervised classification methods identify patterns of interest in an image data. This group of methods does not require training data or prior knowledge of the true class labels. The output of classification is a class label for each pixel. For example, a regression-based information extraction system can provide a quantitative map of biomass or other target properties of interest. In contrast, unsupervised classification results in a class number that can be used as such, or labeled postclassification. The postclassification labeling is supervised process and requires ground data on true labels for sample areas. As in the previous category, unsupervised classification methods could use a set of features selected by the feature selection methods to classify the image into patterns of interest.

4.4.2.1   Clustering

Clustering uses an iterative optimization method to classify pixel vectors to a number of groups based on a similarity measure. The similarity between pixels can be measured using one of the distance measures such as Euclidean distance, Mahalanobis, distance, city block distance, etc. The clustering algorithm can divide pixels into a user-specified number of groups using a k-mean clustering, or to a flexible number of groups within a specified range using the ISODATA clustering method. In both cases, it is possible to delineate clusters in multiple ways. For example, assume that a user is interested in classifying the pixel vectors shown in Figure 4.7 into two classes. All three decision boundaries or lines shown in Figure 4.7 could potentially classify the pixels into two groups. The clustering methods usually adopt a clustering quality indicator such as "sum of squared errors" to select the best decision boundaries.

The clustering algorithm will check the distance between cluster means and a pixel vector that needs to be classified, and will assign a pixel to the closest cluster. In one iteration, all the pixels will be assigned to its nearest cluster and the cluster means will be recalculated. This process will be repeated until it meets one of the user-specified criteria for stopping the process. The user specified criteria includes the number of classes, number of iterations, and percentage of pixels migrating from one cluster to another. More on clustering can be found on literature focusing on multispectral classification such as [61].

4.4.2.2   ICA Mixed Model Classification

The ICA can be used for deriving the independent components that contribute to the mixed signals in a hyperspectral image when the source signals and the mixing information are missing. In a classification problem, the independent components can be viewed as a set of mutually exclusive classes.

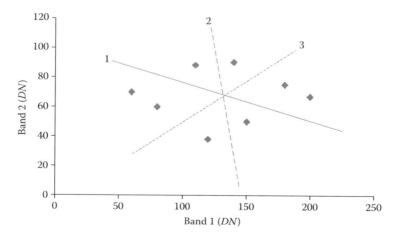

FIGURE 4.7   Three clustering options for a set of pixel vectors in a two-dimensional space. A line represents a decision boundary, the two sides of which represent the two clusters.

The ICA mixed model (ICAMM) considers the hyperspectral image as a mixture of these mutually exclusive classes with non-Gaussian probability density functions. The ICAMM first estimates these independent components as a linear transformation of the original hyperspectral bands using the statistical independency criterion and calculates the class component density [62–63]. It then calculates the class membership probability of each pixel using the Bayes theorem. Each pixel is assigned to a class based on the Bayes decision rule. Additional explanation of the theory behind ICA-based mixed model classification is given by Shah et al. [62–64].

The ICAMM is promulgated as a more appropriate method for unsupervised classification of hyperspectral images transformed with ICA. Because a majority of the classification methods assume normality of data and use second order statistics for classification, they are not considered complementary to ICA that employs higher order statistics [62–63]. Classification using ICAMM has shown to be superior to K-means classification used in combination with various feature selection methods such as PCA [62–63].

4.4.3 Supervised Classification

The first step in supervised classification is obtaining ground data or class signature [61]. Ground data refer to areas or pixels within an image with known class labels. The actual reflectance or absorbance of a class for all wavebands in the image is referred to as the class signature. Surveying the scene is a common method for collecting ground data for a current image scene. Other sources of ground data should be used if a field survey is not possible due to budgetary or time constraints. In such cases, maps, high resolution aerial photographs or images, farm records, archived spectral signatures of the classes of interest, and other historical records can be used [65,66].

Care should be taken to select areas representative of the class. Once ground data are developed, it is often divided into a training set and test set. Training set is used to train the classification algorithm to identify similar pixels in the entire image while test data are used for independent validation of the performance of the classifier.

The next step in supervised classification is to select or extract features from the hyperspectral image that needs to be classified. Either a supervised or unsupervised method can be used to select the features used for classification. If a supervised feature selection/extraction method is used, the ground data under training set are used to select/extract the features.

The next step is to train the classifier using the subset of features selected and the training data. The training process estimates the parameters used in a particular classification algorithm until one of the user-specified classification criteria is achieved for the training set.

The next step is to classify the entire image using the trained classifier. Depending on the type of classification method used, the output could be a hard classification, where each pixel is assigned one class, or a soft classification, where the membership of each pixel to each of the class is the output.

The last step in classification is a postclassification accuracy assessment using the ground data under the independent validation set to ensure that classification result is of acceptable quality. If the validation accuracies, represented by error matrix and kappa coefficient are acceptable, the process stops here. Otherwise, the classification process is repeated.

There are many methods available for supervised classification of hyperspectral data. Most methods available for classifying multispectral images can technically be used on hyperspectral data if the number of features has been reduced to similar levels as the multispectral data. Detailed explanation of various multispectral image classification methods is available from books [21,61,66].

Usually, even after feature selection, the input space for classification of hyperspectral data could be considerably larger than that of multispectral data. Therefore, some of the multispectral classification methods are not optimal for hyperspectral data because of the large number of training samples required for obtaining acceptable classification accuracy. In such cases, methods developed for hyperspectral data classification is better suited. The most commonly used hyperspectral data classification methods are listed in the following.

4.4.3.1 Spectral Angle Mapping

In an N-band (or N-feature) hyperspectral image, each pixel can be considered as an N-dimensional vector. Therefore, each vector defines a set of angles with the coordinates representing the band or features. In spectral angle mapping (SAM) method of classification, the angular distance between pixels is considered as the measure of distance [67]. Each pixel is assigned to the class which is closest to it based on the angular distance. In this respect, the SAM method is similar to the nearest neighbor classification method, except that SAM uses angular distance. This method is developed for classifying hyperspectral data. It can easily handle high-dimensional data and large number of pixels in training set data as it reduces the dimensionality to the axes.

The advantage of SAM is that it is insensitive to the magnitude of the pixel vectors since only the angular distance between vectors are used in establishing class membership. Therefore, it holds special significance for classifying vegetation.

Topographic shading usually interferes with vegetation signals in remote sensing. Since this interference tends to distribute the pixels along the same angular direction with different magnitude, it does not affect the angular orientation of the pixel vectors. Therefore, SAM is a good classification method to use when topographic shading is suspected.

4.4.3.2 Orthogonal Subspace Projection

The orthogonal subspace projection (OSP) is very effective in detecting and classifying constituent materials in a mixed pixel while suppressing undesired signatures [68]. Another benefit of OSP is its ability to reduce data dimensionality. Because of these properties, OSP is especially useful for class signature detection and discrimination and also for subpixel classification [69–71].

In this method, the classifier projects an unlabeled pixel vector onto a particular class vector (or subspace) of interest that is orthogonal to undesired signatures or other class vectors [68]. Since the pixel vector is projected orthogonal to all other class vectors, their effect on the pixel vector under consideration is nullified. The basic idea used in OSP is that if a pixel vector with the unknown label is projected to each of the class vectors while nullifying influences from all other class signatures, it will provide the highest membership with the class where it belongs the best.

If an image is being classified into L classes of interest using supervised classification, there will be L corresponding mean class signatures. Matrix E represents the $K \times L$ class signatures or endmembers. The matrix containing the first $L - 1$ columns or endmembers is called U, and the last column containing the endmember of interest is called d. The OSP classification operator q^T is defined as

$$q^T = d^T(I - UU^\#) \tag{4.14}$$

$$U^\# = (U^T U)^{-1} U^T \tag{4.15}$$

$$\alpha_p = \beta q^T DN \tag{4.16}$$

$$\beta = (d^T P d)^{-1} \tag{4.17}$$

where
    $U^\#$ is the pseudoinverse of U
    $(1 - UU^\#)$ is the projection matrix P
    DN is the pixel vector that needs to be classified
    α_p is the projection of the pixel vector (with unknown label) to the specific endmember d
    β is a scalar normalizing factor

If the pixel vector with the unknown label belongs to the endmember d, the value of α_p will be the highest. Large values of α_p indicate better membership in the class with the signature of d.

This method can be used successfully for pure pixel classification as well as mixed pixel classification [68]. Although OSP is used for supervised classification, it can also be used for unsupervised classification as demonstrated by Ren and Chang [72]. A drawback of this method is that it should have a minimum number of bands that is equal to the number of classes, which is usually not a problem in hyperspectral image classification.

4.4.3.3 Maximum Likelihood Classification

The MLC is by far the most commonly used method of supervised classification of multispectral data when data are Gaussian distributed. The MLC can be performed based on the Bayes' classification rule, which is based on the conditional probabilities of pixel vectors.

Using the training data, an estimate of the conditional probabilities is made and these conditional probabilities are used to develop the maximum likelihood decision rule. If the data distribution is multivariate normal and there are adequate numbers of training samples, MLC can result in high accuracy of classification. However, the number of training pixels required to maintain reasonable accuracy could be quite large in a hyperspectral data set.

The accuracy of MLC depends of the accuracy of the mean vector and covariance matrix estimated for the classes. If N features are used for classification, the training set for each class must contain a minimum of $N + 1$ pixels in order to calculate the sample covariance matrix, although Hoffbeck and Landgrebe [73] have reported a leave-one-out covariance estimation method in case of limited training data. As a rule of thumb, the number of pixels per class per feature is set around 10–100 for obtaining acceptable accurate class statistics [74]. For example, if one wants to classify a hyperspectral image with 100 bands into 10 different classes, a minimum of 1,000–10,000 training pixels per class will be needed for reasonable estimation of mean vector and covariance matrix. In other words, 10,000–100,000 total training pixels will be required. This requirement is without considering the accuracy degradation in hyperspectral images due to Hughes phenomenon. Additionally, the number of training pixels required for achieving a specific accuracy also increases as the pixel variability within a class increases.

In summary, a large number of training samples are required in each class to obtain reasonable accuracy [2,10]. In a hyperspectral image with many features, such large numbers may not be always achievable, especially for classes with limited spatial extent. However, many scientists have utilized feature selection methods to effectively reduce the number of features and then applied MLC successfully [3,63]. A comparison of MLC, SAM, artificial neural network (ANN), and decision tree classifiers found that MLC had the highest accuracy [75].

4.4.3.4 Artificial Neural Network

The ANN is a nonparametric method of classification in that the decision boundaries of the classes are not determined by a deterministic rule. In this method, the decision boundaries are assigned in an iterative fashion, to minimize the error of labeling the training data.

A neural network is a simplified representation of how the human brain works in identifying objects and patterns. It contains data processing elements called neurons communicating through synaptic connections. Although there are many types of neural networks, the most commonly used ANN model for hyperspectral image classification include multi-layer perceptron (MLP), radial basis function (RBF), and AdaBoost models [60,75]. Both RBF and AdaBoost can be used with and without regularization.

A typical MLP neural network for image classification will have one input layer, one output layer, and one or more hidden layers (Figure 4.8). The output layer and hidden layers contains neurons where data are processed. The output layer can have one or more neurons depending on the application. The most common method of training an MLP network is using error back-propagation.

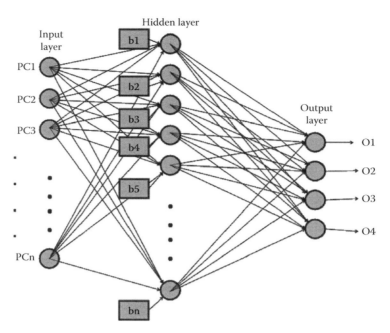

FIGURE 4.8 A example of MLP network showing input, hidden, and output layers, all with multiple neurons. Here principal components (PC) are listed as the input.

In this method, the data are fed from input layer to output layer, where it is processed by each successive neuron, while the error is propagated back from output layer to input layer.

The input of a network for image classification is typically the features selected using one of the feature selection methods. Initially, the model is trained using the training data with known output labels. The model training involves adjusting the weights associated with each synaptic connector and biases associated with each neuron to minimize the error between predicted output and target output. In other words, the network is trained until it learns the input patterns associated with a specific output with a user-specified accuracy. There are also other criteria used for stopping training, which include inability of the network to learn any further, or reaching a maximum number of training epochs specified by the user.

Each processing node contains a summation operator and a transformation operator, which together processes the inputs into a weighted sum and then transforms it into the output (Equations 4.18 through 4.20) using a transformation function called activation function.

$$S_j = \sum_i w_{ji} p_i \tag{4.18}$$

$$O_j = f(S_j) \tag{4.19}$$

$$f(S) = \frac{1}{1 + e^{-S}} \tag{4.20}$$

where
  p_i represents the ith inputs to the jth neuron in a specific layer (either hidden or output)
  w_{ij} represents the weight of the synaptic connection from ith input from the previous layer to jth neuron in the current layer
  O_j is the output from the jth neuron in the current layer
  f represents the transformation function

Although there are many transformation functions that are available for use in a neuron, the most common are sigmoid functions, with the general form indicated in Equation 4.20. The selection of transformation function can affect the rate of convergence in a neural network. Since ANN classifier is capable of developing highly nonlinear and nonparametric relationships, a properly trained network can provide highly accurate classification results. ANN-based classification of land cover types have shown to have higher accuracy than MLC, SAM, and minimum distance classifier [77]. More about ANN classification is given by Camps-Valls et al. [60,76–79]. Also, an overview of the role of ANN in remote sensing is given by Jensen et al. [80].

4.4.3.5 Support Vector Machines

SVM represent a machine learning method that works well with high-dimensional data with limited training samples [81]. The SVM can be used for feature selection, predictive modeling, and classification [82–84]. For linearly separable classes, the SVM tries to find the optimal separation surface between classes based on the training data. If the classes are not linearly separable, then the SVM uses a kernel-based method to find a nonlinear projection of the data where the classes are linearly separable. It is effective in separating classes with means very close to each other.

The SVM method is most commonly used for two-class separation problems although it can be extended to multiclass separation as well. In a two class problem, assume that (y_i, x_i) for $i = 1, 2, ..., N$ represents N training samples, where y_i is the label of the ith observation with values +1 or −1, and x_i is the corresponding feature vector with n features. The hyperplane separating the two classes has to be located such that the class labels +1 and −1 lies on either side of the hyperplane, and the minimum distance of the sample vectors to either side is maximized (Figure 4.9). The hyperplane is defined as

$$wx + b = 0 \qquad (4.21)$$

where w and b are parameters of the hyperplane. Here, the intercept b is a scalar, whereas the feature vector, x is L-dimensional. The vectors on either side of the hyperplane satisfy the condition that $wx + b \gtrless 0$. Therefore, the classifier can be expressed as

$$f(x, \alpha) = \text{sign}(wx + b) \qquad (4.22)$$

Therefore, the support vectors lie on two hyperplanes with equation

$$wx + b = \pm 1 \qquad (4.23)$$

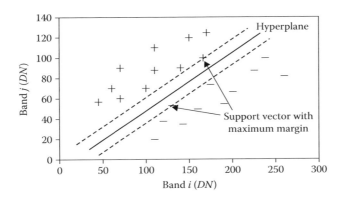

FIGURE 4.9 An example distribution of two-dimensional pixel vectors indicating the support vectors with maximum margin and the hyperplane defined as the decision boundary.

The optimal hyperplane is defined such that the margins (distance to support vector) is maximized. This constraint for optimizing the parameters of hyperplane can be expressed as

$$\min\left\{\frac{1}{2}\parallel w \parallel^2\right\} \quad \text{with } y_i(wx+b)\geq 1, \quad i=1,2,\dots,N \tag{4.24}$$

If the classes are not linearly separable, a regularization parameter C and error variable ε_i are introduced into the constraint in Equation 4.24.

The kernels used in SVM are functions based on the quadratic distance between support vectors. The common kernel functions used include local kernels such as radial basis, kernel with moderate decreasing (KMOD) and inverse multiquadratic, and global kernels such as linear, polynomial and sigmoid. Spectral kernels were also defined, specifically for classifying hyperspectral data that uses the local kernel function with spectral angle as the measure of distance [84]. The spectral kernel function tended to decrease false classification caused by shadows when classical kernel was used. Also, SVM-based classification using different kernels such as linear, polynomial, sigmoid, and RBF performed better than the MLC and a back propagation-based neural network classifier for land use classification [63]. This superior performance was consistent when all original bands were used as inputs as well as with a reduced number of features extracted with discriminant analysis and discriminant boundary feature extraction methods. SVM can also be combined with optimization models such as genetic algorithm to obtain high accuracy of classification [82].

4.5 ACCURACY ASSESSMENT

Information extraction procedures should be followed by an accuracy assessment before they can be utilized for any real applications. In case of decision support models for estimating a numerical variable, the accuracy of the model should be assessed using an independent set of data (data not used for model development) called validation data. In such cases, the model accuracy can be expressed with several performance measures including the coefficient of determination (or R^2 value), root mean square error (RMSE) of prediction and cross-validation, and standard error of prediction.

In case of classification problems, error can be expressed as the error matrix based on an independent validation data set that was not used for training. The error matrix can provide information on total number or percentage of correctly classified pixels as well as the errors of omission and commission for each class. Error of omission for a specific class is the number of pixels in that class that are classified as something else or given a wrong label, while error of commission for a certain class is the number of pixels from other classes labeled as the class under consideration.

In addition to the error matrix, classification accuracy can also be expressed using kappa coefficient. Kappa coefficient compares the classification results with respect to that of random assignment. A value of kappa coefficient of 1 indicates a perfectly accurate classification. More details on accuracy assessment can be obtained from [85].

For unsupervised classification, the common method of accuracy assessment requires postclassification labeling of the classified image based on information available on the actual class types. Once the classified image is labeled, the accuracy of labeling can be assessed in the same fashion as the supervised classification.

4.6 APPLICATIONS

A majority of the studies focused on developing or evaluating data mining techniques have utilized land cover analysis as the target application. Some of the other applications where various hyperspectral data mining methodologies covered in this chapter were applied successfully include estimation of biophysical variables of vegetation such as LAI, biomass, forest stand parameters, vegetation species, pigment concentration, canopy nutrient status, canopy temperature, etc., and physiological processes such as transpiration flux and leaf/canopy moisture content [86–99].

Hyperspectral remote sensing can be also utilized for studying biogeochemical processes by providing information on inventories of vegetation types, primary productivity, and photosynthetic activity. Detailed reviews of hyperspectral remote sensing applications in vegetation monitoring are provided by Treitz et al. [86–88]. Additional examples of selected vegetation monitoring applications utilizing various data mining techniques of feature selection and information extraction on hyperspectral image data are described in Table 4.2.

TABLE 4.2

A Summary of Research Showing of Numerous Applications of Data Mining (Feature Selection/Extraction and Information Extraction) Techniques Applied to Hyperspectral Remote Sensing

References	Application	Feature Selection Method	Information Extraction Methods
[48]	Biomass		Simple or multivariate regression, PLSR
[49]	Forest productivity (wood production, canopy N)—AVIRIS		PLSR
[77]	Land cover classification—OMIS data	PCA	ANN
[81]	Land cover classification—DAIS data	SVM, MNF	SVM, SVM-GA
[60]	Land cover classification—AVIRIS data92AV3C	LDA, SVM, Regularized RBFNN(R-RBFNN), kernel based Fisher's discrimination (k-FD). Regularized AdaBoost (R-Adaboost)	
[89]	Crop type detection—HyMap	R-RBFNN, SVM-RBF kernel, R-AdaBoost	
[91]	Forest pigment (Chl-a) concentration and LAI—CASI	Cross-correlation	simple linear regression
[92]	Forest understory information (canopy structure and pigments)—DAIS and ROSIS		MLP NN model
[93]	Forest LAI, forest species distribution—DAIS, ROSIS, MIVIS	Cross-correlation	Regression model for LAI, SAM for species distribution
[94]	Postfire vegetation recovery—Hyperion	Cross-correlation, PCA, stepwise discriminant analysis	Object oriented nearest-neighbor fuzzy classification
[95]	Crop coverage-ImpSensor V10	ICA	ICA
[96]	Forest Canopy N, LAI—AVIRIS and Hyperion	Derivative analysis	PLSR
[97]	Forest canopy LAI—Hyperion	Cross-correlation	Simple linear regression
[98]	Invasive species mapping-HyMap	MNF, continuum removal	SMA, SAM
[99]	Canopy water content and LAI- AVIRIS and MODIS	Knowledge-based band selection	RTM and simple regression
[100]	Vegetation classification-HyMap and CASI	Transformed divergence, SBS and band width increase	MLC

4.7 DISCUSSIONS AND FUTURE DIRECTIONS

This chapter gives an overview of data mining methods that can be used for extracting useful information from hyperspectral image (or imaging spectrometry) data. Many of the multispectral data processing methods may be unsuitable for hyperspectral data because of the high dimensionality and data redundancy. In the past couple of decades, many data mining techniques for extracting useful information from hyperspectral data have been developed and tested. A characteristic of these data mining methods is the feature reduction or feature extraction step. It is essential to employ one of the feature reduction or extraction methods on hyperspectral data to reduce the dimensionality of the data. Some of the feature selection methods have the added benefit of generating uncorrelated or independent features. Feature selection is usually followed by an information extraction step. If the feature selection method substantially reduces the data dimensionality, the general suit of methods available for information extraction from multispectral data can be applied to the reduced features successfully. However, there are also methods that are specifically suitable for hyperspectral data. This chapter covered both supervised and unsupervised feature selection and information extraction methodologies.

A review of the published literature indicates that for characterizing a target property that can be expressed as a numerical variable (e.g., various biophysical and biochemical properties), empirical models are the most commonly used.

The most widely used empirical models are regression models (simple linear or multivariate or PLSR) on a subset of features selected or extracted with PCA or PLS factors, cross-correlation, or features selected based on knowledge of the absorbance characteristics of the target of interest. Relatively fewer studies utilized ANN and SVM. For characterizing a categorical variable (such as forest species, land cover, etc), the most commonly adopted methods were classification with SAM, MLC, ANN, or SVM on a set of reduced features. The features were selected with PCA, discriminant analysis, derivative analysis, ICA, etc.

In summary, one may choose a specific combination of feature selection/extraction and information extraction methods depending on the application and objectives, the scale of the problem, skill level available, availability of training data, time, and budget constraints. With the proper selection of a suite of data mining techniques, it is possible to reduce data dimensionality and data redundancy, and extract unique information from hyperspectral images that are, often, substantial improvement when compared with multispectral image data.

ACKNOWLEDGMENT

The authors thank Michele Ann Tabler of the Interlibrary Loan Department of Mullins Library at the University of Arkansas for her help in getting copies of many of the papers that are cited in this chapter.

REFERENCES

1. Hughes, G.F., On the mean accuracy of statistical pattern recognizers, *IEEE Transactions on Information Theory*, IT-14, 55–63, 1968.
2. Shahshahani, B.M. and Landgrebe, D.A., The effect of unlabeled samples in reducing the small sample size problem and mitigating the Hughes phenomenon, *IEEE Transactions on Geosciences and Remote Sensing*, 32, 1087–1095, 1994.
3. Tu, T.M., Chen, C.-H., and Chang, C.-I., A fast two-stage classification method for high dimensional remote sensing data, *IEEE Transactions on Geosciences and Remote Sensing*, 36, 171–181, 1998.
4. Arzuaga-Cruz, E., Jimenez-Rodriguez, L.O., and Velez-Reyes, M., Unsupervised feature extraction and band subset selection techniques based on relative entropy criteria for hyperspectral data analysis, *Proceedings of SPIE*, Bellingham, WA, Vol. 5093, pp. 462–473, 2003.
5. Mitra, P., Murthy, C.A., and Pal, S.K., Unsupervised feature selection using feature similarity, *IEEE Transactions on Pattern Analysis and Machine Intelligence*, 24, 301–312, 2002.

6. Kudo, M. and Sklansky, J., Comparison of algorithms that select features for pattern classifiers, *Pattern Recognition*, 33, 25–41, 2000.
7. Rencs, A.N., *Remote Sensing for the Earth Sciences—Manual of Remote Sensing*, 3rd edn., Vol. 3, John Wiley & Sons, New York, 1999.
8. Almeida, T.I.R. and De Souza Filho, C.R., Principal component analysis applied to feature oriented band ratios of hyperspectral data: A tool for vegetation studies, *International Journal of Remote Sensing*, 25, 5005–5023, 2004.
9. Bajwa, S.G., Bajcsy, P., Groves, P., and Tian, L.F., Hyperspectral image data mining for band selection in agricultural applications, *Transactions of the ASABE*, 47, 895–907, 2004.
10. Bajcsy, P. and Groves, P., Methodology for hyperspectral band selection, *Photogrammetric Engineering and Remote Sensing*, 70, 793–802, 2004.
11. Wang, H. and Angelopoulou, E., Sensor band selection for multispectral imaging via average normalized information, *Journal of Real-Time Image Processing*, 1, 109–121, 2007.
12. Jiménez, L. and Landgrebe D.A., Supervised classification in high dimensional space: Geometrical, statistical, and asymptotical properties of multivariate data, *IEEE Transactions on Systems, Man, and Cybernetics-Part C: Applications and Reviews*, 28, 39–53, 1998.
13. Malpica, J.A., Rejas, J.G., and Alonso, M.C., A projection pursuit algorithm for anomaly detection in hyperspectral imagery, *Pattern Recognition*, 41, 3313–3327, 2008.
14. Ifarraguerri, A. and Chang, C., Unsupervised hyperspectral image analysis with projection pursuit, *IEEE Transactions on Geoscience and Remote Sensing*, 38, 2529–2538, 2000.
15. Huber, P.J., Projection pursuit, *Annals of Statistics*, 13, 435–475, 1985.
16. Jones, M.C. and Sibson, R., What is projection pursuit? *Journal of the Royal Statistical Society. A (Statistics in Society)*, 150, 1–36, 1987.
17. Tsai, F., Lin, E.K., and Yoshino, K., Spectrally segmented principal component analysis of hyperspectral imagery for mapping invasive plant species, *International Journal of Remote Sensing*, 28, 1023–1039, 2007.
18. Richards, J.A. and Jia, X., Segmented principal components transformation for efficient hyperspectral remote sensing of image display and classification, *IEEE Transactions on Geoscience and Remote Sensing*, 37, 538–542, 1999.
19. Hyvärinen, A., Karhunen, J., and Oja, E., *Independent Component Analysis*, John Wiley & Sons, New York, 2001.
20. Robila, S.A. and Varshney, P.K., Target detection in hyperspectral images based on independent component analysis, *Proceedings of SPIE Automatic Target Recognition XII*, Vol. 4726, pp. 173–182, 2002.
21. Varshney, P.K. and Arora, M.K., *Advanced Image Processing Techniques for Remotely Sensed Hyperspectral Data*, Springer-Verlag, Heidelberg, Germany, 2004.
22. Hyvärinen, A. and Oja, E., Independent component analysis, algorithms and applications, *Neural Networks*, 13, 411–430, 2000.
23. Petrie, G.M. and Heasler, P.G., Optimal band selection strategies for hyperspectral data sets, *Proceedings of IEEE International Symposium on Geoscience and Remote Sensing*, Seattle, WA, Vol. 3, pp. 1582–1584, 1998.
24. Gomez-Chova, L., Calpe, J., Camps-Valls, G., Martin, J.D., Soria, E., Vila, J., Alonso-Chorda, L., and Moreno, J., Feature selection of hyperspectral data through local correlation and SFFS for crop classification, *Proceedings of IEEE International Symposium on Geoscience and Remote Sensing*, Anchorage, AK, Vol. 1, pp. 555–557, 2004.
25. Bajwa, S.G., Mishra, A.R., and Norman, R.J., Canopy reflectance response to plant nitrogen accumulation in rice, *Precision Agriculture*, 11, 488–506.
26. Thenkabail, P.S., Smith, R.B., and Pauw, E.D., Hyperspectral vegetation indices and their relationships with agricultural crop characteristics, *Remote Sensing of Environment*, 71, 158–182, 2000.
27. Conese, C. and Maselli, F., Selection of optimum bands from TM scenes through mutual information analysis, *ISPRS Journal of Photogrammetry and Remote Sensing*, 48, 2–11, 1993.
28. Guo, B., Gunn, S.R., Damper, R.I., and Nelson, J.D.B., Band selection for hyperspectral image classification using mutual information, *IEEE Geoscience and Remote Sensing Letters*, 3, 522–526, 2006.
29. Tsai, F. and Philpot, W., A derivative-aided hyperspectral image analysis system for land-cover classification, *IEEE Transactions on Geoscience and Remote Sensing*, 40, 416–425, 2002.
30. Becker, B.L., Lusch, D.P., and Qi, J., Identifying optimal spectral bands from in sit measurements of Great Lakes coastal wetlands using second-derivative analysis, *Remote Sensing of Environment*, 97, 238–248, 2005.
31. Jain, A. and Zongkar, D., Feature selection: Evaluation, application and small sample performance, *IEEE Transactions of Pattern Analysis and Machine Intelligence*, 19, 153–158, 1997.

32. Mao, K.Z., Orthogonal forward selection and backward elimination algorithms for feature subset selection, *IEEE Transactions on Systems, Man, and Cybernetics, Part B: Cybernetics*, 34, 629–634, 2004.
33. Pudil, P., Novovicova, J., and Kittler, J., Floating search methods in feature selection, *Pattern Recognition Letters*, 15, 1119–1125, 1994.
34. Siedlecki, W. and Sklansky, J., A note on genetic algorithms for large-scale feature selection, *Pattern Recognition Letters*, 2, 197–222, 1988.
35. Raymer, M.L., Punch, W.F., Goodman, E.D., Kuhn, L.A., and Jain, A.K., Dimensionality reduction using genetic algorithm, *IEEE Transactions on Evolutionary Computation*, 4, 164–171, 2000.
36. Hong, J.-H. and Cho, S.-B., Efficient huge-scale feature selection with speciated genetic algorithm, *Pattern Recognition Letters*, 27, 143–150, 2006.
37. Siedlecki, W. and Sklansky, J., On automatic feature selection, *International Journal of Pattern Recognition*, 10, 335–347, 1989.
38. Chang, Y.L., Fang, J.P., Liu, J.-N., Ren, H., and Liang, W.-Y., A simulated annealing band selection approach for hyperspectral images, *Proceedings of IEEE International Geoscience and Remote Sensing Symposium*, Barcelona, Spain, pp. 3190–3193, 2007.
39. Serpico, S.B. and Bruzzone, L., A new search algorithm for feature selection in hyperspectral remote sensing images, *IEEE Transactions on Geoscience and Remote Sensing*, 1360–1367, 2001.
40. Bruce, L.M. and Li, J., Wavelets for computationally efficient hyperspectral derivative analysis, *IEEE Transactions on Geoscience and Remote Sensing*, 39, 1540–1546, 2001.
41. Bruce, L.M., Koger, C.H., and Li, J., Dimensionality reduction of hyperspectral data using discrete wavelet transform feature extraction, *IEEE Transactions on Geoscience and Remote Sensing*, 40, 2331–2338, 2002.
42. Pu, R. and Gong, P., Wavelet transform applied to EO-1 hyperspectral data for forest LAI and crown closure mapping, *Remote Sensing of Environment*, 91, 212–224, 2004.
43. Jacquemoud, S., Bacour, C., Poilve, H., and Frangi, J.-P., Comparison of four radiative transfer models to simulate plant canopies reflectance: Direct and inverse mode, *Remote Sensing of Environment*, 74, 471–481, 2000.
44. Verhoef, W. and Back, H., Simulation of hyperspectral and directional radiance images using coupled biophysical and atmospheric radiative transfer models, *Remote Sensing of Environment*, 87, 23–41, 2003.
45. Meroni, M., Colombo, R., and Panigada, C., Inversion of a radiative transfer model with hyperspectral observations for LAI mapping in poplar plantations, *Remote Sensing of Environment*, 92, 195–206, 2004.
46. Gong, P., Pu, R., and Miller, J.R., Coniferous forest leaf area index estimation along the Oregon transect using compact airborne spectrographic imager data, *Photogrammetric Engineering and Remote Sensing*, 61, 1107–1117, 1995.
47. Gastellu-Etchegorry, J.P., Zagolski, F., Mougtn, E., Marty, G., and Giordano, G., An assessment of canopy chemistry with AVIRIS—A case study in the Landes forest, South-west France, *International Journal of Remote Sensing*, 16, 487–501, 1995.
48. Bajwa, S.G. and Vories, E.D., Spatial analysis of cotton (*Gossypium hirsutum* L.) canopy responses to irrigation in a moderately humid area, *Irrigation Science,* 25, 429–441, 2007.
49. Smith, M.-L., Ollinger, S.V., Martin, M.E., Aber, J.D., Hallett, R.A., and Goodale, C.L., Direct estimation of aboveground forest productivity through hyperspectral remote sensing of canopy nitrogen, *Ecological Applications*, 12, 1286–1302, 2002.
50. Cho, M.A., Skidmore, A., Corsi, F., van Wieren, S.E., and Sobhan, I., Estimation of green grass/herb biomass from hyperspectral imagery using spectral indices and partial least square regression, *International Journal of Applied Earth Observation and Geoinformation*, 9, 414–424, 2007.
51. Bajwa, S.G., Modeling rice plant nitrogen effect on canopy reflectance with partial least square regression, *Transactions of the ASAE*, 49, 229–237, 2006.
52. Cressie, N.A.C., *Statistics for Spatial Data,* John Wiley & Sons, New York, 1993.
53. Stroup, W.W., Baenziger, P.S., and Mulitze, D.K., Removing spatial variation from wheat yield trials: A comparison of methods, *Crop Science,* 34, 62–66, 1994.
54. Lambert, D.M., Lowenberg-Deboer, J., and Bongiovanni, R., A comparison of four spatial regression models for yield monitor data: A case study from Argentina, *Precision Agriculture,* 5, 579–600, 2004.
55. Lambert, D.M., Lowenberg-DeBoer, J., and Malzer, G.L., Economic analysis of spatial-temporal patterns in corn and soybean response to nitrogen and phosphorus, *Agronomy Journal,* 98, 43–54, 2006.
56. Bajwa, S.G. and Mozaffari, M., Effect of N availability on vegetative index of cotton canopy: A spatial regression approach, *Transactions of the ASABE,* 50, 1883–1892, 2007.
57. Kulkarni, S.S., Bajwa, S.G., Rupe, J., and Kirkpatrick, T., Spatial correlation of crop response to soybean cyst nematode (*Heterodera glycines*), *Transactions of the ASABE,* 51, 1451–1459, 2008.

58. Fukunaga, K., *Introduction to Statistical Pattern Recognition,* 2nd edn., Academic Press, Boston, MA, 1990.
59. Du, Q. and Chang, C.-I., A linear constrained distance-based discriminant analysis for hyperspectral image classification, *Pattern Recognition*, 34, 361–373, 2000.
60. Camps-Valls, G. and Bruzzone, L., Kernel-based methods for hyperspectral image classification, *IEEE Transactions on Geoscience and Remote Sensing*, 43, 1351–1362, 2005.
61. Schowengerdt, R.A., *Remote Sensing—Models and Methods for Image Processing*, Academic Press, San Diego, CA, 1997.
62. Shah, C.A., Varshney, P.K., and Arora, M.K., ICA mixture model algorithm for unsupervised classification of remote sensing imagery, *International Journal of Remote Sensing*, 28, 1711–1731, 2007.
63. Shah, C.A., Watanachaturaporn, P., Varshney, P.K., and Arora, M.K., Some recent results on hyperspectral image classification, *Proceedings of IEEE Workshop on Advances in Techniques for Analysis of Hyperspectral Data*, Greenbelt, MD, pp. 346–353, 2003.
64. Lee, T.-W., Girolami, M., and Sejnowski, T.J., Independent component analysis using an extended infomax algorithm for mixed sub-Gaussian and super-Gaussian sources, *Neural Computation*, 11, 417–441, 1999.
65. Hoffbeck, J.P. and Landgrebe, D.A., Classification of remote sensing images having high spectral resolution, *Remote Sensing of Environment*, 57, 119–126, 1996.
66. Atkinson, P.M. and Tate, N.J., *Advances in Remote Sensing*. John Wiley & Sons, New York, 1999.
67. van der Meer, F., The effectiveness of spectral similarity measures for the analysis of hyperspectral imagery, *International Journal of Applied Earth Observation and Geoinformation*, 8, 3–17, 2006.
68. Harsanyi, J.C. and Chang, C.-I., Hyperspectral image classification and dimensionality reduction: An orthogonal subspace projection, *IEEE Transactions on Geoscience and Remote Sensing*, 32, 779–785, 1994.
69. Farrand, W. and Harsanyi, J.C., Mapping the distribution of mine tailing in the coeur d'Alene river valley, Idaho, through the use of constrained energy minimization technique, *Remote Sensing Environment*, 59, 64–76, 1997.
70. Chang, C.-I., Zhao, X.-L., Althouse, M.L.G., and Pan, J.J., Least square subspace projection approach to mixed pixel classification for hyperspectral images. *IEEE Transactions on Geoscience and Remote Sensing*, 36, 898–912, 1998.
71. Settle, J.J., On the relationship between spectral unmixing and subspace projection, *IEEE Transactions on Geoscience and Remote Sensing*, 34, 1045–1046, 1996.
72. Ren, H. and Chang, C.-I., A generalized orthogonal subspace projection approach to unsupervised multispectral image classification, *IEEE Transactions on Geoscience and Remote Sensing*, 38, 2515–2528, 2000.
73. Hoffbeck, J.P. and Landgrebe, D.A., Covariance matrix estimation and classification with limited training data, *IEEE Transactions on Pattern Analysis and Machine Intelligence*, 18, 763–767, 1996.
74. Swain, P.H. and Davis, S.M. (Eds.), *Remote Sensing: The Quantitative Approach*, McGraw-Hill, New York, 1978.
75. Shafri, H.Z.M., Suhaili, A., and Mansor, S., The performance of maximum likelihood, spectral angle mapper, neural network and decision tree classifiers in hyperspectral image analysis, *Journal of Computer Science*, 3, 419–423, 2007.
76. Goel, P.K., Prasher, S.O., Patel, R.M., Landry, J.A., Bonnell, R.B., and Viau, A.A., Classification of hyperspectral data by decision trees and artificial neural networks to identify weed stress and nitrogen status of corn, *Computers and Electronics in Agriculture*, 39, 67–93, 2003.
77. Du, P., Tan, K., Zhang, W., and Yan, Z., ANN classification of hyperspectral remotely sensed imagery: Experiments and analysis, *Proceedings of IEEE Congress on Image and Signal Processing*, Piscataway, NJ, pp. 692–696, 2008.
78. Benediktsson, J.A. and Kanellopoulos, I., Classification of multisource and hyperspectral data based on decision fusion, *IEEE Transactions on Geoscience and Remote Sensing*, 37, 1367–1377, 1999.
79. Benediktsson, J.A., Palmason, J.A., and Sveinsson, J.R., Classification of hyperspectral data from urban areas based on extended morphological profiles, *IEEE Transactions on Geoscience and Remote Sensing*, 43, 480–491, 2005.
80. Jensen, R.R., Hardin, P.J., and Yu, G., Artificial neural networks and remote sensing, *Geography Compass*, 3, 630–646, 2009.
81. Pal, M. and Mather, P.M., Support vector machines for classification in remote sensing, *International Journal of Remote Sensing*, 26, 1007–1011, 2005.
82. Pal, M., Support vector machine-based feature selection for land cover classification: A case study with DAIS hyperspectral data, *International Journal of Remote Sensing*, 27, 2877–2894, 2006.
83. Melgani, F. and Bruzzone, L., Classification of hyperspectral remote sensing images with support vector machines, *IEEE Transactions on Geoscience and Remote Sensing*, 42, 1778–1790, 2004.

84. Mercier, G. and Lennon, M., Support vector machines for hyperspectral image classification with spectral based kernels, *Proceedings of IEEE International Geoscience and Remote Sensing Symposium*, Toulouse, France, Vol. 1, pp. 288–290, 2003.

85. Congalton, R.G. and Green K., *Assessing the Accuracy of Remotely Sensed Data—Principles and Practices*, CRC Press, Boca Raton, FL, 1999.

86. Treitz, P.M. and Howarth, P.J., Hyperspectral remote sensing for estimating biophysical parameters of forest ecosystems, *Progress in Physical Geography*, 23, 359–390, 1999.

87. Govender, M., Chetty, K., and Bulcock, H., A review of hyperspectral remote sensing and its applications in vegetation and water resource studies, *Water SA*, 33, 145–152, 2007.

88. Im, J. and Jensen, J.R., Hyperspectral remote sensing of vegetation, *Geography Compass*, 2, 1943–1961, 2008.

89. Camps-Valls, G., Serrano-Lopez, A.J., Gomez-Chova, L., Martin-Guerrero, J.D., Calpe-Maravilla, J., and Moreno, J., Regularized RBF networks for hyperspectral data classification, *Image Analysis and Recognition*, 3212, 429–436, 2004.

90. Blackburn, G.A. and Milton, E.J. 1997. An ecological survey of deciduous woodlands using airborne remote sensing and geographical information systems, *International Journal of Remote Sensing*, 18, 1919–1935. 1997.

91. Boschetti, M., Gallo, I., Meroni, M., Brivio, P.A., and Binaghi, E., Retrieval of vegetation understory information fusing hyperspectral and panchromatic airborne data, *Proceedings of EARSel Workshop on Imaging Spectroscopy*, Herrsching, Germany, May 13–16, pp. 483–491, 2003.

92. Boschetti, M., Brivio, A.A., Carnesale, D., and Guardo, D., The contribution of hyperspectral remote sensing to identify vegetation characteristics necessary to assess the fate of persistent organic pollutants (POPs) in the environment, *Annals of Geophysics*, 49, 177–186, 2006.

93. Mitri, G.H. and Gitas, I.Z., Mapping postfire vegetation recovery using EO-1 Hyperion imagery, *IEEE Transactions on Geoscience and Remote Sensing*, 48, 1613–1618, 2010.

94. Kosaka, N., Uto, K., and Kosugi, Y., ICA-aided mixed pixel analysis of hyperspectral data in agricultural land, *IEEE Geoscience and Remote Sensing Letters*, 2, 220–224, 2005.

95. Smith, M.-L., Martin, M.E., Plourde, L., and Ollinger, S.V., Analysis of hyperspectral data for estimating temperate forest canopy nitrogen concentration: Comparison between an Airborne (AVIRIS) and a spaceborne (Hyperion) sensor, *IEEE Transactions on Geoscience and Remote Sensing*, 41, 1332–1337, 2003.

96. Gong, P., Pu, R., Biging, G.S., and Larrieu, M.R., Estimation of forest leaf area index using vegetation indices derived from Hyperion hyperspectral data, *IEEE Transactions on Geoscience and Remote Sensing*, 41, 1355–1362, 2003.

97. Hestir, E., Khanna, S., Andrew, M.E., Santos, M.J., Viers, J.H., Greenberg, J.A., Rajapakse, S.R., and Ustin, S.L., Identification of invasive vegetation using hyperspectral remote sensing in the California delta ecosystem, *Remote Sensing of Environment*, 112, 4034–4047, 2008.

98. Chang, Y.-B., Ustin, S.L., Riano, D., and Vanderbilt, V.C., Water content estimation from hyperspectral images, and MODIS indexes in Southeastern Arizona, *Remote Sensing of Environment*, 112, 363–374, 2004.

99. Riedmann, M. and Milton, E.J., Supervised band selection for optimal use of data from airborne hyperspectral sensors. *Proceedings of IEEE International Geoscience and Remote Sensing Symposium*, Toulouse, France, Vol. 1, pp. 1770–1772, 2003.

5 Hyperspectral Data Processing Algorithms

Antonio Plaza, Javier Plaza, Gabriel Martín, and Sergio Sánchez

CONTENTS

5.1 INTRODUCTION

Hyperspectral imaging is concerned with the measurement, analysis, and interpretation of spectra acquired from a given scene (or specific object) at a short, medium, or long distance by an airborne or satellite sensor [1]. The concept of hyperspectral imaging originated at NASA's Jet Propulsion Laboratory in California with the development of the Airborne visible infrared imaging spectrometer (AVIRIS), able to cover the wavelength region from 400 to 2500 nm using more than 200 spectral channels, at nominal spectral resolution of 10 nm [2]. As a result, each pixel vector collected by a hyperspectral instrument can be seen as a *spectral signature* or *fingerprint* of the underlying materials within the pixel.

The special characteristics of hyperspectral data sets pose different processing problems [3], which must be necessarily tackled under specific mathematical formalisms, such as classification, segmentation, image coding, or spectral mixture analysis [4]. These problems also require specific dedicated processing software and hardware platforms. In most studies, techniques are divided into full-pixel and mixed-pixel techniques, where each pixel vector defines a *spectral signature* or *fingerprint* that uniquely characterizes the underlying materials at each site in a scene [5]. Mostly based on previous efforts in multispectral imaging, full-pixel techniques assume that each pixel vector measures the response of one single underlying material. Often, however, this is not a realistic assumption. If the spatial resolution of the sensor is not fine enough to separate different pure signature classes at a macroscopic level, these can jointly occupy a single pixel, and the resulting spectral signature will be a composite of the individual pure spectra, called *endmembers* in hyperspectral terminology [6]. Mixed pixels can also result when distinct materials are combined into a homogeneous or intimate mixture, which occurs independently of the spatial resolution of the sensor.

To address these issues, spectral unmixing approaches have been developed under the assumption that each pixel vector measures the response of multiple underlying materials [7].

Our main goal in this chapter is to provide a seminal view on recent advances in techniques for full-pixel and mixed-pixel processing of hyperspectral images, taking into account both the spectral and spatial properties of the data. Due to the small number of training samples and the high number of features available in remote sensing applications, reliable estimation of statistical class parameters is a challenging goal [4]. As a result, with a limited training set, classification accuracy (in full-pixel sense) tends to decrease as the number of features increases. This is known as the Hughes effect. Furthermore, high-dimensional spaces are mostly empty, thus making density estimation more difficult. One possible approach to handle the problem of dimensionality is to consider the geometrical properties rather than the statistical properties of the classes. In this regard, it is important to develop techniques able to select the most highly informative training samples from the available training set [8]. The good classification performance already demonstrated by techniques such as kernel methods and support vector machines (SVMs) in remote sensing applications [9], using spectral signatures as input features, has been further increased using intelligent training sample selection algorithms [10].

It should be noted that most available hyperspectral data processing techniques (including both full-pixel and mixed-pixel techniques) focused on analyzing the data without incorporating information on the spatially adjacent data, that is, hyperspectral data are usually not treated as images, but as unordered listings of spectral measurements with no particular spatial arrangement. In certain applications, however, the incorporation of spatial and spectral information is mandatory to achieve sufficiently accurate mapping and/or classification results [11–13]. To address the need for developments able to exploit a priori information about the spatial arrangement of the objects in the scene in order to complement spectral information, this chapter also presents several techniques for spatial–spectral data processing in the context of a mixed-pixel classification scenario.

5.2 SUPPORT VECTOR MACHINES

Supervised classification is one of the most commonly undertaken analyses of remotely sensed hyperspectral data. The output of a supervised classification is effectively a thematic map that provides a snapshot representation of the spatial distribution of a particular theme of interest such as land cover. Recent research has indicated the considerable potential of SVM-based approaches for the supervised classification of remotely sensed hyperspectral data [14]. Comparative studies have shown that classification by an SVM can be more accurate than techniques such as neural networks, decision trees, and probabilistic classifiers such as maximum likelihood classification [9]. This is due to the superior performance of SVMs when analyzing high-dimensional data (particularly in the presence of limited training samples), which generally results in higher relative accuracies than those reported for other classification methods. SVMs were designed for binary classification but various methods exist to extend the binary approach to multiclass classification, such as the *one versus the rest* and the *one versus one* strategies [15].

In essence, the SVM classification is based on fitting an optimal separating hyperplane between classes by focusing on the training samples that lie at the edge of the class distributions, which are the support vectors (Figure 5.1, reproduced from [9]). All of the other training samples are effectively discarded as they do not contribute to the estimation of hyperplane location. In this way not only is an optimal hyperplane fitted, in the sense that it is expected to be generalizable to a large degree, but also a high accuracy may be obtained with the use of a small training set. It should be noted that the SVM used with a kernel function is a nonlinear classifier, where the nonlinear ability is included in the kernel. Different kernels lead to different SVMs. The most used kernels are the polynomial kernel, the Gaussian kernel, or the spectral angle mapper kernel, among many others [9].

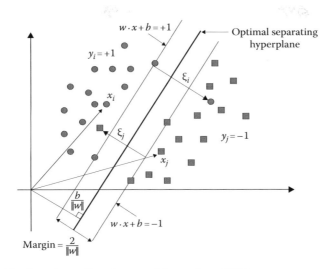

FIGURE 5.1 Classification of a nonlinearly separable case by a SVM.

Recently, innovative kernel-based algorithms with enhanced properties have been developed. These include semisupervised or *transductive* SVMs (TSVMs) learning procedures [16], which are used to exploit both labeled and unlabeled pixels in the training stage, or contextual SVMs [17], in which spatial and spectral information is incorporated by means of the use of proper kernel functions. The capability of semisupervised SVMs to capture the intrinsic information present in the unlabeled data can further mitigate the Hughes phenomenon, and contextual SVMs can address the issues related to the nonstationary behavior of the spectral signatures of classes in the spatial domain.

5.3 SPECTRAL UNMIXING OF HYPERSPECTRAL DATA

Spectral mixture analysis (also called *spectral unmixing*) has been an alluring exploitation goal from the earliest days of hyperspectral imaging [1] to the present [18]. No matter what the spatial resolution is, the spectral signatures collected in natural environments are invariably a mixture of the signatures of the various materials found within the spatial extent of the ground instantaneous field view of the imaging instrument [7]. The availability of hyperspectral imagers with a number of spectral bands that exceeds the number of spectral mixture components [2] has cast the unmixing problem in terms of an over-determined system of equations in which, given a set of pure spectral signatures (called *endmembers*) the actual unmixing to determine apparent pixel *abundance fractions* can be defined in terms of a numerical inversion process.

A standard technique for spectral mixture analysis is *linear* spectral unmixing [19], which assumes that the collected spectra at the spectrometer can be expressed in the form of a linear combination of endmembers weighted by their corresponding abundances. It should be noted that the linear mixture model assumes minimal secondary reflections and/or multiple scattering effects in the data collection procedure, and, hence, the measured spectra can be expressed as a linear combination of the spectral signatures of materials present in the mixed pixel (Figure 5.2a).

Although the linear model has practical advantages such as ease of implementation and flexibility in different applications [3], *nonlinear* spectral unmixing may best characterize the resultant mixed spectra for certain endmember distributions, such as those in which the endmember components are randomly distributed throughout the field of view of the instrument [10,20]. In those cases, the mixed spectra collected at the imaging instrument are better described by assuming that part of the source radiation is multiply scattered before being collected at the sensor (Figure 5.2b).

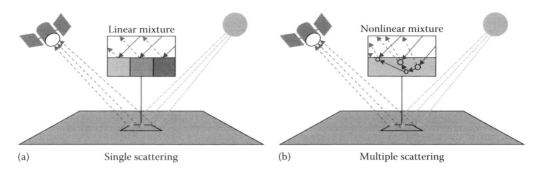

FIGURE 5.2   Graphical interpretation of the linear (a) versus the nonlinear (b) mixture model.

5.3.1 LINEAR SPECTRAL UNMIXING

In order to be able to correctly unmix a hyperspectral data set using the linear model, two requirements are needed:

1. A successful estimation of the number of endmembers (spectrally distinct pure signatures) present in the input hyperspectral scene.
2. The correct determination of a set of endmembers and their correspondent abundance fractions at each pixel.

In order to address the first requirement, two successful techniques in the literature have been the virtual dimensionality (VD) [21] and HySime [22]. The VD concept formulates the issue of whether a distinct signature is present or not in each of the spectral bands as a binary hypothesis testing problem, where a so-called Neyman–Pearson detector is generated to serve as a decision-maker based on a prescribed false alarm probability. In light of this interpretation, the issue of determining an appropriate value for the number of endmembers is further simplified and reduced to setting a specific value of the false alarm probability. In turn, the HySime uses a minimum mean squared error-based approach to determine the signal subspace in hyperspectral imagery.

Regarding the second requirement for successful implementation of the linear mixture model, several algorithms have been developed in recent years for automatic or semiautomatic extraction of spectral endmembers [6]. Classic techniques include the pixel purity index (PPI) [23], N-FINDR [24–26], iterative error analysis (IEA) [27], optical real-time adaptive spectral identification system (ORASIS) [28], convex cone analysis (CCA) [29], vertex component analysis (VCA) [30], and an orthogonal subspace projection (OSP) technique in [31]. Other advanced techniques for endmember extraction have been recently proposed, but few of them consider spatial adjacency. However, one of the distinguishing properties of hyperspectral data is the multivariate information coupled with a two-dimensional (pictorial) representation amenable to image interpretation.

Subsequently, most endmember extraction algorithms listed earlier could benefit from an integrated framework in which both the spectral information and the spatial arrangement of pixel vectors are taken into account. An example is given in Figure 5.3, in which a hyperspectral data cube collected over an urban area (high spatial correlation) is modified by randomly permuting the spatial coordinates of the pixel vectors (i.e., removing the spatial correlation). In both scenes, the application of a spectral-based processing method would yield the same analysis results, while it is clear that a spatial–spectral technique could incorporate the spatial information present in the original scene into the process.

To the best of our knowledge, only a few attempts exist in the literature aimed at including the spatial information in the process of extracting spectral endmembers. Extended morphological operations [13] have been used as a baseline to develop an automatic morphological endmember

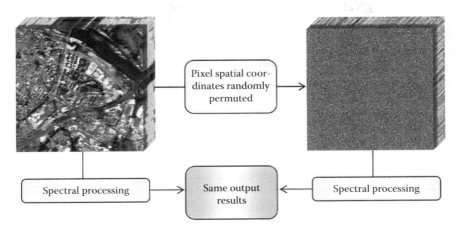

FIGURE 5.3 The importance of including spatial information in hyperspectral data processing.

extraction (AMEE) algorithm [32] for spatial–spectral endmember extraction. Also, spatial averaging of spectrally similar endmember candidates found via singular value decomposition (SVD) was used in the development of the spatial spectral endmember extraction (SSEE) algorithm [33].

Recently, a spatial preprocessing (SPP) algorithm [34] has been proposed. A spatially derived factor is used by this technique to weight the importance of the spectral information associated to each pixel in terms of its spatial context. The SPP is intended as a preprocessing module that can be used in combination with an existing spectral-based endmember extraction algorithm.

Once a set of endmembers have been extracted, their corresponding abundance fractions in a specific pixel vector of the scene can be estimated (in least squares sense) by using the unconstrained and constrained techniques [35]. It should be noted that the fractional abundance estimations obtained in unconstrained fashion do not satisfy the abundance sum-to-one (ASC) and the abundance non-negativity (ANC) constraints that should hold in order for the linear mixture model to be physically meaningful (i.e., the derived endmember set should be complete and negative abundance estimations lack physical interpretation). Imposing the ASC and ANC constraints leads to a more complex optimization problem, which has been solved (in least-squares sense) in the literature [36].

5.3.2 Nonlinear Spectral Unmixing

In a nonlinear model, the interaction between the *endmembers* and their associated *fractional abundances* is given by a nonlinear function, which is not known a priori. Various machine learning techniques have been proposed in the literature to estimate this function. In particular, artificial neural networks have demonstrated great potential to decompose mixed pixels due to their inherent capacity to approximate complex functions [37]. Although many neural network architectures exist, for decomposition of mixed pixels in terms of nonlinear relationships mostly feed-forward networks of various layers, such as the multilayer perceptron (MLP), have been used [10,38,39]. It has been shown in the literature that MLP-based neural models, when trained accordingly, generally outperform other nonlinear models such as regression trees or fuzzy classifiers [40].

A variety of issues have been investigated in order to evaluate the impact of training in mixed pixel classification accuracy, including the size and location of training sites, and the composition of training sets, but most of the attention has been paid to the issue of training set size, that is, the number of training samples required for the learning stage [41]. Sometimes the smallness of a training set represents a major problem. This is especially apparent for analyses using hyperspectral sensor data, where the requirement of large volumes of training sites is a serious limitation [42].

Even if the endmembers participating in mixtures in a certain area are known, proportions of these endmembers on a per-pixel basis are difficult to be estimated a priori. Therefore, one of the most challenging aspects in the design of neural network-based techniques for spectral mixture analysis is to reduce the need for very large training sets. Studies have investigated a range of issues [43], including the use of feature selection and feature extraction methods to reduce the dimensionality of the input data [38], the use of unlabeled and semilabeled samples [42], the accommodation of spatial dependence in the data to define an efficient sampling design [33], or the use of statistics derived on other locations [44].

Our speculation (and that of many thoughtful investigators over the past 40 years [42,43]) is that the problem of mixed pixel interpretation demands intelligent training sample selection algorithms, able to seek for the most informative training samples, thus optimizing the compromise between estimation accuracy (to be maximized) and ground-truth knowledge (to be minimized). In this sense, several efforts in the literature have been oriented toward the selection of mixed (border) training samples using previous work developed by Foody, as well as core (pure) training samples developed by simple *endmember* extraction algorithms.

In our experience, machine learning techniques such as MLP neural networks or SVMs can produce stable results when trained accordingly, a fact that leads us to believe that training can indeed be more important than the choice of a specific network architecture in mixture analysis applications.

5.4   EXPERIMENTAL RESULTS

5.4.1   Analysis of Supervised Hyperspectral Data Classification Using SVMs

The hyperspectral scene used for experiments in this subsection was gathered by AVIRIS over the Indian Pines test site in Northwestern Indiana, a mixed agricultural/forested area, early in the growing season, and consists of 1939×677 pixels and 204 spectral bands in the wavelength range 400–2500 nm (523 MB in size). Twenty AVIRIS bands (151–170) were removed from the original scene prior to analysis due to low signal-to-noise ratio (SNR) in those bands. The AVIRIS Indian Pines data set represents a very challenging classification problem dominated by similar spectral classes and mixed pixels. Specifically, the primary crops of the area, mainly corn and soybeans, were very early in their growth cycle with only about 5% canopy cover. This fact makes most of the scene pixels highly mixed in nature. Discriminating among the major crops under this circumstances can be very difficult, a fact that has made this scene an extensively used benchmark to validate classification accuracy of hyperspectral imaging algorithms. For illustrative purposes, Figure 5.4a shows a randomly selected spectral band (587 nm) of the original scene and Figure 5.4b shows the corresponding ground-truth map, displayed in the form of a class assignment for each labeled pixel, with 30 mutually exclusive ground-truth classes. Part of these data, including the ground-truth, are available online from Purdue University (from http://dynamo.ecn.purdue.edu/~biehl/MultiSpec).

In the following, three types of kernels are used in experiments: polynomial, Gaussian, and spectral angle mapper. Small training sets, composed of 1%, 2%, 4%, 6%, 8%, 10%, and 20% of the ground-truth pixels available per class, were extracted using pure (core) and mixed (border) training sample selection algorithms [10], and also using a random selection procedure. The SVM was trained with each of these training subsets and then evaluated with the remaining test set. Each experiment was repeated five times in order to guarantee statistical significance, and the mean accuracy values were reported. Table 5.1 summarizes the overall classification results obtained using the three considered kernels and training sample selection algorithms.

From Table 5.1, it can be seen that SVMs generalize quite well: with only 1% of training pixels per class, almost 90% overall classification accuracy is reached by all kernels when trained using border training samples. In all cases, classification accuracies decreased when random and pure samples were used for the training site. This confirms the fact that kernel-based methods in general and SVMs in particular are less affected by the Hughes phenomenon. It is also clear from Table 5.1 that the classification accuracy is generally correlated with the training set size.

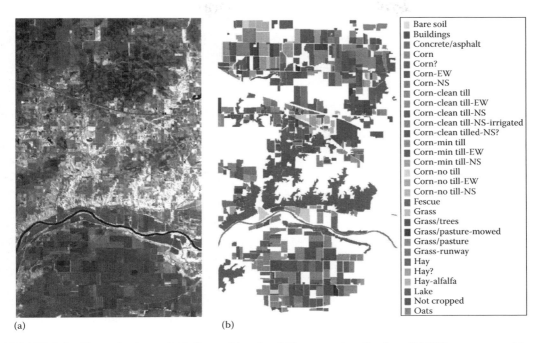

(a)   (b)

FIGURE 5.4 **(See color insert.)** (a) Spectral band at 587 nm wavelength of an AVIRIS scene comprising agricultural and forest features at Indian Pines region. (b) Ground-truth map with 30 mutually exclusive land-cover classes.

TABLE 5.1

Overall Classification Accuracies (in Percentage) Achieved by the SVM Classifier after Applying Polynomial, Gaussian and Spectral Angle Mapper Kernels to the AVIRIS Indian Pines Data Set, Using Different Strategies for Training Sample Selection (Random, Pure, Border Patterns)

Kernel	Training	1%	2%	4%	6%	8%	10%	20%
Polynomial kernel	Random	82.33	82.94	83.21	83.82	85.34	86.12	86.52
	Pure	81.23	82.06	82.80	83.00	84.03	84.45	85.57
	Border	83.44	84.23	84.45	84.96	86.27	87.44	89.96
Gaussian kernel	Random	87.94	88.23	88.78	88.96	89.45	89.48	90.77
	Pure	86.53	87.02	87.64	87.93	88.12	88.26	88.55
	Border	89.45	90.25	91.24	92.08	92.93	93.04	93.67
Spectral angle kernel	Random	85.90	86.22	86.49	87.03	87.56	88.09	88.72
	Pure	85.12	85.67	86.08	86.45	86.97	87.13	87.81
	Border	86.05	86.93	87.57	88.12	89.30	90.12	90.57

However, when border training samples were used, higher classification accuracies were achieved with less training samples. The aforementioned results indicate the importance of including mixed pixels at the border of class boundaries in the training set, as these border patterns are most efficient to determine the hyperplane between two classes.

Finally, it can be seen in Table 5.1 that the best classification scores were generally achieved for the Gaussian kernel, in which the overall accuracy obtained with 1% of the training pixels per class is only 4.22% lower than the overall accuracy obtained with 20% of the training pixels per class (extracted using border training sample selection). On the other hand, the spectral angle mapper kernel gives slightly degraded classification results. However, with accuracies above 85% in a challenging classification problem, this kernel also provides promising results. Finally, the polynomial kernel needs more training samples than the two other kernels to perform appropriately, as can be seen from the relatively poor results obtained by this kernel for a very limited number of training samples.

5.4.2 ANALYSIS OF UNSUPERVISED LINEAR UNMIXING OF HYPERSPECTRAL DATA

The hyperspectral scene used for experiments is the well-known AVIRIS Cuprite data set, available online in reflectance units (from http://aviris.jpl.nasa.gov/html/aviris.freedata.html) after atmospheric correction. This scene has been widely used to validate the performance of endmember extraction algorithms. The portion used in experiments corresponds to a 350 × 350-pixel subset of the sector labeled as f970619t01p02_r02_sc03.a.rfl in the online data. The scene (displayed in Figure 5.5a) comprises 224 spectral bands between 400 and 2500 nm, with full width at half maximum of 10 nm and spatial resolution of 20 m per pixel.

Prior to the analysis, several bands (1–3, 150–170, and 217–224) were removed due to water absorption and low SNR in those bands, leaving a total of 192 reflectance channels to be used in the experiments. The Cuprite site is well understood mineralogically [45,46], and has several exposed minerals of interest included in a spectral library compiled by the U.S. Geological Survey (USGS) available online (from http://speclab.cr.usgs.gov/spectral-lib.html). A few selected spectra from the USGS library, corresponding to several highly representative minerals in the Cuprite mining district (Figure 5.5b), are used in this work to substantiate endmember signature purity.

Two different metrics have been used to compare the performance of endmember extraction and spectral unmixing algorithms in the AVIRIS Cuprite scene. The first metric is the spectral angle [3,19] between each extracted endmember and the set of available USGS ground-truth spectral signatures. Low spectral angle scores mean high spectral similarity between the compared vectors. This spectral similarity measure is invariant in the multiplication of pixel vectors by constants and, consequently, is invariant before unknown multiplicative scalings that may arise due to differences in illumination and angular orientation. In our experiments, the spectral angle allows us to identify the USGS signature that is most similar to each endmember

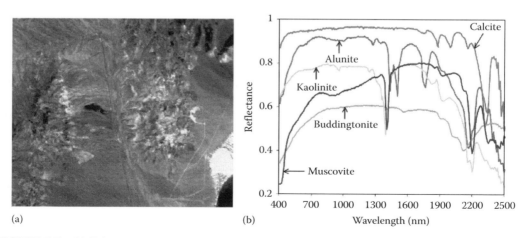

(a)   (b)

FIGURE 5.5   (a) False color composition of the remote sensing image used in experiments. (b) Reference spectral signatures provided by USGS and used for validation purposes.

automatically extracted from the scene by observing the minimum spectral angle reported for such endmember across the entire set of USGS signatures.

A second metric employed to evaluate the goodness of the reconstruction is the root mean square error (RMSE) obtained in the reconstruction of the hyperspectral image (using the derived endmembers and their corresponding abundance fractions). This metric is based on the assumption that a set of high-quality endmembers (and their corresponding estimated abundance fractions) may allow reconstruction of the original hyperspectral scene with higher precision than a set of low-quality endmembers. In this case, the original hyperspectral image is used to measure the fidelity of the reconstructed version of the same scene on a per-pixel basis.

Table 5.2 tabulates the spectral angles (in degrees) obtained after comparing the USGS library spectra of five highly representative minerals in the Cuprite mining district (alunite, buddingtonite, calcite, kaolinite, and muscovite) with the corresponding endmembers extracted by several different algorithms (listed in subsection 5.3.1) from the AVIRIS Cuprite scene. In all cases, the input parameters of the different endmember extraction methods tested have been carefully optimized so that the best performance for each method is reported. Again, the smaller the spectral angles across the five minerals in Table 5.2, the better the results. It should be noted that Table 5.2 only displays the smallest spectral angle scores of all endmembers with respect to each USGS signature for each algorithm.

For reference, the mean spectral angle values across all five USGS signatures are also reported. In all cases, the number of endmembers to be extracted was set to 14 after using a consensus between the VD concept and the HySime method. Table 5.2 reveals that the AMEE provides very good results (all spectral angle values scores below 10°), with the SSEE and the SPP + OSP (where SPP indicates SPP prior to the classic OSP procedure for endmember extraction) are the algorithms that can provide comparable—but slightly worst—results. Table 5.2 also reveals that, in this real example, SPP generally improves the signature purity of the endmembers extracted by spectral-based algorithms.

On the other hand, Figure 5.6 graphically represents the per-pixel RMSE obtained after reconstructing the AVIRIS Cuprite scene using 14 endmembers extracted by different methods. It can be seen that the methods using SPP (SPP + OSP, SPP + N-FINDR, SPP + VCA) improve their respective spectral-based versions in terms of the quality of image reconstruction, while both AMEE and SSEE also provide lower reconstruction errors than OSP, N-FINDR, and VCA. These results suggest the advantages of incorporating spatial information into the automatic extraction of image endmembers from the viewpoint of obtaining more spatially representative spectral signatures, which can be used to describe other mixed signatures in the scene.

TABLE 5.2

Spectral Angle Scores (in Degrees) between the USGS Mineral Spectra and Their Corresponding Endmember Pixels Produced by Several Endmember Extraction Algorithms

Algorithm	Alunite	Buddingtonite	Calcite	Kaolinite	Muscovite	Mean
OSP	4.81	4.16	9.62	11.14	5.41	7.03
N-FINDR	9.96	7.71	12.08	13.27	5.24	9.65
VCA	10.73	9.04	6.36	14.05	5.41	9.12
SPP + OSP	4.95	4.16	9.96	10.90	4.62	6.92
SPP + N-FINDR	12.81	8.33	9.83	10.43	5.28	9.34
SPP + VCA	12.42	4.04	9.37	7.37	6.18	7.98
AMEE	4.81	4.21	9.54	8.74	4.61	6.38
SSEE	4.81	4.16	8.48	11.14	4.62	6.64

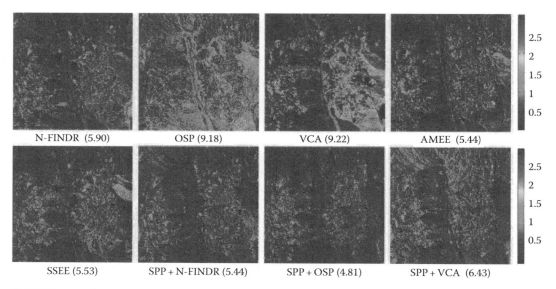

FIGURE 5.6   (See color insert.) RMSE reconstruction errors (in percentage) for various endmember extraction algorithms after reconstructing the AVIRIS Cuprite scene.

5.4.3   ANALYSIS OF SUPERVISED NONLINEAR UNMIXING OF HYPERSPECTRAL DATA USING MLPs

In the Iberian Peninsula, Dehesa systems are used for a combination of livestock, forest, and agriculture activity [47]. The outputs of these systems include meat, milk, wool, charcoal, cork bark, and grain. Around 12%–18% of the area is harvested on a yearly basis. The crops are used for animal feed or for cash cropping, depending on the rainfall of the area. Determination of fractional land-cover using remote sensing techniques may allow for a better monitoring of natural resources in Dehesa agro-ecosystems.

Our choice of this type of landscape for evaluating nonlinear unmixing techniques was made on several accounts. The first one is the availability of hyperspectral image data sets with accurate geo-registration for a real Dehesa test site in Caceres, SW Spain, collected simultaneously in July 2001 by two instruments operating at multiple spatial resolutions: Digital Airborne Imaging Spectrometer (DAIS) 7915 and Reflective Optics Spectrographic Imaging System (ROSIS), operated by the German Aerospace Agency (DLR). A second major reason is the simplicity of the Dehesa landscape, which greatly facilitates the collection of reliable field data for model validation purposes. It is also important to emphasize that the scenes were collected in summertime, so atmospheric interferers were greatly minimized. Before describing our experiments, we first provide a comprehensive description of the data sets used and ground-truth activities in the study area.

The data used in this study consisted of two main components: image data and field measurements of land-cover fractions, collected at the time of image data acquisition. The image data are formed by a ROSIS scene collected at high spatial resolution, with 1.2 m pixels, and its corresponding DAIS 7915 scene, collected at low spatial resolution with 6 m pixels. The spectral range from 504 to 864 nm (consisting of a total of 112 spectral bands) was selected for experiments, not only because it is adequate for analyzing the spectral properties of the landscape under study, but also because this spectral range is well covered by the two considered sensors through narrow spectral bands. Figure 5.7 shows the full flightline of the ROSIS scene, which comprises a Dehesa area located between the facilities of University of Extremadura in Caceres (leftmost part of the flightline) and Guadiloba water reservoir at the center of the flightline. Figure 5.8a shows the Dehesa test site selected for experiments, which corresponds to a highly representative Dehesa area that

FIGURE 5.7   Flightline of a ROSIS hyperspectral scene collected over a Dehesa area in Caceres, Spain.

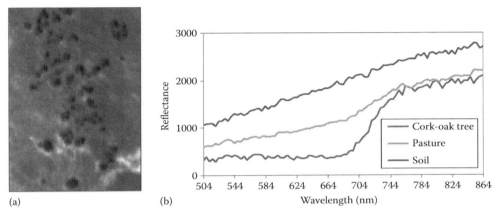

(a)   (b)

FIGURE 5.8   (See color insert.) (a) Spectral band (584 nm) of a ROSIS Dehesa subset selected for experiments. (b) Endmember signatures of soil, pasture, and cork-oak tree extracted by the AMEE algorithm, where scaled reflectance values are multiplied by a constant factor.

contains several cork-oak trees (appearing as dark spots) and several pasture (gray) areas on a bare soil (white) background. Several field techniques were applied to obtain reliable estimates of the fractional land cover for each DAIS 7915 pixel in the considered Dehesa test site:

1. First, the ROSIS image was roughly classified into the three land-cover components discussed earlier using a maximum-likelihood supervised classification approach based on image-derived spectral endmembers, where Figure 5.8b shows the three endmembers used for mapping that were derived using the AMEE algorithm. Our assumption was that the pixels in the ROSIS image were sufficiently small to become spectrally simple to analyze.
2. Then, the classified ROSIS image was registered with the DAIS 7915 image using a ground control point-based method with subpixel accuracy [48].
3. The classification map was then associated with the DAIS 7915 image to provide an initial estimation of land cover classes for each pixel at the DAIS 7915 image scale. For that purpose, a 6 × 6 m grid was overlaid on the 1.2 × 1.2 m classification map derived from the ROSIS scene, where the geographic coordinates of each pixel center point were used to validate the registration with subpixel precision.
4. Next, fractional abundances were calculated within each 6 × 6 m grid as the proportion or ROSIS pixels labeled as cork-oak tree, pasture, and soil located within that grid, respectively.

(a)   (b)   (c)

FIGURE 5.9   Ground measurements in the Dehesa area of study located in Caceres, Spain. (a) Spectral sample collection using an ASD FieldSpec Pro spectroradiometer. (b) High-precision GPS geographic delimitation. (c) Field spectral measurements at different altitudes.

5. Most importantly, the abundance maps at the ROSIS level were thoroughly refined using field measurements (Figure 5.9a) before obtaining the final proportions. Several approaches were developed to refine the initial estimations:
   a. Fractional land cover data were collected on the ground at more than 30 evenly distributed field sites within the test area. These sites were delineated during the field visit as polygons, using high-precision global positioning system (GPS) coordinates (see Figure 5.9b).
   b. Land cover fractions were estimated at each site using a combination of various techniques. For instance, field spectra were collected for several areas using an Analytical Spectral Devices (ASD) FieldSpec Pro spectro-radiometer. Of particular interest were field measurements collected on top of tree crowns (Figure 5.9c), which allowed us to model different levels of tree crown transparency.
   c. On the other hand, the early growth stage of pasture during the summer season allowed us to perform ground estimations of pasture abundance in selected sites of known dimensions, using pasture harvest procedures supported by visual inspection and laboratory analyses.

After following the aforementioned sequence of steps, we obtained a set of approximate fractional abundance labels for each pixel vector in the DAIS 7915 image. Despite our effort to conduct a reliable ground estimation of fractional land-cover in the considered semiarid environment, absolute accuracy is not claimed. We must emphasize, however, that the combined use of imagery data at different resolutions, subpixel ground control-based image registration, and extensive field work including high-precision GPS field work, spectral sample data collection, and expert knowledge represents a novel contribution in the area of spectral mixture analysis validation, in particular, for Dehesa-type ecosystems.

In order to evaluate the accuracy of linear spectral in the considered application, Figure 5.10 shows the scatter plots of measured versus linearly estimated fractional abundances (using linear spectral unmixing with the ASC and ANC constraints imposed) for the three considered land-cover materials in the DAIS 7915 (low spatial resolution) image data set, where the diagonal represents perfect match and the two flanking lines represent plus/minus 20% error bound. Here, the three spectral endmembers were derived using the AMEE algorithm, which incorporates spatial information into the endmember extraction process.

As expected, the flatness of the test site largely removed topographic influences in the remotely sensed response of soil areas. As a result, most linear predictions for the soil endmember fall within the 20% error bound (see Figure 5.10a). On the other hand, the multiple scattering within the pasture and cork-oak tree canopies (and from the underlying surface in the latter case) complicated the spectral mixing in nonlinear fashion, which resulted in a generally higher number of estimations lying

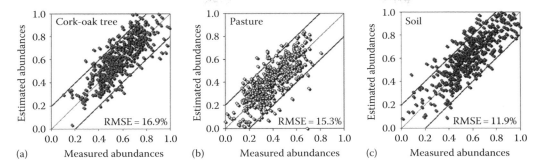

FIGURE 5.10 Abundance estimations of cork-oak tree (a), pasture (b), and soil (c) by the fully constrained linear mixture model from the DAIS 7915 image.

outside the error bound, as illustrated in Figures 5.10b and c. Also, the RMSE scores in abundance estimation for the soil (11.9%), pasture (15.3%), and cork-oak tree (16.9%) were all >10% estimation error in percentage, which suggested that linear mixture modeling was not flexible enough to accommodate the full range of spectral variability throughout the landscape.

In order to characterize the Dehesa ecosystem structure better than linear models do, we used nonlinear spectral unmixing to better characterize nonlinear mixing effects. For this purpose, we applied a mixed (border) training sample selection algorithm to automatically locate highly descriptive training sites in the DAIS 7915 scene and then used the obtained samples (and the ground-truth information associated to those samples) to train the MLP-based neural network model described in subsection 5.3.2. Figure 5.11 shows the scatter plots of measured versus predicted fractional abundances for soil, pasture, and cork-oak tree by the proposed MLP-based model, first trained with the three pure training samples by the AMEE algorithm (Figure 5.11b) plus 40 additional training samples selected by an algorithm designed to seek for the most highly mixed training samples [10]. This represents <1% of the total number of pixels in the DAIS 7915 scene. These samples were excluded from the testing set made up of all remaining pixels in the scene. From Figure 5.11, it is clear that the utilization of intelligently selected training samples resulted in fewer points outside the two 20% difference lines, most notably, for both pasture and cork-oak abundance estimates.

The pattern of the scatter plots obtained for the soil predictions (Figure 5.11a) was similar (in particular, when the soil abundance was high). Most importantly, the RMSE scores in abundance estimation were significantly reduced (with regard to the experiment using fully constrained linear unmixing) for the soil (6.1%), pasture (4%), and cork-oak tree (6.3%). These results confirm our intuition that nonlinear effects in Dehesa landscapes mainly result from multiple scattering effects in vegetation canopies. It is worth noting that, although the ASC and ANC constraints were not imposed in our proposed MLP-based learning stage, negative and/or unrealistic abundance

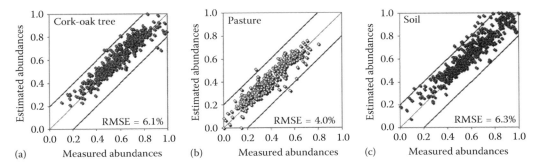

FIGURE 5.11 Abundance estimations of cork-oak tree (a), pasture (b), and soil (c) by the MLP-based mixture model, trained using mixed (border) samples, from the DAIS 7915 image.

estimations (which usually indicate a bad fit of the model and reveal inappropriate endmember/ training data selection) were very rarely found in our experiments.

The experimental validation carried out in this subsection indicates that the intelligent incorporation of mixed training samples can enable a more accurate representation of nonlinearly mixed signatures. It was apparent from experimental results that the proposed neural network-based model was able to generate abundance estimates that were close to abundance values measured in the field, using only a few intelligently generated training samples. The need for mixed training data does, however, require detailed knowledge on abundance fractions for the considered training sites. In practice, these data are likely to be derived from imagery acquired at a finer spatial resolution than the imagery to be classified, for example, using data sets acquired by sensors operating simultaneously at multiple spatial resolutions as it is the case of the DAIS 7915 and ROSIS instruments considered in this experiment. Such multiresolution studies may also incorporate prior knowledge or ancillary information, which can be used to help target the location of training sites, and to focus training site selection activities on regions likely to contain the most informative training samples.

5.5   CONCLUSIONS AND FUTURE PERSPECTIVES

This chapter focused on hyperspectral data processing algorithms that included (a) SVM techniques for supervised classification using limited training samples and (b) development of linear and nonlinear spectral unmixing techniques, some of them integrating the spatial and the spectral information. The special characteristics of hyperspectral images pose new processing problems, not to be found in other types of remotely sensed data:

1. The high-dimensional nature of hyperspectral data introduces important limitations in supervised, full-pixel classifiers, such as the limited availability of training samples or the inherently complex structure of the data (leading to the Hughes phenomenon).
2. There is a need to integrate the spatial and spectral information to take advantage of the complementarities that both sources of information can provide, in particular, for unsupervised mixed-pixel classifiers.

In this regard, the SVM experiments reported in our quantitative assessment demonstrated that, with only 1% of training pixels per class, almost 90% overall classification accuracy is reached by all kernels when trained using border training samples. This highlighted the opportunity of overcoming the Hughes phenomenon using kernel approaches. On the other hand, our unmixing experiments indicate that new trends in algorithm design (such as the joint use of spatial and spectral information in linear spectral unmixing, or the development of nonlinear unmixing models based on machine learning techniques with an appropriate exploitation of limited training samples) can significantly improve the accuracy in the estimation of fractional abundances in real analysis scenarios.

As demonstrated by our experimental results and the determination of the accuracy of these approaches, techniques are rapidly changing from *hard* classifiers to soft classifiers. In this regard, we anticipate that the full adaptation of *soft* classifiers to mixed-pixel classification problems (e.g., via multiregression and robust training sample selection algorithms) may push the frontiers of hyperspectral data classification to new application domains. Further developments on the joint exploitation of the spatial and the spectral information in the input data are also needed to complement initial approximations to the problem of interpreting the data in unsupervised fashion, thus being able to cope with the dramatically enhanced spatial and spectral capabilities expected in the design of future imaging spectrometers. Advances in high performance computing [49], including clusters of computers and distributed grids, as well as specialized hardware modules such as field programmable gate arrays (FPGAs) or graphics processing units (GPUs), will also be crucial to help increase algorithm efficiency and meet timeliness needs in many remote sensing applications.

ACKNOWLEDGMENT

This work has been supported by the European Community's Marie Curie Research Training Networks Programme under reference MRTN-CT-2006-035927, Hyperspectral Imaging Network (HYPER-I-NET). This work has also been supported by the Spanish Ministry of Science and Innovation (HYPERCOMP/EODIX project, reference AYA2008-05965-C04-02). Gabriel Martín and Sergio Sánchez are sponsored by research fellowships with references BES-2009-017737 and PTA2009-2611-P, respectively, both associated to the aforementioned project. Funding from Junta de Extremadura (local government) under project PRI09A110 is also gratefully acknowledged. The authors thank Andreas Mueller for his lead of the DLR project that allowed us to obtain the DAIS 7915 and ROSIS hyperspectral datasets over Dehesa areas in Extremadura, Spain; David Landgrebe at Purdue University for making the AVIRIS Indian Pines scene available to the scientific community, and Robert O. Green at NASA/JPL for also making the AVIRIS Cuprite scene available to the scientific community. Last but not least, the authors would like to take this opportunity to gratefully acknowledge the editors of this volume for their very kind invitation to contribute a chapter and for all their support and encouragement during the different stages of the production process for this monograph.

REFERENCES

1. A. F. H. Goetz, G. Vane, J. E. Solomon, and B. N. Rock, Imaging spectrometry for Earth remote sensing, *Science*, 228, 1147–1153, 1985.
2. R. O. Green, Imaging spectroscopy and the airborne visible-infrared imaging spectrometer (AVIRIS), *Remote Sensing of Environment*, 65, 227–248, 1998.
3. C.-I. Chang, *Hyperspectral Imaging: Techniques for Spectral Detection and Classification*, Kluwer Academic and Plenum Publishers, New York, 2003.
4. D. A. Landgrebe, *Signal Theory Methods in Multispectral Remote Sensing*, John Wiley & Sons, Hoboken, NJ, 2003.
5. J. A. Richards, Analysis of remotely sensed data: The formative decades and the future, *IEEE Transactions on Geoscience and Remote Sensing*, 43, 422–432, 2005.
6. A. Plaza, P. Martinez, R. Perez, and J. Plaza, A quantitative and comparative analysis of endmember extraction algorithms from hyperspectral data, *IEEE Transactions on Geoscience and Remote Sensing*, 42, 650–663, 2004.
7. J. B. Adams, M. O. Smith, and P. E. Johnson, Spectral mixture modeling: A new analysis of rock and soil types at the Viking Lander 1 site, *Journal of Geophysical Research*, 91, 8098–8112, 1986.
8. G. M. Foody and A. Mathur, Toward intelligent training of supervised image classifications: Directing training data acquisition for SVM classification, *Remote Sensing of Environment*, 93, 107–117, 2004.
9. A. Plaza, J. A. Benediktsson, J. Boardman, J. Brazile, L. Bruzzone, G. Camps-Valls, J. Chanussot et al., Recent advances in techniques for hyperspectral image processing, *Remote Sensing of Environment*, 113, 110–122, 2009.
10. J. Plaza, A. Plaza, R. Perez, and P. Martinez, On the use of small training sets for neural network-based characterization of mixed pixels in remotely sensed hyperspectral images, *Pattern Recognition*, 42, 3032–3045, 2009.
11. P. Gamba, F. Dell'Acqua, A. Ferrari, J. A. Palmason, and J. A. Benediktsson, Exploiting spectral and spatial information in hyperspectral urban data with high resolution, *IEEE Geoscience and Remote Sensing Letters*, 1, 322–326, 2004.
12. J. A. Benediktsson, J. A. Palmason, and J. R. Sveinsson, Classification of hyperspectral data from urban areas based on extended morphological profiles, *IEEE Transactions on Geoscience and Remote Sensing*, 42, 480–491, 2005.
13. A. Plaza, P. Martinez, J. Plaza, and R. Perez, Dimensionality reduction and classification of hyperspectral image data using sequences of extended morphological transformations, *IEEE Transactions on Geoscience and Remote Sensing*, 43(3), 466–479, 2005.
14. G. Camps-Valls and L. Bruzzone, Kernel-based methods for hyperspectral image classification, *IEEE Transactions on Geoscience and Remote Sensing*, 43, 1351–1362, 2005.
15. K. R. Muller, S. Mika, G. Ratsch, K. Tsuda, and B. Scholkopf, An introduction to kernel-based learning algorithms, *IEEE Transactions on Neural Networks*, 12, 181–202, 2001.

16. L. Bruzzone, M. Chi, and M. Marconcini, A novel transductive SVM for the semisupervised classification of remote sensing images, *IEEE Transactions on Geoscience and Remote Sensing*, 44, 3363–3373, 2006.
17. G. Camps-Valls, L. Gomez-Chova, J. Munoz-Mari, J. Vila-Frances, and J. Calpe-Maravilla, Composite kernels for hyperspectral image classification, *IEEE Geoscience and Remote Sensing Letters*, 3, 93–97, 2006.
18. M. E. Schaepman, S. L. Ustin, A. Plaza, T. H. Painter, J. Verrelst, and S. Liang, Earth system science related imaging spectroscopy—An assessment, *Remote Sensing of Environment*, 113, 123–137, 2009.
19. N. Keshava and J. F. Mustard, Spectral unmixing, *IEEE Signal Processing Magazine*, 19, 44–57, 2002.
20. K. J. Guilfoyle, M. L. Althouse, and C.-I. Chang, A quantitative and comparative analysis of linear and nonlinear spectral mixture models using radial basis function neural networks, *IEEE Transactions on Geoscience and Remote Sensing*, 39, 2314–2318, 2001.
21. C.-I. Chang and Q. Du, Estimation of number of spectrally distinct signal sources in hyperspectral imagery, *IEEE Transactions on Geoscience and Remote Sensing*, 42, 608–619, 2004.
22. J. M. Bioucas-Dias and J. M. P. Nascimento, Hyperspectral subspace identification, *IEEE Transactions on Geoscience and Remote Sensing*, 46, 2435–2445, 2008.
23. J. W. Boardman, F. A. Kruse, and R. O. Green, Mapping target signatures via partial unmixing of AVIRIS data, *Proceedings JPL Airborne Earth Science Workshop*, JPL Publication, Washington, DC, pp. 23–26, 1995.
24. M. E. Winter, N-FINDR: An algorithm for fast autonomous spectral endmember determination in hyperspectral data, *Proceedings of SPIE*, Big Sky, Montana, Vol. 3753, pp. 266–277, 1999.
25. M. E. Winter, A proof of the N-FINDR algorithm for the automated detection of endmembers in a hyperspectral image, *Proceedings of SPIE Algorithms and Technologies for Multispectral, Hyperspectral, and Ultraspectral Imagery X*, Orlando, Florida, Vol. 5425, pp. 31–41, 2004.
26. M. Zortea and A. Plaza, A quantitative and comparative analysis of different implementations of N-FINDR: A fast endmember extraction algorithm, *IEEE Geoscience and Remote Sensing Letters*, 6, 787–791, 2009.
27. R. A. Neville, K. Staenz, T. Szeredi, J. Lefebvre, and P. Hauff, Automatic endmember extraction from hyperspectral data for mineral exploration, *Proceedings of 21st Canadian Symposium on Remote Sensing*, Ottawa, Ontario, Canada, pp. 21–24, 1999.
28. J. H. Bowles, P. J. Palmadesso, J. A. Antoniades, M. M. Baumback, and L. J. Rickard, Use of filter vectors in hyperspectral data analysis, *Proceedings of SPIE Infrared Spaceborne Remote Sensing III*, San Diego, California, Vol. 2553, pp. 148–157, 1995.
29. A. Ifarraguerri and C.-I. Chang, Multispectral and hyperspectral image analysis with convex cones, *IEEE Transactions on Geoscience and Remote Sensing*, 37(2), 756–770, 1999.
30. J. M. P. Nascimento and J. M. Bioucas-Dias, Vertex component analysis: A fast algorithm to unmix hyperspectral data, *IEEE Transactions on Geoscience and Remote Sensing*, 43(4), 898–910, 2005.
31. J. C. Harsanyi and C.-I. Chang, Hyperspectral image classification and dimensionality reduction: An orthogonal subspace projection, *IEEE Transactions on Geoscience and Remote Sensing*, 32(4), 779–785, 1994.
32. A. Plaza, P. Martinez, R. Perez, and J. Plaza, Spatial/spectral endmember extraction by multidimensional morphological operations, *IEEE Transactions on Geoscience and Remote Sensing*, 40, 2025–2041, 2002.
33. D. M. Rogge, B. Rivard, J. Zhang, A. Sanchez, J. Harris, and J. Feng, Integration of spatial–spectral information for the improved extraction of endmembers, *Remote Sensing of Environment*, 110, 287–303, 2007.
34. M. Zortea and A. Plaza, Spatial preprocessing for endmember extraction, *IEEE Transactions on Geoscience and Remote Sensing*, 47, 2679–2693, 2009.
35. D. Heinz and C.-I. Chang, Fully constrained least squares linear mixture analysis for material quantification in hyperspectral imagery, *IEEE Transactions on Geoscience and Remote Sensing*, 39, 529–545, 2001.
36. C.-I. Chang and D. Heinz, Constrained subpixel target detection for remotely sensed imagery, *IEEE Transactions on Geoscience and Remote Sensing*, 38, 1144–1159, 2000.
37. C. M. Bishop, *Neural Networks for Pattern Recognition*, Oxford University Press, Oxford, U.K., 1995.
38. J. Plaza and A. Plaza, Spectral mixture analysis of hyperspectral scenes using intelligently selected training samples, *IEEE Geoscience and Remote Sensing Letters*, 7, 371–375, 2010.
39. A. Baraldi, E. Binaghi, P. Blonda, P. A. Brivio, and P. Rampini, Comparison of the multilayer perceptron with neuro-fuzzy techniques in the estimation of cover class mixture in remotely sensed data, *IEEE Transactions on Geoscience and Remote Sensing*, 39, 994–1005, 2001.
40. W. Liu and E. Y. Wu, Comparison of non-linear mixture models, *Remote Sensing of Environment*, 18, 1976–2003, 2004.
41. X. Zhuang, B. A. Engel, D. F. Lozano, R. B. Fernndez, and C. J. Johannsen, Optimization of training data required for neuro-classification, *International Journal of Remote Sensing*, 15, 3271–3277, 1999.

42. M. Chi and L. Bruzzone, A semilabeled-sample-driven bagging technique for ill-posed classification problems, *IEEE Geoscience and Remote Sensing Letters*, 2, 69–73, 2005.

43. G. M. Foody, The significance of border training patterns in classification by a feedforward neural network using back propagation learning, *International Journal of Remote Sensing*, 20, 3549–3562, 1999.

44. C. C. Borel and S. A. W. Gerslt, Nonlinear spectral mixing models for vegetative and soil surfaces, *Remote Sensing of Environment*, 47, 403–416, 1994.

45. R. N. Clark, G. A. Swayze, K. E. Livo, R. F. Kokaly, S. J. Sutley, J. B. Dalton, R. R. McDougal, and C. A. Gent, Imaging spectroscopy: Earth and planetary remote sensing with the USGS Tetracorder and expert systems, *Journal of Geophysical Research*, 108, 1–44, 2003.

46. G. Swayze, R. N. Clark, F. Kruse, S. Sutley, and A. Gallagher, Ground-truthing AVIRIS mineral mapping at Cuprite, Nevada, *Proceedings of the JPL Airborne Earth Science Workshop*, JPL Publication, Washington, DC, pp. 47–49, 1992.

47. F. J. Pulido, M. Diaz, and S. J. Hidalgo, Size structure and regeneration of Spanish holm oak Quercus ilex forests and dehesas: Effects of agroforestry use on their long-term sustainability, *Forest Ecology and Management*, 146, 1–13, 2001.

48. A. Plaza, J. L. Moigne, and N. S. Netanyahu, Morphological feature extraction for automatic registration of multispectral scenes, *Proceedings of the IEEE International Geoscience and Remote Sensing Symposium*, Barcelona, Spain, Vol. 1, pp. 421–424, 2007.

49. A. Plaza and C.-I. Chang, *High Performance Computing in Remote Sensing*, CRC Press, Boca Raton, FL, 2007.

Part IV

Leaf and Plant Biophysical and Biochemical Properties

6 Nondestructive Estimation of Foliar Pigment (Chlorophylls, Carotenoids, and Anthocyanins) Contents: Evaluating a Semianalytical Three-Band Model

Anatoly A. Gitelson

CONTENTS

6.1 INTRODUCTION

Quantification of vegetation physiology and phenology, including the rate of gas exchange with the atmosphere, can be achieved by accurate measurements of the pigments present in plant leaves that play very important role in plant photosynthesis and protection. There are three major classes of pigments found in plants: chlorophylls, carotenoids, and anthocyanins. The chlorophyll-*a* and chlorophyll-*b* are essential pigments for the conversion of light energy to stored chemical energy. The amount of solar radiation absorbed by a leaf is a function of the photosynthetic pigment content; thus, chlorophyll content can directly determine photosynthetic potential and primary production [1,2]. In addition, chlorophylls give an indirect estimation of the nutrient status because much of leaf nitrogen is incorporated in chlorophyll [2]. Furthermore, leaf chlorophyll content is indicative and closely related to plant stress and senescence [3–5].

Carotenoids usually are represented by two (α- and β-) carotenes and xanthophylls (lutein, zeaxanthin, violaxanthin, antheraxanthin, and neoxanthin), which exhibit strong light absorption in the blue region of the spectrum and are nonuniformly distributed in photosystems and individual pigment–protein complexes of chloroplasts [6]. Several specific and important physiological functions have been attributed to carotenoids because of their unique physicochemical and photophysical properties: structural role in the organization of photosynthetic membranes, participation in light harvesting, energy transfer, quenching of chlorophyll excited states and singlet oxygen, and interception of deleterious free oxygen and organic radicals [7]. The reversible conversion of violaxanthin to zeaxanthin via antheraxanthin (violaxanthin cycle) is considered to be an important mechanism of excess energy dissipation in chloroplasts [11,13]. The retention of carotenoids in the progress of chlorophyll breakdown [8] has been suggested as a mechanism of photoprotection during leaf senescence [9]. The changes of leaf carotenoids content and their proportion to chlorophyll are widely used for diagnosing the physiological state of plants during development, senescence, acclimation, and adaptation to different environments and stresses [9–14].

The anthocyanins are pigments frequently occurring in higher plants and responsible for their red coloration. In leaves, they localize in vacuoles of epidermal cells or those just below adaxial epidermis, but occasionally, also in the cells of abaxial epidermis, palisade, and spongy mesophyll [15]. The induction of anthocyanins biosynthesis occurs as a result of deficiencies in nitrogen and phosphorus, wounding, pathogen infection, desiccation, low temperature, UV-irradiation, etc. [15]. It is generally accepted that anthocyanins fulfill important physiological functions by being involved in the adaptation to numerous stresses and environmental strain reduction [15]. Some lines of evidence suggest that protective effects of anthocyanins are related to their ability, via screening and/or internal light trapping, to reduce the amount of excessive solar radiation reaching photosynthetic apparatus [16].

Traditionally, high-performance liquid chromatography (HPLC) or leaf extraction with organic solvents and spectrophotometric determination in solution is required for pigment analysis using wet chemistry methods (e.g., Ref. [6]). Recently, alternative solutions of leaf pigment analysis (i.e., chlorophyll, carotenoids, and anthocyanins) with nondestructive optical methods have been developed (e.g., Refs. [1,2,4,5,9–12,17–23]). Monitoring plant physiological status via measuring leaf reflectance offers a number of distinct advantages over traditional destructive and alternative nondestructive (e.g., chlorophyll fluorescence measurement-based) approaches. The most important advantages are simplicity, sensitivity, reliability, and a high throughput. These methods are nondestructive, inexpensive, quick, and possible to use in the field [12,17–22]. Nondestructive techniques save a great deal of manual labor and therefore have a potential for application in studies of plant productivity, physiology, and so on.

Attempts to apply nondestructive methods based on optical spectroscopy for assessment of plant physiological state via measuring pigment content have been undertaken for several decades. The situation has changed drastically during the last few decades when significant amount of research was dedicated to the development of techniques for nondestructive evaluation of leaf pigments. This chapter provides a brief overview of foliar absorbance and reflectance spectral features and the recent developments in the field of reflectance-based techniques for estimating foliar pigments contents and discuss three-band model for estimating foliar total chlorophylls, carotenoids, and anthocyanins contents.

6.2 BACKGROUND

6.2.1 CHLOROPHYLLS

Transmittance and reflectance spectroscopy is applied extensively for nondestructive estimation of foliar chlorophyll [1,18–20,23–29]. Richardson et al. [29] evaluated both nondestructive absorbance and reflectance methods for chlorophyll assessment. They compared the performance of two commercially available handheld chlorophyll absorbance meters with that of several reflectance indices for the estimation of leaf-level chlorophyll and found that indices based on reflectance in the red edge (700–740 nm) region [18,19] were much better indicators of chlorophyll content than some of the more commonly used indices. The best reflectance indices were found to be better indicators of chlorophyll compared with those based on absorbance and transmittance measurements [29].

A large number of reflectance-based methods have been proposed to detect plant pigments, ranging from simple band ratios to radiative transfer models. Analytical radiative transfer models have the potential to produce accurate and consistent prediction of pigment contents because they use the full spectrum rather than individual bands [30–32]. However, the goodness-of-fit of radiative transfer models predicting optical properties of leaves or needles depends on how well all processes affecting reflectance are understood and how they are accounted for in the models [39]. Although radiative transfer models have potential to predict pigment concentrations more consistently and accurately than empirical and semianalytical methods, they require more input parameters, which if not accurate result in poor model performance. Thus, semianalytical and empirical models can be more accurate than analytical models if the components are improperly modeled or the input data are wrong [39].

Empirical and semianalytical models currently employed for estimating chlorophyll using leaf reflectance exploit the differences in reflectance between leaves with different chlorophyll content in the visible and the near infra-red (NIR) regions (e.g., Refs. [18,19,33–35], see also comprehensive reviews of nondestructive estimation of foliar pigments [36–40] and references therein).

6.2.2 CAROTENOIDS

During the last decade, several attempts have been undertaken to develop nondestructive techniques for content assessment of leaf carotenoids [5,21,41]. It was shown that in different plant species, high photon flux induces small reversible changes of reflectance near 530 nm, attributable to the transformation of violaxanthin cycle xanthophylls. The corresponding reflectance indices applicable for nondestructive estimation of the cycle and photosynthetic activities have been suggested [11,36,42].

Chappelle et al. [43] used ratio analysis of reflectance spectra to find a spectral band sensitive to pigment content. A ratio spectrum obtained by dividing the mean reflectance spectrum of soybeans grown at a high nitrogen level by the mean reflectance spectrum of soybeans grown at a medium nitrogen level had a small peak around 500 nm that was attributed to carotenoids absorption. They recommended using a ratio of reflectances at 760 and 500 nm, R_{760}/R_{500}, as a proxy of carotenoids. Blackburn [27] suggested that the optimal individual waveband for carotenoids estimation is located at 470 nm and used so-called pigment-specific ratio R_{800}/R_{470} and a pigment-specific normalized difference, $(R_{800} - R_{470})/(R_{800} + R_{470})$, for carotenoids content assessment.

To retrieve the carotenoid/chlorophyll ratio for a range of individual leaves and conditions, Filella et al. [2] proposed a structure-insensitive pigment index in the form, $(R_{800} - R_{445})/(R_{800} - R_{680})$. Merzlyak et al. [5] found that the difference of reflectances in the green and the blue ranges $(R_{680} - R_{500})$ depends on the pigment composition. The index $(R_{680} - R_{500})/R_{750}$ was found to be sensitive to carotenoid/chlorophyll ratio and used as a quantitative measure of plant senescence.

6.2.3 ANTHOCYANINS

Several vegetation indices were designed for determining leaf anthocyanins content nondestructively [12,22,45]. These indices were based on reflectances in a few spectral bands with varying level of their sensitivity to changes in anthocyanins content as well as to content of other pigments (e.g., chlorophyll). Van den Berg and Perkins [45] suggested using the ratio of absorbance (a) in the near-infrared band (940 nm) to the green band (530 nm) as a proxy of anthocyanins, calling it an anthocyanin content index (ACI):

$$ACI = \frac{a_{green}}{a_{NIR}}$$

The authors claimed that the technique was effective and accurate in estimating anthocyanins content in sugar maple leaves in autumn. ACI was modified (it was called modified ACI, mACI) for using reflectance rather than absorbance for anthocyanins estimation [68]:

$$mACI = \frac{R_{NIR}}{R_{green}} \tag{6.1}$$

Gamon and Surfus [12] suggested using a ratio of reflectances in the red (R_{red}) and the green (R_{green}) spectral bands to estimate anthocyanins content:

$$\frac{Red}{Green} = \frac{R_{red}}{R_{green}} \tag{6.2}$$

Anthocyanins absorb in situ around 540–550 nm [16,22], and the red peak of chlorophyll absorption in situ is around 670–680 nm. Thus, the R_{red}/R_{green} estimates anthocyanins content by comparing reflectance in the red region of the spectrum (where only chlorophyll-a and -b absorb) to reflectance in the green region where both chlorophylls and anthocyanins absorb.

To eliminate the influence of absorption by other pigments (e.g., chlorophyll) on reflectance in the green range, the reflectance in the second spectral band, in red edge region, was used [22]. Anthocyanins do not absorb in the red edge region of the spectrum but chlorophylls do absorb in the same way as in the green [18,19]. The difference $R_{green}^{-1} - R_{red\,edge}^{-1}$ was called anthocyanin reflectance index (ARI) [22]:

$$ARI = (R_{green}^{-1} - R_{red\,edge}^{-1}) \tag{6.3}$$

6.3 SPECTRAL FEATURES OF LEAF REFLECTANCE

A comprehensive understanding of optical properties of leaves is a prerequisite for estimating pigment content. A general approach implies the analysis of reflectance variation in response to variation in pigment content in order to find spectral bands of maximal sensitivity of reflectance to content of pigment of interest. This is combined with minimal sensitivity to content of other pigments as well as to leaf thickness and structure. Remote and nondestructive estimation of vegetation status via pigment content measurement based on relationship between reflectance and leaf inherent optical properties—absorption coefficient (a) and backscattering coefficient (b_b). Absorption of light by plant pigments in the visible spectrum produces a unique spectral signature of reflected light. Figure 6.1 shows the absorbance spectra of chlorophyll-a and -b and carotenoids extracted in acetone. Absorbance of both chlorophyll-a and -b and carotenoids overlapped in the blue range 400–500 nm. Another absorption band of both chlorophylls is in the red range of the spectrum around 650–660 nm. Importantly, absorbance of both chlorophylls and carotenoids in the green (around 550 nm) and the red edge range (beyond 700–740 nm) is much smaller than absorbance in the blue and the red ranges (e.g., Ref. [6]).

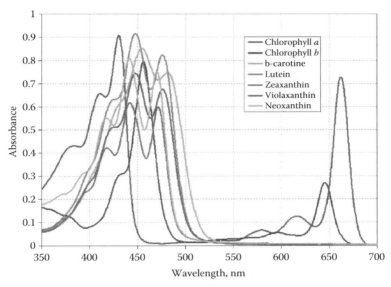

FIGURE 6.1 **(See color insert.)** Absorbance spectra of pigments extracted in 90% acetone. (Merzlyak, M.N., unpublished.)

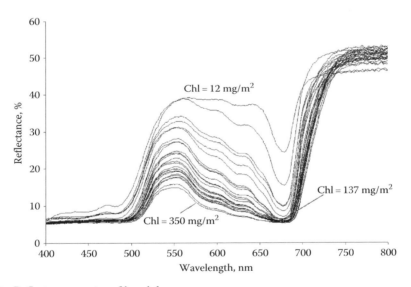

FIGURE 6.2 Reflectance spectra of beech leaves.

Leaf reflectance spectra with different chlorophyll content are presented in Figure 6.2. Reflectance in the blue range (400–500 nm) is very low even for yellow leaves with small chlorophyll content (top spectra). In such leaves, the absorption of carotenoids is very strong, causing the amount of reflected light to be much below 10%. In the green range of the spectrum, reflectance varies widely, decreasing with an increase in chlorophyll content. In the red range, reflectance of yellow leaves (top spectrum in Figure 6.2) is high (above 20%) and sharply drops with increase in chlorophyll reaching about 5% in green leaves. Significant decrease in reflectance with increase in chlorophyll content occurs in the range 700–740 nm. This range is called "red edge"; it locates

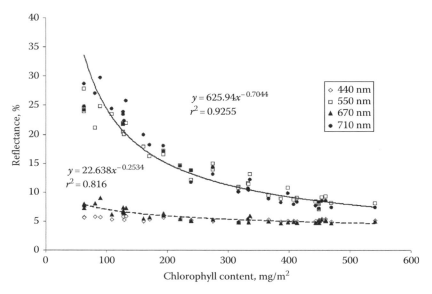

FIGURE 6.3 Reflectance in the blue, green, red, and red edge ranges of the spectrum plotted vs. total chlorophyll content in beech leaves.

between strong chlorophyll absorption in the red and scattering in the NIR range. These spectral features are common for different types of vegetation [5,12,17,19,23,26–30].

Absorption coefficients of pigments are high in the blue and the red ranges (Figure 6.1) [6,46] and the depth of light penetration into the leaf is very low [9,47–49]. As a result, even low amounts of pigments are sufficient to saturate absorption. For yellowish-green leaves, as chlorophyll is around $100 \, mg/m^2$, total absorption in the red and blue regions can exceed 90% and a further increase of chlorophyll content does not cause increase of total absorption. Thus, the relationship of absorption vs. total chlorophyll reaches a plateau, and reflectance becomes virtually insensitive to further chlorophyll increase (Figure 6.3) [12,16–19,43]. The closer the wavelength to the main absorption bands of pigments (the blue or the red), the lower the chlorophyll content at which saturation of the relationship "reflectance vs. chlorophyll" appears. Thus, in the blue neighboring 440 nm and in the red around 670 nm, reflectance of the leaves with chlorophyll above $100 \, mg/m^2$ is small (around 5%) and is not sensitive to chlorophyll content. This effect was detected numerously in different vegetation types [5,12,17,19,23,26–29,34–41,43,49–51] and is also indicated by radiative transfer models [30–32].

As it was mentioned earlier (Figure 6.1), the absorbance of chlorophylls in extract in the green (around 550 nm) and red edge (near 700 nm) is very low and seldom exceeds 4% of that for blue and red (e.g., Ref. [6]). However, green leaves absorb more than 80% of incident light in the green and red edge spectral regions [24,46,50–52]. The depth of light penetration into the leaf in the green and the red edge was found to be four- to sixfold higher than for the blue and the red (e.g., Ref. [53] and Figure 2 in Ref. [9]). Therefore, sensitivity of absorption and reflectance to chlorophyll content is much higher in these spectral regions than in the blue and red; the reflectance decreases hyperbolically as chlorophyll content increases (Figure 6.3) [18,26].

Importantly, the green and the red edge are the only spectral ranges where reflectance is sensitive to wide variation in chlorophyll content (i.e., from yellow to dark green leaves). The relationship between reflectances in these spectral regions is very close in a wide range of chlorophyll content (from yellow to dark green leaves; Figure 6.4) [18,19,43]. This is fundamental spectral feature of anthocyanin-free leaves. The relationship R_{green} vs. $R_{red \, edge}$ is also very indicative of presence of anthocyanins in leaves. For the same chlorophyll content in anthocyanin-containing leaves the green reflectance decreases while red edge reflectance remains the same. Thus, scattering of points from best fit function can be used as sensitive indicator of the presence of anthocyanins [22,67].

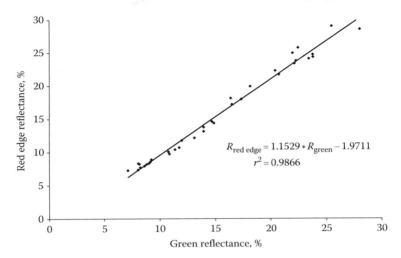

FIGURE 6.4 The red edge reflectance (at 700 nm) vs. the green reflectance (at 560 nm) of dogwood anthocyanin-free leaves. The green and red edge spectral regions are carriers of the same information in anthocyanin-free leaves. Thus, either the green or the red edge bands can be used for chlorophyll estimation. This relationship is very responsive to leaf anthocyanins content: the green reflectance decreases while the red edge reflectance remains the same. This feature can be used as sensitive indicator of anthocyanins presence in leaves.

Thus, spectral features of leaf absorption and reflectance of anthocyanin-free leaves are as follows: (a) minimum sensitivity to pigment content in the blue between 400 and 500 nm and in the NIR; (b) in the red absorption band of chlorophyll-a in situ near 670 nm, both absorption and reflectance are insensitive to chlorophyll content above 150 mg/m^2; (c) reflectances in the green and the red edge regions are related very closely (Figure 6.4); and (d) the highest sensitivity of reflectance and absorption to pigment variation is in the green from 530 to 590 nm and in the red edge around 700 nm (Figures 6.2 and 6.3).

These fundamental spectral features of absorption and reflectance are widely recognized (e.g., Refs. [17–19,24,43,49–52,54]). It was shown that indices composed of reflectance in the bands of maximal chlorophyll absorption saturate even at low chlorophyll content (in slightly green leaves with chlorophyll above 150 mg/m^2). That is why newly developed algorithms for nondestructive chlorophyll retrieval are not based on reflectance in the main chlorophyll absorption bands (in the blue and the red). The use of green and red edge bands avoids saturation and the accompanying loss of sensitivity to chlorophyll, and is usually preferred because reflectances in these ranges remain sensitive to widely variable chlorophyll content. Indices based at these spectral bands were proposed and used to estimate chlorophyll content in the leaves of various plant species [12,17–19,24,29,36,38,43,55].

While carotenoids to chlorophyll ratio varies, generally carotenoids and chlorophyll contents are related closely; thus, it is difficult to find carotenoids specific spectral features in leaf reflectance. Reflectance spectra of leaves with variable carotenoid/chlorophyll ratio are presented in Figure 6.5. With increase in the ratio, reflectance in whole visible spectrum increases. While increase in the red edge is due to decrease in chlorophyll (carotenoids do not absorb in this region), increase in reflectance the region around 500–530 nm is due to variation of both chlorophyll and carotenoids. In Figure 6.6, spectrum of sensitivity of reciprocal reflectance R^{-1} to carotenoids content is plotted for leaves with wide variation of pigment content and composition: from yellow to dark green leaves with chlorophyll from 1.4 to 540 mg/m^2. Sensitivity is defined as a product of slope and determination coefficient (r^2) of the linear relationship "reciprocal reflectance vs. carotenoids content" [21]. The sensitivity has prominent peak at 515 nm. This band around 510 nm is a specific spectral feature of carotenoids that was used for carotenoids content estimation [21,41].

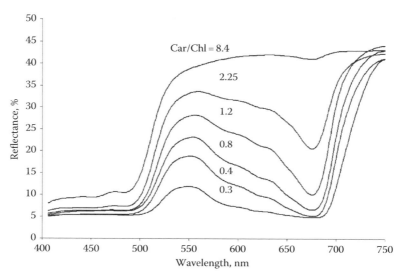

FIGURE 6.5 Reflectance spectra of the chestnut leaves with different carotenoid to chlorophyll ratio. Top spectrum corresponds to a yellow leaf while bottom spectrum corresponds to a dark-green leaf.

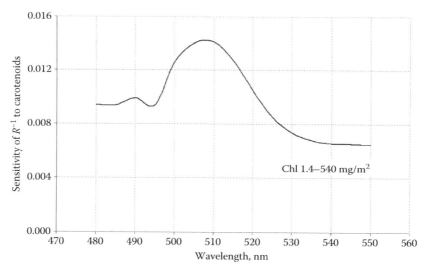

FIGURE 6.6 Sensitivity of reciprocal reflectance to carotenoids content in maple yellow to dark-green leaves. Sensitivity is defined as a product of two characteristics of the linear relationship "R^{-1} vs. Carotenoids": slope and determination coefficient.

Anthocyanin absorption peaks in vivo is between 537 and 542 nm [16]. It was found that anthocyanins absorption in situ has a maximum around 550 nm (Figure 6.7, see also Ref. [22]) and anthocyanins absorbance in leaves is linearly related to Anth content [16]. Thus, Anth absorbs in the green range where chlorophyll also has significant absorption and reflectance closely related to total chlorophyll content (Figure 6.3). Anth content affects leaf absorption spectrum between 450 and 630 nm. In Figure 6.8 spectra of reciprocal of reflectance (R^{-1}) of the leaves with different anthocyanins contents are shown. R^{-1} relates directly to total absorption coefficient of all pigments [54]. The main difference in spectra occurred in the green range around 550 nm.

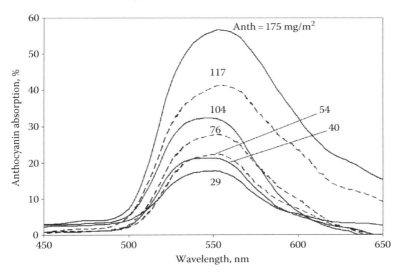

FIGURE 6.7 Spectra of anthocyanin absorption of cotoneaster (solid line) and dogwood (dashed line) leaves.

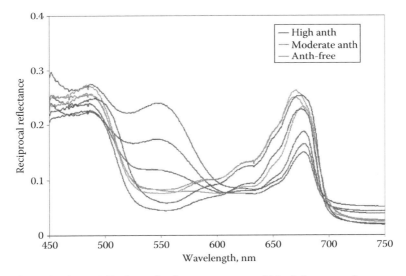

FIGURE 6.8 **(See color insert.)** Reciprocal reflectance spectra of Virginia creeper leaves.

In anthocyanin-free leaves (green curves), absorption depends only on chlorophyll and carotenoids content and R_{green}^{-1} is lowest among leaves presented. In green-reddish leaves with moderate anthocyanins and significant amount of chlorophyll (orange curves), R_{green}^{-1} is much higher. In the red leaves with high anthocyanins and significant amount of chlorophyll content R_{green}^{-1} is maximal (red curves).

While in anthocyanin-free leaves R_{550} and chlorophyll relate closely (Figure 6.3), in anthocyanin-containing leaves with the same chlorophyll content, R_{550} may vary up to 10-fold depending upon anthocyanins content (Figure 6.9). The green reflectance of anthocyanin-containing leaves is affected by both chlorophylls and anthocyanins contents; it presents a great challenge for the retrieval of anthocyanins content from reflectance spectra.

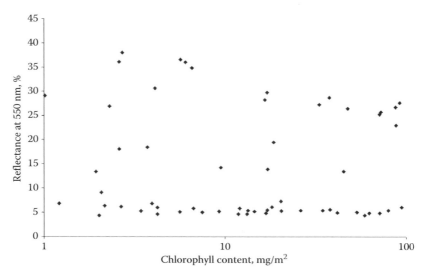

FIGURE 6.9 Reflectance in the green range (at 550nm) plotted vs. chlorophyll content in anthocyanin-containing Virginia creeper leaves. Reflectance in the green range does not follow chlorophyll content as in anthocyanin-free leaves.

6.4 CONCEPTUAL THREE-BAND MODEL

All physically based approaches to estimation of pigment content use the relationship between reflectance and inherent optical properties: absorption coefficients of each pigment a_i and leaf backscattering coefficient b_b. Kubelka-Munk theory established relationship between the infinite reflectance of an ideal layer, R_∞, in which a further increase in thickness results in no noticeable difference, and inherent optical properties as following [56]:

$$f(R_\infty) = \frac{(1 - R_\infty^2)}{2R_\infty} = \frac{a}{b_b} \tag{6.4}$$

To find the relationship between reflectance of a real leaf and its inherent optical properties, one should find a relationship between $f(R_\infty)$ and the measured leaf reflectance. To measure reflectance, a leaf is placed on a black velvet background or other material with reflectance close to zero and the reflectance is measured using an integrating sphere or leaf clip (e.g., Ref. [34]). In terms of the Kubelka-Munk theory, the measured reflectance is defined as R_0 where the index "0" is used to designate the ideal black background with zero reflectance. The relationship between R_0 and $f(R_\infty)$ may be found from the equation [56]:

$$f(R_\infty) = \frac{(1 - T^2 + R_0^2)}{2R_0 - 1} \tag{6.5}$$

where T is leaf transmittance. For reflectance R_0 ranging from 0% to 50%, the relationship between $f(R_\infty)$ and R_0^{-1} was found to be is linear (see Figure 9 in Ref. [54]):

$$f(R_\infty) = 0.4874 \times (R_0)^{-1} - 0.707$$

with $r^2 = 0.9998$. Therefore, the reciprocal reflectance directly relates to absorption coefficient of each pigment and inversely relates to leaf backscattering coefficient [54]:

$$f(R_\infty) \propto (R_0)^{-1} = \frac{a}{b_b} \tag{6.6}$$

To estimate pigment content, one needs to isolate absorption coefficient of each pigment of interest (a_p). To isolate a_p, the conceptual model which uses reflectances at three spectral bands was developed [54]. The model is based on different sensitivities of reflectances in three spectral bands λ_1, λ_2, and λ_3 to absorption by the pigment of interest a_p and to absorption by other pigments a_0 and leaf scattering. Specifically, reflectance in the first band $R(\lambda_1)$ should be maximally sensitive to absorption coefficient of the pigment of interest a_p. However, $R(\lambda_1)$ might also be affected by the absorption of other pigments a_0 and by the variability in backscattering among leaf samples b_b. To remove the effect of absorption by other pigments one needs to find a spectral band λ_2 that meets two requirements:

1. Absorption by the pigment of interest is much lower than at λ_1:

$$a_p(\lambda_2) \ll a_p(\lambda_1) \tag{6.7}$$

2. Absorption by other pigments as well as the leaf backscattering are close to that at λ_1:

$$a_0(\lambda_2) \cong a_0(\lambda_1) \tag{6.8}$$

$$b_b(\lambda_2) \cong b_b(\lambda_1) \tag{6.9}$$

The subtraction of $R(\lambda_2)^{-1}$ from $R(\lambda_1)^{-1}$ gives

$$R(\lambda_1)^{-1} - R(\lambda_2)^{-1} \propto \frac{a_p(\lambda_1)}{b_b} \tag{6.10}$$

Although the effect of other pigments absorption on difference $R(\lambda_1)^{-1} - R(\lambda_2)^{-1}$ has been removed, the difference is still affected by leaf backscattering that depends on leaf thickness and structure. To remove b_b from denominator and isolate a_p, a third spectral band λ_3 should be used. In this spectral band, reflectance has to be controlled by leaf backscattering and minimally depends on any pigment absorption

$$R(\lambda_3) \propto b_b \tag{6.11}$$

Multiplying the difference (Equation 6.10) by $R(\lambda_3)$, we have the model that may isolate a_p:

$$\text{Pigment content} \propto a_p = [R(\lambda_1)^{-1} - R(\lambda_2)^{-1}] \times R(\lambda_3) \tag{6.12}$$

The choice of spectral bands λ_1, λ_2, and λ_3 for chlorophyll, carotenoids, and anthocyanins content estimation is based on fundamental leaf reflectance spectral properties briefly described earlier in Section 6.3. For chlorophyll estimation in anthocyanins-free leaves, to avoid saturation of absorption at moderate to high chlorophyll content, λ_1 should be in the green and red edge ranges of the spectrum where reflectance is sensitive to chlorophyll content (Figure 6.3). However, R_{green}^{-1} and $R_{red\,edge}^{-1}$ are affected also by nonpigment leaf absorption that is almost spectrally independent [54]. Thus, the NIR range beyond 760 nm, where none pigment absorbs, is a candidate for λ_2 [54]. The λ_3 should also be in the NIR spectral region; the reflectance in this range is controlled by leaf scattering (i.e., thickness and density).

For carotenoids estimation, λ_1 should be in the range of maximal sensitivity of reflectance to carotenoids content, thus around 510–520 nm (Figure 6.6). However, both chlorophyll-a and -b also absorb in this region; so, to subtract chlorophyll effect on $R(\lambda_1)$, the λ_2 should be either in the green or the red edge regions of the spectrum where carotenoids influence is minimal. As in the case of chlorophyll estimation, λ_3 should be in the NIR spectral region.

For anthocyanins estimation, λ_1 should be in the range of maximal sensitivity of reflectance to anthocyanins content in the green range around 550 nm (Figures 6.7 and 6.8). However, both chlorophylls-*a* and -*b* absorb in this region and to subtract chlorophyll effect on $R(\lambda_3)$, λ_2 should be in the red edge range where only chlorophylls absorb. As for other pigments, λ_3 should be in the NIR spectral region.

To find the optimal positions of spectral bands λ_1, λ_2, and λ_3, a stepwise technique to minimize root mean square error (RMSE) of the linear relationship between the pigment content and the model (Equation 6.12) was used.

6.5 ESTIMATION OF LEAF PIGMENT CONTENT

6.5.1 CHLOROPHYLL

6.5.1.1 Spectral Bands Tuning

As the first step in tuning of the spectral bands of the model (Equation 6.12) for chlorophyll estimation in anthocyanin-free leaves, we found the optimal position of λ_2 using an initial λ_1^0 and λ_3^0. λ_1^0 was chosen in the red edge range of the spectrum at 700 nm; reflectance in this range is sensitive to chlorophyll content (Figure 6.3). The λ_3^0 was chosen in the NIR range of the spectrum at 800 nm where chlorophyll absorption is negligible ($a_{Chl}(\lambda_3) \cong 0$) and backscattering b_b controls reflectance. RMSE of chlorophyll estimation by the model ($R_{700}^{-1} - R_{\lambda 2}^{-1}) \times R_{800}$ had minimal values at λ_2 beyond 760 nm for all species (Figure 6.10A). We selected $\lambda_2^1 = 790$ nm.

In the second step of the spectral bands tuning, we found the position of λ_3^1 in the model ($R_{700}^{-1} - R_{790}^{-1}) \times R_{\lambda 3}$. As for λ_2^1, minimal RMSE was in the NIR range beyond 760 nm (Figure 6.10B). We selected $\lambda_2^1 = \lambda_3^1 = 790$ nm.

In the third step, we found the optimal position of λ_1^1 in the model ($R_{\lambda 1}^{-1} - R_{790}^{-1}) \times R_{790}$. RMSE had two distinct minima: in the green (around 550 nm) and in the red edge (690–725 nm) ranges

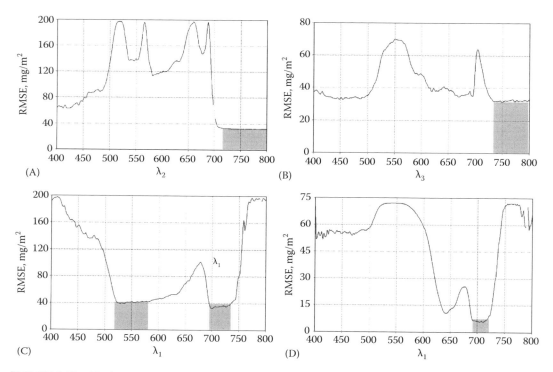

FIGURE 6.10 Tuning spectral bands of the three-band model for chlorophyll retrieval in anthocyanin-free (A–C) and anthocyanin-containing (D) leaves.

(Figure 6.10C). Therefore, for chlorophyll estimation in anthocyanin-free leaves two models with NIR set beyond 760 nm can be used (the so-called chlorophyll index [CI] [54,61]):

$$CI_{green} = [R_{540-560}^{-1} - R_{NIR}^{-1}] \times R_{NIR} = \left(\frac{R_{NIR}}{R_{green}} \right) - 1 \qquad (6.13)$$

$$CI_{red\ edge} = [R_{690-725}^{-1} - R_{NIR}^{-1}] \times R_{NIR} = \left(\frac{R_{NIR}}{R_{red\ edge}} \right) - 1 \qquad (6.14)$$

In Figure 6.11, RMSE of chlorophyll estimation by three indices, $(R_\lambda)^{-1}$, $(R_\lambda)^{-1} - (R_{NIR})^{-1}$, and $(R_{NIR}/R_\lambda) - 1$, is presented. The $(R_\lambda)^{-1}$ is able to produce minimal RMSE (below 60 mg/m^2) in the narrow bands in the green and the red edge. As $(R_{NIR})^{-1}$ was subtracted from $(R_{\lambda 1})^{-1}$, the RMSE dropped to 50 mg/m^2 and spectral regions with minimal RMSE became significantly wider (about 20 nm). Further step, multiplication of the difference $(R_\lambda)^{-1} - (R_{NIR})^{-1}$ to R_{NIR}, significantly decreased uncertainties of chlorophyll estimation (RMSE dropped below 40 mg/m^2) and both spectral regions with minimal RMSE became much wider.

The models R_{740}/R_{720} [34], R_{NIR}/R_{550}, and R_{NIR}/R_{700} [18,19,25] are specific cases of the conceptual models (Equations 6.13 and 6.14) for narrow spectral bands. Importantly, the intercept of the function CI vs. chlorophyll content is very close to zero, which made the chlorophyll estimates linearly proportional to the chlorophyll content.

In anthocyanin-containing leaves, the first and second steps of band tuning gave the same results as for anthocyanin-free leaves: RMSE was minimal for both λ_2 and λ_3 in the NIR range and we selected $\lambda_2^1 = \lambda_3^1 = 790$ nm (not shown). However, in the third step, RMSE for λ_1 had only one minimum in the red edge range from 690 to 725 nm (Figure 6.10D). In contrast to anthocyanin-free leaves, RMSE of chlorophyll estimation in the green range was maximal due to absorption by anthocyanins in this range (Figures 6.7 and 6.8, see also Ref. [22]). Thus, for chlorophyll estimation in anthocyanin-containing leaves, the only model Equation 6.14 should be used.

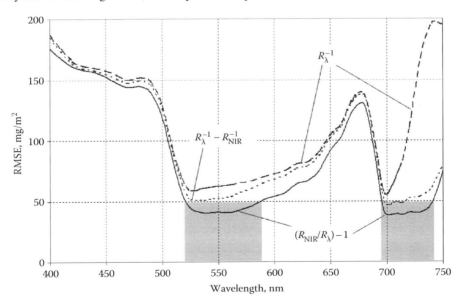

FIGURE 6.11 Spectra of RMSE of chlorophyll estimation by reciprocal reflectance R_λ^{-1}, difference of reciprocal reflectances $R_\lambda^{-1} - R_{NIR}^{-1}$, and chlorophyll index $(R_{NIR}/R_\lambda) - 1$.

6.5.1.2 Model Performance

Equations 6.13 and 6.14 were used for chlorophyll estimation in different unrelated tree and crop species [18,19,25,26,54,55,57–60]. For each species, the linear relationship between the chlorophyll content and the models was found to be the best fit function (Table 1 in Ref. [61]). Using data sets described in Ref. [61], the accuracy of the different indices that had been previously developed for chlorophyll estimation was tested and compared with the performance of the models (Equations 6.13 and 6.14). Relationships between total chlorophyll content and the following indices are presented in Figure 6.12:

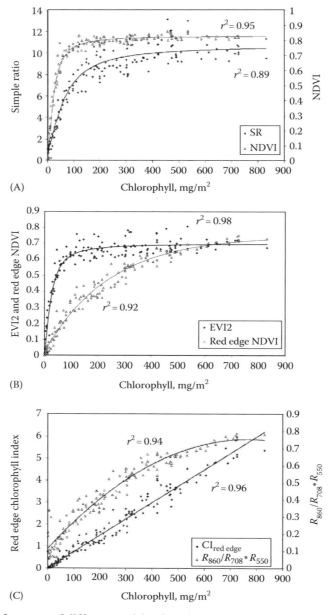

FIGURE 6.12 Performance of different models of total chlorophyll estimation in maple, chestnut, wild grapevine, and beech leaves: (A) Simple Ratio and NDVI; (B) EV12 and Red edge NDVI; (C) $CI_{red\ edge}$ and ECI.

Simple ratio [62] with spectral bands suggested in Ref. [27]—Figure 6.12A:

$$\text{SR} = \frac{R_{800}}{R_{680}}$$

Normalized difference vegetation index [63] with spectral bands suggested in Ref. [27] (Figure 6.12A):

$$\text{NDVI} = \frac{(R_{800} - R_{680})}{(R_{800} + R_{680})}$$

Enhanced vegetation index [64,65] with spectral bands of 250 m resolution MODIS system (Figure 6.12B):

$$\text{EVI2} = 2.5 \times \frac{(R_{\text{NIR}} - R_{\text{red}})}{(1 + R_{\text{NIR}} + 2.4 \times R_{\text{red}})}$$

Red edge NDVI [18,19]—Figure 6.12B:

$$\text{Red edge NDVI} = \frac{(R_{800} - R_{700})}{(R_{800} + R_{700})}$$

Eucalyptus chlorophyll index [66]—Figure 6.12C:

$$\text{ECI} = \left(\frac{R_{860}}{R_{708} \times R_{550}} \right)$$

Red edge chlorophyll index [61]—Figure 6.12C:

$$\text{CI}_{\text{red edge}} = \frac{(R_{750-800})}{(R_{695-740})} - 1$$

Indices that use reflectance in the red range around 680 nm (SR, NDVI, and EVI2) were sensitive to chlorophyll content below 150 mg/m^2 and nonsensitive to moderate to high chlorophyll content (Figure 6.12A). Datt's [28] index [66], which uses red edge and green bands, was a much better indicator of chlorophyll content. It was more sensitive to moderate to high chlorophyll content than the indices that use red band; however, its sensitivity declines considerably as chlorophyll content exceeds 300 mg/m^2 (Figure 6.12C). The red edge NDVI (Figure 6.12B) was shown to be a good predictor of chlorophyll content for a variety of plant species [12,18,19,29,36]; however, the sensitivity of the index to chlorophyll content >400 mg/m^2 was fourfold lower than to low-to-moderate chlorophyll. In the range of total chlorophyll content variation from 1 to 830 mg/m^2, eucalyptus index [66] and the red edge NDVI [18,19] provided estimation of total chlorophyll with a RMSE below 80 mg/m^2. The CI$_{\text{red edge}}$ = [($R_{750-800}$)/($R_{695-740}$) − 1] (Figure 6.12C) and the CI$_{\text{green}}$ = [($R_{750-800}$)/($R_{520-585}$) − 1] (not shown) with broad spectral bands were the best chlorophyll content predictors with a RMSE <39 mg/m^2.

The performance of CI$_{\text{red edge}}$ in anthocyanin-containing grape leaves [58,59] as well as in leaves of four unrelated tree species (European hazel, Siberian dogwood, Norway maple, and Virginia creeper) with widely variable pigment content and composition was also studied [67]. In latter study, the relationship between chlorophyll content and CI$_{\text{red edge}}$ for all species taken together was established (Figure 6.13). The best fit function "chlorophyll vs. CI$_{\text{red edge}}$" was linear with $r^2 > 0.97$. In all four species, with no reparameterization of algorithm, the CI$_{\text{red edge}}$ was able to very accurately estimate chlorophyll content ranged from 0.63 to 539 mg/m^2 with RMSE below 21 mg/m^2.

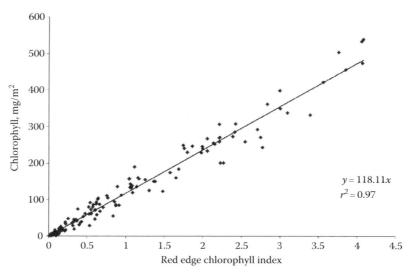

FIGURE 6.13   Total chlorophyll content plotted vs. $CI_{red\ edge}$ in anthocyanin-containing leaves of four species (European hazel, Siberian dogwood, Norway maple, and Virginia creeper).

6.5.2   CAROTENOIDS

6.5.2.1   Spectral Bands Tuning

In many species studied, carotenoids content was related to chlorophyll content very closely (determination coefficient $r^2 > 0.90$) [21,41,44,54,61]; therefore, carotenoids content cannot be treated as an independent variable. However, in tree species (e.g., beech, chestnut, and maple), it was possible to estimate carotenoids content independently from chlorophyll content despite the quite close correlation between chlorophyll and carotenoids (for beech, r^2 was 0.78 in leaves sampled in 1996 and 0.86 in 2000, for chestnut r^2 was 0.69 in 1996–1997, and 0.72 in 2000, for maple r^2 was 0.65 in 1992–1999, and 0.75 in 2000 [21,61]).

The same procedure described earlier was used for spectral bands tuning in anthocyanin-free beech, chestnut, and maple leaves. In the first step, we found the optimal position of λ_2 using an initial $\lambda_1^0 = 500\,nm$ (Figure 6.6, see also Refs. [21,41]) and $\lambda_3^0 = 760\,nm$ ($a_{car}(\lambda_3) \cong 0$ and b_b controls reflectance). For leaves of all three species analyzed together, the model ($R_{500}^{-1} - R_{\lambda 2}^{-1}$) $\times R_{760}$ brought minimal RMSE values at $\lambda_2^1 = 560$–$570\,nm$ and around $700\,nm$ (Figure 6.14A). In these spectral bands the absorption coefficient of carotenoids is much lower than that at $500\,nm$, $a_{Car}(\lambda_2) \ll a_{Chl}(\lambda_1)$ [18,19] and chlorophyll absorption coefficient is much higher than absorption coefficient of carotenoids [18,43].

In the second step, the minimal RMSE and optimal position of λ_3 in the models ($R_{500}^{-1} - R_{560-570}^{-1}$) $\times R_{\lambda 3}$ and ($R_{500}^{-1} - R_{690-710}^{-1}$) $\times R_{\lambda 3}$ was found in the NIR range beyond $760\,nm$ where $a_{Car}(\lambda_2) \cong a_{Chl}(\lambda_1) = 0$ and b_b controls reflectance (Figure 6.14B). For the third step we selected $\lambda_3^1 = 790\,nm$.

In the third step, the optimal position of λ_1 in the models ($R_{\lambda 1}^{-1} - R_{560-570}^{-1}$) $\times R_{790}$ and ($R_{\lambda 1}^{-1} - R_{690-710}^{-1}$) $\times R_{790}$ was found at 510–$520\,nm$ (Figure 6.14C). Thus, two models can be used for carotenoids estimation in anthocyanin-free leaves with NIR set beyond $760\,nm$ (so-called carotenoids reflectance index (CRI) [21]):

$$CRI_{green} \propto [R_{510-520}^{-1} - R_{560-570}^{-1}] \times R_{NIR} \qquad (6.15)$$

$$CRI_{red\ edge} \propto [R_{510-520}^{-1} - R_{690-710}^{-1}] \times R_{NIR} \qquad (6.16)$$

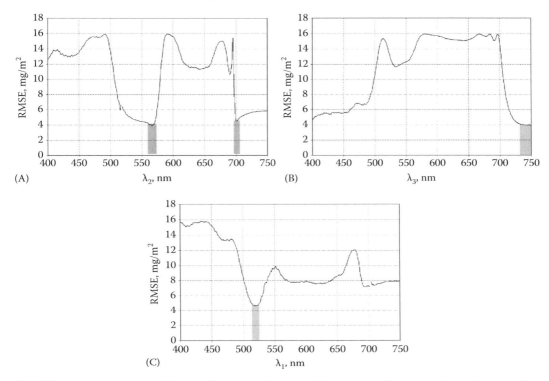

FIGURE 6.14 Tuning spectral bands of the three-band model for carotenoids retrieval in anthocyanin-free maple leaves: (A) λ_2, (B) λ_3, and (C) λ_1.

6.5.2.2 Model Performance

Performance of $CRI_{red\ edge}$ in estimating carotenoids content in maple leaves is shown in Figure 6.15A. The models are able to retrieve carotenoids content quite accurately at the background of very variable chlorophyll content. Performance of the models (Equations 6.15 and 6.16) for carotenoids estimation in maple, beech, and chestnut leaves was investigated in Ref. [61]. It was found that for each species the carotenoids content and the models related almost linearly (although in some cases quadratic function was the best fit function). Importantly, coefficients of the relationships relating carotenoids to models remained almost the same for the independent data sets of each species (Figure 6.15B, Table 2 in Ref. [61]).

To validate the models for carotenoids estimation in different species of the same origin (maple and chestnut collected in Moscow, Russia [21]), the data, which included spectral reflectance of 119 leaves and their carotenoids contents, were analyzed. Samples with odd numbers were used for model calibration, that is, establishment of relationships "carotenoids vs. CRI_{green}" and "carotenoids vs. $CRI_{red\ edge}$." These relationships were validated using data set containing samples with even numbers. Measured reflectances in the validation data set were used to estimate carotenoids content values (Car_{est}), and then Car_{est} were compared with carotenoids content measured in Lab, Car_{meas} (Figure 6.16). In the range of carotenoids of 15–92 mg/m², RMSE of carotenoids estimation was below 10.6 mg/m² and mean normalized bias was less than −5.7%.

In the range 500–520 nm, where reflectance is maximally sensitive to carotenoids content (Figure 6.6), anthocyanins absorption is also significant (Figure 6.8). Thus, in anthocyanin containing leaves, three pigments—chlorophylls, carotenoids, and anthocyanins—contribute to total absorption and affect reflectance in the range 500–520 nm. Therefore, uncertainties of the three-band model in estimating carotenoids content in anthocyanin-containing leaves might increase drastically. I am not aware of any successful attempt to estimate carotenoids content in anthocyanin-containing leaves.

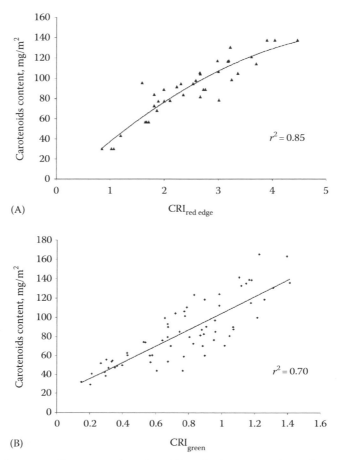

FIGURE 6.15 Total carotenoids content plotted vs. (A) red edge carotenoids reflectance index for maple leaves and (B) green carotenoids reflectance index for chestnut and beech leaves.

6.5.3 ANTHOCYANINS

6.5.3.1 Spectral Bands Tuning

Spectral bands tuning is shown in Figure 6.17 for Virginia creeper [67]. In the first step of the tuning, the optimal position of λ_2 was found using an initial $\lambda_1^0 = 530$ nm, which is close to maximum of anthocyanins absorption in acidic alcohols [16], and $\lambda_3^0 = 760$ nm. RMSE of anthocyanins content estimation by the model $(R_{530}^{-1} - R_{\lambda 2}^{-1}) \times R_{760}$ had minimal values at $\lambda_2^1 = 690$–700 nm (Figure 6.17A). In this spectral band, reflectance is governed mainly by chlorophyll content [54]. The subtraction of $R_{690-700}^{-1}$ from R_{530}^{-1}, made the difference $R_{530}^{-1} - R_{690-700}^{-1}$ closely related to anthocyanins content. This difference was called ARI and successfully used for anthocyanins content estimation [22,67,68]. However, the ARI is also affected by leaf scattering that might vary among samples due to variation in leaf structure, density, and thickness.

In the second step, the optimal position of λ_3 in the model $(R_{530}^{-1} - R_{690-700}^{-1}) \times R_{\lambda 3}$ was found in the NIR range beyond 760 nm where $a_{\text{Anth}}(\lambda_2) \cong a_{\text{Chl}}(\lambda_1) = 0$ and b_b controls reflectance (Figure 6.17B). In the third step, the optimal position of λ_1 in the model $(R_{\lambda 1}^{-1} - R_{690-710}^{-1}) \times R_{790}$ was found in a wide range around 550 nm (Figure 6.17C). The model for anthocyanins estimation, with NIR range beyond 760 nm, had the form (it is called modified anthocyanin reflectance index (mARI) [68]):

$$\text{mARI} = [R_{530-570}^{-1} - R_{690-710}^{-1}] \times R_{\text{NIR}} \tag{6.17}$$

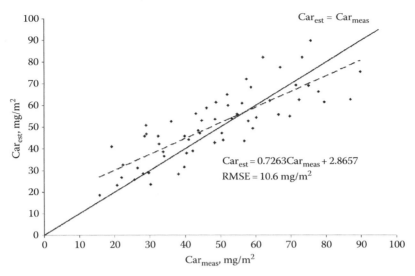

FIGURE 6.16 Validation of CRI$_{red\,edge}$: Estimated total carotenoids content plotted vs. analytically measured in maple and chestnut leaves. Dash line is the best fit function.

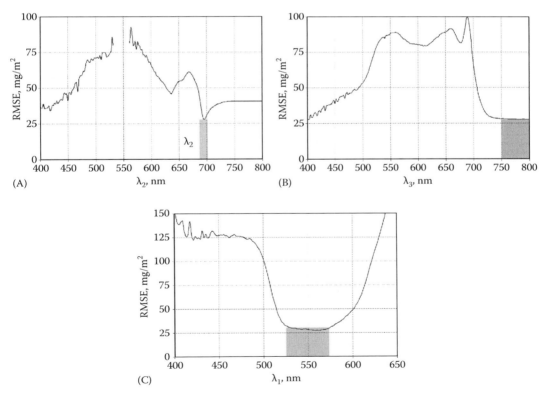

FIGURE 6.17 Tuning spectral bands of the three-band model for anthocyanin retrieval in Virginia creeper leaves: (A) λ_2, (B) λ_3, and (C) λ_1.

6.5.3.2 Model Performance

In an attempt to apply different models to estimate anthocyanins content in four unrelated species (European hazel, Siberian dogwood, Norway maple, and Virginia creeper) with no reparameterization of the model coefficients, the performance of mACI, Red/Green, ARI, and mARI was studied [67]. The mACI (Equation 6.1) is identical to the first term in mARI (Equation 6.17) rewritten as $mARI = R_{NIR}/R_{green} - R_{NIR}/R_{red\ edge}$. Thus, the mACI is affected by absorption of all pigments as well as leaf scattering that may vary with leaf thickness and density. The mACI was closely related to anthocyanins content in hazel leaves ($r^2 = 0.92$) in which chlorophyll content was almost invariable [67]. However, for each of other species with variable chlorophyll content as well as for all four species together the relationships were much weaker ($r^2 < 0.67$) due to strong effect of chlorophyll absorption to reflectance in the green range (Figure 5 in Ref. [67]).

The red/green ratio was accurate in anthocyanins estimations in hazel leaves with virtually one pigment (anthocyanins) governed reflectance (chlorophyll content was almost invariant). In other species, where chlorophyll content varied broadly, the relationships red/green vs. anthocyanins content were weak (r^2 was between 0.34 and 0.51).

In Figure 6.18A and B, the ARI and the mARI were plotted vs. anthocyanins content, measured analytically, for all four species. The ARI and mARI explained more than 92% of anthocyanins content

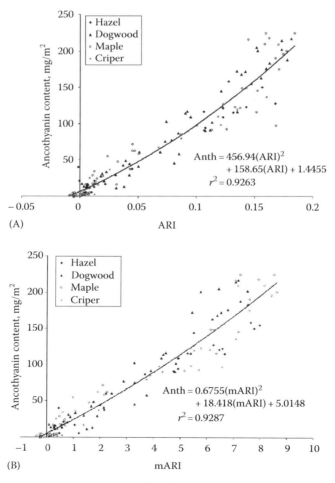

(A)

(B)

FIGURE 6.18 Anthocyanin content plotted vs. (A) ARI and (B) mARI in leaves of four species (European hazel, Siberian dogwood, Norway maple, and Virginia creeper).

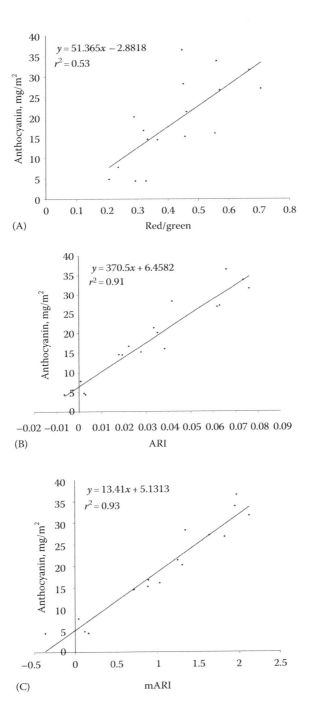

FIGURE 6.19 Performance of (A) red/green, (B) ARI, and (C) mARI in estimating anthocyanin content in grapevine (Saint Croix and Saint Pepin) leaves.

variation. Both indices were very accurate in anthocyanins estimation ranged from 0 to 225 mg/m^2; RMSE was below 19 mg/m^2. Thus, established relationships "anthocyanins vs. ARI" and "anthocyanins vs. mARI" can be used for anthocyanins content retrieval with no reparameterization.

Anthocyanins content was also estimated in two grape cultivars (Saint Croix and Saint Pepin; Figure 6.19) [68]. The relationship "anthocyanins vs. Red/Green" was week with $r^2 < 0.54$, while for the ARI and the mARI relationships were very close with $r^2 = 0.91$ and 0.93, respectively. Both the ARI and the mARI were capable to accurately predict anthocyanins content in grapevine leaves with a RMSE below 3 and 2.3 mg/m^2, respectively. The authors concluded that this technique has potential for developing simple handheld field instrumentation for accurate nondestructive anthocyanins estimation and also for analyzing digital airborne or satellite imagery to assist the agricultural producer in making informed decisions regarding vineyard management.

6.6 CONCLUSIONS AND FUTURE PROSPECTS

In spite of complicated morphological and optical properties and still imprecise understanding of leaf optics, considerable progress has been achieved in the development of nondestructive techniques for sensing of plant physiological status via quantitative estimation of pigment content and composition (chlorophylls, carotenoids, and anthocyanins). The results presented show that reflectance spectroscopy is a useful and efficient tool for quantification of pigment contents. Fundamental spectral features of leaf reflectance and its relation with pigment content and composition, recently revealed and reviewed here, provide a solid basis for the development of reliable technologies for monitoring vegetation status. The developed models are closely related to pigment content and are able to accurately estimate foliar pigments in a wide range of their content. Remarkably, for accurate retrieval of three foliar pigments (chlorophylls, carotenoids, and anthocyanins), one needs reflectances in only four, relatively wide spectral bands: blue-green (around 510 nm), green (540–560 nm), red edge (700–720 nm), and NIR (beyond 760 nm). The three-band model was very accurate also in estimating four pigments in fruits [69] chlorophylls, carotenoids, anthocyanins, and flavonoids.

ACKNOWLEDGMENTS

This chapter is dedicated to late Mark N. Merzlyak, extremely bright and productive scholar. He worked in many fields of biology and physiology. In each of them his contributions were among the highest. Among others were free-radical oxidation of lipids, syndrome of lipid peroxidation in plants, phytoimmunology, and stress- and senescence-induced degradation of plant pigments. He contributed enormously in leaf optics and development techniques for foliar pigment retrieval. He was the best friend of mine and I missed him tremendously. I acknowledge contributions of my former PhD students Drs. Yoav Zur, Robi Stark, Andres Vina, Veronica Ciganda, and Mark Steele. I thank my colleagues Drs. Claus Buschmann, Olga B. Chivkunova, Harmut Lichtenthaler, and Donald C. Rundquist greatly helping me at different stages of this study. The support of Center for Advanced Land Management and Information Technologies at University of Nebraska and J. Blaustein Institute for Desert Research of Ben-Gurion University, Israel, is greatly appreciated.

REFERENCES

1. Curran, P.J., Dungan, J.L., and Gholz, H.L., Exploring the relationship between reflectance red-edge and chlorophyll content in slash pine, *Tree Physiology*, 7, 33–48, 1990.
2. Filella, I., Serrano, I., Serra, J., and Penuelas, J., Evaluating wheat nitrogen status with canopy reflectance indices and discriminant analysis, *Crop Science*, 35, 1400–1405, 1995.
3. Hendry, G.A.F., Houghton, J.D., and Brown, S.B., The degradation of chlorophyll—A biological enigma, *New Phytologist*, 107, 255–302, 1987.

4. Carter, G.A. and Knapp, A.K., Leaf optical properties in higher plants: Linking spectral characteristics to stress and chlorophyll concentration, *American Journal of Botany*, 88, 677–684, 1991.

5. Merzlyak, M.N., Gitelson, A.A., Chivkunova, O.B., and Rakitin, V.Y., Non-destructive optical detection of leaf senescence and fruit ripening, *Physiologia Plantarum*, 106, 135–141, 1999.

6. Lichtenthaler, H.K., Chlorophyll and carotenoids: Pigments of photosynthetic biomembranes, *Methods Enzymology*, 148, 331–382, 1987.

7. Edge, R.D., McGarvey, J., and Truscotte, T.G., The carotenoids as anti-oxidants, *Journal of Photochemistry and Photobiology (B)*, 41, 189–200, 1997.

8. Biswall B., Carotenoid catabolism during leaf senescence and its control by light, *Journal of Photochemistry and Photobiology (B)*, 30, 3–14, 1995.

9. Merzlyak, M.N. and Gitelson, A.A., Why and what for the leaves are yellow in autumn? On the interpretation of optical spectra of senescing leaves (*Acer platanoides* L.), *Journal of Plant Physiology*, 145, 315–320, 1995.

10. Penuelas, J. and Filella, I., Visible and near-infrared reflectance techniques for diagnosing plant physiological status, *Trends in Plant Science*, 3, 151–156, 1998.

11. Gamon, J.A., Field, C.B., Bilger, W., Björkman, O., Fredeen, A.L., and Peñuelas, J., Remote sensing of the xanthophyll cycle and chlorophyll fluorescence in sunflower leaves and canopies, *Oecologia*, 85, 1–7, 1990.

12. Gamon, J.A. and Surfus, J.S., Assessing leaf pigment content and activity with a reflectometer, *New Phytologist*, 143, 105–117, 1999.

13. Young, A. and Britton, G., Carotenoids and stress, In: *Stress Responses in Plants: Adaptation and Acclimation Mechanisms*, R.G. Alscher and J.R. Cumming (Eds.), Wiley-Liss, New York, pp. 87–112, 1990.

14. Merzlyak, M.N. and Chivkunova, O.B., Light stress induced pigment changes and evidence for anthocyanin photoprotection in apple fruit, *Journal of Photochemistry and Photobiology (B)*, 55, 154–162, 2000.

15. Gould, K., Kevin, D., and Winefield, C. (Eds.), *Anthocyanins: Biosynthesis, Functions, and Applications*, Springer, New York, p. 330, 2008.

16. Merzlyak, M.N., Chivkunova, O.B., Solovchenko, A.E., and Naqvi, K.R., Light absorption by anthocyanins in juvenile, stressed and senescing leaves, *Journal of Experimental Botany*, 59, 3903–3911, 2008.

17. Buschmann, C. and Nagel, E., In vivo spectroscopy and internal optics of leaves as basis for remote-sensing of vegetation, *International Journal of Remote Sensing*, 14, 711–722, 1993.

18. Gitelson, A.A. and Merzlyak, M.N., Quantitative estimation of chlorophyll-*a* using reflectance spectra: Experiments with autumn chestnut and maple leaves, *Journal of Photochemistry and Photobiology B: Biology*, 22, 247–252, 1994.

19. Gitelson, A.A. and Merzlyak, M.N., Spectral reflectance changes associated with autumn senescence of *Aesculus hippocastanum* L. and *Acer platanoides* L. leaves. Spectral features and relation to chlorophyll estimation, *Journal of Plant Physiology*, 143, 286–292, 1994.

20. Markwell, J., Osterman, J.C., and Mitchell, J.L., Calibration of the Minolta SPAD-502 leaf chlorophyll meter, *Photosynthesis Research*, 46, 467–472, 1995.

21. Gitelson, A.A., Zur, Y., Chivkunova, O.B., and Merzlyak, M.N., Assessing carotenoid content in plant leaves with reflectance spectroscopy, *Photochemistry and Photobiology*, 75, 272–281, 2002.

22. Gitelson, A.A., Merzlyak, M.N., and Chivkunova, O.B., Optical properties and non-destructive estimation of anthocyanin content in plant leaves, *Photochemistry and Photobiology*, 74, 38–45, 2001.

23. Aoki, M., Yabuki, K., Totsuka, T., and Nishida, M., Remote sensing of chlorophyll content of leaf (I) Effective spectral reflection characteristics of leaf for the evaluation of chlorophyll content in leaves of Dicotyledons, *Environmental Control in Biology*, 24, 21–26, 1986.

24. Gitelson, A.A. and Merzlyak, M., Signature analysis of leaf reflectance spectra: Algorithm development for remote sensing of chlorophyll, *Journal of Plant Physiology*, 148, 495–500, 1996.

25. Gitelson, A.A. and Merzlyak, M.N., Remote estimation of chlorophyll content in higher plant leaves, *International Journal of Remote Sensing*, 18, 2691–2697, 1997.

26. Gitelson, A.A., Merzlyak, M.N., and Lichtenthaler, H., Detection of red edge position and chlorophyll content by reflectance measurements near 700 nm, *Journal of Plant Physiology*, 148, 501–508, 1996.

27. Blackburn, G.A., Quantifying chlorophylls and carotenoids at leaf and canopy scales: An evaluation of some hyperspectral approaches, *Remote Sensing of Environment*, 66, 273–285, 1998.

28. Datt, B., Remote sensing of chlorophyll *a*, chlorophyll *b*, chlorophyll *a+b*, and total carotenoid content in Eucalyptus leaves, *Remote Sensing of Environment*, 66, 111–121, 1998.

29. Richardson, A.D., Duigan, S.P., and Berlyn, G.P., An evaluation of noninvasive methods to estimate foliar chlorophyll content, *New Phytologist*, 153, 185–194, 2002.

30. Jacquemoud, S. and Baret, F., PROSPECT: A model of leaf optical properties, *Remote Sensing of Environment*, 34, 75–91, 1990.

31. Maier, S.W., Lüdeker, W., and Günther, K.P., SLOP: A revised version of the stochastic model for leaf optical properties, *Remote Sensing of Environment*, 68, 273–280, 1999.

32. Feret, J.B., François, C., Asner, G.P., Gitelson, A.A., Martin, R.E., Bidel, L.P.R., Ustin, S.L., le Maire, G., and Jacquemoud, S., PROSPECT-4 and 5: Advances in the leaf optical properties model separating photosynthetic pigments, *Remote Sensing of Environment*, 112, 3030–3043, 2008.

33. Horler, D.N., Dockray, M., and Barber, J., The red edge of plant leaf reflectance, *International Journal of Remote Sensing*, 4, 273–288, 1983.

34. Vogelmann, J.E., Rock, B.N., and Moss, D.M., Red edge spectral measurements from sugar maple leaves, *International Journal of Remote Sensing*, 14, 1563–1575, 1993.

35. Carter, G.A., Ratios of leaf reflectances in narrow wavebands as indicators of plant stress, *International Journal of Remote Sensing*, 15, 3, 697–736, 1994.

36. Sims, D.A. and Gamon, J.A., Relationships between leaf pigment content and spectral reflectance across a wide range of species, leaf structures, and developmental stages, *Remote Sensing of Environment*, 81, 337–354, 2002.

37. Gitelson, A.A., Viña, A., Rundquist, D.C., Ciganda, V., and Arkebauer, T.J., Remote estimation of canopy chlorophyll content in crops, *Geophysical Research Letters*, 32, L08403, 2005. doi:10.1029/2005GL022688.

38. le Maire, G., Francois, C., and Dufrene, E., Towards universal broad leaf chlorophyll indices using PROSPECT simulated database and hyperspectral reflectance measurements, *Remote Sensing of Environment*, 89, 1–28, 2004.

39. Ustin, S.L., Gitelson, A.A., Jacquemoud, S., Schaepman, M., Asner, G.P., Gamon, J.A., and Zarco-Tejada, P., Retrieval of foliar information about plant pigment systems from high resolution spectroscopy, *Remote Sensing of Environment*, 113, S67–S77, 2009. doi:10.1016/j.rse.2008.10.019.

40. Hatfield, J.L., Gitelson, A.A., Schepers, J.S., and Walthall, C.L., Application of spectral remote sensing for agronomic decisions, *Agronomy Journal*, 100, S-117–S-131, 2008. doi:10.2134/agronj2006.0370c.

41. Zur, Y., Gitelson, A.A., Chivkunova, O.B., and Merzlyak, M.N., The spectral contribution of carotenoids to light absorption and reflectance in green leaves, In: *Proceedings of the Second International Conference on Geospatial Information in Agriculture and Forestry*, Lake Buena Vista, FL, Vol. 2, pp. II-17–II-23, 2000.

42. Gamon, J.A., Peñuelas, J., and Field, C.B., A narrow waveband spectral index that tracks diurnal changes in photosynthetic efficiency, *Remote Sensing of Environment*, 41, 35–44, 1992.

43. Chappelle, E.W., Kim, M.S., and McMurtrey, J.E., Ratio analysis of reflectance spectra (RARS): An algorithm for the remote estimation of the concentrations of chlorophyll A, chlorophyll B, and carotenoids in soybean leaves, *Remote Sensing of Environment*, 39, 239–247, 1992.

44. Peñuelas, J., Baret, F., and Filella, I., Semi-empirical indices to assess carotenoids/chlorophyll a ratio from leaf spectral reflectance, *Photosynthetica*, 31, 221–230, 1995.

45. Van den Berg, A.K. and Perkins, T.D., Nondestructive estimation of anthocyanin content in autumn sugar maple leaves, *HortScience*, 40, 685–686, 2005.

46. Heath, O.V.S., *The Physiological Aspects of Photosynthesis*, Stanford University Press, Stanford, CA, 1969.

47. Kumar, R. and Silva, L., Light ray tracing through a leaf cross section, *Applied Optics*, 12, 12, 2950–2954, 1973.

48. Fukshansky, L., Optical properties of plant tissue, In: *Plants and the Daylight Spectrum*, H. Smith (Ed.), Springer, Berlin, Germany, pp. 253–303, 1981.

49. Vogelmann, T.C., Plant tissue optics, *Annual Review of Plant Physiology Plant Molecular Biology*, 44, 231–251, 1993.

50. Gausman, H.W., Evaluation of factors causing reflectance differences between sun and shade leaves, *Remote Sensing of Environment*, 15, 177–181, 1984.

51. Gausman, H.W. and Allen, W.A., Optical parameters of leaves of 30 plant species, *Plant Physiology*, 52, 57–62, 1973.

52. Moss, R.A. and Loomis, W.E., Absorption spectra of leaves. I. The visible spectrum, *Plant Physiology*, 27, 370–391, 1952.

53. Fukshansky, L.A., Remisowsky, A.M., McClendon, J., Ritterbusch, A., Richter, T., and Mohr, H., Absorption spectra of leaves corrected for scattering and distributional error: A radiative transfer and absorption statistics treatment, *Photochemistry and Photobiology*, 57, 538–555, 1993.

54. Gitelson, A.A., Gritz, U., and Merzlyak, M.N., Relationships between leaf chlorophyll content and spectral reflectance and algorithms for non-destructive chlorophyll assessment in higher plant leaves, *Journal of Plant Physiology*, 160, 271–282, 2003.

55. Lichtenthaler, H.K., Gitelson, A.A., and Lang, M., Non-destructive determination of chlorophyll content of leaves of a green and an aurea mutant of tobacco by reflectance measurements, *Journal of Plant Physiology*, 148, 483–493, 1996.

56. Kortum, G., *Reflectance Spectroscopy: Principles, Methods, Applications*, Springer-Verlag New York, Inc., New York, 1969.

57. Ciganda, V., Gitelson, A.A., and Schepers, J., Non-destructive determination of maize leaf and canopy chlorophyll content, *Journal of Plant Physiology*, 166, 157–167, 2009.

58. Steele, M.R., Gitelson A.A., and Rundquist, D.C., Non-destructive estimation of leaf chlorophyll content in grapes, *American Journal of Enology and Viticulture*, 59, 299–305, 2008.

59. Steele, M.R., Gitelson, A.A., and Rundquist, D.C., A Comparison of two techniques for non-destructive measurement of chlorophyll content in grapevine leaves, *Agronomy Journal*, 100, 779–782, 2008. doi:10.2134/agronj2007.0244.

60. Imanishi, J., Nakayama, A., Suzuki, Y., Imanishi, A., Ueda, N., Morimoto, Y., and Yoneda, M., Nondestructive determination of leaf chlorophyll content in two flowering cherries using reflectance and absorptance spectra, *Landscape and Ecological Engineering*, 6, 219–234, 2010.

61. Gitelson, A.A., Keydan, G.P., and Merzlyak, M.N., Three-band model for noninvasive estimation of chlorophyll, carotenoids, and anthocyanin contents in higher plant leaves, *Geophysical Research Letters*, 33, L11402, 2006.

62. Jordan, C.F., Derivation of leaf area index from quality of light on the forest floor, *Ecology*, 50, 663–666, 1969.

63. Rouse, J.W., Haas R.H. Jr., Schell, J.A., and Deering, D.W., Monitoring vegetation systems in the Great Plains with ERTS, NASA SP-351, *Third ERTS-1 Symposium*, NASA, Washington, DC, Vol. 1, pp. 309–317, 1974.

64. Huete, A.R., Liu, H.Q., Batchily, K., and van Leeuwn, W.J.D., A comparison of vegetation indices over a global set of TM images for EOS-MODIS, *Remote Sensing of Environment*, 59, 440–451, 1997.

65. Jiang, Z., Huete, A.R., Didan, K., and Miura, T., Development of a two-band enhanced vegetation index without a blue band, *Remote Sensing of Environment*, 112, 3833–3845, 2008.

66. Datt, B., A new reflectance index for remote sensing of chlorophyll content in higher plants: Tests using Eucalyptus leaves, *Journal of Plant Physiology*, 154, 30–36, 1999.

67. Gitelson, A.A., Chivkunova, O.B., and Merzlyak, M.N., Non-destructive estimation of anthocyanins and chlorophylls in anthocyanic leaves, *American Journal of Botany*, 96, 10, 1861–1868, 2009.

68. Steele, M.R., Gitelson, A.A., and Rundquist, D.C., Non-destructive estimation of anthocyanin content in grapevine leaves, *American Journal of Enology and Viticulture*, 60, 87–92, 2009.

69. Solovchenko, A.E., Chivkunova, O.B., Gitelson, A.A., and Merzlyak, M.N., Non-destructive estimation pigment content, ripening, quality and damage in apple fruit with spectral reflection in the visible range, *Global Science Books, Fresh Produce*, 4 (Special Issue), 91–102, 2010.

7 Forest Leaf Chlorophyll Study Using Hyperspectral Remote Sensing

Yongqin Zhang

CONTENTS

7.1 INTRODUCTION*

A plant's physiological state is governed by its biochemical constituents, including photosynthetic and other enzyme systems, structural and nonstructural carbohydrates, chlorophyll and associated light harvesting complexes, and photoprotective and ancillary pigments. Many biochemical processes, such as photosynthesis, net primary production, and decomposition, are related to the content of biochemicals in leaves [1–3]. Of these biochemicals, leaf chlorophyll content stands out as being both sensitive to environmental conditions and having a very strong influence on leaf optical properties and canopy albedo [4–6]. All green leaves have major absorption features in the 400–700 nm range caused by electron transitions in chlorophyll and carotenoid pigments [7]. Most green vegetation shows absorption peaks near 420, 490, and 670 nm due to the strong absorption peaks of chlorophyll *a* and *b*. The absorption of other pigments in the leaf, such as carotenes and xanthophylls, is usually obscured by the absorption of chlorophyll *a* and *b*. Differences in leaf and canopy reflectance between healthy and stressed vegetation due to changes in chlorophyll levels have been detected in the green peaks and along the red edge (701–740 nm) [8–12]. The changes of red-edge position and slope are associated with vegetation stress [9,13–15]. Canopy reflectance in the green and far-red regions is also sensitive to variations in chlorophyll concentration and can act as an indicator of vegetation stress [10]. Leaf chlorophyll content is of great use for forest health

---

* ©Copyright permission from Canadian Aeronautics and Space Institute and Elsevier.

status evaluation and sustainable forest management [16,17]. Leaf chlorophyll content also serves as an input to photosynthesis and carbon cycle models. Changes in leaf optical properties and chlorophyll content, including responses to rising atmospheric CO_2 and other global change variables, may have important implications to climate forcing as well [18].

Quantitative estimates of leaf chlorophyll content may thus provide a useful indicator of important physiological processes in vegetation that can be readily assessed via hyperspectral remote sensing. The cutting-edge hyperspectral remote sensing technology provides simultaneous acquisition of information in narrow but contiguous spectral bands. Hyperspectral data are almost spectrally "continuous," which are sufficient to detect subtle absorption features in foliar spectra and to study the correlations of vegetation's minor absorption features to biochemical parameters. They are particularly useful for estimating vegetation structure and biochemistry that are important for studying nutrient cycling, productivity, and vegetation stress and for ecosystem modeling [19–22].

In this chapter, methods for estimating leaf and forest canopy chlorophyll content using hyperspectral remote sensing are summarized, and the suitability and limitations of each method are analyzed. In particular, recent advances [23–25] of quantitative methods using physically based algorithms are presented in more detail. These descriptions follow a sequence of estimating broadleaf and needleleaf chlorophyll content from leaf-level hyperspectral measurements, retrieving leaf chlorophyll content of forest canopies from hyperspectral remote sensing imagery, and scaling of leaf-level chlorophyll estimation to forest canopy level. The applications of estimating forest chlorophyll content using hyperspectral remote sensing are discussed as well.

7.2   METHODS FOR ESTIMATING LEAF CHLOROPHYLL CONTENT

Leaf optical properties (and thus leaf chemistry) and canopy structure determine the remote sensing signals originating from vegetated surfaces. Deriving leaf optical and canopy structural properties are the two main domains of remote sensing of terrestrial ecosystems.

Leaf optical properties depend on leaf structure, leaf biochemical composition, distribution of leaf biochemical constituents, and the complex refraction index of these constituents [26–31]. Radiation scattering by a leaf is caused by optical inhomogeneities in the leaf surface, leaf thickness, cell size, and the internal cell structure. Variability in leaf optical properties is wavelength dependent. Spectrally continuous leaf-level hyperspectral measurements are especially useful to detect subtle features and changes in the leaf optical spectra that correlate with leaf chlorophyll content. Accurate estimation of leaf chlorophyll content from leaf optical spectra is an essential step to derive canopy chlorophyll content from remote-sensed imagery.

7.2.1   EMPIRICAL METHOD FOR LEAF CHLOROPHYLL CONTENT ESTIMATION

At the leaf level, empirical relationships between spectral indices and chlorophyll content measurements have been widely exploited for estimating leaf chlorophyll content. They have been proven effective for nondestructive estimations of leaf chlorophyll content from field and laboratory measurements of leaf spectral reflectance [32–37].

Many spectral indices have been developed through identifying relationships between the leaf reflectance and chlorophyll content. le Maire et al. [38] summarized the chlorophyll spectral indices published until 2002. In the red and blue regions of the electromagnetic spectrum, chlorophylls have strong absorbance peaks. Spectral indices employ ratios of narrow bands within spectral ranges that are sensitive to chlorophylls to those not sensitive and/or related to some other control on reflectance. To avoid the saturation of indices under low chlorophyll content, reflectances near, instead of exactly at, the maximum absorption wavelengths are generally selected to develop spectral indices. This method solves the problem of overlapping absorption spectra of different pigments, the effects of leaf surface interactions, and leaf structure. The majority of chlorophyll indices are based on ratios of narrow bands in the visible and near-infrared region [39–41]. Some indices use ratios of narrow bands only in the visible [42], red-edge [9], or in the red-edge and near-infrared shoulder regions [34].

TABLE 7.1
**Estimates of Leaf Chlorophyll Content Using Empirical Indices
in Comparisons to the Laboratory Chlorophyll Measurements**

Spectral Index	Formula	RMSE ($\mu g/cm^2$)	R^2
mSR	$(R_{728} - R_{434})/(R_{720} - R_{434})$	3.94	0.875
mND	$(R_{728} - R_{720})/(R_{728} + R_{720} - 2 * R_{434})$	3.97	0.872
BmSR	$(\delta R_{722} - \delta R_{502})/(\delta R_{701} - \delta R_{502})$	3.98	0.872
NDI	$(R_{750} - R_{705})/(R_{750} + R_{705})$	4.11	0.864
DD	$(R_{749} - R_{720}) - (R_{701} - R_{672})$	4.15	0.861
BmND	$(\delta R_{722} - \delta R_{699})/(\delta R_{722} + \delta R_{699} - 2 * \delta R_{502})$	4.27	0.853

Some indices use three bands [38,43–45] and can differentiate chlorophyll *a* and *b* [46]. Generally, indices using three bands are applicable for leaf-level chlorophyll estimation. Multiple narrow bands, or whole spectrum through principal components transformation [47], factor analysis [48], artificial neural networks [49], and stepwise multiple regression [50,51] have attempted to quantify leaf and canopy chlorophyll concentrations.

le Maire et al. [38] concluded that at leaf-level, simple spectral indices give better estimations than indices related to the red-edge inflection point, derivative-based indices, or indices based on neural network analysis of empirical hyperspectral data. Zhang et al. [23] measured reflectance and transmittance spectral of 255 sugar maples (*Acer saccharum*) from 350 to 2500 nm at 1 nm interval using a portable field spectroradiometer FieldSpec Pro FR (Analytical Spectral Devices, Inc. Boulder, Colorado) attached via a fiber optic to the Li-Cor 1800 integrating sphere (Li-Cor 1800-12S, Li-COR, Inc., Lincoln, Nebraska). Chlorophyll *a* and *b* content of the 255 leaves were measured subsequently to the leaf spectral measurements. Using the leaf spectral and chlorophyll content measurements, Zhang et al. [23] analyzed a number of simple chlorophyll indices for estimating broadleaf chlorophyll content. These indices have previously been shown to produce low deviation from measurements of chlorophyll content. The indices include the modified simple ratio index (mSR) and the modified normalized difference index (mND) of Sims and Gamon [44], the double difference index (DD), the first derivative-based index (BmSR), and the first derivative-based index (BmND) of le Maire et al. [38], and the red-edge normalized difference index (NDI) of Gitelson and Merzlyak [34]. The performances of these indices for chlorophyll content estimations are listed in Table 7.1. Compared to the laboratory measurements of leaf chlorophyll content, these spectral indices estimated broadleaf chlorophyll content very well. Among the indices, the mSR produced the best estimation with a root mean square error (RMSE) of 3.94 $\mu g/cm^2$ and an R^2 of 0.875, and the mND performed nearly as well.

Spectral indices provide nondestructive, efficient, and sensitive measurements of leaf chlorophyll content from leaf spectral reflectance. They can serve as indicators of vegetation stress, senescence, and disease. However, they are generally suitable for broadleaves. They may perform poorly for small and narrow needleleaf plants. Spectral indices are also developed for some specific species. As the size, shape, surface, and internal structure of leaves may vary from species to species, the application of optical indices to other vegetation types or biomes needs to be reinvestigated. Efforts have been made to improve the robustness and generality of chlorophyll indices by testing over a range of species and physiological conditions. Nevertheless, when applied to a specific species, spectral indices need calibration.

7.2.2 Physically Based Model Inversion Method

Considerable progress has been made in physically based models that simulate leaf spectral characteristics based on interactions of incident radiation with foliar medium. These radiometric models take the underlying physics and the complexity of leaf biochemical properties into account and therefore are robust and applicable for different species. Four categories of models now exist, and they include the following: (1) Plate models that assume a leaf as one or several absorbing plates with rough surfaces giving rise to scattering of light [52], typical of which is PROSPECT [53]; (2) N-flux models consider a leaf as a horizontally homogeneous parallel medium with downwelling and upwelling irradiance and isotropic scattering [54,55]; (3) Ray tracing models describe the complexity of leaf internal structure with a detailed description of individual cells and their unique arrangement inside tissues [56–59]; and (4) Stochastic models such as the LEAFMOD simulate leaf optical properties by a Markov chain or basic radiative transfer (RT) equations [60–62].

Numerical inversion of leaf-level RT models, such as PROSPECT and LEAFMOD, has demonstrated success for predicting leaf chlorophyll content [53,61,63–65]. Numerical inversion techniques offer the potential of a generically superior approach to estimate leaf chlorophyll content from hyperspectral data than spectral indices and other approaches that are based on empirical calibrations. Of all the leaf optical models, PROSPECT is a simple but effective model for leaf biochemical content retrieval. The PROSPECT model has been widely validated with various broadleaf species.

7.2.2.1 Modeling Method for Broadleaf Chlorophyll Content Estimation

The PROSPECT model assumes that a leaf is a stack of L identical elementary layers separated by $L - 1$ air spaces. The number of layers mimics the scattering within the leaf. Scattering is described by the refractive index (n) of leaf materials and by a parameter characterizing the leaf mesophyll structure (L). Layers are defined by their refractive index and absorption coefficient K_i. Absorption is the linear summation of the contents of the constitute chemicals and the corresponding specific absorption coefficients.

$$K(\lambda) = K_e(\lambda) + \frac{\sum K_i(\lambda)C_i}{L} \tag{7.1}$$

where
λ is the wavelength
$K_e(\lambda)$ is the absorption coefficient of elementary albino and dry layer
C_i is the content of layer constituent i (chlorophyll$_{a+b}$, water and dry matter) per unit area
$K_i(\lambda)$ is the corresponding specific absorption coefficients of the constituent i
L is the leaf structure parameter, which is the number of compact layers specifying the average number of air/cell walls interfaces within the mesophyll

Through numerical iteration, the chemical contents can be derived from the leaf spectra. First, an initial guess of the structure parameter L and the concentration of three constituents are input in the forward model to calculate the absorption coefficient $K(\lambda)$ and the hemispherical reflectance and transmittance. The estimated hemispherical reflectance and transmittance are then compared with the measured leaf reflectance and transmittance. Using an optimization algorithm, the i constituents can be numerically iterated by minimizing the merit function [66]:

$$\Delta = \sum_{\lambda} \{[R_{mes}(\lambda) - R_{mod}(\lambda)]^2 + [T_{mes}(\lambda) - T_{mod}(\lambda)]^2\} \tag{7.2}$$

where
R_{mes} and T_{mes} are the measured reflectance and transmittance, respectively
R_{mod} and T_{mod} are the estimated reflectance and transmittance from the model

Zhang et al. [23] found that the original PROSPECT model performed well for the overstory leaf samples collected in summer (July and August) that have high chlorophyll content (Figure 7.1a). However, the variations of leaf chlorophyll content across the season and canopy height were not well captured. Specifically, understory leaf samples and samples collected in the early (in May and early June) and

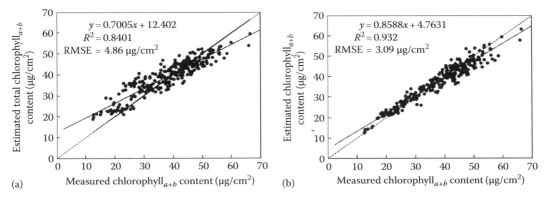

FIGURE 7.1 Comparison of leaf chlorophyll$_{a+b}$ content from measurements and from estimates from (a) the original PROSPECT model and (b) the PROSPECT model taking the thickness of the leaf into consideration. The dotted lines in the figure are 1:1 lines.

late growing season (in middle to late September) have low leaf chlorophyll content. The model predictions tend to overestimate leaf chlorophyll content. The specific absorption coefficients of the biochemical constituents in the PROSPECT were calibrated using the data collected in summer [21,67]. Leaf structure, chlorophyll content, and optical properties vary both seasonally and with respect to canopy position. Assuming that leaf structural variables were across-season averages, the specific absorption coefficient of chlorophyll in summer, or at upper canopies would tend to be low. Using the specific absorption coefficient calibrated in summer (overstory) would thus result in overestimation of leaf chlorophyll content for leaves in other seasons (understory).

To simulate the seasonal and canopy height variations of leaf chlorophyll content, Zhang et al. [23] defined and incorporated a leaf thickness factor in the PROSPECT model to consider the influence of seasonal and canopy-height variability in leaf structure on light absorption. The specific absorption coefficients of all constituents were adjusted using the same leaf thickness factor:

$$K(\lambda) = \frac{K_e(\lambda)}{T} + \frac{\sum K_i(\lambda)C_i}{L * T} \tag{7.3}$$

where T is the thickness factor, which is the ratio of the leaf thickness in summer to that in other growing seasons, or the thickness ratio of overstory leaves to understory leaves. Based on the leaf thickness measurements for sugar maple leaves, the thickness factor was calculated for overstory and understory leaves through the whole season (Table 7.2).

Leaf chlorophyll content was estimated with leaf reflectance and transmittance measurements from 255 sugar maple leaves as input the PROSPECT with consideration of leaf thickness. After incorporating the leaf thickness factor in the model for these samples, the estimation was improved, compared with the estimation from the original PROSPECT, from RMSE = 4.86 μg/cm^2 and R^2 = 0.84 to RMSE = 3.09 μg/cm^2 and R^2 = 0.93 (Figure 7.1b). With the additional input of leaf thickness as a surrogate to capture the seasonal and location variations in leaf structure and

TABLE 7.2

Leaf Thickness Factor Used in the Model for Overstory and Understory Leaves through the Growing Season

	May 27	June 10	July 1	July 27	August 16	August 30	September 11	September 30
Overstory leaves	1.45	1.07	1.02	1.00	1.00	1.00	1.05	1.30
Understory leaves	1.55	1.25	1.20	1.18	1.05	1.05	1.12	1.20

nonchlorophyll light absorption, the model performed better in estimating the low chlorophyll content from understory leaves and overstory leaves in the early and late growing seasons, though there remains some bias with low values of leaf chlorophyll content being slightly, but systematically overestimated by the model inversion.

Féret et al. [68] recalibrate the physical and optical constants in the PROSPECT model using comprehensive datasets compassing hundreds of leaves from various species. The chlorophyll content estimates were improved for leaves with low chlorophyll content.

7.2.2.2 Modeling Method for Needleleaf Chlorophyll Content Estimation

The relative small size, narrow width, and irregular shape of needleleaves impose challenges for both the measurement of leaf optical properties and estimation of chlorophyll content. Leaf-level models mentioned in Section 7.2.2.1 are intended for broadleaf species. An RT model named LIBERTY is developed for estimating leaf optical spectra of Jack pine needles [69]. However, comparative studies show that PROSPECT produces better estimates of needleleaf chlorophyll content [70].

Needleleaves are usually narrow and thick. Needle structures with adaxial and abaxial surfaces are neither parallel nor necessarily flat planes. The validity of the model for needleleaves needs investigation since the assumption of infinite plane layers as for broadleaves is violated. Figure 7.2 schematically shows structural effects of a broadleaf and a needleleaf on light transfer. By assumption, a broadleaf is composed of horizontally infinite layers. As the thickness of a broadleaf is much smaller than the leaf width, the incident light leaves the outer layers as reflected or transmitted light after multiple interactions with the internal layers. Light transmitted through leaf edges is negligible compared to the amount that exits the first (uppermost) and last (lowermost) layer. For thick

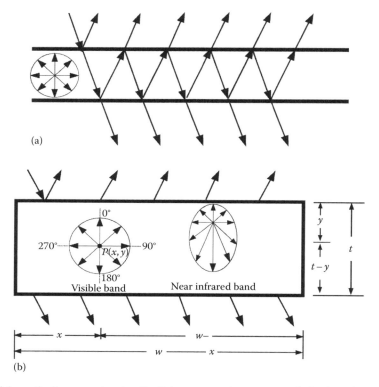

FIGURE 7.2 Schematic diagrams showing the light transport in a monocotyledon broad leaf (a) based on the assumption of the PROSPECT model and (b) in needle. For a broad leaf, the light transport in the leaf interior is assumed to be isotropic. For a needle, a phase function is introduced to characterize the directional scattering in the visible and near-infrared regions.

and narrow black spruce needles (Figure 7.2b), needleleaf length can be assumed to be infinite relative to the leaf thickness and width, but leaf width can significantly affect the measurements of leaf reflectance and transmittance. For example, the needle width of black spruce, which is the dominant species in the boreal ecosystem, is only around two times of the leaf thickness. Therefore, the amount of light that escapes from needle edges, that is, along the direction of leaf width, could be large. This portion of light "loss" is not included in the PROSPECT model for the estimation of reflectance and transmittance.

Moorthy et al. [70] adapted PROSPECT for estimating chlorophyll content of jack pine needles by using a simple geometric correction of needle form. With the consideration of the needle morphology as an equivalent flat plate, the accuracy of estimating chlorophyll content in jack pine needles can be improved, although a significant departure remained in the ability of PROSPECT to quantitatively predict the measured optical properties of the needles as a function of pigment content.

The edge effects of needleleaves on light transfer and measured spectra were considered by incorporating two morphology factors, needleleaf thickness, and width in PROSPECT [24]. Firstly, laboratory experiments were conducted to investigate the effects of needle morphology and needle-holding device on leaf spectral measurements. The effects induced by the leaf morphology and size are evident in leaf optical properties, which demonstrate that the PROSPECT model, as designed for broadleaf species, is not adequate to accurately represent the link between the optical properties and the pigment content of needleleaves. Secondly, to compensate the portion of light "loss" to needleleaf edge that are not considered in PROSPECT simulation of leaf reflectance and transmittance, two leaf biophysical parameters, needle width and thickness, were introduced in the model to take into account the effects of leaf morphology on chlorophyll content retrieval. Light scattering is assumed to be non-isotropic in all directions. The Henyey-Greenstein phase function, which characterizes the angular distribution of scattered light in layered materials such as biological tissues of leaves [71], was used to describe the directional scattering from the incoming direction to the outgoing direction of the light:

$$p(\cos\theta) = \frac{1-g^2}{4(1+g^2-2g\cos\theta)^{3/2}} \cdot \frac{1}{4\pi} \tag{7.4}$$

where
  θ is the angle between the incoming and the outgoing direction
  g is the average cosine of the scattered angle

The phase function is characterized by the parameter g. Light scattering is symmetric about the incident direction. For different values of g, the phase function indicates different scattering:

$$\begin{cases} g = 0, & \text{isotropic scattering} \\ g > 0, & \text{predominantly forward scattering} \\ g < 0, & \text{predominantly backward scattering} \end{cases}$$

The value of g can be determined through an optimization procedure. When the incoming light penetrates through the upper leaf surface, it is scattered at point $P(x,y)$ (Figure 7.2b). Leaf width is denoted as w, leaf thickness as t, and the light scattered toward the leaf width (along 90° and 270°) and thickness (along 0° and 180°) direction as $L_{width}(w)$ and $L_{width}(t)$, respectively. A_g is the global absorption coefficient by chlorophyll a and b, liquid water, and biochemicals. Based on Beer's law, $L_{width}(w)$ and $L_{width}(t)$ can be calculated:

$$L_{width}(x) = \exp[-A_g P(\cos 90°)x] + \exp[-A_g P(\cos 270°)(w-x)] \tag{7.5}$$

$$L_{thick}(y) = \exp[-A_g P(\cos 0°)y] + \exp[-A_g P(\cos 180°)(t-y)] \tag{7.6}$$

Then the total light penetrating through the leaf width and thickness are as follows:

$$L_{width} = \frac{1}{w} \int_0^w \{\exp[-A_g P(\cos 90°)x] + \exp[-A_g P(\cos 270°)(w-x)]\} dx \tag{7.7}$$

$$L_{thick} = \frac{1}{t} \int_0^t \{\exp[-A_g P(\cos 0°)y] + \exp[-A_g P(\cos 180°)(t-y)]\} dy \tag{7.8}$$

These two portions of light are not accounted for in the simulation of leaf reflectance and transmittance by PROSPECT. With the consideration of these two portions of light transfer, the corresponding reflectance R_{width} and transmittance T_{width} losses from the needle width direction can be simply inferred from the measurements:

$$R_{width} = \frac{L_{width}}{1/t \int_0^t \exp[-A_g P(\cos 0°)y] dy} \times R_{mes} \tag{7.9}$$

$$T_{width} = \frac{L_{thick}}{1/t \int_0^t \exp[-A_g P(\cos 180°)y] dy} \times T_{mes} \tag{7.10}$$

Then the following adjustments were given to estimate needle reflectance and transmittance:

$$R_{mod} = R - R_{width} \tag{7.11}$$

$$T_{mod} = T - T_{width} \tag{7.12}$$

where
R_{mod} and T_{mod} are the needle reflectance and transmittance after modifications
R and T are the reflectance and transmittance estimated from the PROSPECT model

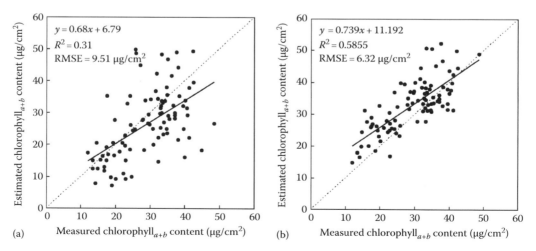

(a)   Measured chlorophyll$_{a+b}$ content (μg/cm^2)   (b)   Measured chlorophyll$_{a+b}$ content (μg/cm^2)

FIGURE 7.3 Relationship between the measured and estimated chlorophyll contents using (a) the original PROSPECT inversion model and (b) the modified PROSPECT model, which takes into account the effects of needle thickness and width. The dotted line is the 1:1 line.

With the modifications to the model for reflectance and transmittance estimation, the chlorophyll content inversion were noticeably improved for small and short black spruce needles ($R^2 = 0.59$ and RMSE = $6.32\,\mu g/cm^2$) in comparison to the original model ($R^2 = 0.31$ and RMSE = $9.51\,\mu g/cm^2$), which tends to underestimate chlorophyll$_{a+b}$ content of black spruce needles (Figure 7.3).

The modified model captures the variation of needle chlorophyll content well from different sites, age classes, and branch orientations. As black spruce needles are significantly small and difficult to measure, the improvements achieved by this new approach are significant. Nevertheless, this improvement in the PROSPECT applicability to needles did not appreciably reduce the departure of the slope of the estimated versus measured total needle chlorophyll content from the 1:1 line. This remaining issue calls for further model refinements specific to the needle internal structure.

7.3 METHODS FOR ESTIMATING FOREST CANOPY CHLOROPHYLL CONTENT

The biochemical composition of the canopy depends strongly on leaf optical properties as well as on canopy structure. Leaf optical properties contribute directly to canopy-level reflectance. In closed canopies with high leaf area index (LAI), leaf optical properties strongly influence canopy-level reflectance [72]. The structure of forest canopies, such as the vertical and horizontal distribution, orientation, and density of foliage, determines light attenuation and thus influences canopy reflectance, distribution of photosynthesis, respiration, transpiration, and nutrient cycling in the canopy [73]. Photon scattering by leaves, stems, and soils also influences the radiation regime of forest canopies. LAI, leaf angle distribution (LAD), and foliage clumping represent the effects of leaf, stem, and soil optical properties on canopy reflectance and thus are dominant controllers of the relationship between leaf and canopy spectral characteristics [4]. The canopy reflectance is also a function of wavelength, soil reflectance, solar illumination conditions, and viewing geometry of the remote sensing instrument [74]. From the leaf to the canopy scale, the complicated perturbations of canopy structure to light transfer need to be carefully considered. For complex forest canopies, it is impossible to derive leaf biochemical parameters from above-canopy hyperspectral measurements without consideration of canopy structural effects.

7.3.1 Empirical and Semiempirical Methods

Statistical estimation of canopy-level biochemical contents is performed through different methods. The simplest way is to directly develop statistical relationships between ground-measured biochemical contents and canopy reflectance measured in the field or by airborne or satellite sensors [75–78]. Alternatively, some leaf-level relationships between optical indices and pigment content are directly applied to the canopy-level estimation [79–81].

Two chlorophyll spectral indices, the transformed chlorophyll absorption in reflectance index/optimized soil-adjusted vegetation index (TCARI/OSVAI) [82] and the MERIS terrestrial chlorophyll index (MTCI) [83], were developed for chlorophyll content estimation. These two spectral indices were based on model spectra, field spectra, and airborne/spaceborne remote sensing data, so may be viewed as semiempirical indices sensitive to chlorophyll content. TCARI/OSVAI minimizes the effects of varying LAI and soil background using narrowband hyperspectral remote sensing data and has been reported to be suitable for crop, as well as open and structured canopies [82,84,85]. MTCI relates a measure of the red-edge position with canopy chlorophyll content [83].

The application of these two indices to forest chlorophyll content estimation was evaluated for nine black spruce sites that have open canopies [25]. It was found that TCARI/OSAVI correlates best with leaf chlorophyll a and b. But with its relative insensitivity to LAI variations, it shows only poor relationship with canopy chlorophyll a and b content (Table 7.3). For open forest canopies, when the forest background has a green and variable vegetation signature instead of bare soil, the relationship of TCARI/OSAVI to canopy chlorophyll a and b content is not as tight as those shown for crop canopies or open canopies [82,84,85]. MTCI shows a good relationship with canopy

TABLE 7.3
**Performance of TCARI/OSAVI and MTCI Index for Estimating Chlorophyll Content
at Leaf and Canopy Scales**

Index	Correlation to Leaf Chlorophyll Content	Correlation to Canopy Chlorophyll Content
TCARI/OSAVI	$y = -12.41\ln(x) + 8.5886$, $R^2 = 0.4025$	$y = -81.998\ln(x) - 48.634$, $R^2 = 0.1915$
MTCI	$y = 53.265x^2 - 204.51x + 226.05$, $R^2 = 0.2005$	$y = 151.29x - 184.35$, $R^2 = 0.6456$

y, leaf/canopy chlorophyll content; x, index.

chlorophyll$_{a+b}$ content, as it is based on a measure of the red-edge position, which is influenced by both leaf chlorophyll content and LAI. However, it shows no measurable correlation with leaf chlorophyll$_{a+b}$ content because of the confounding effect of LAI variations. For both indices, when applied to open boreal canopies, a strong influence of the vegetated understory remains as a perturbing parameter.

Confounding factors that influence the remotely sensed optical properties make it difficult to spatially and temporally extrapolate leaf-level relationships to the canopy level. Statistical relationships are often site- and species-specific and thus cannot be directly applied to other study sites since the canopy structure and viewing geometry may vary from different sites and species.

7.3.2 Physically Based Modeling Method

7.3.2.1 Modeling Method for Closed Forest Canopies

At the canopy level, physically based modeling approaches have been applied to estimate forest chlorophyll content. To derive forest biochemical parameters accurately from hyperspectral remote sensing imagery, one approach is to estimate leaf reflectance spectrum from canopy-level hyperspectral data. Retrieving leaf-level information from canopy-level measurements requires comprehensive consideration of the interactions of radiation with plant canopies. Canopy RT models or tracing models are often coupled with leaf RT models to quantify leaf chlorophyll content from canopy-level remote sensing data [64,70,84,86–88]. Studies using coupled leaf and canopy RT models attempt to understand the effects of controlling factors on leaf reflectance properties at the canopy scale [88]. Coupled modelling methods are also used to scale up the leaf-level relationship between optical indices and pigment content. This method refines the development of spectral indices that are insensitive to factors such as canopy structure, illumination geometry, and background reflectance for estimating foliar chlorophyll concentrations from canopy reflectance [64,89,90].

Canopy RT models often assume that a canopy is composed of horizontal, homogeneous vegetation layers with infinite extent and Lambertian scatterers. Elements of the canopy are assumed to be randomly distributed in space as a turbid medium [91–93]. This type of models is suitable for close and dense canopies such as corn, soybean, and grass canopies where the foliage spatial distribution is close to randomness. As canopy architecture is not considered in canopy RT models, the models are valid only for closed canopy with high LAI (LAI > 3) [94]. Zarco-Tejada et al. [64] estimate broadleaf chlorophyll content through coupled leaf- and canopy-level RT models using the red-edge index (R750/R710) as a merit function to minimize the effects of forest canopy structure, shadows, and openings. This method was extended to coniferous forests for scaling leaf-level pigment estimation to canopy level using high spatial resolution airborne imagery to select only bright crown pixels in the scene for analysis [70,95]. Promising results for leaf chlorophyll content retrieval were obtained using this method. The bright crown methodology is successful in the case where a pixel is completely occupied by sunlit foliage and the shadow effects are small. However, remote sensing pixels, even at submeter resolutions, generally contain both sunlit and shaded fractions. The structural effects imposed by the open forest canopies are not tackled using this method.

7.3.2.2 Modeling Methods for Open Forest Canopies

Open forest canopies, especially heterogeneous, open, and clumped forest canopies, present a big challenge for the retrieval of leaf biochemical parameters. This is due to the compound effects of distinct spatial distribution of trees, the geometry of tree crowns, the structure of tree elements, forest background, and leaf optical properties on canopy reflectance. Efforts have been made to deal with canopy structural effects on the retrieval of forest biochemical parameters. A methodology that is applicable to not only closed but also open canopies is highly desirable.

Models based on radiosity [96,97] and ray tracing [98,99] can simulate the complexity of multiple scattering (MS), but simplifications of the mathematical and canopy architectural descriptions are inevitable due to computational limitations [99,100]. Geometrical optical (GO) models use geometrical shapes, such as cones for conifers and spheres or spheroids for deciduous trees, to simulate the angular and spatial distribution patterns of reflected solar radiance from forests [101–103]. GO models combine the advantages of simplicity, easy implementation, and capability of simulating the effects of canopy structure on the single and MS processes [104]. When considering the complex canopy architectural conditions, geometrical optical-radiative transfer (GO-RT) models are a good solution to solve the MS issues, in which geometrical optics are used to describe the shadowing (the first order scattering) effects, and RT theories are adopted to estimate the second and higher order scattering effects.

Zhang et al. [25,105] developed a look-up-table (LUT) approach to separate structural effects of canopy from leaf optical properties for deriving individual leaf reflectance spectra, and thereby leaf chlorophyll content from hyperspectral remote sensing imagery using a combination of GO model 4-scale and leaf RT model PROSPECT. The estimated leaf chlorophyll content was further scaled up to the landscape scale using forest structural parameter LAI. The method treats deciduous and coniferous trees separately, with the geometrical shapes of cones for conifers and spheres or spheroids for deciduous trees, to simulate the angular and spatial distribution patterns of reflected solar radiance from forests. The impacts of forest structural parameters, LAI, LAD, and clumping index, on canopy reflectance were taken into account. The method is suitable for both closed and open forest canopies.

Figure 7.4 shows the procedures of the forest chlorophyll content estimation and mapping using this method. There are four main procedures:

1. Canopy model forward simulation for canopy reflectance estimation. The forward modeling takes into account spectral scattering/absorbing properties of canopy components (leaf, forest background, canopy structure), canopy architecture, and directions of illumination and view. Forest canopy reflectance is calculated as a linear combination of four contributing components:

$$R = R_T P_T + R_{ZT} P_{ZT} + R_G P_G + R_{ZG} P_{ZG} \qquad (7.13)$$

where
  R is the canopy reflectance
  R_T, R_{ZT}, R_G, and R_{ZG} are the reflectivities of sunlit tree crown, shaded tree crown, sunlit background, and shaded background, respectively
  P_T, P_{ZT}, P_G, and P_{ZG} are the probabilities of a sensor viewing sunlit and shaded tree crowns, sunlit and shaded background, respectively

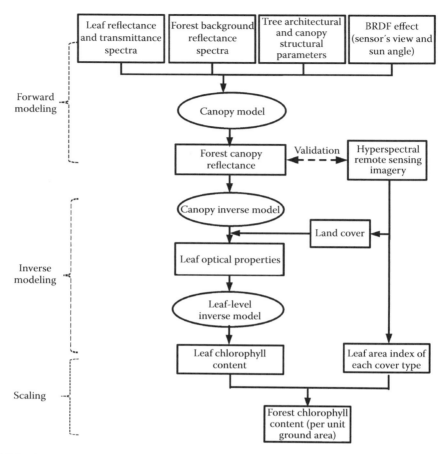

FIGURE 7.4 Relationship between the measured and estimated chlorophyll contents using (a) the original PROSPECT inversion model and (b) the modified PROSPECT model, which takes into account the effects of needle thickness and width. The dotted line is the 1:1 line.

Trees in forest are nonrandom discrete objects and spatially follow a Neyman distribution to simulate patchiness of a forest stand.

The forward modeling estimates the canopy reflectance and four scene components with inputs including leaf optical properties (reflectance and transmittance), forest background reflectance spectra, and general parameters of tree architecture for coniferous and deciduous canopies such as tree height, crown radius, forest density etc., canopy structure (LAI, clumping index), and image view geometry (solar zenith angle [SZA], view zenith angle [VZA], and azimuth angle). The estimated canopy reflectance is then compared with the reflectance measured by the hyperspectral remote sensing sensor for validation.

2. Canopy-level model inversion for estimating the contribution of an individual sunlit leaf to the measured canopy-level reflectance R across the spectrum, and thereby leaf chlorophyll content with the retrieved individual leaf reflectance as input to leaf-level inverse model.

To fulfil the canopy GO model inversion, a MS factor M was introduced to convert R_T, the reflectance of an assemblage of sunlit leaves to individual leaf reflectance R_L. M includes the contributions of the two shaded components to canopy reflectance across spectrum:

$$M = \frac{R - R_G \times P_G}{R_L \times P_T} \tag{7.14}$$

Given that canopy reflectance R can be remotely measured and forest background reflectivity R_G is known, the probabilities of observing the sunlit foliage (P_T) and background (P_G) components and the spectral MS factor (M) need to be estimated to invert the individual leaf reflectivity R_L.

At given SZA, VZA, and azimuthal angle (PHI) between sun and view, variations of R, P_T, and P_G depend on stem density, LAI, tree height, and radius of tree crown. The sensitivity of R, P_T, and P_G to the model input parameters was investigated to develop the LUTs. For simplicity, two LUTs were developed as functions of viewing geometry (VZA, SZA, and PHI) and one dominant contributor LAI for both coniferous and deciduous species. One LUT provides the probabilities of viewing the sunlit foliage P_T and background P_G components, and the other LUT provides the spectral MS factors to convert the average reflectivity of sunlit leaves to the reflectivity of an individual sunlit leaf.

Combining the two LUTs, forest background reflectance measurements, and canopy reflectance derived from hyperspectral remote sensing image, Rimage, the sunlit leaf reflectivity (R_L) can be derived:

$$R_L = \frac{R_{image} - R_G \times P_G}{P_T \times M} \tag{7.15}$$

where R_{image} is the reflectance of the site derived from the hyperspectral remote sensing image.

The retrieved individual leaf reflectance was input to leaf-level inversion model PROSPECT to estimate leaf chlorophyll content. The original and modified PROSPECT models (as described in Section 7.2.2) were applied separately to deciduous, grass, and coniferous species (three vegetation types in the study area) for estimating leaf chlorophyll content of these three cover types.

3. Scaling leaf level chlorophyll content up to the canopy level. LAI is an important forest structural variable to scale up leaf-level biochemical parameters to a canopy scale. When the canopy is open, forest background is often visually greener than the forest canopy and the effect of forest background is large. Using vegetation indices such as simple ratio for LAI mapping, the effect of forest background should be subtracted from the canopy reflectance for the correlation of LAI to vegetation index [25]. Combining leaf chlorophyll content (per unit leaf area) and LAI map, vegetation chlorophyll content (per unit ground area) can then be generated from hyperspectral remote sensing image.

This procedure was performed for forest canopy chlorophyll content estimation and mapping over a study area in boreal forest near Sudbury, Ontario, Canada. Airborne hyperspectral remote-sensing image compact airborne spectrographic imager (CASI), with 72 bands and half bandwidth 4.25–4.36 nm in the visible and near-infrared region and a 2 m spatial resolution, was applied for retrieving forest chlorophyll content. To meet statistical tree distribution pattern defined in the 4-scale model, the original 2 m resolution image was resampled to 20 m for LAI and forest chlorophyll content mapping. Figure 7.5 shows the spatially coarsed leaf and canopy chlorophyll content map using this method. The spatial variability of leaf chlorophyll$_{a+b}$ content, ranging from 16.2 to 43.6 μg/cm^2 (Figure 7.5, left), and canopy chlorophyll content, ranging from 30 to 2170 mg/m^2 (Figure 7.5, right), for vegetated pixels were clearly seen from the map.

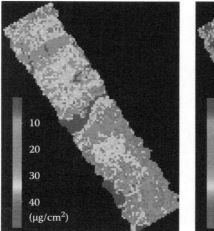

FIGURE 7.5 **(See color insert.)** CASI leaf chlorophyll$_{a+b}$ content (per unit leaf area) distribution (left) and forest chlorophyll content (per unit ground surface area) distribution (right). The images were produced based on the retrieved chlorophyll$_{a+b}$ content for the three vegetated cover types. The spatial resolution of the image is 20×20 m.

7.4 CONCLUSIONS AND APPLICATIONS

Hyperspectral data reveal the subtle spectral responses of leaves to leaf chlorophyll content, which facilitates the radiometrical retrieval of leaf chlorophyll content from remotely sensed leaf and canopy optical properties. This chapter presents methods for retrieving leaf chlorophyll content from hyperspectral remote sensing at different scales for complex forest canopies. Leaf chlorophyll content of both broadleaf and needleleaf species can be quantitatively estimated at leaf and canopy scales through the combination of field measurements on forest structure and architecture, laboratory optical and chemical experiments, hysperspectral remote sensing imagery, and statistical and modeling approaches. The successful demonstration of the physically based approach to complex canopies points to a future direction of retrieving vegetation information from hyperspectral remote sensing imagery.

The leaf-level radiometrical retrieval is a critical step for accurate mapping of the canopy-level chlorophyll content using hyperspectral remote sensing imagery. For broadleaf species, accurate estimation of leaf-level chlorophyll content should rely either on validated algorithms based on empirical indices, or model inversions that take into account leaf thickness change. With leaf thickness as a surrogate to capture the seasonal and locational variations in leaf structure and nonchlorophyll light absorption, the improved PROSPECT model performed better than spectral indices and was capable of deriving the seasonal and canopy height variations in leaf chlorophyll content from leaf reflectance and transmittance spectra. For needleleaf species, the effects of needle width and thickness on light transfer need to be considered for chlorophyll retrieval. With the consideration of these two parameters, the model developed for deciduous leaves can be applied to needleleaf species with greater confidence. Needleleaf chlorophyll content is estimated with a reasonable accuracy. Nevertheless, this improvement to the broadleaf model PROSPECT did not appreciably reduce the departure of the slope of the estimated versus measured total needle chlorophyll content from the 1:1 line. This remaining issue calls for further model refinements specific to the needle internal structure. The current apparatus for measuring needle reflectance and transmittance spectra limits the leaf area exposed to the light source, which can result in a negative bias in the spectra measurements and accordingly an underestimation of chlorophyll content. A new apparatus that can fully expose the needle surface to the light source is desirable for accurate needle spectral measurements.

The transformation of chlorophyll content estimation from leaf scale to canopy scale is a very important issue in contemporary remote sensing. The LUT approach presented in this chapter successfully separates the leaf optical properties and canopy structural effect, and links leaf-level optical properties to the canopy-level hyperspectral measurements. With the LUTs, the effects of canopy structure and forest background are removed for the retrieval of individual leaf reflectance, and thereby leaf chlorophyll content. This physically based approach is developed for retrieving leaf reflectance and thereby leaf chlorophyll content from hyperspectral remote sensing imagery for both open and closed forests, which is in contrast to existing empirically based methods and some highly simplified methods for closed canopies. This approach produced systematically better estimates of chlorophyll content than a widely used spectral index approach.

This complex modeling process requires accurate measurements, as model inputs, of leaf optical properties, forest background reflectance, and forest structural parameter LAI. For simplicity, LUTs are only developed as functions of LAI, and view and sun geometry. The contribution of the two shaded components (shaded foliage and shaded background) was incorporated in the MS factor to reduce the unknown parameters in the model. In future studies, more comprehensive LUTs, which include LAI, sun and view geometry, stem density, and the two shaded components, can be generated to take into account the complexity of forests.

The probability of viewing forest background by the sensor is large in open forests. The LUTs are developed under the assumption that forest background does not vary significantly across the landscape. Due to the heterogeneity and variability of forest background, the effects of forest background on the total remote sensing signals are uneven. To map chlorophyll content more accurately, the spatial variation of forest background needs to be estimated. Multiangle remote sensing data have demonstrated the capability for deriving background reflectance [106]. With the retrieval of spatially explicit forest background reflectance, forest chlorophyll content mapping will be improved.

It is noted that this method also uses a spatially coarsened LAI map for forest chlorophyll content mapping as the GO model 4-scale assumes trees in a forest have certain patchiness. If a pixel is too small, one pixel might not even cover a complete tree crown. To meet the conditions for this statistical tree distribution pattern, the canopy-to-leaf inversion algorithm is applied to coarsened pixels. Further algorithm development is yet needed to retrieve chlorophyll content at the original CASI resolution. This algorithm would have to avoid using any tree distribution assumptions. This is one of the remaining challenges that have not been tackled yet.

Leaf chlorophyll content is potentially one of the most important bioindicators of vegetation stress. Long- or medium-range changes in canopy chlorophyll content can be related to photosynthetic capacity (thus productivity), developmental stage, vegetation phenology, and canopy stresses [107,108]. Chlorophyll content has been used for estimating gross primary productivity [109]. Studies found that leaf chlorophyll content provides an accurate, indirect estimate of plant nutrient status [42,110]. Canopy-level chlorophyll may appear to be most directly relevant for the prediction of productivity [111,112]. Forest chlorophyll content is a bioindicator of forest health and physiological status. Estimates of forest chlorophyll content can reveal forest health status and serve forest management.

Detecting forest health requires a theoretical basis. The algorithms presented in this chapter quantitatively estimate leaf chlorophyll content across tree species and forest canopies from hyperspectral remote sensing. The algorithms would be valuable for various purposes including ecosystem health assessment, natural resources management, and carbon cycle estimation. The spatial distribution of chlorophyll content can potentially be applied for the estimation of spatially explicit carbon budgets. Vegetation biochemical and biophysical parameters enable a linkage between optical remote sensing and the carbon cycle. Estimation of chlorophyll content provides useful insight into plant–environment interactions as chlorophylls change with light [113]. Leaf chlorophyll content can potentially serve as an input to photosynthesis and carbon cycle models.

ACKNOWLEDGMENTS

The author gratefully acknowledges Dr. Jing M. Chen (University of Toronto), Dr. John R. Miller (York University), and Dr. Thomas L. Noland (Ontario Forest Research Institute, Canada) for their insightful advice and constant support for the research.

REFERENCES

1. Running, S.W. and Coughlan, J.C., 1988. A general model of forest ecosystem process for regional applications. 1. Hydrologic balance, canopy gas exchange and primary production processes. *Ecological Modelling* **42**, 125–154.
2. Running, S.W., 1990. Estimating terrestrial primary productivity by combining remote sensing and eco-system simulation. In *Remote Sensing of Biosphere Functioning* (Eds. Hobbs, R.J. and Mooney, H.A.), Springer-Verlag, New York, pp. 65–86.
3. Goetz, S.L. and Prince, S.D., 1996. Remote sensing of net primary production in boreal forest stands. *Agricultural and Forest Meteorology* **78**, 149–179.
4. Asner, G.P., Wessman, C.A., Schimel, D.S., and Archer, S., 1998a. Variability in leaf and litter optical properties: Implications for canopy BRDF model inversions using AVHRR, MODIS, and MISR. *Remote Sensing of Environment* **63**, 243–257.
5. Blackburn, G.A. and Pitman, J.I., 1999. Biophysical controls on the directional spectral reflectance prop-erties of bracken (Pteridium aquilinum) canopies: Results of a field experiment. *International Journal of Remote Sensing* **20**(11), 2265–2282.
6. Baltzer, J.L. and Thomas, S.C., 2005. Leaf optical responses to light and soil nutrient availability in tem-perate deciduous trees. *American Journal of Botany* **92**, 214–223.
7. Belward, A.S., 1991. Spectral characteristics of vegetation, soil and water invisible, near infrared and middle-infrared wavelengths. In *Remote Sensing and Geographical Information Systems of Resource Management in Developing Countries* (Eds. Belward A.S. and Valenzuela, C.R.), ECSC, EEC, EAEC, Brussels and Luxembourg, Germany, pp. 31–53.
8. Rock, B.N., Hoshizaki, T., and Miller, J.R., 1988. Comparison of in situ and airborne spectral measure-ments of the blue shift associated with forest decline. *Remote Sensing of Environment* **24**, 109–127.
9. Vogelmann, J.E., Rock, B.N., and Moss, D.M., 1993. Red-edge spectral measurements from sugar maple leaves. *International Journal of Remote Sensing* **14**, 1563–1575.
10. Carter, G.A., 1994. Ratios of leaf reflectances in narrow wavebands as indicators of plant stress. *International Journal of Remote Sensing* **15**, 697–704.
11. Gitelson, A.A., Merzlyak, M.N., and Lichtenthaler, H.K., 1996. Detection of red-edge position and chlorophyll content by reflectance measurements near 700 nm. *Journal of Plant Physiology* **148**, 501–508.
12. Belanger, M.J., Miller, J.R., and Boyer, M.G., 1995. Comparative relationships between some red-edge parameters and seasonal leaf chlorophyll concentrations. *Canadian Journal of Remote Sensing* **21**, 16–21.
13. Chang, S. and Collins, W., 1983. Confirmation of the airborne biogeophysical mineral exploration tech-nique using laboratory methods. *Economic Geology* **1983**, 723–736.
14. Horler, D.N.H., Barber, J., and Barringer, A.R., 1980. Effects of heavy metals on the absorbance and reflectance spectra of plants. *International Journal of Remote Sensing* **1**, 121–136.
15. Horler, D.N.H., Dockray, M., and Barber, J., 1983. The red-edge of plant leaf reflectance. *International Journal of Remote Sensing* **4**(2), 273–288.
16. Sampson, P.H., Mohammed, G.H., Zarco-Tejada, P.J., Miller, J.R., Noland, T.L., Irving, D., Treitze, P.M., Colombo, S.J., and Freemantle, J., 2000. The bioindicators of forest condition project: A physiological, remote sensing approach. *The Forestry Chronicle* **76**(6), 941–952.
17. Sampson, P.H., Zarco-Tejada, P.J., Mohammed, G.H., Miller, J.R., and Noland, T.L., 2003. Hyperspectral remote sensing of forest condition: Estimating chlorophyll content in tolerant hardwoods. *Forest Science* **49**(3), 381–391.
18. Thomas, S.C., 2005. Increased leaf reflectance in tropical trees under elevated CO_2. *Global Change Biology* **11**, 197–202.
19. Curran, P.J., 1994. Imaging spectrometry. *Progress in Physical Geography* **18**(2), 247–266.
20. Asner, G.P., Braswell, B.H., Schimel, D.S., and Wessman, C.A., 1998. Ecological research needs from multi-angle remote sensing data. *Remote Sensing of Environment* **63**, 155–165.

21. Jacquemoud, S., Ustin, S.L., Verdebout, J., Schmuck, G., Andreoli, G., and Hosgood, B., 1996. Estimating leaf biochemistry using the PROSPECT leaf optical properties model. *Remote Sensing of Environment* **56**, 194–202.

22. Noland, T.L., Miller, J.R., Moorthy, I., Panigada, C., Zarco-Tejada, P.J., Mohammed, G.H., and Sampson, P.H., 2003. Bioindicators of forest sustainability: Using remote sensing to monitor forest condition. In *Meeting Emerging Ecological, Economic, and Social Challenges in the Great Lakes Region: Popular Summaries*. Compiled by Buse, L.J. and A.H. Perera, Ontario Ministry of Natural Resources, Ontario Forest Research Institute, Forest Research Information, Ontario, Canada, **155**, pp. 75–77.

23. Zhang, Y., Chen, J.M., and Thomas, S.C., 2007. Retrieving seasonal variation in chlorophyll content of overstorey and understorey sugar maple leaves from leaf-level hyperspectral data. *Canadian Journal of Remote Sensing* **5**, 406–415.

24. Zhang, Y., Chen, J.M., Miller, J.R., and Noland, T.L., 2008. Retrieving chlorophyll content in conifer needles from hyperspectral measurements. *Canadian Journal of Remote Sensing* **34**(3), 296–310.

25. Zhang, Y., Chen, J.M., Miller, J.R., and Noland, T.L., 2008. Leaf chlorophyll content retrieval from airborne hyperspectral remote sensing imagery. *Remote Sensing of Environment* **112**, 3234–3247.

26. Gates, D.M., Keegan, H.J.J., Schleter, C., and Wiedner, V.R., 1965. Spectral properties of plants. *Applied Optics* **4**, 11–20.

27. Myneni, R.B., Ross, J.K., and Asrar, G., 1989. A review on the theory of photon transport in leaf canopies. *Agricultural Forest Meteorology* **45**, 1–153.

28. Wessman, C.A., 1990. Evaluation of canopy biochemistry. In *Remote Sensing of Biosphere Functioning* (Eds. Hobbs, R.J. and Mooney, H.A.), Springer-Verlag, New York, pp. 135–156.

29. Walter-Shea, E.A. and Norman, J.M., 1991. Leaf optical properties. In *Photon Vegetation Interactions*. (Eds. Myneni, R.B. and Ross, J.), Springer-Verlag, New York, pp. 227–251.

30. Curran, P.J., Dungan, J.L., Macler, B.A., Plummer, S.E., and Peterson, D.L., 1992. Reflectance spectroscopy of fresh whole leaves for the estimation of chemical concentration. *Remote Sensing of Environment* **39**, 153–166.

31. Fourty, T., Baret, F., Jacquemoud, S., Schmuck, G., and Verdebout, J., 1996. Leaf optical properties with explicit description of its biochemical composition: Direct and inverse problems. *Remote Sensing of Environment* **56**, 104–117.

32. Datt, B., 1999. Visible/near infrared reflectance and chlorophyll content in eucalyptus leaves. *International Journal of Remote Sensing* **20**, 2741–2759.

33. Gamon, J.A., Serrano, L., and Surfus, J.S., 1997. The photochemical reflectance index: An optical indicator of photosynthetic radiation-use efficiency across species, functional types, and nutrient levels. *Oecologia* **112**, 492–501.

34. Gitelson, A.A. and Merzlyak, M.N., 1997. Remote estimation of chlorophyll content in higher plant leaves. *International Journal of Remote Sensing* **18**, 2691–2697.

35. Gitelson, A.A., Buschman, C., and Lichtenthaler, H.K., 1999. The chlorophyll fluorescence ratio F735/F700 as an accurate measure of chlorophyll content in plants. *Remote Sensing of Environment* **69**, 296–302.

36. Maccioni, A., Agati, G., and Mazzinghi, P., 2001. New vegetation indices for remote measurement of chlorophylls based on leaf directional reflectance spectra. *Journal of Photochemistry and Photobiology B: Biology* **61**, 52–61.

37. Peñuelas, J., Filella, I., Llusia, J., Siscart, D., and Pinol, J., 1998. Comparative field study of spring and summer leaf gas exchange and photobiology of the mediterranean trees *Quercus ilex* and *Phillyrea latifolia*. *Journal of Experimental Botany* **49**, 229–238.

38. le Maire, G., Francois, C., and Dufrene, E., 2004. Towards universal broad leaf chlorophyll indices using PROSPECT simulated database and hyperspectral reflectance measurements. *Remote Sensing of Environment* **89**, 1–28.

39. Schepers, J.S., Blackmer, T.M., Wilhelm, W.W., and Resende, M., 1996. Transmittance and reflectance measurements of corn leaves from plants with different nitrogen and water supply. *Journal of Plant Physiology* **148**, 523–529.

40. Blackburn, G.A., 1998. Spectral indices for estimating photosynthetic pigment concentrations: A test using senescent tree leaves. *International Journal of Remote Sensing* **19**, 657–675.

41. Blackburn, G.A., 1998. Quantifying chlorophylls and carotenoids from leaf to canopy scales: An evaluation of some hyperspectral approaches. *Remote Sensing of Environment* **66**, 273–285.

42. Filella, I., Serrano, L., Serra, J., and Peñuelas, J., 1995. Evaluating wheat nitrogen status with canopy reflectance indices and discriminant analysis. *Crop Science* **35**, 1400–1405.

43. Chappelle, E.W., Kim, M.S., and McMurtrey, J.E., 1992. Ratio analysis of reflectance spectra (RARS): An algorithm for the remote estimation of the concentrations of chlorophyll A, chlorophyll B and the carotenoids in soybean leaves. *Remote Sensing of Environment* **39**, 239–247.

44. Sims, D.A. and Gamon, J.A., 2002. Relationship between pigment content and spectral reflectance across a wide range of species, leaf structures and developmental stages. *Remote Sensing of Environment* **81**, 337–354.

45. Gitelson, A.A., Gritz, Y., and Merzlyak, M.N., 2003. Relationships between leaf chlorophyll content and spectral reflectance and algorithms for non-destructive chlorophyll assessment in higher plant leaves. *Journal of Plant Physiology* **160**, 271–282.

46. Datt, B., 1998. Remote sensing of chlorophyll a, chlorophyll b, chlorophyll a + b and total carotenoid content in eucalyptus leaves. *Remote Sensing of Environment* **66**, 111–121.

47. Yao, H.B. and Tian, L., 2003. A genetic algorithm-based selective principal component analysis (GA-SPCA) method for high-dimensional data feature extraction. *IEEE Transactions on Geoscience and Remote Sensing* **41**, 1469–1478.

48. Coops, N., Drury, S., Smith, M.L., Martin, M., and Ollinger, S., 2002. Comparison of green leaf eucalypt spectra using spectral decomposition. *Australian Journal of Botany* **50**, 567–576.

49. Tumbo, S.D., Wagner, D.G., and Heinemann, P.H., 2002. Hyperspectral-based neural network for predicting chlorophyll status in corn. *Transactions of the ASAE* **45**, 825–832.

50. O'Neill, A.L., Kupiec, J.A., and Curran, P.J., 2002. Biochemical and reflectance variation throughout a Sitka spruce canopy. *Remote Sensing of Environment* **80**, 134–142.

51. Osborne, S.L., Schepers, J.S., Francis, D.D., and Schlemmer, M.R., 2002. Use of spectral radiance to estimate in-season biomass and grain yield in nitrogen- and water-stressed corn. *Crop Science* **42**, 165–171.

52. Allen, W.A., Gausman, H.W., Richardson, A.J., and Thomas, J.R., 1969. Interaction of isotropic light with a compact plant leaf. *Journal of the Optical Society of America* **59**(10), 1376–1379.

53. Jacquemoud, S. and Baret, F., 1990. PROSPECT: A model of leaf optical properties spectra. *Remote Sensing of Environment* **34**, 75–91.

54. Allen, W.A. and Richardson, A.J., 1968. Interaction of light with a plant canopy. *Journal of the Optical Society of America* **58**(8), 1023–1028.

55. Yamada, N. and Fujimura, S., 1991. Nondestructive measurement of chlorophyll pigment content in plant leaves from three-color reflectance and transmittance. *Applied Optics* **30**, 3964–3973.

56. Allen, W.A., Gausman, H.W., and Richardson, A.J., 1973. Willstätter-Stoll theory of leaf reflectance evaluation by ray tracing. *Applied Optics* **12**(10), 2448–2453.

57. Brakke, T.W. and Smith, J.A., 1987. A ray tracing model for leaf bidirectional scattering studies. In *Proceedings of the 7th International Geoscience and Remote Sensing Symposium (IGARSS'87)*, May 18–21, Ann Arbor, MI, pp. 643–648.

58. Govaerts, Y.M., Jacquemoud, S., Verstraete, M.M., and Ustin, S.L., 1996. Three-dimensional radiation transfer modeling in a dicotyledon leaf. *Applied Optics* **35**(33), 6585–6598.

59. Ustin, S.L., Jacquemoud, S., and Govaerts, Y.M., 2001. Simulation of photon transport in a three-dimensional leaf: Implication for photosynthesis, *Plant Cell Environment* **24**, 1095–1103.

60. Tucker, C.J. and Garratt, M.W., 1977. Leaf optical system modeled as a stochastic process. *Applied Optics* **16**(3), 635–642.

61. Ganapol, B., Johnson, L., Hammer, P., Hlavka, C., and Peterson, D., 1998. LEAFMOD: A new within-leaf radiative transfer model. *Remote Sensing of Environment* **6**, 182–193.

62. Maier, S.W., Lüdeker, W., and Günther, K.P., 1999. SLOP: A revised version of the stochastic model for leaf optical properties. *Remote Sensing of Environment* **68**(3), 273–280.

63. Demarez, V., Gastellu-etchegorry, J.P., Mougin, E., Marty, G., Proisy, C., Duferene, E., and Dantec, V.L.E., 1999. Seasonal variation of leaf chlorophyll content of a temperate forest. Inversion of the PROSPECT model. *International Journal of Remote Sensing* **20**(5), 879–894.

64. Zarco-Tejada, P.J., Miller, J.R., Noland, T.L., Mohammed, G.H., and Sampson, P.H., 2001. Scaling-up and model inversion methods with narrowband optical indices for chlorophyll content estimation in closed forest canopies with hyperspectral data. *IEEE Transactions in Geosciences and Remote Sensing* **39**(7), 1491–1507.

65. Renzullo, L.J., Blanchfield, A.L., Guillermin, R., Powell, K.S., and Held, A.A., 2006. Comparison of PROSPECT and HPLC estimates of leaf chlorophyll contents in a grapevine stress study, *International Journal of Remote Sensing* **27**(4), 817–823.

66. Forsythe, G.E., Malcolm, M.A., and Moler, C.B. 1976. Computer methods for mathematical computations, Prentice-Hall, Englewood Cliffs, N.J.

67. Hosgood, B., Jacquemoud, S., Andreoli, G., Verdebout, J., Pedrini, G., and Schmuck, G., 1995. *Leaf Optical Properties Experiment 93 (LOPEX93) Report EUR-16095-EN*, European Commission, Joint Research Centre, Institute for Remote Sensing Applications, Ispra, Italy.

68. Féret, J.B., François, C., Asner, G.P., Gitelson, A.A., Martin, R.E., Bidel, L.P.R., Ustin, S.L., le Maire, G., and Jacquemoud, S., 2008, PROSPECT-4 and 5: Advances in the leaf optical properties model separating photosynthetic pigments. *Remote Sensing of Environment* **112**(6), 3030–3043.

69. Dawson, T.P., Curran, J.P., and Plummer, S.E., 1998. LIBERTY—Modeling the effects of leaf biochemical concentration on reflectance spectra. *Remote Sensing of Environment* **65**(1), 50–60.

70. Moorthy, I., Miller, J.R., and Noland, T.L., 2008. Estimating chlorophyll concentration in conifer needles: An assessment at the needle and canopy level. *Remote Sensing of Environment* **12**(6), 2824–2838.

71. Hanrahan, P. and Krueger, W., 1993. Reflection from layered surfaces due to subsurface scattering. In *Proceedings of the 20th Annual Conference on Computer Graphics and Interactive Techniques*, New York, pp. 165–174.

72. Myneni, R.B. and Asrar, G., 1993. Radiative transfer in three-dimensional atmosphere-vegetation media. *Journal of Quantitative Spectroscopy and Radiative Transfer* **49**, 585–598.

73. Ross, J.K., 1981. *The Radiation Regime and Architecture of Plant Stands*, Kluwer, Boston, MA.

74. Goel, N.S., 1988. Models of vegetation canopy reflectance and their use in estimation of biophysical parameters from reflectance data. *Remote Sensing Reviews* **4**, 1–212.

75. Johnson, L.F., Hlavka, C.A., and Peterson, D.L., 1994. Multivariate analysis of AVIRIS data for canopy biochemical estimation along the oregon transect. *Remote Sensing of Environment* **47**, 216–230.

76. Matson, P., Johnson, L., Billow, C., Miller, J.R., and Pu, R., 1994. Seasonal patterns and remote spectral estimation of canopy chemistry across the oregon transect. *Ecological Applications* **4**, 280–298.

77. Curran, P.J., Kupiec, J.A., and Smith, G.M., 1997. Remote sensing the biochemical composition of a slash pine canopy. *IEEE Transactions on Geosciences and Remote Sensing* **35**, 415–420.

78. Zarco-Tejada, P.J. and Miller, J.R., 1999. Land cover mapping at BOREAS using red-edge spectral parameters from CASI imagery. *Journal of Geophysics Research* **104**(D22), 27921–27948.

79. Peterson, D.L., Aber, J.D., Matson, P.A., Card, D.H., Swanberg, N.A., Wessman, C.A., and Spanner, M.A., 1988. Remote sensing of forest canopy leaf biochemical contents. *Remote Sensing of Environment* **24**, 85–108.

80. Yoder, B.J. and Pettigrew-Crosby, R.E., 1995. Predicting nitrogen and chlorophyll content and concentrations from reflectance spectra (400–2500 nm) at leaf and canopy scales. *Remote Sensing of Environment* **53**(3), 199–211.

81. Zagolski, F., Pinel, V., Romier, J., Alcayde, D., Fotanari, J., Gastellu-Etchegorry, J.P., Giordano, G., Marty, G., and Joffre, R., 1996. Forest canopy chemistry with high spectral resolution remote sensing. *International Journal of Remote Sensing* **17**, 1107–1128.

82. Haboudane, D., Miller, J.R., Tremblay, N., Zarco-Tejada, P.J., and Dextraze, L., 2002. Integrated narrowband vegetation indices for prediction of crop chlorophyll content for application to precision agriculture. *Remote Sensing of Environment*, **81**(2–3), 416–426.

83. Dash, J. and Curran, P.J., 2004. The MERIS terrestrial chlorophyll index. *International Journal of Remote Sensing* **25**, 5403–5413.

84. Zarco-Tejada, P.J., Miller, J.R., Morales, A., Berjón, A., and Agüera, J., 2004. Hyperspectral indices and model simulation for chlorophyll estimation in open-canopy tree crops. *Remote Sensing of Environment* **90**(4), 463–476.

85. Zarco-Tejada, P.J., Berjón, A., López-Lozano, R., Miller, J.R., Martín, P., Cachorro, V., González, M.R., and Frutos, A., 2005. Assessing vineyard condition with hyperspectral indices: Leaf and canopy reflectance simulation in a row-structured discontinuous canopy. *Remote Sensing of Environment* **99**, 271–287.

86. Jacquemoud, S., Baret, F., Andrieu, B., Danson, F.M., and Jaggard, K., 1995. Extraction of vegetation biophysical parameters by inversion of the PROSPECT + SAIL models on sugar beet canopy reflectance data: Application to TM and AVIRIS sensors. *Remote Sensing of Environment* **52**, 163–172.

87. Demarez, V. and Gastellu-Etchegorry, J.P., 2000. A modeling approach for studying forest chlorophyll content. *Remote Sensing of Environment* **71**, 226–238.

88. Dawson, T.P., Curran, P.J., North, P.R.J., and Plummer, S.E., 1997. The potential for understanding the biochemical signal in the spectra of forest canopies using a coupled leaf and canopy model. In *Physical Measurements and Signatures in Remote Sensing*, Vol. 2 (Eds. Guyot, G. and Phulpin, T.), Balkema, Rotterdam, the Netherlands, pp. 463–470.

89. Broge, N.H. and Leblanc, E., 2000. Comparing prediction power and stability of broadband and hyperspectral vegetation indices for estimation of green leaf area index and canopy chlorophyll density. *Remote Sensing of Environment* **76**, 156–172.

90. Daughtry, C.S.T., Walthall, C.L., Kim, M.S., Brown, de C.E., and McMurtrey, J.E., 2000. Estimating corn leaf chlorophyll concentration from leaf and canopy reflectance. *Remote Sensing of Environment* **74**, 229–239.

91. Verhoef, W., 1984. Light scattering by leaf layers with application to canopy reflectance modelling: The SAIL model. *Remote Sensing of Environment* **16**, 125–141.

92. Verstaete, M.M., Pinty, B., and Dickinson, R.E., 1990. A physical model of the bidirectional reflectance vegetation canopies, I theory. *Journal of Geophysics Research* **95**(D8), 11755–11765.

93. Liang, S. and Strahler, A.H., 1993. Calculation of the angular radiance distribution for a coupled atmosphere and canopy. *IEEE Transactions on Geoscience and Remote Sensing* **31**(2), 491–502.

94. Zarco-Tejada, P.J., 2000. Hyperspectral remote sensing of closed forest canopies: Estimation of chlorophyll fluorescence and pigment content. PhD thesis, York University, Toronto, Ontario, Canada.

95. Zarco-Tejada, P.J., Miller, J.R., Harron, J., Hu, B., Noland, T.L., Goel, N., Mohammed, G.H., and Sampson, P., 2004. Needle chlorophyll content estimation through model inversion using hyperspectral data from boreal conifer forest canopies. *Remote Sensing of Environment* **89**, 189–199.

96. Borel, C.C., Gerstl, S.A.W., and Powers, B.J., 1991. The radiosity method in optical remote sensing of structured 3-D surfaces. *Remote Sensing of Environment* **36**, 13–44.

97. Goel, N.S., Rozehnal, I., and Thompson, R.L., 1991. A computer graphics based model for scattering from objects of arbitrary shapes in the optical region. *Remote Sensing of Environment* **36**(2), 73–104.

98. Myneni, R., Asrar, G., and Gerstl, S., 1990. Radiative transfer in three dimensional leaf canopies. *Transport Theory and Statistical Physics* **19**, 205–250.

99. Gastellu-Etchegorry, J.P., Demarez, V., Pinel, V., and Zagolski, F., 1996. Modelling radiative transfer in heterogeneous 3-D vegetation canopies. *Remote Sensing of Environment* **58**, 131–156.

100. Thompson, R.L. and Goel, N.S., 1999. SPRINT: A universal canopy reflectance model for kilometer level scene. In *Abstract of the Second International Workshop on Multiple-Angle Measurements and Models*, September 15–17, Ispra, Italy.

101. Li, X. and Strahler, A.H., 1988. Modeling the gap probability of discontinuous vegetation canopy. *IEEE Transactions on Geoscience and Remote Sensing* **26**, 161–170.

102. Li, X., Strahler, A.H., and Woodcock, C.E., 1995. A hybrid geometric optical-radiative transfer approach for modeling albedo and directional reflectance of discontinuous canopies. *IEEE Transactions on Geoscience and Remote Sensing* **33**, 466–480.

103. Chen, J.M. and Leblanc, S., 1997. A 4-scale bidirectional reflection model based on canopy architecture. *IEEE Transactions on Geoscience and Remote Sensing* **35**, 1316–1337.

104. Chen, J.M. and Leblanc, S.G., 2001. Multiple-scattering scheme useful for hyperspectral geometrical optical modelling. *IEEE Transactions on Geoscience and Remote Sensing* **39**(5), 1061–1071.

105. Zhang, Y., Hyperspectral remote sensing algorithms for retrieving forest chlorophyll content. PhD thesis, University of Toronto, Toronto, Ontario, Canada.

106. Canisius, F. and Chen, J.M., 2007. Retrieving forest background reflectance in a boreal region from multi-angle imaging spectroradiometer (MISR) data. *Remote Sensing of Environment* **107**, 312–321.

107. Ustin, S.L., Smith, M.O., Jacquemoud, S., Verstraete, M.M., and Govaerts, Y., 1998. GeoBotany: Vegetation mapping for earth sciences. In *Manual of Remote Sensing: Remote Sensing for the Earth Sciences*, 3rd edn., Vol. 3 (Ed. Rencz, A.N.), John Wiley, Hoboken, NJ, pp. 189–248.

108. Zarco-Tejada, P.J., Miller, J.R., Mohammed, G.H., Noland, T.L., and Sampson, P.H., 2002. Vegetation stress detection through chlorophyll a+b estimation and fluorescence effects on hyperspectral imagery. *Journal of Environmental Quality* **31**, 1433–1441.

109. Gitelson, A.A., Vina, A., Verma, S.B., Rundquist, D.C., Arkebauer, T.J., Keydan, G., Leavitt, B., Ciganda, V., Burba, G.G., and Suyker, A.E., 2006. Relationship between gross primary production and chlorophyll content in crops: Implications for the synoptic monitoring of vegetation productivity. *Journal of Geophysical Research–Atmospheres* **111**, Art. No. DO8S11, 13.

110. Moran, J.A., Mitchell, A.K., Goodmanson, G., and Stockburger, K.A., 2000. Differentiation among effects of nitrogen fertilization treatments on conifer seedlings by foliar reflectance: A comparison of methods. *Tree Physiology* **20**, 1113–1120.

111. Whittaker, R.H. and Marks, P.L., 1975. Methods of assessing terrestrial productivity. In *Primary Productivity of the Biosphere, Ecological Studies*, Vol. 14 (Eds. Lieth, H. and Whittaker, R.H.), Springer, New York, pp. 55–118.

112. Dawson, T.P., North, P.R.J., Plummer, S.E., and Curran, P.J., 2003. Forest ecosystem chlorophyll content: Implications for remotely sensed estimates of net primary productivity. *International Journal of Remote Sensing* **24**, 611–617.

113. Fang, Z., Bouwkamp, J., and Solomos, T., 1998. Chlorophyllase activities and chlorophyll degradation during leaf senescence in nonyellowing mutant and wild type of *Phaseolus vulgaris* L. *Journal of Experimental Botany* **49**, 503–510.

8 Estimating Leaf Nitrogen Concentration (LNC) of Cereal Crops with Hyperspectral Data

Yan Zhu, Wei Wang, and Xia Yao

CONTENTS

8.1 INTRODUCTION

Optimization of nitrogen (N) fertilizer management is a key solution to enhance grain yield and quality in agronomic crops. Current N management strategies for cereal crop production have caused excessive N use and low use efficiency (Raun et al., 2002; Shanahan et al., 2008), which resulted in economic loss and environment pollution (Ju et al., 2009). Scientific and rational application of N fertilizer by real-time monitoring of N status in crop plants to balance external N supply and internal N uptake of plant should be an effective technology for improving N use efficiency (Cassman et al., 2002; Shanahan et al., 2008). Traditional measurement method of crop N status normally depends on plant sampling from the field and analytic assay in the laboratory. The results from this protocol are relatively reliable, but are unable to meet the needs of real-time, fast, and nondestructive monitoring and diagnosis of plant N status (Sims and Gamon, 2002). Hyperspectral remote sensing has high resolution and large amount of spectral information (Sims and Gamon, 2003), which can provide a new means for nondestructive, fast, and real-time monitoring of plant physiological parameters (Kokaly, 2001; Dorigo et al., 2007; Zhao et al., 2007; Asner and Martin, 2008) and is exhibiting a promising prospect in crop growth monitoring.

Previous studies have found that monitoring of crop N status has high accuracy at the leaf scale using remote sensing (Kokaly, 2001; Sims and Gamon, 2002, 2003). However, since canopy spectral reflectance reflects the comprehensive information of plant population, including leaf properties, canopy structure, soil background, and atmosphere noise, monitoring of crop N is more challenging at the canopy scale (Dorigo et al., 2007). Effectively removing the interference of soil background from canopy reflectance is helpful for mining the intrinsic characteristics of canopy spectral reflectance and deriving more applicable spectral indices (Asner and Martin, 2008). Several spectral indices can be used for eliminating background noise to some extent (Zhao et al., 2007). Normalized difference vegetation index (NDVI) is one of the most common spectral indices for crop growth monitoring (Huang et al., 2004). Under high canopy coverage, NDVI becomes saturated because of insensitivity of reflectance in red region. If the canopy is too sparse, the value of NDVI changes significantly due to impact of background noise (e.g., soil; Liang, 2004). Soil-adjusted vegetation index (SAVI) is widely used to reduce or eliminate the effects of soil background on canopy reflectance with the parameter L (Huete, 1988; Gitelson et al., 2002; Inoue et al., 2007). Ratio vegetation index (RVI) can slow down the rate of saturation under high canopy coverage, but its resolution ability will decrease when the vegetation coverage is less than 50% (Pearson and Miller, 1972). The application of difference vegetation index (DVI) is much less than NDVI and RVI due to the sensitivity to soil background, and the sensitivity of DVI to vegetation is reduced when the vegetation coverage is larger than 80% (Jordan, 1969).

Vegetation indices have been widely used to estimate plant N status in recent studies with cereal crops (Hansen and Schjoerring, 2003; Zhu et al., 2008; Yao et al., 2009a,b). However, different researchers provided various characteristic wavebands involved in vegetation indices for crop N monitoring (Hansen and Schjoerring, 2003; Xue et al., 2004; Zhu et al., 2007; Yao et al., 2010), which brought some difficulty in developing the portable sensor of N monitoring. Most of previous studies assumed that crop population was the homogeneous entity during the whole growing period for monitoring of crop N status; thus, they generally chose the sensitive bands and constructed a single spectral index to establish the same quantitative model by pooling all the experimental data during the whole growing period (Zhu et al., 2008; Yao et al., 2009). However, the canopy backgrounds as well as spectral reflectance are changing with growing stages, which may result in inconsistent relationships between vegetation index and crop N status during different growing stages. Therefore, it is necessary to carry out a systematic study on selecting the sensitive wavebands of N, constructing the optimal spectral index, and establishing the N monitoring model with high accuracy and solid explanation in cereal crop (Zhao, 2003).

On the basis of eight field experiments with different cultivars, N levels, and water regimes in wheat and rice crops, the present study was targeted to (1) locate the common sensitive wavelength combination for monitoring leaf nitrogen concentration (LNC) of two cereal crops (wheat and rice) by integrating the spectral analysis method, eco-physiological principle, and statistical approach, (2) construct the optimal spectral indices based on the common sensitive wavelength combination by taking full account of characteristics of canopy spectral reflectance during different growing stages in rice and wheat crops, (3) establish the quantitative model for monitoring LNC with strong explanation and high accuracy in cereal crop, and (4) expand the bandwidths of common sensitive wavelength combination with relatively accurate model performance.

The goal was to provide technical support for real-time monitoring of N status and development of portable sensor in crop growth monitoring.

8.2   MATERIALS AND METHODS

8.2.1   EXPERIMENTAL DESIGN

Data used in the present study were obtained from four field experiments (Exp.) of winter wheat (*T. aestivum* L.) and four of wet rice (*O. sativa* L.) across four growing seasons in each crop, involving different eco-sites, cultivars, N rates, and water regimes, and the sampling was conducted at key growing stages from jointing to physiological maturity in each experiment (Table 8.1).

8.2.2   MEASUREMENT AND DATA ANALYSIS

8.2.2.1   Measurement of Canopy Hyperspectral Reflectance

All spectral measurements were made with a portable spectrometer (FieldSpec-FR, ASD). This spectrometer has a spectral sampling interval of 1.4 nm between 350 and 1000 nm, and 2 nm between

TABLE 8.1
The Basic Information for Different Field Experiments

No. of Experiments	Season	Crop	Eco-Site	Cultivar	Treatment	Sampling Date
Exp. 1	2004	Rice	Jiangsu Academy of Agricultural Sciences	Wuxiangjing 9, Huajing 2, Nipponbare	N rate (kg·ha⁻¹): N0 (0), N1 (100), N2 (210), N3 (315)	Heading, initial-filling, mid-filling, late-filling
Exp. 2	2005	Rice	Nanjing Agricultural Bureau	Wuxiangjing 14, 27123	N rate (kg·ha⁻¹): N0 (0), N1 (90), N2 (270), N3 (405)	Jointing, booting, heading, initial-filling, mid-filling, late-filling
Exp. 3	2006	Rice	Nanjing Agricultural Bureau	Wuxiangjing 14, 27123	N rate (kg·ha⁻¹): N0 (0), N1 (90), N2 (240), N3 (360)	Same with Exp. 2
Exp. 4	2009	Rice	Nanjing Pailou Experiment Station	Wuxiangjing 14, Liangyoupeijiu	Water regimes: W1 (−60 kPa), W2 (−40 kPa), W3 (−20 kPa), W4 (5 cm water layer)	Same with Exp. 1
Exp. 5	2003–2004	Wheat	Jiangsu Academy of Agricultural Sciences	Husimai 20, Ningmai 9, Xumai 26, Yangmai 10	N rate (kg·ha⁻¹): N0 (0), N1 (75), N2 (150), N3 (225), N4 (300)	Jointing, booting, anthesis, mid-filling
Exp. 6	2004–2005	Wheat	Jiangsu Academy of Agricultural Sciences	Ningmai 9, Yangmai 12, Yumai 34	N rate (kg·ha⁻¹): N0 (0), N1 (75), N2 (150), N3 (225)	Jointing, booting, anthesis, initial-filling, mid-filling, late-filling
Exp. 7	2005–2006	Wheat	Nanjing Agricultural Bureau	Ningmai 9, Yumai 34	N rate (kg·ha⁻¹): N0 (0), N1 (90), N2 (180), N3 (270)	Same with Exp. 1
Exp. 8	2007–2008	Wheat	Nanjing Jiangpu Experiment Station	Ningmai 9	N rate (kg·ha⁻¹): N0 (0), N1 (45), N2 (90), N3 (135)	Same with Exp. 5

1000 and 2500 nm. The spectral resolution is 3 nm between 350 and 1000 nm, and 10 nm between 1000 and 2500 nm (Ray et al., 2010). The measurements were carried out from a height of 1.0 m above the canopy and 0.44 m view diameter under clear sky conditions between 10:00 h and 14:00 h (Beijing local time) (Analytical Spectral Devices, Inc., 2002). Measurements of vegetation radiance were carried out at 10 sample sites in each plot, with each sample from averaging 20 scans at an optimized integration time. The saved spectrum file contained continuous spectral reflectance at 1 nm step over the band region of 350–2500 nm. A panel radiance measurement was taken before and after the vegetation measurement by two scans each time. In each of the experiments, spectral data were obtained at several major growing stages (Table 8.1).

8.2.2.2 Determination of Leaf Nitrogen Concentration

After measurements of canopy spectral reflectance, 10 plants were randomly selected from each plot for determination of LNC. From each sample, all green leaves were separated from the stems, oven-dried at 70°C to a constant weight, and then weighed. Dried samples were ground to pass through a 1 mm screen and then stored in plastic bags for chemical analysis. Total LNC was determined by the micro-Kjeldahl method. LNC (g/100 g) was expressed based on unit leaf dry weight.

8.2.2.3 Data Analysis

Different spectral indices (Table 8.2) of all available two-band combinations were constructed from original reflectance between 400 and 1000 nm for 1 nm intervals, and they indicated LNC during different growing stages in both rice and wheat crops. Since vegetation is not fully closed before booting stage (defined as the early period), and the influence of soil background on canopy reflectance is relatively large, the SAVI was used to reduce soil noise by adjusting the parameters of L from −2 to 2 in steps by 0.001 increments. From heading to filling stage (defined as the mid-late period), the vegetation was nearly closed, and the soil background had less impact on canopy reflectance; thus, NDVI, RVI, and DVI were used to develop the monitoring models for LNC.

The linear regressions between the aforementioned four types of vegetation indices (SAVI, NDVI, RVI, and DVI) and LNC were performed for two separated periods (the early period and the mid-late period) and two crops (wheat and rice) to inspect the variation in coefficients of determination (R^2). Then, the sensitive wavelength combinations with higher R^2 were explored for each type of spectral index at each period in each crop by sorting the R^2 values of the linear regression equations. Next, by expanding the ranges of sensitive wavelength combinations with top 1%–n% steps by 0.01% R^2 for each type of spectral index at each period of each crop, the common ranges of sensitive wavelength combinations for four types of spectral indices of two crops and two growing periods were identified.

The detailed ranges of each band in sensitive wavelength combination were described by the maximum inscribed rectangle, in which the bandwidths of two sensitive bands can simultaneously be as broad as possible in the regions of common sensitive wavelength combinations. Further, based

TABLE 8.2
Spectral Indices Used in This Study

Abbreviation	Name	Algorithm	Reference
DVI	Difference vegetation index	$R_{\lambda1} - R_{\lambda2}$	Jordan (1969)
RVI	Ratio vegetation index	$R_{\lambda1}/R_{\lambda2}$	Pearson and Miller (1972)
NDVI	Normalized difference vegetation Index	$\dfrac{R_{\lambda1} - R_{\lambda2}}{R_{\lambda1} + R_{\lambda2}}$	Rouse et al. (1973)
SAVI	Soil-adjusted vegetation Index	$\left(\dfrac{R_{\lambda1} - R_{\lambda2}}{R_{\lambda1} - R_{\lambda2} + L}\right) \times (1 + L)$	Huete (1988)

on the common sensitive wavelength combination with the top n% R^2, the quantitative models for LNC monitoring were developed. Finally, the independent datasets were used to test the performance of the linear regression models based on the selected waveband combinations with the relative root mean square error (RRMSE) (Zhu et al., 2007) and the slope between observed and estimated values.

Among eight rice and wheat experiments, the data from rice experiments in 2005 and 2009 and wheat experiments during the growing seasons of 2006–2008 were used for model development, and the other experimental data were used for model testing.

8.3 RESULTS

8.3.1 CANOPY LNC IN RICE AND WHEAT DURING DIFFERENT GROWING PERIODS

The LNC showed a wide range of changes with different field experiments, and the range of LNC during the early period was smaller than during the mid-late growing period. The wheat experiment from 2004 to 2005 had higher soil fertility, which reduced the possible influences of N application rates, so the range of LNC was the smallest during the early period in this growing season. In rice crop, the maximal LNC was 3.64% during the early period in 2005, and the minimum value was 0.71% during the mid-late growing period in 2006. In wheat crop, the maximal LNC was 4.33% during the early period in the growing season from 2004 to 2005, and the minimum value was 0.46% during the mid-late period in the growing season from 2005 to 2006. The mean value of LNC for wheat was greater than that of rice, because the larger biomass was observed in rice than wheat, which resulted in the greater dilution effect on N concentration in rice (Table 8.3).

8.3.2 RELATIONSHIPS OF CANOPY LNC WITH HYPERSPECTRAL REFLECTANCE

The correlation coefficients were obviously different under two crops and two growing periods in the range of 350–2500 nm, with consistent patterns in the range of 400–1000 nm (Figure 8.1). LNC was negatively related to the reflectance in the region between 350 nm and around 725 nm and

TABLE 8.3
Changes of Leaf Nitrogen Content (% Dry Weight) for Different Growing Periods in Rice and Wheat Experiments

Crop	Season	Growing Period	Mean	Min	Max	Standard Deviation	Range	Coefficient of Variation
Rice	2004	Mid-late	2.197	1.271	2.792	0.376	1.521	0.171
	2005	Early	2.855	1.823	3.644	0.532	1.821	0.186
		Mid-late	1.990	0.786	3.186	0.705	2.399	0.354
	2006	Early	2.657	1.806	3.360	0.502	1.554	0.189
		Mid-late	1.770	0.708	2.805	0.567	2.097	0.320
	2009	Mid-late	2.498	1.457	3.407	0.565	1.949	0.226
Wheat	2003–2004	Early	2.492	1.851	3.491	0.446	1.639	0.179
		Mid-late	2.069	0.595	3.766	0.747	3.171	0.361
	2004–2005	Early	3.569	2.239	4.327	0.586	2.088	0.164
		Mid-late	2.532	0.575	3.930	0.915	3.355	0.361
	2005–2006	Early	3.029	1.801	4.218	0.727	2.418	0.240
		Mid-late	2.140	0.455	3.921	1.006	3.466	0.470
	2006–2007	Early	3.386	2.080	4.074	0.631	1.994	0.186
		Mid-late	2.791	0.566	4.168	0.929	3.602	0.333

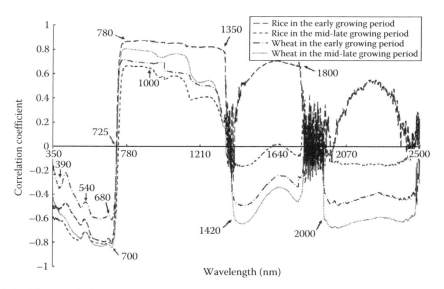

FIGURE 8.1   The correlation coefficients of leaf nitrogen concentration to canopy hyperspectral reflectance in rice and wheat during different growing periods.

positively related to the reflectance in the region between 725–1350 nm. In the region between 350 nm and around 700 nm, there were two obvious peaks at around 540 and 680 nm and a small valley at around 700 nm, and the peak at around 390 nm only existed in the early growing period. In the region between 700 and 780 nm, the correlation coefficients increased quickly with the increasing wavelength. Yet in the region between 780 and 1000 nm, the correlation coefficients were relatively flat. From 1000 to 1350 nm, the correlation coefficients began to decrease quickly except that of rice in the early growing period. In two regions of 1420–1800 and 2000–2500 nm, LNC was positively related to the reflectance during the early growing period of rice, while negatively related to the reflectance during the whole growing period of wheat and the mid-late period of rice.

8.3.3   Relationships of Canopy LNC with Spectral Indices: SAVI during Early Growing Periods

Since the correlation coefficients of LNC to the original reflectance in the range of 400–1000 nm exhibited the consistent patterns under two crops and two growing periods, the analyses on linear regressions between LNC and SAVI with all available two-band combinations in the region from 400 to 1000 nm were performed to inspect variation in the coefficient of determination (R^2).

Figure 8.2 shows contour maps of relative R^2 (ratio of the actual value to the largest one), because the plots with relative R^2 values will be more easy to find the band ranges of top 10% R^2 from linear regressions of SAVI ($R_{\lambda 1}$, $R_{\lambda 2}$) against canopy LNC than with actual R^2 values. It should be noted that SAVI ($R_{\lambda 1}$, $R_{\lambda 2}$) = −SAVI ($R_{\lambda 2}$, $R_{\lambda 1}$); thus, only half of the contour map was displayed. The detailed band ranges of band 1 and band 2 pairs with top 10% R^2 were 700–750 nm (λ_1) and 725–890 nm (λ_2), or 420–520 and 570–675 nm in rice (Figure 8.2A, Table 8.4), and were 695–750 and 735–1000 nm, or 520–575 and 735–1000 nm, or 435–510 and 565–595 nm, or 415–495 and 505–525 nm in wheat (Figure 8.2B, Table 8.4). The best sensitive band combinations for rice and wheat were, respectively, chosen from these band ranges with the highest R^2 for each crop, as shown in Table 8.5. The highest R^2 of LNC monitoring model with SAVI (R_{763}, R_{793}) in rice of 0.887 is slightly higher than that with SAVI (R_{685}, R_{690}) in wheat of 0.863 (Table 8.5).

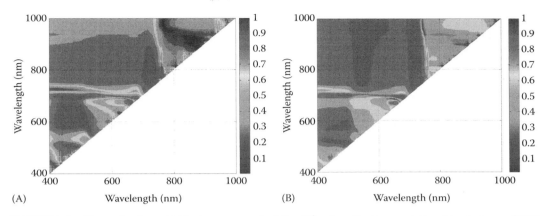

FIGURE 8.2 **(See color insert.)** Contour maps of relative R^2-values for linear relationships between SAVI and leaf nitrogen concentration for rice (A) and wheat (B).

8.3.4 RELATIONSHIPS OF CANOPY LNC WITH SPECTRAL INDICES: DVI, RVI, AND NDVI DURING MID-LATE GROWING PERIOD

Figure 8.3 displays contour maps of relative R^2 from linear regressions of DVI ($R_{\lambda 1}$, $R_{\lambda 2}$), RVI ($R_{\lambda 1}$, $R_{\lambda 2}$), and NDVI ($R_{\lambda 1}$, $R_{\lambda 2}$) against canopy LNC during the mid-late growing period of rice and wheat, respectively. The trends of R^2 from linear regressions of three types of vegetation indices (Table 8.6) were approximately the same for rice and wheat during mid-late growing period, and the detailed band ranges of band 1 and band 2 pairs with the top 10% R^2 in three kinds of vegetation indices (Table 8.4) were mainly in the red-edge region (700–740 nm), green region (500–560 nm), and near-infrared (NIR) region (780–890 nm). In addition, the largest R^2 were 0.875 for NDVI (R_{716}, R_{930}) in rice and 0.901 for NDVI (R_{720}, R_{722}) in wheat, 0.869 for RVI (R_{727}, R_{843}) in rice, and 0.901 for RVI (R_{720}, R_{722}) in wheat, and 0.799 for DVI (R_{722}, R_{927}) in rice and 0.904 for DVI (R_{697}, R_{991}) in wheat during the mid-late period.

8.3.5 RELATIONSHIPS OF CANOPY LNC WITH SPECTRAL INDICES: SAVI AND RVI DURING WHOLE GROWING PERIOD

The sensitive band ranges for LNC monitoring in each crop during the whole growing period were decided by the top 10% R^2 from linear model of SAVI and RVI against LNC, as shown in Table 8.4. Most of the sensitive band ranges of band 1 and band 2 pairs with top 10% R^2 during the whole growing period were both in the visible region in two crops, or in the red-edge and NIR regions in rice, and in the visible and NIR regions in wheat. The largest R^2 during the whole growing period in rice were 0.82 and 0.79 for SAVI (R_{545}, R_{550}) and RVI (R_{538}, R_{562}), respectively, and were 0.876 and 0.864 for SAVI (R_{720}, R_{721}) and RVI (R_{720}, R_{721}) in wheat, respectively (Table 8.5).

8.3.6 COMMON SENSITIVE WAVELENGTH RANGES OF CANOPY LNC

By comprehensively considering the sensitive band ranges in SAVI during the early growing period and in DVI, NDVI, and RVI during the mid-late growing period, we identified the common sensitive wavelength combination (722 and 812 nm) of canopy LNC at about top 5.61% R^2 level for four types of vegetation indices. Figure 8.4 shows the changes of the common wavelength ranges in SAVI during the early growing period and in DVI, NDVI, and RVI during the mid-late growing period by expanding the R^2 level from top 6% to top 10% is step by 0.01%. The regions of the common sensitive wavelength expanded quickly with R^2 from top 5.61% to top 8%, but the

TABLE 8.4

The Wavelength Ranges of Sensitive Band 1 and Band 2 Pairs with Relative Higher R^2 (Top 10% R^2) for Leaf Nitrogen Content Based on Four Types of Spectral Indices in Rice and Wheat during Different Growing Periods

Crop	Growing Period	Spectral Index	Band Pair 1		Band Pair 2		Band Pair 3		Band Pair 4	
			$\lambda 1$ (nm)	$\lambda 2$ (nm)	$\lambda 1$ (nm)	$\lambda 2$ (nm)	$\lambda 1$ (nm)	$\lambda 2$ (nm)	$\lambda 1$ (nm)	$\lambda 2$ (nm)
Rice	Early	SAVI	700–750	725–890	420–520	570–675				
	Mid-late	DVI	700–735	735–1000						
		RVI	465–605	860–1000	520–570	720–860	705–750	735–1000		
		NDVI	450–665	725–1000	695–740	735–1000				
	Whole	SAVI	701–727	718–833	400–520	680–705	400–530	565–680	420–485	490–515
		RVI	417–495	620–695	700–735	705–845	505–525	560–630	615–660	430–510
Wheat	Early	SAVI	695–750	735–1000	520–575	735–1000	435–510	565–595	415–495	505–525
	Mid-late	DVI	400–750	740–1000						
		RVI	460–590	725–1000	695–740	735–1000	415–505	615–690	510–590	555–620
		NDVI	505–740	710–1000	420–505	610–690	510–595	550–660		
	Whole	SAVI	400–470	455–520	400–570	460–665	400–570	680–700	400–720	720–1000
		RVI	515–645	715–1000	685–730	695–1000	615–690	405–475	555–645	515–580

TABLE 8.5

The Performance of Leaf Nitrogen Content Monitoring Models Based on the Best Sensitive Band Combinations in SAVI, DVI, RVI, and NDVI in Rice and Wheat during Different Growing Periods

Crop	Growing Period	Spectral Parameter	Calibration			Validation		
			Regression Equation	R^2	SE	R^2	RRMSE (%)	Slope
Rice	Early	SAVI (R_{763}, R_{793}) $L = 0.355$	$Y = -66.257x + 1.106$	0.887	0.031	0.778	9.3	0.635
	Mid-late	DVI (R_{722}, R_{727})	$Y = -8.908x + 0.377$	0.799	0.097	0.625	24.7	0.968
		RVI (R_{727}, R_{843})	$Y = -4.959x + 4.626$	0.869	0.063	0.854	18.1	1.022
		NDVI (R_{716}, R_{930})	$Y = -4.717x - 0.217$	0.875	0.060	0.844	21.7	1.069
	Whole	SAVI (R_{545}, R_{550})	$Y = 4.115 + 230.25x$	0.820	0.093	0.702	18.2	0.778
		RVI (R_{538}, R_{562})	$Y = -12.778 + 15.798x$	0.790	0.108	0.777	16.6	0.787
Wheat	Early	SAVI (R_{685}, R_{690}) $L = -0.768$	$Y = -1151.304x + 5.376$	0.863	0.068	0.348	26.3	0.603
	Mid-late	DVI (R_{697}, R_{991})	$Y = -12.056x - 0.075$	0.903	0.100	0.793	33.5	1.191
		RVI (R_{720}, R_{722})	$Y = -38.204x + 38.731$	0.901	0.103	0.860	22.7	1.035
		NDVI (R_{720}, R_{722})	$Y = -72.966x + 0.544$	0.901	0.103	0.860	22.8	1.040
	Whole	SAVI (R_{720}, R_{721})	$Y = 0.532 - 250.783x$	0.876	0.127	0.830	28.1	1.014
		RVI (R_{720}, R_{721})	$Y = 75.899 - 75.275x$	0.864	0.140	0.824	27.2	0.984

expanding rate decreased with R^2 from top 8% to top 10%; thus, further analysis was carried out on the common sensitive wavelength regions during the range from top 6% to 8% R^2 level. At top 6% R^2 level, because the shape of the common sensitive band combinations was like a 90° corner, two maximal inscribed rectangular regions of 721–724 nm (λ_1) and 807–813 nm ($\lambda 2$), or 725–728 nm ($\lambda 1$) and 813–821 nm ($\lambda 2$) were selected to precisely locate the regions of common sensitive wavelength combinations. At top 7% and 8% R^2 levels, the maximum inscribed rectangular areas of 719–731 nm ($\lambda 1$) and 797–826 nm ($\lambda 2$), 719–732 nm ($\lambda 1$) and 754–853 nm ($\lambda 2$) were found, respectively. The band ranges were from 4 to 14 nm at band 1 ($\lambda 1$) and from 7 to 100 nm at band 2 ($\lambda 2$) with R^2 from top 6% to top 8% (Table 8.7).

In addition, the best spectral indices with common sensitive band combinations for two crops during the whole growing period were SAVI (R_{722}, R_{724}) or RVI (R_{723}, R_{724}) at top 4.31% and 2.19% R^2 levels, respectively (Table 8.6).

8.3.7 Quantitative Models for Estimating LNC

8.3.7.1 Common Sensitive Wavelength

The common sensitive wavelength pair at about top 5.61% R^2 level (722 and 812 nm) was then used to develop the quantitative models with four types of spectral indices during two different growing periods for monitoring of LNC in two crops. SAVI (R_{722}, R_{812}) (L = 0.216 for rice and L = −0.084 for wheat) was used to estimate canopy LNC during the early period; and DVI (R_{722}, R_{812}), RVI (R_{722}, R_{812}), and NDVI (R_{722}, R_{812}) were used to estimate LNC during the mid-late period. Table 8.6 indicated that the linear regression model in the early period with SAVI (R_{722}, R_{812}) had high R^2 (0.837 in rice and 0.815 in wheat) and low SE (0.045 in rice and 0.092 in wheat). Model test with independent datasets gave the SAVI (R_{722}, R_{812}) a good estimation accuracy for rice with R^2 of 0.928, RRMSE of 8.3%, and slope of 0.965, but a less accurate estimation for wheat with R^2 of 0.443, RRMSE of 29.4%, and slope of 0.544. For LNC monitoring models during the mid-late period, RVI (R_{722}, R_{812}) for linear equation was found to have the best performance for LNC

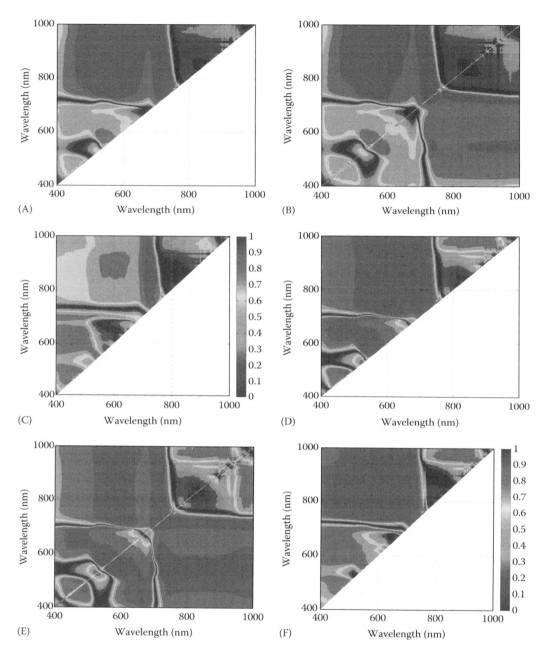

FIGURE 8.3 **(See color insert.)** Contour maps of relative R^2-values for linear relationships of NDVI (A and D), RVI (B and E), and DVI (C and F) against canopy leaf nitrogen concentration during mid-late growing periods of rice (A–C) and wheat (D–F).

prediction, with R^2 of 0.860 and 0.877 and SE of 0.068 and 0.128 for model calibration and with R^2 of 0.860 and 0.877 and RRMSE of 15.7% and 12.8% for model validation in rice and wheat, respectively, and the detailed performance of other models based on DVI and NDVI was displayed in Table 8.6. Figures 8.5 and 8.6 showed the quantitative relationship of SAVI (R_{722}, R_{812}) during the early period and RVI (R_{722}, R_{812}) during the mid-late period against canopy LNC and the 1:1 plotting with the observed and predicted values, respectively.

TABLE 8.6
The Performance of Leaf Nitrogen Content Monitoring Models Based on SAVI, DVI, RVI, and NDVI with Common Sensitive Band Pair during Different Growing Periods in Two Crops

Crop	Growing Period	Spectral Parameter	Calibration			Validation		
			Regression Equation	R^2	SE	R^2	RRMSE (%)	Slope
Rice	Early	SAVI (R_{722}, R_{812}) $L = 0.261$	Y = −4.314x + 1.435	0.837	0.045	0.928	8.3	0.965
	Mid-late	DVI (R_{722}, R_{812})	Y = −7.588x + 0.779	0.756	0.118	0.658	18.2	0.781
		RVI (R_{722}, R_{812})	Y = −4.409x + 4.119	0.860	0.068	0.878	15.7	1.028
		NDVI (R_{722}, R_{812})	Y = −4.656x + 0.303	0.859	0.068	0.860	15.2	0.890
	Whole	SAVI (R_{722}, R_{724}) $L = -0.009$	Y = −49.592x + 0.524	0.775	0.115	0.741	18.4	0.794
		RVI (R_{723}, R_{724})	Y = −52.424x + 52.872	0.774	0.116	0.734	19.4	0.795
Wheat	Early	SAVI (R_{722}, R_{812}) $L = -0.084$	Y = −6.022x + 1.186	0.815	0.092	0.443	29.4	0.544
	Mid-late	DVI (R_{722}, R_{812})	Y = −12.758x + 0.654	0.869	0.136	0.811	32.1	1.219
		RVI (R_{722}, R_{812})	Y = −5.707x + 5.694	0.877	0.128	0.864	20.9	0.984
		NDVI (R_{722}, R_{812})	Y = −7.226x + 0.364	0.874	0.131	0.855	22.3	1.035
	Whole	SAVI (R_{722}, R_{724}) $L = 0.562$	Y = −123.549x + 0.662	0.857	0.147	0.829	27.0	1.026
		RVI (R_{723}, R_{724})	Y = −71.035x + 71.744	0.845	0.159	0.826	26.0	0.986

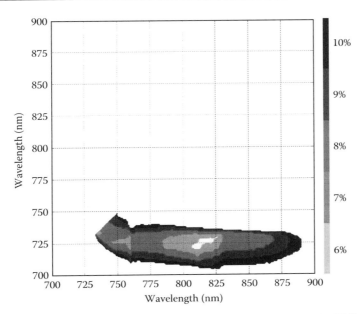

FIGURE 8.4 The common sensitive waveband regions for SAVI, DVI, RVI, and NDVI at different top R^2 levels (each color represents the expanding area with each 1% higher top R^2).

It can be seen that the overall performance was excellent with all monitoring models on LNC (Table 8.6) except for the model based on SAVI (R_{722}, R_{812}) during the early period of wheat. Considering different backgrounds in early periods of different wheat experiments, different L values were tried in different experiments, which provided better performance of SAVI (R_{722}, R_{812}) for LNC estimation during the early growing period, with $R^2 > 0.86$ and RRMSE of 9.4%. The 1:1 plotting with

TABLE 8.7
The Regions of Common Sensitive Wavelength Combinations at Different Top R² Level

Top R² Level	Band Ranges		Center Bands		Band Widths	
	λ1 (nm)	λ2 (nm)	λ1 (nm)	λ2 (nm)	Δλ1 (nm)	Δλ2 (nm)
6%	721–724	807–813	722 or 723	810	4	7
	725–728	813–823	726 or 727	818	4	11
7%	719–731	797–826	725	811 or 812	13	30
8%	719–732	754–853	725 or 726	798 or 799	14	100

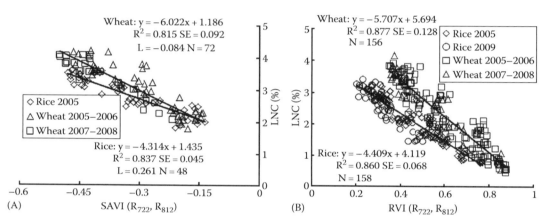

FIGURE 8.5 Quantitative relationships of SAVI (R_{722}, R_{812}) during the early growing period (A) and RVI (R_{722}, R_{812}) during the mid-late growing period (B) against canopy leaf nitrogen concentration of rice and wheat.

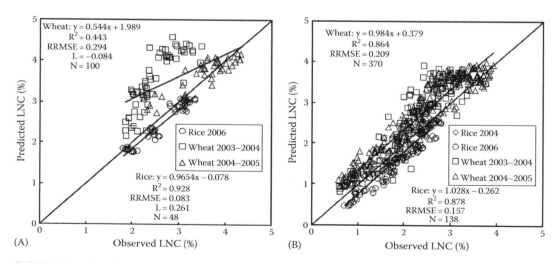

FIGURE 8.6 The 1:1 relationship between the predicted and observed leaf nitrogen concentration based on SAVI (R_{722}, R_{812}) during the early growing period (A) and RVI (R_{722}, R_{812}) during the mid-late growing period (B) in rice and wheat.

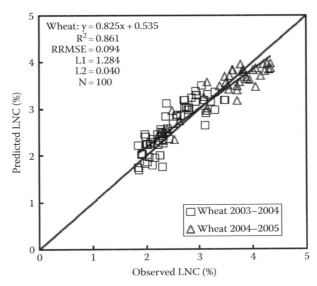

FIGURE 8.7 The 1:1 relationship between the predicted and observed leaf nitrogen concentration based on SAVI (R_{722}, R_{812}) during the early growing period with varied L in different experiments of wheat.

the observed and predicted values exhibited the reliable performance of SAVI (R_{722}, R_{812}) with varied L in different experiments (Figure 8.7).

Compared with the linear regression models against LNC with the best band pairs for each of spectral indices and crops during different growing periods (Table 8.5), the linear regression models with the common sensitive band pair (722 and 812 nm) for four types of spectral indices and two crops had a little lower R^2, but with a better performance of model validation (Table 8.6).

Figure 8.8 showed the quantitative relationships with SAVI (R_{722}, R_{724}) for monitoring of LNC during the whole growing period in two crops. Compared with the models based on two kinds of spectral indices during two growing periods [SAVI (R_{722}, R_{812}) in the early growing period and RVI (R_{722}, R_{812}) in the mid-late growing period], the model with SAVI (R_{722}, R_{724}) during the whole growing period in two crops had lower R^2 and higher RRMSE, indicating a bad accuracy

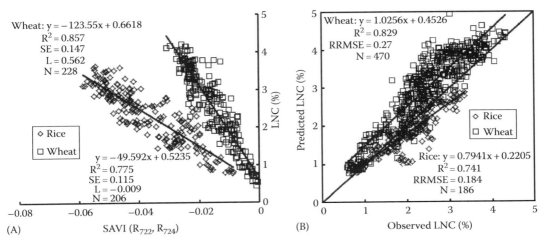

FIGURE 8.8 Quantitative relationships of SAVI (R_{722}, R_{724}) during the whole growing period against canopy leaf nitrogen concentration of rice and wheat (A) and the 1:1 relationship between the predicted and observed leaf nitrogen concentration based on SAVI (R_{722}, R_{724}) in rice and wheat (B).

of LNC estimation. In addition, the two sensitive bands in SAVI (R_{722}, R_{724}) were together during the region of red-edge and were very close; thus, this combination cannot be suggested as the optimum result.

8.3.7.2 Center Bands with Optimum Bandwidths

Table 8.7 showed that, with R^2 from top 6% to top 8%, four combinations of common sensitive waveband regions were found, and the bandwidths of these four combinations varied from 4 to 14 nm at band 1 ($\lambda1$) and from 7 to 100 nm at band 2 ($\lambda2$), with center wavebands varied from 722 to 727 nm at band 1 ($\lambda1$) and from 803 to 818 nm at band 2 ($\lambda2$). Then, the reflectance values in the common sensitive band range for each of four combinations were averaged for construction of SAVI and RVI, and the models were developed based on SAVI and RVI with broadband spectra. The results indicated that the combination with center bands of 724 or 725 nm at band 1 ($\Delta\lambda1 = 13$ nm) and 811 nm at band 2 ($\Delta\lambda2 = 30$ nm) had the best performance in model calibration, but had the worse performance in model validation than the combination with center bands of 725 nm at band 1 ($\Delta\lambda1 = 14$ nm) and 803 nm at band 2 ($\Delta\lambda2 = 100$ nm). The combination of broad bands at top 6% R^2 level did not give better performance than combinations at top 7% and 8% R^2 levels. In addition, no significant differences in model performance were observed with the combinations of broad bands (Table 8.8) and narrow bands (Table 8.6). Since the center bands of the combination at top 7% R^2 level (724 or 725 nm at band 1, and 811 nm at band 2) were close to the wavebands of the combination at top 5.61% R^2 level (722 nm at band 1 and 812 nm at band 2), and the corresponding bandwidths at band 1 ($\Delta\lambda1 = 13$ nm) and band 2 ($\Delta\lambda2 = 30$ nm) were simultaneously larger than 10 nm, the combination with center bands of about 725 nm at band 1 ($\Delta\lambda1 = 13$ nm) and 810 nm at band 2 ($\Delta\lambda2 = 30$ nm) was recommended as the optimum selection for estimating LNC in both rice and wheat. Figure 8.9 showed the quantitative relationships of SAVI (R_{725}, R_{810}) during the early growing period and RVI (R_{725}, R_{810}) during the mid-late growing period with bandwidths of 13 nm at band 1 (725 nm) and

---

TABLE 8.8
The Performance of Leaf Nitrogen Content Monitoring Models Based on Broadband Vegetation Indices

Center Bands (Bandwidths, nm)	Crop	Growing Period	Type of Vegetation Indices	Calibration		Validation		
				R^2	SE	R^2	RRMSE	Slope
$\lambda1$:722 or 723	Rice	Early	SAVI	0.84	0.044	0.928	0.082	0.962
($\Delta\lambda1 = 4$)		Mid-late	RVI	0.86	0.068	0.877	0.156	1.018
$\lambda2$:810	Wheat	Early	SAVI	0.814	0.093	0.859	0.094	0.798
($\Delta\lambda2 = 7$)		Mid-late	RVI	0.876	0.129	0.863	0.21	0.985
$\lambda1$:726 or 727	Rice	Early	SAVI	0.84	0.044	0.928	0.082	0.962
($\Delta\lambda1 = 4$)		Mid-late	RVI	0.86	0.068	0.877	0.156	1.018
$\lambda2$:818	Wheat	Early	SAVI	0.814	0.093	0.859	0.094	0.823
($\Delta\lambda2 = 11$)		Mid-late	RVI	0.876	0.129	0.863	0.21	0.985
$\lambda1$:725	Rice	Early	SAVI	0.84	0.044	0.926	0.078	0.945
($\Delta\lambda1 = 13$)		Mid-late	RVI	0.863	0.066	0.872	0.157	0.996
$\lambda2$:811 or 812	Wheat	Early	SAVI	0.813	0.093	0.841	0.1	0.81
($\Delta\lambda2 = 30$)		Mid-late	RVI	0.873	0.132	0.858	0.214	0.991
$\lambda1$:725 or 726	Rice	Early	SAVI	0.839	0.045	0.931	0.065	0.951
($\Delta\lambda1 = 14$)		Mid-late	RVI	0.862	0.067	0.872	0.143	0.988
$\lambda2$:798 or 799	Wheat	Early	SAVI	0.812	0.094	0.846	0.087	0.813
($\Delta\lambda2 = 100$)		Mid-late	RVI	0.874	0.131	0.861	0.215	0.998

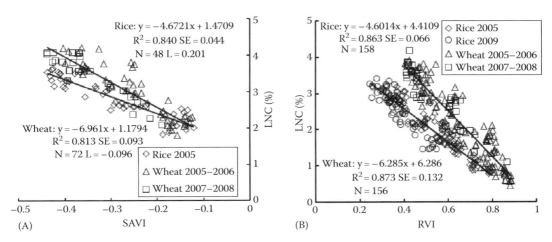

FIGURE 8.9 Quantitative relationships of SAVI (R_{725}, R_{810}) during the early growing period (A) and RVI (R_{725}, R_{810}) during the mid-late growing period (B) with bandwidths of 13 nm at band 1 (725 nm) and 30 nm at band 2 (810 nm) against canopy leaf nitrogen concentration of rice and wheat.

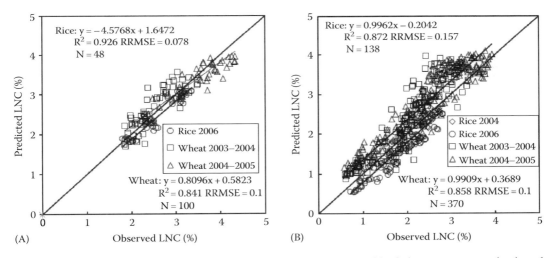

FIGURE 8.10 The 1:1 relationship between the predicted and observed leaf nitrogen concentration based on SAVI (R_{725}, R_{810}) during the early growing period (A) and RVI (R_{725}, R_{810}) during the mid-late growing period (B) with bandwidths of 13 nm at band 1 (725 nm) and 30 nm at band 2 (810 nm) in rice and wheat.

30 nm at band 2 (810 nm) against canopy leaf nitrogen concentration of rice and wheat. Figure 8.10 showed the 1:1 relationship between the predicted and observed leaf nitrogen concentration based on SAVI (R_{725}, R_{810}) during the early growing period and RVI (R_{725}, R_{810}) during the mid-late growing period with bandwidths of 13 nm at band 1 and 30 nm at band 2 in rice and wheat.

8.3.8 PERFORMANCE OF SPECTRAL INDICES IN PREVIOUS STUDIES

The performance of spectral indices for LNC monitoring in wheat and rice in previous studies had been analyzed with our experiment dataset. The results showed that the previous spectral indices for LNC monitoring (Table 8.9) are less precise and stable than SAVI (R_{722}, R_{812}) and RVI (R_{722}, R_{812}) with narrowband reflectance (Table 8.6) or SAVI (R_{725}, R_{810}) and RVI (R_{725}, R_{810}) with broad band reflectance in this study (Figures 8.9 and 8.10).

TABLE 8.9
The Performance of Leaf Nitrogen Content Monitoring Models Based on Other Spectral Indices in Previous Studies with the Data from Eight Experiments

Crop	Spectral Parameter	Calibration Regression Equation	R^2	SE	Validation R^2	RRMSE	Slope	Source of Spectral Index
Rice	NDRE	$Y = 4.069x + 0.437$	0.738	0.135	0.727	0.231	0.705	Barnes et al. (2000)
Wheat		$Y = 6.089x + 0.582$	0.818	0.187	0.806	0.190	0.875	
Rice	CCCI	$Y = 2.436x + 1.010$	0.738	0.135	0.727	0.181	0.892	
Wheat		$Y = 3.724x + 0.910$	0.818	0.187	0.806	0.162	0.837	
Rice	mND	$Y = 3.892x - 0.299$	0.777	0.115	0.800	0.222	0.902	Sims and Gamon
Wheat		$Y = 4.385x + 0.195$	0.855	0.149	0.828	0.196	0.917	(2002)
Rice	mSR	$Y = 0.334x + 1.160$	0.588	0.212	0.597	0.229	0.386	
Wheat		$Y = 0.803x + 0.823$	0.772	0.234	0.765	0.242	0.895	
Rice	NDSI (R_{440}, R_{573})	$Y = 6.104x + 4.934$	0.530	0.242	0.349	0.276	0.497	Hansen and
Wheat		$Y = 16.838x + 9.110$	0.444	0.572	0.521	0.427	0.924	Schjoerring (2003)
Rice	REMI	$Y = 0.722x + 1.222$	0.659	0.176	0.680	0.237	0.412	Gitelson (2005)
Wheat		$Y = 1.780x + 1.052$	0.747	0.260	0.726	0.207	0.785	
Rice	RSI (R_{870}, R_{660})	$Y = 0.053x + 1.479$	0.578	0.217	0.570	0.218	0.374	Zhu et al. (2008)
Wheat		$Y = 0.162x + 1.309$	0.701	0.308	0.734	0.385	1.092	
Rice	RSI (R_{810}, R_{660})	$Y = 0.055x + 1.486$	0.581	0.216	0.577	0.217	0.370	
Wheat		$Y = 0.166x + 1.314$	0.702	0.307	0.738	0.382	1.096	
Rice	NDSI (R_{860}, R_{720})	$Y = 3.929x + 0.541$	0.596	0.208	0.506	0.303	0.680	Yao et al. (2010)
Wheat		$Y = 7.421x + 0.155$	0.824	0.181	0.773	0.190	0.849	
Rice	RSI (R_{990}, R_{720})	$Y = 0.555x + 0.708$	0.608	0.201	0.582	0.255	0.409	
Wheat		$Y = 1.552x - 0.643$	0.776	0.230	0.732	0.199	0.794	

8.4 DISCUSSION

8.4.1 SELECTION OF SPECTRAL INDICES DURING DIFFERENT GROWING PERIODS

Reflectance of vegetation canopy is a kind of synthetic information, mainly influenced by the internal biochemical components of vegetation and external factors such as atmospheric characteristics, soil background, and canopy structure. Vegetation index can reduce the effects of soil background, atmospheric disturbance, and canopy structure to some extent (Zhao, 2003; Dorigo et al., 2007). Thus, considering the influencing factors of canopy reflectance during different growing stages and the advantages of different vegetation indices, different types of vegetation indices should be selected to estimate the LNC during different growing stages. Soil background is the main external factor of canopy reflectance during the early growing period because canopy is not fully closed. The application of SAVI can effectively reduce soil background by adjusting the parameters L (Huete, 1988; Rondeaux et al., 1996).

SAVI (R_{722}, R_{812}) constructed in this paper with L of 0.261 to adjust background of water layer for rice and L of −0.084 to adjust background of soil for wheat performed better for LNC monitoring than NDVI. Different L in SAVI achieved the effective regulation of differential background in rice and wheat crops. Huete (1988) found that the parameter L in SAVI showed negative correlation with leaf area index and recommended the value of parameter L as 0.5. In this study, however, we found that the background impact on canopy reflectance during the early growing period should be regulated with different L in rice and wheat, which may be because the background of rice is water layer, rather than traditional soil background. In addition, the suitable L changed with different experiments in wheat crop, because of different soil backgrounds in different wheat experiments. In contrast, the different backgrounds of water layer in rice have less impact on the canopy reflectance, because reflectance value of water is far less than that of soil and vegetation; thus, the same L in SAVI performed well not only with the dataset for model development, but also with the independent dataset for model testing in rice crop. In this paper, we have only provided the recommended values of L during different growing seasons; thus, the precise values of parameter L during varied growing periods in different experiments still need further research. A possible solution may be development of dynamic relationship between L and canopy coverage during entire growth period.

Comparing the overall relationships of NDVI, RVI, and DVI with all available two-band combinations against canopy LNC during the mid-late growing period of rice and wheat, RVI was the best index for LNC monitoring during the mid-late period, DVI had the worst performance because of insensitivity under high vegetation coverage, and the NDVI performance was slightly worse than RVI because of saturation under high vegetation coverage. Previous studies also showed that RVI performed well on monitoring N status (Jacobsen et al., 1998; Zhu et al., 2008; Yao et al., 2009a,b). However, because RVI does not take into account the impact of atmosphere and soil background, the performance of RVI under airborne or spaceborne situation still need to be tested.

8.4.2 COMMON SENSITIVE WAVELENGTH RANGES OF CANOPY LNC

The characteristic spectral information of N is essentially same during different growing stages of different crops, so the trends of correlations between LNC and reflectance in the range of 400–1000 nm were overall consistent. The obvious common band combinations were found during two different growing periods of rice and wheat, but the correlation of reflectance to LNC could be improved by constructing different vegetation indices during different growing periods because of the differential impacts of the varied backgrounds on canopy spectra. We fully exploited the massive hyperspectral information by analyzing vegetation indices with all available two-band combinations based on eight field experiments under different growing seasons, eco-sites, N rates, and water regimes in rice and wheat crops. The common sensitive two-band combination ranges were identified mainly in the red-edge region and NIR region at different R^2 levels with SAVI during the early growing period and with DVI, RVI, and NDVI during the mid-late growing period. The characteristics of reflectance during red-edge region (680–760 nm) were generated from the combining results of strong chlorophyll absorption in the red region and high reflectance in the NIR platform (Cho and Skidmore, 2006). The existing studies showed a strong correlation between reflectance in the red-edge region and pigment content (Horler et al., 1983; Dawson and Curran, 1998), which was closely related with N (Yoder and Pettigrew-Crosby, 1995). The common sensitive ranges in the NIR region were related to leaf internal scattering, especially the relative thickness of the leaf and cell gap (Zhao, 2003). The precisely common sensitive band ranges were presented at different top R^2 levels of regression models, which can be selected to design the spectral sensor with relatively broad bands, and would help to overcome the shortcomings in practical application of hyperspectral sensors (Thenkabail et al., 2004), such as large data acquisition, transmission, processing, and other issues.

8.4.3 ACCURACY AND UNIVERSALITY OF MONITORING MODELS FOR CANOPY LNC

Physiological parameters and canopy reflectance in agronomic crops differ with cultural conditions, but the intrinsic relationship between physiological parameters and canopy reflectance was essentially stable (Hansen and Schjoerring, 2003; Dorigo et al., 2007). Based on a large amount of experimental datasets involving different crops, N rates, water regimes, eco-sites, and cultivars, four quantitative linear models for LNC monitoring during two growing periods of rice and wheat were constructed, respectively, and the model testing with independent dataset indicated a good fit between estimated and observed LNC. The models with common sensitive band pair (722 and 812 nm) selected from four types of vegetation indices of two crops during two growing periods had higher universality than that with single vegetation index of different sensitive bands for each growing period and each crop. With the increasing of bandwidths, the accuracy of models had not significant change and the models with optimum bandwidths of 14 and 100 nm for each center band still had stable performance for model validation. In our study, the best monitoring models with SAVI (R_{722}, R_{812}) and RVI (R_{722}, R_{812}) still need different linear equations for LNC monitoring during different growing periods of rice and wheat because of varied growth conditions and backgrounds (Huete, 1988). However, intrinsic relationship between N status and spectral reflectance is not changing with crop type and backgrounds, so further studies could be undertaken to refine the LNC and reflectance relations with new data mining method, such as artificial neural network (Huang et al., 2004), genetic algorithm (Whitley et al., 1990), and support vector regression (Ahmad et al., 2010).

8.5 CONCLUSIONS

Through comprehensive analyses on hyperspectral reflectance and canopy leaf nitrogen concentration (LNC) from eight field experiments involving different N rates, water regimes, cultivars, and growing seasons, the present study found some common features in the relationships between canopy LNC and different vegetation indices during two growing periods (early growing period and mid-late growing period) of two cereal crops (rice and wheat) studied here. The common sensitive wavelength ranges for LNC monitoring were identified mainly in specific portions of red-edge (centered at 725 nm) and NIR (centered at 811 nm) at different top R^2 levels in both rice and wheat crops. Based on the reflectance of narrow bands, SAVI (R_{722}, R_{812}) during the early growing period and RVI (R_{722}, R_{812}) during the mid-late growing period were the ideal hyperspectral indices to construct the linear regression models for LNC estimation in both rice and wheat, with different regression equations for two growing periods and two crops. Based on the reflectance of broad bands, the combination with center bands of about 725 nm at band 1 ($\Delta\lambda 1 = 13$ nm) and 810 nm at band 2 ($\Delta\lambda 2 = 30$ nm) were recommended to construct the SAVI during the early growing period and RVI during the mid-late growing period for estimating LNC in both rice and wheat. In addition, the parameter L in SAVI during the early period of wheat somewhat differed with the experiments, and optimization of L with the prior knowledge of vegetation coverage can improve the accuracy of LNC estimation in rice and wheat. These results will provide a technical support for real-time monitoring of N status and development of portable sensor in crop growth monitoring.

REFERENCES

Ahmad, S., Kalra, A., and Stephen, H. (2010) Estimating soil moisture using remote sensing data: A machine learning approach, *Advances in Water Resources*, 33, 69–80.

Analytical Spectral Devices, Inc. (1997) *FieldSpecTM User's Guide*. Analytical Spectral Devices, Inc., Boulder, CO.

Analytical Spectral Devices, Inc. (2002) *FieldSpec Pro User's Guide*.

Asner, G.P. and Martin, R.E. (2008) Spectral and chemical analysis of tropical forests: Scaling from leaf to canopy levels, *Remote Sensing of Environment*, 112, 3958–3970.

Barnes, E.M., Clarke, T.R., Richards, S.E., Colaizzi, P.D., Haberland, J., Kostrzewski, M., Waller, P., Choi, C., Riley, E., Thompson, T., Lascano, R.J., Li, H., and Moran, M.S. (2000) Coincident detection of crop water stress, nitrogen status and canopy density using ground based multispectral data. In: *Proc. 5th Int. Conf. Precis Agric.* (Robert, P.C., Rust, R.H., Larson, W.E., Eds.), Bloomington, MN.

Cassman, K.G., Dobermann, A., and Walters, D.T. (2002) Agroecosystems, nitrogen-use efficiency, and nitrogen management, *AMBIO: A Journal of the Human Environment*, 31, 132–140.

Cho, M.A. and Skidmore, A.K. (2006) A new technique for extracting the red-edge position from hyperspectral data: The linear extrapolation method, *Remote Sensing of Environment*, 101, 181–193.

Dawson, T. and Curran, P. (1998) Technical note A new technique for interpolating the reflectance red-edge position, *International Journal of Remote Sensing*, 19, 2133–2139.

Dorigo, W.A., Zurita-Milla, R., de Wit, A.J.W., Brazile, J., Singh, R., and Schaepman, M.E. (2007) A review on reflective remote sensing and data assimilation techniques for enhanced agroecosystem modeling, *International Journal of Applied Earth Observation and Geoinformation*, 9, 165–193.

Gitelson, A.A., Kaufman, Y.J., Stark, R., and Rundquist, D. (2002) Novel algorithms for remote estimation of vegetation fraction, *Remote Sensing of Environment*, 80, 76–87.

Gitelson, A.A., Vina, A., Ciganda, V., Rundquist, D.C., and Arkebauer, T.J. (2005) Remote estimation of canopy chlorophyll content in crops. *Geophysical Research Letters,* 32, L08403.

Hansen, P.M. and Schjoerring, J.K. (2003) Reflectance measurement of canopy biomass and nitrogen status in wheat crops using normalized difference vegetation indices and partial least squares regression, *Remote Sensing of Environment*, 86, 542–553.

Horler, D.N.H., Dockray, M., and Barber, J. (1983) The red-edge of plant leaf reflectance, *International Journal of Remote Sensing*, 4, 273–288.

Huang, Z., Turner, B.J., Dury, S.J., Wallis, I.R., and Foley, W.J. (2004) Estimating foliage nitrogen concentration from HYMAP data using continuum removal analysis, *Remote Sensing of Environment*, 93, 18–29.

Huete, A.R. (1988) A soil-adjusted vegetation index (SAVI), *Remote Sensing of Environment*, 25, 295–309.

Inoue, Y., Moran, M.S., and Horie, T. (1998) Analysis of spectral measurements in paddy field for predicting rice growth and yield based on a simple crop simulation model, *Plant Production Science*, 1, 269–279.

Jacobsen, S.E., Pedersen, H., and Jensen, C.R. (1998) Reflectance measurements, a quick and nondestructive technique for use in agricultural research. In *International Conference on Sustainable Agriculture in Tropical and Subtropical Highlands with Special Reference to Latin America (SATHLA)* (Janeiro, R.D., Ed.), Condensan, Lima, 1–5.

Jordan, C. (1969) Derivation of leaf-area index from quality of light on the forest floor, *Ecology*, 50, 663–666.

Ju, X.-T., Xing, G.-X., Chen, X.-P., Zhang, S.-L., Zhang, L.-J., Liu, X.-J., Cui, Z.-L., Yin, B., Christie, P., Zhu, Z.-L., and Zhang, F.-S. (2009) Reducing environmental risk by improving N management in intensive Chinese agricultural systems, *Proceedings of the National Academy of Sciences*, 106, 3041–3046.

Kokaly, R.F. (2001) Investigating a physical basis for spectroscopic estimates of leaf nitrogen concentration, *Remote Sensing of Environment*, 75, 153–161.

Liang, S. (2004) *Quantitative Remote Sensing of Land Surfaces*, Wiley-Interscience, Hoboken, NJ.

Pearson, R.L. and Miller, L.D. (1972) Remote mapping of standing crop biomass for estimation of the productivity of the short-grass prairie. In *Proceedings of the Eighth International Symposium on Remote Sensing of Environment* (Asrar, G., Ed.), Pawnee National Grasslands, Grover, CO, 1357–1381.

Raun, W.R., Solie, J.B., Johnson, G.V., Stone, M.L., Mullen, R.W., Freeman, K.W., Thomason, W.E., and Lukina, E.V. (2002) Improving nitrogen use efficiency in cereal grain production with optical sensing and variable rate application, *Agronomy Journal*, 94, 815–820.

Ray, S.S., Jain, N., Miglani, A., Singh, J.P., Singh, A.K., Panigrahy, S., and Parihar, J.S. (2010) Defining optimum spectral narrow bands and bandwidths for agricultural applications, *Current Science*, 98, 1365–1369.

Rondeaux, G., Steven, M., and Baret, F. (1996) Optimization of soil-adjusted vegetation indices, *Remote Sensing of Environment*, 55, 95–107.

Rouse, J.W., Haas, R.H., Schell, J.A., and Deering, D.W. (1973) Monitoring vegetation systems in the great plains with ERTS, *Third ERTS Symposium*, NASA SP-351 I, 309–317.

Shanahan, J.F., Kitchen, N.R., Raun, W.R., and Schepers, J.S. (2008) Responsive in-season nitrogen management for cereals, *Computers and Electronics in Agriculture*, 61, 51–62.

Sims, D.A. and Gamon, J.A. (2002) Relationships between leaf pigment content and spectral reflectance across a wide range of species, leaf structures and developmental stages, *Remote Sensing of Environment*, 81, 337–354.

Sims, D.A. and Gamon, J.A. (2003) Estimation of vegetation water content and photosynthetic tissue area from spectral reflectance: A comparison of indices based on liquid water and chlorophyll absorption features, *Remote Sensing of Environment*, 84, 526–537.

Thenkabail, P.S., Enclona, E.A., Ashton, M.S., and Van Der Meer, B. (2004) Accuracy assessments of hyperspectral waveband performance for vegetation analysis applications, *Remote Sensing of Environment*, 91, 354–376.

Whitley, D., Starkweather, T., and Bogart, C. (1990) Genetic algorithms and neural networks: Optimizing connections and connectivity, *Parallel Computing*, 14, 347–361.

Xue, L., Cao, W., Luo, W., Dai, T., and Zhu, Y. (2004) Monitoring leaf nitrogen status in rice with canopy spectral reflectance, *Agronomy Journal*, 96, 135–142.

Yao, X., Zhu, Y., Feng, W., Tian, Y.-C., and Cao, W.-X. (2009a) Exploring novel hyperspectral band and key index for leaf nitrogen accumulation in wheat, *Guang Pu Xue Yu Guang Pu Fen Xi*, 29, 2191–2195.

Yao, X., Zhu, Y., Tian, Y., Feng, W., and Cao, W. (2009b) Research of the optimum hyperspectral vegetation indices on monitoring the nitrogen content in wheat leaves, *Scientia Agricultura Sinica*, 42, 2716–2725.

Yao, X., Zhu, Y., Tian, Y., Feng, W., and Cao, W. (2010) Exploring hyperspectral bands and estimation indices for leaf nitrogen accumulation in wheat, *International Journal of Applied Earth Observation and Geoinformation*, 12, 89–100.

Yoder, B.J. and Pettigrew-Crosby, R.E. (1995) Predicting nitrogen and chlorophyll content and concentrations from reflectance spectra (400–2500 nm) at leaf and canopy scales, *Remote Sensing of Environment*, 53, 199–211.

Zhao, D., Huang, L., Li, J., and Qi, J. (2007) A comparative analysis of broad-band and narrow-band derived vegetation indices in predicting LAI and CCD of a cotton canopy, *ISPRS Journal of Photogrammetry and Remote Sensing*, 62, 25–33.

Zhao, Y.S. (2003) *Remote Sensing application Analysis of Principle and Methods,* Science Press, Beijing, China.

Zhu, Y., Yao, X., Tian, Y., Liu, X., and Cao, W. (2008) Analysis of common canopy vegetation indices for indicating leaf nitrogen accumulations in wheat and rice, *International Journal of Applied Earth Observation and Geoinformation*, 10, 1–10.

Zhu, Y., Zhou, D., Yao, X., Tian, Y., and Cao, W. (2007) Quantitative relationships of leaf nitrogen status to canopy spectral reflectance in rice, *Australian Journal of Agricultural Research*, 58, 1077–1085.

9 Characterization on Pastures Using Field and Imaging Spectrometers

Izaya Numata

CONTENTS

9.1 INTRODUCTION

Pasturelands cover extensive areas of non-iced land in the world and are distributed across diverse ecosystems [1]. This landcover type is of interest in terms of feed availability, quality for livestock production, and regional ecosystem functioning. Biophysical and biochemical properties of pastures are highly variable and are altered by environmental conditions such as climate, soils, and human management. It is critical to understand changes in pastures under human–environment interactions in order to provide relevant information to management practices as well as estimate the impacts on regional ecosystem functioning. Remote sensing has played an important role in monitoring pasture dynamics and providing estimates of pasture properties using multispectral sensors. From the pasture management perspective, one of the ultimate goals for the use of remote sensing would be the quantification of biophysical and biochemical properties of pasture based solely on remotely sensed data, data that are available at a low cost and do not require time-consuming and costly field sampling and subsequent laboratory analysis. Despite the contributions of previous studies based upon multispectral sensors, the accuracy of broadband remote sensing data for grass estimation is still limited due to their spatial and spectral resolution [2].

Hyperspectral remote sensing provides spectrally rich datasets to measure vegetation properties with many narrowbands and refined techniques for the estimation of pasture properties or develop new approaches to improve our ability of pasture characterization. In the past decade, studies on hyperspectral remote sensing of pasture has made significant progress toward detection and quantification of pasture biophysical (e.g., biomass) and biochemical (e.g., nutrients and water) variables by identifying critical wavebands and exploiting new absorption features. Despite these efforts, there are a limited number of research studies on pasture characterization using hyperspectral remote sensing. As a result, an integrated and comprehensive knowledge-base on hyperspectral remote sensing of pasture does not exist.

Given the above background this chapter first provides the current state-of-the-art on studies pertaining to hyperspectral remote sensing of pasture using field-based and imaging spectrometers. This is followed by an overview of ideas on the future directions of hyperspectral remote sensing of pasture.

9.2   FIELD AND IMAGING SPECTROMETERS FOR PASTURE CHARACTERIZATION

Laboratory or in situ spectrometer measurements are necessary steps in order to identify basic spectral characteristics of different vegetation species and establish relationships between vegetation attributes and hyperspectral measures [3,4]. A number of studies on pasture/grass characterization have been conducted using field spectrometers for a wide range of focuses such as the estimates of biophysical [biomass and leaf area index (LAI)] and biochemical concentrations (pigments and nutrients, water content), fractional cover, litter estimates to pasture degradation, etc. [5–10]. Compared to other land cover types such as forest, field spectrometer–based analysis for pasture characterization is well adopted by a large number of research studies. As most grasslands and pastures are composed of vegetation with height of 1 m or less, field-based analysis of pasture at the canopy level via a field spectrometer is a practical way to assess relationships between field biophysical and biochemical data and remote sensing measurements, while allowing sources of errors and their impacts on the analyzed relationships to be assessed easily [11]. Field-based experiments are applied for broader spatial scales using imaging spectrometers or used in the inversion of radiative transfer models to estimate vegetation properties. Reflectance spectra from distinct grass species or vegetation materials measured in the field can be stored in spectral libraries and used as reference or ideal spectra to calibrate imaging spectrometers and map distinct species at larger spatial scales.

While laboratory analysis is performed under controlled conditions, vegetation spectra measured in situ are governed by many factors such as canopy vertical and spatial structure, presence of live and dead materials, and the diversity of background (discussed in the next section). Thus, in order to have good relationships between grass biophysical and biochemical parameters and field spectra, reliable measurements of both field grass and field spectrometers should be taken. To reduce the cloud interference into field spectral measurement, Kawamura et al. [10] used a canopy pasture probe (CAPP), which has its own artificial light source enclosed within a dark chamber attached to a portable spectroradiomater in New Zealand, where sky is often covered by cloud. The chamber is placed over pasture and excludes sunlight. This is semicontrolled spectral measurement independent from sunlight.

Imaging spectroscopy provides regional biophysical and biochemical measurements of landscapes and offers an avenue to extend field-level relationships between grass productivity and other ecosystem processes to broader spatial scales. Compared to field experiments for pasture, there are still a small number of studies using imaging spectrometers for pasture characterization. Currently, several airborne (e.g., AVIRIS, Hydice, HyMap, and CASI) and satellite (e.g., Hyperion) imaging spectrometers with different spectral and spatial resolutions are available. One of the goals of the use of hyperspectral remote sensing for pastures is to extract biophysical and biochemical attributes (biomass, LAI, and biochemical concentrations) and detect their spatial distributions through quantitative analytical methods. Another goal is to improve discrimination of grasslands from other land covers types that are spectrally ambiguous in the broad spectral band domain. For example, discrimination between dry pasture and bare soil and between green pasture and secondary forest is a real challenge for land cover mapping in the Amazon using broadband sensors like Landsat [12,13]. Spatial patterns of pasture characteristics at landscape scales provide relevant information for land owners to make decisions on management strategy.

9.3 CONTROLLING FACTORS FOR BIOPHYSICAL AND BIOCHEMICAL CHARACTERSTICS OF PASTURE

There are several factors altering grass biophysical and biochemical characteristics that also directly affect grass spectral reflectance signatures. Hill [11] lists several grass physical characteristics important for remote sensing, including (a) height and variation in height, (b) proportion of bare soil, (c) leaf area, (d) leaf orientation or leaf angle distribution, (e) density of reflective or absorptive structures, (f) proportion of live and dead materials, and (g) spatial arrangement of structures. In addition, natural and human-related factors such as climate, soils, species composition in pastureland, variable grazing pressures as well as age of grass all impact on grass biophysical and chemical properties. Here, some of the issues will be highlighted.

9.3.1 STRUCTURE

Asner [14] and Asner and Heidebrecht [15] provide an excellent summary of vegetation spectral properties and their changes as a function of vegetation structure. The authors emphasized the complexity of these factors by using an inversed radiative model. Grass structure influences the interactions between radiation and grass canopy and the responses of radiation (reflected or scattered) to remote sensors. Vertical structure and spatial arrangement of structures of grass vary according to grass species. Figures 9.1 and 9.2 illustrate structures of two predominant grass species used in Brazilian Amazon pastures.

Brachiaria brizantha presents highly heterogeneous surface. This species has stout erect culms and forms bunched crowns. This creates a tufted structure that does not cover the soil surface evenly, which results in the significant effects of soil background on vegetation reflectance (Figures 9.1a and 9.2a and b). *Brachiaria decumbens* is low growing and more decumbent, and forms a dense cover, creating a more homogeneous canopy surface (Figures 9.1b and 9.2c and d). In pastures, grass structures are heavily altered by grazing [12]. Average reflectance signatures and the spectral variability of the two species are strongly related to their structural differences especially at in situ scale (Figure 9.2c and d). Numata et al. [12] observed that the variation in canopy structure within the field of view of a field spectrometer contributes to spectral variability of canopy reflectance even for those areas with the same amount of biomass. In the case of these two species, the heterogeneous and complex canopy structure of *B. brizantha* makes biomass estimation more challenging [9].

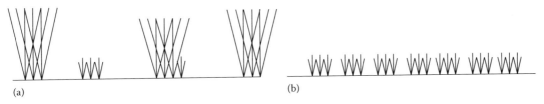

(a)   (b)

FIGURE 9.1   Grass structure: (a) *Brachiaria brizantha*; (b) *Brachiaria decumbens*.

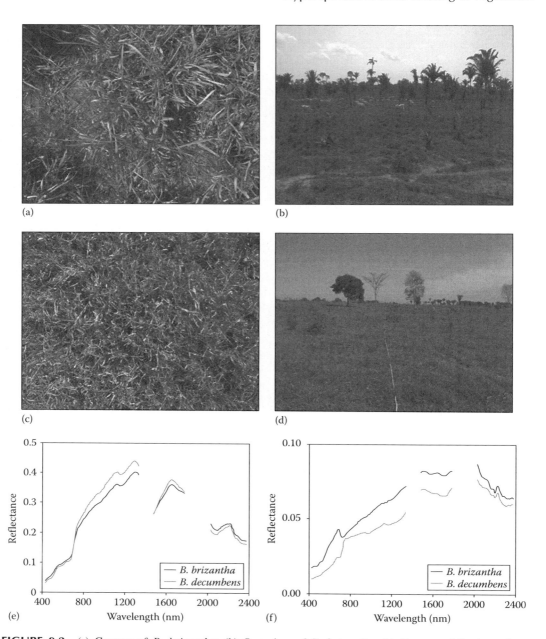

FIGURE 9.2 (a) Canopy of *B. brizantha*; (b) Overview of *B. brizantha*; (c) Canopy of *B. decumbens*; (d) Overview of *B. decumbens*; (e) Averaged reflectance of *B. brizantha* and *B. decumben*; (f) Standard deviation from averaged reflectance of two species.

9.3.2 FOLIAR CHEMICAL COMPOSITION

The concentrations of chemical constituents such as chlorophyll a and b, protein, lignin, cellulose, water, and others have a great influence on vegetation reflectance (Figure 9.3). The absorption features found in the visible region (430–660 nm) such as chlorophyll absorptions are correlated with major nutrients important for animal production, while the shortwave infrared (SWIR) region is highly characterized by lignin-cellulose absorptions [3]. Green leaf reflectance is determined primarily by

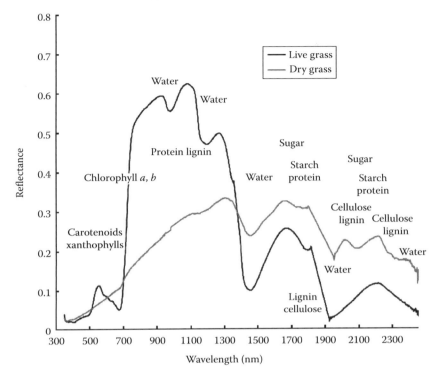

FIGURE 9.3 The reflectance spectra with characteristic absorption features associated with plant chemical constituents for live and dry grass. (Adapted from Hill, M.J., Grazing agriculture: Managed pasture, grassland, and rangeland. In *Manual of Remote Sensing Volume 4. Remote Sensing for Natural Resource Management and Environmental Monitoring*, Ustin, S.L., Ed., John Wiley & Sons, Hoboken, NJ, 2004, pp. 449–530.)

water, pigment, and carbon content, while dry leaf reflectance shows strong signals of lignin and cellulose content. In a green leaf, water absorption obscures lignin, cellulose, and other constituents [16,17]. These chemical constituents determine forage quality and are also responsible for forming characteristic absorption features in the reflectance spectra of forage plants. These spectral characteristics associated with chemical composition can indicate grass physiological conditions and vigor, and also offer information on nutrient status important for animal productivity. These absorption features and shapes have been used to estimate the concentrations of chemical compounds as well as to distinguish grass from different land covers [18] and different grass species [19].

9.3.3 Nonphotosynthetic Vegetation and Background Effects

Most studies dealing with grass/pasture characterization through remote sensing emphasize only green materials to estimate biomass or chemical elements. However, nongreen or nonphotosynthetic vegetation (NPV), referring to litter, wood, and bark, is an important component in pasture especially in dry regions or in dry weather scenarios and plays a critical role in shaping the overall grass spectral reflectance. NPV has its greatest effect in the SWIR region between 2000 and 2400 nm, mainly related to the concentration of lignin-cellulose in dry plant residue as discussed earlier (Figure 9.3). The variation of NPV creates a significant impact on vegetation indexes such as normalized differences vegetation indices (NDVI) and soil adjusted vegetation index (SAVI) [20]. Changes in substrate reflectance beneath the grass canopy including litter on the surface and exposed soil affect the reflectance of grass canopies with lower vegetation coverage. Litter and soil have very high spectral signatures throughout the range 400–2500 nm, and their signals can dominate the reflectance of grass canopies of low coverage [21,22]. Neglecting the effects of background such as litter and soil fractions in grassland may cause erroneous estimation of biophysical characteristics such as grass biomass through remotely sensed data. Hyperspectral data have the potential to detect vegetation covers, such as green vegetation (GV), NPV and soil, individually, as well as their combinations [18,23,24]. Monitoring of changes in these covers can provide better characterization of grass ecosystem change.

9.4 HYPERSPECTRAL APPROACHES FOR PASTURE CHARACTERIZATION

A large number of spectral bands in hyperspectral systems provide an opportunity to develop a range of new measurements and refine conventional approaches for the estimation of grass properties by using narrow bands critical to the target parameters. Here, some typical approaches important for pasture/grass characterization are presented.

9.4.1 VEGETATION INDICES

Vegetation indices such as simple ratios (SRs) and NDVI with two or more bands are widely used to estimate biophysical and biochemical properties of vegetation. These indices are usually based upon the contrast between two spectral bands. In the case of the conventional NDVI, low reflectance in the red, due to chlorophyll absorption, and high reflectance in the near-infrared (NIR), related to multiple scattering effects are used to estimate vegetation greenness [25]. Many research studies of hyperspectral remote sensing of pasture employ vegetation indices to estimate vegetation properties by discovering new combinations of two narrow bands (relevant to target parameters) that are averaged out over the broadbands of multispectral sensors.

A typical approach to determine the best narrowband vegetation indices is to calculate all possible combinations of two bands and identify a combination that has the highest coefficient of determination (R^2) with a target variable [26,27]. Many two-band combinations derived from hyperspectral data for the estimation of a target variable in recent studies have shown much better performance than the traditional red–NIR band combination. For example, Mutanga and Skidmore [27] found that the standard red–NIR based NDVIs derived from a laboratory based hyperspectral analysis performed poorly in estimating dense biomass of tall grass due to the saturation level observed in dense vegetation. However, a modified NDVI with 746 and 755 nm bands had high R^2 (0.78 compared to 0.25 with the standard NDVI), showing the potential of hyperspectral data to overcome saturation problem with high-density grass canopy. Fava et al. [28] analyzed the variability of reflectance and vegetation properties in different pasture growth stages, determined the impact of this variability on VI-based assessment of pasture properties, and evaluated the potential of narrowband NDVI and SR for assessing biomass and LAI as well as canopy N.

9.4.2 RED EDGE

Like all GV covers, green grasses have been characterized by a maximum slope in the red edge between 680 and 740 nm. The chlorophyll concentration is strongly correlated with the point of maximum slope between very low reflectance in the red resulting from chlorophyll absorption and very high reflectance in the NIR due to internal cellular scattering in this region [3]. This region of chlorophyll absorption deepens and expands as chlorophyll concentration increases and consequently the red-edge position moves to a longer wavelength [29].

The structure of the chlorophyll red edge is best identified through the first derivative of vegetation reflectance. Due to the strong relationship between chlorophyll concentration and plant productivity, the location of the red-edge point has been used to estimate vegetation nutritional status [30,31]. Additionally, LAI and biomass have been found to be well correlated to red-edge parameters in the first derivative reflectance curves and vegetation indices with red-edge wavebands [30,32]. Jago et al. [33] generated red-edge position images by a linear equation for chlorophyll estimation and observed high correlation with grassland canopy chlorophyll concentration (r = 0.84).

Cho and Skidmore [32] compared red-edge positions extracted by two methods (the Lagrangian and linear extrapolation) from HyMap images acquired in two different years and found high correlation with field grass biomass ($R^2 > 0.50$). These results indicate that the red-edge-based methods may be widely used for grassland monitoring of vigor, nutritional status, and biomass production as imaging spectrometer data become more available [11].

Red edge has also been used to study plant stress due to nutrient deficiency [34,35] and contamination with pollutants such as gas and metals [36,37]. The shift of red-edge position may be used as an indicator of plant stress. One of typical approaches is to identify nutrient deficiency in grass based upon the shift of the red-edge position. Mutanga and Skidmore [38] related the red-edge position to nitrogen supply to *Cenchus ciliaris* grass in a greenhouse. They observed that the red-edge position of grass canopies was shifted from the control at 703 nm to the high nitrogen treatment at 725 nm.

Kooistra et al. [36] studied the effects of soil metal concentrations on grass and other vegetation based upon red-edge positions derived from the first derivative calculated from the 690 to 720 nm absorption feature and other vegetation indices. Some satisfactory relationships were found between the red-edge position and soil metals such as Pb ($R^2 = 0.61$) and Cu ($R^2 = 0.51$). They also observed that the red-edge positions from grass reflectance increased as soil metal concentrations decreased. Smith et al. [37] used ratios of the magnitude of the derivatives from the red-edge region as an index of plant stress responses to soil–oxygen depletion from natural gas leakage. They found that the ratios of the magnitude derivative at 725 nm to that at 702 nm were less in areas where gas was present. The plant stress responses based upon these ratios were identified for long-term leaks in all studied crops but for short-tem leaks only in grass. These studies demonstrate the potential of hyperspectral remote sensing for plant stress study.

9.4.3 Normalized Absorption Features by Continuum Removal

Across a full spectral range in the optical wavelength (400–2500 nm), vegetation reflectance presents several spectral absorptions associated with biochemical attributes. The depth and the area of these absorptions and indices based upon these features have been increasingly employed for pasture characterization. A paper by Kokaly and Clark [17] has been one of the most important references for many research studies on hyperspectral remote sensing of vegetation in the past decade. Their methodology was developed originally to enhance and standardize known chemical absorption features usually affected by the effects of the factors such as water on carbon-related absorption features in the SWIR region and exposed soil. The approach uses a continuum removal method [39] that normalizes the spectral curves of the absorption features by establishing a common baseline between the edges of the absorption region (Figure 9.4). In this way, differences in absorption strengths are enhanced. Absorption depth is a normalized depth of the absorption feature from the common base line. Band depths within absorption features are divided by the band depth at the

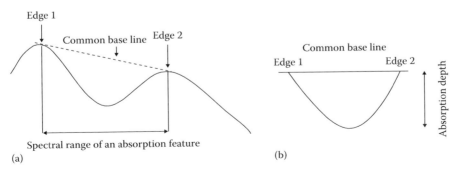

FIGURE 9.4 Illustration of a normalized spectral absorption depth: (a) a spectral absorption feature with established common line and (b) normalized spectral curve by common base line.

center of the feature, called normalized band depths. Then, stepwise multiple linear regression is used to analyze normalized band depths for all wavelengths in continuum-removed absorption features and select the most sensitive wavelengths to a target vegetation parameter in each absorption feature [3,17]. Then, linear equations are developed between band depth and chemical concentration and are used to predict biochemicals such as N, P, lignin, and cellulose [17]. The method has contributed to improving the estimation of biochemicals [6,7,10,40,41], biomass [10,27,28], and grass species discrimination [19]. The use of this methodology for imaging spectrometers for the estimation of biochemical concentration in different vegetation types has been successful [42,43], which indicates the applicability of this technique for grassland and pasture to landscape scales.

9.4.4 SPECTRAL MIXTURE ANALYSIS

Estimating only the amount of GV with vegetation indices such as NDVI limits our ability to characterize pasture biophysical and biochemical properties because other nongreen materials also may have significant impacts on the spectral signatures of pasture. For example, the amount of dry or senesced biomass in vegetation plays an important role in the estimation of carbon storage and plant stress [44,45]. Spectral mixture analysis (SMA) provides fractional cover measurements related to a land surface within a pixel or the field of view of a sensor [46,47]. Mixture modeling is based upon an assumption that a measured spectral signal is the sum of the signals from the components weighted by their fractions [46]. SMA has been found very useful for estimating fractional covers of main components of a landscape including pastures and grasslands [9,13,18,48,49].

For an ecosystem like grazing pasture, usually four components are considered including GV, senesced vegetation or NPV, soil, and shade. SMA involves the basic steps as following: (1) selection of endmembers that represent major materials or components existent in IFOV or pixel and (2) unmixing or solving the mixing equation, linear or nonlinear, for the fractions of the selected endmembers.

This method has been widely used for multispectral data, and the potential of hyperspectral data to accurately estimate vegetation covers using SMA has been evaluated by several researchers. Their results indicate that hyperspectral data provide more accurate estimates of fractional cover measurements compared to multispectral data [9,15]. For example, NPV and soil fractions are not well separable spectrally in the visible and NIR regions in the broadband domain, but these materials can be differentiated based on lignin-cellulose absorption bands in the SWIR [18,23,24]. Using AVIRIS, high-performance hyperspectral airborne sensor with high-signal-to-noise ratios, Asner and Heidebrecht [15] found that the SWIR 2000–2300 nm is a crucial spectral region to estimate accurate fractional covers of PV, NPV, and bare soil for shrub and grass land sites. Numata et al. [9] compared pasture fractional covers including shade-normalized NPV, GV, and Soil estimated by the field reflectance spectra (ASD), the Hyperion and convolved Landsat data from the Hyperion data to field grass covers estimated by CCD data (i.e., NPV, GV, and Soil) as reference, all measured from the same field transects in Rondônia, the Amazon. NPV and GV fractions derived from ASD field spectrometer showed the closest values to the reference fractions and were statically the same as of the reference, while differences between the reference and Hyperion became larger for NPV and GV and but a smaller difference for soil fraction, although no fraction showed statistical difference with the reference data. Landsat-derived fractions such as NPV and GV were statistically different from the same fractions from the reference, overestimating NPV and underestimating GV (Table 9.1). The results indicate that hyperspectral data provide more accurate grass fractional covers than Landsat data. This is especially true for pastures in dry regions and/or degraded pastures where senesced grass and bare soil are present and affect spectral signatures of these pastures [13]. Fraction images derived from EO-1 Hyperion imagery for a pasture area in the Amazon are shown in Figure 9.5. The endmembers are field spectra convolved from the ASD spectrometer to Hyperion.

TABLE 9.1

Shade Normalized Fractional Covers[a] of NPV, GV, and Soil for CCD, ASD, Hyperion/EO-1 and Convolved ETM+/Landsat 7 Spectra from Hyperion

Measurements		Mean	Standard Deviation	p-Value
Field measurements (CCD)	NPV	0.81	0.06	—
(63 measurements)	GV	0.16	0.05	—
	Soil	0.04	0.07	—
ASD field spectrometer	NPV	0.80	0.17	0.8086
(63 measurements)	GV	0.13	0.13	0.3878
	Soil	0.08	0.12	0.2380
Hyperion/EO-1 (12 pixels)	NPV	0.85	0.11	0.3589
	GV	0.10	0.07	0.1241
	Soil	0.05	0.06	0.8419
ETM+/Landsat 7 (12 pixels)	NPV	0.94[b]	0.14	0.0078
	GV	0.07[b]	0.07	0.0095
	Soil	−0.01	0.09	0.1951

Source: Numata, I. et al., *Remote Sens. Environ.*, 112(4), 1569, 2008.

[a] The NPV, GV, and Soil fractions were generated from full optical range reflectance (400–2500 nm) and were normalized by shade fraction to minimize illumination problems [41]. Fraction values vary from 0 to 1.

[b] Statistically significant mean differences relative to CCD fraction determined using a t-test at 0.95 level.

9.4.5 STATISTICAL METHODS

One challenge of hyperspectral data is its high spectral dimentionality and the extraction of critical spectral information from a large volume of spectral data. Statistical models have been employed for better estimation of vegetation parameters using the original spectral bands or transformations from hyperspectral data as independent variables. Multiple linear regression has been used with more than one spectral measurement to predict foliar nutrients and biomass [3,17].

Hyperspectral data often present a high degree of collinearity of neighboring bands. Multiple regression with hyperspectral data suffers from multicollinearity or spectral overfitting when the number of observations is smaller than the number of wavelengths studied and when input data show a high correlation [3,50]. To avoid this problem, the selection of a few contiguous regions such as some known absorption features is recommended [7,17]. The stepwise multiple regression approach has been widely used to select the optimal set of spectral bands for estimating vegetation parameters [17]. The advantage of this technique is that the derived features are easily interpretable from a physical point of view [51].

More recently, partial least square regression (PLSR) [52] has been increasingly used to predict biophysical and chemical properties of vegetation including pasture/grass as an alternative to the previous models [10,51,53,54]. PLSR can deal with all available spectral wavelengths simultaneously with a high degree of collinearity and selects a limited number of latent variables. This method operates similarly to principal component analysis, but instead of first decomposing the spectra into a set of eigenvectors and scores and regressing them against the response variables (vegetation parameters) as a separate step, PLS regression uses the response variable information

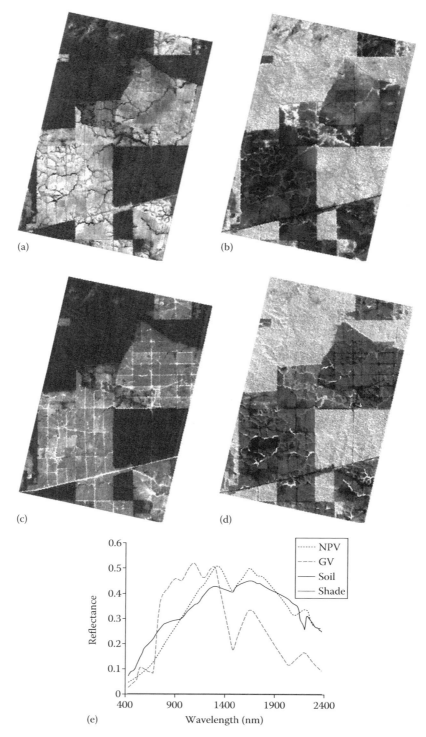

FIGURE 9.5 Fraction images of a pasture property in the Amazon derived from EO-1 Hyperion imagery: (a) NPV; (b) GV; (c) Soil; (d) Shade; (e) Endmembers.

during the decomposition. The selection of spectral subsets, the prediction of all grass variables through PLSR, was further improved [54]. However, the physical interpretation of the latent variables is difficult as with principal components [51].

9.5 APPLICATION OF HYPERSPECTRAL REMOTE SENSING FOR PASTURE ESTIMATION

9.5.1 BIOMASS

Pasture biomass is directly related to pasture and animal productivity (meat and milk), and its quantification is one of the most important, but also challenging applications of remote sensing for pasture research (Table 9.2). Several efforts have been made to estimate biomass in pastures and grassland using multispectral satellite data and most of them have had generally good relationships between field data and remote sensing derived measures [55,56]. The potential of hyperspectral data for the estimation of grass biomass has been evaluated across different spatial scales such as controlled laboratory, in situ, and landscape scales [27,40,51,53,57].

Although reflectance is directly related to LAI, the relationship between reflectance and biomass is indirect. This implies that the same LAI may be representative of different amounts of biomass, depending on the relationships between various canopy structural and density characteristics [11]. Biomass estimation is more problematic particularly for pasture areas with high vegetation density. This is because the conventional NDVI using red and NIR bands reaches a level of saturation above certain LAI level (2–3) [58]. Thus, the conventional NDVI is not appropriate for biomass estimation in dense grass [2,56]. To overcome this problem, new two narrowband combinations for NDVI have been tested from hyperspectral data (Table 9.2). Mutanga and Skidmore [27] found that although the standard red–NIR-based NDVIs derived from the laboratory hyperspectral analysis performed poorly in estimating dense biomass of tall grass

TABLE 9.2
The Results of the Pasture Biomass Estimation Using Hyperspectral-Derived Measures

	Authors	Sensor	Location	Grass Type	Best Band Combination (nm)	R^2	R^2 (NDVI[a])
NDVI	Mutanga and Skidmore [27]	Field	South Africa	Dense canopy grass	745/755	0.78	0.25
	Cho et al. [60]	HyMap	Italy	Mixed grass	771/740	0.7	0.4
	Cho and Skidmore [32]	HyMap	Italy	Mixed grass	695/786 in 2004	0.56 in 2004	
					740/786 in 2005	0.64 in 2005	
SR	Fava et al. [28]	Field	Italy	Mixed grass	920/729	0.77	0.35 (standard SR)
NBD[b]	Mutanga and Skidmore [59]	Field	South Africa	Dense canopy	744, 689, 653, 556 (selected by stepwise linear regression)	0.86	0.31–0.32

[a] R^2, standard NDVI.
[b] NBD, normalized band depth [17].

(*Cenchrus ciliaris*), a modified NDVI with 746 and 755 nm bands had high R^2 (0.78 compared to 0.25 with the standard NDVI). However, they found that SR yielded higher coefficients of determination ($R^2 = 0.80$ on average) with biomass as compared to NDVI (average $R^2 = 0.77$). These results indicate that hyperspectral data provide narrowband-derived vegetation indices more sensitive to canopy biomass compared to conventional vegetation indices such as NDVI [27,56].

In a similar study, Mutanga and Skidmore [59] found that continuum-removed absorption features such as band depth ratio, band depth index, and band area calculated with the bands selected by stepwise linear regression from the red-edge region had much higher coefficients of relationships ($R^2 > 0.80$) with dense grass biomass (*Cenchrus ciliaris*) measured in the laboratory compared to standard red–NIR NDVIs ($R^2 = 0.31$–0.32). Kawamura et al. [10] also found high prediction capabilities of standing biomass ($R^2 > 0.85$) using first derivative reflectance and continuum-removed derivative reflectance in a PLSR model in New Zealand.

Cho et al. [60] used HyMap, an airborne hyperspectral sensor, to identify the best spectral measures for the prediction of grass biomass in the Majella National Park, Italy. Like other laboratory and field based studies, those NDVIs derived from a spectral region between 725 and 800 nm had much higher correlations compared to the traditional NDVI. They also found that PLSR models with the six selected continuum-removed bands produced the highest correlation and lowest standard error for biomass prediction, compared to single variables such as the original reflectance, first derivative, and continuum-removed reflectance.

In the dry season in the Amazon, where a significant amount of dry grass material is found in pasture, Numata et al. [9] evaluated hyperspectral data to estimate live biomass, dead biomass individually, and both combined at the canopy level for two grass species. The results were highly affected by structural differences between two species (Figure 9.2) (see Section 9.3.). *Brachiaria decumbens* (Figure 9.2b) had better coefficients of determination between biomass and hyperspectral data than *Brachiaria brizantha* in general. Continuum removal of water absorption depth and area derived from the water absorption region (1100–1250 nm) showed the highest coefficients of determination for above ground biomass including total biomass ($R^2 = 0.35$–0.57) and live biomass ($R^2 = 0.31$–0.54), whereas lignin-cellulose absorption depth and area (2105–2230 nm) performed best for senesced biomass ($R^2 = 0.25$–0.64) and NDVI performed very poorly. Beeri et al. [57] estimated photosynthetic vegetation (PV) and NPV biomass based upon an accumulated continuum-removal reflectance between 991 and 1306 nm and broad- and narrow band-based NDVIs derived from HyMap in the Northwestern Glaciated Plains and the Northwestern Great Plains. Again, the continuum-removal-based data performed best for PV and NPV combined biomass data and for PV biomass data. The lowest relative error was found when PV live biomass measured alone. They also notice that the performance of NDVIs was affected by the presence of NPV, which masks spectral responses in the red and NIR.

9.5.2 Leaf Area Index

LAI is one of the main drivers of canopy primary productivity and has been a key variable in most ecosystem models. LAI can be directly measurable by optical remote sensing of vegetation including grasslands and pastures. Vegetation indices such as NDVI, SAVI, and enhanced vegetation index (EVI), derived from multispectral sensors, have been widely used to estimate LAI over large regions. Narrow spectral bands derived vegetation indices potentially provide additional improvements over two broadband-based vegetation indices to estimate LAI.

There are a limited number of studies on grass LAI estimation using hyperspectral data. Darvishzadeh et al. [54] tested narrowband vegetation indices and the red-edge inflection point to estimate LAI of a Mediterranean grassland. They found that the best combination of the vegetation indices for LAI was found in the NIR to SWIR regions, 1105/1229 nm for NDVI ($R^2 = 0.61$) and 1998/1402 nm for SAVI ($R^2 = 0.64$). Fava et al. [28] had the best performance for

LAI estimation with SR of 895/730 nm ($R^2 = 0.76$), compared to standard SR, 780/680 nm with ($R^2 = 0.39$). These results indicate the potential of hyperspectral data to improve LAI estimation by remote sensing.

LAI estimate may be very challenging in mixed grassland, where soil and litter effects significantly influence grass reflectance. He et al. [61] evaluated the performance of fifteen different vegetation indices in estimating LAI of a grassland in the semiarid region in Canada. Although the relationships between grassland LAI and studied VIs were statistically significant, their predictive capabilities were low ($R^2 = 0.37–0.44$). A new vegetation index was developed in this study, which incorporates the cellulose absorption index that varies as a function of the proportion of litter as a litter factor in adjusted transformed soil-adjusted vegetation index. This index improved the LAI estimation capability by about 10% ($R^2 = 55\%$). The results indicate the potential contribution of hyperspectral data to improve the LAI estimation by minimizing the effects of litter.

9.5.3 NUTRIENTS

Pasture nutritional quality indicates grass nutrient deficiency and degradation as well as animal grazing distribution patterns [62]. The ability to quantify the concentration of nutrients and determine pasture nutritional quality is a priority for pasture characterization via remote sensing. An accurate estimation of biochemical concentrations from hyperspectral data is one of the most desired applications for pasture management strategies.

The concentrations of nutrients (nitrogen, phosphorus, calcium, potassium, magnesium, and other agronomically important elements) have been found to be well correlated with spectral features and reflectance indices derived from hyperspectral data [7,17,43,50]. Nitrogen is one of the most important elements and is strongly related to chlorophyll activity. This element is also required for the protein synthesis that promotes the photosynthetic process. As chlorophyll determines spectral reflectance in the visible region, strong relationships between visible absorption bands and nitrogen concentration have been identified [7,50].

Phosphorus is a fundamental element to tissue composition as well as being one of the components of the nucleic acids and enzymes. Compared to nitrogen, P has received less attention but is an important nutrient for pasture management. This has been considered a limiting element for forage production in the tropics like Amazonia [63]. Most narrow bands and absorption features highly correlated with nutrients are concentrated in the visible, particularly in the red-edge region [7,10,17,28,64]. Mutanga et al. [63] found that within the red-edge region, those bands selected by stepwise linear regression for prediction of nutrients were most frequently located around 680 nm, a region of pigment absorption. The green reflectance region (550–580 nm) also was another important region for nutrient prediction.

On the other hand, weak correlations between nutrients and hyperspectral data have been observed in the SWIR [7]. For example, in fresh grass canopies, leaf mineral contents such as P, K, and S are usually difficult to be estimated due to the presence of water that masks the biochemical absorption features, particularly, in the SWIR region [17]. Additionally, internal scattering and mixing of spectral signatures obscures the absorption signal of nutrients due to differences in the physical structure of the canopies of different species [10]. Continuum-removal methodology by Kokaly and Clark [17] minimizes these effects on biochemical absorption features and enhances absorption strengths.

Mutanga et al. [7] evaluated four absorption variables derived from continuum-removed absorption features to predict canopy N, P, K, Ca, and Mg concentrations in five African grass species in the field through multiple linear regressions. The continuum-removal derivative reflectance (CRDR) variable yielded highest coefficients of determination of 0.7, 0.8, 0.64, 0.5, and 0.68, with low errors for N, P, K, Ca, and Mg, respectively. Higher R^2 values were obtained (>0.80) from all variables, when data were partitioned into species groups.

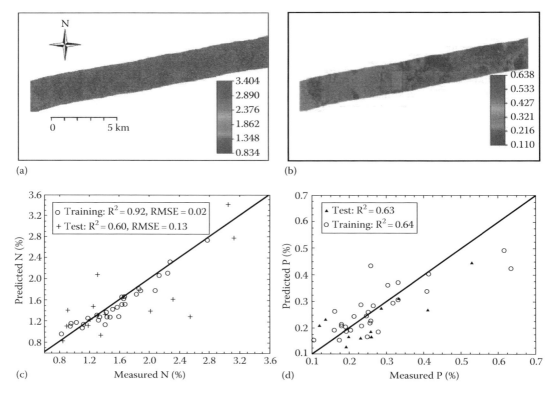

FIGURE 9.6 (See a and b in color insert.) Maps showing spatial distribution of concentration (%) of (a) nitrogen and (b) phosphorus and scatterplots obtained from the best-trained neural network used for mapping. Scatterplots of (c) nitrogen (%) and (d) phosphorus (%). (From Mutanga, O. and Skidmore, A.K., *Remote Sens. Environ.*, 90, 104, 2004; Mutanga, O. and Kumar, L., *Int. J. Remote Sensing.*, 28, 21, 2007.)

In a similar study, Kawamura et al. [10] used PLSR models for the prediction of pasture biochemical concentrations such as N, P, K, and S and biomass from grass with absolute reflectance, first derivative reflectance, and CRDR as input variables for PLSR models. Again, CRDR had the highest R^2 values for all minerals; 0.895, 0.943, 0.809, and 0.943 for N, P, K, and S, respectively.

Despite high prediction capability of hyperspectral data for biochemical concentrations evaluated at laboratory and field levels, the estimates of grass nutritional quality should be spatially explicit. Therefore, further efforts have been made by using imaging spectrometers. Mutanga and Skidmore [6] integrated continuum-removal absorption features from the visible (550–757 nm), the SWIR (2015–2199 nm), and the red-edge position derived from HyMap imagery and neural networks to map grass nitrogen concentration in an African savanna rangeland (Figure 9.6a). While the method used obtained a high coefficient of determination ($R^2 = 0.92$) with RMSE of 0.02 for the training dataset, the predictive capability with the test dataset indicated 60% of the variation in grass nitrogen concentration with RMSE of 0.13 (Figure 9.6c). Using the same method, Mutanga and Kumar [65] estimated and mapped grass phosphorus concentration in the same African rangeland and obtained a coefficient of determination of 0.63 with RMSE of 0.07 for the test dataset (Figure 9.6b and d). They also found that the input of SWIR bands greatly contributed to improving the estimation of grass phosphorus concentration and the prediction errors were drastically reduced when the visible and SWIR bands were used together compared with using the visible input only.

9.5.4 DEGRADATION ANALYSIS

About 20% of the world's pastures and rangelands are in some degradation stages [1], and several factors cause pasture degradation such as overgrazing, compaction and erosion caused by livestock action, soil, and climate. Multispectral satellite sensors have been used to assess the effects of land degradation around watering points, grazing intensity, soil biogeochemistry, and climate on grass biophysical changes [12,56,66,67].

Grazing is one of the main driving factors for pasture biomass change and can lead to pasture degradation. Hyperspectral remote sensing has been utilized to assess the impacts of grazing intensity on vegetation changes at pastureland and ecosystem levels. Most studies of this sort have analyzed impacts on pasture structural changes based upon fractional covers of PV, NPV, bare soil, and, in some studies, SHADE derived from SMA. Elmore and Asner [45] investigated the effects of grazing intensity on soil carbon stocks in Hawaii by estimating plant litter cover based upon NPV derived from AVIRIS. They observed that intensively grazed areas were characterized by higher exposed substrate or soil fraction and lower NPV fraction. As distance from grazing center increases, substrate fraction decreased, whereas NPV fraction increased. Furthermore, NPV had the strongest relationship with grazing intensity. They concluded that high levels of NPV can be used to identify areas of lower grazing intensity in their study area (Hawaii). A similar study conducted by Harris and Asner [68] detected a grazing gradient with fractional covers derived from AVIRIS in a rangeland in Utah and demonstrated the potential of airborne hyperspectral sensors to assess rangeland condition based upon accurately estimated fractional covers sensitive to grazing.

In a field study, Asner et al. [44] characterized the vegetation structures of Amazonian pastures with different ages planted in different soil types based upon NPV and PV derived from photon inverse models applied to grass reflectance measured by a field spectrometer. LAI and nonphotosynthetic vegetation area index (NPVAI) estimated from spectral reflectance through photon transport modeling were highly correlated with field LAI and NPVAI, and these factional covers varied according to grass age and soil texture. Furthermore, the variation of soil biogeochemical elements P and Ca across the sample pastures was well correlated with Canopy LAI + NPVAI inversion calculated from hyperspectral data.

9.5.5 SPECIES DISCRIMINATION

Understanding distribution of different grass species in landscapes is essential to measure ecological characteristics such as plant functional types. Hyperspectral data have been used in species distinction by developing spectral libraries for spectrally distinct species and creating species maps derived from airborne hyperspectral sensors [69]. In a laboratory-based analysis, Schmidt and Skidmore [19] measured spectral reflectance from eight African grass species to compare the reflectance and the continuum-removed reflectance curves for each one of all possible two species pairs to assess whether these species are spectrally separable. They found that bands that maximize the discrimination between species occur in the visible region (550–680 nm) indicating pigment concentrations vary between species.

On the other hand, the NIR section, after the absorption curves normalized by continuum removal, separates species better. Yamano et al. [70] used derivative reflectance in the visible and NIR region for distinction analysis of four predominant grass species in Inner Mongolia. They found that fourth-derivative peaks around 670 and 720 nm were an effective discriminator for distinguishing the grass species *Caragana microphylla* from the others. However, the discrimination capability of hyperspectral data may highly depend on the season and requires ad hoc calibrations to select a specific model and set of bands for species discrimination [71]. Recent study suggests that hyperspectral data have been a useful tool to detect invasive species within pasture and cropland [72–74]. At the landscape scale, species discrimination by hyperspectral data can be further complicated by diversity of background [22].

9.6 CONCLUSIONS

This chapter demonstrated the significant advances made in characterizing and mapping biophysical and biochemical properties of pastures through hyperspectral narrowband data when compared with the broadband data, thus assisting in much improved studies on livestock productivity. Some specific conclusions are possible from this chapter. These were:

1. Narrowband-derived vegetation indices in the red–NIR especially over 680–780 nm have been found more sensitive to canopy biomass when compared with the conventional broadband vegetation indices such as the NDVI [27]. This is because the narrowband indices substantially overcome the saturation issues of broadband indices. As a result, the narrowband SR involving 706/755 nm bands had much higher coefficients of correlations ($R^2 = 0.80$) with dense grass biomass (*Cenchrus ciliaris*) than any broadband indices.
2. Foliar nutrients are especially best estimated with hyperspectral data at the visible region. In the case of nitrogen, high correlations are found around 680 nm, the pigment absorption band, and the green reflectance region (550–580 nm). Accurate estimates of nutrient concentrations such as N, P, and K with high R^2 values (>0.80) are obtained by using continuum-removal derivative reflectance (CRDR) even in the field conditions.
3. The visible region (580–680 nm or longer) highlights the discrimination between species by taking advantage of varying pigment concentrations between species.
4. The ability to accurately estimate nonphotosynthetic vegetation (NPV) and bare soil is one of the greatest advantages of hyperspectral remote sensing of pasture over multispectral remote sensing, especially for pastures in dry regions and/or degraded pastures where senesced grass and bare soil are present and affect spectral signatures of these pastures.

Despite the potential improvement on pasture characterization via hyperspectral data, the results of research studies discussed in this chapter are "snap shot" or site and time specific. Also, currently available imaging spectrometers are not adequate for regularly monitoring of global pastures and grasslands. Therefore, the robustness of the hyperspectral techniques presented in this chapter requires further evaluation.

The ultimate goal for hyperspectral remote sensing of pasture would be the routine detection of different pasture biophysical and biochemical properties. This may be addressed in the near future as the data of new hyperspectral sensors such as HyspIRI, a satellite hyperspectral sensor, will be available regularly. HyspIRI will provide hyperspectral data with 210 spectral bands in the 400–2500 nm region for every 30 days at 45 m spatial resolution. Although the spatial resolution is coarser than most current imaging spectrometers, the HyspIRI data will be acquired more regularly over the same place covering a broader area (90 km SWATH). This will allow us to better monitor pasture dynamics in time at the landscape level. Moreover, new quantitative analytical models need to be developed for time series hyperspectral data in order to estimate in a robust manner biomass, biochemical concentrations, and their changes through space and time.

REFERENCES

1. FAO, *Livestock's Long Shadow: Environmental Issues and Options*, FAO: Rome, Italy, p. 391, 2006.
2. Gao, J., Quantification of grassland properties: How it can benefit from geoinformatic technologies? *International Journal of Remote Sensing*, 27(7):1351–1365, 2006.
3. Curran, P.J., Remote sensing of foliar chemistry. *Remote Sensing of Environment*, 30(3):271–278, 1989.
4. Elvidge, C.D., Visible and near-infrared reflectance characteristics of dry plant materials. *International Journal of Remote Sensing*, 11:1775–1795, 1990.
5. Asner, G.P., Wessman, C.A., and Archer, S., Scale dependence of absorption of photosynthetically active radiation in terrestrial ecosystems. *Ecological Applications*, 8(4):1003–1021, 1998.

6. Mutanga, O. and Skidmore, A.K., Integrating imaging spectroscopy and neural networks to map grass quality in the Kruger National Park, South Africa. *Remote Sensing of Environment*, 90:104–115, 2004.

7. Mutanga, O., Skidmore, A.K., and Prins, H.H.T., Predicting in situ pasture quality in the Kruger National Park, South Africa, using continuum-removed absorption features. *Remote Sensing of Environment*, 89(3):393–408, 2004.

8. He, Y.H. et al., Studying mixed grassland ecosystems II: Optimum pixel size. *Canadian Journal of Remote Sensing*, 32(2):108–115, 2006.

9. Numata, I. et al., Evaluation of hyperspectral data for pasture estimate in the Brazilian Amazon using field and imaging spectrometers. *Remote Sensing of Environment*, 112(4):1569–1583, 2008.

10. Kawamura, K. et al., Field radiometer with canopy pasture probe as a potential tool to estimate and map pasture biomass and mineral components: A case study in the Lake Taupo catchment, New Zealand. *New Zealand Journal of Agricultural Research*, 52(4):417–434, 2009.

11. Hill, M.J., Grazing agriculture: Managed pasture, grassland, and rangeland. In *Manual of Remote Sensing Volume 4. Remote Sensing for Natural Resource Management and Environmental Monitoring*, Ustin, S.L., Ed., John Wiley & Sons: Hoboken, NJ, 2004, pp. 449–530.

12. Numata, I. et al., Characterization of pasture biophysical properties and the impact of grazing intensity using remotely sensed data. *Remote Sensing of Environment*, 109(3):314–327, 2007.

13. Roberts, D.A. et al., Large area mapping of land-cover change in Rondonia using multitemporal spectral mixture analysis and decision tree classifiers. *Journal of Geophysical Research—Atmospheres*, 107(D20):18, 2002.

14. Asner, G.P., Biophysical and biochemical sources of variability in canopy reflectance. *Remote Sensing of Environment*, 64:234–253, 1998.

15. Asner, G.P. and Heidebrecht, K.B., Spectral unmixing of vegetation, soil and dry carbon cover in arid regions: Comparing multispectral and hyperspectral observations. *International Journal of Remote Sensing*, 23(19):3939–3958, 2002.

16. Fourty, T., Baret, F., and Verdebout, J., Leaf optical properties with explicit description of its biochemical composition: Direct and inverse problems. *Remote Sensing of Environment*, 56:104–116, 1996.

17. Kokaly, R.F. and Clark, R.N., Spectroscopic determination of leaf biochemistry using band-depth analysis of absorption features and stepwise multiple linear regression. *Remote Sensing of Environment*, 67(3):267–287, 1999.

18. Roberts, D.A., Smith, M.O., and Adams, J.B., Green vegetation, nonphotosynthetic vegetation, and soils in AVIRIS data. *Remote Sensing of Environment*, 44:255–269, 1993.

19. Schmidt, K.S. and Skidmore, A.K., Exploring spectral discrimination of grass species in African rangelands. *International Journal of Remote Sensing*, 22(17):3421–3434, 2001.

20. van Leeuwen, W.J.D. and Huete, A.R., Effects of standing litter on the biophysical interpretation of plant canopies with spectral indices. *Remote Sensing of Environment*, 55:123–138, 1996.

21. Asner, G.P. et al., Impact of tissue, canopy, and landscape factors on the hyperspectral reflectance variability of arid ecosystems. *Remote Sensing of Environment*, 74(1):69–84, 2000.

22. Okin, G.S., Roberts, D.A., Murray, B., and Okin, W.J., Practical limits on hyperspectral vegetation discrimination in arid and semiarid environments. *Remote Sensing of Environment*, 77:212–225, 2001.

23. Asner, G.P. and Lobell, D.B., A biogeophysical approach for automated SWIR unmixing of soils and vegetation. *Remote Sensing of Environment*, 74:99–112, 2000.

24. Nagler, P.L., Daughtry, C.S.T., and Goward, S.N., Plant litter and soil reflectance. *Remote Sensing of Environment*, 1:207–215, 2000.

25. Rouse, J.W., Haas, R.H., Schell, J.A., and Deering, D.W. Monitoring vegetation systems in the Great Plains with ERTS. In *Proceedings of the Third ERTS Symposium*, NASA, Washington, DC, December 1973.

26. Thenkabail, P.S., Smith, R.B., and De Pauw, E., Hyperspectral vegetation indices and their relationships with agricultural crop characteristics. *Remote Sensing of Environment*, 71:158–182, 2000.

27. Mutanga, O. and Skidmore, A.K., Narrow band vegetation indices overcome the saturation problem in biomass estimation. *International Journal of Remote Sensing*, 25(19):3999–4014, 2004.

28. Fava, F. et al., Identification of hyperspectral vegetation indices for Mediterranean pasture characterization. *International Journal of Applied Earth Observation and Geoinformation*, 11(4):233–243, 2009.

29. Pinar, A. and Curran, P.J., Grass chlorophyll and the reflectance red-edge. *International Journal of Remote Sensing*, 17:351–57, 1996.

30. Filella, I. and Penuelas, J., The red-edge position and shape as indicators of plant chlorophyll content, biomass and hydric status. *International Journal of Remote Sensing*, 15:1459–70, 1994.

31. Lamb, D.W., Steyn-Ross, M., Schaare, P., Hanna, M.M., Silvester, W., and Steyn-Ross, A., Estimating leaf nitrogen concentration in ryegrass (*Lolium* spp) pasture using the chlorophyll red-edge: Theoretical modelling and experimental observations. *International Journal of Remote Sensing*, 23:3619–3648, 2002.

32. Cho, M.A. and Skidmore, A.K., Hyperspectral predictors for monitoring biomass production in Mediterranean mountain grasslands: Majella National Park, Italy. *International Journal of Remote Sensing*, 30(2):499–515, 2009.

33. Jago, R.A., Cutler, M.E.J., and Curran, P.J., Estimating canopy chlorophyll concentration from field and laboratory spectra. *Remote Sensing of Environment*, 68:217–224, 1999.

34. Schut, A.G.T. and Ketelaars, J., Imaging spectroscopy for early detection of nitrogen deficiency in grass swards. *Njas-Wageningen Journal of Life Sciences*, 51(3):297–317, 2003.

35. Schut, A.G.T. et al., Imaging spectroscopy for on-farm measurement of grassland yield and quality. *Agronomy Journal*, 98(5):1318–1325, 2006.

36. Kooistra, L. et al., Exploring field vegetation reflectance as an indicator of soil contamination in river floodplains. *Environmental Pollution*, 127(2):281–290, 2004.

37. Smith, K.L., Steven, M.D., and Colls, J.J., Use of hyperspectral derivative ratios in the red-edge region to identify plant stress responses to gas leaks. *Remote Sensing of Environment*, 92(2):207–217, 2004.

38. Mutanga, O. and Skidmore, A.K., Red edge shift and biochemical content in grass canopies. *ISPRS Journal of Photogrammetry and Remote Sensing*, 62(1):34–42, 2007.

39. Clark, R.N. and Roush, T.L., Reflectance spectroscopy: Quantitative analysis techniques for remote sensing applications. *Journal of Geophysical Research*, 89:6329–6340, 1984.

40. Mutanga, O., Skidmore, A.K., and van Wieren, S., Discriminating tropical grass (*Cenchrus ciliaris*) canopies grown under different nitrogen treatments using spectroradiometry. *ISPRS Journal of Photogrammetry and Remote Sensing*, 57(4):263–272, 2003.

41. Mutanga, O. et al., Explaining grass-nutrient patterns in a savanna rangeland of southern Africa. *Journal of Biogeography*, 31(5):819–829, 2004.

42. Kokaly, R.F. et al., Mapping vegetation in Yellowstone National Park using spectral feature analysis of AVIRIS data. *Remote Sensing of Environment*, 84(3):437–456, 2003.

43. Huang, Z. et al., Estimating foliage nitrogen concentration from HYMAP data using continuum removal analysis. *Remote Sensing of Environment*, 93(1–2):18–29, 2004.

44. Asner, G.P., Townsend, A.R., and Bustamante, M.M.C., Spectrometry of pasture condition and biogeochemistry in the Central Amazon. *Geophysical Research Letters*, 26:2769–2772, 1999.

45. Elmore, A.J. and Asner, G.P., Effects of grazing intensity on soil carbon stocks following deforestation of a Hawaiian dry tropical forest. *Global Change Biology*, 12:1761–1772, 2006.

46. Adams, J.B. et al., Classification of multispectral images based on fractions of endmembers—Application to land-cover change in the Brazilian Amazon. *Remote Sensing of Environment*, 52(2):137–154, 1995.

47. Roberts, D.A., Batista, J.L., Pereira, J.L.G., Waller, E., and Nelson, B., Change identification using multitemporal spectral mixture analysis: Applications in Eastern Amazonia. In *Remote Sensing Change Detection: Environmental Monitoring Applications and Methods*, Elvidge, C.D. and Lunetta, R., Eds., Ann Arbor Press: Ann Arbor, MI, 1998, pp. 137–161.

48. Wessman, C.A., Bateson, C.A., and Benning, T.L., Detecting fire and grazing patterns in tallgrass prairie using spectral mixture analysis. *Ecological Applications*, 7(2):493–511, 1997.

49. Numata, I., Roberts, D.A., Chadwick, O.A., and Hatzel, Y., Spectral characterization of changes in grassland under climatic and soil gradients in Kohala, Hawaii. In *Proceedings of AVIRIS Workshop*, Pasadena, CA, 2004.

50. Curran, P.J., Remote sensing: Using the spatial domain. *Environmental and Ecological Statistics*, 8(4):331–344, 2001.

51. Clevers, J.G.P.W., van der Haijden, G.W.A.M., Verzakov, S., and Schaepman, M.E., Estimating grassland biomass using SVM band shaving of hyperspectral data. *Photogrammetric Engineering and Remote Sensing*, 73:1141–1148, 2007.

52. Geraldi, P. and Kowalski, B.R., Partial least-squares regression: A tutorial. *Analytical Chemistry Acta*, 185:1–17, 1986.

53. Cho, M.A. and Skidmore, A.K., A new technique for extracting the red edge position from hyperspectral data: The linear extrapolation method. *Remote Sensing of Environment*, 101(2):181–193, 2006.

54. Darvishzadeh, R. et al., LAI and chlorophyll estimation for a heterogeneous grassland using hyperspectral measurements. *ISPRS Journal of Photogrammetry and Remote Sensing*, 63(4):409–426, 2008.

55. Field, C.B., Randerson, J.T., and Malmstrom, C.M., Global net primary production: Combining ecology and remote sensing. *Remote Sensing of Environment*, 51:74–88, 1995.

56. Todd, S., Hoffer, R.M., and Milchunas, D.G., Biomass estimation on grazed and ungrazed rangelands using spectral indices. *International Journal of Remote Sensing*, 19:427–438, 1998.
57. Beeri, O. et al., Estimating forage quantity and quality using aerial hyperspectral imagery for northern mixed-grass prairie. *Remote Sensing of Environment*, 110(2):216–225, 2007.
58. Franklin, J., Prince, S.D., Strahler, A.H., Hanan, N.P., and Simonett, D.S., Reflectance and transmission properties of West African savanna trees from ground radiometer measurements. *International Journal of Remote Sensing*, 12:1369–1385, 1991.
59. Mutanga, O. and Skidmore, A.K., Hyperspectral band depth analysis for a better estimation of grass biomass (*Cenchrus ciliaris*) measured under controlled laboratory conditions. *International Journal of Applied Earth Observation and Geoinformation*, 5:87–96, 2004.
60. Cho, M.A., Skidmore, A., Corsi, F., van Wieren, S.E., and Sobhan, I., Estimation of green grass/herb biomass from airborne hyperspectral imagery using spectral indices and partial least squares regression. *International Journal of Applied Earth Observation and Geoinformation*, 9:414–424, 2007.
61. He, Y.H., Guo, X.L., and Wilmshurst, J.F., Comparison of different methods for measuring leaf area index in a mixed grassland. *Canadian Journal of Plant Science*, 87(4):803–813, 2007.
62. McNaughton, S.J. and Banyikwa, F.F., Plant communities and herbivory. In *Serengeti II—Dynamics, Management, and Conservation of an Ecosystem*, Sinclair, A.R.E. and Arcese, P., Eds., Chicago Press: Chicago, IL, 1995, pp. 49–70.
63. Dias Filho, M., Davidson, E.A., and de Carvalho, C.J.R., Linking biogeochemical cycles to cattle pasture management and sustainability in the Amazon Basin. In *Biogeochemistry of the Amazon Basin*, McClain, M., Victoria, R.L., and Ritchey, J.E., Eds., Oxford University: New York, 2000, pp. 84–105.
64. Mutanga, O., Skidmore, A.K., Kumar, L., and Ferwerda, J., Estimating tropical pasture quality at canopy level using band depth analysis with continuum removal in the visible domain. *International Journal of Remote Sensing*, 26:1093–1108, 2005.
65. Asner, G.P. et al., Pasture degradation in the central Amazon: Linking changes in carbon and nutrient cycling with remote sensing. *Global Change Biology*, 10(5):844–862, 2004.
66. Mutanga, O. and Kumar, L., Estimating and mapping grass phosphorus concentration in an African savanna using hyperspectral image data, *International Journal of Remote Sensing*, 28(21):4897–4911.
67. Pickup, G., Bastin, G.N., and Chewings, V.H., Identifying trends in land degradation in non-equilibrium rangelands. *Journal of Applied Ecology*, 35:365–377, 1998.
68. Harris, A.T. and Asner, G.P., Grazing gradient detection with airborne imaging spectroscopy on a semi-arid rangeland. *Journal of Arid Environments*, 55(3):391–404, 2003.
69. Dennison, P.E. and Roberts, D.A., Endmember selection for multiple endmember spectral mixture analysis using Endmember Average RMSE. *Remote Sensing of Environment*, 87(2–3):123–135, 2003.
70. Yamano, H., Chen, J., and Tamura, M., Hyperspectral identification of grassland vegetation in Xilinhot, Inner Mongolia, China. *International Journal of Remote Sensing*, 24(15):3171–3178, 2003.
71. Irisarri, J.G.N. et al., Grass species differentiation through canopy hyperspectral reflectance. *International Journal of Remote Sensing*, 30(22):5959–5975, 2009.
72. Lass, L.W. et al., A review of remote sensing of invasive weeds and example of the early detection of spotted knapweed (*Centaurea maculosa*) and babysbreath (*Gypsophila paniculata*) with a hyperspectral sensor. *Weed Science*, 53(2):242–251, 2005.
73. Lopez-Granados, F. et al., Using remote sensing for identification of late-season grass weed patches in wheat. *Weed Science*, 54(2):346–353, 2006.
74. Wang, C.Z., Zhou, B., and Palm, H.L., Detecting invasive sericea lespedeza (*Lespedeza cuneata*) in Mid-Missouri pastureland using hyperspectral imagery. *Environmental Management*, 41(6):853–862, 2008.

10 Optical Remote Sensing of Vegetation Water Content

*Colombo Roberto, Busetto Lorenzo, Meroni Michele,
Rossini Micol, and Panigada Cinzia*

CONTENTS

10.1 INTRODUCTION

The presence of water within the Earth's physical system sustains the environment and the life that depends on it. Plants, animals, and humans rely on the water in the atmosphere and in the environment, as well as what is in their bodies, to maintain the basic functions of life.

Water is one of the most important factors regulating plant growth and development in ecosystems [1] and it is required for the maintenance of leaf structure and shape, thermal regulation, and photosynthesis. The knowledge of the spatial and temporal variability of vegetation water content is very useful in fire risk assessment (e.g., [2–4]), in detecting vegetation physiological status (e.g., [5–8]) and provides important information for irrigation decisions in agriculture (e.g., [9–11]).

In the last 50 years, several studies have been conducted to provide reliable methods for the estimation of vegetation water content at different scales: leaf, field, airborne, or spaceborne observations. Currently, near real-time measurements of vegetation water content from space are also being developed [12]. Most of the studies exploit spectral data collected in the visible (VIS, 400–750 nm), near infrared (NIR, 750–1300 nm), and short wave infrared (SWIR, 1300–2500 nm) spectral regions, while experiments using thermal and microwave data for the retrieval of vegetation water content are rare [13,14]. In this chapter, different approaches of optical remote sensing (RS) for estimating leaf and canopy water content are presented and discussed.

10.2 LABORATORY AND FIELD MEASUREMENTS OF VEGETATION WATER CONTENT

The amount of water in vegetation (leaves or canopies) can be measured by RS, laboratory analysis, and field surveys. Vegetation water content can be expressed by means of several variables, which are functions of different leaf and canopy parameters.

The equivalent water thickness of a leaf (EWT_L) corresponds to the hypothetical thickness of a single layer of water averaged over the whole leaf area (e.g., [15]) and can be computed in laboratory by measuring fresh and dry weights (FW and DW, respectively, g) and the one-sided leaf area (A, cm^2):

$$EWT_L = \frac{FW - DW}{A} \; (g/cm^2 \text{ or cm})$$

Besides the EWT_L, for RS applications, it is also important to consider the water content per unit of ground area and therefore to measure EWT at canopy level. For grassland and for some crops, canopy EWT (EWT_C) can be directly estimated by measuring fresh and dry weight of vegetation harvested on a known area (kg/m^2) [16,17], or by scaling the plant water content at field level using the areal stand density of crops [18]. The amount of water in forest canopies is instead generally obtained by scaling leaf EWT with leaf area index (LAI, m^2/m^2), which is usually estimated indirectly from canopy gap fraction measurements:

$$EWT_C = LAI \times EWT_L \, (kg/m^2)$$

Sometimes, vegetation fractional cover is also used to scale foliar content to canopy EWT [19].

In the framework of fire risk analysis, leaf water content is frequently expressed as the percentage of leaf fresh weight (range between 0% and 100%) or as a percentage of leaf dry weight, namely, the fuel moisture content (FMC, %) [20], which may assume values above 100%:

$$FMC = \frac{FW - DW}{DW} = \frac{EWT}{LMA} * 100 \; (\%)$$

FMC is related to two different leaf biochemical parameters that affect leaf optical properties: the EWT_L and the specific leaf dry mass area (LMA), which is the weight of dry matter per leaf unit area.

For plant status analysis, relative water content (RWC) and leaf water potential have frequently been investigated from optical RS. Leaf RWC compares the water content of a leaf with the maximum water content at full turgor and can be considered as an indicator of vegetation status [21]. RWC can be obtained from laboratory measurements of leaf weight and leaf turgid weight (TW) according to the following expression [22]:

$$RWC = \frac{FW - DW}{TW - DW} * 100 \; (\%)$$

Both FMC and RWC are not directly or physically linked to the absorption processes and do not scale with LAI from leaf to canopy level. However, leaf RWC is sometimes averaged at canopy level using information of vegetation cover and height to derive a canopy RWC with arbitrary units [23].

10.3 EFFECTS OF WATER CONTENT ON SPECTRAL REFLECTANCE

Water absorption features result from the vibrational processes of OH bonds (i.e., small displacements of the atoms about their resting positions) and appear at the overtones of their fundamental frequencies. Water absorption coefficients are extremely low in the visible part of the electromagnetic spectrum whereas in the NIR and SWIR, four major absorption peaks are present (Figure 10.1). These peaks are located at approximately 975, 1175, 1450, and 1950 nm and increase in magnitude with wavelength.

Reflectance variations due to leaf water content differences can, for example, be appreciated by analyzing simulations conducted with the PROSPECT leaf radiative transfer model [24,25] (see Section 10.4.5). Figure 10.2 shows leaf reflectance and transmittance simulation results for a leaf characterized by six levels of EWT_L.

Visible and NIR reflectance is primarily influenced by chlorophyll content, leaf internal structure, and dry matter content, whereas leaf reflectance decreases with increasing EWT at longer wavelengths of NIR and SWIR spectral regions. Leaf EWT slightly affects reflectance values in the weaker absorption bands of the NIR, and strongly influences reflectance in the SWIR at 1450 and 1950 nm [26–30]. Leaf internal structure and leaf dry matter content also affect reflectance values in the NIR and SWIR spectral regions so that their variations may originate differences in leaf reflectance, which may be unrelated to water content variations (e.g., [15,21,31]).

In general, leaf reflectance is measured by coupling a spectroradiometer with an integrating sphere. This allows measurement of leaf bihemispherical reflectance and transmittance over a selected spectral region. Very often leaf stacks rather than single leaves are measured (e.g., [8,32,33]), to determine the so-called infinite reflectance, which is defined as the maximum reflectance of an optically thick medium. This quantity approximates the reflectance of closed deciduous canopies characterized by high LAI in which the effect of soil background and understory on reflectance is very low. It should however be noted that infinite reflectance may show an excessive magnitude of NIR reflectance with respect to real canopy spectra and is not a direct representation of the whole-canopy spectrum [34,35].

At canopy level, spectral variations due to water content can be well appreciated by analyzing PROSAILH (PROSPECT and SAILH) radiative transfer model simulations [25,36,37] (see Section 10.4.5). Figure 10.3 shows the spectral reflectance simulated by PROSAILH for a canopy with different values of LAI and EWT_L.

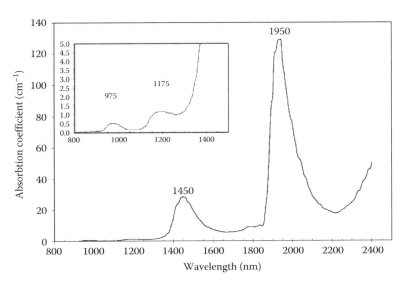

FIGURE 10.1 Specific absorption coefficients of water as used in the PROSPECT model. (Feret, J.B., et al., *Remote Sensing of Environment*, 112, 6, 3030, 2008.)

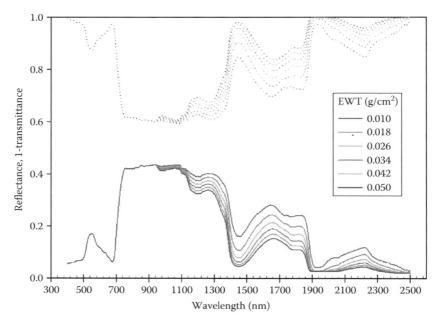

FIGURE 10.2 **(See color insert.)** Example of leaf reflectance (R, continuous lines) and transmittance (plotted as 1-T, dotted lines) spectra, simulated with the PROSPECT model for different levels of EWT.

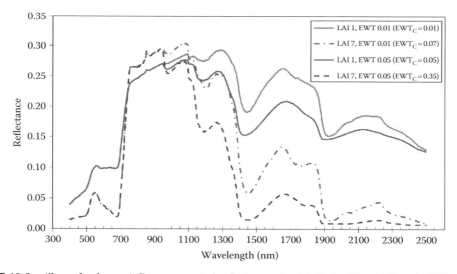

FIGURE 10.3 **(See color insert.)** Canopy spectral reflectance simulated with PROSAILH for different LAI and EWT values.

The visible and the first part of the NIR plateau are affected only by LAI variations, whereas the wavelengths above 900 nm are influenced by both LAI and EWT_L. A major difficulty when estimating canopy water content from reflectance data at landscape level comes from the fact that the effect of LAI and background may vary spatially, temporally, and with sensor view angle, so that these variations may cancel out water-related features in spectral reflectance. Additionally, vegetation water content retrieval from aerial and satellite data requires a very accurate estimation of the reflectance factor starting from the measured at-sensor radiance, so that fine geometric, radiometric, and atmospheric corrections have to be carefully pursued. Since the effect of absorption of water vapor in the atmosphere greatly reduces the energy reaching the ground surface around 1450 and 1950 nm, these water absorption bands are generally not exploited for landscape level studies. Some signal remains around the weaker water absorption peaks located in the NIR spectral region, but any change in canopy reflectance at the 975 and 1175 nm wavelengths due to variations in water content should be decoupled from changes due to the variations of water vapor in the atmosphere. The depth of penetration for radiation within the canopy differs in the NIR and SWIR spectral regions [23,38,39]. Wavelengths that are weakly absorbed penetrate more deeply into canopy and are expected to sense a larger portion of the total water content, whereas wavelengths that are strongly absorbed may be sensitive only to the water in the upper layers of the canopy and become quickly saturated.

The spectral regions most suitable for optical RS of vegetation water content at landscape level are therefore located in the SWIR wavelengths where leaf water shows intermediate absorption (around 1520–1540, 1650, and 2130–2200 nm) and in the weak absorption bands in the NIR region (near 975 and 1150–1260 nm).

10.4 METHODS USED FOR ESTIMATING VEGETATION WATER CONTENT

Several empirical and physically based models have been developed in the previous decades to retrieve vegetation water content from remotely sensed data. The possibility to use the different methods strongly depends on the spectral resolution of the available RS data. For example, the studies based on the use of low spectral resolution satellite data (e.g., Landsat TM) are usually based on the use of broadband spectral indices exploiting the wide water absorption features in the SWIR region. High-resolution field spectrometers and the recently developed hyperspectral aerial or satellite sensors allows instead the acquisition of spectral data in hundreds of contiguous spectral bands so that slight changes in reflectance at specific wavelengths affected by water absorption can be detected and analyzed [40]. This allows the use of narrowband spectral indices exploiting also the weaker absorption regions in the NIR, and also of more sophisticated methods such as continuum removal, derivative analysis, and curve fitting techniques. Moreover, the inversion of physically based models has come to the fore as the most promising technique for retrieving biophysical and biochemical vegetation parameters at leaf and canopy scale starting from high spectral resolution data.

10.4.1 HYPERSPECTRAL VEGETATION INDICES SENSITIVE TO WATER CONTENT

Spectral indices allow estimation of leaf and canopy water content by means of empirical approaches established using regression techniques with field measured water content or related variables. Table 10.1 shows the main vegetation indices sensitive to leaf and canopy water content developed in RS studies.

Table 10.1 reports the band center wavelengths used to compute spectral indices sensitive to water content from hyperspectral data. The devices and bandwidths refer instead to the instrument used in the original study where the spectral index was firstly proposed, and to its nominal bandwidth. In different studies focused to compute and to compare spectral indices, band centers and bandwidths may change because of the use of different devices.

In general, 2.5–5 nm bandwidths were used to develop narrowband indices from spectral data collected from field spectroradiometers (such as GER or ASD FS) or airborne hyperspectral airborne sensors (such as CASI and AISA); 10–50 nm bandwidths were used for indices based on AVIRIS, HYMAP, Probe-1, DAIS, MIVIS, HYPERION, and MODIS sensors while bandwidths larger than 100 nm were used to compute broadband spectral indices starting from satellite data such as Landsat TM, SPOT VEGETATION, ASTER, and NOAA AVHRR. Broadband indices can be in any case computed starting from data acquired by hyperspectral sensors, either by selecting the bands roughly corresponding to the band centers of the sensor used to develop the index or by

TABLE 10.1

Summary of Reflectance-Based Indices Developed for Estimating Vegetation Water Content at Different Scales (Leaf L, Ground G, Airborne A, Spaceborne S)

Index	Full Name	Formulation	Reference	Estimated Variable	Level	Device and Bandwidth
NIR (750–1300 nm)						
WI	Water index	$\dfrac{\rho_{900}}{\rho_{970}}$	[6]	RWC	G	Spectron, 15 nm
NDWI	Normalized difference water index	$\dfrac{(\rho_{858} - \rho_{1240})}{(\rho_{858} + \rho_{1240})}$	[41]	EWT_C	A	MODIS, 20–35 nm
WI/NDVI	Ratio WI normalized difference vegetation index	$\dfrac{(\rho_{900}/\rho_{970})}{(\rho_{800} - \rho_{680})/(\rho_{800} + \rho_{680})}$	[42]	FMC	G	Spectron, 15 nm
RDI	Relative depth index at 1175	$\dfrac{(\rho_{max} - \rho_{min})}{\rho_{max}}$	[16]	EWT_C	G	GER IRIS MK IV, 2, 4 nm
SRWI	Simple ratio water index	$\dfrac{\rho_{858}}{\rho_{1240}}$	[43]	EWT_L	S	MODIS, 20–35 nm
R975	Ratio at 975	$\dfrac{2\bar{\rho}_{960-990}}{(\bar{\rho}_{920-940} + \bar{\rho}_{1090-1100})}$	[44]	FMC	L	ASD FieldSpec Pro Fr, 1.4, 2 nm
SWIR (1300–2500 nm)						
NDII	Normalized difference infrared index	$\dfrac{(\rho_{820} - \rho_{1650})}{(\rho_{820} + \rho_{1650})}$	[45]	EWT_C	S	TM, 140, 195 nm
R5/R7	Ratio of TM band 5 to band 7	$\dfrac{\rho_{1650}}{\rho_{2218}}$	[46]	EWT_L	A	ATM 195, 340 nm
LWCI	Leaf water content index	$\dfrac{-\log(1 - (\rho_{820} - \rho_{1600}))}{-\log(1 - (\rho_{820,FT} - \rho_{1600,FT}))}$	[47]	RWC	L	MK I TM, 140, 195 nm
MSI	Moisture stress index	$\dfrac{\rho_{1600}}{\rho_{820}}$	[21,48]	RWC EWT_L	L	GER VIRIS, 3, 12 nm
DRI	Datt reflectance index	$\dfrac{(\rho_{816} - \rho_{2218})}{(\rho_{816} + \rho_{2218})}$	[33]	EWT_L	L	GER IRIS MK IV, 2, 4 nm
GVMI	Global vegetation moisture index	$\dfrac{(\rho_{820,rect} + 0.1) - (\rho_{1600} + 0.02)}{(\rho_{820,rect} + 0.1) + (\rho_{1600} + 0.02)}$	[28]	EWT_C	S	SPOT VGT 20, 170 nm
RDI1450	Relative depth index at 1450	$\dfrac{(\rho_{max} - \rho_{min})}{\rho_{max}}$	[49]	EWT_L	G	ASD FieldSpec Pro Fr, 1.4, 2 nm

TABLE 10.1 (continued)
Summary of Reflectance-Based Indices Developed for Estimating Vegetation Water Content at Different Scales (Leaf L, Ground G, Airborne A, Spaceborne S)

Index	Full Name	Formulation	Reference	Estimated Variable	Level	Device and Bandwidth
NDWI2130	Normalized difference water index at 2130	$\dfrac{(\rho_{858} - \rho_{2130})}{(\rho_{858} + \rho_{2130})}$	[50]	EWT_C	S	MODIS, 24, 50 nm
NMDI	Normalized multi-band drought index	$\dfrac{\rho_{858} - (\rho_{1640} - \rho_{2130})}{\rho_{858} + (\rho_{1640} - \rho_{2130})}$	[51]	EWT_L	S	MODIS, 24, 50 nm
MSI/SR	Ratio MSI/simple ratio	$\dfrac{\rho_{1600}/\rho_{820}}{\rho_{895}/\rho_{675}}$	[52]	EWT_L	A	MIVIS, 2, 9 nm

resampling the data to a lower spectral resolution. On the contrary, the computation of spectral indices developed from high-spectral resolution sensors on low resolution data (for those indices where the band centers coincide) may lead to erroneous results, particularly when the spectral index considered exploits the use of the weak and narrow water absorption features in the NIR region. More than bandwidth, the main limitations connected with low spectral resolution sensors are related to the reduced number of bands that may prevent the selection of the best spectral regions for indices computation. It is difficult to identify the ideal instrument bandwidth that should be used for the retrieval of vegetation water content from spectral indices, since no studies have explicitly and systematically tested this issue. Some studies show that, in general, spectral indices computed with bandwidths ranging from 2.5 to 50 nm have similar performances for leaf water content and other biochemical compounds estimation, while the use of broader bandwidths may hamper the accuracy of the estimates [32,53,54]. However, spectral resolution may not be critical when the investigated spectral feature is quite broad. Instead, the accuracy of leaf and canopy water content retrieval can be more sensitive to the signal-to-noise ratio and thus resampling hyperspectral data to a lower spectral resolution (e.g., 20 nm) may be beneficial because it suppresses part of the instrument noise.

10.4.1.1 NIR-Based Spectral Indices

The ability of the water index (WI) to evaluate physiological status and water content is documented in several studies exploiting leaf, ground, and airborne measurements [6,55–59]. Peñuelas et al. [6,42] showed that WI closely tracks changes in RWC, especially when there are no important changes in LAI and architectural canopy parameters. To minimize structural effects and therefore to maximize sensitivity to water content, Peñuelas et al. [42] and Piñol et al. [60] successfully tested the ratio of WI with the normalized difference vegetation index (NDVI) to estimate FMC in Mediterranean ecosystems from field spectroradiometer measurements. At the landscape level, by using airborne visible/infrared imaging spectrometer (AVIRIS) imagery, Serrano et al. [23] identified canopy structure as the main source of variability in WI and other spectral indices sensing water content, although they appeared to respond also to canopy RWC. Serrano et al. [23] showed that NIR-based spectral indices such as WI and normalized difference water index (NDWI) were more sensitive to changes in canopy RWC than those formulated using SWIR bands and curve fitting techniques (Section 10.4.4). Sims and Gamon [61] using the GER 2600 instrument at ground level also showed that water spectral indices formulated at 960 and 1180 mm have a strong correlation with EWT_C and provide better correlation than that obtained by using indices based on SWIR wavelengths.

NDWI was originally developed to retrieve vegetation water content from quasi-hyperspectral spaceborne measurements (e.g., MODIS) and was successfully tested on imaging data acquired with AVIRIS [41]. The 860 and 1240 nm wavelengths, which compose NDWI, are located in the high reflectance plateau where the contribution of vegetation scattering to reflectance is similar. NDWI should also be insensitive to foliar dry matter biochemical compounds (lignin, cellulose, and protein), which typically absorb at 1500–2500 nm (e.g., [32]). Several studies have demonstrated the ability of this index to estimate leaf and canopy EWT using airborne hyperspectral sensors and MODIS data (e.g., [23,52, 62]) and for estimating FMC at leaf, ground [63–65], and satellite data (e.g., [66,67]). To estimate EWT_C at ground level, Rollin and Milton [16] developed the relative depth index (RDI) at the water absorption feature centered around 1150 nm, in a way that employed reference wavelengths (at the shoulder and at the bottom of the absorption well) are so close that it is possible to assume that the contribution of other canopy factors is constant. RDI was found well correlated with canopy EWT as the first derivative (see Section 10.4.3) and insensitive to data smoothing.

10.4.1.2   SWIR-Based Spectral Indices

Spectral bands outside the main water vapor absorbing bands in the SWIR have been investigated for a long time, starting from the pioneering studies of Tucker [68] who identified the 1550–1750 nm spectral range as the best suited for monitoring vegetation water content. Leaf water content index (LWCI), moisture stress index (MSI), normalized difference infrared index (NDII), and SWIR ratio were specifically designed for satellite application, mainly exploiting broadbands 5 (1550–1750 nm) and 7 (2080–2350 nm) of the Landsat TM. Several studies exploited broadband indices and narrowband spectroscopy for plant status analysis at different scales. Results of these studies are somewhat controversial because in certain cases spectral indices were found to be strongly related to RWC and leaf water potential, while other studies found no significant statistical evidence of correlation (e.g., [8,11,69–72]). Hunt and Rock [48] and Hunt [73] have shown that the MSI computed at leaf level and landscape level by means of Daedalus Thematic Mapper Airborne Simulator data performed better for the estimation of leaf EWT, EWT_C, and LAI, rather than for the estimation of vegetation status indicators. Satellite RS studies showed that SWIR-based indices, such as NDII, allow successful tracking of seasonal variability of EWT_C in crops and forest ecosystems (e.g., [50,74–76]) and of FMC in grassland and shrubland landscapes [4,13,77]. In the context of soil moisture experiments, Anderson et al. [18] and Jackson et al. [78] tested SWIR-based indices (and greenness indices, such as NDVI) for monitoring seasonal EWT_C in agricultural areas by means of ground reflectance and multitemporal Landsat data, showing that spectral indices sensitive to water content are less susceptible to saturation than NDVI at high levels of LAI.

10.4.1.3   Greenness Indices

Indirect effects of water content were found on using a visible spectral range at 400 nm, in the red-edge region at 680–780 nm [79,80] and on spectral vegetation indices expected to respond to canopy greenness such as NDVI [81]. Greenness indices derived from airborne measurements and coarse resolution satellite data have been successfully used to track seasonality of EWT_C and FMC in grassland and savanna ecosystems, while poor results were in general found for shrubs and forest environments [4,82–85].

In general, NIR- and SWIR-based spectral indices are more efficient than greenness indices in estimating vegetation water content. The correlation between greenness indices and EWT_C or FMC is spurious and due to the strong correlation between NDVI and LAI and to plant chlorophyll content, which are in turn affected by plant water status.

10.4.2   Absorption-Band-Depth Analysis and Continuum-Removed Spectral Indices

These methods have been largely exploited for estimating mineralogical composition in geology and soil science and only fairly recently employed in vegetation analysis. Kokaly and Clark [32] proposed a methodology to standardize laboratory-based NIR spectroscopy measurements to estimate vegetation biochemical parameters. This method benefits from continuum-removal, calculation of band depth, normalization, and statistical methods to derive foliar concentration of various biochemical constituents. Briefly, the values associated with a wavelength are expressed as band depth normalized to the waveband at the center of the absorption feature or as the area of the absorption feature, and then related to water content by means of multiple linear regression. This approach has been found accurate for estimating water content at leaf level; it performs better than standard first derivative analysis and it is promising for RS application at canopy level [86].

Spectral indices computed from continuum-removed spectra have been used to estimate FMC at leaf and ground level [44,87]. Colombo et al. [52] found that continuum-removed spectral indices based on absorption features around 1200 nm allow good estimates of EWT_C at landscape level by means of airborne multispectral infrared and visible imaging spectrometer (MIVIS) data. Similar to relative depth indices or derivative techniques, the good performances of continuum-removed spectral indices based on water absorption features are probably due to their ability to minimize the effects of the background signal and albedo variations on reflectance spectra, thus allowing a better estimation of canopy water content.

10.4.3 DERIVATIVE TECHNIQUES

Derivative analysis of absorption spectra is a well-established technique in analytical chemistry where it is used to suppress background signals and resolve overlapping spectral features [88,89]. These techniques are usually based on the analysis of the relationships between the slope of the reflectance curve in the proximity of the main water absorption bands and leaf or canopy water content. Derivative techniques have been less frequently used than narrowband spectral indices for estimating vegetation water content, although they appear to be less affected by leaf and canopy structure variations and can be advantageous in standardizing for the overall level of reflectance. For example, Danson et al. [15] found that the first derivative of the reflectance spectrum at 1450 nm (measured at leaf level by a GER IRIS Mark IV specroradiometer) is closely correlated with leaf water content and insensitive to leaf structure. More recently, Kumar [53] also found statistically significant correlation between FMC and first derivative spectra at leaf level. At ground level, Rollin and Milton [16] found fairly good correlations between the first derivative at 1156 nm and EWT_C in grasslands. Clevers et al. [90] recently showed that the spectral derivatives for wavelengths positioned on the right slope of the water absorption feature at 970 nm (over the 1015–1050 nm spectral interval) may be successfully used to estimate EWT_C of grassland ecosystems from ASD field measurements and airborne HYMAP reflectance data and provide better results than NIR-based spectral indices and continuum-removal techniques.

10.4.4 FITTING OF THE PURE WATER ABSORPTION SPECTRA

This approach was originally developed to simultaneously retrieve the water vapor and liquid water from AVIRIS data [81,91]. Absorption due to the presence of canopy liquid water can be separated from water vapor absorption in the atmosphere owing to an approximately 40 nm shift toward longer wavelengths for liquid water (e.g., [92]). The equivalent water thickness can be estimated by combining the MODTRAN radiative transfer model and a simple vegetation reflectance model and using spectral fitting techniques (e.g., [81]). Twenty AVIRIS spectral bands between 865 and 1085 nm are generally used to fit measured radiance as a function of the equivalent transmittance spectrum of a water sheet modeled using a Beer-Lambert model modified to account for the scattering contribution [23,81,91–96]. Assuming that reflectance can be substituted for transmittance [26], the EWT is calculated as the slope of the plot of the natural log of reflectance as a function of known absorption coefficients for pure water in a selected spectral region (e.g., [19,61,97]). It is interesting to note that water thickness derived with such a technique has been proved to be sensitive to canopy structure and very useful for estimating LAI, with better performance than those found using traditional NDVI (e.g., [23,98]). Curve fitting has also been implemented for application with Probe-1 airborne hyperspectral sensor over the 850–1200 nm range [99] and with multi-angular HYMAP measurements over the 860–1320 nm range [100].

In general, studies comparing the performance of curve-fitting techniques with spectral indices sensitive to water content and other techniques at ground, aerial, and satellite level have demonstrated good agreement between different methods and with field data (e.g., [19,92,95,96,99,101]). Roberts et al. [101] compared the estimates of EWT from HYPERION and AVIRIS reflectance across two wavelength regions (865–1088 and 1088–1200 nm) found that the measures based on the 1200 nm band are more effective than that based on 980 nm liquid water band to map canopy water content across a wide range of vegetation types.

10.4.5 RADIATIVE TRANSFER MODELS

Radiative transfer models simulate reflectance and transmittance of leaves or canopies by describing the interactions (i.e., absorption and scattering) of solar radiation with the different vegetation elements. At leaf level, reflectance is generally simulated as a function of the main biochemical and structural constituents of the leaf (e.g., chlorophyll, dry matter, and water content). Several leaf optical models exist, and they use different approaches for broadleaf and needleleaf species (e.g., [25,102,103]). At canopy level, reflectance is instead simulated as a function of leaf reflectance and canopy structural and biophysical parameters (e.g., LAI, Fractional Cover), also taking into account factors such as background reflectance and viewing and illumination geometry. Canopy radiative transfer models differ in the description of the radiative regime within the canopy, from one-dimensional formulation where the canopy is described as a sparse homogeneous medium to three-dimensional descriptions of heterogeneous canopies (see [104]). In the framework of vegetation water content studies, leaf and canopy radiative transfer models have been primarily used in performing sensitivity analysis, optimizing/designing spectral vegetation indices and retrieving leaf and canopy EWT from model inversion on RS data.

10.4.5.1 Sensitivity Analysis Studies on Reflectance and Spectral Indices

Sensitivity analysis is used to quantify the contribution of canopy biophysical and biochemical properties to canopy reflectance. This makes it possible to analyze the magnitude of reflectance changes as a function of vegetation water content, to understand its interactions with other leaf and canopy parameters and to select appropriate wavelengths and spectral indices suitable for the retrieval of leaf and canopy water content. For example, the sensitivity studies performed using the PROSPECT model by Ceccato et al. [28] and Bowyer and Danson [29] showed that the stronger SWIR absorption bands around 1450, 1940, and 2480 nm are more sensitive to leaf water content variations than the weaker water absorption bands of the NIR region. Model simulations have underscored the important contribution of dry matter content to reflectance around the water absorption feature at 970 nm (which may hamper the estimation of EWT_L with the WI spectral index) and the rather high stability at 860 nm at changing of water content (this explains why this wavelength is often used as a reference for formulating spectral indices sensitive to water content). Using the SAIL model [36,37], Asner [105] showed the deepening of the two water absorption features within the NIR as LAI increased and their stability at changes of mean leaf angle. The sensitivity study by Bacour et al. [106], performed using the PROSAIL model, showed that EWT_L explained 50% of the total variance in the SWIR spectral region but that this contribution is reduced in favor of LAI in the regions where the specific absorption of pure liquid water is higher (i.e., 1950 and 2500 nm). Bowyer and Danson [29] carried out a canopy sensitivity analysis using PROSAILH and PROGEOSAIL (PROSPECT and GEOSAIL, [107]) showing that EWT_L have a strong influence only at the longer wavelengths of the NIR and in the SWIR spectral region, although the importance of EWT_L and LAI differs depending upon which model is used.

Canopy radiative transfer models have also been used to evaluate the performance of known or proposed spectral vegetation indices sensitive to water content. This process involves the use of physical models to simulate a set of reflectance spectra corresponding to canopies characterized by different water content and other leaf and canopy parameters. For example, Ceccato et al. [28] optimized a new water spectral index by exploiting the SWIR spectral region able to minimize various biophysical factors, geometry of observation, and atmospheric conditions, which was recently successfully applied in the estimation of EWT_C from SPOT-VEGETATION data [108]. Dawson et al. [19] showed the potential of LIBERTY [102] and FLIGHT [109] models in simulating leaf and canopy reflectance and found that NIR-based spectral indices were closely related to leaf and canopy EWT. The modeling analysis conducted by Zarco-Tejada et al. [62] also demonstrated that NIR-based spectral indices (MODIS NDWI and the Simple Ratio Water Index, SRWI) efficiently respond to the change in EWT in the LAI range 4 and $10 \, m^2/m^2$, while at lower and higher values LAI, background and saturation effects dominate the spectral indices. Colombo et al. [52] used PROSAILH to evaluate the sensitivity of double ratio indices to leaf EWT and their capability to minimize LAI effects at landscape level with airborne hyperspectral data. Simulation results underlined that NIR- and SWIR-based double ratio indices are sensitive to leaf EWT but are strongly influenced by LAI and soil background in sparse canopy conditions. Finally, other studies used PROSAILH to develop spectral indices sensitive to water content, which minimize disturbing influences of soil background in sparse coverage conditions [110,111].

10.4.5.2 Retrieval of Vegetation Water Content from Model Inversion

The inversion of a radiative transfer model consists of finding the model parameters (i.e., model input) that provide a "good match" between the modeled (i.e., model output) and observed reflectance (for a review of radiative transfer model inversion on RS data see Kimes et al. [112]). Model inversion allows the estimation of both leaf and canopy properties from RS data and thus allows estimation of EWT_L and EWT_C (e.g., [52,62,95,108,113–120]).

At leaf level, uncertainties in the estimation of leaf EWT have been demonstrated to be caused by the difficulties of radiative transfer models to completely decouple leaf EWT from leaf structure and leaf dry matter parameters [25,121,122]. At canopy level, the estimation of leaf water content and canopy structural parameters is also complicated by the coupling between the quantity of leaf absorbing material and vegetation structure/architecture, so that some studies suggest that EWT_C may be more conveniently estimated from canopy reflectance data than from leaf EWT. Among the few studies that inverted radiative transfer models starting from reflectance computed from airborne hyperspectral sensors, Kötz et al. [115] found reasonable accuracy in both canopy structure and foliage water content estimates in heterogeneous coniferous forest canopies using the GEOSAIL and the FLIGHT models on DAIS data. More recently, Colombo et al. [52] also showed that PROSAILH can be successfully inverted for the retrieval of both leaf and canopy EWT in poplar plantations starting from MIVIS data. New approaches that combine hyperspectral data and light detection and ranging (LiDAR) technologies or microwave information should make it possible to obtain detailed observations of vegetation and to provide accurate estimation of biophysical/biochemical parameters, canopy height, and other 3D structural parameters useful in different applications. For example, LiDAR data can be used to estimate canopy structural parameters that may help to stabilize the radiative transfer model inversion to improve the retrieval of leaf EWT with imaging spectroscopy [123].

It is important to underline that for applications at regional and global level, the inversion of radiative transfer models for vegetation water content estimation requires an appropriate parameterization of the structural parameters of the different ecosystems and the use of appropriate regularization techniques to solve the ill-posed problem in order to obtain stable and reliable solutions [96,124,125]. Detailed land cover classifications, which make it possible to account for mixed vegetation composition, are moreover necessary to accurately determine vegetation water content, especially when using medium to coarse spatial resolution satellite imageries.

10.4.6 OTHER TECHNIQUES

Besides the methods described earlier, other techniques have been developed to estimate vegetation water content from remotely sensed data. For example, artificial neural networks have been successfully used to estimate EWT from leaf spectral data, AVIRIS, and MODIS imageries in different ecosystems (e.g., [126–130]). Dawson et al. [130] used an artificial neural network to estimate the concentration of biochemical parameters in stacked fresh slash pine needles. Five wavelengths, located in correspondence to absorption features due to leaf water, lignin–cellulose, and nitrogen concentrations were selected as input for training an artificial neural network, which was then used to successfully estimate FMC and the other biochemical compounds. Pinzon et al. [131] proposed the use of hierarchical Foreground/Background Analysis to detect leaf water content. This approach is based on a nonlinear spectral mixing technique that makes it possible to highlight sharp absorption features that may be directly related to the desired property (i.e., water content). More recently, Li et al. [132] proposed a new technique based on genetic algorithm and partial least squares regression to estimate vegetation water content from leaf spectral data and AVIRIS imageries.

10.5 SUMMARY AND CONCLUSIONS

Measuring vegetation water content remotely is not only an appealing prospect but also a challenging one. In this chapter, we have reviewed progress toward this goal and discussed the main approaches and techniques used to estimate vegetation water content from remotely sensed data at leaf, ground, aircraft, and satellite level. The methods for retrieval of vegetation water content from optical remote sensing have made much progress in the last decades. In particular, the increasing availability of high resolution field spectrometers and of hysperspectral imaging offer the possibility of exploiting several techniques to accurately estimate vegetation water content and evaluate its spatial and temporal variability for different ecosystems.

Remote sensing data are better suited for the estimation of leaf and canopy equivalent water thickness (EWT), rather than fuel moisture content (FMC) and relative water content (RWC). At leaf level, studies that explicitly tested the performances of different vegetation indices or employed radiative transfer models for performing sensitivity analysis have shown that the stronger absorption bands in the SWIR are more responsive to water content variations than the weaker water absorption bands of the NIR spectral region. At canopy level, it appears that spectral indices based on reflectance around 1175 nm and other techniques exploiting contiguous spectral bands within this water absorption peak are better suited for the prediction of water content than that based on 975 nm and on the longer wavelengths in the SWIR. Beside other methods, canopy reflectance models represent an essential tool in the analysis of optical data and their inversion has been demonstrated to be useful in producing spatial and temporal maps of both leaf and canopy water content.

Most studies indicate that the reflectance signal near the main water absorption bands is largely influenced by leaf structure and LAI variations and, over partially vegetated areas, also by soil and background reflectance. Therefore, prediction of vegetation water content remains a challenge when both leaf structure and canopy architecture vary within the image. Issues associated with scaling measurements from leaf to canopy levels and the existence of major covariances between vegetation structure and water content make the deconvolution of these vegetation properties quite challenging. Two developments show great potential for minimizing canopy effects, such as the availability of physically based reflectance models that contain explicit descriptions of canopy structure and leaf biochemical compounds and the possibility of simultaneous LiDAR measurements that could be used to measure and minimize the influence of canopy structure and architecture on the acquired spectra.

In summary, remote sensing is a powerful tool that provides accurate estimations of vegetation water content and offers the only possibility of generating maps at different spatial and temporal scales that can be assimilated in ecological studies that include vegetation as a dynamic component.

REFERENCES

1. Kramer, P. J. and Boyer, J. S., *Water Relations in Plants and Soils.* Academic Press, San Diego, CA, 1995.
2. Nelson, R. M., Water relations of forest fuels. In: Johnson, E. A. and Miyanishi, K. (Eds.), *Forest Fires: Behavior and Ecological Effects.* Academic Press, San Diego, CA, pp. 79–149, 2001.
3. Carlson, J. D. and Burgan, R. E., Review of user needs in operational fire danger estimation: The Oklahoma example. *International Journal of Remote Sensing,* 24, 1601–1620, 2003.
4. Chuvieco, E., Cocero, D., Aguado, I., Palacios-Orueta, A., and Prado, E., Improving burning efficiency estimates through satellite assessment of fuel moisture content. *Journal of Geophysical Research—Atmospheres,* 109, 1–8, 2004.
5. Carter, G. A., Responses of leaf spectral reflectance to plant stress. *American Journal of Botany,* 80, 239–243, 1993.
6. Peñuelas, J., Filella, I., Biel, C., Serrano, L., and Save, R., The reflectance at the 950–970 nm region as an indicator of plant water status. *International Journal of Remote Sensing,* 14, 1887–1905, 1993.
7. Peñuelas, J., Gamon, J. A., Fredeen, A. L., Merino, J., and Field, C. B., Reflectance indices associated with physiological changes in nitrogen and water limited sunflower leaves. *Remote Sensing of Environment,* 48, 135–146, 1994.
8. Stimson, H. C., Breshears, D. D., Ustin, S. L., and Kefauver, S. C., Spectral sensing of foliar water conditions in two co-occurring conifer species: *Pinus edulis* and *Juniperus monosperma. Remote Sensing of Environment,* 96, 108–118, 2005.
9. Begg, J. E. and Turner, N. C., Crop water deficits. *Advances in Agronomy,* 28, 161–217, 1976.
10. Strachan, I. B., Pattey, E., and Boisvert, J. B., Impact of nitrogen and environmental conditions on corn as detected by hyperspectral reflectance. *Remote Sensing of Environment,* 80, 213–224, 2002.
11. Dzikiti, S., Verreynne, J. S., Stuckens, J., Strever, A., Verstraeten, W. W., Swennen, R., and Coppin, P., Determining the water status of Satsuma mandarin trees [Citrus Unshiu Marcovitch] using spectral indices and by combining hyperspectral and physiological data. *Agricultural and Forest Meteorology,* 150, 3, 369–379, 2010.
12. Hao, X. and Qu, J. J., Retrieval of real-time live fuel moisture content using MODIS measurements. *Remote Sensing of Environment,* 108, 2, 130–137, 2007.
13. Chuvieco, E., Riaño, D., Aguado, I., and Cocero, D., Estimation of fuel moisture content from multitemporal analysis of Landsat Thematic Mapper reflectance data: Applications in fire danger assessment. *International Journal of Remote Sensing,* 23, 11, 2145–2162, 2002.
14. Notarnicola, C. and Posa, F., Inferring vegetation water content from C and L band images. *IEEE Transactions on Geoscience and Remote Sensing,* 45, 10, 3165–3171, 2007.
15. Danson, F. M., Steven, M. D., Malthus, T. J., and Clark, J. A., High-spectral resolution data for determining leaf water content. *International Journal of Remote Sensing,* 13, 3, 461–470, 1992.
16. Rollin, E. M. and Milton, E. J., Processing of high spectral resolution reflectance data for the retrieval of canopy water content information. *Remote Sensing of Environment,* 65, 86–92, 1998.
17. Clevers, J. G. P. W., Kooistra, L., and Schaepman, M. E., Using spectral information from the NIR water absorption features for the retrieval of canopy water content. *International Journal of Applied Earth Observation and Geoinformation,* 10, 3, 388–397, 2008.
18. Anderson, M. C., Neale, C. M. U., Li, F., Norman, J. M., Kustas, W. P., Jayanthi, H. et al. Upscaling ground observations of vegetation water content, canopy height, and leaf area index during SMEX02 using aircraft and Landsat imagery. *Remote Sensing of Environment,* 92, 447–464, 2004.
19. Dawson, T. P., Curran, P. J., North, P. R. J., and Plummer, S. E., The propagation of foliar biochemical absorption features in forest canopy reflectance: A theoretical analysis. *Remote Sensing of Environment,* 67, 147–159, 1999.
20. Burgan, R. E., Use of remotely sensed data for fire danger estimation. *EARSeL Advances in Remote Sensing,* 4, 4, 1–8, 1996.
21. Ceccato, P., Flasse, S., Tarantola, S., Jacquemoud, S., and Gregoire, J. M., Detecting vegetation leaf water content using reflectance in the optical domain. *Remote Sensing of Environment,* 77, 22–33, 2001.
22. Kramer, P. J., *Plant and Soil Water Relationships: A Modern Synthesis.* McGraw-Hill, Inc. New York, p. 482, 1969.
23. Serrano, L., Ustin, S. L., Roberts, D. A., Gamon, J. A., and Peñuelas, J., Deriving water content of chaparral vegetation from AVIRIS data. *Remote Sensing of Environment,* 74, 570–581, 2000.

24. Feret, J. B., François, C., Asner, G. P., Gitelson, A. A., Martin, R. E., Bidel, L. P. R., Ustin, S. L., le Maire, G., and Jacquemoud, S., PROSPECT-4 and -5: Advances in the leaf optical properties model separating photosynthetic pigments. *Remote Sensing of Environment*, 112, 6, 3030–3043, 2008.

25. Jacquemoud, S. and Baret, F., PROSPECT: A model of leaf optical properties spectra. *Remote Sensing of Environment*, 34, 75–91, 1990.

26. Knipling, E. B., Physical and physiological basis for the reflectance of visible and near-infrared radiation from vegetation. *Remote Sensing of Environment*, 1, 155–159, 1970.

27. Thomas, J. R., Namken L. N., Oerther, G. F., and Brown, R. G., Estimating leaf water content by reflectance measurements. *Agronomy Journal*, 63, 845–847, 1971.

28. Ceccato, P., Gobron, N., Flasse, S., Pinty, B., and Tarantola, S., Designing a spectral index to estimate vegetation water content from remote sensing data: Part 1, Theoretical approach. *Remote Sensing of Environment*, 82, 188–197, 2002.

29. Bowyer, P. and Danson, F. M., Sensitivity of remotely sensed spectral reflectance to variation in live fuel moisture content. *Remote Sensing of Environment*, 92, 297–308, 2004.

30. Seelig, H. D., Adams, W. W., Hoehn, A., Stodieck L. S., Klaus D. M., and Emery W. J., Extraneous variables and their influence on reflectance-based measurements of leaf water content. *Irrigation Science*, 26, 5, 407–414, 2008.

31. Aldakheel, Y. Y. and Danson, F. M., Spectral reflectance of dehydrating leaves: Measurements and modelling. *International Journal of Remote Sensing*, 18, 3683–3690, 1997.

32. Kokaly, R. F. and Clark, R. N., Spectroscopic determination of leaf biochemistry using band-depth analysis of absorption features and stepwise multiple linear regression. *Remote Sensing of Environment*, 67, 267–287, 1999.

33. Datt, B., Remote sensing of water content in eucalyptus leaves. *Australian Journal of Botany*, 47, 909–923, 1999.

34. Myneni, R. B., Ross, J., and Asrar, G., A review on the theory of photon transport in leaf canopies in slab geometry. *Agricultural and Forest Meteorology*, 45, 1–153, 1989.

35. Kokaly, R. F., Asner, G. P., Ollinger, S. V., Martin, M. E., and Wessman, C. A., Characterizing canopy biochemistry from imaging spectroscopy and its application to ecosystem studies. *Remote Sensing of Environment*, 113, 78–91, 2009.

36. Verhoef, W., Light scattering by leaf layers with application to canopy reflectance modeling: The SAIL model. *Remote Sensing of Environment*, 16, 125–141, 1984.

37. Kuusk, A., A fast, invertible canopy reflectance model. *Remote Sensing of Environment*, 51, 342–350, 1995.

38. Bull, C. R., Wavelength selection for near-infrared reflectance. *Journal of Agricultural Engineering Research*, 49, 113–125, 1991.

39. Lillesaeter, O., Spectral reflectance of partly transmitting leaves: Laboratory measurements and mathematical modeling. *Remote Sensing of Environment*, 12, 247–254, 1982.

40. Goetz, A. F. H., Three decades of hyperspectral remote sensing of the Earth: A personal view. *Remote Sensing of Environment*, 113, 5–16, 2009.

41. Gao, B. C., NDWI, a normalized difference water index for remote sensing of vegetation liquid water from space. *Remote Sensing of Environment*, 58, 257–266, 1996.

42. Peñuelas, J., Piñol, J., Ogaya, R., and Filella, I., Estimation of plant water concentration by the reflectance water index (R900/R970). *International Journal of Remote Sensing*, 18, 2869–2875, 1997.

43. Zarco-Tejada, P. L. J. and Ustin, S. L., Modeling canopy water content for carbon estimates from MODIS data at land EOS validation sites. *International Geoscience and Remote Sensing Symposium*, 342–344, 2001.

44. Pu, R., Ge, S., Kelly, N. M., and Gong, P., Spectral absorption features as indicators of water status in coast live oak (Quercus agrifolia) leaves. *International Journal of Remote Sensing*, 24, 1799–1810, 2003.

45. Hardisky, M. A., Lemas, V., and Smart, R. M., The influence of soil salinity, growth form, and leaf moisture on the spectral reflectance of Spartina alternifolia canopies. *Photogrammetric Engineering and Remote Sensing*, 49, 77–83, 1983.

46. Elvidge, C. D. and Lyon, R. J. P., Estimation of the vegetation contribution to the 1.65/2.22 mm ratio in air-borne thematic-mapper imagery of the Virginia Range, Nevada. *International Journal of Remote Sensing*, 6, 75–88, 1985.

47. Hunt, E. R., Jr., Rock, B. N., and Nobel, P. S., Measurement of leaf relative water content by infrared reflectance. *Remote Sensing of Environment*, 22, 429–435, 1987.

48. Hunt, E. R. and Rock, B. N., Detection of changes in leaf water content using near and middle-infrared reflectances. *Remote Sensing of Environment*, 30, 43–54, 1989.

49. Zhao, C. J., Zhou, Q. F., Wang, J. H., and Huang, W. J., Band selection for relative depth indices (RDI) in analyzing wheat water status under field conditions. *International Journal of Remote Sensing*, 25, 2575–2584, 2004.

50. Chen, D., Huang, J. F., and Jackson, T. J., Vegetation water content estimation for corn and soybeans using spectral indices derived from MODIS near- and short-wave infrared bands. *Remote Sensing of Environment*, 98, 222–236, 2005.

51. Wang, L. and Qu, J. J., NMDI: A normalized multi-band drought index for monitoring soil and vegetation moisture with satellite remote sensing. *Geophysical Research Letter*, 34, L20405, doi:10.1029/2007GL031021, 2007.

52. Colombo, R., Meroni, M., Marchesi, A., Busetto, L., Rossini, M., Giardino, C., and Panigada, C., Estimation of leaf and canopy water content in poplar plantations by means of hyperspectral indices and inverse modeling. *Remote Sensing of Environment*, 112, 4, 1820–1834, 2008.

53. Kumar, L., High-spectral resolution data for determining leaf water content in Eucalyptus species: Leaf level experiments. *Geocarto International*, 22, 1, 3–16, 2007.

54. Fava, F., Colombo, R., Bocchi, S., Meroni, M., Sitzia, M., Fois, N., and Zucca, C., Identification of hyperspectral vegetation indices for Mediterranean pasture characterization. *International Journal of Applied Earth Information and Geoinformation*, 11, 233–243, 2009.

55. Peñuelas, J. and Inoue, Y., Reflectance indices indicative of changes in water and pigment contents of peanut and wheat leaves. *Photosynthetica*, 36, 3, 355–360, 1999.

56. Gamon, J. A. and Qiu, H.-L., Ecological applications of remote sensing at multiple scales. In: Pugnaire, F. I. and Valladares, F. (Eds.), *Handbook of Functional Plant Ecology*. Marcel Dekker, Inc., New York, pp. 805–846, 1999.

57. Gamon, J. A., Qiu, H.-L., Roberts, D. A., Ustin, S. L., Fuentes, D. A., Rahman, A., Sims, D., and Stylinski, C., Water expressions from hyperspectral reflectance: Implications for ecosystem flux modeling. In: Green, R. O. (Ed.), *Summaries of the Eighth JPL Airborne Earth Science Workshop*. Pasadena, CA, 1999.

58. Claudio, H. C., Cheng, Y., Fuentes, D. A., Gamon, J. A., Luo, H., Oechel, W., Qiu, H. L., Rahman, A. F., and Sims, D. A., Monitoring drought effects on vegetation water content and fluxes in chaparral with the 970 nm water band index. *Remote Sensing of Environment*, 103, 3, 304–311, 2006.

59. Eitel, J. U. H., Gessler, P. E., Smith, A. M. S., and Ronbrecht, R., Suitability of existing and Novel spectral indices to remotely detect water stress in *Populus* spp. *Forest Ecology and Management*, 229, 170–182, 2006.

60. Piñol, J., Filella, I., Ogaya, R., and Peñuelas, J., Ground-based spectroradiometric estimation of life fine fuel moisture of Mediterranean plants. *Agricultural Forest Meteorology*, 90, 173–186, 1998.

61. Sims, D. A. and Gamon, J. A., Estimation of vegetation water content and photosynthetic tissue area from spectral reflectance: A comparison of indices based on liquid water and chlorophyll absorption features. *Remote Sensing of Environment*, 84, 526–537, 2003.

62. Zarco-Tejada, P. J., Rueda, C. A., and Ustin, S. L., Water content estimation in vegetation with MODIS reflectance data and model inversion methods. *Remote Sensing of Environment*, 85, 109–124, 2003.

63. De Santis, A., Vaughan, P., and Chuvieco, E., Foliage moisture content estimation from one-dimensional and two-dimensional spectroradiometry for fire danger assessment. *Journal of Geophysical Research*, 111, G04S03, 2006.

64. Wu, C., Niu, Z., Tang, Q., and Huang, W., Predicting vegetation water content in wheat using normalized difference water indices derived from ground measurements. *Journal of Plant Research*, 122, 3, 317–326, 2009.

65. Zhang, J., Xu, Y., Yao, F., Wang, P., Guo, W., Li, L., and Yang, L., Advances in estimation methods of vegetation water content based on optical remote sensing techniques. *Science China Technological Science*, 53, 5, 2010.

66. Dennison, P. E., Roberts, D. A., Peterson, S. H., and Rechel, J., Use of normalized difference water index for monitoring live fuel moisture. *International Journal of Remote Sensing*, 26, 1035–1042, 2005.

67. Dasgupta, S., Qu, J. J., Hao, X., and Bhoi, S., Evaluating remotely sensed live fuel moisture estimations for fire behavior predictions in Georgia. *Remote Sensing of Environment*, 108, 138–150, 2007.

68. Tucker, C. J., Remote sensing of leaf water content in the near infrared. *Remote Sensing of Environment*, 10, 23–32, 1980.

69. Ripple, W. J., Spectral reflectance relationships to leaf water stress. *Photogrammetric Engineering and Remote Sensing*, 52, 10, 1669–1675, 1986.

70. Cohen, W. B., Temporal versus spatial variation in leaf reflectance under changing water-stress conditions. *International Journal of Remote Sensing*, 12, 1865–1876, 1991.

71. Riggs, G. A. and Running, S. W., Detection of canopy water stress in conifers using the airborne imaging spectrometer. *Remote Sensing of Environment*, 35, 51–68, 1991.
72. Cibula, W. G., Zetka, E. F., and Rickman, D. L., Response of thematic mapper bands to plant water stress. *International Journal of Remote Sensing*, 13, 1869–1880, 1992.
73. Hunt, E. R., Jr., Airborne remote sensing of canopy water thickness scaled from leaf spectrometer data. *International Journal of Remote Sensing*, 12, 643–649, 1991.
74. Yilmaz, M. T., Hunt, E. R., Goins, L. D., Ustin, S. L., Vanderbilt, V. C., and Jackson, T. J., Vegetation water content during SMEX04 from ground data and Landsat 5 Thematic Mapper imagery. *Remote Sensing of Environment*, 112, 2, 350–362, 2008.
75. Yilmaz, M. T., Hunt, E. R., and Jackson, T. J., Remote sensing of vegetation water content from equivalent water thickness using satellite imagery. *Remote Sensing of Environment*, 112, 2514–2522, 2008.
76. Maki, M., Ishiahra, M., and Tamura, M., Estimation of leaf water status to monitor the risk of forest fires by using remotely sensed data. *Remote Sensing of Environment*, 90, 441–450, 2004.
77. Davidson, A., Wang, S., and Wilmshurst, J., Remote sensing of grassland–shrubland vegetation water content in the shortwave domain. *International Journal of Applied Earth Observation and Geoinformation*, 8, 225–236, 2006.
78. Jackson, T. J., Chen, D., Cosh, M., Li, F., Anderson, M., Walthall, C. et al., Vegetation water content mapping using Landsat data derived normalized difference water index for corn and soybeans. *Remote Sensing of Environment*, 92, 475–482, 2004.
79. Filella, I. and Peñuelas, J., The red edge position and shape as indicators of plant chlorophyll content, biomass and hydric status. *International Journal of Remote Sensing*, 15, 1459–1470, 1994.
80. Liu, L., Wang, J., Huang, W., Zhao, C., Zhang, B., and Long, Q., Estimating winter wheat plant water content using red edge parameters. *International Journal of Remote Sensing*, 25, 3331–3342, 2004.
81. Roberts, D. A., Green, R. O., and Adams, J. B., Temporal and spatial patterns in vegetation and atmospheric properties from AVIRIS. *Remote Sensing of Environment*, 62, 223–240, 1997.
82. Hardy, C. C. and Burgan, R. E., Evaluation of NDVI for monitoring live moisture in three vegetation types of the western US. *Photogrammetric Engineering and Remote Sensing*, 65, 603–610, 1999.
83. Dilley, A. C., Millie, S., O'Brien, D. M., and Edwards, M., The relation between Normalized Difference Vegetation Index and vegetation moisture content at three grassland locations in Victoria, Australia. *International Journal of Remote Sensing*, 25, 3913–3928, 2004.
84. Verbesselt, J., Somers, B., Lhermitte, S., Jonckheere, I., Aardt, J. V., and Coppin, P., Monitoring herbaceous fuel moisture content with SPOT VEGETATION time-series for fire risk prediction in savanna ecosystems. *Remote Sensing of Environment*, 108, 357–368, 2007.
85. Cheng, Y. B., Wharton, S., Ustin, S. L., Zarco-Tejada, P. J., Falk, M., and Paw U. K. T., Relationships between Moderate Resolution Imaging Spectroradiometer water indexes and tower flux data in an old growth conifer forest. *Journal of Applied Remote Sensing*, 1, 013513, 2007.
86. Curran, P. J., Dungan, J. L., and Peterson, D. L., Estimating the foliar biochemical concentration of leaves with reflectance spectrometry: Testing the Kokaly and Clark methodologies. *Remote Sensing of Environment*, 76, 349–359, 2001.
87. Tian, Q., Tong, Q., Pu, R., Guo, X., and Zhao, C., Spectroscopic determinations of wheat water status using 1650–1850 nm spectral absorption features. *International Journal of Remote Sensing*, 22, 2329–2338, 2001.
88. Butler, W. L. and Hopkins, W., Higher derivative analysis of complex absorption spectra. *Photochemistry and Photobiology*, 12, 439–450, 1970.
89. O'Haver, T. C., Derivative spectroscopy: Theoretical aspects. *Analytical Proceedings, Proceedings of the Analytical Division of the Royal Society of Chemistry*, 19, 22–28, 1982.
90. Clevers, J. G. P. W., Kooistra, L., and Schaepman, M. E., Estimating canopy water content using hyperspectral remote sensing data. *International Journal of Applied Earth Observations and Geoinformation*, 12, 2, 119–125, 2010.
91. Green, R. O., Conel, J. E., Margolis, J. S., Bruegge, C. J., and Hoover, G. L., An inversion algorithm for retrieval of atmospheric and leaf water absorption from AVIRIS radiance with compensation for atmospheric scattering, In: Green, R. O. (Ed.), *Proceedings of the 3rd Airborne Visible/Infrared Imaging Spectrometer Workshop*, 93–26, JPL Publication, Pasadena, CA, pp. 51–61, 1991.
92. Cheng, Y. B., Ustin, S. L., Riaño, D., and Vanderbilt, V. C., Water content estimation from hyperspectral images and MODIS indexes in Southeastern Arizona. *Remote Sensing of Environment*, 112, 2, 363–374, 2008.
93. Gao, B. C. and Goetz, A. F. H., Column atmospheric water vapor and vegetation liquid water retrievals from airborne imaging spectrometer data. *Journal of Geophysical Research*, 95, 3549–3564, 1990.

94. Gao, B. C. and Goetz, A. F. H., Retrieval of equivalent water thickness and information related to bio-chemical components of vegetation canopies from AVIRIS data. *Remote Sensing of Environment*, 52, 155–162, 1995.

95. Ustin, S. L., Roberts, D. A., Pinzón, J., Jacquemoud, S., Gardner, M., Scheer, G. C., Castaneda, M., and Palacios-Orueta, A., Estimating canopy water content of chaparral shrubs using optical methods. *Remote Sensing of Environment*, 65, 280–291, 1998.

96. Cheng, Y. B., Zarco-Tejada, P. J., Riaño, D., Carlos, A., and Ustin, S. L., Estimating vegetation water content with hyperspectral data for different canopy scenarios: Relationships between AVIRIS and MODIS indexes. *Remote Sensing of Environment* 105, 4, 354–366, 2006.

97. Roberts, D. A., Brown, K. J., Green, R., Ustin, S. L., and Hinckley, T., Investigating the relationship between liquid water and leaf area in clonal populus. In: *Proceedings of the 7th AVIRIS Earth Science Workshop JPL 97–21*, Pasadena, CA, p. 10, 1998.

98. Roberts, D. A., Ustin, S. L., Ogunjemiyo, S., Greenberg, J., Dobrowski, S. Z., and Chen, J. et al., Spectral and structural measures of northwest forest landscapes at leaf to landscape scales. *Ecosystems*, 7, 545–562, 2004.

99. Champagne, C. M., Staenz, K., Bannari, A., Mcnairn, H., and Deguise, J. C., Validation of a hyperspectral curve-fitting model for the estimation of plant water content of agricultural canopies. *Remote Sensing of Environment*, 87, 148–160, 2003.

100. Moreno, J. F., Baret, F., Leroy, M., Menenti, M., Rast, M., and Shaepman, M., Retrieval of vegetation properties from combined hyperspectral/multiangular optical measurements: Results from the DAISEX campaigns. In: *Geoscience and Remote Sensing Symposium. Proceedings of the IGARSS '03 IEEE International*, Piscataway, NJ, 2003.

101. Roberts, D. A., Dennison, P. E., Gardner, M. E., Hetzel, Y., Ustin, S. L., and Lee, C. T., Evaluation of the potential of Hyperion for fire danger assessment by comparison to the airborne visible/infrared imaging spectrometer. *IEEE Transactions on Geoscience and Remote Sensing*, 41, 1297–1310, 2003.

102. Dawson, T. P., Curran, P. J., and Plummer, S. E., LIBERTY—Modeling the effects of leaf biochemical concentration on reflectance spectra. *Remote Sensing of Environment*, 65, 50–60, 1998.

103. Ganapol, B. D., Johnson, L. F., Hammer, P. D., Hlavka, C. A., and Peterson, D. L., LEAFMOD: A new within-leaf radiative transfer model. *Remote Sensing of Environment*, 63, 182–193, 1998.

104. Liang, S., *Quantitative Remote Sensing of Land Surfaces*, John Wiley and Sons, Inc., River Street Hoboken, NJ, p. 534, 2004.

105. Asner, G. P., Biophysical and biochemical sources of variability in canopy reflectance. *Remote Sensing of Environment*, 64, 234–253, 1998.

106. Bacour, C., Jacquemoud, S., Tourbier, Y., Dechambre, M., and Frangi, J.-P., Design and analysis of numerical experiments to compare four canopy reflectance models. *Remote Sensing of Environment*, 79, 72–83, 2002.

107. Huemmrich, K. F., The GeoSAIL model: A simple addition to the SAIL model to describe discontinuous canopy reflectance. *Remote Sensing of Environment*, 75, 3, 423–431, 2001.

108. Ceccato, P., Flasse, S., and Gregoire, J. M., Designing a spectral index to estimate vegetation water content from remote sensing data: Part 2. Validation and applications. *Remote Sensing of Environment*, 82, 198–207, 2002.

109. North, P. R. J., Three-dimensional forest light interaction model using a Monte Carlo method. *IEEE Transaction on Geoscience and Remote Sensing*, 34, 946–956, 1996.

110. Shen, Y., Shi, R., Niu, Z., and Yan, C., Estimation models for vegetation water content at both leaf and canopy levels. In: *Geoscience and Remote Sensing Symposium, Proceedings of the IEEE International*, Seoul, Korea, 2005.

111. Ghulam, A., Li, Z. L., Qin, Q. M., Yimit, H., and Wang, J., Estimating crop water stress with ETM + NIR and SWIR data. *Agricultural Forest Meteorology*, 148, 1679–1695, 2008.

112. Kimes, D. S., Knyazikhin, Y., Privette, J. L., Abuelgasim, A. A., and Gao, F., Inversion of physically-based models. *Remote Sensing Reviews*, 18, 381–439, 2000.

113. Jacquemoud, S. and Baret F., Inversion of the PROSPECT + SAILH canopy reflectance model from AVIRIS equivalent spectra: Theoretical study. *Remote Sensing of Environment*, 44, 281–292. 1993.

114. Jacquemoud, S., Baret, F., Andrieu, B., Danson F. M., and Jaggard, K., Extraction of vegetation biophysical parameters by inversion of the PROSPECT + SAIL models on sugar beet canopy reflectance data. Application to TM and AVIRIS sensors. *Remote Sensing of Environment*, 52, 163–172, 1995.

115. Jacquemoud, S., Ustin, S. L., Verdebout, J., Schmuck, G., Andreoli, G., and Hosgood, B., Estimating leaf biochemistry using the PROSPECT leaf optical properties model. *Remote Sensing of Environment*, 56, 194–202, 1996.

116. Fourty, T. and Baret, F., On spectral estimates of fresh leaf biochemistry. *International Journal of Remote Sensing*, 19, 1283–1297, 1997.

117. Kötz, B., Schaepman, M., Morsdorf, F., Itten, K., and Allgöwer, B., Radiative transfer modeling within a heterogeneous canopy for estimation of forest fire fuel properties. *Remote Sensing of Environment*, 92, 332–344, 2004.

118. Asner, G. P. and Vitousek, P. M., Remote analysis of biological invasion and biogeochemical change. *Proceedings of the National Academy of Sciences U. S. A.*, 102, 4383–4386, 2005.

119. Li, J. and Goodenough, D. G., Mapping relative water content in douglas-fir with AVIRIS and a canopy model. In: *Proceedings of the IGARSS'05*, Seoul, Korea, pp. 3572–3574, 2005.

120. Toomey, M. P. and Vierling, L. A., Estimating equivalent water thickness in a conifer forest using Landsat TM and ASTER data: A comparison study. *Canadian Journal of Remote Sensing*, 32, 288–299, 2006.

121. Combal, B., Baret, F., Weiss, M., Trubuil, A., Macé, D., Pragnère, A., Myneni, R., Knyazikhin, Y., and Wang, L., Retrieval of canopy biophysical variables from bidirectional reflectance using prior information to solve the ill-posed inverse problem. *Remote Sensing of Environment*, 84, 1–15, 2002.

122. Riaño, D., Vaughan, P., Chuvieco, E., Zarco-Tejada, P. J., and Ustin, S. L., Estimation of fuel moisture content by inversion of radiative transfer models to simulate equivalent water thickness and dry matter content: Analysis at leaf and canopy level. *IEEE Transaction on Geoscience and Remote Sensing*, 43, 819–826, 2005.

123. Kötz, B., Sun, G., Morsdorf, F., Ranson, K. J., Kneubühler, M., Itten, K., and Allgöwe, B., Fusion of imaging spectrometer and LIDAR data over combined radiative transfer models for forest canopy characterization. *Remote Sensing of Environment*, 106, 4, 449–459, 2007.

124. Danson, F. M. and Bowyer, P., Estimating live fuel moisture content from remotely sensed reflectance. *Remote Sensing of Environment*, 92, 309–321, 2004.

125. Yebra, M. and Chuvieco, E., Linking ecological information and radiative transfer models to estimate fuel moisture content in the Mediterranean region of Spain: Solving the ill-posed inverse problem. *Remote Sensing of Environment*, 113, 2403–41, 2009.

126. Fourty, T. and Baret F., Vegetation water and dry matter contents estimated from top-of-the-atmosphere reflectance data: A simulation. *Remote Sensing of Environment*, 61, 34–45, 1997.

127. Riaño, D., Ustin, S., Usero, L., and Patricio, M. A., Estimation of fuel moisture content using neural networks. *Lecture Notes in Computer Science*, 3562, 489–498, 2005.

128. Rubio, M. A., Riaño, D., Cheng, Y. B., and Ustin, S. L., *Estimation of Canopy Water Content from MODIS Using Artificial Neural Networks Trained with Radiative Transfer Models*. ECAC, European Meteorological Society, Ljubljana, Slovenia, 2006.

129. Trombetti, M., Riaño, D., Rubio, M. A., Cheng, Y. B., and Ustin, S. L., Multi-temporal vegetation canopy water content retrieval and interpretation using artificial neural networks for the continental USA. *Remote Sensing of Environment*, 112, 203–215, 2008.

130. Dawson, T. P., Curran, P. J., and Plummer, S. E., The biochemical decomposition of slash pine needles from reflectance spectra using neural networks. *International Journal of Remote Sensing*, 19, 1433–1438, 1998.

131. Pinzon, J. E., Ustin, S. L., Canstenada, C. M., and Smith, M. O., Investigation of leaf biochemistry by hierarchical foreground/background analysis. *IEEE Transaction Geoscience and Remote Sensing*, 36, 1–15, 1998.

132. Li, L., Cheng, Y. B., Ustin, S., Hu, X. T., and Riaño, D., Retrieval of vegetation equivalent water thickness from reflectance using genetic algorithm (GA)-partial least squares (PLS) regression. *Advances in Space Research*, 41, 11, 1755–1763, 2008.

11 Estimation of Nitrogen Content in Crops and Pastures Using Hyperspectral Vegetation Indices

Daniela Stroppiana, F. Fava, M. Boschetti, and P.A. Brivio

CONTENTS

11.1 INTRODUCTION

The nitrogen cycle is of particular interest to ecologists because nitrogen availability can affect the rate of key ecosystem processes, including primary production. However, the excessive use of artificial nitrogen fertilizers, together with fossil fuel combustion and release of nitrogen in wastewater, has dramatically altered the global nitrogen cycle [1]. The extent and effects of the anthropogenically induced doubling of biologically available nitrogen in the soils, waters, and air of the Earth during the past century is still poorly understood. Nitrogen is an essential nutrient for crop and pasture growth. Crop nitrogen status is key information for the application of Variable Rate Technology in precision agriculture, which aims to maximize productivity and, at the same time, to limit the environmental impact of excessive fertilization. For pastures, nitrogen content is a key determinant of the nutritive value, being directly related to the crude protein concentration in herbage or forage [2]. Traditionally, nitrogen content has been estimated through soil testing and plant tissue analysis [3], which involves destructive field sampling and laboratory chemical analyses. Besides the effort required, destructive measurements can prevent the study of successive developmental stages. Moreover, they create a lag between sampling, laboratory analyses, and nitrogen treatments [4] or livestock grazing management [5].

Improved technology, such as the chlorophyll soil and plant analyzer development (SPAD) meter (Minolta Osaka Co., Ltd, Japan), is based on measuring leaf transmittance in two wavebands centered at 650 and 940 nm (e.g., [6]), which well correlate to leaf chlorophyll content. Yet SPAD readings still require laborious field measurements on single leaves and these may not really represent the great field variability unless they are made on an impracticable number of leaves and plants. Moreover, SPAD measurements have been shown to be affected by growth stage, nutrient deficiency other than nitrogen, environmental conditions, and measurement position on leaves [4]. Ideally, a method is required that is accurate, nondestructive, simple to use, and from the canopy as opposed to the leaf point of view [7]. Remote sensing techniques satisfy these requirements by measuring the optical properties of leaves and canopies that are driven by the biophysical and biochemical properties of the plant.

Early work laying the foundations of remote sensing applications described the basic theory of relationships between the optical and morphological properties of the leaf (e.g., [8]) and investigated the effects of environmental stresses (e.g., water stress, nutrient deficiency and pests) on leaf spectra [9,10]. Laboratory applications of reflectance spectroscopy on leaf and plant samples clearly showed the potential of using optical properties for assessing nitrogen content [11]. The next step was to scale this knowledge up to the canopy level [10] where this information could find practical applications by describing the spatial and temporal variability of canopy characteristics.

Radiometric canopy data have in fact the potential of measuring reflected radiation from many plants within the field of view of the sensor [3], thus making nitrogen canopy assessment feasible, non-destructive, and cost-effective. Precursory field experiments at leaf and canopy level with multispectral sensors demonstrated the sensitivity of visible bands and vegetation indices (VIs) in the visible and near-infrared (NIR) to pigment and nitrogen content [12,13]. However, the development and diffusion of hyperspectral instruments was what really boosted the use of remote sensing data for nitrogen estimation since the most important reflectance features related to nitrogen content can only be measured by using contiguous and narrow bands [14,15]. The most widely used approach with hyperspectral data is based on empirical relationships between nitrogen content and reflectance in narrow bands and VIs [15–21]. Empirical approaches do not attempt to directly relate changes in reflectance to nitrogen variability through the description of a causal relationship, but they measure the strength of the statistical correlation observed.

This chapter will review the use of hyperspectral vegetation indices (HVIs) for estimating nitrogen content in herbaceous plants (crops and pastures) by focusing on the most common/recent advancements, describing and discussing their theoretical basis and the methodological approaches.

11.2 STATE OF THE ART

Most of the recent literature for crops (Table 11.1) focuses on nitrogen status assessment of rice (*Oryza sativa*), maize (*Zea mays L. ssp*), and wheat (*Triticum ssp.*) as they are the most commonly produced cereals worldwide [15,16,22–26] although cotton [27], sorghum [28], sunflowers [29], and sugar cane [30], among others, have also been studied. Leaf scale measurements have been used to understand the physiological processes governing plant growth and to estimate leaf nitrogen content as an indicator of crop nutrition status [4,11,23,27,29,31]. The use of canopy reflectance has shown promise for deriving crop nitrogen status for precision agriculture and yield estimation [16,24,25].

Some authors have more specifically addressed the issue of estimating photosynthetic pigment concentrations as indicators of physiological vegetation conditions due to their direct relationship with reflectance [32–34]. Principally, empirical statistical approaches, such as simple linear and nonlinear regression [20,24,27] and multiple linear regression (MLR) [4], have been used with VIs. Partial least square (PLS) regression has also been shown to be promising by Hansen and Schjoerring [15] and Nguyen and Lee [22]. Finally, Artificial Neural Networks (ANN) were found suitable for crop nitrogen assessment by Yi et al. [4].

Although most studies relied on proximal sensing techniques (i.e., field spectrometers), airborne hyperspectral sensors, such as AISA for rice [35] and Compact Airborne Spectrographic Imager (CASI) for corn [36], have been tested.

Finally, only a few studies have used simulated data to investigate the influence of biophysical and biochemical parameters on leaf and canopy reflectance in relation to nutrient stress [33]. Simulations

TABLE 11.1

Herbaceous Vegetation	Sensor	Range (nm)	Scale		Parameter		Spectral Transform			HVI							Reference
			L	C	PP	N	CR	De	PC	R$_\lambda$	SR	ND	Ch	S	RE	Ot	
Crop																	
Cotton	FS	300–1100								a,b					a,b		[97]
Cotton	FS	350–1050									c	c					[27]
Rice	FS	447–1752								a	c	a		a			[16]
Rice	FS	300–1100															[22]
Rice	FS	350–2500								d,e							[4]
Rice	FS	447–1752								c	c	c					[24]
Rice	AS	400–1000								a							[35]
Sorghum	FS	350–2500									c	c					[67]
Sugarcane	FS	350–2500									c						[30]
Sunflowers	FS	390–1100								c	c	a					[29]
Wheat	FS	400–900											c	c			[15]
Wheat	FS	350–2500										c	c	c		b	[81]
Wheat	FS	350–2500									b	b				c	[83]
Wheat	FS	350–2500									c	c				c	[23]
Wheat	FS	350–2500									c	a				a	[25]
Wheat	FS	350–2500								c	c	c	c	c		c	[82]
Wheat, corn	FS,AS	325–1075								b							[26]
Corn	AS	407–949															[36]
Pasture																	
Bermudagrass (M)	FS	280–1100								c	c	c				a	[37]
Bermudagrass (M)	FS	368–1100															[5]
Bermudagrass (M)	FS	350–2500								a,d	c						[38]
Bermudagrass (M)	FS	350–2500								c,d	c						[19]
Bermudagrass (M)	FS	350–2500								a,d	c					c	[38]
Bermudagrass (M) and mixed grasslands (MG)	FS	350–2500															[39]

(*continued*)

TABLE 11.1 (continued)

Herbaceous Vegetation	Sensor	Range (nm)	Scale		Parameter		Spectral Transform			HVI							Reference
			L	C	PP	N	CR	De	PC	R_λ	SR	ND	Ch	S	RE	Ot	
Pasture																	
Ryegrass (M)	FS	400–900													d		[41]
Ryegrass (M)	FS	404–1650															[42]
Ryegrass (M)	FS	350–2500															[40]
Blue grama and Sandberg bluegrass (M)	FS	350–2500															[98]
Ryegrass (M), Clover (M) and mixed grasslands (MG)	FS	350–2500															[43]
Sainfoin pastures (M)	FS	325–1150								a							[18]
Blue Buffalo Grass (G)	FS	350–2500								c,d	c	c					[45]
Blue Buffalo Grass (G)	FS	350–2500								c	c				c		[46]
Tropical grass (G)	FS	350–2500								d							[47]
Mixed grassland (MG)	AS	423–2507								a,d	c	c					[52]
Natural prairie (MG)	FS,AS	350–2500															[53]
Alpine meadow (MG)	FS	325–1075								a	c	c					[17]
Mediterranean pastures (MG)	FS	325–1075									c						[21]
Meadow (MG)	FS	360–1010														a	[99]
Temperate pastures (MG)	FS	350–2500								a,d						a,d	[54]
Mixed grasslands (MG) and Ryegrass	FS	350–2500								c,d					c		[45]
Savanna (MG)	FS	350–2500								c,d							[48]
Savanna (MG)	AS	500–2450								d,e					a,d		[49]
Savanna (MG)	AS	500–2400								e							[50]

M, monoculture; G, greenhouse; MG, mixed grassland; FS, field sensor; AS, airborne sensor; L, leaf spectra; C, canopy spectra; PP, photosynthetic pigments; N, nitrogen; CR, continuum removal; De, derivative spectra; PC, principal components; R, Band reflectance; SR, simple ratio; ND, normalized difference; Ch, chlorophyll sensitive indices; S, soil indices; RE, red edge; Ot, other indices.

a   Partial least square regression.
b   Discriminant analysis.
c   Simple linear regression.
d   Multiple linear regression.
e   Artificial neural networks.

certainly offer the advantage of exploring a wide range of conditions that could not simultaneously occur in experimental fields where crops are grown under controlled conditions.

Concerning pastures, the majority of studies in the last 15 years (Table 11.1) have been performed in monoculture pastures, mainly bermudagrass (*Cynodon dactylon* L.) [5,19,37–39], and ryegrass (*Lolium* sp.) [40–43], or in greenhouse experiments at leaf and canopy level [44–47]. Beside their economic relevance, monoculture pastures are particularly suited for reflectance spectroscopy experiments, since their canopy structure is vertically and horizontally homogeneous compared to mixed species canopies.

Estimating nitrogen content in natural and seminatural pastures, composed of mixed species and with a complex canopy structure, is more challenging. Nevertheless, in the last few years, a number of studies have been conducted in savannas [48–50], mountain and Mediterranean pastures [17,21,51], and north American grasslands [52,53], demonstrating increasing scientific interest in these ecosystems, which play a key role in providing several ecosystem goods and services.

Leaf level experiments for nitrogen assessment in pastures are limited to just a few examples [41,45], as studies on leaves have been conducted mainly on crops. Most of the research has been done at canopy level, mainly by means of field spectrometric data (Table 11.1). Airborne sensors have been little used for nitrogen assessment at landscape scale on pastures. Mutanga and Skidmore [49] and Skidmore et al. [50] analyzed airborne HYMAP data (450–2500 nm) to map nitrogen concentration (nitrogen on a dry weight basis [mg N g^{-1} dry weight] and generally expressed as [%]) in African savannas. Beeri et al. [53] used HYMAP to accurately map carbon:nitrogen ratios in the American Northwestern Great Plains, while Mirik et al. [52] acquired PROBE-1 (423–2507 nm) data from a helicopter to map forage quality in different grassland systems in the Yellowstone National Park.

As far as nitrogen retrieval is concerned, most research has been based on empirical statistical approaches. Studies differ either according to the type of spectral transformation applied (e.g., derivative, continuum removal, band depth, VIs, etc.) or/and to the statistical technique used (e.g., ordinary least square regression, MLR, PLS regression, ANN, etc.). A positive tendency is that an increasing number of studies compared different techniques for pastures and discussed the main advantages and constraints [17–19,38,39,43–45]. Research done by Mutanga et al. [40,44,48,49], for example, emphasize the advantages of using continuum removal [55] as a spectral transformation to enhance nitrogen absorption features, minimizing the spectral variability, which was independent of the nitrogen content. The use of VIs and red-edge indices has been also widely explored and compared to other methodologies.

Finally, to our knowledge, satellite instruments have not yet been used to map nitrogen in either crops or pasturelands, probably due to the limited spatial and/or spectral resolution of most satellite sensors.

11.3 PHYSICAL AND PHYSIOLOGICAL BASIS

The domain of optical remote sensing used in vegetation monitoring (400–2500 nm) is usually divided into three spectral regions (VIS-NIR 400–700 nm; NIR 700–1100 nm; and short wave infrared [SWIR] 1100–2500 nm) characterized by specific light-vegetation interactions influenced by leaf biochemicals and structural canopy characteristics. The amount of radiation reflected by vegetation in the photosynthetically active radiation (PAR) region of the electromagnetic spectrum (400–700 nm) is regulated by pigment absorption within leaves. Chlorophyll-a and chlorophyll-b absorb the greatest proportion of radiation and provide energy for the reactions of photosynthesis, while carotenoids protect the reaction centers from excess light and help intercept PAR as auxiliary pigments of chlorophyll-a [56]. Chlorophylls absorb radiation mainly in the blue (~450 nm) and red (~680 nm) wavelengths, whereas carotenoids have an absorption feature in the blue overlapping with chlorophyll. The red absorption peak is solely due to the presence of chlorophylls but low concentrations might saturate the 660–680 nm region, thus making it poorly sensitive to high chlorophyll contents [10,32]. Longer (~700 nm, red edge) or shorter (~550 nm, green) wavelengths are therefore preferred because reflectance is more sensitive to moderate-to-high chlorophyll content [10].

In the NIR wavelengths (700–1100 nm), the high reflectance factor is caused by the internal leaf cellular structure [8]. Between the red and NIR wavelengths, leaf reflectance is associated with the transition from chlorophyll absorption to leaf scattering; this position, referred to as red edge, is identified by the wavelength of the maximum slope of the reflectance spectrum between 650 and 800 nm [57]. An increase in the amount of chlorophyll in the canopy, either due to increases in the chlorophyll concentration or to Leaf Area Index (LAI), results in the broadening of the red absorption feature, and, consequently, in the shift of the red-edge position (REP) toward longer wavelengths [58].

The SWIR (1100–2500 nm) is dominated by water absorption and minor absorption features related to other foliar biochemicals, including nitrogen [11]. However, determining the nitrogen content from reflectance measurements on fresh leaf or canopies is extremely complex due, among other things, to the strong effect of water and to the impossibility of unequivocally associating a specific absorption feature with the chemical abundance of the biochemical of interest [55].

The relationship between nitrogen supply and chlorophyll formation has long been observed [59]. Since part of leaf nitrogen is contained in chlorophyll molecules, the amount of available nitrogen largely determines the amount of chlorophyll formed in plants, provided that other requirements for chlorophyll formation, such as light, iron supply, and magnesium, are present in sufficient quantities [60]. However, the nitrogen/chlorophyll relation can be influenced by environmental conditions (nutrients and water stress), leaf position in the canopy, genotype, temperature, and leaf growth stage [10,61,62]. Since nitrogen stress induces a physiological change (pigment concentration), which in turn produces changes in leaf spectra, reflectance can be used to assess nitrogen status [29,63].

Also the relation between leaf nitrogen and canopy spectra is indirectly due to its association with chlorophyll [64] since canopy spectra are determined by optical leaf properties besides density and geometry of the canopy (LAI and Leaf Angle Distribution [LDA]) and background reflectivity [65]. Canopy spectra can change dramatically during the season as a consequence of changes in the architecture and arrangement of plant components and changes in the proportion of soil and vegetation [9]. Reflectance from a canopy is considerably less than that from an individual leaf, although in the NIR wavelengths attenuation is less pronounced. In fact, the radiation transmitted through the upper leaves is reflected by the lower strata and transmitted up to enhance the reflectivity of the upper leaves [8].

In conclusion, the use of leaf and canopy spectra for nitrogen assessment generally relies on the close relation between nitrogen and chlorophylls in the cell metabolism although the experimental relationship established at the canopy scale remains purely empirical. Kokaly et al. [67] in fact state that the two variables are only moderately correlated within and across ecosystems.

11.4 RETRIEVAL APPROACHES

The most widely used approaches to estimate nitrogen content are based on regressive models relating in situ measurements and VIs. Simple linear and nonlinear regressions have been widely applied [20,24,27,67] although multivariate techniques, such as MLR or PLS regression, have generally allowed slightly better nitrogen prediction [17,19,22,39]. PLS regression, with respect to MLR, is able to reduce the large number of collinear spectral variables measured to a few noncorrelated principal components (PCs) [15]. Stepwise linear regression has been exploited to identify wavebands for MLR where correlation between reflectance and chemical concentrations is high. This technique generally provided good results for nitrogen estimation [4,19,39] although the selected wavelengths were found to be only indirectly related to nitrogen content. Some major drawbacks to multivariate techniques can, however, be pointed out: limited exportability of statistical models that use many predictors, overfitting of wavebands in the calibration equation, intercorrelation of leaf chemicals and omission of expected wavelengths where absorption features are known to be present [11].

ANN have also been applied to derive biochemical characteristics of crop canopies [4,40,50]. These methods are suitable to describe the nonlinear relationships that may exist between observations. However, training ANN presents limitations similar to regression analysis due to the low generalization power.

Physically based canopy reflectance models also deserve to be mentioned. By means of radiative transfer equations, they describe the interactions between the incident radiation and the biophysical and biochemical vegetation parameters. They can be numerically inverted to retrieve canopy parameters from leaf/canopy radiometric measurements. However, models can be applied only to retrieve those parameters that are directly involved in the physical processes of radiative transfer; in this case photosynthetic pigments rather than nitrogen. Since models can be very complex, numerical inversion could be computationally intensive, especially when applied at the pixel scale. Moreover, since no universally applicable canopy reflectance model for all vegetation types has yet been defined, model selection is a compromise between complexity, invertibility, and computational efficiency [10,68]. Neither should we forget issues related to numerical inversion such as lack of convergence, sensitivity to parameter initialization, and the difficulty of estimating all model parameters. However, radiative transfer models could help us to gain further insight into the spectral features directly influenced by nitrogen and to describe the influence of structural canopy parameters on nitrogen estimates more effectively.

In general, statistical approaches are the simplest way to predict nitrogen content but they provide relationships that are significantly space, time, and species dependent. The regression equations established by different studies do not extrapolate to other sites and years as they depend on viewing and radiation geometries, canopy morphology, soil background, and the spectral characteristics of plant parts. The simplified approach of regressive techniques is suitable when the goal is prediction of in situ quantities rather than understanding radiative transfer processes. In this framework, regressive models based on spectral VIs may be preferable to physically based models that are complex to design and parameterize [69].

11.5   REVIEW OF COMMON VEGETATION INDICES

VIs in more or less complex and physically driven ways enhance the signal from vegetation while minimizing the effect of factors such as solar irradiance, canopy architecture, and background [33,71]. VIs can be classified on the basis of the spectral attributes employed, including: reflectance in individual narrow bands, band ratio and (normalized) difference, derivative spectra, and chlorophyll/soil sensitive indices [56]. Some of these indices have been designed for broadband sensors while others have been specifically developed for hyperspectral data such as the use of derivative analysis, which requires continuous spectra. This review can be integrated by reading Hatfield et al. [10], Pinter et al. [9], and Zarco-Tejada et al. [72].

11.5.1   Individual/Multiple Waveband Reflectance

Nitrogen availability and growth stage significantly affect canopy reflectance (R_λ) although the correlation between R_λ and nitrogen content is a function of the wavelength (Figure 11.1). According to Stroppiana et al. [20], in the visible wavelengths R_λ has significant negative correlation to nitrogen concentration: an increase of LAI decreases nitrogen concentration and hence reflectance due to the greater proportion of absorbed solar radiation (Figure 11.1). The opposite can be observed for NIR wavelengths while correlation drastically drops to zero in coincidence with the red edge; it is in fact the position of the red edge rather than the reflectance that is sensitive to nitrogen stress [25].

Visible wavelengths appear to be more suitable for estimating nitrogen concentration although the high extinction coefficients of photosynthetic pigments result in very low reflectance, especially in coincidence with pigment absorption spectra (<5%), leading to a loss of detail. To enhance

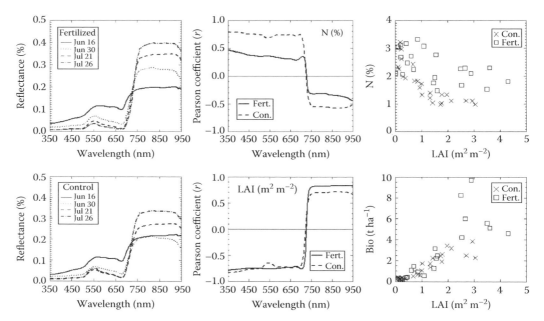

FIGURE 11.1   From the left: changes in canopy spectra with rice growth for fertilized (Fert.) and controlled conditions (Con.), Pearson correlation (r) of rice nitrogen concentration and LAI to canopy reflectance (R_λ) in the VIS-NIR, scatters of nitrogen-LAI and Biomass-LAI measures. Canopy spectra were acquired with a FieldSpec FR PRO spectroradiometer over experimental rice fields in Italy. (From Stroppiana, D. et al., *Field Crops Res.*, 111, 119, 2009.)

features of pigment absorption, Chapelle et al. [31] used the Ratio Analysis of Reflectance Spectra (RARS) and Mutanga et al. [44] successfully tested the continuum removal technique [55].

Other authors have investigated the correlation between R_λ and nitrogen content and results are inconsistent. Some studies have found negative correlation in the visible wavelengths and positive in the NIR wavelengths [4,17,23–25]. Fava et al. [21] found low positive correlation (<0.5) for visible and NIR wavelengths. Hansen and Schjoerring [15] found low (either positive or negative) correlation coefficients in the blue, red, and NIR regions for chlorophyll and nitrogen concentration and high correlation (either positive or negative in the NIR and visible regions, respectively) between reflectance and area-dependent variables such as LAI and chlorophyll/nitrogen density (nitrogen on an area basis [g N m^{-2} soil]). For corn leaves, Alchanatis et al. [72] found positive correlation in the visible and no correlation in the NIR range. This discrepancy may arise from the different variables used to quantify nitrogen status in plants: if nitrogen content is expressed on a unit area basis (nitrogen density) and therefore correlated to LAI, then the expected correlation is very similar to the R_λ-LAI correlation shown in Figure 11.1.

Clearly, common protocols for field measurements are necessary to make studies comparable. Although correlated to nitrogen content, the use of reflectance in individual bands has limitations due to the overlap of the effect of different nutrients [73] and the influence of leaf and structural canopy parameters; hence, the scientific focus has moved toward the use of VIs [56] or multivariate approaches [11].

11.5.2   Simple Ratio and Normalized Difference Indices

Simple ratio (SR, Equation 11.1) and normalized difference (ND, Equation 11.2) indices have been used for almost 40 years and are still widely used for many applications mainly due to their simple formalization:

$$SR = \frac{R_{\lambda 2}}{R_{\lambda 1}} \tag{11.1}$$

$$ND = \frac{(R_{\lambda 2} - R_{\lambda 1})}{(R_{\lambda 2} + R_{\lambda 1})} \tag{11.2}$$

where $R_{\lambda 1}$, $R_{\lambda 2}$ is the spectral reflectance measured at wavelengths $\lambda 1$ and $\lambda 2$. Initially designed for multispectral instruments, with $\lambda 1$ = red and $\lambda 2$ = NIR, they have been tested with almost all hyperspectral narrow bands [15,20–22,24,38]. The SR combines nutrient stress sensitive and insensitive bands to normalize it for variations in exogenous factors such as irradiance, leaf orientation, irradiance angles, and shadowing. To estimate chlorophyll and nitrogen contents, Kim et al. [74] proposed $\lambda 1$ = 550 nm and $\lambda 2$ = 700 nm since the use of a green wavelength, more sensitive to nitrogen rate, instead of red, could improve the precision and accuracy of the predictions [16,17,76]. For rice, Xue et al. [16] suggested R_{810}/R_{560} as the best index and Zhu et al. [24] found R_{950}/R_{660} and R_{950}/R_{680} the best predictors of leaf nitrogen accumulation (i.e., the product of leaf nitrogen concentration per unit dry weight and leaf dry weight per unit ground area).

Zhao et al. [28] found R_{405}/R_{715} and R_{1075}/R_{735} well correlated to sorghum leaf nitrogen concentration. Fava et al. [21] found that SR involving NIR bands (780–820 nm) and longer wavelengths of the red edge (740–770 nm) yielded the best correlation with nitrogen concentration in pasture (Figure 11.2). In a study on sugarcane leaf nitrogen concentration, Abdel-Rahman et al. [30] concluded that R_{743}/R_{1316} calculated from first-order derivatives of the reflectance yielded the best correlation. Since SR can be influenced by cloud and soil [77], the Modified Simple Ratio (MSR) was introduced with $\lambda 2$ = 800 nm and $\lambda 1$ = 670 nm (Equation 11.3) [78].

$$MSR = \frac{(R_{\lambda 2}/R_{\lambda 1} - 1)}{\sqrt{(R_{\lambda 2}/R_{\lambda 1}) + 1}} \tag{11.3}$$

The Normalized Difference Vegetation Index (NDVI) is certainly a good indicator of vegetation greenness but, in its initial form with $\lambda 1$ = red ($\sim$670 nm) and $\lambda 2$ = NIR ($\sim$800 nm) [79], it cannot follow physiological changes determined by stress conditions that produce changes at specific wavebands [29]. Changes in photosynthetic activity do not always determine changes

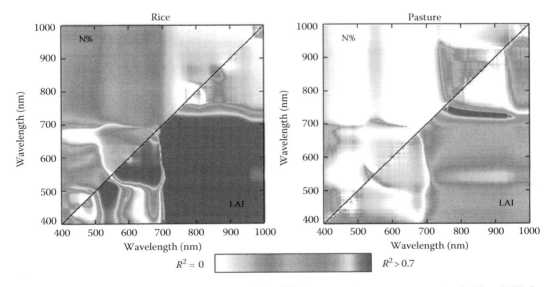

FIGURE 11.2 (See color insert.) Linear correlation (R^2) between nitrogen concentration/LAI and ND for rice (left; From Stroppiana, D. et al., *Field Crops Res.*, 111, 119, 2009) and SR for pasture (right; From Fava, F. et al., *Int. J. Appl. Earth Obs. Geoinf.*, 11, 233, 2009. With permission.) for field canopy spectra acquired with a FieldSpec FR PRO spectroradiometer.

in canopy structure [80] so indices sensitive to canopy characteristics may not be the best indicators of nutrient deficiency.

However, hyperspectral narrow bands sensitive to physiological changes can be used for defining a ND-type index: for example, the Photochemical Reflectance Index (PRI = $(R_{550} - R_{530})/R_{550} + R_{530})$) [81] and the Normalized Pigments Chlorophyll Ratio Index (NPCI = $(R_{680} - R_{430})/(R_{680} + R_{430})$) were found sensitive to physiological leaf status and nitrogen stress in sunflower leaves [29]. Wang et al. [82] indicated that a chlorophyll-based index, the Plant Pigment Ratio (PPR = $(R_{550} - R_{450})/(R_{550} + R_{450})$), yielded the best predictions of leaf nitrogen concentration from canopy spectra of winter wheat. Hansen and Schjoerring [15] found the greatest correlation with chlorophyll/nitrogen concentration when wavebands were selected in the visible domain and, in particular, in the blue (440–501 nm), characterized by strong light absorption due to chlorophyll-a and chlorophyll-b, combined with a green (573–586 nm) or red (692 nm) band.

The same authors found the red edge almost absent from band combinations in contrast with Gianelle and Guastella [17] and Fava et al. [21] who found red-edge bands in most of the best indices obtained for nitrogen concentration retrieval in pastures. Note that, due to low signal-to-noise ratio (SNR) at both ends of the spectrum, Hansen and Schjoerring [15] restricted the analyses to the 438–883 nm range, discarding the longer NIR wavelengths. Zhu et al. [24] found that the best NDVI for leaf nitrogen content estimation involved NIR wavelengths (1220, 710). On the contrary, Stroppiana et al. [20] found that, for paddy rice, the combination of reflectance measured with a field spectrometer (ASD FieldSpec FR PRO) in two visible wavelengths ($\lambda1 = 483$ nm, $\lambda2 = 503$ nm) provided the highest correlation with field measurements of nitrogen concentration ($R^2 = 0.65$, ***$p < 0.001$), whereas visible/NIR combinations determine a high correlation with LAI rather than nitrogen concentration (Figure 11.2). The Canopy Chlorophyll Content Index (CCCI) is derived from the Normalized Difference Red Edge (NDRE = $(R_{790} - R_{720})/(R_{790} + R_{720})$) and has been proved to relate to plant nitrogen in wheat [83,84].

11.5.3 Chlorophyll and Soil Sensitive Indices

The chlorophyll reflectance indices have been developed to enhance sensitivity to photosynthetic pigments and to reduce the influence of canopy architecture and the background [71]. The Chlorophyll Absorption in Reflectance Index (CARI) was proposed by Kim et al. [74] to reduce the influence of nonphotosynthetic parts of the plants and incorporates the green and red-edge regions of the spectrum.

The Modified Chlorophyll Absorption in Reflectance Index (MCARI) was introduced by Daughtry et al. [3] (Equation 11.4) and it measures the depth of chlorophyll absorption at 670 nm relative to the reflectance at 550 and 700 nm. To reduce the combined effect of nonphotosynthetic materials and soil background, they propose the use of the (R_{700}/R_{670}) ratio: the slope of the spectrum when the canopy contains no green biomass. However, MCARI is not very predictive at low LAI due to the influence of the background [3]. Haboudane et al. [33] found it is still influenced by nonphotosynthetic elements at low chlorophyll concentrations and proposed the Transformed Chlorophyll Absorption in Reflectance Index (TCARI) (Equation 11.5) where the ratio (R_{700}/R_{670}) is used to compensate for the influence of the background only on the difference ($R_{700} - R_{550}$), since the change of background reflectance mainly influences the slope between 550 and 700 nm [74].

$$\text{MCARI} = \left[(R_{700} - R_{670}) - 0.2 * (R_{700} - R_{550}) \right] \left(\frac{R_{700}}{R_{670}} \right) \tag{11.4}$$

$$\text{TCARI} = 3 * \left[(R_{700} - R_{670}) - 0.2 * (R_{700} - R_{550}) \left(\frac{R_{700}}{R_{670}} \right) \right] \tag{11.5}$$

The family of Soil Adjusted Vegetation Indices (SAVI, [85]) should further reduce the contribution of background reflectance and includes the Optimized Soil Adjusted Vegetation Indices (OSAVI) [85]) (Equation 11.6) and the Transformed Soil Adjusted Vegetation Index (TSAVI) [86]) (Equation 11.8). The major difference between these two indices is that OSAVI requires no information on optical soil properties and therefore has a more direct application, whereas TSAVI requires knowledge of the soil line (constants a and b in Equation 11.7) which in practise can be derived from measuring an unsown area within the field.

$$OSAVI = \frac{(1+0.16)(R_{800}R_{670})}{(R_{800} + R_{670} + 0.16)} \quad (11.6)$$

$$TSAVI = \frac{a\left[R_{800} - (a*R_{670}) - b\right]}{\left[R_{670} + (a*R_{800}) - (a*b)\right]^3} \quad (11.7)$$

Since neither sensitivity to chlorophyll nor insensitivity to the background seemed to be achievable with either of the aforementioned models, Daughtry et al. [3] and Haboudane et al. [33] proposed MCARI/OSAVI and TCARI/OSAVI. Using model simulations of the reflectance of open-canopy tree crops, Zarco-Tejada et al. [72] found, however, that MCARI/OSAVI was less affected by soil background variations than TCARI/OSAVI for total chlorophyll content estimation; tree crops are a challenge for remote sensing due to the effect of background and shadowing of a complex canopy architecture.

Figure 11.3 shows scatter plots of the correlation between nitrogen concentration/biomass and example HVIs ($NDI_{483,503}$, $NDVI_{670,800}$, $SR_{670,800}$, MCARI, and TCARI/OSAVI) derived for rice canopy reflectance data acquired with a field spectrometer (FieldSpec FR PRO). The coefficients of determination (R^2) for both linear and exponential regressive models are given in the figure. Interestingly, $NDI_{483,503}$ shows a good correlation with nitrogen concentration and, at the same time, a poor correlation with aboveground biomass whilst the other indices are correlated with biomass rather than nitrogen [20]. On the other hand, $NDVI_{670,800}$ and $SR_{670,800}$ are not informative for nitrogen assessment since they are correlated to biomass. The same indices derived with different red/NIR wavelength combinations provided similar results when applied to the same dataset. Soil sensitive indices were insufficiently investigated for nitrogen assessment in pastures probably because they are often characterized by complete canopy cover in late stage of growth, unless high grazing pressure occurred.

11.5.4 The Red Edge

The red edge identifies the steep transition between the reflectance absorption feature in red wavelengths (~680 nm) and the high NIR reflectance (~740 nm) [87]; the point of the maximum slope (or inflection point) is defined as the REP or red-edge wavelength. The red edge has been extensively used to estimate plant nitrogen content and nutritional status in crops and pastures [25,41,46,88] and several techniques have been proposed to locate the REP. The linear interpolation technique assumes that the red-edge curve can be modeled as a straight line and the REP is estimated by fitting a linear equation using the slope of the line.

The algorithm proposed by Guyot and Baret [90] uses four wavebands (670, 700, 740, and 780 nm) and calculates the REP as:

$$\lambda_{REP} = 700 + 40\left(\frac{R_{RE} - R_{700}}{R_{740} - R_{700}}\right) \quad (11.8)$$

where
700 and 40 are constants derived from the interpolation in the 700–740 nm range
$R_{RE} = (R_{670} + R_{780})/2$ is the reflectance at the inflection point

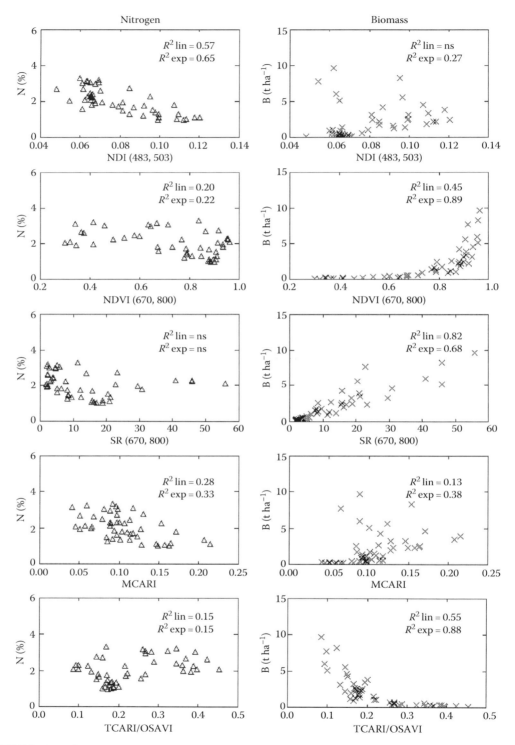

FIGURE 11.3 Scatter plots of VI-N (%) and VI-Biomass (t ha⁻¹) for a subset of HVIs described earlier. Field spectra data were acquired with FieldSpec FR PRO spectroradiometer for experimental paddy fields in Italy. (From Stroppiana, D. et al., *Field Crops Res.*, 111, 119, 2009.) The coefficient of determination (R^2) for linear and exponential regressive models is given in the panels.

The Inverted Gaussian (IG) technique [91,92] consists of fitting a Gaussian model to the reflectance red edge by means of an iterative fitting optimization procedure (Equation 11.9). The REP is calculated as the midpoint of the ascending edge of the Gaussian curve (REP = $\lambda_0 + \sigma$).

$$R(\lambda) = R_s - (R_s - R_0)\exp\left(-\frac{(\lambda_0 - \lambda)^2}{2\sigma^2}\right) \qquad (11.9)$$

where
R_s is the maximum red-edge reflectance (NIR shoulder)
R_0 and λ_0 are the minimum spectral reflectance and corresponding wavelength
σ is the Gaussian function variance

A third group of methods uses fitting of continuous high-order curves, such as cubic-spline or high-order polynomial, to the first derivative of the reflectance spectrum: REP is calculated as the maximum value [87,92,93].

The main drawback to the techniques based on first derivative spectrum is that they do not account for the possibility of more than one maximum in the derivative of the red edge [94]. In contrast, several studies evidenced a bimodal distribution of the derivative reflectance curve in the red edge, characterized by two maxima around 700 and 725 nm [41,45,87]. As a consequence, the REP could represent whichever maximum was dominant, and, therefore, a red-edge shift could involve a jump between the two maxima, creating a discontinuity in the REP [94].

To mitigate the destabilizing effect of the double peak feature, and to predict leaf nitrogen content with high accuracy in herbaceous vegetation, Cho and Skidmore [45] have proposed an alternative methodology called the linear extrapolation technique. This technique is based on the linear extrapolation of two straight lines through two points on the far-red (680–700 nm) and two points on the NIR (725–760 nm) flanks of the first derivative reflectance spectrum of the red edge. The REP is identified by the wavelength value at the intersection of the straight lines.

Overall, comparative studies showed that the accuracy of different techniques to estimate the red edge is dependent on several factors related to the biological characteristics of the plant material [45], and no single extraction method could be considered superior in general [95]. Despite the good performances of red-edge indices [25,45,46], the effect of varying LAI on the red edge is still an open issue. Lamb et al. [41] demonstrated by radiative transfer model simulations that only high LAI values determine low sensitivity of the red edge to LAI, thus allowing accurate chlorophyll/nitrogen estimation.

11.6 CONCLUSIONS

With the advent of hyperspectral sensors, remote sensing techniques have become particularly attractive for assessing crop and pasture nitrogen status. Compared to multispectral sensors, the availability of narrow and contiguous bands is fundamental for retrieving nitrogen content [14,17,21,26]. Indeed spectral resolution becomes even more important than spatial resolution [97]. A vast literature is now available, most of which has focused on the use of field spectrometers. Few studies have used airborne sensors, whereas space-borne instruments still suffer from numerous problems for this type of application: atmospheric correction, spectral and spatial resolution, and signal to noise ratio (SNR) not always adequate for identifying the absorption features of the chemical compounds. There is a clear need for further investigations into the optimal characteristics of a space-borne hyperspectral sensors for a wide range of applications related to vegetation.

The high correlation observed between chlorophylls and nitrogen is the basis for using spectra for nitrogen assessment although the relation between spectral changes and physiological processes has to be further investigated. Statistical regression methods, based on the use of vegetation indices (VIs), have been the most widely exploited approaches for retrieving nitrogen content from leaf and canopy spectra.

There is still little agreement over which hyperspectral vegetation indices (HIVs) have the strongest relationship with nitrogen. Ratio and normalized difference (ND) indices have been shown to normalize for the influence of exogenous factors (e.g., canopy, background signal) but little consensus has been reached on the narrow wavebands that should be used in their formulation. The visible and red-edge wavelengths are often reported as important spectral regions for nitrogen assessment and could be better suited than near infrared (NIR) wavebands, which are strongly influenced by structural canopy parameters. The de-correlation of the optimal HVIs from factors other than nitrogen concentration has proved to be rather difficult to achieve. Canopy architecture can significantly influence changes in canopy spectra and some indices have shown a correlation with nitrogen concentration only as a consequence of the nitrogen–LAI correlation. Since NIR wavelengths are mainly influenced by canopy structure, HVIs that use visible and red-edge wavebands should be preferred [20,21]. The role of vegetation phenology and growth stage in the retrieval of nitrogen content has also been insufficiently investigated [21,41]. Not to forget that other nutrient deficiencies might produce spectral changes that overlap with nitrogen stress features. Hence, the ability of predicting nitrogen status will depend not only on the knowledge of nitrogen itself, but also on the knowledge of other nutrients and other crop growing influences such as moisture, soils, and irrigation.

Several studies have, however, shown that the performance of the VIs is often a function of site and vegetation characteristics and none of the indices appears to be robust enough across sites. Empirical models remain therefore applicable mainly at local/regional scale. Moreover, there is a need for comparison of not only different VIs, which has been extensively done, but also different methodological approaches in order to identify those that are able to provide consistent results. Yet comparisons are often hampered by the different field conditions/parameters used in the experiments (e.g., nitrogen concentration/density/accumulation).

In summary, there are some priorities to be addressed by future research:

1. Further understanding of the physiological processes involved in the relation between plant nitrogen content and leaf/canopy spectra by also exploiting simulations from radiative transfer models.
2. Comparison of methodologies on datasets acquired with common protocols of measurements to evaluate consistency and robustness.
3. Set up of field experiments designed to evaluate the influence on VIs of canopy architecture, other nutrients stresses and other crop growing conditions (e.g., moisture, soils, irrigation).
4. The potential of SWIR wavelengths for nitrogen content monitoring should be further explored.

REFERENCES

1. Holland, E.A., Dentener, F.J., Braswell, B.H., Sulzman, J.M., Contemporary and pre-industrial global reactive nitrogen budgets, *Biogeochemistry*, 46, 7–43, 1999.
2. Pearson, C.J., Ison, R.L., *Agronomy of Grassland Systems*, Cambridge University Press, Cambridge, U.K., 1997.
3. Daughtry, C.S.T., Waithall, C.L., Kim, M.S., de Colstoun, E.B., McMurtrey III, J.E., Estimating corn leaf chlorophyll concentration from leaf and canopy reflectance, *Remote Sensing of Environment*, 74, 229–239, 2000.
4. Yi, Q.-X., Huang, J.-F., Wang, F.-M., Wang, X.-Z., Liu, Z.-Y., Monitoring rice nitrogen status using hyperspectral reflectance and Artificial Neural Networks, *Environmental Science and Technology*, 41, 6770–6775, 2007.
5. Starks, P.J., Coleman, S.W., Phillips, W.A., Determination of forage chemical composition using remote sensing, *Journal of Range Management*, 57, 635–640, 2004.
6. Wood, C.W., Reeves, D.W., Himelrick, D.G., Relationships between chlorophyll meter reading and leaf chlorophyll concentration, N status, and crop yield: A review, *Proceedings of Agronomy Society New Zealand* 23, 1–9, 1993.

7. Curran, P.J., Dungan, J.L., Gholz, H.L., Exploring the relationship between reflectance red edge and chlorophyll content in slash pine, *Tree Physiology*, 7, 33–48, 1990.
8. Knipling, E.B., Physical and physiological basis for the reflectance of visible and near-infrared radiation from vegetation, *Remote Sensing of Environment*, 1, 155–159, 1970.
9. Pinter, P.J., Jr., Hatfield, L., Schepers, J.S., Barnes, E.M., Moran, M.S., Daughtry, C.S.T., Upchurch, D.R., Remote Sensing for crop management, *Photogrammetric Engineering and Remote Sensing*, 69, 6, 647–664, 2003.
10. Hatfield, J.L., Gitelson, A.A., Schepers, J.S., Walthall, C.L., Application of spectral remote sensing for agronomic decisions, *Agronomy Journal*, 100, 117–131, 2008.
11. Curran, P.J., Remote sensing of foliar chemistry, *Remote Sensing of Environment*, 30, 271–278, 1989.
12. Richardson, A.J., Everitt, J.H., Gausman, H.W., Radiometric estimation of biomass and nitrogen-content of Alicia grass, *Remote Sensing of Environment*, 13, 179–184, 1983.
13. Everitt, J.H., Richardson, A.J., Gausman, H.W., Leaf reflectance-nitrogen-chlorophyll relations in buffelgrass, *Photogrammetric Engineering and Remote Sensing*, 51, 463–466, 1985.
14. Thenkabail, P.S., Smith, R.B., Pauw, E.D., Hyperspectral vegetation indices and their relationships with agricultural crop characteristics, *Remote Sensing of Environment*, 71, 158–182, 2000.
15. Hansen, P.M., Schjoerring, J.K., Reflectance measurement of canopy biomass and nitrogen status in wheat crops using normalized difference vegetation indices and partial least squares regression, *Remote Sensing of Environment*, 86, 542–553, 2003.
16. Xue, L., Cao, W., Luo, W., Dai, T., Zhu, Y., Monitoring leaf nitrogen status in rice with canopy spectral reflectance, *Agronomy Journal*, 96, 135–142, 2004.
17. Gianelle, D., Guastella, F., Nadir and off-nadir hyperspectral field data: Strengths and limitations in estimating grassland biophysical characteristics, *International Journal of Remote Sensing*, 28, 1547–1560, 2007.
18. Albayrak, S., Use of reflectance measurements for the detection of N, P, K, ADF and NDF contents in sainfoin pasture, *Sensors*, 8, 7275–7286, 2008.
19. Starks, P.J., Zhao, D., Brown, M.A., Estimation of nitrogen concentration and in vitro dry matter digestibility of herbage of warm-season grass pastures from canopy hyperspectral reflectance measurements, *Grass and Forage Science*, 63, 168–178, 2008.
20. Stroppiana, D., Boschetti, M., Brivio, P.A., Bocchi, S., Plant nitrogen concentration in paddy rice from field canopy hyperspectral radiometry, *Field Crops Research*, 111, 119–129, 2009.
21. Fava, F., Colombo, R., Bocchi, S., Meroni, M., Sitzia, M., Fois, N., Zucca, C., Identification of hyperspectral vegetation indices for Mediterranean pasture characterization, *International Journal of Applied Earth Observation and Geoinformation*, 11, 233–243, 2009.
22. Nguyen, H.T., Lee, B-W., Assessment of rice leaf growth and nitrogen status by hyperspectral canopy reflectance and partial least square regression, *European Journal of Agronomy*, 24, 349–356, 2006.
23. Yao, X., Feng, W., Zhu, Y., Tian, Y.C., Cao, W.X., A non-destructive and real-time method of monitoring leaf nitrogen status in wheat, *New Zealand Journal of Agricultural Research*, 50, 935–942, 2007.
24. Zhu, Y., Zhou, D., Yao, X., Tian, Y., Cao, W., Quantitative relationships of leaf nitrogen status to canopy spectral reflectance in rice, *Australian Journal of Agricultural Research*, 58, 1077–1085, 2007.
25. Feng, W., Yao, X., Zhu, Y., Tian, Y.C., Cao, W.X., Monitoring leaf nitrogen status with hyperspectral reflectance in wheat, *European Journal of Agronomy*, 28, 394–404, 2008.
26. Chen, P., Haboudane, D., Trembaly, N., Wang, J., Vigneault, P., Li, B., New spectral indicator assessing the efficiency of crop nitrogen treatment in corn and wheat, *Remote Sensing of Environment*, 114, 1987–1997, 2010.
27. Tarpley, L., Reddy, K.R., Sassenrath-Cole, G.F., Reflectance indices with precision and accuracy in predicting cotton leaf nitrogen concentration, *Crop Science*, 40, 1814–1819, 2000.
28. Zhao, D., Reddy, K.R., Kakani, V.G., Reddy, V.R., Nitrogen deficiency effects on plant growth, leaf photosynthesis, and hyperspectral reflectance properties of sorghum, *European Journal of Agronomy*, 22, 391–403, 2005.
29. Peñuelas, J., Gamon, J.A., Fredeen, A.L., Merino, J., Field, C.B., Reflectance indices associated with physiological changes in nitrogen- and water-limited sunflower leaves, *Remote Sensing of Environment*, 48, 135–146, 1994.
30. Abdel-Rahman, E.M., Ahmed, F.B., van der Berg, M., Estimation of sugarcane leaf nitrogen concentration using in situ spectroscopy, *International Journal of Applied Earth Observation and Geoinformation*, 12S, S52–S57, 2010.
31. Chapelle, E.W., Kim, M.S., McMurtrey, J.E., III, Ratio analysis of reflectance spectra (RARS): An algorithm for the remote estimation of the concentration of chlorophyll a, chlorophyll b, and carotenoids in soybean leaves, *Remote Sensing of Environment*, 39, 239–247, 1992.

32. Sims, D., Gamon, J.A., Relationship between leaf pigment content and spectral reflectance across a wide range of species, leaf structure and development stages, *Remote Sensing of Environment*, 81, 337–354, 2002.

33. Haboudane, D., Miller, J.R., Tremblay, N., Zarco-Tejada, P.J., Dextraze, L., Integrated narrow-band vegetation indices for prediction of crop chlorophyll content for application to precision agriculture, *Remote Sensing of Environment*, 81, 416–426, 2002.

34. Gitelson, A.A., Keydan, G.P., Merzlyak, M.N., Three-band model for noninvasive estimation of chlorophyll, carotenoids, and anthocyanin contents in higher plant leaves, *Geophysical Research Letters*, 33, 111402, 2006, doi:10.1029/2006Gl026457.

35. Ryu, C., Suguri, M., Umeda, M., Model for predicting the nitrogen content of rice at panicle initiation stage using data from airborne hyperspectral remote sensing, *Biosystems Engineering*, 104, 465–475.

36. Goel, P.K., Prasher, S.O., Landry, J.A., Patel, R.M., Bonnell, R.B., Viau, A.A., Miller, J.R., Potential of airborne hyperspectral remote sensing to detect nitrogen deficiency and weed infestation in corn, *Computers and Electronics in Agriculture*, 38, 99–124, 2003.

37. Sembiring, H., Raun, W.R., Johnson, G.V., Stone, M.L., Solie, J.B., Phillips, S.B., Detection of nitrogen and phosphorus nutrient status in bermudagrass using spectral radiance, *Journal of Plant Nutrition*, 21, 1189–1206, 1998.

38. Starks, P.J., Zhao, D.L., Phillips, W.A., Coleman, S.W., Development of canopy reflectance algorithms for real-time prediction of bermudagrass pasture biomass and nutritive values, *Crop Science*, 46, 927–934, 2006.

39. Zhao, D., Starks, P.J., Brown, M.A., Phillips, W.A., Coleman, S.W., Assessment of forage biomass and quality parameters of bermudagrass using proximal sensing of pasture canopy reflectance, *Grassland Science*, 53, 39–49, 2007.

40. Mutanga, O., Skidmore, A.K., van Wieren, S., Discriminating tropical grass (*Cenchrus ciliaris*) canopies grown under different nitrogen treatments using spectroradiometry, *ISPRS Journal of Photogrammetry and Remote Sensing*, 57, 263–272, 2003.

41. Lamb, D.W., Steyn-Ross, M., Schaare, P., Hanna, M.M., Silvester, W., Steyn-Ross, A., Estimating leaf nitrogen concentration in ryegrass (*Lolium* spp.) pasture using the chlorophyll red-edge: Theoretical modelling and experimental observations, *International Journal of Remote Sensing*, 23, 3619–3648, 2002.

42. Schut, A.G.T., Lokhorst, C., Hendriks, M., Kornet, J.G., Kasper, G., Potential of imaging spectroscopy as tool for pasture management, *Grass and Forage Science*, 60, 34–45, 2005.

43. Biewer, S., Fricke, T., Wachendorf, M., Development of canopy reflectance models to predict forage quality of legume-grass mixtures, *Crop Science*, 49, 1917–1926, 2009.

44. Mutanga, O., Skidmore, A.K., Kumar, L., Ferwerda, J., Estimating tropical pasture quality at canopy level using band depth analysis with continuum removal in the visible domain, *International Journal of Remote Sensing*, 26, 1093–1108, 2005.

45. Cho, M.A., Skidmore, A.K., A new technique for extracting the red edge position from hyperspectral data: The linear extrapolation method, *Remote Sensing of Environment*, 101, 181–193, 2006.

46. Mutanga, O., Skidmore, A.K., Red edge shift and biochemical content in grass canopies, *ISPRS Journal of Photogrammetry and Remote Sensing*, 62, 34–42, 2007.

47. Knox, N.M., Skidmore, A.K., Schlerf, M., de Boer, W.F., van Wieren, S.E., van der Waal, C., Prins, H.H.T., Slotow, R., Nitrogen prediction in grasses: Effect of bandwidth and plant material state on absorption feature selection, *International Journal of Remote Sensing*, 31, 691–704, 2010.

48. Mutanga, O., Skidmore, A.K., Prins, H.H.T., Predicting in situ pasture quality in the Kruger National Park, South Africa, using continuum-removed absorption features, *Remote Sensing of Environment*, 89, 393–408, 2004.

49. Mutanga, O., Skidmore, A.K., Integrating imaging spectroscopy and neural networks to map grass quality in the Kruger National Park, South Africa, *Remote Sensing of Environment*, 90, 104–115, 2004.

50. Skidmore, A.K., Ferwerda, J.G., Mutanga, O., Van Wieren, S.E., Peel, M., Grant, R.C., Prins, H.H.T., Balcik, F.B., Venus, V., Forage quality of savannas—Simultaneously mapping foliar protein and polyphenols for trees and grass using hyperspectral imagery, *Remote Sensing of Environment*, 114, 64–72, 2010.

51. Boschetti, M., Bocchi, S., Brivio, P.A., Assessment of pasture production in the Italian Alps using spectrometric and remote sensing information, *Agriculture Ecosystems and Environment*, 118, 267–272, 2007.

52. Mirik, M., Norland, J.E., Crabtree, R.L., Biondini, M.E., Hyperspectral one-meter-resolution remote sensing in Yellowstone National Park, Wyoming: I. Forage nutritional values, *Rangeland Ecology and Management*, 58, 452–458, 2005.

53. Beeri, O., Phillips, R., Hendrickson, J., Frank, A.B., Kronberg, S., Estimating forage quantity and quality using aerial hyperspectral imagery for northern mixed-grass prairie, *Remote Sensing of Environment*, 110, 216–225, 2007.

54. Thulin, S.M., Hill, M.J., Held, A.A., Spectral sensitivity to carbon and nitrogen content in diverse temperate pastures of Australia. *IGARSS 2004: IEEE International Geoscience and Remote Sensing Symposium Proceedings*, Vols. 1–7, 1459–1462, 2004.

55. Kokaly, R.F., Clark, R.N., Spectroscopic determination of leaf biochemistry using band-depth analysis of absorption features and stepwise multiple linear regression, *Remote Sensing of Environment*, 67, 267–287, 1999.

56. Blackburn, G.A., Relationship between spectral reflectance and pigment concentrations in stacks of deciduous broadleaves, *Remote Sensing of Environment*, 70, 224–237, 1999.

57. Curran, P.J., Dungan, J.L., Macler, B.A., Plummer, S.E., The effect of a red leaf pigment on the relationship between red edge and chlorophyll concentration, *Remote Sensing of Environment*, 35, 69–76, 1991.

58. Dawson, T.P., Curran, P.J., A new technique for interpolating red edge position, *International Journal of Remote Sensing*, 19, 11, 2133–2139, 1998.

59. Schertz, F.M., A chemical and physiological study of mottling of leaves, *Botanical Gazette*, 71, 81–130, 1921.

60. Tam, R.K., Magistad, O.C., Relationship between nitrogen fertilization and chlorophyll content in pineapple plants, *Plant Physiology*, 10, 159–168, 1935.

61. Houlès V., Guérif, M., Mary, B., Elaboration of a nitrogen nutrition indicator for winter wheat based on leaf area index and chlorophyll content for making nitrogen recomendations, *European Journal of Agronomy*, 27, 1–11, 2007.

62. Filella, I., Serrano, I., Serra, J., Penuelas, J., Evaluating wheat nitrogen status with canopy reflectance indices and discriminant analysis, *Crop Science*, 35, 1400–1405, 1995.

63. Thomas, J.R., Oerther, G.F., Estimating nitrogen content of sweet pepper leaves by reflectance measurements, *Agronomy Journal*, 64, 11–13, 1972.

64. Yoder, B.J., Pettigrew-Crosby, R.E., Predicting nitrogen and chlorophyll content and concentrations from reflectance spectra (400–2500 nm) at leaf and canopy scales, *Remote Sensing of Environment*, 53, 199–211, 1995.

65. Gausman, H.W., Gerbermann, A.H., Wiegand, C.L., Leamer, R.W., Rodriguez, R.R., Noriega, J.R., Reflectance differences between crop residues and bare soils, *Soil Science Society of American Journal*, 39, 752–755, 1975.

66. Kokaly R.F., Asner, G.P., Ollinger, S.V., Martin, M.E., Wessman, C.A., Characterizing canopy biochemistry from imaging spectroscopy and its application to ecosystem studies, *Remote sensing of Environment*, 113, 578–591, 2009.

67. Zhao, D.H., Li, J.L., Qi, J.G., Identification of red and NIR spectral regions and vegetative indices for discrimination of cotton nitrogen stress and growth stage, *Computers and Electronics in Agriculture*, 48, 155–169, 2005.

68. Jacquemoud, S., Baret, F., Andrieu, B., Danson, F.M., Jaggard, K., Extraction of vegetation biophysical parameters by inversion of the PORSPECT+SAIL models on sugar beet canopy reflectance data. Application to TM and AVIRIS sensors, *Remote Sensing of Environment*, 52, 163–172, 1995.

69. Fernandes, R., Leblanc, S.G., Parametric (modified least square) and non-parametric (Theil-Sen) linear regression for predicting biophysical parameters in the presence of measurements errors, *Remote Sensing of Environment*, 95, 301–316, 2005.

70. Jackson, R.D., Huete, A.R., Interpreting vegetation indices, *Preventive Veterinary Medicine*, 11, 185–200, 1991.

71. Zarco-Tejada, P.J., Miller, J.R., Morales, A., Berjón, A., Agüera, J., Hyperspectral indices and model simulation for chlorophyll estimation in open-canopy tree crops, *Remote Sensing of Environment*, 90, 463–476, 2004.

72. Alchanatis, V., Schmilvitch, Z., In-field assessment of single leaf nitrogen status by spectral response measurements, *Precision Agriculture*, 6, 25–39, 2005.

73. Masoni, A., Ercoli, L., Mariotti, M., Spectral properties of leaves deficient in iron, sulfur, magnesium, and manganese, *Agronomy Journal*, 88, 937–943, 1996.

74. Kim, M.S., Daughtry, C.S.T., Chapelle, E.W., McMurtrey, J.E., The use of high spectral resolution bands for estimating absorbed photosynthetically active radiation (APAR). *Proceedings of the 6th International Symposium on Physical Measurements and Signatures in Remote Sensing*, Val D'Isere, France, 1994, pp. 229–306.

75. Hinzman, L.D., Bauer, M.E., Daughtry, C.S.T., Effects of N fertilization on growth and reflectance characteristics of winter wheat, *Remote Sensing of Environment*, 19, 47–61, 1986.

76. Slater, P.N., Jackson, R.D., Atmospheric effects on radiation reflected from soil and vegetation as measured by orbit sensors using various scanning directions, *Applied Optics*, 21, 3923, 1982.

77. Chen, J., Evaluation of vegetation indices and modified simple ratio for boreal applications, *Canadian Journal of Remote Sensing*, 22, 229–242, 1996.

78. Rouse, J.W., Haas, R.H., Jr., Schell, J.A., Deering, D.W., Monitoring vegetation systems in the Great Plains with ERTS, NASA SP-351. *3rd ERTS-1 Symposium*, Washington, DC, pp. 309–317, 1974.

79. Running, S.R., Nemani, R.R., Relating seasonal patterns of the AVHRR vegetation index to simulated photosynthesis and transpiration of forests in different climates, *Remote Sensing of Environment*, 24, 347–367, 1998.

80. Gamon, J.A., Peñuelas, J., Field, C.B., A narrow-waveband spectral index that tracks diurnal changes in photosynthetic efficiency, *Remote Sensing of Environment*, 41, 35–44, 1992.

81. Wang, Z.J., Wang, J.H., Liu, L.Y., Huang, W.J., Zhao, C.J., Wang, C.Z., Prediction of grain protein content in winter wheat (*Triticum aestivum* L.) using plant pigment ratio (PPR), *Field Crops Research*, 90, 311–321, 2004.

82. Fitzgerald, G.J., Rodriguez, D., Christensen, L.K., Belford, R., Sadras, V.O., Clarke, T.R., Spectral and thermal sensing for nitrogen and water status in rainfed and irrigated wheat environments, *Precision Agriculture*, 7, 233–248, 2007.

83. Tilling, A.K., O'leary, G.J., Ferwerda, J.G., Jones, S.D., Fitzgerald, G.J., Rodriguez, D., Belford, R., Remote sensing of nitrogen and water stress in wheat, *Field Crops Research*, 104, 77–85, 2007.

84. Huete, A.R., Jackson, R.D., Post, D.F., Spectral response of a plant canopy with different soil backgrounds, *Remote Sensing of Environment*, 17, 37–53, 1985.

85. Rondeaux, G., Steven, M., Baret, F., Optimization of soil-adjusted vegetation indices, *Remote Sensing of Environment*, 55, 95–107, 1996.

86. Baret, F., Guyot, G., Major, D.J., TSAVI: A vegetation index which minimizes soil brightness effects on LAI and APAR estimations. In *Quantitative Remote Sensing for the Nineties*, Proceedings of the IGARSS'89, Vancouver, Canada, Vol. 3, 1989, pp. 1355–1358.

87. Horler, D.N.H., Dockray, M., Barber, J., The red edge of plant leaf reflectance, *International Journal of Remote Sensing*, 4, 273–288, 1983.

88. Jongschaap, R.E.E., Booij, R., Spectral measurements at different spatial scales in potato: Relating leaf, plant and canopy nitrogen status, *International Journal of Applied Earth Observation and Geoinformation*, 5, 3, 205–218, 2004.

89. Guyot, G. and Baret, F., Utilisation de la haute résolution spectrale pour suivre l'état des couverts végétaux. *Proceedings of the 4th International Colloquium on Spectral Signatures of Objects in Remote Sensing*. ESA SP-287, Aussois, France, 1988, pp. 279–286.

90. Bonham-Carter, G.F., Numerical procedures and computer program for fitting an inverted Gaussian model to vegetation reflectance data, *Computers and Geosciences*, 14, 3, 339–356, 1988.

91. Miller, J.R., Hare, E.W., Wu, J., Quantitative characterization of the red edge reflectance. An inverted-Gaussian reflectance model, *International Journal of Remote Sensing*, 11, 10, 1755–1773, 1990.

92. Savitzky, A., Golay, M.J.E., Smoothing and differentiation of data by simplified least-squares procedures, *Analytical Chemistry*, 36, 8, 1627–1639, 1964.

93. Demetriades-Shah, T.H., Steven, M.D., Clark, J.A., High resolution derivative spectra in remote sensing, *Remote Sensing of Environment*, 33, 55–64, 1990.

94. Clevers, J.G.P.W., De Jong, S.M., Epema, G.F., Van der Meer, F., Bakker, W.H., Skidmore, A.K., Sholte, K.H., Derivation of the red edge index using MERIS standard band setting, *International Journal of Remote Sensing*, 23, 16, 3169–3184, 2002.

95. Baranoski, G.V.G., Rokne, J.G., A practical approach for estimating the red edge position of plant leaf reflectance, *International Journal of Remote Sensing*, 26, 3, 503–521, 2005.

96. Thenkabail, P.S., Enclona, E.A., Ashton, M.S., Van der Meer, B., Accuracy assessment of hyperspectral wavebands performance for vegetation analysis applications, *Remote Sensing of Environment*, 91, 354–376, 2004.

97. Fridgen, J.L., Varco, J.J., Dependency of cotton leaf nitrogen, chlorophyll, and reflectance on nitrogen and potassium availability, *Agronomy Journal*, 96, 63–93, 2004.

98. Phillips, R.L., Beeri, O., Liebig, M., Landscape estimation of canopy C: N ratios under variable drought stress in Northern Great Plains rangelands, *Journal of Geophysical Research-Biogeosciences*, 111, G02015, 2006, doi:10.1029/2005JG000135.

99. Suzuki, Y., Tanaka, K., Kato, W., Okamoto, H., Kataoka, T., Shimada, H., Sugiura, T., Shima, E., Field mapping of chemical composition of forage using hyperspectral imaging in a grass meadow, *Grassland Science*, 54, 179–188, 2008.

Part V

Vegetation Biophysical Properties

12 Spectral Bioindicators of Photosynthetic Efficiency and Vegetation Stress

*Elizabeth M. Middleton, K. Fred Huemmrich,
Yen-Ben Cheng, and Hank A. Margolis*

CONTENTS

12.1 INTRODUCTION

The term "spectral bioindicators" refers to a developing scientific field that uses remote sensing to assess the health and physiology of ecosystems. It places particular emphasis on responses to environmentally imposed stresses that affect photosynthetic efficiency at the scale of ecosystems and vegetation canopies. These spectral indices express new information about plant physiological processes that are not available from broadband satellite data that have been collected over more than three decades. These historical data sets, primarily utilizing spectral vegetation indices such as the Normalized Difference Vegetation Index (NDVI), have been foundational in demonstrating the potential of satellite remote sensing to study ecosystem processes [1]. Based on the reflectance characteristics of green vegetation, satellite observations have made it possible to track interannual trends and to infer increased photosynthetic activity and longer growing seasons in Northern Hemisphere land regions [2,3], However, current and future narrowband sensors and spectrometers give us potent new tools to examine ecosystem function and health from space, supplementing the existing capability for monitoring seasonality and phenology of terrestrial ecosystems. Furthermore, these new measurements push us beyond *inference* of photosynthetic activity into more direct observations closely related to *measurement* of photosynthetic function, with less dependence on supporting meteorological data for interpretation. This is critical to enable a global monitoring capability for detecting and monitoring the dynamics in time and space of our Earth's ecosystems, in the face of increasing human populations and decline in natural biodiversity. The Earth's climate is closely linked to biogeophysical and biogeochemical processes in ecosystems. With climate-induced changes globally, there is a pressing need to apply the best technologies to bear on monitoring our home planet, Earth.

12.2 DESCRIBING AND MEASURING ECOSYSTEM PHOTOSYNTHESIS AND RELATED PROCESSES

Vegetation forms in assemblages that grow under a specific range of environmental conditions, characterized as biomes, and thrive when optimal conditions for that particular assemblage occur. However, vegetation is subjected to "stress" when environmental conditions differ significantly from optimal conditions, such as when drought or unfavorable temperatures occur, or insufficient or excess nutrients are present [4–6]. When stress factors impair physiological function through biochemical mechanisms, this impacts the exchange of carbon, water, and energy between vegetation and the atmosphere. Since nonoptimal conditions are more common than optimal conditions over extensive periods of time, assumptions based on optimal conditions are typically not reasonable, either in interpreting observations or for model retrievals.

It is imperative to understand how, where, and when environmental stresses invoke limiting conditions on terrestrial ecosystems. Although there are multiple ways that plants manage their exposure to suboptimal environmental stresses, we focus on the subset of physiological responses that affect photosynthetic function and efficiency. Consequently, we will discuss photosynthetic light use efficiency (LUE) in conjunction with a primary photoprotective mechanism that regulates it, the xanthophyll cycle and its connection to a related energy dissipation pathway for chlorophyll fluorescence (ChlF), as well as spectral bioindicators that capture these processes.

12.2.1 RELEVANCE OF ECOSYSTEM PHOTOSYNTHETIC PROCESSES TO CLIMATE CHANGE

Detecting vegetation stress at landscape scales is important for assessing the physiological status of ecosystems, to obtain a synoptic view in order to quantify impacts from climate change and human activities and to predict impacts on future human societies. Photosynthesis represents the largest carbon flux between the biosphere and the atmosphere and thus is of critical importance to climate change. Based on direct observations, gross global terrestrial photosynthesis has recently been estimated to be 123 ± 8 Pg C year^{-1} (1 Pg = peta gram = 10^{15} g) [7] (Figure 12.1), while photosynthesis in oceans has been estimated to be 93 ± 19 Pg C year^{-1} [8,9].

Photosynthetic activity by the large land area in the Northern Hemisphere during the northern growing season is clearly manifested in the seasonal drawdown of atmospheric CO_2 concentrations that have been measured for more than a half century at Mauna Loa, Hawaii [7], and provide important evidence underlying concerns about increased greenhouse gas warming and climate change attributed to fossil fuel combustion and land use change [10–12]. Canadell et al. [10] proposed that terrestrial and ocean carbon sinks were becoming less efficient over time and that this decreased efficiency accounted for 35% ($\pm$16%) of the recent increases (2000–2006 versus 1970–1990) in the atmospheric growth rate of CO_2. The near disappearance of a terrestrial carbon sink during

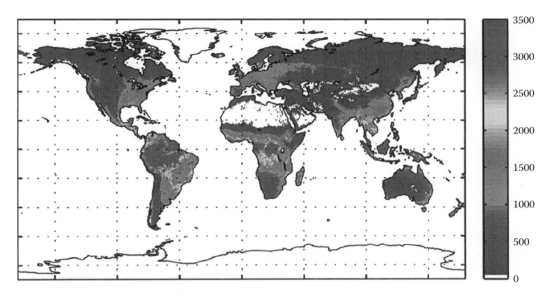

FIGURE 12.1 **(See color insert.)** Spatial variations of the global median annual GPP ($gC/m^2/a$) from various spatially explicit approaches. (From Beer, C. et al., Terrestrial gross carbon dioxide uptake: Global distribution and covariation with climate, *Science*, 329 (5993), 834–838, 2010. Reprinted with permission of AAAS.)

the 1998 El Niño event was thought to be due to a combination of decreased photosynthesis and increased respiration [11,13], although Van der Werf et al. [14] argued that increased combustion from fires was also important.

Present approaches depend on predicting vegetation responses to environmental conditions using models driven by meteorological data. Environmental changes, such as temperature, precipitation, atmospheric CO_2 concentration, and nitrogen deposition, may introduce uncertainties into the model results. Additionally, responses of terrestrial systems to climate-ecosystem feedback factors are not well understood and therefore contribute important uncertainties in climate change studies. Spectral bioindicators observe reflectance (or apparent reflectance) changes that are directly related to vegetation physiological stress responses, thus providing an independent check on the models.

12.2.2 Ecosystem Photosynthesis, Photosynthetic Efficiency, and Plant Stress

Photosynthesis in plants is a fundamental biochemical process that occurs within chloroplasts in cells of leaf mesophyll tissue but sometimes also in stem and bark tissue [15]. During photosynthesis, CO_2 is removed from the atmosphere and fixed into carbohydrates to provide metabolic energy that fuels biochemical processes for growth, maintenance, reproduction, and biomass accumulation. However, in terrestrial ecosystems, CO_2 is also returned to the atmosphere as a by-product of respiratory processes by plants or soil microbes. An ecosystem is referred to as a net carbon sink if the CO_2 influx from photosynthesis exceeds the efflux from respiration, but is referred to as a net carbon source if more carbon is lost than is gained [16]. In the aggregate across terrestrial ecosystems, small-scale physiological processes within plants can affect CO_2 exchange with large-scale global impacts on the Earth's climate, atmosphere, and biogeochemical cycles [7,17].

The photosynthetic efficiency of terrestrial ecosystems, and associated ecosystem services (e.g., carbon sequestration, climate regulation, hydrological function, and wildlife habitat) [18,19], can be negatively impacted by stress factors such as climate change (e.g., those related to temperature, irradiance levels, and atmospheric CO_2 concentrations), land use change, chemical pollution, biodiversity loss, and an overabundance or imbalance of nitrogen, phosphorus, and other essential minerals. Rockström et al. [20] made a first attempt to define the boundary conditions beyond which various stress factors could push terrestrial ecosystems to the point where significant negative impacts accrue on ecosystems and human well-being. While the response to these stress factors can be abrupt and irreversible, the responses can also be gradual and thus more difficult to detect in their initial stages.

At the terrestrial ecosystem and landscape scales, three primary approaches have been utilized to examine the dynamics of photosynthesis: measurements from instrumented towers; remote sensing observations from various platforms; and process models. By any of these methods, it should be borne in mind that ecosystem health and physiological status are evaluated indirectly, based on responses of a biological process (e.g., photosynthesis, growth, reproduction) over time.

12.2.3 How Do We Measure Canopy/Ecosystem Photosynthesis?

The eddy covariance flux tower technique developed in the 1980s–1990s and now deployed at approximately 400 instrumented flux towers around the world, measures the net exchange of carbon, water, and energy between an ecosystem and the atmosphere, and offers a powerful means for obtaining near-continuous time series of ecosystem-level photosynthesis over multiple years [21]. Flux towers provide the data to develop, calibrate, and validate models driven by spectral bioindicators. Data from tower sites supply direct, 30 min measurements of the net CO_2 flux for ecosystems representing a large range of different climatic zones, plant functional types, and ecological disturbances [7,22]. The net flux is determined from the covariance of the vertical wind speed and the CO_2 concentration, measured using sonic anemometers and infrared gas analyzers, respectively. For towers of different heights, the measured flux footprint typically averages between 0.5 and 1.0 km^2 [23,24] over a period of several weeks or months, varying with wind direction, turbulence, and surface characteristics. To interpret and analyze the net flux, it is important to make precise simultaneous meteorological measurements for air and soil temperatures, soil moisture, atmospheric vapor pressure deficit (VPD), and precipitation. The net CO_2 flux is referred to as net ecosystem productivity (NEP), which can be partitioned into two streams: gross ecosystem productivity (GEP), the carbon taken up by photosynthesis; and ecosystem respiration (R_{eco}), the carbon flux exiting the ecosystem due to metabolic activity by microbial and plant cells [25,26].

Recently, there has been increasing interest in the use of spectral bioindicators to help make predictions about biological processes, and to measure and model large areas with some understanding of those underlying processes. The bioindicators approach offers promise, and the photosynthetic LUE model offers an example of a new capability that combines remote sensing and in situ measures.

12.2.4 The Photosynthetic Light Use Efficiency Model

A widely used approach for modeling canopy or ecosystem photosynthesis is based on the LUE concept. An LUE model, rather than describing complex biochemical, canopy structure, and meteorological information, attempts to describe the overall ability of plants/canopies/ecosystems to convert solar energy into useable carbohydrates and biomass [27,28]. This type of model assumes that gross maximum photosynthesis (under ideal conditions) is proportional to the available Photosynthetically Active Radiation (PAR) that is absorbed by vegetation (or APAR), adjusted by an efficiency term (ε). Typically, a single efficiency term attempts to account for different vegetation types as well as all nonoptimal or limiting environmental stress factors (e.g., air and soil temperature, soil moisture, nutrient availability, VPD, etc.) that affect the absorption and utilization of PAR, and for APAR losses through photoprotection mechanisms such as heat dissipation via the xanthophylls pigment cycle and emission as ChlF. Therefore, the basic LUE model describes GEP in terms of the physiologically active APAR fraction (fAPAR) and a photosynthetic efficiency term [29]:

$$\int GEP(t)dt = \int \varepsilon(t) * fAPAR(t) * Q_{in}(t)dt, \qquad (12.1)$$

where

Q_{in} is the incident PAR

APAR is the product of fAPAR and Q_{in}

ε is the LUE parameter at time t

GEP accumulated over time, as for an hour, a day, or a growing season, represents the integral of the instantaneous GEP

As conceived by Monteith [27,29], the original LUE model was used for crop dry biomass production at harvest [30,31] and for wheat yields [32]; thus, the start and end times for the integration in Equation 12.1 were the start and end of the growing season. However, such determinations for natural ecosystems (e.g., forests) cannot practically be done using the harvest approach used in agriculture. Rather, a noninvasive method was needed.

With the advent of instrumented towers to measure carbon exchange, ecosystem LUE can be determined from tower-based measurements through incorporation of measured fluxes and meteorological information into models. It should be realized that net ecosystem CO_2 and LUE determinations from a tower include not only the amount fixed by the overstory foliage, but also include the net storage by understory plants and the respiratory contributions from nonphotosynthetic material (i.e., bark, standing dead foliage), litterfall, and soil components. In addition to the use of towers, the maturing of optical sensors and techniques has enabled the adaption of the LUE concept from land and space-based remote sensing platforms for monthly, weekly, daily, and even instantaneous ecosystem measurements throughout the growing season, as the ratio of GPP/APAR at midday, at the time of satellite overpass observation(s), and/or as a temporal (e.g., daily, monthly) average. These two methods are compatible and provide a straightforward approach to describing ecosystem carbon uptake using both flux and spectral information.

When the original LUE concept was formulated, some suggested that the full growing season LUE could be a conservative value, not varying widely [29,33,34]. However, further field studies have clearly indicated that seasonally averaged LUE varies substantially among vegetation types (Figure 12.2) and in response to environmental conditions [35–38]. It has also been observed that LUE is more variable over short time periods, since it is affected by many such factors: temperature, atmospheric humidity, soil type, water availability, disease, nutrient availability, plant type, and plant age [35,39,40].

This new ability to track photosynthetic activity and LUE from a dense temporal dataset (using tower and/or remote sensing information) has enabled researchers to appreciate and quantify the daily, seasonal and interannual LUE variability across ecosystems (Figure 12.3). Consequently, variations on the basic LUE model have been developed to estimate LUE using models that incorporate environmental information. Typically, these draw upon values from a look-up table for maximum *unstressed* LUE (ε^*) [41], and apply modifying factors ranging between 0 and 1 that explicitly describe the expected effects of environmental variables such as soil water deficits, low temperatures, and VPD [41–43]. One such LUE-based regional model is the Vegetation Photosynthesis and Respiration Model (VPRM) [44] that has been used to calculate GEP, and which can explain 60%–80% of hourly variability in fluxes at test sites. Matross et al. [45] have shown how atmospheric CO_2 concentrations can be used in an inverse modeling framework to derive the LUE efficiency parameters in VPRM. Furthermore, Lin et al. [46] have recently demonstrated that biases in estimating LUE can accumulate to extremely large values at seasonal to annual time scales.

It is clear that additional information is required to constrain modeled estimates of LUE because the current approaches have the disadvantage of relying on meteorological measurements that often represent a large area average rather than the local conditions to which a plant canopy/ecosystem is subjected. Furthermore, some variables affecting LUE, such as soil moisture, can have significant spatial variations and are difficult to measure over large areas. Developing approaches that use spectral information to directly determine LUE can reduce uncertainties in this key ecosystem descriptor as it varies over the landscape and over time.

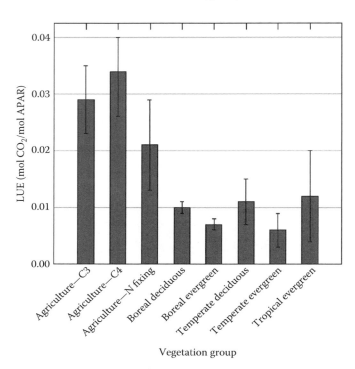

FIGURE 12.2 The variability in annual values for photosynthetic light use efficiency (LUE, mol CO_2/mol APAR) is shown for a range of vegetation groups. (Reproduced from *Remote Sensing of Environment*, 70, Gower, S.T., Kucharik, C.J., and Norman, J.M., Direct and indirect estimation of leaf area index, *f*APAR, and net primary production of terrestrial ecosystems, 29–51, Copyright (1999), with permission from Elsevier; *Agricultural and Forest Meteorology*, 101, Anderson, M.C. et al., An analytical model for estimating canopy transpiration and carbon assimilation fluxes based on canopy light-use efficiency, 265–289, Copyright (2000), with permission from Elsevier.) Sample sizes vary, $n \geq 50$.

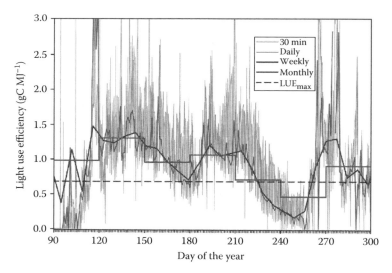

FIGURE 12.3 **(See color insert.)** Variations in LUE at different temporal resolutions for Shindler, Oklahoma, C4 grassland from green-up in April through senescence in October in 1998. LUE calculated using GEP and incident PAR from flux tower, with green *f*PAR estimated from broadband NDVI using PAR and shortwave radiation sensors. (Flux data from Verma, S., University of Nebraska, Lincoln, NE.)

12.3 PHYSIOLOGICAL BASIS FOR SPECTRAL OBSERVATIONS OF STRESS RESPONSES

Plants often absorb more energy from the sun than they can use for photosynthesis. Therefore, various protection mechanisms have been developed so that plants can utilize sunlight while disposing of dangerous excess energy. These include biochemical mechanisms related to the light reactions of Photosystem II (PSII) and the production of screening pigments in the external epidermal leaf layer. Without these protective processes, excess energy could result in lethal photooxidations, endangering the organisms [47].

Absorbed radiation (e.g., APAR) has three primary pathways: through the electron transport chain to the dark cycle photosynthetic processes; dissipation as heat via the xanthophyll system; and emission as ChlF. Plants continually adjust energy flow to support photosynthesis, while avoiding photooxidation under stressful conditions. Thus, these three pathways are interrelated (Figure 12.4), and stress responses and photosynthetic function can be inferred from bioindicators related to the xanthophyll pigment cycle pigment and ChlF. In addition to these physiological mechanisms, plants have a number of other ways to adjust to environmental stress, such as altering phenology and/or canopy structural components (leaf angle distributions, leaf area, epidermal screening pigments, plant form, and root depth, etc.), but we will not address those here.

12.3.1 XANTHOPHYLL PIGMENT CYCLE

The xanthophyll pigment cycle serves as a photoprotective mechanism underlying reflectance responses produced by physiologically induced spectral changes at 531 nm captured by a spectral bioindicator, referred to as the Photochemical Reflectance Index (PRI) [discussed in the following].

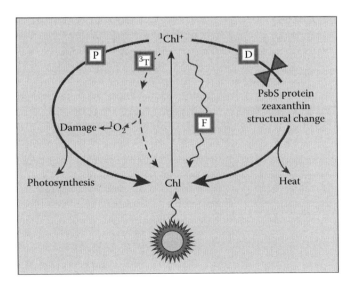

FIGURE 12.4 (See color insert.) A schematic showing the interrelationship of photosynthesis and two photoprotective mechanisms—the xanthophylls cycle and chlorophyll fluorescence. Sunlight is absorbed by chlorophyll (Chl) in the light harvesting complexes, producing an excited singlet chlorophyll (^{1}Chl*). Energy is used either for: (i) photochemistry (P) via electron transport to yield photosynthesis (left circuit); (ii) safe dissipation of excess excitation energy as heat (D) via the xanthophyll cycle (right circuit); or (iii) fluorescence emission (F), wavy line. (T3) is the triplet pathway leading to the formation of singlet oxygen ($^1O_2$*) and photooxidative damage. (Reprinted by permission from Macmillan Publishers Ltd. [*Nature*] (Demmig-Adams, B. and Adams, W.W., Photosynthesis: Harvesting sunlight safely, 403, 371), Copyright (2000).)

The xanthophyll cycle's primary function is the prevention of oxidative damage to the photosynthetic apparatus imbedded in the chloroplasts' thylakoid membranes [48–52]. The xanthophyll cycle operates as a light- and pH-induced interconversion of carotenoid pigments that ultimately sheds excess energy through heat dissipation. This nonphotochemical energy reduction (or quenching) mechanism involving the interconversion of the xanthophylls violaxanthin (V) to zeaxanthin (Z) under light and dark regimes was described by Müller et al. [53].

In the dark, the V form predominates but it is converted first to antheraxanthin (A) and then to Z in the light. This reversible biochemical cycle begins when sufficiently high light intensities induce a lower chloroplast lumen pH, initiating the sequential removal of double-bonded oxygen groups (or deepoxidation) to form A and Z. Z binds to PSII proteins causing a conformational change that quenches excess excitation energy [54]. In the dark, Z reverts to V through two epoxidation steps. The dependence of xanthophyll deepoxidation state on lumen pH [55] and the fraction of PSII antennae complexes bound to Z [56] have been described quantitatively. Since A and Z have higher absorption coefficients than V at ~531 nm [57], reflectance decreases and a lower PRI value is produced when A and Z accumulate in the thylakoid membrane under high light conditions. PRI is defined in Equation 12.2.

12.3.2 Chlorophyll Fluorescence

Another pathway that plants use to dissipate excess energy is through ChlF, primarily emitted from chlorophyll *a* associated with the PSII antennae system. We include a brief discussion of ChlF here because thermal energy dissipation can be assessed from changes in ChlF rate [47,58], with a known photosynthetic rate. This is possible due to the interrelationship of electron transport to support photosynthesis versus the two pathways that dissipate energy, the xanthophylls cycle and ChlF.

The emission spectrum of ChlF is characterized by two broad peaks between 600 and 800 nm, centered at ~685 nm (Red) and ~740 nm (Far Red). For a given chlorophyll content, ChlF declines when plants are not under environmental stress; and, in general, ChlF and photosynthesis rate are negatively correlated. However, ChlF can also decline when the system shifts toward increasing dissipation of excess energy as heat [59] by invoking the xanthophyll cycle. Most of the foundational research on ChlF was accomplished with active systems including lasers and other illumination sources in laboratory and field studies [60–62]. Laboratory research until recently relied upon dark-adapted leaves that were given an ultraviolet or visible radiation laser pulse to stimulate fluorescence (F), during which it was discovered that photochemical efficiency was strongly correlated with the magnitude of the normalized fluorescence response, or $\Delta F/F$maximum. The advent of new field fluorometers that can measure ChlF under ambient light, coupled with software that computes the full set of fluorescence and photochemical parameters, has broadened the application of ChlF measurements in crops and ecosystems.

Furthermore, exciting new methods have emerged in the past decade to retrieve ChlF from hyperspectral reflectance, especially very high resolution spectra (<1 nm), using various Fraunhofer Line Depth (FLD) methods. These typically utilize the atmospheric absorption features located in the vicinity of the two telluric oxygen (O_2) lines centered at 687.0 nm (O_2-B) and 760.4 nm (O_2-A) where telluric refers to contamination of spectra by the Earth's atmosphere [61]. The quantity retrieved is the passively solar-induced steady-state fluorescence, or Fs. The FLD methods enable separation of the "apparent reflectance" into two components: the weak ChlF emission signal (Fs, e.g., <5% of the apparent reflectance) and the much greater reflected contribution [59,61]. Due to the fine structure of Fraunhofer lines, high spectral resolution (i.e., 0.025–5 nm) must be used. The original FLD method utilized two narrowbands inside and outside the Fraunhofer feature, while other methods require additional narrowbands. For example, the 3FLD approach uses three wavelengths and linear interpolation—the wavelength at the deepest absorption (e.g., inside the feature) and wavelengths on both shoulders of the feature (outside) [63]. Nonlinear spectral fitting methods require all of the wavelengths within the Fraunhofer line region [61,64]. These radiance-based methods have the advantage of retrieving ChlF with actual physical units (e.g., mW m^{-2} nm^{-2} sr^{-2}) and are less sensitive to other changes in leaf biochemistry.

Reflectance-based approaches examine the effects of ChlF on apparent reflectance of plants in the red edge spectral region (~650 to 800 nm) and have been developed from either reflectance or derivative reflectance spectra. These indices are usually combinations of narrowbands located at the ChlF signal peaks (685 or 740 nm) and unaffected reference bands. Reflectance-based methods do not yield ChlF in physical units, but rather provide a relative influence of ChlF on apparent reflectance, and may be contaminated by changes in other biochemical properties. Nonetheless, these reflectance-based methods provide a relatively easier alternative to capture ChlF information and open a way for further investigation and progress [59,61].

12.3.3 Influence of PAR Levels on Vegetation Responses

If the xanthophyll photoprotection mechanism is the dominant process driving instantaneous LUE, then it should vary with PAR intensity and time of day, and exhibit different LUE for sunlit versus shaded leaves, as successfully demonstrated by several studies [35,65,66]. Optimal PAR levels for photosynthesis in foliage of most species of higher plants typically range from 1200 to 2000 µmol m^{-2} s^{-1} for "sun" leaves that grow in the exposed light environment found in the upper and outer foliage of a canopy [67]. Optimal PAR levels are lower for "shade" leaves that grow in the lower or inner portions of a canopy, where transmitted PAR dominates and shaded foliage receives a higher proportion of PAR as diffuse radiation [68].

In addition to high PAR intensity levels, it has been realized that LUE is strongly affected by the direct/diffuse irradiance ratio, increasing under diffuse solar radiation [68–70]. At the canopy level, this direct/diffuse ratio determines the instantaneous fraction of shaded versus sunlit foliage. For scaling up photosynthetic activities from leaf to canopy level, the importance of separating sunlit and shaded leaves has been recognized because of the nonlinear response of leaf carbon assimilation to light intensity [71–75]. Carbon uptake rates of shaded leaves are likely to have a linear response to irradiance, whereas this response often saturates for leaves exposed to direct sunlight [72,76]. Moreover, since sunlit leaves might be much warmer than shaded leaves on a clear and sunny day, ignoring the temperature difference could also bias estimates among carbon, water, and heat fluxes [73].

The physiological and morphological differences of sun and shade foliage have been well documented, especially concerning leaf anatomy and pigment concentrations [75,77–81]. As a result, several modeling methods separating canopy into sunlit and shaded fractions were developed [71,73,76,82–84]. However, it should be noted that unfavorable environmental conditions such as insufficient water and nutrient levels, and low/high temperatures also produce lower LUE, possibly also contributing to responses mediated through the xanthophyll cycle. High light intensities can be damaging to the photosystems of both sun and shade leaves, especially if water or nutrients are not sufficiently available, or temperatures are unfavorable.

12.4 REMOTE SENSING OBSERVATIONS OF PHOTOSYNTHESIS AND LUE

12.4.1 Remote Sensing Estimates of APAR

APAR is a key variable in LUE models, representing the integrated effect of all physiologically active absorbing pigments present in a vegetated canopy. Incident PAR is variable over time periods from minutes to days, due to predictable daily and season changes in solar elevation angle as well as less-predictable variations in atmospheric scattering. The use of remote sensing for the measurement of the components of APAR, incident PAR and *f*APAR, is well established (e.g., [85]). A number of approaches have been developed to estimate atmospheric scattering from satellite to determine incident PAR at the surface [86–88].

The *f*APAR parameter is also temporally variable, but as it is related to vegetation structure and solar angle its change is generally smoother over time. Some techniques for estimating *f*APAR are more sensitive to green vegetation rather than to total *f*APAR absorbed by the whole canopy [89,90]. The most commonly employed method utilizes the NDVI as a surrogate for *f*APAR, utilizing relatively broadband (~20 to 50 nm) observations in the red and near-infrared (NIR) spectrum [91–94]. In the field, values for total *f*APAR can be easily collected, however, accurate measures of the fraction of green material in the canopy are more difficult to obtain when a significant amount of material is nonphotosynthetic, such as dead leaves and branches, requiring an adjustment to the total *f*APAR in both field and remotely acquired measurements.

12.4.2 HYPERSPECTRAL BIOINDICATORS OF VEGETATION STRESS

Physical and chemical changes occur within leaves when they are subjected to stress conditions. The developing science of spectral bioindicators endeavors to extract information about the health and physiology of ecosystems, putting emphasis on spectroscopic techniques to examine the fine details captured with continuous narrow spectral bands (i.e., hyperspectral sensors), or a collection of well-chosen physiologically responsive narrowbands having spectral resolution between 1 and 10 nm. Spectroscopic techniques are based on optical properties of leaf tissues and their constituents and are especially valuable for elucidating alterations in spectral properties due to different amounts of these constituents [2]. This concept applies at the leaf level as well as for integrated values representing a canopy or ecosystem.

Many spectral vegetation indices have been published that relate reflectance characteristics to vegetation state and composition, and a large percentage of them are in the form of two-band normalized difference indices (i.e., magnitude of the reflectance difference between two spectral bands divided by their sum) (review, [95]). Most broadband indices of this type are sensitive to changes in total pigment content or to canopy structure such as Leaf Area Index (LAI) and leaf angle distribution. Attempts using broadband indices to capture stress-induced changes in photosynthetic activity (e.g., unfavorable temperature and irradiance levels) have met with mixed success [96]. In contrast, the photosynthetic LUE of vegetation has been monitored successfully with hyperspectral bioindicators, tracking leaf/canopy changes based on established spectroscopy methods [35,39,40].

Hyperspectral remote sensing requires different instruments with specific spectral configurations and radiometric accuracy than those used, for example, to produce satellite images from well-known broadband satellite sensors, such as the National Oceanic and Atmospheric Administration (NOAA) Advanced Very High Resolution Radiometer (AVHRR) series, the NASA and U.S. Geological Survey (USGS) Landsat series, and the NASA Moderate-Resolution Imaging Spectroradiometer (MODIS). However, one drawback of hyperspectral sensors and imaging spectrometers is that large amounts of data are generated, requiring special considerations for collection, documentation, management, storage, and powerful state-of-the-art data processing and information systems. Technological advances (see Chapter 4) help to alleviate this impediment to widespread adoption of spaceborne spectrometers for ecosystem studies [97].

12.4.3 ADVANTAGES OF REMOTE SENSING OBSERVATIONS

A flux tower can provide a continuous temporal profile of whole ecosystem carbon exchange, but only for a single location. In contrast, aircraft and satellite remote sensing provide information on spatial variability, thereby extending observations related to carbon fluxes over the entire landscape. Another advantage of remote sensing is that it does not disturb vegetation being measured and allows efficient repeated measurements over time. Consequently, remote sensing technologies offer one of the main tools for monitoring ecosystem responses to environmental stresses for large land areas over time [35,98,99]. But this information is predominantly associated with the gross production of foliage in the exposed, upper layer of the ecosystem, not the whole system sampled by towers.

Another limitation is that with sun-synchronous polar orbiting satellites, measurements are often limited to a single observation per day at a fixed time of day. From these data, a typical measurement series represents samples acquired on clear days composited over periods of several days, although limited daily observations are possible from some platforms. Most commonly, remote sensing observations from aircraft or satellite sensors provide snapshots of vegetation at variable time intervals across a growing season. Nevertheless, this information is extremely useful for examining the spatial and temporal patterns of ecosystem health and physiological status.

The developing field of spectral bioindicators benefits greatly from the valuable GEP time series acquired from flux towers providing an essential way to validate remote sensing estimates of photosynthesis. Moreover, the in situ and space-based observations can be used together to develop and test different methods of analyzing and modeling the spectral data against the "ground truth" of the flux tower measurements. For example, time series from flux towers have been used to validate the MODIS satellite sensor algorithm that computes global GEP [100].

12.5 PHOTOCHEMICAL REFLECTANCE INDEX AND LUE

The PRI is a normalized difference spectral index in form, but its original formulation is based on two narrow (3–10 nm) wavelengths in the green spectrum to capture physiological responses [101–103]. The PRI expresses the relative down-regulation of photosynthesis induced primarily by high light intensities via the xanthophyll pigment cycle, but it also is affected by secondary or compounding factors such as drought [104,105]. The general form for the PRI is:

$$PRI_{REF\lambda} = \frac{[\rho 531 - \rho(REF\lambda)]}{[\rho 531 + \rho(REF\lambda)]}. \tag{12.2}$$

where

$\rho 531$ is the reflectance for the physiologically active 531 nm spectral band

$\rho(REF\lambda)$ is the reflectance in a second non-responsive wavelength band, which in the original formulation of the PRI used a REF λ at 570 nm, or PRI_{570}

Several other reference wavelengths have been used to provide a PRI, depending on available sensor bands. These include 551 and 488 nm, which provided options that have enabled the use of MODIS data acquired since 2000 for PRI determinations [104,106–108].

With both PRI_{570} and PRI_{551}, short-term responses to high light conditions are indicated by relatively lower and typically negative values, so that the greatest stress-induced relative photosynthetic downregulation is indicated by the most negative values observed. The reverse situation occurs with PRI_{488}, for which increasingly greater downregulation is indicated with larger positive values.

The use of the PRI for LUE monitoring was first documented at the leaf level by Gamon et al. [101–103] and substantiated by many others [109–112]. Although the PRI signal is relatively weak compared with spectral indices using red versus NIR wavelengths, it has successfully been observed remotely over forests from aircraft and towers [105,113–116] and from satellites [106,107].

12.5.1 LEAF, WHOLE PLANT, AND CANOPY PRI STUDIES

The original demonstration of the PRI and its relationship to LUE was made in sunflower leaves by Gamon et al. [101,103], who linked the PRI response to the xanthophyll cycle's role in photosynthetic downregulation and to photosynthetic function [102] (Figure 12.5). Subsequently, several groups investigated the capability of the PRI to track variations in plant physiological status and photosynthetic activities induced by various environmental stressors including excessive sunlight, water limitations, and nitrogen availability. From measurements of individual plants, Trotter et al. [117]

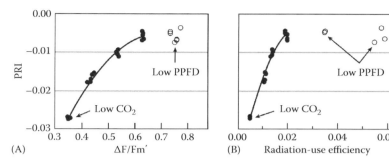

(A)

(B)

FIGURE 12.5 Measurements on leaves of *G. barbadense* (cotton) exposed to light and CO_2 treatments. PRI is plotted against (A) the photochemical efficiency, expressed by the relative change in chlorophyll fluorescence in the light, or $\Delta F/Fm'$ ($r^2 = 0.93$ for ●) and (B) photosynthetic radiation use efficiency (RUE or LUE, $r^2 = 0.94$ for ●). Low-light points (○; 0–100 μmol m^{-2} s^{-1}); medium to high-light points (●; 500–1500 μmol m^{-2} s^{-1}). (With kind permission from Springer Science+Business Media: *Oecologia*, The photochemical reflectance index: An optical indicator of photosynthetic radiation use efficiency across species, functional types, and nutrient levels, 112, 1997, 492–501, Gamon, J.A., Serrano, L., and Surfus, J.S.)

reported a linear relationship between PRI_{570} and LUE among eight different species that exhibited a range of physiological properties and nitrogen content. These early studies were based on in situ observations acquired using handheld spectroradiometers at the leaf level [101,102,110,118], and were supported by other studies, including the estimations of leaf-level LUE of soybean plants under different soil moisture conditions [118].

These successful leaf-level results led to localized canopy-level studies [39,103,111] For crops, PRI from handheld and pole-mounted spectrometer data have shown clear diurnal and seasonal dynamics related to variations in GEP determined from eddy covariance measurements [119–121] (Figure 12.6). These studies established a correlation between PRI and vegetation physiological condition that strongly influence photosynthetic activities. They also support the hypothesis

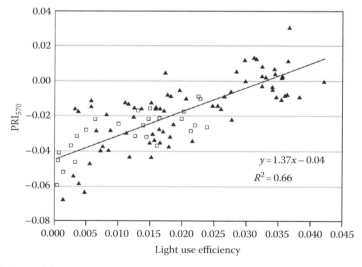

Light use efficiency

FIGURE 12.6 Relationship between LUE and PRI_{570} for cornfield in Beltsville, Maryland, United States, collected at multiple times during selected clear days during the 2007 (□) and 2008 (▲) growing seasons. PRI_{570} values from averages of nadir spectral reflectance collected along a 100 m transect in field. LUE (with units of mol C mol^{-1} APAR) are hourly values calculated using GEP and incident PAR from flux tower and green fPAR. (Estimated from NDVI from reflectance data; Flux data from W.P. Kustas; USDA/Beltsville Agricultural Research Service, Beltsville, MD.)

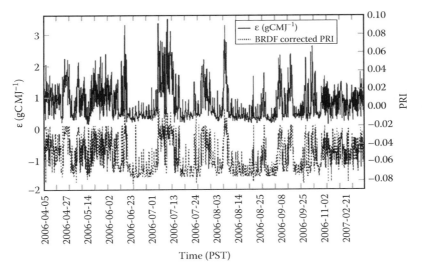

FIGURE 12.7 Temporal variability of half hourly LUE (ε; gC MJ^{-1}) determined from eddy flux measurements between April 1, 2006 and March 31, 2007 at a Canadian Carbon Program instrumented tower at a Douglas fir forest in British Columbia. The tower-derived LUE pattern was followed closely by directionally corrected PRI. (From *Remote Sensing of Environment*, 112, Hilker, T. et al., Separating physiologically and directionally induced changes in PRI using BRDF models, 2777–2788, Copyright (2008), with permission from Elsevier.)

that PRI is an indicator of PSII photochemical efficiency (as determined by ChlF measurements, Figure 12.5). Consistent with this hypothesis, Guo and Trotter [122] found increased LUE occurred for C$_3$ plants with elevated CO$_2$ without affecting PRI, while low temperatures decreased both LUE and PRI proportionally.

Canopy studies have benefited from recent advances in instrumentation such as the development of automated near surface instruments [123]. This allows frequent observations of a canopy as it experiences a range of environmental conditions, including variations in temperature, light quality, humidity, and soil moisture at the temporal and spatial scales relevant to flux measurements [124]. From a forested tower site in Canada, Hilker et al. [115] demonstrated similar temporal trends for the PRI and LUE (Figure 12.7). The challenge for these systems is the development of methods to observe an area large enough to be representative of a flux tower footprint. Multiple approaches have been tried including: Mounting a spectroradiometer on a tram that moves on a track above the canopy [125]; having a spectroradiometer mounted on a tower to scan the canopy [123,126]; using an imaging spectrometer from a fixed platform [127]; or having multiple narrowband sensors that observe different parts of the canopy [128]. By sampling a number of smaller areas within the flux tower footprint frequently, near surface measurements provide rich datasets for the study of spectral bioindicators.

12.5.2 Canopy Airborne PRI Observations

The need to capture vegetation under a range of conditions presents a challenge for studying spectral bioindicators, accomplished either by viewing multiple sites and/or capturing varying conditions at different times [124]. The BOReal Ecosystem and Atmosphere Study (BOREAS), with multiple flux towers that were frequently observed by airborne hyperspectral sensors, provided such an opportunity, leading to important advances in using PRI. Nichol et al. [113] demonstrated a significant linear correlation across four types of boreal forests between LUE and PRI$_{570}$. LUE was determined from tower-based eddy covariance and radiometer data, while the PRI$_{570}$ was derived from narrow ($\sim$10 nm) waveband reflectances measured by a helicopter-mounted spectroradiometer. Additionally, Rahman et al. [129] used Airborne Visible/Infrared Imaging Spectrometer (AVIRIS) aircraft data and a scaled version of PRI (combined with the NDVI) to map carbon fluxes over the BOREAS study area. Since then, AVIRIS aircraft data have also been used in PRI studies of chaparral [130].

Inoue and Peñuelas [118] found that the correlation to LUE appeared stronger when PRI was derived from narrower spectroradiometer measurements (~3 nm bandwidth) than from simulated AVIRIS bandwidth of ~10 nm. Strachan et al. [121] used PRI from aircraft to determine gross photosynthesis, showing that it identified stressed areas that were correlated with lower yield in corn and wheat fields. Suárez et al. [121] found airborne PRI was sensitive to diurnal changes in water stress in orchards.

12.5.3 PRI FROM SATELLITE

The MODIS instruments flying on the NASA Aqua and Terra satellites [131] have several relatively narrow (10 nm) spectral bands intended for ocean studies in the region needed for PRI, in particular band 11 centered at 531 nm (and reference band 13, 550 nm), but these MODIS bands have generally not been used over land. Several studies have shown relationships between a MODIS-measured PRI, using the ocean bands or ocean/land band combinations, and LUE from flux towers [106–108,132]. In Rahman et al. [108] and Drolet et al. [106], single site relationships between MODIS PRI and LUE were shown for deciduous forests. Subsequently, Drolet et al. [107] found consistent relationships between LUE and PRI (Figure 12.8) in a chronosequence of boreal conifer sites consisting of multiple sites dominated by the same species, and extended those results to produce a MODIS-PRI-based LUE map at the scale of a few kilometers to demonstrate LUE variability for a 5300 km^2 region in Saskatchewan, Canada. Huemmrich et al. [132] examined four widely separated forest sites, and identified significant differences in the PRI:LUE relationships for deciduous versus conifer forests.

These studies have shown that PRI from MODIS can be used to determine ecosystem LUE, although an optimum approach for calculating MODIS-PRI has not yet emerged because the MODIS sensor lacks the commonly used PRI reference band at 570 nm. Therefore, these MODIS-based PRI studies used a number of different reference bands, including narrow (10 nm) "ocean" bands centered at 488, 551, 667, and 678 nm (all at 1 km spatial scale), as well as the broad (50 nm) land band centered at 645 nm (500 m resolution) [106–108,133]. Another important issue in the analysis of the MODIS data is the correction for atmospheric effects on the reflectance, since currently, the MODIS ocean bands are not atmospherically corrected to surface reflectance over land operationally. Thus, all of the studies have either used top of atmosphere reflectance [134] or developed their own implementation of the MODIS atmospheric correction algorithm [106,108,133]. Hilker et al. [135] processed MODIS data using both the 6S model [136] and the Multi-Angle Implementation of Atmospheric Correction (MAIAC) algorithm [137], which were compared to the PRI measurements from a tower-mounted spectrometer, and found better MODIS-PRI relationships to tower measurements when corrected with MAIAC versus 6S.

Other satellite studies were conducted using the Earth Observing 1 (EO-1) hysperspectral spaceborne imager, Hyperion, to calculate PRI, which successfully detected drought stress responses in Amazon forests [138] and biochemical differences among native and invasive species in Hawaiian rainforests [139,140]. The effects of viewing geometry on PRI were further examined and verified using data from European Space Agency's (ESA) CHRIS/PROBA (the Compact High Resolution Imaging Spectrometer flying on the Project for On Board Autonomy platform) satellite sensor [141].

12.5.4 CONFOUNDING EFFECTS ON PRI

Various studies have demonstrated that a number of factors influence the PRI:LUE relationship at the canopy/ecosystem level, including sun angle [142], view angle and canopy structure [35,66,106,114,115,143], soil background [144], or chlorophyll or carotenoid pigment levels

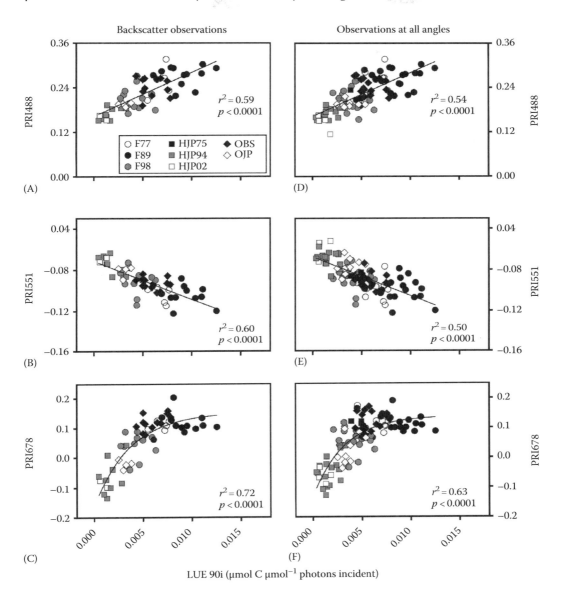

FIGURE 12.8 Relationships between PRI and LUE for different PRI versions using MODIS satellite data at the physiological active band, 531 nm, and reference bands at either 488, 551, or 678 nm [denoted as PRI488, PRI551, and PRI678]. LUE is computed as 90 min averages from flux towers, where (A–C) observations are close to the backscattering direction, and (D–F) observations are for all view angles. The different symbols represent different Boreal Ecosystem Research and Monitoring Sites (BERMS) in the Canadian Carbon Program. (From *Remote Sensing of Environment*, 112, Drolet, G.G. et al., Regional mapping of gross light-use efficiency using MODIS spectral indices, 3064–3078, Copyright (2008), with permission from Elsevier.)

[39,142,145,146]. A positive correlation was demonstrated between PRI and LUE at the canopy level when these factors were evaluated with a one-dimensional ray tracing radiative transfer modeling study [144], but the PRI was also sensitive to variations in viewing angle, LAI, canopy structure, shadow fraction, and background reflectance [105,114,144].

These view angle and canopy architecture factors have been studied further. Drolet et al. [106,107] found that viewing direction (back vs. forward scattering) was an important factor influencing the PRI value obtained from MODIS imagery, with sunlit canopy displaying lower PRI than

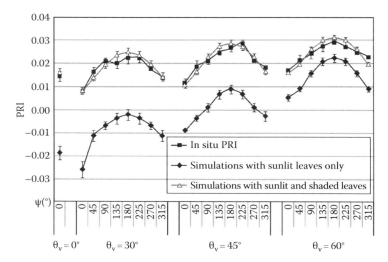

FIGURE 12.9 Effects of viewing geometry on PRI values for canopy measurements in a USDA cornfield in Beltsville, MD. Comparisons show in situ observations (■) and radiative transfer modeling simulations using sunlit leaves only (♦) versus both sunlit and shaded leaves (△). (From Cheng, Y.-B. et al., *Ecol. Inf.*, 5(5), 330, 2010.)

shadowed canopy. Sunlit foliage has lower LUE and thus lower PRI values, due to increased light-induced stress at high light intensities. This finding revealed canopy three-dimensional structure was affecting the canopy PRI:LUE relationships, consistent with earlier modeling studies [144]; aircraft or tower measurements are also influenced by canopy structure due to varying amounts of shadowed versus sunlit foliage in the instrument field of view [35,66,114]. In a further study in a cornfield, where in situ spectrometer measurements were made in 15° increments around the full azimuth range with off-nadir view angles, the PRI values exhibited the expected stress effects in sunlit foliage as compared to shaded foliage, in agreement with simulations with a radiative transfer model that could adequately account for these effects (Figure 12.9).

The dynamics of PRI variability within canopies has been addressed using automated tower-mounted spectrometers that collect measurements for known angular sectors of a canopy over time

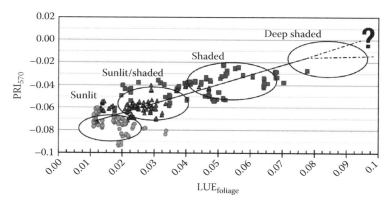

FIGURE 12.10 (See color insert.) This figure describes the relationship between the Photochemical Reflectance Index (PRI) and photosynthetic light use efficiency ($LUE_{foliage}$, μmol C μmol^{-1} APAR) for foliage exposed to a range of illumination conditions in a Douglas-fir forest in Canada. The lowest PRI and $LUE_{foliage}$ values are associated with sunlit foliage throughout the 2006 growing season. The highest PRI and $LUE_{foliage}$ values measured were associated with shaded foliage, but high values are also expected for foliage residing in the deeply shaded canopy sectors that could not be measured. (From Middleton, E.M. et al., *Can. J. Remote Sens.*, 35(2), 166, 2009.).

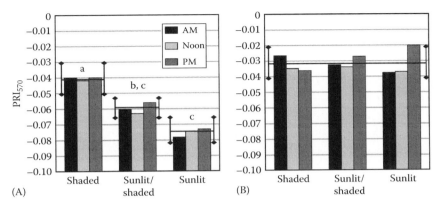

FIGURE 12.11 For three foliage groups exposed to a range of illumination conditions, the seasonal PRI_{570} averages at three times of day are shown (morning [■], noon [■], and afternoon [■]) with mean and SD indicated with horizontal and vertical bars: (A) on sunny days, displayed significant diurnal differences; (B) on cloudy days, no foliage group differences were significant across the season at a Douglas-fir stand during the 2006 growing season. (From Middleton, E.M. et al., *Can. J. Remote Sens.,* 35(2), 166, 2009.)

and over a range of illumination conditions, as was first undertaken for a conifer forest in 2006 [115]. For that same forest, these initial results were intensely studied across the growing season by Middleton et al. [35] and Cheng et al. [66], clearly showing different PRI behavior associated with sunlit versus shaded canopy sectors. When corrected for PAR differences in sunlit and shaded canopy sectors of this Douglas fir forest (Figures 12.10 and 12.11), PRI_{570} was linearly related to LUE.

12.6 FUTURE REMOTE SENSING OPPORTUNITIES

Implementation of a remote sensing approach for regional views to study LUE is appealing considering the relative simplicity of the approach, and applicability to mapping. A number of airborne spectrometers are available for select transect studies, including the AVIRIS aircraft, and have proved especially useful for agricultural studies because of their flexibility in deployment and ability to provide detailed observation at high spatial resolution. However, currently, only a limited set of satellite sensors in space have the correct set of narrow and/or continuous high spectral resolution bands, which include a narrowband centered at 531 nm along with an appropriate reference band, to obtain viable instantaneous estimates of ecosystem stress responses, including (ε) needed to implement the LUE model.

Our expectation is that improved and newly implemented spectral bioindicators will be able to detect photosynthetic stress at regional and global scales from satellites, aircraft, and ground-based sensors over the coming decades. Advances in instruments that will decrease costs and increase availability for ground-based and aircraft operation are expected to make it easier to use spectral bioindicators locally. Unlike satellites, imaging spectrometers on aircraft can provide observations over specific areas at high spatial resolutions on demand. Utilizing unmanned aerial vehicles (UAV) may further decrease the cost and increase availability of spectral bioindicator data. Further, radiometers mounted on towers or within canopies can be integrated into sensor networks for continuous monitoring of vegetation.

Presently, civilian U.S. satellite instruments with adequate spectral characteristics to track LUE and vegetation stress include MODIS on Terra and Aqua (using ocean bands having 1 km+ coarse spatial resolution) and Hyperion on EO-1, a relatively high spatial resolution sensor (30 m), but since these sensors have already greatly exceeded their life expectancies, we cannot expect them to be available very far into the future. The ESA has sponsored narrowband satellite sensors such as MEdium Resolution Imaging Spectrometer (MERIS) [147] and CHRIS/PROBA; several other countries also have satellite spectrometers (e.g., Japan's Greenhouse gases Observing Satellite, or GOSAT launched January 2009, with a 10.5 km nadir footprint) or plan to launch spectrometers in the next decade (e.g., Germany, Italy), but many of these are regional sampler missions.

Unfortunately, none of the future instruments planned by the United States, such as Landsat-8, National Polar-orbiting Operational Environmental Satellite System (NPOESS), NPOESS Preparatory Project (NPP), or National Oceanic and Atmospheric Administration-Worldwide Repeat Data (NOAA-R) have included the physiologically essential band at ~530 nm in their multiband sensor designs, while the U.S. Tier 2 Decadal Survey spectrometer/thermal global mission, HyspIRI, is now delayed until the 2020s. ESA is presently evaluating a possible future mission, the FLuourescence EXplorer (FLEX), carrying a high spectral resolution imaging spectrometer to measure fluorescence within the oxygen bands. An exciting new development is the demonstration that ChlF can in fact be retrieved from space using the FLD approach, as shown for global terrestrial maps derived from the GOSAT [148]. Unfortunately, the progress made by scientists over the past decade to utilize spectral bioindicators for photosynthetic efficiency and stress indicators has not yet been widely incorporated into satellite remote sensing sensor design plans.

12.7   CONCLUSIONS

Environmental pressures on terrestrial vegetation, both from natural and human causes, induce physiological stress responses that result in changes in ecosystem processes, such as carbon exchange. These stress responses also affect the optical properties of plants. The new field of study addressing spectral bioindicators links observations of plant optical properties to ecosystem physiology. This approach to monitor ecosystem health at a range of scales with spectral bioindicators allows the connection of surface, aircraft, and satellite observations, and has enabled the great strides in showing clear connections with light use efficiency (LUE). Spectral bioindicators have the capacity to provide unique information about ecosystem health and physiology that can be incorporated into process models and/or coupled climate carbon models to better assess the feedbacks between ecosystems and the climate system. Narrow and/or continuous high spectral resolution bands, including a narrowband centered at 531 nm along with an appropriate reference band, are essential to obtaining viable instantaneous estimates of ecosystem stress responses, such as LUE, for use in carbon cycle models. Nevertheless, studies have also shown how viewing geometry, soil background, and satellite data preprocessing can significantly affect the retrieved measurements, for which generalized approaches for mitigating these effects have not yet been developed. Advancement in this field and utilization of these approaches will require technical developments in sensor systems with narrow spectral bands to provide frequent and consistent observations from satellite, aircraft, and near surface sensor networks.

ACKNOWLEDGMENTS

We thank NASA's Terrestrial Ecology Program for support of E.M.'s spectral bioindicator project. Support for writing this chapter was also provided by an NSERC Discovery grant awarded to HM.

REFERENCES

1. Tucker, C.J. et al., An extended AVHRR 8-km NDVI dataset compatible with MODIS and SPOT vegetation NDVI data, *International Journal of Remote Sensing, 26* (20), 4485–4498, 2005.
2. Goetz, A.F.H. and Curtiss, B., Hyperspectral imaging of the earth: Remote analytical chemistry in an uncontrolled environment, *Field Analytical Chemistry and Technology, 1* (2), 67–76, 1996.
3. Myneni, R.B. et al., Increased plant growth in the northern high latitudes from 1981 to 1991, *Nature, 386* (6626), 698–702, 1997.
4. Margolis, H.A. and Brand, D.G., An ecophysiological basis for understanding plantation establishment, *Canadian Journal of Forest Research, 20*, 375–390, 1990.

5. Cech, J.J.J., Wilson, B.W., and Crosby, D.G. (Eds.), *Multiple Stresses in Ecosystems*. Boca Raton, Florida: CRC Press, 1998.
6. Mooney, H.A., Winner, W.E., and Pell, E.J. (Eds.), *Response of Plants to Multiple Stresses*. New York: Academic Press, 1991.
7. Beer, C. et al., Terrestrial gross carbon dioxide uptake: Global distribution and covariation with climate, *Science, 329* (5993), 834–838, 2010.
8. Denman, K.L. et al., Couplings between changes in the climate system and biogeochemistry. In S. Solomon, D. Qin, M. Manning, Z. Chen, M. Marquis, K.B. Avery, M. Tignor, and H.L. Miller (Eds.), *Climate Change 2007: The Physical Science Basis. Contribution of Working Group 1 to the Fourth Assessment Report of the Intergovernmental Panel on Climate Change*. Cambridge, U.K.: Cambridge University Press, 2007.
9. Falkowski, P.G., Barber, R.T., and Smetacek, V., Biogeochemical controls and feedbacks on ocean primary production, *Science, 281* (5374), 200–206, 1998.
10. Canadell, J.G. et al., Contributions to accelerating atmospheric CO_2 growth from economic activity, carbon intensity, and efficiency of natural sinks, *Proceedings of the National Academy of Sciences of the United States of America, 104* (47), 18866–18870, 2007.
11. Nemani, R.R. et al., Climate-driven increases in global terrestrial net primary production from 1982 to 1999, *Science, 300* (5625), 1560–1563, 2003.
12. Piao, S. et al., The carbon balance of terrestrial ecosystems in China, *Nature, 458* (7241), 1009–1013, 2009.
13. Bousquet, P. et al., Regional changes in carbon dioxide fluxes of land and oceans since 1980, *Science, 290* (5495), 1342–1346, 2000.
14. Van der Werf, G.R. et al., Continental-scale partitioning of fire emissions during the 1997 to 2001 El Nino/La Nina period, *Science, 303* (5654), 73–76, 2004.
15. Kharouk, V.I. et al., Aspen bark photosynthesis and its significance to remote sensing and carbon budget estimates in the boreal ecosystem, *Water, Air, and Soil Pollution, 82* (1), 483–497, 1995.
16. Goulden, M.L. et al., Patterns of NPP, GPP, respiration, and NEP during boreal forest succession, *Global Change Biology*, 2010 (Accepted).
17. Jung, M. and Reichstein, M., Recent decline in the global land evapotranspiration trend due to limited moisture supply, *Nature,* 2010. In press.
18. Costanza, R. et al., The value of the world's ecosystem services and natural capital, *Nature, 387* (6630), 253–260, 1997.
19. De Groot, R.S., Wilson, M.A., and Boumans, R.M.J., A typology for the classification, description and valuation of ecosystem functions, goods and services, *Ecological Economics, 41* (3), 393–408, 2002.
20. Rockström, J. et al., A safe operating space for humanity, *Nature, 461* (7263), 472–475, 2009.
21. Baldocchi, D., TURNER REVIEW No. 15. 'Breathing' of the terrestrial biosphere: Lessons learned from a global network of carbon dioxide flux measurement systems, *Australian Journal of Botany, 56* (1), 1–26, 2008.
22. Mahecha, M.D. et al., Global convergence in the temperature sensitivity of respiration at ecosystem level, *Science, 329* (5993), 838–840, 2010.
23. Kljun, N. et al., A simple parameterisation for flux footprint predictions, *Boundary-Layer Meteorology, 112* (3), 503–523, 2004.
24. Falge, E. et al., Gap filling strategies for defensible annual sums of net ecosystem exchange, *Agricultural and Forest Meteorology, 107* (1), 43–69, 2001.
25. Desai, A.R. et al., Cross-site evaluation of eddy covariance GPP and RE decomposition techniques, *Agricultural and Forest Meteorology, 148* (6–7), 821–838, 2008.
26. Reichstein, M. et al., On the separation of net ecosystem exchange into assimilation and ecosystem respiration: Review and improved algorithm, *Global Change Biology, 11* (9), 1424–1439, 2005.
27. Monteith, J., Climate and efficiency of crop production in Britain, *Philosophical Transaction of the Royal Society of London B: Biological Sciences, 281*, 271–294, 1977.
28. Russell, G., Jarvis, P.G., and Monteith, J.L., Absorption of radiation by canopies and stand growth. In G. Russell, B. Marshall, and P.G. Jarvis (Eds.), *Plant Canopies: Their Growth, Form and Function* (pp. 21–40). Cambridge, U.K.: Cambridge University Press, 1989.
29. Monteith, J., Solar-radiation and productivity in tropical ecosystems, *Journal of Applied Ecology, 9*, 747–766, 1972.
30. Kumar, M. and Monteith, J.L., Remote sensing of crop growth. In H. Smith (Ed.), *Plants and the Daylight Spectrum* (pp. 134–144). London, U.K.: Academic Press, 1981.
31. Daughtry, C.S.T. et al., Spectral estimates of absorbed radiation and phytomass production in corn and soybean canopies, *Remote Sensing of Environment, 39* (2), 141–152, 1992.

32. Wiegand, C.L. and Richardson, A.J., Leaf area, light interception and yield estimates from spectral components analysis, *Agronomy Journal, 76*, 543–548, 1984.

33. Goetz, S.J. and Prince, S.D., Modelling terrestrial carbon exchange and storage: Evidence and implications of functional convergence in light-use efficiency. In A.H. Fitter and D. Raffaelli (Eds.), *Advances in Ecological Research* (pp. 57–92). San Diego, CA: Academic Press, 1999.

34. Field, C.B., Ecological scaling of carbon gain to stress and resource availability. In H.A. Mooney, W.E. Winner, and E.J. Pell (Eds.), *Integrated Responses of Plants to Stress*. San Diego, CA: Academic Press, 1991.

35. Middleton, E.M. et al., Linking foliage spectral responses to canopy level ecosystem photosynthetic light use efficiency at a Douglas-fir forest in Canada, *Canadian Journal of Remote Sensing, 35* (2), 166–188, 2009.

36. Gower, S.T., Kucharik, C.J., and Norman, J.M., Direct and indirect estimation of leaf area index, fAPAR, and net primary production of terrestrial ecosystems, *Remote Sensing of Environment, 70* (1), 29–51, 1999.

37. Green, D.S., Erickson, J.E., and Kruger, E.L., Foliar morphology and canopy nitrogen as predictors of light-use efficiency in terrestrial vegetation, *Agricultural and Forest Meteorology, 115* (3–4), 163–171, 2003.

38. Anderson, M.C. et al., An analytical model for estimating canopy transpiration and carbon assimilation fluxes based on canopy light-use efficiency, *Agricultural and Forest Meteorology, 101* (4), 265–289, 2000.

39. Gamon, J. et al., Assessing photosynthetic downregulation in sunflower stands with an optically-based model, *Photosynthesis Research, 67* (1), 113–125, 2001.

40. Prince, S.D., A model of regional primary production for use with coarse resolution satellite data, *International Journal of Remote Sensing, 12* (6), 1313–1330, 1991.

41. Heinsch, F.A. et al., *User's Guide, GPP and NPP (MOD17A2/A3) Products, NASA MODIS Land Algorithm*, 2003.

42. Law, B.E. and Waring, R.H., Combining remote sensing and climatic data to estimate net primary production across Oregon, *Ecological Applications, 4* (4), 717–728, 1994.

43. Prince, S.D. and Goward, S.N., Global primary production: A remote sensing approach, *Journal of Biogeography, 22* (4–5), 815–835, 1995.

44. Mahadevan, P. et al., A satellite-based biosphere parameterization for net ecosystem CO_2 exchange: Vegetation photosynthesis and respiration model (VPRM), *Global Biogeochemical Cycles, 22* (2), GB2005, 2008.

45. Matross, D.M. et al., Estimating regional carbon exchange in New England and Quebec by combining atmospheric, ground-based and satellite data, *Tellus, 58B*, 344–358, 2006.

46. Lin, J.C. et al., Attributing uncertainties in simulated biospheric carbon fluxes to different error sources, *Global Biogeochemical Cycles*, 2011. In press.

47. Demmig-Adams, B. and Adams, W.W., Photosynthesis: Harvesting sunlight safely, *Nature, 403* (6768), 371–374, 2000.

48. Pfündel, E.E. and Bilger, W., Regulation and possible function of the violaxanthin cycle, *Photosynthesis Research, 42* (2), 89–109, 1994.

49. Demmig-Adams, B., Linking the xanthophyll cycle with thermal energy dissipation, *Photosynthesis Research, 76* (1), 73–80, 2003.

50. Demmig-Adams, B. and Adams, W.W., III, The role of xanthophyll cycle carotenoids in the protection of photosynthesis, *Trends in Plant Science, 1* (1), 21–26, 1996.

51. Öquist, G. and Huner, N.P.A., Photosynthesis of overwintering evergreen plants, *Annual Review of Plant Biology, 54* (1), 329–355, 2003.

52. Štroch, M. et al., Dynamics of the xanthophyll cycle and non-radiative dissipation of absorbed light energy during exposure of Norway spruce to high irradiance, *Journal of Plant Physiology, 165* (6), 612–622, 2008.

53. Müller, P., Li, X.-P., and Niyogi, K.K., Non-photochemical quenching. A response to excess light energy, *Plant Physiology, 125* (4), 1558–1566, 2001.

54. Havaux, M. and Niyogi, K.K., The violaxanthin cycle protects plants from photooxidative damage by more than one mechanism, *Proceedings of the National Academy of Sciences of the United States of America, 96* (15), 8762–8767, 1999.

55. Pfundel, E.E. and Dilley, R.A., The pH dependence of violaxanthin deepoxidation in isolated pea chloroplasts, *Plant Physiology, 101* (1), 65–71, 1993.

56. Gilmore, A.M. and Yamasaki, H., 9-Aminoacridine and dibucaine exhibit competitive interactions and complicated inhibitory effects that interfere with measurements of ΔpH and xanthophyll cycle-dependent photosystem II energy dissipation, *Photosynthesis Research, 57* (2), 159–174, 1998.

57. Bilger, W. and Björkman, O., Role of the xanthophyll cycle in photoprotection elucidated by measurements of light-induced absorbance changes, fluorescence and photosynthesis in leaves of *Hedera canariensis*, *Photosynthesis Research, 25* (3), 173–185, 1990.

58. Li, X.-P. et al., A pigment-binding protein essential for regulation of photosynthetic light harvesting, *Nature, 403* (6768), 391–395, 2000.

59. Zarco-Tejada, P.J. et al., Steady-state chlorophyll a fluorescence detection from canopy derivative reflectance and double-peak red-edge effects, *Remote Sensing of Environment, 84* (2), 283–294, 2003.

60. Cerovic, Z.G. et al., Ultraviolet-induced fluorescence for plant monitoring: Present state and prospects, *Agronomie, 19* (7), 543–578, 1999.

61. Meroni, M. et al., Remote sensing of solar-induced chlorophyll fluorescence: Review of methods and applications, *Remote Sensing of Environment, 113* (10), 2037–2051, 2009.

62. Mohammed, G.H., Binder, W.D., and Gillies, S.L., Chlorophyll fluorescence: A review of its practical forestry applications and instrumentation, *Scandinavian Journal of Forest Research, 10* (1), 383–410, 1995.

63. Maier, S.W., Günther, K.P., and Stellmes, M., Sun-induced fluorescence: A new tool for precision farming. In T. VanToai, D. Major, M. McDonald, J. Schepers, and L. Tarpley (Eds.), *Digital Imaging and Spectral Techniques: Applications to Precision Agriculture and Crop Physiology* (pp. 209–222). Madison, WI: American Society of Agronomy, 2003.

64. Alonso, L. et al., Improved Fraunhofer line discrimination method for vegetation fluorescence quantification, *Geoscience and Remote Sensing Letters, IEEE, 5* (4), 620–624, 2008.

65. Cheng, Y.-B. et al., Utilizing in situ directional hyperspectral measurements to validate bio-indicator simulations for a corn crop canopy, *Ecological Informatics, 5* (5), 330–338, 2010.

66. Cheng, Y.-B. et al., Dynamics of spectral bio-indicators and their correlations with light use efficiency using directional observations at a Douglas-fir forest, *Measurement Science and Technology, 20* (9), 095107, 2009.

67. Jones, H.G., *Plants and Microclimate, a Quantitative Approach to Environmental Plant Physiology.* Cambridge, U.K.: Cambridge University Press, 1983.

68. Alton, P.B., North, P.R., and Los, S.O., The impact of diffuse sunlight on canopy light-use efficiency, gross photosynthetic product and net ecosystem exchange in three forest biomes, *Global Change Biology, 13* (4), 776–787, 2007.

69. Gu, L. et al., Advantages of diffuse radiation for terrestrial ecosystem productivity, *Journal of Geophysical Research, 107* (D6), 4050, 2002.

70. Hollinger, D.Y. et al., Carbon dioxide exchange between an undisturbed old-growth temperate forest and the atmosphere, *Ecology, 75* (1), 134–150, 1994.

71. Chen, J.M. et al., Daily canopy photosynthesis model through temporal and spatial scaling for remote sensing applications, *Ecological Modelling, 124* (2–3), 99–119, 1999.

72. Gurevitch, J., Scheiner, S.M., and Fox, G.A., *The Ecology of Plants.* Sunderland, MA: Sinauer Associates, Inc., 2002.

73. Wang, Y.P. and Leuning, R., A two-leaf model for canopy conductance, photosynthesis and partitioning of available energy I: Model description and comparison with a multi-layered model, *Agricultural and Forest Meteorology, 91* (1–2), 89–111, 1998.

74. Sellers, P.J. et al., Canopy reflectance, photosynthesis, and transpiration. III. A reanalysis using improved leaf models and a new canopy integration scheme, *Remote Sensing of Environment, 42* (3), 187–216, 1992.

75. Lichtenthaler, H.K. et al., Differences in pigment composition, photosynthetic rates and chlorophyll fluorescence images of sun and shade leaves of four tree species, *Plant Physiology and Biochemistry, 45* (8), 577–588, 2007.

76. De Pury, D.G.G. and Farquhar, G.D., Simple scaling of photosynthesis from leaves to canopies without the errors of big-leaf models, *Plant, Cell and Environment, 20* (5), 537–557, 1997.

77. Boardman, N.K., Comparative photosynthesis of sun and shade plants, *Annual Review of Plant Physiology, 28* (1), 355–377, 1977.

78. Thayer, S.S. and Björkman, O., Leaf xanthophyll content and composition in sun and shade determined by HPLC, *Photosynthesis Research, 23* (3), 331–343, 1990.

79. Zhang, H., Sharifi, M., and Nobel, P., Photosynthetic characteristics of sun versus shade plants of *Encelia farinosa* as affected by photosynthetic photon flux density, intercellular CO_2 concentration, leaf water potential, and leaf temperature, *Functional Plant Biology, 22* (5), 833–841, 1995.

80. Demmig-Adams, B., Survey of thermal energy dissipation and pigment composition in sun and shade leaves, *Plant and Cell Physiology, 39* (5), 474–482, 1998.

81. Sarijeva, G., Knapp, M., and Lichtenthaler, H.K., Differences in photosynthetic activity, chlorophyll and carotenoid levels, and in chlorophyll fluorescence parameters in green sun and shade leaves of Ginkgo and Fagus, *Journal of Plant Physiology, 164* (7), 950–955, 2007.

82. Spitters, C.J.T., Separating the diffuse and direct component of global radiation and its implications for modeling canopy photosynthesis. Part II. Calculation of canopy photosynthesis, *Agricultural and Forest Meteorology, 38* (1–3), 231–242, 1986.

83. Spitters, C.J.T., Toussaint, H.A.J.M., and Goudriaan, J., Separating the diffuse and direct component of global radiation and its implications for modeling canopy photosynthesis. Part I. Components of incoming radiation, *Agricultural and Forest Meteorology, 38* (1–3), 217–229, 1986.

84. White, H.P., Miller, J.R., and Chen, J.M., Four-scale linear model for anisotropic reflectance (FLAIR) for plant canopies. I. Model description and partial validation, *IEEE Transactions on Geoscience and Remote Sensing, 39* (5), 1072–1083, 2001.

85. Myneni, R.B. et al., Estimation of global leaf area index and absorbed par using radiative transfer models, *IEEE Transactions on Geoscience and Remote Sensing, 35* (6), 1380–1393, 1997.

86. Liang, S. et al., Estimation of incident photosynthetically active radiation from moderate resolution imaging spectrometer data, *Journal of Geophysical Research, 111* (D15), D15208, 2006.

87. Pinker, R.T. et al., Surface radiation budgets in support of the GEWEX Continental-Scale International Project (GCIP) and the GEWEX Americas Prediction Project (GAPP), including the North American Land Data Assimilation System (NLDAS) project, *Journal of Geophysical Research, 108* (D22), 8844, 2003.

88. Frouin, R. and Pinker, R.T., Estimating photosynthetically active radiation (PAR) at the earth's surface from satellite observations, *Remote Sensing of Environment, 51* (1), 98–107, 1995.

89. Hall, F.G. et al., Satellite remote sensing of surface energy balance: Success, failures, and unresolved issues in FIFE, *Journal of Geophysical Research, 97* (D17), 19061–19089, 1992.

90. Huemmrich, K.F. et al., Remote sensing of tundra gross ecosystem productivity and light use efficiency under varying temperature and moisture conditions, *Remote Sensing of Environment, 114* (3), 481–489, 2010.

91. Daughtry, C.S.T., Gallo, K.D., and Bauer, M.E., Spectral estimates of solar radiation intercepted by corn canopies, *AgRISTARS Technical Report sr-PZ-04236*, Purdue University, West Lafayette, IN, 1982.

92. Demetriades-Shah, T.H. et al., Comparison of ground- and satellite-based measurements of the fraction of photosynthetically active radiation intercepted by tallgrass prairie, *Journal of Geophysical Research, 97* (D17), 18947–18950, 1992.

93. Walter-Shea, E.A. et al., Biophysical properties affecting vegetative canopy reflectance and absorbed photosynthetically active radiation at the FIFE site, *Journal of Geophysical Research, 97* (D17), 18925–18934, 1992.

94. Goward, S.N. and Huemmrich, K.F., Vegetation canopy PAR absorptance and the normalized difference vegetation index: An assessment using the SAIL model, *Remote Sensing of Environment, 39* (2), 119–140, 1992.

95. Bannari, A. et al., A review of vegetation indices, *Remote Sensing Reviews, 13* (1), 95–120, 1995.

96. Dobrowski, S.Z. et al., Simple reflectance indices track heat and water stress-induced changes in steady-state chlorophyll fluorescence at the canopy scale, *Remote Sensing of Environment, 97* (3), 403–414, 2005.

97. HyspIRI Group, NASA 2010 HyspIRI Science Workshop Report, Pasadena, CA: Jet Propulsion Laboratory, National Aeronautics and Space Administration, 2010.

98. Grace, J. et al., Can we measure terrestrial photosynthesis from space directly, using spectral reflectance and fluorescence? *Global Change Biology, 13* (7), 1484–1497, 2007.

99. Running, S.W. et al., A continuous satellite-derived measure of global terrestrial primary production, *Bioscience, 54* (6), 547–560, 2004.

100. Turner, D.P. et al., Site-level evaluation of satellite-based global terrestrial gross primary production and net primary production monitoring, *Global Change Biology, 11* (4), 666–684, 2005.

101. Gamon, J.A. et al., Remote sensing of the xanthophyll cycle and chlorophyll fluorescence in sunflower leaves and canopies, *Oecologia, 85* (1), 1–7, 1990.

102. Gamon, J.A., Serrano, L., and Surfus, J.S., The photochemical reflectance index: An optical indicator of photosynthetic radiation use efficiency across species, functional types, and nutrient levels, *Oecologia, 112* (4), 492–501, 1997.

103. Gamon, J.A., Penuelas, J., and Field, C.B., A narrow-waveband spectral index that tracks diurnal changes in photosynthetic efficiency, *Remote Sensing of Environment, 41* (1), 35–44, 1992.

104. Peñuelas, J. et al., Reflectance indices associated with physiological changes in nitrogen- and water-limited sunflower leaves, *Remote Sensing of Environment, 48*, 135–146, 1994.

105. Suárez, L. et al., Assessing canopy PRI for water stress detection with diurnal airborne imagery, *Remote Sensing of Environment, 112* (2), 560–575, 2008.
106. Drolet, G.G. et al., A MODIS-derived photochemical reflectance index to detect inter-annual variations in the photosynthetic light-use efficiency of a boreal deciduous forest, *Remote Sensing of Environment, 98* (2–3), 212–224, 2005.
107. Drolet, G.G. et al., Regional mapping of gross light-use efficiency using MODIS spectral indices, *Remote Sensing of Environment, 112* (6), 3064–3078, 2008.
108. Rahman, A.F. et al., Potential of MODIS ocean bands for estimating CO_2 flux from terrestrial vegetation: A novel approach, *Geophysical Research Letters, 31* (10), L10503, 2004.
109. Inoue, Y. et al., Normalized difference spectral indices for estimating photosynthetic efficiency and capacity at a canopy scale derived from hyperspectral and CO_2 flux measurements in rice, *Remote Sensing of Environment, 112* (1), 156–172, 2008.
110. Peñuelas, J. et al., Photochemical reflectance index and leaf photosynthetic radiation-use-efficiency assessment in Mediterranean trees, *International Journal of Remote Sensing, 18* (13), 2863–2868, 1997.
111. Filella, I. et al., Relationship between photosynthetic radiation-use efficiency of barley canopies and the photochemical reflectance index (PRI), *Physiologia Plantarum, 96* (2), 211–216, 1996.
112. Meroni, M. et al., Leaf level early assessment of ozone injuries by passive fluorescence and photochemical reflectance index, *International Journal of Remote Sensing, 29* (17), 5409–5422, 2008.
113. Nichol, C.J. et al., Remote sensing of photosynthetic-light-use efficiency of boreal forest, *Agricultural and Forest Meteorology, 101* (2–3), 131–142, 2000.
114. Hall, F.G. et al., Multi-angle remote sensing of forest light use efficiency by observing PRI variation with canopy shadow fraction, *Remote Sensing of Environment, 112* (7), 3201–3211, 2008.
115. Hilker, T. et al., Separating physiologically and directionally induced changes in PRI using BRDF models, *Remote Sensing of Environment, 112* (6), 2777–2788, 2008.
116. Nichol, C.J. et al., Remote sensing of photosynthetic-light-use efficiency of a Siberian boreal forest, *Tellus B, 54* (5), 677–687, 2002.
117. Trotter, G.M., Whitehead, D., and Pinkney, E.J., The photochemical reflectance index as a measure of photosynthetic light use efficiency for plants with varying foliar nitrogen contents, *International Journal of Remote Sensing, 23* (6), 1207–1212, 2002.
118. Inoue, Y. and Peñuelas, J., Relationship between light use efficiency and photochemical reflectance index in soybean leaves as affected by soil water content, *International Journal of Remote Sensing, 27* (22), 5109–5114, 2006.
119. Huemmrich, K.F. et al., Using reflectance measurements to determine light use efficiency in corn. In *IEEE International Geoscience and Remote Sensing Symposium (IGARSS)*, Boston, MA, 2008.
120. Middleton, E.M. et al., Diurnal and seasonal dynamics of canopy-level solar-induced chlorophyll fluorescence and spectral reflectance indices in a cornfield. In *6th EARSeL SIG IS Workshop, European Association of Remote Sensing Laboratories Workshop*, Tel Aviv, Israel, 2009.
121. Strachan, I.B. et al., Use of hyperspectral remote sensing to estimate the gross photosynthesis of agricultural fields, *Canadian Journal of Remote Sensing, 34* (3), 333–341, 2008.
122. Guo, J.M. and Trotter, C.M., Estimating photosynthetic light-use efficiency using the photochemical reflectance index: The effects of short-term exposure to elevated CO_2 and low temperature, *International Journal of Remote Sensing, 27* (20), 4677–4684, 2006.
123. Hilker, T. et al., Instrumentation and approach for unattended year round tower based measurements of spectral reflectance, *Computers and Electronics in Agriculture, 56* (1), 72–84, 2007.
124. Gamon, J.A. et al., Spectral Network (SpecNet)—What is it and why do we need it? *Remote Sensing of Environment, 103* (3), 227–235, 2006.
125. Gamon, J.A. et al., A mobile tram system for systematic sampling of ecosystem optical properties, *Remote Sensing of Environment, 103* (3), 246–254, 2006.
126. Leuning, R. et al., A multi-angle spectrometer for automatic measurement of plant canopy reflectance spectra, *Remote Sensing of Environment, 103* (3), 236–245, 2006.
127. Corp, L.A. et al., FUSION: A fully ultraportable system for imaging objects in nature. In *IEEE International Geoscience and Remote Sensing Symposium (IGARSS)*, Honolulu, HI, 2010.
128. Garrity, S.R., Vierling, L.A., and Bickford, K., A simple filtered photodiode instrument for continuous measurement of narrowband NDVI and PRI over vegetated canopies, *Agricultural and Forest Meteorology, 150* (3), 489–496, 2010.
129. Rahman, A.F. et al., Modeling spatially distributed ecosystem flux of boreal forest using hyperspectral indices from AVIRIS imagery, *Journal of Geophysical Research, 106* (D24), 33579–33591, 2001.

130. Fuentes, D.A. et al., Mapping carbon and water vapor fluxes in a chaparral ecosystem using vegetation indices derived from AVIRIS, *Remote Sensing of Environment, 103* (3), 312–323, 2006.

131. Justice, C.O. et al., The moderate resolution imaging spectroradiometer (MODIS): Land remote sensing for global change research, *IEEE Transactions on Geoscience and Remote Sensing, 36* (4), 1228–1249, 1998.

132. Huemmrich, K.F. et al., Remote sensing of light use efficiency. In *Proceedings of the 30th Canadian Symposium on Remote Sensing*, Lethbridge, Alberta, Canada, 2009.

133. Goerner, A., Reichstein, M., and Rambal, S., Tracking seasonal drought effects on ecosystem light use efficiency with satellite-based PRI in a Mediterranean forest, *Remote Sensing of Environment, 113* (5), 1101–1111, 2009.

134. Garbulsky, M.F. et al., Remote estimation of carbon dioxide uptake by a Mediterranean forest, *Global Change Biology, 14* (12), 2860–2867, 2008.

135. Hilker, T. et al., An assessment of photosynthetic light use efficiency from space: Modeling the atmospheric and directional impacts on PRI reflectance, *Remote Sensing of Environment, 113* (11), 2463–2475, 2009.

136. Vermote, E.F. et al., Second simulation of the satellite signal in the solar spectrum, 6S: An overview, *IEEE Transactions on Geoscience and Remote Sensing, 35* (3), 675–686, 1997.

137. Lyapustin, A. and Wang, Y., The time series technique for aerosol retrievals over land from MODIS. In A.A. Kokhanovsky and G. Leeuw (Eds.), *Satellite Aerosol Remote Sensing over Land* (pp. 69–99). Berlin, Germany: Springer, 2009.

138. Asner, G.P. et al., Drought stress and carbon uptake in an Amazon forest measured with spaceborne imaging spectroscopy, *Proceedings of the National Academy of Sciences of the United States of America, 101* (16), 6039–6044, 2004.

139. Asner, G. et al., Vegetation–climate interactions among native and invasive species in Hawaiian rainforest, *Ecosystems, 9* (7), 1106–1117, 2006.

140. Asner, G.P., Carlson, K.M., and Martin, R.E., Substrate age and precipitation effects on Hawaiian forest canopies from spaceborne imaging spectroscopy, *Remote Sensing of Environment, 98* (4), 457–467, 2005.

141. Verrelst, J. et al., Angular sensitivity analysis of vegetation indices derived from CHRIS/PROBA data, *Remote Sensing of Environment, 112* (5), 2341–2353, 2008.

142. Sims, D.A. et al., Parallel adjustments in vegetation greenness and ecosystem CO_2 exchange in response to drought in a Southern California chaparral ecosystem, *Remote Sensing of Environment, 103* (3), 289–303, 2006.

143. Hilker, T. et al., A modeling approach for upscaling gross ecosystem production to the landscape scale using remote sensing data, *Journal of Geophysical Research— Biogeosciences, 113*, G03006, 2008.

144. Barton, C.V.M. and North, P.R.J., Remote sensing of canopy light use efficiency using the photochemical reflectance index: Model and sensitivity analysis, *Remote Sensing of Environment, 78* (3), 264–273, 2001.

145. Sims, D.A. and Gamon, J.A., Relationships between leaf pigment content and spectral reflectance across a wide range of species, leaf structures and developmental stages, *Remote Sensing of Environment, 81* (2–3), 337–354, 2002.

146. Stylinski, C.D., Gamon, J.A., and Oechel, W.C., Seasonal patterns of reflectance indices, carotenoid pigments and photosynthesis of evergreen chaparral species, *Oecologia, 131* (3), 366–374, 2002.

147. Guanter, L. et al., Estimation of solar-induced vegetation fluorescence from space measurements, *Geophysical Research Letters, 34* (8), L08401, 2007.

148. Joiner, J. et al., First observations of global and seasonal terrestrial chlorophyll fluorescence from space, *Biogeosciences Discuss*, 2010 (Submitted).

13 Spectral and Spatial Methods of Hyperspectral Image Analysis for Estimation of Biophysical and Biochemical Properties of Agricultural Crops

Victor Alchanatis and Yafit Cohen

CONTENTS

13.1 INTRODUCTION

The ultimate goal of hyperspectral (HS) sensing in this area is to delineate and characterize homogeneous management zones for optimal agricultural management like fertilization, irrigation, or other agro-technical operations. This work concentrates on three characteristics of HS images: First, their unique *spectral* properties, namely, the narrowband widths and the plethora of the bands, as opposed to wider and limited number of bands in other broadband spectral sensing systems; second, the *spatial* attribute of HS images, as opposed to point measurements of other spectral systems; and third, the state-of-the art algorithms for HS image processing that show the added value of spatial information when combined with spectral information for mapping plant biophysical and biochemical properties of agricultural crops (BB-PACs—phonetically pronounced bee-bee-pax).

Modern agricultural crop production relies on close monitoring of the crop status. This enables efficient management of available resources for profitable and environmentally friendly agricultural practice. Broadly used monitoring tools are mainly based on point sampling of biophysical and biochemical properties of the crop. Numerous crop properties have been studied over the years and act as indicators of the crop condition. Local and global growth protocols have been developed based on these measured biophysical and biochemical properties. For example, irrigation management of cotton is widely based on the height measurement of the plants in selected points, a biophysical property that can be easily measured by simple means, but it is labor intensive and based on selected sampled spots. Another example is fertilization management in potatoes, where nitrate content in the petiole is used as an indication for the fertilizer need. Table 13.1 presents a list of some important biophysical and biochemical properties that are used in growing protocols of agricultural crops.

These examples illustrate the high importance of monitoring biophysical and biochemical properties of agricultural crops. The desire to upgrade from point measurements to maps with high density of data has brought remote sensing to the front of the technologies that can provide such a mission.

A number of other crop health conditions related to crop protection, like pests damage, plant diseases, and weeds infestation are also expressed through changes in the biophysical and biochemical properties of the crop. Several reports in the literature show the contribution of remote sensing techniques on the detection of plant diseases [29,31], pests damage [38,47], and weeds infestation [30,40,44]. All studies report that HS remote sensing can detect the phenomena assuming that they are the factor that causes the anomalies in the field. This chapter will focus on sensing plant

---

TABLE 13.1

Biophysical and Biochemical Properties of Crops That Serve as Indicators for Agricultural Crop Management

	Property (BB-PAC)	Example Crops	Agro-Technical Management Parameter
Biophysical	Biomass (kg m⁻¹)	Wheat, rice, corn	Fertilization
	LAI/Crop cover (No units/%)	Wheat, soybean, corn, cotton	Fertilization
	Crop height (m)	Cotton, wheat	Irrigation, application of growth regulators
	Canopy volume (m³)	Orchards, wheat	Irrigation, fertilization
	Yield (kg m⁻¹)	Wheat, corn, cotton	—
	Stomata conductance (mmol s⁻¹)	Vineyards	Irrigation
	Leaf/stem water potential (MPa)	Cotton, orchards, vineyards	Irrigation
	Flowering intensity (relative units)	Orchards	Growth regulators, mechanical thinning
Biochemical	Nitrogen content (%N)	Corn, wheat, potatoes	Fertilization
	Chlorophyll content (µg cm⁻²)	Corn, wheat, cotton	Fertilization
	Salinity (mg L⁻¹)	Cotton	Water quality management, not used in practice
	Leaf water content (%)	Wheat, potato	Irrigation
	Leaf macro-elements like phosphorus (P) and potassium (K) (mg kg⁻¹)	Olives	Fertilization, not used in practice

properties related to manageable agricultural resources like irrigation and fertilization and will not discuss in extent the issues with sensing plant properties related to plant diseases, pests damage, and weeds infestation.

First, we review the most prominent methods of HS data processing to model and enhance quantification of BB-PACs. Description of the methods is followed by their application. In the second part of the chapter, we present approaches to integrate HS data with hyperspatial data which exploit the spatial attribute of the HS images.

13.2 SPECTRAL METHODS

A number of methods are commonly used for analyzing spectral data to extract BB-PACs. The source of the data may be a point spectral sensor, as well as a HS imaging camera. In the latter case, each pixel is regarded as single point measurement. In both cases, there are hundreds of narrow spectral bands, with bandwidth around 1–10 nm. There are three main methods for spectral analysis: (a) band selection, (b) use of spectral indices, and (c) linear and nonlinear multivariate statistics and models.

Individual bands selection and spectral indices are mainly developed in the field of remote sensing, whereas multivariate statistical methods are mainly developed in the chemometrics field.

13.2.1 SPECTRAL BAND SELECTION

Spectral bands selection comprises a methodology for choosing HS bands that provide sufficient, but not redundant information to classification or prediction algorithms, using a practical amount of computational resources. There are two conceptually different approaches of band selection: unsupervised and supervised. Unsupervised methods order the spectral bands without training based on generic information evaluation approaches. They are usually very fast and computationally efficient, and can provide information for clustering an image to classes of common spectral signatures. Supervised methods require training data in order to build an internal predictive model. They are usually more computationally intensive than unsupervised methods and can provide quantitative models for predicting BB-PACs [2].

Unsupervised methods for spectral band selection include the use of methods like principal components analysis [20] and band–band correlation [41]. Supervised methods include the use of methods like correlation of the spectral bands with the BB-PAC studied [43] and stepwise discriminant analysis [20] to extract the number of independent wavelengths that can explain the variability of the measured BB-PAC. Both methods result in an optimum number of spectral bands that contain unique information.

13.2.2 SPECTRAL INDICES

Spectral indices assume that the combined interaction between a small number of wavelengths is enough to describe the biochemical or biophysical interaction between light and matter. The simplest form of index is a simple ratio (SR), where the ratio between two wavelengths is indicative for a BB-PAC under investigation. The typical form of an SR index is

$$I = \frac{R_{\lambda i}}{R_{\lambda j}}$$

where

I is the index value

$R_{\lambda i}$ and $R_{\lambda j}$ are the reflectance values in wavelength λi and λj, respectively

Enhanced SRs are the normalized difference vegetation indices (NDVI), which also exploit the difference between two distinct wavelengths, but normalize it using the following equation:

$$I = \frac{R_{\lambda i} - R_{\lambda j}}{R_{\lambda i} + R_{\lambda j}}$$

Another category of spectral indices comprise integrated indices (or derivative indices), where more than two wavelengths are combined together to produce a value that is correlated with BB-PACs. Integrated indices are usually specific to a certain BB-PAC and sometimes to the crop that they were developed for. An extensive compilation of all three index categories can be found in Li et al. [27].

13.2.3 MULTIVARIATE STATISTICS

Spectral indices that are based on a small number of bands are indicators of irregular conditions and provide evidence that an anomaly is present. Despite their widespread use, it has not been possible to design an index that is sensitive only to a desired variable and totally insensitive to all other vegetation parameters [14]. Thus, if the factor or the cause of the anomaly in the field is known, then some of the spectral indices may be able to quantify the level or the severity of the anomaly. The advantage of the whole spectral signature of the crop is that it contains information that can be used to identify the cause for the spectral changes in the light reflected from the canopy as well as to quantify it.

Multivariate statistics assumes that there is an underlying relation between the spectral signature of the crop and its biochemical or biophysical properties. Statistical tools extract this underlying relationship as a model, which is often a linear model. The large number of independent variables (wavelengths) together with the high colinearity between the variables (spectral bands) do not permit the use of common multivariate methods, like multivariate regression based on least squares (MLR), before prior selection of the most indicative independent wavelengths. Therefore, methods that overcome these constraints are used and taken from the field of chemometrics. Two main methods are common; principal components regression (PCR), which has a core of unsupervised data extraction, and partial least squares regression (PLSR), which is a supervised method. Both methods produce a linear model.

PLS regression [46] is related to both PCR and MLR, and can be thought of as occupying a middle ground between them. PCR finds factors that capture the greatest amount of variance in the predictor variables (spectra). MLR seeks to find a single factor that best correlates predictor variables with predicted variables (BB-PACs). PLS attempts to find factors that both capture variance and achieve correlation while avoiding the colinearity. We commonly say that PLS attempts to maximize covariance and this is the explicit objective of the algorithm for PLS. PLS attempts to find factors (called Latent Variables) that maximize the amount of variation explained in the spectra that is relevant for predicting the BB-PAC. This is in contrast to PCR, where the factors (called Principal Components) are selected solely based on the amount of variation that they explain in spectra. In mathematical terms, the difference between them is the objective function that is used to optimize during the calculation of the regression coefficients. Unsupervised methods tend to minimize only the interclass (between classes) variance based on the spectral curves of the samples. Supervised methods either minimize the variance of the intraclass (within the class) variance or a combination of the interclass and intraclass variance.

Recently, nonlinear methods have also been developed for processing of HS data. Spectral Angle Mapper (SAM) [34,38], a novel method for spectral similarity measure, comprises a nonlinear transformation and removes high colinearity among bands [34].

Wavelets are a group of functions that vary in complexity and mathematical properties that are used to dissect data into different frequency components and then to characterize each component with a resolution appropriate to its scale. Wavelet analysis of a reflectance spectrum is performed by scaling and shifting the wavelet function to produce wavelet coefficients that are assigned to different frequency components. By selecting appropriate wavelet coefficients, a spectral model can be established between the coefficients and biochemical concentrations. Hence, wavelet analysis has the potential to capture much more of the information contained within high-resolution spectra than previous approaches and offers the prospect of developing robust, generic methods for pigment determinations [4].

Wavelet methods have also been developed for detection of vegetation stress using HS data. Compared to traditional vegetation stress indexes, the proposed approach uses the complete reflectance spectrum and its wavelet representation. The detection strategy is formulated as a classification problem, showing the superior performance of the proposed strategy and demonstrating its generic nature [17].

Most of these methods are confined to classification and detection problems, and are not used for quantitative estimation crop characteristics from HS data.

13.3   SENSING OF AGRICULTURAL CROP PROPERTIES

Spectral characteristics of green vegetation have very prominent features: two valleys in the visible portion of the spectrum are determined by the pigments contained in the plant. Chlorophyll absorbs strongly in the blue (450 nm) and red (680 nm) regions, also known as the chlorophyll absorption bands. Chlorophyll is the primary photosynthetic pigment in green plants. This is the reason for the human eye perceiving healthy vegetation as green. When the plant is subjected to stress that hinders normal growth and chlorophyll production, there is less absorption in the red and blue regions and the amount of reflection in the red waveband increases. In some cases where stress is severe, the stress can be sensed by the human eyes.

The spectral reflectance signature has a dramatic increase in the reflection for healthy vegetation at around 700 nm. In the near infrared (NIR) between 700 and 1300 nm, a plant leaf will typically reflect between 40% and 60%, the rest is transmitted, with only about 5% being adsorbed. For comparison, the reflectance in the green range reaches 15%–20%.

This high reflectance in the NIR is due to scattering of the light in the intercellular volume of the leaves mesophyll. Structural variability in leaves in this range allows one to differentiate between species, even though they might look the same in the visible region. Beyond 1300 nm, the incident energy upon the vegetation is largely absorbed or reflected with very little transmittance of energy. Three strong water absorption bands are noted at around 1400, 1900, and 2700 nm and can be used for plant water content estimation.

13.3.1   Prediction of Biophysical Properties

13.3.1.1   LAI

Green leaf area index (LAI) is a key variable used by crop physiologists and modelers for estimating foliage cover, as well as forecasting crop growth and yield. The exposed area of living leaves plays a key role in various biophysical processes such as plant transpiration and CO_2 exchange. Because LAI is functionally linked to the canopy spectral reflectance, its retrieval from remote sensing data has prompted many investigations and studies over the years [1,4,13,15,18,25]. Most of these studies have relied on empirical relationships between the ground-measured LAI and observed spectral responses.

The most common index to estimate LAI and its counterparts of the crop cover and biomass is the NDVI [45], which expresses the normalized ratio between the reflected energy in the red chlorophyll absorption region and the reflected energy in the NIR mesophyll scattering region. Yet, it is well documented that the NDVI approaches saturation asymptotically under conditions of moderate-to-high aboveground biomass [13] and therefore it may be a good predictor only for low to medium LAIs (0–4).

Linear regression analysis of single bands and two-band combinations of pseudo NDVIs have shown the importance of the red edge spectral region (700–740 nm), the short wave infrared (SWIR) spectral region, and the advantage of narrowbands over traditional broadband in LAI prediction [10,18,25,42,43]. A major problem in the use of indices to estimate LAI arises from the fact that canopy reflectance, in the visible and NIR, is strongly dependent also on chlorophyll of the canopy (e.g., [48]). Moreover, both variables have similar effects on canopy reflectance particularly in the spectral region from the green (550 nm) to the red edge (740 nm). To uncouple the LAI effect, Haboudane et al. [15] developed two indices (MTVI2 and MCARI2). Prediction algorithms based on these two indices were applied for compact airborne spectrographic imager (CASI) HS image over fields of soybean, corn, and wheat and showed excellent agreements between modeled and measured LAI.

Other studies exploit wider range of the spectra or even the whole spectra to improve LAI prediction. Delegido et al. [11] have shown that the spectra between 500 and 750 nm can be adjusted with good precision to third degree polynomials and that there was strong correlation between one of its coefficients and LAI values that ranged from 0 to 7. This is a significant improvement over other methods since it covers the whole range of LAI (0–7) and is not limited to low (0–2) and medium (2–4) LAI. Multivariate and PLS regression models based on selected narrowbands or the whole spectra, respectively, have shown to be comparable or better LAI predictors than narrowband normalized difference indices (NDIs) [18,25]. While narrowband NDVI had strong correlation in LAI range of 0–3 and explained 80% of the LAI variability, the multivariate regression of wider range had a very high correlation in LAI range of 0–6 and explained 90% of the variability [25].

13.3.1.2 Biomass

Forecasting and estimating crop production using remote sensing have great consequence on food provisions management and is fundamental to applications of precision agriculture. In-season biomass estimation from remote sensing for yield forecasting and variable rate applications have been a challenge for various studies.

Biomass and LAI have similar effects on spectral characteristics and studies have shown similarities as well as some differences in the estimation of both crop properties using spectral measurements and HS images. Correlation coefficients between spectral reflectance in discrete narrowbands and LAI and biomass in various crops presented similar shape [42,43]. Spectral bands that are best suited for characterizing LAI and biomass were determined by Thenkabail et al. [42,43] and no significant differences were found. Yet, while no improvement was achieved by PLSR models for LAI estimation, PLS models significantly improved the prediction of biomass by lowering the root mean square error (RMSE) by 22%, compared to the best narrowband indices [18]. Correspondingly, PLS models using the spectral range of 350–2500 nm were found to better predict wheat dry biomass compared to common vegetation indices (VIs): R^2 of 0.80 and 0.50, respectively [36].

13.3.1.3 Water Status

Crop water status is a key biophysical property that is used to manage irrigation, as well as to evaluate crop health. In most cases, it is directly associated with water availability in the soil, and when this is not the case (i.e., water availability is not the limiting factor), water status becomes an indicator of crop health. For example, when salinity is a limiting factor of water uptake, crop water status becomes an indicator of salinity stress. Similarly, plant diseases that damage water flow in the plant affect the crop water status, which becomes an indicator of the disease presence or its severity.

Crop water status can be quantified at the canopy and leaf level by measuring leaf or stem water potential, or leaf water content. Leaf and stem water potential are important biophysical parameters that indicate the crop condition. The reports in the literature show limited ability to remotely detect them using HS sensing methods in the VIS/NIR region since they express the physical status of water potential in the plant tissue [39]. Nevertheless, they affect the status of the leaves' stomata, which control the evapotranspiration process and affect leaf temperature. Sensing methods in the thermal IR region have been reported as capable of predicting leaf and stem water potential by measuring leaf temperature [33], but this is beyond the scope of this section.

Leaf water content, on the other hand, can also serve as an indicator of crop water status. The spectral characteristics of water can then be used to quantify the water content in the leaves. For wavelengths sensitive to water absorption (760, 970, 1450, 1940, and 2950 nm), leaf reflectance decreases as water content increases. Attempts to use indices as algebraic expressions of reflectance values for specific wavelengths do not yield significant relationships at canopy level [39]. Nevertheless, when methods that use the whole spectrum were analyzed at the canopy level as well, water content could be predicted from remote sensed data. Namely, PLS models based on the first derivative of the spectrum in the range 350–2500 nm predicted water content with R^2 of 0.87, while spectral indices with exponential model achieved R^2 0.2 [36]. In addition, when the water absorbance band at 970 nm was considered, leaf water content was successfully predicted based on the slope (first derivative) of the spectral curve at 1015–1050 nm ($R^2 = 0.97$) [8]. Other methods that consider the entire wavelength spectrum between 700 and 1300 nm showed that nonlinear models based on functional radial basis functions produce considerably better results than functional linear regression models (relative error of 4% and 17%, respectively) [35]. This outcome might indicate the existence of a complex dependency relationship between reflectance and leaf water content. It might also explain the poor results obtained by some methods based on indices in other studies.

13.3.2 PREDICTION OF BIOCHEMICAL PROPERTIES

13.3.2.1 Chlorophyll Content

The most commonly used biochemical property of crops is chlorophyll content. It reflects the general condition of the crop, since chlorophyll is the producing factory of the crop. Changes in chlorophyll may indicate limited availability of important elements, among a wide possibility of options or other biotic or abiotic stresses. Chlorophyll deficiency can be detected by remote sensing, using specific spectral indices. Nevertheless, detection of chlorophyll deficiency is not an indicator of the cause that induced the deficiency.

Chlorophyll-specific spectral indices can be divided into two categories: (a) indices based on chlorophyll absorption in the blue (around 450 nm) and red (around 680 nm) spectral region and (b) indices that are based on the displacement of the red edge inflection point (700–740 nm). Several reports in the literature describe the use of simple and combined spectral indices for leaf chlorophyll estimation [3,17]. Among the set of indices tested, index combinations like Modified Chlorophyll Absorption Ratio Index/Optimized Soil-Adjusted Vegetation Index (OSAVI), Triangular Chlorophyll Index/OSAVI, Moderate Resolution Imaging Spectrometer Terrestrial Chlorophyll Index/Improved Soil-Adjusted Vegetation Index (MSAVI), and Red Edge Model/MSAVI seem to be relatively consistent and more stable as estimators of crop chlorophyll content [17].

Chlorophyll content was also estimated using wavelet decomposition on the HS data. In the context of remote sensing of foliar chlorophyll, wavelet analysis has the potential to capture much more of the information contained with reflectance spectra than previous analytical approaches, which uses a small number of optimal wavebands. This approach was found to be more reliable than simple linear regression analysis when linking chlorophyll to the reflectance measured. This was observed both for leaf level measurements as well as top of canopy measurements (peach trees) [22]. The wavelet-based approach outperformed models based on untransformed spectra (like stepwise derivative) and a range of existing spectral indices. While wavelet-based models yielded 1:1 relationships between measured and predicted chlorophyll content in the range of 0–60 μg cm^{-2} (with R^2 of 0.88), other methods (including indices and first derivative) saturated above 30 μg cm^{-2} [7]. These findings indicate that wavelet analysis warrants further investigation as a method for extracting meaningful quantitative information from HS data.

Refinements in the technique for quantifying chlorophyll could explore the use of new wavelet functions or combinations of functions, multiple scales of wavelet coefficients, alternative methods for calculating derivatives prior to wavelet decomposition, and different approaches to the selection of wavelet coefficients during model calibration. The value of wavelet analysis of spectra for quantifying leaf chlorophyll in principle has been demonstrated; it is now important that this is tested in practice and that the generality of the technique for HS remote sensing of vegetation is explored, particularly at the canopy and landscape scales [7].

The approach providing the highest predictive accuracy was the multiple regression models based on wavelet coefficient energy feature vectors. This was closely followed by multiple regression models derived from the energy feature vectors of the nth largest wavelet coefficients, which in turn was closely followed by stepwise regression models based on wavelet coefficients. The predictive accuracy of the stepwise regression models derived from narrowband reflectance was substantially lower than that of the wavelet-based approaches and the SR and normalized difference ratio spectral indices had the poorest performance by some margin [6].

13.3.2.2 Nitrogen Content

Nitrogen deficiency is one of the most important conditions to be detected, since it affects directly the productivity of the crop. An additional reason that makes the detection of nitrogen deficiency very important is the fact that nitrogen leaches under the root zone when irrigation or water management is not appropriate, creating conditions that are suboptimal for the crop growth.

Nitrogen indices can be divided to indices that are based on wavelengths in the visible and the NIR region, and indices that include specific nitrogen absorption wavelengths in the SWIR spectral region. The additional value of using SWIR-based indices has been shown in studies on wheat where a firm advantage was revealed for the proposed SWIR-based indices in their ability and sensitivity to predict nitrogen content in potato leaves [19].

Many HS VIs have been developed to estimate crop nitrogen status at leaf and canopy levels. They have been evaluated for different growth stages and years using data from both nitrogen experiments and farmers' fields. Furthermore, to identify alternative promising HS VIs, evaluation of all possible two-band combinations of SRs and NDIs has been performed. The results indicated that best performing published and newly identified VIs included SRs in the red edge region and in the blue region [20,27]. Red edge and NIR bands were more effective for nitrogen estimation at early growing stage, but visible bands, especially ultraviolet, violet, and blue bands, were more sensitive at later growing stage.

Across site years, cultivars, and growth stages, the combination of wavelengths in the blue range (370–400 nm) as either SR or an NDI performed most consistently in both experimental and field data for wheat. Therefore, growth stage had a significant influence on the performance of different VIs and on the selection of sensitive wavelengths for leaf nitrogen estimation [27].

The observed interchangeability of wavelengths and indices along growth stages and cultivars is addressed by multivariate methods, which make use of the whole spectrum and not only selected wavelengths. For instance, multivariate methods were used to estimate leaf nitrogen content based on narrowband spectral data in potatoes. PLSR analysis has resulted in a stronger correlation between predicted and measured leaf nitrogen content ($R^2 = 0.95$) than the nitrogen-specific index transformed chlorophyll absorption reflectance index (TCARI) ($R^2 = 0.82$), even though in both models, data from narrowbands was used. Moreover, the improved PLS correlation was achieved with a single model for both the vegetative and the tuber-bulking periods while the TCARI yielded a different model for each period [9]. In the same study, when the number of wavelengths was reduced from 400 to 11, and the bands' bandwidth was broadened from 1.3 to 20–40 nm, in order to simulate future satellite data, the accuracy of the spectral model was decreased ($R^2 = 0.78$). Similar results were obtained for nitrogen prediction in winter wheat: models based on NDVI had an exponential characteristic, which implies saturation for high nitrogen values, and low coefficient of determination ($R^2 = 0.15$). When the derivative of the spectrum between 350 and 2500 nm was used in conjunction with PLSR models, the coefficient of determination was significantly better ($R^2 = 0.82$).

13.4 SPATIAL METHODS

Most available HS data processing techniques focused on analyzing the spectral data without incorporating information on the spatially adjacent data. In other words, HS data are usually not treated as images, but as unordered listings of spectral measurements with no particular spatial arrangement [37]. The importance of analyzing both spectral and spatial patterns has been identified as a desired goal by many scientists devoted to multidimensional data analysis. This type of processing has been approached from various points of view representing different levels of combination between spectral and spatial information.

Nearly all of the methods combining spectral and spatial information were developed for land-cover classification. Several methods are presented and illustrated in [37]. There is a basic difference between land-use/cover classification and estimation of crop properties. Land-use/cover types are discrete elements with relatively well-defined borders. Moreover, most of them have relatively distinct spectral signature. In comparison, biophysical and biochemical crop properties are continuous variables, with smooth differences in spectral signature and with amorphous shapes. Most of the methods that were developed for classification are less suitable for partitioning a field into homogenous zones with similar levels of continuous properties.

Methods combining spectral and spatial information were not reported in studies designated to estimate levels of crop biophysical and biochemical properties and to divide them into homogenous zones. Indeed, spectral models manipulated over HS images were used to create maps of biophysical and biochemical crop properties [15,16,32] or further to partition fields into management zones on spectral-based properties [28]. For the creation of these maps smoothing operations have been applied for reducing the speckle effect. Yet, all of the maps were created merely based on a pixel-by-pixel spectral data without incorporating information on the neighboring pixels. In this section we shortly describe potential approaches and illustrate methods for combining spectral and spatial information for segmentation of HS images based on spectral-based crop properties.

13.4.1 HYPERSPECTRAL DATA SET

For illustrating the potential that lies in some of the described methods, we use a sample from an aerial HS image taken over a potato plot under different nitrogen treatments. The HS image was acquired on May 25, 2007, using a push-broom AISA system in the range of 400–1000 nm, with 420 bands with spectral resolution of 1.3 nm. The image was acquired from a 500 m height and has 1 m spatial resolution. The nitrogen treatments are listed in Table 13.2. More details on the multiyear study can be found in [9]. Figure 13.1 is a red, green, and blue (RGB) (670, 550, and 420 nm) image of the experimental plot derived from the narrowband HS image. Figure 13.2 presents a combination of the bands 750 nm (IR), 670 nm (red), and 550 nm (green) of the experimental plot overlaid by the nitrogen treatment borders.

TABLE 13.2
Nitrogen Treatments Applied in the Potato Field in Spring 2007

Nitrogen Treatment	N Rate (kg ha⁻¹)	Percentage N Rate Relative to Commercial Rate	Application Type
T100%	400	100	Commercial (urea); fertigation
T75%	300	75	Multigro® 43-0-0 SRN; base
T50%	200	50	Multigro 43-0-0 SRN; base
T25%	100	25	Multigro 43-0-0 SRN; base
T0%	0	0	Base

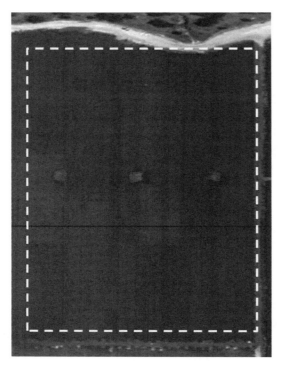

FIGURE 13.1 **(See color insert.)** RGB (670, 550, and 420 nm) image of the experimental plot.

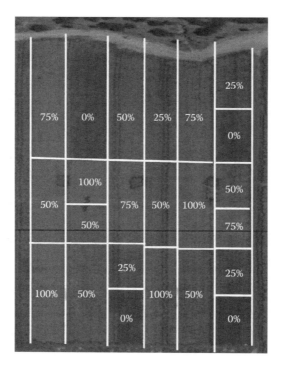

FIGURE 13.2 **(See color insert.)** Combination of IR (750 nm), Red (670 nm), and Green (550 nm) bands of the experimental plot overlaid by the borders of the N treatments.

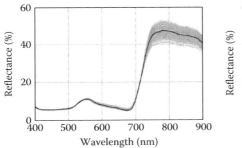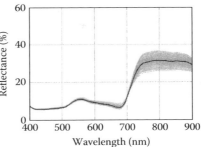

FIGURE 13.3 Individual spectra for ROIs taken from subplots of T75% (left) and T0% (right) along with the mean spectra (black thick line).

13.4.2 SPATIAL INFORMATION AS A PREPROCESSING TOOL

Individual spectra of the same object or property taken from neighboring pixels in the HS image present relatively high variability. Figure 13.3 shows individual spectra for region of interest (ROIs) taken from subplots T75% and T0% along with the mean spectra (black-thick line). These HS data are rather noisy in comparison to spectra collected using a spectrometer [24]. This is primarily a result of how HS data are collected. In most spectrometers, a single measurement is actually the mean of several independent spectra that were collected over a small area, which greatly reduces the noise in the spectra.

Reduction of the noise is essential for calibrating a spectral-based model. However, in an aerial HS image, each pixel of a hypercube is a single spectra of a relatively wide area. To reduce the spectral noise in calibrating models Lawrence et al. [24] suggest using a spatially averaged ROI spectra and then apply the model on a pixel-by-pixel basis. Lawrence et al. [24] used manual selection of ROI spectra from a close range HS image to calibrate a PLS spectral model for contaminant detection on poultry carcasses. Manual selection of homogenous ROI from an aerial HS image of a field is problematic and might suffer from subjectivity. In the next paragraph, we describe spatial methods for classification that can be used also for automatic or semiautomatic selection of homogenous ROI for model calibration.

13.4.3 SPATIAL INFORMATION TO IMPROVE SPECTRAL CLASSIFICATION

Spatial context was suggested as a second step for the refinement of results obtained by spectral-based techniques. This approach consists of three parts: (1) a pixel-by-pixel spectral classification; (2) definition of a pixel neighborhood (surrounding each pixel); and (3) performance of a local operation; so if there is a strong evidence that individual spectra of pixels in a neighborhood are spectrally homogenous they are enforced to the same cluster. This approach was developed for multispectral images and enhanced by Jimenes et al. [21] to airborne HS sensors. The developed classifier is an unsupervised modification of the supervised extraction and classification of homogenous objects (ECHO). Based on the dataset, the developed algorithm, called UnECHO, automatically estimates the required threshold of homogeneity level of the entire neighborhood without input from the human analyst [21]. When applied to urban and rural areas, the UnECHO successfully uncovered spatial structures and significantly improved spectral classifications (C-means or ML).

Another example for this approach is the Markov random field (MRF) in which spatial characterization is performed by modeling the spatial neighborhood of a pixel as a spatially distributed random process. The MRF attempts to make regularization via the minimization of an energy function using known land covers and their prior probabilities. Similar to Jimenes et al. [21], Plaza et al. [37] developed an unsupervised version of this methodology. They used a neuro-fuzzy classifier to perform classification in the spectral domain and to compute a first approximation of the posterior probabilities of classes. The output of this step is then fed to the MRF spatial analysis stage, performed using a maximum likelihood probabilistic reclassification. The performance of the MRF in classifying urban land-cover types was compared with the results of the first stage, that is, a neuro-fuzzy classifier. Similar classification accuracies were achieved mainly because the spatial analysis stage reassigned only border pixels to different classes.

In both classification methods, the UnECHO and the modified MRF, the neighborhoods are determined in advance and do not account for the real size and shape of the objects in the image. This kind of division might not be suitable for the gradual change of biophysical and biochemical properties of crops over the field. Instead of a deterministic definition of the neighborhood, a geo-statistical approach can be adopted [23,34]. Lark [23] suggested a spatially weighted averaging of the class memberships within a local neighborhood based on the variogram. Although the generation of Spatially Coherent Regions (SCR) developed by [23] was initially applied for limited dimensionality of nonspectral data (multitemporal yield data), it can be adopted to HS images. To initially investigate the ability to generate SCRs in the spectral domain, we applied it for the HS image of the experimental plot. Preprocessing included selection of every second band of the original 420 bands and smoothing of the 210-band spectra of the new cube, with a 15-points window. Then, a fuzzy C-means classification into seven classes was applied (Figure 13.4). The fuzzy C-means is an unsupervised classification. Assuming that in the experimental plot the main effect on the reflectance is the nitrogen level, the classes were labeled with a nitrogen level based on a visual inspection. Despite the noisy result, the fuzzy classification captured differences in nitrogen levels: similar nitrogen levels were assigned with similar classes. Yet, the classes do not fully match with nitrogen treatments. For example, treatments 75%, 50%, and 100% in the western part of the plot were all classified as having the highest nitrogen level. In the eastern part, their parallels are distinguished from each other and classified as having lower nitrogen levels. These differences can be explained by several reasons: (1) there are other crop properties other than nitrogen level that affected the reflectance; (2) there are differences in actual nitrogen levels between the replicates of each treatment; (3) since it is located in the plot margin, the eastern side is affected by dust or other marginal effects that change the reflectance. No matter what the reason is, the accuracy of the spectral classification is not the main topic in this part of the chapter but the added value that can be achieved by utilizing the spatial aspect.

Following the spectral classification, the variogram range was calculated and used as the neighborhood radius for refining the spectral classification (Figure 13.5). The resulted SCR significantly

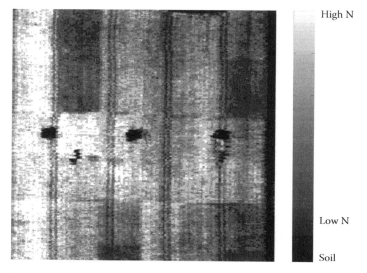

FIGURE 13.4 A fuzzy C-means classification of the 210-band HS image of the experimental plot, seven classes.

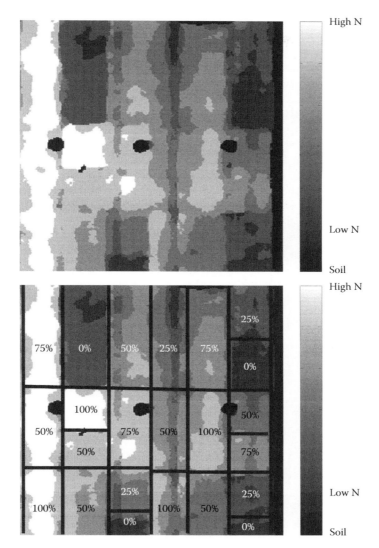

FIGURE 13.5 SCR of the fuzzy C-means.

reduced the speckle effect by uncovering most of the spatial structures of the subplots with different nitrogen N levels. Since the neighborhood is not determined by geometrical shapes, the borders between subplots are not crisp but rather fuzzy. This result implies that this type of flexible neighborhood definition is more suitable for the real situation in the field where changes in nitrogen levels are gradual and not sharp. If the affecting factor for the variability in the field is known in advance, the resulted SCRs can be used as management zones for variable rate application. If not, they can be used both for selecting spectra free of noise for calibrating spectral model [24] or for implementing a validated model for crop properties estimation.

13.4.4 FUSION OF SPECTRAL AND SPATIAL INFORMATION

The previous approaches separate spatial from spectral information, and, thus, the two types of information are not treated simultaneously. Plaza et al. [37] suggested incorporating spatial context into the support vector machine (SVM) spectral classifier. In this method, a pixel entity is redefined simultaneously both in the spectral and spatial domains by applying some feature extraction to its surrounding area, which yields spatial (contextual) features like the mean or standard deviation per spectral band. These separated entities lead to two different kernel matrices that can be summed up and introduce cross-information features in the formulation. When applied for land-cover classification in an agricultural area, the contextual SVM obtained classification accuracy of 95%. It outperformed spectral classifier based on Euclidean distance and performed much better than other methods like the ECHO, which use spectral and spatial information to classify homogeneous objects. Similar to the UnECHO and the modified MRF, the contextual SVM is based on a predefined neighborhood of N × N windows.

Another approach for fusing spectral and spatial information is a multiscale or hierarchical segmentation. Hierarchical segmentation is based on sequential optimization to produce a hierarchical data-driven decomposition of the picture with no restriction on segment shapes [5]. Beamlet analysis

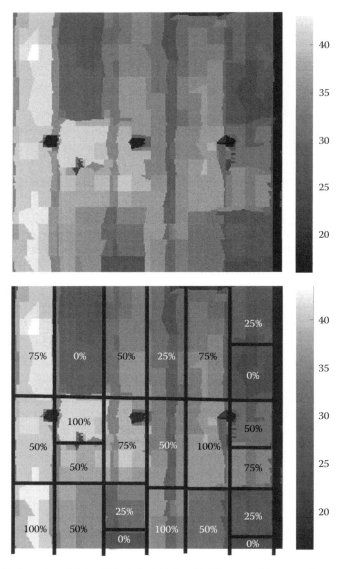

FIGURE 13.6   Multidimensional BD-RDP of the 210-band HS image of the experimental plot. The value of each segment is the average of the reflectance in an IR band (750 nm).

is a framework for multiscale image analysis in which line segments play a role analogous to the role played by points in wavelet analysis [12]. The beamlet-decorated recursive-dyadic-partitioning (BD-RDP) is one realization of the beamlet analysis. While partitioning with basic RDP is limited to square elements, the BD-RDP allows that some of its squares (optionally) are decorated by a beamlet, that is, it can be partitioned not only by squares but also by other geometrical shapes. In comparison to the basic RDP, this additional flexibility allows the BD-RDP to approximate an image more accurately with much fewer segments. The BD-RDP was originally designated to one-dimensional images by implementing two main steps: a spreading phase, where the image is partitioned into its smallest parts according to a quad-tree structure and a folding phase, where the tree is folded up according to a target function. The target function has to serve the idea of minimum variation between the original and reassigned values with a penalty entails for the number of different segments. The BD-RDP does not require a priori knowledge of the number of segments.

We introduce two enhancements of the algorithm [26]. First, it was modified to suit multidimensional images based on the Euclidian distance between vectors and second, a merging neighbor's phase was added that checks the possibility of merging segments that belong to different dyadic squares using the target function. The multidimensional three-step BD-RDP was applied to the 210-band HS image of the experimental plot and the segmentation result is shown in Figure 13.6. As it is not a classification, the color of each segment was determined by its average reflectance in an IR band (750 nm) to partially demonstrate the differences in nitrogen levels. In general, reflectance increases with increasing nitrogen levels.

The multiscale segmentation successfully uncovered the spatial structures in the image according to differences in nitrogen levels. Unlike the SCR classification, the beamlet analysis results in homogeneous segments and needs further analysis to classify the segments according to their nitrogen level. For that, a fuzzy C-means or a calibrated PLS model can be applied.

13.5 DISCUSSION AND FUTURE DIRECTIONS

Hyperspectral (HS) remote sensing (imaging spectroscopy) systems enable the collection of spectral data in hundreds or even thousands of narrowbands continuously over a spectral range (e.g., 400–2500 nm). Key biophysical and biochemical properties of agricultural crops (BB-PACs) like left area index (LAI), biomass, chlorophyll level, and water status have major effect in the specific narrowband transition zones of the spectral reflectance curve like the red edge and water absorption bands. Thus, the narrowband widths of HS data allow for better estimation of crop properties when compared with the relatively coarse bandwidths acquired with multispectral scanners. However, practical issues related to data volumes and data processing need to be considered when using hyperspectral data in applications over large areas. The processing complexity and the statistical concerns of colinearity and over-fitting entailed with spectral analysis have led to the widespread adoption of dimensionality reduction approaches. Various narrowband indices were developed and were shown to improve over the broadband indices. Stepwise discriminant analysis was used in many studies to select few optimal bands for characterizing agricultural crop variables. In general, this type of analysis has demonstrated the importance of the red edge and the SWIR regions and for lesser extent, the blue, green, and NIR regions. These findings together with the high cost of HS systems and the analysis complexity promted the development of hyperspatial platforms like the Rapid-Eye, World-View2, and the Venus. The selected bands in literature were neither identical for the same crop property in different studies nor to different crop properties in the same study. Figure 13.7 shows single bands and band ranges that were used in the reviewed studies for estimating LAI and biomass, water content, and nitrogen and chlorophyll. This overview strongly demonstrates the necessity of HS systems that provide contiguous spectra for the estimation of key BB-PACs. The PLS chemometric method enable the analysis of the whole spectra avoiding the colinearity or over-fitting complexities. Recent studies have confirmed the potential of partial least squares regression (PLSR) analysis to interpret HS remote sensing data for the estimation of crop properties. Other methods that use

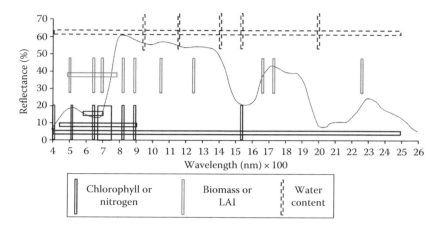

FIGURE 13.7 Spectral bands and spectral ranges that were used in various studies to estimate key BB-PACs. Vertical sticks refer to single bands and horizontal lines refer to bands range.

the whole spectra like wavelet analysis, continuum removal, SAM, and the area under the spectra (integral) were also suggested and proved to be effective.

The canopy reflectance, in the visible and NIR, is intensely dependent on both biophysical and biochemical properties of the canopy. Moreover, several properties have similar effects on canopy reflectance. Very few studies have introduced new spectral indices and algorithms and demonstrated that leaf chlorophyll content can be estimated with minimal confounding effects due to LAI and vice versa. Fusion of the thermal region and the VIS/NIR spectrum was used to decouple the effect of water status and nitrogen level. Still, most of the ongoing studies concentrate on a single factor effect and an in-depth research is required to address the challenge of discriminating between combined effects of various crop variables.

HS images are distinguished from point spectral measurements by their added spatial aspect. Yet, HS images are usually not treated as images, but as lists of spectral measurements with no particular spatial arrangement. Estimation of crop properties using HS images was based merely on spectral information. Precision agriculture necessitates the partition of the field into homogeneous zones. Similar to land-use segmentation and classification, delineation of homogeneous zones would benefit from incorporating spectral and spatial information. In this chapter, we have introduced different levels of integration of spatial information in estimating crop properties. The ability of integrating and fusing spatial analysis with spectral information was initially demonstrated using an HS image of an experimental potato plot.

In general, the suggested methods were effective in classifying nitrogen levels and uncovered spatial structures that coincide with the blocks of the different treatments but no quantitative evaluation was done. In other studies, spatial information derived from HS images was found to be valuable in land-use classification. The contribution of spatial analysis for BB-PACs estimation is yet to be studied. For that, research is needed to investigate existing methods, adjust them or develop new methodologies to incorporate spatial and spectral information.

While integrated spectral/spatial algorithms hold great promise for in-season management zones delineation using HS images, they also introduce computational challenges. Since agro-technical decisions are made routinely by the farmer once or twice a week, a temporal aspect should also be taken into consideration. With the rapid developments of satellites, it seems that in the near future frequent HS images will be available. In order to fully exploit HS images, processing methods that can take advantage of their enhanced spectral, spatial, and temporal features are required. Parallel processing hardware has necessarily become a requirement to speed up processing performance and to satisfy high computational requirements. As a result, the future potential of HS image-processing methods will also be largely defined by their suitability for being implemented in parallel [37].

The hierarchical segmentation approach that was presented in this chapter is suitable for parallel processing hardware and thus the routine of in-season management zones delineation utilizing this approach may be accelerated to meet the timeline requirements of the agro-technical applications.

REFERENCES

1. Aparicio, N., Villegas, D., Casadesus, J., Araus, J. L., and Royo, C. 2000. Spectral vegetation indices as nondestructive tools for determining durum wheat yield. *Agronomy Journal* 92 (1):83–91.
2. Bajcsy, P. and Groves, P. 2004. Methodology for hyperspectral band selection. *Photogrammetric Engineering and Remote Sensing* 70 (7):793–802.
3. Bannari, A., Khurshid, K. S., Staenz, K., and Schwarz, J. 2008. Potential of hyperion EO-1 hyperspectral data for wheat crop chlorophyll content estimation. *Canadian Journal of Remote Sensing* 34:S139–S157.
4. Baret, F. and Guyot, G. 1991. Potentials and limits of vegetation indexes for LAI and APAR assessment. *Remote Sensing of Environment* 35 (2–3):161–173.
5. Beaulieu, J. M. and Goldberg, M. 1989. Hierarchy in picture segmentation—A stepwise optimization approach. *IEEE Transactions on Pattern Analysis and Machine Intelligence* 11 (2):150–163.
6. Blackburn, G. A. 2007. Wavelet decomposition of hyperspectral data: A novel approach to quantifying pigment concentrations in vegetation. *International Journal of Remote Sensing* 28 (12):2831–2855.
7. Blackburn, G. A. and Ferwerda, J. G. 2008. Retrieval of chlorophyll concentration from leaf reflectance spectra using wavelet analysis. *Remote Sensing of Environment* 112 (4):1614–1632.
8. Clevers, J., Kooistra, L., and Schaepman, M. E. 2010. Estimating canopy water content using hyperspectral remote sensing data. *International Journal of Applied Earth Observation and Geoinformation* 12 (2):119–125.
9. Cohen, Y., Alchanatis, V., Zusman, Y., Dar, Z., Bonfil, D. J., Karnieli, A., Zilberman, A. et al. Leaf nitrogen estimation in potato based on spectral data and on simulated bands of the VEN mu S satellite. *Precision Agriculture* 11 (5):520–537.
10. Darvishzadeh, R., Atzberger, C., Skidmore, A. K., and Abkar, A. A. 2009. Leaf area index derivation from hyperspectral vegetation indices and the red edge position. *International Journal of Remote Sensing* 30 (23):6199–6218.
11. Delegido, J., Fernandez, G., Gandia, S., and Moreno, J. 2008. Retrieval of chlorophyll content and LAI of crops using hyperspectral techniques: Application to PROBA/CHRIS data. *International Journal of Remote Sensing* 29 (24):7107–7127.
12. Donoho, D. and Huo, X. 2002. Beamlets and multiscale image analysis, pp. 149–196. In: *Multiscale and Multiresolution Methods*, T. J. Barth, T. Chan, and R. Haimes, eds. Springer Lecture Notes in Computational Science and Engineering.
13. Gitelson, A. A. 2004. Wide dynamic range vegetation index for remote quantification of biophysical characteristics of vegetation. *Journal of Plant Physiology* 161 (2):165–173.
14. Govaerts, Y. M., Verstraete, M. M., Pinty, B., and Gobron, N. 1999. Designing optimal spectral indices: A feasibility and proof of concept study. *International Journal of Remote Sensing* 20 (9):1853–1873.
15. Haboudane, D., Miller, J. R., Pattey, E., Zarco-Tejada, P. J., and Strachan, I. B. 2004. Hyperspectral vegetation indices and novel algorithms for predicting green LAI of crop canopies: Modeling and validation in the context of precision agriculture. *Remote Sensing of Environment* 90 (3):337–352.
16. Haboudane, D., Miller, J. R., Tremblay, N., Zarco-Tejada, P. J., and Dextraze, L. 2002. Integrated narrow-band vegetation indices for prediction of crop chlorophyll content for application to precision agriculture. *Remote Sensing of Environment* 81 (2–3):416–426.
17. Haboudane, D., Tremblay, N., Miller, J. R., and Vigneault, P. 2008. Remote estimation of crop chlorophyll content using spectral indices derived from hyperspectral data. *IEEE Transactions on Geoscience and Remote Sensing* 46 (2):423–437.
18. Hansen, P. M. and Schjoerring, J. K. 2003. Reflectance measurement of canopy biomass and nitrogen status in wheat crops using normalized difference vegetation indices and partial least squares regression. *Remote Sensing of Environment* 86 (4):542–553.
19. Herrmann, I., Karnieli, A., Bonfil, D. J., Cohen, Y., and Alchanatis, V. SWIR-based spectral indices for assessing nitrogen content in potato fields. *International Journal of Remote Sensing* 31 (19):5127–5143.
20. Jain, N., Ray, S. S., Singh, J. P., and Panigrahy, S. 2007. Use of hyperspectral data to assess the effects of different nitrogen applications on a potato crop. *Precision Agriculture* 8 (4–5):225–239.

21. Jimenez, L. O., Rivera-Medina, J. L., Rodriguez-Diaz, E., Arzuaga-Cruz, E., and Ramirez-Velez, M. 2005. Integration of spatial and spectral information by means of unsupervised extraction and classification for homogenous objects applied to multispectral and hyperspectral data. *IEEE Transactions on Geoscience and Remote Sensing* 43 (4):844–851.

22. Kempeneers, P., de Backer, S., Debruyn, W., Coppin, P., and Scheunders, P. 2005. Generic wavelet-based hyperspectral classification applied to vegetation stress detection. *IEEE Transactions on Geoscience and Remote Sensing* 43 (3):610–614.

23. Lark, R. M. 1998. Forming spatially coherent regions by classification of multi-variate data: An example from the analysis of maps of crop yield. *International Journal of Geographical Information Science* 12 (1):83–98.

24. Lawrence, K. C., Windham, W. R., Park, B., Heitschmidt, G. W., Smith, D. P., and Feldner, P. 2006. Partial least squares regression of hyperspectral images for contaminant detection on poultry carcasses. *Journal of Near Infrared Spectroscopy* 14 (4):223–230.

25. Lee, K. S., Cohen, W. B., Kennedy, R. E., Maiersperger, T. K., and Gower, S. T. 2004. Hyperspectral versus multispectral data for estimating leaf area index in four different biomes. *Remote Sensing of Environment* 91 (3–4):508–520.

26. Levi, O., Cohen, S., and Mharaby, Z. 2010. Effective hyper-spectral image segmentation using multi-scale geometric analysis. In: *IADIS Multi Conference on Computer Science and Information Systems 2010*, Freiburg, Germany.

27. Li, F., Miao, Y. X., Hennig, S. D., Gnyp, M. L., Chen, X. P., Jia, L. L., and Bareth, G. 2010. Evaluating hyperspectral vegetation indices for estimating nitrogen concentration of winter wheat at different growth stages. *Precision Agriculture* 11 (4):335–357.

28. Liu, J. G., Miller, J. R., Haboudane, D., Pattey, E., and Nolin, M. C. 2005. Variability of seasonal CASI image data products and potential application for management zone delineation for precision agriculture. *Canadian Journal of Remote Sensing* 31 (5):400–411.

29. Liu, Z. Y., Wu, H. F., and Huang, J. F. Application of neural networks to discriminate fungal infection levels in rice panicles using hyperspectral reflectance and principal components analysis. *Computers and Electronics in Agriculture* 72 (2):99–106.

30. López-Granados, F. 2011. Weed detection for site-specific weed management: Mapping and real-time approaches. *Weed Research* 51 (1):1–11.

31. Mahlein, A. K., Steiner, U., Dehne, H. W., and Oerke, E. C. Spectral signatures of sugar beet leaves for the detection and differentiation of diseases. *Precision Agriculture* 11 (4):413–431.

32. Miao, Y. X., Mulla, D., Randall, G., Vetsch, J., and Vintila, R. 2009. Combining chlorophyll meter readings and high spatial resolution remote sensing images for in-season site-specific nitrogen management of corn. *Precision Agriculture* 10 (1):45–62.

33. Moller, M., Alchanatis, V., Cohen, Y., Meron, M., Tsipris, J., Naor, A., Ostrovsky, V., Sprintsin, M., and Cohen, S. 2007. Use of thermal and visible imagery for estimating crop water status of irrigated grapevine. *Journal of Experimental Botany* 58 (4):827–838.

34. Nansen, C., Sidumo, A. J., and Capareda, S. 2010. Variogram analysis of hyperspectral data to characterize the impact of biotic and abiotic stress of maize plants and to estimate biofuel potential. *Applied Spectroscopy* 64 (6):627–636.

35. Ordonez, C., Martinez, J., Matias, J. M., Reyes, A. N., and Rodriguez-Perez, J. R. 2010. Functional statistical techniques applied to vine leaf water content determination. *Mathematical and Computer Modelling* 52 (7–8):1116–1122.

36. Pimstein, A., Karnieli, A., and Bonfil, D. J. 2007. Wheat and maize monitoring based on ground spectral measurements and multivariate data analysis. *Journal of Applied Remote Sensing* 1:16.

37. Plaza, A., Benediktsson, J. A., Boardman, J. W., Brazile, J., Bruzzone, L., Camps-Valls, G., Chanussot, J. et al. 2009. Recent advances in techniques for hyperspectral image processing. *Remote Sensing of Environment* 113 (Supplement 1):S110–S122.

38. Reisig, D. D. and Godfrey, L. D. 2010. Remotely sensing arthropod and nutrient stressed plants: A case study with nitrogen and cotton aphid (Hemiptera: Aphididae). *Environmental Entomology* 39 (4):1255–1263.

39. Rodriguez-Perez, J. R., Riano, D., Carlisle, E., Ustin, S., and Smart, D. R. 2007. Evaluation of hyperspectral reflectance indexes to detect grapevine water status in vineyards. *American Journal of Enology and Viticulture* 58 (3):302–317.

40. Shapira, U., Herrmann, I., Karnieli, A., and Bonfil, D. J. 2010. Weeds detection by ground-level hyperspectral imaging. In: *10th International Conference on Precision Agriculture*, July 18–21, 2010, Hyatt Regency Tech Center, Denver, CO.

41. Thenkabail, P. S., Enclona, E. A., Ashton, M. S., and Van der Meer, B. 2004. Accuracy assessments of hyperspectral waveband performance for vegetation analysis applications. *Remote Sensing of Environment* 91 (3–4):354–376.

42. Thenkabail, P. S., Smith, R. B., and De Pauw, E. 2000. Hyperspectral vegetation indices and their relationships with agricultural crop characteristics. *Remote Sensing of Environment* 71 (2):158–182.

43. Thenkabail, P. S., Smith, R. B., and De Pauw, E. 2002. Evaluation of narrowband and broadband vegetation indices for determining optimal hyperspectral wavebands for agricultural crop characterization. *Photogrammetric Engineering and Remote Sensing* 68 (6):607–621.

44. Thorp, K. R. and Tian, L. F. 2004. A review on remote sensing of weeds in agriculture. *Precision Agriculture* 5 (5):477–508.

45. Tucker, C. J. 1979. Red and photographic infrared linear combinations for monitoring vegetation. *Remote Sensing of Environment* 8:127–150.

46. Wold, S., Ruhe, A., Wold, H., and Dunn, W. J. 1984. The colinearity problem in linear-regression—The partial least-squares (PLS) approach to generalized inverses. *Siam Journal on Scientific and Statistical Computing* 5 (3):735–743.

47. Yang, Z., Rao, M. N., Elliott, N. C., Kindler, S. D., and Popham, T. W. 2009. Differentiating stress induced by greenbugs and Russian wheat aphids in wheat using remote sensing. *Computers and Electronics in Agriculture* 67 (1–2):64–70.

48. Zarco-Tejada, P. J., Berjon, A., Lopez-Lozano, R., Miller, J. R., Martin, P., Cachorro, V., Gonzalez, M. R., and de Frutos, A. 2005. Assessing vineyard condition with hyperspectral indices: Leaf and canopy reflectance simulation in a row-structured discontinuous canopy. *Remote Sensing of Environment* 99 (3):271–287.

14 Hyperspectral Vegetation Indices

Dar A. Roberts, Keely L. Roth, and Ryan L. Perroy

CONTENTS

14.1 INTRODUCTION

Vegetation properties are often measured by converting a reflectance spectrum into a single number value or vegetation index (VI). Hyperspectral, or narrowband [1], vegetation indices (HVIs) include narrower band features or wavelengths only captured by hyperspectral instruments (e.g., [2]). Vegetation properties measured with HVIs can be divided into three main categories: (1) structure; (2) biochemistry; and (3) plant physiology/stress. Measured structural properties include fractional cover, green leaf biomass, leaf area index (LAI), senesced biomass, and fraction absorbed photosynthetically active radiation (FPAR) [3–5]. A majority of the indices developed for structural analysis were formulated for broadband systems and have narrowband, hyperspectral equivalents. Biochemical properties include water, pigments (chlorophyll, carotenoids, anthocyanins), other nitrogen-rich compounds (e.g., proteins), and plant structural materials (lignin and cellulose) [6–8]. Physiological and stress indices measure subtle changes due to a stress-induced change in the state of xanthophylls [9], changes in chlorophyll content [10], fluorescence [11], or changes in leaf moisture [12]. In general, biochemical and physiological/stress indices were formulated using laboratory or field instruments ($\leq$10 nm spectral sampling) and are targeted at very fine spectral features. As a result they are strictly hyperspectral. The one exception is indices developed for water.

Many structurally oriented VIs rely on some combination of near-infrared (NIR) to red reflectance, such as the NIR to red ratio, or simple ratio (SR) [3]. This is because increases in LAI correspond with increases in chlorophyll absorption and NIR-scattering and decreases in exposed substrate, resulting in decreasing red and increasing NIR reflectance (Figure 14.1). Thus, equations for these VIs often compare reflectance at an absorbing wavelength to a nonabsorbing wavelength. However, more subtle changes also occur with an increase in LAI, including increasing green reflectance [13] and increases in the absorption of liquid water [14].

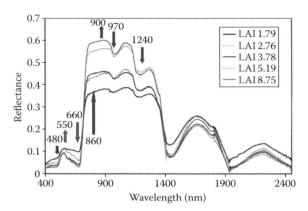

FIGURE 14.1 Reflectance spectra of *Populus trichocarpa* hybrids over a range in LAI. Wavelengths labeled refer to absorption features (480, 660, 970, and 1240 nm) or NIR scattering regions (860 and 900 nm) typically used in combination to quantify structure. Arrows mark regions of decreasing reflectance due to absorption (down) or increasing reflectance due to scattering (up).

Biochemical and stress-related indices rely on a similar comparison of absorbing and nonabsorbing wavelengths, varying the absorbing wavelength by biochemical (Figure 14.2). For example, canopy moisture/moisture stress indices include wavelengths associated with liquid water absorption (e.g., 970 and 1200 nm), while ligno-cellulose content indices utilize the short-wave-infrared (SWIR) and wavelengths from 1500 to 1800 and 2000 to 2350 nm (Figure 14.2). By contrast, pigments (carotenoids, anthocyanins, and chlorophylls) absorb the visible and ultraviolet, with distinct, but overlapping absorption features. Chlorophyll primarily absorbs blue and red light (*AcerLf*, Figure 14.2). Anthocyanins absorb all but red light (*Fagus*, Figure 14.2); and many carotenoids are yellow due to strong blue light absorption (*Betula*, Figure 14.2). Thus, pigment-sensitive VIs frequently include a combination of visible bands.

In this chapter, we provide an overview of common HVIs associated with vegetation structure (Section 14.2.1), canopy biochemistry (Section 14.2.2), and plant physiology (Section 14.2.3). Canopy biochemistry is further divided into pigments (Section 14.2.2.1), moisture (Section 14.2.2.2), and plant residues (Section 14.2.2.3). Many of the most commonly used HVIs, with their equations and key citations, are found in Table 14.1. We conclude with two applied examples: one study examining the relationship between LAI and HVIs for *populus* (Section 14.3.1), and another evaluating the relationship between HVIs and seasonal environmental changes for two plant species (Section 14.3.2).

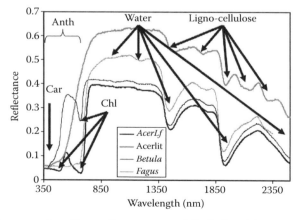

FIGURE 14.2 Reflectance spectra of leaves from a senesced birch (*Betula*), ornamental beech (*Fagus*), and healthy and fully senesced maple (*AcerLf*, Acerlit) illustrating carotenoid (Car), anthocyanin (Anth), chlorophyll (Chl), water, and ligno-cellulose absorptions.

TABLE 14.1
List of the Major Hyperspectral Vegetation Indices, Including Relevant Formulas and Key Citations

Index	Equation	Reference
Structure (LAI, Green Biomass, Fraction)		
NDVI[a]	$(R_{NIR} - R_{red})/(R_{NIR} + R_{red})$	Rouse et al. [15]
SR[a]	R_{NIR}/R_{red}	Jordan [3]
EVI[a]	$2.5 * (R_{NIR} - R_{red})/(R_{NIR} + 6 * R_{red} - 7.5 * R_{blue} + 1)$	Huete et al. [23]
NDWI[a]	$(R_{857} - R_{1241})/(R_{857} + R_{1241})$	Gao [29]
WBI[b]	R_{900}/R_{970}	Peñuelas et al. [28]
ARVI[a]	$(R_{NIR} - [R_{red} - \gamma * \{R_{blue} - R_{red}\}])/(R_{NIR} + [R_{red} - \gamma * (R_{blue} - R_{red})])$	Kaufman and Tanré [22]
SAVI[a]	$[(R_{NIR} - R_{red})/(R_{NIR} + R_{red} + L)] * (1 + L)$	Huete [21]
1DL_DGVI[b]	$\sum_{\lambda_{626nm}}^{\lambda_{795nm}} \mid R'(\lambda_i) - R'(\lambda_{626nm}) \mid \Delta\lambda_i$	Elvidge and Chen [1]
1DZ_DGVI[b]	$\sum_{\lambda_{626nm}}^{\lambda_{795nm}} \mid R'(\lambda_i) \mid \Delta\lambda_i$	Elvidge and Chen [1]
VARI[a]	$(R_{green} - R_{red})/(R_{green} + R_{red} - R_{blue})$	Gitelson et al. [13]
VIgreen[a]	$(R_{green} - R_{red})/(R_{green} + R_{red})$	Gitelson et al. [13]
Biochemical		
Pigments		
SIPI[b]	$(R_{800} - R_{445})/(R_{800} - R_{680})$	Peñuelas et al. [31]
PSSR[b]	$(R_{800}/R_{675}); (R_{800}/R_{650})$	Blackburn [30]
PSND[b]	$[(R_{800} - R_{675})/(R_{800} + R_{675})]; [(R_{800} - R_{650})/(R_{800} + R_{650})]$	Blackburn [32]
PSRI[b]	$(R_{680} - R_{500})/R_{750}$	Merzlyak et al. [33]
Chlorophyll		
CARI[b]	$[(R_{700} - R_{670}) - 0.2 * (R_{700} - R_{550})]$	Kim [34]
MCARI[b]	$[(R_{700} - R_{670}) - 0.2 * (R_{700} - R_{550})] * (R_{700}/R_{670})$	Daughtry et al. [35]
$CI_{red edge}$[b]	$R_{NIR}/R_{red edge} - 1$	Gitelson et al. [36]
Anthocyanins		
ARI[b]	$(1/R_{green}) - (1/R_{red edge})$	Gitelson et al. [40]
mARI[b]	$[(1/R_{green}) - (1/R_{red edge})] * R_{NIR}$	Gitelson et al. [36]
RGRI[b]	R_{red}/R_{green}	Gamon and Surfus [7]
ACI[b]	R_{green}/R_{NIR}	Van den Berg and Perkins [41]
Carotenoids		
CRI1[b]	$(1/R_{510}) - (1/R_{550})$	Gitelson et al. [42]
CRI2[b]	$(1/R_{510}) - (1/R_{700})$	Gitelson et al. [42]
Water		
NDII[a]	$(R_{NIR} - R_{SWIR})/(R_{NIR} + R_{SWIR})$	Hunt and Rock [12]
NDWI[a], WBI[b]	See above	See above
MSI[a]	R_{SWIR}/R_{NIR}	Rock et al. [43]
Lignin and cellulose/residues		
CAI[b]	$100 * [0.5 * (R2031 + R2211) - R2101]$	Daughtry [47]
NDLI[b]	$[\log(1/R_{1754}) - \log(1/R_{1680})]/[\log(1/R_{1754}) + \log(1/R_{1680})]$	Serrano et al. [48]
Nitrogen		
NDNI[b]	$[\log(1/R_{1510}) - \log(1/R_{1680})]/[\log(1/R_{1510}) + \log(1/R_{1680})]$	Serrano et al. [48]

(continued)

TABLE 14.1 (continued)
List of the Major Hyperspectral Vegetation Indices, Including Relevant Formulas and Key Citations

Index	Equation	Reference
Physiology		
Light use efficiency		
RGRI[b], SIPI[b]	See above	See above
PRI[b]	$(R_{531} - R_{570})/(R_{531} + R_{570})$	Gamon et al. [9]
Stress		
MSI[a]	See above	See above
REP[b]	L(max first derivative: 680–750 nm)	Horler et al. [10]
RVSI[b]	$[(R_{714} + R_{752})/2 - R_{733}$	Merton and Huntington [52]

Note: Indices marked in gray are ones that serve multiple purposes.
[a] Narrow band equivalent of a broad band index.
[b] Strictly narrow band/hyperspectral.

14.2 APPLICATIONS OF HYPERSPECTRAL VEGETATION INDICES

Most structural indices were developed for broadband systems but have narrowband (≤10 nm) equivalents (Table 14.1). Exceptions include indices based on first derivatives and the water band index (WBI) [28], which includes wavelengths that are not sampled by broadband systems (Table 14.1). In contrast, most biochemical/physiological indices are strictly hyperspectral, requiring narrow bands (≤10 nm) and specific band centers that are not sampled by broadband systems. Band centers and spectral sampling were typically defined by field or laboratory instrumentation used in the original research, but can be modified for alternate systems.

14.2.1 Vegetation Structure (i.e., LAI, FPAR)

Using VIs for vegetation analysis dates back to Jordan [3], who proposed the SR to estimate over-story LAI. Normalized forms of the SR, such as the normalized difference vegetation index (NDVI) [15], were proposed soon after and were designed to reduce the impact of atmospheric scattering by using a normalized difference between two bands. Both the SR and NDVI are good predictors of wet and dry green biomass [4], LAI [3], FPAR [5], and fractional cover [16–17]. However, NDVI saturates at high LAI values [14] and varies with viewing geometry [18] and substrate reflectance [19]. Narrowband versions of the NDVI have been proposed by Galvao et al. [2] and Thenkabail et al. [20], in which wavelength selection was optimized to reduce sensitivity to nonphotosynthetic vegetation [2] or to improve crop-specific estimates of LAI, wet biomass, and canopy height [20].

Other indices based on NDVI have been proposed to improve canopy structure estimates and minimize the impact of the atmosphere and substrate. For example, the soil adjusted vegetation index (SAVI) [21] includes an offset in the denominator designed to force the NDVI to radiate from the origin of an NIR and red scatterplot, independent of a change in substrate reflectance. The atmospherically resistant vegetation index (ARVI) [22] includes a blue band in the numerator and denominator and a weighting factor to compensate for enhanced atmospheric scattering in red wavelengths. The enhanced vegetation index (EVI) [23] incorporates both a substrate reflectance correction (i.e., SAVI) and a blue band to compensate for the atmosphere (i.e., ARVI). EVI has largely replaced NDVI as a primary global product because of improved resistance to the atmosphere and less evidence of saturation at high LAI [24].

Visible-reflectance variants of the NDVI and ARVI have also been proposed, with the NIR band replaced by a green band, such as the VI Green and the vegetation atmospherically resistant index (VARI) [13]. These indices respond more linearly to changes in vegetation cover fraction than does NDVI [13]. VARI has also proven to be highly effective for estimating LAI and moisture stress in maize [25,26] and live fuel moisture (LFM) in shrub lands [27].

Changes in LAI also impact the shape and position of the red edge and the expression of liquid water in canopy spectra [1,14]. Elvidge and Chen [1] proposed several indices based on the red edge, using first derivative spectra between 626 and 796 nm (1DZ_DGVI and 1DL_DGVI). The relationship between these indices and LAI is significantly improved (compared to other normalized indices) because the slope of the red edge is more sensitive to changes in LAI and the first derivative spectrum is less sensitive to changes in albedo. Although Elvidge and Chen [1] proposed a range between 626 and 796 nm, hyperspectral systems offer numerous possibilities for alternate, narrower or broader spectral ranges for integration that could be explored. Indices that contrast liquid water absorption, either at 970 or 1200 nm to NIR reflectance outside of these wavelengths such as the WBI [28] and the normalized difference water index (NDWI) [29], are also sensitive to LAI. Perry and Roberts [26] found WBI and NDWI to be the most sensitive of 15 indices to a change in maize biomass.

14.2.2 Canopy Biochemistry

14.2.2.1 Plant Pigments

A number of indices have been developed to measure overall pigment concentrations or to quantify specific pigments in plant leaves and canopies. Three types of plant pigments contribute significantly to visible reflectance in leaves and canopies: chlorophylls (a and b), carotenoids, and anthocyanin (Figure 14.2). Many indices are formulated either as an SR, such as the pigment specific spectral ratio (PSSR) [30] or a normalized ratio, such as the structurally insensitive pigment index (SIPI) [31] or the pigment sensitive normalized difference (PSND) [32]. These indices respond to either a single pigment or combinations, such as the ratio of carotenoids to chlorophyll a (SIPI). Similar to SIPI, the plant senescence reflectance index (PSRI) [33] also changes in response to a change in the ratio of carotenoids to chlorophyll as plants senesce.

Indices developed to estimate chlorophyll content include the chlorophyll absorption in reflectance index (CARI) [34], modified CARI (MCARI) [35], and chlorophyll red-edge index ($CI_{red\,edge}$) [36]. CARI quantifies the 670 nm chlorophyll absorption feature as the mathematical difference between 700 and 670 nm reflectance, adjusted by a weighted difference between 700 and 550 nm to compensate for nonphotosynthesizing materials [37]. MCARI further adjusts the soil compensating component by the ratio of NIR to red reflectance. It has proven effective in identifying nutrient stress in maize [26] and drought stress in Amazonian forests [37]. $CI_{red\,edge}$ is based on a three-band generalized model for quantifying pigments [25]. In this model, the concentration of an absorber is quantified as the mathematical difference in reciprocal reflectance, $R_{\lambda 1}^{-1}$, within an absorption region and reciprocal reflectance at a second wavelength, $R_{\lambda 2}^{-1}$, outside of the main absorption region but with similar backscattering. This quantity is multiplied by NIR reflectance to compensate for backscatter-dependent variation in brightness. $CI_{red\,edge}$ simplifies to the form shown in Table 14.1 because $R_{\lambda 2}^{-1}$ and R_{NIR} are the same. $CI_{red\,edge}$ has shown a near linear relationship to chlorophyll content over a diversity of broadleaf tree species [38]. In addition to chlorophyll-specific indices, a number of the structural indices have also been used effectively to estimate chlorophyll at leaf scales, such as the SR [39].

Anthocyanins are plant pigments that can increase in response to environmental stress and may play a role in minimizing photoinhibition [40]. Two indices proposed by Gitelson et al. [36,40] are based on the same concept of reciprocal reflectance developed for chlorophyll, but adjusted for anthocyanin. These include the anthocynanin reflectance index (ARI), calculated as the difference between reciprocal green reflectance (540–560 nm) and reciprocal red edge reflectance (690–710 nm), and the modified ARI, which weights ARI by NIR reflectance (760–800 nm). Additional indices built upon the SR model include the red/green ratio (RGRI) [7] and the anthocyanin content index (ACI) [41], calculated as the ratio of green to NIR reflectance. RGRI is based on the concept that high anthocyanin content, which results in red leaves, will increase the green to red ratio, while ACI should also increase in response to increased anthocyanins as green leaf reflectance drops.

Carotenoids aid in the process of light harvesting for photosynthesis and protect chlorophyll from photooxidation via the reversible conversion of the xanthophylls violaxanthin to zeaxanthin [9,42]. They are most readily apparent in leaves during senescence, as they are retained while chlorophyll breaks down. Carotenoid indices include SIPI [31] and two reciprocal reflectance models proposed by Gitelson et al. [42]; one using the difference between reciprocal reflectance at 510 and 550 nm, the other replacing the 550 nm band with a band at 700 nm (carotenoid reflectance index: CRI1 and 2) [42]. The most widely used caretonoid index, however, is the photochemical reflectance index (PRI) [9], which is designed to capture the shift from violaxanthin to zeaxanthin. This transition results in a subtle (<1%) decrease in reflectance at 531 nm that can be quantified using a normalized difference index and 570 nm as the reference band [9]. In this form, increasingly negative PRIs will occur with increasing plant stress. The PRI is discussed in more detail in Section 14.2.3.

14.2.2.2 Canopy Moisture

Plant canopy moisture varies as a function of the number of leaves within a crown (Figure 14.1) and the water content of individual leaves (Figure 14.2). SR and normalized indices have been proposed to compare the expression of subtle or strong liquid water bands relative to a reference nonabsorbing wavelength. SRs include the moisture stress index (MSI) [43] and the WBI [28]. The MSI is calculated as the ratio of an SWIR band (1650 nm) to NIR band (830 nm), while the WBI is calculated as the reflectance ratio of 900–970 nm. Increases in water content correspond with decreases in the MSI and increases in the WBI. Normalized versions of the MSI include the normalized difference infrared index (NDII) [12], in which the SWIR band can either be at a short wavelength (1650 nm) or long wavelength (2200 nm). The NDWI [29] is roughly equivalent to the normalized version of the WBI, although the 1240 nm water band takes the place of the 970 nm band.

Several studies have evaluated the relationship between moisture indices and relative water content (RWC) [44], plant water content [29], LFM [27], and moisture stress [12,26,43]. Serrano et al. [44] found WBI and NDWI performed better than NDVI and NDII in shrublands, and Penuelas et al. [29] found a strong correlation between the WBI and plant water content across a wide range of Mediterranean ecosystems. Roberts et al. [27] found WBI and NDWI superior to NDVI, EVI, and NDII for estimating LFM. Several forms of the WBI, including water absorption features at 960, 1180, and 1450 nm were tested to determine their ability to estimate water content for leaves, thin tissues (i.e., stems) and leaves and thin tissues [45]. The 960 and 1180 nm WBI were the most strongly correlated with thin tissue water ($r^2 > 0.75$) over a wide range of plant functional types with slightly higher correlations found for 1180 nm. It should be noted that two greenness measures, the VIG and VARI, have consistently shown stronger relationships to moisture stress than other greenness or water based indices [26,27], potentially because water stress also manifests as an increase in green reflectance [46].

14.2.2.3 Lignin and Cellulose/Plant Residues

Specific absorption bands associated with proteins, starch, sugars, lignin, and cellulose, make HVIs especially well suited for measuring the biochemistry of branches or senesced plant materials in the absence of water and pigments [6] (Figure 14.2). Of these biochemicals, two of the most evident in reflectance spectra are lignin and cellulose, which produce many prominent absorption features in the SWIR. Several HVIs have been developed specifically to estimate the ligno-cellulose content or mass of senesced plant materials, including the cellulose absorption index (CAI) [47] and the normalized difference lignin index (NDLI) [48]. The CAI is a band-depth measure, calculated as the difference in reflectance within a strong cellulose absorption band at 2101 nm and average reflectance for two bands outside of this absorption feature, at 2031 and 2211 nm. The NDLI targets a prominent lignin absorption band at 1754 nm and uses a normalized difference index formula but with the natural logarithm of reciprocal reflectance, a common transform used in biochemical spectroscopy (e.g., [49]). Serrano et al. [48] also proposed a similar index targeted at the 1510 nm nitrogen absorption feature, called the normalized difference nitrogen index (NDNI). Both normalized indices use the nonabsorbing 1680 nm band as a reference band.

Daughtry et al. [17] evaluated the performance of CAI and NDVI for estimating fractional cover of crop residues and green cover for corn, soybean, and wheat over several soils for cases of dry and wet residual biomass. They noted that the CAI and NDVI were essentially uncorrelated, with CAI strongly linearly correlated to changes in the fraction of crop residues and NDVI strongly correlated to changes in green cover for all crops and conditions. When applied to AVIRIS data the combination of CAI and NDVI enabled them to accurately discriminate conservation, reduced, and intensively tilled soils [50], which is critical in assessing soils for potential erosion or carbon uptake. Serrano et al. [48] analyzed AVIRIS data acquired from drought deciduous and evergreen chaparral species, finding a strong linear relationship between NDLI and NDNI for bulk lignin and nitrogen, respectively.

14.2.3 PLANT PHYSIOLOGY

Leaf physiology also impacts reflectance spectra. A good example is how the position and shape of the red edge shifts in response to plant stress, either toward shorter wavelengths (blueshift) or longer wavelengths (redshift) [10]. Blueshifts were reported by Rock et al. [43] and Horler et al. [10], in response to heavy metal stress in *Picea* [43] and *Pisum* [10]. Redshifts typically occur during chlorophyll development, and nutrient stress sometimes decreases this shift [51]. A quantitative measure of the position of the red edge is the maximum of the first derivative of reflectance between 650 and 750 nm [10]. Merton and Huntington [52] proposed the red edge vegetation stress index (RVSI), an index that captures variation in shape of the red edge associated with plant stress. The RVSI is calculated as the average canopy reflectance at 714 and 752 nm, minus reflectance at 733 nm. A concave upward red edge and slightly negative or positive RVSI is found in stressed plants, while a concave downward red edge, and strongly negative RVSI occurs in unstressed plants [52]. Naidu et al. [53], studying leafroll-infected grape vines, found stressed leaves had slightly less negative RVSI values than healthy leaves. In contrast, Perry and Roberts [26] found RVSI to become less negative with higher leaf nitrogen.

The PRI has proven to be one of the most effective stress/physiology-oriented HVIs. As plants become progressively more stressed and are unable to utilize light absorbed by chlorophyll, reflectance at 531 nm drops as violazanthin shifts to zeaxanthin, producing an increasingly negative PRI. As a result, the PRI provides a viable surrogate for measuring light use efficiency (LUE). Gamon et al. [54] evaluated carbon uptake by dry tropical forest species using canopy reflectance. They found NDVI and SR responded primarily to changes in FPAR, while PRI responded to photosynthetic downregulation, becoming increasingly negative at midday. Rahman et al. [55] combined strengths of the NDVI for FPAR and PRI for LUE to develop a multiplicative model for estimating carbon uptake by boreal forests. This model showed a near-linear relationship between NDVI*PRI-predicted carbon uptake, and CO_2 flux measured by several flux towers [55].

14.3 APPLICATIONS

14.3.1 ESTIMATING LAI USING HYPERSPECTRAL VEGETATION INDICES

We evaluate several structurally oriented HVIs for estimating LAI of hybrid poplars (*Populus trichocarpa*). The original study, described in Roberts et al. [56], was designed to evaluate the relationship between LAI and the expression of liquid water in canopies. Here we focus strictly on HVIs, adding several that were not evaluated in the original study.

The study site is located near Wallula, Washington (46°4′N, 118°54′W), at the Boise-Cascade Wallula fiber farm. Field work was conducted between July 20 and 25, 1997. Seventy-six young stump-sprouting plants, ranging between 10 and 60 cm in height were sampled in a 6-year-old stand that had been recently harvested. Reflectance spectra were measured above each plant using an analytical spectral devices (ASD) full range instrument (Analytical Spectral Devices, Boulder, Colorado) and standardized to reflectance with a spectralon panel (Labsphere Inc., North Sutton, New Hampshire) measured at approximately 10 min intervals. At least three replicates were measured for each plant. One to four sets of spectra were measured per plant depending on the size of the resprout at a height of 0.5 m above the canopies. In order to determine LAI, plants were destructively harvested, with five plants randomly sampled in each of five height classes designed to ensure a range in LAI. Plant height and diameter along the major and minor axes were measured to determine ground area cover and plant volume for each plant. Leaf area was calculated for each plant using a linear equation relating stem diameter to leaf area. Stem diameter was measured for every stem on the sampled plants using calipers. In order to develop the linear equation relating leaf area to stem diameter, one out of every 10 stems was stored in a plastic bag, cooled, and then transported to the laboratory for analysis. Leaves from each stem were harvested, measured for leaf area, and then regressed against stem diameter. This relationship was then combined with the resprout stem data to calculate total leaf area for each resprout, and LAI was calculated as leaf area divided by the areal projection of each resprout. The analysis included three soil spectra and 24 plant spectra. For more details see [56].

LAI increases led to dramatic changes in canopy reflectance, including decreasing visible light reflectance, increasing NIR reflectance, and the increased expression of weak and strong absorption features (Figure 14.1). Eight indices were calculated: two that respond to canopy water (WBI, NDWI), five to greenness (SR, NDVI, EVI, VIG, VARI) and one to the red edge (1DZ_DGVI). All the indices were highly correlated with LAI (Table 14.2, Figure 14.3), and all but NDVI showed a linear relationship, producing r^2 values between 0.69 (WBI) and 0.78 (VARI; Figure 14.3). The NDVI-LAI scatter plot (Figure 14.3b) illustrates NDVI saturating at high LAI, beginning around an LAI of 4. The water-based indices had the lowest r^2 values, the NIR to red combinations (SR, NDVI, and EVI) intermediate, and VIG and VARI were the highest. The red edge–based index (1DZ_DGVI) outperformed the SR and linear NDVI model, but had a slightly lower r^2 than a nonlinear fit for NDVI or the linear relationship for the EVI.

TABLE 14.2
Linear and Nonlinear Relationship between LAI and Eight HVIs

Fit Metric	WBI	NDWI	SR	NDVI	NDVI(P)	EVI	VIG	VARI	1DZDG
Slope	44.51	34.31	0.38	7.23	8.12	7.28	11.32	8.11	10.15
Intercept	−43.19	1.67	0.27	−1.11	0	−0.75	1.18	1.24	−0.03
Exponent					2.87				
r^2	0.69	0.73	0.74	0.72	0.77	0.77	0.78	0.78	0.76
RMS	0.21	0.2	0.19	0.2	0.18	0.18	0.18	0.18	0.19
MAE	0.9	0.82	0.8	0.77	0.67	0.69	0.65	0.64	0.71
MaxAE	2.29	2.57	2.17	3.47	3.03	3.57	3.13	3.24	3.65
MinAE	0.09	0.01	0.03	0.07	0	0.05	0.03	0.02	0.06
Soil error	0.23	0.18	0.72	0.42	0	0.25	0.07	0.14	0.21

Note: Fit metrics include r^2, root mean squared error (RMS), mean absolute error (MAE) and variants on the MAE to quantify LAI prediction errors at high LAI (MaxAE), and over bare soils (soil error). Only the NDVI-LAI relationship was improved by a nonlinear model.

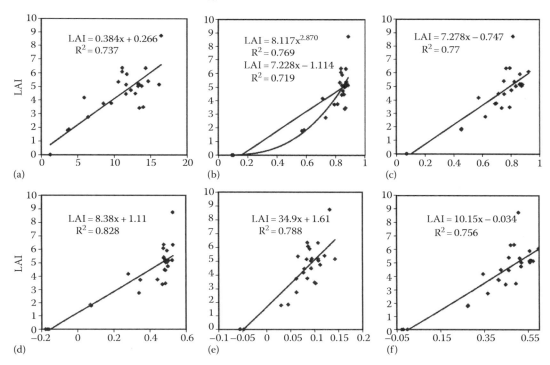

FIGURE 14.3 Showing scatter plots between LAI (y) and a subset of indices including (a) the SR; (b) NDVI; (c) EVI; (d) VARI; (e) NDWI; and (f) 1DZ_DGVI.

Error metrics were calculated to examine model differences. The maximum absolute error (MaxAE) captures a model's over or under-prediction across the range of LAI values. Based on this error metric, the WBI, NDWI, and SR all produced considerably lower errors than the greenness measures, demonstrating that though the greenness measures fit the total population better, they under-predicted high LAI values, suggesting water-based indices and the SR are likely to perform better in high LAI forests. Soil error was calculated as the average mean absolute error for predictions of LAI over bare soils. A high value indicates that an index will produce an error in LAI when cover is sparse or absent. Based on this metric, the lowest error occurred for a power function fit of the NDVI, followed closely by VIG and VARI. The highest soil error observed was for the SR, equal to nearly one LAI, suggesting that the SR is not effective for sparsely covered areas. The second highest error was for the NDVI linear model, equal to 0.5 LAI. EVI had a lower soil error than NDVI as would be expected given the formulation of EVI.

14.3.2 Soil Moisture and AET

In the second example, we focus on how seasonal changes in soil moisture impact the reflectance spectra and HVI values for two invasive species. We extend the analysis to evaluate the relationship between several HVIs and measures of potential (PET) and actual (AET) evapotranspiration, calculated using Penman Monteith [57] and the Bowen Ratio [58], respectively. This study was conducted at coal oil point reserve (COPR), California, one of three areas sampled by the innovative datasets for environmental analysis by students (IDEAS) network (www.geog.ucsb.edu/ideas; [59]). Additional data shown include webcam imagery used to track changes in canopy greenness at solar noon for annual and perennial plants [60].

COPR is located at 34.41386°N and 119.8802°W at an elevation of 6 m and is dominated by a mixture of native and invasive annual grasses and forbs and perennial shrubs. The micrometeorological tower at COPR measures all variables needed to calculate PET and AET, including wind speed and direction, air temperature, and relative humidity (at 0.75 and 2.85 m), and net radiation using a four-channel net radiometer. Additional instrumentation includes a tipping bucket rain gauge, fog collector, and leaf wetness sensor. Three belowground sensors measure soil temperature and volumetric water content at 10, 20, and 50 cm. Soil heat flux, G, was estimated from a combination of soil temperature at 10 cm and surface temperature using an assumed thermal conductivity of 0.4 Wm^{-1} K^{-1}. Soils at COPR are clay loams with clay content increasing from 29.6% to 35.7% from 10 to 50 cm. For more details on instrumentation or site properties see www.geog.ucsb.edu/ideas.

Seasonal environmental data, plotted over 2 years starting in November, 2007, illustrate a typical Mediterranean climate, defined by winter precipitation and summer drought (Figure 14.4a). Rainfall was highly variable between years, with almost all of the rainfall in the 2007–2008 hydrological year falling in a single month. In 2008–2009, the rains started earlier and persisted longer, but totaled less. Soil moisture responded rapidly to precipitation, with the greatest seasonal fluctuations at the shallowest depths and a rapid increase followed by a more gradual dry down (Figure 14.4c). Overall, soil moisture remained relatively high due to the high clay content of soils in the study region.

A webcam, deployed in the late spring of 2008 shows the pronounced, but distinct seasonal cycle of annual and perennial vegetation (Figure 14.4b). Solar noon images were analyzed using software described in [60] to generate estimates of percent greenness for uniform regions of annuals and perennials. PET, AET, and the ratio of AET to PET also show pronounced seasonal cycles as would be expected (Figure 14.4d). PET, to a large extent, is driven by a combination of net radiation and vapor pressure deficit (VPD) and thus peaks in the summer when the skies are clear and the air is dry. AET, by contrast, is controlled by available moisture, as well as net radiation and VPD and thus peaks in the

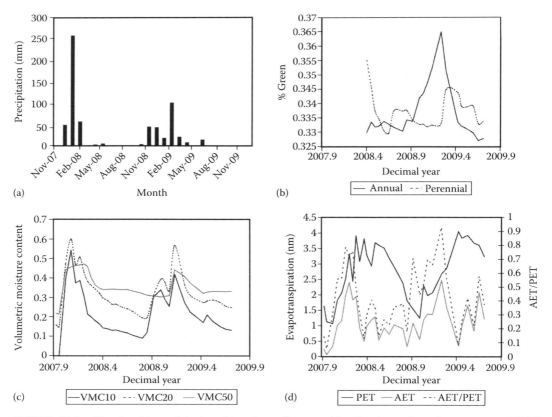

FIGURE 14.4   Showing (a) precipitation; (b) webcam % green; (c) soil volumetric moisture content (VMC) at 10, 20, and 50 cm; and (d) PET, AET, and AET/PET.

spring, when soil water is plentiful and can either be directly evaporated from the surface, or transpired by green plants. Based on the ratio of AET to PET, between 80% and 90% of the atmospheric moisture demand was met by ET in the spring, yet typically less than 20% over the summer.

Reflectance spectra of two invasive plants species were collected in the area over two hydrological years starting in October, 2007. Species measured included *Brachypodium distachyon* (BRDI: Purple false brome) and *Carduus pycnocephalus* (CAPY: Italian thistle). Spectra were collected over five individuals of each species by measuring 25 spectra for each individual plant, five at the plant center and in the four cardinal directions from approximately 0.5–1 m above the canopy. Spectra were acquired within 2 h of solar noon at 2–4 week intervals using an ASD spectrometer and a spectralon standard.

Biweekly reflectance spectra demonstrate significant variation in plant responses to changes in moisture, temperature, and net radiation (Figure 14.5). BRDI started the year highly senesced, beginning to green up by Julian day (JD)77 (Figure 14.5a). Peak greenness was reached quickly by JD137 followed by a rapid senescence. While chlorophyll and water absorption features are evident, they were never expressed as clearly as might be expected for an individual leaf. Furthermore, ligno-cellulose absorptions were evident throughout the year, becoming the dominant absorption feature by JD167. After senescence, BRDI spectra continued to evolve, showing a gradual decrease in reflectance and change in convexity of the visible-NIR region as stems became decomposed.

Unlike BRDI, CAPY showed very pronounced chlorophyll and liquid water absorptions and had a well-defined, steeply sloping red edge (Figure 14.5b). Furthermore, CAPY greened up earlier than BRDI and remained active far longer into the growing season. For example, on JD167, when

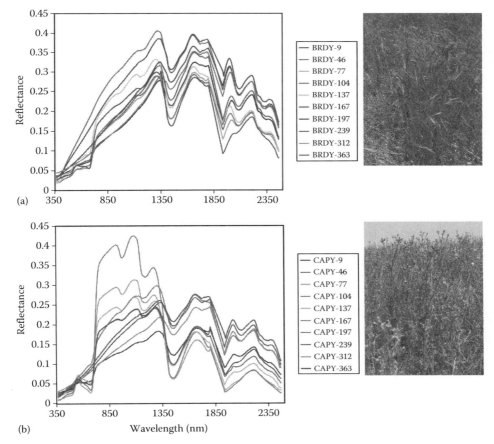

FIGURE 14.5 **(See color insert.)** Reflectance spectra of *Brachypodium distachyon* (BRDI) and *Carduus pycnocephalus* (CAPY) for 2009. The number to the right on the legend reports Julian day.

BRDI was fully senesced, CAPY still showed a pronounced red edge and chlorophyll absorptions, although ligno-cellulose bands also suggest it was senescing. By JD197 CAPY was senesced, but had lower reflectance than BDRI because of its taller canopy.

A subset of HVIs were calculated from reflectance spectra to explore the relationship between HVIs and soil moisture and ET (Figure 14.6). In this figure, BRDI is shown on the left and CAPY on the right. Temporal plots for three indices, $CI_{red\ edge}$ (chlorophyll), WBI (Water), and RVSI (Stress) are plotted with volumetric moisture content (VMC)20 and the ratio of AET to PET. All indices show a clear seasonal pattern that corresponds well to soil moisture, illustrating the strong dependence of plant growth on soil moisture in a Mediterranean climate. However, VMC20 and the HVIs are not perfectly aligned. For example, in 2008, the WBI significantly lags the peak in soil moisture, while $CI_{red\ edge}$ lacks VMC20 asymmetry. RVSI, interestingly, appears to capture some of the asymmetry in soil moisture in 2008. In 2009, all three indices were better aligned with VMC20. Furthermore, indices calculated for CAPY tend to better track seasonal changes in VMC20, particularly in 2008.

The HVIs align far better with the ratio of AET to PET than with VMC20. The peak in $CI_{red\ edge}$ and WBI, and trough in RVSI align very well with peak AET/PET for CAPY. For BRDI,

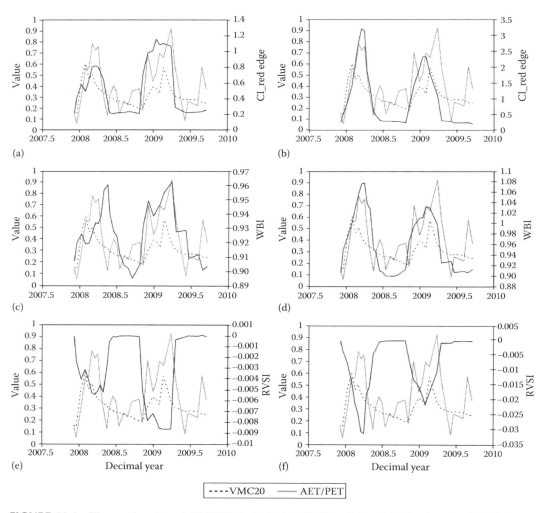

FIGURE 14.6 Time series plots of VMC20 (dashed), AET/PET (solid), and HVI values for BRDY (a, c, e) and CAPY (b, d, f) for CI_red edge (a, b), WBI (c, d), and RVSI (e, f).

CI$_{red\ edge}$, and RVSI match AET/PET, but WBI is clearly lagged in 2008, in which peak AET occurs well before peak WBI. Better alignment for CAPY, suggests that AET at the site may be largely controlled by the seasonal response in this species, which was already photosynthetically active by January and thus was able to take advantage of early winter rains and remain green as long as enough moisture was available to support high AET. Better alignment for AET than VMC20 is not surprising, given that VMC can be high when plants are completely senesced, yet AET will only be high when plants are green if transpiration is a major contributor to ET.

Statistical relationships between HVIs and environmental measures were evaluated by regressing a selection of HVIs against each of the measures over a 2 year period for each species (Figure 14.7; Tables 14.3 and 14.4). Many HVIs proved to show statistically significant relationships to VMC at 10 or 20 cm depth, PET, or AET/PET. For example, CRI1, had the highest r^2 value with soil moisture, followed closely by NDVI for BRDI (Table 14.3).

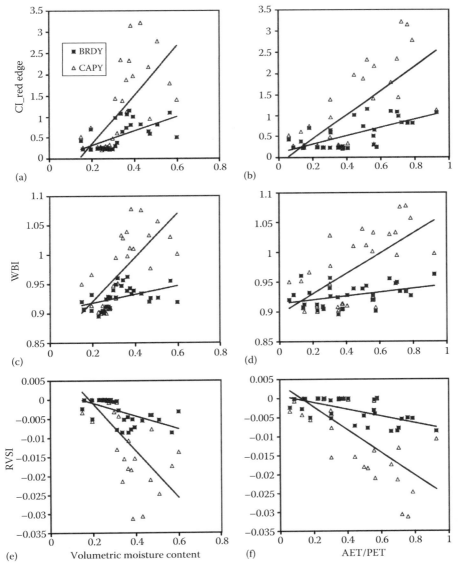

FIGURE 14.7 Scatterplots between VMC (a, c, e) and AET/PET (b, d, f) for three HVIs, CI_red edge, WBI and RVSI for BRDY (solid squares) and CAPY (open triangles).

TABLE 14.3

Statistical Relationships Using Linear Regression for BRDI for 21 HVSIs[a]

	VMC10	VMC20	PET	AET	AET/PET	VMC10	VMC20	PET	AET	AET/PET
			r^2					F-Stat		
Greenness										
DL1DGVI	0.302	0.295	0.043	**0.136**	0.342	13**	12.5**	1.3	4.7*	15.6***
DZ1DGVI	0.363	0.31	0.268	0.096	0.402	17.1***	13.9***	1.1**	3.1	20.2***
EVI	0.372	0.334	0.101	0.123	0.411	17.8***	15***	5.8*	4.2*	21***
NDVI	**0.414**	**0.367**	0.257	0.123	0.444	21.2***	17.4***	10.4**	4.2*	23.9***
SR	0.37	0.312	0.242	0.121	**0.448**	17.6***	13.6***	9.6**	4.2	24.4***
VARI	0.295	0.208	**0.51**	0.01	0.239	12.6**	7.9**	32***	0.3	9.4**
Pigments										
ARI2	0.308	0.321	0.018	**0.259**	0.454	13.3***	14.2***	0.6	10.5**	24.9***
CI_red edge	0.377	0.335	0.242	0.139	0.463	18.2***	15.1***	9.6**	4.9*	25.8***
CRI1	**0.429**	**0.397**	0.209	0.157	**0.477**	22.6***	19.8***	7.9**	5.6*	27.4***
MCARI	0.352	0.284	0.251	0.065	0.357	10.3***	11.9**	10**	2.1	10.7***
SIPI	0.332	0.291	**0.404**	0.055	0.311	14.9***	12.3**	20.4***	1.7	13.6***
Nitrogen										
NDNI	0.391	0.318	0.571	0.003	0.221	19.2***	14***	40***	0.1	8.5**
Water										
MSI	**0.173**	**0.194**	0.004	**0.147**	0.221	6.3*	7.2*	0.1	5.2*	8.5**
NDII	0.136	0.154	0.004	0.146	**0.222**	4.7*	5.5*	0.1	5.1*	8.6**
NDWI	0.073	0.086	0.014	0.076	0.116	2.4	2.8	0.4	2.5	3.9
WBI	0.155	0.185	**0.027**	0.048	0.15	5.5*	6.8*	0.8	1.5	5.3*
Cellulose/residues										
CAI	0.299	0.195	**0.558**	0.007	0.249	12.8**	7.2*	37.9***	0.2	9.9**
NDLI	**0.328**	**0.343**	0.02	**0.178**	**0.347**	14.6***	15.6***	0.6	6.5*	15.9***
Plant stress/physiology										
PRI	0.054	0.096	0.206	**0.12**	0.056	1.7	3.2	7.8**	4.1	1.8
RVSI	**0.371**	**0.332**	**0.259**	0.104	**0.413**	17.7***	14.9***	10.5**	3.5	21.1***

Note: Statistical significance is reported as (* = 0.05, ** = 0.01, *** = 0.001).

[a] r^2 values in bold highlight indices with the highest correlation for a specific measure and category of index.

For this shallow-rooted species, higher correlations were observed for 10 cm than 20 cm. Other indices that showed a strong correlation with soil moisture included all of the pigment and greenness measures, the NDNI, NDLI, and RVSI. The water-based indices showed a poorer relationship and no significant relationship was observed for PRI. Most indices were positively correlated to VMC and AET, although RVSI was negatively correlated, consistent with increasingly positive RVSI with increasing plant stress [52].

Several indices were strongly correlated to PET, most likely driven by a strong relationship to net radiation. These included VARI, SIPI, NDNI, and CAI. The poorest relationships were observed for AET. By contrast, the highest r^2 values were observed for AET/PET, with most greenness measures producing r^2 values greater than 0.4, and CRI1 producing the highest r^2 value of 0.477, followed closely by $CI_{red\ edge}$ and ARI2.

Similar correlations were observed for CAPY, although r^2 values tended to be considerably higher, as might be expected based on the better temporal match between VMC, AET, and reflectance changes in CAPY. The highest r^2 value for this species was 0.803, between the CRI1 and PET.

TABLE 14.4

Statistical Relationships Using Linear Regression for CAPY for 21 HVSIs[a]

	VMC10	VMC20	PET	AET	AET/ PET	VMC10	VMC20	PET	AET	AET/ PET
			r^2					F-Stat		
Greenness										
DL1DGVI	0.348	0.452	0.027	**0.336**	**0.506**	15.5***	23.9***	0.8	14.7***	29.7***
DZ1DGVI	0.369	0.448	0.084	0.259	0.48	16.9***	23.6***	2.7	10.2**	26.7***
EVI	0.388	0.476	0.057	0.289	0.501	18.4***	26.3***	1.8	11.8**	29.1***
NDVI	**0.48**	**0.534**	0.143	0.191	0.464	26.8***	33.2***	4.9*	6.8*	25.1***
SR	0.326	0.401	0.054	0.291	0.476	14***	19.4***	1.7	11.9**	26.4***
VARI	0.432	0.452	**0.213**	0.13	0.389	22***	23.9***	7.9**	4.3*	18.5***
Pigments										
ARI2	0.208	0.329	0.007	**0.433**	0.472	7.6**	14.2***	0.2	22.1***	25.9***
CI_red edge	0.387	**0.451**	0.091	0.246	**0.476**	18.3***	23.8***	2.9	9.5**	26.3***
CRI1	**0.542**	0.439	**0.803**	0.038	0.37	9.5*	6.3*	32.6***	0.3	4.7
MCARI	0.327	0.423	0.043	0.323	0.5	14.1***	21.3***	1.3	13.8***	29***
SIPI	0.397	0.423	0.3	0.059	0.293	19.1***	21.2***	12.4**	1.8	12**
Nitrogen										
NDNI	0.491	0.507	0.143	0.196	0.482	28***	29.8***	4.8*	7.1*	27***
Water										
MSI	**0.443**	0.517	**0.109**	0.163	0.396	23.1***	31.1***	3.6	5.7*	19***
NDII	0.404	0.504	0.064	**0.253**	**0.464**	19.7***	29.5***	2	9.8**	25.1***
NDWI	0.421	0.508	0.081	0.228	0.451	21.1***	29.9***	2.5	8.6**	23.8***
WBI	0.42	0.501	0.109	0.212	0.438	21***	29.1***	3.5	7.8**	22.6***
Cellulose/residues										
CAI	0.335	0.393	**0.21**	0.125	0.376	14.6***	18.7***	7.7**	4.1	17.4***
NDLI	**0.41**	**0.517**	0.044	**0.261**	**0.457**	20.2***	31.1***	1.3	10.2**	24.4***
Plant stress/physiology										
PRI	0.11	0.056	**0.598**	0.036	0.026	3.6	1.7	43.1***	1.1	0.8
RVSI	**0.391**	**0.47**	0.083	**0.256**	**0.48**	18.6***	25.7***	2.6	10**	26.8***

Note: Statistical significance is reported as (* = 0.05, ** = 0.01, *** = 0.001).

[a] r^2 values in bold highlight indices with the highest correlation for a specific measure and category of index.

This pigment index differed starkly from all other pigment indices, which showed typically poor relationships with PET. Very high r^2 values were also observed between greenness measures and VMC, with VMC at 20 cm showing a slightly stronger relationship (in contrast to BRDI). Pigment measures were similarly highly correlated with VMC, although the highest correlation, with an r^2 value of 0.542 was observed for CRI1 with VMC10. Contrary to BRDI, HVIs for CAPY were also highly correlated with AET and water-based indices were significantly correlated with soil moisture. Finally, r^2 values between HVIs and AET/PET tended to be higher for CAPY than BRDI.

Overall, we conclude that (1) HVIs are better predictors of AET/PET than soil moisture, (2) these relationships vary significantly between plant species with different phenologies, and (3) while pigment-based and greenness-based indices were effective across species, the performance of water-based indices was more species-dependent. These results show that HVIs can be of great use in applications ranging from individual species identification to quantifying environmental variables where field data is sparse or absent.

14.4   DISCUSSION

Most hyperspectral systems collect a large volume of data in wavelengths that are highly correlated. HVIs are one means by which the wealth of information captured in a spectrum can be distilled to a few, physically meaningful variables. In one form, HVIs take advantage of finer spectral sampling to generate a narrowband equivalent of a broadband VI. While narrowband equivalents of broadband indices have not necessarily improved performance (e.g., [1]), a change in the wavelength position of one or more bands within the index, better tuned for a specific absorption, has been shown to significantly improve performance [2,20].

A potentially greater contribution of hyperspectral systems is their ability to create new indices that incorporate wavelengths not sampled by any broadband system and to quantify absorptions that are specific to important biochemical and/or biophysical quantities of vegetation. Examples include most of the pigment-oriented indices, all indices formulated for the red edge, several water absorption indices, and indices that use three or more wavelengths. In many cases, these indices have either proven to be less sensitive to saturation, such as water-based indices for LAI, or less sensitive to changes in lighting/viewing geometry such as VARI [27]. NIR and SWIR based indices, because they sample spectral regions with reduced atmospheric scattering, would be expected to be less sensitive to atmospheric contamination. Some of the most promising HVIs are designed to quantify key plant physiological responses such as the PRI used as a proxy for LUE and RVSI, which is highly sensitive to seasonal changes in environmental stress. There is also considerable potential in combining several indices through multivariate regression to improve a biophysical or biochemical retrieval or as inputs into classification [61]. Such an approach takes advantage of the greater diversity of spectral features available through hyperspectral sensors to improve a retrieval based on multiple measures of the same variable and offers the potential of classifying vegetation based on inferred biochemistry and structure as in [61].

In this chapter, we provided a review of many of the most commonly used and best-adopted HVIs reported in the literature. We strongly linked the formulation of the HVI to its physical basis, typically including at least one strong absorption band and a nonabsorbing reference band. To illustrate some of the potential of these indices, we supplied two example case studies. As the number and amount of hyperspectral sensors and data increases, HVIs will become ever more important and prevalent.

REFERENCES

1. Elvidge, C.D. and Chen, Z., Comparison of broad-band and narrow-band red and near-infrared vegetation indices, *Remote Sensing of Environment*, 54:38–48, 1995.
2. Galvao, L.S., Viterello, I., and Almeida Filho, R., Effects of band positioning and bandwidth on NDVI measurements of tropical savannas, *Remote Sensing of Environment*, 67:181–193, 1999.
3. Jordan, C.F., Leaf-area index from quality of light on the forest floor, *Ecology*, 50(4):663–666, 1969.
4. Tucker, C.J., Red and photographic infrared linear combinations for monitoring vegetation, *Remote Sensing of Environment*, 8:127–150, 1979.
5. Sellers, P.J., Canopy reflectance, photosynthesis and transpiration, *International Journal of Remote Sensing*, 6:1335–1372, 1985.
6. Curran, P.J., Remote sensing of foliar chemistry, *Remote Sensing of Environment*, 30:271–278, 1989.
7. Gamon, J.A. and Surfus, J.S., Assessing leaf pigment content and activity with a reflectometer, *New Phytologist*, 143:105–117, 1999.
8. Ustin, S.L., Gitelson, A.A., Jacquemoud, S., Schaepman, M., Asner, G.P., Gamon, J.A., and Zarco-Tejada, P., Retrieval of foliar information about plant pigment systems from high resolution spectroscopy, *Remote Sensing of Environment*, 113:S67–S77, 2009.
9. Gamon, J.A., Serrano, L., and Surfus, J.S., The photochemical reflectance index: An optical indicator of photosynthetic radiation-use efficiency across species, functional types, and nutrient levels, *Oecologia*, 112:492–501, 1997.

10. Horler, D.N.H., Dockray, M., and Barber, J., The red-edge of plant leaf reflectance, *International Journal of Remote Sensing*, 4:273–288, 1983.

11. Zarco-Tejada, P.J., Miller, J.R., Mohammed, G.H., and Noland, T.L., Chlorophyll fluorescence effects on vegetation Apparent reflectance: I Leaf-level measurements and model simulation, *Remote Sensing of Environment*, 74:582–595, 2000.

12. Hunt, E.R. Jr., and Rock, B.N., Detection of changes in leaf water content using near- and middle-infrared reflectances, *Remote Sensing of Environment*, 30:43–54, 1989.

13. Gitelson, A.A., Kaufman, Y.J., Stark, R., and Rundquist, D., Novel algorithms for remote estimation of vegetation fraction, *Remote Sensing of Environment*, 80:76–87, 2002.

14. Roberts, D.A., Ustin, S.L., Ogunjemiyo, S., Greenberg, J., Dobrowski, S.Z., Chen, J., and Hinckley, T.M., Spectral and structural measures of Northwest forest vegetation at leaf to landscape scales, *Ecosystems*, 7:545–562, 2004.

15. Rouse, J.W., Haas, R.H., Schell, J.A., and Deering, D.W., Monitoring vegetation systems in the great plains with ERTS, in *Third ERTS Symposium*, NASA SP-351, NASA, Washington, DC, Vol. 1, pp. 309–317, 1973.

16. Carlson, T.N. and Ripley, D.A., On the relation between NDVI, fractional vegetation cover and leaf area index, *Remote Sensing of Environment*, 62:241–252, 1997.

17. Daughtry, C.S.T., Hunt, E.R. Jr., and McMurtrey III, J.E., Assessing crop residue cover using shortwave infrared reflectance, *Remote Sensing of Environment*, 90:126–134, 2004.

18. Deering, D.W., Middleton, E.M., and Eck, T.F., Reflectance anisotropy for a Spruce-Hemlock forest canopy, *Remote Sensing of Environment*, 47:242–260, 1994.

19. Huete, A.R., Jackson, R.D., and Post, D.F., Spectral response of a plant canopy with different soft backgrounds, *Remote Sensing of Environment*, 17:37–53, 1984.

20. Thenkabail, P.S., Smith, R.B., and De-Pauw, E., Hyperspectral vegetation indices for determining agricultural crop characteristics, *Remote Sensing of Environment*, 71:158–182, 2000.

21. Huete, A.R., A soil adjusted vegetation index (SAVI), *Remote Sensing of Environment*, 25:295–309, 1988.

22. Kaufman, Y.J. and Tanier, D., Atmospherically resistant vegetation index (ARVI) for EOS-MODIS, *IEEE Transactions on Geoscience and Remote Sensing*, 30(2):261–270, 1992.

23. Huete, A.R., Liu, H.Q., Batchily, K., and van Leeuwen, W., A comparison of vegetation indices over a global set of TM images for EOS-MODIS, *Remote Sensing of Environment*, 59:440–451, 1997.

24. Huete, A.R., Didan, K., Miura, T., Rodriguez, E.P., Gao, X., and Ferreira, L., Overview of the radiometric and biophysical performance of the MODIS vegetation indices, *Remote Sensing of Environment*, 83:195–213, 2002.

25. Gitelson, A.A., Viña, A., Arkebauer, T.J., Rundquist, D.C., Keydan, G., and Leavitt, B., Remote estimation of leaf area index and green leaf biomass in maize canopies, *Geophysical Research Letters*, 30(5):1248, 2003, doi:10.1029/2002GL016450.

26. Perry, E.M. and Roberts, D.A., Sensitivity of narrow-band and broad-band indices for assessing nitrogen availability and water stress in annual crop, *Agronomy Journal*, 100(4):1211–1219, 2008.

27. Roberts, D.A., Dennison, P.E., Peterson, S., Sweeney, S., and Rechel, J., Evaluation of AVIRIS and MODIS measures of live fuel moisture and fuel condition in a shrubland ecosystem in Southern California, *Journal of Geophysical Research Biogeosciences*, 111:G04S02, 2006, doi:10.1029/2005JG000113.

28. Peñuelas, J., Pinol, J., Ogaya, R., and Lilella, I., Estimation of plant water content by the reflectance water index WI (R900/R970), *International Journal of Remote Sensing*, 18:2869–2875, 1997.

29. Gao, B., NDWI: A normalized difference water index for remote sensing of vegetation liquid water from space, *Remote Sensing of Environment*, 58:257–266, 1996.

30. Blackburn, G.A., Spectral indices for estimating photosynthetic pigment concentrations: A test using senescent tree leaves, *International Journal of Remote Sensing*, 19:657–675, 1998.

31. Penuelas, J., Baret, F., and Filella, I., Semi-empirical indices to assess carotenoids/chlorophyll a ratio from leaf spectral reflectance, *Photosynthetica*, 31:221–230, 1995.

32. Blackburn, G.A., Quantifying chlorophylls and carotenoids from leaf to canopy scale: An evaluation of some hyperspectral approaches, *Remote Sensing of Environment*, 66:273–285, 1998.

33. Merzlyak, M.N., Gitelson, A.A., Chivkunova, O.B., and Rakitin, Y., Non-destructive optical detection of pigment changes during leaf senescence and fruit ripening, *Physiologia Plantarum*, 105:135–141, 1999.

34. Kim, M.S., The use of narrow spectral bands for improving remote sensing estimation of fractionally absorbed photosynthetically active radiation (fAPAR), Masters Thesis, Department of Geography, University of Maryland, College Park, MD, 1994.

35. Daughtry, C.S.T., Walthall, C.L., Kim, M.S., de Colstoun, E.B., and McMurtrey, J.E., Estimating corn leaf chlorophyll concentration from leaf and canopy reflectance, *Remote Sensing of Environment*, 74:229–239, 2000.

36. Gitelson, A.A., Keydan, G.P., and Merzlyak, M.N., Three band model for noninvasive estimation of chlorophyll, carotenoids, and anthocyanin contents in higher plant leaves, *Geophysical Research Letters*, 33:L11402, 2006.

37. Asner, G.P., Nepstad, D., Cardinot, G., and Ray, D., Drought stress and carbon uptake in an Amazon forest measured with spaceborne imaging spectroscopy, *Proceedings of the National Academy of Sciences of the United States of America*, 101(16):6039–6044, 2004.

38. Gitelson, A.A., Chivkunova, O.B., and Merzlyak, M.N., Nondestructive estimation of anthocyanins and chlorophylls in anthocyanic leaves, *American Journal of Botany*, 96(10):1861–1868, 2009.

39. Sims, D.A. and Gamon, J.A., Relationships between leaf pigment content and spectral reflectance across a wide range of species, leaf structures and developmental stages, *Remote Sensing of Environment*, 81:337–354, 2002.

40. Gitelson, A.A., Merzlyak, M.N., and Chivkunova, O.B., Optical properties and non-destructive estimation of anthocyanin content in plant leaves, *Photochemistry and Photobiology*, 74(1):38–45, 2001.

41. Van den Berg, A.K. and Perkins, T.D., Non-destructive estimation of anthocyanin content in autumn augar maple leaves, *Horticultural Science*, 40(3):685–685, 2005.

42. Gitelson, A.A., Zur, Y., Chivkunova, O.B., and Merzlyak, M.N., Assessing carotenoid content in plant leaves with reflectance spectroscopy, *Photochemistry and Photobiology*, 75(3):272–281, 2002.

43. Rock, B.N., Vogelmann, J.E., Williams, D.L., Vogelmann, A.F., and Hoshizaki, T., Detection of forest damage, *BioScience*, 36(7):439–445, 1986.

44. Serrano, L., Ustin, S.L., Roberts, D.A., Gamon, J.A., and Penuelas, J., Deriving water content of chaparral vegetation from AVIRIS data, *Remote Sensing of Environment*, 74:570–581, 2000.

45. Sims, D.A. and Gamon, J.A., Estimation of vegetation water content and photosynthetic tissue area from spectral reflectance: A comparison of indices based on liquid water and chlorophyll absorption, *Remote Sensing of Environment,* 84:526–537, 2003.

46. Zygielbaum, A.I., Gitelson, A.A., Arkebauer, T.J., and Rundquist, D.C., Non-destructive detection of water stress and estimation of relative water content in maize, *Geophysical Research Letters*, 36:L12403, 2009, doi:10.1029/2009GL038906.

47. Daughtry, C.S.T., Discriminating crop residues from soil by shortwave infrared reflectance, *Agronomy Journal*, 93:125–131, 2001.

48. Serrano, L., Penuelas, J., and Ustin, S.L., Remote sensing of nitrogen and lignin in Mediterranean vegetation from AVIRIS data: Decomposing biochemical from structural signals, *Remote Sensing of Environment*, 81:355–364, 2002.

49. Card, D.H., Peterson, D.L., Matson, P.A., and Aber, J.D., Prediction of leaf chemistry by the use of visible and near infrared reflectance spectroscopy, *Remote Sensing of Environment,* 26:123–147, 1988.

50. Daughtry, C.S.T., Hunt, E.R. Jr., Doraiswamy, P.C., and McMurtrey, J.E. III, Remote sensing the spatial distribution of crop residues, *Agronomy Journal*, 97:864–871, 2005.

51. Milton, N.M., Eiswerth, B.A., and Ager, C.M., Effect of phosphorus deficiency on spectral reflectance and morphology of soybean plants, *Remote Sensing of Environment*, 36:121–127, 1991.

52. Merton, R. and Huntington, J., Early simulation results of the ARIES-1 satellite sensor for multi-temporal vegetation research derived from AVIRIS, NASA Jet Propulsion Laboratory, Pasadena, CA, 1999. Available at ftp://popo.jpl.nasa.gov/pub/docs/workshops/ 99_docs/41.pdf

53. Naidu, R.A., Perry, E.M., Pierce, F.J., and Mekuria, T., The potential of spectral reflectance technique for the detection of Grapevine leafroll-associated virus-3 in two red-berried wine grape cultivars, *Computers and Electronics in Agriculture*, 66, 38–45, 2009.

54. Gamon, J.A., Kitajima, K., Mulkey, S.S., Serrano, L., and Wright, S.J., Diverse optical and photosynthetic properties in a neotropical dry forest during the dry season: Implications for remote estimation of photosynthesis, *Biotropica*, 37(4):547–560, 2005.

55. Rahman, A.F., Gamon, J.A., Fuentes, D.A., Roberts, D.A., and Prentiss, D., Modeling spatial distributed ecosystem flux of boreal forests using hyperspectral indices from AVIRIS imagery, *Journal of Geophysical Research Atmospheres*, 106(d24):33579–33591, 2001.

56. Roberts, D.A., Brown, K.J., Green, R., Ustin, S., and Hinckley, T., Investigating the relationship between liquid water and leaf area in clonal Populus, in *Proceedings of the 7th AVIRIS Earth Science Workshop JPL 97–21*, Pasadena, CA, pp. 335–344, 1998.

57. Monteith, J.L. and Unsworth, M.H., *Principles of Environmental Physics*, 2nd edn., Edward Arnold, London, U.K., 1990.

58. Bowen, I.S., The ratio of heat losses by conduction and by evaporation from any water surface, *Physical Review*, 27:779, 1926.
59. Roberts, D.A., Bradley, E.S., Roth, K., Eckmann, T., and Still, C., Linking physical geography education and research through the development of an environmental sensing network and project-based learning, *Journal of Geoscience Education*, 58:262–275, 2010.
60. Bradley, E., Still, C., and Roberts, D., Design of an image analysis website for phenological and meteorological monitoring, *Environmental Modelling and Software,* 25:107–116, 2010.
61. Fuentes, D., Gamon, J.A., Qiu, H.-L., Sims, D., and Roberts, D., Mapping Canadian Boreal forest vegetation using pigment and water absorption features derived from the AVIRIS sensor, *Journal of Geophysical Research—Atmospheres*, 106(D24):33565–33577, 2001.

15 Remote Sensing Estimation of Crop Biophysical Characteristics at Various Scales

Anatoly A. Gitelson

CONTENTS

15.1 INTRODUCTION

Remote sensing has provided valuable insights into agronomic management over the past few decades. The use of remote sensing for determining crop physiological and phenological status has its roots in the pioneering work by William Allen, Harold Gausman, and Joseph Woolley [1–3], who provided much of the basic theory relating morphological characteristics of crop plants to their optical properties. These pioneering works have led to the understanding of how leaf reflectance changes in response to leaf thickness, species, canopy shape, leaf age, nutrient status, and water status. Leaf chlorophyll content and its absorption in the visible spectrum provide the basis for utilizing reflectance as a tool either with broadband radiometers typical of current satellite systems or hyperspectral sensors that measure reflectance at narrowbands. The basic understanding of leaf reflectance has led to the development of various vegetation indices (VIs) that have been extended to crop canopies and have been used to quantify various agronomic parameters (e.g., leaf area, crop cover, biomass, crop type, nutrient status, and yield). These tools are still being developed as we learn more about how to use the information contained in reflectances from a range of different sensors.

A summary of the progress in applying remote sensing to agriculture was recently published in a collection of articles in *Photogrammetric Engineering and Remote Sensing* (volume 69) [4–8]. Other recent reviews of the application of remote sensing methods to crops were developed by Hatfield et al. [9,10]. These articles provide a summary of the multispectral and hyperspectral remote sensing efforts in more detail and the reader is referred to these articles for a more thorough understanding.

This chapter contains a summary of experiences and advances made in the last 10 years in understanding how remote sensing at a close range and at satellite level can be used to quantitatively assess crop biophysical characteristics (BPCs). In what follows, the performance of VIs in estimating crop vegetation fraction (VF), fraction of photosynthetically active radiation (PAR) absorbed by photo-synthetically active ("green") vegetation, total canopy chlorophyll content, green leaf area index (LAI), and gross primary production is tested at close range (6 m above the top of canopy [TOC]). All spectral VIs discussed here were calculated in the spectral bands of the Moderate Resolution Imaging Spectrometer (MODIS) and the Medium Resolution Imaging Spectrometer (MERIS): near-infrared (NIR) (841–876 nm), red edge (707–717 nm), red (620–670 nm), and green (545–565 nm). Examples of green LAI estimation using MODIS data and gross primary production estimation employing enhanced thematic mapper (ETM) Landsat data are also presented.

15.2 VEGETATION FRACTION

One of the principal variables in the growth of crops is the fraction of the solar radiation intercepted by foliage. The productivity of crops may be analyzed as the product of the solar energy intercepted over a season and the efficiency with which that energy is converted to biomass. In many crops the relationship between radiation interception and green foliage cover/fraction is sufficiently close for the latter to be used as a substitute for more elaborate measurements of light interception [11]. Thus, the VF is an important quantity that helps determine crop productivity.

There are three basic approaches for estimating green VF from remotely sensed data: spectral mixture analysis (e.g., [12,13]), neural networks (e.g., [14]), and VIs. For proximal sensing, VIs are commonly used. Spectral mixing with the use of reference end-members has been employed to model reflectance data as mixtures of green vegetation, nonphotosynthetic vegetation, soils, and shade [15]. It is an effective way to monitor VF. Linear spectral mixture modeling combined with principal component analysis has allowed for retrieving agriculture and agronomic information from an image using 10–20 spectral channels [16]. However, this technique requires ancillary ground measurements of pertinent biophysical variables in order to determine the empirical relationship between those variables and unmixing fractions for each different set of hyperspectral data. Other problems with spectral mixing have been described by Price [17].

Using neural networks VF can be accurately estimated from the NIR and red reflectances, with the only ancillary information being the vegetation type and soil line characteristics [14]. This "black box" technique, which is dependent on the data set used in the training process, was found to perform better than the VIs examined. Any neural network model is constrained by the assumptions and approximations that define the model and may fail when applied to data sets that do not conform to these assumptions and approximations. The main limitation of this approach results from the approximations and assumptions made in the modeling of the radiative transfer, and in the distribution of the input variables of the model. Thus, the VF derived from the mixture modeling as well as from neural networks is not a straightforward estimation of the vegetation amount and requires careful analysis [18].

Spectral VIs are widely used as indicators of temporal and spatial variations in vegetation structure and biophysical parameters (e.g., reviews [10,19–21] and references therein). This approach has already proven to be relevant to many requirements of vegetation status monitoring, including the assessment and monitoring of changes in BPCs, such as VF, LAI, fraction of absorbed photosynthetically active radiation (fAPAR), and net primary production (NPP) [22–26]. Most VIs combine reflectance (ρ) in two spectral bands, the red and NIR. Simple Ratio (SR) [27] and Normalized Difference Vegetation Index (NDVI) [28], were among the first used for estimating vegetation BPCs:

$$SR = \frac{\rho_{NIR}}{\rho_{red}} \tag{15.1}$$

$$NDVI = \frac{(\rho_{NIR} - \rho_{red})}{(\rho_{NIR} + \rho_{red})} \tag{15.2}$$

The relationship between NDVI and VF for wheat is shown in Figure 15.1. For low-to-moderate VF, NDVI is sensitive to VF. However, as VF exceeds 50%, NDVI levels off and its sensitivity to VF drops drastically. The reason for that is the very low variability of the red reflectance [29,30] and the very mathematical formulation of NDVI, which makes the index insensitive to variation in ρ_{red} when $\rho_{NIR} \gg \rho_{red}$; for crops with moderate-to-high biomass, $\rho_{red} \sim 2\%$ to 5% and $\rho_{NIR} > 40\%$ [31].

Considerable effort has been expended in improving NDVI and in developing new indices to compensate for the atmosphere [32,33] and the canopy background such as soil and crop residue [34–37]. Nevertheless, the indices have limitations, some of which are due either to the choices of band location and bandwidth [24,25,38–40] or to the fact that NIR reflectance levels off or even decreases with an increase in VF during later stages of growth [41–45].

Huete et al. introduced Enhanced Vegetation Index (EVI) [46], which has a higher sensitivity to moderate-to-high vegetation biomass and is widely used as a product of the MODIS system. EVI2 [47] containing only the red and NIR bands was formulated in the form:

$$EVI2 = \frac{2.5 \times (\rho_{NIR} - \rho_{red})}{(1 + \rho_{NIR} + 2.4 \times \rho_{red})} \tag{15.3}$$

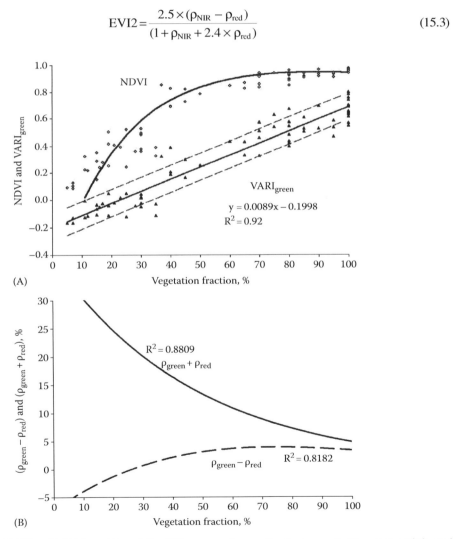

(A)

(B)

FIGURE 15.1 NDVI and VARI$_{green}$ (A) and the difference and sum of reflectances in the green and the red ranges of the spectrum (B), plotted against the VF in wheat; Israel 1998. Spectral bands correspond to the spectral bands of MODIS.

Wide Dynamic Range Vegetation Index (WDRVI) [31], is a nonlinear transformation of NDVI in the form:

$$WDRVI = \frac{(\alpha \times \rho_{NIR} - \rho_{red})}{(\alpha \times \rho_{NIR} + \rho_{red})} \qquad (15.4)$$

The weighting coefficient, α, is introduced to attenuate the contribution of the NIR reflectance at moderate-to-high green biomass, and to make it comparable to that of the red reflectance [31,48].

A few authors have considered using only the visible range of the spectrum for VF estimation. Kanemasu [43] recommended using the ratio of the reflectance at 545 nm to the reflectance at 655 nm as a measure of VF regardless of the crop type. Pickup et al. [49] developed a green vegetation cover index in the green and red spectral regions and applied it to Australia's rangelands [49]. This index is similar in form to the perpendicular vegetation index [50], although only the visible range of the spectrum was used in the green vegetation cover index.

To estimate VF it was suggested to use atmospherically resistant VIs in the form [51]:

$$VARI_{green} = \frac{(\rho_{green} - \rho_{red})}{(\rho_{green} + \rho_{red} - \rho_{blue})} \qquad (15.5)$$

$$VARI_{red\ edge} = \frac{(\rho_{red\ edge} - 1.7 \times \rho_{red} + 0.7 \times \rho_{blue})}{(\rho_{red\ edge} + 2.3 \times \rho_{red} - 1.3 \times \rho_{blue})} \qquad (15.6)$$

These indices are based on the fundamental spectral properties of vegetation: (a) high sensitivity of the red reflectance to low-to-moderate VF and low sensitivity to moderate-to-high VF; and (b) high sensitivity of the green and the red edge reflectances to moderate-to-high VF [29,52,53]. For VF ranging from 0% to 60%, the difference ($\rho_{green} - \rho_{red}$) increases (Figure 15.1B) due to a sharp decline in ρ_{red}, which is associated with chlorophyll absorption by crops during early-season growth, and a moderate decrease in ρ_{green}. As VF increases beyond 60%, the difference tends to level off: ρ_{green} decreases slightly while ρ_{red} is virtually invariant. The relationship between the sum ($\rho_{green} + \rho_{red}$) and VF tends to become hyperbolic with the sum decreasing significantly even as VF increases above 60%. As a result, $VARI_{green}$ and $VARI_{red\ edge}$ do increase and their relationship with VF is linear. Therefore, the normalization of the reflectance difference ($\rho_{green} - \rho_{red}$) or ($\rho_{red\ edge} - \rho_{red}$) to the sum of these reflectances results in a linear relationship between the index and VF (Figure 15.1A).

Using hyperspectral radiometry at a distance of 6 m above the TOC (details are in [51,54,55]), the performances of seven general types of VIs (summarized in Table 5 in [21]) in estimating VF of maize and soybeans were tested. VF was retrieved from digital camera images acquired using a Kodak DC-40 system. The camera, mounted adjacent to the hyperspectral radiometer SE-590, provided above-canopy images of the crops. A image was acquired concurrently with spectral data collection over each of the 16 plots. The images were imported into ERDAS Imagine (ver. 8.3.1) for processing. The area (size) and location of the field of view of the SE-590 radiometer within each image was determined and a model was designed to exclude from the image the data from outside radiometer's field of view. The model also separated nonvegetation (soil) pixels from vegetation pixels by subtracting the green band from the red band. The images were saved as files containing two classes: nonvegetation and vegetation (details are in [51]).

TABLE 15.1
Slope, Intercept, and Determination Coefficient, R^2, of the Linear Relationship, VIs versus VF, Established Using the Calibration Data Set, for Maize in Three Test Sites in Nebraska

VI	$VARI_{green}$	$VARI_{red\ edge}$	WDRVI	Green NDVI	EVI2	Red edge NDVI
Slope	196.02	246.36	72.30	172.64	144.48	125.67
Intercept	44.32	−13.12	65.82	−63.99	−12.38	−14.46
R^2	0.911	0.913	0.928	0.936	0.939	0.955

The performance of the vegetation indices was investigated. In addition to the already-mentioned EVI2, $VARI_{green}$, and $VARI_{red\ edge}$, the indices that performed the best were Green NDVI and Red edge NDVI, which use the same NDVI formulation, but with the green (around 550 nm) and the red edge (around 700 nm) bands, respectively [57]:

$$Green\ NDVI = \frac{(\rho_{NIR} - \rho_{green})}{(\rho_{NIR} + \rho_{green})} \tag{15.7}$$

$$Red\ edge\ NDVI = \frac{(\rho_{NIR} - \rho_{red\ edge})}{(\rho_{NIR} + \rho_{red\ edge})} \tag{15.8}$$

As stated earlier, all indices were calculated with reflectances in the simulated discrete spectral bands of MODIS and MERIS.

The relationships between VF and VIs were established using data collected during 1 year (33 field campaigns) at three irrigated and rainfed maize sites (Table 15.1) and validated using data collected at three irrigated and rainfed maize sites (35 campaigns) during another year. In Figure 15.2 the relationships of VI versus VF are presented for two independent data sets, one for calibration (2005) and the other for validation (2003). For all six VIs, the relationships were significantly close (p < 0.001) with $R^2 > 0.9$ with the red edge NDVI performing slightly better (Table 15.1).

Validation of the established relationships (Figure 15.3, Table 15.2) showed that all of the aforementioned VIs, which were selected from a pool of 30 indices, were able to estimate VF ranging from 0% to 95% with a root mean square error (RMSE) below 11%. The best index, the red edge NDVI, yielded a highly accurate estimation, with the RMSE below 7% and the coefficient of variation (CV) <12% (Table 15.2). WDRVI and EVI are recommended for estimating VF using data from MODIS with a 250 m spatial resolution, which has spectral bands in the red and NIR regions. The red edge NDVI is recommended for estimating VF from MERIS data with a spatial resolution of 300 m.

15.3 FRACTION OF ABSORBED PHOTOSYNTHETICALLY ACTIVE RADIATION

The fAPAR is one of the most important crop BPCs. It is one of the main factors used in the formulation of production efficiency models (PEM). Numerous studies (e.g., [22,25]) have found that under specified canopy reflectance properties, fAPAR can be estimated remotely using the NDVI (Equation 15.2). Several PEMs have been developed that use synoptic NDVI data to estimate NPP (e.g., [59]). Roujean and Breon [60] underscored the fact that the linear relationship between fAPAR

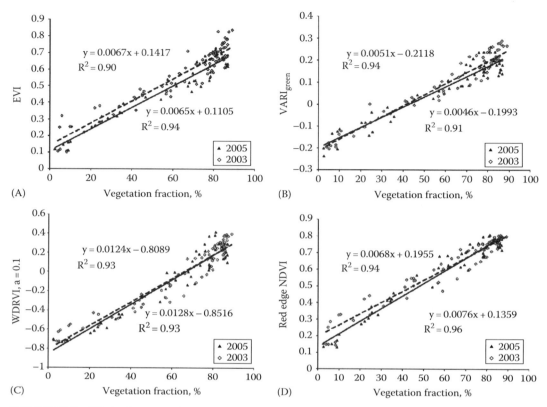

FIGURE 15.2 VIs in the spectral bands of MODIS (EVI, WDRVI, and VARI$_{green}$) and MERIS (Red edge NDVI) plotted against the VF in three maize sites in Nebraska: (A) EVI, (B) VARI$_{green}$, (C) WDRVI, and (D) Red edge NDVI.

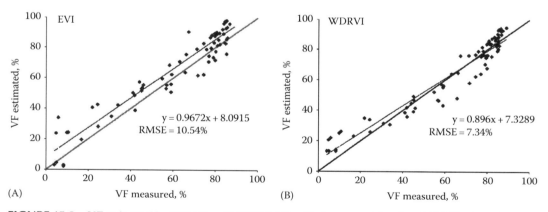

FIGURE 15.3 VF estimated by EVI (A) and WDRVI (B), with the spectral bands of MODIS, plotted against the measured VF for maize and soybean. Thick solid lines are VF$_{pred}$ = VF$_{meas}$ thin lines are best fit functions.

and NDVI is an approximation, which is only valid during the growing stage. A likely explanation for this is that during the reproductive and senescence stages in crops, the canopy still intercepts the incoming radiation, but the leaves contain less photosynthetic pigments, and, hence, only a fraction of the absorbed light is used for photosynthesis [61,62]. Therefore, since only the green parts of the canopy are used for photosynthesis, fAPAR needs to be separated into its photosynthetically

TABLE 15.2

Slope and Intercept of the Linear Relationship between VF Measured and VF Predicted for Three Maize Validation Test Sites in Nebraska by Algorithms Presented in Table 15.1

VI	EVI2	Green NDVI	VARI$_{red\ edge}$	VARI$_{green}$	WDRVI	Red Edge NDVI
Slope	0.93	1.21	1.07	0.94	1.03	1.10
Intercept	−1.45	−14.28	−6.92	1.18	−3.05	−7.49
MNB (%)	22.91	24.52	22.14	8.34	19.69	13.43
RMSE	10.54	9.32	8.72	7.38	7.34	6.95
CV (%)	17.06	15.09	14.11	11.94	11.88	11.37

The mean normalized bias (MNB), the RMSE of VF prediction, and the CV (CV = RMSE/mean VF) are also presented.

(fAPAR$_{green}$) and nonphotosynthetically active components, in order to improve the estimation of vegetation productivity over time [63]. Thus, fAPAR$_{green}$ is defined as

$$fAPAR_{green} = fAPAR \times \left(\frac{LAI_{green}}{LAI_{total}} \right) \tag{15.9}$$

where LAI$_{green}$ and LAI$_{total}$ are the green and total LAI, respectively.

A significant decrease in the sensitivity of NDVI has been observed when fAPAR exceeds 0.7 [22,64] as the NDVI does not capture the changes in vegetation with moderate-to-high biomass. In this section, the NDVI versus fAPAR$_{green}$ relationship in maize and soybean, grown under irrigated and rainfed conditions, was evaluated and other spectral VIs were tested in order to overcome the limitation due to the decreased sensitivity of NDVI to fAPAR at moderate-to-high biomass (details in [65]). In both maize and soybean, fAPAR showed a progressive increase during the vegetative stage until maximum canopy development, and, importantly, remained virtually invariant during the reproductive stage, with a decrease during the senescence stage (Figure 15.4). In soybean, (Figure 15.4B) this decrease was particularly conspicuous due to a drastic loss of leaf cover. During the vegetative stage, the increase in fAPAR coincided with an increase in LAI up to a point at which further increase in LAI (above 4 m²/m²) virtually did not induce an increase in fAPAR. In contrast,

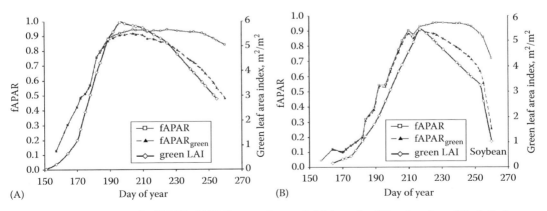

FIGURE 15.4 Three BPCs, fAPAR, fAPAR$_{green}$, and green LAI, in maize (A) and soybean (B) plotted versus the day of year.

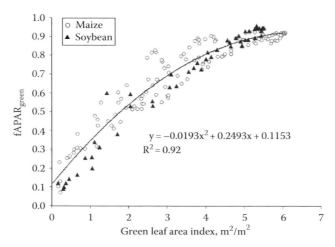

FIGURE 15.5 fAPAR$_{green}$ plotted versus the green LAI in maize and soybean; three sites, 3 years of observation.

during the reproductive and senescence stages, fAPAR remained almost insensitive to decrease in LAI. During these stages, the crops absorb PAR by photosynthetic and nonphotosynthetic components, but the contribution of the photosynthetic component decreases considerably toward the end of the season. Thus, although the canopy was still intercepting PAR, it was progressively used less for photosynthesis.

The fAPAR$_{green}$ (Equation 15.9) is a measure of the fAPAR absorbed only by the photosynthetic component of the vegetation [63,66]. In the reproductive and senescence stages, the behavior of fAPAR$_{green}$ was very different from that of fAPAR. It decreases significantly during the reproductive stage and drops drastically in senescence. The temporal behavior of fAPAR$_{green}$ generally coincided with that of green LAI (Figure 15.4). However, the relationship fAPAR$_{green}$ versus green LAI for both maize and soybean was asymptotic, with a considerable decrease in the sensitivity of fAPAR$_{green}$ to green LAI exceeding $4\,m^2/m^2$ (Figure 15.5). Thus, increase in green LAI beyond $4\,m^2/m^2$ virtually did not cause an increase in the absorption of PAR by photosynthetically active vegetation.

The relationship between NDVI and fAPAR$_{green}$ was almost nonspecies specific (Figure 15.6). Thus, NDVI can be thought of as a proxy of fAPAR$_{green}$, although the relationship was asymptotic,

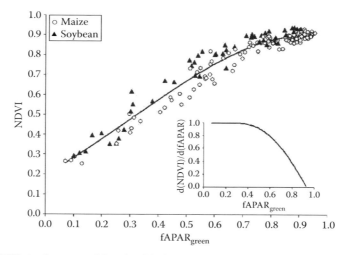

FIGURE 15.6 NDVI, in the spectral bands of MODIS, plotted versus fAPAR$_{green}$ in maize and soybean. Insert: the first derivative of NDVI with respect to fAPAR$_{green}$.

with a significant decrease in the slope as $fAPAR_{green}$ exceeds 0.5 (see first derivative d(NDVI)/d($fAPAR_{green}$) in the insert in Figure 15.6). This has been also observed by several authors (e.g., [24,43,67]).

Thus, NDVI exhibits limitations at moderate-to-high vegetation density. Its sensitivity to $fAPAR_{green}$ drops twofold for $fAPAR_{green}$ = 0.74 and 10-fold for $fAPAR_{green}$ = 0.89. As $fAPAR_{green}$ exceeds 0.7, the noise equivalent of $fAPAR_{green}$ estimation by NDVI grows exponentially, reaching 0.25 for $fAPAR_{green}$ = 0.8 [65]. It means that for more than 2 months during the growing season, as $fAPAR_{green}$ exceeds 0.7 (Figure 15.4A and B). NDVI does not bring reliable information about $fAPAR_{green}$. This limitation is due to: (a) choices of band location and bandwidth [e.g., 25,38,40]; and (b) the very mathematical formulation of the NDVI [31], as mentioned earlier.

To correct for this significant loss of sensitivity, different spectral bands have been incorporated into the mathematical formulation of NDVI and new indices have been also developed. EVI2 (Equation 15.3) [46,47] uses the same spectral bands as NDVI. However, due to the increased weight of the NIR band, it is more sensitive than NDVI to moderate-to-high biomass [68].

The performances of seven general types of VIs (summarized in Table 5 in [21]) were tested for estimating $fAPAR_{green}$ using hyperspectral radiometric data taken from 6m above the top of the crop canopy (details in [54,65]). Daily measurements of PAR were obtained using the following procedures. The incoming PAR (PAR_{inc}) was measured with Li-Cor (Lincoln, Nebraska) point quantum sensors pointed to the sky, and placed at 6m above the ground. PAR reflected by the canopy and soil (PAR_{out}) was measured with Li-Cor point quantum sensors pointed down, and placed at 6m above the ground. PAR transmitted through the canopy (PAR_{transm}) was measured with Li-Cor line quantum sensors placed at about 2cm above the ground, and pointed upward; PAR reflected by the soil (PAR_{soil}) was measured with Li-Cor line quantum sensors placed about 12cm above the ground, and pointed downward (details in [66]):

$$APAR = PAR_{inc} - PAR_{out} - PAR_{transm} + PAR_{soil}$$

fAPAR was calculated as ($APAR/PAR_{inc}$) and $fAPAR_{green}$ as fAPAR × (green LAI/total LAI) [63]. Measurement of LAI is described in Section 15.5.

All indices were calculated with reflectances simulated in the discrete spectral bands of MODIS and MERIS. Six VIs were found to have close relationships with $fAPAR_{green}$ (Table 15.3 and Figure 15.7). The relationships of $fAPAR_{green}$ with EVI2, WDRVI, $VARI_{green}$, and $VARI_{red\ edge}$ were significantly close ($p < 0.001$) although asymptotic, especially for EVI, which lost sensitivity to $fAPAR_{green}$

TABLE 15.3
Relationships between VIs and $fAPAR_{green}$
and Established for Maize in Three Test Sites
during 3 Years of Observations in Nebraska

VI	VI versus $fAPAR_{green}$	R^2
$VARI_{green}$	$y = -5.567x^2 + 5.159x - 0.344$	0.79
$VARI_{red\ edge}$	$y = -1.211x^2 + 1.571x + 0.409$	0.88
EVI2	$y = -1.063x^2 + 2.285x - 0.264$	0.88
Green NDVI	$y = 1.6891x - 0.5271$	0.92
WDRVI	$y = -0.316x^2 + 0.576x + 0.719$	0.93
Red edge NDVI	$y = 1.2531x - 0.1035$	0.95

Source: Viña, A. and Gitelson, A.A., *Geophys. Res. Lett.*, 32, L17403, 2005.

R^2 is the coefficient of determination.

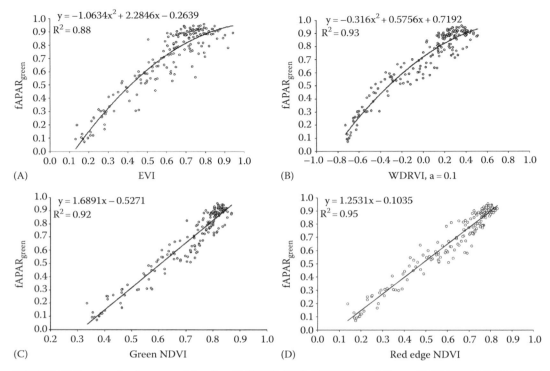

FIGURE 15.7 VIs, in the spectral bands of MODIS (EVI, WDRVI, and Green NDVI) and MERIS (Red edge NDVI), plotted versus fAPAR$_{green}$ in maize and soybean; three sites, 3 years of observation. (A) EVI, (B) WDRVI, (C) Green NDVI, and (D) Red edge NDVI.

above 0.6. The shape of the relationships fAPAR$_{green}$ versus EVI and fAPAR$_{green}$ versus WDRVI was almost the same as the shape of the relationship fAPAR$_{green}$ versus LAI (compare Figure 15.5 with Figure 15.7A and B). It means that EVI and WDRVI are better proxies of LAI than fAPAR$_{green}$. In contrast, both Green NDVI and Red edge NDVI relate closely and linearly with fAPAR$_{green}$ (R^2 is above 0.92) and are able to accurately estimate fAPAR$_{green}$.

It is important to note that the determination coefficient, R^2, represents the dispersion of the points from the best fit regression line. It constitutes a measure of how good the regression model is in capturing the relationship between the variables of interest. However, when the regression functions are nonlinear, the R^2 value might be misleading. For example, although EVI2 versus fAPAR$_{green}$ shows high values of R^2 and EVI2 has a high sensitivity to fAPAR$_{green}$ below 0.6, the sensitivity drops by five to six times when fAPAR$_{green}$ exceeds 0.6 (Figure 15.7A). Although the decrease in the sensitivities of WDRVI, VARI$_{green}$, and VARI$_{red\,edge}$ to high fAPAR$_{green}$ is less than that of EVI2, this factor should be taken into account while choosing a model for fAPAR$_{green}$ estimation.

To estimate how sensitive each of the VIs is to changes in BPC, it was recommended to use the noise equivalent of BPC (e.g., fAPAR) estimation (NE ΔBPC), calculated as following [65]:

$$NE\triangle(BPC) = \frac{RMSE(VI\ vs.\ BPC)}{d(VI)/d(BPC)}$$

where
   BPC is the biophysical characteristic of interest
   RMSE (VI versus BPC) is the RMSE of the relationship VI versus BPC
   d(VI)/d(BPC) is the first derivative of VI with respect to BPC

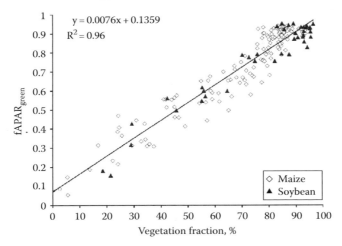

FIGURE 15.8 fAPAR$_{green}$ versus the VF in maize (three sites, 3 years of observation) and soybean (two sites, 1 year of observation).

The noise equivalent of BPC estimation should be calculated in each case when a nonlinear relationship of VI versus BPC is observed. The noise equivalent of BPC estimation allows making direct quantitative comparisons among different indices with different scales and dynamic ranges.

All six indices, which were found to be the best for VF estimation, performed much better than the other indices tested for estimating fAPAR$_{green}$. Moreover, the two best indices for VF estimation (Red edge NDVI and WDRVI) were also the best for estimating fAPAR$_{green}$. This is not surprising because the relationship between fAPAR$_{green}$ and VF is very close, with R^2 higher than 0.96. Figure 15.8 illustrates this using the data from 3 years of observation at three test sites, including rainfed and irrigated maize and soybean. Thus, the best indices for VF estimation can be used for estimating fAPAR$_{green}$. Nevertheless the question as to whether VF can be used as an accurate proxy of fAPAR$_{green}$ and vice versa requires further investigation.

Thus, to obtain accurate estimates of fAPAR$_{green}$ at moderate-to-high vegetation biomass densities, where NDVI loses its sensitivity, alternative indices can be used. Red edge NDVI appears to be the best index for such estimation. It can be used in satellite systems with spectral bands in the red edge region (e.g., Hyperion and MERIS). The WDRVI exhibited high sensitivity to fAPAR$_{green}$ over the full range of fAPAR$_{green}$ variation. It can be employed to estimate fAPAR$_{green}$ using sensors such as Landsat, MODIS, and Advanced Very High Resolution Radiometer (AVHRR) among others.

15.4 CHLOROPHYLL CONTENT

The importance of studying chlorophyll content in vegetation has been well recognized (e.g., [69]). Long- or medium-term changes in chlorophyll can be related to photosynthetic capacity (thus, productivity), developmental stage, and canopy stresses (e.g., [70]). It was suggested that chlorophyll content may be the plant community property that is most directly relevant to the prediction of productivity [71].

There are two different approaches to the remote estimation of chlorophyll content in crops. One of them is by the assessment of the *leaf chlorophyll content* [72,73] and the other approach is by the assessment of the *total crop chlorophyll content*, which is defined as the product of leaf chlorophyll content and total LAI [55,74]. Monitoring *leaf chlorophyll content* is a crucial component of farm management [73]. *Canopy chlorophyll content* relates closely to crop primary production [e.g., 71,75–77] and, thus, is a good indicator of the physiological status and the carbon sequestration potential of crops [58]. It is also a good predictor of crop yield [78,79].

Changes in leaf chlorophyll content produce large differences in canopy reflectance and transmittance spectra. However, canopy reflectance is also strongly affected by other factors (e.g., canopy architecture, Chl distribution within the canopy, LAI, and soil background) that mask and confound the changes in canopy reflectance caused by variations in leaf chlorophyll content. Daughtry et al. [72] found that VIs that combined NIR reflectance and red reflectance minimized the contributions of background reflectance, while VIs that combined reflectances in NIR and other visible bands (green and/or red edge) were responsive to both leaf chlorophyll content and back-ground reflectance. One example of such an approach is the use of two VIs, namely, the Modified Chlorophyll Absorption Ratio Index, (MCARI) [72] and the Optimized Soil-Adjusted Vegetation Index (OSAVI) [80] to estimate leaf chlorophyll content.

The indices were formulated as follows:

$$MCARI = (\rho_{700} - \rho_{670}) - 0.2 \times (\rho_{700} - \rho_{550}) \times \left(\frac{\rho_{700}}{\rho_{670}} \right) \tag{15.10}$$

$$OSAVI = \frac{(1+0.16) \times (\rho_{800} - \rho_{670})}{(\rho_{800} + \rho_{670} + 0.16)} \tag{15.11}$$

A test with measured canopy reflectance and leaf chlorophyll data confirmed that this technique can be used for *leaf chlorophyll content* retrieval [72]. Later, combination of indices based on the Transformed Chlorophyll Absorption Reflectance Index (TCARI) [73], the MCARI [72], and the OSAVI, such as, TCARI/OSAVI and MCARI/OSAVI, has been shown to accurately estimate *leaf chlorophyll content*, minimizing the effects of the soil background and the LAI variation in crops [72,73].

Several remote sensing techniques have been proposed to estimate the *total crop chlorophyll* content. NDVI was found to be a good indicator of low-to-moderate total chlorophyll content. However, the saturation of the red reflectance and the much higher NIR reflectance than red reflec-tance ($\rho_{NIR} \gg \rho_{red}$) at moderate-to-high total chlorophyll levels [e.g., 29,31,43] limit the applica-bility of NDVI for estimating moderate-to-high crop chlorophyll content. It has been shown that reflectances in the green and red edge regions are sensitive to a wide range of total chlorophyll content [29,38,40,81,82, see also Chapter 6 of this book]. VIs that are based on these spectral regions have been developed and used successfully to estimate the total crop chlorophyll content (e.g., [31,38,46,47,55,74,83]). Dash and Curran, [74] introduced MERIS Terrestrial Chlorophyll Index (MTCI) with spectral bands of MERIS system in the form:

$$MTCI = \frac{(\rho_{753.75} - \rho_{708.75})}{(\rho_{753.75} - \rho_{681.25})} \tag{15.12}$$

Recently, a conceptual model was developed, which relates reflectance with the pigment content (chlorophyll, carotenoids and anthocyanins) in leaves ([84–86], Chapter 6 of this book), LAI [87], and the chlorophyll content in maize and soybean canopy [55]. The rationale of this model is described in Chapter 6. Special cases of the conceptual model for the estimation of the total chloro-phyll content are the so-called "Chlorophyll Indices," CI, in the forms:

$$CI_{green} = \frac{\rho_{NIR}}{\rho_{green}} - 1 \tag{15.13}$$

$$CI_{red\ edge} = \frac{\rho_{NIR}}{\rho_{red\ edge}} - 1 \tag{15.14}$$

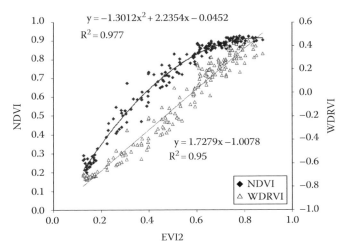

FIGURE 15.9 NDVI and WDRVI, in the spectral bands of MODIS, plotted versus EVI2 for maize in Nebraska.

In Figure 15.9, relationships between three indices, EVI2, NDVI, and WDRVI, are presented for maize canopy with widely varying chlorophyll content. The data were taken from 6 m above the top of the canopy using hyperspectral radiometers. For low-to-moderate chlorophyll content, the relationship NDVI versus EVI2 is linear. As NDVI exceeds 0.8, it levels off, while EVI2 continues to increase. Thus, EVI2 is really more sensitive than NDVI to moderate-to-high chlorophyll content. The relationship EVI2 versus WDRVI is very close and linear throughout the whole range of index values. Thus, these two indices (EVI2 and WDRVI) are comparable in their sensitivity to crop chlorophyll content. It has to be noted, however, that the sensitivity of EVI2 to low chlorophyll content is a little bit higher than that of WDRVI.

The performances of seven general types of VIs (summarized in Table 5 in [21]) were tested for estimating total chlorophyll content at close range (details in [54,55]). The total canopy chlorophyll content was calculated as the product of leaf chlorophyll content and total LAI [55]. The leaf pigment content was measured analytically and was also retrieved from leaf reflectance spectra (details are in [86]). The relationships between the measured leaf chlorophyll content and the chlorophyll content retrieved from leaf reflectance spectra were established for maize and soybean and then used for nondestructive measurements of leaf chlorophyll content [86,89]. Measurement of LAI is described in Section 15.5.

Among the VIs tested, six indices were closely related to chlorophyll content: EVI2, WDRVI, red edge NDVI, CI_{green}, MTCI, and $CI_{red\ edge}$ (Table 15.4). In Figure 15.10, the relationships of VI versus crop chlorophyll content are presented. EVI2 relates to total chlorophyll content closely, although the relation is asymptotic, with a significant decrease in the sensitivity of EVI2 to chlorophyll content above 1 g/m² for soybean and above 1.5 g/m² for maize. WDRVI with $\alpha = 0.1$ has quite a similar relationship with chlorophyll content as EVI2 for maize, although WDRVI performed better for soybean.

CI_{green} relates closely and linearly with total chlorophyll content, with $R^2 > 0.91$ for both maize and soybean. As in the case of EVI2 and WDRVI, CI_{green} is species specific; the sensitivity of CI_{green} to soybean chlorophyll content is more than two times higher than that to maize chlorophyll content. This can be explained by the differences in leaf structure and canopy architecture. For the same leaf chlorophyll content, the chlorophyll content on the adaxial leaf surface is much higher in soybean than in maize (so, ρ_{green} in soybean is lower than ρ_{green} in maize). Due to the different crop canopy structure, for the same LAI, ρ_{NIR} in soybean was higher than ρ_{NIR} in maize [55]. Therefore, for the same total chlorophyll content, the ratio ρ_{NIR}/ρ_{green} in soybean was higher than in maize.

TABLE 15.4

Relationships between VIs and Total Chlorophyll Content (Chl), Established for Maize (a) and Soybean (b) in Three Test Sites during 3 Years of Observations in Nebraska

VI	VI versus Chl	R^2
a. Maize		
EVI	$y = 0.4525x^{0.423}$	0.88
Red edge NDVI	$y = 0.195\text{Ln}(x) + 0.538$	0.92
WDRVI, a = 0.1	$y = 0.39 + 1.39/(1 + \exp(x - 1.1)/0.87)$	0.92
CI_{green}	$y = 2.311x + 1.293$	0.92
$CI_{red\ edge}$	$y = 2.080x + 0.191$	0.92
MTCI	$y = 3.189x + 2.449$	0.93
b. Soybean		
WDRVI, a = 0.1	$y = -0.2778x^2 + 1.1518x - 0.7432$	0.95
EVI2	$y = 0.6871x^{0.396}$	0.88
Red edge NDVI	$y = 0.163\text{Ln}(x) + 0.576$	0.92
CI_{green}	$y = 5.319x + 1.122$	0.91
MTCI	$y = 3.917x + 2.254$	0.89
$CI_{red\ edge}$	$y = 3.398x + 0.362$	0.94

Source: Gitelson, A.A. et al., *Geophys. Res. Lett.*, 32, L08403, 2005.
R^2 is the coefficient of determination.

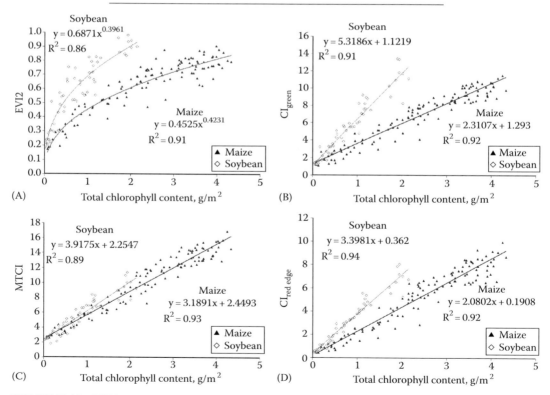

FIGURE 15.10   EVI2 and CI_{green}, in the spectral bands of MODIS, and MTCI, and $CI_{red\ edge}$, in the spectral bands of MERIS, plotted versus the total chlorophyll content in maize and soybean (three sites, 3 years of observation). (A) EVI2, (B) CI_{green}, (C) MTCI, and (D) $CI_{red\ edge}$.

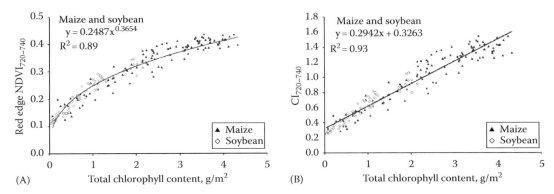

FIGURE 15.11 Red edge NDVI$_{720-740}$ (A) and red edge chlorophyll index CI$_{720-740}$ (B) with red edge spectral band 720–740 nm and MODIS NIR band plotted versus the total chlorophyll content in maize and soybean.

CI$_{red\ edge}$ was able to very accurately estimate chlorophyll content with $R^2 > 0.92$ for both maize and soybean; however, it was species specific. The MTCI was found to be an accurate proxy of total chlorophyll content in both maize and soybean. While MTCI was also species specific ($p < 0.0001$ and F-factor = 77), the difference between the slopes of the relationship MTCI versus chlorophyll content for maize and soybean was smaller than in EVI2, CI$_{green}$ and CI$_{red\ edge}$.

It was shown that CI$_{red\ edge}$ with the red edge band located at longer wavelengths (around 720–740) was nonspecies specific between maize and soybean [55]. In Figure 15.11A and B, Red edge NDVI and CI$_{red\ edge}$, with the red edge band at 720–740 nm, were plotted against the total chlorophyll content for rainfed and irrigated maize and soybean in Nebraska [55]. Red edge NDVI versus chlorophyll content bears a significantly close polynomial relationship when data from both species are plotted together. There is no statistically significant difference between the relationships for maize and soybean; $p > 0.74$ and F-factor = 0.1. The relationship CI$_{red\ edge}$ versus chlorophyll content was linear with $R^2 = 0.93$ (Figure 15.11B) and there was no difference between the relationships for maize and soybean: $p > 0.65$ and F-factor = 0.2. Despite a great difference in the leaf structure and the canopy architecture between these two species, CI$_{red\ edge}$ and Red edge NDVI with the red edge band at 720–740 nm remained nonspecies specific. This indicated that, at least for maize and soybean, there was no need to reparameterize the algorithms derived from these VIs, and prior crop mapping was not required when applying these models to estimate total chlorophyll content from satellite data. This is likely the case for other crops also.

15.5 GREEN LEAF AREA INDEX

One of the key variables required in estimating primary production and in global climate studies is the green LAI (LAI$_{green}$), which is the ratio of the one-sided green leaf area to the ground area underneath. Myneni et al. [24] developed a physically based algorithm for the estimation of LAI$_{green}$ from NDVI observations. As the authors noted, "the algorithm must be viewed within a framework dominated largely by practical consideration and to a lesser extent by accuracy." The relationship between NDVI and LAI$_{green}$ is essentially nonlinear and exhibits significant variations among various vegetation cover types. When LAI$_{green}$ exceeds 2, NDVI is generally insensitive to LAI$_{green}$ in forest canopies with a dense understory and also in grasses, cereal crops, and broadleaf crops (Figure 15.12 for maize and soybean; see also Figure 5 in [24] and Figure 1 in [87]).

The study was carried out in three large production fields (each 65 ha). Within each of the three study sites, six small (20 m × 20 m) plot areas were established for detailed process-level studies [87]. These intensive measurement zones (IMZ) represented all major occurrences of soil and crop production zones within each site. Plant populations were determined (by counting plants) for each IMZ. On each sampling date, plants from a 1 m length of either of two rows within each

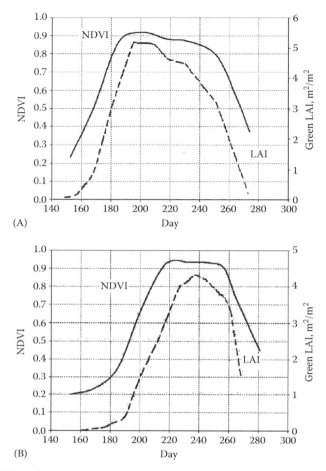

FIGURE 15.12 NDVI, in the spectral bands of MODIS, and the green LAI plotted versus the day of year for maize (A) and soybean (B).

IMZ were collected and total number of plants recorded. Collection rows were alternated on successive dates to minimize edge effects on subsequent plant growth. Plants were transported on ice to the laboratory. In the lab, plants were dissected into green leaves, dead leaves, stems, and reproductive organs. The green leaves were run through an area meter (Model LI-3100, Li-Cor, Inc., Lincoln, Nebraska) and the green leaf area per plant was determined. For each IMZ, the green leaf area per plant was multiplied by the plant population to obtain LAI_{green}. LAI_{green} at the six IMZs were averaged to obtain a site-level value.

The relationship between LAI_{green} and the total canopy chlorophyll content is very close, although with some hysteresis (Figure 15.13 for maize). These BPCs are related in different ways during the vegetative stage and the reproductive and senescence stages [88]. During the vegetative period, the increase in LAI was followed by a corresponding increase in chlorophyll content; at this stage, chlorophyll content and LAI increased synchronously until the LAI_{green} values reached $5\,m^2/m^2$, and the chlorophyll content was about $2.5\,g/m^2$. During the reproductive and senescence stages, however, a significant decrease in chlorophyll content occurred while LAI_{green} decreased only slightly. Thus, a potential bias will be introduced when measuring LAI_{green}, because it is somewhat subjective to decide whether a leaf is green or nongreen in crops that are at either the reproductive or senescence stages [89]. Moreover, in practice, mature dark green leaves with high chlorophyll content during the green-up stage and leaves with much lower chlorophyll content during the reproductive and

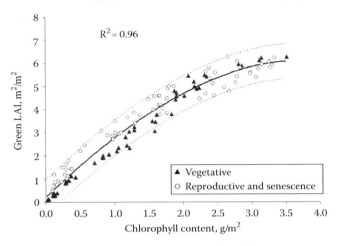

FIGURE 15.13 Green LAI versus the total chlorophyll content in maize.

senescing stages are both designated as "green" leaves. For the same LAI_{green}, the chlorophyll content in a leaf taken during the green-up stage might be more than two times higher than the chlorophyll content in a leaf taken during the reproductive and senescence stages [89] (Figure 15.13). Thus, the total chlorophyll content is a much more objective parameter than LAI_{green} in quantifying vegetation "greenness" or the amount of absorbed radiation.

Nevertheless, the close relationship between LAI_{green} and chlorophyll content allows the use of VIs, developed originally for estimating chlorophyll content, for the remote assessment of LAI_{green}. In Table 15.5a and b, the relationships between six VIs, which were previously tested for chlorophyll content estimation (Section 15.4), and LAI_{green} are presented. Only one VI, the Red edge NDVI, had

TABLE 15.5

Relationships between VIs and the Green LAI (LAI_{green}), Established for Maize (a) and Soybean (b) in Three Test Sites during 3 Years of Observations in Nebraska

VI	VI versus LAI_{green}	R^2
a. Maize		
Red edge NDVI	$y = 0.383x^{0.428}$	0.89
EVI2	$y = 0.098x + 0.248$	0.83
WDRVI, a = 0.1	$y = 0.182x - 0.665$	0.90
MTCI	$y = 2.143x + 1.767$	0.90
$CI_{red\ edge}$	$y = 1.405x - 0.255$	0.91
CI_{green}	$y = 1.604x + 0.590$	0.94
b. Soybean		
Red edge NDVI	$y = -0.025x^2 + 0.251x + 0.134$	0.95
WDRVI, a = 0.1	$y = -0.027x^2 + 0.387x - 0.794$	0.94
EVI2	$y = -0.018x^2 + 0.241x + 0.153$	0.93
MTCI	$y = 0.916x + 1.415$	0.76
CI_{green}	$y = 2.177x + 0.299$	0.88
$CI_{red\ edge}$	$y = 1.370x - 0.112$	0.88

Source: Gitelson, A.A. et al., *Geophys. Res. Lett.*, 30, 5, 1248, 2003.
R^2 is the coefficient of determination.

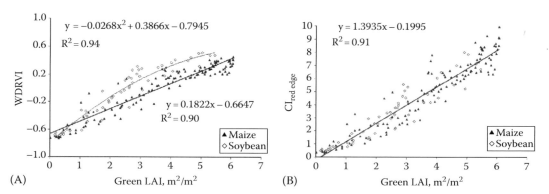

FIGURE 15.14   WDRVI, in the spectral bands of MODIS (A) and $CI_{red\,edge}$, in the spectral bands of MERIS (B), plotted versus the green LAI in maize and soybean; three sites, 3 years of observation.

a nonlinear relationship with LAI_{green} in maize, with decreased sensitivity to $LAI_{green} > 3.5\,m^2/m^2$. Other indices had fairly close linear relationships with LAI_{green} of maize (Table 15.5a). Three VIs, namely, Red edge NDVI, EVI2, and WDRVI, had nonlinear relationships with LAI_{green} of soybean. The sensitivity of Red edge NDVI to LAI_{green} decreased threefold for LAI_{green} above $3.5\,m^2/m^2$. WDRVI and EVI2 showed only a slight decrease in sensitivity to high LAI_{green} (Figure 15.14A for WDRVI). Three VIs, namely, MTCI, CI_{green}, and $CI_{red\,edge}$ were closely and linearly related to LAI_{green} of both maize and soybean (Table 15.5 and Figure 15.14B for $CI_{red\,edge}$). $CI_{red\,edge}$ was the only nonspecies specific (with $R^2 = 0.91$ for both crops taken together); thus, it can be used with no reparameterization for estimating green LAI in both crops.

The performance of the MODIS-derived NDVI and WDRVI were evaluated across three growing seasons (2001–2004) over a wide range of LAI_{green} and also compared with the performances of NDVI and WDRVI derived from reflectance data collected at close range across the same field locations (Table 15.5) [56]. A time series of 16-day composite MODIS 250 m NDVI data (MOD13Q1 V004), spanning from May to October (10 composite periods), was acquired over the three sites for each year of the study. The NDVI data for MODIS were extracted for each composite period, reprojected to the Lambert Azimuthal Equal Area projection, and sequentially stacked to create NDVI time series for each year. Each field was then geolocated on the MODIS imagery and the time-series NDVI data were extracted for a 3×3 pixel window that was centered near the middle of the field. Given the fields' large size and MODIS' 250 m spatial resolution, a block of nine pixels located completely within each field's boundaries could be selected. The median NDVI value of the fields' nine pixels was then calculated for each composite period to produce the time series of MODIS NDVI data for the three growing seasons analyzed in the study [56]. A comparable time series of WDRVI values were also calculated from the median NDVI data sets using the following equation [65] with $\alpha = 0.2$:

$$WDRVI = \frac{[(\alpha+1)\,NDVI + (\alpha-1)]}{[(\alpha-1)\,NDVI + (\alpha+1)]} \quad (15.15)$$

The LAI_{green} varied widely across the growing season for the three fields, reaching a maximum of $6\,m^2/m^2$ in maize and $5.1\,m^2/m^2$ in soybean. Thus, the remote sensing technique used for LAI_{green} estimation must have a wide dynamic range across the highly variable crop conditions encountered during the growing season. The series of MODIS images provided observations for all physiological stages (green-up, reproduction, and senescence) of maize and soybean.

The relationship between the TOC NDVI acquired at close range and LAI_{green} was essentially nonlinear for both crops (Figures 15.15A and 15.16A). The slope of the NDVI versus

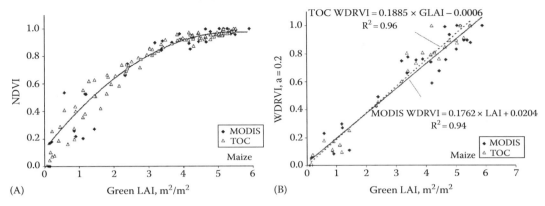

FIGURE 15.15 NDVI (A) and scaled WDRVI (B), retrieved from reflectance measurements taken from 6 m above the TOC and from MODIS data, plotted versus the green LAI in maize. WDRVI was scaled as $(WDRVI-WDRVI_{min})/(WDRVI_{max}-WDRVI_{min})$, where $WDRVI_{max}$ and $WDRVI_{min}$ are maximal and minimal WDRVI values, respectively.

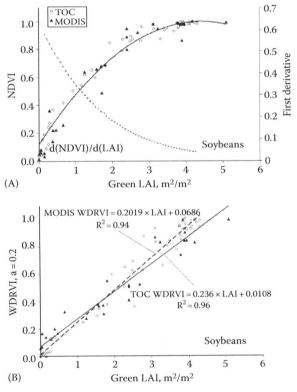

FIGURE 15.16 NDVI (A) and scaled WDRVI (B), retrieved from reflectance measurements taken from 6 m above the TOC and from MODIS data, plotted versus the green LAI in soybean.

LAI_{green} relationship for maize $[d(NDVI)/d(LAI)]$ decreased twofold for $LAI_{green} = 2.5\,m^2/m^2$ and ninefold for $LAI_{green} = 5\,m^2/m^2$. NDVI exhibited a similar behavior for soybean (Figure 15.16A); the sensitivity of NDVI to LAI_{green} decreased threefold at $LAI_{green} = 2\,m^2/m^2$ and 11-fold for $LAI_{green} = 5\,m^2/m^2$. Thus, NDVI cannot be used to accurately estimate $LAI_{green} > 2\,m^2/m^2$, which occurs for more than 2 months of the growing season for maize and 1½ months for soybean.

Relationships between TOC WDRVI and LAI_{green} for both crops were linear, with $R^2 > 0.95$ for maize (Figure 15.15B) and $R^2 > 0.96$ for soybean (Figure 15.16B). Relationships between the MODIS-retrieved VIs and LAI_{green} were very similar to those for the VIs derived from the close-range in situ reflectance data (Figures 15.15 and 15.16). MODIS-retrieved WDRVI accounted for more than 93% of the variation in LAI_{green} in both crops. The RMSE of the MODIS-retrieved LAI_{green} estimation was $0.49\,m^2/m^2$ for maize and $0.41\,m^2/m^2$ for soybean. Importantly, the coefficients of the close-range TOC WDRVI versus LAI_{green} equation and the MODIS-retrieved WDRVI versus LAI_{green} equation were very close to each other (Figures 15.15 and 15.16).

The noise equivalent (NE ΔLAI_{green}) of LAI_{green} estimation by WDRVI remained constant for varying LAI_{green} and equaled $0.46\,m^2/m^2$ for maize and $0.34\,m^2/m^2$ for soybean [56]. The ratio of the noise equivalent of LAI_{green} estimation by the MODIS-retrieved NDVI to that by the MODIS-retrieved WDRVI provides a quantitative measure of the reliability of these two indices in LAI_{green} estimation. When the ratio is <1, NDVI is a more accurate index; when the ratio is >1, WDRVI is more accurate in estimating LAI_{green}. The ratio was slightly <1 for $LAI_{green} < 1.5\,m^2/m^2$ in maize, and $LAI_{green} < 1.4\,m^2/m^2$ in soybean, thus showing that NDVI was more effective in estimating LAI_{green} below $1.5\,m^2/m^2$ [56]. However, for $LAI > 2\,m^2/m^2$ the ratio increased exponentially for both crops. Thus, for $LAI_{green} > 1.5\,m^2/m^2$, WDRVI was much more accurate than NDVI for estimating LAI_{green} using MODIS 250 m data. Since the relationship between LAI_{green} and WDRVI was linear, it was straightforward to invert the relationship between WDRVI and LAI and obtain a measure of LAI_{green}.

15.6 GROSS PRIMARY PRODUCTION

Vegetation productivity is the basis of all the biospheric functions on the land surface and is simply defined as the production of organic matter through photosynthesis. The total amount of organic matter produced through photosynthesis is termed "the gross photosynthesis," and if expressed as the integral of the organic matter produced by all the individual plants in a defined area per unit of time, is termed "the gross primary productivity" (GPP).

The vegetation productivity is commonly assessed through the use of micrometeorological approaches, by means of studying whole-community gas exchange (e.g., [90,92–94]). Micrometeorological methods measure the entire net carbon flux between the land surface and the atmosphere, or what is called the net ecosystem carbon dioxide exchange, NEE. Many field studies in different ecosystems across the globe have used tower-based eddy covariance techniques to provide information on seasonal dynamics and interannual variation of NEE (e.g., [90] and references therein).

Given that the vegetation productivity is directly related to the interaction of solar radiation with the plant canopy [91], remote sensing techniques are used to measure vegetation productivity. Since December 1999, the NASA Earth Observing System has produced a regular global estimate of NPP for the entire globe (e.g., [59]). This estimate is based on the original logic of Monteith [92,93], who suggested that the GPP of stress-free (i.e., well-watered and fertilized) annual crops was linearly related to the amount of the PAR they absorbed, following the expression

$$GPP \propto \varepsilon \times \sum APAR \qquad (15.16)$$

where
   APAR is the absorbed PAR
   ε is the light use efficiency (LUE)

As stated earlier, the linear relationship between fAPAR and NDVI is an approximation, and it is valid only for low to moderate vegetation biomass. In fact, a comparison of models revealed that the NDVI-derived APAR was significantly lower than an independently modeled APAR by a consistent global discrepancy of 28% [94]. As shown in Section 15.2, the relationship NDVI versus $fAPAR_{green}$ was essentially nonlinear, with an extremely low sensitivity of NDVI to $fAPAR_{green} > 0.7$.

The EVI was shown to be much more accurate than NDVI in estimating GPP in different vegetation types, including crops (e.g., [95,97]). EVI is used as a proxy of $fAPAR_{green}$ for estimating GPP [95,96]. According to Monteith [92,93], LUE is a relatively conservative value among plant formations of the same metabolic type. However, it can vary with phenological stage, climatic condition, temperature, and water stress (e.g., [99]). Gamon et al. [100] suggested using the photochemical reflectance index (PRI) as a proxy of LUE. PRI was defined as

$$PRI = \frac{(\rho_{531} - \rho_{570})}{(\rho_{531} + \rho_{570})}$$

where ρ_{531} and ρ_{570} are the reflectances at 531 and 570 nm, respectively. Reflectance changes at 531 nm are associated with the thylakoid energization and zeaxanthin-antheraxanthin-violaxanthin interconversion. The relative concentration of xanthophylls cycle pigments is closely related to plant photochemical efficiency. Several studies have shown linear relationships between PRI and LUE in different vegetation types (e.g., [100]). However, Barton and North [101] showed that PRI was most sensitive to changes in LAI and concluded that the potential use of this index to predict LUE will require an independent estimate of LAI. The attempt to use PRI for crops showed that no major improvement has been achieved in GPP estimation [75]. Therefore, the uncertainty of estimating terrestrial GPP using Monteith's logic might be considerable when no information about the factors influencing LUE is included.

The GPP is a function of the amount of APAR and the capacity of the leaves to export or utilize the product of photosynthesis (i.e., LUE). The product of APAR and LUE depends on the amount and the distribution of photosynthetic biomass; thus, it depends on the amount of chlorophyll content and leaf physiology [75,102]. Therefore, both APAR and LUE are related to the amount of chlorophyll in the canopy. Low frequency (day-to-day) variation in GPP is associated with crop phenological stage and physiological status. It has been also shown that this low frequency variation in GPP is closely related to LAI_{green} (i.e., chlorophyll) in maize [103,104].

A new approach to estimate GPP in crops was proposed by Gitelson et al. [75]. It does not depend on the NDVI/fAPAR linearity assumption, and it does not depend on the constancy, the preestablished variability or the biome/species specificity of LUE. The approach to estimate GPP is based on the hypothesis that total chlorophyll content in crops relates closely to the low frequency variation in GPP. Since long- or medium-term changes in canopy chlorophyll are related to crop phenology, canopy stresses, and the photosynthetic capacity of the vegetation (e.g., [105,106]), it can also be related to GPP. It was shown that canopy chlorophyll appears to be the plant community property that is most relevant for the prediction of productivity [105,107]. Close significant nonspecies-specific relationship of GPP with the product of chlorophyll content and incident PAR has been found. Therefore, it was suggested to estimate GPP as follows [75]:

$$GPP = Chl \times PAR_{in} \qquad (15.17)$$

It has been shown that chlorophyll-related indices have a potential to estimate GPP in crops at close range [75,109] and satellite levels [108] and thus it was suggested to estimate GPP in crops using techniques developed for chlorophyll content assessment [55].

The performances of different VIs in estimating GPP in irrigated and rainfed maize and soybean, using data sets collected in three AmeriFlux sites in Nebraska ([58,75,109], http://public.ornl.gov/ameriflux/site-select.cfm), were studied. Micrometeorological eddy covariance data collection began in 2001 at these three test sites. To have sufficient upwind fetch (in all directions), eddy covariance sensors were mounted at 3 m above the ground while the canopy was shorter than 1 m, and later moved to a height of 6.2 m until harvest. The study sites represented approximately 90%–95% of the flux footprint during daytime and 70%–90% during nighttime. Eddy covariance measurements of fluxes of CO_2, water vapor, sensible heat, and momentum were made [75,109]. More details of the measurements and calculations are given elsewhere [58,103,109]. Daytime estimates of respiration were obtained from the nighttime CO_2 exchange–temperature relationship. The midday GPP (in $gC/m^2/s$, hereafter $g/m^2/s$) was then obtained by subtracting the daily respiration from the daily NEE.

In Figure 15.17 and Table 15.6 the relationships established between the product (VI × PAR_{in}) and GPP are shown for maize and soybean. VIs that were the best in GPP estimation, namely, EVI2, WDRVI, Red Edge NDVI, CI_{green}, MTCI, and $CI_{red\ edge}$, were those found to be the best predictors of crop chlorophyll content (Table 15.4). All these indices have close significant (p < 0.001) relationships with GPP (Table 15.6). NDVI and Green NDVI were also included for comparison; the former was accurate in estimating $fAPAR_{green} < 0.7$, and the latter was closely and linearly related

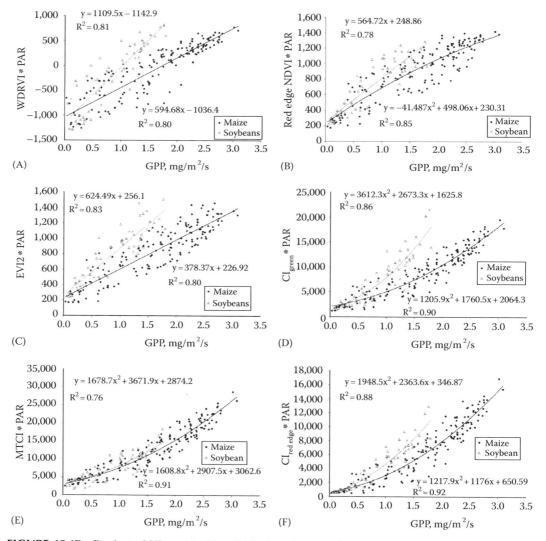

FIGURE 15.17 Product of VIs and incident PAR plotted versus the midday primary production in maize (three sites, 3 years) and soybean (two sites, 2 years). EVI2, WDRVI, and CI_{green} were calculated with the spectral bands of MODIS; Red edge NDVI, MTCI, and $CI_{red\ edge}$ with the spectral bands of MERIS. (A) WDRVI, (B) Red edge NDVI, (C) EVI2, (D) CI_{green}, (E) MTCI, and (F) $CI_{red\ edge}$.

TABLE 15.6

Relationships between the Product of Vegetation Index and the Incident Photosynthetically Active Radiation (PAR$_{in}$) and the Gross Primary Production (GPP), Established for Maize (a) and Soybean (b) in Three Test Sites during 3 Years of Observations in Nebraska

VI	VI × PAR versus GPP	R^2
a. Maize		
Green NDVI	$y = 278.04x + 581.93$	0.69
NDVI	$y = -76.261x^2 + 600.3x + 368.38$	0.79
EVI2	$y = 378.37x + 226.92$	0.80
WDRVI, a = 0.1	$y = 594.68x - 1036.4$	0.80
Red edge NDVI	$y = -41.49x^2 + 498.06x + 230.31$	0.85
CI$_{green}$	$y = 1205.9x^2 + 1760.5x + 2064.3$	0.90
MTCI	$y = 1135x^2 + 1514.1x + 2620.4$	0.90
CI$_{red\ edge}$	$y = 1217.9x^2 + 1176x + 650.59$	0.92
b. Soybean		
Green NDVI	$y = 442.9x + 545.49$	0.62
MTCI	$y = 1678.7x^2 + 3671.9x + 2874.2$	0.76
NDVI	$y = -376.44x^2 + 1237.6x + 292.43$	0.78
Red edge NDVI	$y = 514.8x + 213.78$	0.78
WDRVI, a = 0.1	$y = 1109.5x - 1142.9$	0.81
EVI2	$y = 624.49x + 256.1$	0.82
CI$_{green}$	$y = 3612.3x^2 + 2673.3x + 1625.8$	0.86
CI$_{red\ edge}$	$y = 1948.5x^2 + 2363.6x + 346.87$	0.88

Source: Gitelson, A.A. et al., *J. Geophys. Res.*, 111, D08S11, 2006.
R^2 is the coefficient of determination.

to fAPAR$_{green}$ (Table 15.3). In maize, as it was with chlorophyll content estimation, the best indices were CI$_{green}$, MTCI, and CI$_{red\ edge}$, with R^2 above 0.9. EVI, WDRVI, and Red edge NDVI behave almost similarly, with R^2 around 0.8. In soybean, CI$_{green}$ and CI$_{red\ edge}$ were still the best indices and MTCI was less accurate than for maize with $R^2 = 0.76$. The other indices (EVI2, WDRVI, and Red edge NDVI) performed similar to how they performed for chlorophyll content estimation. Importantly, MTCI calculated using the spectral bands of MERIS was the only nonspecies-specific index ($p > 0.8$ and F-factor = 0.06).

In Figure 15.18A and B the relationships of CI$_{red\ edge}$ versus GPP and Red edge NDVI versus GPP, with the red edge band at 720–740 nm, are presented. There is no statistically significant difference between the relationships for maize and soybean: $p > 0.5$ and F-factor < 4. Thus, these indices can be used for estimating GPP in maize and soybean with no reparameterization of the algorithms.

The comparison of the performances of VIs in estimating fAPAR$_{green}$ and GPP gives an understanding of which BPC each vegetation index represents. Following Monteith's logic and assuming that LUE is a conservative value among plant formations of the same metabolic type, one can expect to get an accurate estimation of GPP using indices that are the best in estimating fAPAR$_{green}$. However, the best predictors of fAPAR$_{green}$ (Green NDVI, Red edge NDVI and WDRVI) are not the best in estimating GPP. In contrast, the best indices for estimating GPP were those that were the best in estimating the chlorophyll content: MTCI, CI$_{red\ edge}$, and CI$_{green}$ [111]. As noted earlier, EVI2 is more closely related to LAI$_{green}$ (i.e., chlorophyll content) than to fAPAR$_{green}$, as it was originally suggested by its authors [46,47,68], and it is quite accurate in estimating GPP. It means

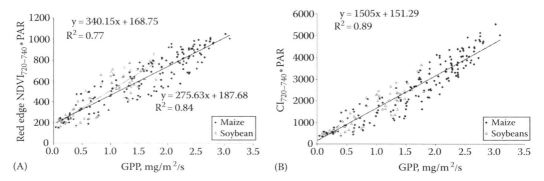

FIGURE 15.18 Products of red edge NDVI (A) and $CI_{red\ edge}$ (B) and incident PAR plotted versus the mid-day gross primary production in maize (three sites, 3 years) and soybeans (two sites, 3 years). Indices were calculated with the red edge band at 720–740 nm and the MODIS NIR band.

that in studies for estimating GPP (e.g., [95–97]), the EVI2 is used as a surrogate of total chlorophyll content, not $fAPAR_{green}$ as it was suggested [96,97].

It is also important to note that a high determination coefficient of the relationship of NDVI × PAR versus GPP indicates a small dispersion of the points from the essentially nonlinear best fit regression line. The product, NDVI × PAR, becomes insensitive to GPP exceeding 1.5 mg/m²/s in maize and 1 mg/m²/s in soybean. Thus, its noise equivalent increases exponentially as GPP increases. In contrast, while the relationships of GPP with $CI_{red\ edge}$, CI_{green}, and MTCI are also nonlinear, the sensitivity of the indices to moderate-to-high GPP is even higher than to low GPP.

Gitelson et al. [108] investigated the performances of four VIs, namely, NDVI, EVI, WDRVI, and CI_{green}, in estimating daily GPP, using satellite data, specifically the Landsat-7 ETM+. Four Landsat-7 ETM+ images (WRS-2 path 28, row 31) acquired throughout the growing season in 2001 (June 10, August 13 and 29, and September 30) and one image acquired in July 15, 2002 were used in that study. These images were first coregistered manually by picking ground control points using the ENvironment for Visualizing Images (ENVI) image processing package, and then resampled to a common grid (30 × 30 m/pixel). The images were atmospherically corrected to the TOC reflectance using the Landsat Ecosystem Disturbance Adaptive Processing System [110].

NDVI was sensitive to low GPP values below 5–7 gC/m²/day, and then its sensitivity to GPP above 10 gC/m²/day dropped drastically (Figure 15.19). Three indices, EVI, WDRVI, and CI_{green}, were significantly closely related to GPP. The relationships of GPP with EVI and WDRVI were nonlinear, with a substantial decrease in sensitivity to GPP exceeding 15 gC/m²/day. CI_{green} was found to be linearly related with GPP ranging from 1.88 to 23.1 gC/m²/day, with $R^2 > 0.96$ and

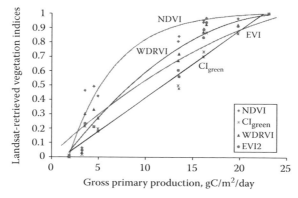

FIGURE 15.19 VIs retrieved from atmospherically corrected Landsat-7 ETM+ imagery plotted versus the daily gross primary production in maize and soybean.

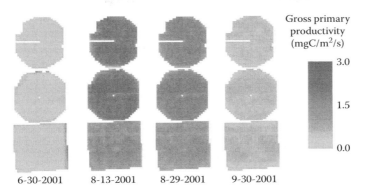

Gross primary
productivity
(mgC/m²/s)

3.0

1.5

0.0

6-30-2001   8-13-2001   8-29-2001   9-30-2001

FIGURE 15.20   (See color insert.) Midday gross primary production in maize (first and second rows) and soybean (bottom row) retrieved from atmospherically corrected Landsat-7 ETM+ imagery taken over Nebraska in 2001.

RMSE $< 1.58\,\text{gC/m}^2/\text{day}$. It therefore presents an accurate surrogate measure for GPP estimation. Due to its high spatial resolution (i.e., $30 \times 30\,\text{m/pixel}$), the Landsat-7 ETM+ satellite system is particularly appropriate for detecting not only between but also within crop field GPP variability during the growing season (Figure 15.20).

15.7   CONCLUSIONS

Remote estimation of five crop BPCs, namely, VF, fraction of PAR absorbed by photosynthetically active vegetation, total crop chlorophyll content, green LAI and gross primary production, were discussed in detail in this chapter. All techniques were tested using reflectances acquired from 6 m above the top of the canopy. The VIs were calculated using reflectances simulated in the discrete spectral bands of MODIS and MERIS systems. It is shown that the aforementioned crop BPCs can be estimated accurately using remotely sensed data. MODIS-retrieved estimates of the green LAI and Landsat-7 ETM+-retrieved estimates of the gross primary production were very close to in situ measured values. Importantly, it was found that the relationships of WDRVI versus LAI and CI_{green} versus GPP, retrieved from MODIS data with a 250 m spatial resolution and Landsat-7 ETM+ data with a 30 m spatial resolution, respectively, were quantitatively very close to the relationships established using reflectance data taken in situ several meters above the top of the crop canopy. It shows that these indices are likely to yield similar accuracies when applied to satellite data as they do when applied to in situ measured data.

Tables 15.2 through 15.5 summarize the accuracy of each index in estimating different BPCs. The choice of the index depends on the spectral characteristics of the radiometer or the satellite sensor being used. The indices employing red edge spectral bands, namely, $CI_{red\,edge}$, MTCI, $VARI_{red\,edge}$, and Red edge NDVI, can be used for satellite systems with spectral band in the red edge region (MERIS, Hyperion, and/or future systems as EnMap, HyspIRI, HISUI). The indices employing green spectral band, namely, green NDVI, $VARI_{green}$, and CI_{green}, can be used in satellite systems with spectral bands in the green region (e.g., Landsat, Hyperion, MODIS 500 m and 1 km spatial resolution, and MERIS). The indices using only red and NIR spectral bands, namely, NDVI, EVI2, and WDRVI, can be used for crop monitoring by satellite systems such as AVHRR, Landsat, MODIS (250 m spatial resolution), and MERIS.

The implications of these findings are far reaching since the described techniques open a new possibility for an accurate estimation of crop BPCs at different scales, from close range to satellite altitudes. Some of the techniques based on the red, green, and NIR bands allow using the extensive archive of Landsat and AVHRR imagery acquired since the early 1970s and the 250 m spatial resolution MODIS imagery acquired since 2001. With these techniques, it is now possible to obtain global synoptic estimates of crop BPCs at the 30 m spatial resolution of Landsat TM and ETM+ and the

250–290 m resolution of MODIS and MERIS. The performances of the VIs were tested for maize, soybean, and wheat. These crops have very different canopy architecture and leaf structure. Still, the remote sensing techniques yielded accurate results, which indicate that these techniques are likely applicable to other crops as well. However, the techniques should be tested for different soil types because among the indices that were tested, only EVI2 is resistant to variability in soil reflectance.

ACKNOWLEDGMENTS

I owe this chapter to Donald C. Rundquist who created the best system for collecting and processing remotely sensed data using hyperspectral radiometers. I was honored to use this system for studies presented here. His help at different stages of this study is greatly appreciated. I also acknowledge contributions of my former PhD students Drs. Veronica Ciganda, Robi Stark, Andres Vina, Wesley Moses, and Arthur Zygielbaum. The supports from the Center for Advanced Land Management and Information Technologies, Carbon Sequestration Program at University of Nebraska, Lincoln, Nebraska, and the J. Blaustein Institute for Desert Research of Ben-Gurion University, Israel, are greatly appreciated.

REFERENCES

1. Allen, W.A., Gausman, H.W., and Richardson, A.J., Willstatter-Stoll theory of leaf reflectance evaluated by ray tracing, *Applied Optics*, 12, 10, 2448–2453, 1973.
2. Gausman, H.W., Allen, W.A., Myers, V.I., and Cardenas, R., Reflectance and internal structure of cotton leaves *Gossypium hirsutum* L., *Agronomy Journal*, 61, 3, 374–376, 1969.
3. Woolley, J.T., Reflectance and transmittance of light by leaves, *Plant Physiology*, 47, 656–662, 1971.
4. Barnes, E.M., Sudduth, K.A., Hummel, J.W., Leach, S.M., Corwin, D.L., Yeng, C.C., Daughtry, S.T., and Bausch, W.C., Remote- and ground-based sensor techniques to map soil properties, *Photogrammetric Engineering and Remote Sensing*, 69, 619–630, 2003.
5. Kustas, W.P., French, A.N., Hatfield, J.L., Jackson, T.J., Moran, M.S., Rango, A., Ritchie, J.C., and Schmugge, T.J., Remote sensing research in hydrometeorology, *Photogrammetric Engineering and Remote Sensing*, 69, 631–646, 2003.
6. Pinter, P.J., Hatfield, J.L., Schepers, J.S., Barnes, E.M., Moran, M.S., Daughtry, C.S., and Upchurch, D.R., Remote sensing for crop management, *Photogrammetric Engineering and Remote Sensing*, 69, 647–664, 2003.
7. Doraiswamy, P.C., Moulin, S., and Cook, P.W., Crop yield assessment from remote sensing, *Photogrammetric Engineering and Remote Sensing*, 69, 665–674, 2003.
8. Moran, S., Fitzgerald, G., Rango, A., Walthall, C., Barnes, E., Bausch, W., Clarke, T. et al., Sensor development and radiometric correction for agricultural applications, *Photogrammetric Engineering and Remote Sensing*, 69, 705–718, 2003.
9. Hatfield, J.L., Prueger, J.H., and Kustas, W.P., Remote sensing of dryland crops. In: *Remote Sensing for Natural Resource Management and Environmental Monitoring: Manual of Remote Sensing*, S.L. Ustin (ed.), 3rd Edn., John Wiley, Hoboken, NJ, Vol. 4, Chapter 10, pp. 531–568, 2004.
10. Hatfield, J.L., Gitelson, A.A., Schepers, J.S., and Walthall, C.L., Application of spectral remote sensing for agronomic decisions, *Agronomy Journal*, 100, S-117–S-131, 2008, doi:10.2134/agronj2006.0370c
11. Steven, M.D., Biscoe, P.V., Jaggard, K.W., and Paruntu, J., Foliage cover and radiation interception, *Field Crops Research*, 13, 75–87, 1986.
12. Adams, J.B. and Smith, M.O., Spectral mixture modeling: A new analysis of rock and soil types at the Viking Lander 1 site, *Journal of Geophysical Research*, 91, 8098–8112, 1986.
13. Ustin, S.L., Hart, Q.J., Duan, L., and Scheer, G., Vegetation mapping on hardwood rangelands in California, *International Journal of Remote Sensing*, 17, 3015–3036, 1996.
14. Baret, F., Clevers, J.G.P.W., and Steven, M.D., The robustness of canopy gap fraction estimates from red and near-infrared reflectances: A comparison of approaches, *Remote Sensing of Environment*, 54, 141–151, 1995.
15. Roberts, D.A., Smith, M.O., and Adams, J.B., Green vegetation, nonphotosynthetic vegetation, and soils in AVIRIS data, *Remote Sensing of Environment*, 44, 255–269, 1993.

16. Lelong, C.C.D., Pinet, P.C., and Poilve, H., Hyperspectral imaging and stress mapping in agriculture: A case study on wheat in Beauce (France), *Remote Sensing of Environment*, 66, 179–191, 1998.

17. Price, J.C., How unique are spectral signatures? *Remote Sensing of Environment*, 49, 181–186, 1994.

18. Garcia-Haro, F.J., Gilabert, M.A., and Melia, J., Linear spectral mixture modeling to estimate vegetation amount from optical spectral data, *International Journal of Remote Sensing*, 17, 3373–3400, 1996.

19. Verstraete, M.M., Pinty, B., and Myneni, R.B., Potential and limitations of information extraction on the terrestrial biosphere from satellite remote sensing, *Remote Sensing of Environment*, 58, 201–214, 1996.

20. Moran, M.S., Inoue, Y., and Barnes, E.M., Opportunities and limitations for image-based remote sensing in precision crop management, *Remote Sensing of Environment*, 61, 319–346, 1997.

21. le Maire, G., François, C., Soudani, K., Berveiller, D., Pontailler, J.Y., and Bréda, N., Calibration and validation of hyperspectral indices for the estimation of broadleaved forest leaf chlorophyll content, leaf mass per area, leaf area index and leaf canopy biomass, *Remote Sensing of Environment*, 112, 3846–3864, 2008.

22. Asrar, G., Fuchs, M., Kanemasu, E.T., and Hatfield, J.L., Estimating absorbed photosynthetic radiation and leaf area index from spectral reflectance in wheat, *Agronomy Journal*, 76, 300–306, 1984.

23. Myneni, R.B., Hall, F.G., Sellers, P.S., and Marshak, A.L., The interpretation of spectral vegetation indexes, *IEEE Transactions on Geoscience and Remote Sensing*, 33, 481–486, 1995.

24. Myneni, R.B., Nemani, R.R., and Running, S.W., Estimation of global leaf area index and absorbed par using radiative transfer models, *IEEE Transactions on Geoscience and Remote Sensing*, 35, 1380–1393, 1997.

25. Sellers, P.J., Canopy reflectance, photosynthesis and transpiration, *International Journal of Remote Sensing*, 6, 1335–1372, 1985.

26. Tucker, J.C., Red and photographic infrared linear combination for monitoring vegetation, *Remote Sensing of Environment*, 8, 127–150, 1979.

27. Jordan, C.F., Derivation of leaf area index from quality of light on the forest floor, *Ecology*, 50, 663–666, 1969.

28. Rouse, J.W., Haas, R.H., Jr., Schell, J.A., and Deering, D.W., Monitoring vegetation systems in the Great Plains with ERTS, NASA SP-351, *Third ERTS-1 Symposium*, NASA, Washington, DC, Vol. 1, pp. 309–317, 1974.

29. Buschmann, C. and Nagel, E., In vivo spectroscopy and internal optics of leaves as basis for remote sensing of vegetation, *International Journal of Remote Sensing*, 14, 711–722, 1993.

30. Gitelson, A.A. and Merzlyak, M.N., Quantitative estimation of chlorophyll-*a* using reflectance spectra: Experiments with autumn chestnut and maple leaves, *Journal of Photochemistry and Photobiology B: Biology*, 22, 247–252, 1994.

31. Gitelson, A.A., Wide dynamic range vegetation index for remote quantification of crop biophysical characteristics, *Journal of Plant Physiology*, 161, 165–173, 2004.

32. Kaufman, Y.J., The atmospheric effect on remote sensing and its corrections, In: *Theory and Application of Optical Remote Sensing*, G. Asrar (ed.), John Wiley & Sons, Inc., New York, pp. 336–428, 1989.

33. Kaufman, Y.J. and Tanre, D., Atmospherically resistant vegetation index (ARVI) for EOS-MODIS, *IEEE Transactions on Geoscience and Remote Sensing*, 30, 261–270, 1992.

34. Baret, F., Guyot, G., and Major, D., TSAVI: A vegetation index which minimizes soil brightness effects on LAI and APAR estimation, *12th Canadian Symposium on Remote Sensing and IGARSS'90*, Vancouver, Canada, July 10–14, 1989.

35. Baret, F., Jacquemoud, S., and Hanocq, J.F., The soil line concept in remote sensing, *Remote Sensing Reviews*, 7, 65–82, 1993.

36. Huete, A.R., A soil-adjusted vegetation index (SAVI), *Remote Sensing of Environment*, 25, 295–309, 1988.

37. Huete, A.R., Justice, C., and Liu, H., Development of vegetation and soil indices for MODIS-EOS, *Remote Sensing of Environment*, 49, 224–234, 1994.

38. Gitelson, A., Kaufman, Y.J., and Merzlyak, M.N., Use of a green channel in remote sensing of global vegetation from EOS-MODIS, *Remote Sensing of Environment*, 58, 289–298, 1996.

39. Roberts, D.A., Green, R.O., and Adams, J.B., Temporal and spatial patterns in vegetation and atmospheric properties from AVIRIS, *Remote Sensing of Environment*, 62, 223–240, 1997.

40. Yoder, B.J. and Waring, R.H., The normalized difference vegetation index of small Douglas-fir canopies with varying chlorophyll concentrations, *Remote Sensing of Environment*, 49, 81–91, 1994.

41. Colwell, J.E., Vegetation canopy reflectance, *Remote Sensing of Environment*, 3, 175–183, 1974.

42. Daughtry, C.S.T., Bauer, M.E., Crecelius, D.W., and Hixson, M.M., Effects of management practices on reflectance of spring wheat canopies, *Agronomy Journal*, 72, 1055–1060, 1980.

43. Kanemasu, E.T., Seasonal canopy reflectance patterns of wheat, sorghum, and soybean, *Remote Sensing of Environment*, 3, 43–47, 1974.

44. Jackson, R.D. and Ezra, C.E., Spectral response of cotton to suddenly induced water stress, *International Journal of Remote Sensing*, 6, 177–185, 1985.

45. Tucker, C.J., Holben, B.N., Elgin, J.H., Jr., and McMurtrey, III, J.E., Remote sensing of total dry-matter accumulation in winter wheat, *Remote Sensing of Environment*, 11, 171–189, 1981.

46. Huete, A.R., Liu, H.Q., Batchily, K., and van Leeuwen, W.J.D., A comparison of vegetation indices over a global set of TM images for EOS-MODIS, *Remote Sensing of Environment*, 59, 440–451, 1997.

47. Jiang, Z., Huete, A.R., Didan, K., and Miura, T., Development of a two-band enhanced vegetation index without a blue band, *Remote Sensing of Environment*, 112, 3833–3845, 2008.

48. Viña, A., Henebry, G.M., and Gitelson, A.A., Satellite monitoring of vegetation dynamics: Sensitivity enhancement by the wide dynamic range vegetation index, *Geophysical Research Letters*, 31, L04503, 2004, doi:10.1029/2003GL019034.

49. Pickup, G., Chewings, V.H., and Nelson, D.J., Estimating changes in vegetation cover over time in arid rangelands using Landsat MSS data, *Remote Sensing of Environment*, 43, 243–263, 1993.

50. Richardson, A.J. and Wiegand, C.L., Distinguishing vegetation from soil background information, *Photogrammetric Engineering and Remote Sensing*, 43, 1541–1552, 1977.

51. Gitelson, A.A., Kaufman, Y.J., Stark, R., and Rundquist, D., Novel algorithms for remote estimation of vegetation fraction, *Remote Sensing of Environment*, 80, 76–87, 2002.

52. Gitelson, A. and Merzlyak, M.N., Remote estimation of chlorophyll content in higher plant leaves, *International Journal of Remote Sensing*, 18, 291–298, 1997.

53. Gitelson, A. and Merzlyak, M.N., Signature analysis of leaf reflectance spectra: Algorithm development for remote sensing of chlorophyll, *Journal of Plant Physiology*, 148, 494–500, 1996.

54. Rundquist, D.C., Perk, R., Leavitt, B., Keydan, G.P., and Gitelson, A.A., Collecting spectral data over cropland vegetation using machine positioning versus hand-positioning of the sensor, *Computers and Electronics in Agriculture*, 43, 173–178, 2004.

55. Gitelson, A.A., Viña, A., Rundquist, D.C., Ciganda, V., and Arkebauer, T.J., Remote estimation of canopy chlorophyll content in crops, *Geophysical Research Letters*, 32, L08403, 2005, doi:10.1029/2005GL022688.

56. Gitelson, A.A., Wardlow, B.D., Keydan, G.P., and Leavitt, B., Evaluation of MODIS 250-m data for green LAI estimation in crops, *Geophysical Research Letters*, 34, L20403, 2007, doi:10.1029/2007GL031620.

57. Gitelson, A.A. and Merzlyak, M., Spectral reflectance changes associated with autumn senescence of *Aesculus hippocastanum* L. and *Acer platanoides* L. leaves: Spectral features and relation to chlorophyll estimation, *Journal of Plant Physiology*, 143, 286–292, 1994.

58. Verma, S.B., Dobermann, A., Cassman, K.G., Walters, D.T., Knops, J.M., Arkebauer, T.J., Suyker, A.E. et al., Annual carbon dioxide exchange in irrigated and rainfed maize-based agroecosystems, *Agricultural and Forest Meteorology*, 131, 77–96, 2005.

59. Running, S.W., Nemani, R.R., Heinsch, F.A., Zhao, M., Reeves, M., and Hashimoto, H., A continuous satellite-derived measure of global terrestrial primary production, *Bioscience*, 54, 547–560, 2004.

60. Roujean, J.L. and Breon, F.M., Estimating PAR absorbed by vegetation from bidirectional reflectance measurements, *Remote Sensing of Environment*, 51, 375–384, 1995.

61. Hatfield, J.L., Asrar, G., and Kanemasu, E.T., Intercepted photosynthetically active radiation estimated by spectral reflectance, *Remote Sensing of Environment*, 14, 65–75, 1984.

62. Gallo, K.P., Daughtry, C.S.T., and Bauer M.E., Spectral estimation on absorbed photosynthetically active radiation in corn canopies, *Remote Sensing of Environment*, 17, 221–232, 1985.

63. Hall, F.G., Huemmrich, K.F., Goetz, S.J., Sellers, P.J., and Nickeson, J.E., Satellite remote sensing of surface energy balance: Success, failures and unresolved issues in FIFE, *Journal of Geophysical Research*, 97, 19061–19089, 1992.

64. Goward, S.M. and Huemmerich, K.E., Vegetation canopy PAR absorptance and the normalized difference vegetation index: An assessment using SAIL model, *Remote Sensing of Environment*, 39, 119–140, 1992.

65. Viña, A. and Gitelson, A.A., New developments in the remote estimation of the fraction of absorbed photosynthetically active radiation in crops, *Geophysical Research Letters*, 32, L17403, 2005, doi:10.1029/2005Gl023647.

66. Hanan, N.P., Burba, G., Verma, S.B., Berry, J.A., Suyker A., and Walter-Shea, E.A., Inversion of net ecosystem CO_2 flux measurements for estimation of canopy PAR absorption, *Global Change Biology*, 8, 563–574, 2002.

67. Baret, F. and Guyot, G., Potentials and limits of vegetation indices for LAI and APAR assessment, *Remote Sensing of Environment*, 35, 161–173, 1991.

68. Huete, A.R., Didan, K., Miura, T., Rodriguez, E.P., Gao, X., and Ferreira, L.G., Overview of the radiometric and biophysical performance of the MODIS vegetation indices, *Remote Sensing of Environment*, 83, 195–213, 2002.

69. Danks, S.M., Evans, E.H., and Whittaker P.A., *Photosynthetic Systems: Structure, Function and Assembly*, John Wiley & Sons, New York, 1984.

70. Ustin, S.L., Smith, M.O., Jacquemoud, S., Verstraete, M.M., and Govaerts, Y., Geobotany: Vegetation mapping for Earth sciences. In: *Manual of Remote Sensing: Remote Sensing for the Earth Sciences*, A.N. Rencz (ed.), 3rd Edn., Wiley, New York, Vol. 3, pp. 189–248, 1998.

71. Lieth, H. and Whittaker, R.H., *Primary Production of the Biosphere*, Springer-Verlag, New York, 1975.

72. Daughtry, C.S.T., Walthall, C.L., Kim, M.S., Brown de Colstoun, E., and McMurtrey, III, J.E., Estimating corn leaf chlorophyll concentration from leaf and canopy reflectance, *Remote Sensing of the Environment*, 74, 229–239, 2000.

73. Haboudane, D., Miller, J.R., Tremblay, N., Zarco-Tejada, P.J., and Dextraze, L., Integrated narrow-band vegetation indices for prediction of crop chlorophyll content for application to precision agriculture, *Remote Sensing of Environment*, 81, 416–426, 2002.

74. Dash, J. and Curran, P.J., The MERIS terrestrial chlorophyll index, *International Journal of Remote Sensing*, 25, 5403–5413, 2004.

75. Gitelson, A.A., Viña, A., Verma, S.B., Rundquist, D.C., Arkebauer, T.J., Keydan, G., Leavitt, B., Ciganda, V., Burba, G.G., and Suyker, A.E., Relationship between gross primary production and chlorophyll content in crops: Implications for the synoptic monitoring of vegetation productivity, *Journal of Geophysical Research*, 111, D08S11, 2006, doi:10.1029/2005JD006017.

76. Wu, C., Niu, Z., Tang, Q., Huang, W., Rivard, B., and Feng, J., Remote estimation of gross primary production in wheat using chlorophyll-related vegetation indices, *Agricultural and Forest Meteorology*, 149, 1015–1021, 2009.

77. Wu, C., Niu, Z., and Gao, S., Gross primary production estimation from MODIS data with vegetation index and photosynthetically active radiation in maize, *Journal of Geophysical Research*, 115, D12127, 2010, doi:10.1029/2009JD013023.

78. Walters, D.T., Diagnosis of nitrogen deficiency in maize and the influence of hybrid and plant density. In: *North Central Extension-Industry Soil Fertility Conference*, Des Moines, IA, Vol. 19, 2003.

79. Solari, F., Shanahan, J., Ferguson, R., Schepers, J., and Gitelson, A., Active sensor reflectance measurements of corn nitrogen status and yield potential, *Agronomy Journal*, 100, 3, 571–579, 2008, doi:10.2134/agronj2007.0244.

80. Rondeaux, G., Steven, M., and Baret, F., Optimization of soil-adjusted vegetation indices, *Remote Sensing of Environment*, 55, 95–107, 1996.

81. Thomas, J.R. and Gaussman, H.W., Leaf reflectance vs. leaf chlorophyll and carotenoid concentration for eight crops, *Agronomy Journal*, 69, 799–802, 1977.

82. Lichtenthaler, H.K., Gitelson, A.A., and Lang, M., Non-destructive determination of chlorophyll content of leaves of a green and an aurea mutant of tobacco by reflectance measurements, *Journal of Plant Physiology*, 148, 483–493, 1996.

83. Broge, N.H. and Mortensen, J.V., Deriving green crop area index and canopy chlorophyll density of winter wheat from spectral reflectance data, *Remote Sensing of Environment*, 81, 45–57, 2002.

84. Gitelson, A.A., Merzlyak, M.N., and Chivkunova, O.B., Optical properties and non-destructive estimation of anthocyanin content in plant leaves, *Photochemistry and Photobiology*, 74, 38–45, 2001.

85. Gitelson, A.A., Zur, Y., Chivkunova, O.B., and Merzlyak, M.N., Assessing carotenoid content in plant leaves with reflectance spectroscopy, *Photochemistry and Photobiology*, 75, 272–281, 2002.

86. Gitelson, A.A., Gritz, U., and Merzlyak, M.N., Relationships between leaf chlorophyll content and spectral reflectance and algorithms for non-destructive chlorophyll assessment in higher plant leaves, *Journal of Plant Physiology*, 160, 3, 271–282, 2003.

87. Gitelson, A.A., Viña, A., Arkebauer, T.J., Rundquist, D.C., Keydan, G., and Leavitt, B., Remote estimation of leaf area index and green leaf biomass in maize canopies, *Geophysical Research Letters*, 30, 5, 1248, 2003, doi:10.1029/2002Gl016450.

88. Ciganda, V., Gitelson, A.A., and Schepers, J., Vertical profile and temporal variation of chlorophyll in maize canopy: Quantitative "crop vigor" indicator by means of reflectance-based techniques, *Agronomy Journal*, 100, 1409–1417, 2008, doi:10.2134/agronj2007.0322.

89. Ciganda, V., Gitelson, A.A., and Schepers, J., Non-destructive determination of maize leaf and canopy chlorophyll content, *Journal of Plant Physiology*, 166, 157–167, 2009.

90. Baldocchi, D.D., Assessing the eddy covariance technique for evaluating carbon dioxide exchange rates of ecosystems: Past, present and future, *Global Change Biology*, 9, 479–492, 2003.

91. Knipling, E.B., Physical and physiological bases for the reflectance of visible and near-infrared radiation from vegetation, *Remote Sensing of Environment*, 1, 155–159, 1970.

92. Monteith, J.L., Solar radiation and productivity in tropical ecosystems, *Journal of Applied Ecology*, 9, 744–766, 1972.

93. Monteith, J.L., Climate and the efficiency of crop production in Britain, *Philosophical Transactions of the Royal Society of London*, 281, 277–294, 1977.

94. Ruimy, A., Kergoat, L., and Bondeau, A., Comparing global models of terrestrial net primary productivity (NPP): Analysis of differences in light absorption and light-use efficiency, *Global Change Biology*, 5, 56–64, 1999.

95. Sims, D.A., Rahman, A.F., Cordova, V.D., El-Masri, B.Z., Baldocchi, D.D., and Flanagan, L.B., On the use of MODIS EVI to assess gross primary productivity of North American ecosystems, *Journal of Geophysical Research*, 111, G04015, 2006, doi:10.1029/2006JG000162.

96. Xiao, X., Zhang, Q., Braswell, B., Urbanski, S., Boles, S., and Wofsy, S., Modeling gross primary production of temperate deciduous broadleaf forest using satellite images and climate data, *Remote Sensing of Environment*, 91, 256–270, 2004.

97. Wang, Z., Xiao, X., and Yana, X., Modeling gross primary production of maize cropland and degraded grassland in northeastern China, *Agricultural and Forest Meteorology*, 150, 1160–1167, 2010.

98. Jarvis, P.G. and Leverenz, J.W., Productivity of temperature, deciduous and evergreen forests. In: *Encyclopedia of Plant Physiology*, O.L. Lange, P.S. Nobel, C.B. Osmond, and H. Ziegler (eds.), New Series, Vol. 12d., Springer-Verlag, New York, pp. 233–280, 1983.

99. Gamon, J.A., Penuelas, J., and Field, C.B., A narrow waveband spectral index that tracks diurnal changes in photosynthetic efficiency, *Remote Sensing of Environment*, 41, 35–44, 1992.

100. Nichol, C.J., Huemmrich, K.F., Black, T.A., Jarvis, P.G., Walthall, C.L., Grace, J., and Hall, F.G., Remote sensing of photosynthetic-light-use efficiency of boreal forest, *Agricultural and Forest Meteorology*, 101, 131–142, 2000.

101. Barton, C.V.M. and North, P.R.J., Remote sensing of canopy light use efficiency using the photochemical reflectance index. Model and sensitivity analysis. *Remote Sensing of Environment*, 78, 264–273, 2001.

102. Sellers, P.J., Berry, J.A., Collatz, G.J., Field, C.B., and Hall, F.G., Canopy reflectance, photosynthesis and transpiration. III. A reanalysis using improved leaf models and a new canopy integration scheme, *Remote Sensing of Environment*, 42, 187–216, 1992.

103. Suyker, A.E., Verma, S.B., Burba, G.G., Arkebauer, T.J., Walters, D.T., and Hubbard, K.G., Growing season carbon dioxide exchange in irrigated and rainfed maize, *Agricultural and Forest Meteorology*, 124, 1–13, 2004.

104. Suyker, A.E., Verma, S.B., Burba, G.G., and Arkebauer, T.J., Gross primary production and ecosystem respiration of irrigated maize and irrigated soybean during a growing season, *Agricultural and Forest Meteorology*, 131, 180–190, 2005.

105. Whittaker, R.H. and Marks, P.L., Methods of assessing terrestrial productivity. In: *Primary Productivity of the Biosphere. Ecological Studies*, H. Lieth and R. H. Whittaker (eds.), Vol. 14. Springer-Verlag, New York, pp. 55–118, 1975.

106. Zarco-Tejada, P.J., Miller, J.R., Mohammed, G.H., Noland, T.L., and Sampson, P.H., Vegetation stress detection through chlorophyll a + b estimation and fluorescence effects on hyperspectral imagery, *Journal of Environmental Quality*, 31, 1433–1441, 2002.

107. Dawson, T.P., North, P.R.J., Plummer, S.E., and Curran, P.J., Forest ecosystem chlorophyll content: Implications for remotely sensed estimates of net primary productivity, *International Journal of Remote Sensing*, 24, 611–617, 2003.

108. Gitelson, A.A., Viña, A., Masek, J.G., Verma, S.B., and Suyker, A.E., Synoptic monitoring of gross primary productivity of maize using Landsat data, *IEEE Geoscience and Remote Sensing Letters*, 5, 2, 2008, doi:10.1109/LGRS.2008.915598.

109. Gitelson, A.A., Verma, S.B., Vina, A., Rundquist, D.C., Keydan, G., Leavitt, B., Arkebauer, T.J., Burba, G.G., and Suyker, A.E., Novel technique for remote estimation of CO_2 flux in maize, *Geophysical Research Letters*, 30, 9, 1486, doi:10.1029/2002GL016543, 2003.

110. Masek, J.G., Vermote, E.F., Saleous, N., Wolfe, R., Hall, F.G., Huemmrich, F., Gao, F., Kutler, J., and Lim, T.K., A Landsat surface reflectance data set for North America, 1990–2000, *Geoscience and Remote Sensing Letters*, 3, 68–72, 2006.

111. Peng, Y., Gitelson, A.A., Keydan, G., Rundquist, D.D., and Moses, W., Remote estimation of gross primary production in maize and support for a new paradigm based on total crop chlorophyll content, *Remote Sensing of Environment*, 115, 978–989, 2011.

Part VI

Vegetation Processes and Function (ET, Water Use, GPP, LUE, Phenology)

16 Hyperspectral Remote Sensing Tools for Quantifying Plant Litter and Invasive Species in Arid Ecosystems

Pamela Lynn Nagler, B.B. Maruthi Sridhar,
Aaryn Dyami Olsson, Willem J.D. van Leeuwen,
and Edward P. Glenn

CONTENTS

Any use of trade, product, or firm names is for descriptive purposes only and does not imply endorsement by the U.S. Government.

16.1 INTRODUCTION—HYPERSPECTRAL REMOTE SENSING OF LANDSCAPE COMPONENTS

Green vegetation can be monitored and distinguished using visible and infrared multiband and hyperspectral remote sensing methods. The problem has been in identifying and distinguishing the nonphotosynthetically active radiation (PAR) landscape components, such as litters and soils, from green vegetation [35–38]. Additionally, distinguishing different species of green vegetation is challenging using the relatively few bands available on most satellite sensors. This chapter focuses both on previously published work by Nagler et al. [35–38] that identified hyperspectral remote sensing characteristics that distinguish between green vegetation, soil, and litter (or senescent vegetation), and on new research conducted to aid in distinguishing invasive species from the mixed landcover surface.

The main message from the previously published work is that the shortwave infrared (SWIR) wavelength range can be used to distinguish plant litter from soils using the cellulose absorption feature at 2100 nm exhibited by litter [1]. A three-band SWIR index, which incorporates wavelengths that capture unique absorption differences, may prove more useful than the visible and near-infrared (VIS-NIR) range in discriminating plant litter from soils.

Quantifying litter by remote sensing methods is important in constructing carbon budgets of natural and agricultural ecosystems. Distinguishing between plant types is important in tracking the spread of invasive species. Green leaves of different species usually have similar spectra, making it difficult to distinguish between species. However, in this chapter we show that phenological differences between species can be used to detect some invasive species by their distinct patterns of greenness and dormancy over an annual cycle based on hyperspectral data. Both applications require methods to quantify the nongreen cellulosic fractions of plant tissues by remote sensing, even in the presence of soil and green plant cover. We explore these methods and offer three case studies. The first concerns distinguishing surface litter from soil using the cellulose absorption index (CAI), as applied to no-till farming practices where plant litter is left on the soil after harvest. The second involves using different band combinations to distinguish invasive tamarisk from agricultural and native riparian plants along the Lower Colorado River (LCR). The third illustrates the use of the CAI and normalized difference vegetation index (NDVI) time-series analyses to distinguish between invasive buffelgrass (*Pennisetum ciliare*) and native plants in a desert environment in Arizona. Together the results show how hyperspectral imagery can be used to inform applications and solve problems that are not amenable to solution by the simple band combinations normally used in remote sensing.

16.1.1 Distinguishing between Green Vegetation, Soil, and Litter Using the CAI in Agricultural Systems

16.1.1.1 Plant Litter

Litter is dead plant material that begins as green leaves, stems, fruits, etc., and then falls from the canopy to the surface; it gradually decomposes into soils over time. Senescent or dormant leaves on plants can have the same spectral properties as litter and are classified with litter in this chapter. In agricultural systems, litter is the straw fraction of annual crops left behind after harvest, as in no-till farming. In the context of invasive species, seasonal senescence is a distinguishing feature of some invasive species such as buffelgrass and methods to detect litter can in theory be used to track the spread of these species.

16.1.1.2 Importance of Litter to the Soil System

Litter contributes greatly to soil nutrient and energy cycles. The amount and composition of litter changes spatially (by region) and temporally (by season) [1,2], and rates of decomposition depend upon species physiology, climate/environmental conditions, and microbial activity [3]. Litter decomposition is more closely correlated with nutrient release than with energy flows [4,5]. Eventually decomposition of litter results in humification and mineralization of the recalcitrant carbon fraction. Initially, worms and other macro-animals break the litter into smaller fractions with greater surface area [6]. Then bacteria and fungi break down complex biochemicals to molecular constituents [7]. The remaining litter contains celluloses, hemicelluloses, lignins, and many other materials including organic nitrogen [5]. Finally, organic material is broken down into carbon dioxide (CO_2), water, and minerals; nitrogen, phosphorus, calcium, magnesium, and potassium are released [8]. The decomposition of litter contributes to atmospheric CO_2 concentrations and contributes to nitrogen and oxygen cycles [1].

16.1.1.3 Benefits of Litter Left in Agriculture Systems

Part of the interest in quantifying soil litter cover by remote sensing is the recent interest in no-till agriculture, a method used to manage residue cover to protect soils from erosion. McMurtrey et al. [9] cited agricultural statistics in the United States (U.S.) as follows: 330 million acres of arable land is tilled, of which 123 million acres are classified as highly erodible land (HEL); the result is the annual loss of 1.25 billion tons of soil. Leaving organic residues on bare soil also affects water infiltration, evaporation, porosity, and soil temperatures [10]. The decay of litter adds nutrients to the soil, improves soil structure, and facilitates tilling, thereby reducing soil erosion, runoff volumes, sediment transport, and movement of pesticides [11]. Maintaining crop residue on the soil surface is frequently the most cost-effective method of reducing soil erosion and complying with Federal regulations [12]. Federal erosion prevention legislation is defined in two acts: the 1985 Food Security Act, specifically the Conservation Compliance Provision (Public Law 99–198), and the 1990 Food, Agriculture, Conservation and Trade Act (Public Law 101–624) [9]. In response to these laws, farmers must implement erosion control practices on HEL.

16.1.1.4 Importance of Quantifying Litter

Erosion prediction models (i.e., the Universal Soil Loss Equation [USLE] and the Water Erosion Prediction Project [WEPP]) incorporate crop residue cover estimates, but there is considerable error in these estimates [13,14]. Usually line-transect methods are used to measure litter cover in the field, but these methods are subject to human error and are time-consuming [12,14,15]. More rapid and more accurate spectral measurement techniques are needed to improve litter quantification methods [9,12,13,16].

16.1.1.5 Senescent Leaves in the Life Cycle of Invasive Species

Tracking the spread of invasive species, particularly introduced range grasses, has become a priority goal for lands managers. Many of these species have distinct dormant periods in which leaves are dry and brown and have the same spectral properties as litter. While native plants may also have a dormant period, it is sometimes possible to distinguish between species by their phenology, focusing on the timing of their senescent periods through time series of hyperspectral imagery [60,62]. We explore the use of hyperspectral imagery for this purpose as well as for quantifying plant litter and crop residues in this chapter.

16.1.2 Reflectance Spectra (400–2400 nm) Used in a Laboratory to Distinguish Green Vegetation, Soil, and Litter in the Landscape

16.1.2.1 Distinguishing between Pure Scenes of Plant Litter and Green Vegetation

Remote sensing studies have typically focused on green vegetation because plant canopies are dominated by green leaf spectral features and because key biophysiological processes, such as photosynthesis, occur in green leaves [1]. Photosynthetically active vegetation has a very distinctive spectral reflectance signature in the landscape, with strong absorption in the visible bands contrasted with strong reflectance in the NIR (see Figure 16.3). Ratio and difference indices based on energy from a band in the VIS (400–700 nm) and in the NIR (700–1100 nm) wavelength range were first developed by Jordan [17] to assess spectral features in green vegetation for estimating energy accumulation in plant canopies, biomass, and the leaf area per unit ground (LAI, leaf area index). The green and nongreen components of a canopy can be separately identified using two wavebands because their spectral reflectance curves have uniquely different shapes. In green leaves, pigment concentrations, water content, and structure affect leaf optical properties [18]. Chlorophylls and other pigments in green vegetation absorb in the Blue (450 nm) and Red (650 nm) wavelengths and cell structure and thickness control NIR optical properties; reflectance in the Green (550 nm) and NIR wavelengths thereby produces a step-function reflectance curve [19–22]. The ability to discriminate green and nongreen component types by differences in their signatures with these two bands alone is the basis for quantifying vegetation parameters for landscape models [23,24].

16.1.2.2 Distinguishing between Pure Scenes of Plant Litter and Soils

The absorption and scattering properties of leaves change as they senesce and decompose [9]. Previous work by Woolley [18] has shown that during senescence, leaves lose moisture and air spaces between cells increase. Spectral changes in the VIS wavelengths occur due to the loss of moisture, pigments, and structure. Celluloses and lignins do not readily compost, resulting in high NIR reflectance [18]. Woolley [18] showed that dried or senescent plant material has higher reflectance than green vegetation at all wavelengths, while Daughtry and Biehl [25] found that litter shows reduced NIR scattering and thus lower values for reflectance and transmittance. Unfortunately, soil spectra are generally similar to those of plant litter, making quantifying litter with remote sensing techniques challenging [26–28].

16.1.2.3 Remote Sensing Techniques (and Their Limitations) to Discriminate Litter from Soils and Green Vegetation

To distinguish litter from green leaves and soil, the best waveband regions and resolutions of reflectance spectra must be chosen to distinguish plant litter and soils from an integrated scene. This entails finding differences between litter hyperspectral reflectance data for a variety of species, decay stages, and moisture levels [1,26]. It is important to note wavelength band permutations such as minimum/maximum, greater than/less than, and concave/convex relationships without emphasizing the absolute magnitude of spectral reflectance. For this study, reflectance was measured by sensors that collect data from VIS (400–700 nm), NIR (700–1100 nm), and SWIR (1100–2500 nm) wavelength bands. The appropriate wavelength range and resolution were examined by looking for differences in the spectral curves of the target types; places where litter cellulose and soil minerals absorb energy were seen in the reflectance signatures of nongreen components. We explored methods to develop a robust index that distinguishes plant litter from soils based on differences in their reflectance curve shapes. Several wavebands, including not only the two bands that are commonly used in vegetation studies (VIS and NIR), but also new combinations, were examined to find a diagnostic feature so that an index could be devised to separate plant litter from soils.

16.1.2.3.1 VIS-NIR Wavelength Range

Although the spectral reflectance of a scene is affected by all included components, such as soil, green vegetation, shadow, surface roughness, and nonsoil residue [29], the spectral reflectance curves of plant litter and soils are often assumed to have the same generally featureless shape in the VIS-NIR (400–1100 nm) [11,12]. There is a problem in discriminating these photosynthetically inactive materials using the VIS-NIR wavelength range because there are generally limited unique features to show differences between the various ground components [30]. An exception to this ability to discriminate contrasting soil and litter signatures is the VIS and infrared (IR) wavelength range which does depend somewhat on background soil color, such as if you had an organic or iron rich soil with overlying yellow-colored litter. Spectral vegetation indices produce values that not only vary within component type, but also vary between types; for instance, the NDVI values for soils (0.08–0.16) and for litter (0.14 to as high as 0.45 for freshly deposited, still green litter) [9] share a common range. Consequently, one difficulty in discriminating plant litter from soils is that wavelengths in the VIS-NIR range do not provide sufficient separability between soils of varying moisture and plant litter of different moisture and ages because their spectral curves are similar and thus are indistinguishable at any one wavelength [9].

Reflectance in two wavebands has been of some use in discriminating litter from soil. For example, McMurtrey et al. [9] measured a separation of soil and crop residue NDVI values. The separability or variability of these spectral reflectance curves indicates that the ground components spectra are not constant; but because their spectra could not be consistently distinguished and were not statistically significant at the least-significant-difference (LSD) (0.05) level, they concluded that the VIS-NIR wavelengths (and thus NDVI) do not produce absorption peaks that can be used to discriminate soils from litter. The two band index commonly used in vegetation studies (i.e., NDVI) does not provide statistically reliable results for detecting differences amongst the three classes of photosynthetically inactive plant material, soils, and photosynthetically active green vegetation. Reflectance (400–1100 nm) in three bands has been used in an attempt to distinguish plant litter from soils. McMurtrey et al. [9] found that a third band in the blue (450 nm) range, used in conjunction with the VIS and NIR bands, appeared to capture major differences in the background components, but this band combination was not tested further. This previously published research was undertaken to determine whether key diagnostic differences in nongreen component spectra are revealed in other wavelengths and/or multiple band combinations.

16.1.2.3.2 UV Wavelength Range

Other wavelength regions were investigated. Fluorescence techniques were tested to distinguish plant litter from soils. Daughtry et al. [16] found that plant litter produced greater fluorescence than most soils when illuminated with ultraviolet (UV, 320–400 nm) radiation. This method was less ambiguous and better suited for discriminating litter from soils than the VIS-NIR reflectance methods, but several potential problems inhibit the implementation of the fluorescence technique. For instance, (i) excitation energy must be supplied to induce fluorescence, and (ii) the fluorescence signal is small relative to normal, ambient sunlight [16].

16.1.2.3.3 SWIR Wavelength Range

Few studies have investigated differences in the SWIR reflectance curves of plant litter and soils [23]; however, several studies have noted spectral features that are unique to each component in the SWIR region [1,31]. A common spectral feature to both litter and soils are two water absorption peaks at 1400 and 1900 nm. In the SWIR spectra of dried plants, a cellulose/lignin absorption peak (a reflectance trough) was noted at 2100 nm [1]. Work with spectral reflectance indicated that the ligno–cellulose absorption feature at 2100 nm and shoulder peaks at 2000 and 2200 nm were useful for discriminating litter from soils [12]. This feature is absent in the spectra of soils, which show no cellulose absorption, but rather a clay mineral absorption feature at 2200 nm [28,31].

Two different experiments using two instruments were carried out in this work: (1) forest litter and soils samples in the VIS-NIR (400–1100 nm) (Figure 16.1a) were first examined with an SE-590 spectroradiometer and (2) forest litter, crop residue, senescent grass, and soils samples in the VIS-SWIR (400–2500 nm) were then measured with an IRIS Mark IV spectroradiometer (Figure 16.1b). The litter types were a general representation of the litter surface beneath a canopy. In total, the spectral reflectance of 82 samples of litter (52 forest litter, 24 residues, and 6 grasses) and 7 soils were measured. Five types of litter (coniferous and deciduous forest litter, soybean and corn crop residue, and senescent grasses) were considered for multiple ages.

16.1.2.4   A Diagnostic Feature—Cellulose Absorption Index

The best wavelength range and resolution for discriminating plant litter from soils were defined using three bands in the CAI. Mean spectral reflectance from each sample for three 50 nm wide bands were used to calculate CAI as follows:

$$CAI = 0.5(R_{2023nm} + R_{2215nm}) - R_{2100nm} \tag{16.1}$$

where R_{2023nm}, R_{2100nm}, and R_{2215nm} are the wavebands centered at 2023, 2100, and 2215 nm, respectively, with a bandwidth of 10 nm.

CAI was defined by the relative depth of the spectral absorption at 2100 nm because dry litter exhibited this spectral ligno-cellulose absorption, as demonstrated first by Elvidge [1].

When contrasted with dry soils spectrums from Stoner and Baumgardner [28], the lack of the ligno-cellulose absorption was evident. The spectral data from soils and plant litter in this work and that of [12] showed that both wet and dry soils and plant litter could be distinguished when their different absorptions at 2100 nm, due to the effects of water, were calculated and indexed according to Equation 16.1. The 50 nm bands appeared to be the most useful for discriminating all types of soils from all types of litter. However, the usefulness of the index in discriminating wet soils from wet litter may possibly be improved with smaller bandwidths, which would capture less of the absorption from the water band at 1900 nm.

16.1.2.5   Effects of Water on CAI

To evaluate the effect of sample water content on the index, CAI was plotted as a function of reflectance in the water absorption band (1900–1950 nm) (Figure 16.2). This band ca. 1900 nm is sensitive to sample moisture content [32]; we use the reflectance in this band to monitor moisture in the plant litter and soils samples in the form of a scatterplot. Wet samples (litter and soils) had reflectance spectra at 1900 nm that were <25%, while dry samples were generally >25%. For wet soils alone, reflectance spectra were <10%, with the exception of the sand, which was very bright even when wet. Dry soils samples had reflectance values in the water absorption band >25% with the exception of Houston Black Clay, which held more moisture when air-dried than other soils.

The presence of water reduced the reflectance of all samples at all wavelengths and made discrimination of litter and soils difficult. Although water absorption dominated the spectral properties of both soils and residues in the SWIR, it was possible to discriminate wet litter from wet soil using CAI. More than 90% of the wet plant litter samples had positive CAI values. However, five wet litter samples also have negative CAI values. The cellulose absorption feature was negative for three wet deciduous samples and two coniferous samples that were all more than 1 year old. All five samples were sufficiently decomposed so that the absorption due to cellulose or lignin fibers was easily masked by moisture. Positive CAI values represented a presence of the cellulose

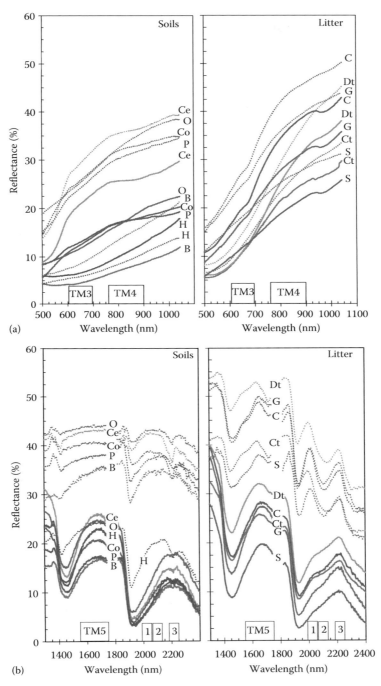

FIGURE 16.1 (See color insert.) (a) VIS-NIR spectral reflectance (500–1100 nm) of dry (dashed lines) and wet (solid lines) soils and litter types. (b) SWIR spectral reflectance (1300–2400 nm) of dry (dashed lines) and wet (solid lines) soils and litters. The soils are Othello (O, o), Cecil (E, e), Codorus (C, c), Portneuf (P, p), Barnes (B, b), and Houston Black Clay (H, h). The plant litters are corn (M, m), soybean (S, s), deciduous tree (D, d), coniferous tree (C, c), and grass (G, g). (From *Remote Sensing of Environment*, 71, Nagler, P.L., Daughtry, C.S.T., Goward, S.N., Plant litter and soil reflectance, 207–215, Copyright (2000), with permission from Elsevier.)

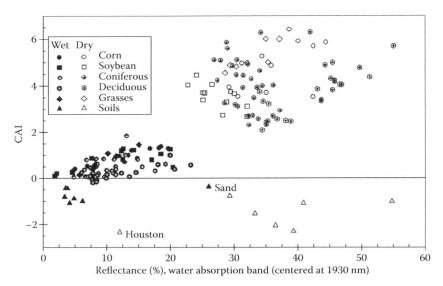

FIGURE 16.2　Plot of the CAI as a function of reflectance in a water absorption band from 1900 to 1950 nm. (From *Remote Sensing of Environment*, 71, Nagler, P.L., Daughtry, C.S.T., Goward, S.N., Plant litter and soil reflectance, 207–215, Copyright (2000), with permission from Elsevier.)

spectral feature. In the spectra for all the soils, the cellulose feature was absent and thus produced CAI values between 0 and −5. Although negative CAI values represented an absence of the cellulose spectral feature, they do not necessarily indicate the absence of cellulose, for the ligno–cellulose feature in plant litter samples was present; however, sometimes it was masked by water and negative CAI values were produced.

The effect of green vegetation on CAI was also determined in previously published work [36]. Although wet and dry forest litter, crop residues, senesced grass, and soils spectra provided data to represent various ground component spectral reflectance values in the VIS-SWIR range, the original study [36] lacked fresh and dry green-leaf reflectance measurements in this range because the authors were focusing on nonphotosynthetically active targets prior to producing [37]. Lignin, cellulose, and organic compounds found in litter showed distinct absorption peaks in the SWIR wavelengths, which were not present in the signatures of most green vegetation or yellowed leaves [1]. Hence, it was expected that green vegetation index values would not confound the use of the index in discriminating litter from soils. Information about the spectral behavior of the cellulose feature in the SWIR wavelengths provided a way to index the reflectance values and distinguish most litter from soils, but whether spectra of green vegetation inhibits the usefulness of CAI to distinguish litter from soils was not determined until we looked at mixtures of litter on soils as well as pure green vegetation in the next step of this study (see Section 1.3.1, in the following). With the new mixed study, we repeated the experiment with pure samples and then proceeded with mixtures. Figure 16.3 shows the typical reflectance spectra (400–2400 nm) of pure scenes of black, red, and gray soils and green vegetation (top figure) and spectra of four crop residues and two tree litters (bottom figure).

16.1.2.6　Benefits (and Limitations) of CAI

The spectral resolution requirements (i.e., the spectral, bandwidths, position or center of the spectral bands and number of spectral bands) and sensor proximity to the target (related to the field of view [FOV]) are some considerations in choosing a sensor for a particular application [33]. A few of the current satellite sensors used are Landsat's Thematic Mapper (TM) and the Moderate resolution Imaging Spectroradiometer (MODIS). The limited spectral range and resolution of most satellite instruments currently inhibit their use in discriminating nongreen canopy components using the CAI.

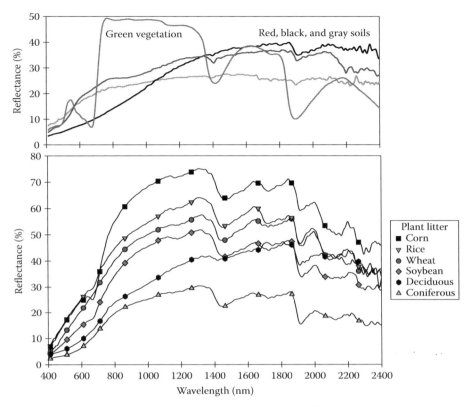

FIGURE 16.3 (See color insert.) Typical reflectance spectra (400–2400 nm) of black, red, and gray soils and green vegetation (top) and spectra of four crop residues and two tree litters (bottom). The symbols in the bottom figure do not represent data sampling points, but rather are placed on the spectral lines for clarity. (From *Remote Sensing of Environment*, 87, Nagler, P.L., Inoue, Y., Glenn, E.P., Russ, A., Daughtry, C.S.T., Cellulose absorption index (CAI) to quantify mixed soil-plant litter scenes, 310–325, Copyright (2003), with permission from Elsevier.)

The bandpasses are too broad for measuring the absorption features of dry plant materials; they lack the ability to spectrally discriminate between plant litter types [1]. However, a high spectral resolution sensor such as the Advanced Visible Infra Red Imaging Spectrometer (AVIRIS) can detect reflectance throughout the 400–2500 nm range in continuous narrow bands (10 nm) [34]. Distinguishing plant litter from soils using AVIRIS and other sensor systems designed with narrow bandwidths in the SWIR region has not yet been tested. Minimizing atmospheric signals due to aerosols and water vapor will be an important prerequisite to quantify CAI.

16.1.3 Summary of Pure Scenes of Soils and Litter

This study supports research that it is not possible to consistently distinguish plant litter from soils using reflectance spectra in the VIS-NIR wavelength range exclusively. The CAI, which we developed using reflectance data from the SWIR wavelength region, is effective at distinguishing litter from soils and may improve quantification estimates of plant litter in a scene by making the measurements more objective and accurate. Further work could serve to calibrate the CAI for quantifying phytomass to improve estimates of productivity and energy balance. Plant litter and soils, regardless of moisture content, were distinguishable from each other using spectral reflectance data acquired in the VIS-SWIR (400–2500 nm) wavelength region. The developed CAI can be used to successfully discriminate litter from soil.

16.1.4 Mixed Scenes of Plant Litter and Soils

The ability to discriminate plant litter from soils using CAI allows ground components to be identified using laboratory spectra of pure samples. In the research described for pure scenes, mixed laboratory samples were not measured and discrimination of mixed components using CAI was not tested. Additionally, CAI was not tested using field samples.

The pure scene research indicates that it would be useful to evaluate the value of this remote sensing method for field systems like natural canopies or agricultural lands to distinguish crop residues from underlying soils and estimate the quantity of litter in field conditions. This can be done by incorporating the experimental CAI values into existing models to derive theoretical estimates of field conditions. However, the hypothetical estimates may not be useful if CAI is incorporated into a model applied outside its intended use, such as forest systems where too many unknown variables exist, to get a reasonable estimate. Future work to test this methodology in agricultural systems alone is recommended to see if CAI is flexible enough to use in noncanopy field circumstances. If successful, it may then replace the current tedious methods employed in quantifying residues as part of conservation tillage efforts. We used spectral measurements of laboratory samples with varying percent cover (mixed targets) to obtain CAI and assess its usefulness as a function of litter weight and litter cover. We employed both (i) photographs of percent cover and (ii) SWIR video/images of percent cover to determine if either the photographs or CAI images were necessary to distinguish varying fractions of litter from underlying soils. The use of SWIR imaging techniques to estimate percent cover and replace time-consuming SWIR spectral measurement and manipulation techniques is being explored. These imaging techniques can be made using a Vidicon/CCD camera with the three CAI bands; a CAI image of the different fractions of percent residue cover over varying soil backgrounds can be produced.

In the field, soils are rarely completely bare (0% litter cover) or completely covered with plant litter (100% cover), except in some no-till cropping systems. Daughtry [35] varied the moisture content of soils and litter samples but only simulated the effect of mixed scenes; in the present work, the reflectance spectra of wet and dry scenes with different proportions of soil and litter were measured. Figure 16.4 shows the dry (upper graph) and wet (lower graph) reflectance spectra for various amounts of wheat litter on the surface of the black soil. As the coverage of plant litter increased in the dry samples, the prominence of the 2100 nm absorption feature also increased. Moisture reduced reflectance and masked the absorption feature at 2100 nm in all the wet, mixed samples. Nagler et al. [36] also showed that discrimination of wet, pure soils from wet, pure litter was possible using CAI, but here, the wet, mixed samples with >20% litter cover did not show negative CAI values as was seen in the dry, mixed samples. Regardless of moisture, adding wheat litter to the black soil increased reflectance at all wavelengths. On the other hand, adding soybean residue to the gray soil reduced reflectance in the visible wavelength region, but increased reflectance at other wavelengths.

The CAI spectral variable describes the average depth of the cellulose absorption feature at 2100 nm. Positive values of CAI represent the presence of cellulose, and thus, plant litters typically had positive CAI values. Negative values of CAI indicate the absence of cellulose. CAI of soils is typically negative [37,36]. Daughtry et al. [38] observed that in the wet samples, absorption by water dominated the reflectance spectra and nearly obscured the differences in their CAI values. The CAI of each mixed scene of plant litter, with seven different levels of cover including the full litter cover, and green vegetation (also the scenes of pure soils [three types]), was plotted as a function of reflectance in the water absorption band at 1910–1950 nm (Figure 16.5). Green vegetation is shown here as being very negative. Mean CAI increased significantly from bare soils (CAI = −0.2) as the amount of plant litter on the soil increased to 100% cover (CAI = 5.2). The plant litter had positive values of CAI and the soils had negative values. The CAI values of green leaves from Inoue et al. [39] were also large negative values, which indicated that the cellulose absorption feature was obscured by the abundance of water in green leaves. CAI can be used to distinguish green canopy cover from underlying nongreen landscape components, but it is also possible—given CAI as a function of reflectance in the water absorption band (1910–1950 nm)—that the nongreen components, litter, and soils are also

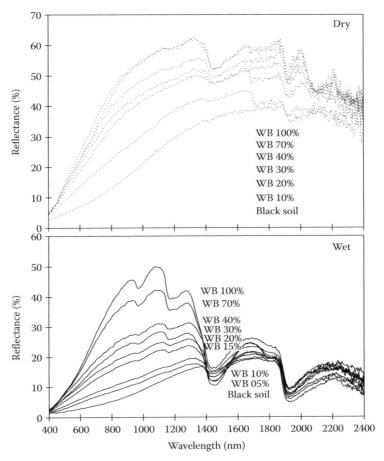

FIGURE 16.4 Mean reflectance spectra for a series of wet and dry mixed scenes of wheat residue with underlying black soil. The litter level, estimated from its weight, of wheat residue in the dry scenes (top figure) were 0% (black soil), 10%, 20%, 30%, 40%, 70%, and 100%. The levels in the wet scenes (bottom figure) were 0% (black soil), 5%, 10%, 15%, 20%, 30%, 40%, 70%, and 100%. (From *Remote Sensing of Environment*, 87, Nagler, P.L., Inoue, Y., Glenn, E.P., Russ, A., Daughtry, C.S.T., Cellulose absorption Index (CAI) to quantify mixed soil-plant litter scenes, 310–325, Copyright (2003), with permission from Elsevier.)

differentiated. A multispectral approach may also be employed; for example, the simple ratio (reflectance in the 760–900 nm band divided by reflectance in the 630–690 nm band [30]) could be used to distinguish green vegetation from bare soil and the CAI could then be used to separate plant litter from soil. Thus, the CAI is relevant to situations where it is important to distinguish residues from soils (agricultural systems) and to discern green vegetation canopies from underlying nongreen vegetation components (natural systems). In this mixed scene work, the CAI of all three soils was negative, but as the amount of litter on the soil surface increased, CAI of the mixed scenes also increased (Figure 16.6). All four residue types showed that mixed scenes with 0% and 10% residue level by weight and black soil underneath were negative, showing that small amounts of residue on black soil could not be discriminated from bare soil. However, for gray soils, the mixed scene litter limit varied depending on the litter type. For corn and soybean residues, the mixed scenes with 0% and 10% residue level and gray soil underneath were negative, as was the case with black soil underneath, showing that small amounts of residue on gray soil could not be discriminated from bare soil. However, for wheat and rice residue, the mixed scenes with 10% residue level and gray soil underneath were positive, showing that these could be discriminated from bare soil. For red soil, for wheat and soybean

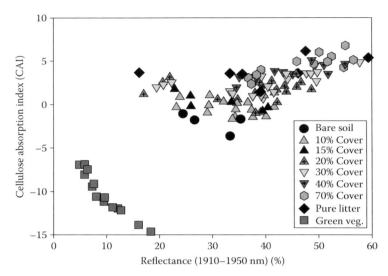

FIGURE 16.5   (See color insert.) CAI of all the pure soils, mixed scenes of plant litter with different levels of cover, and green vegetation as a function of percent reflectance (%) in the water absorption band (1910–1950 nm). (From *Remote Sensing of Environment*, 87, Nagler, P.L., Inoue, Y., Glenn, E.P., Russ, A., Daughtry, C.S.T., Cellulose absorption index (CAI) to quantify mixed soil-plant litter scenes, 310–325, Copyright (2003), with permission from Elsevier.)

residue, the mixed scenes at the 0% and 10% levels were negative, but were positive with corn and rice residue at these percent cover levels. All four crop residue types had positive CAI values for mixed scenes of more than 20% residue level. These were significantly different from the CAI values of the soils. For the mixed experiment with the tree litters, both types showed that mixed scenes with 0% litter level for the black soil were negative, but that any amount of litter (10%, 15%, and 20% litter level or higher) could be discriminated from the black soils using CAI. The situation was different for the gray and red soils. Deciduous, broadleaf tree litter at 10%, 15%, and 20% litter level had negative CAI values and could not be discriminated from the underlying gray or red soils. The mixed scene CAI values only became positive at levels greater than 30% litter level. For coniferous tree litter over gray soil, the CAI values were positive for litter levels greater than 10%, showing that this residue could be easily discriminated from a gray background soil. For coniferous tree litter over red soil, the CAI was negative for 10% and 15% residue levels, but was positive at a 20% residue level.

Residue level by relative weight or relative percent cover (Rel.%C), averaged over three soils, is shown for crop and forest litter levels and is linearly related to CAI (Figure 16.7). Coniferous tree litter had the most variability and lowest correlation ($r^2 = 0.84$). Deciduous tree litter had the lowest CAI values and a high correlation ($r^2 = 0.98$). Although the discrimination of background soils from litter at low densities or residue levels of <10% (crop residues) or <20% (tree litters) may be difficult based on these results, CAI is a very good predictor of the percent of plant litter cover in mixed scenes.

16.1.5   Conclusions for Mixed and Pure Scenes of Soils and Litter

Reflectance spectra of pure and mixed scenes of six plant litter types and three soils were measured and the CAI was calculated using the spectral feature at 2100 nm. The CAI values of pure plant litter were significantly larger than the CAI value of the pure soils. For the mixed scenes, as plant litter cover increased, CAI increased linearly. The results showed that CAI was successful in distinguishing fractions of litter from underlying soils in mixed laboratory samples. In some soil types, such as the red soil in this study, a complication arises with using the depth of the cellulose absorption feature at 2100 nm, because the width of the clay mineral absorption feature at 2200 nm matches the minor reflectance peak of cellulose at 2200 nm that has been induced by absorptions at 2100 and 2300 nm

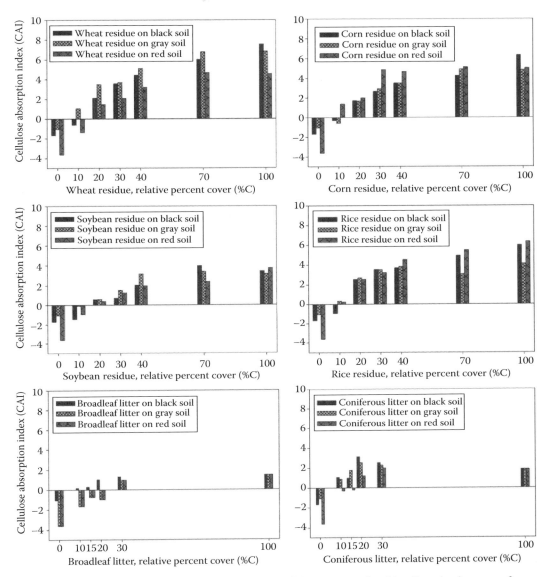

FIGURE 16.6 (See color insert.) CAI as a function of the amount of residue for mixed scenes of varying amounts of each crop residue and tree litter, shown for each of the three soils. (From *Remote Sensing of Environment*, 87, Nagler, P.L., Inoue, Y., Glenn, E.P., Russ, A., Daughtry, C.S.T., Cellulose absorption index (CAI) to quantify mixed soil-plant litter scenes, 310–325, Copyright (2003), with permission from Elsevier.)

in plant material. This leads to lower values of CAI than with either the black or gray soils in this study. Therefore, it is recommended that special attention be given to the shoulder of the absorption feature at 2200 nm before utilizing the countered absorption and reflectance features to calculate CAI. Using a two-way ANOVA for crop residues, soils were found not to be significantly different from one another, although when the statistics were run for tree litters, the red soil was indeed found to be slightly significant. Thus, only one soil type in this study inhibited the detection of tree litter and/or the ability to quantify litter cover. Because the relationship between CAI and litter level did not saturate at low levels of cover in these experiments, this spectral variable was useful over nearly the whole range (>70% cover) of mixed soil–litter scenes and was generally not affected by soil type. Furthermore, the strong linear relationship between the crop residues/tree litters and CAI

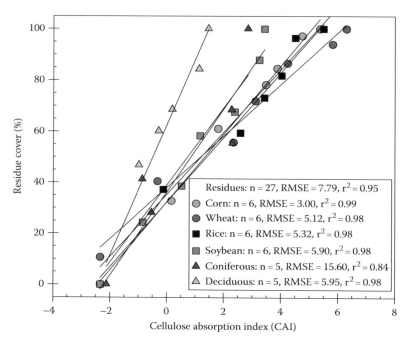

FIGURE 16.7   (See color insert.) Crop and forest litter levels estimated by their weight (Rel.% cover) averaged over three soils shown as a function of CAI. (From *Remote Sensing of Environment*, 87, Nagler, P.L., Inoue, Y., Glenn, E.P., Russ, A., Daughtry, C.S.T., Cellulose absorption index (CAI) to quantify mixed soil-plant litter scenes, 310–325, Copyright (2003), with permission from Elsevier.)

promotes the idea of extrapolating these findings to other residue and litter species, although new experimental data would first have to be obtained.

The relationships between CAI and percent cover were also determined for each plant litter by image analysis of color slides (%C by video) and residue level by weight (Rel.%); the polynomial relationship between CAI and the level by weight was more useful than %C by video for all litters, except corn residue. When CAI was regressed with the average percent cover by image analysis (%C by video) and average %C by weight (residue level) for each experiment, crop residues, and tree litters, the crop residues had more robust coefficients of determination (r^2) values than the tree litter across all three soil types. However, %C by video (image analysis) was a more effective method of discriminating crop residues, while %C by weight (residue level) was a more effective method of discriminating tree litters. The red soil showed a more promising polynomial relationship between CAI and percent cover than the other soils for both methods of estimating percent cover (video and residue level by weight); percent cover was averaged across crop residues and across tree litters. Residue density (g/m^2) can be compared with stacked leaves (weight per unit area) similar to LAI; this may warrant a new study in which the effect of a range of residue densities, all at 100% cover, on CAI is determined. An instrument based on measuring CAI could replace tedious, manual methods of quantifying plant litter cover.

16.2   APPLICATIONS OF HYPERSPECTRAL REMOTE SENSING TO INVASIVE SPECIES—RESEARCH APPROACH AND CASE STUDIES WITH TAMARISK AND BUFFELGRASS

16.2.1   ECOLOGICAL IMPORTANCE OF DISCRIMINATING INVASIVE PLANT SPECIES IN THE LANDSCAPE

Invasive plant species are considered a major threat to global diversity and ecosystem functioning [40]. Plant invasions are known to cause undesirable alterations to the native plant and animal populations, and their community structure and functioning [40]. They also can cause economic loss [1,80]. Of the myriad invaders of arid lands, riparian shrubs (e.g., *Tamarisk ramosissima* and *Eleagnus* sp.) and fire-promoting grasses (e.g., *Pennisetum* spp., *Schismus* spp., and *Bromus rubens*) elicit the greatest cause for concern.

Tamarisk (*Tamarisk ramosissima*) is an invasive shrub introduced to the U.S. from Asia for ornamental and erosion prevention purposes [41,42]. In due course, tamarisk emerged as a dominant woody species in many riparian sites in the western U.S. and northern Mexico, where it forms dense, low thickets that displace native vegetation, impede water flow, and increase sedimentation [42]. Tamarisk has greater salt, drought, and fire tolerance, and resistance to water stress than native plant species [43,44]. At higher salinities, tamarisk has a clear advantage over native trees. Robinson [42] reported that tamarisk covered about 900,000 acres during the 1960s in the U.S. and then it spread and occupied about 1.5 million acres by 1987 [44]. However, these estimates have not been confirmed by actual surveys, in part due to the difficulty in distinguishing between tamarisk and native species by remote sensing methods.

Invasive grasses threaten desert ecosystems by introducing a grass-fire cycle [45,46]. By increasing the abundance and connectivity of fine fuels, invasive grasses facilitate increases in fire frequency and extent in areas poorly adapted to fire. Invasive grasses recover rapidly following fire, whereas most native vegetation does. Both C3 and C4 grasses have altered fire frequencies to the detriment of natives vegetation in North America. The C3 grasses *Bromus rubens* (Mohave and Sonoran Deserts), and *Schismus arabicus* (Mohave and Sonoran Deserts) are seasonally abundant and depend on winter precipitation, which varies over interannual and multidecadal time frames (e.g., in accordance with phases of El Niño Southern Oscillation and Pacific Decadal Oscillation) [46,47]. C4 grasses are common invaders of subtropical and semiarid systems worldwide and exhibit strong dependence on warm season precipitation which, in North America, is coupled tightly with the influence of the North American monsoon. The Sonoran Desert is particularly prone to C4 grass invasion and currently supports various stages of invasion by *Pennisetum ciliare*, *Pennisetum setaceum*, *Eragrostis Lehmanniana*, *Melinis repens*, *Enneapogon cenchroides*, among others. Of particular concern is buffelgrass (*P. ciliare*), which has been planted extensively in pastures in Mexico and is invading southern Arizona but has not yet altered fire regimes in the U.S. Monitoring the spread of buffelgrass in the southwestern U.S. is a key research priority, but at present there are no remote sensing methods to distinguish between buffelgrass and native grasses, cacti, shrubs, and trees.

16.2.1.1 Hyperspectral Reflectance Data as a Monitoring Tool for Invasive Plants

Remote sensing technology has been widely used to monitor weedy and invasive plant species in agriculture and forest environments. Several studies reported the use of aerial photography [48,49], multispectral airborne digital imagery [50], hyperspectral imagery [51–53], and multispectral satellite imagery like advanced very high resolution radiometer (AVHRR) [54], Landsat TM [55], and Satellite Pour l'Observation de la Terre (SPOT) [56] for detecting the spread of different invasive plant species. Application of remote sensing for monitoring invasive plants has proved to be helpful in assessing the extent of infestations, development of management strategies, and evaluation of control measures for the spread of these unwanted plant populations. Studies have been conducted that included the mapping of different species of *Tamarisk* using remote sensing imagery. Aerial photography was used for mapping *Tamarisk chinensis*, specifically during early winter months [48], when their leaves turn to orange-brown colors prior to the leaf drop. *Tamarisk parviflora* was mapped through texture analysis of aerial photographs [49], during the time when the trees were without leaves and had pink flowers, making them distinct from other vegetation. A high spatial resolution (0.5 m) airborne hyperspectral imager was used for mapping the *Tamarisk* species in riparian habitats of Southern California [57] during the time when the trees begin to senesce. A combination of single band and vegetation indices derived from Landsat enhanced thematic mapper plus (ETM+) images [58] were also used for mapping the *Tamarisk* spp. The advanced spaceborne thermal emission reflection radiometer

(ASTER) and MODerate resolution imaging spectroradiometer (MODIS) data [59] were also used to map *Tamarisk* spp., based on vegetation indices. However, most of these studies depend on the distinct visual characteristics of either tamarisk foliage or flowers seen at only one particular time of the year.

The quest to identify invasive grasses in western deserts has led to similar findings, namely, that mapping success is highest when the process is highly manual (e.g., digitization of high resolution aerial imagery) [60–63] or dependent on a phenological opportunity that inconsistently emerges from year to year.

16.2.2 Reflectance Spectra (0.4–2.4 μm) Used Outdoors in Natural Settings

16.2.2.1 Remote Sensing

This chapter documents two studies that identify the utility of hyperspectral data and imagery to map and monitor invasive species. The first study focuses on an invasive riparian tree, salt cedar (*Tamarisk* spp.), in the LCR valley and the second focuses on an invasive C4 grass, buffelgrass (*Pennisetum ciliare*) in the Arizona Upland (AU) zone of the Sonoran Desert. These studies highlight different methodologies for utilizing hyperspectral remote sensing to identify vegetation in highly dynamic and sensitive environments.

16.2.3 Tamarisk Study

The goal of the tamarisk study was to develop a cost-effective, multiseasonal monitoring approach through satellite remote sensing to identify and map tamarisk and other vegetation types growing in the study area. The specific objectives were to (1) identify the spectral characteristics of the major riparian and agricultural vegetation types in the LCR region and (2) determine if Landsat TM data can be used to map tamarisk (*Tamarisk ramosissima*) infestations in this region.

16.2.3.1 Description of Study Area and Vegetation

This study was conducted in two areas, the Palo Verde Irrigation District (PVID) and Cibola National Wildlife Refuge (CWR, Figure 16.8). The study sites are located in Southern California and Northern Arizona, respectively, along the Colorado River. The PVID has a consistent, year-round crop cycle growing crops such as alfalfa, cotton, melons, corn, wheat, and other grasses. Among these crops, alfalfa and cotton cover the largest acreage in the study area. Within the riparian areas of the Colorado River and the CWR, there are several native plants, such as honey mesquite (*Prosopis glandulosa*), cottonwood (*Populus fremontii*), quail bush (*Atriplex lentiformis*), arrow weed (*Pulchea sericea*), palo verde (*Parkinsonia microphylla*), and creosote bush (*Larrea tridentata*). Most of these native plant communities have been invaded by and are being replaced by tamarisk, such that extensive areas of the CWR are covered with tamarisk.

16.2.3.2 Spectral Reflectance and Image Analysis

Ground truth observations were collected from 79 sampling locations across the study area were collected. Observational data were the spectral reflectance measurements of vegetation, type of plant species, plant heights, soil samples, and GPS coordinates for the locations. The study area, overlaid with the sampling locations, is shown in Figure 16.8.

Field measurements of the canopy level spectral reflectance of vegetation at various locations within the study area were designed to coincide with the acquisition of the Landsat TM imagery that was obtained on June 9, 2007. A Fieldspec Pro spectroradiometer (ASD Inc., Boulder, Colorado) with a spectral range of 350–2500 nm was used to collect the field reflectance spectra. The fore-optics of the spectroradiometer were aligned vertically and placed at 1 m above the surface of the

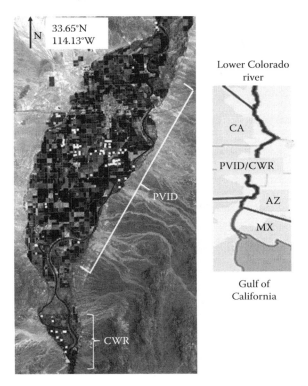

FIGURE 16.8 The Landsat TM image obtained on June 9, 2007 showing the PVID and Cibola Wildlife Refuge (CWR) on the Lower Colorado River. The CWR is seen toward the southern side (bottom) of the image. Ground truth sampling locations of the study area are shown as white squares over the dark vegetated surfaces.

plant canopy, and the instrument was adjusted such that only the reflectance from the targeted area filled the FOV of the instrument. The fiber optic input device was held approximately 1 m above the ground, such that the FOV covered a circle of approximately 50 cm in diameter. The calibration spectra of a white spectralon panel (Labsphere Inc., North Sutton, New Hampshire) were acquired before recording the field spectra. All the spectra were obtained on cloud-free days, with sunlight as the source of illumination. A total of five spectra were collected for each of the sampling locations shown in Figure 16.8. The ground level spectral reflectances acquired at each location were averaged to obtain a mean reflectance spectrum for that location. These means were plotted and analyzed to determine the spectral differences among locations. The leaves of the selected plants were harvested and the reflectance spectra were recorded with a quartz-tungsten-halogen (QTH) lamp as a light source. Diffused light from the 100 W Lowell Pro-Light was used to illuminate the dorsal side of leaf surfaces at 45° angles when spectra were collected in the laboratory. The foreoptics were aligned vertically and the height of the foreoptics was adjusted so that only the leaf surface filled the FOV of the instrument. The height of the foreoptics was kept constant throughout the experiment. The same experimental setup was used to obtain the spectra of all the leaf samples. Spectra were collected from five leaf samples of each selected plant type and then averaged to get a representative leaf spectrum of the plant. The spectral recording and analysis procedure was similar to that of the canopy-level reflectance described earlier.

The Landsat TM image obtained on June 9, 2007 was used in this study. The georeferenced and terrain corrected Landsat TM images were downloaded from the United States Geological Survey (USGS) Earth Resources Observation Systems (EROS) Data Center. The Landsat TM image was processed using the ERMapper image processing software, a commercial product of

Earth Resources Mapping, Inc. (now part of ERDAS). Based on the locations of the 79 sampling points, dark object subtracted (DOS) pixel values were derived corresponding to Landsat TM bands 1–5 and 7. The spectral range of these Landsat TM bands are as follows: Band 1: 450–520 nm; Band 2: 520–600 nm; Band 3: 630–690 nm; Band 4: 760–900 nm; Band 5: 1550–1750 nm; and Band 7: 2080–2350 nm. The dark object of each spectral band is defined as one value less than the minimum digital number (DN) found in all the pixels of the image [64]. The detailed procedure for DOS and its effects on removal of atmospheric haze was given elsewhere [64,65,66]. From the DOS-corrected DN values of the six 30 m resolution Landsat single bands, all the spectral ratio combinations and the NDVI were calculated.

The spectral ratios calculated are: $R_{2,1}$; $R_{3,1}$; $R_{3,2}$; $R_{4,1}$; $R_{4,2}$; $R_{4,3}$; $R_{5,1}$; $R_{5,2}$; $R_{5,3}$; $R_{5,4}$; $R_{7,1}$; $R_{7,2}$; $R_{7,3}$; $R_{7,4}$; $R_{7,5}$; and all their inverse ratios, where R represents the ratio and the numbers represent the Landsat TM band numbers [65]. The NDVI was calculated using the formula NDVI = ((Band4 − Band3)/(Band4 + Band3)) [67]. The spectral ratios and vegetation indices were calculated using the MINITAB statistical software (MINITAB Inc., State College, Pennsylvania).

16.2.3.3 Spectral Characteristics of Riparian and Other Vegetation

16.2.3.3.1 Canopy and Leaf Level Spectral Reflectance

The averaged canopy level spectral reflectance of the various native and invasive plant species obtained at the riparian sites are shown in Figure 16.9 and the spectra of major crop plants obtained in the fields are shown in Figure 16.10. In general, the spectral reflectance of the plants is relatively low in the visible region (400–700 nm) where light absorption by leaf pigments (primarily due to chlorophyll) is the determining factor. The absorption maxima of leaf pigments occur in the blue and red at 470 and 680 nm, respectively, while the familiar green reflectance peak occurs at 550 nm. In the near and middle IR regions, these pigments are transparent and internal leaf structure and biochemical composition control reflectance. The reflectance spectrum of principal biological interest occurs in the NIR between 700 and 1300 nm, where reflectance is high and absorption is minimal (with two minor water absorption bands at 975 and 1175 nm); beyond 1300 nm, major water absorption bands (at 1450 and 1950 nm) become significant.

Among the spectra of native and invasive plant species, the reflectance of quail bush was higher in the NIR region of 700–1300 nm, followed by *Tamarisk*, mesquite, and arrow weed, respectively (Figure 16.9). The reflectance values of alfalfa and cotton were higher in the entire spectral range from 350 to 2500 nm, compared to that of melons (Figure 16.10).

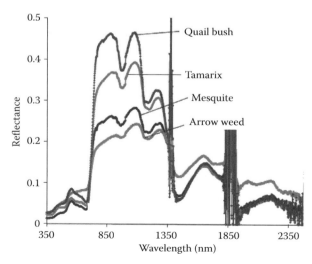

FIGURE 16.9 Averaged (n = 5) canopy spectral reflectance of the native and invasive plant species located in the Cibola National Wildlife Refuge (CWR) and riparian areas of the Lower Colorado River.

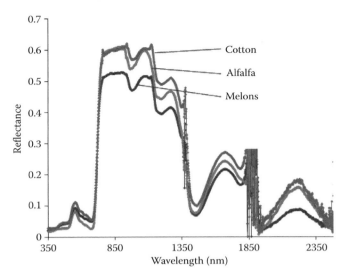

FIGURE 16.10 Averaged (n = 5) canopy spectral reflectance of the agricultural crop plants located in the PVID.

Among the leaf-level spectra of native and invasive plant species, the reflectance of cottonwood was higher in the NIR region of 700–1300 nm, followed by quail bush, salt cedar, arrow weed, creosote, and palo verde, respectively (Figure 16.11). The reflectance in the visible region of 400–700 nm shows a clear chlorophyll peak at 550 nm for cottonwood compared to other plants (Figure 16.11).

The reflectance values of grass and alfalfa were higher in the entire spectral range from 350 to 2500 nm, compared to that of cotton (Figure 16.12). All the plants show a clear chlorophyll peak at 550 nm (Figure 16.12).

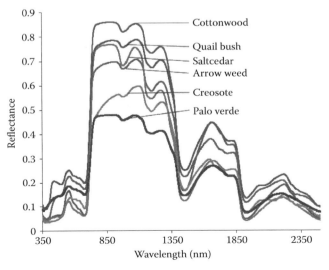

FIGURE 16.11 Averaged (n = 5) leaf-level spectral reflectance of the native and invasive plant species located in the Cibola National Wildlife Refuge (CWR) and riparian areas of the Lower Colorado River.

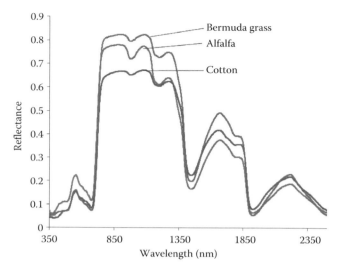

FIGURE 16.12 Averaged (n = 5) leaf-level spectral reflectance of the agricultural crop plants located in the PVID.

16.2.3.4 Landsat TM Spectral Ratios and Image Interpretation

Among all the spectral ratios and vegetation indices calculated from the DOS DN values corresponding to the six Landsat TM bands, the NDVI, $R_{1,5}$, and $R_{1,7}$ were chosen as the three best spectral ratios to differentiate major types of vegetation in the region (Figures 16.13 and 16.14a and b). NDVI values of the alfalfa and melons were significantly ($p < 0.05$) higher, compared to cotton and *Tamarisk*, as shown in Figure 16.13. The dry or senescent grass and soils had NDVI values of <0.2 (Figure 16.13). The ratios $R_{1,7}$ and $R_{1,5}$ were significantly ($p < 0.05$) higher for *Tamarisk*, compared to the rest of the vegetation (Figure 16.14a and b). Alfalfa showed significantly lower values of $R_{1,5}$ ($p < 0.05$) compared to other plants (Figure 16.14b).

The Landsat TM color-composite spectral ratio image, consisting of the band ratios $R_{1,5}$, NDVI, and $R_{1,7}$ assigned to the colors blue, green, and red, respectively, is shown in Figure 16.13. The fully grown alfalfa fields, which were in flowering condition, appear in dark green color, and the alfalfa fields that were of medium growth appear in light green color (Figure 16.15a). The alfalfa fields in

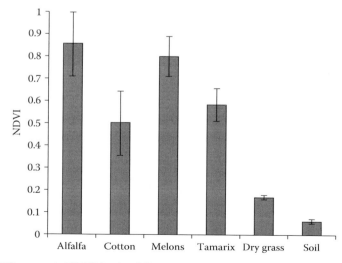

FIGURE 16.13 Differences in NDVI for the different vegetation types and soil in the Lower Colorado River Region. Bars are ± one standard error from 10 replicates.

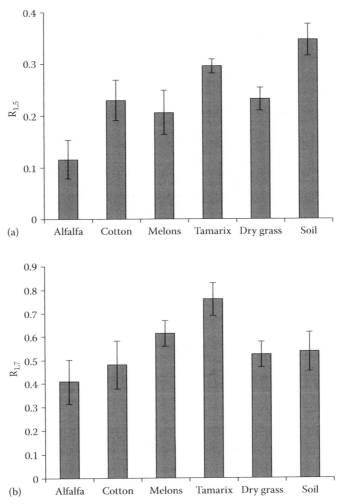

FIGURE 16.14 Differences in $R_{1,5}$ (a) and $R_{1,7}$ (b) for the different vegetation types and soil in the Lower Colorado River Region. Bars are ± one standard error from ten replicates.

flowering condition were of greater than 40cm in height, while alfalfa fields of medium growth were in the range of 10–40cm in height. The melon and cotton fields appear in shades of greenish yellow and bluish green, respectively (Figure 16.15a). The *Tamarisk* plants appear in light yellow, and all other vegetation appears in different shades of red (Figure 16.15a). Masks were created to limit image processing to the areas of vegetation, which were identified as pixels with an NDVI of greater than 0.2.

The unsupervised classification image, where the Landsat TM band ratios $R_{1,5}$, NDVI, and $R_{1,7}$ were used as selected inputs for classification, is shown in Figure 16.15b. The different vegetation classes were clearly distinguished in the classified image and the *Tamarisk*-infested areas of the CWR appears distinctly in the image (Figure 16.15b). An accuracy assessment of the classification was performed using the ground truth data obtained over 76 sampling locations during the period of Landsat overpass. Each sampling location represents an area of about 30–40 acre field size. The accuracy results given in Table 16.1 shows that an overall accuracy of 76% (kappa = 0.71) was obtained by using the three selected band ratios, compared to the low accuracies obtained by the Landsat single band inputs (data not shown). This implies that the selected band ratios are better suited for detecting species with distinct characteristics, such as *Tamarisk*, with its distinctive saline and arid adoptive characteristics.

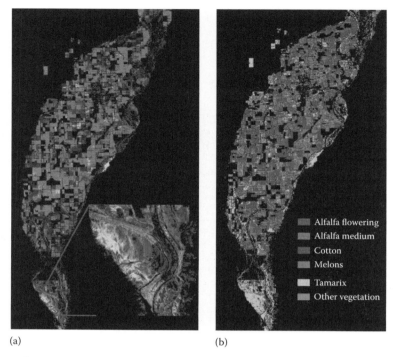

(a)   (b)

FIGURE 16.15   (See color insert.) Landsat TM color composite spectral ratio image (NDVI, $R_{1,5}$ and $R_{1,7}$ displayed as BGR, respectively) of Lower Colorado River Region (a). The Cibola National Wildlife Refuge (CWR) was shown in the insert image. Results of unsupervised classification where NDVI, $R_{1,5}$ and $R_{1,7}$ were used as ratio inputs (b).

TABLE 16.1
Confusion Matrix for Unsupervised Classification of Spectral Ratio (NDVI, $R_{1,5}$ and $R_{1,7}$) Image

Class	Alfalfa Flowering	Alfalfa Medium	Cotton	Melons	*Tamarix*	Other Vegetation	Total
Alfalfa flowering	22	1	0	0	0	0	23
Alfalfa medium	1	5	7	0	0	0	13
Cotton	0	0	6	0	0	2	8
Melons	2	0	3	6	0	0	11
Tamarix	0	0	1	1	11	0	13
Other vegetation	0	0	0	0	0	8	8
Total	25	6	17	7	11	10	76

Class	User's Accuracy (%)	Producer's Accuracy (%)
Alfalfa flowering	95.65	88
Alfalfa medium	38.46	83.33
Cotton	75	35.29
Melons	54.54	85.71
Tamarix	84.61	100
Other vegetation	100	80

Overall accuracy = 76.32%; kappa coefficient = 0.71.

A visual comparison of all the spectral ratio composite images obtained from July 2007 to June 2008 shows that the *Tamarisk* plants can be clearly mapped separately from other vegetation and are reported elsewhere [68]. High densities of *Tamarisk* in the CWR can be clearly seen in all the images and the changes in level of *Tamarisk* density through spring, summer, and fall seasons are distinctly evident and these results were published elsewhere [68].

16.2.3.5 Discussion and Future Directions

The spectral reflectance of the agricultural plants (alfalfa, cotton, and melons) and *Tamarisk* were higher in the NIR (800–1300 nm) region, compared to other plants. The reason for this is that all these four plant types have higher green cover and plant biomass than the rest of the vegetation types present. Spectral reflectance in the NIR region increases with increased percent green cover, plant biomass [69], and LAI [70]. The spectral reflectance of all the agricultural plants from 1300 to 2500 nm regions was higher than that of *Tamarisk*. It was reported that NIR wavelengths can best distinguish the *Tamarisk* from other vegetation types [71]. However, they employed the *Tamarisk* spectra from 400 to 900 nm only [71], whereas our study employs spectra covering the wavelength range of 400–2500 nm (Figure 16.9).

The leaf level spectra shows less atmospheric noise in the 1400 and 1900 nm regions compared to the canopy level reflectance. The minor water absorption bands at 975 and 1175 nm are stronger for the canopy reflectance than for the leaf reflectance. This is due to multipath reflectance in the case of canopy spectra. The canopy level reflectance differs from the leaf level spectra because the canopy spectra are affected by factors, such as leaf orientation, canopy height, diameter, and leaf density etc. Variations in leaf area and leaf angle have a dominant effect on canopy reflectance in a full canopy [72]. The interaction of photons with vegetation components in vertical space is known to be highly nonlinear. The scattering behavior is defined by the bidirectional reflectance distribution function (BRDF) and is beyond the scope of this chapter.

The three best spectral ratios selected in this study discriminate the different vegetation types based on their biophysical and biochemical properties. The NDVI of the alfalfa and melons were higher compared to the cotton and *Tamarisk* (Figure 16.13). The NDVI is a vegetation index [73] that has been widely used to estimate vegetation biomass, LAI, photosynthetic activity, and chlorophyll content. The ratios R1,5 and R1,7 were significantly higher for *Tamarisk* compared to the rest of the vegetation types (Figure 16.14a and b). This can be attributed to the differences in the physical and chemical composition of the *Tamarisk* plants, compared to the other vegetation. *Tamarisk* that has adapted to many different saline soil types is known to secrete a variety of ions such as sodium, chlorine, potassium, calcium, magnesium, and sulfate [74–76] through its salt glands. During the process of evapotranspiration, these salt glands on *Tamarisk* leaves release ions into the transpirational stream, thereby coating all the plant leaves with salt [76]. During the field studies in CWR, we noticed that all the *Tamarisk* plants had a white powdery salt coating over all their leaves. All the salt accumulation on *Tamarisk* leaves makes the leaves appear visually as a bluish green color [77].

The selected Landsat TM spectral ratios (NDVI, $R_{1,5}$ and $R_{1,7}$) emphasize the biophysical and biochemical differences among selected plants, as well as between the selected plants and other vegetation. Assignment of these three ratios to the primary colors green, blue, and red reveals (Figure 16.15a) important information that is not apparent in the single band image (Figure 16.8) made from the same Landsat TM data set. This methodology is particularly useful in discriminating and mapping vegetation types that do not show strong visible color contrasts. Also, the spectral ratios are more robust and can be applied to multiple satellite overpasses for continuous monitoring of the vegetation.

Application of the spectral ratio, color composite technique for all the Landsat TM images obtained from June, 2007 through June, 2008 (Figures 16.13 and 16.15) reveals the seasonal progression of the tamarisk growth. Starting in October, the tamarisk gradually loses leaves until

January and remains leaf-less from January to March, then start producing leaves again from March onward until leaf production peaks in summer. This seasonal progression also reveals that the spectral ratios are robust and can be applied to multiple satellite overpasses to monitor the spread of tamarisk. For mapping the *Tamarisk* along the Arkansas River in Colorado [58], different combinations of single bands, NDVI, and tasseled cap transformations for each of the six Landsat ETM+ images obtained through April to October months. In contrast to the previous study [58], where different Landsat single band combinations were chosen for different Landsat image analysis, the spectral ratios developed in our study are based on DOS-corrected spectral ratios that are more robust compared to the combination of single spectral bands; hence, the same spectral ratio combination can be used on all dates of Landsat TM data for mapping tamarisk.

In conclusion, the spectral ratio, color composite, Landsat TM images can be used to detect and map tamarisk-infested areas in the LCR region. These results show that multispectral and multi-temporal Landsat TM data can be a valuable and cost-effective tool with which natural resource managers can develop regional maps depicting where tamarisk infestations occur over large, poorly accessible areas. Generating tamarisk distribution maps over time with Landsat TM data and combining that information with existing GIS data bases can create models for studying several other problems, such as soil salinity, evapotranspiration, forest fire potential, and displacement of native vegetation and wildlife caused by the spread of tamarisk.

16.2.4 BUFFELGRASS STUDY

The goal of the buffelgrass study was to identify remote sensing strategies that could be effectively utilized to map and monitor buffelgrass invasion in the Sonoran Desert. The specific objectives were to (1) identify the spectral characteristics of buffelgrass and native vegetation in the AU zone of the Sonoran Desert as they varied throughout a single year and (2) identify optimal timing for discriminating between buffelgrass and native plants.

16.2.4.1 Study Area

This study focused on a sensitive habitat dominated by saguaro cactus (*Carnegiea gigantea*) and palo verde trees (*Parkinsonia microphylla*) in the AU zone of the Sonoran Desert in the Santa Catalina Mountains (Catalinas) just north of Tucson, Arizona (Figure 16.16). The lower piedmont of the Catalinas supports some of the most abundant stands of giant saguaro cactus in the world [78], yet is currently threatened by invasion by buffelgrass invasion. The Catalina Mountains are a sky island, with a forested summit but surrounded by a sea of desert. The saguaro–palo verde association forms a ring at the base of the mountain and has the potential to link high elevation fuels with urban ignition sources in the suburbs of Tucson.

16.2.4.2 Measurements of Community Composition

Field data collection was performed to (1) characterize the community composition of the habitat buffelgrass has invaded, is invading, and has not yet invaded, and (2) measure the canopy-level reflectance of dominant species and cover types found in these communities throughout the year. Ten medium-to-large buffelgrass patches were identified in the study area at elevations ranging from 883 to 1097 m. Fieldwork occurred between December 2008 and March 2009. At each patch, a transect of contiguous 10 × 10 m plots was oriented such that one end started at the center of the patch and the other extended beyond the patch edge by at least 20 m. Transects were randomly oriented but confined to similar slope, aspect, and geomorphology. Within each plot, species-level projected canopy cover was measured using a point-intercept method along a regular 1 m grid.

16.2.4.3 Field Spectroscopy of Dominant Cover

On six dates between March and October 2007, we recorded canopy-level reflectance of each major species found in AU habitat at Tumamoc Hill, a small hill dominated by saguaros and palo verdes

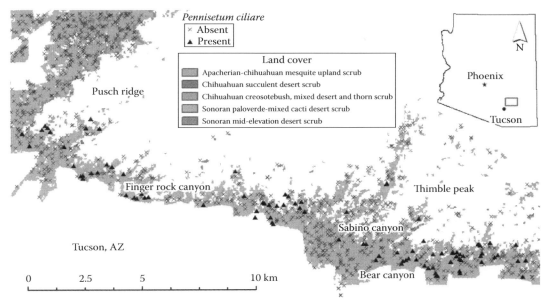

FIGURE 16.16   (See color insert.) Buffelgrass study area.

situated about 15–30 km from the field plots. Tumamoc Hill is at a similar elevation and supports vegetation to that found in the Santa Catalina Mountains. Hyperspectral reflectance data at 10 nm intervals between 0.4 and 2.5 μm were measured using an Analytical Spectral Devices (ASD) Fieldspec Pro III with an 8° FOV held at nadir above the target and adjusted in height to maintain the target in its FOV. Measurements were made between 10 am and noon local time on cloud-free days. The ASD spectrometer was periodically calibrated with a calibrated spectralon reference panel to avoid saturation; spectralon panel reference measurements were made before and after target measurements to account for solar illumination changes and to calculate surface reflectance values. Each reflectance reading was the average of five readings that were taken while the instrument was slightly moved over the target to capture its variability. The same targets were measured on each of the six dates. Throughout the year, approximately 50 targets were measured. The spectra of the most abundant species found in the community characterization step are given in Figure 16.17.

16.2.4.4   Spectral Analysis of Sonoran Desert Vegetation

Spectral separability analysis was performed to assess temporal variations in discriminability between pure buffelgrass and other cover types as well as to distinguish mixed pixels containing different amounts of buffelgrass from pixels without buffelgrass. Due to the low signal-to-noise ratio in the SWIR region of the spectrum, a moving average filter was applied to the SWIR bands. We calculated Pearson's r correlation of spectral signatures of all targets with buffelgrass for all six acquisition dates. While the landscape we are interested in is highly mixed, combinations of several cover types typically comprise greater than 50% of uninvaded cover: rock, soil, *Encelia farinosa*, *Parkinsonia microphylla*, and *Prosopis glandulosa*. We investigated the discriminability of *P. ciliare* from these dominant cover types in more detail, identifying the wavelengths that generate maximum differentiability for each season. To assess magnitude differences, we calculated the difference between the curves representing the reflectance of each cover type vs. *P. ciliare*.

16.2.4.5   Spectral Separability of Mixed Fractional Cover of Buffelgrass

We utilized plot-level cover measurements and predicted hyperspectral reflectance for all 53 plots for the six ASD reflectance data collection dates. We divided plots into low (<5% *P. ciliare* cover),

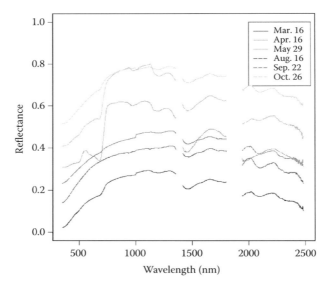

FIGURE 16.17 Reflectance of buffelgrass on six dates in 2007. Curves for acquisitions after March 16, 2007 are offset in sequence of 10% for each subsequent date on the y-axis to increase visibility of the curves.

medium (5%–50%), and high (≥50%) *P. ciliare* classes and performed a student's t-test to examine differences between reflectance for high vs. medium and high vs. low cover at each wavelength.

16.2.4.6 Results

Two species comprised 28.4% cover of the low buffelgrass cover plots (48.6% of the total vegetation cover): *Encelia farinosa* and *Parkinsonia microphylla* (Table 16.2) with no other species comprising more than 10% of the vegetation cover. Plots were typically dominated by bare ground with mineral soil (mean = 24.8%, sd = 8.3%) and rock outcrops (mean = 16.4%, sd = 6.9%) averaging over 40% cover on the uninvaded plots. Therefore, we chose to focus on the relationships between buffelgrass and these four cover types for field-based spectral endmember comparisons. The three plant species (*Encelia farinosa*, *Parkinsonia microphylla*, and *Pennisetum ciliare*) exhibited different spectral responses to phenological changes during the year, varying in their differentiability from each other and from background soil and rock outcrops.

In March 2007, the buffelgrass plants sampled were composed primarily of senesced blades and rachii from previous years' growth, but exhibited some minimal greenup in the form of new leaves arising from the base of the plant. This partial greenup, likely the plants' response to available soil moisture from winter precipitation and rising spring temperatures, was short-lived as the plants became senescent for the hot, arid fore-summer preceding the onset of the North American monsoon. By mid-August, the monsoon had manifested in southern Arizona, bringing abundant precipitation that stimulated greenup and leaf elongation in buffelgrass plants. The photosynthetically active phase diminished significantly by September when the plant was yellow, leaves were curled, and had set seed. The plant was almost completely senesced by October and had taken on an orange hue that differed from native plants.

Spectral characteristics followed buffelgrass phenology and mirrored changes in photosynthetic activity, water content, and cellulose/lignin absorption (Figure 16.17). Absorption at 675 nm was noticeable in March, August, September, and October, although the August absorption was exemplary and the September absorption being noteworthy. August was typical of photosynthetically active vegetation across the VNIR, SWIR1, and SWIR2 wavelength regions, exhibiting additional absorption features near 950, 1150, and 1450 nm. SWIR1 and SWIR2 reflectance was comparatively low during August. March and September had somewhat lower SWIR reflectance, likely due

TABLE 16.2
Percent Cover of Species Found in 15 Plots with <5%
Buffelgrass Cover on Rocky Slopes in the Santa Catalina
Mountains

Species	Type	Uninvaded Mean Cover % (SE)
Encelia farinosa	Shrub	16.14 (1.46)
Parkinsonia microphylla	Tree	12.56 (2.24)
Prosopis glandulosa	Tree	3.80 (2.10)
Janusia gracilis	Vine/Shrub	3.20 (1.08)
Jatropha cardiophylla	Shrub	2.98 (1.00)
Lycium berlandieri	Shrub	2.26 (0.68)
Fouquieria splendens	Succulent	1.93 (0.43)
Calliandra eriophylla	Shrub	1.76 (0.52)
Eysenhardtia orthocarpa	Shrub	1.76 (1.08)
Jacquemontia pringlei	Vine/Shrub	1.65 (0.85)
Abutilon incanum	Forb	1.60 (0.60)
Cylindropuntia versicolor	Succulent	1.38 (0.55)
Evolvulus arizonica	Forb	1.32 (0.56)
Opuntia engelmannii	Succulent	1.16 (0.95)
Trixis californica	Shrub	0.88 (0.35)
All succulents		5.5 (1.3)
All grasses		2.1 (0.43)
All forbs		3.1 (1.0)
All shrubs		31.7 (1.7)
All trees		16.7 (2.8)

to foliar water of green basal leaves (March) or stressed green leaves (September). A cellulose/lignin absorption feature at 2050 nm was one of the most characteristic features of the buffelgrass spectra on five of the six acquisition dates (all dates excluding August 16).

Encelia farinosa followed a bimodal phenological pattern that included spring greenup, floral production, mid-summer senescence, and monsoon greenup (Figure 16.18). Greenup had started by March 16 but was close to full spring production by April. April was also characterized by a showy display of yellow composite flowers. Senescence during the arid fore-summer resulted in shriveled, desiccated leaves. *Encelia* responded strongly to the summer monsoon and remained photosynthetically active between mid-August and late September. By late October, most leaves had desiccated and dehisced, leaving behind only the skeleton of this mostly <1 m shrub. The spectral reflectance curves show the bimodal growing season and the extended (and more intense) response to the monsoon. This is shown by the chlorophyll feature with a peak at 550 nm and absorption at 675 nm contrasted with strong NIR reflectance and a steep red edge. Additionally, lower SWIR reflectance is demonstrated as well as a strongly peaked reflectance at 1650 nm. Floral reflectance is evident in the April 16 spectra in the form of a shoulder between 550 nm peak and 675 nm trough. During the arid fore-summer, *Encelia* mostly lost the chlorophyll/carotenoid absorption at 675 nm while the cellulose/lignin absorption feature manifested slightly (more so in May than April). The reflectance of *Encelia* in the NIR August was >0.6, significantly higher than other vegetation sampled in this study (data not shown). The higher reflectance was mirrored across the VNIR wavelength region with a reflectance of >0.1 at a strong 675 nm absorption feature and >0.15 at the 550 nm chlorophyll peak. Other strongly photosynthetic vegetation we sampled in August reflected <0.1 at 520 nm and close to 0.05 at 675 nm. This is likely caused by the production of reflective trichomes across the surface of the *Encelia* leaves that act to lower leaf-level temperatures and increase water

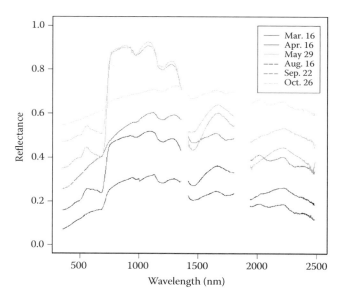

FIGURE 16.18 Reflectance of *Encelia farinosa* on six dates during 2007. Reflectance spectra following March 16 are offset on the y-axis to increase readability. Note that the NIR reflectance (800 nm) for August 16 is actually 10% greater than reflectance on September 22.

use efficiency [79]. Foliar water content of leaves was evidently high in August and September as evidenced by relatively low SWIR reflectance values. After September, plant deconstruction was rapid, with leaves desiccating, falling off, and blowing away. Remaining leaves in October were brown, but the dominant plant cover was composed of dried stems. This is reflected in slight absorption feature at 2050 nm.

Parkinsonia microphylla is a deciduous leguminous tree with green photosynthetic bark and small, compound leaves with very small leaflets. The evergreen bark allows it to photosynthesize at all times of the year while the small leaves minimize potential for water loss from transpiration during the hot, dry summer. *Parkinsonia* produces leaves in spring as temperatures increase. Leaf-out occurs in March and April and floral production is prolific in April. Leaves remain during most summers, although activity is typically suppressed during the arid fore-summer because, like many members of Fabaceae, *Parkinsonia* leaves have pulvini at the base of the rachii that retract leaves and minimize their exposure to sun and wind during times of low water availability and/or humidity. *Parkinsonia* responds to summer precipitation and remains photosynthetically active throughout the summer and fall. *Parkinsonia* reflectance exhibits signs of photosynthetic activity on all six dates (Figure 16.19). While the chlorophyll peak at 550 nm was only apparent from August through late October, the absorption feature at 675 nm was evident on all dates. August and September have the strongest photosynthetic response as evidenced by the clear chlorophyll absorption feature in 550–675 nm and the higher NIR reflectance (750 nm). The slope of the NIR reflectance from 750 to 1250 nm was also noticeably negative in the August spectra. Cellulose absorption features were barely evident in the March, April, and May spectra.

Buffelgrass most strongly resembled *Encelia* during the March 16 acquisition date (Figure 16.18). Only *Encelia* had a cellulose/lignin absorption feature in our first date of the year. Both plants had slight absorption at 675 nm but largely lacked the chlorophyll absorption feature at 550–675 nm. Senesced buffelgrass had a slight orange hue, manifested in the fact that *Encelia* had a lower slope from 450 to 650 nm than buffelgrass. *Parkinsonia microphylla* resembled buffelgrass to a lesser extent but exhibited stronger absorption at 675 nm, strong absorption features in the NIR, less cellulose/lignin feature in SWIR2, and overall less reflectance in SWIR1 and SWIR2. Soil and rock differed by virtue of their steep slope from 450 to 700 nm and flat response curves across the SWIR1

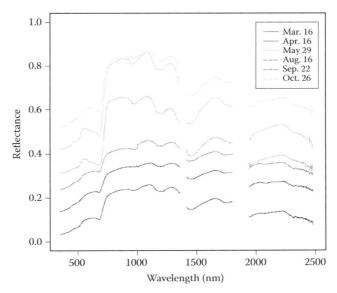

FIGURE 16.19 Spectral reflectance of *Parkinsonia microphylla* on six dates in 2007. Curves following March 16 are sequentially offset by 10% to increase readability.

and SWIR2. In April, buffelgrass was no longer photosynthetically active and resembled both rock and soil in the VNIR wavelength region (Figure 16.17). In the SWIR2, buffelgrass was the only target with a strong absorption feature at 2050 nm. By late May, all vegetation had slowed photosynthetic activity or stopped altogether. *Encelia* resembled buffelgrass in all parts of the spectrum, including the SWIR2 cellulose/lignin absorption wavelengths. Buffelgrass did lack the absorption feature found at 1200 nm in both *Encelia* and *Parkinsonia*. All plants had greened up by August 16 in response to the monsoon. *Encelia* was exemplary in having increased reflectance across the board. The likely cause is the production of reflective trichomes that lower leaf surface temperature and increase water use efficiency [79]. *Parkinsonia* and buffelgrass were almost indistinguishable with only slight differences in the shape of the chlorophyll absorption and slight differences in the overall reflectance in the NIR and depth of the 1200 nm absorption feature. By September, buffelgrass photosynthesis had slowed, differentiating it from both *Parkinsonia* and *Encelia* in the loss of the strong 675 nm absorption and the reappearance of a 2050 nm cellulose/lignin feature. Based on similar SWIR1/SWIR2 reflectance magnitudes, water content between the three plants appeared similar. *Encelia* had dessicated by October 26 while *Parkinsonia* remained photosynthetically active. While *Encelia* reflectance shape in the SWIR1/SWIR2 regions was similar to buffelgrass, buffelgrass reflectance in the SWIR1/SWIR2 was higher overall, possibly due to lower residual water content. Furthermore, the slope from 450 to 750 nm of buffelgrass was quite steep compared to that of *Encelia*. This was also reflected in a strong orange hue to senesced buffelgrass in October that contrasted with that of natives. *Encelia* remains, for example, were mostly grey and light brown.

Overall, buffelgrass exhibited a phenological pattern distinct from that of other dominant native plants and background cover. While some of these differences relate to changes in ecosystem function (e.g., earlier shut-down of photosynthetic activity in the arid fore-summer and following the summer monsoon), others relate to changes in ecosystem structure (e.g., the persistence of highly flammable grasses year-round is evident in a strong cellulose/lignin absorption feature at 2050 nm that is not reflected in the native species or uninvaded soil background spectra). The spread of buffelgrass across a region that is typically characterized by abundant bare ground surrounding islands of shrubs, trees, cacti, and native grasses results in significant changes in both structure and function. Hyperspectral remote sensing offers both the means to map and monitor the invasion and to measure the changes in ecosystem structure and function associated with the invasion.

16.2.4.7 Conclusions

The selection of specific hyperspectral narrowbands is very important to the observation of litter or standing senescent vegetation. CAI, involving hyperspectral narrowbands centered at 2023, 2100, and 2215 nm, was developed for the purpose of discriminating litter from soils. The phenology of invasive plants that have litter phases as well as agricultural crops and other plant litters can be observed with the CAI. Two of the three case studies presented in this chapter demonstrate the utility and strengths of CAI in plant litter and invasive species studies. The CAI distinguishes between soil, green vegetation, and cellulose-lignin plant litter, even in mixed scenes, across different soil types and in soils with varying moisture content. It was first developed to quantify plant litter left on the soil as a result of no-till farming. However, the case study with buffelgrass shows that the CAI combined with time-series NDVI imagery can be used to detect the spread of this invasive species based on its phenology. The third case study with Tamarisk showed that conventional broad-band indices are useful in detecting invasive species.

REFERENCES

1. Elvidge, C.D. 1990. Visible and near infrared reflectance characteristics of dry plant materials. *International Journal of Remote Sensing* 11: 1775–1795.
2. van Leeuwen, W.J.D. and Huete, A.R. 1996. Effects of standing litter on the biophysical interpretation of plant canopies with spectral indices. *Remote Sensing of Environment* 55: 123–138.
3. Swift, M.J., Heal, O.W., and Anderson, J.M. 1979. *Decomposition in Terrestrial Ecosystems*. Oxford, U.K.: Blackwell Scientific Publications.
4. DeAngelis, D.L. 1992. Nutrient interactions of detritus and decomposers. In: *Population and Community Biology: Dynamics of Nutrient Cycling and Food Webs*, Usher, M.B., Rosenzweig M.L., and Kitching R.L. (Eds.), London, U.K.: Chapman & Hall.
5. Mason, C.F. 1976. *Decomposition*. Southampton, U.K.: Camelot Press Ltd.
6. Stalfelt, M.G. 1972. *Stalfelt's Plant Ecology: Plants, the Soil and Man*. Translated by Jarvis, M.A. and Jarvis, P.G., London, U.K.: William Clowes and Sons.
7. Richards, B.N. 1974. *Introduction to the Soil Ecosystem*. New York: Longman, Inc.
8. Remmert, H. 1980. *Ecology: A Textbook*. 2nd Edn., Translated by Brederman-Thorson, M.A., Berlin, Germany: Springer-Verlag.
9. McMurtrey, III, J.E., Chappelle, E.W., Daughtry, C.S.T., and Kim, M.S. 1993. Fluorescence and reflectance of crop residue and soil. *Journal of Soil and Water Conservation* 48(3): 207–213.
10. Skidmore, L. and Siddoway, F.H. 1978. Crop residue requirements to control wind erosion. In *Crop Residue Management Systems*. Oschwald, W. (Ed.), ASA Special Publication 31, Madison, WI: ASA, CSSA, and SSSA.
11. Aase, J.K. and Tanaka, D.L. 1991. Reflectances from four wheat residue cover densities as influenced by three soil backgrounds. *Agronomy Journal* 83: 753–757.
12. Daughtry, C.S.T., McMurtrey, III, J.E., Nagler, P.L., Kim, M.S., and Chappelle E.W. 1996. Spectral reflectance of soils and crop residues. In: *Near Infrared Spectroscopy: The Future Waves*, A.M.C. Davies and P. Williams (Eds.), Chichester, U.K.: NIR Publications, pp. 505–511.
13. Ahn, C.W., Baumgardner, M.F., and Biehl, L.L. 1996. Performance of AVIRIS, adjusted AVIRIS, and simulated TM data for classifying crop residue, PECORA 13 Symposium, Sioux Falls, South Dakota, 20–22 August.
14. Morrison, J.E. Jr., Huang, C., Lightle, D.T., and Daughtry, C.S.T. 1993. Residue measurement techniques. *Journal of Soil and Water Conservation* 48: 479.
15. Shelton, D.P. and Dickey, E.C. 1995. Conservation tillage. In: *Estimating Percent Residue Cover Using the Line-Transect Method*. Conservation Technology Information Center, Cooperative Extension Service.
16. Daughtry, C.S.T., McMurtrey, III, J.E., Chappelle, E.W., Dulaney, W.P., Irons, J.R., and Satterwhite, M.B. 1995. Potential for discriminating crop residues from soil by reflectance and fluorescence. *Agronomy Journal* 87: 165–171.
17. Jordan, C.F. 1969. Derivation of leaf-area index from quality of light on the floor. *Ecology* 50: 663–666.
18. Woolley, J.T. 1971. Reflectance and transmittance of light by leaves. *Plant Physiology* 47: 656–662.
19. Gates D.M., Keegan, H.J., Schleter, J.C., and Weidner, V.R. 1965. Spectral properties of plants. *Applied Optics* 4(1): 11–20.

20. Tucker, C.J. 1979. Red and photographic infrared linear combinations for monitoring vegetation. *Remote Sensing of Environment* 8: 127–150.
21. Bracher, G.A. and Murtha, P.A. 1994. Estimation of foliar macro-nutrients and chlorophyll in douglas-fir seedlings by leaf reflectance. *Canadian Journal of Remote Sensing* 20: 102–115.
22. Myneni, R.B., Hall, F.G., Sellers, P.J., and Marshak, A.L. 1995. The interpretation of spectral vegetation indexes. *IEEE Transactions on Geoscience and Remote Sensing* 35(2): 481–486.
23. Goward, S.N. and Huemmrich, K.F. 1992. Vegetation canopy PAR absorptance and the normalized difference vegetation index: An assessment using the SAIL model. *Remote Sensing of Environment* 39: 119–140.
24. Goward, S.N., Huemmrich, K.F., and Waring, R.H. 1994. Visible-near infrared spectral reflectance of landscape components in western Oregon. *Remote Sensing of Environment* 47: 190–203.
25. Daughtry, C.S.T. and Biehl, L.L. 1985. Changes in spectral properties of detached birch leaves. *Remote Sensing of Environment* 17: 281–289.
26. Kimes, D.S. 1991. Radiative transfer in homogeneous and heterogeneous vegetation canopies. In: *Photon-Vegetation Interactions: Applications in Optical Remote Sensing and Plant Ecology*. R.B. Myneni and J. Ross (Eds.), New York: Springer-Verlag, pp. 339–388.
27. Irons, J.R., Weismiller, R.A., and Peterson, G.W. 1989. Soils reflectance. In: *Theory and Applications of Optical Remote Sensing*. G. Asrar (Ed.), New York: Wiley, pp. 66–110.
28. Stoner, E.R. and Baumgardner, M.F. 1981. Characteristic variations in reflectance of surface soils. *Soil Science Society of American Journals* 45(6): 1161–1165.
29. Stoner, E.R., Baumgardner, M.F., Weismiller, R.A., Biehl, L.L., and Robinson, B.F. 1980. Extension of laboratory-measured soil spectra to field conditions. *Soil Science Society of American Journals* 44: 572–574.
30. Wiegand, C.L. and Richardson, A.J. 1992. Relating spectral observations of the agricultural landscape to crop yield. *Food Structure* 11: 249–258.
31. Henderson, T.L., Baumgardner, M.F., Franzmeier, D.P., Stott, D.E., and Coster, D.C. 1992. High dimensional reflectance analysis of soil organic matter. *Soil Science American Journals* 56: 865–872.
32. Murray, I. and Williams, P.C. 1988. Chemical principles of near infrared technology. In: *Near Infrared Technology in the Agricultural and Food Industries*, P. Williams and K. Norris (Eds.), St. Paul, MN: American Association of Cereal Chemists, pp. 17–34.
33. Sudduth, K.A. and Hummel, J.W. 1993. Soil organic matter, CEC, and moisture sensing with a portable NIR spectrophotometer. *Transactions of the ASAE* 36(6): 1571–1582.
34. Price, J.C. 1992. Variability of high resolution crop reflectance spectra. *International Journal Remote Sensing* 13: 2593–2610.
35. Daughtry, C.S.T. 2001. Discriminating crop residues from soil by shortwave infrared reflectance. *Agronomy Journal* 93: 125–131.
36. Nagler, P.L., Daughtry, C.S.T., and Goward, S.N. 2000. Plant litter and soil reflectance. *Remote Sensing of Environment* 71(2): 207–215.
37. Nagler, P.L., Inoue, Y., Glenn, E.P., Russ, A., and Daughtry, C.S.T. 2003. Cellulose absorption Index (CAI) to quantify mixed soil-plant litter scenes. *Remote Sensing of Environment* 87: 310–325.
38. Inoue, Y., Morinaga, S., and Shibayama, M. 1993. Non-destructive estimation of water status of intact crop leaves based on spectral reflectance measurements. *Japanese Journal of Crop Science* 62(3): 462–469.
39. Mooney, H.A. and Cleland, E.E. 2001. The evolutionary impact of invasive species. *Procedures of National Academy of Science* 98: 5446–5451.
40. Baum, B.R. 1967. Introduced and naturalized tamarisks in the United States and Canada. *Baileya* 15: 19–25.
41. Robinson, T.W., Introduction, spread and areal extent of tamarisk (*Tamarisk*) in the western states, *US Geological Survey Professional Paper 491-A*, Washington, DC, 1965.
42. Horton, J.S. and Campbell, C.J., Measurement of phreatophyte and riparian vegetation for maximum multiple use values, *USDA Forest Service Paper RM117*, 1974.
43. Brotherson, J.D. and Field, D. 1987. Tamarisk: Impacts of a successful weed. *Rangelands* 9: 110–112.
44. D'Antonio, C.M. and Vitousek, P.M. 1992. Biological invasion by exotic grass, the grass/fire cycle, and global change. *Annual Review of Ecology and Systematics* 23: 63–87.
45. Brooks, M.L., D'Antonio, C.M., Richardson, D.M., Grace, J.B., Keeley, J.E., diTomaso, J.M., Hobbs, R.J., Pelland M., and Pyke, D. 2004. Effects of invasive alien plants on fire regimes. *Bioscience* 54(7): 677–688.
46. Swetnam, T.W. and Betancourt, J.L. 1990. Fire-southern oscillation relations in the southwestern United States. *Science* 249: 1017–1021.

47. Everitt, J.H. and Deloach, C.J. 1990. Remote sensing of Chinese Tamarisk (*Tamarisk chinensis*) and associated vegetation. *Weed Science* 38: 273–278.

48. Ge, S., Carruthers, R., Gong, P., and Herrera, A. 2006. Texture analysis for mapping *Tamarisk parviflora* using aerial photographs along the Cache creek, California. *Environmental Monitoring and Assessment* 114: 65–83.

49. Akasheh, O.Z., Neale, C.M.U., and Jayanthi, H. 2008. Detailed mapping of riparian vegetation in the middle Rio Grande River using high resolution multi-spectral airborne remote sensing. *Journal of Arid Environments* 72: 1734–1744.

50. Underwood, E., Ustin, S., and DiPietro, D. 2003. Mapping native plants using hyperspectral imagery. *Remote Sensing of Environment* 46: 150–161.

51. Narumalani, S. et al. 2006. A comparative evaluation of ISODATA and spectral angle mapping for the detection of tamarisk using airborne hyperspectral imagery. *Geocarto International* 21: 59–66.

52. Narumalani, S., Mishra, D.R., Wilson, R., Reece, P., and Kohler, A. 2009. Detecting and mapping four invasive species along the floodplain of North Platte River, Nebraska. *Weed Technology* 23: 99–107.

53. Peters, A.J., Reed, B.C., Eve, M.D., and McDaniel, K.C. 1992. Remote sensing of broom snake weed (*Gutierrezia sarothrae*) with NOAA-10 spectral image processing. *Weed Technology* 6: 1015–1020.

54. Dewey, S.A., Price, K.P., and Ramsey, D. 1991. Satellite remote sensing to predict potential distribution of dyers woad (*Isatis tinctoria*). *Weed Technology* 5: 479–484.

55. Everitt, J.H., Escobar, D.E., Villarreal, R., Alaniz, M.A., and Davis, M.R. 1993. Canopy light reflectance and remote sensing of shin oak (*Quercus havardii*) and associated vegetation. *Weed Science* 41: 291–297.

56. Hamada, Y., Stow, D.A., Coulter, L.L., Jafolla, J.C., and Hendricks, L.W. 2007. Detecting Tamarisk species (*Tamarisk* spp.) in riparian habitats of Southern California using high spatial resolution hyperspectral imagery. *Remote Sensing of Environment* 109: 237–248.

57. Evangelista, P.H., Stohlgren, T.J., Morisette, J.T., and Kumar, S. 2009. Mapping invasive tamarisk (*Tamarisk*): A comparison of single-scene and time series analyses of remotely sensed data. *Remote Sensing* 1: 519–533.

58. Dennison, P.E., Nagler, P.L., Hultine, K.R., Glenn, E.P., and Ehleringer, J.R. 2009. Remote monitoring of tamarisk defoliation and evapotranspiration following tamarisk leaf beetle attack. *Remote Sensing of Environment* 113: 1462–1472.

59. Franklin, K.A., Lyons, K., Nagler, P.L., Lampkin, D., and Glenn. E.P. 2006. Buffelgrass (*Pennisetum ciliare*) land conversion and productivity in the plains of Sonora, Mexico. *Biological Conservation* 127: 62–71.

60. Brenner, Jacob. C. 2010. What drives the conversion of native rangeland to buffelgrass (*Pennisetum ciliare*) pasture in Mexico's Sonoran Desert? The social dimensions of a biological invasion. *Human Ecology* 38(4): 495–505.

61. Olsson, A. 2010. Ecosystem transformation by buffelgrass: Climatology of invastion, effects on Arizona Upland diversity, and remote sensing tools for managers. Dissertation for Arid Lands Resource Sciences, University of Arizona, Tucson, AZ.

62. Huang, C.-Y. and Geiger, E.L. 2008. Climate anomalies provide opportunities for large-scale mapping of non-native plant abundance in desert grasslands. *Diversity and distributions* 14: 875–884.

63. Vincent, R.K., Qin, X., McKay, R.M.L., Miner, J., Czajkowski, K., Savino, J., and Bridgeman, T. 2004. Phycocyanin detection from LANDSAT TM data for mapping cyanobacterial blooms in Lake Erie, *Remote Sensing of Environment*, 89: 381–392.

64. Vincent, R.K. 1997. *Fundamentals of Geological and Environmental Remote Sensing*, Upper Saddle River, NJ: Prentice Hall.

65. Maruthi Sridhar, B.B., Vincent, R.K., Witter, J.D., and Spongberg, A.L. 2009. Mapping the total phosphorus concentration of biosolid amended surface soils using LANDSAT TM data. *Science of the Total Environment* 407: 2894–2899.

66. Rouse, J.W. et al. 1974. Monitoring the vernal advancement and retrogradation (green wave effect) of natural vegetation, *Type III Final Report*, NASA Goddard Space Flight Center, Green belt, MD.

67. Maruthi Sridhar, B.B., Vincent, R.K., Clapaham, W.B., Sritharan, S.I., Osterberg, J., Neale, C.M.U., and Watts, D.R., Mapping saltcedar (*Tamarisk ramosissima*) and other riparian and agricultural vegetation in the Lower Colorado River region using multi spectral LandsatTM imagery, *GeoCarto International* (In Press).

68. Gamon, J.A., Field, C.B., Goulden, K.L., Griffin, A.E., Hartley, G., Joel, G., Penuelas, J., and Vallentini, R. 1995. Relationships between NDVI canopy structure, and photosynthesis in three California vegetation types. *Ecological Applications* 4: 28–41.

69. Qi, J., Cabot, F., Moran, M.S., and Dedieu, G. 1995. Biophysical parameter estimations using multidirectional spectral measurements. *Remote Sensing of Environment* 54: 71–83.

70. Randquist, B.C. and Brookman, D.A. 2007. Spectral characterization of the invasive shrub tamarisk (*Tamarisk* spp.) in North Dakota. *Geocarto International* 22: 63–72.
71. Asner, G.P. 1998. Biophysical and biochemical sources of variability in canopy reflectance. *Remote Sensing of Environment* 64: 234–253.
72. Curran, P.J. 1989. Remote sensing of foliar chemistry. *Remote Sensing of Environment* 30: 271–278.
73. Waisel, Y. 1961. Ecological studies on *Tamarisk aphylla* (L.) Karst. III. The salt economy. *Plant and Soil* 13: 356–364.
74. Berry, W.L. 1970. Characteristics of salts secreted by *Tamarisk aphylla*. *American Journal of Botany* 57: 1226–1230.
75. Storey, R. and Thompson, W.W. 1994. An X-ray microanalysis study of the salt glands and intercellular calcium crystals of tamarisk. *Annals of Botany* 73: 307–313.
76. Nagler, P.L., Glenn, E.P., Thompson, T.L., and Huete, A. 2004. Leaf area index and normalized difference vegetation index as predictors of canopy characteristics and light interception by riparian species on the Lower Colorado River. *Agriculture and Forest Meteorology* 125: 1–17.
77. Whittaker, R.H. and W.A. Niering. 1973. Vegetation of the Santa Catalina Mountains, Arizona. V. Biomass, production, and diversity along the elevation gradient. *Ecology* 56(4): 771–790.
78. Ehleringer, J.R. and Björkman, O. 1978. Pubescence and leaf spectral characteristics in a desert shrub, *Encelia farinosa*. *Oecologia* 36(2): 151–162.
79. Pimentel, D., McNair, S., Janecka, J., Wightman, J., Simmonds, C., O'Connell, C., Wong, E. et al. 2001. Economic and environmental threats of alien plant, animal, and microbe invasions. *Agriculture, Ecosystems & Environment* 84(1): 1–20.

Part VII

Species Identification

17 Crop Type Discrimination Using Hyperspectral Data

*Lênio Soares Galvão, José Carlos Neves Epiphanio,
Fábio Marcelo Breunig, and Antônio Roberto Formaggio*

CONTENTS

17.1 INTRODUCTION

The original definition of hyperspectral remote sensing, also known as imaging spectrometry/ spectroscopy, refers to the acquisition of images in hundreds of contiguous spectral bands to obtain a high-spectral-resolution data for each pixel of the scene [1]. In reality, this concept is much more related to the ability of the sensors to measure narrow absorption bands rather than the number of bands. For example, despite the difference in the number of bands, the Airborne Visible Infrared Imaging Spectrometer (AVIRIS) with 224 bands (400–2500 nm) and the Compact High Resolution Imaging Spectrometer (CHRIS)/PROBA with 62 bands (410–1000 nm) are both hyperspectral sensors because they have contiguous bands in each spectral range with bandwidths of approximately 10 nm. This bandwidth allows adequate measurement of most of the narrow spectral features that appear in surface component spectra.

Hyperspectral sensors enable the calculation of several narrowband vegetation indices (NVIs) and absorption band parameters on a per pixel basis, and the resultant images are useful for agriculture. For crop type discrimination, the use of narrowbands usually results in increased classification accuracy when compared to broadbands of the multispectral sensors [2,3]. However, signal-to-noise ratio (SNR) is an important factor in this comparison because it affects classification accuracy [4]. Unfortunately, several hyperspectral sensors have poor SNR, especially in the shortwave infrared (SWIR). Classification of hyperspectral data requires also feature selection to avoid the Hughes effect: the loss of accuracy when data dimensionality increases while the training sample size remains fixed [5]. The selected set of narrowbands for crop type discrimination is not universally applicable and should be determined for each study area. NVIs are useful for crop type discrimination as well as to estimate biophysical (e.g., leaf area index—LAI, biomass) and biochemical (e.g., chlorophyll, leaf nitrogen) crop attributes [6–9]. At local scale, bands and derived vegetation indices can be correlated with crop yield [10–12]. Using images showing variation in absorption band parameters, one can detect changes in canopy constituents such as chlorophyll and leaf water content within and between agricultural fields. For a given crop type and reproductive stage, information on the causes (e.g., water stress, diseases, or nutrient deficiency) of such variation can be obtained.

Examples of crop type discrimination studies using imaging spectrometers include the following: (1) discrimination of sugarcane varieties with Hyperion/Earth Observing-1 (EO-1) data [13–15]; (2) use of a spectral library generated from field spectroradiometric and Hyperion data to provide reference spectra for spectral angle mapper (SAM) classification of different crop varieties [16]; (3) use of support vector machine (SVM) and linear discriminant analysis for crop classification using different hyperspectral datasets [17,18]; (4) SAM classification of crops using an SVM-based algorithm for endmember extraction [19]; (5) evaluation of multiangular information to improve crop classification accuracy, when compared to a single nadir acquisition, using hyperspectral CHRIS/PROBA data [20]; and (6) use of ground-based hyperspectral imagery for crop–weed species discrimination and oriented herbicide application [21].

Hyperion, on board the EO-1 spacecraft, launched in 2000, is a pushbroom hyperspectral instrument capable of acquiring images in 196 calibrated bands (10 nm of bandwidth) in the visible, near-infrared (NIR), and SWIR (400–2400 nm range) with a spatial resolution of 30 m and a swath width of 7.7 km. The 16 day revisit time of the sensor can be reduced by off-nadir pointing. Despite the poor SNR in the SWIR, Hyperion provides an excellent opportunity to evaluate the hyperspectral technology from orbital level for crop type discrimination.

Here, we first review the factors that affect crop type discrimination by remote sensing. Then, we use different Brazilian Hyperion datasets to show the spectral discrimination between flooded rice, coffee, sugarcane, bean, corn, cultivated pasture, and soybean. The role played by reflectance, reflectance ratios, NVIs, and spectral feature parameters to differentiate crops is also demonstrated. Using Hyperion data to simulate the spectral response of selected multispectral sensors, we discuss the spectral resolution influence on cultivar discrimination from multispectral to hyperspectral remote sensing.

17.2 FACTORS AFFECTING CROP TYPE DISCRIMINATION BY REMOTE SENSING

Measurements of crop area are important for government and economic players. However, the first step for such measurements using remote sensing—the crop identification—is not necessarily an easy task. When compared to field observations, such difficulties increase when measuring the radiance of crops from satellites because the spectral similarities between the crops can be even higher. Remote sensing scientists are always developing methods to improve crop type discrimination. Factors affecting discrimination may be grouped into five general categories: *biophysical, crop development, management, crop calendar*, and *regional aspects*.

In terms of *biophysical aspects,* each crop belongs to a family, species, and so forth. In the lowest scale, each crop belongs to a variety or cultivar. At this scale, plants are expected to be homogeneous concerning their genotypes and phenotypes. As we generalize from cultivar and variety to species and family, it is expected that plant characteristics depart from homogeneity and impose to crops increasing differences. With regard to remote sensing, the differences are related to many aspects such as leaf pigments, leaf structure, duration of life cycle, plant structure, height, and physiology [22]. In this classical review, Gausman [22] showed how some of these factors could affect plant reflectance. At the variety or cultivar scale, leaf pigments affect spectral reflectance and are supposed to be very similar for some specific crops. However, even at this level, we can find differences between cultivars [23,24]. Because some cultivars have different structures (e.g., leaf thickness, leaf angle), this parameter affects the way that light interacts with the canopy [25]. When we look at higher level of species or class, differences between crops are even larger. For instance, leaf mesophyll layers of dicotyledonous have more well-defined palisade and spongy cells than those of the monocotyledonous. For crop type discrimination, it is well known that reflectance is influenced by the mesophyll structure [26].

Crops are very distinct in their *development aspects* or phenological stages. In general, major cultivated crops can be classified into three main groups according to the duration of the life cycle: annual, perennial, and semiperennial crops. Annual crops are planted once or even three times a year. A perennial crop can stay in field for many years, while a semiperennial crop remains in field only for a few years. The duration of the crop cycle impacts on the chances of acquiring cloud-free images using optical remote sensing, which are obviously higher for perennial crops. Especially for annual crops, due to their short cycle of life, another important aspect for remote sensing is how they are split into development stages. Different crops present distinct phenological characteristics and timings according to their nature: germination, tillering, flowering, ball formation (e.g., cotton), ripening, and so forth. Even for the same crop and growing season, the duration and magnitude of each phenological stage can vary between the varieties, which introduce data variability for crop type discrimination with imaging systems.

The farmers use distinct *management practices* depending on the cultivated crop. For instance, tillage (e.g., soybean and corn), burning prior to harvest (sugarcane), type of pruning (coffee and orchards), and harvest timing (e.g., sugarcane can be harvested almost all year long) can introduce strong variability in spectral reflectance between agricultural fields of the same crop type. In addition, technological improvements (e.g., mechanical harvesting of sugarcane) are not adopted by farmers at the same time and place. Management practices in perennial crops have a special impact on crop discrimination. In general, the row or tree spacing is larger in perennial crops than in annual crops, thus exposing the soil background, which influences the canopy reflectance [27,28].

Dry-land crops, which predominate in global production, depend on rain to develop. As a result, cloud cover in the rainy season is an important issue for their discrimination using optical remote sensing. For annual crops, *the crop calendar* is determined mainly by the water availability for dry-land crops. Irrigated crops have a complement of water supplied by mechanized systems. For irrigated cultivations, the crop calendar is not as fixed as it is for dry-land crops, since the water availability is a minor problem for them. The main limitations are temperature and light availability. If irrigation is used, farmers can cultivate more crop types in more times during the year. In general, there is a preference for some crops to be irrigated, and there is a dependence on the irrigation method. For instance, rice is cultivated by flooding rather than by sprinkler systems. So, irrigation management can help crop discrimination in some cases (e.g., rice paddies) or can pose difficulties in others (e.g., various crops under the same center pivot system).

Finally, *regional aspects* can control the crop calendar. For instance, low, humid, and flat lands are prone to grow flooded rice. Intermediate temperatures are favorable for coffee. Gentle topography and fertile soils are preferred for corn, cotton, oats, soybeans, sugarcane, and wheat. However, sugarcane, vineyards, apple tree, and coffee can be cultivated in undulated topography.

When using hyperspectral remote sensing in agriculture, one should consider the influence of these factors and take advantage of them to facilitate crop type discrimination. For example, the water background influence on canopy reflectance can facilitate discrimination of flooded rice if two images are acquired in distinct reproductive stages or in flooded and nonflooded phases of crop development.

17.3 CROP TYPE DISCRIMINATION USING HYPERION DATA

Here, we discuss crop type discrimination using Hyperion data acquired over different Brazilian agricultural areas as well as multispectral sensor-simulated images from this instrument. Obviously, this discussion is a simplification of the "real world," in which the spectral variability between and within crops is much stronger due to the factors mentioned in the previous section. However, it serves to illustrate major differences in reflectance, reflectance ratios, NVIs, and absorption band parameters between the crops. To facilitate the discussion, we use the following nomenclature for the spectral ranges: visible (400–700 nm), red edge (701–760 nm), NIR-1 (761–900 nm), NIR-2 (901–1400 nm), SWIR-1 (1401–1900 nm), and SWIR-2 (1901–2500 nm).

17.3.1 SELECTED CROPS

Six crop types were selected for analysis: coffee (*Coffea arabica* L.), sugarcane (*Saccharum* spp.), flooded rice (*Oryza sativa* L.), common bean (*Phaseolus vulgaris* L.), corn (*Zea mays* L.), and soybean (*Glycine max* (L.) Merrill). Coffee is a perennial crop and sugarcane a semiperennial crop. The others are annual crops. Cultivated pasture was also included in the analysis because of its importance for Brazilian animal production as the main feed resource for cattle. Brazil is one of the largest world producers of these crops, especially of sugarcane and coffee (world leader), and soybean (second in rank). Sugarcane has an important environmental role providing ethanol—a renewable fuel for the great number of Brazilian flexible-fuel vehicles. Coffee has had an historical participation in the Brazilian economy, and soybeans have experienced an impressive rise in production in the last few decades.

From a remote sensing perspective, the selected crops represent completely distinct canopies. Leguminous plants like soybean and bean have broader leaves and a more planophile canopy architecture than grass plants like rice, sugarcane, and corn with erectophile architecture. Among the erectophile crops, rice has the largest number of individuals per linear meter (around 20). Naturally, structural or physiognomic characteristic and differences often influence reflectance.

In contrast to the other crops that have less than 1 m of maximum height, sugarcane and corn are the tallest crops and usually reach more than 2 m height. Such characteristics affect the scattering and absorption interactions of radiation by the canopy components, especially the leaves. They affect also the possibilities for foliar chemistry estimates that are better with planophile architecture due to reduced shadow influences.

Furthermore, annual crops like rice and soybean are generally planted in October–November and harvested in March–April in southern and central Brazil, respectively. Because the peak of crop development is also coincident with the peak of cloud cover in Brazil, optical remote sensing of these crops is sometimes difficult, as demonstrated by Sugawara et al. [29] when mapping soybean in southern Brazil with Landsat. Much better revisit times than that provided by Landsat (16 days) would be required to increase the chances of acquiring cloud-free data for these crops.

17.3.2 HYPERION DATASETS AND PREPROCESSING

In this study, we used six Hyperion images acquired with solar elevation angles that ranged from 36° to 54° (Table 17.1). Data acquisition for pasture and soybean was performed under off-nadir viewing and in the backscattering direction with pointing angles of −12° and −26°, respectively. Off-nadir pointing is an alternative to reduce the 16 day revisit time of the Hyperion, as mentioned before. However, it may result in strong reflectance variation for a given crop type, when compared to nadir data acquisition, depending on the view angle and view direction used in data collection [12].

The 196 radiometrically calibrated Hyperion bands, sampled at approximately 10 nm intervals in the 426–2395 nm range, were used in the analysis. An algorithm to identify bad pixels and to reduce striping effects was applied to the Hyperion images replacing abnormal vertical lines by the average response of adjacent columns. The Hyperion radiance values from the 196 bands were converted into surface reflectance images using the Fast Line-of-Sight Atmospheric Analysis of Spectral Hypercubes (FLAASH) algorithm, a MODTRAN4-based approach to remove atmospheric scattering and absorption effects [30]. The 2-Band (K-T) method, which uses a dark pixel reflectance ratio approach with bands placed around 660 and 2100 nm [31], was selected to estimate the amount of aerosols and the scene average visibility. Precipitable water vapor was derived on a per pixel basis from the 1140 nm spectral feature. A correction for adjacency effects was also applied to data.

TABLE 17.1

Crops and Hyperion Datasets Attributes

Crop	Location in Brazil	Latitude[a]	Longitude[a]	Acquisition Date	Sun Azimuth	Sun Elevation	Pointing Angle
Rice	State of Rio Grande do Sul	29°45′37″S	55°46′57″W	04/10/2003	47°	40°	+1°
Coffee	State of Minas Gerais	20°54′18″S	47°03′51″W	08/21/2008	48°	43°	+2°
Sugarcane	State of São Paulo	20°32′16″S	47°25′09″W	07/16/2002	40°	36°	+1°
Bean and corn	State of São Paulo	23°06′00″S	49°20′23″W	02/19/2004	77°	54°	+2°
Pasture	State of São Paulo	21°40′41″S	52°22′19″W	05/27/2002	39°	36°	−12°
Soybean	State of Mato Grosso	12°45′06″S	52°22′19″W	01/14/2006	108°	54°	−26°

[a] Center coordinates of the study area.

Model parameters included tropical (soybean dataset) and midlatitude summer (remaining datasets) atmospheres with a rural aerosol model. Bands around 1400 and 1900 nm were not useful, even after atmospheric correction, due to the strong atmospheric water vapor absorption. For crop type discrimination, only 150 bands were considered in data analysis.

17.3.3 HYPERION COLOR COMPOSITES

Hyperion color composites of the six study areas are illustrated in Figure 17.1. For annual crops like rice and soybean, knowledge of the phenological stages in the period of image acquisition is

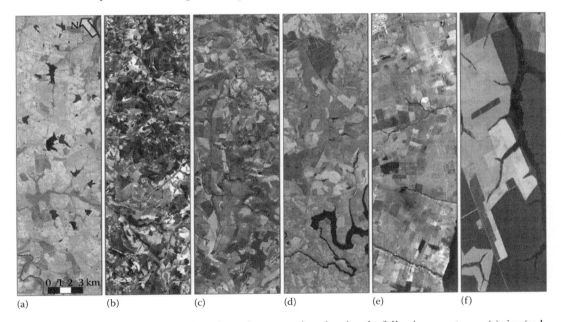

FIGURE 17.1 **(See color insert.)** Hyperion color composites showing the following crop types: (a) rice (reddish shades); (b) coffee (dark green color); (c) sugarcane (reddish shades); (d) corn and bean (pivots with dark and bright reddish shades, respectively); (e) pasture; and (f) soybean. In (a), (c), (d), and (f), bands centered at 864 nm (red), 1649 nm (green), and 671 nm (blue) were used in the false color composites. In (b) and (e), the true color composites refer to bands at 671 nm (red), 569 nm (green), and 487 nm (blue).

essential for the correct interpretation of the data. In southern Brazil, rice is flooded and water-seeded in October–November, flowers in January–February and is harvested in March–April. Thus, the Hyperion image of the Brazilian rice fields (April 10, 2003) comprises the harvest period, as indicated by several harvested crop fields in cyan color (Figure 17.1a). At this period, the spectral influence of the background water to reduce the visible and especially the NIR and SWIR reflectance is greatly reduced. The dominant response is from the senesced mature leaves associated with the ripening growth stage. Because the reflectance of the rice fields changes dramatically with crop development, or with background water modifications, inspection of images from two dates (flooded and nonflooded rice stages) is helpful in case of eventual classification problems between rice and other scene components [32]. Even a single image from the establishment phase (vegetative stage), in which the spectral response of the background water is stronger, can be used to discriminate flooded rice from other crops and to provide an early estimate of its cultivated area [33].

Well-developed coffee canopies in the central and upper portions of the Hyperion scene are characterized by a low reflectance in the visible region (dark green leaves plus shadow effects within and between canopies) or by dark-green canopy shades in the true color composite (Figure 17.1b). Observed spectral variability between the fields is probably introduced by factors such as row spacing, relief, age, cultivar, and management practices [34]. Emerging coffee canopies appear in dark reddish shades due to the strong soil background influence.

Most of the reddish shades in Figure 17.1c are associated with sugarcane. Color variability is resultant from a combination of interrelated factors such as canopy closure, variety, age, successive ratoon harvests (shoot sprouting), leaf width and size, erectness or canopy architecture, degree of lodging, soil type and fertility, local water stress, pests, and diseases. Harvested fields with the predominance of nonphotosynthetic vegetation over the surface (crop residues) occur in cyan. Harvesting may be mechanical or manual. In manual harvesting, sugarcane is burned prior to cutting the stems in order to eliminate the straw and facilitate harvesting [35]. The presence of ash over the soil surface results in dark blue shades in the false color composite (Figure 17.1c), which is, thus, an indicator of manual harvesting. Remote sensing is then used in a Brazilian agro-environmental program to detect areas of biomass burning and support a protocol that establishes deadlines to cease straw burning practice depending on terrain slope [35].

Selected agricultural fields to represent bean and corn came from the same Hyperion image (Figure 17.1d). In the study area, they were cultivated under center pivot irrigation systems. Center pivots with bean have brighter reddish shades than those with corn due to the large amounts of radiation scattered in the NIR by the planophile canopies of bean. Pivots with nonphotosynthetic vegetation appear in cyan.

Pasture predominates in the true color composite of Figure 17.1e. Hyperion acquired the image at the end of the rainy season, and pastures with different vigor were observed. Variation in green color indicates changes in live to senescent biomass ratio. The magnitude of these changes is also dependent on the grass species under analysis, as pointed out by Numata et al. [36] when using Hyperion data to estimate biophysical parameters of grazed pasture in the Amazon region.

Soybean is the highest reflective studied crop, which facilitates its discrimination in false color composites (Figure 17.1f). Hyperion acquired data on January 14, 2006 over seven soybean varieties at reproductive stages that ranged from R1 (beginning bloom) to R3 (beginning pod). The observed color variation in the southern portion of the farm (brighter yellowish shades) is associated with one variety (Monsoy 9010), which was sensed by Hyperion at an earlier reproductive stage [12].

17.3.4 REFLECTANCE AND BAND RATIO DIFFERENCES BETWEEN THE CROPS

We randomly selected 300 pixels per crop from different agricultural fields to illustrate reflectance differences between them. Average reflectance spectra (Figure 17.2) showed that the lowest reflectance was displayed by coffee due to the combined influence of its dark green leaves on the visible and of shadows within and between canopies over the entire wavelength region. Depending on the

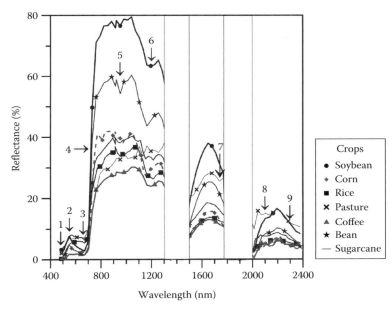

FIGURE 17.2 Average Hyperion reflectance spectra of the crop types under analysis (300 pixels per crop). The major spectral features are as follows: 1 and 3, blue–red chlorophyll absorption bands; 2, green reflectance peak; 4, red edge; 5 and 6, leaf water absorption bands; 7–9, lignin–cellulose spectral features. The two major intervals (1400 and 1900 nm) of strong atmospheric water vapor absorption are indicated.

row orientation, the topographic position of the plantations (slope aspect and gradient) and the viewing-illumination geometry, shadow effects can be dominant in the spectral response of coffee. Such effects may affect the relationships between reflectance and biophysical parameters. On the other hand, the largest reflectance was observed for soybean and bean because of the efficiency of their planophile canopy architecture to scatter radiation toward the sensor. Furthermore, Hyperion data were acquired over the soybean farm with off-nadir viewing in the backscattering direction, which resulted in increased reflectance due to the predominance of sunlit canopy components toward the sensor. Senesced pasture presented high reflectance in the red, NIR-2, SWIR-1, and SWIR-2 spectral intervals due to decreasing amounts of chlorophyll (live biomass) and leaf water and to the resultant predominance of nonphotosynthetic vegetation over live biomass. The lowest variability was observed for soybean that presented the smallest deviation values in all Hyperion bands, whereas the highest variability was observed for sugarcane, corn, and coffee (results not shown).

For the present dataset, the scatter plot of the relationship between red (e.g., 660 nm; maximum chlorophyll absorption) and NIR (e.g., 864 nm; canopy scattering in a well-defined atmospheric window) bands can be thought as a triangle whose vertices have different crop positioning (Figure 17.3). Soybean is placed at the first corner of the triangle because of the chlorophyll absorption in the red and the strong scattering of radiation in the NIR by the planophile canopy. Coffee is located at the second corner (low NIR and red reflectance) because of the dark green leaves and shadow effects mentioned before. Finally, senesced pasture and sugarcane occurred close to the third corner due to decreasing amounts of chlorophyll (presence of nonphotosynthetic vegetation) over the substrate and within canopy, and the resultant higher red reflectance. Inside the triangle, rice occurred together with pasture and sugarcane because Hyperion sensed this crop in the mature grain stage (ripening), in which the red reflectance increased due to senescing vegetation. In this stage, the filled spikelets change in color from green to yellow.

Besides the differences in overall brightness, the crops presented also variation in reflectance ratios, especially for NIR/red and SWIR-1/green ratios. For example, pasture, sugarcane, and rice with senesced foliage had lower 864/660 and 1649/569 nm reflectance ratios than soybean, bean,

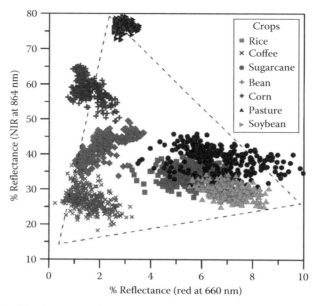

FIGURE 17.3 Relationships between red and NIR Hyperion bands for the studied crop types. The triangle is discussed in the text.

corn, and coffee (Figure 17.4). The same behavior was verified for NIR/SWIR-1 reflectance ratios (results not shown). Inspection of the Hyperion crop spectra (Figure 17.2) highlighted the importance of the NIR-1, SWIR-1, and red bands for crop type discrimination.

The role of each band can be also studied using multiple discriminant analysis (MDA) with a single-variable run procedure. MDA performs linear discriminant analysis for multiple groups with the final objective of classifying samples into one of the groups. MDA deals simultaneously with maximal separability between the groups, feature selection, and classification. Feature selection

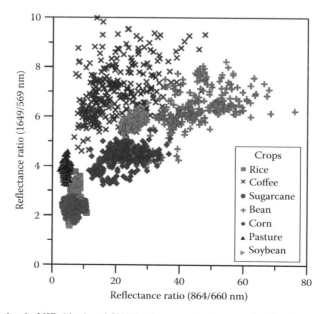

FIGURE 17.4 Variation in NIR-1/red and SWIR-1/green reflectance ratios for the crop types under study.

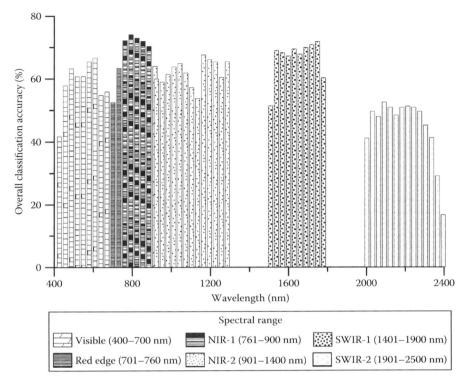

FIGURE 17.5 MDA-derived classification accuracy of the Hyperion bands for a subset of 700 pixels representing the seven crop types. Spectral intervals are indicated to facilitate discussion.

will be discussed later (Section 17.3.7). Data distribution is generally assumed as normal but statistical procedures (e.g., the Kolmogorov–Smirnov and Shapiro–Wilk tests) should be used to evaluate data normality, which is usually dependent on the wavelength and on the groups under analysis.

In this context, it is important to use a subset of samples to test the discriminatory power of the functions by classifying them. In the present example, the 2100 crop pixel spectra were used for training and a separate subset of 700 pixel spectra (100 per crop) was used for validation. We tested the reflectance values of each one of the 150 Hyperion bands placed between 426 and 2395 nm as input variables for MDA. Figure 17.5 confirmed the importance of the NIR-1 and SWIR-1 bands for crop type discrimination. Other important bands were located in the red and green regions, in the NIR-2 interval, around the 1205 nm leaf water absorption, and in the red edge region around 720 nm.

17.3.5 NARROWBAND VEGETATION INDICES

Using hyperspectral remote sensing, one can calculate many NVIs [12]. Care is necessary when using them because some indices are simply reflectance ratios. Others have local applicability and were proposed empirically to solve for specific problems. Several indices are also correlated with each other. In fact, only a few indices have been elaborated on a rigorous scientific basis.

Table 17.2 shows 21 NVIs that can be calculated with Hyperion data. Equations and references [4,13,38–52] are indicated there. We generated images for the following indices: (1) Atmospherically Resistant Vegetation Index (ARVI); (2) Enhanced Vegetation Index (EVI); (3) Normalized Difference Vegetation Index (NDVI); (4) Simple Ratio (SR); (5) Sum Green Index (SGI); (6) Normalized Difference Infrared Index (NDII); (7) Normalized Difference Water Index (NDWI); (8) Water Band Index (WBI); (9) Leaf Water Vegetation Index (LWVI-2); (10) Disease Water Stress Index (DWSI); (11) Moisture Stress Index (MSI); (12) Plant Senescence Reflectance

TABLE 17.2

Narrowband Vegetation Indices Calculated from Hyperion Data

Vegetation Index	Formula[a]	References
ARVI	$(\rho864 - [2 \times \rho671 - \rho467])/(\rho864 + [2 \times \rho671 - \rho467])$	Kaufman and Tanré [37]
EVI	$2.5([\rho864 - \rho671]/[\rho864 + 6 \times \rho671 - 7.5 \times \rho467 + 1])$	Huete et al. [38]
NDVI	$(\rho864 - \rho671)/(\rho864 + \rho671)$	Rouse et al. [39]
SR	$\rho864/\rho671$	Rouse et al. [39]
SGI	$(\rho508 + \rho518 + \rho528 + \rho538 + \rho549 + \rho559 + \rho569$ $+ \rho579 + \rho590 + \rho600)/10$	Lobell and Asner [4]
NDII	$(\rho823 - \rho1649)/(\rho823 + \rho1649)$	Hunt and Rock [40]
NDWI	$(\rho854 - \rho1245)/(\rho854 + \rho1245)$	Gao [41]
WBI	$\rho905/\rho973$	Penuelas et al. [42]
LWVI-2	$(\rho1094 - \rho1205)/(\rho1094 + \rho1205)$	Galvão et al. [13]
DWSI	$\rho803/\rho1598$	Apan et al. [43]
MSI	$\rho1598/\rho823$	Hunt and Rock [40]
PSRI	$(\rho681 - \rho498)/\rho752$	Merzlyak et al. [44]
CRI	$(1/\rho508) - (1/\rho701)$	Gitelson et al. [45]
ARI	$(1/\rho549) - (1/\rho701)$	Gitelson et al. [46]
PRI	$(\rho529 - \rho569)/(\rho529 + \rho569)$	Gamon et al. [47]
SIPI	$(\rho803 - \rho467)/(\rho803 + \rho681)$	Penuelas et al. [48]
RENDVI	$(\rho752 - \rho701)/(\rho752 + \rho701)$	Gitelson et al. [49]
REP	$(\rho n + 1 - \rho n)/10$ in the 690–750 nm interval	Curran et al. [50]
VOG-1	$\rho742/\rho722$	Vogelmann et al. [51]
VARI	$(\rho559 - \rho640)/(\rho559 + \rho640 - \rho467)$	Gitelson et al. [52]
VIg	$(\rho559 - \rho640)/(\rho559 + \rho640)$	Gitelson et al. [52]

[a] ρ is the reflectance of the closest Hyperion bands (n, center in nanometers) to the original wavelength formulations.

Index (PSRI); (13) Carotenoid Reflectance Index (CRI); (14) Anthocyanin Reflectance Index (ARI); (15) Photochemical Reflectance Index (PRI); (16) Structure Insensitive Pigment Index (SIPI); (17) Red Edge Normalized Difference Vegetation Index (RENDVI); (18) Red Edge Position (REP); (19) Vogelmann Red Edge Index 1 (VOG-1); (20) Visible Atmospherically Resistant Index (VARI); and (21) Visible Green Index (VIg).

In general, ARVI, EVI, NDVI, SR, ARI, CRI, SGI, VARI, and VIg are indices closely related to greenness and leaf pigments (e.g., chlorophyll and carotenoids). MSI, NDII, NDWI, LWVI-2, WBI, and DWSI are much more associated with leaf/canopy water content. PSRI indicates canopy stress (pigment changes), whereas PRI and SIPI express light use efficiency. RENDVI, REP, and VOG-1 show spectral variations associated with the red edge wavelength position, which may be affected by changes in chlorophyll concentration or water stress.

Narrowband vegetation indices can be used as potential variables for crop type discrimination. Classification of the training dataset using discriminant analysis showed that the 10 best vegetation indices to discriminate the seven crop types were ARVI, EVI, NDVI, and SGI (greenness/leaf pigment indices); RENDVI and VOG-1 (chlorophyll red edge); SIPI and PRI (light use efficiency); and DWSI and NDWI (leaf water) (Figure 17.6). Examples of the relationships between pairs of indices are shown in Figures 17.7 and 17.8. Pasture presented the lowest NDWI value, as expected. Pasture, sugarcane, and rice displayed lower NDVI values than the other crops because of the influence of the nonphotosynthetic vegetation over the substrate and within the canopies (Figure 17.7). In agreement with these results, decreasing amounts of chlorophyll (presence of

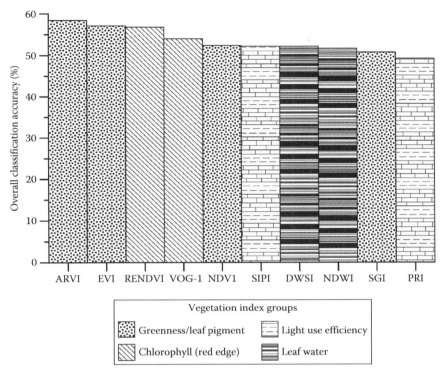

FIGURE 17.6 MDA-derived classification accuracy of the 10 best narrowband vegetation indices for a subset of 700 pixels representing the seven crop types. Indices were grouped according to their meaning.

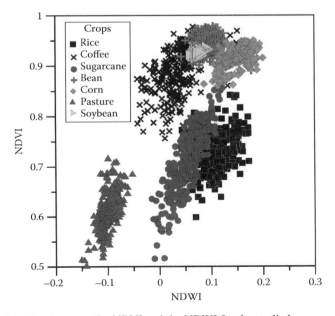

FIGURE 17.7 Relationships between the NDVI and the NDWI for the studied crops.

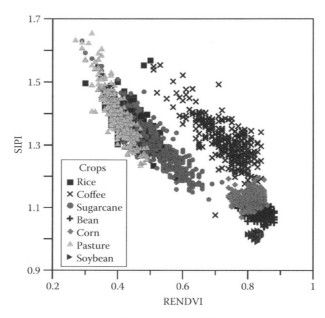

FIGURE 17.8   Relationships between the SIPI and the RENDVI for the studied crops.

senesced canopy foliage) resulted in larger SIPI and lower RENDVI values (canopy stress) for the three mentioned crops (Figure 17.8).

Although useful for crop type discrimination, the major value of the vegetation indices for agriculture is to provide information on the variability of biophysical and biochemical canopy attributes within a given crop rather than between crops. For example, some indices (e.g., NDVI and EVI) are usually correlated with LAI and others with leaf water content (e.g., NDWI). Variation in LAI and leaf water content for a given crop may be produced by several factors such as water stress, nutrient deficiency, pests, and diseases, or even by local problems in soil adaptation. Especially for annual crops, such variation may be also produced by differences in reproductive stages of the cultivars.

An example is presented in Figure 17.9 for seven soybean varieties (300 pixels per variety extracted from different fields) planted in the same growing season and farm (Figure 17.1f) and for two indices (MSI and VOG-1). By definition, higher MSI and VOG-1 values indicate greater water stress and chlorophyll content, respectively. Using a single date (January 14, 2006) and without any agronomical information, one can wrongly conclude that the fields of Monsoy 9010 (brighter yellowish shades in the southern portion of the farm—Figure 17.1f) were under water stress during Hyperion image acquisition. In reality, Monsoy 9010 was sensed by Hyperion at an earlier reproductive stage than the other varieties, which is the major cause to explain the observed higher MSI and lower VOG-1 values (Figure 17.9). On the other hand, some pixels (agricultural fields) of Monsoy 8914 and Perdiz displayed comparatively larger MSI values (right side of Figure 17.9), which would have required attention for potential canopy stress.

Thus, interpretation of the causes of the vegetation index variability for annual crops should be performed relatively to crop development. It is greatly facilitated with image acquisition in different dates along the growing season.

17.3.6   Spectral Features

We can determine spectral feature parameters and test them for crop type discrimination. In Figure 17.2, major spectral features detected by Hyperion were indicated by numbers. Most of the crop spectra displayed well-defined chlorophyll absorption bands in the blue (447 nm) and

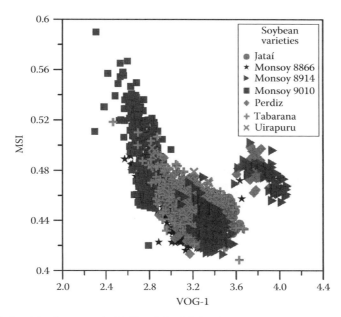

FIGURE 17.9 Relationships between the MSI and the VOG-1 for seven soybean varieties.

red (671 nm) wavelengths (e.g., soybean) and also the characteristic green reflectance peak at 569 nm. Exceptions were pasture, rice, and sugarcane that had comparatively higher red reflectance due to the spectral influence of senesced canopy components.

Senesced pasture showed also lignin–cellulose spectral features in the SWIR-1 and SWIR-2 ranges (1750, 2103, and 2304 nm). Leaf water absorption bands at 983 and 1205 nm, especially the second feature, were observed in all spectra, except for pasture. Unfortunately, the deepest leaf water absorption bands at 1400 and 1900 nm, usually detected in laboratory spectra, cannot be measured by airborne or orbital hyperspectral sensors because of the strong atmospheric water vapor absorption. Even in field data acquisition, if the Sun is used as the illumination source, measurement of these features is difficult because of the small amount of available energy that can be detected by the sensor after it passes through the atmosphere.

Another spectral feature indicated in Figure 17.2 is the red edge, which varies between the studied crops. This term has been used in the literature with two different meanings: (1) to characterize the spectral region of rapid reflectance change between the visible and NIR (701–760 nm); and (2) as the wavelength position (inflection point) equivalent to the maximum rate of reflectance change (slope). The position of the inflection point is usually associated with changes in chlorophyll content. The red edge changes also for the same crop with canopy development (results not shown). At a fixed phenological stage of a given crop, shifts in red edge in agricultural fields may be associated with canopy stress.

Quantitatively, detection of the red edge (inflection point) requires derivative analysis applied to the mentioned interval preceded by some smoothing procedure to reduce noise effects on results. This is critical when using instruments with poor SNR. In the present example, we applied derivative analysis to the 2100 crop spectra in the 701–760 nm interval and used the Savitzky–Golay smoothing procedure to obtain the first-order derivative spectra and the wavelength position of the red edge. First-order derivative results (not presented) showed that senesced pasture, rice, and sugarcane had red edge values positioned at shorter wavelengths (lower than 725 nm) than those observed for the other crops, which was consistent with RENDVI results of Figure 17.8 and with the spectral influence of nonphotosynthetic vegetation. It is important to note that red edge shifting occurred over a 30 nm wavelength range, which is equivalent only to three Hyperion bandwidths.

Quantitatively, we can describe each absorption band by calculating several parameters (e.g., depth, width, and area). Band depth is the most frequently used parameter and is usually correlated with the abundance of the absorber. However, there are cases in which this correlation is affected by the presence of other substances. For example, opaque minerals (magnetite and ilmenite) in tropical soils decrease absorption band depths of other constituents at different wavelengths.

A simple way to calculate absorption band depth is to select bands at the edges and center of the feature and compute the normalized difference between them. For example, NDVI and LVWI-2 (Table 17.2) are indirect measurements of the red chlorophyll and NIR leaf water absorption bands at 671 and 1205 nm, respectively. A more elaborated procedure is to use the continuum removal method to normalize the curves, to isolate and filter the features from spectra, and to allow their comparison from a common baseline [53]. In practice, bands of a given hyperspectral sensor placed at the edges (reflectance maxima) of an absorption feature are selected to represent the limits of straight line segments (continuum). The depth (D) of each absorption band is 1 minus Rb/Rc, where Rb is the reflectance at the center of the absorption band (Rb) and Rc is the reflectance of the continuum at the same wavelength as Rb.

Using the different Hyperion datasets, we generated band depth images for the following absorption bands: chlorophyll at 671 nm (edges at 569 and 763 nm); leaf water at 983 nm (edges at 933 and 1094 nm) and 1205 nm (edges at 1094 and 1286 nm); and lignin–cellulose at 2103 nm (edges at 2052 and 2214 nm) and 2304 nm (edges at 2214 and 2385 nm). Figure 17.10 shows an example of the 671 nm chlorophyll absorption band depth derived from the Hyperion image (corn and bean dataset). Bright pixels indicate green vegetation with deep 671 nm chlorophyll absorption bands in the corresponding Hyperion spectra. Dark pixels are related to nonvegetated surfaces such as bare soils or water.

FIGURE 17.10   Band depth image for the 671 nm chlorophyll spectral feature derived from Hyperion and the continuum removal method. Bright pixels indicate green vegetation with deep chlorophyll absorption bands in their spectra. Dark pixels refer to nonvegetated surfaces (e.g., bare soils or water).

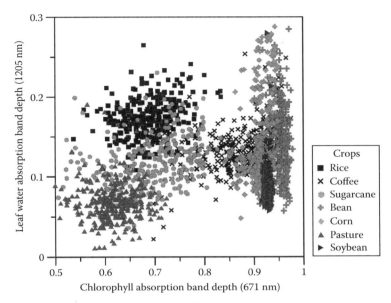

FIGURE 17.11 Relationship between the chlorophyll and leaf water absorption bands with band depth values calculated from the continuum removal method (2100 training pixels).

Figure 17.11 shows crop type discrimination based on the relationship between the 671 nm chlorophyll absorption band and the 1205 nm leaf water spectral feature for the 2100 crop spectra under study. Pasture, rice, and sugarcane presented the shallowest chlorophyll absorption bands due to senesced foliage, whereas bean showed the deepest chlorophyll feature. Flooded rice displayed deeper leaf water features when compared to sugarcane and pasture, which may reflect in some extent the influence of water over the leaf surface and the great number of individuals per linear meter.

17.3.7 Cultivar Discrimination

Because of the greater number of narrowbands and contiguous bands, hyperspectral sensors provide an enhanced level of information that increases the chances to discriminate materials with very small reflectance differences. The example presented here is based on the work by Galvão et al. [13], who used Hyperion data to discriminate five sugarcane varieties in Southeastern Brazil (Figure 17.1c). The main hypothesis was that cultivar-related agronomical differences produced different spectral responses that could be detected by hyperspectral sensors. Detailed sugarcane information was provided by an owner company and included variety, date of planting, frequency and date of cutting, soil types, and farms. Sugarcane age ranged from 1 to 4 years.

Reflectance in the Hyperion bands, narrowband ratios, NVIs, and spectral features parameters, which were discussed in the previous items, were considered as potential variables for MDA. The objective was to look for the optimum discriminant function to differentiate the varieties or to maximize the Mahalanobis distance for the two most similar groups. A training dataset of 200 pixels and a subset of 100 pixels were used to obtain the functions and test their discriminatory power. Final validation was performed by comparing the MDA classification results with the available ground information. Galvão et al. [13] used a stepwise method to select the best variables (feature selection). The probability of F was used as a criterion to include (0.05) and to remove (0.10) variables in forward and backward steps. In classification of hyperspectral data, feature selection is necessary to avoid the Hughes effect or phenomenon [54]. In other words, depending on the classifier, the use of the whole set of highly correlated Hyperion bands may actually

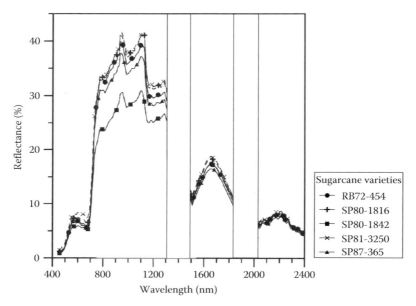

FIGURE 17.12 Average Hyperion surface reflectance spectra of the five studied sugarcane varieties. (Adapted from Galvão, L.S. et al., *Remote Sens. Environ.*, 94, 523, 2005.)

produce worse classification results than the use of a selected set of bands with good discriminatory power. In the literature, several procedures have been reported for feature selection [55–57] but discussion on them is beyond the scope of this chapter. Classification methods such as SVM are pointed out by some researchers as insensitive to the Hughes effect. However, according to Pal and Foody [5], SVM is also affected by the Hughes phenomenon, and feature selection is still recommended before SVM classification.

Figure 17.12 shows the average Hyperion reflectance spectra of the five sugarcane varieties. The transition from the low (SP80-1842) to the high reflectance (SP81-3250) varieties represented the change from erect (SP80-1842) to medium arch foliage (SP81-3250). Such differences in canopy structure affect sunlight penetration and reflectance, resulting in a higher reflectance for planophile than erectophile plants [58]. In comparison with the other sugarcane varieties, SP80-1842 presented also deeper lignin–cellulose absorption bands at 2103 and 2304 nm and shallower leaf liquid water absorption bands at 983 and 1205 nm. The transition from the variety SP80-1842 to SP81-3250 was characterized by an increase in the LWVI-2 values and by a decrease in the depth of the 2304 nm absorption band, both due to the larger amounts of nonphotosynthetic constituents within the canopy viewed by the sensor in the SP80-1842 variety (Figure 17.13).

The variety SP80-1842 was easily discriminated from the other four varieties due to its lower NIR reflectance (Figure 17.12). Thus, the simplest way to perform such discrimination was to use a band threshold in the NIR interval (e.g., pixels with reflectance values lower than 30% at 864 nm). Discrimination between the remaining four varieties (RB72-454, SP80-1816, SP81-3250, and SP87-365) was much more difficult due to the similarity of their average reflectance spectra. It required discriminant analysis.

The predictive power of the reflectance of the Hyperion narrowbands, the reflectance ratios, and of some spectral parameters tested by Galvão et al. [13] to discriminate the four sugarcane varieties is shown in Figures 17.14 through 17.16, respectively. The Hyperion bands were considered individually for the subsequent calculation of the average classification accuracy of the discriminant functions at different spectral intervals. A single-variable run procedure was used also for reflectance ratios and spectral parameters. Results refer to the training dataset of pixels. The best spectral intervals of Hyperion narrowband positioning to discriminate the four Brazilian sugarcane varieties were

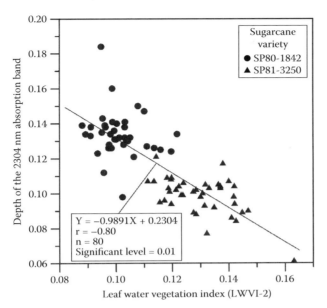

FIGURE 17.13 Relationships between the depth of the 2304 nm absorption band and the LWVI-2 for the two sugarcane varieties with extreme reflectance response. (Adapted from Galvão, L.S. et al., *Remote Sens. Environ.*, 94, 523, 2005.)

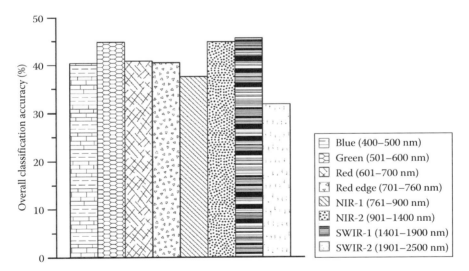

FIGURE 17.14 The potential of selected spectral intervals of Hyperion narrowband positioning to discriminate the four sugarcane varieties (RB72-454, SP80-1816, SP81-3250, and SP87-365). Results refer to average classification accuracy values and the training dataset. (Adapted from Galvão, L.S. et al., *Remote Sens. Environ.*, 94, 523, 2005.)

the SWIR-1 (1498–1780 nm), the NIR-2 (915–1296 nm), and the green (508–600 nm) (Figure 17.14). Hyperion narrowbands located in these intervals showed the largest classification accuracy values.

For Hyperion narrowband ratios, the best results (36%–40% of correct classification of the pixels) were obtained for (1) SWIR-1/green and SWIR-2/green ratios; (2) SWIR-1/NIR-1 and SWIR-1/NIR-2 ratios; and (3) NIR-2/NIR-1 and NIR-2/NIR-2 ratios (Figure 17.15). Finally, the spectral indices that exhibited the best discriminatory power were associated with the leaf liquid water (LWVI-2

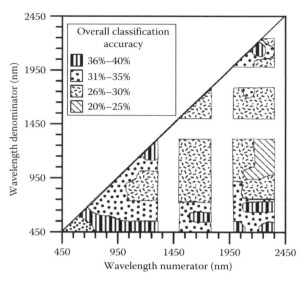

FIGURE 17.15 The potential of the Hyperion narrowband ratios to discriminate sugarcane varieties. Results around 1400 and 1900 nm were omitted due to atmospheric water vapor absorption. (Adapted from Galvão, L.S. et al., *Remote Sens. Environ.*, 94, 523, 2005.)

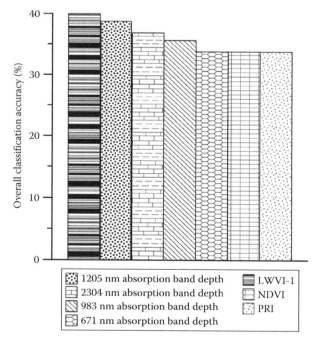

FIGURE 17.16 The potential of some spectral parameters to discriminate sugarcane varieties. (Adapted from Galvão, L.S. et al., *Remote Sens. Environ.*, 94, 523, 2005.)

and depth of the 1205 and 983 nm absorption bands) and lignin–cellulose (depth of the 2304 nm absorption band) spectral features (Figure 17.16).

Using the stepwise procedure, Galvão et al. [13] selected the following set of variables to compose the final discriminant model: (1) the Hyperion bands placed at 651, 722, 813, 1084, 1124, 1649, and 2002 nm; (2) the reflectance ratios 2355/2052, 1750/478, 1750/569, and 1255/478 nm; and (3) the

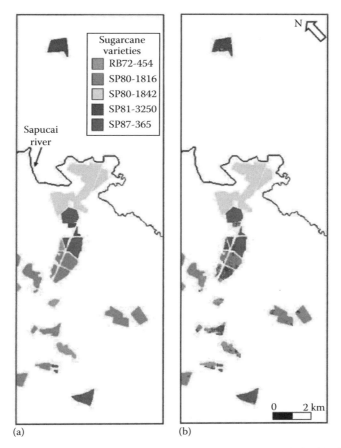

FIGURE 17.17 **(See color insert.)** (a) Ground truth image showing the spatial distribution of the five sugarcane varieties under study in the central portion of the study area. (b) Classification image derived from MDA. (Adapted from Galvão, L.S. et al., *Remote Sens. Environ.*, 94, 523, 2005.)

depth of the absorption bands centered at 671 nm (chlorophyll), 983 nm (leaf liquid water), and 2304 nm (lignin–cellulose); the NDWI, and the DSWI. When applied to the subset of pixels, this discriminant model reached 87.5% of classification accuracy. Comparison between the ground truth data (Figure 17.17a) and the MDA-derived classified image (Figure 17.17b) in the central portion of the study area confirmed the good performance of the discriminant model that presented the best results for the cultivars SP87-365 and RB72-454.

17.3.8 MULTISPECTRAL VERSUS HYPERSPECTRAL DISCRIMINATION

We used the Hyperion soybean image (Figure 17.1f) to evaluate the spectral resolution influence to discriminate seven cultivars: Jataí; Monsoy 8866, 8914, and 9010; Perdiz; Tabarana; and Uirapuru. Filter functions were used to simulate the spectral response of the following multispectral instruments (Table 17.3): Advanced Very High Resolution Radiometer (AVHRR/NOAA-17); CCD Camera on Board the China–Brazil Earth Resources Satellite (CCD/CBERS-2); High Geometric Resolution Instrument (HRG/SPOT-5); Enhanced Thematic Mapper Plus (ETM+/Landsat-7); Moderate Resolution Imaging Spectroradiometer (MODIS/Terra); and Advanced Space Borne Thermal Emission and Reflection Radiometer (ASTER/Terra). MODIS Bands 8 (405–420 nm) and

TABLE 17.3
Simulated Multispectral Sensor Bands from Hyperion Data Using the Filter Functions

Sensor	Blue Band (nm)	Green Band (nm)	Red Band (nm)	Red Edge Band (nm)	NIR-1 Band (nm)	NIR-2 Band (nm)	SWIR-1 Band (nm)	SWIR-2 Band (nm)
AVHRR/NOAA-17	—	—	580–680	—	725–1000	—	1580–1640	—
CCD/CBERS-2	450–520	520–590	630–690	—	770–890	—	—	—
HRG/SPOT-5	—	500–590	610–680	—	780–890	—	1580–1750	—
ETM + /Landsat-7	450–515	525–605	630–690	—	775–900	—	1550–1750	2090–2350
MODIS/Terra	438–448	526–536	620–670	743–753	841–876	1230–1250	1628–1652	2105–2155
	459–479	545–565	662–672	—	862–877	931–941		
	483–493	546–556	673–683	—	890–920	915–965		
ASTER/Terra	—	520–600	630–690	—	760–860	—	1600–1700	2145–2185
								2185–2225
								2235–2285
								2295–2365

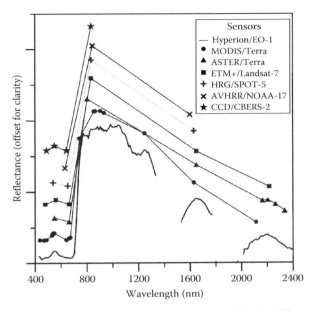

FIGURE 17.18 Simulated spectral response of the cultivar Monsoy 8914 for different multispectral sensors.

20–36 (3660–14385 nm) were not simulated from Hyperion because they were out of the spectral range of operability. The same occurred for ASTER bands 9 (2360–2430 nm) and 10–14 (thermal). The spatial resolution was not simulated.

The discrimination between the soybean varieties was tested using reflectance values of the Hyperion and sensor-simulated images extracted from training (300 pixels per variety) and subset samples (100 pixels per cultivar) as input variables for MDA. The stepwise procedure described before was used for band selection.

Inspection of the average surface reflectance spectra measured by Hyperion (not presented) showed a better discrimination for cultivar Monsoy 9010, which was sensed by Hyperion in an earlier reproductive stage. Considering only the spectral resolution, MODIS was the best multispectral sensor when compared to Hyperion (Figure 17.18). MODIS allows the measurement of the chlorophyll features in the visible and of the 1205 nm leaf water absorption in the NIR-2. It has also red edge (743–753 nm) and SWIR-1 and SWIR-2 bands. On the other hand, CCD/CBERS-2 does not acquire data in the SWIR-1 interval, and AVHRR/NOAA-17 acquires data in a very broad NIR band (Table 17.3). From an agricultural point of view, these are two undesirable specifications for crop type discrimination, as deduced from previous sections.

Among the multispectral sensors, the lowest classification accuracies were observed for AVHRR (53.9%) and CCD (54.7%), and the best accuracy was verified for MODIS (75.4% in Figure 17.19). In comparison with the simulated multispectral sensors, Hyperion had the best result. The overall classification accuracy was 84.1% for the subset of pixels using 36 narrowbands centered at the following wavelengths: 477, 528, 548, 569, 589, 609, 640, 660, 681, 701, 721, 742, 762, 793, 823, 864, 884, 925, 972, 1023, 1063, 1083, 1104, 1194, 1205, 1245, 1517, 1548, 1568, 1649, 1699, 1749, 2133, 2203, 2244, and 2284 nm. Results from the sensor simulation (Figure 17.19) are in agreement with those obtained by Galvão et al. [14] when studying sugarcane varieties with Hyperion. Discriminant scores plotted for four of the seven soybean varieties confirmed the poor discrimination observed for AVHRR (Figure 17.20a) and the good discrimination provided by Hyperion (Figure 17.20b).

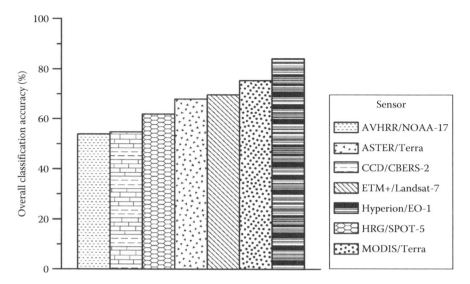

FIGURE 17.19 Overall classification accuracy for Hyperion and simulated multispectral sensors. Results refer to the subset of 700 pixels and seven soybean varieties.

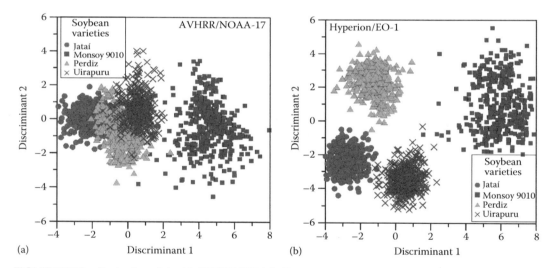

FIGURE 17.20 Projection of the (a) AVHRR/NOAA-17 and (b) Hyperion/EO-1 discriminant scores for four of the seven studied soybean cultivars.

17.4 CONCLUSIONS

In general, hyperspectral sensors provide significantly better classification results than multispectral sensors but their classification performance depends on other factors such as signal to noise ratio (SNR) and adequate feature selection. The most important broad spectral intervals for crop type discrimination using imaging spectrometers are the NIR (760–900 nm), the SWIR (1500–1750 nm), the red (600–700 nm), and the green (500–600 nm). A number of focused hyperspectral narrowbands help distinctly separate crop types and/or crop species based on their biophysical and/or biochemical properties. Some examples demonstrated in this research are (a) spectral reflectance plot involving two narrowbands centered at 864 and 660 nm, (b) leaf water absorption bands such as 983 and 1205 nm, (c) numerous possibilities of computing narrowband vegetation indices (e.g., Table 17.2), and (d) waveband reflectivity or indices derived using data from the red-edge (701–740 nm) portion of the spectrum. These numerous opportunities offered by hyperspectral data are of great advantage when compared with possibilities offered by broadband data. However, narrowband selection in these broad spectral intervals depends on each study area or on the set of crops/cultivars tested for hyperspectral classification. Considering the great data variability introduced by the factors discussed in Section 17.2, applicable set of narrowbands to provide maximum crop discrimination requires large number of studies conducted in wide array of locations, crop types, and crop species in distinct agroecosystems from around the world. By acquiring data in contiguous bands, hyperspectral sensors provide distinct possibilities for narrowband selection aimed at optimizing crop type discrimination and crop classification within and across locations. Furthermore, besides being useful for differentiating crops, NVIs and absorption band parameters, calculated on a per pixel basis, can provide very important information on biophysical and biochemical crop attributes as well as the potential causes of spectral variability between agricultural fields of a given crop (e.g., canopy stress, diseases, and nutrient deficiency).

ACKNOWLEDGMENTS

The authors are grateful to Drs. Prasad Thenkabail, John Lyon, and Alfredo Huete for the invitation to contribute a book chapter and for revising the manuscript. Thanks are also due to *Conselho Nacional de Desenvolvimento Científico e Tecnológico* (CNPq) (302732/2009-8) and to *Fundação de Amparo à Pesquisa do Estado de São Paulo* (FAPESP) (2008/11499-8).

REFERENCES

1. Goetz, A.F.H., Vane, G., Solomon, J., and Rock, B.N., Imaging spectrometry for Earth remote sensing, *Science*, 228, 1147–1153, 1985.
2. Thenkabail, P.S., Enclona, E.A., Ashton, M.S., and van der Meer, B., Accuracy assessments of hyperspectral waveband performance for vegetation analysis applications, *Remote Sensing of Environment*, 91, 354–376, 2004.
3. Thenkabail, P.S., Smith, R.B., and De Pauw, E., Evaluation of narrowband and broadband vegetation indices for determining optimal hyperspectral wavebands for agricultural crop characterization, *Photogrammetric Engineering and Remote Sensing*, 68, 607–621, 2002.
4. Lobell, D.B. and Asner, G.P., Comparison of Earth Observing-1 ALI and Landsat ETM+ for crop identification and yield prediction in Mexico, *IEEE Transactions on Geoscience and Remote Sensing*, 41, 1277–1282, 2003.
5. Pal, M. and Foody, G.M., Feature selection for classification of hyperspectral data for SVM, *IEEE Transactions on Geoscience and Remote Sensing*, 48, 2297–2307, 2010.
6. Bannari, A., Khurshid, K.S., Staenz, K., and Schwarz, J., Potential of hyperion EO-1 hyperspectral data for wheat crop chlorophyll content estimation, *Canadian Journal of Remote Sensing*, 34, 139–157, 2008.
7. Li, Q.M., Hu, B.X., and Pattey, E., A scale-wise model inversion method to retrieve canopy biophysical parameters from hyperspectral remote sensing data, *Canadian Journal of Remote Sensing*, 34, 311–319, 2008.
8. Haboudane, D., Tremblay, N., Miller, J.R., and Vigneault, P., Remote estimation of crop chlorophyll content using spectral indices derived from hyperspectral data, *IEEE Transactions on Geoscience and Remote Sensing*, 46, 423–437, 2008.
9. Smith, A.M., Bourgeois, G., Teillet, P.M., Freemantle, J., and Nadeau, C., A comparison of NDVI and MTVI2 for estimating LAI using CHRIS imagery: A case study in wheat, *Canadian Journal of Remote Sensing*, 34, 539–548, 2008.
10. Yang, C., Everitt, J.H., Bradford, J.M., and Murden, D., Comparison of airborne multispectral and hyperspectral imagery for estimating grain sorghum yield, *Transactions of the ASABE*, 52, 641–649, 2009.
11. Zarco-Tejada, P.J., Ustin, S.L., and Whiting, M.L., Temporal and spatial relationships between within-field yield variability in cotton and high-spatial hyperspectral remote sensing imagery, *Agronomy Journal*, 97, 641–653, 2005.
12. Galvão, L.S., Roberts, D.A., Formaggio, A.R., Numata, I., and Breunig, F.M., View angle effects on the discrimination of soybean varieties and on the relationships between vegetation indices and yield using off-nadir Hyperion data, *Remote Sensing of Environment*, 113, 846–856, 2009.

13. Galvão, L.S., Formaggio, A.R., and Tisot, D.A., Discrimination of sugarcane varieties in southeastern Brazil with EO-1 hyperion data, *Remote Sensing of Environment*, 94, 523–534, 2005.

14. Galvão, L.S., Formaggio, A.R., and Tisot, D.A., The influence of spectral resolution on discriminating Brazilian sugarcane varieties, *International Journal of Remote Sensing*, 27, 769–777, 2006.

15. Everingham, Y., Lowe, K.H., Donald, D., Coomans, D., and Markley, J., Advanced satellite imagery to classify sugarcane crop characteristics, *Agronomy for Sustainable Development*, 27, 111–117, 2007.

16. Rao, N.R., Development of a crop-specific spectral library and discrimination of various agricultural crop varieties using hyperspectral imagery, *International Journal of Remote Sensing*, 29, 131–144, 2008.

17. Camps-Valls, G., Gomez-Chova, L., Calpe-Maravilla, J., Soria-Olivas, E., Martin-Guerrero, J.D., and Moreno, J., Support vector machines for crop classification using hyperspectral data, *Lecture Notes in Computer Science*, 2652, 134–141, 2003.

18. Bandos, T.V., Bruzzone, L., and Camps-Valls, G., Classification of hyperspectral images with regularized linear discriminant analysis, *IEEE Transactions on Geoscience and Remote Sensing*, 47, 862–873, 2009.

19. Filippi, A.M., Archibald, R., Bhaduri, B.L., and Bright, E.A., Hyperspectral agricultural mapping using support vector machine-based endmember extraction (SVM-BEE), *Optics Express*, 17, 23823–23842, 2009.

20. Duca, R. and Del Frate, F., Hyperspectral and Multiangle CHRIS-PROBA images for the generation of land cover maps, *IEEE Transactions on Geoscience and Remote Sensing*, 46, 2857–2866, 2009.

21. Eddy, P.R., Smith, A.M., Hill, B.D., Peddle, D.R., Coburn, C.A., and Blackshaw, R.E., Hybrid segmentation: Artificial neural network classification of high resolution hyperspectral imagery for site-specific herbicide management in agriculture, *Photogrammetric Engineering and Remote Sensing*, 74, 1249–1257, 2008.

22. Gausman, H.W., *Plant Leaf Optical Properties in Visible and Near-Infrared Light*, Graduate Studies Texas No. 29, Texas Tech Press, Lubbock, TX, 78 p., 1985.

23. Yanqun, Z., Yuan, L., Haiyan, C., and Jianjun, C., Intraspecific differences in physiological response of 20 soybean cultivars to enhanced ultraviolet-B radiation under field conditions, *Environmental and Experimental Botany*, 50, 87–97, 2003.

24. Falqueto, A.R., Cassol, D., Magalhães Júnior, A.M., Oliveira, A.C., and Bacarin, M.A., Physiological analysis of leaf senescence of two rice cultivars with different yield potential, *Pesquisa Agropecuária Brasileira*, 44, 695–700, 2009.

25. Gonçalves, B., Correia, C.M., Silva, A.P., Bacelar, E.A., Santos, A., and Moutinho-Pereira, J.M., Leaf structure and function of sweet cherry tree (*Prunus avium* L.) cultivars with open and dense canopies, *Scientia Horticulturae*, 116, 381–387, 2008.

26. Bauer, M.E., The role of remote sensing in determining the distribution and yield of crops, *Advances in Agronomy*, 27, 271–304, 1975.

27. Heilman, J.L. and Boyd, W.E., Soil background effects on the spectral response of a three-component rangeland scene, *Remote Sensing of Environment*, 19, 127–137, 1986.

28. Gilabert, M.A., Segarra, D., and Melia, J., Simulation of citrus orchard reflectance by means of a geometrical canopy model, *International Journal of Remote Sensing*, 15, 2559–2582, 1994.

29. Sugawara, L.M., Rudorff, B.F.T., and Adami, M., Feasibility of the use of Landsat imagery to map soybean crop areas in Parana, Brazil, *Pesquisa Agropecuária Brasileira*, 43, 1777–1783, 2008. (In Portuguese.)

30. ITT Visual Information Solutions, Atmospheric correction module: QUAC and FLAASH user's guide. Version 4.7, ITT Visual Information Solutions, Boulder, CO, 2009.

31. Kaufman, Y.J., Tanré, D., Remer, L.A., Vermote, E.F., Chu, A., and Holben, B.N., Operational remote sensing of tropospheric aerosol over land from EOS moderate resolution imaging spectroradiometer, *Journal of Geophysical Research*, 102, 17051–17067, 1997.

32. D'Arco, E., Alvarenga, B.S., Rizzi, R., Rudorff, B.F.T., Moreira, M.A., and Adami, M., Geotechnologies to estimate flooded rice crop area, *Revista Brasileira de Cartografia*, 58, 247–253, 2006.

33. Van Niel, T.G. and McVicar, T.R., Current and potential uses of optical remote sensing in rice-based irrigation systems: A review, *Australian Journal of Agricultural Research*, 55, 155–185, 2004.

34. Moreira, M.A., Adami, M., and Rudorff, B.F.T., Spectral and temporal behavior analysis of coffee crop in Landsat images, *Pesquisa Agropecuária Brasileira*, 39, 223–231, 2004.

35. Rudorff, B.F.T., Aguiar, D.A., Silva, W.F., Sugawara, L.M., Adami, M., and Moreira, M.A., Studies on the rapid expansion of sugarcane for ethanol production in São Paulo state (Brazil) using Landsat data, *Remote Sensing*, 2, 1057–1076, 2010.

36. Numata, I., Roberts, D.A., Chadwick, O.A., Schimel, J.P., Galvão, L.S., and Soares, J.V., Evaluation of hyperspectral data for pasture estimate in the Brazilian Amazon using field and imaging spectrometers, *Remote Sensing of Environment*, 112, 1569–1583, 2008.

37. Kaufman, Y.J. and Tanré, D., Atmospherically resistant vegetation index (ARVI) for EOS-MODIS, *IEEE Transactions on Geoscience and Remote Sensing*, 30, 261–270, 1992.

38. Huete, A.R., Didan, K., Miura, T., Rodriguez, E.P., Gao, X., and Ferreira, L.G., Overview of the radiometric and biophysical performance of the MODIS vegetation indices, *Remote Sensing of Environment*, 83, 195–213, 2002.

39. Rouse, J.W., Haas, R.H., Schell, J.A., and Deering, D.W., Monitoring vegetation systems in the Great Plains with ERTS, *Proceedings of Third ERTS-1 Symposium*, Washington, DC, December 10–14, NASA, SP-351, Vol. 1, pp. 309–317, 1973.

40. Hunt E.R. Jr. and Rock, B.N., Detection of changes in leaf water content using near- and middle-infrared reflectances, *Remote Sensing of Environment*, 30, 43–54, 1989.

41. Gao, B.C., NDWI: A normalized difference water index for remote sensing of vegetation liquid water from space, *Remote Sensing of Environment*, 58, 257–266, 1996.

42. Penuelas, J., Pinol, J., Ogaya, R., and Filella, I., Estimation of plant water concentration by the reflectance water index WI (R900/R970), *International Journal of Remote Sensing*, 18, 2869–2875, 1997.

43. Apan, A., Held, A., Phinn, S., and Markley, J., Detecting sugarcane 'orange rust' disease using EO-1 Hyperion hyperspectral imagery, *International Journal of Remote Sensing*, 25, 489–498, 2004.

44. Merzlyak, M.N., Gitelson, A.A., Chivkunova, O.B., and Rakitin, V.Y., Non-destructive optical detection of pigment changes during leaf senescence and fruit ripening, *Physiologia Plantarum*, 106, 135–141, 1999.

45. Gitelson, A.A., Zur, A., Chivkunova, O.B., and Merzlyak, M.N., Assessing carotenoid content in plant leaves with reflectance spectroscopy, *Photochemistry and Photobiology*, 75, 272–281, 2002.

46. Gitelson, A.A., Merzlyak, M.N., and Chivkunova, O.B., Optical properties and nondestructive estimation of anthocyanin content in plant leaves, *Photochemistry and Photobiology*, 71, 38–45, 2001.

47. Gamon, J.A., Serrano, L., and Surfus, J.S., The photochemical reflectance index: An optical indicator of photosynthetic radiation use efficiency across species, functional types and nutrient levels, *Oecologia*, 112, 492–501, 1997.

48. Penuelas, J., Baret, F., and Filella, I., Semi-empirical indices to assess carotenoids/chlorophyll-a ratio from leaf spectral reflectance, *Photosynthetica*, 31, 221–230, 1995.

49. Gitelson, A.A., Merzlyak, M.N., and Lichtenthaler, H.K., Detection of red edge position and chlorophyll content by reflectance measurements near 700 nm, *Journal of Plant Physiology*, 148, 501–508, 1996.

50. Curran, P.J., Windham, W.R., and Gholz, H.L., Exploring the relationship between reflectance red edge and chlorophyll concentration in Slash Pine leaves, *Tree Physiology*, 15, 203–206, 1995.

51. Vogelmann, J.E., Rock, B.N., and Moss, D.M., Red edge spectral measurements from sugar maple leaves, *International Journal of Remote Sensing*, 14, 1563–1575, 1993.

52. Gitelson, A.A., Kaufman, Y.J., Stark, R., and Rundquist, D., Novel algorithms for remote estimation of vegetation fraction, *Remote Sensing of Environment*, 80, 76–87, 2002.

53. Clark, R.N. and Roush, T.L., Reflectance spectroscopy: Quantitative analysis techniques for remote sensing applications, *Journal of Geophysical Research*, 89, 6329–6340, 1984.

54. Hughes, G.F., On the mean accuracy of statistical pattern recognizers, *IEEE Transactions on Information Theory*, 14, 55–63, 1968.

55. Bajcsy, P. and Groves, P., Methodology for hyperspectral band selection, *Photogrammetric Engineering and Remote Sensing*, 70, 793–802, 2004.

56. Lu, S., Oki, K., Shimizu, Y., and Omasa, K., Comparison between several feature extraction/classification methods for mapping complicated agricultural land use patches using airborne hyperspectral data, *International Journal of Remote Sensing*, 28, 963–984, 2007.

57. Pal, M., Margin-based feature selection for hyperspectral data, *International Journal of Applied Earth Observation and Geoinformation*, 11, 212–220, 2009.

58. Jackson, R.D. and Pinter, P.J., Spectral response of architecturally different wheat canopies, *Remote Sensing of Environment*, 20, 43–56, 1986.

18 Identification of Canopy Species in Tropical Forests Using Hyperspectral Data

Matthew L. Clark

CONTENTS

18.1 INTRODUCTION

Tropical forests (TFs) are globally important due to their extremely high species diversity and major role in biogeochemical cycles. Spatial and temporal sampling of these forests in the field is greatly restricted due to prohibitive costs and inaccessibility at the ground and canopy level, and extrapolating plot data to broader spatial scales is problematic as species distributions and ecosystem processes vary at different spatial scales due to overlapping factors, such as climate, soil, and past disturbance. Remote sensing plays an important role in mapping TFs over larger areas than available from the ground, especially with the recent use of new sensors, processing and modeling techniques, and integration with field data [1]. In particular, a new generation of hyperspectral, hyperspatial, and hypertemporal sensors is making rapid and innovative advances in understanding TF composition, structure, and function over broad spatial and temporal scales [1]. These advances include canopy chemistry by hyperspectral sensors [2–6]; analysis of individual tree crowns (ITCs) [7,8] and canopy components, such as lianas [9] by hyperspatial (<4 m) sensors; and new insights into canopy chemistry, physiology, and phenology by hypertemporal sensors [4,10,11].

This chapter reviews hyperspectral and hyperspatial remote sensing of TF canopy species, with an emphasis on trees in wetter forests but excluding mangroves. The fundamental biochemical, structural (biophysical), phenological, and site-specific factors that control plant spectral properties across the visible (VIS: 400–700 nm), near-infrared (NIR: 700–1300 nm), and shortwave-infrared (SWIR: 1500–2500 nm) regions of the electromagnetic spectrum are reviewed, beginning with fine-scale canopy photosynthetic and nonphotosynthetic components, and then broadening the scope to the 3D scale of crowns in the canopy. Applications with hyperspectral and hyperspatial imagery from TFs have taken a pixel-based view of analysis, leading to wall-to-wall maps of canopy floristic diversity or fractional species abundance [12,13], or an object-based analysis of ITCs [7,14,15]. This chapter concentrates on ITC species mapping with hyperspectral data, although advances from other research will be discussed in order to provide a holistic understanding of the factors influencing species discrimination using hyperspectral technology. For reference, Table 18.1 provides a summary of hyperspectral research from TFs discussed in this chapter, and Table 18.2 lists the important properties of sensors used in existing, and potentially future, studies.

18.2 DRIVERS OF SPECTRAL VARIATION IN TROPICAL FOREST CANOPIES

18.2.1 LEAVES, BARK, AND OTHER FINE-SCALE CANOPY COMPONENTS

Our ability to remotely detect TF canopy species using hyperspectral imagery hinges upon species having interspecies (among species) spectral differences that are detectable and consistently measured, despite intraspecific (within species) spectral variation. Humid TF canopies are characterized by hemispherical, broadleaf, multilayered crowns with high leaf area index (LAI) and perforated by gaps covered by understory vegetation (Figure 18.1). Spectral variation over a TF is thus primarily determined by the biochemistry (e.g., plant pigments, water, and structural carbohydrates) and structure (e.g., leaf thickness and air spaces) of leaves, and the scaling of these spectral properties due to volumetric scattering of photons in the canopy. However, nonphotosynthetic tissues (e.g., bark, flowers, and seeds) and other photosynthetic canopy organisms (e.g., vines, epiphytes, and epiphylls) can mix in the photon signal and vary depending on a complex interplay of species, structure, phenology, and site differences, none of which are well understood. Research on the spectral connection to biochemical and structural properties of fine-scale canopy tissues—mainly leaves—has accelerated in the last 5 years and now includes several sites [6,9,16–24], thereby permitting a foundation for understanding how these properties scale to the crown or canopy level.

Leaf reflectance in the visible spectrum is dominated by absorption features created by plant pigments, such as chlorophyll a (chl-a) and chlorophyll b (chl-b), carotenoids (e.g., β-carotene and lutein), and anthocyanins [25]. Photosynthetic pigments chl-a absorbs in 410–430 nm and 600–690 nm and chl-b absorbs in 450–470 nm [25,26]. Other accessory pigments increase absorption of light and act in plant defense functions, like avoiding damage from ultraviolet light. Carotenoids have peak absorption in lower wavelengths <500 nm that overlaps chl-a and chl-b absorption features [25,26]. Reflectance in the NIR is relatively high, with absorption features created mainly by water around 970 and 1200 nm, and to a lesser extent by broad absorption from structural carbohydrates such as lignin and cellulose [26,27]. In photosynthetic leaves, the SWIR region is characterized by relatively low reflectance and strong absorption by water that masks other absorption features [26,28]. However, dry leaves do not have strong water absorption and reveal overlapping absorptions by carbon compounds, such as lignin and cellulose, and other plant biochemicals, including protein nitrogen, starch, and sugars [26,28].

Leaf structure also plays a role in a species reflectance signature. Surface reflectance can be influenced by surface topography caused by features such as leaf hairs and waxes [29]. Reflectance in NIR is high due to multiple scattering of photons inside the leaf caused by internal structures, air spaces, and air–cell interfaces, such as in spongy mesophyll [29]. An important summary variable of leaf structure is specific leaf area (SLA—projected leaf area per unit leaf dry mass), which can be thought of the amount of leaf biomass spread over an area. Leaf design scales as a stoichiometric balance of photosynthetic (e.g., chl-a and chl-b) and other plant biochemicals (e.g., water, nitrogen, and cellulose), and different chemical constituents are correlated with structural metrics, such as SLA [5,6,20].

Do species have unique biochemical–structural properties that translate into hyperspectral signatures that can discriminate canopy species? We are only in the beginning stages of answering this question. A pioneering study in Pará, Brazil [16] acquired laboratory leaf spectra (450–950 nm) for 11 TF tree species, and averaged these spectra to analyze differences in species discrimination at simulated branch and crown scales. Using a novel shape filter technique, species discrimination was possible at crown scales and declined at branch and leaf scales. The distributions of red edge wavelength and position, calculated from derivatives of reflectance, were not normally distributed within

TABLE 18.1

Tropical Forest Research Using Hyperspectral Data Reviewed in This Chapter

Reference (Year)	Life Forms (Forest)	Objective	Location	Spectral Region (Sensor)
Laboratory spectrometer				
[16] (2000)	Trees (wet forest)	Explore spectral separability of 11 species	Amazon, Brazil	VIS, NIR (FieldSpec, analytical spectral devices [ASD])
[22] (2004)	Trees, lianas (dry and wet)	Discrimination of lianas and trees with nine classifiers; leaf chlorophyll	PNM, FS Panama	VIS, NIR (UniSpec)
[7,14] (2005)	Trees (wet forest)	Discriminate seven tree species based on LDA, DT, and two other classifiers	LSBS, Costa Rica	VIS, NIR, SWIR (FieldSpec, ASD)
[17] (2006)	Trees (dry and wet)	Spectral separability and spectral–biochemical links across sites	Mexico, Panama, Costa Rica	VIS, NIR (UniSpec, FieldSpec, ASD)
[9] (2007)	Trees, lianas (dry forest)	Discrimination of lianas and trees with data reduction and nine classifiers	PNM, Panama	VIS, NIR, SWIR (FieldSpec)
[6] (2008)	Trees (wet forest)	Link between spectral, biochemical and structural properties	Amazon, Brazil	VIS, NIR, SWIR (FieldSpec, ASD)
[18] (2008)	Trees (wide range)	Discriminate 20 tree species based on LDA classifier	Costa Rica	VIS, NIR, SWIR (FieldSpec, ASD)
[24] (2008)	Trees (wet forest)	Spectral and biochemical differences between sex of two species	LSBS, Costa Rica	VIS, NIR, SWIR (FieldSpec, ASD)
[20] (2008)	Trees (wet/ elev. grad.)	Link between canopy reflectance and leaf properties with RT based variation in crown structure	Queensland Australia	VIS, NIR, SWIR (FieldSpec, ASD)
[19] (2009)	Trees (wet/elev. grad.)	Link between spectral, biochemical, and structural properties	Queensland Australia	VIS, NIR, SWIR (FieldSpec, ASD)
[21] (2009)	Trees, lianas (dry and wet)	Differences in leaf reflectance, biochemistry, and structure for trees and lianas	PNM, FS Panama	VIS, NIR (UniSpec)
[23] (2009)	Trees (wet forest)	Influence of epiphylls on canopy reflectance from RT; link to indices	Amazon, Brazil	VIS, NIR (PS2, ASD)
Airborne hyperspatial and hyperspectral				
[7,14,15] (2005)	Trees (wet forest)	Discriminate emergent tree species based on LDA, DT, and three other classifiers (DT and LDA with lidar)	LSBS, Costa Rica	VIS, NIR, SWIR (HYDICE, FLIMAP lidar)
[3] (2005)	Trees (wet forest)	How biological invasion alters forest canopy chemistry	Hawai'i	VIS, NIR, SWIR (AVIRIS)
[35] (2006)	Trees (wet forest)	Explore spectral separability of five species; data reduction	LSBS, Costa Rica	VIS, NIR (HYDICE)
[9] (2007)	Trees, lianas (dry forest)	Discrimination of lianas and trees with data reduction and nine classifiers	PNM, Panama	VIS, NIR, SWIR (HYDICE)

(continued)

TABLE 18.1 (continued)
Tropical Forest Research Using Hyperspectral Data Reviewed in This Chapter

Reference (Year)	Life Forms (Forest)	Objective	Location	Spectral Region (Sensor)
[12] (2007)	Trees (wet forest)	Link biochemistry to spectral diversity; map species richness	Hawai'i	VIS, NIR, SWIR (AVIRIS)
[5] (2008)	Trees (wet forest)	Canopy spectral, biochemical, and structural differences between native and invasive species	Hawai'i	VIS, NIR, SWIR (AVIRIS)
[13] (2008)	Trees (wet forest)	Mapping the fractional abundance of invasive species	Hawai'i	VIS, NIR, SWIR (AVIRIS—CAO-beta)
[34] (2008)	Trees (dry forest)	Quantify nonphotosynthetic veg. with spectral mixture analysis	PNM, Panama	VIS, NIR, SWIR (HYDICE)
Spaceborne medium resolution (30 m) hyperspectral				
[2] (2004)	Trees (wet forest)	Quantify differences in canopy water and productivity from drought	Amazon, Brazil	VIS, NIR, SWIR (Hyperion)
[4] (2006)	Trees (wet forest)	Track canopy physiology and biochemistry through time	Hawai'i	VIS, NIR, SWIR (Hyperion)
[11] (2008)	Trees (wet forest)	Spectral seasonality related to canopy phenology	Amazon, Brazil	VIS, NIR, SWIR (Hyperion)
[38] (2010)	Trees (wet forest)	Discriminate five tree species based on LDA classifier	Amazon, Peru	VIS, NIR, SWIR (Hyperion)

a species, indicating intraspecific spectral variability, yet there was potential to separate some species with these metrics. Clark et al. [7] measured laboratory reflectance for leaves from seven tree species in a wet TF at the La Selva Biological Station (LSBS), Costa Rica. Interspecific variability was included in the analysis by measuring bidirectional reflectance (i.e., not hemispherical reflectance in an integrating sphere) and leaves with epiphylls, herbivory, and galls (effects discussed later). A linear discriminant analysis (LDA) classifier with a stepwise band selection procedure showed that just 10 reflectance bands could classify species with 90% overall accuracy, while accuracy increased to 100% when using 40 bands. The 10-band classifier included mostly bands in NIR and SWIR, where variability was largely controlled by leaf structure, water content, and possibly other biochemicals (e.g., lignin and cellulose), while VIS bands were in blue (BE), green-peak (GP), and red edge wavelengths. Castro-Esau et al. [17] collected leaf reflectance data, chlorophyll content, and structural (mesophyll attributes and thickness) data from a range of species across seven sites within wet and dry TF in Mexico and Costa Rica, and Panama. Species could be accurately classified with leaf spectra within sites, with overall accuracy declining with species richness (85% for 20 species and 80% for 40 species). The best classifier and wavelengths used varied among sites, and spectral signatures from one site were not useful for classifying the same species at another site. The most frequently-selected bands were from blue, blue–green edge, and red to red edge, while NIR bands were not as important in leaf-scale species discrimination (SWIR not analyzed). Rivard et al. [18] focused on discriminating species of 20 tropical tree species from two sites in Costa Rica using LDA and stepwise selection of reflectance bands and narrowband indices. Accuracies ranged from 70% to 97%. Bands and indices that correlated with leaf water content, followed by pigment properties, were most important in discriminating these species, with indices having an advantage over individual bands. Both Clark et al. [7] and Rivard et al. [18] found that the highest overall accuracy was achieved when information from across-the-visible to SWIR spectrum (i.e., full range) was

TABLE 18.2
Summary of Hyperspectral Sensors Discussed in This Chapter

Name	Wavelength Range (nm)	No. of Bands	Signal-to-Noise Ratio[a]	Ground Resolution (m)	Developer
Laboratory spectrometer					
FieldSpec 3	350–2,500	2,151 (interpolated)	160,000 VIS 1,200 SWIR	<1	ASD
FieldSpec FR	350–2,500	2,151 (interpolated)	87,000 VIS 805 SWIR	<1	ASD
UniSpec	350–1,100	256	~250 VIS	<1	PP Systems
Airborne hyperspatial and hyperspectral					
HYDICE	400–2,500	210	~300 VIS ~300 SWIR at 5,000 ft	≥1	U.S. Naval Research Laboratory
AVIRIS[b]	400–2,500	224	~1000 VIS ~500 SWIR	≥3	NASA [36]
CAO-alpha (CASI-1500)	369–1,052	≤288	~400 VIS	0.45–1.35	CAO [37]; ITRES
HyMap	450–2,480	128	>500	3–10	HyVista
Spaceborne medium resolution (30 m) hyperspectral					
Hyperion	400–2,500	220	~60 VIS ~30 SWIR	30	NASA
HyspIRI[c]	380–2,500	212	~900 VIS ~400 SWIR	60	NASA
Enmap[d]	420–2,450	249	~500 VIS ~150 SWIR	30	German Consortium

[a] VIS = 550 nm, SWIR = 2,100 nm.
[b] Also used in Carnegie Airborne Observatory beta (CAO-beta) system with lidar sensor [37].
[c] Estimated launch date ≥ 2016; hyspiri.jpl.nasa.gov (accessed on September 2010).
[d] Estimated launch date ≥ 2013; www.enmap.org (accessed on September 2010).

FIGURE 18.1   (**See color insert.**) Image of old-growth forest at the La Selva Biological Station, Costa Rica.

used. Taken as whole, these studies represent the initial steps toward statistically demonstrating that for tree leaves, intraspecific variability can be less than interspecific variability, thereby permitting tree species discrimination.

Recent groundbreaking research from humid TF of Australia [19] provides a framework for linking leaf spectral, biochemical, and structural properties from many species together in a unified analysis, and thereby helps understand the success of these earlier leaf-scale studies. Asner et al. [19] measured sunlit leaf chemistry (nitrogen [N], phosphorous [P], chl-a, chl-b, carotenoids, anthocyanins, and water), structure (SLA), and full-range laboratory spectrometer data for 162 canopy species from sites spanning a lowland to montane climate gradient. A partial least squares (PLS) regression technique was used to estimate leaf chemicals and SLA, as well as to assess the relative contribution of wavelengths (i.e., weightings) from the entire spectrum to the model. A cluster analysis was used to sort related species based on their spectral or chemical signatures. Species were found to have high variation in leaf properties at site and regional scales. Leaf chemicals and SLA were weakly to moderately correlated, except for highly correlated chl-a and chl-b that are functionally linked. In particular, lowland forests had the highest variation in chemicals and SLA among species, attributed to a wider range of chemical and physiological adaptive strategies among species. It was also found that certain families could dominate chemical diversity at any site. The PLS weightings revealed that multiple parts of the spectrum are related to the various leaf biochemical and structural properties, often due to scaling between leaf SLA and a stoichiometric balance of chemical constituents. For example, carotenoids and chl-a and chl-b were important contributors to visible reflectance, which is not surprising given their absorption properties, but also in SWIR where SLA was best predicted, indicating correlations between SLA and pigments. There was weak clustering of species based on their reflectance spectra as species tended to have unique signatures. In particular, clustering was lowest in the lowland site, which matched the trends in chemical diversity. These results indicate that species spectral diversity is linked to chemical diversity in the canopy, a finding that the authors then used to ask if spectral and chemical variation tracks species diversity. With a Monte Carlo simulation, as more species were added to a model "virtual" forest (i.e., more richness), both chemical and spectral diversity increased without saturation across all sites. Spectral–chemical diversity increased most rapidly with inclusion of more species in the lowland forest. Similar patterns of increasing spectral and chemical diversity with species richness were also seen in a similar simulation with leaf properties from 150 species in the Brazilian Amazon [6]. It is important to note that these results are based on leaf-level spectral and chemical/structural analyses, but they indicate that these properties can be linked to species richness. The scaling of these results to actual humid TF has also made progress [5,12,20] and will be discussed in the next section.

All of these studies discussed so far have involved controlled laboratory conditions and do not include the multitude of factors (i.e., variable view angle, poor atmospheric conditions, and radiometric scattering) that are confronted when using airborne or spaceborne hyperspectral data over TF canopies. In addition, these studies included a limited subset of tree species from TF canopies, which may include hundreds of tree species, and they generally ignore other canopy components in TFs that ultimately mix in the radiance received by a hyperspectral sensor.

One characteristic life form in TFs are lianas, or vines, which are woody climbers that use trees for support and form a monolayer of leaves that adds to canopy-level species diversity (Figure 18.2a). In neotropical forests, reports of trees covered with lianas have ranged from 43% to 86% (cited in [21]). Our knowledge of how the spectral–biochemical properties of liana and tree leaves contrast is still limited, yet several recent studies from Central American wet and dry TF have focused intensively on this topic [9,21,22]. One study focused on biochemicals, structure, and VIS–NIR optical properties for liana and tree leaves sampled from canopy cranes at tropical dry (Parque Natural Metropolitano—PNM) and wet forest sites (Fort Sherman—FS) in Panama [21]. Lianas in the dry forest had significantly lower chlorophyll and carotenoid concentrations relative to their host trees, and these differences were detected by spectral indices and the red edge. Lianas had higher water content than trees at both sites, although these differences were not

FIGURE 18.2 **(See color insert.)** (a) Liana (darker green) on a tree crown, (b) flowering *D. panamensis,* and (c) deciduous tree with epiphytes.

detected by water indices. Leaf thickness was low, and SLA was high for lianas relative to trees, but there was no difference in mesophyll air spaces due to high variability between life forms, which led to no differences in NIR reflectance. Supervised parametric and nonparametric classifiers applied to VIS–NIR reflectance spectra [22] could separate tree and liana leaves only at the dry forest site, which was attributed to differences in drought stress and phenological strategies between the two groups under dry season conditions. A follow-up study [9] at PNM measured full-range laboratory reflectance, and using the reduction techniques and classifiers in [22], there was a >96% accuracy in distinguishing tree and liana spectra. Important bands were mostly in the VIS and SWIR regions, attributed to greater differences of chlorophyll and other pigments between lianas and tree leaves, rather than differences in internal leaf structure.

Life cycle changes in trees and liana tissues, or phenology, will also add to intraspecific spectral variation. Important phenological factors to consider are leaf drop, period without leaves, leaf flush, flowering, and fruit maturation. In wetter TF, a relatively constant growing season fosters a range of phenological traits, and leaf turnover and reproductive events may follow annual to irregular cycles, with synchronous to asynchronous timing of events among individuals of a species (cited in [7]). Drier TFs typically have a more characteristic community-wide syndrome, with more synchronized flowering events and many trees in a leaf-off (deciduous) stage in the dry season [30]. As leaves age, they can become more susceptible to epiphyll infestations (e.g., lichens, liverworts, fungi, algae, and bacteria), internal galls, herbivory, and necrosis. Epiphyll infestation can be high. For example, it has been reported to occur on 16% of overstory plants at Barro Colorado Island, a dry forest in Panama, with cover exceeding 50% (cited in [23]). A recent study on Amazonian *caatinga* and *terra firme* forests indicates that as epiphyll cover increases, leaves tend to have lower green-peak reflectance (GP-Refl) and a red edge shifted to longer wavelengths, while leaf NIR transmittance is reduced [23].

Examples of factors that can increase intraspecific spectral variation in tree crowns are shown in Figure 18.3 (photographs) and Figures 18.4 through 18.6 (reflectance graphs), with data from my research at LSBS, Costa Rica [14]. Hyperspectral metrics have been computed from these reflectance spectra to illustrate the power of having contiguous, narrowband data (Table 18.3). These metrics, calculated following methods in [14, adapted from 31 and 32], describe absorption features and scattering properties across the VIS to SWIR, including the center wavelength, depth and width of absorptions in blue, red, NIR, and SWIR, and the wavelength and magnitude of derivative (Mag) inflection points in the blue, yellow, and red edges. The narrowband equivalents of the normalized difference vegetation index (NDVI) and normalized difference water index (NDWI) are also shown in Table 18.3.

The effect of epiphylls on leaf reflectance can be seen for *Hymenolobium mesoamericanum* leaves (Figure 18.4; Table 18.3—HYME), where relative to an epiphyll-free leaf (Figure 18.3a), there is lower GP-Refl and less steep yellow and red edges (YE-Mag and RE-Mag) for a leaf completely covered with a lichen (Figure 18.3b). Herbivory affected 4% of leaves from canopy-emergent trees sampled at LSBS. As seen for *Lecythis ampla* leaves (Figure 18.3e and f; Table 18.3—LEAM), a leaf miner insect removes photosynthetic pigments, thereby reducing blue and red absorption depth and width (Blue-D,W and Red-D,W) and increasing GP-Refl; SWIR reflectance increases due to lower

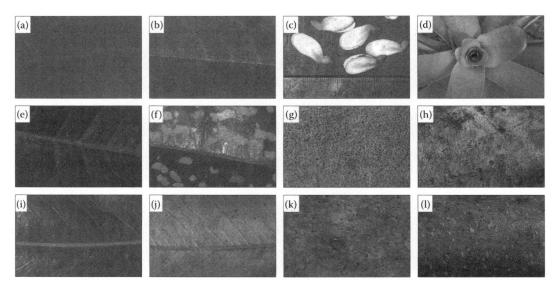

FIGURE 18.3 **(See color insert.)** (a) *H. mesoamericanum* leaf without surface lichen, (b) *H. mesoamericanum* leaf without surface lichen, (c) *D. panamensis* flowers, (d) canopy epiphytic bromeliad, (e) *L. ampla* without herbivory, (f) *L. ampla* with herbivory, (g) *B. elegans* trunk bark, (h) *B. elegans* branch bark, (i) *D. panamensis* mature leaf, (j) *D. panamensis* senescing leaf, (k) *D. panamensis* trunk bark, and (l) *D. panamensis* branch bark.

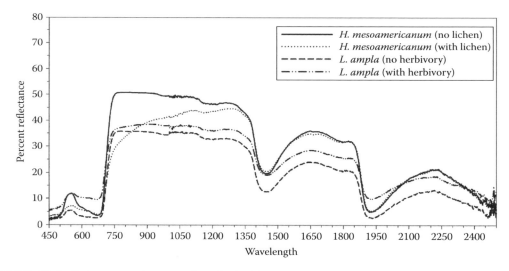

FIGURE 18.4 Laboratory reflectance for a leaf with and without cover by a surface lichen and herbivory from the trees *Hymenolobium mesoamericanum* and *Lecythis ampla*. The spectral data in Figures 18.4 through 18.7 are from an ASD FieldSpec spectrometer (Table 18.2).

water content and more exposed, dry leaf structure, possibly allowing deeper expression of SWIR features at near 1762 and 2301 nm (Figure 18.4; Table 18.3—SWIR1,3-D). Leaves from *Dipteryx panamensis* reveal that as a leaf senesces (Figures 18.3i and j and 18.5; Table 18.3—DIPA), lower concentrations of chlorophyll reduce blue and red absorption depth (Blue-D and Red-D), increase the GP-Refl, shift the blue and red edge inflection points to lower wavelengths (BE-λ and RE-λ) while shift the yellow edge to longer wavelengths (YE-λ). Canopy tree flowers can be very conspicuous and possibly help discriminate species or detect richness at canopy scales. For example, canopy-emergent *Dipteryx panamensis* trees in Costa Rica have conspicuous purple–pink

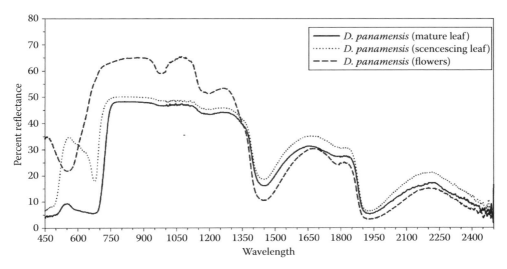

FIGURE 18.5 Laboratory reflectance for flowers and a mature and senescing leaf from the tree *Dipteryx panamensis.*

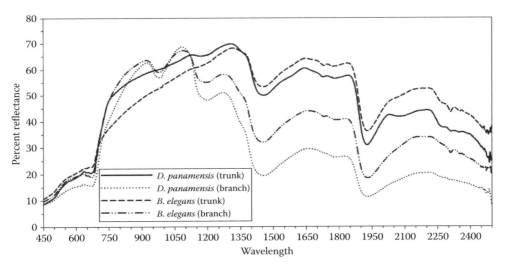

FIGURE 18.6 Laboratory bark reflectance for two tree species at La Selva, *Dipteryx panamensis* and *Balizia elegans.*

flowers (Figures 18.2b and 18.3c) that can be accurately identified with automated processing of color aerial photographs [33], and have distinct hyperspectral reflectance properties (Figure 18.5; Table 18.3—DIPA), such as absence of chlorophyll red absorption, low NDVI, high NIR reflectance, and deeply expressed NIR water absorption (NDWI and NIR1,2-D). Other canopy components such as understory vegetation, bark, and canopy epiphytes may have considerable influence on canopy spectral variation when trees have lower to no LAI due to seasonal phenological cycles (e.g., deciduousness and leaf exchange) or stress [7,14,34]. An example of a tree at LSBS without leaves is shown in Figure 18.2c. Note that most of the exposed bark is on the trunk and outer branches, while larger lateral branches are covered with epiphytes. This is not atypical for wet TF. Reflectance data from LSBS show that tree bark on branches and trunks can have considerable intraspecific reflectance variation [14]. Branches taken from the upper canopy of younger trees (Figure 18.3h and l) show photosynthetic pigment absorption, seen as relatively deep red (Red-D) and NIR (NIR1,2-D)

TABLE 18.3

Hyperspectral Metrics for Components in Figures 18.4 through 18.7

Metric	HYME No Lichen	HYME Lichen	LEAM No Herbivory	LEAM Herbivory	DIPA Green Leaf	DIPA Senesced Leaf	DIPA Flowers	DIPA Trunk	DIPA Branch	Field Epiphyte	Field PEMA
NDVI	0.86	0.79	0.85	0.58	0.79	0.47	0.06	0.42	0.51	0.86	0.93
NDWI	0.04	−0.08	0.04	0.02	0.05	0.04	0.10	−0.09	0.08	0.09	0.03
Blue-λ	492	496	502	503	505	496	528	499	494	493	498
Blue-D	54.1	23.9	34.4	22.5	27.0	50.0	3.0	9.5	5.1	55.7	59.5
Blue-W	51.4	36.3	33.3	46.2	46.9	46.3	23.6	45.0	42.5	48.5	50.0
BE-λ	521	520	523	522	524	520	n/a	520	519	521	523
BE-Mag	0.3	0.1	0.1	0.2	0.2	0.8	n/a	0.1	0.1	0.3	0.1
GP-λ	549	553	548	554	550	560	553	n/a	n/a	555	552
GP-Refl	11.9	7.1	5.4	12.0	9.3	34.7	21.8	7.1	7.1	11.2	5.1
YE-λ	569	570	568	571	570	576	n/a	574	573	572	570
YE-Mag	−0.20	−0.06	−0.09	−0.08	−0.10	−0.12	n/a	0.03	0.03	−0.13	−0.09
Red-Wvl	678	679	674	679	682	678	n/a	676	677	677	676
Red-D	88.2	78.9	86.7	60.3	80.6	57.3	n/a	39.3	45.5	88.4	93.9
Red-W	104.6	98.8	111.2	90.1	112.4	49.8	n/a	72.5	81.5	88.1	114.6
RE-λ	710	708	711	704	724	688	n/a	694	702	702	720

RE-Mag	1.108	0.578	0.781	0.649	0.954	0.982	n/a	0.483	0.536	0.726	0.621
NIR1-λ	1018	1037	996	1015	990	1007	980	996	989	977	965
NIR1-D	2.1	1.2	2.8	4.3	2.1	2.7	9.2	0.9	12.3	13.7	5.5
NIR1-W	50.5	3.9	48.0	14.1	101.9	41.1	59.4	40.9	70.7	70.3	75.4
NIR2-λ	1158	1158	1168	1178	1167	1170	1162	1193	1167	1155	1159
NIR2-D	3.3	2.3	4.3	3.2	4.3	3.4	12.3	2.2	17.3	18.3	10.2
NIR2-W	82.3	72.5	70.1	78.4	77.0	70.7	78.3	73.2	83.1	80.2	78.9
SWIR1-λ	1770	1768	1776	1762	1770	1776	n/a	1725	1771	1777	1776
SWIR1-D	0.2	0.4	0.1	0.5	0.5	0.2	n/a	1.8	0.4	0.0	0.5
SWIR1-W	5.5	14.3	1.5	9.6	19.5	1.9	n/a	25.4	5.3	0.7	3.8
SWIR2-λ	n/a	2215	2202	n/a	2217	2047	n/a	2093	n/a	2049	n/a
SWIR2-D	n/a	0.7	0.0	n/a	0.4	0.1	n/a	2.0	n/a	2.3	n/a
SWIR2-W	n/a	3.7	0.0	n/a	1.8	3.1	n/a	62.9	n/a	3.9	n/a
SWIR3-λ	2354	2303	2321	2301	2344	2338	2350	2266	2309	2257	2300
SWIR3-D	5.2	2.7	5.8	6.7	8.1	3.4	0.7	6.3	3.4	2.4	6.4
SWIR3-W	7.1	47.4	29.4	14.6	33.9	7.8	15.3	86.3	22.5	109.1	60.6

Refl, % reflectance; λ, wavelength (nm); D, depth of absorption feature (% reflectance); W, width of absorption feature (nm); n/a, absorption feature faint or nonexistent; Mag, magnitude of the derivative.

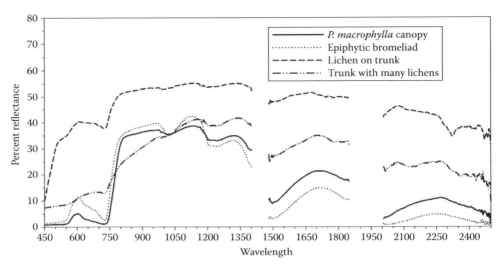

FIGURE 18.7 Field reflectance spectra of canopy components acquired from a bridge at La Selva.

absorption features, related to chlorophyll and water, respectively (Figure 18.6; Table 18.3—DIPA). In contrast, bark from old-growth tree trunks (Figure 18.3g and k), taken lower in the canopy where branches split from the main bole, have shallower red and NIR absorption features, red edge shifted toward longer wavelengths (RE-λ), and deeper expression of SWIR features (SWIR1,2,3-D). Examples of field reflectance spectra of canopy components taken from a suspension bridge at LSBS are shown in Figure 18.7. Lichens tend to have high reflectance throughout the spectrum, with a deep absorption at 2300 nm possibly related to protein nitrogen. These surface organisms can add to tree bark intraspecific spectral variability. An example canopy spectrum for the tree *Pentaclethra mac-rophylla* is shown as a contrast to a spectrum from an epiphytic tank bromeliad (Figure 18.3d—note contains water), which has higher GP-Refl, steeper yellow edge (YE-Mag), and deeper NIR water absorption and higher NDWI than *Pentaclethra* (Figure 18.7; Table 18.3—Epiphyte, PEMA).

18.2.2 PIXEL TO CANOPY SCALES

Operational mapping of individual tree species or overall species diversity in TF canopies using airborne or spaceborne hyperspectral sensors is a grand challenge for remote sensing. As in all remote sensing, moving from the controlled laboratory environment to the canopy scale introduces additional noise to the radiance signal, such as atmospheric moisture absorption and poorer radiometric calibration. Beyond the linkages between leaf reflectance and biochemical–structural properties, canopy reflectance is heavily influenced by species-level differences in crown biophysical structure, which determines the level of 3D scattering of photons among the various tissues within the canopy, including leaves, but also bark, flowers, fruits, lianas, and epiphytes.

Much of our understanding of how tissue-scale biochemical/structure is expressed in canopy-scale reflectance is through radiative transfer (RT) models [3,6,12,20,26]. Canopy-scale reflectance is best understood with the concept of effective photon penetration depth (EPPD, Figure 18.8), which is the canopy depth to which a downward (nadir) viewing hyperspectral sensor is most sensitive—conceptualized as the number of leaf layers (reported in LAI units) from which the sensor can detect biochemicals due to wavelength-specific absorption and scattering among canopy components [6]. In VIS, strong absorption by leaf pigments translates into a very short path length of photons scattering within the canopy, especially in blue and red (Figure 18.8, EPPD LAI < 2). In contrast, photons are heavily scattered in NIR, creating increasing reflectance as canopy LAI increases until the "NIR plateau" is saturated, yielding an EPPD LAI < 4 (Figure 18.8); however, water absorption features at 980 and 1190 nm will continue to broaden and deepen with increasing canopy LAI, making these features highly sensitive

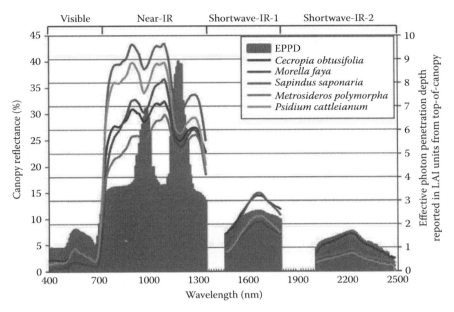

FIGURE 18.8 Effective photon penetration depth (EPPD) of a tropical forest canopy in Hawaii. This EPPD estimate was derived by combining field measurements of leaf hyperspectral optical properties, canopy shape, architectural data, and LAI in a three-dimensional canopy radiative transfer model. EPPD was calculated as the depth from the top of the canopy to which photons no longer contribute to the top-of-canopy reflectance. (From Asner, G.P., Hyperspectral remote sensing of canopy chemistry, physiology and biodiversity in tropical rainforests. In: *Hyperspectral Remote Sensing of Tropical and Sub-Tropical Forests*, Kalacska, M. and Sánchez-Azofeifa, G.A. (Eds.), CRC Press, Boca Raton, FL, pp. 261–296, 2008. With permission.)

to canopy water content deep into the canopy (Figure 18.8, EPPD LAI < 8). In SWIR, the photon path length is relatively shallow due to strong water absorption, and so this region returns mostly information from the upper canopy (Figure 18.8, EPPD LAI < 3).

In a recent study, Asner and Martin [20] modeled canopy reflectance using their leaf reflectance and transmission spectra from Queensland, Australia [19], with a goal of estimating biochemicals and SLA at leaf and canopy scales with PLS regression. As reviewed in the last section, analysis of this leaf-scale dataset demonstrated a link between species-level spectral variation and inherent biochemical–structural diversity [19]. The RT model considered an area at the pixel scale (<3 m), not an entire crown, and included several structural parameters, the most important of which was LAI. The model also considered variable sensor and view zenith and azimuth angles, but not intracrown gaps and shadows, crown density, nor spectral properties from other canopy tissues (e.g., bark and lianas). Foliar chlorophyll, carotenoids, and SLA were highly correlated with leaf reflectance measured in the laboratory (r = 0.89–0.91), with higher PLS regression weights for VIS wavelengths for pigments and NIR to SWIR wavelengths for SLA. Foliar N, P, and water were moderately correlated (r = 0.79–0.85). For N, high weights were for VIS and SWIR wavelengths, associated with chlorophyll and proteins, respectively, while for water higher weights were for SWIR wavelengths. The correlation of P with reflectance was likely due to a stoichiometric link with N, rather than expression of absorption features. At the canopy scale, with the RT model held constant at a crown LAI of 5.0 (an average level for humid TF), correlations of reflectance and pigments, N, P, and SLA were 3%–4% higher, and water was 18% higher than when using leaf-scale reflectance, attesting to the power of the canopy to amplify leaf biochemical and structural properties. In other simulations, pigments and SLA remained highly correlated to canopy reflectance with variable LAI conditions (LAI 1.5–8.0), which can be attributed to low EPPD in VIS (Figure 18.8), and subsequent insensitivity to variation in upper-canopy leaf layering [19,20]. In contrast, foliar water, and particularly N and P, had less predictive capacity with increasing variation in modeled LAI. These chemicals (especially water) are strongly expressed in the NIR, the region with the highest EPPD due to volumetric scattering (Figure 18.8); thus, variations in crown LAI at greater depths will lead to more variable expression of chemical absorptions, thereby weakening predictive models.

Actual airborne hyperspectral data were used by Asner et al. [5] to explore the spectral, biochemical, and structural differences of native, introduced and invasive, N-fixing and nonfixing tree species on Hawai'i island. The sensor used was the airborne visible/infrared imaging spectrometer (AVIRIS), which has very high fidelity (i.e., high signal-to-noise ratio, SNR; Table 18.2). Groups of trees were found to be spectrally unique for natives vs. introduced, N-fixers vs. nonfixers, and introduced vs. invasive. These differences between groups in reflectance and derivative spectra were related to variation in leaf pigments, water, N, P, and SLA and canopy LAI, which caused reflectance and absorption features at both specific and multiple spectral regions. For example, PLS regression weightings showed that N and SLA were well predicted by SWIR wavelengths. Pigments heavily influenced VIS reflectance (e.g., 670 nm of chlorophyll absorption), but also NIR and SWIR reflectance due to an indirect relationship with SLA as all are expressed on a per area basis. The main message of this paper was that no single band or spectral region, nor underlying biochemical or structural property, could distinguish the groups of species, and so full-range measurements are needed to separate groups; in this case, PLS regression was used to analyze the full spectrum simultaneously to target the important predictive features.

Can linkages between canopy reflectance and leaf biochemistry and structural properties help us achieve species mapping with an airborne or spaceborne hyperspectral sensor? There are two important studies from Hawai'i that show that this may be possible. In a pioneering study, Asner et al. [13] took the evidence that invasive and native trees have unique hyperspectral properties [5], and sought to map the fractional abundance of representative species from these groups using AVIRIS imagery (spatial resolution: 3 m) acquired in tandem with small-footprint lidar. Lidar was used to mask forest gaps, intra- and intercrown shadows, and low vegetation in the hyperspectral imagery, while spectral mixture analysis (SMA) based on SWIR was used to mask nonphotosynthetic vegetation. The remaining pixels in the imagery thus had the highest information content. Full-range spectra from two native and three invasive species were used as endmembers in another SMA to map the per-pixel, fractional abundance of each species, with a <7% error rate in detection of invasive species for pixels with 75% cover.

In another Hawai'i study, Carlson et al. [12] sought to link leaf biochemistry to woody species richness through AVIRIS hyperspectral data (spatial resolution: 3.3–3.6 m). Woody species richness was measured from plots in 17 lowland TF sites across the Hawaiian islands. Leaf biochemicals (total chlorophyll, water, N) sampled from species in plots were used in a Monte Carlo simulation to show that as species were progressively added to a model forest community, biochemical diversity also had a nonlinear increase without saturation—a result later confirmed with a much larger dataset from Australia, reviewed in Section 18.2.1 [19]. Carlson et al. [12] next set out to map per-pixel woody species richness from AVIRIS imagery. The range of reflectance and reflectance derivative were highly correlated with plot-based species richness across many wavelengths, but the derivative had fewer inter-correlated bands and so was selected for modeling. Linear regression was used to find the best wavelength in each of four regions determined to add to biochemical diversity: green edge (500–550 nm) and red edge (700–800 nm) affected by total chlorophyll, the 1140–1250 nm feature affected by water absorption, and the 1500–1700 nm feature affected by water and N absorption. The final regression model with four optimal derivative bands had a strong linear relationship to species richness ($r^2 = 0.85$, $p < 0.01$). The greatest predictor of species richness was a SWIR band (1525 nm), although the bands from other regions helped strengthen the model. It should be noted that with 44 species sampled across all sites, these Hawaiian forests are relatively species poor compared to other humid TF, and so we do not yet know how the findings and techniques in this study will scale to other sites.

Tree phenology also has an important role in determining canopy reflectance. For one, structural properties of upper-canopy tree or liana tissues will vary through time—older leaves may accumulate epiphylls, and at certain times of the year, there will be senesced leaves, flowers, and fruits (see Section 18.2.1). Little is known about how these fine-scale tissues will change the crown photon scattering environment and alter the biochemical properties that contribute to overall radiance measured by the sensor.

Toomey et al. [23] used a GeoSAIL RT model to investigate the effect of epiphylls on canopy reflectance. Amazonian *caatinga* forests had a 23%–35% decrease in canopy NIR reflectance and 11%–20% decrease in green reflectance with infestations at 50%–100% leaf area cover, while *terra firme* forests had lower infestation rates and exhibited a 6%–11% decrease in NIR and 5%–10% decrease in green reflectance, respectively. Although only seven species were investigated in their simulations, there were differences in epiphyll effects among species related to factors such as leaf longevity and differences in transmittance [23].

If crown LAI is low due to crown leaf drop (e.g., deciduousness) or general crown architecture (e.g., sparse leaves and branches), photons can penetrate deeper into the canopy, and the biochemical–structural properties of bark, canopy components (e.g., epiphytes and lianas), and understory vegetation can contribute to canopy-scale reflectance [7,9,14]. Clark [14] and Bohlman [34], working in tropical wet (LSBS, Costa Rica) and dry (PNM, Panama) forests, respectively, explored SMA applied to hyperspectral imagery (spatial resolution <1.6 m) to estimate the relative fractional contribution of nonphotosynthetic tissues to canopy reflectance. Both studies used data from the HYperspectral Digital Imagery Collection (HYDICE; Table 18.2) sensor acquired in March 1998 during the dry season. These data are some of the earliest hyperspatial, hyperspectral imagery available from humid TFs for research use. Pixels were unmixed to provide fractional abundance of green vegetation (GV: leaves from trees, lianas, epiphytes, and understory vegetation), nonphotosynthetic vegetation (NPV: bark), and shade. At LSBS, Clark [14] investigated the same seven species analyzed at tissue scales [7] using pixel- and crown-scale reflectance spectra from HYDICE. Mean values of SMA fractions from pixels within ITCs across the seven species were roughly 40% GV, 15% NPV, and 44% shade. Canopies of deciduous tree species (*D. panamensis* and *L. ampla*) had relatively high fractions of NPV and low fractions of GV (Figures 18.9 and 18.10). In contrast, leaf-on, broadleaf species (*Ceiba pentandra, Hyeronima alchorneoides,* and *Terminalia oblonga*) had relatively high fractions of GV and low fractions of NPV (Figures 18.9 and 18.10). There was variability in the crown-level proportions of GV, NPV, and shade among individuals of the same

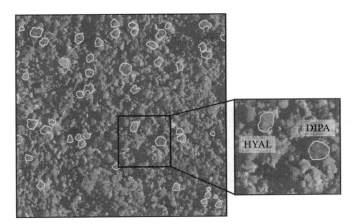

FIGURE 18.9 Fractional abundance of green vegetation (green), non-photosynthetic vegetation (red), and photometric shade (blue) from a spectral mixture analysis. Individual tree crowns delineated with visual interpretation [7] are outlined in yellow. The crowns for *Dipteryx panamensis* (DIPA) and *Hyeronima alchorneoides* (HYAL) are shown in the zoom inset.

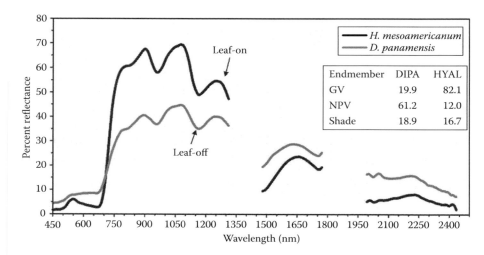

FIGURE 18.10 A reflectance spectrum and associated spectral mixture analysis (SMA) fractions for a single pixel from the *D. panamensis* (DIPA) and *H. alchorneoides* (HYAL) crowns delineated in Figure 18.9. SMA fractions are GV: green vegetation, NPV: non-photosynthetic vegetation, and shade: photometric shade. These data are from the HYDICE airborne sensor (Table 18.2).

species (i.e., intraspecific variability). For example, the tree species *B. elegans* and *H. mesoamericanum* were fully leaved, yet they had fine compound leaves and sparse branching architecture; as a consequence, some individuals of these species had NPV and GV fractions similar to deciduous crowns, while other individuals had fractions that resembled the broadleaf, leaf-on species. At the drier PNM forest site, which had more distinct tree leaf phenology and fewer species, Bohlman [34] used SMA at pixel (1 m) to canopy scales (30 m) using image GV, NPV, and shade endmembers. For ITCs, the percentage of GV increased and percentage of NPV decreased with higher levels of deciduousness, as measured in the field. The model also accurately determined the percent cover of nondeciduous (GV, 53%), deciduous (NPV, 32%), and intercrown shade/shadow (shade, 15%) over a 5 ha test area at these three scales. Deciduous species had a dominant NPV fraction, yet there was still GV cover (<40%), attributed to lianas and exposed understory vegetation. There was also variation in GV and NPV fractions for leaf-on species. For example, *Luehea seemannii* had relatively high NPV despite being fully leaved, possibly due to lower LAI and the presence of senescing leaves and high load of upper-canopy, woody seeds that had spectral properties of NPV. Taken together, these studies at LSBS and PNM show that TF canopies should not be viewed as a continuous blanket of green leaves with randomly mixed structure. Tree phenology and site characteristics will lead to important species-level variation in the relative exposure of differently aged leaves, bark, and other canopy components that form the crown's reflectance signature; these factors vary through time and even within an individual crown, and may provide an important signal for species discrimination that goes beyond just leaf-scale spectral–biochemical–structural properties.

As discussed in the last section, TFs are particularly influenced by lianas and epiphytes, which can contribute to both canopy species richness and overall spectral diversity. Zhang et al. [35], working with HYDICE data at the LSBS, found that pixel-scale reflectance from a single *T. oblonga* crown with heavy epiphyte load had distinct spectral properties than other individuals of the species, thereby contributing to higher intraspecific spectral variation. Kalacska et al. [9] used HYDICE imagery at the PNM dry forest to scale their hyperspectral discrimination of liana and trees from the leaves to crowns. Tree crowns with <40% liana cover could be separated from those with ≥40% liana cover with 93% overall accuracy, indicating that lianas have a profound impact on ITC spectral response, which may be independent of the tree's species. Top wavelengths for discriminating crowns from lianas were in SWIR (>2000 nm), which possibly indicates that lianas affect the water absorption characteristics of the upper canopy. Leaf-level analysis revealed that lianas tend to have more water than trees [21], especially in the dry season when the image was acquired; and in this season, reflectance from deciduous crowns may come mostly from lianas, which tend to lose leaves after host leaves drop or retain them through the year.

18.3   MAPPING CANOPY SPECIES OVER BROAD SPATIAL SCALES

The operational mapping of individual species or species diversity of TF canopy plants is only in the initial stages of development. Hyperspatial, hyperspectral data with high fidelity are now available from commercial (e.g., HyMap [www.hyvista.com], CASI [www.itres.com], and SpecTIR [www.spectir.com]), and experimental (e.g., AVIRIS [36], Carnegie Airborne Observatory, CAO [37]) airborne sensors (Table 18.2), yet to date there have been relatively few published studies using these datasets for mapping canopy species of humid TF, with most work in Hawai'i [3,5,12,13]—a site with relatively low species diversity.

Hyperion is the only spaceborne sensor offering publically accessible hyperspectral data over the large spatial extents needed for regional species mapping; however, the sensor has relatively low fidelity (Table 18.2), and the 30 m resolution of the imagery makes it difficult to identify pure spectra from smaller tree crowns. To date, there has been one attempt to use Hyperion for species-level mapping in humid TF. Working in the Peruvian Amazon, Papeş et al. [38] used hyperspatial imagery from the Quickbird satellite to identify Hyperion pixels that were >50% covered by the crown of a single study species (called the "clean" dataset) relative to those pixels that included more spectral mixtures from multiple crowns. Five genera were then classified with an LDA classifier with 5–25 optimally selected bands, with wet and dry season images. The "clean" dataset achieved 100% overall accuracy with 25 bands for both seasons, while the raw dataset never reached more than 50% overall accuracy. This result indicates that when the crown from a single species covers a large portion of a moderate resolution pixel, its spectral response may be dominant enough for accurate pixel-scale classification. However, in practice, this study's landscape-scale maps of study genera were difficult to assess, accuracy was not quantified, and some distributions of taxa were obviously related to sensor noise. Given the high species diversity and range of crown sizes in humid TFs, it is inevitable that pixels will be spectrally mixed with medium-resolution pixels, even when screened with hyperspatial imagery. Although not yet tested, a subpixel multiple-endmember spectral mixture analysis (MESMA)/Auto-MCU-S [13,14,39] approach that maps fractional abundance of species within a pixel may be more fruitful.

As seen in [38], ITCs in hyperspatial imagery can be resolved as groups of image pixels, or objects, especially for larger crowns (e.g., canopy emergents) [7,14,15,40–44]. This object-based perspective may permit the mapping of species at a finer spatial scale than offered by medium-resolution sensors, especially when hyperspectral data are available from the objects [7,14]. Operational mapping of ITCs to species could allow broadscale delineation of community types [41,43], measures of species richness, and monitoring of important keystone, endemic, rare or commercial tree species [1,7,33]. As found in [38], panchromatic or multispectral hyperspatial imagery (e.g., Quickbird, IKONOS) is also useful for masking pixels from medium-resolution spaceborne hyperspectral imagery (e.g., Hyperion) that have many small crowns, and are thus more spectrally mixed—a step which can greatly improve classification accuracy for the remaining pixels.

There are two important steps in the automated mapping of ITC species with hyperspectral data and an object-based perspective: (1) location and delineation of canopy objects, that is, image segmentation, and (2) classification of species based on predictor variables, that is, spectral reflectance, metrics, texture. Both of these steps are areas of active research, and few techniques have been applied to hyperspectral imagery from TFs.

18.3.1 Automated Delineation of Individual Tree Crowns or Crown Clusters

Image segmentation can be accomplished with manual digitization [7,40]; however, for consistency and reduction of production costs, object-based species inventory over large spatial extents will require automated algorithms for crown detection and delineation. Algorithms are reviewed more extensively in [41,42,44]. Techniques generally involve using panchromatic bands or collapsing spectral data into one illumination/albedo band to form a "hillscape," where radiometric maxima are from upper-crown areas, or "hills," and radiometric minima are assumed to be intercrown shadows, or "valleys." Algorithms specifically tailored to tree crown delineation include valley following, bright point expansion, and template matching [41,42,44]. High spatial resolution, multispectral to panchromatic imagery are generally used to develop techniques, and most algorithms have been designed for forests with fewer species and more regular crown structure (e.g., conifers) than found in subtropical to TFs.

A novel study with multilayered, mixed species subtropical woodland in Australia had more success delineating crowns with the popular eCognition image segmentation and classification software (www.ecognition.com) than with existing crown delineation techniques [42]. The technique used multispectral CASI imagery (14 bands, 446–840 nm) to first segment imagery into forest/no forest, then worked on forest types for iterative delineation and splitting of crown clusters using both spectral and shape properties. Characteristics of the red edge were found to highlight illuminated foliage and help separate crowns from the open understory [41,42]. Accuracies in delineation were >70% within scenes with sparse trees, but accuracy dropped below 50% for forests with dense, multilayered trees. All automated crown delineation studies indicate that TFs are the ultimate challenge, as trees may intertwine and overlap, intercrown shadows are narrow, intracrown shadows are prevalent, and lianas connect multiple crowns, causing segmentation schemes to identify clusters of trees rather than ITCs. Working in wet tropical forests, Palace et al. [44] designed a hybrid algorithm that performs an iterative local maximum filtering (e.g., finding the bright apex of tree) and local minima-finding (e.g., crown edges) radiating in transects from the apex. When applied to IKONOS panchromatic data in forest stands in the Brazilian Amazon, crown widths estimated from the technique were within 3% of field measurements and estimates were better than manual crown delineation [40,44]. The algorithm was applied to Quickbird imagery over a moist tropical semideciduous forest of lowland Bolivia and analyzed at the level of individual trees rather than stands [8]. Emergent tree crowns were generally segmented into smaller crowns, thus appearing as multiple subcrowns. There is still much work needed to refine crown delineation algorithms for TFs, and this will require a greater understanding of tree and liana diversity, incorporation of spectral data to differentiate adjacent tree crowns, and possibly inclusion of biophysical information from small-footprint lidar sensors [8,13,15,37].

18.3.2 Classification Schemes

An object-based approach to mapping individual canopy tree species allows considerable flexibility in choosing a species-level classification scheme. There are many (sometimes hundreds) of pixels within crowns, allowing selection of well-lit pixels with potentially more biochemical–structural signal [7,13,41,43]. Reflectance spectra can be analyzed at the pixel scale or averaged to form crown-scale spectra, and bands can be smoothed with filters to remove noise [7,22,35], transformed to derivatives to reduce the effect of illumination and highlight spectral shape [12,16,22,35], and processed with various data reduction techniques that seek to isolate spectral features that maximize species discrimination and reduce error. In tropical species and life-form classification applications using field spectrometers or hyperspectral sensors, data reduction or "feature selection" techniques have included PCA [9,22], wavelet transforms [9,22,35], stepwise band selection [7,17,18], narrowband indices [14,18,24,43], and hyperspectral metrics based on derivatives, absorption fitting and SMA [14]. In all of these applications, predictor variables are used in supervised classifiers, including spectral angle mapper (SAM) [7,16], MESMA [14], maximum likelihood (ML) [7], LDA [7,14,15,18,38,43], decision trees (DTs) [9,14,15,17,22,24], and neural networks, log linear, quadratic, and k-nearest neighbor [9,17,22,24]. These studies have focused on discriminating different sets of tree and liana species, with hyperspectral data from leaf to crown scales acquired at different seasons and forest types, in concert with multiple data reduction/feature selection techniques—making intercomparison and generalization of findings difficult; however, a recent in-depth review of these applications with a focus on classifiers and feature selection is available in [45].

My research in the species-rich, tropical rain forests of LSBS, Costa Rica using airborne HYDICE reflectance data (437–2434 nm, 161 bands, 1.6 m) provides a unique exploration of different techniques for automated, object-based ITC species discrimination, including ML, SAM, LDA, DT, and MESMA classifiers [7,14]. The hyperspectral data were smoothed with a filter and included the full suite of factors that compound operational species mapping, such as variable illumination and viewing geometry, noise from atmospheric moisture and spectral mixing with nontarget canopy (e.g., lianas, epiphytes) and understory components. As the focus was on species classification with hyperspectral techniques, I bypassed experimenting with automated crown delineation by manually digitizing 214 canopy-emergent ITCs from seven species.

In a first analysis [7], all reflectance bands were used to classify species using SAM, ML, and LDA, in combination with a stepwise band selection procedure. Classifiers were applied to all and sunlit pixel spectra within ITCs and to crown-scale spectra from averaged pixel spectra. ITCs were also classified using a pixel-majority approach in which a species label was assigned according to the majority class of classified pixels within the crown. Overall classification accuracy decreased from leaf scales measured in the laboratory (reviewed in Section 18.2.1) to pixel and crown scales measured from the airborne sensor, but reasonable accuracy was still achieved. The highest ITC classification accuracies for crown-scale spectra and pixel-majority techniques were 92% and 86%, respectively, with LDA and 30 bands. The optimal bands for classifying species at pixel and crown scales were concentrated in the yellow edge to red edge, around NIR water absorption features and throughout the SWIR, indicating that the full-range data are important to identify species at these scales. Furthermore, classifications were significantly more accurate with 10 narrow bands from hyperspectral data relative to using simulated multispectral, broadband data.

In a second analysis [14], classification variables included hyperspectral metrics that responded to crown structure and absorption features from photosynthetic pigments, water, and other biochemicals. The metrics included narrowband indices, derivative-based metrics, absorption-fitting metrics, and SMA fractions (GV, NPV, and shade). Differences in spectral metrics among species at pixel and crown scales were largely dependent on tree leaf phenology and structure, which controlled the relative amounts of leaf and bark tissues within a crown. For example, leaf-off trees had fewer canopy leaves and less overall canopy water, and so those species had lower values of water indices and higher values of metrics responding to overlapping SWIR absorption features. A DT classifier was used to discriminate tree species using crown-scale spectra and pixel spectra with a majority vote, an approach similar to [7]. Some of the most heavily used metrics were related to water absorption in NIR and other nonpigment features in SWIR. The best classification scheme was with crown-scale metrics and had 70.1% overall accuracy, which was lower than with LDA and reflectance bands from the first analysis [7]. However, hyperspectral metrics and the DT classifier were instructive for identifying key spectral reflectance properties for tropical tree species discrimination, which helped establish linkages to underlying biochemical properties in tissues that warrant analysis with chemical assays. A third analysis [14] with MESMA [39] as a classifier only reached accuracies of LDA and reflectance bands [7] when hundreds of endmembers per species were included, similar to the MESMA approach in [13] (i.e., Auto-MCU-S); however, I considered my result to be too optimistic given model endmembers came from the same crowns being classified.

Crown structure metrics derived from small-footprint lidar were also analyzed in concert with reflectance bands and hyperspectral metrics in classifying ITC species, with 6 target and 18 "other" species grouped together [14,15]. There were significant differences in the majority of lidar-derived metrics among the six target species, indicating that species have unique crown structural properties. Crown leaf cover, especially in deciduous leaf-off trees, was the primary factor controlling variation in lidar metrics. Both DT and LDA classifiers were compared for classifying the species of ITCs with lidar and hyperspectral metrics and reflectance bands. As with [7], the best classifier was stepwise selection LDA applied to reflectance bands, with an overall accuracy of 88.9% and 81.5% when classifying six target species alone or with the "other" species class, respectively. The addition of lidar-derived structure information to the classifier did not improve overall species classification accuracy, highlighting the importance of hyperspectral data for tropical tree species discrimination.

In summary, important findings from these series of analyses at LSBS showed that (1) elaborate methods and labor involved in implementing the DT classifier with hyperspectral metrics or the MESMA classifier with optimal endmembers did not translate into improved species classification accuracy relative to using reflectance bands with a more traditional LDA classifier; (2) crown-scale spectra provided more accurate ITC species classification, with both LDA and DT classifiers, than the pixel-majority approach; (3) there were no major improvements in accuracy by isolating sunlit pixels in crowns, suggesting that image spatial resolution can be coarse, as long as it avoids many mixed pixels and can isolate crowns (see [38]); (4) full-spectrum data provided optimal species discrimination; (5) an advantage of hyperspectral data is they are over-sampled, and information that does not optimize separation among species can be discarded; and (6) crown-scale LAI and its changes with leaf phenology are important considerations in operational mapping of tree species, as they altered the volume-scattering and spectral mixing properties at pixel to crown scales. For further details of "feature" extraction methods, please refer to Chapter 4.

18.4   CONCLUSIONS AND FUTURE CHALLENGES

The operational mapping of individual species and canopy diversity in TFs of the world from remote sensing is still a future goal as we are in the initial stages of this research frontier. This chapter reviewed a growing body of evidence—most of it published in the last 5 years—on the advances made and promise shown by hyperspectral sensors in mapping species in TFs.

Half of the research reviewed here (Table 18.1) worked with leaf spectral reflectance from tree and liana species in neotropical and Australian forests, with the goal of establishing statistical separation among species with data reduction, selection, and classification techniques [7,9,14, 16–18,22,24] and connecting spectral differences to their underlying generating mechanisms—leaf biochemical and structural properties [6,17,21,24]. Although limited in breadth of sites and species, these analyses have laid the foundation of basic principles, showing that species have unique biochemical and structural properties at the tissue scale that translate into species-level spectral diversity across the entire VIS to SWIR range (i.e., full range); and, these spectral properties, when expressed at the canopy scale, have potential for discriminating individual species or species diversity despite the myriad of factors that cause intraspecific variation. Some of the most important advances in understanding the link of spectral response to biochemistry and structure have been accomplished with high-fidelity, full-range hyperspectral data from laboratory spectrometers, taken in conjunction with foliar chemical assays and physical measurements [6,17,19,20]. These types of analyses are critical to replicate at other field sites in order to cover the extremely broad range of site conditions and species diversity that exist in the TFs. Our understanding of how leaf spectral properties scale to canopies, where the time-varying, volumetric scattering of photons is critically important, has been advanced with RT modeling [6,20,23,26]. Future RT modeling of TF canopies should incorporate additional spectral (e.g., bark, epiphytes, and lianas) and temporal diversity (e.g., older leaves with epiphylls and herbivory, deciduousness) to better understand variability in canopy-scale reflectance. It is difficult to undertake these types of research with limited funding, and field sites must have adequate access (e.g., roads, rivers, trails, sampling, and equipment permits), logistical support (e.g., transport, supplies, botanists, climbers/shooters, and field assistants), and infrastructure (e.g., shelter, electricity, and canopy access). There are few TF sites that meet these requirements, thus partially explaining the limited sampling and analysis that has been achieved to date.

In order to map canopy plant species across soil, climate and diversity gradients, and track changes through time, we will need more hyperspectral datasets from airborne or spaceborne sensors. The applications using hyperspatial, hyperspectral data reviewed in this chapter indicate some general conclusions: (1) full-range data are best for discriminating species and detecting diversity; (2) high-fidelity spectral data are most useful, especially when trying to understand spectral connections to subtle biochemical absorption properties, such as those found in SWIR; (3) an advantage of narrowband, hyperspectral data is that the most important spectral features can be utilized through analytical techniques, without *a priori* information, and other redundant or noisy data can be discarded or minimized; (4) hyperspatial imagery allows crown delineation, screening of pixels with high information content (e.g., sunlit), and analysis of intracrown spectral variability; (5) there are a wide range of data processing and analytical techniques being used, with most research teams settling on a unique approach—there is no existing synthetic analysis that compares the relative advantages/disadvantages of these methods across site conditions and species assemblages with a common dataset; (6) there is a paucity of detailed field and hyperspectral datasets from TFs, greatly limiting synthesis and advances in the field; and (7) the potential of the temporal domain, which contains valuable information on plant phenological cycles, has not been exploited for tropical species identification and diversity mapping using hyperspectral imagery. The hyperspectral AVIRIS [36] and CAO [37] sensors offer cutting-edge performance, spectral range, and spatial resolution for researchers (Table 18.2). In terms of producing results with greater ecological understanding, these airborne hyperspectral sensors, and coupled small-footprint waveform lidar, have made the greatest strides over the last several years [3,5,6,12,13]. High-fidelity hyperspectral data can detect detailed biochemical information, while lidar peers deeper into the canopy and helps in analyzing the hyperspectral data by interpreting its structural signal and isolating information-rich content. Commercial hyperspectral and lidar sensors are already being combined for joint flights, and improvements in performance and capabilities are also advancing rapidly. These types of sensors need to fly over more sites where detailed field data can be collected, and just as important, these datasets need to reach a broader research community to be fully exploited.

Spaceborne sensors allow mapping at broader spatial and temporal scales, with much less monetary cost and restrictions than airborne sensors, albeit with lower spatial resolution and signal to noise. Hyperion has provided multitemporal, full-range hyperspectral datasets at 30 m for the last decade (Table 18.2). Although not the focus on this chapter, there has been some research on species-level mapping [38] and tracking of landscape-scale biochemical patterns and shifts due to aggregated canopy phenology and climate [2,4,11]. A new generation of future spaceborne hyperspectral sensors, HyspIRI (hyspiri.jpl.nasa.gov, 60 m) and EnMap (www.enmap.org, 30 m), will have higher fidelity (Table 18.2) and will undoubtedly increase our understanding of variability in biochemical, structural, and phenological properties of TF canopies across a range of sites; this, in turn, will both inform and help scale results from species-mapping applications conducted at finer spatial scales.

Our lack of knowledge of tropical species distributions has important implications for understanding changes in biodiversity due to climate and land change, locating and designing representative protected areas, planning extractive timber operations, and implementing global carbon agreements that require species information (e.g., United Nations REDD+, www.un-redd.org [46]). The important goal of having species-level or species diversity maps of TFs over broad spatial and temporal scales is not yet a reality, but hyperspectral remote sensing technology has a bright and exciting future in this endeavor.

REFERENCES

1. Chambers, J.Q. et al., Regional ecosystem structure and function: Ecological insights from remote sensing of tropical forests, *Trends in Ecology and Evolution*, 22, 414–423, 2007.
2. Asner, G.P. et al., Drought stress and carbon uptake in an Amazon forest measured with spaceborne imaging spectroscopy, *Proceedings of the National Academy of Sciences of the United States of America*, 101, 6039–6044, 2004.
3. Asner, G.P. and Vitousek, P.M., Remote analysis of biological invasion and biogeochemical change, *Proceedings of the National Academy of Sciences of the United States of America*, 102, 4383–4386, 2005.
4. Asner, G.P. et al., Vegetation-climate interactions among native and invasive species in Hawaiian rainforest, *Ecosystems*, 9, 1106–1117, 2006.
5. Asner, G.P. et al., Remote sensing of native and invasive species in Hawaiian forests, *Remote Sensing of Environment*, 112, 1912–1926, 2008.
6. Asner, G.P., Hyperspectral remote sensing of canopy chemistry, physiology and biodiversity in tropical rainforests. In: *Hyperspectral Remote Sensing of Tropical and Sub-Tropical Forests,* Kalacska, M. and Sánchez-Azofeifa, G.A. (Eds.), CRC Press, Boca Raton, FL, pp. 261–296, 2008.
7. Clark, M.L., Roberts, D.A., and Clark, D.B., Hyperspectral discrimination of tropical rain forest tree species at leaf to crown scales, *Remote Sensing of Environment*, 96, 375–398, 2005.
8. Broadbent, E.B. et al., Spatial partitioning of biomass and diversity in a lowland Bolivian forest: Linking field and remote sensing measurements, *Forest Ecology and Management*, 255, 2602–2616, 2008.
9. Kalacska, M. et al., Hyperspectral discrimination of tropical dry forest lianas and trees: Comparative data reduction approaches at the leaf and canopy levels, *Remote Sensing of Environment*, 109, 406–415, 2007.
10. Brando, P.M. et al., Seasonal and interannual variability of climate and vegetation indices across the Amazon, *Proceedings of the National Academy of Sciences of the United States of America*, 107(33), 14685–14690, 2010.
11. Huete, A.R. et al., Assessment of phenological variability in Amazon tropical rainforests using hyperspectral Hyperion and MODIS data. In: *Hyperspectral Remote Sensing of Tropical and Sub-Tropical Forests*, Kalacska, M. and Sánchez-Azofeifa, G.A. (Eds.), CRC Press, Boca Raton, FL, pp. 233–259, 2008.
12. Carlson, K.M. et al., Hyperspectral remote sensing of canopy biodiversity in Hawaiian lowland rainforests, *Ecosystems*, 10, 536–549, 2007.
13. Asner, G.P. et al., Invasive species detection in Hawaiian rainforests using airborne imaging spectroscopy and LiDAR, *Remote Sensing of Environment*, 112, 1942–1955, 2008.
14. Clark, M.L., An assessment of hyperspectral and lidar remote sensing for the monitoring of tropical rain forest trees, Doctoral Dissertation, University of California, Santa Barbara, CA, 2005.
15. Clark, M.L., Relative advantages of airborne lidar and hyperspectral data for individual tropical tree classification, *Proceedings of the 32nd International Symposium on Remote Sensing of Environment*, San Jose, Costa Rica, June 25–29, 2007.
16. Cochrane, M.A., Using vegetation reflectance variability for species level classification of hyperspectral data, *International Journal of Remote Sensing*, 21, 2075–2087, 2000.
17. Castro-Esau, K.L. et al., Variability in leaf optical properties of Mesoamerican trees and the potential for species classification, *American Journal of Botany*, 93, 517–530, 2006.
18. Rivard, B. et al., Species classification of tropical tree leaf reflectance and dependence on selection of spectral bands. In: *Hyperspectral Remote Sensing of Tropical and Sub-Tropical Forests,* Kalacska, M. and Sánchez-Azofeifa, G.A. (Eds.), CRC Press, Boca Raton, FL, pp. 261–296, 2008.
19. Asner, G.P. et al., Leaf chemical and spectral diversity in Australian tropical forests. *Ecological Applications,* 19, 236–253, 2009.
20. Asner, G.P. and Martin, R.E., Spectral and chemical analysis of tropical forests: Scaling from leaf to canopy levels, *Remote Sensing of Environment*, 112, 3958–3970, 2008.
21. Sánchez-Azofeifa, G.A. et al., Differences in leaf traits, leaf internal structure, and spectral reflectance between two communities of lianas and trees: Implications for remote sensing in tropical environments, *Remote Sensing of Environment,* 113, 2076–2088, 2009.
22. Castro-Esau, K.L., Sánchez-Azofeifa, G.A., and Caelli, T., Discrimination of lianas and trees with leaf-level hyperspectral data, *Remote Sensing of Environment*, 90, 353–372, 2004.
23. Toomey, M., Roberts, D., and Nelson, B., The influence of epiphylls on remote sensing of humid forests, *Remote Sensing of Environment*, 113, 1787–1798, 2009.

24. Arroyo-Mora, J.P. et al., Spectral expression of gender: A pilot study with two dioecious neotropical tree species. In: *Hyperspectral Remote Sensing of Tropical and Sub-Tropical Forests*, Kalacska, M. and Sánchez-Azofeifa, G.A. (Eds.), CRC Press, Boca Raton, FL, pp. 125–140, 2008.

25. Ustin, S.L. et al., Retrieval of foliar information about plant pigment systems from high resolution spectroscopy, *Remote Sensing of Environment*, 113, S67–S77, 2009.

26. Asner, G.P., Biophysical and biochemical sources of variability in canopy reflectance. *Remote Sensing of Environment*, 64, 234–253, 1998.

27. Ustin, S.L. et al., Using imaging spectroscopy to study ecosystem processes and properties. *Bioscience*, 54, 523–534, 2004.

28. Kokaly, R.F. et al., Characterizing canopy biochemistry from imaging spectroscopy and its application to ecosystem studies, *Remote Sensing of Environment*, 113, S78–S91, 2009.

29. Grant, L., Diffuse and specular characteristics of leaf reflectance. *Remote Sensing of Environment*, 22, 309–322, 1987.

30. Castro-Esau, K.L. and Kalacka, M., Tropical dry forest phenology and discrimination of tropical tree species using hyperspectral data. In: *Hyperspectral Remote Sensing of Tropical and Sub-Tropical Forests*, Kalacska, M. and Sánchez-Azofeifa, G.A. (Eds.), CRC Press, Boca Raton, FL, pp. 1–25, 2008.

31. Pu, R., Ge, S., Kelly, N.M., and Gong, P., Spectral absorption features as indicators of water status in coast live oak (*Quercus agrifolia*) leaves. *International Journal of Remote Sensing*, 24, 1799–1810, 2003.

32. Pu, R.L., Gong, P., Biging, G.S., and Larrieu, M.R., Extraction of red edge optical parameters from Hyperion data for estimation of forest leaf area index. *IEEE Transactions on Geoscience and Remote Sensing*, 41, 916–921, 2003.

33. Chun, S., The utility of digital aerial surveys in censusing *Dipteryx panamensis*, the key food and nesting tree of the endangered great green macaw (*Ara ambigua*) in Costa Rica, Doctoral Dissertation, Duke University, Durham, NC, 2008.

34. Bohlman, S., Hyperspectral remote sensing of exposed wood and deciduous trees in seasonal tropical forests. In: *Hyperspectral Remote Sensing of Tropical and Sub-Tropical Forests*, Kalacska, M. and Sánchez-Azofeifa, G.A. (Eds.), CRC Press, Boca Raton, FL, pp. 177–192, 2008.

35. Zhang, J. et al., Intra- and inter-class spectral variability of tropical tree species at La Selva, Costa Rica: Implications for species identification using HYDICE imagery, *Remote Sensing of Environment*, 105, 129–141, 2006.

36. Green, R.O. et al., Imaging spectroscopy and the airborne visible infrared imaging spectrometer (AVIRIS), *Remote Sensing of Environment*, 65, 227–248, 1998.

37. Asner, G.P. et al., Carnegie airborne observatory: In-flight fusion of hyperspectral imaging and waveform light detection and ranging (LiDAR) for three-dimensional studies of ecosystems, *Journal of Applied Remote Sensing*, 1, 013536, 2007.

38. Papeş, M. et al., Using hyperspectral satellite imagery for regional inventories: A test with tropical emergent trees in the Amazon Basin, *Journal of Vegetation Science*, 21, 342–354, 2010.

39. Dennison, P.E. and Roberts, D.A., Endmember selection for multiple endmember spectral mixture analysis using endmember average RMSE. *Remote Sensing of Environment*, 87, 123–135, 2003.

40. Asner, G.P., Estimation canopy structure in an Amazon forest from laser range finder and IKONOS satellite observations, *Biotropica*, 34, 483–492, 2002.

41. Lucas, R., Mitchell, A., and Bunting, P., Hyperspectral data for assessing carbon dynamics and biodiversity of forests. In: *Hyperspectral Remote Sensing of Tropical and Sub-Tropical Forests*, Kalacska, M. and Sánchez-Azofeifa, G.A. (Eds.), CRC Press, Boca Raton, FL, pp. 47–86, 2008.

42. Bunting, P. and Lucas, R., The delineation of tree crowns in Australian mixed species forests using hyperspectral compact airborne spectrographic imager (CASI) data, *Remote Sensing of Environment*, 101, 230–248, 2006.

43. Lucas, R. et al., Classification of Australian forest communities using aerial photography, CASI and HyMap data, *Remote Sensing of Environment*, 112, 2088–2103, 2008.

44. Palace, M. et al., Amazon forest structure from IKONOS satellite data and the automated characterization of forest canopy properties, *Biotropica*, 40, 141–150, 2008.

45. Ghiyamat, A. and Shafri, H.Z.M., A review on hyperspectral remote sensing for homogeneous and heterogeneous forest biodiversity assessment, *International Journal of Remote Sensing*, 31, 1837–1856, 2010.

46. Stickler, C.M. et al., The potential ecological costs and cobenefits of REDD: A critical review and case study from the Amazon region, *Global Change Biology*, 15, 2803–2824, 2009.

19 Detecting and Mapping Invasive Plant Species by Using Hyperspectral Data

Ruiliang Pu

CONTENTS

19.1 INTRODUCTION

Invasions of plant species have caused significant changes in structure and function of ecosystems. Biological invasions have been identified as a major non-climatic driver of global change [1,2]. Such changes usually have negative impacts on ecological functions of natural ecosystems at various scales and become a serious problem to environment friendly sustainable ecosystems all over the world [3–6]. Invasive plant species (IPS) have caused costs associated with environment, economy, and culture. For example, in the United States, the estimated cost of environmental damages and associated management and control of IPS is about $137 billion per year; the total amount could be several times more if one considers native species extinctions, biodiversity reduction, ecosystem services, and aesthetics [7]. Therefore, monitoring the invasion extent and speed of the IPS and eradicating IPS across their invaded areas is an important task. Traditionally, this task relies heavily on field-based investigations and on methods that are usually expensive and time-consuming [6].

However, remote sensing techniques, especially hyperspectral remote sensing, offer an effective and economical alternative for gathering spatially distributed data over a large area and have potential for detecting and mapping IPS [8–13]. With remote sensing techniques, a large area of coverage can be acquired in a short period of time. Hyperspectral remote sensing, with its sufficient spectral information associated with phenological and structural characteristics of specific invasive species, allows for the species-level detection necessary to map invasive species [10,14,15]. During the last two decades, researchers have had an increasing interest in studies on the hyperspectral detection and mapping of invasions of aquatic species (e.g., [16,17]), grasses and weeds (e.g., [9,18–21]), scrubs and shrubs (e.g., [10,12,13,15,22,23]), and trees (e.g., [6,24–28]).

Based on the existing literature review, this chapter provides an overview on hyperspectral remote sensing techniques for detecting and mapping IPS. The main objectives of this chapter are as follows:

- To review suitable techniques/methods and their applications in detecting and mapping IPS with hyperspectral data
- To address relevant considerations for detecting and mapping IPS
- To discuss challenges and point out future directions for identifying and mapping invasive species using hyperspectral remote sensing techniques

19.2 POTENTIAL OF DETECTING AND MAPPING INVASIVE PLANT SPECIES

19.2.1 Physiological and Phenological Characteristics of IPS

The plant infestation can make several impacts to various ecosystems, such as (1) enhancing fire frequency and severity [29]; (2) increasing soil salinity, which reduces productivity of native plants and results in the loss of natural habitat [7]; (3) consuming soil water to such an extent that it can dry up streams and reduce water levels of rivers and lakes [30]; (4) depleting nutrients and making changes in microclimate and alterations in vegetation succession [31]; and (5) increasing nitrogen (N) deposition [32]. These impacts caused by IPS can directly (e.g., fire frequency and intensity) or indirectly (e.g., depleting nutrients and altering vegetation succession) mirror spectral difference compared to native species.

Compared to native species, most IPS usually have different phenological characteristics. Remote sensing methods have potential to detect such phenological differences, such as peak bloom, bud break, or senescence. Numerous researchers have investigated relationships between plant phenology and environmental factors, which could be exploited for species identification. For example, Ramsey et al. [25,26] applied the five characteristic spectra (senescing foliage, canopy shadow, green vegetation, yellow foliage, and red tallow) to the corrected and normalized Hyperion image data to identify an invasive species, Chinese tallow. Phenological timing information extracted from multitemporal hyperspectral data (two seasonal CASI and AVIRIS image data) helps detect tamarisk [12] via classification and cheatgrass [18] via color changes and altered seasonality relative to native species, respectively.

19.2.2 Canopy Structure and Biochemistry of IPS

Recent work has shown that invasive tree species often express biochemical and physiological properties unique from those of native trees [27,33,34]. Resolving these particular leaf and canopy characteristics in remotely sensed imagery may provide a way to map and monitor invaders at the regional scale. For example, Asner et al. [27,28] found that invasive tree species have unique hyperspectral reflectance signatures from those of native tree species. In their work, canopy reflectance properties in the 400–2500 nm wavelength range, collected from an airborne hyperspectral sensor, Airborne Visible and Infrared Imaging Spectrometer (AVIRIS), demonstrated spectral separability of native, introduced and highly invasive species. Their further analyses showed systematic, wavelength-dependent spectral reflectance differences between plant functional types, such as nitrogen-fixing (*Morella faya*) and nonfixing trees (native species). Most importantly, they presented that the spectral separability of species was tightly linked to their biochemical composition associated with relative differences in measured leaf pigment (chlorophyll and carotenoids), nutrient (N and P), and structural (specific leaf area) properties, as well as to canopy leaf area index [27].

Others have found diagnostic characteristics of IPS. With AVIRIS hyperspectral data, Underwood et al. [10] found that quantification of the water absorption feature (via the continuum removal approach) was sufficient to map the invasive succulent iceplant. Hunt et al. [35,36] detected invasive species leafy spurge through estimating its pigments' content of leaves and floral bracts from simulated, *in situ* and hyperspectral image (AVIRIS) data.

19.3 TECHNIQUES AND METHODS

The goal of detecting and mapping IPS using hyperspectral data is employing suitable analysis techniques and methods to compare spectral signatures characterized in spatial, spectral, and temporal domains, derived from hyperspectral data, between invasive species and native (background) species to separate them spatially. To focus more on application, the section will review the techniques/methods that currently only appear in existing literature and consist of seven techniques/ methods: spectral derivative analysis, spectral matching, vegetation index (VI) analysis, absorption analysis, hyperspectral transformation, spectral mixture analysis (SMA), and classification. To be convenient, each technique/method will be reviewed from describing the characteristics of the technique/method itself, briefly summarizing advantages/disadvantages of the technique/method applied in detecting and mapping IPS with hyperspectral data with a few of typical application cases and listing several major factors to be considered when applied in practice. For the seven techniques/ methods, the main characteristics, advantages and disadvantages, major factors affecting detecting and mapping results, and some application examples are provided in Table 19.1. The hyperspectral data used for implementing the techniques/methods include *in situ* spectral measurements taken with various spectrometers and image data acquired with airborne and spaceborne systems/sensors.

19.3.1 DERIVATIVE ANALYSIS

In situ or imaging hyperspectral data obtained in the field are rarely from a single object. They are contaminated by illumination variations caused by terrain relief, cloud, and viewing geometry [37]. For the first- and second-order derivative spectra, a finite approximation [38] can be applied to calculate them from hyperspectral data. The derivative spectrum is the normalized spectral difference of two continuous/neighbor narrowbands with their wavelength interval. Spectral derivative analysis has been considered a desirable tool to remove or compress the effect of illumination variations with low frequency on target spectra but it is sensitive to the signal-to-noise ratio (SNR) of hyperspectral data and higher order spectral derivative processing is susceptible to the noise [39]. In other words, lower order derivatives (e.g., the first-order derivative) are less sensitive to noise and hence more effective in operational remote sensing. When implementing the spectral derivative analysis, the spectral resolution is required narrower than 10 nm and spectral bands are continuous.

Derivative spectra have been successfully employed in hyperspectral data analysis for estimating biophysical and biochemical parameters, thus helping detecting and mapping IPS. For example, Asner et al. [27] conducted the spectral separability analysis between Hawaiian native and introduced (invasive) tree species with AVIRIS hyperspectral image data for detecting and assessing invasive species. They observed that the spectral differences (measured in reflectance, first- and second-order derivative spectra, see Figure 19.1) in canopy spectral signatures are linked to relative differences in leaf pigment (chlorophyll and carotenoids), nutrient (N and P), and structural (specific leaf area) properties, as well as to canopy LAI. These relative differences associated with leaf and canopy properties of trees are helpful to separate invasive species from its background (native) species. Figure 19.1 shows that there is a greater derivative spectral difference between the invasive and introduced species than that of their reflectance spectra at those gray bars. Combining *in situ* spectral measurements with airborne imaging spectroradiometer for application (AISA) data, Wang et al. [19] mapped an invasive weed (*Sericea lespedeza*) in a public grass field in mid-Missouri. The maximal first-order derivative in red-near infrared region (650–800 nm) was derived to separate

TABLE 19.1

Summary of Techniques Suitable for Detecting and Mapping Invasive Plant Species Using Hyperspectral Data

Technique	Characteristic and Description	Advantage and Disadvantage	Major Factor	Typical Example
1. Derivative analysis	Normalized spectral difference of two continuous/neighbor narrowbands with their wavelength interval	Remove or compress the effect of illumination variations with low frequency on target spectra but sensitive to the SNR of hyperspectral data and higher order spectral derivative processing is susceptible to the noise	Spectral resolution <10nm and also continuous, right threshold	[19,27]
2. Spectral matching	With n-dimensional angles (distances and correlations) to match pixels to reference spectra with smaller angles (shorter distances and higher correlations) representing closer matches to the reference spectrum, otherwise representing no matches to the reference spectrum	It is a physically based spectral classification and is less sensitive to differences in curve magnitude caused by variation in lighting across a scene, but it is sensitive to noise in any particular band	Determination of threshold of angle (distance and correlation)	[9,41,23,45]
3. Vegetation index analysis	Calculate ratio of images of two bands or normalized difference of two or more than two bands	Easy to use and reduce impact of sun angle, atmosphere, shadow, and topography. However, VI image is not normal usually	Identify suitable bands to construct VIs	See Table 20.2
4. Absorption feature analysis	Caused by a combination of factors inside and outside matter surface including electronic processes, molecular vibrations, abundance of chemicals, granular size and physical structure, and surface roughness relative to electromagnetic wavelength	Absorption features directly linked to matter structure, constituents, and concentration/content. Require images with higher SNR	Spectral resolution <10nm and also continuous	[10,17,22]
5. Hyperspectral transformation	A linear or nonlinear combination of raw data to reduce dimensionality and preserve variance contained in raw data as much as possible in the first several components images	Dimension reduction and usefully informative feature extraction; not easy to identify which are more signal components	Identify informative features/components	[6,10,12,18]

TABLE 19.1 (continued)
Summary of Techniques Suitable for Detecting and Mapping Invasive Plant Species Using Hyperspectral Data

Technique	Characteristic and Description	Advantage and Disadvantage	Major Factor	Typical Example
6. Spectral mixture analysis	Spectral reflectances from different materials (>1 material) within a pixel are recorded as one spectral response (mixed spectrum). Using linear or nonlinear spectral mixture model to derive fraction (endmember) images from mixed pixels	The fraction representing the areal proportion of each endmember, but do not know where the proportioned areas locate within a mixed pixel and sometime it is difficult to obtain endmember spectra and know all endmembers in a scene	Identify suitable/pure endmembers and extract their individual spectra	[20,26,28,75]
7. Classification	Using supervised/unsupervised method, parametric/nonparametric algorithms to assign one pixel or image object into one of classes (species). Usually for classifying hyperspectral data, dimension reduction and feature extraction are first considered	There is a basis of statistic/probability or a rule for classification; usually, it is difficult to obtain adequate training samples for supervise method and labeling unsupervised spectral clusters	Identify suitable method/classifier for a specific task and gather adequate training samples	[6,17,22,61]

the invasive species from the target grass in pastures in Missouri. With a simple threshold approach for the maximum first-order derivative spectrum, Sericea with various sizes was successfully identified in the study area.

19.3.2 Spectral Matching

Two measures of spectral similarity widely used are the spectral angle that is calculated between two curve vectors and Euclidean distance (ED) that calculates the vector distance. The algorithm that uses spectral angle is called spectral angle mapper (SAM). It is commonly believed that SAM is a physically based spectral classification and is less sensitive to differences in curve magnitude caused by variation in lighting across a scene. As long as the shapes of spectral curves are similar, it will result in high value. ED is a widely used measure that detects an absolute difference between two spectral curves. It is sensitive to the noise in any particular band spectra.

A third measure of spectral similarity, named spectral correlation measure (SCM, also called a cross correlogram spectral matching (CCSM)), as a modified version of SMA and ED, is calculated with the vector crosscorrelation between a known reference spectrum and an unknown target spectrum [40,41]. Spectral matching uses n-dimensional angles (distances and crosscorrelations) to match pixels to reference spectra with smaller angles (shorter distances and higher crosscorrelations) representing closer matches to the reference spectrum, otherwise representing no matches to the reference spectrum. Usually, a key factor to apply the technique for detecting and mapping IPS with hyperspectral data is how to determine a suitable threshold of angle (or distance or correlation). Spectral matching techniques to identify and label vegetation categories have also been discussed in detail elsewhere [42,43].

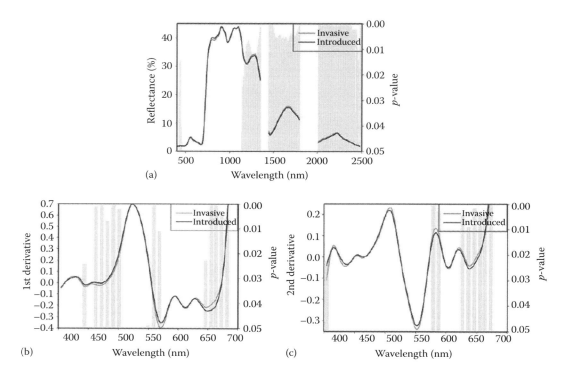

FIGURE 19.1 (a) Reflectance, (b) first-order derivative, and (c) second-order derivative spectra of highly invasive versus other introduced tree species of Hawaiian tropical and subtropical forests, with band-by-band t-tests showing significant differences in gray bars (*p*-values < 0.05). (Reprinted from *Remote Sensing of Environment*, 112, Asner, G.P., Jones, M.O., Martin, R.E., Knapp, D.E., and Hughes, R.F., Remote sensing of native and invasive species in Hawaiian forests, 1912–1926, Copyright (2008), with permission of Elsevier.)

Many researchers have employed spectral matching method to identify and map IPS. For example, with AISA airborne hyperspectral image data and SAM analysis method, Narumalani et al. [23] quantified and mapped four dominant IPS, including saltcedar, Russian olive, Canada thistle, and musk thistle, along the floodplain of the North Platte River, Nebraska. Validation procedures confirmed an overall map accuracy of 74%. Also with the same analysis method (SAM) and airborne hyperspectral image data, Lass and Prather [44] detected the location of Brazilian pepper trees in the Everglades, and Hirano et al. [45] mapped wetland vegetation with an invasive species lather leaf (*Colubrina asiatica*). In addition, Pengra et al. [41] used the SCM measure to map an invasive plant, *Phragmites australis*, in coastal wetland using the Hyperion hyperspectral sensor's data to show a good overall accuracy of 81.4%.

19.3.3 VEGETATION INDEX ANALYSIS

When using hyperspectral data to conduct spectral VI analysis, we can make use of the advantage of increasing chance and flexibility to choose spectral bands. With multispectral data, one may only have one choice of using the only red and NIR bands. However, with hyperspectral data, one can choose many of such red and NIR narrowband combinations [46,47–50]. Accordingly, spectral VIs applied to hyperspectral data are called narrowband VIs [51–53]. Table 19.2 lists a set of 21 VIs that are developed from hyperspectral data and suitable for detecting and mapping IPS. These VIs have appeared in existing literature. To be convenient to locate a (or group) VI(s) for readers, the total 21 VIs are organized into three categories based on the characteristics and functions of the VIs: multiple bioparameters, pigments, and foliar chemistry. Within individual categories, the VIs are

TABLE 19.2

Summary of 21 Spectral Indices Extracted from Hyperspectral Data, Appearing in Detecting and Mapping IPS

Spectral Index	Characteristics and Functions	Definition	References
Multiple Bioparameters			
LI, *Lepidium* index	To be sensitive to the uniformly bright reflectance displayed by *Lepidium* in the visible range	R_{630}/R_{586}	[20]
NDVI, normalized difference vegetation index	Respond to change in the amount of green biomass and more efficiently in vegetation with low to moderate density	$(R_{NIR} - R_R)/(R_{NIR} + R_R)$	[81]
PSND, pigment-specific normalized difference	Estimate LAI and carotenoids (Cars) at leaf or canopy level	$(R_{800} - R_{470})/(R_{800} + R_{470})$	[99]
SR, simple ratio	Same as NDVI	R_{NIR}/R_R	[82,83]
Pigments			
Chl_{green}, chlorophyll index using green reflectance	Estimate chlorophylls (Chls) content in anthocyanin-free leaves if NIR is set beyond 760 nm	$(R_{760-800}/R_{540-560}) - 1$	[84]
$Chl_{red-edge}$, chlorophyll index using red edge reflectance	Estimate Chls content in anthocyanin-free leaves if NIR is set	$(R_{760-800}/R_{690-720}) - 1$	[84]
LCI, leaf chlorophyll index	Estimate Chl content in higher plants, sensitive to variation in reflectance caused by Chl absorption	$(R_{850} - R_{710})/(R_{850} + R_{680})$	[85]
mND_{680}, modified normalized difference	Quantify Chl content and sensitive to low content at leaf level	$(R_{800} - R_{680})/(R_{800} + R_{680} - 2R_{445})$	[86]
mND_{705}, modified normalized difference	Quantify Chl content and sensitive to low content at leaf level. mND_{705} performance better than mND_{680}	$(R_{750} - R_{705})/(R_{750} + R_{705} - 2R_{445})$	[86,87]
mSR_{705}, modified simple ratio	Quantify Chl content and sensitive to low content at leaf level	$(R_{750} - R_{445})/(R_{705} - R_{445})$	[86]
NPCI, normalized pigment chlorophyll ratio index	Assess Cars/Chl ratio at leaf level	$(R_{680} - R_{430})/(R_{680} + R_{430})$	[88]
PBI, plant biochemical index	Retrieve leaf total Chl and nitrogen concentrations from satellite hyperspectral data	R_{810}/R_{560}	[89]
PRI, photochemical/ physiological reflectance index	Estimate car pigment contents in foliage	$(R_{531} - R_{570})/(R_{531} + R_{570})$	[90]
PI2, pigment index 2	Estimate pigment content in foliage	R_{695}/R_{760}	[91]
RGR, red:green ratio	Estimate anthocyanin content with a green and a red band	R_{683}/R_{510}	[86,92]
SGR, summed green reflectance	Quantify Chl content	Sum of reflectances from 500 to 599 nm	[87]
Foliar Chemistry			
CAI, cellulose absorption index	Cellulose and lignin absorption features, discriminates plant litter from soils	$0.5(R_{2020} + R_{2220}) - R_{2100}$	[93]
NDLI, normalized difference lignin index	Quantify variation of canopy lignin concentration in native shrub vegetation	$[\log(1/R_{1754}) - \log(1/R_{1680})]/[\log(1/R_{1754}) + \log(1/R_{1680})]$	[94]

(continued)

TABLE 19.2 (continued)
Summary of 21 Spectral Indices Extracted from Hyperspectral Data,
Appearing in Detecting and Mapping IPS

Spectral Index	Characteristics and Functions	Definition	References
NDWI, ND water index	Improving the accuracy in retrieving the vegetation water content at both leaf and canopy levels	$(R_{860} - R_{1240})/(R_{860} + R_{1240})$	[95,96]
RVI$_{hyp}$, hyperspectral ratio VI	Quantify LAI and water content at canopy level	R_{1088}/R_{1148}	[97]
WI, water index	Quantify relative water content at leaf level	R_{900}/R_{970}	[98]

arranged in an alphabetical order. The explicit advantages of VIs are easy to use and can also reduce impact of sun angle, atmosphere, shadow, and topography on target spectra. However, the probability distribution of VI image for different land cover types usually is not normal. A key factor to determine the usefulness of a VI depends on identifying suitable wave bands.

Developing various VIs (Table 19.2) from hyperspectral data helps detect and map invasive species. For example, Hestir et al. [17] used mSR$_{705}$ VI to map invasive species (three weeds: perennial pepperweed, water hyacinth, and Brazilian waterweed) with airborne hyperspectral data (HyMap). They achieved a moderate to high mapping accuracy. Andrew and Ustin [20,21] used 13 VIs, derived from HyMap data, together with a minimum noise fraction (MNF) transformation, mixture tuned matched filtering (MTMF), and decision tree (CART) classification method to identify and map the perennial pepperweed from the three sites of California's San Francisco Bay/ Sacramento–San Joaquin Delta Estuary. Their approach was sufficiently flexible and robust to detect the invasive species with similar accuracies (~90%) at both Rush Ranch and Jepson Prairie sites, but was unsuccessful at Cosumnes River Preserve due to different environmental context. Also using VIs derived from airborne hyperspectral image data (AVIRIS and CASI), Underwood et al. [10] detected and mapped iceplant and jubata grass in California coastal habitat, and Pu et al. [13] mapped change of saltcedar after a biocontrol measure was taken by USDA. Both studies show high mapping accuracy. Compiling a time series of Hyperion image data, Asner et al. [32] studied the variations in upper-canopy leaf chlorophyll and carotenoid content during a climatological transition with the remotely sensed photochemical and carotenoid reflectance indices (PRI and CRI). They found that the PRI and CRI were related to differences in light-use efficiency between invasive and native tree species, and thus helped separating invasive species (*Myrica faya*) from native species (*Metrosideros polymorpha*).

19.3.4 ABSORPTION FEATURES ANALYSIS

Quantitative characterization of absorption features allows for abundance estimation of materials from hyperspectral data. Spectral absorption features mirror diagnostic spectral features of materials [54] and are caused by a combination of factors inside and outside matter surface including electronic processes, molecular vibrations, abundance of chemical constituents, granular size and physical structure, and surface roughness relative to electromagnetic wavelength. Therefore, the absorption features are directly linked to matter structure, constituents, concentration, and content. For example, the central wavelengths of *in situ* chlorophyll-a absorption are at both 0.45 and 0.67 μm; the central wavelengths of water absorption are near 0.97, 1.20, 1.40, and 1.94 μm; the central wavelengths of N absorption are near 1.51, 2.06, 2.18, 2.30, and 2.35 μm; the central wavelengths of lignin absorption features are near 1.12, 1.42, 1.69, and 1.94 μm; and the central wavelengths of cellulose absorption features are near 1.20, 1.49, 1.78, 1.82, 2.27, 2.34, and 2.35 μm [55].

Extraction of these absorption features requires hyperspectral data with higher SNR and spectral resolution <10 nm and neighbor bands continuous.

In order to analyze the absorption features of a spectral reflectance curve, one needs to normalize the spectral curve, so that only the spectral values inside the absorption features will be less than 1 (100%). This can be done using a continuum removal technique proposed by Clark and Roush [56]. Quantitative measures can be determined from each absorption peak after the normalization of the raw spectral reflectance curve. The quantitative measures can be used to determine abundances of certain compounds in a pixel, thus linked to characteristics of plant species. For example, Hamada et al. [22] explored the effectiveness of depth of chlorophyll absorption feature of Tamarisk species from airborne hyperspectral image data with the continuum removal technique. They combined the absorption feature and other spectral variables extracted from the hyperspectral data with parallelepiped classifier to yield the most accurate and reliable Tamarisk classification products. Underwood et al. [10] successfully used the continuum removal technique to extract water absorption feature from AVIRIS hyperspectral image data to map nonnative species, iceplant, and jubata grass in California's coastal habitat. With the technique, Hestir et al. [17] also estimated foliar water absorption feature from hyperspectral image data HyMap to successfully identify and map invasive species in the California Delta ecosystem.

19.3.5 HYPERSPECTRAL TRANSFORMATION

Hyperspectral transformation is a linear or nonlinear combination of raw data to reduce dimensionality and preserve variance as much as possible in the first several components images. In detecting and mapping IPS with hyperspectral data, the most popular transformation techniques are principal component analysis (PCA) and its modified version: MNF. PCA technique has been applied to reduce the data dimension and feature extraction from hyperspectral data for assessing leaf or canopy parameters (e.g., [57,58]). With a covariance (or correlation) matrix calculated from vegetated pixels only, it is commonly believed that the eigenvalues and corresponding eigenvectors of the first several PC images, computed from the covariance (or correlation) matrix, are expected to be able to enhance vegetation variation. Green et al. [59] developed one transform method called MNF transform to maximize the SNR when choosing principal components with increasing component number. Then several MNFs with maximum SNR are selected for further analysis of hyperspectral data, such as for determining endmember spectra for SMA [60,61]. Although canonical discriminant analysis (CDA) and the wavelet transform (WT) are more attractive for dimension reduction and feature extraction for classification recently (e.g., [58,62]), they were not seen to be used in detecting and mapping IPS with hyperspectral data currently. Therefore, their applications as transformation techniques will not be reviewed here. In general, for all the transformation methods, identification of informative features/components still is a difficult task because some subtle and useful features may not be included in first several component images [17,20].

During the last 2 decades, PCA and MNF techniques have been extensively applied for extracting spectral features for detecting and mapping IPS with either SMA or other supervised classifiers such as maximum likelihood classifier (MLC) and artificial neural networks (ANN). For example, Pu et al. [12] and Tsai et al. [6] used PCA (segmented PCA) transformation method for extracting several important component images from CASI and Hyperion hyperspectral image data to successfully detect and map invasive species, saltcedar and *Leucaena leucocephala* in Lovelock, Nevada and in southern Taiwan, respectively. For detecting and mapping four invasive species: saltcedar, Russian olive, Canada thistle, and musk thistle, along the floodplain of the North Platte River, Nebraska Narumalani et al. [23] transformed AISA airborne hyperspectral image data with the MNF technique to extract first several MNF images that were used as input for running SAM to separate the four invasive species. The overall accuracy of mapping the four species was 74%. Underwood et al. [10,15] employed several MNFs transformed from AVIRIS hyperspectral image data in coastal California to map nonnative plant species (iceplant, jubata grass, and blue gum). They used a standard supervised classifier, MLC with the several MNFs as an input to map the three invasive species, which resulted in a higher accuracy. For mapping some invasive species in the California Delta ecosystem, Hestir et al. [17] also transformed HyMap hyperspectral image to extract several MNFs, together with other extracted spectral variables, as inputs to run a few of supervised classifiers to map the invasive species. They achieved moderate to high success for the task.

19.3.6 SPECTRAL MIXTURE ANALYSIS

Spectral reflectances measured from different materials within a pixel are recorded as one spectral response or as a mixed spectrum. A large portion of remotely sensed data is spectrally mixed because the spatial resolution (pixel size) of image data cannot resolve individual materials. In order to identify various "pure materials" and to determine their spatial proportions from the remotely sensed data, the spectral mixing process has to be properly modeled. Then the model can be inverted to derive the spatial proportions and spectral properties of those "pure materials."

There are two types of spectral mixing, linear spectral mixing, and nonlinear spectral mixing. Linear spectral mixing modeling and its inversion have been widely used since the late 1980s. A linear SMA was extensively applied to extract the abundance of various components within mixed pixels. Nonlinear spectral mixture model can be found in Sasaki et al. [63] and Zhang et al. [64].

In a linear spectral mixture model analysis, there are two solutions: a linear least square solution and a nonlinear solution (e.g., a neural network-based, nonlinear, subpixel classifier by Walsh et al. [61]). At present, since the SMA method is easy to use, it has been widely and successfully applied for mapping abundance of a certain number of invasive species with hyperspectral image data. For instance, with airborne and spaceborne hyperspectral image data (AVIRIS, HyMap, and Hyperion), the invasive species that were detected and mapped include Brazilian pepper [44], perennial pepperweed, Water hyacinth and Brazilian waterweed [16,17], Chinese tallow [25,26], and *Psidium guajava* [61].

In addition, MTMF is an advanced spectral unmixing algorithm that does not require that all materials within a scene are known and have been identified as endmembers [65]. This method represents an improved alternative to SMA analysis for cases where the number of similar spectra are large or where it is problematic to collect spectra of all potential endmember components within the scene. MTMF treats each endmember independently and, at each pixel for each endmember, models the pixel as a mixture of the endmember and an undefined background material. It outputs a matched filter (MF) score and an infeasibility value for each endmember. The MF score is analogous to the fraction value from simple SMA, and the infeasibility is a measure of how likely a pixel is to contain the material of interest. Pixels are likely to contain materials for which they receive high MF scores and low infeasibilities [20]. MTMF has proven to be a very powerful tool to detect specific materials that differ slightly from the background. For example, with airborne hyperspectral image data (e.g., AVIRIS and HyMap), MTMF has successfully mapped a variety of invasive species, including Tamarisk [22], perennial pepperweed [20,21], and cheatgrass [18]. Figure 19.2 presents a map of the *Lepidium* infestation detected at Rush Ranch, California, overlaid on a true-color image of the site, created with MTMF outputs. With other classification algorithm, the MTMF outputs were the most important variables for *Lepidium* detection [20].

As a result of spectral unmixing, the fractions represent the areal proportions of endmembers, but we do not know where the proportioned areas of the endmembers locate within a mixed pixel. A key factor for unmixing mixed spectra is to identify suitable/pure endmembers and extract their individual spectra for training and test purposes.

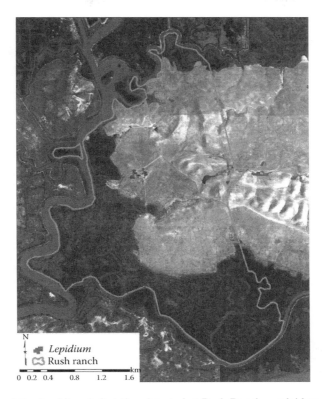

FIGURE 19.2 Map of the *Lepidium* infestation detected at Rush Ranch overlaid on a true-color image of the site, created with MTMF outputs. (Reprinted from *Remote Sensing of Environment*, 112, Andrew, M.E., and Ustin, S.L., The role of environmental context in mapping invasive plants with hyperspectral image data, 4301–4317, Copyright (2008), with permission of Elsevier.)

19.3.7 HYPERSPECTRAL IMAGE CLASSIFICATION

To overcome difficulties faced by traditional classifiers, caused by high dimensionality of hyperspectral data and high correlation of adjacent bands with limited number of training samples, it is necessary for us to reduce dimension and extract features for classifying IPS. Based on existing literature review, there are several hyperspectral transformation techniques/methods (see Section 19.3.5) that, currently, have been successfully applied in dimension reduction and feature extraction for classifying plant species. These transformation techniques/methods include PCA, MNF, linear discriminant analysis (LDA), CDA, and WT, etc. Theoretically, after features/component images in a relatively lower dimension are produced by running appropriate transformation algorithms, some supervised algorithms, such as MLC, LDA, ANN, and classification and regression tree (CART), and unsupervised methods, such as ISODATA, are utilized for detecting and mapping a variety of invasive species. For example, after extracting several MNFs from the AVIRIS hyperspectral image data, Underwood et al. [10,15] used MLC classifier to classify three nonnative plant species (iceplant, jubata grass, and blue gum) in California' coastal areas. With several PC images transformed from CASI and Hyperion hyperspectral image data, Pu et al. [12, Figure 4] utilized ANN and LDA algorithms to map saltcedar invasive species, and Tsai et al. [6] used MLC to map *Leucaena leucocephala*. Figure 19.3 presents change maps of saltcedar invasive species from July, 2002 (JUL02) to August, 2002 (AUG02), from AUG02 to September, 2003 (SEP03), and from JUL02 to SEP03 with multitemporal CASI hyperspectral image data. In addition, there are some researchers who applied a rule-based algorithm, CART, for detecting and mapping a few of IPS, including some aquatic species (water hyacinth and Brazilian waterweed) with inputs of several

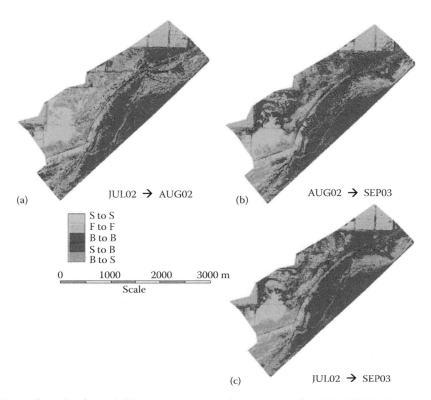

FIGURE 19.3   (See color insert.) Change detection resultant maps produced by ANN with principal components extracted from the CASI data (a) from JUL02 to AUG02, (b) from AUG02 to SEP03, and (c) from JUL02 to SEP03. In legend, "S to S," "F to F," and "B to B" represent no-changes of saltcedar, farmland, and bare/wildland, respectively; "S to B" means saltcedar changed to bare/wildland while "B to S" means bare/wildland changed to saltcedar. (Reprinted with kind permission from Springer Science + Business Media: *Environmental Monitoring and Assessment*, Invasive species change detection using artificial neural networks and CASI hyperspectral imagery, 140, 2008, 15–32, Pu, R., Gong, P., Tian, Y., Miao X., Carruthers, R., and Anderson, G.L.)

MNFs and other spectral variables, derived from HyMap hyperspectral image data, and achieved a varying degree of success [17,20,21,66].

In practice, for the hyperspectral image classification, sometimes it is difficult to obtain adequate training samples for supervise method and labeling unsupervised spectral clusters due to requirement of a large amount of field work. The major factors for detecting and classifying invasive species with the various hyperspectral data are to identify suitable classification method and gather adequate training samples for a specific task.

19.4   CONSIDERATIONS

Successfully implementing detecting and mapping IPS using hyperspectral data requires careful considerations of hyperspectral sensors/systems, image preprocessing methods, vegetation types and characteristics, phenological change, and environmental context before starting a project. This is because the temporal, spatial, spectral, and radiometric resolutions of hyperspectral data have a significant impact on the success of a remote sensing detecting and mapping IPS.

In general, atmospheric correction for hyperspectral data is optimal for conversion of radiance to reflectance [67]. Of the requirements of image preprocessing for detecting and mapping IPS, conversion of digital numbers to radiance or surface reflectance is a most important task for quantitative analysis of plant spectra. This is because the radiometrically corrected and calibrated image data can enhance spectral distinguishability between different plant species due to removing or compressing atmospheric effects on target spectra. A variety of methods [68]: Atmospheric CORrection Now program (ACORN), ATmospheric CORrection program (ATCOR), ATmospheric REMoval program (ATREM), ENVI's Fast Line-of-sight Atmospheric Analysis of Spectral Hypercubes (FLAASH) [69], Second Simulation of the Satellite Signal in the Solar Spectrum (6S), and High-accuracy ATmosphere Correction for Hyperspectral data (HATCH), have been developed [27,67,70,71]. In the rugged or mountainous areas, correcting for slope and aspect effects may be necessary. More detailed information about topographic correction can be found in Meyer et al. [72], Allen [73], and Adler-Golden et al. [74].

Different vegetation types comprised by individual invasive species should be first considered. The vegetation type information may help selecting appropriate hyperspectral sensor data and image analysis techniques. For the aquatic vegetation type, since its water background may significantly absorb and scatter a certain portion of radiometric energy to make reflected signal very weak compared to most terrestrial vegetation types. This requires hyperspectral data with a high SNR, such as HyMap and AVIRIS. Due to relatively small size of individual water plant, the suitable analysis techniques may include MNF transformation to extract several features/component images, SMA/CART for mapping abundance of IPS (e.g., [16,17,45]). For terrestrial vegetation types being comprised of grass, weed, shrub, and scrub species, due to their individual sizes being relatively small and frequently several species growing together, this requires the hyperspectral data with high spectral (<5 nm)/spatial (<5 m) resolutions besides a high SNR (e.g., CASI, HyMap, and AVIRIS). The corresponding analysis techniques/methods may consider using data transformation methods, PCA, MNF, and CDA, and mapping and classification methods, SAM, SMA, MTMF, CART, MLC, ANN (e.g., [9,12,15,18,19–23,61,75]). For trees and forest types, due to their individual sizes being relatively large and spectral difference between different species relatively distinct, we may consider using high altitude AVIRIS data and satellite Hyperion image data. The suitable analysis techniques/methods for data transformation and invasive species detecting and mapping may use those for terrestrial vegetation types for identifying and mapping grass, weed, shrub, and scrub species (e.g., [6,25–28,76]). Finally, the concept of the plant functional types may also be cited here in consideration of a few of invasive species that distribute in the same areas and may have similar structural, functional, and/or phenological properties. After considering both ecology and remote sensing characteristics and the capabilities of new remote sensing instruments (e.g., hyperspectral sensor/system), Ustin and Gamon [77] proposed a new concept of optically distinguishable plant functional types ("optical types") as a unique way to address ecological and remote sensing properties for plant functional types. The proposed concept of "optical type" is based on the assessment of vegetation structure, physiology, and phenology. All the three ecological variables affect vegetation optical properties and contribute to the definition of the concept of "optical type." This would ensure more direct relationships between ecological information and remote sensing observations [77].

Species life history stages or phenological change have different spectral characteristics. When collecting hyperspectral data (either *in situ* measurements or image data), we should consider the life history of the target (invasive) species and find the appropriate phenology stage when spectral difference between invasive species and native species is maximum. For example, the three species (Brazilian waterweed, perennial pepperweed, and water hyacinth) in Hestir et al.'s [17] case studies all differ at different phenological states: flowering, growing peaks, fruiting, and senescing. The flowering and fruiting phenologies of *Lepidium* can be spectrally distinct between its phonologies and spectral different from co-occurring species [20,78] (Andrew and Ustin, 2006 [add it in reference list], 2008) and other general green plants (Figure 19.4). The figure presents the field spectral difference of flowering and fruiting phenologies of *Lepidium*, also showing a typical reflectance spectrum of green vegetation for reference. When mapping downy brome using multidate AVIRIS image data, Noujdina and Ustin [18] found that the temporal offset in phenology of the downy brome relative to native vegetation provided a basis for spectral differences.

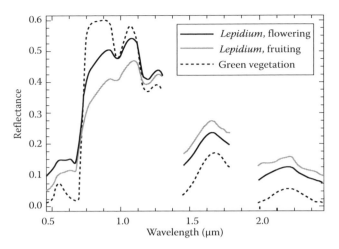

FIGURE 19.4 Field spectra of flowering and fruiting phenologies of *Lepidium*, along with a typical reflectance spectrum of green vegetation for reference. (Reprinted from *Remote Sensing of Environment*, 112, Andrew, M.E. and Ustin, S.L., The role of environmental context in mapping invasive plants with hyperspectral image data, 4301–4317, Copyright (2008), with permission of Elsevier.)

19.5 CHALLENGES AND FUTURE DIRECTIONS

The largest challenge to successfully detect invasive species from their background species perhaps is that all plants are spectrally similar because they are composed of the same spectrally active materials such as pigments, water, and cellulose [79]. The spectral uniqueness requisite for hyperspectral detection most often occurs when invaders possess physiological traits or phenological characteristics that are novel to the invaded ecosystems [20]. Adequate spectral information unique for invasive plant species (IPS) depends on spatial and temporal resolutions of hyperspectral sensors/systems. Currently, most of the spatial and temporal resolutions from airborne or spaceborne hyperspectral data are insufficient to decipher the complex characteristics of invasive species in natural environments. Although high spatial resolution (0.5–5 m) of most airborne hyperspectral sensors/systems (e.g., AVIRIS, HyMap, and CASI) has been proven to be a valid method to map these invasive plants, there are still several drawbacks to these approaches. One of them involves aircraft scheduling. This would limit the flexibility of data collection that catches the chance of maximizing spectral difference between invasive and native species across their phenology states. As a result, detecting and mapping IPS from native plants cannot effectively utilize the maximum difference information of phenology of the different plants. Although this issue (i.e., low temporal resolution) can be resolved by utilizing spaceborne hyperspectral sensors/systems, such as Hyperion (30 m resolution) onboard EO-1 satellite, Compact High Resolution Imaging Spectrometer (CHRIS; 18 m resolution) onboard the ESA's PROBA satellite, and HyspIRI (60 m resolution) [80] planned mission, their spatial resolutions are relatively low compared with individual size of many IPS. Therefore, many nonnative species are still not discernable, especially grasses, weeds, and shrubs. Such challenges of the spatial and temporal resolutions, related to hyperspectral sensors/systems, are not easy to solve because they are related to technical and economic issues, as well as huge data volume storage and processing.

Another challenge to detect an invasive species is its biological heterogeneity. Such a biological heterogeneity may lead to intra-species variation, thus causes overlapping spectral features between co-occurring species in the same study area. Hyperspectral image data acquired with suitable phenologic states may provide sufficient information to overcome these challenges, allowing the applications of more complex spectral analyses and spectral unmixing techniques [17] for detecting and mapping IPS. In other words, using multitemporal hyperspectral image data, the accuracy of detecting and mapping such invasive species can be expected to increase.

The application of multisensor data, which can synergize the high spatial resolution (airborne hyperspectral data) and high temporal resolution (satellite hyperspectral data), provides the potential to more accurately detect IPS through integration of different features of sensor data. The disadvantage of using multisensor data for detecting IPS is the difficulty in various image acquisition/processing and use of appropriate detecting and mapping techniques. In the future, after satellite hyperspectral sensor HyspIRI is in operation, the high frequent hyperspectral data are available globally. Application of multisensor data will become increasingly important in future study of detecting and mapping IPS, and thus, more advanced image processing and invasive species detection and mapping techniques are needed. Accurately detecting and mapping IPS with hyperspectral data remains an active research topic and new techniques continue to be developed. For an advanced invasive species mapping/detecting technique, it is required to be easy to use and it can provide accurate detecting and mapping results.

ACKNOWLEDGMENTS

Three anonymous reviewers' comments and suggestions were greatly valuable to improve the chapter. Author sincerely appreciates their efforts.

REFERENCES

1. Beck, K.G., Zimmerman, K., Schardt, J.D., Stone, J., Lukens, R.R., Reichard, S., Randall, J., Cangelosi, A.A., Cooper, D., and Thompson, J.P., Invasive species defined in a policy context: Recommendations from the federal invasive species advisory committee, *Invasive Plant Science and Management*, 1, 414–421, 2008.
2. Huang, C.-Y. and Asner, G.P., Applications of remote sensing to alien invasive plant studies, *Sensors*, 9, 4869–4889, 2009.
3. Mack, R.N., Simberloff, D., Lonsdale, W.M., Evans, H., Clout, M., and Bazzaz, F., Biotic invasions: Causes, epidemiology, global consequences and control, *Ecological Application*, 10, 689–710, 2000.
4. Simberloff, D., Biological invasions: How are they affecting us and what can we do about them? *Western North American Naturalist*, 61, 308–315, 2001.
5. Jackson, R.B., Banner, J.L., Jobbagy, E.G., Pockman, W.T., and Wall, D.H., Ecosystem carbon loss with woody plant invasive of grasslands, *Nature*, 418, 623–626, 2002.
6. Tsai, F., Lin, E.E., and Yoshino, K., Spectrally segmented principal component analysis of hyperspectral imagery for mapping invasive plant species, *International Journal of Remote Sensing*, 28(5), 1023–1039, 2007.
7. Pimentel, D., Lach, L., Zuniga, R., and Morrison, D., Environmental and economic costs associated with non-indigenous species in the United States, *BioScience*, 50, 53–65, 2000.
8. Ustin, S.L., Scheer, G., DiPietro, D., Underwood, E., and Olmstead, K., Hyperspectral remote sensing for invasive species detection and mapping, *Abstracts of Papers of the American Chemical Society*, 221, U50, 2001.
9. Lass, L.W., Thill, D.C., Shafii, B., and Prather, T.S., Detecting spotted knapweed (*Centaurea maculosa*) with hyperspectral remote sensing technology, *Weed Technology*, 16, 426–432, 2002.
10. Underwood, E.C., Ustin, S.L., and DiPietro, D., Mapping nonnative plants using hyperspectral imagery, *Remote Sensing of Environment*, 86, 150–161, 2003.
11. Miao, X., Gong, P., Swope, S., Pu, R., Carruthers, R., Anderson, G.L. et al., Estimation of yellow starthistle abundance through CASI-2 hyperspectral imagery using linear spectral mixture models, *Remote Sensing of Environment*, 101, 329–341, 2006.
12. Pu, R., Gong, P., Tian, Y., Miao X., Carruthers, R., and Anderson, G.L., Invasive species change detection using artificial neural networks and CASI hyperspectral imagery, *Environmental Monitoring and Assessment*, 140, 15–32, 2008.
13. Pu, R., Gong, P., Tian, Y., Miao, X., Carruthers, R., and Anderson, G.L., Using classification and NDVI differencing methods for monitoring sparse vegetation coverage: A case study of saltcedar in Nevada, USA, *International Journal of Remote Sensing*, 29(14), 1987–4011, 2008.
14. Clark, M.L., Roberts, D.A., and Clark, D.B., Hyperspectral discrimination of tropical rain forest tree species at leaf to crown scales, *Remote Sensing of Environment*, 96, 375–398, 2005.

15. Underwood, E.C., Ustin, S.L., and Ramirez, C.M., A comparison of spatial and spectral image resolution for mapping invasive plants in coastal California, *Environmental Management*, 39, 63–83, 2007.

16. Underwood, E.C., Mulitsch1, M.J., Greenberg, J.A., Whiting, M.L., Ustin, S.L., and Kefauver, S.C., Mapping invasive aquatic vegetation in the Sacramento-San Joaquin Delta using hyperspectral imagery, *Environmental Monitoring and Assessment*, 121, 47–64, 2006.

17. Hestir, E.L., Khanna, S., Andrew, M.E., Santos, M.J., Viers, J.H., Greenberg, J.A., Rajapakse, S.S., and Ustin, S.L., Identification of invasive vegetation using hyperspectral remote sensing in the California Delta ecosystem, *Remote Sensing of Environment*, 112, 4034–4047, 2008.

18. Noujdina, N.V. and Ustin, S.L., Mapping downy brome (*Bromus tectorum*) using multidate AVIRIS data, *Weed Science*, 56, 173–179, 2008.

19. Wang, C., Zhou, B., and Palm, H.L., Detecting invasive Sericea lespedeza (*Lespedeza cuneata*) in mid-Missouri pastureland using hyperspectral imagery, *Environmental Management*, 41, 853–862, 2008.

20. Andrew, M.E. and Ustin, S.L., The role of environmental context in mapping invasive plants with hyperspectral image data, *Remote Sensing of Environment*, 112, 4301–4317, 2008.

21. Andrew, M.E. and Ustin, S.L., The effects of temporally variable dispersal and landscape structure on invasive species spread, *Ecological Applications*, 20(3), 593–608, 2010.

22. Hamada, Y., Stow, D.A., Coulter, L.L., Jafolla, J.C., and Hendricks, L.W., Detecting Tamarisk species (*Tamarix* spp.) in riparian habitats of Southern California using high spatial resolution hyperspectral imagery, *Remote Sensing of Environment*, 109, 237–248, 2007.

23. Narumalani, S., Mishra, D.R., Wilson, R., Reece, P., and Kohler, A., Detecting and mapping four invasive species along the floodplain of North Platte River, Nebraska, *Weed Technology*, 23, 99–107, 2009.

24. Asner, G.P. and Vitousek, P.M., Remote analysis of biological invasion and biogeochemical change, *Proceedings of the National Academy Science of the United States of America*, 102, 4383–4386, 2005.

25. Ramsey III, E., Rangoonwala, A., Nelson, G., and Ehrlich, R., Mapping the invasive species, Chinese tallow, with EO1 satellite Hyperion hyperspectral image data and relating tallow occurrences to a classified Landsat Thematic Mapper land cover map, *International Journal of Remote Sensing*, 26(8), 1637–1657, 2005.

26. Ramsey III, E., Rangoonwala, A., Nelson, G., Ehrlich, R., and Martella, K., Generation and validation of characteristic spectra from EO1 Hyperion image data for detecting the occurrence of the invasive species, Chinese tallow, *International Journal of Remote Sensing*, 26(8), 1611–1636, 2005.

27. Asner, G.P., Jones, M.O., Martin, R.E., Knapp, D.E., and Hughes, R.F., Remote sensing of native and invasive species in Hawaiian forests, *Remote Sensing of Environment*, 112, 1912–1926, 2008.

28. Asner, G.P., Knapp, D.E., Kennedy-Bowdoin, T., Jones, M.O., Martin, R.E., Boardman, J., and Hughes, R.F., Invasive species detection in Hawaiian rainforests using airborne imaging spectroscopy and LiDAR, *Remote Sensing of Environment*, 112, 1942–1955, 2008.

29. Brooks, M.L., D'Antonio, C.M., Richardson, D.M., Grace, J.B., and Keeley, J.E., Effects of invasive alien plants on fire regimes, *BioScience*, 54(7), 677–688, 2004.

30. Friederici, P., The alien saltcedar, *American Forests*, 101, 45–47, 1995.

31. D'Antonio, C.M. and Vitousek, P.M., Biological invasions by exotic grasses, the grass/fire cycle, and global change. *Annual Review of Ecology Systematics*, 23, 63–87, 1992.

32. Asner, G.P., Martin, R., Carlson, K., Rascher, U., and Vitousek, P., Vegetation-climate interactions among native and invasive species in Hawaiian rainforest, *Ecosystems*, 9, 1106–1117, 2006.

33. Hughes, F.R. and Denslow, J.S., Invasion by a N2-fixing tree alters function and structure in wet lowland forests of Hawaii, *Ecological Applications*, 15, 1615–1628, 2005.

34. Funk, J.L. and Vitousek, P.M., Resource-use efficiency and plant invasion in low-resource systems, *Nature*, 446, 1079–1081, 2007.

35. Hunt, E.R., McMurtrey, J.E., Parker, A.E., and Corp, L.A., Spectral characteristics of leafy spurge (*Euphorbia esula*) leaves and flower bracts, *Weed Science*, 52, 492–497, 2004.

36. Hunt, E.R., Daughtry, C.S., Kim, M.S.S., and Williams, A.E.P., Using canopy reflectance models and spectral angles to assess potential of remote sensing to detect invasive weeds, *Journal of Applied Remote Sensing*, 1, 013506–013519, 2007.

37. Pu, R. and Gong, P., Chapter 5: Hyperspectral remote sensing of vegetation bioparameters, *Advances in Environmental Remote Sensing: Sensors, Algorithms, and Applications* (Q. Weng, Ed.), CRC Press/Taylor & Francis Group, pp. 101–142, 2011.

38. Tsai, F. and Philpot, W., Derivative analysis of hyperspectral data, *Remote Sensing of Environment*, 66(1), 41–51, 1998.

39. Cloutis, E.A., Hyperspectral geological remote sensing: Evaluation of analytical techniques, *International Journal of Remote Sensing*, 17(12), 2215–2242, 1996.

40. van der Meer, F., The effectiveness of spectral similarity measures for the analysis of hyperspectral imagery, *International Journal of Applied Earth Observation and Geoinformation*, 8(1), 3–17, 2006.

41. Pengra, B.W., Johnston, C.A., and Loveland, T.R., Mapping an invasive plant, Phragmites australis, in coastal wetlands using the EO-1 Hyperion hyperspectral sensor, *Remote Sensing of Environment*, 108, 74–81, 2007.

42. Thenkabail, P.S., GangadharaRao, P., Biggs, T., Krishna, M., and Turral, H., Spectral matching techniques to determine historical land use/land cover (LULC) and irrigated areas using time-series AVHRR pathfinder datasets in the Krishna River Basin, India, *Photogrammetric Engineering and Remote Sensing*, 73(9), 1029–1040, 2007.

43. Thenkabail, P.S., Lyon, G.J., Turral, H., and Biradar, C.M., *Remote Sensing of Global Croplands for Food Security*, CRC Press/Taylor & Francis Group, Boca Raton, FL, 556 pp. (48 pages in color), June 2009.

44. Lass, L.W. and Prather, T.S., Detecting the locations of Brazilian pepper trees in the everglades with a hyperspectral sensor, *Weed Technology*, 18, 437–442, 2004.

45. Hirano, A., Madden, M., and Welch, R., Hyperspectral image data for mapping wetland vegetation, *Wetlands*, 23(2), 436–448, 2003.

46. Gong, P., Pu, R., Biging, G.S., and Larrieu, M., Estimation of forest leaf area index using vegetation indices derived from Hyperion hyperspectral data, *IEEE Transactions on Geoscience and Remote Sensing*, 41(6), 1355–1362, 2003.

47. Thenkabail, P.S., Smith, R.B., and De-Pauw, E., Hyperspectral vegetation indices for determining agricultural crop characteristics, *Remote sensing of Environment*, 71, 158–182, 2000.

48. Thenkabail, P.S., Smith, R.B., and De-Pauw, E., Evaluation of narrowband and broadband vegetation indices for determining optimal hyperspectral wavebands for agricultural crop characterization, *Photogrammetric Engineering and Remote Sensing*, 68(6), 607–621, 2002.

49. Thenkabail, P.S., Enclona, E.A., Ashton, M.S., and Van Der Meer, V., Accuracy assessments of hyperspectral waveband performance for vegetation analysis applications, *Remote Sensing of Environment*, 91(2–3), 354–376, 2004.

50. Thenkabail, P.S., Enclona, E.A., Ashton, M.S., Legg, C., and Jean De Dieu, M., Hyperion, IKONOS, ALI, and ETM+ sensors in the study of African rainforests, *Remote Sensing of Environment*, 90, 23–43, 2004.

51. Zarco-Tejada, P.J., Miller, J.R., Noland, T.L., Mohammed, G.H., and Sampson, P.H., Scaling-up and model inversion methods with narrowband optical indices for chlorophyll content estimation in closed forest canopies with hyperspectral data, *IEEE Transactions on Geoscience and Remote Sensing*, 39(7), 1491–1507, 2001.

52. Eitel, J.U.H., Gessler, P.E., Smith, A.M.S., and Robberecht, R., Suitability of existing and novel spectral indices to remotely detect water stress in *Populus* spp., *Forest Ecology and Management*, 229(1–3), 170–182, 2006.

53. He, Y., Guo, X., and Wilmshurst, J., Studying mixed grassland ecosystems I: Suitable hyperspectral vegetation indices, *Canadian Journal of Remote Sensing*, 32(2), 98–107, 2006.

54. Hunt, G.R., Electromagnetic radiation: The communication link in remote sensing, in *Remote Sensing in Geology* (B. Siegal and A. Gillespia, Eds.), Wiley, New York, 702 pages, 1980.

55. Curran, P.J., Remote sensing of foliar chemistry, *Remote Sensing of Environment*, 30, 271–278, 1989.

56. Clark, R.N. and Roush, T.L., Reflectance spectroscopy: Quantitative analysis techniques for remote sensing applications, *Journal of Geophysical Research*, 89, 6329–6340, 1984.

57. Gong, P., Pu, R., and Heald, R.C., Analysis of *in situ* hyperspectral data for nutrient estimation of giant sequoia, *International Journal of Remote Sensing*, 23(9), 1827–1850, 2002.

58. Pu, R. and Gong, P., Wavelet transform applied to EO-1 hyperspectral data for forest LAI and crown closure mapping, *Remote Sensing of Environment*, 91, 212–224, 2004.

59. Green, A.A., Berman, M., Switzer, P., and Craig, M.D., A transformation for ordering multispectral data in terms of image quality with implications for noise removal, *IEEE Transactions on Geoscience and Remote Sensing*, 26, 65–74, 1988.

60. Pu, R., Gong, P., Michishita, R., and Sasagawa, T., Spectral mixture analysis for mapping abundance of urban surface components from the Terra/ASTER data, *Remote Sensing of Environment*, 112, 939–954, 2008.

61. Walsh, S.J., McCleary, A.L., Mena, C.F., Shao, Y., Tuttle, J.P., González, A., and Atkinson, R., QuickBird and Hyperion data analysis of an invasive plant species in the Galapagos Islands of Ecuador: Implications for control and land use management, *Remote Sensing of Environment*, 112, 1927–1941, 2008.

62. van Aardt, J.A.N. and Wynne, R.H., Examining pine spectral separability using hyperspectral data from an airborne sensor: An extension of field-based results, *International Journal of Remote Sensing*, 28(2), 431–436, 2007.

63. Sasaki, K., Kawata, S., and Minami, S., Estimation of component spectral curves from unknown mixture spectra, *Applied Optics*, 23, 1955–1959, 1984.

64. Zhang, L., Li, D., Tong, Q., and Zheng, L., Study of the spectral mixture model of soil and vegetation in Poyang Lake area, China, *International Journal of Remote Sensing*, 19, 2077–2084, 1998.

65. Boardman, J.W., Kruse, F.A., and Green, R.O., Mapping target signatures via partial unmixing of AVIRIS data, *Summaries of the Fifth JPL Airborne Geoscience Workshop JPL Publication*, 95–1 (pp. 23–26), Pasadena, CA: NASA Jet Propulsion Laboratory, 1995.

66. Andrew, M.E. and Ustin, S.L., Habitat suitability modeling of an invasive plant with advanced remote sensing data, *Diversity and Distributions*, 15, 627–640, 2009.

67. Goetz, A.F.H., Ferri, M., Kindel, B., and Qu, Z., Atmospheric correction of Hyperion data and techniques for dynamic scene correction, *2002 IEEE International Geoscience and Remote Sensing Symposium and the 24th Canadian Symposium on Remote Sensing*, Toronto, Canada, June 24–28, 2002.

68. Jensen, J.R., Electromagnetic radiation principles and radiometric correction, *Introductory Digital Image Processing: A Remote Sensing Perspective*, 3rd Edn., Pearson Prentice Hall, Upper Saddle River, NJ, pp. 175–222, 2005.

69. *FLAASH User's Guide*, ENVI FLAASH Version 4.1, September, 2004 Edition, Research Systems, Inc. 1-80, 2004.

70. Gao, B.-C., Heidebrecht, K.B., and Goetz, A.F.H., Derivation of scaled surface reflectance from AVIRIS data, *Remote Sensing of Environment*, 44, 165–178, 1993.

71. Qu, Z., Kindel, B.C., and Goetz, A.F.H., The high accuracy atmospheric correction for hyperspectral data (HATCH) model, *IEEE Transactions on Geoscience and Remote Sensing*, 41, 1223–1231, 2003.

72. Meyer, P., Itten, K.I., Kellenberger, T., Sandmeier, S., and Sandmeier, R., Radiometric corrections of topographically induced effects on Landsat TM data in alpine environment, *ISPRS Journal of Photogrammetry and Remote Sensing*, 48, 17–28, 1993.

73. Allen, T.R., Topographic normalization of Landsat thematic mapper data in three mountain environments, *Geocarto International*, 15(2), 13–19, 2000.

74. Adler-Golden, S.M., Matthew, M.W., Anderson, G.P., Felde, G.W., and Gardner, J.A., An algorithm for de-shadowing spectral imagery, in *Proceedings of the 11th JPL Airborne Earth Science Workshop*, 5–8, March, 2002 (Pasadena, CA: JPL Publication 03-04), 2002.

75. Judd, C., Steinberg, S., Shaughnessy, F., and Crawford, G., Mapping salt marsh vegetation using aerial hyperspectral imagery and linear unmixing in Humboldt Bay, California, *Wetlands*, 27(4), 1144–1152, 2007.

76. Asner, G.P., Martin, R.E., David E., Knapp, D.E., and Kennedy-Bowdoin, T., Effects of Morella faya tree invasion on aboveground carbon storage in Hawaii, *Biological Invasions*, 12, 477–494, 2010.

77. Ustin, S.L. and Gamon, J.A., Remote sensing of plant functional types, *New Phytologist*, 186, 795–816, 2010.

78. Andrew, M.E. and Ustin, S.L., Spectral and physiological uniqueness of perennial pepperweed (*Lepidium latifolium*), *Weed Science*, 54, 1051–1062, 2006.

79. Jacquemoud, S. and Baret, F., Prospect—A model of leaf optical properties spectra, *Remote Sensing of Environment*, 34, 75–91, 1990.

80. NRC's Decadal Survey report, *Earth Science and Applications from Space: National Imperatives for the Next Decade and Beyond*, http://www.nap.edu/catalog/11819.html, 2007.

81. Rouse, J.W., Haas, R.H., Schell, J.A., and Deering, D.W., Monitoring vegetation systems in the Great Plains with ERTS, in *Proceedings, Third ERTS Symposium*, Greenbelt, Maryland, USA. vol. 1, pp. 48–62, 1973.

82. Jordan, C.F., Derivation of leaf area index from quality of light on the forest floor, *Ecology*, 50, 663–666, 1969.

83. Tucker, C.J., Red and photographic infrared linear combinations for monitoring vegetation, *Remote Sensing of Environment*, 8, 127–150, 1979.

84. Gitelson, A.A., Keydan, G.P., and Merzlyak, M.M., Three-band model for non-invasive estimation of chlorophyll, carotenoids and anthocyanin contents in higher plant leaves, *Geophysical Research Letters*, 33, L11402, 2006.

85. Datt, B., A new reflectance index for remote sensing of chlorophyll content in higher plants: Tests using Eucalyptus leaves, *Journal of Plant Physiology*, 154, 30–36, 1999.

86. Sims, D.A. and Gamon, J.A., Relationships between leaf pigment content and spectral reflectance across a wide range of species, leaf structures and developmental stages, *Remote Sensing of Environment*, 81, 337–354, 2002.

87. Fuentes, D.A., Gamon, J.A., Qiu, H.-L., Sims, D.A., and Roberts, D.A., Mapping Canadian boreal forest vegetation using pigment and water absorption features derived from the AVIRIS sensor, *Journal of Geophysical Research*, 106, 33565–33577, 2001.

88. Peñuelas, J., Gamon, J.A., Fredeen, A.L., Merino, J., and Field, C.B., Reflectance indices associated with physiological changes in nitrogen- and water-limited sunflower leaves, *Remote Sensing of Environment*, 48, 135–146, 1994.
89. Rama Rao, N., Garg, P.K., Ghosh, S.K., and Dadhwal, V.K., Estimation of leaf total chlorophyll and nitrogen concentrations using hyperspectral satellite imagery, *Journal of Agricultural Science*, 146, 65–75, 2008.
90. Gamon, J.A., Peñuelas, J., and Field, C.B., A narrow waveband spectral index that tracks diurnal changes in photosynthetic efficiency, *Remote Sensing of Environment*, 41, 35–44, 1992.
91. Zarco-Tejada, P.J., Optical indices as bioindicators of forest sustainability, *Report to the Graduate Programme in Earth and Space Science Toronto*, York University, Toronto, Ontario, Canada, 1998.
92. Gamon, J.A. and Surfus, J.S., Assessing leaf pigment content and activity with a reflectometer, *New Phytologist*, 143, 105–117, 1999.
93. Nagler, P.L., Daughtry, C.S.T., and Goward, S.N., Plant litter and soil reflectance, *Remote Sensing of Environment*, 71, 207–215, 2000.
94. Serrano, L., Peñuelas, J., and Ustin, S.L., Remote sensing of nitrogen and lignin in Mediterranean vegetation from AVIRIS data: Decomposing biochemical from structural signals, *Remote Sensing of Environment*, 81, 355–364, 2002.
95. Datt, B., McVicar, T.R., Van Niel, T.G., Jupp, D.L.B., and Pearlman, J.S., Preprocessing EO-1 Hyperion hyperspectral data to support the application of agricultural indexes, *IEEE Transactions on Geoscience and Remote Sensing*, 41, 1246–1259, 2003.
96. Gao, B.C., NDWI—A normalized difference water index for remote sensing of vegetation liquid water from space, *Remote Sensing of Environment*, 58, 257–266, 1996.
97. Schlerf, M., Atzberger, C., and Hill, J., Remote sensing of forest biophysical variables using HyMap imaging spectrometer data, *Remote Sensing of Environment*, 95, 177–194, 2005.
98. Peñuelas, J., Piñol, J., Ogaya, R., and Filella, I., Estimation of plant water concentration by the reflectance water index WI (R900/R970), *International Journal of Remote Sensing*, 18, 2869–2875, 1997.
99. Blackburn, G.A., Quantifying chlorophylls and carotenoids at leaf and canopy scales: An evaluation of some hyperspectral approaches, *Remote Sensing of Environment*, 66, 273–285, 1998.

Part VIII

Land Cover Applications

20 Hyperspectral Remote Sensing for Forest Management

Valerie Thomas

CONTENTS

20.1 INTRODUCTION

Forests cover about 30% of the Earth's terrestrial surface (FAO 2006), play a significant role in the climate system, and are integral to numerous ecosystem, cultural, and economic services (Figure 20.1). As such, there is considerable interest in sustainable management to conserve forest resources, while balancing the competing interests for their use. Forest management is a multifaceted endeavor, defined as the "application of biological, physical, quantitative, managerial, economic, social, and policy principles to the regeneration, management, utilization, and conservation of forests to meet specified goals and objectives while maintaining the productivity of the forest (Helms 1998, p. 71)." In practical terms, this often means some type of silviculture practice, protection activity, and/or forest regulation.

Given the large, and often remote, land area covered by forests, remote sensing technologies have been widely adopted as part of operational forest management portfolios, mainly to monitor the location, type, and amount of forests, as well as changes over time. To date, the use of remote sensing for forest management has been largely driven by aerial photography and multispectral satellite imagery (such as Landsat or SPOT). Hyperspectral technology may allow us to expand our remote characterizations to examine species, forest health and condition, stand structure, and possibly forest ecosystem function (Carter 1994, Goetz 1995, Lichtenhaler et al. 1996, Martin and Aber 1996, Green et al. 1998, Merton 1998, Ustin and Trabucco 2000, Curran 2001, Treitz et al. 2010). Indeed, most of the vegetation applications of hyperspectral remote sensing discussed in Chapters 4 through 7 of this book could be applicable to forest management. However, despite the many potential benefits of this technology, the use of hyperspectral imagery for operational forest management applications has been extremely limited. Most of the work has been in the realm of scientific research or pilot/demonstration studies. This reluctance by practicing forest managers to the

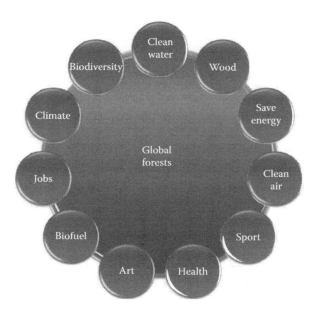

FIGURE 20.1 **(See color insert.)** Ecosystem, economic, and cultural services of forests.

large-scale adoption of hyperspectral remote sensing stems from the size and complexity of both the hyperspectral datasets (Chapter 3) and the forest ecosystems.

20.2 COMPLEXITIES OF FOREST ECOSYSTEMS

There are many factors that affect the reflectance of all vegetation and confound remote sensing analysis. These include phenology, insolation, illumination geometry, soil characteristics, the nutrient regime, hydrology, and spectral similarities between many species. In fact, it has been demonstrated that the inherent within-species variability in reflectance, combined with spectral similarities across species, has made some species indistinguishable using reflectance alone (Hoffbeck and Landgrebe 1996). Analysis of forest ecosystems is further confounded by highly variable three-dimensional structure, particularly in mixedwood canopies (Figure 20.2). Forests often have complex horizontal and vertical species mixtures (causing layering in the canopy), complex canopy

(A)   (B)

FIGURE 20.2 **(See color insert.)** Complex forest canopy architecture. (A) Evident of variable species, canopy height, and layering in a boreal mixedwood forest. (B) Below-canopy variability of a boreal mixedwood forest.

architecture, variable height and biomass, within- and across-species variability in leaf area and in foliar biochemistry. Variability in leaf morphology and foliar biochemistry has also been demonstrated within a single canopy, particularly between sunlit leaves near the top of the canopy and shaded leaves below (Gholz et al. 1991, Vose et al. 1994, Demarez et al. 1999, O'Neill et al. 2002). Complex canopy architecture, particularly the size and location of gaps within the canopy, also influences the light regime (Hardy et al. 2004), and can influence the rate of photosynthesis and absorption of light (Todd et al. 2003, Thomas et al. 2006a).

When analyzing remote sensing data, the inherent heterogeneity of a forest ecosystem can be further magnified by the pixel resolution of the sensor. If the size of the pixel is greater than the size of a single tree canopy, the reflectance signal may contain mixed effects of shadow, nonleaf reflectance, species mixtures, etc. (Woodcock and Strahler 1987, St-Onge and Cavayas 1995, Curran and Atkinson 1999, Treitz and Howarth 2000, Treitz 2001). This has resulted in considerable research effort in canopy reflectance modeling, in an attempt to quantify and/or remove some of the nonleaf reflectance from the signal.

Of particular relevance is the body of research in geometric modeling, which assumes that a forest is composed of multiple objects with quantifiable dimensions, shapes, and arrangements. The simplest of these models, originally developed for crop canopies, are designed to simulate the effect of shadow on canopy reflectance, where the fractions of vegetation, soil, and their shadowed components are quantified (e.g., Jahnke and Lawrence 1965, Terjung and Louie 1972, Jackson et al. 1979). More complex models combine canopy structural geometry and the principles of radiative transfer (RT) within crowns (such as the geometric optical and radiative transfer model [GORT] [Li et al. 1995]). These models, referred to as hybrid radiative transfer models, have been used to characterize canopy structure and foliar biochemistry in a variety of heterogeneous forest environments (Kuusk 1998, Dawson et al. 1999, Demarez and Gastellu-Etchegorry 2000, Hu et al. 2000, Gastellu-Etchegorry and Bruniquel-Pinel 2001, Zarco-Tejada et al. 2001, Gemmell et al. 2002, Kimes et al. 2002, Kötz et al. 2004).

More recently, scientists are examining the possibility of hyperspectral fusion with light detection and ranging (LIDAR) data. LIDAR has been shown to be the ideal technology to characterize canopy structure, including height, crown shape, leaf area, biomass, and basal area (e.g., Næsset 1997, Magnussen and Boudewyn 1998, Means et al. 2000, Lim and Treitz 2004, Hopkinson et al. 2005, Thomas et al. 2006b). Fusing LIDAR with hyperspectral data in a hybrid RT modeling approach could provide better information on canopy structure (LIDAR) and canopy biochemistry (hyperspectral). This approach has been applied on synthetic data to characterize such parameters as height, canopy cover, leaf area, canopy chlorophyll content, and canopy water content (e.g., Koetz et al. 2006). However, this work is very much at the research stage, and has not yet been successfully implemented for real operational forest management scenarios.

20.3 FOREST MANAGEMENT APPLICATIONS OF HYPERSPECTRAL REMOTE SENSING

There are a number of applications for hyperspectral remote sensing that can directly benefit forest management programs that have national and local implementations. These include, but are not limited to, information to supplement forest inventories (e.g., species and biophysical variables), improved estimation of biomass and carbon (of interest for a variety of purposes, including climate modeling), and mapping of wildfire fuels for forest fire risk detection.

20.3.1 FOREST INVENTORIES

Forest inventories involve a systematic collection of data about forest stands that are used for a variety of local and national purposes. Many countries have their own well-developed and long-standing system of standardized data collection, which is usually very manually intensive. For example, in the

United States, the Forest Inventory and Analysis national program has been run by the USDA Forest Service since 1930. Finland has a National Forest Inventory program that has been operating since 1920. Although the specific methods of data collection vary by country, most involve a systematic sampling of trees to attain representative data for to characterize a forest stand. At minimum, this includes species, height, and diameter at breast height (DBH). From these, stand basal area can be calculated and allometric equations can be used to model biomass, volume, and carbon. At the stand level, some measurement is also usually made of stage age and quality. Despite the complexity of forest environments discussed above, hyperspectral remote sensing can provide information that benefits forest inventories, particularly for improved species classifications and mapping certain biophysical variables.

20.3.1.1 Forest Species Mapping

Numerous authors have demonstrated the potential of hyperspectral remote sensing to improve species mapping. A number of authors have used hyperspectral data and indices known to be related to foliar pigments to discriminate between canopy species across the landscape (Wessman et al. 1988, Fuentes et al. 2001, Clark et al. 2003, Kokaly et al. 2003, Townsend et al. 2003, Plourde et al. 2007). Others have developed techniques to use fused LIDAR and hyperspectral data for species mapping, which enables the spectral and structural characteristics of the canopy to be examined concurrently (Asner et al. 2008a,b, Thomas et al. 2009). There have also been numerous studies in which the spectral reflectance of the hyperspectral data has been matched directly to the species of interest. Spectral libraries have been developed for many species, under a variety of phonological conditions. This has been done at the leaf scale, using a portable handheld spectroradiometer (e.g., Clark et al. 2005), and at the canopy scale by locating "pure" pixels (i.e., pixels that contain only one species, also referred to as an endmember) within the hyperspectral image (e.g., Kokaly et al. 2003). Spectral matching algorithms, such as the spectral angle mapper (SAM), can be used to match the spectral pattern of pixels in the image to the endmembers by determining the cosine of the angle in spectral space, where a low angle implies a close match (Buddenham et al. 2005). This approach assumes that the spectral response in a pixel is "pure," and tends to work better at higher spatial resolutions where there is less mixture of species within a pixel, or for broader classes that include multiple species with a similar spectral response. In cases where there is known mixture of desired classes within pixels, spectral mixture analysis has been used to quantify the fraction of each species within pixels (e.g., Darvishsefat et al. 2002).

20.3.1.2 Forest Biophysical Variables

There are numerous forest biophysical variables that are of interest for forest inventories, including leaf area index (LAI), crown closure or crown gaps, tree or canopy height, stem density, crown depth, biomass or volume, DBH, and amount of dead trees in a stand (Olthof and King 1998, Treitz and Howarth 1999, Sampson 2000). Although there has been considerable effort to monitor these variables using remote sensing, most of the work to date has been through the use of broadband sensors, such as Landsat, or through the use of active systems, such as synthetic aperture radar (SAR) and LIDAR. Research has generally followed three paths: (1) the use of broadband indices, such as the calculation of the normalized difference vegetation index (NDVI) for the prediction of LAI (e.g., Peterson et al. 1987, Herwitz et al. 1989, Spanner et al. 1990, Curran et al. 1992, Shippert et al. 1995, Chen and Cihlar 1996, Green et al. 1997, White et al. 1997, Wang et al. 2005); (2) analysis of spatial statistics, texture, or multivariate statistical techniques (e.g., Yuan et al. 1991, Hershey et al. 1998, Olthof and King 1998, Davison et al. 1999, Phinn et al. 1999, Seed et al. 1999, Pellikka et al. 2000); (3) use of active-sensor metrics to develop predictive regression models (e.g., Næsset 1997, Magnussen and Boudewyn 1998, Means et al. 2000, Lim and Treitz 2004, Hopkinson et al. 2005, Thomas et al. 2006b).

Hyperspectral research in this area has generally focused either on the use of narrowband indices, partial regression analysis, spectral mixture analysis, or canopy reflectance modeling to predict biophysical variables of interest. In the case of canopy reflectance modeling (mentioned in Section 20.2 above), the models predict reflectance based on canopy characteristics (e.g., foliar biochemistry, water, leaf morphology, and transmittance characteristics), solar-target-sensor geometry, and soil information. For example, Chen et al. (1999) used the canopy reflectance model 4-Scale to improve hyperspectral predictions of LAI and crown closure in a boreal forest ecosystem. When reflectance is the input, these models can be inverted to predict the desired structural information (referred to as inversion modeling). This was demonstrated in a coniferous ecosystem for LAI, canopy cover, water content, and dry material (Schaepman et al. 2005).

Although there is a known sensitivity of hyperspectral reflectance and indices to structure and leaf area/morphology, use of hyperspectral indices to predict forest biophysical variables has been relatively limited. Notable exceptions to this include Gong et al. (2003a,b), Lee et al. (2004), and Schlerf et al. (2005) for LAI and/or volume prediction. More recently, Thomas et al. (2010) used hyperspectral indices to predict LAI and clumping, and found strong relationships between the derivative chlorophyll index (DCI) and clumping. Clumping is a variable that describes nonrandom foliage distribution within canopies (Chen et al. 2005), and can be thought of as a descriptor of canopy gap distribution. It partially drives processes that are strongly affected by the canopy light regime, such as photosynthesis, evapotranspiration, and the distribution of canopy foliar nutrients (Stenberg 1998, Chen et al. 1999, Alt et al. 2000, Dreccer et al. 2000, Bernier et al. 2001, Palmroth and Hari 2001, Baldocchi et al. 2002, Liu et al. 2002, Thomas et al. 2006a, 2009). Thomas et al. (2010) demonstrated strong relationships between clumping and several canopy nutrients, including nitrogen, chlorophyll, carotenoids, phosphorus, and magnesium.

The work of Schlerf et al. (2005) and Thomas et al. (2010) suggests that some hyperspectral indices may provide direct insight into volume, canopy height, above ground biomass/carbon, and crown closure. Expanding upon the work of Thomas et al. (2010), the relationship between airborne hyperspectral indices and plot-based stand variables was assessed using simple bivariate regression. For most variables, the relationship was shown to be logarithmic (Figure 20.3) and significant, with 50%–75% of the variance explained by the models (Table 20.1). As expected, given the relationship to clumping, crown closure was well predicted by the DCI. These results suggest that hyperspectral data is currently being underutilitilzed for the purpose of stand inventories, and may offer more than just species or foliar biochemistry.

Goodenough et al. (2006, 2008) demonstrated the ability to map biomass and carbon from the Airborne Visible/Infrared Imaging Spectrometer (AVIRIS) data using partial least square regression analysis on reflectance data and the first and second derivatives of reflectance ($r^2 = 0.82$, Goodenough et al. 2006). In their work, they also compared the performance of two airborne hyperspectral

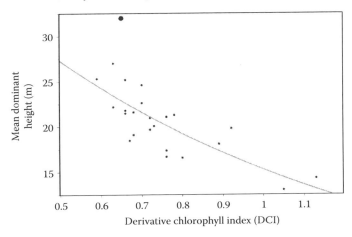

FIGURE 20.3 Logarithmic relationship between the hyperspectral DCI and mean dominant height. Regression statistics shown in Table 20.1.

TABLE 20.1

Logarithmic Bivariate Models to Predict Field Metrics from the Mean Derivative Chlorophyll Index

Field Metric versus DCI (n = 24)	r^2	r^2_{adj}	RMSE	Plot Mean
Mean dominant height (m)[a]	0.59 (0.64)	0.57 (0.62)	2.16 m	20.4 m
Quadratic DBH (cm)	0.55	0.53	3.51 cm	19.3 cm
Total above ground biomass (v)[a]	0.63, (0.67)	0.61, (0.65)	1740	4426.5
Total above ground carbon (v)[a]	0.63, (0.67)	0.61, (0.65)	870.5	2208.3
Crown closure	0.75	0.74	0.48	0.98
Stem density (#/ha)	No significant relationship			

Assumption of normality tested and satisfied for all regression residuals.

[a] The r^2 and r^2_{adj} were also calculated after removal of a significant outlier, shown as a larger circle in Figure 20.3.

sensors with significantly different signal-to-noise ratios. The AVIRIS sensor, designed and operated by NASA, is known to produce high quality images with a high signal-to-noise ratio. These data were compared to imagery from the airborne imaging spectrometer for applications (AISA) sensor, which had very comparable signal-to-noise ratios in the visible and near infrared wavelengths, but significantly lower values in the short wave infrared wavelengths (3.43 versus 4.50 relative signal-to-noise values for AISA SWIR and AVIRIS SWIR, respectively (Goodenough et al. 2008). They attribute the superior performance of AVIRIS to the higher signal-to-noise ratio, as well as to solar geometry effects in the AISA data. Their findings highlight one of the limiting factors to the use of hyperspectral at the operational forest management applications. That is, there is significant preprocessing of hyperspectral data required to derive high quality reflectance data, which requires considerable user expertise.

20.3.2 Carbon Exchange

Forests are the largest terrestrial carbon stores and make a significant contribution to the global carbon cycle, which is widely felt to have a fundamental role in regulating the climate of the Earth. In North America, the combined boreal and temperate forests cover a significant portion of our landscape and, in addition to the major role they play in the carbon cycle, also have a significant influence on the water and energy balance, animal habitat, and the economic function of many regions. Despite the importance of our forests to climate change and other processes, our understanding of and ability to model and predict the carbon cycle remains weak in some areas, and there are numerous ongoing research efforts to improve our understanding of these systems. Part of the challenge exists because carbon sequestration processes occur at many spatial and temporal scales and it is difficult to envision a measurement/modeling scheme that can adequately represent this variability.

Carbon cycling is often described in terms of net ecosystem productivity (NEP) or net primary productivity (NPP). NEP and NPP are fundamental ecological concepts, which are not only critical to regional, national, and global-scale carbon budgets and climate models, but also indicate terrestrial land surface condition and ecological processes (Ciais et al. 2005, Kashian et al. 2006, Bo et al. 2007). Unfortunately, there are significant problems and inaccuracies with current global NEP and NPP models, often resulting in poor correlation between measured NEP at micrometeorological flux stations and national/global-scale model predictions. A large number of national/global-scale carbon models calculate NPP from the fraction of photosynthetically active radiation absorbed by the canopy (fPAR) and a light use efficiency (LUE) term that is considered constant for a given vegetation type and is typically biome-specific (Running et al. 1999, Ahl et al. 2004). However, growing evidence suggests that LUE varies according to many factors across and within ecosystems, including ecosystem type, stand age, species composition, nutrient availability, and vegetation stress (e.g., Goetz and Prince 1996, Goetz and Prince 1998, Medlyn 1998, Turner et al. 2003a,b, Ahl et al. 2004, Drolet et al. 2005, Jenkins et al. 2007). To improve global terrestrial models of NEP, it is essential to develop a deeper understanding of the factors that control LUE.

Research at satellite and airborne scales suggest that hyperspectral remote sensing has much to offer toward the improved accuracy of global carbon exchange models, though it is underutilized in this context. These data have a direct role to play for: improved species mapping (e.g., Wessman et al. 1988, Fuentes et al. 2001, Clark et al. 2003, Kokaly et al. 2003, Townsend et al. 2003, Plourde et al. 2007, Asner et al. 2008a,b, Thomas et al. 2009), characterization of vegetation structure (e.g., Schlerf et al. 2005, Thomas et al. 2010), quantification of above-ground carbon and biomass (e.g., Schlerf et al. 2005), mapping of fPAR and LUE (Thomas et al. 2009), nutrient availability (e.g., Zarco-Tejada et al. 2001, 2004), vegetation stress, vegetation water content, and wildfire regimes (discussed below).

20.3.3 WILDFIRE FUEL

Another forest management application for which hyperspectral remote sensing can make a direct contribution is the mapping of forest fire risk, particularly with regard to the type and quantity of fire fuel, which can affect the behavior and intensity of fires. This topic is drawing increasing attention and urgency, with observed increases in the frequency and intensity of fires due to climate change (Tymstra et al. 2007, Flannigan et al. 2009) and exurban encroachment into forested areas (Syphard et al. 2007a,b). There is an identified need for a better understanding of the spatial distribution of different wildfire fuel types and amounts, as well as areas that have been previously burned (Ustin et al. 2004, Jia et al. 2006).

Wildfire fuels include all dead or living vegetation that can be ignited (Miller 2001). Their arrangement and availability, including the structure of the canopy itself, will determine whether the burn will be a ground surface, or crown fire (Miller 2001, Cruz et al. 2003). Hyperspectral remote sensing can be used to map the spatial distribution of wildfire fuel properties, providing much-needed information on the type and arrangement, as well as the location and severity of previous burns. Most of the work in this area has focused on the use of narrowband vegetation indices or RT modeling to map fuel properties within the canopy. This may include canopy moisture content (Jacquemoud and Ustin 2003, Li et al. 2005, Riano et al. 2005), amount of dry or senescent carbon or degree of fuel curing (Melillo et al. 1982, Fourty et al. 1996, Serrano et al. 2002), canopy stress detection and foliar biochemistry (Ustin and Trabucco 2000, Smith et al. 2003), and relative amount of photosynthetic verses nonphotosynthetic materials (Asner et al. 1998).

Another approach to fire fuel mapping follows the species classification work discussed above. Some authors have collected the field reflectance spectra of a number of fuel attributes and used those spectra as endmembers (i.e., pure spectra). Authors have attempted to quantify the relative fraction of endmembers at the subpixel level using spectral mixture analysis (e.g., Green et al. 1998, Clark et al. 2003, Ustin et al. 2004, Jia et al. 2006) and at the pure-pixel level, using spectral matching algorithms (e.g., Jia et al. 2006).

Although fire fuel attribute mapping is really in the research phase, there has been some attempt to move toward operational capabilities in this area. To assist forest planners, Environment for Visualizing Images (ENVI), one of the software packages used to analyze hyperspectral data, has combined a number of vegetation indices that detect canopy water content, greenness, and dry carbon into a "Fire Fuel Tool," which produces a map of relative fire risk based on the combination of these three attributes (RSI 2009). The tool requires limited user expertise, both in its operation and interpretation of results, setting the stage for the adoption of the technique by users beyond the research community.

In addition to wildfire fuel mapping, hyperspectral remote sensing has also been used as an input to fire behavioral models. Fire behavior is largely controlled by topography, weather, and the availability of fuel (Miller 2001). Topography and weather are generally well known and accurately characterized across space and time. Fire behavior models, which range in complexity from simple empirical approaches (Andrews 1986, Finney 1998) to complex physics models that are based upon thermodynamics and the conservation of mass, momentum, and energy (e.g., Linn 1997, Dupuy and Larini 1999, Porterie et al. 2000, Grishin 2001a,b). For many of these models, fire fuel attributes are averaged over large areas, or broad ecological types. With advances in the ability to map complex wildfire fuel characteristics from hyperspectral data, fuel maps can now be directly input into some models (e.g., FIRETEC, Linn 1997) to provide more accurate simulations of fire behavior (e.g., Bossert et al. 2000, Dennison et al. 2000).

20.4 POTENTIAL FUTURE APPLICATIONS

As briefly mentioned in Section 20.2, a very promising form of data fusion is the integration of LIDAR data with hyperspectral data, which would provide insight into canopy architecture, the canopy light regime, and it influence on reflectance (in other words, insight into both structure and function of a forest environment). This type of data fusion has the potential to advance research in numerous areas, including bidirectional reflectance modeling, species mapping, GORT modeling, ecosystem modeling, prediction of canopy biochemistry and photosynthesis parameters, and modeling of canopy photosynthesis and carbon exchange.

Asner et al. (2008a,b) have made significant contributions toward developing more robust techniques for species mapping using a combination of hyperspectral and LIDAR data. They developed techniques to identify invasive species in Hawaiian forests, with the hope of eventually being able to map forest spread over time. Their approach, which builds upon previous work by Blackburn (2002), is to develop masks of sunlit tree crowns (at predetermined heights), which are clearly visible to an airborne or satellite sensor, thereby eliminating the confounding effect of gaps, shadows, and canopy architecture. Superimposing this mask on a hyperspectral image, Asner et al. then used a two-stage spectral mixture analysis approach to determine (1) the fraction of live and dead canopy within the sunlit areas and (2) the fraction of individual species within the sunlit crown.

Thomas et al. (2008) used fused LIDAR and hyperspectral data to scale estimates of chlorophyll from leaf to canopy, to improve species classifications, and to model the spatial variability of the fPAR for a boreal mixedwood ecosystem (Thomas et al. 2006a). These data have been ingested into models of photosynthesis and carbon exchange that, in conjunction with meteorological data being collected continuously at a flux tower site, allow for the analysis of variability within the footprint of the flux tower (Thomas et al. 2009), providing insight into the importance of canopy architecture on canopy function.

The fusion of LIDAR and hyperspectral data can also be used to model the vertical light profile of the canopy and its relationship to foliar biochemistry and photosynthesis parameters through the use of GORTs. There have been a number of studies at the leaf scale that have successfully predicted foliar biochemistry based on leaf reflectance and transmittance using leaf optical models (e.g., PROSPECT; Jacquemoud et al. 1996, Baret and Fourty 1997, Fourty and Baret 1998). At the canopy scale, RT models are used to model the interaction of solar radiation with vegetation elements and can describe the spectral reflectance of a forest stand. The simplest of these models assume that canopies are homogenous and can be described by simple geometric shapes (e.g., SAIL). More complex models, the GORT models, attempt to represent canopy geometry by modeling gaps and shadows within the canopy (e.g., GeoSAIL, GORT). By coupling RT models with leaf optical models (e.g., PROSPECT-SAIL, PROSPECT-GeoSAIL) and running backwards simulations, numerous authors have predicted canopy biochemistry from remotely sensed reflectance data (Jacquemoud et al. 2000, Zarco-Tejada et al. 2004, Malenovsky et al. 2006). Unfortunately, due to the complexity of forest canopy architecture and the confounding factors highlighted above, validation of these inversion model predictions with leaf samples has proven difficult. There is an exciting potential to incorporate detailed information about the vertical and horizontal distribution of canopy available from airborne LIDAR data directly into the GORT models. Some work has been done to use RT models to simulate LIDAR waveforms and to predict tree height, spacing, gap distribution, foliar chlorophyll, and water content from LIDAR data (e.g., Peterson et al. 2001, Kotchenova et al. 2003, Koetz et al. 2006, Koetz et al. 2007).

A better understanding of canopy structure and function across the landscape, such as might be gleaned though the fusion of LIDAR and hyperspectral data, opens the possibility to advance the research beyond the stage of mapping biochemistry, species, and biophysical variables and into the realm of predicting parameters of photosynthesis and ecosystem function. A few authors have attempted to link foliar nutrients (particularly chlorophyll and nitrogen content) to photosynthesis parameters that can be measured with an LI-6400 portable photosynthesis unit (i.e., LUE, maximum rate of light saturated photosynthesis, quantum yield efficiency, Jmax, and Vcmax) with the objective of relating photosynthesis to leaf or canopy reflectance (e.g., Richardson and Beryln 2002, Zhao et al. 2003). However, given the logistic difficulty in taking these measurements in the field, there have been relatively few studies to characterize natural forest canopies (with one notable exception being Richardson and Beryln 2002). As a result, there are many unanswered questions with regard to the potential of a leaf or canopy to photosynthesis under different environmental conditions. For example, although we know that within-species variable of foliar chlorophyll is usually relatively low compared to across-species variability (unless a large environmental gradient exists within your study area) (Carleson et al. 2007, Thomas et al. 2008), we cannot say anything with confidence about the within-species versus across-species variability in photosynthesis parameters for most forested ecosystems. Also, although we know that these parameters change throughout the growing season, there is uncertainty about how well these temporal patterns follow the temporal variability of foliar biochemistry as viewed at the canopy scale.

20.4.1 THE NEED FOR REPEATED GLOBAL MEASUREMENTS

One of the major limitations to the use of hyperspectral remote sensing in a forest management context is the lack of large-area measurements taken with consistent sensor settings and repeated over time, such as that which would be possible with satellite-based hyperspectral sensors. This has resulted in work done from airborne platforms, in localized contexts, primarily for scientific objectives.

Satellite-based hyperspectral remote sensing offers the potential for more accurate derivations of geochemical, biochemical, and biophysical variables that drive regional and global ecosystem process models. For instance, satellite hyperspectral data could be used to improve the accuracy of maps of species (or communities of species), to derive foliar pigment and nutrient concentrations for large areas, and to estimate LAI, fPAR, and parameters of photosynthesis. Satellite hyperspectral data, operated continuously, would provide this information across landscapes continually altered by disturbance and recovery processes, land use management and change, and climate change. This would improve our understanding of biogeochemical cycling and the impact of land management decisions on these processes.

Although several satellite-based hyperspectral sensors have been proposed in the past decade, there are only two in operation, neither of which were designed to provide global mapping. NASA's Hyperion sensor provides 220 bands of data across visible, near infrared, and shortwave infrared wavelengths (i.e., 0.4–2.5 µm) at 30 m resolution (Folkman et al. 2001). Hyperion is part of the Earth observing (EO-1) mission, originally intended to validate several instrument and spacecraft bus technologies. The satellite was launched in 2000, with a design life of 18 months. It is still operating, and is now in an extended mission phase. Data are not collected continuously from this sensor, but must be specifically acquired according to a user request. This makes large-area forest management applications challenging, because a single scene is relatively small (i.e., 7.7 km wide by

42 or 185 km long), and the dates of acquisition may be in conflict with other users, in which case acquisitions would be prioritized. The other satellite-based option is the Compact High Resolution Imaging Spectrometer (CHRIS) sensor on the European Space Agency's Proba-1 mission. CHRIS has a higher spatial resolution (i.e., 17 m) but fewer bands (up to 62). It was launched in 2001, with a 2 year lifespan, and is also still in operation (ESA 2008).

Research success at the satellite scale lends credence to the potential use of satellite hyperspectral for forest management in ways that are simply not possible with the existing suite of technologies. In forest contexts, Hyperion has been used for improved mapping of forest species (Goodenough et al. 2003), species communities (Thenkabail et al. 2004), various invasive species (Asner et al. 2006), crown closure and LAI (Gong et al. 2003, Pu et al. 2005), drought stress (Asner et al. 2004), biomass and carbon mapping (Asner et al. 2004, Thenkabail et al. 2004), and canopy nitrogen (Smith et al. 2003). Similarly, CHRIS has been used to map forest successional stages (Galvão et al. 2009), canopy structure and heterogeneity (Koetz et al. 2005), chlorophyll fluorescence (Raddi et al. 2005), foliar biochemistry and water content (Kneubühler et al. 2008), and wildfires (Valencia et al. 2005).

There are three new satellite-based hyperspectral sensors planned for launch within the next 5 years, two of which could be used as operational instruments. The Italian Space Agency (ASI) plans to launch the PRISMA hyperspectral instrument by 2012. This sensor is conceived as a pre-operational hyperspectral technology demonstration (ASI 2010), which will restrict the applications to being primarily research-driven. However, Germany plans to launch the EnMAP mission by 2014 and NASA the HyspIRI mission by 2015, both of which will provide repeated global coverage, enabling forest planners to study the impacts of management decisions on the function of biogeochemical cycling in a way never before possible.

20.5 CONCLUSIONS

Although the applications for hyperspectral remote sensing in a forest management context are numerous, the technology is currently being underutilized for this purpose. It is evident that hyperspectral imaging is part of the solution to global monitoring and predictions for biogeochemical cycling, forest biophysical variables, and forest physiology and function. Many successes have been demonstrated at the laboratory and localized scales, but consistent repeatable results are necessary at national and global scales (i.e., through use of satellite-based technologies) before the full potential of this technology can be realized for forest management and policy. With rising social, political, and scientific concerns surrounding the role of the world's forests in the global carbon cycle and climate, satellite-based hyperspectral data will likely become a critical tool to inform global modeling efforts, as well as our understanding of the function of remote boreal and tropical ecosystems. Even greater advances in studies of forest biochemistry, structure and species are possible when hyperspectral data is used in fusion with LIDAR.

REFERENCES

Ahl, D., S.T. Gower, D.S. Mackay, S.N. Burrows, J.M. Norman, and G.R. Diak, 2004. Heterogeneity of light use efficiency in a northern Wisconsin forest: Implication for modeling net primary production using remote sensing. *Remote Sensing of Environment*, 93: 168–178.

Alt., C., H. Stutzel, and H. Kage, 2000. Optimal nitrogen content and photosynthesis in cauliflower (*Brassica oleracea* L. *botrytis*). Scaling leaf to plant. *Annals of Botany*, 85: 779–787.

Andrews, P.L., 1986. BEHAVE: Fire behavior prediction and fuel modeling system—BURN Subsystem, Part 1. Gen. Tech. Rep. INT-194. Ogden, UT: U.S. Department of Agriculture, Forest Service Intermountain Forest and Range Experiment Station. p. 130.

ASI 2010. PRISMA. http://www.asi.it/en/activity/earth_observation/prisma_ [last cited, October 8, 2010].

Asner, G.P., M.O. Jones, R.E. Martin, D.E. Knapp, and R.F. Hughes, 2008a. Remote sensing of native and invasive species in Hawaiin forest. *Remote Sensing of Environment*, 112: 1912–1926.

Asner, G.P., D.E. Knapp, T. Kennedy-Bowdoin, M.O. Jones, R.E. Martin, J. Boardman, and R.F. Hughes, 2008b. Invasive species detection in Hawaiian rainforests using airborne imaging spectroscopy and LiDAR. *Remote Sensing of Environment*, 112: 1942–1955.

Asner, G.P., R.E. Martin, K.M. Carlson, U. Rascher, and P.M. Vitousek, 2006. Vegetation-climate interactions among native and invasive species in Hawaiian rainforest. *Ecosystems*, 9: 1106–1117.

Asner, G.P., D. Nepstad, G. Cardinot, and D. Ray, 2004. Drought stress and carbon uptake in an Amazon forest measured with spaceborne imaging spectroscopy. *PNAS*, 101(16): 6039–6044.

Asner, G.P., C.A. Wessman, and D.S. Shimel, 1998. Heterogeneity of savannah canopy structure and function from imaging spectrometer and inverse modeling. *Ecological Applications* 8: 1022–1036.

Baldocchi, D.D., K.B. Wilson, and L. Gu, 2002. How the environment, canopy structure and canopy physicalogical functioning influence carbon, water and energy fluxes of a temperate broad-leaved deciduous forest—An assessment with the biophysical model CANOAK. *Tree Physiology*, 22: 1065–1077.

Baret, F., and T. Fourty, 1997. Estimation of leaf water content and specific leaf weight from reflectance and transmittance measurements. *Agronomie*, 17: 455–464.

Bernier, P.Y., F. Raulier, P. Stenberg, and C.H. Ung, 2001. Importance of needle age and shoot structure on canopy net photosynthesis of balsam fir (*Abies balsamea*): A spatially inexplicit modeling analysis. *Tree Physiology*, 21: 815–830.

Blackburn, G.A., 2002. Remote sensing of forest pigments using airborne imaging spectrometer and LIDAR imagery. *Remote Sensing of Environment*, 82: 311–320.

Bo, T., C. MingKui, L. KeRang, G. FengXue, J. JinJun, H. Mei, and Z. LeiMing, 2007. Spatial patterns of terrestrial net ecosystem productivity in China during 1981–2000. *Science in China Series D: Earth Sciences*, 50(5): 745–753.

Bossert, J.E., R.R. Linn, J.M. Reisner, J.L. Winterkamp, P. Dennison, and D. Roberts, 2000. Coupled atmosphere–fire behavior model sensitivity to spatial fuels characterization. *Proceedings of Third Symposium on Fire and Forest Meteorology of the American Meteorological Society*, January 9–14, 2000, Long Beach, CA, pp. 21–26.

Buddenbaum, H., M. Schlerf, and J. Hill, 2005. Classification of coniferous tree species and age classes using hyperspectral data and geostatistical methods. *International Journal of Remote Sensing*, 26(24): 5453–5465.

Carleson, K.M., G.P. Asner, R.F. Hughes, R. Ostertag, and R.E. Martin, 2007. Hyperspectral remote sensing of canopy biodiversity in Hawaiian lowland rainforests. *Ecosystems*, 10: 536–549.

Carter, G.A., 1994. Ratios of leaf reflectances in narrow wavebands as indicators of plant stress. *International Journal of Remote Sensing*, 15: 697–703.

Chen, J.M. and J. Cihlar, 1996. Retrieving leaf area index of boreal conifer forests using landsat TM images. *Remote Sensing of Environment*, 55: 153–162.

Chen, J.M., S.G. Leblanc, J.R. Miller, J. Freemantle, S.E. Loechel, C.L. Walthall, K.A. Innanen, and H.P. White, 1999. Compact airborne spectrographic imager (CASI) used for mapping biophysical parameters of boreal forests. *Journal of Geophysical Research*, 104: 927–945.

Chen, J.M., C.H. Menges, and S.G. Leblanc, 2005. Global mapping of foliage clumping index using multiangular satellite data. *Remote Sensing of Environment*, 97: 447–457.

Ciais, P., M. Reichstein, N. Viovy, A. Granier, J. Ogee, V. Allard, M. Aubinet et al., 2005. Europe-wide reduction in primary productivity caused by the heat and drought in 2003. *Nature*, 437: 529–533.

Clark, M.L., D.A. Roberts, and D.B. Clark, 2005. Hyperspectral discrimination of tropical rain forest tree species at leaf to crown scales. *Remote Sensing of Environment*, 96: 375–398.

Clark, R.N., G.A. Swayze, K.E. Livo, R.F. Kokaly, S.J. Sutley, J.B. Dalton, R.R. McDougal, and C.A. Gent, 2003. Imaging spectroscopy: Earth and planetary remote sensing with the USGS Tetracorder and expert systems. *Journal of Geophysical Research*, 108(E12): 5131–5146.

Cruz, M., M. Alexander, and R. Wakimoto, 2003. Assessing canopy fuel stratum characteristics in crown fire prone fuel types of western North America. *International Journal of Wildland Fire*, 12: 39–50.

Curran, P.J., 2001. Imaging spectrometry for ecological applications. *International Journal of Applied Earth Observation and Geoinformation*, 3: 305–312.

Curran, P.J. and P.M. Atkinson, 1999. Issues of scale and optimal pixel size. *Spatial Statistics for Remote Sensing*. A. Stein, F. Van Der Meer, and B. Gorte (eds.), Kluwer Academic Publishers, Dordrecht, the Netherlands, pp. 115–133.

Curran, P.J., J.L. Dungan, and H.L. Gholz, 1992. Seasonal LAI in slash pine estimated with Landsat TM. *Remote Sensing of Environment*, 39: 3–13.

Darvishefat, A., T. Kellenburger, and K. Itten, 2002. Application of hyperspectral data for forest stand mapping. *Proceedings of Symposium on Geospatial Theory, Processing, and Applictions, ISPRS Commission IV*, Symposium 2002 Ottawa, Canada, July 9–12, 2002. IAPRS, Vol. XXXIV, part 4.

Davison, D., S. Achal, S. Mah, R. Gauvin, M. Kerr, A. Tam, and S. Preiss, 1999. Determination of tree species and tree stem densities in Northern Ontario forests using airborne CASI data. *Proceedings of the Fourth International Airborne Remote Sensing Conference and Exhibition II*, June 21–24, 1999, Ottawa, ON, Canada, pp. 187–196.

Dawson, T.P., P.R.J. North, and P.J. Curran, 1999. The propagation of foliar biochemical absorption features in forest canopy reflectance: A theoretical analysis. *Remote Sensing of Environment*, 67: 147–159.

Demarez, V. and J.P. Gastellu-Etchegorry, 2000. A modeling approach for studying forest chlorophyll content. *Remote Sensing of Environment*, 71: 226–238.

Demarez, V., J.P. Gastellu-Etchegorry, E. Mougin, G. Marty, C. Proisy, E. Duferene, and V.L.E. Dantec, 1999. Seasonal variation of leaf chlorophyll content of a temperate forest. Inverstion of the PROSPECT model. *International Journal of Remote Sensing*, 20: 879–894.

Dennison, P.E., D.A. Roberts, and J.C. Regelbrugge, 2000. Characterizing chaparral fuels using combined hyperspectral and synthetic aperture radar data. *Proceedings of the Ninth AVIRIS Earth Science Workshop*, vol. 6, February 23–25, 2000, Pasadena, CA, pp. 119–124.

Dreccer, M.F., M. van Oijen, A.H. Schapendonk, C.S. Pot, and R. Rabbinge, 2000. Dynamics of vertical leaf nitrogen distribution in a vegetative wheat canopy. Impact on canopy photosynthesis. *Annals of Botany*, 86: 821–831.

Drolet, G.G., K.F. Huemmrich, F.G. Hall, E.M. Middleton, T.A. Black, A.G. Barr, and H.A. Margolis, 2005. A MODIS-derived photochemical reflectance index to detect inter-annual variations in the photosynthetic light-use efficiency of a boreal deciduous forest. *Remote Sensing of Environment*, 98: 212–224.

Dupuy, J.L. and M. Larini, 1999. Fire spread through a porous forest fuel bed: A radiative and convective model including fire-induced flow effects. *International Journal of Wildland Fire*, 9: 155–172.

ESA, 2008. ESA Earthnet: Proba. http://earth.esa.int/missions/thirdpartymission/proba.html [last cited, October 8, 2010].

FAO, 2006. *Global Forest Resources Assessment 2005, Main Report. Progress towards Sustainable Forest Management. FAO Forestry Paper*, Rome, 147 pp.

Finney, M.A., 1998. FARSITE: Fire Area Simulator—Model development and evaluation. Res. Pap. RMRSRP-4. Fort Collins, CO: U.S. Department of Agriculture, Forest Service, Rocky Mountain Research Station. 47 pp.

Flannigan, M., B. Stocks, M. Turetsky, and M. Wotton, 2009. Impacts of climate change on fire activity and fire management in the circumboreal forest. *Global Change Biology*, 15: 549–560.

Folkman, M.A., P. Lee, P.J. Jarecke, S.L. Carman, and J. Pearlman, 2001. EO-1/Hyperion hyperspectral imager design, development, characterization, and calibration. *Proceedings of SPIE*, vol. 4151, 2000, 40–51.

Fourty, T. and F. Baret, 1998. On spectral estimates of fresh leaf biochemistry. *International Journal of Remote Sensing*, 19: 1283–1297.

Fourty, T., F. Baret, S. Jacquemoud, G. Schmuck, and J. Verdebout, 1996. Leaf optical properties with explicit description of its biochemical composition: Direct and inverse problems. *Remote Sensing of Environment*, 56: 104–117.

Fuentes, D.A., J.A. Gamon, H.L. Qui, D.A. Sims, and D.A. Roberts, 2001. Mapping Canadian boreal forest vegetation using pigment and water absorption features derived from the AVIRIS sensor. *Journal of Geophysical Research—Atmospheres*, 106: 33565–33577.

Galvão, L.S., F.J. Ponzoni, V. Liesenberg, and J.R. dos Santos, 2009. Possibilities of discriminating tropical secondary succession in Amazônia using hyperspectral and multiangular CHRIS/PROBA data. *International Journal of Applied Earth Observation and Geoinformation*, 11: 8–14.

Gastellu-Etchegorry, J.P. and V. Bruniquel-Pinel, 2001. A modeling approach to assess the robustness of spectrometric predictive equations for canopy chemistry. *Remote Sensing of Environment*, 76: 1–15.

Gemmell, F., J. Varjo, M. Strandstrom, and A. Kuusk, 2002. Comparison of measured boreal forest characteristics with estimates from TM data and limited ancillary information using reflectance model inversion. *Remote Sensing of Environment*, 81: 365–377.

Gholz, H.L., S.A. Vogel, W.P. Cropper, K. McKelvey, K.C. Ewel, R.O. Teskey, and P.J. Curran, 1991. Dynamics of canopy structure and light interception in Pinus elliottii stands, north Florida. *Ecological Monographs*, 61: 33–51.

Goetz, A.F.H., 1995. Imaging spectrometry for remote sensing: Vision to reality in 15 years. *Proceedings of the SPIE International Society for Optical Engineers*, Bellingham, WA, 2480, p. 2–13.

Goetz, S.J. and S.D. Prince, 1996. Remote sensing of net primary production in boreal forest stands. *Agricultural and Forest Meteorology*, 78: 149–179.

Goetz, S.J. and S.D. Prince, 1998. Variability in carbon exchange and light utilization among boreal forest stands: Implications for remote sensing of net primary production. *Canadian Journal of Forest Research*, 28: 375–389.

Gong, P., R. Pu, G. Biging, and M. Larrieu, 2003a. Estimation of forest leaf area index using vegetation indices derived from Hyperion hyperspectral data, *IEEE Transactions on Geoscience and Remote Sensing*, 41: 1355–1362.

Gong, P., R. Pu, G.S. Biging, and M.R. Larrieu, 2003b. Extraction of red edge optical parameters from hyperion data for estimation of forest leaf area index. *IEEE Transactions on Geoscience and Remote Sensing*, 41: 916–920.

Goodenough, D.G., A. Dyk, K.O. Niemann, J.S. Pearlman, H. Chen, T. Han, M. Murdoch, and C. West, 2003. Processing hyperion and ALI for forest classification. *IEEE Transactions on Geoscience and Remote Sensing*, 41: 1321–1331.

Goodenough, D.G., J.Y. Li, G.P. Asner, M.E. Schaepman, S.L. Ustin, and A. Dyk, 2006. Combining hyperspectral remote sensing and physical modeling for applications in land ecosystems. *IEEE International Geoscience and Remote Sensing Symposium (IGARSS)*, Denver, CO, 5 pp.

Goodenough, D.G., K.O. Niemann, A. Dyk, G. Hobart, P. Gordon, M. Loisel, and H. Chen, 2008. Comparison of AVIRIS and AISA airborne Hyperspectral sensing for above-ground forest carbon mapping. *IGARSS 2008*, II: 129–132.

Green, E.P., P.J. Mumbyb, A.J. Edwards, C.D. Clark, and A.C. Ellis, 1997. Estimating leaf area index of mangroves from satellite data. *Aquatic Botany*, 58: 11–19.

Green, R.O., M.L. Eastwood, and O. Williams, 1998. Imaging spectroscopy and the airborne visible/infrared imaging spectrometer (AVIRIS). *Remote Sensing of Environment*, 65: 227–240.

Grishin, A.M., 2001a. Heat and mass transfer and modeling and prediction of environmental catastrophes. *Journal of Engineering Physics and Thermophysics*, 74: 895–903.

Grishin, A.M., 2001b. Conjugate problems of heat and mass exchange and the physicomathematical theory of forest fires. *Journal of Engineering Physics and Thermophysics*, 74: 904–911.

Hardy. J.P., R. Melloh, G. Koenig, D. Marks, A. Winstral, J.W. Pomeroy, and T. Link, 2004. Solar radiation transmission through conifer canopies. *Agricultural and Forest Meteorology*, 126: 257–270.

Helms, J.A. (ed.), 1998. *Terminology of Forest Science, Technology, Practice, and Products. The Dictionary of Forestry*. Society of American Foresters, Bethesda, MD, 224 pp.

Hershey, R.R., W.H. McWilliams, and G.C. Reese, 1998. Utilizing the spatial structure available: Creating maps of forest attributes from forest inventory data. *Proceedings of the First International Conference on Geospatial Information in Agriculture and Forestry I*, June 1–3, 1998, Lake Buena Vista, FL, pp. 64–71.

Herwitz, S.R., D.L. Peterson, and J.R. Eastman, 1989. Thematic mapper detection of change in the leaf area index of closed canopy pine plantations in Central Massachusetts. *Remote Sensing of Environment*, 29: 129–140.

Hoffbeck, J.P. and D.A. Landgrebe, 1996. Classification of remote sensing images having high spectral resolution. *Remote Sensing of Environment*, 57: 119–126.

Hopkinson, C., L.E. Chasmer, G. Sass, I.F. Creed, M. Sitar, W. Kalbfleisch, and P. Treitz, 2005. Vegetation class dependent errors in lidar ground elevation and canopy height estimates in a boreal wetland environment. *Canadian Journal of Remote Sensing*, 31: 191–206.

Hu, B.X., K. Inannen, and J.R. Miller, 2000. Retrieval of leaf area index and canopy closure from CASI data over the BOREAS flux tower sites. *Remote Sensing of Environment*, 74: 255–274.

Jackson, R.D., R.T. Reginato, P.J. Printer, and S.B. Idso, 1979. Plant canopy information extraction from composite scene reflectance of row crops. *Applied Optics*, 18: 3775–3782.

Jacquemoud, S., C. Bacour, H. Poilvé, and J.P. Frangi, 2000. Comparison of four radiative transfer models to simulate plant canopies reflectance—Direct and inverse mode. *Remote Sensing of Environment*, 74: 471–481.

Jacquemoud, S. and S.L. Ustin, 2003. Application of radiative transfer models to moisture content estimation and burned land mapping. *Proceedings of the Fourth International Workshop on Remote Sensing and GIS Applications to Forest Fire Management*, Ghent, Belgium, pp. 3–12.

Jacquemoud, S., S.L. Ustin, J. Verdebout, G. Schmuck, G. Andreoli, and B. Hosgood, 1996. Estimating leaf biochemistry using the PROSPECT leaf optical properties model. *Remote Sensing of Environment*, 56: 194–202.

Jahnke, L.S. and D.B. Lawrence, 1965. Influence of photosynthetic crown structure on potential productivity of vegetation, based primarily on mathematical models, *Ecology*, 46: 319–326.

Jenkins, J.P., A.D. Richardson, B.H. Braswell, S.V. Ollinger, D.Y. Hollinger, and M.-L. Smith, 2007. Refining light use efficiency calculations for a deciduous forest canopy using simultaneous tower based carbon flux and radiometric measurements. *Agriculture and Forest Meteorology*, 143: 64–79.

Jia, G.J., I.C. Burke, M.R. Kaufmann, A.F.H. Goetz, B.C. Kindel, and Y. Pu, 2006. Estimates of forest canopy fuel attributes using hyperspectral data. *Forest Ecology and Management*, 229: 27–38.

Kashian, D., W.H. Romme, D.B. Tinker, M.G. Turner, and M.G. Ryan. 2006. Carbon storage on landscapes with stand-replacing fires. *Bioscience*, 56(7): 598–605.

Kimes, D., J. Gastellu-Etchegorry, and P. Esteve, 2002. Recovery of forest canopy characteristics through inversion of a complex 3D model. *Remote Sensing of Environment*, 79: 320–328.

Kneubühler, M., B. Koetz, S. Huber, N.E. Zimmermann, and M.E. Schaepman, 2008. Spectro-directional CHRIS/Proba data over two Swiss test sites for improved estimation of biophysical and chemical variables— Five years of activities. The International Archives of the Photogrammetry, Remote Sensing and Spatial Information Sciences. Vol. XXXVII. Part B7. Beijing 2008, p. 6.

Koetz, B., M. Kneubühler, J.L. Widlowski, F. Morsdorf, M. Schaepman, and K. Itten, 2005. Assessment of canopy structure and heterogeneity from multi-angular CHRIS-PROBA data. *Proceedings of the Ninth International Symposium on Physical Measurements and Signatures in Remote Sensing (ISPMSRS)*, Beijing, China XXXVI, pp. 73–78.

Koetz, B., F. Morsdorf, G. Sun, K.J. Ranson, K. Itten, and B. Allgöwer, 2006. Inversion of a lidar waveform model for forest biophysical parameter estimation. *IEEE Geoscience and Remote Sensing Letters*, 3(1): 49–53.

Koetz, B., G. Sun, F. Morsdorf, K.J. Ranson, M. Kneubühler, K. Itten, and B. Allgöwer, 2007. Fusion of imaging spectrometer and LIDAR data over combined radiative transfer models for forest canopy characterization. *Remote Sensing of Environment*, 106: 449–459.

Kokaly, R.F., D.G. Despain, R.N. Clark, and K.E. Livo, 2003. Mapping vegetation in Yellowstone National Park using spectral feature analysis of AVIRIS data. *Remote Sensing of Environment*, 84: 437–456.

Kotchenova, S.Y., N.V. Shabanov, Y. Knyazikhin, A.B. Davis, R. Dubayah, and R.B. Myneni, 2003. Modeling lidar waveforms with time-dependent stochastic radiative transfer theory for remote estimations of forest structure. *Journal of Geophysical Research*, 108(D15, 4484, ACL 12): 1–13.

Kötz, B., M. Schaepman, F. Morsdorf, P. Bowyer, K. Itten, and B. Allgöwer, 2004. Radiative transfer modeling within a heterogeneous canopy for estimation of forest fire fuel properties. *Remote Sensing of Environment*, 92: 332–344.

Kuusk, A., 1998. Monitoring of vegetation parameters on large areas by the inversion of a canopy reflectance model. *International Journal of Remote Sensing*, 19: 2893–2905.

Lee, K.-S., W.B. Cohen, R.E. Kennedy, T.K. Maiersperger, and S.T. Gower, 2004. Hyperspectral versus multispectral data for estimating leaf area index in four different biomes. *Remote Sensing of Environment*, 91: 508–520.

Li, J.Y., D.G. Goodenough, and A. Dyk, 2005. Mapping relative water content in douglas-fir with AVIRIS and a canopy model. *Proceedings of IGARSS 2005*, Seoul, Korea, vol. V, pp. 3572–3574, 2005.

Li, X., A.H., Strahler, and C.E. Woodock, 1995. A hybrid geometric optical radiative transfer approach for modeling albedo and directional reflectance of discontinuous canopies. *IEEE Transactions on Geosciences and Remote Sensing*, 33: 466–480.

Lichtenhaler, H.K., M. Lang, M. Sowinska, F. Heisel, and J.A. Mieh, 1996. Detection of vegetation stress via a new high resolution fluorescence imaging system. *Journal of Plant Physiology*, 148: 599–612.

Lim, K.S. and P.M. Treitz, 2004. Estimation of above ground forest biomass from airborne discrete return laser scanner data using canopy-based quantile estimators. *Scandinavian Journal of Forest Research*, 19: 558–570.

Linn, R.R., 1997. A transport model for prediction of wildfire behavior. *Los Alamos National Laboratory Scientific Report*, LA 13334-T.

Liu, J., J.M. Chen, J. Chilar, and W. Chen, 2002. Remote sensing-based estimation of net primary productivity over Canadian landmass. *Global Ecology and Biogeography*, 11: 115–129.

Magnussen, S. and P. Boudewyn, 1998. Derivations of stand heights from airborne laser scanner data with canopy-based quantile estimators. *Canadian Journal of Forest Research*, 28: 1016–1031.

Malenovský, Z., J. Albrechtova, Z. Lhotáková, R. Zurita-Milla, J.G.P.W. Clevers, M.E. Schaepman, and P. Cudlín, 2006. Applicability of the PROSPECT model for Norway spruce needles. *International Journal of Remote Sensing*, 27: 5315–5340.

Martin, M.E. and J.D. Aber, 1996. Estimating canopy characteristics as inputs for models of forest carbon exchange by high spectral resolution remote sensing. *The Use of Remote Sensing in the Modeling of Forest Productivity*. H.G. Gholz, K. Nakane, and H. Shimoda (eds.), Kluwer Academic, Dordrecht, the Netherlands, pp. 61–72.

Means, J.E., S.A. Acker, J.F. Brandon, M. Renslow, L. Emerson, and C.J. Hendrix, 2000. Predicting forest stand characteristics with airborne scanning LiDAR. *Photogrammetric. Engineering Remote Sensing*, 66: 1367–1371.

Medlyn, B.E., 1998. Physiological basis of the light use efficiency model. *Tree Physiology*, 18: 167–176.

Melillo, J.M., J.D. Aber, and J.F. Muratore, 1982. Nitrogen and lignin control of hardwood leaf litter decomposition dynamics. *Ecology*, 63: 621–626.

Merton, R., 1998. Monitoring community hysteresis using spectral shift analysis and the red-edge vegetation stress index. *Proceedings of the Seventh Annual JPL Airborne Earth Science Workshop*, January 12–16, 1998, NASA, Jet Propulsion Laboratory, Pasadena, CA.

Miller, M., 2001. *Fire Effects Guide, Chapter III—Fuels. National Wildfire Coordinating Group.* NEES 2394. National Interagency Fire Center, Boise, http://www.nwcg.gov/pms/RxFire/FEG.pdf [cited September 30, 2010].

Næsset, E., 1997. Determination of mean tree height of forest stands using airborne laser scanner data. *ISPRS Journal of Photogrammetry and Remote Sensing*, 52: 49–56.

Olthof, I. and D.J. King, 1998. Determination of soil property and forest structure relations with airborne digital camera images spectral and spatial information. *Proceedings of the 19th Canadian Symposium on Remote Sensing*, May 1997, Ottawa, ON, pp. 103–106.

O'Neil, A.L., J.A. Kupiec, and P.J. Curran, 2002. Biochemical and reflectance variation throughout the canopy of a Sitka spruce plantation. *Remote Sensing of Environment*, 80: 134–142.

Palmroth, S. and P. Hari, 2001. Evaluation of the importance of acclimation of needle structure, photosynthesis, and respiration to available photosynthetically active radiation in a Scots pine canopy. *Canadian Journal of Forest Research*, 31: 1235–1243.

Pellikka, P.K.E., E.D. Seed, and D.J. King, 2000. Modeling deciduous forest ice storm damage using CIR aerial imagery and hemispheric photography. *Canadian Journal of Remote Sensing*, 26: 394–405.

Peterson, B., W. Ni-Meister, J. Blair, M. Hofton, P. Hyde, and R. Dubayah, 2001. Modeling lidar waveforms using a radiative transfer model. *International Archives of Photogrammetry, Remote Sensing and Spatial Information Sciences*, 34: 121–124.

Peterson, D.L., Spanner, M.A., Running, S.W., and Teuber, K.B., 1987. Relationship of Thematic Mapper simulator data to leaf area index of temperate coniferous forests. *Remote Sensing of Environment*, 22: 323–341.

Phinn, S.R., P. Scarth, and D. Mitchell, 1999. Estimation of forest structural parameters for forestry and koala habitat monitoring in South-East Queensland, Australia. *Proceedings of the Fourth International Airborne Remote Sensing Conference and Exhibition II*, June 21–24, 1999, Ottawa, ON, pp. 179–186.

Plourde, L.C., S.V. Ollinger, M.-L. Smith, and M.E. Martin, 2007. Estimating species abundance in a northern temperate forest using spectral mixture analysis. *Photogrammetric Engineering & Remote Sensing*, 73: 829–840.

Porterie B., D. Morvan, J.-C. Loraud, and M. Larini, 2000. Firespread through fuel beds: Modeling of wind-aided fires and induced hydrodynamics. *Physics of Fluids*, 12: 1762–1782.

Pu, R., Q. Yu, P. Gong, and G.S. Biging, 2005. EO-1 Hyperion, ALI, and Landsat 7 ETM+ data comparison for estimating forest crown closure and leaf area index. *International Journal of Remote Sensing*, 26: 457–474.

Raddi, S., S. Cortes, I. Pippi, and F. Magnani, 2005. Estimation of vegetation photochemical processes: An application of the photochemical reflectance index at the San Rossore test site. *Proceedings of the Third ESA CHRIS/Proba Workshop*, March 21–23, ESRIN, Frascati, Italy (ESA SP-593, June 2005).

Riano, D., P. Vaughan, E. Chuvieco, P.J. Zarco-Tejada, and S.L. Ustin, 2005. Estimation of fuel moisture content by inversion of radiative transfer models to simulate equivalent water thickness and dry matter content: Analysis at leaf and canopy level. *IEEE Transactions on Geoscience and Remote Sensing*, 43: 819–825.

Richardson, A.D. and G.P. Berlyn, 2002. Spectral reflectance and photosynthetic properties of *Betula Papyrifera* (Betulaceae) leaves along an elevational gradient on Mt. Mansfield, Vermont, USA. *American Journal of Botany*, 89: 88–94.

Running, S.W., D.D. Baldocchi, D.P. Turner, P.S. Bakwin, and K.A. Hibbard, 1999. A global terrestrial monitoring network integrating tower fluxes, flask sampling, ecosystem modeling and EOS satellite data. *Remote Sensing of Environment*, 70: 108–128.

RSI, 2009. Fire Fuel Tool. ENVI User's Guide, Version 4.7. ITT Visual Information Solutions, www.ittvis.com

Sampson, P.H., 2000. *Forest Condition Assessment: An Examination of Scale, Structure, and Function Using High Spatial Resolution Remote Sensing Data*, MSc Thesis, York University, Toronto, Canada, 157 pp.

Schaepman, M.E., B. Koetz, G. Schaepman-Strub, and K.I. Itten, 2005. Spectrodirectional remote sensing for the improved estimation of biophysical and chemical variables: Two case studies. *International Journal of Applied Earth Observation and Geoinformation*, 6: 271–282.

Schlerf, M., C. Atzberger, and J. Hill, 2005. Remote sensing of forest biophysical variables using HyMap imaging spectrometer data. *Remote Sensing of Environment*, 95: 177–194.

Seed, E.D., D.J. King, and P.K.E. Pellikka, 1999. Multivariate analysis of low cost airborne CIR imagery for the determination of forest canopy structure. *Proceedings of the Fourth International Airborne Remote Sensing Conference and Exhibition II*, June 21–24, 1999, Ottawa, ON, pp. 139–146.

Serrano, L., J. Penuelas, and S.L. Ustin, 2002. Remote sensing of nitrogen and lignin in mediterranean vegetation from AVIRIS data: Decomposing biochemical from structural signals. *Remote Sensing of Environment*, 81: 355–364.

Shippert, M.M., D.A. Walker, N.A. Auerbach, and B.E. Lewis, 1995. Biomass and leaf-area index maps derived from SPOT images for Toolik Lake and Imnavait Creek areas, Alaska. *Polar Record*, 31: 147–54.

Smith, M.-L., M.E. Martin, L. Plourde, and S.V. Ollinger, 2003. Analysis of hyperspectral data for the estimation of temperate forest canopy nitrogen concentration: Comparison between an airborne (AVIRIS) and a spaceborne (Hyperion) sensor. *IEEE Transactions on Geoscience and Remote Sensing*, 41: 1332–1337.

Spanner, M.A., L.L. Pierce, D.L. Peterson, and S.W. Running, 1990. Remote sensing of temperate coniferous forest leaf area index: The influence of canopy closure, understory vegetation, and background reflectance. *Remote Sensing of Environment*, 33: 97–112.

Stenberg, P., 1998. Implications of shoot structure on the rate of photosynthesis at different levels in a coniferous canopy using a model incorporating grouping and penumbra. *Functional Ecology*, 12: 82–91.

St-Onge, B.A. and F. Cavayas, 1995. Estimating forest stand structure from high resolution imagery using the directional variogram. *International Journal of Remote Sensing*, 16: 1999–2020.

Syphard, A.D., K.C. Clarke, and J. Franklin, 2007a. Simulating fire frequency and urban growth in southern California coastal shrublands, USA. *Landscape Ecology*, 22: 431–445.

Syphard, A.D., V.C. Radeloff, J.E. Keeley, T.J. Hawbaker, M.K. Clayton, S.I. Stewart, and R.B. Hammer, 2007b. Human Influence on California Fire Regimes. *Ecological Applications*, 17: 1388–1402.

Terjung, W.H. and S.S.F. Louie, 1972. Potential solar radiation on plant shapes. *International Journal of Biometeorology*, 16: 25–43.

Thenkabail, P.S., E.A. Enclona, M.S. Ashton, C. Legg, and M.J. De Dieu, 2004. Hyperion, IKONOS, ALI, and ETM+ sensors in the study of African rainforests. *Remote Sensing of Environment*, 90: 23–43.

Thomas, V., D.A. Finch, J.H. McCaughey, T. Noland, L. Rich, and P. Treitz, 2006a. Spatial modeling of the fraction of photosynthetically active radiation absorbed by a boreal mixedwood forest using a lidar-hyperspectral approach. *Agricultural and Forest Meteorology*, 140: 287–307.

Thomas, V., J.H. McCaughey, P. Treitz, D.A. Finch, T. Noland, and L. Rich, 2009. Spatial modeling of photosynthesis for a boreal mixedwood forest by integrating micrometeorological, lidar and hyperspectral remote sensing data. *Agricultural and Forest Meteorology*, 149: 639–654.

Thomas, V., T. Noland, P. Treitz, and J.H. McCaughey, 2010. Leaf area and clumping indices for a boreal mixedwood forest: Lidar, hyperspectral, and Landsat models. *International Journal of Remote Sensing* (in press).

Thomas, V., P. Treitz, J.H. McCaughey, and I. Morrison, 2006b. Mapping stand-level forest biophysical variables for a mixedwood boreal forest using LiDAR: An examination of scanning density, *Canadian Journal of Forest Research*, 36: 34–47.

Thomas, V., P. Treitz, J.H. Mccaughey, T. Noland, and L. Rich, 2008. Canopy chlorophyll concentration estimation using hyperspectral and lidar data for a boreal mixedwood forest in northern Ontario, Canada. *International Journal of Remote Sensing*, 29: 1029–1052.

Todd, K.W., F. Csillag, and P.M. Atkinson, 2003. Three-dimensional mapping of light transmittance and foliage distribution using lidar, *Canadian Journal of Remote Sensing*, 29(5): 544–555.

Townsend, P.A., J.R. Foster, R.A. Chastain, and W.S. Currie, 2003. Application of imaging spectroscopy to mapping canopy nitrogen in the forests of the central Appalachian Mountains using Hyperion and AVIRIS. *IEEE Transactions on Geoscience and Remote Sensing*, 41: 1347.

Treitz, P.M., 2001. Variogram analysis of high spatial resolution remote sensing data: An examination of boreal forest ecosystems. *International Journal of Remote Sensing*, 22: 3895–3900.

Treitz, P.M. and P.J. Howarth, 1999. Hyperspectral remote sensing for estimating biophysical parameters of forest ecosystems. *Progress in Physical Geography*, 23(3): 359–390.

Treitz, P.M. and P.J. Howarth, 2000. High spatial resolution remote sensing data for forest ecosystem classification: An examination of spatial scale. *Remote Sensing of Environment*, 72: 268–289.

Treitz, P.M., V. Thomas, P.J. Zarco-Tejada, P. Gong, and P.J. Curran, 2010. ASPRS monograph-hyperspectral remote sensing for forestry. *ASPRS Monograph Series*, ISBN 1-57083-093-2, 107 pages.

Turner, D.P., W.D. Ritts, W.B Cohen, S.T. Gower, M. Zhao, S.W. Running, S.C. Wofsy, S. Urbanski, A.L. Dunn, and J.W. Munger, 2003a. Scaling gross primary production (GPP) over boreal and deciduous forest landscapes in support of MODIS GPP product validation. *Remote Sensing of Environment*, 88: 256–270.

Turner, D.P., S. Urbanski, D. Bremer, S.C. Wofsy, T. Meyers, S.W. Gower, and M. Gregory, 2003b. A cross-biome comparison of daily light use efficiency for gross primary production. *Global Change Biology*, 9: 383–395.

Tymstra, C., M.D. Flannigan, O.B. Armitage, and K. Logan, 2007. Impact of climate change on area burned in Alberta's boreal forest. *International Journal of Wildland Fire*, 16: 153–160.

Ustin, S.L., D.A. Roberts, J.A. Gamon, G.P. Asner, and R.O. Green, 2004. Using imaging spectroscopy to study ecosystem processes and properties. *BioScience*, 54: 523–534.

Ustin, S.L. and A. Trabucco, 2000. Using hyperspectral data to assess forest structure, *Journal of Forestry*, 98: 47–49.

Valencia, D., P. Martínez, J. Plaza, R.M. Pérez, M.C. Cantero, and R. Paniagua, 2005. Pre-evaluation of wild fires in Monfragüe Regional Park using CHRIS imagery. *Proceedings of the Third ESA CHRIS/Proba Workshop*, March 21–23, ESRIN, Frascati, Italy (ESA SP-593, June 2005), p. 4.

Vose, J.M., P.M. Dougherty, J.N. Long, F.W. Smith, H.L. Gholy, and P.J. Curran, 1994. Factors influencing the amount and distribution of leaf area of pine stands. *Ecological Bulletins*, Copenhagen, 43: 102–114.

Wang, Q., S. Adiku, J. Tenhunen, and A. Granier, 2005. On the relationship of NDVI with leaf area index in a deciduous forest site. *Remote Sensing of Environment*, 94: 244–255.

Wessman, C.A., J.D. Aber, D.L. Peterson, and J. Melillo, 1988. Remote sensing of canopy chemistry and nitrogen cycling in temperate forest ecosystems. *Nature*, 335: 154–156.

White, J.D., S.W. Running, R. Nemani, R.E. Keane, and K.C. Ryan, 1997. Measurement and remote sensing of LAI in rocky mountain montane ecosystems. *Canadian Journal of Forest Research*, 27: 1714–1727.

Woodcock, C.E. and A.H. Strahler, 1987. The factor of scale in remote sensing, *Remote Sensing of Environment*, 21: 333–339.

Yuan, X., D. King, and J. Vlcek, 1991. Sugar Maple decline assessment based on spectral and textural analysis of multispectral aerial videography. *Remote Sensing of Environment*, 37: 47–54.

Zarco-Tejada, P.J., J.R. Miller, D. Haboudane, N. Tremblay, and S. Apostol, 2004. Detection of chlorophyll fluorescence in vegetation from airborne hyperspectral CASI imagery in the red edge spectral region. *International Geoscience and Remote Sensing Symposium, IGARSS'03*, I, Toulouse, France, pp. 598–600.

Zarco-Tejada, P.J., J.R. Miller, T.L. Noland, G.H. Mohammed, and P.H. Sampson, 2001. Scaling-up and model inversion methods with narrowband optical indices for chlorophyll content estimation in closed forest canopies with hyperspectral data. *IEEE Transactions on Geoscience and Remote Sensing*, 39: 1491–1507.

Zhao, D., K.R. Reddy, V.G. Kakani, J.J. Read, and G.A. Carter, 2003. Corn (*Zea mays* L.) growth, leaf pigment concentration, photosynthesis and leaf hyperspectral reflectance properties as affected by nitrogen supply. *Plant and Soil*, 257: 205–217.

21 Hyperspectral Remote Sensing of Wetland Vegetation

Elijah Ramsey III and Amina Rangoonwala

CONTENTS

21.1 INTRODUCTION

Wetlands proportionally exert a higher influence on biogeochemical fluxes among the land, the atmosphere, and hydrologic systems than their 1% worldwide occurrence suggests [1]. Although their frequency of occurrence is low and their importance is high, wetlands continue to face high detrimental pressures from natural and human-induced forces [2]. Remote sensing offers the single best source of timely, synoptic wetland status and trends information at a variety of spatial and temporal scales [3].

The remote sensing of wetlands does not generally differ in technique or process from remote sensing-based mapping of other terrestrial features (e.g., [4]). Differences exist, because wetlands occupy a unique interface, or ecotone, between aquatic and upland ecosystems [5]. Although there are environmental factors that affect all vegetation, such as climate, soils, and geology, the uniqueness of wetland vegetation stems from the biophysical features that define the wetland ecotone. Infrequent to near constant inundation by fresh to saline waters promotes adaptations that set wetland plants uniquely apart from all other terrestrial plants. As the spatial and temporal complexities in flushing strength and salinity increase, so do the variety and complexity of wetland species, forms, and associations. Lacustrine and riparian wetlands reveal that aforementioned uniqueness primarily in response to seasonal and longer term cycles of hydrology or changing water inputs into the systems. Coastal wetlands experience complexities common to lacustrine and riparian wetlands, but their locations add to their complexities [5].

Tidal flushing that carries pulses of elevated salinity waters dominates the dynamics of coastal wetlands. Tidal periods and amplitudes vary in time and space; however, all coastal regions experience up to two low and high tides daily and a 28 day cycle exhibiting neap (low) to spring (high) amplitudes. Storms can augment the tidal amplitudes because of associated wind and storm surges. In many areas, coastal wetlands experience freshwater flooding, sometimes alternating with tidal flooding. Because of these dynamics, wetlands at the confluence of river and overland freshwater flow and tidal flooding can support a variety of species and structural forms.

Storm surges carrying water with elevated salinity can invade the normally fresher wetland zones impacting less salt tolerant wetland species. Excessive flood duration, or water logging, can also adversely impact coastal wetlands. The overall results of these spatially and temporally varying forces converging on broad transitional coastal areas exhibiting low-shore-normal topographic gradients are highly diverse wetland ecotones that transition from the coastal ocean to the upland ecosystem. These types of wetlands support highly diverse vegetation that dominates and codominates narrow niches in the flushing and salinity regime. Drought, fire, invasives, and human development are additional forces that contribute to the spatial complexity of these coastal wetland systems. These expansive, highly diverse, and valuable coastal wetlands can benefit from a variety of remote sensing mapping techniques, including those applications uniquely available when using airborne and satellite hyperspectral image data.

21.1.1 Benefits of Hyperspectral Data

All remote sensing is based on linking spatial to spectral features. In terrestrial mapping, wetlands can be represented simply as a land cover surface composed of soil, litter, rocks, and water, and a vegetation canopy component including spectral properties of plant leaves and structure of the plant canopy. Components of the plant canopy structure broadly consist of the total standing plant live (green and nonphotosynthetic) and dead biomass and the three-dimensional orientation of that biomass. If the onset of change from normal healthy condition is to be recognized in the plant canopy, the spectral changes in the plant leaves must be detectable before such vegetative features can be effectively mapped.

Leaf spectral properties have been directly related to vegetation type and state of vigor and are general indicators of the leaf chlorophyll, water content, and leaf biomass [6]. If the mapping objective is vegetation classification, then spectral discernment of the targeted classes should be analyzed at the plant leaf, plant canopy, and background levels. If the mapping objective is to identify changes primarily occurring in the leaf, then spectral uniqueness at the plant leaf level must be present. Furthermore, the ability to separate or maximize the plant leaf variability from the plant canopy (i.e., foreground) and background variability in the pixel reflectance must be possible. On the other hand, if the sought change is related to the plant canopy structure, the pixel reflectance component linked to this physical feature should be isolatable. Numerous studies have related the canopy structure variable of leaf area index (LAI) to vegetation type, health, and phenology.

The benefits of hyperspectral imaging (HSI) narrowband data compared to broadband multispectral (MS) data depend on the spectral resolutions of the HSI and MS sensors. Both sensor types can record in the same reflected electromagnetic (EM) energy range from about 400 to about 3000 nm; the MS sensors record that EM information in fewer spectral bands (about 3–16 bands) than HSI sensors (often hundreds of bands). In addition, to accommodate this high number of spectral bands, the HSI sensor bands are much narrower (typically around or less than 10 nm) than those used in common MS sensors (Figure 21.1, [7]). In essence, the HSI sensor records a

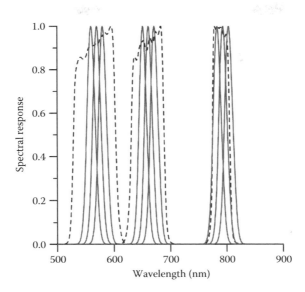

FIGURE 21.1 The spectral response curves of the EO-1 ALI sensor bands 2 (525–605 nm), 3 (630–690 nm), and 4 (775–805 nm) (shown in dashed lines) and similar curves for the co-located EO-1 Hyperion sensor (solid lines) approximated as Gaussian distributions [7]. Three Hyperion bands are plotted for each ALI band.

more continuous spectral record, whereas the MS records only selected broad areas of the reflected EM range. MS bands normally include at a minimum the blue, green, and red visible bands, one or more near infrared (NIR) bands, and similarly one or more shortwave infrared bands. Well-known MS satellite sensors are carried by the Landsat, Quickbird, and SPOT satellites. Heavily used airborne HSI sensors include the National Aeronautics and Space Administration's (NASA) airborne visible infrared imaging spectrometer (AVIRIS) and the privately owned compact airborne spectrographic imager (CASI). The EO-1 satellite carries both the Hyperion hyperspectral sensor and the advanced land imager (ALI) broadband MS sensor. The Hyperion and ALI sensor combination illustrates the trade-off in the quantity of spectral information extracted from each ground-resolution element (or pixel) and the imaged area (e.g., Figure 21.1). Both sensors have 30 m pixel resolutions; however, the Hyperion sensor with 224 bands has a nominal 7.6 km swath width, whereas the ALI with nine MS bands and one panchromatic band collects data over a much larger 37 km swath width.

By using a high density, multi-narrowband spectral recording, the HSI sensors provide spectral discrimination similar to laboratory spectrophotometers (e.g., Figure 21.1). If the HSI sensor normally on board an aircraft or satellite was moved within a few hundred meters of the Earth's surface, the HSI record would simulate a spectrophotometer. The main difference would be the variable illumination and the diffuse nature of the surface target. Because there is a higher number of spectral channels (or bands) and the higher spectral resolution, many mapping applications operationally performed with broadband sensor systems have been enhanced with HSI image data. Some of these classical broadband applications include land cover type mapping, change detection, and biomass determination.

Regarding hyperspectral imagery, the near-continuous spectra enhance the use of specialized image processing techniques that map spectral variations within a pixel. This specialized processing can partition each pixel into percent occurrences of various target compositions on the basis of determined spectral differences. By applying linear processing to HSI data, the classical broadband point classifications are transformed into continuous classification of percent occurrences (i.e., fractional abundances) of land cover features. Analogous to spectral analyses in a laboratory, the HSI sensor data can also be used to detect subtle variation in the reflectances of plant leaves if the canopy structure and background influences can be removed from the recorded reflectance.

Although it is not always necessary for the spectral changes in the plant leaves to be linked to the abnormal plant canopy change, it is always advantageous and many times critical. By defining the biophysical indicator of the onset of adverse change at the plant leaf level, a more targeted monitoring strategy can be constructed. If the onset and progression of vegetation change cannot be defined and characterized at the plant leaf level, the ability to document the onset and progression at the canopy level via synoptic remote sensing is hindered, especially for stands with less dense foliage and more open canopies. The use of reflectance is of paramount importance in the successful application of the HSI sensor for detecting subtle changes of biophysical and biochemical properties at the plant leaf level.

The advantage of HSI spectral alignment over that of MS sensors is realized only if the spectral autocorrelation within the EM response from the terrestrial target does not eliminate the more effective spectral isolation available from HSI sensors (e.g., [8]). Unlike mineral reflectance spectra from nonvegetated landscapes that often exhibit well-defined spectral features, vegetation spectra tend to have subtle varying spectral absorption and reflectance features. In the visible wavelengths, the wide and skewed reflectance features tend to imitate pigment absorptions; however, image processing techniques are available that may remove or at least diminish the spectral autocorrelation from the HSI data before subsequent effective classification of the image reflectance data (e.g., [8]). In all vegetation type and biophysical classifications, careful site-specific analyses should also be conducted to understand how changes in the HSI data are linked to the landscape features of interest.

The near spectrally continuous HSI data can be used to rate single bands and combinations of spectral bands that best produce the mapping objective [9]. If successful mapping can be isolated to a small number of spectral bands, then the mapping can be greatly simplified. If these selected bands are replicated on operational broadband systems, the mapping process can be further simplified by transfer to that operational remote sensing system. The HSI data analyses provide the unique ability to define the most appropriate spectral bands (or regions) for any mapping exercise. This unique, advanced spectral analysis function cannot be replicated by spectral analyses that are based on broadband MS data.

21.1.2 Chapter Outline

Most remote sensing technologies applied to wetlands are commonly used for terrestrial mapping; however, our focus is centered on wetland mapping, which is uniquely a function of the variety of wetland plant species, forms, and associations. Within that focus, the application of hyperspectral remote sensing to mapping and monitoring wetlands is demonstrated in temporally and spatially complex coastal forests and marshes that occupy the northern Gulf of Mexico but also share a wide variety of similarities with wetland vegetation types worldwide. The conveyance of HSI applications is facilitated by partitioning these wetlands into three broad components: plant leaf, plant canopy, and the background. The plant canopy is further partitioned into preferred canopy orientation as the leaf angle distribution (LAD) and the foliar biomass as observed through the LAI.

This chapter focuses on those applications of HSI that extend the more common canopy compositional classification (characteristic spectra mapping) and inferential prediction analyses. Descriptions of mangrove biophysical mapping, invasive detection and mapping, marsh dieback onset and progression mapping, and other applied research projects present how hyperspectral information can elucidate relations observed with broadband spectral analyses, transfer high spatial resolution broadband information to spatial resolutions amenable for regional mapping, and determine what pertinent diagnostic information may be obtained from broadband data and what is more suited to hyperspectral data. In each project description, the plant leaf and plant canopy reflectance are coupled. In all cases, the canopy structure is considered as a necessary component for interpreting the vegetation canopy reflectance; in two cases, this coupling is directly shown. This chapter illustrates the importance of understanding the structure and compositional characteristics of the target canopy as well as its background. These examples incorporate the application of light-interaction models (based on radiative transfer [RT] equations), direct measurements, and above the top of canopy (TOC) reflectance and photography.

21.2 HYPERSPECTRAL REMOTE SENSING OF WETLAND FORESTS

Three wetland forest types occupy different flood and salinity regimes in the coastal wetland. Of the three swamp types, only mangroves are evergreen and tolerant of saline waters. Mangroves occupy 75% of the world's coastlines between 25°N and 25°S latitudes [10]. Although new world mangroves (red mangrove [*Rhizophora mangle L.*], black mangrove [*Avicennia germinans L.*], and white mangrove [*Laguncularia racemosa*]) thrive in the more southern portions of the Gulf of Mexico (hereafter Gulf), black mangroves have tenaciously established footholds as far north as 30° latitude. Situated at the ocean and land transition, mangrove swamps directly experience frequent tidal flooding and the many perturbations of the coastal ocean.

Baldcypress (*Taxodium distichum*) and bottomland hardwood (BLH) dominated forests can make up major parts of deciduous freshwater swamps, and in contrast to mangrove forests, these swamps extend well into the temperate zones. Baldcypress swamps can occupy more permanently flooded portions of a wetland forest or isolated oxbows that constitute the remnant of former meanders in riparian flood plains [5]. BLH make up intermittently flooded floodplain forests that occur along rivers and streams throughout the central and southern United States [11]. These forests can occur on an elevation gradient between drier upland hardwood forests and more persistently flooded swamps. Canopy closure tends to be high in mature, nondisturbed BLH canopies.

21.2.1 MANGROVE FORESTS

In a study of mangroves located in the southern Gulf off the southwest coast of Florida (Figure 21.2), ground-based measurements of canopy closure, laboratory leaf spectral measurements, and helicopter-based spectroradiometer measurements were obtained in the five mangrove types: basin, overwash, fringe, riverine, and dwarf [12]. The study objective was to determine whether or not the black, red, and white mangroves were spectrally separable, and, therefore, discernable with satellite optical remote sensing mapping [10,13]. In addition, LAI was calculated on the basis of ground-based measurements of observed vegetation canopies and acquired optical spectral data.

Leaves from near the TOC were collected at 23 nondisturbed mangrove sites throughout the northern 10,000 islands off the southwest coast of Florida [10,13]; these leaves were used to determine whether or not the three mangrove species were spectrally separable. Spectral properties of the leaves were obtained by using flat-plate methods. The multiple addition of reflected light from lower leaves in the stack and the flat-plate was accounted for by using a simple adaption of a method introduced by Lillesaeter [13,14]. Spectral measurements were collected over the visible and near infrared (VNIR) wavelengths (about 400–1000 nm as applied here).

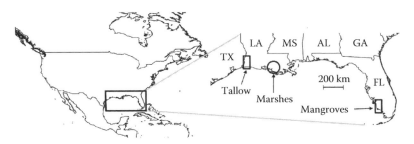

FIGURE 21.2 General locations of projects that used hyperspectral remote sensing techniques.

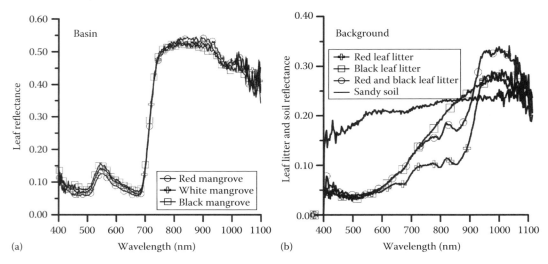

(a)   (b)

FIGURE 21.3 (a) (BASIN) Mangrove top of canopy leaf spectra acquired by using a stacked-plate reflectance design. The basin site contained about 40% red, 30% white, and 30% black mangroves. (b) (BACKGROUND) Spectra of litter samples obtained at three mangrove sites and sand spectra obtained at a dwarf mangrove site. Figure 21.2 shows the location of the mangrove sites in southwest Florida. (Adapted from Ramsey, E., III and Jensen, J., *Photogramm. Eng. Rem. Sens.*, 62(8), 939, 1996.)

Results of these leaf spectral analyses indicated that although black and red mangrove leaves had slightly different mean reflectance magnitudes (0.02 on average), relatively high variances of about ±0.04 in the visible (VIS) and ±0.06 in the NIR eliminated the between-species spectral discernment. Spectral reflectance of leaves obtained from the TOC was not a unique identifier of species (an example is given in Figure 21.3a). Another result of the flat-plate spectral analyses was related to the leaf-litter spectra. Interestingly, even though spectra of the TOC leaves could not separate mangrove species, the leaf litter collected within stands dominated by each of the three single species spectrally differed throughout the VNIR wavelengths (Figure 21.3b, [10]). Although no linkage to the length of decomposition in each of the litter samples was considered, these spectral differences illustrate the importance of considering the changing background (leaf litter in this case) reflectance as a spectrally complex component of the canopy reflectance.

Even though leaf spectral properties would not singularly separate the three mangrove species, canopy VNIR spectra representing a 20 m instantaneous field of view (IFOV) of the TOC were highly variable (Figure 21.4, [10]). The canopy spectra were collected from a helicopter platform with a handheld radiometer at the 23 mangrove sites. To understand what caused the canopy reflectance differences, a RT model was implemented by following the construction outlined in Goudriaan [15]. Particulars of the model, assumptions, inputs, and validations are described in Ramsey and Jensen [13]. RT model results with predictions >97% showed that the LAD of black, red, and white mangrove canopies were predominantly spherical. This consistency implied that LAD did not control the high spectral variance observed in the canopy reflectance spectra. In contrast, the LAI, as an indicator of the number of canopy leaf layers, although exhibiting low variance, was highly correlated to canopy reflectance but not to the species of mangrove. These LAI values predicted with the RT model were validated with ground-based measurements (Figure 21.5, [10,13]).

The study conducted to determine whether or not the three mangrove species were spectrally separable resulted in three major findings. First, although the spectral reflectance of the TOC leaves did exhibit differences in the mean, the spectral variance for each of the three mangrove species removed that spectral differentiability. Second, the RT model predicted that all mangrove species LADs largely followed a spherical distribution. The overall differences in the TOC

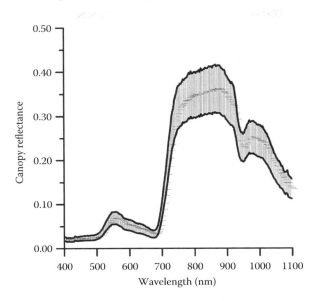

FIGURE 21.4 Range of mangrove canopy reflectance spectra calculated from top of canopy site-specific upwelling (~20 m instantaneous field of view) and surface downwelling light measurements (mean ± one standard deviation, n = 23). The 23 sites included basin, overwash, fringe, riverine, and dwarf forest sites containing mixtures of black, red, and white mangroves. Location of mangrove sites shown in Figure 21.2. (Adapted from Ramsey, E., III and Jensen, J., *Photogramm. Eng. Rem. Sens.*, 62(8), 939, 1996.)

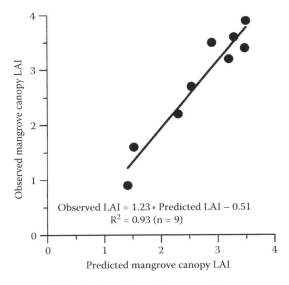

FIGURE 21.5 Mangrove canopy LAI calculated from site canopy closure measurements and predicted with the radiative transfer canopy model by using measured background (e.g., Figure 21.3b BACKGROUND), leaf (e.g., Figure 21.3a BASIN), and canopy reflectance hyperspectral reflectance spectra (e.g., Figure 21.4) inputs.

reflectance spectra were not explainable by leaf-spectral differences or variability in the canopy LADs. Third, although the LAI variance was fairly low, the canopy structural differences related to the canopy LAI had the strongest relation to the canopy-reflectance spectral variance. Canopy closure tended to dominate the overall mangrove canopy variance. With the methods applied, hyperspectral analyses in conjunction with canopy light interaction modeling found no consistent discernable differences between mangrove species at the plant leaf and plant canopy levels.

21.2.2 BALDCYPRESS FORESTS

The cypress family includes 142 conifer species collectively exhibiting a near-global distribution (Cupressaceae, http://www.conifers.org/cu/Cupressaceae.php [Accessed December 27, 2010]). Baldcypress (*Taxodium distichum* [L.] Rich) is a species of the cypress family that is widely distributed in the east and southeast United States (Plants Profile, http://plants.usda.gov/java/profile?symbol = TADI2 [Accessed September 21, 2010]). These trees grow in a wide variety of coastal and inland environments; however, in southeastern wetland forests, baldcypress trees commonly occupy the more frequently flooded areas where it often occurs with water tupelo (*Nyssa aquatica*) trees in the northern Gulf. In the current era, the baldcypress deciduous forests typically exhibit more open canopies, which allow a higher penetration of sunlight onto the understory surface than typically observed in mature, minimally disturbed BLH forests. Baldcypress leaves are needle-like, whereas tupelo leaves are somewhat narrow and elliptical. There are two conditions in these forests that cause dramatic changes in the canopy spectral reflectance: winter leaf-off and the presence and absence of subcanopy flooding. During leaf-off conditions in the winter months, baldcypress forest canopies become highly transparent to sunlight.

The spectral detection of baldcypress swamps is enhanced by its unique environment and canopy structures. Baldcypress mapping is not usually problematic for broadband optical sensors; however, monitoring changes in these swamps that may indicate detrimental and abnormal stresses could be enhanced with hyperspectral mapping. Saltwater intrusion related to relative sea-level rise is causing widespread diebacks of coastal baldcypress swamps, commonly termed "ghost forests" [16]. HSI would be the most appropriate tool to survey broad areas to detect abnormal change in these forests before irreversible change becomes imminent. Hindrances to consistently detecting subtle changes in the more vulnerable baldcypress swamp are the typically open canopies and the persistent subcanopy flooding. These two physical factors can combine to create high variability in the canopy reflectance that could overwhelm leaf spectral responses to salt stress. Although this is not necessarily a problem in the more closed canopy stands, another obstruction to detection of abnormal change is the linear and featherlike leaves of baldcypress.

The fine structure of the baldcypress leaves poses a challenge for leaf level observation and characterization with remote sensing. Methods developed by Daughtry et al. [17] describe measurement of the reflectance and transmittance of thin and short leaves; however, these methods are time-consuming and are not amenable for frequent and spatially distributed sampling and timely analysis. More rapid methods that analyze very thin single leaves may be appropriate; however, no literature was found describing these types of analytical procedures. Development and implementation of appropriate plant-leaf spectral monitoring coupled with hyperspectral monitoring would be useful for conservation of these at-risk coastal forests.

21.2.3 BOTTOMLAND HARDWOOD FORESTS

BLH forests occupy about 2.8 million ha in the Lower Mississippi River Alluvial Valley of the United States and comprise a variety of deciduous species. As in the mangrove mapping, the ability to separately monitor the different bottomland communities would greatly enhance the value of the information obtainable from remote sensing data. As an example, White [18] classified the BLH forests occupying a wetland forest on the southern Louisiana and Mississippi border into two community classes. The classification represented two different BLH forests in terms of function and species composition. The cause of the stand composition differences of the two BLH forest classes was related to different hydroperiods (i.e., hydrologic regimes) for these two types, each dominating physically separate portions of the wetland landscape [18]. In a study of hurricane impacts to the same wetland forest, the most severe BLH stand damage was generally associated with one of the two BLH classes [19]. In essence, the composition and functional differences ascribed to each BLH class resulted in disparate sensitivity to wind damage. This disparity highlights the need for more detailed BLH classifications.

As part of the hurricane impact study, attempts were made to separately classify the two BLH forest classes with prehurricane TM and SAR 25 m image data [20]. These classification attempts were unsuccessful. In a separate study that combined the 25 m damage classes with 3 years of daily moderate-resolution imaging spectroradiometer (MODIS) data, a temporal difference in foliage onset of the two BLH classes was observed in the early spring [21]. Even though the daily temporal resolution of the MODIS product provided discrimination of the two classes, the spatial scale is not conducive to the needed stand-level monitoring and the temporal frequency unreasonable for these nominally 25 m spatial resolution sensor systems. A more appropriate mapping strategy would apply hyperspectral TOC reflectance data and spectral analyses to discriminate the two BLH forests.

21.3 HYPERSPECTRAL REMOTE SENSING OF INVASIVE PLANTS

An applied research study of mapping Chinese tallow (*Triadica sebifer*), an invasive tree species, was undertaken in the north-central Gulf coastal region [22–24] (Figure 21.2). The wetland environments within the study area included baldcypress and BLH forests along with palustrine and estuarine marshes.

The initial mapping of Chinese tallow was conducted with color-infrared aerial photography captured in the late fall when leaf senescence most often results in red leaves that provide a spectral contrast between tallow trees and most adjacent vegetation [25]. Converted to digital images, the mapping successfully detected tallow forests whose senescing leaves had turned red in upland and BLH forest and in marshes where topographic highs existed. The mapping was successful because the red leaves were unique in the CIR image and contrasted highly with all surroundings, and the 1 m CIR spatial resolution was fine enough to capture small clumps of red leaves. The mapping also elucidated complications and limits of the tallow mapping. First, the tallow had to be visible from above the TOC. Second, in forests where tallow existed, most often, they were scattered or in clumps making up less than 15% of the overstory in a 30 m × 30 m area. Third, the tallow distribution was not spatially uniform. In forests, higher occurrences seemed spotty; however, there tended to be consistent occurrences along fence rows, in fallow fields, along levees, and anywhere herbaceous plants would grow in the marsh. Fourth, tallows contained a wide variety of leaf colors, including green, yellow, red, and brown and every gradation and mixture of those colors, even on the same tree (Figure 21.6); some tallows were mostly defoliated. Only red or possibly yellow tallow leaves contrasted with the surrounding vegetation. Within these limitations, the CIR mapping was successful; however, the mapping was costly and spatially limited.

To build a regional monitoring system, late fall hyperspectral data collected by the Hyperion sensor on board the EO-1 satellite was used. The CIR mapping determined that success required the detection of TOC occurrences of 10%. To be detected consistently at that level, the VIS data must be reproducible to <1% and the NIR to <5%. To obtain that level of reproducibility, the Hyperion data were transformed to TOC reflectance estimates [22].

Transformation entailed the establishment of sites spatially distributed throughout the approximately 8 km × 50 km coverage area. The coverage area included widely varying percent occurrences of tallow that were located in upland (pine and hardwood forests, agriculture, and grasslands) and wetland (baldcypress and BLH) land covers. As outlined in Ramsey and Nelson [22], TOC reflectance spectra were created for each of 34 sites, which included six baldcypress and six BLH sites. The helicopter-based upwelling radiance and downwelling irradiance spectra were collected the same day as the Hyperion data (Figure 21.7). Similar to the RT construction used in the mangrove canopy reflectance simulation, programs built on equations derived by Turner and Spencer [26] were enclosed within an optimization procedure [22,27]. The optimization minimized the difference between the calculated and predicted reflectance spectra of the site TOC. The optimized atmospheric variables were used to transform the Hyperion data to TOC reflectance data. Validation confirmed that the TOC maximum reflectance error was <1% in the VIS and <5% in the NIR (Figure 21.8) [22].

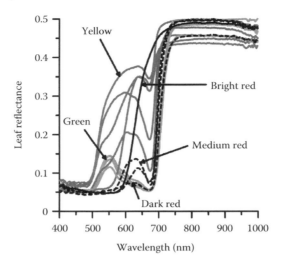

FIGURE 21.6 Leaf reflectance of tallow leaves collected from a few trees in fall senescence period. Yellow (dashed grey), green (solid grey), and bright (circles), medium (dashed black), and dark red (solid black) refer to the mix of different leaf colors exhibited by tallow at one site and at one time. The location of the tallow sites is shown in Figure 21.2. (Adapted from Ramsey, E., III et al., *Int. J. Remote Sens.*, 26, 1611, 2005.)

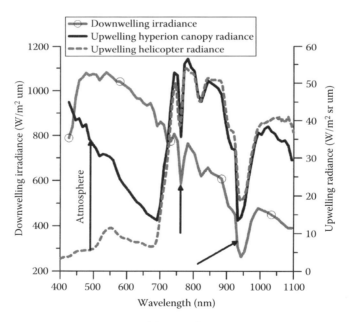

FIGURE 21.7 Near concurrent measurements of downwelling irradiance and upwelling reflected radiance as measured by the Hyperion sensor and the helicopter-based radiometer over a forest site in southwest Louisiana (a slight bias was added to the helicopter spectra for clarity). The forest site location is shown in Figure 21.2.

The CIR mapping also determined that leaf senescence and tallow occurrence patterns were extremely varied and that large (>30 m) stands of tallow likely did not exist within the region. Even if possible, a detection procedure based on a leaf spectral library or uniform stand of tallow trees was unfeasible. A more appropriate approach was provided by the multivariate analysis technique, polytopic vector analysis (PVA). PVA extracted characteristic spectra directly from the input canopy reflectance without requiring that spectra form and type be defined *a priori* as in spectral libraries

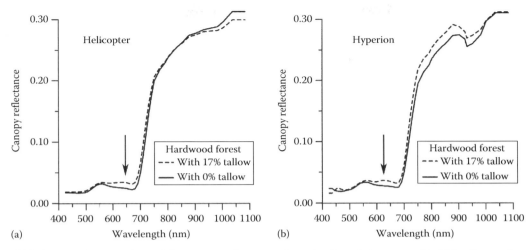

FIGURE 21.8 A comparison of the canopy reflectance of two hardwood sites, one with 17% tallow and one without tallow (general location of tallow sites shown in Figure 21.2). (a) Helicopter-based canopy reflectance derived by dividing the helicopter-based reflected radiance by the measured downwelling irradiance (radiance and irradiance of a similar forest site is shown in Figure 21.7). (b) Similarly, the Hyperion reflected radiance of the same two hardwood sites corrected and normalized to canopy reflectance. The arrows point out the slight differences in reflectance related to the tallow percent occurrence differences at the two sites. (Adapted from Ramsey, E., III and Nelson, G., *Int. J. Remote Sens.*, 26, 1589, 2005.)

or as extracted from homogeneous imaged areas [23,28]. The extracted characteristic spectra corresponded to percent occurrences per 30 m Hyperion pixel of tallow, live vegetation, and senescent vegetation. The senescent vegetation was in marshes, some agriculture fields, and most cypress and tupelo forests. Live vegetation included canopy shadows. A fourth characteristic spectrum was created in the PVA that exhibited a form similar to yellow vegetation; however, this spectrum seemed primarily to correspond to noise. The fourth spectrum was not used in the Hyperion classification.

Loadings representing the tendency of each TOC reflectance spectra to align with the tallow, live, and senescent characteristic spectra were compared to classifications of each of the 34 sites [23]. The loading represented the percent occurrences of tallow, live, and senescent vegetations. The classifications were prepared from photography collected at each site at the time of the helicopter-based upwelling radiance recordings [23]. Correspondence demonstrated that 78% of the tallow percent occurrences were correctly mapped. Calculated confidence limits for individual predicted values indicated tallow occurrences making up 10% (<10 m × 10 m) of the 30 m pixel were detected 68% of the time and 15% (<12 m × 12 m) of the occurrences were detected 85% of the time. In addition, 92% of live and 82% of senescent vegetation occurrences were accurately mapped. Next, the three characteristic spectra were applied to the Hyperion reflectance image.

The application of the characteristic spectra transformed the Hyperion image into three images that represented continuous percent occurrences per 30 m pixel of tallow, live, and senescent vegetation [24]. In contrast to a point classifier, the continuous classifier ranged from 0% to 100% (Figure 21.9). The final validation was performed with oblique photography to confirm the predicted occurrences of tallow outside the selected 34 sites. The validation showed that a high correspondence existed between the Hyperion tallow percent occurrences and the actual spatial distribution. The final step was to associate the mapped tallow percent occurrences with land cover types and possible land-based activities that promoted the establishment of tallow. The latter information relied on TM land cover classifications and knowledge of land cover activities drawn from successive map production every 3 years over a 9 year period. The land cover mapping followed protocols outlined by the National Oceanic and Atmospheric Administration's (NOAA) Coastal Change Analysis Program (C-CAP) [29,30].

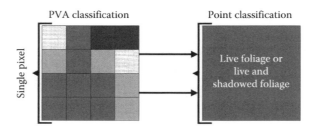

FIGURE 21.9 An illustration of classifications of the same target based on hyperspectral and broadband sensors. (Left) A PVA percent composition classification (■ 50% live vegetation, ☐ 12.5% red tallow, ▨ senescing foliage, and ■ 12.5% shadowed foliage) based on hyperspectral data. (Right) A point classification (single class per pixel) based on broadband data (i.e., Landsat TM).

Land cover type association was accomplished by spatially cross tabulating occurrences of tallow with each of the defined C-CAP land covers [24]. The cross tabulations were based solely on red tallow occurrences and therefore were conservative estimates. Although advanced HSI and TM integrations are available [31], this postclassification hyperspectral and TM integration fulfilled the purposes of the study. Restricted to wetland land covers, results suggested that the highest occurrences of tallow were in BLH forests. Also, tallow was persistent in the flooded baldcypress forests, a testament to the invasiveness of the exotic Chinese tallow. Cross-tabulation suggested that relatively high occurrences of tallow within the coastal marsh were located on the numerous cheniers and scattered topographic highs. These tallow occurrences were not identified as woody wetlands in the TM C-CAP classification; however, the TM collection time was not considered in that comparison. Additionally, even though percent occurrences were low, tallow was associated with the C-CAP water class. Most likely, these associations were linked to tallow occurring on levees lining the ubiquitous canals within the marsh. These narrow and linear features are mixed with the surrounding water or marsh in the TM pixel and hidden within the C-CAP land cover classification. By linking tallow occurrences determined from satellite HSI to land cover type produced from broadband TM data, we were able to take the first steps toward determining the relationship of various land covers and activities to the establishment of tallow.

21.4 HYPERSPECTRAL REMOTE SENSING OF MARSH WETLANDS

Multiple marsh types occupy estuarine to lacustrine coastal wetlands; however, two dominant types were used to encapsulate hyperspectral methods most relatable to marshes (general marsh locations shown in Figure 21.2). Within saline marshes, smooth cordgrass (*Spartina alterniflora*) and black needlerush (*Juncus romerianus*) produce nearly constant rates of live and dead turnover showing no clear seasonal trends after reaching maturity [32]. Needlerush and cordgrass (needlerush is comparatively more vertical and less leafy) exhibit more vertical than horizontal canopy orientations; however, dominant leaf orientation can change from top to bottom. The brackish salt-hay or salt-meadow cordgrass (*Spartina patens*) and fresh maidencane (*Panicum hemitomon*) marshes occupy the more interior coastal marshes of the Gulf coast [32]. Salt-hay marshes tend to be hummocky with vertical shoots rising above a layer of thick and lodged (nearly horizontal) dead material. As in needlerush and cordgrass marshes, salt-hay marshes appear to have low turnover with little seasonal pattern in live and dead composition. Maidencane marsh canopies exhibit yearly turnover. Beginning with nearly vertical shoots in the early spring, the canopy gains height and increasingly adds mixed orientations and density through the late spring to summer and then begins fall senescence. The maidencane seasonal turnover and its implications in interpreting canopy hyperspectral reflectance spectra will be presented.

In addition, bacteria, cyanobacteria, diatoms, and other microbes, which are intermittently exposed on intertidal surfaces and in degrading marshes, often as microbial mats, mediate important processes in marsh ecosystems, particularly the primary production and degradation of below-ground organic matter [33]. Because of the spectral complexities of the various microbes, the high variability of sediment moisture content, and high spatial composition variability, remote sensing mapping of these microbial marsh components require high spectral detail. Although not discussed here, HSI techniques pertinent to these important marsh components are being developed [33].

21.4.1 CANOPY REFLECTANCE AND STRUCTURE

Although spectral and structural variability is exemplified in canopies within and between the four marsh types, observed patterns suggest more commonality of these variables within each marsh. To best encapsulate the range and covariance of canopy spectra and structure profiles, seasonal changes of a maidencane marsh are described with respect to canopy hyperspectral reflectance differences. Outside of external factors that result in death and regrowth, particularly burns [34], the only marsh within the four introduced that completes a full senescence and regrowth cycle is maidencane (see Figure 21.2 for general locations).

To document canopy structure, up to 22 light attenuation profiles (LAPs) were obtained at each maidencane marsh site in coordination with helicopter-based radiometer upwelling light recordings from the same sites. Collection techniques and analyses of these data have been fully described [32]. The canopy LAPs were represented as a function of canopy LAI and LAD as

$$L_z = L_{sun} \cdot EXP - (LAI \cdot LAD) \cdot Z$$

where L_z is the measured photosynthetically active radiation (PAR) (sunlight from 400 to 700 nm) at a height $= Z$ above the ground surface (0 cm) remaining of the L_{sun} PAR illuminating the TOC. The L_z/L_{sun} ratio is the fraction of PAR that penetrates to any depth below the TOC.

Canopy reflectance was estimated by dividing the helicopter-based recorded upwelling light with simultaneous downwelling recordings of a separate radiometrically and spectrally aligned radiometer or by using before and after recordings of downwelling sunlight reflected from diffuse cards of constant reflectance [22,32]. The upwelling radiometer recordings ranged from about 380–1100 nm with nominal band center spacings of 2.6 nm and an estimated spectral bandpass of 10 nm [35].

Canopy LAPs and reflectance spectra of a single maidencane marsh site illustrate the covarying changes in these two variables (Figure 21.10). No surface flooding was observed during the collection of the LAP and radiance recordings. The combined LAP and reflectance information shown in the graphics indicate that the canopy comprised standing (LAP) dead (reflectance) stalks in winter (December). By early spring (April), the canopy had added new green shoots but had lost dead plant material. The February reflectance spectrum was added to demonstrate that the addition of new plant material occurred even before the end of winter. The summer (July) canopy LAP indicates a denser (or higher LAI) and less vertical (mixed orientation components or changing LAD) canopy. The reflectance spectrum indicates that the canopy contained a higher percent of live biomass than either the winter or spring canopies. The dramatic change in the early fall (October) reflectance signified a high decrease in live plant material. The concurrent LAP suggests that the decrease in live material did not represent a large loss of canopy plant material. Instead, the combined information of the LAP and reflectance suggests that the live material senesced in place.

The maidencane example (Figure 21.10) helps illustrate the need for canopy structure data as well as canopy hyperspectral data to correctly and fully interpret changes in the marsh condition. For this example, the references to canopy orientation and density were generalized, because the coupled response of the marsh canopy LAI and LAD were not separated. This coupling suited the purpose of the example; however, decoupling of these canopy structure variables is a part of ongoing applied research studies and has been partially automated in some field instruments. In addition, key aspects of the canopy reflectance were not described, for example, the increase in VIS reflectance amplitudes and the movement of the red-edge (far red, at around 700 nm) from higher to

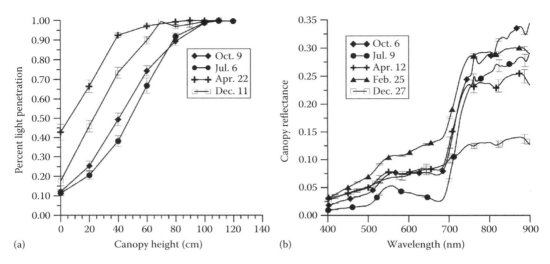

FIGURE 21.10 Light penetration and canopy reflectance of the same single 30 m × 30 m maidencane marsh (*Panicum hemitomon*) site (location of marsh shown in Figure 21.2). The error bars on the graphs represent variance of replicates. Note: Light penetration represents the combined LAD and LAI as a canopy structure indicator. (a) The light penetration curves illustrate the winter dieback, the spring turnover of dead material, the summer regrowth, and the beginning of the fall senescence. (b) The canopy reflectance spectra show the complementary changes in green leaf material. A completely dead canopy in the winter begins to add green material in the late winter and into spring. The full growth summer canopy that includes the maximum green leaf material exhibits the most defined spectral peaks and the early fall senescence reflectance spectrum illustrates the initial loss of green leaf material.

lower wavelengths with the loss of live plant material. These changes in the canopy reflectance are used to foretell adverse changes in the plant canopy condition, which is more fully illustrated in the smooth cordgrass example.

21.4.2 Detecting Subtle Changes

In contrast to maidencane marshes, cordgrass marshes maintain a more stable canopy reflectance and structure throughout the year. Although reflectance and structure are more stable temporally, they are not uniform spatially. Like many marsh types, cordgrass exists in a variety of forms that have adapted to the differences in flushing and salinities. Canopy structures are short to tall and dense to moderately sparse. In expansive coastal marshes with low-shore-normal inland elevation gradients and microtides, changes in form can occur in response to changes in topography as small as a centimeter [36]. The rapid and spatially heterogeneous changes in microelevations create a complex tapestry of cordgrass structures over the broad coastal zone. Interlaced within these spatially complex changes of flushing strength and form is the varying exposure of the background; primarily mud. Furthermore, because flushing strength is one causal agent of the varying background exposures, the moisture content of the mud mixed with litter also affects the spectral nature of the reflectance intensity (Figure 21.11). This complexity of smooth cordgrass forms in response to changes in flushing also occurs in "homogeneous" black needlerush marshes along the northern Gulf [36]; however, these monotypic marshes are anything but spectrally and spatially homogeneous.

Overlain on this tapestry of structural forms, a spatially distributed and seemingly spatially heterogeneous coastal cordgrass marsh dieback was discovered (see Figure 21.2 for general location). The reason for the sudden dieback was unknown. It spread rapidly but without obvious patterns. Only its predominant association with the cordgrass marsh was somewhat unifying. Reconnaissance of the dieback distribution was started soon after the recognition that the phenomenon was occurring

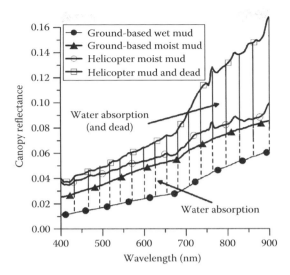

FIGURE 21.11 Reflectance as a function of mud background and water content. Wet and moist terms refer to high and moderate mud-water contents, respectively, that were based on visual observations and were not quantitatively determined. "Mud and dead" refers to relatively low mud-water content and the presence of dead marsh vegetation. Ground-based refers to measurements obtained about 2 m above the ground level. Note the abrupt reflectance increase around 700 nm in the "mud and dead" spectra indicating live leaf material was present even though visually the marsh was brown. (Adapted from Ramsey, E., III and Rangoonwala, A., Mapping the onset and progression of marsh dieback, in *Remote Sensing of Coastal Environments*, Wang, Y. (Ed.), CRC Press, Remote Sensing Applications Series, Boca Raton, FL, 2009.)

at the landscape scale. Visually, the occurrences of dieback seemed to be fairly certain; however, the myriad of dieback severities were difficult to describe. Complex classifications were used to characterize the visual progression of the dieback, such as yellow-green, or brown-green, or in the extreme case, removed (Figure 21.12). There was a need to better quantify the detection of dieback and the dieback progression.

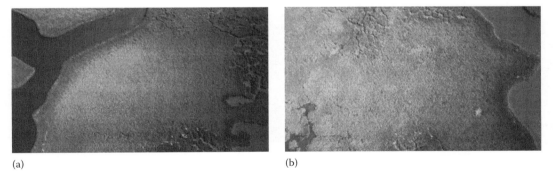

(a)   (b)

FIGURE 21.12 (See color insert.) Two marsh sites that could be visually classified as (a) brown-green and (b) green-brown dieback marshes. Although the descriptive visual aerial interpretation of dieback progression is helpful, to provide comparability and repeatability needed for regional mapping and monitoring, additional quantitative measures of the marsh condition are needed. Hyperspectral canopy reflectance provided more quantitative assessments of dieback progression and determination of dieback onset in the highly temporally and spatially complex coastal marshes. The shown marsh sites were located in coastal Louisiana (Figure 21.2). (Adapted from Ramsey, E., III and Rangoonwala, A., Mapping the onset and progression of marsh dieback, in *Remote Sensing of Coastal Environments*, Wang, Y. (Ed.), CRC Press, Remote Sensing Applications Series, Boca Raton, FL, 2009.)

A remote sensing strategy was developed that would determine whether spectral indicators of dieback onset and progression existed at the leaf level, and if so, transfer these indicators to a satellite hyperspectral or broadband sensor system. The strategy first established a metric for assessing change and the relative change magnitude [6,37,38]. The metric was based on a concept that each isolated dieback occurrence began in a localized area and spread outward. A transect from the most severe portion of the dieback through progressively less-impacted marsh and finally to the local healthy marsh simulated the temporal dieback progression. The inclusion of the local healthy marsh attempted to account for the naturally occurring spatial complexity in cordgrass by compensating for disparities in the progression from dieback site to dieback site with the differences in the local healthy marsh. The conceptual model was tested at four dieback sites scattered within a 12 km × 12 km area of cordgrass marsh experiencing sudden dieback [6].

Plant leaf reflectance and transmittance spectra were obtained from plants collected every 5 m along transects spanning the dead to local healthy marsh [6]. Transects were established at two fairly extensive diebacks and two smaller diebacks. Only the greenest leaves on the plants were used. The reflectance spectra from the most extensive dieback site showed a nearly progressive change in blue (400–500 nm) and red (600–700 nm) magnitudes (Figure 21.13). In both wavelength regions, plant leaf reflectance decreased from the severest dieback region to the healthiest marsh. Similar but more variable correspondences were exhibited in the green (500–600 nm) and red-edge (around 770 nm) spectral bands. The progressive changes of these VIS reflectance bands with transect distance supported the concept of dieback onset at a localized area as well as its progressive outward spread.

The progressive change also confirmed that a spectral indicator of the dieback progression was obtainable [6]. The direction of changes and their similar but different correspondences of the blue-red and green-red-edge bands to the transect progression also supported the existence of a spectral indicator of dieback onset and progression. Considering only the most common and usually dominant green leaf pigments, carotene (CAR) and chlorophyll a and b (CHL), a reason for the similarities and differences of leaf reflectance at the four VIS bands with respect to the dieback was proposed (Figure 21.14, [39]). Changes in the absorption magnitudes are logarithmically related

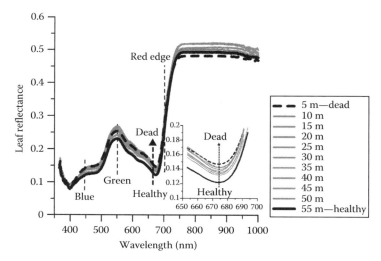

FIGURE 21.13 The dieback transect site was located in coastal Louisiana (Figure 21.2). Leaf reflectance spectra depicting changes in pigment concentrations (see Figure 21.14) from the dead (dashed line) to local healthy (solid line) marsh every 5 m at a dieback site. The insert exhibits the increase in reflectance with dieback progression in the red band. The same progression was shown in the blue band and similar but less monotonic in the green and red-edge bands (vertical lines and labels locate band centers noted on Figure 21.14). (Adapted from Ramsey, E., III and Rangoonwala, A., *Photogramm. Eng. Rem. Sens.*, 71, 299, 2005.)

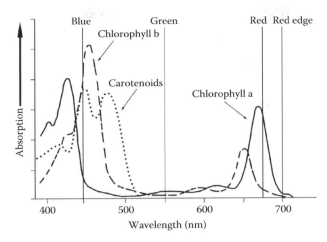

FIGURE 21.14 Absorption spectra of chlorophyll and carotene. Note that the blue and red bands are located near peak absorptions, and the green and red edge bands are located at the tails of these absorptions. (Adapted from Kirk, J., *Light and Photosynthesis in Aquatic Ecosystems*, Cambridge University Press, Cambridge, U.K., 1994, pp. 229–233.)

to the leaf pigment concentrations. CAR and CHL pigments exhibit high absorptions in the blue band with a maximum between 440 and 460 nm. Secondary CHL absorption peaks are exhibited between about 600 and 700 nm and the main CHL-a peak near 670 nm. Green and red-edge bands also experience CHL absorption, although situated in the tails of the pigment absorption spectra, changes would be relatively lower. The spectrally varying strengths of the pigment absorptions indicated that small changes in pigment concentrations would be represented as dramatic changes in the green and red-edge bands but by relatively small changes in the blue and red bands. In fact, this difference in spectral sensitivity as suggested by Gitelson and Kaufman [40] was used to provide greater discernment of the dieback onset and progression.

The changes in the leaf pigment absorptions are related to changes in the leaf reflectance and transmittance via Kirchoff's radiation law, as

$$\% \text{ absorption } (\lambda) = 1 - [\% \text{ reflectance } (\lambda) + \% \text{ transmittance } (\lambda)]$$

where λ is the wavelength. This relation is also illustrated by combining the three spectral variables in the same graphic (Figure 21.15). Inspection of all leaf spectra representing the four dieback transects found that narrow wavelength bands located in the blue, green, red, red-edge, and NIR maximized the spectral information obtainable as related to changes in the CHL and CAR pigment concentrations. Even without the need to use the whole spectrum (400–1100 nm), as was the case in this study, the availability of hyperspectral information provided the ability to closely define the spectral positions and widths best suited for the dieback onset and progression mapping. TM bands are overlain on the same leaf spectra to illustrate their general lack of specificity (Figure 21.15). The TM overlay also points out the lack of a red-edge band, considered an important spectral region in this study as in many others.

Leaf spectral analyses indicated that the red and blue bands best represented the larger and presumably older diebacks, and the green and red-edge bands best represented the presumably younger diebacks. It is also noteworthy that trends in CHL and CAR concentrations estimated from a selection of narrow VIS reflectance bands [41] indicated that the younger diebacks were in fact in transition [42], further supporting the older to younger dieback depiction drawn from the blue-red band and green-red-edge band comparisons. HSI provided the bands and bandwidths appropriate for the algorithm [41].

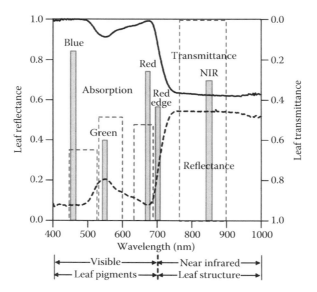

FIGURE 21.15 The reflectance and transmittance spectra calculated from direct measurements of smooth cordgrass leaves. The solid bars represent the bands selected to determine changes in the plant-leaf that were indicative of dieback progression. The dashed-line bars depict visible and near infrared bands available on the TM sensor. Note the differences in the selected and TM bandwidths and locations and that lack of a TM red-edge band. The leaf sample was obtained from a dieback site in coastal Louisiana (Figure 21.2). (Adapted from Ramsey, E., III and Rangoonwala, A., *Photogramm. Eng. Rem. Sens.*, 71, 299, 2005.)

To provide a single indicator conducive to most operational broadband sensors, an NIR/green simple ratio was used [40]. Overall, the NIR/green ratio performed better than the more common NIR/red simple ratio and provided an indication of the dieback onset at each location [6]. Both ratios helped desensitize dieback onset and progression mapping to atmospheric influences, variable soil background reflectance, and changing canopy compositions (plant and background) (e.g., [43]). The next step was to transfer this spectral information from the plant leaf level to the plant canopy level.

The method used to transfer the plant leaf results to the plant canopy was helicopter-based radiance collections [37,38]. This technique allowed targeted and controlled collections with little atmospheric influences. As illustrated in Figure 21.12, the ability to detect and better quantify the progression of dieback was needed. The circles on Figure 21.16a depict the location and spatial extent of the canopy reflectance spectra in Figure 21.16b. As expected from the plant leaf spectral analyses, the canopy VIS reflectance increases as the impact becomes more severe. The canopy reflectance associated with the dead transect portion is dominated by the mud background, whereas the canopy reflectance at the healthy end represents the local healthy marsh reflectance.

As in the invasive mapping with Hyperion data, PVA was used to extract the healthy and dead characteristic reflectance spectra [37]. The characteristic spectra defined from canopy reflectance spectra of the transect sites and a more inland reference cordgrass site provided a method to deemphasize the background variability and to isolate those influences related to plant leaf reflectance within the marsh canopy [37]. Subsequently, these two characteristic spectra refined through comparison to the plant leaf dieback onset and progression results were applied to 13 transect canopy reflectance spectra and 10 additional sites depicting cordgrass marsh of various visual hues.

Loadings representing the tendency of the canopy reflectance to align with healthy or dead spectra were compared to marsh classifications of each of the 23 sites [37]. In addition, NIR/green and NIR/red ratios calculated for each site from the pertinent canopy reflectance

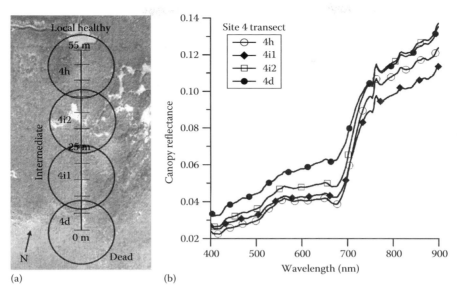

(a)   (b)

FIGURE 21.16   Marsh canopy sampling strategy and transect canopy reflectance spectra. In the left portion (a), circles along the transect overlain onto photography taken during the helicopter overflight represent the marsh areas targeted by the helicopter radiometer, dead (4d); intermediate-1 (4i1); intermediate-2 (4i2); and local healthy (4h). On the right (b), canopy reflectance associated with those marsh areas is depicted by circles on the transect. Note that the progressively higher VIS canopy reflectance spectra simulate the leaf transect reflectance spectra shown in Figure 21.13. The dieback transect site was located in coastal Louisiana (Figure 21.2). (Adapted from Ramsey, E., III and Rangoonwala A., *Photogramm. Eng. Rem. Sens.*, 72, 641, 2006.)

spectrum were compared to the marsh classifications. The marsh classifications provided generalized amounts of green to yellow to brown marsh and visible background. The loadings based on the PVA hyperspectral analyses better aligned with the plant leaf transect results and better corresponded with the marsh classifications than did NIR/green or NIR/red indicators of marsh dieback progression [37]. Of the two ratios, the NIR/green ratio exhibited higher alignment with the classifications than NIR/red and had a higher correspondence with the PVA results [38]. Mapping based on the whole canopy reflectance spectra (400–1000 nm) produced the best depiction of marsh dieback progression and the NIR/green band ratio produced lower but acceptable results.

Classification of the PVA results was accomplished with a distance-clustering procedure [44]. The classification defined three broad categories: healthy, impacted, and severely impacted (including dead) [37]. The local healthy marshes at the most advanced dieback transect sites were designated as impacted marshes, whereas the local healthy marshes at the transitional transect sites were designated as healthy marshes. The reference site was also classified as healthy. Surprisingly, the marshes shown in Figure 21.12 were both designated as healthy. Canopy reflectance spectra of these two sites exhibited the expected spectral differences in the VIS bands; however, the difference is low (Figure 21.17). Overall, based on hyperspectral data analyses, differences in severity of dieback impact were indicated in the impacted and dead marsh categories. In addition, results based on the whole spectra analyses captured the spectral variability in the non-impacted or healthy cordgrass marshes that was expected given the high variability in physical controls (e.g., Figures 21.12 and 21.17). Based on that natural spectral variability, classification of marsh status based on more traditional visual inspection and even broadband data in these highly complex coastal marsh systems may not provide a consistent strategy for monitoring the status or trends of these coastal resources.

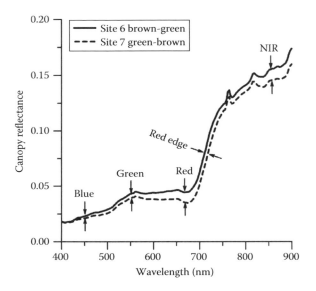

FIGURE 21.17 Canopy reflectance associated with the two marsh sites shown in Figure 21.12 (located in coastal Louisiana, Figure 21.2). The position of the bands used in the study (as depicted on Figures 21.13 through 21.15) is included. Note the small change in canopy reflectance (600–700 nm) that caused the perceived visual change in Figure 21.12. (Adapted from Ramsey, E., III and Rangoonwala, A., Mapping the onset and progression of marsh dieback, in *Remote Sensing of Coastal Environments*, Wang, Y. (Ed.), Remote Sensing Applications Series, CRC Press, Boca Raton, FL, 2009.)

21.5 SUMMARY

The benefits of HSI to remote sensing mapping of wetlands are not based on specific image processing techniques but on the uniqueness of vegetation (at the species or cover type level) and hydrodynamics and phenodynamics of the wetland system. The mixtures of vegetation types, densities, heights and forms, high spatial complexity, and dominant role of flooding in controlling the wetland landscape lead to unique remote sensing solutions to detect, map, and monitor these often dynamic systems. This unique use of remote sensing is particularly true in coastal wetlands where the spatial and temporal frequency of change is even more dynamic than in interior wetlands. As demonstrated in this chapter, the ability to overcome these complexities relies on targeted applications of HSI that are enhanced when supplemented with pertinent plant leaf spectral and canopy structure and background information.

21.5.1 Hyperspectral Imaging Enhancing Broadband Mapping

HSI was used to determine if the successful application of broadband data to mangrove canopy LAI determination could be extended to mangrove species mapping. Hyperspectral leaf analyses led to the conclusion that spectral variability within species overwhelmed mean differences between species; leaf spectral properties lacked species specificity. Light interaction modeling based on field data and hyperspectral TOC spectra found that even though LAI variability was low, the near stability of the canopy LAD and nonspecies specificity of the leaf spectra allowed LAI to fully dominate the broadband variance and remain consistent across all species. Although HSI was not successful in mapping mangroves at the species level, the hyperspectral analyses provided a more complete understanding of how spectral information was uniquely linked to the individual features making up the mangrove canopy. In providing that direct linkage, species mapping based on broadband data were determined to be most likely a function of canopy LAI variability or possibly background variability in more open canopies than canopy leaf spectral differences associated with species composition.

Although species separation was not successful, more refined leaf spectral measurements (i.e., single leaves) and the use of more precise canopy closure measurements might improve the species-specific relations with canopy reflectance. The addition of radar could provide a more direct measure of canopy structure variability. That structure information could be used to remove canopy structural influences from the canopy reflectance and thereby enhance the leaf reflectance component in the hyperspectral canopy reflectance, possibly improving species mapping.

21.5.2 HYPERSPECTRAL MAPPING OF INVASIVE PLANT OCCURRENCES AND BROADBAND FUSION FOR REGIONAL RISK ASSESSMENT

An invasive species mapping project on evaluating Chinese tallow invasions illustrated how broadband mapping requiring a high spatial resolution (1 m) could be regionalized with moderate spatial resolution (30 m) hyperspectral data (EO-1 Hyperion) and image processing. The invasive occurred as single trees or in small clumps dispersed throughout BLH and cypress-tupelo wetland forests and occupying topographic highs in coastal marshes. The tallow tree occurred in varieties of canopy structures and compositions and exhibited a high mix of leaf colors in progressive stages of senescence. These extreme physical heterogeneities and the lack of large tallow stands in the Hyperion scene excluded the more conventional creation of hyperspectral characteristic spectra from homogeneous stands within the image or from leaf spectral libraries. The use of a novel technique that identified the underlying characteristic directly from the image data was applied to overcome these limitations in creating a characteristic spectrum for tallow detection. Application of this technique provided tallow detection to <10% of the 30 m Hyperion pixel and a high 78% compositional mapping accuracy inclusive of all wetlands. Most notably, the successful mapping relied on highly accurate canopy reflectance input and targeted field sites of known canopy compositions. The regional hyperspectral mapping provided not only tallow detection as the broadband 1 m mapping but also provided a continuous classification of tallow percent occurrence and the live and senescing vegetation compositions within the 30 m Hyperion pixel. The continuous tallow percent occurrence classification was used to create an added value product for resource managers. An integration of land resource information created from broadband Landsat TM data and tallow occurrences created from the hyperspectral Hyperion data provided new information about what land cover types were most associated with tallow occurrences and what activities promoted or led to more frequent tallow occurrences.

21.5.3 HYPERSPECTRAL IMAGING AND CANOPY STRUCTURE INFLUENCES

Another study illustrated the high level of coupling between marsh canopy structure and canopy hyperspectral reflectance. The study was undertaken to determine what information about the marsh condition could be extracted from canopy reflectance spectra and to determine the best collection times were. To emphasize the canopy structure and reflectance linkage, a marsh that exhibited total seasonal turnover was studied. The demonstration showed that correct interpretation of the marsh trends required information describing the canopy structure (e.g., LAI and LAD). Canopy reflectance spectra depicted the seasonal changes in live canopy biomass; however, if solely based on that information, impressions about changes in the marsh could be misinterpreted. In fact, the structural changes in the marsh were offset from the changes in live and dead proportions. For example, although fully senesced by fall, the canopy structure remained largely intact until new growth in late winter.

As a discovery and evaluation tool, the full canopy hyperspectral reflectance spectra clarified the VNIR spectral response; the spectral response outside of the selected broadband regions did not have to be inferred. As illustrated in the mapping of marsh dieback onset, further analyses of the canopy spectra should provide more accurate measurement of the fall senescence and spring regrowth patterns. In addition, hyperspectral data extending into the SWIR region can provide dead material estimates of biomass. This added information should improve the canopy live and dead determinations, however, to fully understand changes in marsh function as related to biomass and structure requires the coupling of hyperspectral data with other remote sensing data such as radar and light detection and ranging (LiDAR) data that are more revealing about the canopy structure.

21.5.4 HYPERSPECTRAL IMAGING FOR DETECTING SUBTLE AND ABNORMAL LANDSCAPE CHANGE

The final application demonstrated the ability of hyperspectral data processing to capture subtle leaf and canopy spectral changes related to the onset and progression of marsh dieback in a spectrally complex, monotypic marsh. In this case, the enhanced mapping performance compared to broadband sensor systems was illustrated even when the hyperspectral data were transformed to user-specified bands of defined bandwidths. The bands were placed at leaf reflectance features exhibiting high variability and closely aligned with the major leaf pigment spectral absorption regions in the VIS. The primary outcome of the leaf spectral analyses was that red-edge reflectance, lacking from most operational broadband remote sensing systems, and green reflectance were effective indicators of marsh dieback onset and early progression. Blue and red bands were better indicators of later stages in the dieback progression. Translated to the plant canopy level, the same hyperspectral bands and band transforms produced fairly broad divisions of marsh dieback. The most detailed accounting of marsh onset and dieback progression, however, was determined from analyses of the full plant canopy hyperspectral reflectance spectra (400–1000 nm). The full spectra analyses replicated the broadband analyses on the basis of user specified bands and added dieback onset and progression details within the broad divisions of severe impact, intermediate impact, and healthy marsh. In fact, the full reflectance analyses seemed to separate natural variability in the plant canopy from variability because of the dieback progression. In the plant canopy reflectance, structure and background contributed to the reflectance suppressing the plant leaf reflectance, sometimes overwhelmingly. The heightened information captured by the full hyperspectral reflectance spectra helped compensate for the structure and background spectral influences, thus increasing the performance of hyperspectral versus broadband dieback mapping.

21.5.5 HYPERSPECTRAL METHODS SUMMARY

HSI provided increased information about the status or condition of the wetland canopy beyond what was obtainable from broadband sensors in all described comparisons. In each case, the hyperspectral mapping advantage was based on the use of high-quality canopy reflectance spectra free from atmospheric distortions and illumination variability's. Once these high fidelity reflectance datasets were obtained, specialized tools were required that could transform the reflectance data into information about the wetland condition or composition with enough precision, so that subtle changes could be detected. Each of these transforms to reflectance and to spectral characteristics is often arduous and always detailed. As demonstrated in the described applied research studies, however, the obtainment of the mapping objective often depends on the use of HSI; broadband optical sensing or other sensors operating in alternate wavelength regions will not provide the necessary data. Also introduced was the need at times to incorporate new and more detailed field collection and analyses techniques. Often, particularly in investigative works, mapping success relies on targeted and well-structured field campaigns that provide consistent and pertinent biophysical data that is relatable to the hyperspectral reflectance data. Those biophysical and reflectance relationships provide quantitative interpretation of the reflectance data and the ultimate transformation of the reflectance data into meaningful products and information that can only be provided by HSI.

21.6 FUTURE DIRECTIONS

There are two specific forest types where hyperspectral imaging (HSI) could be focused: cypress and BLHs. A remote sensing method that uses hyperspectral detection of the onset of cypress impact could be developed. As stated, the disappearance of these forests in subsiding coastal areas is alarming, and an early detection tool is needed to identify impact before visible change (e.g., conversion of wetland forest to marsh) has occurred. The BLHs are commonly grouped into a single map class because of the inability to spectrally separate different communities by using traditional broadband multispectral (MS) approaches. Hyperspectral mapping could provide an appropriate basis for determining whether or not these hardwood communities can be mapped separately.

In all wetland mapping projects, applied remote sensing researchers should endeavor to obtain hyperspectral canopy reflectance of meaningful sites within the project area. In addition, leaf hyperspectral data should be obtained and complemented with canopy structure data where possible. The interrelations among the plant leaf and plant canopy reflectance spectra variances, features, and canopy structure should be examined with respect to the mapping objective. Can features in the hyperspectral data be isolated that are relatable to the mapping objective? If so, what wetland type and condition information is extractable based on operational broadband data and what information is more amenable to analysis of hyperspectral data? Can the information obtainable from the broadband be improved through integration with the hyperspectral data? If structural information is available, is the wetland vegetation separable into groups that are more or less structurally stable and those that exhibit high structural variability spatially or temporally? In vegetation exhibiting high structure variability, and most likely variable background influences in the plant canopy reflectance, would integration of the hyperspectral mapping with a more structurally appropriate sensors such as radar or lidar improve the success of the mapping project or capability for such work?

In addition to using hyperspectral data collections as the first step in the wetland mapping project, applied research should be directed at integration of hyperspectral data with broadband optical, radar, and lidar data sources. The broadband and HSI integration can be as simple as demonstrated in the invasive tallow mapping in this chapter. More refined and complete integrations would use the hyperspectral data to enhance the mapping performance of the broadband data and extend that enhanced mapping performance throughout the broadband swath extent, thus providing wider coverage and thereby higher repeat frequency. HSI integration with radar would help account for the plant canopy and background variability that can obscure the interpretation of the plant canopy reflectance. Radar data could also be useful for monitoring the hydrologic state of observed wetlands, such as flooding within closed canopy swamps. Similar to what was discovered with broadband and HSI integration, current research is pursuing integration of optical and radar with HSI in an effort to provide a more consistent and on-demand land cover monitoring system. The full integration of hyperspectral, broadband, lidar, and radar image data would advance the relevance of remote sensing for the unique spatial, compositional, and functional setting of dynamic coastal wetlands.

ACKNOWLEDGMENTS

We thank Joseph Spruce (Science Systems and Applications, Inc.) and Beth Middleton (U.S. Geological Survey) for their help in preparing this manuscript for publication. Partial funding for the described works were provided by the Louisiana Department of Natural Resources (Agreement Number 2512-01-11) and the National Aeronautics and Space Administration (Grant Number EO-1-0100-0042). Mention of trade names or commercial products is not an endorsement or recommendation for use by the U.S. Government.

REFERENCES

1. Sahagian, D. and Melack, J., *Global Wetland Distribution and Functional Characterization: Trace Gases and the Hydrologic Cycle*, Report of the Joint GAIM-DIS-BAHC-IGAC-LUCC Workshop held in Santa Barbara CA, on May 16–20, 1996.

2. Ramsey, E., III, Radar remote sensing of wetlands, in *Remote Sensing Change Detection: Environmental Monitoring Methods and Applications*, R. Lunetta and C. Elvidge (Eds.), Ann Arbor Press, Inc., Ann Arbor, MI, pp. 211–243, 1998.

3. Wickland, D., Mission to planet earth: The ecological perspective, *Ecology*, 72, 1923–1933, 1991.

4. Ramsey, E., III, Remote sensing of coastal environments, in *Encyclopedia of Coastal Science*, M. L. Schwartz (Ed.), Kluwer Academic Publishers, Dordrecht, the Netherlands, pp. 797–803, 2005.

5. Mitsch, W. and Gosselink, J., *Wetlands*, 3rd edn., John Wiley & Sons, New York, 2000.

6. Ramsey, E., III and Rangoonwala, A., Leaf optical property changes associated with the occurrence of *Spartina alterniflora* dieback in coastal Louisiana related to remote sensing mapping, *Photogrammetry Engineering and Remote Sensing*, 71, 299–311, 2005.

7. Bo Liu, B., Zhang, L., Zhang, X., Zhang, B., and Tong, Q., Simulation of EO-1 Hyperion data from ALI multispectral data based on the spectral reconstruction approach, *Sensors*, 9, 3090–3108, 2009.

8. Warner, T. and Shank, M., Spatial autocorrelation analysis of hyperspectral imagery for feature selection, *Remote Sensing of Environment*, 60, 58–70, 1997.

9. Thenkabail, P., Smith, R., and Pauw, E., Hyperspectral vegetation indices and their relationships with agricultural crop characteristics, *Remote Sensing of Environment*, 71, 158–182, 2000.

10. Ramsey, E., III and Jensen, J., Remote sensing of mangroves: Relating canopy spectra to site-specific data, *Photogrammetric Engineering and Remote Sensing*, 62(8), 939–948, 1996.

11. King, S. and Keeland, B., Evaluation of reforestation in the Lower Mississippi River Alluvial Valley, *Restoration Ecology*, 7, 348–359, 1999.

12. Lugo, A. and Snedaker, S., The ecology of mangroves, *Annual Review of Ecological Systems*, 5, 39–64, 1974.

13. Ramsey, E., III and Jensen, J., Modeling mangrove canopy reflectance using a light interaction model and an optimization technique, in J. Lyon and J. McCarthy (Eds.), *Wetland and Environmental Applications of GIS*, CRC Press, Boca Raton, FL, pp. 61–81, 1995.

14. Lillesaeter, O., Spectral reflectance of partly transmitting leaves: Laboratory measurements and mathematical modeling, *Remote Sensing of Environment*, 12, 247–254, 1982.

15. Goudriaan, J., *Crop Micrometeorology: A Simulation Study*, Netherlands Centre for Agricultural Publishing and Documentation, Wageningen, the Netherlands, 249 pp., 1977.

16. Krauss, K., Chambers, J., and Allen, J., Salinity effects and differential germination of several half-sib families of baldcypress from different seed sources, *New Forests*, 15(1), 53–68, 1998.

17. Daughtry, C., Ranson, K., and Biehl, L., A new technique to measure the spectral properties of conifer needles, *Remote Sensing of Environment*, 27, 81–91, 1989.

18. White, D., Plant communities of the lower Pearl River basin, Louisiana, *American Midland Naturalist*, 110, 381–396, 1983.

19. Ramsey, E., III and Rangoonwala, A., Bottomland hardwood type related to hurricane impact severity, manuscript submitted for publication.

20. Ramsey, E., III, Rangoonwala, A., Middleton, B., and Lu, Z., Satellite optical and radar image data of forested wetland impact on and short-term recovery from Hurricane Katrina in the lower Pearl River flood plain of Louisiana, USA, *Wetlands*, 29, 66–79, 2009.

21. Ramsey, E., III, Spruce, J., Rangoonwala, A., Suzuoki, Y., Smoot, J., Gasser, J., and Bannister, T., Monitoring wetland forest recovery along the lower Pearl River with daily MODIS satellite data, *Photogrammetric Engineering and Remote Sensing*, in press.

22. Ramsey, E., III and Nelson, G., A whole image approach for transforming EO1 Hyperion hyperspectral data into highly accurate reflectance data with site-specific measurements, *International Journal of Remote Sensing*, 26, 1589–1610, 2005.

23. Ramsey, E., III, Rangoonwala, A., Nelson, G., Ehrlich, R., and Martella, K., Generation and validation of characteristic spectra from EO1 Hyperion image data for detecting the percent occurrence of invasive species, specifically Chinese tallow, *International Journal of Remote Sensing*, 26, 1611–1636, 2005.

24. Ramsey, E., III, Rangoonwala, A., Nelson, G., and Ehrlich, R., Mapping the invasive species, Chinese tallow with EO1 satellite Hyperion hyperspectral image data and relating tallow percent occurrences to a classified Landsat Thematic Mapper landcover map, *International Journal of Remote Sensing*, 26, 1637–1657, 2005.

25. Ramsey, E., III, Nelson, G., Sapkota, S., Seeger, E., and Martella, K., Mapping Chinese tallow with color-infrared photography, *Photogrammetric Engineering and Remote Sensing*, 68(3), 251–255, 2002.
26. Turner, R. and Spencer, M., Atmospheric model for correction of spacecraft data, in *Proceedings of Eighth International Symposium of Remote Sensing of Environment, Environmental Research Institute of Michigan*, Ann Arbor, MI, pp. 895–934, 1972.
27. Himmelblau, D.M., *Applied Nonlinear Programming*, McGraw-Hill, New York, 1972.
28. Ehrlich, R. and Crabtree, S., *The PVA Multivariate Unmixing System, Self-Training Classification*, Tramontane, Inc., and C & E Enterprises, Salt Lake City, UT, 2000.
29. Klemas, V., Dobson, J., Ferguson, R., and Haddad, K., A coastal land cover classification system for the NOAA coast watch change analysis project, *Journal of Coastal Research*, 9, 862–872, 1993.
30. Ramsey, E., III, Nelson, G., and Sapkota, S., Coastal change analysis program implemented in Louisiana, *Journal of Coastal Research*, 17, 55–71, 2001.
31. Gomez, R., Jazaeri, A., and Kafatos, M., Wavelet-based hyperspectral and multi-spectral image fusion, *Proceedings of SPIE*, 4383, 36–42, 2001.
32. Ramsey, E., III, Nelson, G., Baarnes, F., and Spell, R., Light attenuation profiling as an indicator of structural changes in coastal marshes, in *Remote Sensing and GIS Accuracy Assessment*, R. Lunetta and J. Lyon (Eds.), CRC Press, Boca Raton, FL, pp. 59–73, 2004.
33. Ramsey, E., III, Rangoonwala, A., Christian, R., and Marsh, A., Spectral reflectance of purple bacteria occurring in an upper saline marsh, manuscript submitted for publication.
34. Ramsey, E., III, Rangoonwala, A., Baarnes, F., and Spell, R., Mapping fire scars and marsh recovery with remote sensing image data, in *Remote Sensing and GIS for Coastal Ecosystem Assessment and Management*, X. Yang (Ed.), Springer Lecture Notes in Geoinformation and Cartography, Berlin, Germany, pp. 415–438, 2009.
35. Markham, B., Williams, D., Schafer, J., Wood, F., and Kim, M., Radiometric characterization of diode-array field spectroradiometers, *Remote Sensing of Environment*, 51, 317–330, 1995.
36. Stout, J.P., The ecology of irregularly flooded salt marshes of the northeastern Gulf of Mexico: A community profile, U.S. Fish Wildlife Service Biology Report 85, 1984.
37. Ramsey, E., III and Rangoonwala, A., Site-specific canopy reflectance related to marsh dieback onset and progression in coastal Louisiana, *Photogrammetric Engineering and Remote Sensing*, 72, 641–652, 2006.
38. Ramsey, E., III and Rangoonwala, A., Mapping the onset and progression of marsh dieback, in *Remote Sensing of Coastal Environments*, Y. Wang (Ed.), Remote Sensing Applications Series, CRC Press, Boca Raton, FL, 2009.
39. Kirk, J., in *Light and Photosynthesis in Aquatic Ecosystems*, Cambridge University Press, Cambridge, U.K., pp. 229–233, 1994.
40. Gitelson, A. and Kaufman, J., MODIS NDVI optimization to fit the AVHRR data series-spectral considerations, *Remote Sensing of Environment*, 66, 343–350, 1998.
41. Gitelson, A., Zur, Y., Chivkunova, O., and Merzlyak, M., Assessing carotenoid content in plant leaves with reflectance spectroscopy, *Photochemistry and Photobiology*, 75, 272–281, 2002.
42. Mendelssohn, I. and McKee, K., Spartina *alterniflora* die-back in Louisiana: Time-course investigation of soil waterlogging effects, *Journal of Ecology*, 76, 509–521, 1988.
43. Ehrlich, D., Estes, J., and Singh, A., Applications of NOAA-AVHRR 1 km data for environmental monitoring, *International Journal of Remote Sensing*, 15, 145–161, 1994.
44. SAS Institute Inc., *SAS/STAT User's Guide: Version 6*, 4th edn., vol. 1, SAS Institute Inc., Cary, NC, 1989.

22 Characterization of Soil Properties Using Reflectance Spectroscopy

E. Ben-Dor

CONTENTS

22.1 INTRODUCTION

Reflectance spectroscopy has become a very useful tool for the soil sciences in the past 20 years. This technique enables the extraction of quantitative and qualitative information on many soil attributes in real time and can shed light on the soil's composition without the need for labor-intensive wet-chemistry analyses (e.g., [1–3]). Reflectance spectroscopy information is acquired in both point and image arrangements in the laboratory, field, and, recently, also from air and space domains. Whereas in the laboratory, soil-reflectance measurements are performed under controlled conditions using standard protocols (and hence with minimum interference), in the field, reflectance measurements are fraught with a variety of problems, such as variations in viewing angle, changes in illumination, soil roughness, and soil sealing (Ben-Dor et al. [4]). Acquiring soil reflectance data from air and space involves additional difficulties, resulting from, for example, relatively low signal-to-noise ratio sensors and atmospheric attenuation. Laboratory-based measurements enable an understanding of the chemical and physical principles of soil reflectance and are widely used for practical applications requiring a quantitative approach. As the sensitivity of portable field spectrometers increases, field soil spectroscopy is becoming a promising tool for rapid point-by-point monitoring of the soil environment. Recently, considerable effort has been invested in commercializing field spectroscopy–based sensors for agricultural applications (e.g., VERIS Technologies http://www.veristech.com/index.aspx), NovoSpec (www.novospec.com). Several papers have described the possibility of using soil reflectance in the field for precision agriculture applications (e.g., [5,6]). In this regard, the future looks bright for the development of a new spectral technology, termed imaging spectroscopy (IS), which combines the spectral and spatial domains [7]. Due to the large number of airborne IS sensors operating today for many terrestrial applications, this technology is slowly but surely entering the field of soil science, where its use will rely heavily on the spectral foundation generated over the past two decades in soil analysis laboratories [8]. Understanding the principles and limitations of soil spectra is crucial to the use of the forthcoming soil-IS technology. Information about soils from reflectance spectra in the visible-near-infrared (VIS–NIR) (400–1000 nm) and short-wave-infrared (1000–2500 nm) spectral regions constitutes almost all of the data that passive solar sensors can provide and therefore this chapter will only cover these regions. We provide a historical overview of soil reflectance spectroscopy and a general overview of the chemical–physical principles of the soil-reflectance spectrum in this spectral region. We also discuss the basic interactive processes between soils and electromagnetic radiation and shed light on the principle of quantitative soil-spectral approaches. This chapter will provide recent examples on how reflectance spectroscopy of soil is used in a modern remote sensing arena, using both point and imaging sensors, as well as future notes on the potential of this methodology.

22.2 SOIL

22.2.1 GENERAL BACKGROUND

Soil is a complex material that is extremely variable in physical and chemical composition. It is formed from exposed masses of partially weathered rocks and minerals of the Earth's crust. Soil formation, or genesis, is strongly dependent upon the environmental conditions of both the atmosphere and lithosphere. Soils are a product of five factors: climate, vegetation, fauna, topography, and parent materials. The great variability in soils is the result of interactions among these factors and their influence on the formation of different soil profiles [9]. Whereas soil taxonomy and soil mapping require knowledge of the entire soil profile, optical remote sensing measures only the thin (about 50,000 nm) upper surface layer [4]. Thus, reflectance remote sensing cannot be an effective tool for classifying soils from a pedological standpoint where examination of the entire soil profile is required. This limitation is becoming more salient as the natural surface becomes altered, for example, by agricultural activities, and therefore less of the natural soil body exists [10,11]. Another limiting factor in the use of optical remote sensing as a petrological tool is the masking effect of vegetation and snow, which both affect radiant flux at the surface.

22.2.2 SOIL COMPOSITIONS

Any given soil mixture is made up of all three phases of matter: solid, liquid, and gas. A typical soil may consist of about 50% pore space, containing spatially and temporally varying proportions of gas and liquid.

22.2.2.1 Solid Phase

This phase contains organic and inorganic matter in a complicated and generic mixture of primary and secondary minerals, organic components, and salts. The solid phase consists of three main particle size fractions—sand (2–0.2 mm), silt (0.2–0.002 mm), and clay (<0.002 mm), which together govern two major soil properties: texture and structure. Soil texture is a function of the distribution of these three main components and is generally described in terms of quantities of gravel, sand, silt, and clay. Soil structure (a function of adhesive forces between generally fine particles) describes the aggregation characteristics of a soil. These two properties play a major role in soil behavior and influence some major soil characteristics such as drainage, fertility, moisture, and erosion. The inorganic portion of the solid phase consists of soil minerals, which are generally categorized as either primary or secondary. Primary minerals are derived directly from the weathering of parent materials that were formed under much higher temperatures and pressures than those found at the Earth's surface. Secondary minerals are formed by geochemical weathering of the primary minerals. An extensive description of minerals in the soil environment is given by Dixon and Weed [12], and readers who wish to expand their knowledge in this area are referred to that text. In general, the dominant primary minerals are quartz, feldspar, orthoclase, and plagioclase. Some layer silicate minerals, such as mica and chlorite, and ferromagnesian silicates, such as amphibole, peroxide, and olivine, also exist. The secondary minerals in soils—often termed clay minerals—are aluminosilicates such as smectite, illite, vermiculite, sepiolite, kaolinite, and gibbsite. The type of clay mineral present is strongly dependent on the weathering stage of the soil and can be a significant indicator of the environmental conditions under which the soil was formed. Other secondary minerals in soils are aluminum and iron oxides and hydroxides, carbonates (calcite and dolomite), sulfates (gypsum), and phosphates (apatite). Most of these minerals are relatively insoluble in water and maintain equilibrium with the water solution. Soluble salts such as halite may also be found in soil but they are mobile in water. Clay minerals are most likely found in the fine-sized soil particles (<2 mm; clay fraction) and are characterized by relatively high specific surface areas (50–800 m^2/g). The primary minerals and other nonclay minerals are usually found in both the sand and silt portions, and consist of relatively small specific surface areas (<1 m²/g). In addition to the inorganic components in the solid phase, organic components also exist. Although the organic matter content in mineral soils does not exceed 15% (and is usually less), it plays a major role in the soil's chemical and physical behavior [13]. Organic matter is composed of decaying tissues from vegetation and the bodies of micro- and macrofauna. Soil organic matter can be found in various stages of degradation, from coarse dead to complex fine components called humus [14]. Its content is naturally higher in the upper soil horizon, making consideration of organic matter essential for remote sensing applications, where only the upper thin layer is detected.

22.2.2.2 Liquid and Gas Phases

These phases in soils are complementary to the solid phase and occupy about 50% of the soil's total volume. The liquid consists of components of water and dissolved anions in various amounts and positions. The water molecules either fill the entire pore volume in the soil "saturated," occupy a portion of its pore volume "wet," or are adsorbed on the surface areas "air dry." The composition of the soil's gaseous phase is normally very similar to that of the atmosphere, except for the concentrations of oxygen and carbon dioxide, which vary according to the biochemical activity in the root zone.

22.3 SOIL SPECTROSCOPY

22.3.1 DEFINITIONS AND LIMITATIONS

A soil spectrum is a collection of discrete energies, covering a wide spectral range, of photons that travel along the sun (or other equivalent source)-surface-sensor pathways after removing the atmospheric and solar (source) effects. The soil reflectance spectrum is a collection of values obtained from the ratio of radiance (E) and irradiance (L) fluxes across most of the spectral region of the solar emittance function. The values are traditionally described, from a practical standpoint, by a relative ratio against a perfect reflector spectrum measured for the same soil geometry and position [10,15,16]. To illustrate the reflectance product of the calculation discussed earlier, Figure 22.1 provides a spectra of five representative soil types in semiarid environment (taken from RSL-TAU collection). The electromagnetic energy in question covers the VIS (400–700 nm), NIR (700–1000 nm), and SWIR (1000–2500 nm) spectra. Soil reflectance is an inherent soil property that should not be affected by external conditions such as radiation intensity and the instrument used. Observational capabilities can be extended by the use of spectrometers and radiometers that can quantify the characteristics of radiation scattered from the soil. Soil reflectance data have been acquired with such instruments in a substantial number of remote sensing studies performed in both the field and

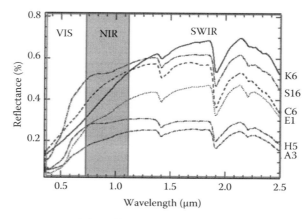

FIGURE 22.1 Reflectance spectra of six different mineral soils from arid and semi-arid area from Israel. The spectral were taken from TAU spectral library (http://www.tau.ac.il/~rslweb/links.html).

the laboratory [10]. Although most of these studies have focused on the spectral distribution of the scattered radiation, some data on the directional distribution and polarization state of radiation scattered by soils are also available in the literature. The studies generally demonstrate relationships between spectral reflectance data and certain soil properties that correspond to the well-known relationships with soil color. Recent studies have also made use of new sensitive sensors from either point or image domains that can provide an immediate spatial view of the spectral information. To provide a better understanding of soil spectra, this section will provide some background information on the electromagnetic spectrum, radiation interactions with soil, and soil attributes that affect the spectral response. This discussion is mainly limited to the 400–2500 nm region of the electromagnetic spectrum.

22.3.2 Spectral Measurements

Soil-reflectance measurements can be acquired in the laboratory or in the field, as well as from both the air and orbit. Whereas in the laboratory soil reflectance is measured under controlled conditions (geometry, illumination, no atmospheric interference, standard protocols), in the field, reflectance measurements are encumbered by a number of interfering factors such as variations in viewing angle, changes in illumination, soil roughness, soil sealing, and differences in measured areas [11,17]. The acquisition of soil reflectance data from the air or orbit involves additional difficulties, such as low signal-to-noise ratio and atmospheric interference. In this regard, laboratory-based measurements enable an understanding of the chemical and physical principles of soil reflectance, but all of the laboratory applications cannot be adopted for the remote sensing of soils. An extensive body of work has been applied to soil-laboratory analyses (see later in this chapter), whereas only a few (but growing) number of studies have been conducted for soil–field (i.e., using a portable spectrometer [18] or soil–air (or space, see later in this chapter) (i.e., using IS technique) applications. In general, field measurements are characterized by better spectral performance than that provided by IS sensors (see later discussion), and, thus, they are preferable for extracting soil properties in the NIR–SWIR region. However, using a field (point) spectrometer rather than an image (point-by-point) spectrometer reduces the accuracy of the first mapping process. This is because interpolation of selected points to a thematic soil map is less accurate than information obtained from a mosaic of hundreds of individual pixels having chemical-physical information. Some common portable point spectrometers are available for remote sensing, such as ASD (http://www.asdi.com/prod/ps2.html), PIMA (http://www.intspec.com/pima/pima.htmt), GER (http://www.ger.com), LICOR (http://licor.alcavia.net/), and PSR-2500 (www.spectraevolution.com/portable_spectrometer.html). A comprehensive technical review of field spectrometers can be found at http://www.themap.com.au/overview_spectrometer.htm. Point spectrometers are characterized by a high signal-to-noise ratio and good stability. Their quick response time can make them operable from aircraft, where they can spectrally track kinetic processes (e.g., soil and vegetation drying processes). Based on this ability, Karnieli et al. [19] mounted an analytical spectral device (ASD) spectrometer on a light aircraft equipped with a video camera and acquired measurements along a climate cross section in Israel (personal communication). They showed for the first time that a laboratory/field point spectrometer can be used as an airborne tool for assessing soil and vegetation status with very high signal-to-noise standards. Recently, Fiemstein et al. [14] observed that even in a single well-calibrated spectrometer, instabilities can occur and consequently they suggested a standard protocol and correction factor to normalize internal and external variations of the spectrometer being used. It should be pointed out that measuring soil reflectance in the field using sun illumination is still problematic since it involves atmospheric attenuation (e.g., at 1400 and 1900 nm, there is strong absorption of water vapor) as well soil sealing. To solve the first problem, an artificial light source is used to directly illuminate the target. For example, in the ASD spectrometer, an external device with a closed chamber (contact probe) is used, whereas in the PIMA spectrometer, a built-in illumination source is in effect. To solve the second problem, a prototype device was recently developed (named 3S-HEAD) to acquire reflectance measurements from soil boreholes, and it provides excellent information on the soil profile without the need to open trenchs [20]. A new version of this device, named SpectralTool©, furnished with a video camera and a unique interface for field work, is now commercially available through NovoSpec Ltd (www.novospec.com). This device can also acquire soil reflectance from a soil surface that has not been offended in the field as well as several soil attributes based on the spectroscopy (see later discussion). This is an important issue that prevents the effects of external parameters such as sun illumination and atmospheric conditions on the final results. Whereas in the laboratory, a standard protocol is used and the samples are properly prepared for measurement (sieved to 2 mm and well mixed), while the illumination and fore optics geometry remain constant, in the field these factors vary and produce uncertain variations in the spectral response. The geometries of both irradiance and radiance play a major role in deriving the soil spectrum. The bidirectional reflectance distribution function (BRDF, see later) assumes that the radiation source, the target, and the sensor are all points in the measurement space and that the ratio calculated between absolute values of radiance and irradiance is strongly dependent upon the geometry of their positions.

22.3.3 Spectral Chromophores

A chromophore is a parameter or substance (chemical or physical) that significantly affects the shape and nature of a soil spectrum via its attenuation of incident radiation. A given soil sample consists of a variety of chromophores, which vary with environmental conditions. In many cases, the spectral signals related to a given chromophore overlap with the signals of other chromophores, hindering the assessment of a given chromophore's effect. Because of the complexity of the chromophores in soil, it is important to understand their physical activity as well as their origin and nature. The spectra of pure minerals are extensively discussed elsewhere and readers are referred to those sources (e.g., [16,21–27]). In the following section, our discussion focuses primarily on factors affecting soil spectra, directly or indirectly, from both chemical and physical chromophores.

22.3.3.1 Chemical Chromophores

Chemical chromophores are those materials that absorb incident radiation at discrete energy levels. The absorption process usually appears on the reflectance spectrum as troughs whose positions are attributed to specific chemical groups in various structural configurations. All features in the VIS–NIR–SWIR spectral regions have a clearly identifiable physical basis. In soils, three major chemical chromophores can be roughly categorized as follows: minerals (mostly clay and iron oxides), organic matter (living and decomposing), and water (solid, liquid, and gas phases).

22.3.3.1.1 Clay Minerals

Clay minerals (also referred to as phyllosilicate minerals) are crystalline aluminosilicates organized in a layered structure. The crystal structure consists of two basic units: the Si tetrahedron, which is formed by a Si^{4+} ion surrounded by four O^{2-} ions in a tetrahedral configuration, and the Al octahedron formed by an Al^{3+} ion surrounded by four O^{2-} and two OH^- ions in an octahedral configuration. These structural units are joined together into tetrahedral and octahedral sheets, respectively, by adjacent Si tetrahedrons sharing all three basal corners and by Al octahedrons sharing edges. These sheets, in turn, form the clay mineral layer by sharing the optical O of the tetrahedral sheet. Layer silicates are classified into eight groups according to layer type, layer charge, and type of interlayer cations. The layer type designated 1:1 is organized with one octahedral and one tetrahedral sheet, whereas the 2:1 layer type is organized with two octahedral and one tetrahedral sheet. A one-layer octahedral sheet [1] is also found in highly leached acid soils. The layer silicate charge is a function of isomorphic substitution that occurs in both the tetrahedral and octahedral positions during the weathering process. Charge density is one of the major factors governing soil behavior and therefore mineral species and composition are considered to be key to understanding soil behavior.

22.3.3.1.1.1 Origin of Layer Minerals in Soils   All clay minerals are derived from the weathering of primary minerals. The occurrence of smectite, vermiculite, illite, or kaolinite is related to the degree of weathering and the chemical nature of the soil environment. Muscovite tends to produce illite, whereas biotite tends to produce vermiculite. Both illite and vermiculite are associated with slightly weathered materials. Vermiculite requires large amounts of magnesium during clay formation, which is most likely to occur in neutral to slightly alkaline soils. Illite occurs to a greater extent than vermiculite in soils of moderate acidity. Illite tends to form smectite as surface potassium ions are removed by the weathering processes and new cation substitution occurs. Smectite minerals (2:1 configuration) are an important component of slightly to moderately weathered soils, which are formed under relatively high pH values and specific Si and Al concentrations in the soil solution. Kaolinite minerals (1:1 configuration) are predominant in highly weathered, leached soils that turn, under stronger weathering and acid conditions, into gibbsite (1 configuration). Whereas gibbsite is quite rare, smectite and kaolinite are more commonly found in soils. Kaolinite may be formed from a 2:1 mineral during the weathering process and requires an environment in which both silica and alumina are accumulated in a ratio that favors its formation. Illite and vermiculite are associated with youthful materials. Smectite is formed in the middle stages of the weathering process and is therefore most likely to be found in many soils as the major or secondary mineral. Similarly, the kaolinite component in soils tends to increase with increasing stages of weathering. Soils of warm temperate regions have a high percentage of kaolinite in the clay fraction, whereas cold areas tend to form more illitic and smectitic type minerals. Of all of the soil minerals discussed earlier, smectite is thought to be the most active, because of its high specific surface area and electrochemical reactivity. These characteristics are known to affect many of the soil's properties, as reported by Banin and Amiel [28] and others.

22.3.3.1.2   Nonclay Minerals

Figure 22.2 provides pure spectra of select nonclay minerals found in soils. The most common of these are divided into five groups: silicates, phosphates, oxides and hydroxides, carbonates, and sulfides and sulfates. The fraction of each mineral in soils depends upon the environmental conditions and the parent materials. Primary minerals will most likely be found in young soils, where the weathering process is weak. Whereas silicates such as feldspars are rarely found in mature soils, quartz may also be found in some developed soils, depending on their environmental conditions and parent material. In general, the quartz mineral is spectrally inactive in the VIS–NIR–SWIR region and therefore diminishes other spectral features in the soil mixture. Other nonclay silicate minerals such as feldspars may have some diagnostic absorption features that make the soil spectrum less monotonous.

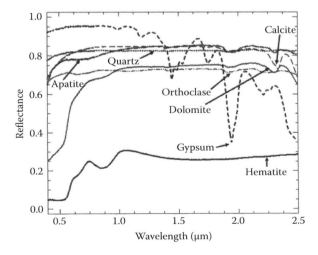

FIGURE 22.2   Reflectance spectra of representative non-clay minerals in soils.

Oxide-group minerals occur in highly weathered areas such as those associated with slopes, highly leached profiles, or "mature" soils. Phosphate and sulfate minerals can be found in soils as apatite and gypsum, respectively. Although both minerals have unique spectral features, their occurrence in soils may be relatively rare and even undetectable. Other oxides, such as iron, are strongly spectrally active, mostly in the VIS region, because of their crystal field and charge-transfer mechanism. The content of free oxides (both iron and aluminum) is low in young soils but increases gradually as the soil ages, similar to organic matter.

22.4 MECHANISMS OF SOIL–RADIATION INTERACTIONS

A comprehensive description of the physical mechanisms governing the electromagnetic radiation of diverse minerals and rocks is provided by Clark [27]. This section focuses on the most common chromophores in the soil environment and their relationship with electromagnetic radiation across the VIS–NIR–SWIR spectral region (taken from [4]).

22.4.1 Chemical Process

22.4.1.1 Clay Minerals

Basically, the spectral features of clay minerals in the NIR–SWIR region are associated with overtone and combination modes of fundamental vibrations of functional groups in the IR region. Of all of the clay mineral elements, only the hydroxide group is spectrally active in the VIS–NIR–SWIR region. The OH group can be found as part of either the mineral structure (mostly in the octahedral position, termed lattice water) or a thin water molecule directly or indirectly attached to the mineral surfaces (termed adsorbed water). Three major spectral regions are active for clay minerals in general and for smectite minerals in particular: around 1300–1400, 1800–1900, and 2200–2500 nm. For Ca-montmorillonite (SCa-2), a common clay mineral in the soil environment, the lattice OH features are found at 1410 nm (assigned $2\upsilon OH$, where υOH symbolizes the stretching vibration at around $3630\,cm^{-1}$) and at 2206 nm (assigned $\upsilon OH + \delta OH$ where δOH symbolizes the bending vibration at around $915\,cm^{-1}$). In comparison, OH features of free water (W) are found at 1456 nm (assigned $\upsilon W + 2\delta W$, where υW symbolizes the stretching vibration at around $3420\,cm^{-1}$, and δW the bending vibration at around $1635\,cm^{-1}$), 1.910 μm (assigned $\upsilon'W + \delta W$ where $\upsilon'W$ symbolizes the high-frequency stretching vibration at around $3630\,cm^{-1}$), and 1978 nm (assigned $\upsilon W + \delta W$). Note that these assigned positions can change slightly from one smectite to the next, depending upon their chemical composition and surface activity. The spectra of three smectite endmembers are given in Figure 22.3 as follows: montmorillonite (dioctahedral, aluminous), nontronite (dioctahedral, ferruginous), and hectroite (trioctahedral, manganese). The OH absorption feature of the $\upsilon OH + \delta OH$ in combination mode at around 2200 nm is slightly, but significantly shifted for each endmember. In highly enriched Al smectite, (montmorillonite) the Al–OH bond is spectrally active at 2160–2170 nm. In highly enriched iron smectite (nontronite), the Fe–OH bond is spectrally active at 2210–2240 nm and in highly enriched magnesium smectite (hectorite), the Mg–OH bond is spectrally active at 2300 nm. Based on these wavelengths, Ben-Dor and Banin [29] found a significant correlation between the absorbance values derived from the reflectance spectra and the total content of Al_2O_3, MgO, and Fe_2O_3. Except for a significant lattice OH absorption feature at around 2200 μm in smectite, invaluable information about OH in free water molecules can be culled at around 1400 μm and 1900 nm. Because smectite minerals contribute relatively high specific surface areas to the soils, and these are covered by free and hydrated water molecules, these absorption features can be significant indicators of soil water content.

Kaolinite and illite minerals are also spectrally active in the SWIR region as they both consist of octahedral OH sheets. From Figure 22.4, which presents pure spectra of nonsmectite layer clay minerals (kaolinite, chlorite, vermiculite and illite), one can see that different positions and spectral shapes of the lattice OH in the layer minerals affect soil spectra across the SWIR region. These changes are a result of the different structures and chemical compositions of the minerals. In the case of kaolinite,

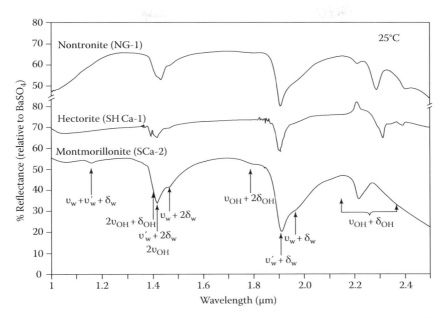

FIGURE 22.3 Reflectance spectra of three pure smectite endmembers across the NIR-SWIR region (nontronite = Fe-smectite; hectorite = Mg-smectite; montmorillonite Al-smectite). Also given are possible combination and overtone modes for explaining each of the spectral feature. (After Ben-Dor, E., et al., Soil spectroscopy. In: *Manual of Remote Sensing*, 3rd Edn., A. Rencz (ed.). John Wiley & Sons Inc., New York, pp. 111–189, 1998.)

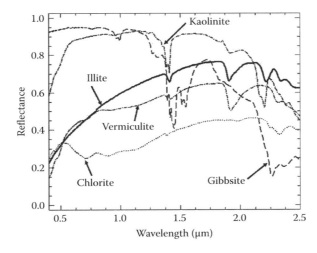

FIGURE 22.4 Reflectance spectra of representative pure non-smectite clay minerals.

a 1:1 mineral (one octahedron and one tetrahedron), the fraction of the OH group is higher than in 1:2 minerals (one octahedron and two tetrahedrons), and, hence, the lattice OH signals at around 1400 and 2200 nm are relatively strong, whereas the signal at 1900 nm is very weak (because of relatively low surface areas and adsorbed water molecules). In the case of gibbsite, an octahedral aluminum structure [1], the 1400 nm signal is even stronger, but the signal at 2200 nm is shifted significantly to the IR region relative to kaolinite. Note that under relatively high signal-to-noise conditions, a second overtone feature of the structural OH (3υOH) can be observed at around 950 nm in layer OH-bearing minerals as well [30]. Based on the aforementioned spectral features, Chabrillat and Goetz [31,32] used Airborne Visible InfraRed Imaging Spectrometer (AVIRIS) sensor data to assess and map expansive clay soils in Colorado for urban planning and environmental applications.

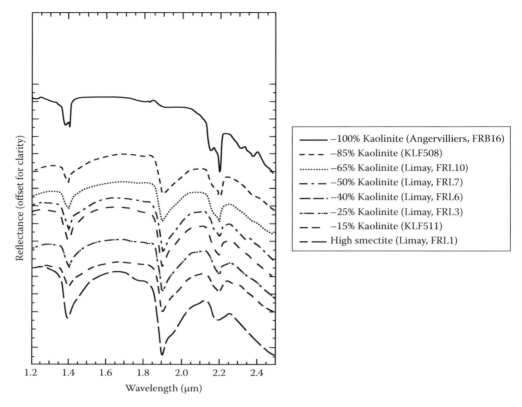

FIGURE 22.5 Reflectance spectra of kaolinite mixed-layer kaolinite/smectite from Paris Basin, halloysite from Tintic Utha and Ca-smectite. (After Kruse, F.A., et al., *Proceedings of the 5th International Colloquium, Physical Measurements and Signatures in Remote Sensing*, Courchevel, France, 181–184, 1991.)

The affinity of water molecules to clay mineral surfaces is correlated to their specific surface area. The specific surface area sequence of the aforementioned minerals is smectite > vermiculite > illite > kaolinite > chlorite > gibbsite, which usually provide a similar spectral sequence at the water absorption feature near 1800 nm (area and intensity). As smectite and kaolinite are often found in soils, they can also appear in a mixed-layer formation that overlaps spectrally. Kruse et al. [33] described a specific case in Paris Basin, France, where interstratifications of smectite/kaolinite (a result of the alkaline weathering process of the flint-bearing chalk) was identified. Figure 22.5 presents the spectra of smectite, kaolinite, and halloysite (hydrated kaolinite) endmembers with the two representative spectra from the basin area soils examined by Kruse et al. [33]. The noticeable asymmetrical OH absorption feature at 2200 nm was further examined by those authors to yield a graph that predicts the relative amount of kaolinite in the mixture (Figure 22.6).

22.4.1.2 Carbonates

Carbonates, and especially calcite and dolomite, are found in soils that are formed from carbonic parent materials, or in a chemical environment that permits calcite and dolomite precipitation. Carbonates, especially those of fine particle size, play a major role in many of the soil chemical processes most likely to occur in the root zone. A relatively high concentration of fine carbonate particles may cause fixation of iron ions in the soil and consequently inhibition of chlorophyll production. On the other hand, the absence of carbonate in soils may affect the soil's buffering capacity and hence negatively affect biochemical and physicochemical processes. The C–O bond, part of the –CO_3 radical in carbonate, is the spectrally active chromophore. Hunt et al. [16,21] indicated the availability of five major overtones and combination modes to describe the C–O bond in the SWIR region. In their table,

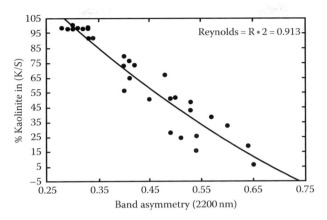

FIGURE 22.6 A correlation between the asymmetry of the 2.2 mm (2200 nm) absorption band and percentage kaolinite form Paris Basin soil samples consisting of interstratification of kaolinite/smectite. (After Kruse, F.A., et al., *Proceedings of the 5th International Colloquium, Physical Measurements and Signatures in Remote Sensing*, Courchevel, France, 181–184, 1991.)

υ1 accounts for the symmetric C–O stretching mode, υ2 for the out-of-plane bending mode, υ3 for the asymmetric stretching mode, and υ4 for the in-plane bending mode in the IR region. Gaffey [34] added two additional significant bands centered at 2230–2270 nm (moderate) and at 1750–1800 nm (very weak), whereas Van-der-Meer [35] summarized the seven possible calcite and dolomite absorption features with their spectral widths. It is evident that significant differences occur between the two minerals. This enabled Kruse et al. [36], Ben-Dor and Kruse [37] and others to differentiate between calcite and dolomite formations using airborne spectrometer data with bandwidths of 10 nm. Aside from the seven major C–O bands, Gaffey and Reed [38] were able to detect copper impurities in the calcite minerals, as indicated by the broad band between 903 and 979 nm. However, such impurities are difficult to detect in soils, because overlap with other strong chromophores may occur in this region. Gaffey [39] showed that Fe impurities in dolomite shift the carbonate's absorption band toward longer wavelengths, whereas Mg in calcite shifts the band toward shorter wavelengths. As carbonates in soils are quite likely to be impure, it is only reasonable to expect that the carbonates' absorption feature positions will differ slightly from one soil to the next.

A correlation between reflectance spectra and soil carbonate concentration was found by Ben-Dor and Banin [40]. Those authors used a calibration set of soil spectra and their chemical data to find three wavelengths that best predict the calcite content in arid soil samples (1800, 2350, and 2360 nm). They concluded that the strong and sharp absorption features of the C–O bands in the examined soils provide an ideal tool for studying soil carbonate content solely from reflectance spectra. The best performance obtained for quantifying soil carbonate content ranged between 10% and 60%.

22.4.1.3 Organic Matter

Organic matter plays a major role in many chemical and physical processes in the soil environment, and has a strong influence on soil reflectance characteristics. Soil organic matter is a mixture of decomposing tissues of plants, animals, and secreted substances. The sequence of organic matter decomposition in soils is strongly determined by the soil microorganism activity. In the initial stages of the decomposition process, only marginal changes occur in the chemistry of the parent organic material. The mature stage refers to the final stage of microorganism activity, in which new, complex compounds, often called humus, are formed. The most important factors affecting the amount of soil organic matter are those involved in soil formation, that is, topography, climate, time, type of vegetation, and oxidation state. Organic matter, particularly humus, plays an important role in many soil properties, such as aggregation, fertility, water retention, ion transformation, and color.

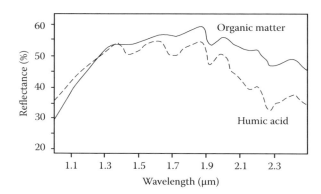

FIGURE 22.7   The spectral reflectance curves of pure organic matter isolated from Alfisol and its extracted humic acid. (After Ben-Dor, E., et al., Soil spectroscopy. In: *Manual of Remote Sensing*, 3rd Edn., A. Rencz (ed.). John Wiley & Sons Inc., New York, pp. 111–189, 1998.)

Because organic matter exhibits spectral activity throughout the entire VIS–NIR–SWIR region, especially in the (VIS-NIR) VIS region, workers have extensively studied organic matter from a remote sensing standpoint (e.g., [41,42]) and have noted that if the organic matter in soils drops below 2%, there is only a minimal effect on the reflectance property. Montgomery [43] indicated that organic matter content as high as 9% does not appear to mask the contribution of other soil parameters to soil reflectance. In another study, Schreier [44] indicated that the relation of organic matter content to soil reflectance follows a curvilinear exponential function. Mathews et al. [45] found that organic matter correlates with reflectance values in the 500–1200 nm range, whereas Beck et al. [46] suggested that the 900–1220 nm region is suited for mapping organic matter in soils. Krishnan et al. [47] used a slope parameter at around 800 nm μm to predict organic matter content and Da-Costa [48] found that simulated Landsat channels (bands 4–6) yield reflectance readings that are significantly correlated with organic carbon content in soils. Downey and Byrne [49] showed that it is possible to predict both moisture and bulk density of milled peat using spectral information.

The wide spectral range found by different workers for assessing organic matter content suggests that organic matter has important chromophores across the entire spectral region. Figure 22.7 shows the reflectance spectra of coarse organic matter (in the NIR–SWIR region), isolated from an Alfisol, and of the humus compounds extracted from this organic matter. Numerous absorption features exist that relate to the high number of functional groups in the organic matter. These can all be spectrally explained by combination and vibration modes of organic functional groups [50]. Vinogradov [51] developed an exponential model to predict the humus content in the upper horizon of plowed forest soils by using reflectance parameters between 600 and 700 nm for two extreme endmembers (humus-free parent material and humus-enriched soil). Schreier [44] found an exponential function that accounts for soil organic matter content in reflectance spectra. Al-Abbas et al. [52] used a multispectral scanner, with 12 spectral bands covering the 400 nm to 2600 μm range, from an altitude of 1200 m and showed that a polynomial equation will predict the organic matter content from only five channels. They implemented the equation on a pixel-by-pixel basis to generate an organic content map of a 25 ha field. Dalal and Henry [53] were able to predict the organic matter and total organic nitrogen content in Australian soils using wavelengths in the SWIR region (1702–2052 nm), combined with chemical parameters derived from the soils. Using a similar methodology, Morra et al. [32] showed that the SWIR region is suitable for identification of organic matter composition between 1726 and 2426 nm. Evidence that organic matter assessment from soil reflectance properties is related to soil texture, and more likely to the soil's clay, was provided by Leger et al. [54] and Al-Abbas et al. [52]. Aber et al. [55] noted that the organic matter, including its decomposition stage, affects the reflectance properties of mineral soil.

Baumgardner et al. [10] showed that three organic soils at different levels of decomposition yield different spectral patterns. Morra et al. [56] showed that using reflectance spectroscopy carbon and nitrogen

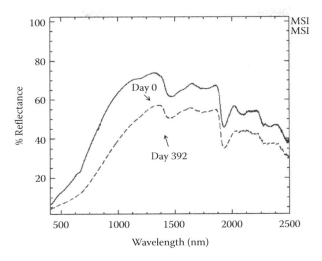

FIGURE 22.8 The reflectance spectra of two endmembers that represent two extreme compositing stage t0 = 0 days and t8 = 378 days for grape MARC material (CGM).

analysis can be done in the laboratory. A study by Ben-Dor et al. [57], using a controlled decomposition process over more than 1 year, revealed significant spectral changes across the entire VIS–NIR–SWIR region as the organic matter aged. Figure 22.8 shows a typical spectrum of grape MARC (CGM) organic matter during 392 days of decomposition. Significant changes can be seen in the slope values across the VIS–NIR region and in the spectral features across the entire spectrum. Ben-Dor et al. [58] postulated that some of the analyses traditionally used to assess organic matter content in soils from reflectance spectra may be biased by the age factor. As many soils consist of dry vegetation in various different stages of degradation, assessment of organic matter using reflectance spectra should consider the vegetation's aging status. Although mineral soil has relatively low organic matter content (around 0%–4%), its accurate assessment requires high spectral resolution data across the entire VIS–NIR–SWIR region. Recently, a comprehensive attention is given to develop spectral assessment tool to map organic matter in soil remotely using IS devices that are based on laboratory and field models [59,60].

22.4.1.4 Water

The various forms of water in soils are all active in the VIS–NIR–SWIR region (based on the vibration activity of the OH group) and can be classified into three major categories: (1) *hydration water*, which is incorporated into the lattice of the mineral, for example, limonite ($Fe_2O_3 \cdot 3H_2O$) and gypsum ($CaSO_4 \cdot 4H_2O$), (2) *hygroscopic water*, which is adsorbed on the soil surface as a thin layer, and (3) *free water*, which occupies the soil pores. Each of these categories influences soil spectra differently, providing the ability to identify the soil's water status (this is discussed separately further on). Three basic fundamentals in the IR regions exist for water molecules, particularly the OH group: υw1—asymmetric stretching, δw—bending, and υw3—symmetric stretching vibrations. Theoretically, in a mixed system of water and minerals, combination modes of these vibrations can yield OH absorption features at around 0.95 μm (very weak), 1200 nm (weak), 1400 nm (strong), and 1900 nm (very strong) related to 2W1 + υW3, υW1 + υW3 + υW, υW3 + 2υW, and υW3 + υW, respectively.

(1) *Hydration water* can be seen in minerals such as gypsum as strong OH absorption features at around 1400 and 1900 nm [21].
(2) *Hygroscopic (adsorbed) water* is adsorbed on the surface areas of clay minerals (especially smectite) and organic matter (especially humus). Early results by Obukhov and Orlov [61] in the VIS region showed that the slope of the spectral curve for soils is not affected by wetting and that the ratio of the reflectance of moist soil to that of dry soil remains practically constant. Sheilds et al. [62] also pointed out that "moisture has no significant effect

on the hue or chroma of several soils." Peterson [63] observed linear relationships between bidirectional reflectance factors at 0.71 μm in oven-dried soil samples that consisted of water tensions between 15 and 0.33 bar. These findings actually suggest that soil albedo is the first factor in the soil spectrum that is altered upon soil wetting [64]. The primary reason for this is the change in the medium surrounding the particles from air to water, which decreases their relative refractive index [65,66]. Based on this idea, Ishida et al. [66] developed a quantitative theoretical model to estimate the effect of soil moisture on soil reflection. The shape of soil reflectance curves is strongly affected by the presence of water absorption bands at around 1400 and 1900 nm, and occasionally weaker absorption bands at around 950 and 1200 nm. Because the amount of hygroscopic water in soil is governed by atmospheric conditions (i.e., relative humidity), the significant spectral changes are related to changes in the adsorbed water molecules on the mineral surfaces. It is interesting to note that a similar observation was already made years ago by Bowers and Hanks [67] with soils that consisted of different moisture values (ranging from 0.8% to 20.2%). This observation demonstrates that the gas phase (water vapor in this case) in the soil environment plays a major role in the quantitative assessment of both structural and free water OH. Further insight into this problem was provided by Montgomery and Baumgardner [68] and Montgomery [43], who indicated that it is not possible to quantitatively assess water content in soils because of the different states of dryness under which the soils were measured. Using reflectance spectra of several treated smectite minerals, Cariati et al. [69] examined shifts in the OH absorption features at 1400, 1900, and 2200 nm. They found that vibration properties of the adsorbed water strongly depend upon the composition of the smectite structure. In another study, Cariati et al. [70] indicated that several kinds of interactions are responsible for the vibration properties of the hygroscopic molecules, and that these may even change with water content. Because smectite is the most effective clay mineral in the soil environment at affecting the reflectance spectrum in the major water absorption features, Cariati et al.'s [70] observations may help us understand the spectral activity of hygroscopic moisture in soils. However, further work is still required to implement the results obtained for pure smectite in the complex soil system.

(3) *Free pore water* (*wet condition*) is water that is not in either the hygroscopic phase or filling the entire pore volume size (saturated condition). The rate of movement of this water into the plant is governed by water tension or water potential gradients in the plant–soil system. Water potential is a measure of the water's ability to do work compared to pure free water, which has zero energy. In soils, water potential is less than that of pure free water due in part to the presence of dissolved salts and the attraction between soil particles and water. Water will flow from areas of high potential to lower potential and hence flow from the soil to the root and up the plant occurs along potential gradients. In agricultural systems, plant growth occurs at soil water potentials between 15 and 0.3 bar (note these are actually negative water potentials); however, water tensions in desert environments are far greater. Baumgardner et al. [10] studied the reflectance spectra of a representative soil (typic Hapludalf by the USDA) with various water tensions (Figure 22.9). As expected, when water tension decreased (and, hence, water content increased), the general albedo decreased and the area under the strong 1.4 and 1.9 μm water absorption peaks also decreased. Clark [71] examined the reflectance of montmorillonite at room temperature for two different water conditions (Figure 22.10) and showed a dramatic decrease in albedo from dry to wet material. Other changes related to water and lattice OH can be observed across the entire spectrum as well. Some of these changes are directly related to the total amount of free and adsorbed water and some to the increase in the spectral reflectance fraction of the soil (wet) surface. In kaolinite minerals, a similar trend was observed under two moisture conditions; however, the changes around the water OH absorption features were less pronounced than in montmorillonite. In the latter, adding water to the

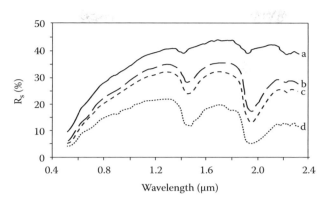

FIGURE 22.9 Spectra curve of Typic Hapludalf soil at four different moisture tensions: oven dry (a), 15 bar (b), 0.3 bar (c), 0.1 bar (d). (After Baumgardner, M.F., et al., *Advances in Agronomy*, 38, 1, 1985.)

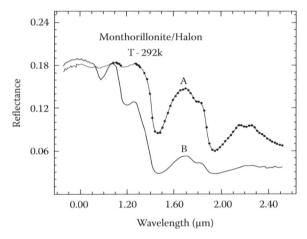

FIGURE 22.10 Reflectance spectra of montmorillonite clay mineral with 50% (A) and 90% (B) water mixed in the sample (by weigh) at room temperature. (After Clark, R.N., *Journal of Geophysical Research*, 86, 3074, 1981.)

sample enhanced the water OH features at 940, 1200, 1400, and 1900 nm, because of the relatively high surface area and the correspondingly high content of adsorbed water. In kaolinite, the relatively low specific surface area obscured a similar response and hence only small changes were noticeable. In the montmorillonite, the lattice-OH features at 2200 nm decreased, suggesting that hygroscopic moisture is a major factor affecting the clay minerals' (and soil's) spectra. In soils in which the entire pore volume size (or more) is filled with water (under saturated [or flooded] conditions, respectively), the soil reflectance is more likely to consist of more secular than Lambertian components. It should be noted that under remote sensing conditions, the water vapor absorptions overlap with the soil water signals, putting use of the aforementioned relationship into question.

22.4.1.5 Iron

Iron is the most abundant element on the Earth's surface and the fourth most-abundant element in the Earth's crust. The average Fe concentration in the Earth's crust is 5.09 mass %, and the average Fe^{3+}/Fe^{2+} ratio is 530 nm [72]. The geochemical behavior of iron in the weathering environment is largely determined by its significantly higher mobility in the divalent vs. trivalent state. Changes in its oxidation state, and consequently in its mobility, tend to occur under different soil conditions. The major Fe-bearing minerals in the Earth's crust are the mafic silicates, Fe-sulfides, carbonates, oxides, and smectite clay minerals. All Fe^{3+} oxides have striking colors, ranging from red and

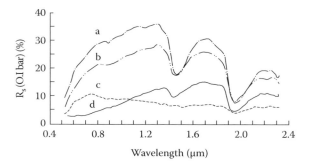

FIGURE 22.11 Reflectance spectra of soils consisting of different textures but exhibiting iron absorption bands: Fine sand, 0.20% Fe_2O_3; sandy loam, 0.64% Fe_2O_3; silty loam, 0.76% Fe_2O_3; clay, 25.6% Fe_2O_3. (After Baumgardner, M.F., et al., *Advances in Agronomy*, 38, 1, 1985.)

yellow to brown, due to selective light absorption in the VIS range caused by transitions in the electron shell. It is well known that even a small amount of iron oxides can change the soil's color significantly. The red, brown, and yellow "hue" values, all caused by iron, are widely used in soil classification systems in almost all countries and languages.

Representative soil spectra with various amounts of total Fe_2O_3 are presented in Figure 22.11. The iron's feature assignments in the VIS–NIR region result from the electronic transition of iron cations (3+, 2+), either as the main constituent (as in iron oxides) or as impurities (as in iron smectite). Hunt et al. [21] summarized the physical mechanisms responsible for Fe^{2+} (ferrous) and Fe^{3+} (ferric) spectral activity in the VIS–NIR region as follows: the ferrous ion typically produces a common band at around 1 μm due to the spin allowed during transition between the E_g and T_{2g} quintet levels into which the D ground state splits into an octahedral crystal field. Other ferrous bands are produced by transitions from the $5T_{2g}$ to $3T_{1g}$ states at 0.55 μm, to $1A_{1g}$ at around 0.51 μm, to $3T_{2g}$ at 0.45 μm and to $3T_{1g}$ at 430 nm. For the ferric ion, the major bands produced in the spectrum are the result of the transition from the $6A_{1g}$ ground state to $4T_{1g}$ at 870 nm, to $4T_{2g}$ at 700 nm and to either $4A_{1g}$ or $4E_g$ at 400 nm.

Just as organic matter is an important indicator for soils, iron oxides provide significant evidence that soil is being formed in a given area of the Earth's crust [73]. Iron oxide content and species are strongly correlated with short- and long-term soil-weathering processes. Iron oxide transformation in the soil often occurs under natural soil conditions. Hematite and goethite are common iron oxides in soils and their relative content is strongly controlled by soil temperature, water, organic matter, and annual precipitation. Hematitic soils are reddish and goethitic soils are yellowish brown. Their reflectance spectra also differ, as can be seen in Figure 22.12. Hematite (α-Fe_2O_3) has Fe^{3+} ions in

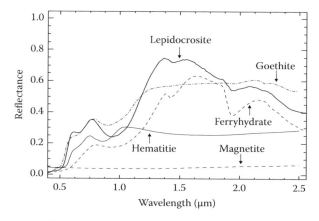

FIGURE 22.12 Reflectance of iron oxide in the soil that can be found under natural soil conditions.

octahedral coordination with oxygen. Goethite (α-FeOOH) also has Fe^{3+} in octahedral coordination, but different site distortions along with the oxygen ligand (OH) provide the main absorption features that appear near 900 nm. Leidocrocite (α-FeOOH), which is associated with goethite but rarely with hematite, is another common unstable iron oxide found in soils.

It appears mostly in subtropical regions and is often found in the upper subsoil position [73]. Maghemite (α-Fe_2O_3) is also found in soils, mostly in subtropical and tropical regions, and it has occasionally been identified in soils in humid temperate areas. Ferrihydrite is a highly disordered Fe^{3+} oxide mineral found in soils in cool or temperate, moist climates, characterized by young iron oxide formations and soil environments that are relatively rich in other compounds (e.g., organic matter, silica, etc.). Iron associated with clay mineral structures is also an active chromophore in both the VIS–NIR and SWIR spectral regions. This can be seen in the nontronite-type mineral presented in Figure 22.3. Based on the structural OH-Fe features of smectite in the SWIR region, Ben-Dor and Banin [29] generated a prediction equation to account for the total iron content in a series of smectite minerals. The wavelengths selected automatically by their method were 2294, 2259, 2291, and 1266 nm. Stoner [74] also observed a higher correlation between reflectance in the 1550–2320 nm region and iron content in soils, whereas Coyne et al. [75] found a linear relationship between total iron content in montmorillonite and the absorbance measured in the 600–1100 nm spectral region. Ben-Dor and Banin (used spectra of 91 arid soils to show that their total iron content (both free and structural iron) can be predicted by multiple linear regression analysis and wavelengths 1075, 1025, and 425 nm. Obukhov and Orlov [61] generated a linear relationship between reflectance values at 640 nm and the total percentage of Fe_2O_3 in other soils. Taranik and Kruse [76] showed that a binary encoding technique for the spectral slope values across the VIS–NIR spectral region is capable of differentiating a hematite mineral from a mixture of hematite-goethite-jarosite.

It is important to mention that iron can often have an indirect influence on the overall spectral characteristics of soils. In the case of free iron oxides, it is well known that soil particle size is strongly related to absolute iron oxide content [77–79]. As iron oxide content increases, the size fraction of the soil particles increases as well, because of the cementing effect of the free iron oxides. As a result, problems resulting from different scattering effects are introduced into the soil analysis. Moreover, free iron oxides, mostly in their amorphous state, may coat the soil particles with a film that prevents natural interaction between the soil particle (clay or nonclay minerals) and the sun's photons. Karamanova [80] found that well-crystallized iron compounds have the strongest effect on the spectral reflectance of soil and that removal of nonsilicate iron (mostly iron oxides) helps enhance other chromophores in the soil. In this respect, Kosmas et al. [81] demonstrated a second-derivative technique in the VIS region as a feasible approach for differentiating even small features of synthetic goethite from clays, and they suggested that such a method may be adopted to assess quantities of iron oxide in mixtures. Based on these spectral characteristics, Dematte et al. [82] showed the possibility of spectrally assessing the alteration of soil properties and Gerbermann and Neher [83] showed that soil mixtures of clay and sand can be predicted from reflectance spectra. Ben-Dor et al. [84] modeled iron oxides' absorption features in a sand dune and were able to account for the rubification of the soil formed over the dune. Recent studies by Lugassi et al. [85] have shown that iron oxide spectra of soils that have burned can be used as a quantitative indicator to assess the temperature of the fire. Iron oxide alteration during the fire events played a major role in the spectral domains that were used by the authors to study the fire days after it had occurred. Other reflectance changes within the soil undergo fire have described by Kokaly et al. [86].

It can thus be concluded that iron is a very strong chromophore in soil, and that a determination of its content in clay and soil from reflectance spectra in the entire VIS–NIR–SWIR region is feasible. Based on the complexity of the iron component in the soil environment, as well as on the intercorrelation between iron and other soil components, sophisticated methods and relatively high spectral resolution data are absolutely required to determine iron content from reflectance spectra.

22.4.1.6 Soil Salinity

Soil salinity is one of the major factors affecting biomass production and is the principal cause of soil degradation [87]. Salt-affected areas cover about 7% of Earth's land surface [88] and are located mostly in arid and semiarid regions [89]. However, salt-affected soils can also be found in subhumid and coastal areas associated with hydrogeological structures. Soil salts have been reported to be in the form of Na_2CO_3, $NaHCO_3$, and $NaCl$, which are very soluble and mobile components of the soil environment. Typically saline soils have a poor structure, are highly erosive, have low fertility, low microbial activity, and other attributes that are not conducive to plant growth.

The spectral signature of saline soils can be a result of the salt itself, or indirectly, of other chromophores related to the presence of the salt (e.g., organic matter, particle-size distribution). Hunt et al. [22] reported an almost featureless spectrum for halite (NaCl 433B from Kansas). Although salt is spectrally featureless, Hick and Russell [90] raised the hypothesis that there are certain wavelengths in the VIS–NIR–SWIR region that can provide more accurate information about saline-affected areas. Dwivedi and Sreenivas [91] applied an image manipulation tool to the study of soil salinity by remote sensing means while Rao et al. [92] investigated the spectral reflectance of salt-affected soils and found some spectral variations.

Vegetation is an indirect factor that can facilitate the detection of saline soils from reflectance measurements [93,94]. Gausman et al. [95], for example, showed that cotton leaves grown in saline soils have a higher chlorophyll content than those grown in low-salt soil. Hardisky et al. [93], using the spectral reflectance of a *Spartina alterniflora* canopy, showed a negative correlation between soil salinity and spectral vegetation indices. In the absence of vegetation, the major influence of salt is on the structure of the upper soil surface.

Because no direct significant spectral features are found in the VIS–NIR–SWIR region for sodic soil, indirect techniques are thought to be more suitable for classifying salt-affected areas [89,96]. Salt in water is most likely to affect the hydrogen bond in water molecules, causing subtle spectral changes; based on this, Hirschfeld [97] suggested that high spectral resolution data are required. Support for this idea was provided by Szilagyi and Baumgardner [98] who reported on the feasibility of characterizing soil salinity status with high-resolution laboratory spectra. A relatively high number of spectral channels are also important in indentifying indirect relationships between salinity and other soil properties that appear to involve chromophores in the VIS–NIR–SWIR regions. Csillag et al. [87] analyzed high-resolution spectra taken from about 90 soils in the United States and Hungary for their chemical parameters, including clay and organic matter content, pH, and salt. They claimed that because salinity is such a complex phenomenon, it cannot be attributed to a single soil property. While studying the capability of commercially available Earth-observing optical sensors, they were able to show that six broad bands in the VIS–NIR–SWIR region best discriminate soil salinity. These six channels were selected solely on the basis of their overall spectral distribution, which provided complete information about salinity status. In another study, Metternicht and Zinck [209] showed that by using six combined Landsat bands [1,2,4–7], it is possible to discriminate salt and sodium-affected soil with varying confident limits. They discussed the nondirect salt effect on soil spectral responses and suggested the addition of more electromagnetic radiation to shed more light on this problem. Thus, it can be concluded that the entire spectral region needs to be considered in evaluating salinity levels in different environments and unknown soil systems. Mougenot et al. [100] noted that in addition to an increase in reflectance with salt content, high salt content may mask ferric ion absorption in the VIS region. Those authors concluded that salts are not easily identified in proportions below 10% or 15%.

Another important factor in saline soils is the fact that in modern agriculture, farmers add gypsum to sodic soils for soil reclamation [101]. The artificial increase in gypsum content in such soils may alter the soil reflectance spectra significantly and this therefore requires attention. It hence should be remembered that although salt is not a strong or direct chromophore, its interaction with other soil components (water, structure, iron, and organic matter) makes its assessment possible but complicated. Farifteh et al. [102] studied the reflectance spectra of soils affected by salt, both

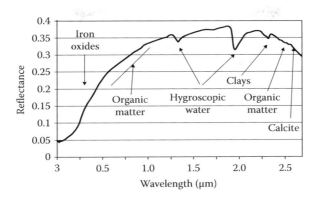

FIGURE 22.13 A soil spectrum of Haploxeralf that represents the major chromophors in soils (see text for more details.)

artificially and naturally, and established some interesting findings toward understanding the spectral features of salt-affected soils. Ben-Dor et al. [103] and Metternich et al. [104] reviewed the spectral-based studies of remote sensing of soil salinity using data combined with other electromagnetic means. Recent studies by Livne et al. [105] describe a robust spectral model to account for soil salinity that has been used with both Israeli and Uzbekistani soils. It is anticipated that more development in soil spectroscopy and salinity interactions will be done in the near future. This is mainly based on the potential of soil reflectance to spot shed on soil salinity and the increasing need to assess soil salinity in the field rapidly and frequently.

22.4.1.7 Chemical Chromophores: Summary

To provide an overview of chemical chromophore activity in soils, Figure 22.13 gives spectrum illustration of a selected soil from Israel (Haploxeralf) with the positions of all possible chromophores.

Also given is Figure 22.14, which summarizes the chromophores associated with soil and geological matter as collected from the literature and summarized by Ben-Dor et al. [4]. It also lists the intensities of each chromophore in the VIS–NIR–SWIR spectral regions as they appear in those studies. The current review demonstrates that high-resolution spectral data can provide additional, sometimes quantitative information about soil properties that are strongly correlated to those chromphores, that is, primary and secondary minerals, organic matter, iron oxides, water, and salt.

22.5 PHYSICAL PROCESSES

22.5.1 GENERAL

In addition to chemical processes, the reflectance of light from the soil surface is dependent upon numerous physical processes. Reflection, or scattering, is clearly described by Fresnel's equation and depends upon the angle of the incident radiation and the refraction index of the materials in question. Generally, physical factors are those parameters that affect soil spectra with regard to Fresnel's equation but do not cause changes in the position of the specific chemical absorption. These parameters include particle size, sample geometry, viewing angle, radiation intensity, incident angle, and azimuth angle of the source. Changes in these parameters are most likely to affect the shape of the spectral curve through changes in baseline height and absorption feature intensities. In the laboratory, measurement conditions can be maintained constant. In the field, several of these parameters are unknown and may hinder an accurate assessment of soil spectra.

Many studies covering a wide range of materials have shown that differences in particle size alter the shape of soil spectra [10,16,106]. Specifically, Hunt and Salisbury [16] quantified effects

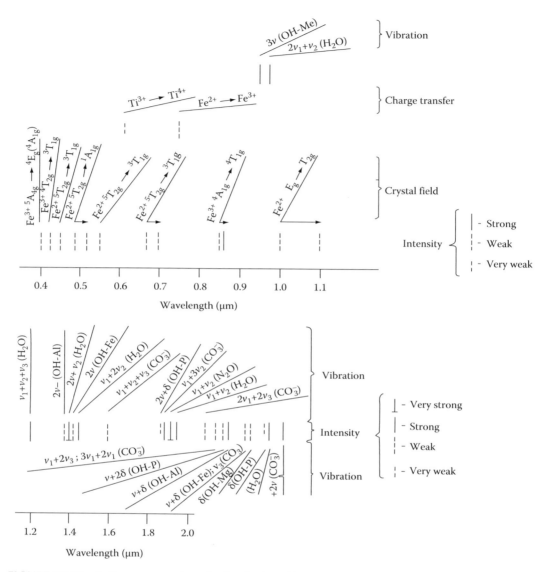

FIGURE 22.14 Active groups and mechanism in the soil chromophors. For each possible group the wavelength range and absorption feature intensity is given. (After Ben-Dor, E., et al., Soil spectroscopy. In: *Manual of Remote Sensing*, 3rd Edn., A. Rencz (ed.). John Wiley & Sons Inc., New York, pp. 111–189, 1998.)

of about 5% in absolute reflectance due to particle size differences, and these changes occurred without altering the position of the diagnostic spectral features. Under field conditions, aggregate size rather than particle size distributions may be more important in altering soil spectra [10,107]. In the field, aggregate size may change over a short time due to tillage, soil erosion, aeolian accumulation, or physical crust formation (e.g., [108,109]). Basically, the aggregate size, or more accurately roughness, plays a major role in the shape of field and airborne soil spectra (e.g., [110,111]). Escadafal and Hute [112] showed strong anisotropy reflectance properties in five soils with a rough surface.

A practical solution for evaluating the effects of physical parameters is to evaluate the reflectance of a given target relative to a perfect reflector measured at the same geometry and viewing angle of the target in question. In reality, such conditions are impossible to achieve in the field, and complex effects, such as those of particle size, cannot be completely eliminated by this method. It is postulated that more effort should be expended in the more precise accounting of physical effects under field conditions (both from a spectroscopy and IS point of view), such as Pinty et al. [113] tried to do by simulating the bidirectional effect over bare soils (see the next section).

22.5.2 MODELS OF RADIATION SCATTERING BY SOILS

The geometry of both irradiance and radiance plays a major role in deriving the soil spectrum. The BRDF assumes that the radiation source, the target, and the sensor are all points in the measurement space and that the ratio calculated between absolute values of radiance and irradiance is strongly dependent upon the geometry of their positions. Theories and models explaining the BRDF phenomenon in relation to soil components are widely discussed and covered in the literature [114–121]. The following equation describes the basic BRDF value:

$$fr(q, q', f, f') = \frac{dL(q, q', f, f', E)}{dE(q, f)}$$

where
E is the radiance
L is the irradiance
q, q', f, f' are source and sensor zenith angles, and source and target azimuth angles, respectively

Whereas the BRDF is better suited to remote sensing applications, hemispheric and bihemispheric reflectance factors are also used in the laboratory [10]. To reduce the effects of geometry and to eliminate systematic and nonsystematic measurement interference, reflectance standards such as MgO, $BaSO_4$, and Halon are often used to correct the relative reflectance spectrum [62,122,123]. Another factor that affects soil spectra is the sensor's field of view (FOV) and sun target geometry. If the soil is homogeneous, a small FOV may be sufficient. However, where some variation occurs in the soil, the FOV should be adjusted to cover a representative portion of that soil. Recent studies by Feingersh et al. [124] have shown that BRDF measurements of selected soils in the laboratory using a controlled geometry setup. Using such a device enable Fiengresh et al. [124] to develop a correction scheme for correcting soil urban material and vegetation BRDF effect that appear on IS images.

22.6 RELATIONSHIP BETWEEN SOIL CHROMOPHORE AND PROPERTIES

22.6.1 QUALITATIVE ASPECTS

Color as a human's eyes and brain property serve as a day-by-day spectrometer and analyzer. Pedologists have long used soil color to describe soils, help classify them, and infer their characteristics [9,62,125]. As Baumgardner stated, "ever since soil science evolved into an important discipline for study and research, color has been one of the most useful soil variables in characterizing and describing a particular soil." Certain qualitative relationships between color and soil properties are well recognized by pedologists on the basis of their collective observations and of a conceptual understanding of the interaction of visible light with soil material. Today's instruments can convert soil reflectance curves into color parameters [125,126], soil color continues to play a major role in modern soil classification. Models that formulate color as mathematical functions of soil properties, however, are not well established in the pedological community. As soil color is a spectral phenomenon and today, spectrometers are replacing the eye and computers, the brain, the following section discusses various models describing soil's reflectance properties in the VIS through SWIR regions of the electromagnetic spectrum.

22.6.2　Quantitative Aspects of Proximal Soil Spectroscopy

22.6.2.1　Historical Notes

Today, quantitative soil spectroscopy is a mature discipline that has come quite a long way since the mid-1960s, when Bowers and Hanks [67] published their paper on the correlation between soil reflectance and soil moisture content. That pioneering study, followed by a series of papers by Hunt et al. [16,21–25] and Hunt [26], proved that water and minerals in the soil environment have unique spectral fingerprints that can be further used for specific recognition. In parallel to this development, Ben-Gera and Norris [127] published a paper showing a correlation between the reflectance reading and moisture content in soybeans. Based on this finding, the new discipline of near-infrared analysis (NIRS) emerged, which focused on extracting quantitative information from reflectance data, first in the food sciences and then in other disciplines. The first conference to gather together scientists in this field, the first international NIRS conference, was held in 1987 in Norwich, U.K. Today, many conferences, workshops, and scientific meetings are dedicated to the NIRS concept, along with a specialized scientific journal dedicated to the NIRS method which was established in 1993 (*The Journal of Near Infrared Spectroscopy*). In 1987, Davies published an article in *European Spectroscopy News* entitled "near infrared spectroscopy analysis: Time for the giant to wake up" [128]. He appealed to potential users at the time to use reflectance spectroscopy across the NIR wavelength region for chemical analysis of powders. A decade later, Davies [129] published another article entitled "the history of near infrared spectroscopic analysis: Past, present and future 'from sleeping technique to the morning star of spectroscopy,'" which showed that the NIRS technique had come a long way and that the quantitative optical approach (mostly for food products) was mature, successful, and applicable. The disciplines that made use of NIRS were the food sciences, pharmacology, the textile, tobacco and oil industries, agriculture, art, the paper industry, and more. Learning from these sectors' successes, Dalal and Henry [32] applied the NIRS approach to soils in 1986. This pioneering study captured the attention of many researchers who realized the potential of soil reflectance spectroscopy. The first scientists to systematically gather soil spectral information and publish it in the form of a soil spectral atlas were Stoner et al. [130]. Their soil spectral library very soon became a classic tool that soil scientists came to rely on. Later, when laboratory and portable field spectrometers were introduced into the market (around 1993), more scientists realized the potential of soil spectroscopy, and consequently more spectral libraries were assembled (e.g., [131]). A comprehensive summary of the quantitative applications of the soil reflectance spectroscopy was provided by Ben-Dor [57] whereas in April 2008, a world soil spectroscopy group was established by Viscarra Rossel (http://groups.google.com/group/soil-spectroscopy), who gathered soil spectra and corresponding attributes from more than 80 countries worldwide in order to generate a global soil spectral/attribute database providing soil-NIRS capability to all [132]. This initiative was based on the idea that the NIRS approach in soil sciences had become well-established and applicable and should be more collaborative. It was only an obvious step after understanding that only sharing information together will help to forward soil quantitative spectroscopy (e.g., [3,133,134]). Comprehensive reviews on NIRS applications for soils can be found in Malley et al. [135] and Viscarra Rossel et al. [136], and other important reviews focusing on soil reflectance theory and applications can be found in Clark and Roush [27], Irons et al. [137], Ben-Dor et al. [4], and Ben-Dor [57].

22.6.2.2　Quantitative Applications

In general, soil reflectance spectra are directly affected by chemical and physical chromomophores, as already discussed. The spectral response is also a product of the interaction between these parameters, calling for a precise understanding of all chemical and physical reactions in soils. For example, even in a simple mixture of iron oxide, clay and organic matter, the spectral response cannot be judged simply by linear mixing models of the three endmembers. Strong chemical interactions between these components are, in most cases, nonlinear and rather complex. For instance, organic components, mostly humus, affect soil clay minerals in chemical and physical ways. Likewise, free iron oxides may coat soil particles and mask photons that interact with the real mineral components or the iron oxides themselves (and organic matter as well). In addition, the coating material may collate fine particles into coarse aggregates that may physically change the soil's spectral behavior from a physical standpoint. Karmanova [81] selectively removed the iron oxides from soil samples and concluded that the effects of various iron compounds on the spectral reflectance and color of soils were not proportional to their relative contents. Another example of the strong relationship between chromophores is given by Bedidi et al. [138,139], who showed that the normally accepted view of decreasing soil baseline height with increasing moisture content (VIS region) does not hold for lateritic (highly leached low-pH) soils. They concluded that the spectral behavior of such soils under various moisture conditions is more complex than originally thought. In this context, Galvao et al. [140] showed spectra from laterite soils (VIS–NIR region) consisting of complex spectral features that appeared to deviate from other soils. Al-Abbas et al. [52] found a correlation between clay content and reflectance data in the VIS–NIR–SWIR region and suggested that this was not a direct but an indirect relationship, strongly controlled by the organic matter chromophore. Another anomaly that relates to the interactions between soil chromophores was identified by Gerbermann and Neherc [83]. They carefully measured the reflectance properties in the VIS region of a clay–sand mixture extracted from the upper horizon of a montmorillonic soil and found that "adding of sand to a clay soil decreases the percent of soil reflectance." This observation stands in contrast to what is traditionally expected from adding coarse (sand) to fine (clay) particles in a mixture (soil), that would have to increase soil reflectance. Likewise, Ben-Dor and Banin [141–144] concluded that intercorrelations between feature and featureless properties play a major role in assessing unexpected information about soil solely from their reflectance spectra in either the VIS–NIR or SWIR regions. Ben-Dor and Banin [143] examined arid and semiarid soils from Israel and showed that "featureless" soil properties (i.e., properties without direct chromophores such as K_2O, total SiO_2, and Al_2O_3) can be predicted from the reflectance curves due to their strong correlation with "feature" soil properties (i.e., properties with direct chromophores). Csilage et al. [87] best described the effect of multiple factors indirectly affecting soil spectra in their discussion on soil salinity, which can be considered a featureless property. They stated that "salinity is a complex phenomenon and therefore variation in the (soil) reflectance spectra cannot be attributed to a single (chromophoric) soil property." To get the most out of soil spectra, they examined chromophoric properties of organic matter and clay content, among others, and ran a principal component analysis to fully account for the salinity status culled from the soil reflectance spectra. Recently Whiting et al. [145] has demonstrated that water content in soil can be quantitative assessed by using spectral features away from the central water peak in the TIR region. Another recent innovative work has been done by Schwartz et al. [146] who demonstrated a capability to detect carbohydrate contamination status (content and species) in soil using neural network on both soil spectroscopy and chemical analysis spectroscopy. These examples show that soil chromophores do not stand alone in the soil matrix and that spectral anomalies are often found in the soil environment. Examining all available information on a soil's population (spectral and chemical) is key to understanding soil reflectance spectra and their relationship to soil properties. To some extent, this suggests that soil spectra should be judged and examined with caution in order to obtain quantitative information about the soil, despite the fact that the chemical and physical mechanisms in the VIS–NIR–SWIR region, taken separately, are well understood.

It is interesting to note that the soil spectra community is growing rapidly on a global scale, with concomitant development and use of commercial applications and methods to assess soil attributes from reflectance spectroscopy. Recently, NIRS has also been incorporated in IS data and used to map soil surface properties (see later discussion in this chapter). Over the past decade, many users have discovered the potential of soil spectroscopy and much work has therefore been published. Mulley [135] summarized all of the quantitative applications of soil spectroscopy prior to this surge in activity, and later, Viscarra Rossel [136] published a review that covered the applications that had been added by the growing community of soil spectral users. MacBarthy et al. [147,148] has also shed light on this technology through his pioneering work over the years. Brown et al. [3] had concluded that NIRS technique is soil has the potential to replace or augment standard soil characterization techniques, and based their conclusion in 3768 soil samples from the United States. Recently, in view of the growing soil spectral community, Viscarra Rossel [132] generated an initiative (Soil World Spectral Group, http://groups.google.com/group/soil-spectroscopy) in which all members of the soil spectral community were asked to join together and contribute their local spectral library in order to generate a worldwide spectral library that would be accessible to all. The world spectral library is composed (at the time of this writing, 2010) of about 10,000 soil spectra with their chemical attributes. This initiative, beside being the first attempt to gather spectral information on the world's soils, is an important step toward establishing a standard protocol and quality indicators that will be accepted by all members of the growing soil spectral community. To that end, it is important to mention that special sessions dealing with soil spectra have been organized in several leading conferences for both earth material and soil sciences (eg., EGU 2007, 2008, WSC 2010), along with specific workshops (e.g., EUFAR 2009, 2010), and many soil scientists who were unaware of this technology are now being exposed to it.

22.7   FACTORS AFFECTING SOIL REFLECTANCE

22.7.1   GENERAL

In the laboratory, where soil spectra are recorded under controlled conditions, it was thought that the data could be simply analyzed. This assumption was recently reexamined by Fimestein et al. [149] who realized that even under controlled conditions, it is important to maintain a strict protocol and a standard internal procedure to enable comparisons between users. In the laboratory, samples are mixed and homogenized prior to measurement. In the field, natural soil is affected by different factors, such as dust accumulation and soil crust (both biogenic and physical) that prevent sensing the "real" soil surface that is measured in the laboratory. In the field, the soil might also be only air dried, which can significantly affect the soil spectral signatures. Vegetation also plays a major role in masking the "real" soil signals and can be classified as biospheric interference. *The FAO Production Yearbook* [150] states that about 56% of the land area is covered by green vegetation, such as forest, pasture, and crops, whereas the rest is bare or covered by dry vegetation, snow, or urban development. Within the nonvegetated areas, only a portion of the soils are characterized by an unaltered surface layer (e.g., not tilled or not having undergone natural soil sealing) and hence even partial sensing of the natural soil surface is difficult. This effect can be termed surface-coverage interference. Another important problem in acquiring accurate soil reflectance spectra from air and space is atmospheric interference. Electromagnetic energy interacts with atmospheric gas molecules and aerosol particles that may cause misinterpretation of the "soil spectrum" derived from airborne sensors. The measurement geometry (referred to earlier as BRDF) also plays a major role in affecting soil reflectance. It is also important to mention other factors that can change the soil spectrum by natural incidence, such as fire. This latter occurrence can significantly change the mineralogy of the upper surface of the soil and bias exact identification of the soil entity. Here, we provide a brief overview of these interferences.

22.7.2   BIOSPHERE

22.7.2.1   Higher Vegetation

Soil is a growth environment for green plants (natural and agricultural) and a sink for decomposing tissues of vegetation and fauna. Because large parts of the world's soils are vegetated (green or dry), the problem of deriving soil spectra from the mixture of soil and vegetation signals is complex. Siegal and Goetz [151] postulated that "the effect of naturally occurring vegetation on spectral reflectance of earth materials is a subject that deserves attention." At one extreme are situations in which the canopy cover is so dense that reflectance from soils is too difficult to interpret. Where the vegetation cover is only partial, a mixed signal from soil and vegetation is received and to some extent, the chemical and physical components can be resolved [152]. In a soil–vegetation mixture, nonlinear models are typically used to resolve the soil spectral components [153,154]. Otterman et al. [155] noted that the relationship between the amount, type, and architecture of a vegetation cover and the reflectance properties of the underlying soil is an important issue (e.g., low-albedo soils are those most significantly affected by vegetation). The 680–1300 nm spectral region of soils is the region most strongly affected by green vegetation as a result of the steep rise in reflectance that it causes (e.g., [156]). Dry vegetation does not alter the spectrum in the VIS–NIR region, aside from changing the albedo, whereas in the SWIR region, significant vegetation effects are related to cellulose, lignin, and water. The low reflectance of green vegetation beyond 1.4 μm indicates that if a soil–vegetation mixture exists, most of the spectral information will relate to rock and soil types [150]. Two chromophores that exist in both plant and soil material—water and organic matter—can complicate interpretation of the spectra, particularly in the SWIR region. In the green vegetation–soil mixture, liquid water of green and dry vegetation may overlap with the soil water forms. Signatures of lignin, cellulose, and protein can also significantly affect the soil components in the soil–vegetation mixture. Murphy and Wadge [152] showed that in one case, although live vegetation had a greater impact on the SWIR region of the soil spectra, dead vegetation had a greater impact on the 2200 nm absorption features (see, e.g., the reflectance spectra of pure organic matter in Figures 22.7 and 22.8). Murphy and Wadge [152] concluded that dead vegetative tissue has a greater impact on soil spectra than live vegetation, and they suggested that workers consider this effect more seriously. Pinty et al. [157] studied the effect of soil anisotropy as affected from vegetation canopies.

From a vegetation point of view, Tucker and Miller [158] postulated that "remotely sensed data of vegetated surfaces could be analyzed more accurately if the contribution of the underlying soils spectra are known." Tuller [159] and Smith et al. [160] noted that it is difficult to extract vegetation information when its coverage consists of less than 30%–40%. The normalized differential vegetation index (NDVI) is a parameter that is commonly used to estimate the green vegetation cover in satellite and airborne data. The index, which is based on the normalized difference between the NIR and VIS reflectance values, is very sensitive to soil background, atmosphere, and sunangle conditions. Based on that background, Huete [161] developed a new index called soil-adjusted vegetation index (SAVI), which accounts for soil brightness and shadows, and more recently, Liu and Huete [162] presented another index, the modified NDVI (MNDVI), which accounts for atmospheric attenuation as well. The SAVI has been shown to significantly minimize soil-related problems in nadir measurements over a variety of plant canopies and densities and in data derived from canopy radiant transfer models [163] noted that the optimal correction factor is achieved at the point at which dark and light SAVI values are the same. More precise models take into account the vegetation architecture [155] or contain additional correction factors [164]. Richardson et al. [165] developed three plant canopy models for extracting plant, soil, and shadow reflectance components of a cropped field. Using such models, Murphy and Wadge [152] were able to separate soil and vegetation spectra by using GER 63-channel IS data [37]. Roberts et al. [166] also incorporated an unmixing procedure to discriminate vegetation, litter, and soils using AVIRIS 224-channel IS data [163] and were able to account for different soil types using a residual spectrum technique. It can be concluded that soil spectral signatures can be extracted from areas that are partially covered by decaying or live vegetation; however, caution should be exercised when assessing the "true" soil reflectance spectra in a vegetation–soil mixture.

22.7.2.2 Lower Vegetation

A major vegetation component in arid soil areas that is usually ignored by workers is the biogenic crust. Recently, however, this issue has been receiving more attention, and its importance to explaining anomalies in field soil spectra and satellite data has been demonstrated [167]. The biogenic crust consists mainly of lower, nonvascular (microphytic) plants covering the upper soil surface in a thin

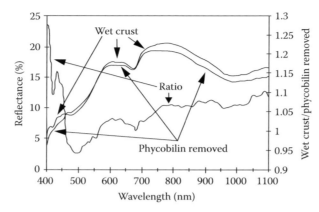

FIGURE 22.15 Spectra of wetted cyanobacteria crust, phycobilin extracted crust, and the ratio between them. It is shown that relative high reflectivity of the crust in the blue region is due to the spectral characteristics of the phycobilin pigments. (After Karnieli, A., et al., *Remote Sensing of Environment*, 69, 67, 1999.)

layer [168,169]. The microphytic community consists of mosses, lichens, algae, fungi, cyanobacteria, and bacteria. Each of these groups have pigments that are spectrally active in the VIS region under certain environmental conditions (Figure 22.15) and thus may mask soil features, or more seriously, may be interpreted as part of the soil signature [170]. O'Neill [171] showed that some soil spectral features (between 2080 and 2100 µm) could be attributed to the microphytic crust and speculated that this was due to cellulose. Karnieli and Tsoar [172] showed that the microphytic crust causes a decrease in the soil's overall albedo, leading to the false identification of anomalies in arid soils. The spectral response related to the biogenic crust permits linear mixing models, unlike the complex architecture of higher vegetation, which requires nonlinear models to analyze mixed signals. During a long-term study, Zaady et al. [173] have demonstrated that several indices are valid for assessing the crust based on its recovery condition. It is however important to note that in addition to more basic research and consideration of the biogenic crust issues, more quantitative studies are still needed to fully account for its effects on soil spectra.

22.7.3 LITHOSPHERE

22.7.3.1 Soil Cover and Crust

Soil crust and cover can be formed by different processes. The biogenic crust, as outlined earlier, is one example of such interference. Aeolian material and desert varnish are others. A lithosphere crust that is often found in soil is the "rain crust." This crust is formed by raindrops [174] that cause segregation of fine particle sizes at the surface of the soil. This can increase runoff and lead to soil erosion. The crusting effect is more pronounced in saline soils and well studied with relation to the mineralogical and chemical changes in the soil surface [175]. The immediate observation after a rainstorm is an enhancement of "hue" and "value" of the soil color because of an increase in the fine fraction on the surface. One can assume that the reflectance spectrum of the "rain crust" will be totally different from that of the original soil, because it contains a greater clay fraction with a different textural component. In the literature, the issue of "rain crust" as it affects the spectral signature of soils has not received considerable attention; we therefore encourage workers to consider this problem in their studies. The spectral changes observed at the soil surface are caused by changes in the soil's texture (clay-fraction enrichment), structure (from loose to compact), and roughness. Several innovative studies have shown a significant relationship between spectral information and the infiltration rate of water into the soil profile as measured in the laboratory (e.g., [176,177]). The next requisite step to the rain-simulation studies was to test the use of an AIS to characterize a structural crust in the field. Ben-Dor et al. [178], used an IS sensor that covered the

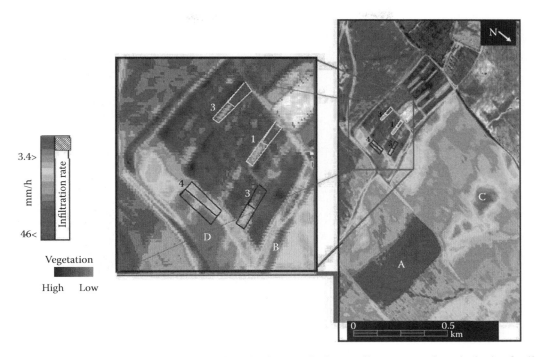

FIGURE 22.16 **(See color insert.)** The infiltration image of a loess soil as generated on the basis of soil reflectance information and rain simulator measurements. (After Ben-Dor, E. et al., *Soil Science Society of American Journal*, 72, 1, 2008.)

VIS–NIR region (AISA Eagle hyper spectral IS) over highly crusted areas in southern Israel. Using a spectral-based index (the normalized spectral area [NSA]), which is the area under the ratio curve generated by using a tested spectrum against a standard reference spectrum, they were able to generate a possible erosion hazard map of the soil area (Figure 22.16). An important question based on that finding is whether a generic spectral model can describe the crust status rather than the kinetics of the formation process. A study by Goldshleger et al. [176] showed that the spectral model used to predict crust status might be more robust than originally thought. By using four soils from Israel and three soils from the United States subjected to rain events in a rain simulator, promising results were obtained using a combined prediction equation for infiltration rate with a root mean square error (RMSE) of cross validation of 15.2% and a ratio of prediction to deviation (RPD) of 1.98. In another study, Chappell et al. [109] also investigated the effect of soil-structure changes due to rain and wind-tunnel events. Their results showed that the spectral information can shed more light on the soil composition and structure generated by these two factors (rain crust and Aeolian abrasion).

22.7.3.2 Surface Affected by Fire

Fire affects a variety of physical and chemical soil properties, including loss of structure and soil organic matter, reduced soil porosity, increased pH, and alteration of the soil minerals. Soil degradation by fire plays a major role in the indirect cost of wildfires worldwide. Therefore, a large effort is being invested in the development of monitoring tools for fires. The interaction between heat and soil minerals has been extensively studied in the last decades by thermal-analysis methods such as differential thermal analysis, differential scanning calorimetric, and thermal gravimetric analysis during a controlled heating process. The integration of the knowledge gained from soil spectroscopy (both imaging and point) with that on changes in soil properties that might be induced by the heating process could serve as a powerful tool for studying postfire consequences for the environment. In a recent study by Lugassi et al. [85], a heterogeneous natural

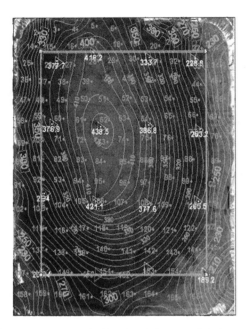

FIGURE 22.17 **(See color insert.)** The extended isothermal maps of the burned soil based on spectral measurements and prediction models to reconstruct the surface temperature. (After Lugassi, R., *Remote Sensing of Environment*, 114, 322, 2010.)

fire simulation that lasted 10 min at 190°C–450°C was applied to a uniform area of loess soil (Xeric Torriorthent). The burned soil (after cooling) exhibited a higher albedo along the entire solar illumination region relative to the original soil. Common spectral analysis, using continuum removal, showed that heating the soil caused a shift in the absorption bands of iron oxides and a gradual disappearance of illite absorption. The significant spectral and mineralogical changes with heat enabled these authors to relate the maximum soil surface temperatures measured with a net of thermocouples to the associated soil spectral reflectance (350–2500 nm) measured after the soil had cooled down, and thus to understand the new mineralogy and degradation potential of the burned soil. Figure 22.17 shows the reconstructed soil temperature map as derived from the spectral information. An important finding was that the estimation of the maximum surface temperature using soil reflectance information was not affected by the thin layer of scorched organic matter present on the burned surface; however, the interference of larger amounts of ash on the soil surface in this analysis has not yet been tested and further investigations are therefore very important. We hypothesize that we can further use IS to assess and map fire areas and to generate a hazard map indicating the potential dangers and ways of preventing consequent soil degradation. Further study to this end is highly requested.

22.7.3.3 Soil Moisture

Water is considered to be one of the most significant chromophores in the soil system [10,64,78,179]. Muller and Décamps [180] determined that the impact of soil moisture on reflectance could be greater than the differences in reflectance between soil categories; hence, they stressed its importance in the previously discussed applications. Soil moisture affects the baseline height (albedo) as well as several spectral features across the entire spectral range (see, e.g., [181]). Figure 22.18 [182] shows the features directly associated with the OH group in the water molecule (at 1400 and 1900 nm), and some that are indirectly associated with the strong OH group in the TIR region (around 2750–3000 nm), which affect the lattice OH in clay (at 2200 nm) and CO_3 in carbonates (at 2330 nm).

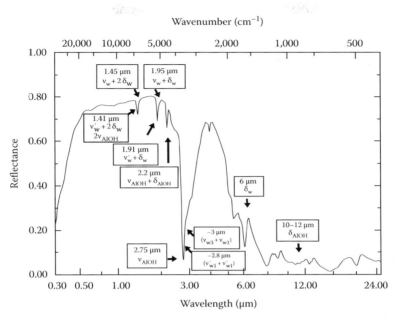

FIGURE 22.18 Bond stretching and bending vibrations in montmorillonite clay mineral due water and aluminium hydroxyl. (After Bishop, J.L., et al., *Clays and Clay Minerals*, 42, 702, 1994.)

Bowers and Hanks's [67] frequently quoted work defined a loss of albedo and spread of absorptions in the 1440 and 1900 μm regions, whereas a clay–OH band at 2200 nm diminished with increasing water content. Figure 22.19 shows the loss in albedo from the VIS through SWIR regions (400–2500 nm) with increasing moisture, increasing water band depths, and decreasing band depth of the 2200 nm region (the OH–lattice band), described by Bowers and Hanks [67] in a sample with high clay content [183].

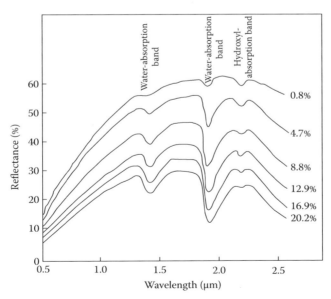

FIGURE 22.19 Spectral reflectance curves for Newton silt loam at various moisture contents. (After Bowers, S., and R.J. Hanks, *Soil Science*, 100, 130, 1965.)

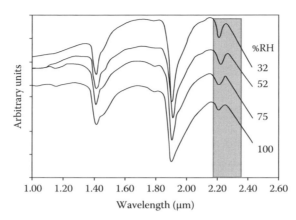

FIGURE 22.20 Reflectance spectra (in the NIR-SWIR region) of Ca-montmorillonite clay mineral at various relative humidity (RH) conditions. (After Ben-Dor, E., et al., Soil spectroscopy. In: *Manual of Remote Sensing*, 3rd Edn., A. Rencz (ed.). John Wiley & Sons Inc., New York, pp. 111–189, 1998.)

Ben-Dor et al. [4] noted a decrease in the 2200 nm absorption feature in Ca-montmorillonite at various relative humidities (Figure 22.20). In the highly sensitive 1900 nm region, a water OH combination band showed excellent nonlinear fit to the increase in water content.

Recent work by Demattê et al. [2] applied this feature and others to practical field use. They found that the best interpretation of water content emerges when both dry and wet soil samples are spectrally measured. Dalal and Henry [32] isolated the main differences in absorbance (log 1/reflectance), and found them to be related to the variation in moisture contents across the 1100–2500 nm SWIR spectral region. In this region, they determined that the correlation coefficient was greater than 0.92, when the 1926, 1954, and 2150 nm wavelengths were used, in a near-infrared analysis (NIRS) approach using gravimetric moisture ranging from air-dry (ca. 4%) to intermediate moisture (ca. 13%), with a standard error of prediction of 0.58% water content with finely ground samples (<0.25 mm), and greater error for coarse ground samples (<2 mm). Lobell and Asner [181] also showed that the SWIR region is much more sensitive than the VIS region for assessing soil moisture, and suggested that this region be used for practical purposes in the field. However, for the four soils they examined, different exponential decay rates between the volumetric content and spectral parameters were noted. This suggests that their method cannot be considered generic, and that special attention must be given to every soil group examined. Recently, a robust spectral technique to estimate soil moisture content was developed by Whiting et al. [185] using a broad range of soils. In their approach, they isolated the influence of the fundamental water band from a sequence of gravimetric moisture contents in two distinctly different soils in the California Central Valley (high clay content, low carbonate) and La Mancha, Spain (low clay content, high carbonate). They fitted an inverted Gaussian function centered on the assigned fundamental water absorption region at 2800 nm, beyond the limit of commonly used instruments, over the logarithmic soil spectrum continuum found with convex hull boundary points (Figure 22.21). The area of the inverted function, the soil moisture Gaussian model (SMGM), accurately estimated the water content within an RMSE of 2.7% and coefficient of determination (R^2) of 0.94 among both soil regions, and an RMSE of 1.7%–2.5% with R^2 ranging from 0.94 to 0.98 when samples were separated according to landform position (Spain) and salinity (USA). Using AVIRIS hyperspectral images of these soil regions in an air-dried state, they improved the abundance estimates by 10% of the regression mean by including the SMGM area as a parameter in the empirical determination of clay–OH and carbonate abundance based on the continuum-removed mineral-band depth (Whiting et al. [145]).

This method is novel since it uses the entire SWIR region, it is not directly affected by the atmospheric water vapor, and it works in the real IS domain. Based on the aforementioned method, the authors were also able to present a processed AVIRIS image that provides soil moisture content (Figure 22.22) [185].

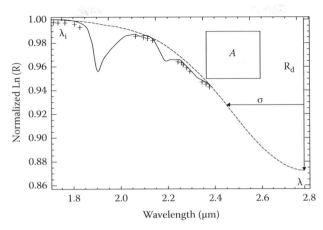

FIGURE 22.21 Inverted Gaussian function is fitted to the fundamental water absorption center at 2800 nm to the convex hull boundary points of a lograthmic transformed SWIR region, where λ_i, wavelength value at maximum reflectance; λ, wavelength value of 2800; σ, distance at inflexion; R_d, depth of Gaussian, and A, area above the continuum. (After Whiting, M.L. et al., *Remote Sensing of Environment*, 89, 535, 2004.)

FIGURE 22.22 **(See color insert.)** Surface water content (gravimetric) from AVIRIS data (May 3, 2003, near Lemoore, California) as estimated with the SMGM. (After Whiting, M.L. et al., *Remote Sensing of Environment*, 89, 535, 2004.)

Working on the similar goal of developing a novel approach to estimating soil moisture content solely from spectral readings not affected by atmospheric attenuations, Haubrock et al. [186,187] developed and successfully tested a new model for determining soil moisture by means of remote sensing techniques. This model was based on combining multitemporal high-spatial-resolution IS observations with field and laboratory spectral studies, along with hydrological measurements. This method, termed normalized soil moisture index (NSMI), was tested for the best spectral prediction of soil moisture content in the field using the 400–2450 nm spectral region. R^2 was 0.61 (up to 0.71) for natural field samples, taking into account the influence of different environmental factors: heterogeneous soil types and related field moisture content, variable soil water profiles, and the presence of soil crust and vegetation cover. Moisture is an integral part of a soil's reflectance, and future modeling attempts may support the contribution of soil background to vegetation radiative transfer models. Jacquemoud et al. [119] using a modification of Hapke's single-scattering albedo model [114], separated the surface geometry component in a radiative transfer model for soil reflectance, SOILSPECT. They also noticed "quasihomothetic variations" in the VIS–NIR and SWIR regions with moisture content, though the moisture dataset was limited [119]. Future investigations to account for this decline in albedo may help resolve this modeling problem [113]. In summary, it can be concluded that soil surface water content should be estimated with caution, and due to its effect on other soil components, its spectral absorptions require proper attention. Soil moisture is an important property, not only for assessing the water content available for plant utilization, but also for assessing the direct exchange of soil water with the atmosphere (i.e., evaporation). This innovative direction has not yet been fully studied and developed for use in IS, though it appears very promising and highly necessary.

22.7.4 ATMOSPHERE

22.7.4.1 Gases and Aerosols

When reflectance is measured from air- or space-borne sensors, the atmosphere's gases and aerosols play a major role in the VIS–NIR–SWIR spectral regions and thus may attenuate soil reflectance. Absorption and scattering of electromagnetic radiation takes place across these regions. Water vapor, oxygen, carbon dioxide, methane, ozone, nitrous oxides, and carbon monoxide are the spectrally active components across approximately half of the VIS–NIR–SWIR regions. Some good models for retrieving gas and aerosol interference exist and are widely used by many workers (e.g., LOWTRAN-7 [188]), 5S and 6S codes [189], and recently ATOCR [190]. Although a discussion of these models is beyond the scope of this chapter (see, e.g., [178]) one should be aware that in many cases, the models do not perfectly remove all atmospheric attenuation and may alter the soil spectrum. This problem is most likely to appear in hyperchannel data, where discrete absorption features are more pronounced relative to the multichannel data, which practically averages small features into one wide value.

To illustrate the spectral regions under which atmospheric attenuation can affect the soil spectrum we provide Figure 22.22. This figure shows the reflectance spectrum of a E-7 soil from Isreal (Haploxeralf, taken from TAU spectral library) ovelean to its simulated (soil) radiance as calculated by MODTRAN. The last is normalized to the Plank sun function on top at the atmosphere in order to illustrate only the atmosphere transmittance. As the atmospheric attenuations remain it can be clearly seen the most affected spectral regions: The VIS region is affected by aerosol scattering (monotonous decay from 400 to 800 nm) and absorption of ozone (around 600 nm), water vapor (730, 820 nm) and oxygen (760 nm). The NIR–SWIR regions are affected by absorption of water vapor (940, 1140, 1380, and 1880 nm), oxygen (around 1300 nm) carbon dioxides (around 1560, 2010, and 2080 nm), and methane (2350 nm).

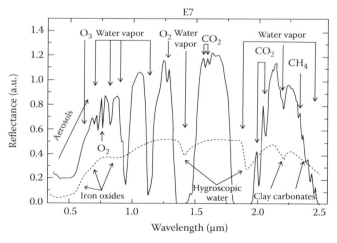

FIGURE 22.23 An simulated AVIRIS spectrum of a Haploxeralf soil from semi arid environment from Israel (E7 from TAU spectral library), after removing the solar effect. Across the spectrum the major gas absorption's absorption features are annotated to show area where atmospheric attenuation's might overlap with soil features. The reflectance spectrum of this soil is also given in dots line.

Also seen are the absorption peaks of the soil chromophres at 2330 nm (carbonates, 2200 nm (clay), 1900 and 1400 nm (hygroscopic water), and 500, 600, and 900 nm (iron oxides) that are overlapped with the aforementioned atmospheric chromophores. As already mentioned, even weak spectral features in the soil spectrum can contain very useful information. Therefore, great caution must be taken before applying any quantitative models to soil reflectance spectra derived from air- or space-borne hyperchannel sensors. Validation of the (atmospherically) corrected data is an essential step in ensuring that the reflectance spectrum consists of reliable soil information (Figure 22.23).

22.8 SOIL REFLECTANCE AND REMOTE SENSING

22.8.1 GENERAL

A comprehensive description of soil spectral remote sensing can be found in Ben-Dor et al. [8]. In general, many studies have been conducted with the intention of classifying soils and their properties using optical sensors on board orbital satellites, such as Landsat MSS and TM, SPOT and NOAA-AVHRR (e.g., [191–194]). Qualitative classification approaches have traditionally been used to analyze multichannel data in cases where the spectral information was relatively scarce. Nevertheless, it has also been possible to obtain useable sets of information on soil type, soil degradation, and soil conditions from "broad" channel sensors by applying sophisticated classification approaches [144,195]. Over the years, soil spectra have been collected and analyzed in the laboratory both quantitatively and qualitatively, by many workers (e.g., [196,197]). Although the use of these libraries involves many limitations, it is understood that the spectral domain is very important for soil mapping. The last 25 years have seen the development of a new remote sensing technique, termed hyperspectral remote sensing (HSR) (also termed IS in this chapter). This is an advanced tool that provides high-spectral-resolution data in an image, with the aim of providing near-laboratory quality reflectance or emittance for each individual picture element (pixel) from far or near distances [198]. This information enables the identification of objects based on the spectral absorption features of their chromophores and has found many uses in terrestrial and marine applications [198–200]. Figure 22.24 illustrates the concept, in which the spectral information of a given pixel shows a new dimension that cannot be obtained by traditional point

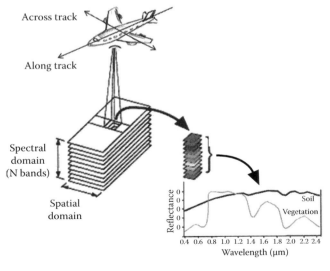

FIGURE 22.24 The IS concept: An IS image is composed of N spectral bands that generates a spectral cube. For each pixel a spectrum can be extracted representing a spectral foot print of the object.

spectroscopy, air photography, or other multiband images. IS can thus be described as an "expert" geographical information system (GIS) in which layers are built on a pixel-by-pixel basis, rather physical-chemical based than on a selected group of points [147]. This enables spatial recognition of the phenomenon in question with a precise spatial view and use of the traditional GIS-interpolation technique in precise thematic images. Since the spatial-spectral-based view may provide better information than viewing either the spatial or spectral views separately, IS serves as a powerful and promising tool in the modern remote sensing arena. Since 1983, when the first AIS, [148] ushered in the IS era [148], this technique has been used mostly for geology, water, and vegetation applications. It appears that its application does not yet extend to soils because these present a complex matrix: it is only recently, with the advent of better signal-to-noise sensors, the manufacture of less expensive IS sensors and the development many soil (point) spectroscopy applications, that soil-IS activity has progressed somewhat.

22.8.2 IMAGING SPECTROSCOPY APPLICATIONS IN SOILS: HISTORICAL NOTES

Whereas in 2000 only a limited number of sensors were available worldwide and the applications varied from vegetation to water, geology, and atmosphere, in 2010, many sensors have entered the field of remote sensing, some even designed to work underground rather than in the aerial domain. As a consequence, many new users have discovered the technology and the number of papers on the subject, along with internet searches, has grown exponentially [201]. It seemed obvious that the proven spectral information from soils and the gathered soil-NIRS knowledge would direct potential users toward employing IS for soil applications in a quantitative rather than qualitative manner. However, this step was not being taken, as the scientific community continued to treat IS and NIRS as separate entities. This was the situation until 1998, when Palacios-Orueta and Ustin [202] showed that a multivariate analysis specifically a principal component analysis and canonical discriminate analysis, as well as a band-depth analysis, could capture quantitative information on organic matter, iron content, and texture in an arid soil environment. One of the first papers that applied the NIRS concept to the IS domain for soil applications was published by Ben-Dor et al. [57]. They used the airborne DAIS 7915 IS sensor to map soil salinity, organic matter, hygroscopic moisture, and soil pH from a distance of 10 km. At around the same time, Kemper and Sommer [203] also applied a complete NIRS chain to airborne IS data (HyMap) and were able to show that heavy metals (such as As, Cd, Cu, Fe, Hg, Pb, S, Sb, and Zn) can be quantitatively mapped in soil using an AIS sensor. Finally, as more and more studies adopted the NIRS approach in the IS domain, it was implemented in soil-surface mapping with an effort to use more quantitative spectral approaches. The soil properties that have been quantitatively mapped using IS data combined with spectral analytical tools are iron oxides, organic matter, salinity, texture, soil contamination, crust formation, pH, soil moisture, soil infiltration, soil degradation, and soil classification (see examples in Ben-Dor et al. [15]). Recent efforts have been made by several authors to account Stevens et al. [59], Gomez et al. [204], and Bartholomeus [60]. Nonetheless, all of the soil-based IS studies still suffer from the fact that the soil profile cannot be sensed directly from afar and more work to that end must be applied. Other studies have shown fairly good results with iron oxides [84,205]. A new approach to using both IS data and NIRS at the field level was suggested by Ben-Dor et al. [20], who developed a penetrating optical sensor (POS) that fits into a small drilled hole and uses spectral models to describe the soil profile in situ and further map soil according to USDA (or other) classification systems.

22.8.3 LIMITATIONS OF IMAGING SPECTROSCOPY FOR SOIL MAPPING

It appears that the limited number of studies in soil-IS applications (relative to other disciplines such as geology, vegetation and water) is due to the difficulties encountered on the journey from point spectroscopy to a cognitive (imaging) spectral view of soils. The IS approach is still a costly method, it is difficult to process, it is operated by only a few sensors worldwide, and it has not yet been recognized by many end users. The relatively low signal-to-noise ratio, atmospheric attenuation, varying FOV for every pixel, spectral instability, low integration time for a given pixel, spectral mixing problems, optical shifts from one pixel to another, and BRDF effects are only a few of the problems. The relatively low spatial resolution from air and space domains with low geolocation accuracy also hinders accurate mapping with the IS approach. IS data require fine preprocessing of the raw data prior to any advanced correction. Highly skilled personnel are needed to meet all of these requirements, with experience, knowledge, and a well-equipped infrastructure (software and field measurements). In addition, it should be remembered that the soil surface is not always flat, smooth, or homogeneous, and, therefore, sample preparation (as performed in the laboratory) is almost impossible. This leads to problems such as variations in particle size, adjacency, BRDF effects, and the need to develop methodologies that will represent a pixel on the ground and in the IS sensor from both the chemical and spectral perspectives. In the face of all of these obstacles, one should remember that although optical remote sensing does not go beneath the surface, it can eventually produce precise soil (profile) maps, by combining more electromagnetic methods with IS, and by using smart approaches, such as the spectral penetrating probe assembly presented by Ben-Dor et al. [15]. Another important problem is the validation-stage assessment. Since the pixel size cannot really represent point measurements, at least 3×3 pixels have to be averaged for both true spectral ground measurements and chemical analysis. Since a field may generate a nonhomogeneous presentation, this can cause problems that, if not estimated properly, will bias the final "spectral-based" map.

22.9 SPECTRAL PROXYMATION OF SOIL USING POINT AND IMAGE DOMAINS: FUTURE NOTES

For soil applications, air- and space-borne imaging spectrometers should consist of a reasonable number of spectral channels across the entire VIS–NIR–SWIR region that will cover the spectrally active regions of all chromophores with a reasonable bandwidth and sampling interval. Price [195] believes that a relatively low number of spectral channels [15,16,21–24] with bandwidths of 40–100 nm and high signal-to-noise ratio promise better remote sensing capabilities for soils. Goetz and Herring [206] prefer more spectral channels [146] but a narrow bandwidth (about 10 nm) to permit diagnostic evaluation of specific features across the entire VIS–NIR–SWIR region. We believe that for quantitative analysis of soil spectra, the optimal bandwidth, and number of channels may be strongly dependent upon the soil population and the property being examined. There is no doubt, however, that high signal-to-noise ratio is a crucial factor in quantitative analysis of soil spectra derived from both air and space measurements. It is also important that the airborne data be well calibrated and the atmosphere attenuation be accurately removed. Brook and Ben-Dor [207] demonstrated a practical way of retrieving accurate reflectance information from IS data, by minimizing major uncertainty factors via application of a "supervised vicarious calibration" method. This method permits better reflectance retrieval from the soil surface. As the IS technology holds promising capability for soil mapping, especially if accurate reflectance is extracted, the use of all know how in the laboratory can be adopted with small modifications, driving soil spectroscopy forward [207]. The innovative progress in quantitative analyses of soils from a spectral perspective, and the recent advances in IS sensors, will create a valuable environment for innovative studies and practical applications in the soil sciences. The remarkable achievements in IS sensor manufacturing are evidenced by the many relatively low-cost IS sensors which are now commercially available (e.g., SPECIM, a well-known IS producer, has sold more than 80 sensors over the past 5 years (Hyvärinen, 2010 Personal communication). The introduction of unmanned airborne vehicles with simple operational capabilities and ground-sensor availability has become a driving force in IS technology. Separation between preprocessing of IS data (including atmospheric rectification) and quantitative analysis is recommended. This can be achieved by collaboration or by purchasing the former service in advance. Likewise, developing full-chain capability from raw reflectance data is strongly recommended if IS is to become a major tool for the potential user. The current drawback of high operational costs is diminishing as the technology develops. Software to analyze IS data is available today for simple or complex applications and for spectral and spatial analyses, atmospheric correction, BRDF, and geometrical rectification. The problem of sensing the soil profile can be solved by merging IS technology with other remote sensing techniques, as was done by Ben-Dor et al. [15] for soil salinity (using ground-penetrating radar [GPR] and frequent domain electromagnetic [FDEM] antennas, as well as by developing new approaches, such as combining the IS information with point spectroscopy in the field). Another point of note is that the spectral range available today (VIS–NIR–SWIR) can be expanded to both the UV (<300 nm) and thermal (TIR 2500–14000 nm) spectral regions, providing more quantitative information on the soils. Moving from airborne to ground IS sensors may also create new applications for soil mapping. A new initiative to put an IS sensor into orbit (e.g., EnMAP, [208]) is more good news, as it will generate high signal-to-noise spectral information on soils with high temporal resolution and wide spatial coverage. Ground minimization is optics end up with light IS camera that would be carried on UMV are also a (not far) view to the future. In summary, it can be said that although IS for soil applications is still in its infancy, it holds great promise as an innovative vehicle to study soils from afar in a quantitative domain.

22.10   GENERAL SUMMARY AND CONCLUDING REMARKS

Based on the accumulated knowledge presented in this chapter, it is obvious that soil-spectral information is a treasure chest for mapping and recognizing soils remotely. Over the past decade, a strong foundation has been established for the practical use of soil spectroscopy, despite the complexity of the soil matrix. Many researchers have generated intersecting studies that utilized a chromometric analytical approach to retrieve many soil properties solely from the reflectance measurements. Significant success has been achieved under controlled laboratory conditions and the methods have become well accepted by the soil science community. It is very clear that moving from the point to spatial remote sensing domain will advance the impact of spectral capability for practical applications. In the imaging spectroscopy (IS) domain, more cognitive information is being generated, opening new frontiers in the field of soil science. However, this technology, especially when used from the air and space domains, has encountered severe difficulties in providing accurate soil reflectance values. These problems include atmospheric attenuation, mixed pixels, low signal-to-noise ratios, geometric and optical distortions, bidirectional reflectance distribution function (BRDF) effects, and sensing only the first 50,000 nm of the soil body.

Attempts to overcome these problems are being explored scientifically and it is believed that soon, spectral information from the air and space domain will be as accurate as that obtained in the laboratory. This will permit the implementation of spectral models originally developed for point spectrometry in the spatial (IS) domain. Several reports have shown innovative results using both point and image spectrometry to map soils, such as on soil degradation (salinity and erosion), soil genesis and formation, soil water content, soil contamination, soil formation, soil sealing, and soil expansion. Soil mapping and classification, which require soil profile information, have also become possible with the combined use of point and image spectrometry. The new (penetrating spectral) device, which can provide the soil profile composition within a short time (using reflectance spectroscopy and NIRS), may replace the traditional trencher soil survey scheme. Although there is no doubt that soil spectroscopy in general and IS in particular harbor high potential for soil applications, so far, not many users have adopted this technology on a routine basis. This is mainly because of the relatively high price of operating AISs and the need for skilled personnel to process the data; it is also because of failure to disseminate the technology to other end users. The new generation of IS sensors that can work on the ground may also open up a new frontier in soil spectroscopy applications. All of this leads to the promise of a bright future for soil spectroscopy technique. Electro-optic technology is now well developed and can offer infrastructure and never-before-seen capabilities. As progress continues, it is estimated that soil spectral technology will soon be fully commercialized. The success of soil spectroscopy relies on active chromophores in the soil matrix and on the development of sophisticated analytical approaches to extract the highly correlated wavelength of the attribute in question. The quantitative approach using NIRS technology (as developed in other disciplines) enables the extraction of spectral models that can be used simply and at low cost. NIRS is a very-well-established technique in soil science and is a successful means of determining several soil attributes, such as total C, organic C, total N, cation-exchange capacity, and moisture (in "as is" soil), clay content, free iron oxides, carbonates, and more. The field of soil NIRS is dynamic, with on-going developments in field portability, software capability, sample presentation, and interfacing of instruments with GPS/GIS for mapping and monitoring soil. Soil reflectance spectroscopy has commercial applications and has been adopted by several companies. Nonetheless, soil spectroscopy still requires marketing and educating farmers, decision makers, and other end users (e.g., soil scientists). It is believed that as soon as a good IS sensor becomes operational from orbit, a new era in soil mapping will begin. In summary, it can be concluded that reflectance spectroscopy is a powerful tool for quantitative and qualitative applications with soil material in the laboratory, field and far-distance domains and is moving toward practical usage. We strongly feel that if all spectral domains (point and image) are thoroughly researched and other active remote sensing methods will be merged and combined, the applications are nearly unlimited. We therefore believe that this review will shed light on future activities of imaging spectroscopy for soil discipline. We hope that this will spark the imagination of both scientists and end users on how best to utilize this imaging spectroscopy for effective management of one of the World's greatest resource-soils.

REFERENCES

1. Dunn B.W., Batten G.D., Beecher H.G., and S. Ciavarella, 2002. The potential of near-infrared reflectance spectroscopy for soil analysis—A case study from the Riverine Plain of south-eastern Australi. *Australian Journal of Experimental Agriculture* 42(5):607–614.
2. Nanni M.R. and J.A.M. Dematte, 2006. Spectral reflectance methodology in comparison to traditional soil analysis. *Soil Science Society of American Journal* 70:393–407.
3. Brown D.J., Shepherd K.D., Walsh M.G., Dewayne Mays M., and T.G. Reinsch, 2006. Global soil characterization with VNIR diffuse reflectance spectroscopy. *Geoderma* 132:273–290.
4. Ben-Dor E., Irons J.A., and A. Epema, 1998. Soil spectroscopy. In: *Manual of Remote Sensing*, 3rd Edn. A. Rencz (ed.). John Wiley & Sons Inc., New York, pp. 111–189.
5. Demattê J.A.M., Morelli J., Nelly E., and R. Negrão, January 10–12, 2000. Precision agriculture applied to sugar cane cultivation in São Paulo, Brazil. In: *Second International Conference on Geospatial Information in Agriculture and Forestry, Proceedings*, v. II, Lake Buena Vista, FL, pp. 388–394.
6. Colin C., Collings K., Drummond P., and E. Lund, 2004. A mobile sensor platform for measurement of soil pH and buffering. Paper #041042, ASAE annual meeting.
7. Ben-Dor E., Chabrillat S., Demattê J.A.M., Taylor G.R., Hill J., Whiting M.L., and S. Sommer, 2009. Using imaging spectroscopy to study soil properties. *Remote Sensing of Environment* 113:38–55.
8. Ben-Dor E., Taylor R.G., Hill J., Demattê J.A.M., Whiting M.L., Chabrillat S., and S. Sommer, 2007. Imaging spectrometry for soil applications, *Agronomy Journal* 97:323–381.
9. Buol S.W., Hole F.D., and R.J. McCracken, 1973. *Soil Genesis and Classification*. The Lowa State University Press, Ames, IA, 360p.
10. Baumgardner M.F., Silva L.F., Biehl L.L., and E.R. Stoner, 1985. Reflectance properties of soils. *Advances in Agronomy* 38:1–44.
11. Aase J.K. and D.L. Tanaka, 1983. Effect of tillage practices on soil and wheat spectral reflectance. *Journal of Agronomy* 76:814–818.
12. Dixon J.B. and S.B. Weed, 1989. *Minerals in Soil Environments*, Soil Science Society of America Publishing, Madison, WI.

13. Odlare M., Svensson K., and M. Pell, 2005. Near infrared reflectance spectroscopy for assessment of spatial soil variation in an agricultural field. *Geoderma* 126:193–202.
14. Stevenson F.J., 1982. *Humus Chemistry*. John Wiley & Sons Inc., New York.
15. Ben-Dor E., Heller D., and A. Chudnovsky, 2008. A novel method of classifying soil profiles in the field using optical means. *Soil Science Society of American Journal* 72:1–13.
16. Hunt, G.R. and J.W. Salisbury, 1970. Visible and near infrared spectra of minerals and rocks: I: Silicate minerals. *Modern Geology* 1:283–300.
17. Karnieli A. and H. Tsoar, 1994. Spectral reflectance of biogenic crust developed on desert dune sand along the Israel–Egypt border. *International Journal of Remote Sensing* 16:369–374.
18. Milton E.J., 1987. Review article principles of field spectroscopy 1987. *International Journal of Remote Sensing* 8:1807–1827.
19. Karnieli A., Kaufman Y.J., Remer L., and A. Aldc, 2001. AFRI—Aerosol free vegetation index. *Remote Sensing of Environment* 77(2001):10–21.
20. Schnitzer M. and S.U. Khan, 1978. *Soil Organic Matter*. Elsevier Publication, Amsterdam, the Netherlands.
21. Hunt G.R., Salisbury J.W., and A. Lenhoff, 1971. Visible and near-infrared spectra of minerals and rocks: III Oxides and hydroxides. *Modern Geology* 2:195–205.
22. Hunt G.R., Salisbury J.W., and C.J. Lenhoff, 1971. Visible and near-infrared spectra of minerals and rocks: Sulfides and sulfates. *Modern Geology* 3:1–14.
23. Hunt G.R., Salisbury J.W., and C.J. Lenhoff, 1971. Visible and near-infrared spectra of minerals and rocks: Halides, phosphates, arsenates, vandates and borates. *Modern Geology* 3:121–132.
24. Hunt G.R. and J.W. Salisbury, 1971. Visible and near infrared spectra of minerals and rocks: Carbonates. *Modern Geology* 2:23–30.
25. Hunt G.R. and J.W. Salisbury, 1976. Visible and near infrared spectra of minerals and rocks: XI Sedimentary rocks. *Modern Geology* 5:211–217.
26. Hunt G.R., 1980. Spectoscopic properties of rock and minerals. In: *Handbook of Physical Properties Rocks*. Stewart C.R (ed.). CRC Press, Boca Raton, FL, p. 295.
27. Clark R.N., 1998. Spectroscopy of rocks and minerals, and principles of spectrocopy. In: *Remote Sensing for the Earth Sciences: Manual of Remote Sensing*, 3rd Edn., Vol. 3. A.N. Rencz (ed.). John Wiley & Sons Inc., New York.
28. Banin A. and A. Amiel, 1970. A correlation of the chemical physical properties of a group of natural soils of Israel. *Geoderma* 3:185–198.
29. Ben-Dor E. and Banin A. 1990. Diffuse reflectance spectral of smectitte minerals in the near infrared and their relation to chemical composition. *Science Geological Bulletin* 43(2–4):117–128.
30. Goetz F.A.H., Hauff P., Shippert M., and A.G. Maecher, 1991. Rapid detection and identification of OH-bearing minerals in the 0.9–1.0 mm region using new portable field spectrometer. *Proceedings of the 8th Thematic Conference on Geologic Remote Sensing*, Denver, CO I:1–11.
31. Chabrillat S. and A.F.H. Goetz, 1999. The search for sweeling clays along the colorado front range: The role of AVIRIS resolutoin in detection. *Summaries of the 8th JPL Airborne Earth*. JPL Publication, 99-17: 69–78.
32. Morra M.J., Hall M.H., and L.L. Freeborn, 1991. Carbon and nitrogen analysis of soil fractions using near-infrared reflectance spectroscopy. *Soil Science Society of American Journal* 55:288–291.
33. Kruse F.A., Thiry M., and P.L. Hauff, 1991. Spectral identification (1.2–2.5 mm) and characterization of Paris basin kaolinite/smectite clays using a field spectrometer. *Proceedings of the 5th International Colloquium, Physical Measurements and Signatures in Remote Sensing*, Courchevel, France I:181–184.
34. Gaffey S.J., 1986. Spectral reflectance of carbonate minerals in the visible and near infrared (0.35–2.55 μm): Calcite, aragonite and dolomite. *American Mineralogist* 71:151–162.
35. Van-der-Meer F., 1995. Spectral reflectance of carbonate mineral mixture and bidirectional reflectance theory: Quantitative analysis techniques for application in remote sensing. *Remote Sensing Reviews* 13:67–94.
36. Kruse F.A., Kierein-Young., and J.W. Boardman, 1990. Mineral mapping of Cuprite, Nevada with a 63-channel imaging spectrometer. *Photogrammetric Engineering and Remote Sensing* 56:83–92.
37. Ben-Dor E. and F.A. Kruse, 1995. Surface mineral mapping of Makhtesh Ramon Negev, Israel using GER 63 channel scanner data. *International Journal of Remote Sensing* 18:3529–3553.
38. Gaffey S.J. and K.L. Reed, 1987. Copper in calcite: Detection by visible and near infra-red reflectance. *Economic Geology* 82:195–200.
39. Gaffey S.J., 1985. Reflectance spectroscopy in the visible and near infrared (0.35–2.55 μm): Applications in carbonate petrology. *Geology* 13:270–273.
40. Ben-Dor E. and A. Banin, 1990. Near infrared reflectance analysis of carbonate concentration in soils. *Applied Spectroscopy* 44(6):1064–1069.

41. Kristof S.F., Baumgardner M.F., and C.J. Johannsen, 1971. Spectral mapping of soil organic matter. *Journal Paper* No.5390, *Agricultural Experiment Station*, Purdue University, West Lafayette, IN.

42. Baumgardner M.F., Kristof S.J., Johannsen C.J., and A.L. Zachary, 1970. Effects of organic matter on multispectral properties of soils. *Proceedings of the Indian Academy of Science* 79:413–422.

43. Montgomery O.L., 1976. An investigation of the relationship between spectral reflectance and the chemical, physical and genetic characteristics of soils. PhD Thesis, Purdue University, West Lafayette, IN (Libr. Congr. no 79–32236).

44. Schreier H., 1977. *Proceedings of the 4th Canadian Symposium on Remote Sensing*, Quebec, Canada I:106–112.

45. Mathews H.L., Cunningham R.L., and G.W. Peterson, 1973. Spectral reflectance of selected Pennsylvania soils. *Proceedings of the Soil Science Society of American Journal* 37:421–424.

46. Beck R.H., Robinson B.F., McFee W.H., and J.B. Peterson, 1976. Information note 081176. *Laboratory Application of Remote Sensing*, Purdue University, West Lafayette, IN.

47. Krishnan P., Alexander J.D., Bulter B.J., and J.W. Hummel, 1980. Reflectance technique for predicting soil organic matter. *Soil Science Society of American Journal* 44:1282–1285.

48. Da-Costa L.M., 1979. Surface soil color and reflectance as related to physicochemical and mineralogical soil properties. PhD dissertation, University of Missouri, Columbia.

49. Downey G. and P. Byrne, 1986. Prediction of moisture and bulk density in milled peat by near infrared reflectance. *Journal of Food and Agriculture Science* 37:231–238.

50. Chen Y. and Y. Inbar, 1994. Chemical and spectrscopical analysis of organic matter transformation during composting in relation to compost maturity. In: *Science and Engineering of Composting: Design, Environmental, Microbiology and Utilization Aspects*. Hoitink H.A.J. and Keener H.M. (eds.). Renaissance Publications, Worthington, OH, pp. 551–600.

51. Vinogradov B.V., 1981. Remote sensing of the humus content of soils. *Soviet Soil Science* 11:114–122.

52. Al-Abbas H.H., Swain H.H., and M.F. Baumgardner, 1972. Relating organic matter and clay content to multispectral radiance of soils. *Soils Science* 114:477–485.

53. Dalal, R.C. and R.J. Henry, 1986. Simultaneous determination of moisture, organic carbon and total nitrogen by near infrared reflectance spectroscopy. *Soil Science Society of American Journal* 50:120–122.

54. Leger R.G., Millette G.J.F., and S. Chomchan, 1979. The effects of organic matter, iron oxides and moisture on the color of two agricultural soils of Quebec. *Canadian Journal of Soil Science* 59:191–202.

55. Aber J., Wessman C.A., Peterson D.L., Mellilo J.M., and J.H. Fownes, 1990. Remote sensing of litter and soil organic matter decomposition in forest ecosystems. In: *Remote Sensing of Biosphere Functioning*. Hobbs R.J. (ed.). Springer-Verlag, NY, pp. 87–101.

56. Pinty B., Verstraete M.M., and N. Gobron, 1998. The effect of soil anisotropy on the radiance field emerging from vegetation canopies. *Geophysical Research Letters* 25:797–800.

57. Ben-Dor E., 2002. Quantitative remote sensing of soil properties. *Advances in Agronomy* 75:173–243.

58. Bellinaso H., Demattê J.A.M., and S. Araújo, 2010. Soil spectral libraray and its use in soil classification. *Revista Brasileira de Ciência do Solo* 34:861–870.

59. Stevens A., Wesemael B.V., Bartholomeus H., Rosillon D., Tychon B., and E. Ben-Dor, 2008. Laboratory, field and airborne spectroscopy for monitoring organic carbon content in agricultural soils. *Geoderma* 144:395–40.

60. Condit H.R., 1970. The spectral reflectance of American soils. *Photogrammetric Engineering* 36:955–966.

61. Obukhov A.I. and D.S. Orlov, 1964. Spectral reflectance of the major soil groups and the possibility of using diffuse reflection in soil investigations. *Soviet Soil Science* 2:174–184.

62. Shields J.A., Paui E.A., Arnaud R.J., and W.K. Head, 1968. Spectrophtometric measurement of soil color and its relation to moisture and organic matter. *Canadian Journal of Soil Science* 48:271–280.

63. Peterson J.B., 1980. Use of spectral data to estimate the relationship between soil moisture tension and their corresponding reflectance. *Annual Report* OWRT, Purdue University, West Lafayette, IN, pp. 1–18.

64. Idso S.B., Jackson R.D., Reginato R.J., Kimball B.A., and F.S. Nakama, 1975. The dependence of bare soil albedo on soil water content. *Journal of Applied Meteorology* 14:109–113.

65. Twomey S.A., Bohren C.F., and J.L. Mergenthaler, 1986. Reflectance and albedo differences between wet and dry surfaces. *Applied Optics* 25:431–437.

66. Ishida T., Ando H., and M. Fukuhara, 1991. Estimation of complex refractive index of soil particles and its dependence on soil chemical properties. *Remote Sensing of Environment* 38:173–182.

67. Bowers S. and R.J. Hanks, 1965. Reflectance of radiant energy from soils. *Soil Science* 100:130–138.

68. Montgomery O.L. and M.F. Baumgardner, 1974. The effects of the physical and chemical properties of soil and the spectral reflectance of soils. Information note 1125. *Laboratory for Applications of Remote Sensing*, Purdue University, West Lafayette, IN.

69. Cariati F., Erre L., Micera G., Piu P., and C. Gessa, 1983. Polarization of water molecules in phyllosylicates in relation to exchange cations as studied by near infrared spectroscopy. *Clays and Clay Minerals* 31:155–157.

70. Cariati F., Erre L., Micera G., Piu P., and C. Gessa, 1981. Water molecules and hydroxyl groups in montmorillonites as studied by near infrared spectroscopy. *Clays and Clay Minerals* 29:157–159.

71. Clark R.N., 1981. The reflectance of water-mineral mixtures at low temperatures. *Journal of Geophysical Research* 86:3074–3086.

72. Ronov A.A. and A.A. Yaroshevsky, 1971. Chemical composition of the earth's crust. In: *The Earth's Crust and Upper Mantle*. Hart P.J. (ed.). American Geophysical Union, Washington, DC, pp. 37–57.

73. Schwertmann U., 1988. Occurrence and formation of iron oxides in various pedoenvironment. In: *Iron in Soils and Clay Minerals*. Stucki J.W., Goodman B.A., and U. Schwertmann (eds.). NATO ASI Series, Reidel Publishing Company, Dordrecht, the Netherlands, pp. 267–308.

74. Stoner E.R., 1979. Physicochemical, site and bidirectional reflectance factor characteristics of uniformly-moist soils. PhD Thesis, Purdue University, West Lafayette, IN.

75. Coyne, L.M., Bishop J.L., Sacttergood T., Banin A., Carle G., and J. Orenberg, 1989. Near-infrared correlation spectroscopy. Quantifying iron and surface water in series of variably cation-exchanged montmorillonite clays. In: *Spectroscopic Characterization of Minerals and Their Surfaces*. ACS Symposium Series No. 415:407–429, ISSN 0097-6156.

76. Taranik D.L. and F.A. Kruse, 1989. Iron minerals reflectance in geophysical and environmental research imaging spectrometer (GERIS) data. *Proceedings of the 7th Thematic Conference on Remote Sensing for Exploration Geology*, Calgary, Alberta I:445–458.

77. Soileau J.M. and R.J. McCraken, 1967. Free iron and coloration in certain well-drained costal plain soils in relation to their other properties and classification. *Soil Science Society of American Proceedings* 31:248–255.

78. Stoner E.R. and M.F. Baumgardner, 1981. Characteristic variations in reflectance of surface soils. *Soil Science Society of American Journal* 45:1161–1165.

79. Ben-Dor E. and A. Singer, 1987. Optical density of vertisol clays suspensions in relation to sediment volume and dithionite-citrate-bicarbonate extractable iron. *Clays and Clay Minerals* 35:311–317.

80. Karmanova L.A., 1981. Effect of various iron compounds on the spectral reflectance and color of soils. *Soviet Soil Science* 13:63–60.

81. Kosmas C.S., Curi N., Bryant R.B., and D.P. Franzmeier, 1984. Characterization of iron oxide minerals by second derivative visible spectroscopy. *Soil Science Society of American Journal* 48:401–405.

82. Dematte A.J.M. and G.J. Garcia, 1999. Alteration of soil properties through a weathering sequence as evaluated by spectral reflectance. *Soil Science Society of American Journal* 63:327–342.

83. Gerbermann A.H. and D.D. Neher, 1979. Reflectance of varying mixtures of a clay soil and sand. *Photgrammertic Engineering and Remote Sensing* 45:1145–1151.

84. Ben-Dor E., Levin N., Singer A., Karnieli A., Braun O., and G.J. Kidron, 2005. Quantitative mapping of the soil rubification process on sand dunes using an airborne hyperspectral sensor *Geoderma* 131:1–21.

85. Lugassi R., Ben-Dor E., and G. Eshel, 2010. A spectral-based method for reconstructing spatial distributions of soil surface temperature during simulated fire events. *Remote Sensing of Environment* 114:322–331.

86. Kokaly R., Rockwell B.W.M., Haire S., and T.V.V King, 2007. Characterization of post fire surface cover, soils and burn severity at the Cerro Grande Fire, New Mexico, using hyperspectral and multispectral remote sensing. *Remote Sensing of Environment* 106:305–325.

87. Csillag F., Pasztor L., and L.L. Biehl, 1993. Spectral band selection for the characterization of salinity status of soils. *Remote Sensing of Environment* 43:231–242.

88. Toth T., Csillag F., Biehl L.L., and E. Micheli, 1991. Characterization of semivegetated salt-affected soil by means of field remote sensing. *Remote Sensing of Environment* 37:167–180.

89. Verma K.S., Saeena R.K., Barthwal A.K., and S.N. Deshmukh, 1994. Remote sensing technique for mapping salt affected soils. *International Journal for Remote Sensing* 15:1901–1914.

90. Hick R.T. and W.G.R. Russell, 1990. Some spectral considerations for remote sensing of soil salinity. *Australian Journal of Soil Research* 28:417–431.

91. Dwivedi R.S. and K. Sreenivas, 1998. Image transforms as a tool for the study of soil salinity and alkalinity dynamics. *International Journal of Remote Sensing* 19:605–619.

92. Rao B.R.M., Ravi Sankar T., Dwivedi R.S., Thammappa S.S., and L. Venkataratnam, 1995. Spectral behavior of salt-affected soils. *International Journal of Remote Sensing* 16:2125–2136.

93. Hardisky M.A., Klemas V., and R.M. Smart, 1983. The influence of soil salinity, growth form and leaf moisture on the spectral radiance of Spartina Alterniflora canopies. *Photogrammetric Engineering and Remote Sensing* 49:77–83.

94. Wiegand C.L., Rhoades J.D., Escobar D.E., and J.H. Everitt, 1994. Photographic and videographic observations for determining and mapping the response of cotton to soil salinity. *Remote Sensing of Environment* 49:212–222.

95. Gausman H.W., Allen W.A., Cardenas R., and R.L. Bowen, 1970. Color photos, cotton leaves and soil salinity. *Photogrammertic Engineering and Remote Sensing* 36:454–459.

96. Sharma R.C. and G.P. Bhargava, 1988. Landsat imagery for mapping saline soils and wetlands in northwest India. *International Journal of Remote Sensing* 9:39–44.

97. Hirschfeld T., 1985. Salinity determination using NIRA. *Applied Spectroscopy* 39:740–741.

98. Szilagyi A. and M.F. Baumgardner, 1991. Salinity and spectral reflectance of soils. *Proceedings of ASPRS Annual Convention*, Baltimore, MD, pp. 430–438.

99. Roberts D.A., Smith M.O., and J.B. Adams, 1993. Green vegetation, nonphotosynthetic vegetation, and soils in AVIRIS data. *Remote Sensing of Environment* 44:255–269.

100. Mougenot B., Epema G.F., and M. Pouget, 1993. Remote sensing of salt-affected soils. *Remote Sensing Review* 7:241–259.

101. Singh A.N., 1994. Monitoring change in the extent of salt-affected soils in northern India. *International Journal of Remote Sensing* 16:3173–3182.

102. Farifteh J., Farshad A., and R.J. George, 2006. Assessing salt-affected soils using remote sensing, solute modeling, and geophysics. *Geoderma* 130:191–206.

103. Ben-Dor E., Goldshleger N., Eshel M., Mirablis V., and U. Bason, 2008. Combined active and passive remote sensing methods for assessing soil salinity. In: *Remote Sensing of Soil Stalinization: Impact and Land Management*. Metternicht G. and A. Zinck (eds.). CRC Press, Boca Raton, FL.

104. Metternich G., Ben-Dor E., Goldshleger N., and U. Basson, 2008. Sensors/platforms and popular classification algorithms. In: *Remote Sensing of Soil Stalinization: Impact and Land Management*. Metternicht G. and A. Zinck (eds.). CRC Press, Boca Raton, FL.

105. Livne I., Goldshleger N., Ben-Dor E., Mirlas V., and R. Ben-Binyamin, March 16–19, 2009. Monitoring soil salinity in agricultural lands using combined hyperspectral data and chemical measurements. In: *EARSeL SIG Imaging Spectroscopy Workshop*, 6th Edn., Tel Aviv University, Tel Aviv, Israel.

106. Pieters C.M., 1983. Strength of mineral absorption features in the transmitted component of near-infrared reflected light. First results from RELAB. *Journal of Geophysical Research* 88:9534–9544.

107. Orlov D.C., 1966. Quantitative patterns of light reflectance on soils I: Influence of particles (aggregate) size on reflectivity. *Soviet Soil Science* 13:1495–1498.

108. Jackson R.D., Teillet P.M., Slater P.N., Fedosjsvs G., Jasinski M.F., Aase J.K., and M.S. Moran, 1990. Bidirectional measurements of surface reflectance for view angle corrections of oblique imagery. *Remote Sensing of Environment* 32:189–202.

109. Chappell A., Zobeck T.M., and G. Brunner, 2005. Using on-nadir spectral reflectance to detect soil surface changes induced by simulated rainfall and wind tunnel abrasion. *Earth Surface Processes and Landforms*, 30:489–511.

110. Cierniewski J., 1987. A model for soil surface roughness influence on the spectral response of bare soils in the visible and near infrared range. *Remote Sensing of Environment* 23:98–115.

111. Cierniewski J., 1989. The influence of the viewing geometry of bare soil surfaces on their spectral response in the visible and near infrared. *Remote Sensing of Environment* 27:135–142.

112. Escadafal R. and A.R. Hute, 1991. Influence of the viewing geometry on the spectral properties (high resolution visible and NIR) of selected soils from Arizona. *Proceedings of the 5th International Colloquium, Physical Measurements and Signatures in Remote Sensing*. Courchevel, France I:401–404.

113. Pinty B., Verstraete M.M., and R.E. Dickson, 1989. A physical model for prediction biderctional reflectance over bare soil. *Remote Sensing of Environment* 27:273–288.

114. Hapke B.W., 1981. Bidirectional reflectance spectroscopy I. Theory. *Journal of Geophysical Research* 86:3039–3054.

115. Hapke B.W., 1981. Bidirectional reflectance spectroscopy: 2 experiments and observation. *Journal of Geophysical Research* 86:3055–3060.

116. Hapke B.W., 1984. Bidirectional reflectance spectroscopy: Correction for macroscopic roughens. *Icarus* 59:41–59.

117. Hapke B.W., 1986. Bidirectional reflectance spectroscopy 4: The extinction coefficient and the opposition effect. *Icarus* 67:264–280.

118. Hapke B.W., 1993. *Theory of Reflectance and Emittance Spectroscopy*, Cambridge University Press, New York.

119. Jacquemoud S., Baret F., and J.F. Hanocq, 1992. Modeling spectral and bidirectional soil reflectance. *Remote Sensing of Environment* 41:123–132.

120. Liang S. and R.G. Townshend, 1996. A modified Hapke model for soil biderctional reflectance. *Remote Sensing of Environment* 55:1–10.
121. Chabrillat S., Goetz A.F.A., Krosley L., and H.W. Olsen, 1999. Use of hyperspectral images in the identification and mapping of expansive clay soils and the role of spatial resolution. *Remote Sensing of Environment* 82:431–445.
122. Tkachuk R. and D.P. Law, 1978. Near infrared diffuse reflectance standards. *Cereal Chemistry* 55:981–995.
123. Young E.R., Clark K.C., Bennett R.B., and T.L. Houk, 1980. Measurements and parameterization of the bidirectional reflectance feature of BaSO4 paint. *Applied Optics* 19(20):3500–3505.
124. Feingresh T., Ben-Dor E., and S. Filin, 2010. Correction of reflectance anisotropy: A multi-sensor approach. *International Journal of Remote Sensing* 31:49–74.
125. Escadafal R., 1993. Remote sensing of soil color: Principles and applications. *Remote Sensing Reviews* 7:261–279.
126. Escadafal R., Girard M., and D. Courault, 1989. Munsell soil color and soil reflectance in the visible spectral bands of Landsat MSS and TM data. *Remote Sensing of Environment* 27:37–46.
127. Ben-Gera I. and K.H. Norris, 1968. Determination of moisture content in soybeans by direct spectrophotometry. *Israeli Journal of Agriculture Research* 18:124–132.
128. Davies A.M., 1987. *Near Infrared Spectroscopy: Time for the Giant to Wake Up! European Spectroscopy News* 73, ISSN 0307-0026.
129. Davies T., 1998. The history of near infrared spectroscopy analysis: Past, present and future—From sleeping technology to morning star of spectrosocopy. *Analusis Magazin* 26:4.
130. Palmer J.M., 1982. Field standards of reflectance. *Photogrametirc Engineerings and Remote Sensing* 48:1623–1625.
131. Stoner E.R., Baumgardner M.F., Weismiller R.A., Biehl L.L., and F. Robinson, 1980. Extension of laboratory soil spectra to field conditions. *Soil Science Society of American Journal* 44:572–574.
132. Viscarra Rossel R., 2009. The soil spectroscopy group and the development of a global soil spectral library. *Geophysical Research Abstracts*, Vol. 11, EGU2009–14021.
133. Shepherd K.D. and M.G. Walsh, 2002. Development of reflectance spectra libraries for characterization of soil properties. *Soil Science Society of American Journal* 66:988–998.
134. Ben-Dor E., Inbar Y., and Y. Chen, 1997. The reflectance spectra of organic matter in the visible near infrared and short wave infrared region (400–2,500 nm) during a control decomposition process. *Remote Sensing of Environment* 61:1–45.
135. Malley D. and E. Ben-Dor, 2004. Application in analysis of soils. Chapter 26. In: *Near Infrared Spectroscopy in Agriculture*. Craig R., Windham R., and J. Workman (eds.). A Three Societies Monograph (ASA, SSSA, CSSA) 44:729–784. Soil Science Society of America Inc. Publisher, Madison, Wisconsin, USA.
136. Viscarra-rossel R.A., Walvoort D.J.J., Mcbratney A.B., Janik L.J., and J.O. Skjemstad, 2006. Visible, near infrared, mid infrared or combined diffuse reflectance spectroscopy for simultaneous assessment of various soil properties. *Geoderma* 131:59–75.
137. Irons J.R., Weismiller R.A., and G.W. Petersen, 1989. Soil reflectance. In: *Theory and Aapplication of Optical Remote Sensing*. Asrar G. (ed.). Willey Series in Remote Sensing, John Wiley & Sons, New York, pp. 66–106.
138. Bedidi A., Cervelle B., Madeira J., and M. Pouget, 1990. Moisture effects on spectral characteristics (visible) of lateritic soils. *Soil Science* 153:129–141.
139. Beidid A., Cervelle B., and J. Madeira, 1991. Moisture effects on spectral signatures and CIE-color of lateritic soils. *Proceedings of the 5th International Colloquium, Physical Measurements and Signatures in Remote Sensing*. Courchevel, France I:209–212.
140. Galavo L.S., Vitroello I., and W.R. Paradella, 1995. Spectroradiometric discrimination of laterites with principle components analysis and additive modeling. *Remote Sensing of Environment* 53:70–75.
141. Ben-Dor E. and A. Banin, 1994. Visible and near infrared (0.4–1.1 mm) analysis of arid and semiarid soils. *Remote Sensing of Environment* 48:261–274.
142. Ben-Dor E. and A. Banin, 1995. Near infrared analysis (NIRA) as a rapid method to simultaneously evaluate, several soil properties. *Soil Science Society of American Journal* 59:364–372.
143. Ben-Dor E. and A. Banin, 1995. Near infrared analysis (NIRA) as a simultaneously method to evaluate spectral featureless constituents in soils. *Soil Science* 159:259–269.
144. Ben-Dor E. and A. Banin, 1995. Quantitative analysis of convolved TM spectra of soils in the visible, near infrared and short-wave infrared spectral regions (0.4–2.5 mm). *International Journal of Remote Sensing* 18:3509–3528.

145. Whiting M.L., Palacios-Orueta A., Li L., and S.L. Ustin, October 23–27, 2005. Light absorption model for water content to improve soil mineral estimates in hyperspectral imagery. In: *Pecora 16: Global Priorities in Land Remote Sensing*, American Society of Photogrammetry and Remote Sensing, Sioux Falls, SD.

146. Weindner V.R. and J.J. Hsia, 1981. Reflection properties of presses plytetrafluorothylene powder. *Journal of Optical Society of America* 71:856–862.

147. McBratney A.B., Mendonca Santos M.L., and B. Minasny, 2003. On digital soil mapping. *Geoderma* 117:2–52.

148. McBratney A.B., Minasny B., and R. Viscara Rossel, 2006. Spectral soil analysis and inference sustme: A powerful combination for solving soil data crisis. *Geoderma* 136:272–278.

149. Feimstein A., Nutesko G., and E. Ben-Dor, 2010. Performance of three identical spectrometers in retrieving soil reflectance under laboratory conditions. *Soil Sceince Society of America Journal* 75:S0110–S0174.

150. FAO Year Book, 1994. Production. Vol. 48, p. 3, *FAO Statistical Series* #125, Food and agriculture organization of the Untied Nations, Rome.

151. Bartholomeusa H., Kooistraa L., Stevens A., Leeuwen M.V., Wesemael B.V., Ben-Dor E., and B. Tychon, 2010. Soil organic carbon mapping of partially vegetated agricultural fields. *International Journal of Applied Earth Observation and Geoinformation* 13:81–88.

152. Murphy R.J. and G. Wadge, 1994. The effects of vegetation on the ability to map soils using imaging spectrometer data. *International Journal of Remote Sensing* 15:63–86.

153. Ray T.W. and B.C. Murray, 1996. Nonlinear spectral mixing in desert vegetation. *Remote Sensing of Environment* 55:59–79.

154. Goetz A.F.A., 1992. Principles of narrow band spectrometry in the visible and IR: Instruments and data analysis. In: *Imaging Spectroscopy: Fundamentals and Prospective Applications*. Toselli F. and Bodechtel (eds.). ECSE, EEC, EAEC, Brusels and Luxembourg, pp. 21–32.

155. Otterman J., Brakke T., and A. Marshak, 1995. Scattering by Lamberian-leaves canopy: Dependents of leaf-area projections. *International Journal of Remote Sensing* 16:1107–1125.

156. Ammer U., Koch B., Schneider T., and H. Wittmeier, 1991. High resolution spectral measurements of vegetation and soil in field and laboratory. *Proceedings of the 5th International Colloquium, Physical Measurements and Signatories in Remote Sensing*, Courchevel, France I:213–218.

157. Schwartz G., Eshel G., Ben-Haim M., and E. Ben-Dor, 2009. Reflectance spectroscopy as a rapid tool for qualitative mapping and classification of hydrocarbons soil contamination. Available at http://www.earsel6th.tau. ac.il/~earsel6/CD/PDF/earsel-PROCEEDINGS/3080%20Schwartz.

158. Tucker C.J. and L.D. Miller, 1977. Soil spectra contributions to grass canopy spectral reflectance. *Photogrammetric Engineering and Remote Sensing* 43:721–726.

159. Tueller P.T., 1987. Remote sensing science application in arid environment. *Remote Sensing of Environment* 23:143–154.

160. Smith M.O., Ustin S.L., Adams J.B., and A.R. Gillespie, 1990. Vegetation in desert: I A regional measure of abundances from multispectral images. *Remote Sensing of Environment* 31:1–26.

161. Huete A.R., 1988. Soil adjusted vegetation index (SAVI). *Remote Sensing of Environment* 25:47–57. *Proceedings of the 5th International Colloquium, Physical Measurements and Signatures in Remote Sensing*, Courchevel, France I:419–422.

162. Liu H.Q. and A. Huete, 1995. A feedback based modification of the NDVI to minimize canopy background and atmospheric noise. *IEEE* 33:457–465.

163. Vane G., Reimer J.H., Chrien T.G., Enmark H.T., Hansen E.G., and W.M. Porter, 1993. Airborne visible/ infrared imaging spectrometer (AVIRIS). *Remote sensing of environment* 44:127–143.

164. Siegal B.S. and A.F.H. Goetz, 1977. Effect of vegetation on rock and soil type discrimination. *Photogrammetric Engineering and Remote Sensing* 43:191–196.

165. Richardson A.J., Wiegand C.L., Gausman H.W., Cullar J.A., and A.H. Gerbermann, 1975. Plant, soil and shadow reflectance components of raw crops. *Photogammetric Engineering and Remote Sensing* 41:1401–1407.

166. Rondeaux G., Steven M., and F. Baret, 1996. Optimization of soil-adjusted vegetation indices. *Remote Sensing of Environment* 55:95–107.

167. Pinker and A. Karnieli, 1995. Characteristics spectral reflectance of semi-arid environment. *International Journal of Soil Science* 16:1341–1363.

168. Rogers R.W. and R.T. Langer, 1972. Soil surface lichens in arid and subarid south-eastern Australia. Introduction and floristics. *Australian Journal of Botany* 20:197–213.

169. West N.E., 1990. Structure and function of microphytic soil crust in wildland ecosystems of arid to semi-arid regions. *Advances in Ecological Research* 20:179–222.

170. Chappell A., Strong C., McTainsh G., and J. Leys, 2006. Detecting induced in situ erodobility of dust-producing playa in Australia using a bi-directional soil spectral reflectance model. *Remote Sensing of Environment* 106:508–524.

171. O'neill A.L., 1994. Reflectance spectra of microphytic soil crusts in semi-arid Australia. *International Journal of Remote Sensing* 15:675–681.

172. Karnieli A., Kidron G.J., Glaesser C., and E. Ben-Dor, 1999. Spectral characteristics of cyanobacteria soil crust in semiarid environments. *Remote Sensing of Environment* 69:67–75.

173. Zaady E., Karnieli A., and M. Shachak, 2007. Applying a field spectroscopy technique for assessing successional trends of biological soil crusts in a semi-arid environment. *Journal of Arid Environments* 70:463–477.

174. Morin Y., Benyamini Y., and A. Michaeli, 1981. The dynamics of soil crusting by rainfall impact and the water movement in the soil profile. *Journal of Hydrology* 52:321–335.

175. Shainberg I., 1992. Chemical and mineralogical components of crusting. In: *Soil Crusting*. Sumenr M.E. and B.A. Stewart (eds.). Lewis Publications, Ann Arbor, MI.

176. Goldshlager N., Ben-Dor E., Chudnovsky A., and M. Agassi, 2009. Soil reflectance as a generic tool for assessing infiltration rate induced by structural crust for heterogeneous soils. *European Journal of Soil Science* 60:1038–1051.

177. Ben-Dor E., Goldshalager N., Braun O., Kindel B., Goetz A.F.H., Bonfil D., Agassi M., Margalit N., Binayminy Y., and A. Karnieli, 2004. Monitoring of infiltration rate in semiarid soils using airborne hyperspectral technology. *International Journal of Remote Sensing* 25:1–18.

178. Ben-Dor E., Kindel B., and A.F.H. Goetz, 2004. Quality assessment of several methods to recover surface reflectance I using synthetic imaging spectroscopy (IS) data. *Remote Sensing of Environment* 90:389–404.

179. Hummel J.W., Sudduth K.A., and S.E. Hollinger, 2001. Soil moisture and organic matter prediction of surface and subsurface soils using an NIR soil sensor. *Computers and Electronics in Agriculture* 32:149–165.

180. Muller E. and H. Decamps, 2001. *Remote Sensing of Environment* 76:173–180.

181. Lobell D.B. and G.P. Asner, 2002. Moisture effects on soil reflectance. *Soil Science Society of American Journal* 66:722–727.

182. Bishop J.L., Pieters C.M., and J.O. Edwards, 1994. Infrared spectroscopic analyses on the nature of water in montmorillonite. *Clays and Clay Minerals* 42:702–716.

183. Whiting M.L., Li L., and S.L. Ustin, March 30, April 3, 2004. Correcting mineral abundance estimates for soil moisture. In: *13th Annual JPL Airborne Earth Science Workshop*, Pasadena, CA. R.O. Green (ed.). JPL Publication 05-3-1.

184. Lekner J. and M.C. Dorf, 1988. Why some things are darker when wet. *Applied Optics* 27:1278–1280.

185. Whiting M.L., Li L., and S.L. Ustin, 2004. Predicting water content using Gaussian model on soil spectra. *Remote Sensing of Environment* 89:535–552.

186. Haubrock S., Chabrillat S., Lemmnitz C., and H. Kaufmann, 2008. Surface soil moisture quantification models from reflectance data under field conditions. *International Journal of Remote Sensing* 29:3–29.

187. Haubrock S., Chabrillat S., and H. Kaufmann, 2004. Application of hyperspectral imaging and laser scanning for the monitoring and assessment of soil erosion in a recultivation mining area. In: *Remote Sensing and GIS for in Geography* (Göttinger geographische Abhandlungen; 113). Erasmi S., Cyffka B., and M. Kappas (eds.). Goltze, Germany, pp. 230–237.

188. Kneizys F.X., Abdersen G.P., Shettle E.P., Gallery W.O., Abreu L.W., Selby J.E.A., Chetwynd J.H., and S.A. Clough, 1988. Users guide to LOWTRAN-7. *Air Force Geophysics Laboratory*, Hanscom AFB, Massachusetts AFGL-TR-88-0177.

189. Tanre D., Deroo C., Duhaut P., Herman M., Morcrette J.J., Perbos J., and P.Y. Deschamps, 1986. *Simulation of the Satellite Signal in the Solar Spectrum (5S), User Gide (UST de Lille)*. 59655 Villenueve D'asc, France: Laboratory d'Optique Atmospherique.

190. Richter R. and D. Schläpfer, 2002. Geo-atmospheric processing of airborne imaging spectrometry data. Part 2: atmospheric/topographic correction. *International Journal of Remote Sensing*, 23:2631–2649.

191. Cipra J.E., Franzmeir D.P., Bauer M.E., and R.K. Boyd, 1980. Comparison of multispectral measurements from some nonvegetated soils using landsat digital data and a spectroradiometer. *Soil Science Society of American Journal* 44:80–84.

192. Frazier B.E. and Y. Cheng, 1989. Remote sensing of soils in the Eastern Palouse region with landsat thematic mapper. *Remote Sensing of Environment* 28:317–325.

193. Kierein-Young K. and F.A. Kruse, 1989. Comparison of landsat thematic mapper images and geophysical and environmental reassert imaging spectrometer data for alteration mapping. *Proceedings of the 7th Thematic Conference on Remote Sensing for Exploration Geology*, Calgary, Alberta, Canada I:349–359.

194. Morran S.M., Jackson R.D., Slater P.N., and P.M. Teillet, 1992. Evaluation of simplified procedures for retrieval of land surface reflectance factors from satellite sensor output. *Remote Sensing of Environment* 41:169–184.

195. Price J.C., 1990. On the information content of soil reflectance spectra. *Remote Sensing of Environment* 33:113–121.

196. Latz K., Weismiller R.A., and G.E. Van Scoyoc, 1981. A study of the spectral reflectance of selected eroded soils of Indiana in relationship to their chemical and physical properties. LARS Technical Report 082181.

197. Price J.C., 1995. Examples of high resolution visible to near-infrared reflectance spectra and a standardized collection for remote sensing studies. *International Journal of Remote Sensing* 16:993–1000.

198. Vane G., Goetz A.F.H., and J.B. Wellman, 1984. Airborne imaging spectrometer: A new tool for remote sensing. *IEEE Trans Geosciences Remote Sensing* 22(6):546–549.

199. Clark R.N. and T.L. Roush, 1984. Reflectance spectroscopy: Quantitative analysis techniques for remote sensing applications. *Journal of Geophysical Research* 89:6329–6340.

200. Dekker A.G., Brando V.E., Anstee J.M., Pinnel N., Kutser T., Hoogenboom H.J., Pasterkamp R., Peters S.W.M., Vos R.J., Olbert C., and T.J. Malthus, 2001. Imaging spectrometry of water, Chapter 11. In: *Imaging Spectrometry: Basic Principles and Prospective Applications: Remote Sensing and Digital Image Processing*, v. IV: Kluwer Academic Publishers, Dordrecht, the Netherlands, pp. 307–335.

201. Ustin S.L. and M.E. Schaepman, 2009. Imaging spectroscopy: Special issue. *Remote Sensing of Environment* 113:1–3.

202. Palacios-Orueta A. and S.L. Ustin, 1998. Remote sensing of soil properties in the Santa Monica mountains: I. Spectral analysis. *Remote Sensing of Environment* 65:170–183.

203. Kemper T. and S. Sommer, 2002. Estimate of heavy metal contamination in soils after a mining accident using reflectance spectroscopy. *Environmental Science & Technology* 36:2742–2747.

204. Gomez C., Viscarra Rossel R.A., and A.B. McBratney, 2008. Soil organic carbon prediction by hyperspectral remote sensing and field vis-NIR spectroscopy: An Australian case study. *Geoderma* 146:403–411.

205. Bartholomeus H., Epema G., and M. Schaepman, 2007. Determining iron content in mediterranean soils in partly vegetated areas, using spectral reflectance and imaging spectroscopy. *International Journal of Applied Earth Observation and Geoinformation* 9:194–203.

206. Goetz A.F.H. and M. Herring, 1989. A high resolution imaging spectrometer (HIRIS) for EOS. *IEEE Transaction on Geoscience and Remote Sensing* 27:136–144.

207. Ben-Dor E. and A. Brook, 2011. Supervised vicarious calibration (SVC) of hyperspectral remote-sensing data. *Remote Sensing of Environment* 115:1543–1555.

208. Stuffler T., Förster K., Hofer S., Leipold M., Sang B., Kaufmann H., Penné B., Mueller A., and C. Chlebek, 2009. Hyperspectral imaging—An advanced instrument concept for the EnMAP mission (environmental mapping and analysis programme). *Acta Astronautica* 65:7–8, 1107–1112.

209. Metternicht G.I. and J.A. Zink, 1997. Spatial distribution of saline and sodium affected soil surfaces. *International Journal of Remote Sensing* 18:2571–2586.

Part IX

Detecting Crop Management,
Plant Stress, and Disease

23 Analysis of the Effects of Heavy Metals on Vegetation Hyperspectral Reflectance Properties

E. Terrence Slonecker

CONTENTS

23.1 INTRODUCTION

Absolute definitions of "heavy metals" are elusive in modern science. Many different definitions have been proposed. Some are based on density, some on atomic number or atomic weight, and some on chemical properties or toxicity [1]. One definition holds that they are elements with a specific weight higher than $6\,g/cm^3$ [2] (see Figure 23.1). But no single definition fits well in modern usage. The term "toxic metals" has become to some extent synonymous with heavy metals, but that term is equally problematic because levels of toxicity are highly variable between different metals and vegetation species. At best, heavy metals can be classified as a poorly defined subset of elements that exhibit metallic properties and are toxic to living organisms at some level of concentration or exposure. The term "heavy metals" itself has been criticized as functionally meaningless [1,3].

Metals in the environment, however, are a real concern for a variety of reasons, including their commercial and industrial value, medicinal applications, use in agricultural chemicals, and their toxic effects on human and ecological resources as chemical weapons or as fugitive, uncontrolled, anthropogenic releases into the environment. Some metals such as selenium, copper, and zinc are micronutrients that are actually required by most plant and animal life forms in very small doses while others such as mercury and lead are toxic and have no known benefit to living organisms.

Although the toxicity and ecotoxicity of many heavy metals can vary widely, the term has evolved to connote a pejorative meaning synonymous with anthropogenic pollution. Heavy metals can occur naturally and can arise from many anthropogenic sources such as mining and processing of other metals, the smelting of copper, processing of gold, steel, iron, and coal, the preparation of nuclear fuels, and the production of industrial construction materials. In addition, many computer parts and chips contain heavy metals or have production that results in waste products with heavy metals.

Hydrogen 1 H 1.0079																		Helium 2 He 4.0026
Lithium 3 Li 6.941	Beryllium 4 Be 9.0122												Boron 5 B 10.811	Carbon 6 C 12.011	Nitrogen 7 N 14.007	Oxygen 8 O 15.999	Fluorine 9 F 18.998	Neon 10 Ne 20.180
Sodium 11 Na 22.990	Magnesium 12 Mg 24.305												Aluminium 13 Ai 26.982	Silicon 14 Si 28.086	Phosphorus 15 P 30.974	Sulfur 16 S 32.065	Chlorine 17 Cl 35.453	Argon 18 Ar 39.948
Potassium 19 K 39.098	Calcium 20 Ca 40.078		Scandium 21 Sc 44.956	Titanium 22 Ti 47.867	Vanadium 23 V 50.942	Chromium 24 Cr 51.996	Manganese 25 Mn 54.938	Iron 26 Fe 55.845	Cobalt 27 Co 58.933	Nickel 28 Ni 58.693	Copper 29 Cu 63.546	Zinc 30 Zn 65.39	Gallium 31 Ga 69.723	Germanium 32 Ge 72.61	Arsenic 33 As 74.033	Selenium 34 Se 78.06	Bromine 35 Br 79.904	Krypton 36 Kr 83.90
Rubidium 37 Rb 85.468	Strontium 38 Sr 87.62		Yttrium 39 Y 88.906	Zirconium 40 Zr 91.224	Niobium 41 Nb 92.906	Molybdenum 42 Mo 95.94	Technetium 43 Tc [98]	Ruthenium 44 Ru 101.07	Rhodium 45 Rh 102.91	Palladium 46 Pd 106.42	Silver 47 Ag 107.87	Cadmium 48 Cd 112.41	Indium 49 In 114.82	Tin 50 Sn 118.71	Antimony 51 Sb 121.76	Tellurium 52 Te 127.60	Iodine 53 I 126.90	Xenon 54 Xe 131.29
Caesium 55 Cs 132.91	Barium 56 Ba 137.33	57–70 *	Lutetium 71 Lu 174.97	Hafnium 72 Hf 178.49	Tantalum 73 Ta 180.95	Tungsten 74 W 183.84	Rhenium 75 Re 186.21	Osmium 76 Os 190.23	Iridium 77 Ir 192.22	Platinum 78 Pt 195.08	Gold 79 Au 196.97	Mercury 80 Hg 200.59	Thallium 81 Ti 204.38	Lead 82 Pb 207.2	Bismuth 83 Bi 208.08	Polonium 84 Po (209)	Astatine 85 At (210)	Radon 86 Rn (222)
Francium 87 Fr (223)	Radium 88 Ra (226)	89–102 * *	Lawrencium 103 Lr (262)	Rutherfordium 104 Rf (261)	Dubnium 105 Db (262)	Seaborgium 106 Sg (266)	Bohrium 107 Bh (264)	Hassium 108 Hs (269)	Meitnerium 109 Mt (268)	Ununnilium 110 Uun (271)	Unununium 111 Uuu (272)	Ununbiun 112 Uub (277)		Ununquadium 114 Uuq [289]				

FIGURE 23.1 The periodic table showing the elements generally considered heavy metals. Lanthanides and actinides are not shown. (Modified from Shaw, B. et al., Heavy metal induced oxidative damage in terrestrial plants, in *Heavy Metal Stress in Plants—From Biomolecules to Ecosystems*, Vol. 2, Springer, Berlin, Germany, 2004, pp. 84–126.)

Electroplating is a primary source of chromium and cadmium pollution. Arsenic has been used extensively in pesticides and in wood treating [4] and, because of its toxicity, has been used for years as a base compound for chemical warfare weapons such as Lewisite gas [5].

Hyperspectral remote sensing (HRS), also known as imaging spectroscopy and to a greater extent traditional field and laboratory spectroscopy, has a long history of investigations into the identification of metals and their effects on vegetation in the environment. However, fugitive metals in the environment do not usually exist in their pure form but rather in a soil–water–vegetation matrix as waste rock materials, sediments, or as a result of soil deposition. Besides detecting the minerals themselves, spectroscopy and imaging spectroscopy can also be used to detect the composition and condition of vegetation, which can then be used to interpret the mineral deposits or metal composition of the soil in the area of the vegetation growth. It has long been acknowledged by scientists that a relation exists between vegetation, soils, and underlying mineral deposits [6]. In several studies, airborne spectroscopy was used to detect "hidden" mineral deposits through forest-covered areas by revealing subtle variations in the reflected spectrum of vegetation under stress due to the presence of heavy metals [7–10]. In addition, a growing body of spectroscopic literature has been involved with the identification of environmental hazards, many of which are heavy metals. Another area that has received recent attention in the area of spectroscopy of metal stress in vegetation is that of vegetation indices (VIs). This chapter reviews the scientific background of spectroscopy and imaging spectroscopy with respect to the effects of heavy metals on vegetation reflectance.

23.2 PHYSIOLOGY OF METAL STRESS IN PLANTS

Plants are generally more exposed to pollution risks in the environment because they are stationary and cannot avoid interacting with environmental pollutants such as metals. Plants have evolved various complex strategies for adapting to heavy metals pollution in soil or water mediums. Plants respond to exposure to heavy metals in several different ways. Metals usually interfere with basic plant metabolism, and enzyme activity is often negatively affected. Metals present in the plant tissues can cause the plant to form chelate structures, molecules that enclose and isolate metal ions and cause them to lose the functional properties in metabolic cycles, such as the citric acid cycle.

Plants generally fall into two categories with respect to strategies for dealing with exposure to heavy metals: *Accumulators* uptake metal ions and process them in some manner, storing them in internal tissues or reducing or processing them in biochemical reactions; *excluders* generally restrict the uptake of metals by preventing their uptake of metals into plant tissues. This is often accomplished by trapping metal ions in the cell walls of the root tissue.

Whether the plants are accumulators or excluders, excess metals in the soil or in the plant tissues tend to have negative effects on plant health, growth, and biomass accumulation, and can cause visual symptoms at toxic levels. Table 23.1 shows examples of the visual injuries to various flowering plants from metal exposures. These visual symptoms also affect reflectance characteristics of the typical vegetation spectra. Figure 23.2a and b shows increasing visual damage to plant health and the corresponding changes in the blue and red energy absorption troughs at 480 and 680 nm, respectively, seen as increasing reflectance and a blueshift.

Excess metal exposure negatively affects photosynthetic processes and typically induces a general "stress" reaction in plants. In some cases, the absorbed metal ion will replace the central magnesium atom in the chlorophyll molecule, which generally causes oxidative stress in the plant. This substitution reduces or prevents photosynthetic light-harvesting and results in a breakdown of photosynthesis [11].

Heavy metal exposure can also interfere with plant–water relations. Metals may alter plasma membrane properties, affect enzyme activities, inhibit root growth and elongation, affect osmotic potential, and generally inhibit the ability of the plant to acquire water [12]. This may be manifested as a general drought-stress response but is actually caused primarily by the interference of heavy metals and not simply the lack of water availability.

TABLE 23.1

Examples of Visual Symptoms of Metals Stress in Plants

Metal	Characteristics	Reference(s)
Arsenic	Red/brown necrotic spots on old leaves, yellow/brown roots, reduced growth	[36,37]
Aluminum	Stunted growth, inhibition of root elongation, purple coloration, curling, and yellowing of leaf tips	[72,73]
Cadmium	Brown edges to leaves, chlorosis, necrosis, curled leaves, stunted roots	[74,75]
Copper	Chlorosis, yellow and purple coloration, decreased root growth, and leaf biomass	[76–78]
Lead	Dark green leaves, stunted growth, chlorosis, and blackening of root system	[79]
Mercury	Severe stunting of seedlings and roots, chlorosis, reduced biomass	[80]
Nickel	Chlorosis. necrosis, stunting, reduced root, and leaf growth	[81]
Selenium	Interveined chlorosis, black spots, bleaching and yellowing of young leaves, pink spots on roots	[17]
Zinc	Chlorosis, stunting, reduced root elongation	[82]

Source: Modified from Shaw, B. et al., Heavy metal induced oxidative damage in terrestrial plants, in *Heavy Metal Stress in Plants—From Biomolecules to Ecosystems*, Vol. 2, Springer, Berlin, Germany, 2004, pp. 84–126.

In general, many different photosynthetic reactions and physiological processes are negatively affected by plant exposure to heavy metals. These vary widely among different species and metals, but in many cases, both light and dark photosynthetic reactions are generally inhibited [13].

23.3 BASIC SPECTROSCOPY OF VEGETATION

Spectroscopy is the study of the interaction between energy and matter as a function of either wavelength (λ) or frequency (v). Historically, spectroscopy refers to the use of visible light dispersed by a prism according to its wavelength and is the parent science to all visible and near infrared (VNIR) HRS. Dating from the nineteenth century [14], spectroscopic techniques have been used widely in analytical chemistry and astronomy to identify many elemental substances, minerals, and organic compounds.

The use of spectral reflectance methods to gain an understanding of photosynthesis and related vegetative processes is a field of scientific study that has been ongoing for decades [15,16]. Laboratory instruments called spectrometers, spectrophotometers, spectrographs, or spectroradiometers are all different names for instruments that essentially use some type of prism to separate light into its component parts and measure the reflectance and absorption of each of those individual component parts from a target surface. Early instruments separated light into the basic colors of the spectrum. Modern instruments separate light into individual nanometers of reflectance energy.

In this review, "hyperspectral" remote sensing technology is afforded the broadest possible definition. The papers reviewed here represent a variety of spectroscopic remote sensing systems and approaches that include individual leaf-level and plant-level analysis under controlled conditions in the laboratory to spectroscopic measurements of plants in the field to overhead aircraft and satellite systems. The common thread is that multiple bands of energy reflectance are being recorded and analyzed with spectroscopic methods.

Different spectroscopic collection perspectives also contain inherent advantages and disadvantages that include complications involving the detection and analysis of the reflected energy signal. Outside of a pure laboratory setting, field collections generally involve variable solar lighting, background effects from soil and other materials, and effects from bidirectional reflection distribution function (BRDF). Aircraft and especially satellite sensors contain increasingly significant signal noise from atmospheric moisture and constituent gasses.

(a)

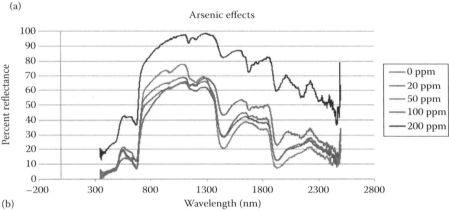

(b)

FIGURE 23.2 **(See color insert.)** (a) Visual effects of arsenic stress on *Nephrolepis exaltata* (Boston fern). Ferns are planted in clean sand amended with, from left to right, 0, 20, 50, 100, and 200 ppm sodium arsenate. (From Slonecker, E., *Remote Sensing Investigations of Fugitive Soil Arsenic and Its Effects on Vegetation Reflectance*, George Mason University, Fairfax, VA, 2007.) (b) Laboratory reflectance spectra of arsenic-affected ferns in (a). Spectra were collected with an ASD full range spectrometer from 6 in. above the canopy of each plant. Note the loss of photosynthetic absorption at 680 nm, causing higher reflectance, the blueshift, and the general increase in reflectance in SWIR (due to loss of water) with increasing soil arsenic. (From Barcelo, J. and Poschenrieder, C., *J. Plant Nutr.*, 13, 1, 1990.)

The majority of papers and research studies reviewed here involve spectrometers used in either a laboratory or field setting. There are some that utilize aircraft and satellite systems and a few that represent multiscale data collection from the laboratory to field to aircraft or satellite sensor. While the availability and applications of aircraft and satellite systems is growing significantly and this will be a prime focus area of future research, hyperspectral research in the laboratory and field represents a critical first step in developing and understanding the repeatable spectral measurement of heavy metal effects in plant reflectance.

23.4 SPECTROSCOPY AND IMAGING SPECTROSCOPY OF METAL INTERACTIONS WITH PLANTS

Early spectroscopic analysis of vegetation–metal interactions from both laboratory and aircraft sensors can be traced to the late 1970s and early 1980s to researchers, such as Collins, Milton, and Horler who demonstrated repeated shifts in the "red edge" of typical vegetation reflectance-based

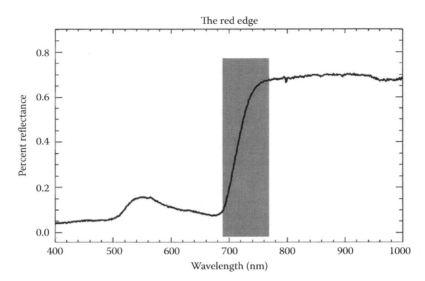

FIGURE 23.3 The red edge. An important region of vegetation spectra known as the red edge. Much research has focused on measuring shifts in this region corresponding to stress or the enhancement of chlorophyll. (From Slonecker, T. et al., *Remote Sens.*, 1, 644, 2009.)

stress or enhanced growth caused by excessive exposure to metals in the soil [7,17,18]. This has evolved into a fundamental spectroscopic-plant principle that is still widely used today. The red edge of vegetation reflectance is an area usually centered around 720 nm and represented by the typical sharp rise in reflectance in the 680–760 nm range of the classic vegetation spectral signature. Figure 23.3 shows the classic red edge area of vegetation spectra.

Although the general concept of the red edge is easily understood as the area of the sharp rise in reflectance, a variety of definitions and quantitative methods for computing the red edge are found in the literature. Ray et al. [19] defined the red edge as the sharp transition between absorption by chlorophyll in the visible wavelengths and the strong scattering in the near infrared from the cellular structure of leaves. The red edge is defined by Horler et al. [18] as the wavelength of maximum $\Delta R/\Delta \lambda$ where R is the reflectance and λ is the specific wavelength. Guyot et al. [20] defines the red edge as an inflection in the sharp rise in reflectance between 670 and 760 nm. Although variable in the literature, most modern definitions of the red edge involve the peak of the first derivative [21]. Additional red edge related measurements include a ratio of R_{740}/R_{720} and a ratio of first derivative values D_{715}/D_{705} [22].

The general movement of the spectral features in the red edge area is one of the keys to its analytical strength. When plants are healthy and are producing more chlorophyll, the red edge tends to shift toward the right to longer wavelengths. This is also usually accompanied by an increase in the absorption trough at 680 nm as the plant absorbs more energy in the photosynthetic process. When the plant is stressed, such as in the case of excessive heavy metals in the soil, the spectra tend to shift toward the left and shorter wavelengths. Stress also tends to produce an increase in reflectance at the 680 nm absorption trough because less light is being utilized for photosynthesis and chlorophyll production. Figure 23.4 shows an example of this stress based on a laboratory experiment with varying levels of copper sulfate in the soil.

Horler et al. [18] studied the feasibility of utilizing a red edge measurement as an indication of plant chlorophyll status. Using derivative reflectance spectroscopy in the laboratory, plant chlorophyll status and red edge measurements were acquired from single leaves of several different species under heavy metal stress. By using spectroscopic and laboratory methods to measure the chlorophyll content of the same leaf samples, direct evidence of the red edge–chlorophyll correlation was obtained. Measuring *in situ* vegetation using a field spectrometer, Ray et al. [19] discovered

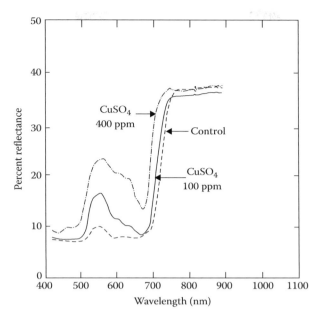

FIGURE 23.4 The "blueshift" in the red edge in laboratory-grown sorghum exposed to different levels of copper sulfate in the soil. (From Chang, S. and Collins, W., *Econ. Geol.*, 78, 723, 1983.)

significant differences in the size and shape of the red edge in different types of arid vegetation and found for a common yellow grass species, that there was no chlorophyll "bump" at the green peak and no detectable red edge.

A critical component of spectral analysis of vegetation is the shift in absorption and reflectance features that occur as a result of chemical and nutrient exposures. A general relation between increases in chlorophyll concentration and a "redshift" toward longer wavelengths has been established by several researchers utilizing both laboratory and field spectrographic methods. Gates et al. [23] showed the basic relationship between increased chlorophyll and plant health and the shift of the red edge toward longer wavelengths. Guyot et al. [20] similarly showed that the red edge inflection point shifts to longer red wavelengths as chlorophyll concentrations increase. This general correlation between chlorophyll content and redshift was confirmed by Horler et al. [24] and Baret et al. [25] for different crop species.

More important to this specific research topic, however, is the "blueshift" (i.e., shift toward shorter wavelengths) of the red edge that occurs when vegetation has undergone stress from some mineral or chemical agent. The blue or redshift toward shorter or longer wavelengths, respectively, is one of the keys to detection of stress and growth in all green vegetation. The blueshift is usually accompanied by a general increase in overall reflectance and an increase in the 680 nm absorption feature showing that less light energy is being utilized for photosynthesis.

In some of the first applications comparing field and airborne spectroscopic measurements of metal stress, Collins et al. [8] and Chang and Collins [10] showed a blueshift in the 700–780 nm region of reflectance spectra from conifers affected by metal sulfide (see Figure 23.4). Similar blueshift results have been reported by Schwaller and Tkach from field applications and aerial photographs [26] and Milton et al. in the laboratory [17,27]. In a seminal remote sensing research application using both *in situ* and airborne measurements, Rock et al. [28] demonstrated a 5 nm blueshift in spruce and fir species in Vermont and Germany as a result of stress caused by airborne pollutant deposition.

Although the underlying physiology is not completely understood, the uptake of heavy metals has the effect of reducing photosynthetic activity and the concentration of chlorophyll. One mechanism of heavy metal induced damage in plants that leads to reduction in photosynthesis involves the *in vivo* replacement of the central Mg^2 ion in the chlorophyll molecule by a heavy metal ion. This replacement is generally toxic to the plant depending on the metal and at the very least, inhibits the overall ability of the plant to conduct photosynthesis. In general, the magnesium–chlorophyll molecule has a much higher capacity to release electrons than other metals and replacement by other metals quenches, or reduces the ability of the plant to regulate excess light energy and protect the plant from damage [11,29,30].

In another classic paper utilizing both lab and field spectral measurements, Horler et al. [18] studied the effects of heavy metals on the reflectance spectra of plants. Utilizing both natural vegetation growing in known areas of metal concentrations and specific greenhouse experiments, relationships were established between metal stress, total chlorophyll, chlorophyll a/b ratios, and reduced reflectance at specific wavelengths. Controlled experiments with pea plants and other species showed that the general effect of exposure to cadmium (Cd), copper (Cu), lead (Pb), and zinc (Zn) was growth inhibition. Also, the pea plants showed changes in the leaf chlorophyll a/b ratios for exposure to Cd and Cu but showed no changes for Pb and Zn. Metal-treated plants in both controlled and natural environments showed a decrease in reflectance at 850, 1650, and 2200 nm and an increase at 660 nm. Metal concentration in the soil has strong negative correlations to reflectance at 1650 and 2200 nm and strong positive correlations at 660 nm. In general, the ability to measure stress effects from heavy metals is dependent on species, the phase of the growth cycle, and the environment.

Kooistra et al. [31] conducted a study to examine the possibilities for *in situ* evaluation of soil properties in river floodplains using field reflectance spectroscopy of cover vegetation. Results determined that a combination of field spectroscopy and multivariate calibration results in a qualitative relation between organic matter and clay content, which are intercorrelated with levels of Cd and Zn. The study indicated the potential for these multivariate methods for mapping soil properties using HRS techniques. Kooistra et al. [32,33] conducted two additional studies to investigate the relation between vegetation reflectance and soil characteristics including elevated concentrations of the metals Ni, Cd, Cu, Zn, and Pb found in the floodplain soils along the rivers Rhine and Meuse in the Netherlands. These studies obtained high-resolution vegetation reflectance spectra in the visible to near-infrared using a field radiometer [32]. The relationships were evaluated using simple linear regression in combination with two spectral VIs: the Difference Vegetation Index (DVI) and the red edge position (REP). The r^2 values between metal concentrations and vegetation reflectance ranged from 0.50 to 0.73. The results of the study demonstrated the potential of remote sensing data to contribute to the survey of spatially distributed soil contaminants in floodplains under natural grasslands, using the spectral response of the vegetation as an indicator. Modeling the relationship between soil contamination and vegetation reflectance resulted in similar results for DVI, REP, and the multivariate approach using partial least squares (PLS) regression [32,33].

Similar studies were conducted by Clevers et al. [34,35] in contaminated floodplains in the Netherlands. Analysis of field spectrometer measurements of reflectance found that REP and the first derivative peaks around 705 and 725 nm were the best predictors of heavy metal contamination. Similarly, Slonecker [36,37] showed the spectral relationship between arsenic uptake and spectral reflectance in arsenic-hyperaccumulating *Pteris* ferns, using PLS regression. Rosso et al. successfully detected plant stress due to metal pollution at the leaf level, and reiterate that more investigations need to take place that link their results to canopy-level reflectance [38].

Slonecker [36] used both laboratory spectra and HyMAP imagery spectra of arsenic stress in common lawn grasses to map the distribution of fugitive arsenic and other metals in household lawns in an urban setting. The hyperspectral imagery was processed with a linear spectral unmixing algorithm and mapped with a maximum likelihood classifier. Classes included grass, arsenic-affected grass, trees, buildings, soil, asphalt, and concrete and showed an overall accuracy of 82.9%. Critical spectral parameters for identifying arsenic stress were located in the

Laboratory and imagery hyperspectral signatures of arsenic stress in grass

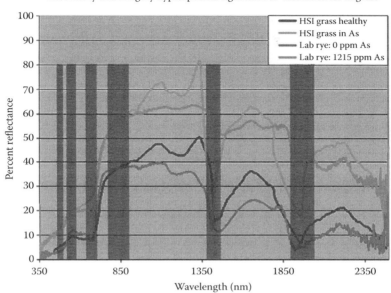

FIGURE 23.5 (See color insert.) Healthy and stressed grass signatures from both the laboratory and hyperspectral imagery. The same critical areas in the green, red, near-infrared, and SWIR show the patterns of spectral separation between the healthy and stressed grass that enable the image processing algorithm to separate, identify, and map arsenic stressed grasses. (From Slonecker, E., *Remote Sensing Investigations of Fugitive Soil Arsenic and Its Effects on Vegetation Reflectance*, George Mason University, Fairfax, VA, 2007.)

green, red, near-infrared plateau, and water-absorption bands in both the laboratory and imagery spectra. Validated against comprehensive ground sampling efforts, final maps of the arsenic-affected grass showed an overall producer's accuracy of 55.8% and an overall user's accuracy of 82.7% (see Figure 23.5).

Gallagher et al. [39] utilized field spectrometry and Ikonos multispectral satellite measurements to assess basal area, plant productivity, and chlorophyll content of gray birch growing in soils containing elevated metals in a New Jersey Brownfields site. Biomass production, measured by a red/green ratio index, showed an inverse relationship ($r^2 = 0.46 - 0.81$) to soil zinc concentration. The relationship was stronger when the total metal levels (TMLs) were higher. Threshold TMLs were established for several species beyond which the normalized difference vegetation index (NDVI) decreased at both the assemblage and individual tree level.

Mars and Crowley [40] utilized AVIRIS and the digital elevation model (DEM) data to evaluate hazardous waste contamination in southeastern Idaho including mine waste dumps, wetlands vegetation, and other relevant vegetation types. With the mapped information and the DEM, delineation of mine dump morphologies, catchment watershed areas above each mine dump, flow directions from the dumps, stream gradients, and the extent of downstream wetlands available for selenium absorption were determined. Compared to ground truth maps, the AVIRIS imagery correctly identified 76% of all mine waste pixels. Additionally, Mars and Crowley were able to characterize the physical settings of mine dumps and test hypotheses concerning the causes of selenium contamination in the area [40].

Ren et al. [41] found that rice exposed to lead in the soil weakened the photosynthetic process of rice as measured by field spectral measurements. Lead concentrations in rice could be reliably predicted by changes in the normalized band absorption depth, blueshifts in the red edge region, and the distance of the shift.

23.5 VEGETATION INDICES

One area that has received recent attention in the area of spectroscopy of metal stress in vegetation is that of VIs. VIs are mathematical manipulations of digital number values of two or more bands of data and have been a fundamental part of the remote sensing analysis of vegetation for decades. VIs typically stretch or enhance a particular part of the reflected electromagnetic spectrum (EMS) known to relate to specific vegetation qualities, such as chlorophyll content, leaf moisture, pigment ratios, and stress level. The search for stressed or unusual growth patterns in cover vegetation, such as potential metal stress patterns, has been enhanced by the use of one or more VIs that have been reported in the scientific literature.

The most widely known and used VI is the NDVI, which is calculated by the following general band formula:

$$NDVI = \frac{NIR - Red}{NIR + Red}$$

where
  NIR is the reflectance from the near-infrared band
  R is the reflectance from the red visible band

The NDVI was first proposed by Pearson and Miller [42] and has been widely utilized as a general measure of vegetation condition and has both broadband and narrowband formulae for its computation. Although the NDVI has been the most widely used vegetation index, it has clear limitations. NDVI saturates in areas of multilayered canopy and shows nonlinear relationships with critical vegetation parameters, such as leaf area index (LAI). As a result, there has been substantial effort to develop new indices that improve on the shortcomings of NDVI [43].

VIs have often been developed for specific purposes and optimized to assess a specific condition or process. Also, the emergence and increasing availability of hyperspectral data and imagery has resulted in a new class of VIs, known as "narrowband" indices that capitalize on the increased spectral resolution of hyperspectral data.

For example, Peñuelas et al. [44] proposed a structurally insensitive pigment index (SIPI), which incorporates an NIR band (800 nm) to minimize internal leaf structure effects such as increased scattering due to refractive index discontinuities between air and cell walls inside the leaf. Gamon et al. [45] developed the photochemical reflectance index (PRI) to estimate physiological parameters of sunflowers undergoing nitrogen stress. Huete [46] developed a vegetation index that accounts for, and minimizes, the effect of soil background conditions. The soil-adjusted vegetation index (SAVI) equation introduces a soil-brightness-dependent correction factor, L that compensates for the difference in soil background conditions. NIR is the reflectance from the near-infrared band, and R is the reflectance from the red visible band. Applying the correction for the soil provides more accurate information on the condition of the vegetation itself. The triangular vegetation index (TVI) was developed as very precise measures of chlorophyll concentration and absorption and depends on very specific narrow wavelengths [47].

Agricultural vegetation applications of both field and airborne hyperspectral data analysis have been conducted by several researchers showing the promise of this technology in monitoring plant production for food supplies. Strachan et al. [48] and Daughtry et al. [49] both showed that very-narrow, crop-specific VIs could be developed and utilized from hyperspectral data and applied to the assessment of agricultural productivity. In general, the use of VIs has seen a significant increase with the development and availability of hyperspectral data. Elvidge and Chen [50], Blackburn [51,52], and Thenkabail et al. [53,54] have demonstrated the effectiveness of narrowband VIs, which continues as one of the most important analytical approaches in the area of spectroscopic analysis of vegetation. Table 23.2 shows the several VIs that are mentioned in this paper along with the spectral calculation and literature source.

TABLE 23.2
VIs Specifically Referenced in This Paper

Name	Acronym	Formula	Reference(s)
Anthocyanin reflectance index	ARI	$(1/R_{550}) - (1/R_{700})$	[83]
Difference vegetation index	DVI	$2.4 * MSS7 - MSS5$	[84]
Modified triangular vegetation index 2	MTVI2	$1.5 [1.2(R_{800} - R_{550}) - 1.3(R_{670} - R_{550})]/$ $SQRT[(2 * (R_{800} + 1) 2) - (6 * R_{800} - 5 *$ $SQRT(R_{670})) - 0.5]$	[85]
Moisture stress index	MSI	$(R_{1599} - R_{819})$	[86]
Normalized difference vegetation index (broadband)	NDVI	$(NIR - RED)/(NIR + RED)$	[87]
Normalized difference vegetation index (narrowband)	NDVI	$(R_{800} - R_{670})/(R_{800} + R_{670})$	[88]
Normalized pigment chlorophyll index	NPCI	$(R_{680} - R_{430})/(R_{680} + R_{430})$	[44]
Photochemical reflectance index	PRI	$(R_{531} - R_{570})/(R_{531} + R_{570})$	[45]
Red edge position	REP	R1Dmax: (R1D690 – R1D740)	[89]
Red edge vegetation stress index	RVSI	$((R_{714} - R_{752})/2) - R_{733}$	[90]
Soil adjusted vegetation index	SAVI	$(1 + 0.5) (R_{800} - R_{670})/(R_{800} + R_{670} + 0.5)$	[46]
Structure insensitive pigment index	SIPI	$(R_{800} - R_{445})/(R_{800} - R_{680})$	[44]
Triangular vegetation index	TVI	$0.5 * [120 * (R_{750} - R_{550}) - 200 *$ $(R_{670} - R_{550})]$	[47]

VIs have also played an important role in the detection and analysis of stress due to heavy metals (Table 23.3). Reusen et al. [55] successfully mapped heavy metal contamination in Belgium through the expressions of vegetation stress in conifers near abandoned zinc smelting facilities. Utilizing imaging data from an airborne hyperspectral sensor (CASI), they utilized a spectral angle mapper (SAM) classification to build a mask for pine trees and then computed 18 separate VIs of stress.

TABLE 23.3
Some Key Spectral Features and VIs Related to Metal Stress in the Literature

Spectral Feature	Metal(s)	Vegetation Type	Sensor	Reference(s)
DVI, REP	Ni, Cd, Cu, Pb, Zn	Floodplain, ryegrass	ASD	[32]
EGFN	Zn	Conifers	CASI	[55]
NDVI, RGI	Cr, Pb, Zn, V	Gray birch	ASD Ikonos	[39]
NDVI	Ni, Cd, Cu, Pb, Zn	Rice	Landsat TM	[91]
PRI	General HM	Floodplain	ASD	[56]
PRI	As	Ferns	ASD	[36,37]
REP	Pb	Rice	ASD	[41]
REP	Cu, Zn	Peas, maize, sunflower	PE 554	[61]
REP	General HM	Floodplain bluegrass, ryegrass	ASD	[34,35]
RVI NDVI, REP	Hg	Mustard spinach	ASD	[59]
NPCI, PRI, REP	General HM	Stinging nettles, reed canarygrass, meadow foxtail	ASD	[56]
R_{850}	Cd, Cu, Pb, Zn, As	Peas	PE 554	[18]
R_{1650}	Cd, Cu, Pb, Zn, As	Peas	PE 554	[18]
CR_{1730}	General HM	Floodplain	ASD	[56]
R_{2200}	Cd, Cu, Pb, Zn, As	Peas	PE 554	[18]

The edge green first derivative normalized difference (EGFN) vegetation index proved to be the best indicator of zinc stress in the pine trees in the surrounding area [55].

Götze et al. [56] used reflectance spectroscopic methods in both the laboratory and in the field to quantify and separate heavy metal stress in floodplain vegetation. Testing a series of VIs, they showed that metal stress could be uniquely separated from other forms of stress, such as water or nutrient stress. The indices that proved to be most sensitive to the stress from heavy metals in the soil were the PRI, the REP, the normalized pigment chlorophyll index (NPCI), and the continuum removed band depth at 1730 nm (CR1730) [56].

Using both filed and laboratory measurements, Slonecker [36] showed that the PRI was sensitive to metal stress in the form of inorganic arsenic. Thorhaug et al. [57] showed that the PRI was sensitive to the effects of low salinity in seagrass health. Gallagher et al. [39] showed that a red/green ratio index had an inverse relationship with zinc concentrations in gray birch trees.

Several VIs seem to dominate the literature with respect to metal stress in vegetation. The REP described earlier is the most dominant spectral feature used to assess plant stress. It has been used by many researchers to evaluate decreases in plant chlorophyll, biomass, or physiological health with respect to metal stress [7–10,17,24,34,35,39,41,56,58–63].

The PRI was developed by Gamon et al. [45] as a narrowband hyperspectral indicator of changes in the pigment balance of plants due to photosynthetic stress. Originally designed to track diurnal changes in photosynthetic efficiency, the PRI is sensitive to changes in carotenoid pigments and the epoxidation state of the xanthophyll cycle. This is a measure of photosynthetic light use efficiency and the rate of carbon dioxide uptake. The PRI measures the relative reflectance on either side of the green maxima around 550 nm and compares reflectance parameters in both the red and green regions simultaneously. Because the change in pigment concentrations due to metal stress in most vascular plants is similar, the PRI index has been shown to be a successful indicator of a variety of stress conditions, including stress from soil metals. Slonecker [36] computed a suite of 67 broadband and hyperspectral VIs and used a PLS and stepwise linear regression (SLR) analysis to isolate the best VIs for explaining arsenic stress in Boston ferns and arsenic hyperaccumulating *Pteris* ferns. The results for the control Boston ferns showed that the PRI, along with the moisture stress index, the red edge vegetation stress index, and the modified triangular vegetation index 2, provided the best model for explaining the level of arsenic uptake. These indices measure plant stress in one form or another, which generally increases with higher concentrations of soil arsenic. The best indices for the hyperaccumulating *Pteris* ferns were the broadband green index (GI), the sum green index (SGI), and the carotenoid reflectance index (CRI), all relating to the green part of the spectrum. Although not fully understood, the different indices for stressed and hyperaccumulating species reflect key differences in internal plant physiology [36].

Götze et al. [56] found that four indices were highly correlated between heavy metal content and chlorophyll content. The r^2 values for the NCPI (0.91), PRI (0.75), REP (0.80), and the value of the continuum-removed spectra at 1730 nm (0.74) were all sensitive to metal stress in plants. Although the underlying physiology is not fully understood, the authors speculate that the correlation could be related to lignin or protein production in plant synthesis. Further, this study shows promising results for using these values to separate heavy metal stress from water and nutrient stress [56].

23.6 EMERGING STATISTICAL METHODS

A wide variety of analytical methods can be noted in the review of the hyperspectral analysis of vegetation and vegetation stress. One of the fundamental issues can be attributed to the fact that the analysis of hyperspectral data presents unique analytical problems for standard multivariate techniques because of the highly correlative and overlapping nature of data. The large numbers of independent variables (>1500 spectral bands) and the highly correlated nature of those variables stem from the fact that each individual spectral band is only a few nanometers away from the spectral bands above and below it and the result is that each spectral band records an energy pattern that is similar to its neighboring bands. Highly correlated independent variables create a condition known as collinearity, which violates the assumptions of linear regression. To develop a predictive and effective linear model, variables must be independent. The overall result of a collinearity condition is that correlated independent variables have unstable coefficients, and although the model developed may have a high r^2 value and low residuals, it will perform poorly outside of the immediate data set that was used to develop it.

In recent years, a special statistical technique has emerged that addressed the problems of numerous, highly correlated variables. The technique, known as PLS was first introduced in 1966 by Swedish mathematician Herman Wold as an exploratory analysis technique in the field of econometrics [64]. It was specifically designed to help researchers in situations of small, non-normally distributed data sets with numerous but highly correlated explanatory variables. General PLS and all of its variants consist of a set of regression and classification tasks as well as dimension reduction techniques and modeling tools. Sometimes called a "soft" modeling technique, the strength of PLS resides in its relaxation, or "softening" of the distribution, normality, and collinearity restrictions that are inherent in standard multiple linear regression techniques [65,66].

The underlying assumption of all PLS methods is that the observed data are generated by a system or process that is driven by a small number of latent (not directly observed or intuitive) variables. Projection of the observed data to its latent structure by means of PLS is a variation of principal component analysis (PCA). PLS generalizes and combines features from PCA and multiple regression and is similar to canonical correlation analysis in that it can also relate the set of independent variables to a set of multiple dependent response variables and extract latent vectors with maximum correlation [67,68].

The overall goal of PLS processing of laboratory spectral data is the reduction of 2151 variables (bands 350–2500 spectrometer data) down to a manageable number of variables (~100) that have a high probability of significance in a predictive model. PLS regression produces a number of significant factors using a "leave-one-out" cross-validation method [60]. At several stages in the PLS process, diagnostic checks are performed, sometimes graphically, to help isolate variables for deletion in the model that do not have any significant predictive value or are outliers. The end result of a PLS run is a variable importance for projection (VIP) table. The VIP represents the value of each variable in fitting the PLS model for both predictors and responses. The VIP for each factor is defined as the square root of the weighted average times the number of predictors. If a predictor has a relatively small coefficient (in absolute value) and a small value of VIP, then it is a prime candidate for deletion. Variables with VIP values <0.8 and outliers are dropped from the variable list. The VIP table results are then typically divided into four to nine groups. The PLS analysis process is then repeated on the individual groups of variables. Typically, the process is iterated two to five times until a manageable subset of variables can be identified based on the top VIP scores in each group and some a priori knowledge of the process being modeled. PLS can itself be used to construct a predictive model but has some drawbacks. One of the strengths of PLS is its relaxation of collinearity and distribution assumptions, but this also can result in a set of collinear or redundant independent variables. Also, the best combinations of variables are not necessarily reflected in the VIP table values.

In spectral applications, a common practice is to take the final subset of variables and then place them into an SLR model. The stepwise method is a modification of the forward variable selection technique and differs in that variables already in the model do not necessarily stay there. The SLR model computes the F-statistic for each variable and contains parameters for significance levels for variables to *enter* and to *stay* in the model. The SLR process computes all possible combinations of linear variables and ends when none of the variables outside the model has significance (p-value) at or below the *entry* level, and every variable in the model is significant at the *stay* level. Using these sigma-restricted parameterization and general linear model methods, the SLR process simply regresses all possible combinations of input variables and returns the model with the best regression coefficient and the lowest residuals [36,65].

PLS is also used as an exploratory/data mining and analysis tool in remote sensing. As a relatively new technique, the full utilization of PLS is still evolving, but it is clear that it has a major application in several types of spectral, remote sensing analyses, due to the large numbers of potential predictive variables and the highly correlated nature of hyperspectral reflectance and hyperspectral imaging data.

23.7   SUMMARY AND CONCLUSIONS

This chapter has reviewed the hyperspectral applications of detecting the effects on vegetation of heavy metals in the soil. Most of the spectral applications have been in the form of laboratory or field studies with portable spectrometers as opposed to hyperspectral imagery applications. But because field spectrometers and hyperspectral remote sensing (HRS) instruments essentially measure the same phenomenon at high spatial and spectral resolutions, these studies serve as a form of benchmark for airborne and/or spaceborne remote sensing development. However, several studies with airborne and/or spaceborne HRS instruments, such as AVIRIS [40], CASI [55], and HyMAP [36] have been successfully demonstrated, and it is only a matter of time until metal-specific vegetation applications of hyperspectral imagery make their way into the scientific literature.

The metals involved included a wide range of elements, including general heavy metal contamination as might be expected in industrial or urban floodplains [31–35,56] and metal-specific applications such as arsenic [17,36,37,69], lead [41], zinc [55], and selenium [17]. Vegetation targets included general forest canopy, general floodplain, common grasses, and species-specific applications. Hyperspectral methods included standard applications of the NDVI and red edge and newer methods that included vegetation indices (VIs), such as the photochemical reflectance index (PRI), normalized pigment chlorophyll index (NPCI), edge green first derivative normalized difference (EGFN), and a very interesting application of continuum removed band depth at 1730 nm (CR1730) [56,70].

Research on hyperspectral detection of heavy metals and their effects on vegetation is in its infancy. Although much research has been accomplished on other forms of vegetation condition, such as stress and/or agricultural productivity, specific attention to metals is currently a primary scientific gap that needs research attention.

One of the direct needs for hyperspectral research is developing the ability to differentiate metal-induced stress from other types of stress, such as a drought or nutrient stress. Greenhouse experiments, where stress levels are controlled and then measured with a field spectrometer, could be extremely valuable in determining where metal stress can be confidently and uniquely identified in spectra and for establishing underlying mechanisms causing spectral variation. Götze et al. [56] provided a breakthrough in the identification of specific stress agents, and additional work in this area is encouraged.

Further, controlled experiments could be developed to determine if stress from specific metals can be uniquely identified using hyperspectral methods. As various metals interact differently with plant biochemistry and photosynthetic processes, it is feasible that stress patterns due to specific metals could be indentified and utilized effectively. There could also be specific indicator species that identify the presence of metals in the soil, and the development of this line of research would have commercial as well as ecological value.

Additional studies that utilize both field and overhead instruments, and scale-up the spectral responses as a function of spatial scale are needed and represent a critical gap in the current state of the science. Lastly, data-mining efforts such as partial least square (PLS) regression, which systematically considers thousands or even millions of possible band combinations and compute their statistical relevance against a known data set, would be a valuable approach to teasing out very narrow and specific spectral parameters that are not fully understood.

23.8   FUTURE APPLICATIONS

A better understanding of the spectral response to metals in the soil has three primary and valuable applications. First, economic prospecting for metal deposits was one of the early applications and remains just as viable today. Second, metals are often a hindrance to agricultural productivity and a remote method of monitoring their presence would have immediate application to food production throughout much of the world. Third, the problem of fugitive hazardous wastes in the environment is not one that is likely to diminish in the future. As the global population grows, the need for natural resource exploitation will increase dramatically along with the negative side effects of mining, industrial by-products, and both controlled and fugitive wastes. As this review has indicated, there have been numerous successful hyperspectral applications of remote sensing for the location and monitoring of hazardous metals in the environment. Unlike earlier systems, hyperspectral remote sensing (HRS) has the potential to identify specific materials based on molecular structure, and although there is considerable laboratory research, overhead aircraft and satellite remote sensing applications are still in their infancy due to complex atmospheric interferences, cost, and data availability. But all of these factors are steadily improving, and there is opportunity for considerable research in the area of hyperspectral monitoring of metal effects on vegetation.

REFERENCES

1. Duffus, J., Heavy metals—A meaningless term. *Pure and Applied Chemistry* 2002, *74*, 793–807.
2. Alloway, B., *Heavy Metals in Soils*. Springer: New York, 1995.
3. Nieboer, E.; Richardson, D., The replacement of the nondescript term 'heavy metals' by a biologically and chemically significant classification of metal ions. *Environmental Pollution Series B, Chemical and Physical* 1980, *1*, 3–26.
4. Nriagu, J. O., *Arsenic in the Environment: Cycling and Characterization*. John Wiley & Sons, Inc.: New York, 1994, Vol. 1.
5. Albright, R., *Cleanup of Chemical and Explosive Munitions: Locating, Identifying Contaminants, and Planning for Environmental Remediation of Land and Sea Military Ranges and Ordnance Dumpsites*. William Andrew Publishing: New York, 2008, p. 267.
6. Sabins, F., Remote sensing for mineral exploration. *Ore Geology Reviews* 1999, *14*, 157–183.
7. Collins, W., Spectroradiometric detection and mapping of areas enriched in ferric iron minerals using airborne and orbiting instruments: Unpub. PhD dissert., Columbia University, 1978, p. 120, Remote sensing of crop type and maturity: *Photo-Grammetric Engineering and Remote Sensing*, 1976.
8. Collins, W.; Chang, S.; Kuo, J., Detection of hidden mineral deposits by airborne spectral analysis of forest canopies. NASA Contract NSG-5222, Final Report 1981, p. 61.
9. Collins, W.; Chang, S.; Raines, G.; Canney, F.; Ashley, R., Airborne biogeophysical mapping of hidden mineral deposits. *Economic Geology* 1983, *78*, 737.
10. Chang, S.; Collins, W., Confirmation of the airborne biogeophysical mineral exploration technique using laboratory methods. *Economic Geology* 1983, *78*, 723.
11. Küpper, H.; Küpper, F.; Spiller, M., Environmental relevance of heavy metal-substituted chlorophylls using the example of water plants. *Journal of Experimental Botany* 1996, *47*, 259.
12. Barcelo, J.; Poschenrieder, C., Plant water relations as affected by heavy metal stress: A review. *Journal of Plant Nutrition* 1990, *13*, 1–37.
13. Mysliwa-Kurdziel, B.; Prasad, M.; Strzalka, K., Photosynthesis in heavy metal stressed plants. In *Heavy Metal Stress in Plants: From Biomolecules to Ecosystems*. Springer: Berlin, Germany, 2004, p. 198.
14. Rood, J. J., *Modern Chromatics with Application to Art and Industry*. D. Appleton and Company: New York, 1879.
15. Shull, C., A spectrophotometric study of reflection of light from leaf surfaces. *Botanical Gazette* 1929, *87*, 583–607.
16. Willstatter, R.; Stoll, A., Investigations on Chlorophyll, 1913. (Translation by Schertz and Merz.) Science Press: Lancaster, PA, 1928, pp. 290–291.
17. Milton, N.; Ager, C.; Eiswerth, B.; Power, M., Arsenic- and selenium-induced changes in spectral reflectance and morphology of soybean plants. *Remote Sensing of Environment* 1989, *30*, 263–269.
18. Horler, D.; Barber, J.; Barringer, A., Effects of heavy metals on the absorbance and reflectance spectra of plants. *International Journal of Remote Sensing* 1980, *1*, 121–136.
19. Ray, T.; Murray, B.; Chehbouni, A.; Njoku, E., The red edge in arid region vegetation: 340–1060 nm spectra. In *Summaries of the 4th Annual JPL Airborne Geoscience Workshop*. Vol. 1: AVIRIS Workshop, Jet Propulsion Laboratory: Pasadena, CA, 1993, pp. 149–152.

20. Guyot, G.; Baret, F.; Jacquemoud, S., Imaging spectroscopy for vegetation studies. *Imaging Spectroscopy: Fundamentals and Prospective Application*. Kluwer Academic Press: Dordrecht, the Netherlands, 1992, pp. 145–165.
21. Curran, P.; Dungan, J.; Gholz, H., Exploring the relationship between reflectance red edge and chlorophyll content in slash pine. *Tree Physiology* 1990, *7*, 33.
22. Vogelmann, J.; Rock, B.; Moss, D., Red edge spectral measurements from sugar maple leaves. *International Journal of Remote Sensing* 1993, *14*, 1563–1575.
23. Gates, D.; Keegan, H.; Schleter, J.; Weidner, V., Spectral properties of plants. *Applied Optics* 1965, *4*, 11–20.
24. Horler, D.; Dockray, M.; Barber, J., The red edge of plant leaf reflectance. *International Journal of Remote Sensing* 1983, *4*, 273–288.
25. Baret, F.; Champion, I.; Guyot, G.; Podaire, A., Monitoring wheat canopies with a high spectral resolution radiometer. *Remote Sensing of Environment* 1987, *22*, 367–378.
26. Schwaller, M.; Tkach, S., Premature leaf senescence: Remote-sensing detection and utility for geobotanical prospecting. *Economic Geology* 1985, *80*, 250.
27. Milton, N.; Eiswerth, B.; Ager, C., Effect of phosphorus deficiency on spectral reflectance and morphology of soybean plants. *Remote Sensing of Environment* 1991, *36*, 121–127.
28. Rock, B.; Hoshizaki, T.; Miller, J., Comparison of *in situ* and airborne spectral measurements of the blue shift associated with forest decline. *Remote Sensing of Environment* 1988, *24*, 109–127.
29. Küpper, H.; Küpper, F.; Spiller, M., In situ detection of heavy metal substituted chlorophylls in water plants. *Photosynthesis Research* 1998, *58*, 123–133.
30. Küpper, H.; Šetlík, I.; Spiller, M.; Küpper, F.; Prášil, O., Heavy metal-induced inhibition of photosynthesis: Targets of *in vivo* heavy metal chlorophyll formation1. *Journal of Phycology* 2002, *38*, 429–441.
31. Kooistra, L.; Wehrens, R.; Leuven, R.; Buydens, L., Possibilities of visible-near-infrared spectroscopy for the assessment of soil contamination in river floodplains. *Analytica Chimica Acta* 2001, *446*, 97–105.
32. Kooistra, L.; Salas, E.; Clevers, J.; Wehrens, R.; Leuven, R.; Nienhuis, P.; Buydens, L., Exploring field vegetation reflectance as an indicator of soil contamination in river floodplains. *Environmental Pollution* 2004, *127*, 281–290.
33. Kooistra, L.; Wanders, J.; Epema, G.; Leuven, R.; Wehrens, R.; Buydens, L., The potential of field spectroscopy for the assessment of sediment properties in river floodplains. *Analytica Chimica Acta* 2003, *484*, 189–200.
34. Clevers, J.; Kooistra, L., Assessment of heavy metal contamination in river floodplains by using the red-edge index. *Chemical Analysis*. In *Proceedings of the 3rd EARSeL Workshop on Imaging Spectroscopy*, Herrsching, Germany, 13–16, 2003. 2001.
35. Clevers, J.; Kooistra, L.; Salas, E., Study of heavy metal contamination in river floodplains using the red-edge position in spectroscopic data. *International Journal of Remote Sensing* 2004, *25*, 3883–3895.
36. Slonecker, E., *Remote Sensing Investigations of Fugitive Soil Arsenic and Its Effects on Vegetation Reflectance*. George Mason University: Fairfax, VA, 2007.
37. Slonecker, T.; Haack, B.; Price, S., Spectroscopic analysis of arsenic uptake in *Pteris* ferns. *Remote Sensing* 2009, *1*, 644.
38. Rosso, P.; Pushnik, J.; Lay, M.; Ustin, S., Reflectance properties and physiological responses of *Salicornia virginica* to heavy metal and petroleum contamination. *Environmental Pollution* 2005, *137*, 241–252.
39. Gallagher, F.; Pechmann, I.; Bogden, J.; Grabosky, J.; Weis, P., Soil metal concentrations and productivity of Betula populifolia (gray birch) as measured by field spectrometry and incremental annual growth in an abandoned urban Brownfield in New Jersey. *Environmental Pollution* 2008, *156*, 699–706.
40. Mars, J.; Crowley, J., Mapping mine wastes and analyzing areas affected by selenium-rich water runoff in southeast Idaho using AVIRIS imagery and digital elevation data. *Remote Sensing of Environment* 2003, *84*, 422–436.
41. Ren, H.; Zhuang, D.; Pan, J.; Shi, X.; Wang, H., Hyper-spectral remote sensing to monitor vegetation stress. *Journal of Soils and Sediments* 2008, *8*, 323–326.
42. Pearson, R.; Miller, L. Remote mapping of standing crop biomass for estimation of the productivity of the shortgrass prairie. In *Proceedings of the Eighth International Symposium on Remote Sensing of Environment*. Willow Run Laboratories, Environmental Research Institute of Michigan: Ann Arbor, MI, 1972, p. 1355.
43. Carlson, T.; Ripley, D., On the relation between NDVI, fractional vegetation cover, and leaf area index. *Remote Sensing of Environment* 1997, *62*, 241–252.

44. Peñuelas, J.; Gamon, J.; Fredeen, A.; Merino, J.; Field, C., Reflectance indices associated with physiological changes in nitrogen- and water-limited sunflower leaves. *Remote Sensing of Environment* 1994, *48*, 135–146.

45. Gamon, J.; Peñuelas, J.; Field, C., A narrow-waveband spectral index that tracks diurnal changes in photosynthetic efficiency. *Remote Sensing of Environment* 1992, *41*, 35–44.

46. Huete, A., A soil-adjusted vegetation index (SAVI). *Remote Sensing of Environment* 1988, *25*, 295–309.

47. Broge, N.; Leblanc, E., Comparing prediction power and stability of broadband and hyperspectral vegetation indices for estimation of green leaf area index and canopy chlorophyll density. *Remote Sensing of Environment* 2001, *76*, 156–172.

48. Strachan, I.; Pattey, E.; Boisvert, J., Impact of nitrogen and environmental conditions on corn as detected by hyperspectral reflectance. *Remote Sensing of Environment* 2002, *80*, 213–224.

49. Daughtry, C.; Walthall, C.; Kim, M.; De Colstoun, E.; McMurtrey III, J., Estimating corn leaf chlorophyll concentration from leaf and canopy reflectance. *Remote Sensing of Environment* 2000, *74*, 229–239.

50. Elvidge, C.; Chen, Z., Comparison of broad-band and narrow-band red and near-infrared vegetation indices. *Remote Sensing of Environment* 1995, *54*, 38–48.

51. Blackburn, G., Quantifying chlorophylls and carotenoids at leaf and canopy scales: An evaluation of some hyperspectral approaches. *Remote Sensing of Environment* 1998, *66*, 273–285.

52. Blackburn, G., Hyperspectral remote sensing of plant pigments. *Journal of Experimental Botany* 2007, *58*, 855.

53. Thenkabail, P.; Smith, R.; De Pauw, E., Hyperspectral vegetation indices and their relationships with agricultural crop characteristics. *Remote Sensing of Environment* 2000, *71*, 158–182.

54. Thenkabail, P.; Smith, R.; De Pauw, E., Evaluation of narrowband and broadband vegetation indices for determining optimal hyperspectral wavebands for agricultural crop characterization. *Photogrammetric Engineering and Remote Sensing* 2002, *68*, 607–622.

55. Reusen, I.; Bertels, L.; Debruyn, W.; Deronde, B.; Fransaer, D.; Sterckx, S., Species Identification and Stress Detection of Heavy-Metal Contaminated Trees. VITO—Flemish Institute for Technological Research: Mol, Belgium, 2003.

56. Götze, C.; Jung, A.; Merbach, I.; Wennrich, R.; Gläßer, C., Spectrometric analyses in comparison to the physiological condition of heavy metal stressed floodplain vegetation in a standardised experiment. *Central European Journal of Geosciences* 2010, *2*, 132–137.

57. Thorhaug, A.; Richardson, A.; Berlyn, G., Spectral reflectance of *Thalassia testudinum* (Hydrocharitaceae) seagrass: Low salinity effects. *American Journal of Botany* 2006, *93*, 110.

58. Choe, E.; van der Meer, F.; van Ruitenbeek, F.; van der Werff, H.; de Smeth, B.; Kim, K., Mapping of heavy metal pollution in stream sediments using combined geochemistry, field spectroscopy, and hyperspectral remote sensing: A case study of the Rodalquilar mining area, SE Spain. *Remote Sensing of Environment* 2008, *112*, 3222–3233.

59. Dunagan, S.; Gilmore, M.; Varekamp, J., Effects of mercury on visible/near-infrared reflectance spectra of mustard spinach plants (*Brassica rapa* P.). *Environmental Pollution* 2007, *148*, 301–311.

60. Guang-yu, C.; Xin-hui, L.; Su-hong, L.; Zhi-feng, Y., Spectral characteristics of vegetation in environment pollution monitoring [J]. *Environmental Science and Technology* 2005, *1*, 173–178.

61. Horler, D.; Barber, J.; Darch, J.; Ferns, D.; Barringer, A., Approaches to detection of geochemical stress in vegetation. *Advances in Space Research* 1983, *3*, 175–179.

62. Wickham, J.; Chesley, M.; Lancaster, J.; Mouat, D., *Remote Sensing for the Geobotanical and Biogeochemical Assessment of Environmental Contamination*. Desert Research Institute, Nevada University: Reno, NV, 1993, DOE/NV/10845-27.

63. Xia, L.; Shoo-feng, L.; Zheng, L., High spectral resolution data applied to identify plant stress response to heavy metal in mine site [J]. *Science of Surveying and Mapping* 2007, *2*, 1–6.

64. Wold, H., Estimation of principal components and related models by iterative least squares. *Multivariate Analysis* 1966, *1*, 391–420.

65. Tobias, R., *An Introduction to Partial Least Squares Regression*. SAS Institute Inc.: Cary, NC, 1995, pp. 1250–1257.

66. Abdi, H., Partial least squares (PLS) regression. *Encyclopedia of Social Sciences Research Methods*, M. Lewis-Beck, A. Bryman, and T. Futing (eds.). Sage: Thousand Oaks, CA, 2003, pp. 1–7.

67. Höskuldsson, A., PLS regression methods. *Journal of Chemometrics* 1988, *2*, 211–228.

68. Rosipal, R.; Krämer, N., Overview and recent advances in partial least squares. In *Subspace, Latent Structure and Feature Selection: Statistical and Optimization Perspectives Workshop (SLSFS 2005)*, C. Craig Saunders, M. Grobelnik, J. Gunn, and J. Shawe-Taylor (eds.). Springer-Verlag: New York, 2006, pp. 34–51.

69. Sridhar, B.; Han, F.; Diehl, S.; Monts, D.; Su, Y., Spectral reflectance and leaf internal structure changes of barley plants due to phytoextraction of zinc and cadmium. *International Journal of Remote Sensing* 2007, *28*, 1041–1054.

70. Clark, R.; Roush, T., Reflectance spectroscopy: Quantitative analysis techniques for remote sensing applications. *Journal of Geophysical Research* 1984, *89*, 6329–6340.

71. Shaw, B.; Sahu, S.; Mishra, R., Heavy metal induced oxidative damage in terrestrial plants. In *Heavy Metal Stress in Plants—From Biomolecules to Ecosystems*. Springer: Berlin, Germany, 2004; Vol. 2, pp. 84–126.

72. Delhaize, E.; Ryan, P., Aluminum toxicity and tolerance in plants. *Plant Physiology* 1995, *107*, 315.

73. Roy, A.; Sharma, A.; Talukder, G., Some aspects of aluminum toxicity in plants. *The Botanical Review* 1988, *54*, 145–178.

74. Jastrow, J.; Koeppe, D., *Uptake and Effects of Cadmium in Higher Plants*. John Wiley & Sons: New York, 1980, pp. 607–638.

75. Das, P.; Samantaray, S.; Rout, G., Studies on cadmium toxicity in plants: A review. *Environmental Pollution* 1997, *98*, 29–36.

76. Kukkola, E.; Rautio, P.; Huttunen, S., Stress indications in copper- and nickel-exposed Scots pine seedlings. *Environmental and Experimental Botany* 2000, *43*, 197–210.

77. Mocquot, B.; Vangronsveld, J.; Clijsters, H.; Mench, M., Copper toxicity in young maize (*Zea mays* L.) plants: Effects on growth, mineral and chlorophyll contents, and enzyme activities. *Plant and Soil* 1996, *182*, 287–300.

78. Masarovicová, E.; Cicák, A.; Štefan ík, I., Plant responses to air pollution and heavy metal stress. In *Handbook of Plant and Crop Stress*, M. Pessaraki (ed.). Marcel Dekker, Inc.: New York, 1999, pp. 569–598.

79. Sharma, P.; Dubey, R., Lead toxicity in plants. *Brazilian Journal of Plant Physiology* 2005, *17*, 35–52.

80. Patra, M.; Sharma, A., Mercury toxicity in plants. *The Botanical Review* 2000, *66*, 379–422.

81. Khalid, B.; Tinsley, J., Some effects of nickel toxicity on rye grass. *Plant and Soil* 1980, *55*, 139–144.

82. Bonnet, M.; Camares, O.; Veisseire, P., Effects of zinc and influence of *Acremonium lolii* on growth parameters, chlorophyll *a* fluorescence and antioxidant enzyme activities of ryegrass (*Lolium perenne* L. cv Apollo). *Journal of Experimental Botany* 2000, *51*, 945.

83. Gitelson, A.; Merzlyak, M.; Chivkunova, O., Optical properties and nondestructive estimation of anthocyanin content in plant leaves. *Photochemistry and Photobiology* 2001, *74*, 38–45.

84. Richardson, A.; Wiegand, C., Distinguishing vegetation from soil background information (by gray mapping of Landsat MSS data). *Photogrammetric Engineering and Remote Sensing* 1977, *43*, 1541–1552.

85. Haboudane, D.; Miller, J.; Pattey, E.; Zarco-Tejada, P.; Strachan, I., Hyperspectral vegetation indices and novel algorithms for predicting green LAI of crop canopies: Modeling and validation in the context of precision agriculture. *Remote Sensing of Environment* 2004, *90*, 337–352.

86. Hunt, Jr., E.; Rock, B., Detection of changes in leaf water content using near- and middle-infrared reflectances. *Remote Sensing of Environment* 1989, *30*, 43–54.

87. Rouse, J., Monitoring vegetation systems in the Great Plains with ERTS. In *3rd ERTS-1 Symposium*. NASA Goddard Space Flight Center: Greenbelt, MD, 1974.

88. Sims, D.; Gamon, J., Relationships between leaf pigment content and spectral reflectance across a wide range of species, leaf structures and developmental stages. *Remote Sensing of Environment* 2002, *81*, 337–354.

89. Curran, P.; Dungan, J.; Macler, B.; Plummer, S., The effect of a red leaf pigment on the relationship between red edge and chlorophyll concentration. *Remote Sensing of Environment* 1991, *35*, 69–76.

90. Merton, R. Monitoring community hysteresis using spectral shift analysis and the red-edge vegetation stress index. In *Proceedings of the Seventh Annual JPL Airborne Earth Science Workshop*. NASA, Jet Propulsion Laboratory: Pasadena, CA, 1998.

91. Boluda, R.; Andreu, V.; Gilabert, M.; Sobrino, P., Relation between reflectance of rice crop and indices of pollution by heavy metals in soils of albufera natural park (Valencia, Spain). *Soil Technology* 1993, *6*, 351–363.

24 Hyperspectral Narrowbands and Their Indices on Assessing Nitrogen Contents of Cotton Crop Applications

Jianlong Li, Cherry Li, Dehua Zhao, and Chengcheng Gang

CONTENTS

24.1 INTRODUCTION

The use of remote sensing for precision agriculture applications is very popular and a popular use is the characterization of nitrogen (N) conditions. N is one of the most important fertilizer elements for crop production, and is an essential element for crop growth, development, and yield formulation. Nitrogen content deficiency will bring a series of changes to crop leaves, such as color, thickness, water content, form, and structure. A lack of nitrogen will directly affect the composition of amino acid, protein, nucleic acid, and other materials, which will lead to the reduction of crops photosynthesis capacity and the final yields. Therefore, N management is a very important management measure in agriculture produce for obtaining high yields and good quality. At the same time, overapplication of nitrogen will pollute underground water and also get into streams. To overcome this and other similar problems, precision agriculture (also referred to as precision farming or site-specific farming) has been put forward. Precision N management is a key content of precision agriculture, which reduces pollution of the water resources and yet results in sustained high yields over space and time.

Given the importance of nitrogen in crop growth and yield as well as the need to maintain environmentally acceptable levels of N application, many studies in precision farming are focused on N application rate and timing for high yield, crop quality, and pollution control [1,2]. Under normal conditions, N fertilizer influences chlorophyll concentration in green leaves. Since chlorophyll, a key indicator of crop physiological status, has a strong absorbance peak in the red spectral region, empirical models of predicting chlorophyll status from spectral reflectance are largely based on red spectra [3–5]. Arrangement of cells within the mesophyll is affected with chlorophyll status by N supply and is an important factor determining canopy near-infrared (NIR) reflectance [6–8].

Traditional methods to determine plant tissue nutrient concentrations in a laboratory can be time-consuming and costly. Remote sensing has a great potential to assess and manage timely crop stress affected by environment from leaf to landscape scales of crop physiology [9,10]. Recent studies have realized this potential and found close relationships between plant physiological parameters and spectral reflectance. Several studies have documented that N status of field crops can be assessed by spectral reflectance data of crops' leaf or canopy.

Red–NIR-based vegetation indices (VIs) could also be used to estimate crop N stress [11–13]. The red and NIR reflectance data, used to generate the popular VIs and monitor crop growth conditions, are acquired from two kinds of sensors—broadband and narrowband sensors. Broadband spectral reflectance, currently the popular remotely sensed data, is obtained from the current generation of earth-orbiting satellites carrying multispectral sensors such as advanced spaceborne thermal emission and reflection radiometer (ASTER), moderate resolution imaging spectroradiometer (MODIS), and landsat-7 enhanced thematic mapper plus (ETM+). Most of these sensors have several channels among which the red and NIR are the most popular bands. Narrowband spectral data used to monitor crop condition are generated from imaging sensors such as airborne visible/infrared imaging spectrometer (AVIRIS) and Compact High Resolution Imaging Spectrometer (CHRIS) Project for on Board Autonomy (PROBA).

Hyperspectral narrowband remote sensing provides a perfect opportunity to characterize and advance the study of N content in plants [2,13]. A number of authors [15] have discussed the superiority of narrowband/hyperspectral imaging sensors over broadband/multispectral instruments. Multispectral imaging sensors gather spectral data in large, noncontiguous ranges of the electromagnetic spectrum; thus, a single band represents the average of a relatively large portion of the spectrum. When comparing the predictive powers and stability of broadband and narrowband VIs for deriving crop growth variables, there are some other opinions [3,14]. Besides the indicators of crop growth variables, VIs were applied to detect N stress [13,15]. The selection of optimum wavebands in hyperspectral data has focused mainly on how to improve the correlation between VIs and crop biophysical/biochemical variables. But few studies have been focused on how to increase the sensitivity of the VIs to N stress.

Hence, this study was to analyze hyperspectral remote sensing capability in detecting characteristic difference of cotton crops under different N application rates and different growing stages. In order to address the overall objective, three specific sub-objectives were identified, which were to

1. Identify sensitive hyperspectral wavelengths to different N treatment
2. Evaluate if the continuum-removal method improves the ability to recognize different N status in the spectral wavebands of absorbing chlorophyll (550–750 nm) at full green canopy coverage period
3. Test canonical, discriminant analysis

24.2   MATERIALS AND METHODS

To enhance the red–NIR-based VIs to green vegetation spectral signals and to reduce external effects such as noise-related soil and atmospheric influences, many VIs have been developed in the past three decades. These VIs can be divided into four groups: (1) Ratio-based VIs based on the ratio between red and NIR reflectance. The normalized difference vegetation index (NDVI) and ratio vegetation index (RVI) are the most commonly used ratio-based VIs. (2) Orthogonal VIs defined by a line in spectral space for identification at bare soils. The transformed soil-adjusted vegetation index (TSAVI), second soil-adjusted vegetation index (SAVI2), and modified second soil-adjusted vegetation index (MSAVI2) are examples of orthogonal VIs. (3) Derivative VIs: First and second-order derivative green VIs. (4) Atmospheric corrected indices, such as the visible atmospherically resistant index (VARI) [4]. These VIs have been shown to be quantitatively and functionally related with canopy parameters such as the leaf area index (LAI), aboveground biomass, and chlorophyll content (CC) and vegetation fraction. Research results indicate that these VIs have potential applications in agriculture for forecasting and estimating crop productions, monitoring crop conditions, classifying and mapping crops, and directing precision farming activities [6,13,14].

24.2.1 Experiment Designs and Treatments

Experimental 1 (E1): A completely randomized design experiment containing three replicates was conducted in a cotton (Gossypium hirsutum L. cv. Sumian 12) field at Zhejiang University, Zhejiang Province, China (30°42 N, 120°102 E). Treatments included three N application rates of 0, 60, and 120 kg N ha^{-1} (termed LN, MN, and HN, respectively). The soil of experiment field is sandy soil, which contains 0.95 g kg^{-1} total-N, 148.5 mg kg^{-1} available-N, 1.21 g kg^{-1} available-P, 72.7 mg kg^{-1} available-K and 9.96 g kg^{-1} organic matters. Each sampling plot consisted of two rows of 0.3 m apart, 3.7 m wide, and 5.0 m long (3.7 × 5.0 = 17.5 m^2) with a density of 50,000 plants/ha. Cotton was sown on April 29 directly in fields with north/south row orientation. Phosphorous and potassium fertilizers were supplied in adequate amounts according to the general nutrient status of the field as determined by soil samples: 80 kg ha^{-1} P$_2$O$_5$ and 160 kg ha^{-1} K$_2$O.

Experimental 2 (E2): A completely randomized design experiment containing three replicates was conducted in a cotton (Gossypium hirsutum L. cv. Sumian 3) field at Zhangjiagang, Jiangsu Province, China (31°502 N, 120°492 E). The soil was of sandy texture, which contains 41.6 mg kg^{-1} available-N, 47.2 mg kg^{-1} total-P, 63.9 mg kg^{-1} total-K, and 13.2 g kg^{-1} organic matters. Treatments included three N application rates of 90, 180, and 360 kg N ha^{-1} (termed LN, MN, and HN, respectively). Each sampling plot consisted of two rows of 0.8 m apart, 0.4 m wide, and 14 m long (2.4 × 14 = 33.6 m^2) with a density of 45,000 plants/ha. Cotton was sown on April 12 in greenhouses and later transplanted on May 28 to fields with north/south row orientation. Phosphorous and potassium fertilizers were also supplied: 180 kg ha^{-1} P$_2$O$_5$ and 240 kg ha^{-1} K$_2$O. Irrigation was not used due to the high rainfall (above 1200 mm) and high ground water table of the soil at the study site.

24.2.2 Observed Dates

In accordance with canopy structure and leaf function of cotton plants, the cotton growth cycle was divided into three stages: (1) rapid growth period (early stage when the soil was partially covered by cotton, and, therefore, its contribution to spectral signals was significant), (2) full green canopy coverage period (middle stage when the canopy reached almost 100% cover), and (3) senescent period (late stage when cotton boll opened and part of the leaves were senesced). Timing of growth stages corresponded to sampling dates of July 15, August 14, and October 1, 2002 in experiment 1 and of July 12, August 22, and September 29, 2002 in experiment 2 respectively, when agronomic and hyperspectral data were collected.

24.2.3 Canopy Hyperspectral Reflectance Measurements

A 512-channel spectroradiometer (300–1100 nm) by Analytical Spectral Devices™ (FieldSpec FR) was used to acquire cotton canopy spectral data. Noise at both ends of the spectrum limited the useful data range between 400 and 1000 nm for the analyses. Data were colleted on cloudless days with solar elevations ranging from 50° to 55° to minimize external effects from atmospheric conditions and changes in solar position. Prior to the cotton planting, spectral reflectance measurements of the bare soil surface were made. Spectral reflectance was calculated as the ratio of measured radiance to radiance from a white standard reference panel. Immediately after the white standard radiance measurement, two spectra of the cotton canopy were obtained—one with the sensor located directly over the center of two rows on a ridge, and the other with the sensor located directly over the furrow. Then the two spectra were averaged to represent a single mean field spectrum of ridge. The measurements were repeated 10 times for each plot.

For experiment 1 (E1), reflectance measurements were obtained three times on July 12, August 22, and September 29 by the spectroradiometer with 25° field of view and 1.0 m nadir orientation above the canopy which resulted in a sensor field of view of 45 cm diameter. For experiment 2 (E2), reflectance measurements were obtained three times on July 15, August 14, and December 1 by the spectroradiometer with 15° field of view and 2.3 m nadir orientation above the canopy which resulting in a sensor field of view of 60 cm diameter.

24.2.4 BIOMASS MEASUREMENTS

After spectral measurement, 10 cotton samples were selected to analyze biophysical variables immediately in the same place. The samples were dried at 70°C in an oven for 48 h to constant weight, and then acquired dry weight biomass was determined.

24.2.5 AGRONOMIC VARIABLE MEASUREMENTS

Six cotton plants were harvested on the same days that the canopy spectral measurements were made. Each plant was separated into leaves, branches, and stems and then weighed for leaf biomass calculations (g m^{-2}). The green leaves from two plants (thus decreasing the workload) were measured with a leaf area meter (CI-203, CID) to estimate the total leaf area per sample plot (1.333 m^2) in order to calculate LAI. The LAI of cotton was computed as the ratio of green leaf area per sampled area (m^2 m^{-2}). CC was measured from 0.15 g leaf samples that were ground in 3 mL cold acetone/Tris buffer solution (80:20 Vol/Vol, pH = 7.8), centrifuged to remove particulates, and the supernatant diluted to a final volume of 15 mL with additional acetone/Tris buffer. The absorbance of the extract solutions was measured with a U-3000 spectrophotometer at 663, 647, and 537 nm. The chlorophyll concentration was calculated using the following statistically derived equations [5]:

$$Chla = 0.01373A663 - 0.000897A537 - 0.0030464A647$$

$$Chlb = 0.120405A647 - 0.004305A537 - 0.005507A663$$

where Ax was the absorbance of the extract solution in a 1 cm path length cuvette at wavelength x. The units for all the equations are micromoles per milliliter (μ mol mL^{-1}). Canopy chlorophyll density (g m^{-2}) was computed by multiplying CC by total leaf weights.

24.2.6 DATA PROCESS AND ANALYSIS

The research hypothesis was whether the mean values of the reflectance between the three treatments were significantly different at each wavelength. This was tested using one-way analysis of variance (ANOVA), and it was concluded that there are differences between the groups. The statistical tests were done at different time periods (rapid growth period, full green canopy coverage period, senescent period) in order to assess the spectral differences between treatments at different stages of the plants' physiological status. Specially, we tested the utility of the visible absorption feature (R550–R750 nm) to discriminate different levels of N concentration after continuum removal. This red absorption feature was selected since it has consistently proved to be an indicator of vegetation condition and was not affected by water absorption in fresh plants. This was in contrast to the mid-infrared bands where chemical absorption is largely masked by water. Continuum removal normalizes reflectance spectra to allow comparison of individual absorption features from a common baseline [16]. The continuum is a convex hull fitted over the top of a spectrum utilizing straight-line segments that connect local spectra maxima. The continuum is calculated by dividing the reflectance value for each point in the absorption pit by the reflectance level of the continuum line (convex hull) at the corresponding wavelength. The first and last spectral data values are on the hull and therefore the first and last bands in the output continuum-removed data file are equal to 1. The output curves have values between 0 and 1, in which the absorption pits are enhanced and the absolute variance removed [17].

Initially, this method was applied to identify mineral components in geology, then applied to vegetation science by Kokaly [16] to analyze chemical component of several plant dry leaves. In recent years, continuum-removal measure has been applied to vegetation canopy for measuring biochemical content of plants [8].

One-way ANOVA method is often used to assess significant degree of the spectral reflectance difference between different N treatment during four growth stages with statistical product and service solutions (SPSS11.0) software.

24.3 RESULTS AND ANALYSIS

24.3.1 BIOMASS ANALYSIS UNDER DIFFERENT NITROGEN TREATMENTS

Biomass is an important agriculture parameter in crop canopy structure reflectance. In this study, dry weight of aboveground biomass (DWAB) was selected as the assessment standard. As expected, different N treatments resulted in significantly different DWABs in experiments 1 and 2 (Figure 24.1). In general, nitrogen increased cotton DWAB. For these two experiments, DWAB experienced significant difference between LN- and HN-treatments at their respective three observed dates ($p < 0.01$).

24.3.2 DIFFERENCE OF CANOPY SPECTRAL REFLECTANCE
UNDER DIFFERENT NITROGEN TREATMENTS

Figures 24.2 and 24.3 were canopy reflectance spectra and results of one-way ANOVA of canopy reflectance among three N treatments at different wavelengths. In general, canopy spectral reflectance showed no significant difference between N treatments in visible light. But in NIR regions, canopy spectral reflectance showed significant difference between N treatments ($p < 0.05$).

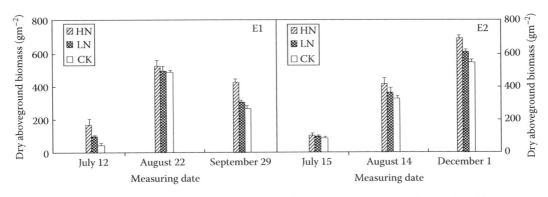

FIGURE 24.1 The variation of cotton biomass under different treatments at experiments 1 and 2.

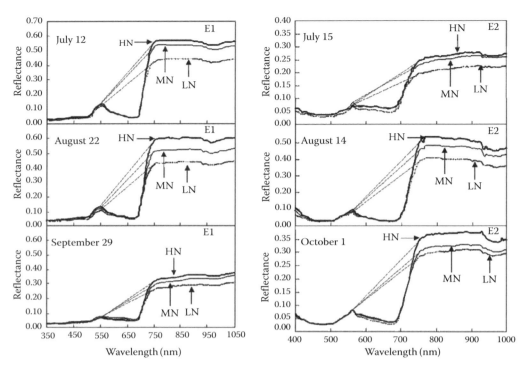

FIGURE 24.2 Mean canopy spectral reflectance with continuum-removed line under different treatments.

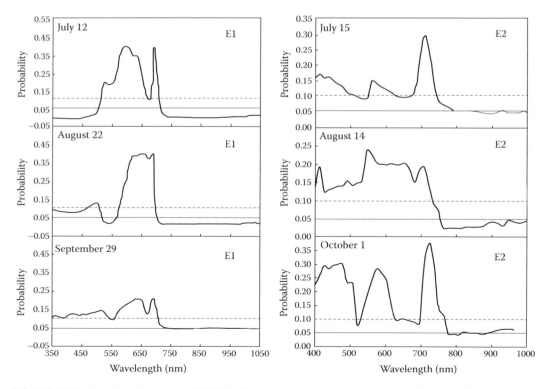

FIGURE 24.3 Results of one-way ANOVA of cotton canopy reflectance among the three N treatments at different wavelengths.

Figure 24.2 show that canopy reflectance spectra under HN, MN, and LN treatments decreased in turn in NIR region, especially in full green canopy coverage period, which presented a significant difference between different N treatments. This vigorous growth stage is not only nutrition-driven growth but also the development growth of young cotton buds and bolls. In the senescent period, all of canopy spectra reflectance under three different N applications were lower and spectral reflectance curve presents identical trend in different N treatments, because most cotton leaves had been defoliated and only a few withered leaves and unpicked bolls were left.

24.3.3 Changes of Normalized Difference Spectra Characteristic

Figures 24.3 and 24.4 were continuum-removed mean reflectance spectra and results of one-way ANOVA of continuum-removed reflectance among three N treatments at the wavelengths of chlorophyll maximal absorptance (550–750 nm) in experiments 1 and 2. Results suggested that the difference of spectral reflectance between 550 and 750 nm was improved by using continuum-removed technology. Nitrogen could enlarge spectral vale between 550 and 750 nm especially in the full green canopy coverage period. One-way ANOVA results showed that there always existed some wavelengths at which the canopy reflectance showed significant difference between N treatments ($p < 0.05$). The most sensitive reflectance to N rate was located at two sides of the chlorophyll maximal absorptance (680 nm).

24.3.4 Multiple Variable Comparison Analysis under Different Treatments

Most of the red–NIR VIs were established for the purpose of estimating plant biophysical/biochemical variables, and have been related to crop variables such as biomass, LAI, and chlorophyll [3,5,6,14]. In general, since N conditions result in a significant variation in these

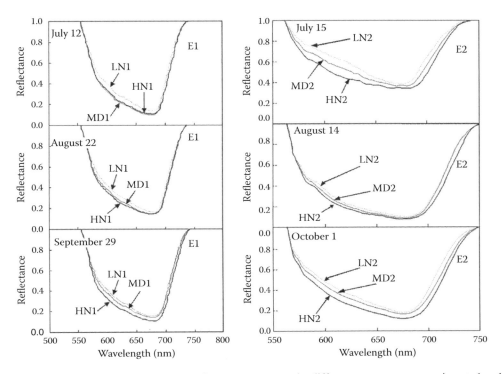

FIGURE 24.4 Continuum-removed mean reflectance spectra under different treatments at experiments 1 and 2.

TABLE 24.1

Multiple Comparisons of Mean Values of Three Cotton Variables Observed on July 15, August 14, and October 1 and Seed Cotton Yield under Different Nitrogen Treatments (at 95% Confidence Level)

| Variables | Treatments | Means[a] | | | Seed Cotton Yield (kg h m^{-2}) |
		July 14	August 15	October 1	
LAI (m^2 m^{-2})	N0	0.90a	2.23a	1.34a	
	N90	0.97ab	2.75b	1.92b	
	N180	1.06bc	3.12c	2.37c	
N360		1.12c	3.25c	2.33c	
CC (%)	N30	1.25a	0.85a	0.53a	
	N90	1.31ab	1.23b	1.07b	
	N180	1.37ab	1.37bc	1.26c	
	N360	1.41b	1.52c	1.37c	
Aboveground dry	N0	118.7a	424.8a	636.2a	
Biomass (g m^{-2})	N90	130.2ab	488.7b	816.9b	
	N180	139.4bc	548.6bc	911.7c	
	N360	153.3c	625.2c	1032.3d	
Seed cotton	N0				2908.5a
Yield (kg h m^{-2})	N90				3914.9b
	N180				4474.9c
	N360				4592.3c

[a] Means within columns followed by the same letter (a–d) are not significantly different based on ANOVA at 95% confidence level ($p \leq 0.05$).

variables, which has been proved by this experiment, it is also possible to discriminate canopies grown under different N treatments using these VIs [8,13].

As expected, the N fertilizer treatments resulted in broad variations in the three variables (Table 24.1). Generally, variable values increased with the N application rates. With cotton growth, the differences between N treatments were greater. Multiple comparison analysis was used to test if the mean values of the three variables were significantly different between N treatments. The results showed that the differences between N1 (no nitrogen applied) and N360 treatments (the highest N rate) were significant ($p < 0.05$). At middle and late growth stages, the N90 treatment also was significantly different from other treatments ($p < 0.05$). At late stages, each difference in above ground dry biomass (ADB) between two N treatments was significant at the 95% level.

24.4   DISCUSSIONS

The earlier mentioned results showed that the N nutrition difference of crops greatly affected the canopy spectral reflectance (Figures 24.3 and 24.4). In the visible portion of the spectrum, around 550 nm, showed obvious difference tendency on statistics, especially at the senescent period, due to the absorption difference of chlorophyll, which was caused by different N application, and led to the difference of canopy spectral reflectance between different treatments due to the close positive linear correlation between nitrogen and chlorophyll.

In this study, canopy reflectance showed a stable and significant difference in NIR spectra region, which is sustainable from rapid growth period to senescent period. The spectra reflectance of plant in the NIR is mainly affected by leaf and canopy structures [6]. The biomass and LAI of plant is

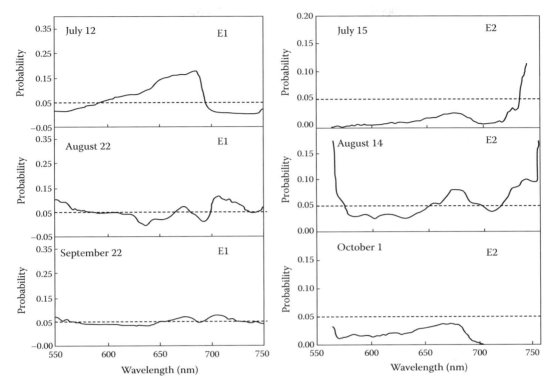

FIGURE 24.5 Results of one-way ANOVA showing wavelengths where continuum-removed reflectance differences between the three treatments are significant. Horizontal dashed lines showing 95% confidence limits.

added with the increasing of N amount, which leads to the evident difference of reflection spectra in different treatments. This study also shows that in NIR spectra region the spectral difference between different N applications could reach a 5% significant level. The reflected characteristic in NIR spectra provided a possibility for distinguishing the nutrition condition of plant nitrogen in different N treatments.

The study results of absorption characteristic in red valley absorption area between 550 and 750 nm in normalized difference method shows that the width and depth of absorption characteristic will increase with N application rate increasing. In general, it is believed that the absorption center of the chlorophyll is 680 nm. This study on two experiments showed the difference of spectra statistics at 680 nm is not significant; the best spectra difference among N treatments existed on the absorption valley side of slope (Figures 24.4 and 24.5). The reason might be that higher level of CCs made the red valley wider. Therefore, the normalized difference technology can improve the difference of reflected spectra in visible light region under different N treatments; the best-distinguished spectra band existed around the 620–640 and 690–710 nm. The study results confirmed the conclusions of [8,9] which hold that the sensitive spectra area exists at the 535–640 and 685–700 nm. The reason that the normalized difference method improved the difference of reflected spectra in different N treatments might be that this method can eliminate the absolute error by reflectance and can strengthen the absorption valley. Normalized difference method can successfully identify the characteristic of cotton canopies without being affected by spectra reflectance of cotton canopy structure and background. The index and the ratio operations also potentially reduced systematic atmospheric influences. This research result has an important significance to evaluated N content in visible spectra area.

24.5 CONCLUSIONS

Increasing the nitrogen (N) application will increase the content of chlorophyll and biomass of crops, which typically decreases the spectral reflectance in the visible light region. But this study has shown that the differences of canopy spectral reflectance were not stable in visible light and are dependent on factors such as crop growth stages and soil background effects. The difference in reflectance, for example, between 10% difference level from spectral band 660 and 680 nm is as high as 10% or greater during the rapid growth period as a result of significant soil background effects (as a result of <100% canopy cover) or during senescent period, as a result of fast loosing chlorophyll (lower reflectance or higher absorption associated with healthier plants with greater canopy cover, plant moisture, and biomass). But in the full green canopy coverage period the difference was not significant because the canopy coverage in three different treatments could all reach almost 100% cover and show no obvious difference between each other. These factors highlight the need to use specific narrowbands from targeted portions of the spectrum to better characterize and study vegetation.

Overall, the results of this study offer a possibility to estimate cotton canopy quality at the field level. The results trigger the need to investigate band depths and slopes, particularly the red edge, and to estimate cotton nutrition quality at canopy level. The development of models that can manipulate the influence of factors such as the atmosphere, species mix, and non-photosynthetic vegetation (standing litter, woody stems, etc.) at different times will be important for hyperspectral remote sensing of cotton quality.

REFERENCES

1. Weisz, R., Crozier, C.R., and Heiniger, R.W., 2001. Optimizing nitrogen application timing in no-till soft red winter wheat. *Agron. J.* 93, 435–442.
2. Pattey, E., Strachan, I.B., Boisvert, J.B., Desjardins, R.L., and McLaughlin, N.B., 2001. Detecting effects of nitrogen rate and weather on corn growth using micrometeorological and hyperspectral reflectance measurements. *Agric. Forest Meteorol.* 108, 85–99.
3. Broge, N.H. and Leblanc, E., 2001. Comparing prediction power and stability of broad-band and hyperspectral vegetation indices for estimation of green leaf area index and canopy chlorophyll density. *Remote Sens. Environ.* 76, 156–172.
4. Gitelson, A.A., Kaufman, Y.J., Stark, R., and Rundquist, D., 2002. Novel algorithms for remote estimation of vegetation fraction. *Remote Sens. Environ.* 80, 76–87.
5. Sims, D.A. and Gamon, J.A., 2002. Relationships between leaf pigment content and spectral reflectance across a wide range of species, leaf structures and developmental stages. *Remote Sens. Environ.* 81, 331–354; Smil, V., 1997. Global population and nitrogen cycle. *Sci. Am.* 277, 76–81.
6. Serrano, L., Filella, L., and Peñuelas, J., 2000. Remote sensing of biomass and yield of winter wheat under different nitrogen supplies. *Crop Sci.* 40, 723–731.
7. Kumar, L., Schmidt, K.S., Dury, S., and Skidmore, A.K., 2001. Imaging spectrometry and vegetation science. In: Van der Meer, F., De Jong, S.M. (Eds.), *Imaging Spectrometry*. Kluwer Academic Publishers, Dordrecht, the Netherlands, pp. 111–155.
8. Mutanga, O., Skidmore, A.K., and Wieren, S., 2003. Discrimination tropical grass (*Cenchrus ciliaris*) canopies grown under different nitrogen treatments using spectroradiometry. *J. Photogram. Remote Sens.* 57, 263–272.
9. Daughtry, C.S.T., Walthall, C.L., Kim, M.S., Brown de Colstoun, E., and McMurtrey III, J.E., 2000. Estimating corn leaf chlorophyll concentration from leaf and canopy reflectance. *Remote Sens. Environ.* 74, 229–239.
10. Zarco-Tejada, P.J., Miller, J.R., Mohammed, G.H., Noland, T.L., and Sampson, P.H., 2000. Chlorophyll fluorescence effects on vegetation apparent reflectance: II. Laboratory and airborne canopy-level measurements with hyperspectral data. *Romote Sens. Environ.* 74, 596–608.
11. Boegh, E., Soegaarda, H., Broge, N., Hasagerc, C.B., Jensenc, N.O., Scheldeb, K., and Thomsen, A., 2002. Airborne multispectral data for quantifying leaf area index, nitrogen concentration, and photosynthetic efficiency in agriculture. *Remote Sens. Environ.* 81, 179–193.

12. Hansen, P.M. and Schjoerring, J.K., 2003. Reflectance measurement of canopy biomass and nitrogen status in wheat crops using normalized difference vegetation indices and partial least squares regression. *Remote Sens. Environ.* 86, 542–553.

13. Strachan, I.B., Pattey, E., and Boisvert, J.B., 2002. Impact of nitrogen and environment conditions on corn as detected by hyperspectral reflectance. *Remote Sens. Environ.* 80, 213–224.

14. Broge, N.H. and Mortensen, J.V., 2002. Deriving crop area index and canopy chlorophyll density of winter wheat from spectral reflectance data. *Remote Sens. Environ.* 81, 45–57.

15. Craig, J.C., 2001. Multi-scale remote sensing techniques for vegetation stress detection. PhD dissertation. University of Florida, Gainesville, Florida; Datt, B., 1998. Remote sensing of chlorophyll a, chlorophyll b, chlorophyll a + b, and total carotenoid content in Eucalyptus leaves. *Remote Sens. Environ.* 66, 111–121.

16. Kokaly, R.F., 2001. Investigating a physical basis for spectroscopic estimates of leaf nitrogen concentration. *Remote Sens. Environ.* 75, 153–161.

17. Schmidt, K.S. and Skidmore, A.K., 2001. Exploring spectral discrimination of vegetation types in African rangelands. *Int. J. Remote Sens.* 22, 3421–3434.

25 Using Hyperspectral Data in Precision Farming Applications

Haibo Yao, Lie Tang, Lei Tian, Robert L. Brown,
Deepak Bhatnagar, and Thomas E. Cleveland

CONTENTS

25.1 INTRODUCTION

25.1.1 PRECISION FARMING

Rather than managing the entire field as one uniform unit, a crop field can be handled site specifically based on local field needs. This is the concept behind using precision agriculture for in-field variability management. The goals of precision agriculture can be described as the following and are based on economic, productivity, and environmental considerations:

- Greater yield than traditional farming with the same amount of input
- The same yield with reduced input
- Greater yield than traditional farming with reduced input

The precision agriculture concept has drawn much attention from farmers and researchers all over the world [1,2]. A complete precision agriculture system can be described in terms of four indispensable parts. They are (1) field variability sensing and information extraction, (2) decision making, (3) precision field control, and (4) operation and result assessment. The success of any precision agriculture system depends on the correct implementation of these four parts. Among the four parts, the decision making is the central component [3]. The decision making process involves making the right management decisions based on the variability information derived from data collected in the crop field.

To make correct decisions, the most important step is to obtain the right information about in-field variabilities. Agricultural engineers put a significant amount of effort into field variability sensing and information extraction, as well as into precision field control and operation. Sensing and information extraction are crucial parts, which require obtaining desired information at the right location at the right time. Sensing and information extraction involve using various sensors to capture data associated with field conditions. Once the raw data are obtained, appropriate algorithms can be used to extract the field information. Remote sensing from either close distance (ground) or remote distance such as from airborne or spaceborne sensors is an import source for field data acquisition [4,5]. Agricultural remote sensing typically uses surface reflectance information in the visible and near infrared (NIR) regions of the electromagnetic spectrum. It provides a fast and economical way to acquire detailed field data in a short period of time. Remote sensing has thus been used in a broad range of applications in the farm industry.

25.1.2 HYPERSPECTRAL DATA

Traditionally, agricultural remote sensing used multispectral imagery. With advances in sensor technology over the past two decades, the introduction of hyperspectral remote sensing imagery to agriculture provided more opportunities for field level information extraction. Figure 25.1 presents a system approach to using hyperspectral imagery for precision agriculture applications. A hyperspectral image has more bands (tens to hundreds or even thousands) with a narrow bandwidth (one to several nanometers) in the same spectral range (e.g., 400–2500nm) as a multispectral image. When presenting hyperspectral imagery, each pixel within the image is typically described as a data vector and the entire image as an image cube. Due to the high data volume of a hyperspectral image, it is expected that hyperspectral imagery could potentially provide more information for precision agriculture. On the other hand, the increased number of data dimensions in a hyperspectral image also increases complexity in image processing and might impact the accuracies. One example of the influence of data dimensionality on accuracy is the Hughes phenomenon [6], which shows that classification accuracy decreases as data dimensions increase, especially when a large number of

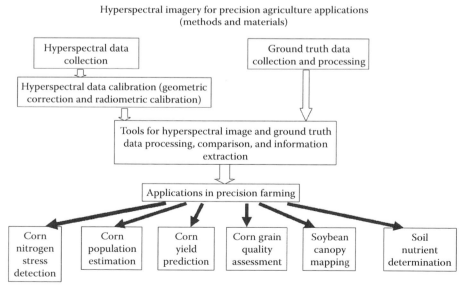

FIGURE 25.1 System diagram of using hyperspectral data in precision farming applications. (Adapted from Yao, H., Hyperspectral imagery for precision agriculture, PhD dissertation, University of Illinois at Urbana-Champaign, 2004.)

wavebands are involved. In order to reduce image-processing complexity and to increase image-interpretation accuracy, it is desirable to reduce the original image dimensionality through a feature reduction process.

There are two major types of feature reduction methods [7]. One is feature selection and the other is feature extraction. The purpose of feature selection is to remove the least effective features (image bands) and select the most effective features. Feature selection is an evaluation of the existing set of features for the hyperspectral image in order to select the most discriminating features and discard the rest. Feature extraction involves transforming the pixel vector into a new set of coordinates in which the basis for feature selection is more evident. Common feature extraction techniques used in remote sensing include the linear combination of image bands such as in principal component transformation and canonical analysis, and arithmetic transformation such as vegetation indices. Vegetation indices and their application in precision farming will be discussed later in this chapter.

25.2 APPLICATIONS OF HYPERSPECTRAL DATA IN PRECISION AGRICULTURE

25.2.1 Precision Farming Management Considerations

Crop production in the agricultural industry has relied heavily on the development and implementation of technologies. Crop yield can be regarded as the single most important output from the crop production systems. Other aspects related to crop production such as field topography, soil characteristics and fertility, tillage practice, fertilizer application, crop rotation, seeding, weed and pest control, irrigation, and weather can all be regarded as inputs for the crop production systems. Remote sensing provides field variability information of the manageable inputs in a map-driven approach for precision farming practices. For example, one of the most successful precision farming technologies is variable rate technology [2]. The map-driven approach would provide a prescription map based on field variability measured by remote sensing. Subsequent variable rate application of fertilizer, herbicide, or other agriculture chemicals can be implemented using the prescription map. In this process, the use of GPS (Global Positioning System) and GIS (Geographic Information System) is also necessary. Additionally, the concept of "management zone" is an important topic in precision farming. The management zones are smaller sections of a large field where the field properties of interest are regarded as relatively homogeneous. Remote sensing has proven to be quite a useful tool for management zone delineation.

The following section first discusses the spatial, spectral, and temporal aspects of using hyperspectral imagery in precision farming. The second part is a discussion of hyperspectral vegetation indices. Lastly, applications of hyperspectral data in precision farming, including weed mapping and control, soil property and fertility sensing, crop nitrogen (N) stress detection, crop yield estimation, and insect/pest infestation identification, are discussed.

25.2.2 Spatial, Spectral, and Temporal Considerations

There are three issues related to using hyperspectral remote sensing imagery in agricultural applications. They are the spatial, spectral, and temporal issues of the image. One of the advantages of airborne or spaceborne remote sensing images is the large spatial coverage. Aerial or space remote sensing data can cover a large area in a short period of time. Thus, they can provide a fast, accurate, and economical method for precision applications. Spatial resolution is another important factor, which varies drastically depending on the sensor platform. For proper data interpretation, the spatial resolution of remote sensing data and ground truth should be matched. Sometimes spatial resampling on one data is necessary to meet this requirement.

The second issue is image spectral range and resolution. The spectral range normally is in the visible NIR region from 400 to 1000 nm. This is the region where plants show distinct spectral signatures under different conditions. Some applications such as soil characterizations extend the spectral region to short wave infrared such as from 1000 to 2500 nm. For spectral resolution, agricultural remote sensing has traditionally used multispectral images with a spectral resolution (or bandwidth) of several hundred nanometers. Multispectral images are sometimes called broadband images, with each broadband covering a specific wavelength range such as blue, green, red, or NIR. On the other hand, a hyperspectral image has a bandwidth of one to several nanometers and thus provides significant fine image spectral resolution. Hyperspectral imagery thus provides the potential for more detailed information extraction in agricultural applications.

The third issue is related to temporal hyperspectral data acquisition. It pertains to time of acquisition of each image and the time interval between image acquisitions. For example, identifying the temporal relationship between image and yield is helpful for yield estimation and management. The spatial yield pattern does not appear immediately before harvest. Rather, the yield pattern is built up gradually during the growing season. One study found that the spectral reflectance of plants has both a temporal and a spatial aspect [8]. Because this variation of crop spectral reflectance during the growing season can be used to relate to yield, it could help growers estimate yield during the growing season.

25.2.3 HYPERSPECTRAL NARROWBAND VEGETATION INDICES

Vegetation indices have been used widely in remote sensing. The most important vegetation index is the normalized difference vegetation index (NDVI) calculated by using the red and NIR wavelengths. The use of hyperspectral images makes it possible to build more refined vegetation indices by using distinct narrowbands and improving the indices for the correction of the effects of soil background [9]. Many hyperspectral vegetation indices have been developed for different applications. The simplest vegetation index is based on individual bands. Filella et al. [10] used individual image bands located at 430, 550, 680, and 780 nm to build different indices for wheat N status evaluation. Blackburn [11] also used individual spectral bands for developing different hyperspectral indices for estimating chlorophyll concentrations.

Vegetation indices could be calculated based on band ratio and combinations. Elvidge and Chen [12] used narrowbands with a 4 nm bandwidth at 674 and 755 nm to calculate several narrowband indices for the leaf area index (LAI) and percent green cover and compared the results with the corresponding broadband indices. Hurcom and Harrison [13] used the NDVI calculated from 677 to 833 nm to measure vegetation cover in a semi-arid area. Serrano et al. [14] used two image bands at 680 and 900 nm to compute vegetation indices, including the NDVI for estimating the biomass and yield of winter wheat. Broge and Leblanc [15] calculated narrowband vegetation indices from spectral bands centered at 670 and 800 nm and having a 10 nm bandwidth. Daughtry et al. [16] used discrete bands at 550, 670, and 801 nm to develop narrowband indices for N stress estimation in corn.

Vegetation indices were also studied based on data from different platforms. Broge and Mortensen [17] utilized field spectrometer data and spectral bands centered at 550, 650, and 800 nm and having a 10 nm bandwidth to calculate various hyperspectral vegetation indices. The authors also used an aerial hyperspectral image (CASI—compact airborne spectral imager) and chose, based on their sensitivity to chlorophyll, several individual image bands for vegetation indices calculation. These indices were used for LAI and N prediction over different types of crops. Another study using CASI images [18] calculated several vegetation indices using image bands centered at 550, 670, 700, and 800 nm for crop chlorophyll content prediction. The reason for choosing 700 nm was because it is located at the edge between the region where vegetation reflectance is dominated by pigment absorption and the beginning of the red edge region where reflectance is more influenced by the structural characteristics of the vegetation. Hyperspectral imagery acquired from spaceborne Hyperion sensor was also used for the calculation of vegetation indices [9]. This study evaluated 12 vegetation indices using 168 bands selected from the image after removing the water absorption bands and noise bands. These indices were two-band indices and were constructed using all possible two-band combinations.

A common trend in the earlier mentioned hyperspectral indices is the use of individual image bands, where, most of the time, one specific image band pair is selected based on crop characteristics. One reason for doing this is that it is simple to construct such indices. Another reason is that it is complicated to construct a multiple narrowband-based index if one follows the traditional construction and comparison approach for all possible solutions. For example, even though many vegetation indices have been designed in the past studies, these studies generally only test several indices for result comparison and best index identification. This practice may miss some important indices in the vast vegetation index database. Yao [19] presented a generic approach to automate the process for vegetation index selection and generation with hyperspectral data. The study first established a collection of available vegetation indices. A genetic algorithm–based method was then used to select the best vegetation index and spectral band combination for a specific application. Reader is also referred to a much broader discussion on wavebands and indices to study crop N using hyperspectral wavebands and indices in Chapter 11.

25.2.4 APPLICATION 1: SOIL MANAGEMENT ZONING

Research has been carried out using remotely sensed data for soil property mapping. It is expected that soil surface spectral reflectance could be used for soil constituent and nutrient content discrimination. As discussed before, the soil property maps are used for prescribing variable rate applications. For example, a soil pH map is a good source for decision making in variable rate lime application. A general concept for utilizing remote sensing in soil nutrient mapping as recommended by Moran et al. [20] can be stated as "Measurements of soil and crop properties at sample sites combined with multispectral imagery could produce accurate, timely maps of soil and crop characteristics for defining precision management units." Some ground and lab-based studies have focused on using reflectance in the visible and NIR region for looking at soil nutrients. It was found that there are different sensitive regions in the electromagnetic spectrum for different soil nutrient properties under controlled lab conditions. Ben-Dor and Banin [21] used reflectance curves in the infrared region to study six soil properties, including soil organic matter (OM) content. The results showed that the optimum prediction performance of each property required a different number of bands ranging from 25 to 3113. Palacios-Orueta and Ustin [22] found that the total iron and OM contents were the main factors affecting soil spectral shape, and concluded that the levels of iron and OM could be identified from Advanced Visible/Infrared Imaging Spectrometer (AVIRIS) images. Thomasson et al. [23] found that the spectral regions from 400 to 800 and from 950 to 1500 nm are sensitive to soil nutrient composition.

The earlier mentioned results, that is, there were different sensitive regions in the image spectral data for different nutrient properties, would be a viable source for soil nutrient content classification and mapping. In this case, a single hyperspectral image would provide the opportunity for different nutrient classifications using various sensitive ranges. With a 79-band hyperspectral image ($0.4–1.4\,\mu m$) Ben-Dor [24] was able to build a multiple regression model for each of four soil properties, OM, soil field moisture, soil saturated moisture, and soil salinity, respectively, each with different bands. Because the sensitive spectral region is valuable for soil nutrient identification and mapping using hyperspectral images, identification of such sensitive regions remains a major task in hyperspectral remote sensing research.

A more common and traditional way in this application is to explore spectral information using only the hyperspectral data. Bajwa and Tian [25] used first derivatives from aerial hyperspectral data and PLSR (partial least square regression) to model soil fertility factors including pH, OM, Ca, Mg, P, K, and soil electrical conductivity. The conclusion was that some wavebands explained a high degree of variability in the model. Ge and Thomasson [26] incorporated wavelet analysis with conventional regression methods with field spectrometer measurements for soil property determination. It was found that Ca, Mg, P, and Zn could be predicted with reasonable R^2 values (>0.5). DeTar et al. [27] pointed out that some soil properties could be accurately detected using aerial hyperspectral data

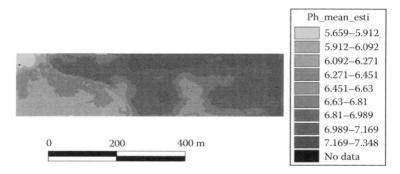

FIGURE 25.2 **(See color insert.)** In-field variability of soil pH as indicated by the sequential Gaussian cosimulation method using aerial hyperspectral data. (From Yao, H., Hyperspectral imagery for precision agriculture, PhD dissertation, University of Illinois at Urbana-Champaign, 2004.)

over nearly bare fields. The best regression R^2 (0.806) was for percentage of sand. Other properties such as silt, clay, chlorides, electric conductivity, and P had slightly lower R^2 (0.66–0.76).

Geostatistical techniques were also used with hyperspectral data to incorporate spatial information for soil property determination. Yao [19] applied two geostatistical approaches, colocated ordinary cokriging and sequential Gaussian cosimulation to predict soil nutrient factors. It was found that the cosimulation method yielded the best estimation ($R^2 = 0.71$) for K prediction. Figure 25.2 is an in-field pH map generated from the cosimulation process ($R^2 = 0.58$). It shows that the pH zones can be divided into two regions along the grass waterway located in the middle-left of the field. To the left of the waterway, the soil is acidic with low pH value estimations. To the right of the waterway the soil varies from acidic to basic. This analysis thus provides important information to help decision making on variable rate lime applications. Ge et al. [28] worked on regression-kriging method to analyze soil sampling data and reflectance measurements. The regression-kriging model R^2 was 0.65 for Na, which was much better than a principal component regression approach. Volkan Bilgili et al. [29] also concluded that cokriging and regression-kriging improved the predictions of soil properties with reflectance data. Ladoni et al. [30] reviewed statistical techniques including simple regression, the "soil line" approach, principal component analysis (PCA), and geostatistics for soil OM estimation using remote sensing data. The study pointed out that remote sensing data could help the design of the soil sampling strategy.

25.2.5 APPLICATION 2: WEED SENSING AND CONTROL

Effective weed management is of vital importance for ensuring the profitability of agricultural and horticultural crops. However, weed control has been heavily relying on herbicide application, leading to an increased environmental contamination concerning to the general public. This situation calls for more effective use of herbicide, that is, applying a minimal dosage of herbicide to only the weeds. In recent years, there has been a growing trend of shifting to organic farming, particularly in vegetable crop production. Weed control in organic farming practice excludes the use of synthetic chemicals and often requires mechanical means to control weeds with or without minimal collateral damage to crops. For both, reduced herbicide application and organic farming scenarios, it is self-evident that there must be an effective and reliable weed sensing system that allows for differentiation of weeds from crops and soils. Numerous sensing technologies have been investigated and developed for weed detection, among which optical and machine vision systems dominate. When compared with conventional machine vision systems, hyperspectral sensors capture more than a hundred spectral bands simultaneously, providing a detailed spectrum of light for each pixel. Thus, hyperspectral images are considered to be of superior potential in revealing the physical nature of different materials and leading to a more accurate classification. This information enrichment offered by hyperspectral sensors has a direct implication to weed detection, which has long been a very challenging task imposed by the biological complexity of a rather large number of weed species and their similarities to crop plants in the visible color domain as well as in the morphological feature space. To this end, a significant amount of research evident in the literature has been spent in developing hyperspectral-based weed sensors.

When using hyperspectral images for vegetation analysis, Thenkabail et al. [31] identified 22 optimal bands (in 400–2500 nm spectral range) that best characterize and classify vegetation and agricultural crops. Accuracies of over 90% were attained when classifying shrubs, weeds, grasses, and agricultural crop species. Hyperspectral imaging for weed sensing can be categorized into the following typical application areas: mapping invasive weed species [32,33], weed stress characterization [34,35], and weed species identification [36,37]. For instance, Hestir et al. [32] used remotely sensed hyperspectral images to map invasive weeds in wetland systems. They reported a moderate mapping accuracy primarily due to significant spectral variation of the mapped weed species. Goel et al. [34] conducted research using hyperspectral data acquired by a compact airborne spectrographic imager to classify the results of four different weed management strategies in corn fields where three different N application rates were also employed. Satisfactory classification results were obtained when one factor (weed or nitrogen) was considered at a time. As for weed species identification, Vrindts et al. [38] used reflectance spectra to classify sugar beet, maize, and seven weed species (Figure 25.3). When tested under controlled laboratory conditions, crop and weeds were separated with more than 97% accuracy using a limited number of wavelength band ratios. When testing in field conditions, over 90% of crop and weed spectra can be classified correctly when the model is specific to the prevailing light conditions. Nieuwenhuizen et al. [39] investigated the use of two spectral sensors that used spectra of 450–900 and 900–1650 nm for differentiating volunteer potato plants from sugar beet. They found the best classification accuracy was produced when coupling 10 wavebands in the NIR range with an artificial neural network (ANN) algorithm.

Though the fine spectral resolution of the hyperspectral images provides an invaluable source of information for more accurate classification, it also implies that the high dimensionality of the data presents challenges in image analysis and classification. Without the development of effective image processing and pattern recognition algorithms, the advantages presented by the rich information in the spectral dimension cannot be utilized. Furthermore, these algorithms

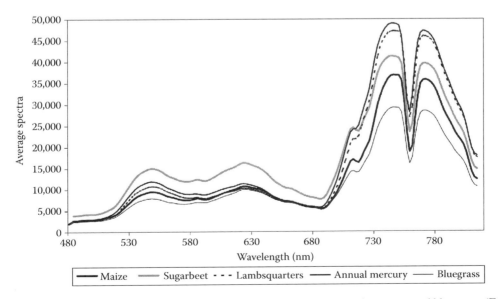

FIGURE 25.3 Reflectance spectra of maize, sugarbeet, lambsquarter, annual mercury, and bluegrass. (From Vrindts, E. et al., *Prec. Agr.*, 3, 63, 2002.)

are often application specific and require substantial efforts of exploration and testing. To cope with the high dimensionality of hyperspectral data, multivariate analysis, and computational intelligence techniques have been widely used and reported in literature. When processing the raw hyperspectral data to delineate spectral characteristics of weeds, techniques such as PCA [31,36], wavelet transforms [36], and spectral angle mapping [32] are often used. Commonly found classification algorithms include Linear or Stepwise Discriminant Analysis [31,39], Spectral Mixture Analysis [32,36], ANN [34,35,39], Support Vector Machines [35], and Mixture Tuned Matched Filtering [33].

25.2.6 APPLICATION 3: HYPERSPECTRAL IMAGERY FOR CROP NITROGEN STRESS DETECTION

In season crop nutrient management such as side dressing application of nitrogen for corn plants requires crop N content estimation (N stress detection) for application map (or prescription map) generation. Many studies have suggested the use of remotely sensed canopy reflectance for crop N detection. Different N levels in plants affect crop chlorophyll concentration and results in different canopy reflectance spectra [40]. For ground truth estimation, actual plant N level could be obtained through (1) using as applied plot N level; (2) chemical measurement of plant tissues; or (3) estimation of leaf chlorophyll concentration with a field instrument such as the Minolta SPAD meter.

For corn plant N estimation, Zara et al. [41] found that the slope of the reflectance spectra between 560 and 580 nm generated the best results for corn N stress detection with AVIRIS aerial hyperspectral images. Three traditional indices [42], NDVI, Photosynthetic Reflectance Index (PRI), and Red Edge Vegetation Stress Index (RVSI), were computed using certain bands from AVIRIS images. It was found that RVSI had the highest correlation with both applied nitrogen and measured chlorophyll levels in corn. Boegh et al. [43] concluded that the CASI image green and NIR bands, which are the maximum reflectance bands of chlorophyll, were the most important predictors. Haboudane et al. [18] worked on developing a combined modeling-based and indices-based approach to predict corn chlorophyll content using CASI imagery. This method used the ratio of an index sensitive to low chlorophyll values to a soil adjusted index, both calculated from distinct narrow image bands, to build the prediction model. Yao [19] developed a generic vegetation index generation algorithm and obtained an R^2 of 0.79 for corn plant N estimation.

Similar studies for plant N estimation have been implemented for other crops. Christensen et al. [44] concluded that N content in barley could be predicted with 81% of accuracy. Min and Lee [45] reported that the R^2 for N content in citrus tree leaves was 0.839. Zhao et al. [46] obtained an accuracy of 62.4% in discriminating N in cotton plants. Jain et al. [47] developed a regression model for N estimation in potato plants with R^2 equal to 0.551. Li et al. [48] studied winter wheat for N content estimation. The R^2 was 0.58 in an experimental field and 0.51 in a farmer's field. Nguyen et al. [49] developed regression models for rice plant N estimation with R^2 from 0.76 to 0.87 for validation data. The in-field variation maps from this study are presented in Figure 25.4. Bajwa [50] stated that the PLSR models could explain 47%–71% of the variability in rice plant N. Additionally, in a 3 year study a regression model was reported to have an $r = 0.938$ for N estimation in rice [51].

25.2.7 APPLICATION 4: CROP YIELD ESTIMATION

Crop yield prediction or estimation is one of the most important aspects in the farm industry. Traditionally, farmers have estimated crop yield for a whole field or for large parts of a field. This approach involves manually counting the number of ears per acre and the number of kernels per ear after the kernel number is established. The total yield of a field can also be predicted with a yield model, using information such as weather, crop, and up to 170 soil parameters [52]. By utilizing a yield sensor and the GPS, crop yields can be measured and stored with

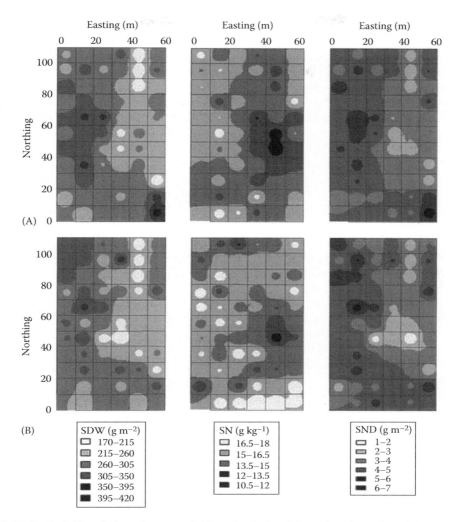

FIGURE 25.4 In-field variation of measured (A) and calculated (B) values by PLSR of shoot dry weight (SDW), shoot N concentration (SN), and shoot N density (SND) in year 2004. The wavebands selected by a multiple stepwise linear regression for an N nutrition index (NNI) calculation are 977, 583, 702, and 725 nm [49]. The NNI is a function of SN and SDW. (From Yang, C. et al., *Prec. Agr.*, 5, 445, 2004.)

detailed location information on-the-go. In this way, yield maps with both high accuracy and high spatial resolution can be generated [53] to be incorporated in the farm GIS database. For management practices, yield maps can be used for giving input prescriptions for future growing seasons. Although a yield monitor can output yield data in great detail, its yield estimation is only a "post-harvest" estimation.

One commonly used method for in-season yield monitoring is based on remote sensing images [54,55]. Different crop yields can cause different crop canopy reflectance. Such reflectance difference can be recorded by images captured from a close distance or by aerial or satellite remote sensing systems. Based on satellite data, regression models [56,57] could be developed for whole field yield estimation. Satellite images were also used for an in-field yield variability study [58]. Yang et al. [59] summarized practical procedures for yield estimation using aerial remote sensing images. The following paragraphs will describe the use of hyperspectral data for yield estimation of different crops.

- *Corn yield estimation*

  Yao [19] worked on using aerial hyperspectral images for corn yield estimation. The ground truth yield ranging from 128 to 259 bushels/acre was obtained with a plot harvester for each 50 m² experiment plot. The total number of harvested plots was 570. The R^2 value for yield estimation with a vegetation index was 0.59. Uno et al. [60] also worked on corn yield with aerial hyperspectral imaging. The ground truth yield was obtained from four 1 m² subplots for each 400 m² treatment plot (48 total treatment plots). The highest yield prediction using ANN was r = 0.76. The study concluded there was no clear difference between ANNs and stepwise multiple linear regression models.

- *Grain sorghum yield estimation*

  Yang et al. [61] used aerial hyperspectral images for grain sorghum yield estimation over two Texas fields. The ground truth yield was collected with a grain combine. It was reported that the sorghum yield was significantly correlated to the visible NIR bands. Regression analysis using both principal component transformation and stepwise band selection produced similar results. The yield estimation R^2 was from 0.69 to 0.82 for the two fields. In addition, linear spectral unmixing techniques were employed for sorghum yield variability estimation using aerial multispectral [62] and hyperspectral images [63]. These techniques are based on the assumption that canopy reflectance is a linear mixture of different spectral components (end members) such as soil and sorghum plants. The multispectral image study gave the best correlation coefficient of 0.90. The two study fields with hyperspectral data had the best fraction-based r-values (0.67 and 0.82). Yang et al. [64] further applied Spectral Angle Mapper (SAM) algorithm for sorghum yield estimation. He concluded that the SAM technique can be used alone or with other vegetation indices for yield estimation with hyperspectral imagery. Yang et al. [65] also pointed out that hyperspectral imagery had the potential to significantly improve yield estimation accuracy when compared with multispectral imagery.

- *Wheat yield estimation*

  In a study using ground-based canopy reflectance for winter wheat yield estimation, it was found that spectral indices, such as the infrared/red, the normalized difference (ND), the transformed vegetation index, and the greenness index obtained between flowering to milking stages, gave the best results [66]. The indices were calculated by integrating the reflectance data taken by a spectrometer into broadband images. Reyniers et al. [67] used a line scanner mounted on a tractor for hyperspectral data acquisition. In this study, winter wheat was planted in 60 plots of 12 m × 16 m each. The ground truth yield data was collected through a plot harvester. Narrowband NDVI calculated at wavelength 630 and 750 nm was used for crop coverage measurement. The optically measured crop coverage was positively correlated with grain yield. The correlation coefficient was 0.74 between yield and coverage data collected on April 27, 2000. Migdall et al. [68] also provided a modeling-based approach to simulate winter wheat yield using airborne and spaceborne hyperspectral imagery.

- *Other crops*

  Rasmussen [69] used AVHRR (Advanced Very High Resolution Radiometer) images for millet yield forecasting. The NDVI was used to build the yield regression model. The conclusion was that millet yield could be measured 1 month before harvest. Yang et al. [70] used aerial hyperspectral images for cotton yield estimation over two fields in Texas. Cotton yield ground truth data was generated by using a cotton yield monitor mounted on the cotton picker. To compensate for image resolution and geo-registration errors, both the image data and yield data were aggregated into 8 m × 8 m cells. The stepwise regression analysis produced yield estimation R^2 of 0.61 and 0.69 for the two experiment fields. The yield variation map of one field is presented in Figure 25.5. Pettersson et al. [71] used a hand-held

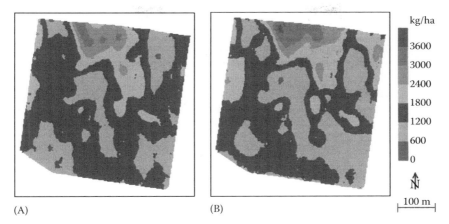

(A)   (B)

FIGURE 25.5 **(See color insert.)** In-field yield variability maps generated from (A) an airborne hyperspectral image using a nine-band regression model and (B) yield monitor data for a 16 ha cotton field [70]. The nine bands are 499, 546, 601, 702, 717, 738, 771, 778, and 826 nm. (From Yao, H., Hyperspectral imagery for precision agriculture, PhD dissertation, University of Illinois at Urbana-Champaign, 2004.)

multispectral scanner to collect canopy reflectance data of malting barley. The reflectance data was then used to generate nine vegetation indices. It was reported that all the vegetation indices were significantly correlated with grain yield. The correlation coefficient was 0.9 for the regression models.

- *Hyperspectral imagery for temporal yield analysis*
  Research has been carried out to find the temporal relationship between remote sensing imagery and yield. Moran et al. [20] suggested that remote sensing images from the late growing season have the best results for preharvest crop yields prediction. Vellidis et al. [72] used an unsupervised classification method and found that a cotton yield pattern could be identified in the early growth stage, within 10 weeks of crop growth, from multispectral CIR (color infrared) aerial photos. Although the aforementioned multispectral image–based research showed that images obtained from different times and different vegetation indices could be used for yield estimation, problems like when to acquire the images and how to properly correlate spectral information from hyperspectral imagery with yield, still require much more research.

The potential of using hyperspectral image data for yield estimation varied based on the date of image data acquisition during the growing season. From the decision making point of view, it is desirable to know when the best time for image data collection would be. The best time could be determined by estimation accuracy, economical consideration, and other factors. Among them, the ability to accurately estimate crop yield is of utmost importance. There were several studies conducted on using hyperspectral imagery for temporal yield analysis. Goel et al. [34] showed that the largest correlation between reflectance spectra and crop yield occurred at tasseling stage. In a study using ground spectrometer data for temporal yield analysis, an ANN model was used [73]. The conclusion was that ANN could be used to predict yield in the early planting stage using ground spectrometer-based hyperspectral data. Reyniers et al. [67] pointed out that spectral data from mid-season were more related to wheat yield measurements at harvest. Yao [19] used an index-based approach for corn temporal yield estimation. The aerial hyperspectral image data collected at five different dates were correlated with corn yield. It was found that image bands at around 700 nm (the red edge region) including 694, 700, and 706 nm were strong indicators for yield estimation among all five images. The late season images show better correlation with the measured corn yield than do the early season images (Figure 25.6).

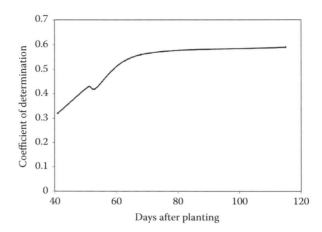

FIGURE 25.6 R^2 values between corn yield and the calculated vegetation indices, plotted on days after planting (DAP). (From Yao, H., Hyperspectral imagery for precision agriculture, PhD dissertation, University of Illinois at Urbana-Champaign, 2004.)

25.2.8 Application 5: Pest and Disease Detection

The changes in plant canopy reflectance due to pest invasion or disease infestation could be used for pest and disease detection. The following studies summarize this potential. The cotton plant feeding strawberry spider mite causes leaf puckering and reddish discoloration in early stages of infestation followed by leaf drop. Fitzgerald [74] successfully distinguished between adjacent mite-free and mite-infested cotton field areas through implementing a spectral unmixing process on AVIRIS imagery. Bacterial leaf blight is a vascular disease of irrigated rice. Its serious infestations might result in up to 50% yield loss. Yang [75] found that changes in leaf color and appearance caused by different levels of disease severity, exhibited reflectance spectra differences. The model for infestation area estimation of the highly susceptible cultivar had R^2 equal to 0.978 based on a single spectral band (745 nm).

Another important disease for many agriculture crops is fungal pathogen infection. Muhammed [76] used spectral classification to estimate the fungal disease severity of a spring winter wheat field. The spectral analysis included data normalization and a nearest-neighbor classification. The resulting R^2 was generally >0.93. Zhang et al. [77] investigated another fungal pathogen causing disease, late blight, in tomato fields. The study used a vegetation index- based approach. The results showed that the diseased tomatoes could be separated from healthy ones before economic damage happened. Huang et al. [78] used wavelengths of 531 and 570 nm to calculate a normalized vegetation index for yellow rust detection in wheat. The infestation ground truth was interpreted as disease index through an independent field measurement and assessment. The resulting R^2 was 0.91 between the disease index and the vegetation index generated from airborne data. Mahlein et al. [79] explored the potential of using hyperspectral data to detect and differentiate three fungal leaf diseases in sugar beet. A vegetation index-based approach was used in the study and concluded that a distinctive differentiation of the three diseases was possible using a combination of several indices.

25.3 CONCLUSIONS

Precision farming practices such as variable rate applications of fertilizer and agricultural chemicals require accurate field variability mapping. This chapter investigated the value of hyperspectral remote sensing in providing useful information for five applications of precision farming: (1) Soil management zoning, (2) weed control, (3) N stress detection, (4) crop yield estimation, and (5) pest and disease control. When using remotely sensed hyperspectral data for soil management zone delineation, it was found that there were different sensitive regions in the electromagnetic spectrum (0.4–1.4 μm) for different soil nutrient properties. In addition, the combination of geostatistical techniques and remote sensing data holds great potential for soil nutrient mapping. For selective weed control, canopy reflectance in the spectral region from 450 to 900 nm, with emphasis on the region from red to NIR is important for weed and crop differentiation. Successful classification between broadleaf plants and grasses using reflectance spectra has been reported in literature. This can potentially lead to smart herbicide application systems which can correctly deliver broadleaf specific or grass specific herbicide to correct weed patch targets, resulting in improved weed control efficacy and reduction of herbicide usage.

Canopy reflectance has also been used in plant N stress sensing and yield estimation, as well as pest and disease detection. In plant N stress sensing applications, the most significant spectral region is the visible to NIR region. Many vegetation indices were developed for N stress detection. Among them the most frequently cited index was the NDVI. The prediction accuracy varied between different crops, and even for the same crop. For example, the coefficient of determination could be as low as 0.47 or as high as 0.87 for nitrogen in rice plants. The sources for such variability could come from many different sources such as plant cultivar, planting date and growth stages, local environment, weather conditions, sensor platforms, as well as sensor calibration and data processing. For crop yield estimation, it was generally regarded that canopy reflectance measured in the middle to late growing season gave the best prediction result. Many studies used vegetation indices and explored the visible NIR region for different crops. The subsequently generated yield map is one of the most important maps for precision farming practices. Lastly, this chapter summarized another potential precision farming application using hyperspectral data, which includes some research in detection and assessment of insect invasion, onset of disease, and fungal pathogen infection.

In summary, hyperspectral remotely sensed data could be an important data source for field variability sensing and information extraction. This is a crucial step for the implementation of precision farming technology, which also consists of management decision making, precision field operation control, and result assessment. A great deal of research has been focused on the use of canopy reflectance for N stress detection and crop yield estimation. The red edge region where the spectrum shifts from red to NIR region was found to be important in plant canopy reflectance, as the reflectance differences at these wavelengths reflect the differences of internal leaf tissue (mesophyll) structures. It was also concluded that three aspects of hyperspectral data including spatial, spectral, and temporal issues are all important to precision farming applications.

The use of relative reflectance was reported in many studies. Employment of relative reflectance values can minimize the influence of variable lighting conditions, and the effects of relative positioning variation between the sensor and the object. However, outdoor variable lighting (in particular the direct sunlight conditions) still pose challenges for developing practical spectral systems for robust outdoor plant sensing. It was obvious that hyperspectral data obtained from different ground-based, airborne, and spaceborne sensors for precision farming was generally calibrated in different ways. For this reason, it is important that all future studies involving hyperspectral imaging should calibrate the image sensors to produce data in standard radiometric units in order to establish a calibrating standard across all sensors. More applications of hyperspectral data in precision farming are expected with better data availability and improved data quality.

Finally, one common issue related to hyperspectral remote sensing is the spectral signature mixture of different target components. The implementation of different unmixing techniques would help to extract canopy spectral reflectance information for better data analysis. Another potential direction is the use of geostatistical techniques. This chapter reviewed studies on combining such techniques with soil reflectance for soil nutrient mapping. Similar approaches could be extended to plant reflectance to improve data interpretation.

REFERENCES

1. National Research Council, *Precision Agriculture in the 21st Century: Geospatial and Information Technologies in Crop Management*, National Academy Press, Washington, DC, 1997.
2. Zhang, N., Wang, M., and Wang, N., Precision agriculture—A worldwide overview, *Computers and Electronics in Agriculture*, 36, 2–3, 113–132, 2002.
3. Stafford, J. V., Implementing precision agriculture in the 21st century, *Journal of Agricultural Engineering Research*, 76, 3, 267–275, 2000.
4. Scotford, I. M. and Miller, P. C. H., Applications of spectral reflectance techniques in northern European cereal production: A review, *Biosystems Engineering*, 90, 3, 235–250, 2005.
5. Larson, J. A., Roberts, R. K., English, B. C., Larkin, S. L., Marra, M. C., Martin, S. W., Paxton, K. W., and Reeves J. M., Factors affecting farmer adoption of remotely sensed imagery for precision management in cotton production, *Precision Agriculture*, 9, 4, 195–208, 2008.
6. Hughes, G. F., On the mean accuracy of statistical pattern recognition, *IEEE Transactions on Information Theory*, 14, 55–63, 1968.
7. Richard, J. A. and Jia, X., *Remote Sensing Digital Image Analysis: An Introduction*, 3rd edn. Springer-Verlag, Berlin, Germany, 1999.
8. Zwiggelaar, R., A review of spectral properties of plants and their potential use for crop/weed discrimination in row-crops, *Crop Protection*, 17, 189–206, 1998.
9. Gong, P., Ru, R. L., and Biging, G. S., Estimation of forest leaf area index using vegetation indices derived from Hyperion hyperspectral data, *IEEE Transactions on Geoscience and Remote Sensing*, 41, 6, 1355–1362, 2003.
10. Filella, I., Serrano, L., Serra, J., and Penuelas, J., Evaluating wheat nitrogen status with canopy reflectance indices and discriminant analysis, *Crop Science*, 35, 1400–1405, 1995.
11. Blackburn, G. A., Quantifying chlorophylls and caroteniods at leaf and canopy scales: An evaluation of some hyperspectral approaches, *Remote Sensing of Environment*, 66, 3, 273–285, 1998.
12. Elvidge, C. D. and Chen, Z., Comparison of broad-band and narrow-band red and near-infrared vegetation indices, *Remote Sensing of Environment*, 54, 38–48, 1994.
13. Hurcom, S. J. and Harrison, A. R., The NDVI and spectral decomposition for semi-arid vegetation abundance estimation, *International Journal of Remote Sensing*, 19, 16, 3109–3125, 1998.
14. Serrano, L., Filella, I., and Peuelas, J., Remote sensing of biomass and yield of winter wheat under different nitrogen supplies, *Crop Science*, 40, 3, 723–731, 2000.
15. Broge, N. H. and Leblanc, E., Comparing prediction power and stability of broad-band and hyperspectral vegetation indices for estimation of green leaf area index and canopy chlorophyll density, *Remote Sensing of Environment*, 76, 156–172, 2000.
16. Daughtry, C. S. T., Walthall, C. L., Kim, M. S., Colstoun, E. B. D., and McMurtrey, J. E., Estimating corn leaf chlorophyll concentration from leaf and canopy reflectance, *Remote Sensing of Environment*, 74, 2, 229–239, 2000.
17. Broge, N. H. and Mortensen, J. V., Deriving green crop area index and canopy chlorophyll density of winter wheat from spectral reflectance data, *Remote Sensing of Environment*, 81, 1, 45–57, 2002.
18. Haboudane, D., Miller, J. R., Tremblay, N., Zarco-Tejada, P. J., and Dextraze, L., Integrated narrow-band vegetation indices for prediction of crop chlorophyll content for application to precision agriculture, *Remote Sensing of Environment*, 81, 2–3, 416–426, 2002.
19. Yao, H., Hyperspectral imagery for precision agriculture, PhD dissertation, University of Illinois at Urbana-Champaign, 2004.
20. Moran, M., Inoue, Y., and Barnes, E. M., Opportunities and limitations for image-based remote sensing in precision crop management, *Remote Sensing for Environment*, 61, 3, 319–346, 1997.
21. Ben-Dor, E. and Banin, A., Near-infrared analysis as a rapid method to simultaneously evaluate several soil properties, *Soil Science Society of America Journal*, 59, 2, 364–372, 1995.
22. Palacios-Orueta, A. and Ustin, S. L., Remote sensing of soil properties in the Santa Monica mountains I. Spectral analysis, *Remote Sensing for Environment*, 65, 2, 170–183, 1998.
23. Thomasson, J. A., Sui, R., Cox, M. S., and Al-Rajehy, A., Soil reflectance sensing for determining soil properties in precision agriculture, *Transactions of the ASAE*, 44, 6, 1445–1453, 2001.
24. Ben-Dor, E., Quantitative remote sensing of soil properties, *Advances in Agronomy*, 75, 173–243, 2002.
25. Bajwa, S. G. and Tian, L. F., Soil fertility characterization in agricultural fields using hyperspectral remote sensing, *Transactions of the ASABE*, 48, 6, 2399–2406, 2005.
26. Ge, Y. and Thomasson, J. A., Wavelet incorporated spectral analysis for soil property determination, *Transactions of the ASABE*, 49, 4, 1193–1201, 2006.

27. DeTar, W. R., Chesson, J. H., Penner, J. V., and Ojala, J. C., Detection of soil properties with airborne hyperspectral measurements of bare fields, *Transactions of the ASABE*, 51, 2, 463–470, 2008.

28. Ge, Y., Thomasson, J. A., Morgan, C. L., and Searcy, S. W., VNIR diffuse reflectance spectroscopy for agricultural soil property determination based on regression-kriging, *Transactions of the ASABE*, 50, 3, 1081–1092, 2007.

29. Volkan Bilgili, A., Akbas, F., and van Es, H. M., Combined use of hyperspectral VNIR reflectance spectroscopy and kriging to predict soil variables spatially, *Precision Agriculture*, 2010.

30. Ladoni, M., Bahrami, H. A., Alavipanah, S. K., and Norouzi, A. A., Estimating soil organic carbon from soil reflectance: A review, *Precision Agriculture*, 11, 1, 82–99, 2010.

31. Thenkabail, P. S., Enclona, E. A., Ashton, M. S., and Van Der Meer, B., Accuracy assessments of hyperspectral waveband performance for vegetation analysis applications, *Remote Sensing of Environment*, 91, 3–4, 354–376, 2004.

32. Hestir, E. L., Khanna, S., Andrew, M. E., Santos, M. J., Viers, J. H., Greenberg, J. A., Rajapakse, S. S., and Ustin, S. L., Identification of invasive vegetation using hyperspectral remote sensing in the California Delta ecosystem, *Remote Sensing of Environment*, 112, 11, 4034–4047, 2008.

33. Glenn, N. F., Mundt, J. T., Weber, K. T., Prather, T. S., Lass, L. W., and Pettingill, J., Hyperspectral data processing for repeat detection of small infestations of leafy spurge, *Remote Sensing of Environment*, 95, 3, 399–412, 2005.

34. Goel, P. K., Prasher, S. O., Patel, R. M., Landry, J. A., Bonnell, R. B., Viau, A. A., and Miller, J. R., Potential of airborne hyperspectral remote sensing to detect nitrogen deficiency and weed infestation in corn, *Computers and Electronics in Agriculture*, 38, 2, 99–124, 2003.

35. Karimi, Y., Prasher, S. O., Patel, R. M., and Kim, S. H., Application of support vector machine technology for weed and nitrogen stress detection in corn, *Computers and Electronics in Agriculture*, 51, 1–2, 99–109, 2006.

36. Koger, C. H., Bruce, L. M., Shaw, D. R., and Reddy, K. N., Wavelet analysis of hyperspectral reflectance data for detecting pitted morningglory (Ipomoea lacunosa) in soybean (Glycine max), *Remote Sensing of Environment*, 86, 1, 108–119, 2003.

37. Piron, A., Leemans, V., Kleynen, O., Lebeau, F., and Destain, M.-F., Selection of the most efficient wavelength bands for discriminating weeds from crop, *Computers and Electronics in Agriculture*, 62, 2, 141–148, 2008.

38. Vrindts, E., De Baerademaeker, J., and Ramon, H., Weed detection using canopy reflection, *Precision Agriculture*, 3, 63–80, 2002.

39. Nieuwenhuizen, A. T., Hofstee, J. W., van de Zande, J. C., Meuleman, J., and van Henten, E. J., Classification of sugar beet and volunteer potato reflection spectra with a neural network and statistical discriminant analysis to select discriminative wavelengths, *Computers and Electronics in Agriculture*, 73, 2, 146–153, 2010.

40. Walburg, G., Bauer, M. E., Daughtry, C. S. T., and Housle, Y. T. L., Effects of nitrogen nutrition on the growth, yield, and reflectance characteristics of corn canopies, *Agronomy Journal*, 74, 677–683, 1982.

41. Zara, P. M., Doraiswamy, P. C., and McMutrey, J., Assessing variability of nitrogen status in corn plants with hyperspectral remote sensing, *Proceedings of ASPRS Meeting*, 2000.

42. Cassady, P. E., Perry, E. M., Gardner, M. E., and Roberts, D. A., Airborne hyperspectral imagery for the detection of agricultural crop stress, *Proceedings of SPIE. Hyperspectral Remote Sensing of the Land and Atmosphere*, 4151, 197–204, 2000.

43. Boegh, E., Soegaard, H., Broge, N., Hasager, C. B., Jensen, N. O., Schelde, K., and Thomsen, A., Airborne multispectral data for quantifying leaf area index, nitrogen concentration, and photosynthetic efficiency in agriculture, *Remote Sensing of Environment*, 81, 2–3, 179–193, 2002.

44. Christensen, L. K., Bennedsen, B. S., Jørgensen, R. N., and Nielsen, H., Modelling nitrogen and phosphorus content at early growth stages in spring barley using hyperspectral line scanning, *Biosystems Engineering*, 88, 1, 19–24, 2004.

45. Min, M. and Lee, W. S., Determination of significant wavelengths and prediction of nitrogen content for citrus, *Transactions of the ASAE*, 48, 2, 455–461, 2005.

46. Zhao, D. H., Li, J. L., and Qi, J. G., Identification of red and NIR spectral regions and vegetative indices for discrimination of cotton nitrogen stress and growth stage, *Computers and Electronics in Agriculture*, 48, 155–169, 2005.

47. Jain, N., Ray, S. S., Singh, J. P., and Panigrahy, S., Use of hyperspectral data to assess the effects of different nitrogen applications on a potato crop, *Precision Agriculture*, 8, 225–239, 2007.

48. Li, F., Miao, Y., Hennig, S. D., Gnyp, M. L., Chen, X., Jia, L., and Bareth, G., Evaluating hyperspectral vegetation indices for estimating nitrogen concentration of winter wheat at different growth stages, *Precision Agriculture*, 11, 335–357, 2010.

49. Nguyen, H. T., Kim, J. H., Nguyen, A. T., Nguyen, L. T., Shin, J. C., and Lee, B., Using canopy reflectance and partial least squares regression to calculate within-field statistical variation in crop growth and nitrogen status of rice, *Precision Agriculture*, 7, 249–264, 2006.

50. Bajwa, S. G., Modeling rice plant nitrogen effect on canopy reflectance with partial least square regression (PLSR), *Transactions of the ASABE*, 49, 1, 229–237, 2006.

51. Ryu, C., Suguri, M., and Umeda, M., Model for predicting the nitrogen content of rice at panicle initiation stage using data from airborne hyperspectral remote sensing, *Biosystems Engineering*, 104, 465–475, 2009.

52. Villalobos, F. J., Hall, A. J., Ritchie, J. T., and Orgaz, F., OILCROP-SUN: A development, growth, and yield model of sunflower crop, *Agronomy Journal*, 88, 3, 403–415, 1996.

53. Taylor, R. K., Kluitenberg, G. J., Schrock, A. D., Zhang, N., Schmidt, J. P., and Havlin, J. L., Using yield monitor data to determine spatial crop production potential Source, *Transactions of the ASAE*, 44, 6, 1409–1414, 2001.

54. Ma, B. L., Morrison, M. J., and Dwyer, L. M., Canopy light reflectance and field greenness to assess nitrogen fertilization and yield of maize, *Agronomy Journal*, 88, 6, 915–920, 1996.

55. Taylor, J. C., Wood, G. A., and Thomas, G., Mapping yield potential with remote sensing, *Precision Agriculture-Soil and Crop Modeling*, 2, 713–720, 1997.

56. Thenkabail, P. S., Ward, A. D., and Lyon, J. G., LANDSAT-5 Thematic Mapper models of soybean and corn crop characteristics, *International Journal of Remote Sensing*, 15, 1, 49–61, 1994.

57. Hammar, D., Ferencz, C., Lichtenberger, J., Tarcsai, G., and Ferencz-Arkos, I., Yield estimation for corn and wheat in the Hungarian great plain using LANDSAT MSS data, *International Journal of Remote Sensing*, 17, 9, 1689–1699, 1997.

58. Hayes, M. J. and Decker, W. L., Using NOAA AVHRR data to estimate maize production in the United States corn belt, *International Journal of Remote Sensing*, 17, 16, 3189–3200, 1996.

59. Yang, C., Bradford, J. M., and Wiegand, C. L., Airborne multispectral imagery for mapping variable growing conditions and yields of cotton, grain sorghum, and corn, *Transactions of the ASAE*, 44, 6, 1983–1994, 2001.

60. Uno, Y., Prasher, S. O., Lacroix, R., Goel, P. K., Karimi, Y., Viau, A., and Pate, R. M., Artificial neural networks to predict corn yield from compact airborne spectrographic imager data, *Computers and Electronics in Agriculture*, 47, 2, 149–161, 2005.

61. Yang, C., Everitt, J. H., and Bradford, J. M., Airborne hyperspectral imagery and yield monitor data for estimating grain sorghum yield variability, *Transactions of the ASAE*, 47, 3, 915–924, 2004.

62. Yang, C., Everitt, J. H., and Bradford, J. M., Using multispectral imagery and linear spectral unmixing techniques for estimating crop yield variability, *Transactions of the ASAE*, 50, 2, 667–674, 2007.

63. Yang, C., Everitt, J. H., and Bradford, J. M., Airborne hyperspectral imagery and linear spectral unmixing for mapping variation in crop yield, *Precision Agriculture*, 8, 279–296, 2007.

64. Yang, C., Everitt, J. H., and Bradford, J. M., Yield estimation from hyperspectral imagery using spectral angle mapper (SAM), *Transactions of the ASAE*, 51, 2, 729–737, 2008.

65. Yang, C., Everitt, J. H., Bradford, J. M., and Murden, D., Comparison of airborne multispectral and hyperspectral imagery for estimating grain sorghum yield, *Transactions of the ASAE*, 52, 2, 641–649, 2009.

66. Das, D. K., Mishra, K. K., and Kalra, N., Assessing growth and yield of wheat using remotely-sensed canopy temperature and spectral indices, *International Journal of Remote Sensing*, 14, 17, 3081–3092, 1993.

67. Reyniers, M., Vrindts, E., and De Baerdemaeker, J., Optical measurement of crop cover for yield prediction of wheat, *Biosystems Engineering*, 89, 4, 383–394, 2004.

68. Migdall, S., Bach, H., Bobert, J., Wehrhan, M., and Mauser, W., Inversion of a canopy reflectance model using hyperspectral imagery for monitoring wheat growth and estimating yield, *Precision Agriculture*, 10, 508–524, 2009.

69. Rasmussen, M. S., Operational yield forecasting using AVHRR NDVI data: Reduction of environmental and inter-annual variability, *International Journal of Remote Sensing*, 18, 5, 1059–1077, 1997.

70. Yang, C., Everitt, J. H., Bradford, J. M., and Murden, D., Airborne hyperspectral imagery and yield monitor data for mapping cotton yield variability, *Precision Agriculture*, 5, 445–461, 2004.

71. Pettersson, C.G., Soderstrom, M., and Eckersten, H., Canopy reflectance, thermal stress, and apparent soil electrical conductivity as predictors of within-field variability in grain yield and grain protein of malting barley, *Precision Agriculture*, 7, 343–359, 2006.

72. Vellidis, G., Thomas, D., Wells, T., and Kvien, C., Cotton yield maps created from aerial photographs, *ASAE Meeting Paper No. 991139*, St. Joseph, MI, ASAE, 1999.

73. Yang, C.-C., Prahser, S. O., and Whalen, J., In-season yield prediction of corn and soybean with hyperspectral imagery, *ASAE Meeting Paper No. 023139*, 2002.

74. Fitzgerald, G. J., Spider mite detection and canopy component mapping in cotton using hyperspectral imagery and spectral mixture analysis, *Precision Agriculture*, 5, 275–289, 2004.
75. Yang, C.-M., Assessment of the severity of bacterial leaf blight in rice using canopy hyperspectral reflectance, *Precision Agriculture,* 11, 61–81, 2010.
76. Muhammed, H. H., Hyperspectral crop reflectance data for characterizing and estimating fungal disease severity in wheat, *Biosystems Engineering*, 91, 1, 9–20, 2005.
77. Zhang, M., Zhang, Q., and Liu, X., Remote sensed spectral imagery to detect late blight in field tomatoes, *Precision Agriculture*, 6, 489–508, 2005.
78. Huang, W., Lamb, D. W., Niu, Z., Zhang, Y., Liu, L., and Wang, J., Identification of yellow rust in wheat using in-situ spectral reflectance measurements and airborne hyperspectral imaging, *Precision Agriculture*, 8, 187–197, 2007.
79. Mahlein, A.-K., Steiner, U., Dehne, H.-W., and Oerke, E.-C., Spectral signatures of sugar beet leaves for the detection and differentiation of diseases, *Precision Agriculture*, 11, 413–431, 2010.

Part X

Hyperspectral Data in Global Change Studies

26 Hyperspectral Data in Long-Term, Cross-Sensor Continuity Studies

Tomoaki Miura and Hiroki Yoshioka

CONTENTS

26.1 INTRODUCTION

Numerous satellite optical sensors have been launched and planned for launch for monitoring and characterization of the Earth system and its behaviors. These sensors have been providing and will continue to provide systematic observations of terrestrial vegetation at various spatial, spectral, and temporal resolutions. Spectral vegetation indices (VIs) are among the most widely used satellite data products in monitoring temporal and spatial variations of vegetation photosynthetic activities and biophysical properties. VIs are optical measures of vegetation canopy "greenness," a direct measure of photosynthetic potential resulting from the composite property of total leaf chlorophyll, leaf area, canopy cover, and structure [1]. Although they are not intrinsic physical quantities, VIs are widely used as proxies in the assessments of many canopy state and biophysical process variables, including leaf area index, fraction of absorbed photosynthetically active radiation, vegetation fraction, and gross primary production (e.g., Refs. [2,3]).

Utilities of these observations greatly increase when datasets from multiple sensors are combined, for example, multi-decadal land cover characterization and change detection via multi-sensor data sources (e.g., Refs. [4–6]), synergistic applications of multi-resolution remote sensing for forest and rangeland inventory (e.g., Refs. [7,8]), and a development of multi-sensor, long-term data records for climate studies [9–12].

Applications of multi-sensor observations, however, require consideration and account of continuity and compatibility due to differences in sensor/platform characteristics that include band position, spatial resolution, and overpass time [13,14]. Multi-sensor VI continuity becomes a critical and complicated issue because it involves consideration of differences in both sensor/platform characteristics and product generation algorithms, a requirement that needs to be addressed [15]. The underlying issue in multi-sensor VI continuity is that VI values for the same targets will not be directly comparable because input reflectance values differ from sensor to sensor [13,16].

Swinnen and Veroustraete [15] provide a comprehensive list of factors to be taken into consideration for extending the SPOT VEGETATION NDVI time series back in time with NOAA AVHRR data.

Hyperspectral remote sensing, in particular imaging sensors, has great potential in addressing several key issues of multi-sensor VI continuity and providing deeper insights and understanding of the issues. An ultimate advantage of using hyperspectral remote sensing for multi-sensor continuity studies is that it allows analyzing the effects of multiple factors simultaneously [13]. Specific issues of multi-sensor VI continuity that can be addressed with hyperspectral remote sensing include the following:

1. Spectral—a large number of narrow spectral bands that continuously cover the visible-nearinfrared-shortwaveinfrared wavelength regions can be spectrally convolved to simulate spectral responses of virtually any broadband sensors. The simulated data can be used for multi-sensor comparisons devoid of mis-registration (e.g., Ref. [17]). It should be noted that, although the word, "simulation," is used here, the resultant, spectrally aggregated values are actual observations.
2. Spatial—current and future hyperspectral sensors provide medium resolution images (3–100 m spatial resolution with 30 m being typical) with swaths of 3–150 km with 30 km being typical. These resolutions are fine enough and these swath widths are wide enough to allow simulation of various pixel footprint sizes via spatial aggregation. The aggregated data can be used to examine VI compatibility across multiple resolutions [18].
3. Algorithmic—the effects of algorithmic differences (e.g., atmospheric correction schemes) can be examined on spectrally and/or spatially aggregated data from hyperspectral imagery, although limited in types of algorithms that could be tested.
4. Angular—most of current and future satellite hyperspectral sensor systems have a cross-track pointing capability (see Table 26.2). In addition, the CHRIS sensor acquires data at five discrete view angles along track. Although limited in a range of possible observation geometry, multi-angular hyperspectral observations could be used to address bidirectional reflectance distribution function (BRDF) effects on multi-sensor VI continuity.

The purpose of this chapter is to discuss the potential utilities of hyperspectral remote sensing data in long-term VI continuity for global change studies. We present analysis results obtained from a regional set of Earth Observing-One (EO-1) hyperspectral Hyperion images [19,20] over the conterminous United States along with literature reviews for this purpose.

26.2　MATERIALS

Five sites within the conterminous United States were selected based upon availability of nearly cloud-free Hyperion scenes, availability of *in situ* atmospheric measurements from the Aerosol Robotic Network (AERONET) [21], and a diversity of land cover types. Level 1R Hyperion scenes were obtained for the five sites for the dates listed in Table 26.1. For each Hyperion scene, Level 2 AERONET data were acquired for a 2 h time period bracketing the image acquisition time (±1 h) (Table 26.1).

Hyperion images were spectrally convolved to spectral bandpasses of various satellite sensors described in Section 26.3. The spectral response curves of these satellite sensors were splined to Hyperion band center wavelengths for each Hyperion pixel [17] because each pixel had a slightly different spectral calibration (spectral smile) [19].

The convolved images were first converted to top-of-atmosphere (TOA) reflectances and then corrected for atmosphere with the "6S" radiative transfer code [22]. The 6S radiative transfer code was constrained with scene specific geometric conditions extracted from the corresponding image metadata and *in situ* AERONET atmospheric data (Table 26.1). We performed three types of atmospheric corrections: (1) partial correction for molecular scattering and ozone absorption, (2) partial

TABLE 26.1
List of Study Sites, Hyperion Image Properties, and *In Situ* Atmospheric Properties

Geographic Location	Latitude/Longitude (Degrees)	Elevation[a] (m)	Biome Type	Date (yyyy/mm/dd)	θ_s/θ_v[b] (Degrees)	Ozone[c] (Dobson)	W.V.[c,d] (cm-atm)	AOT[c,e] (550 nm)
Harvard Forest, MA	42.532/−72.188	322	Broadleaf forest	2001/09/05	40.5/3.6	302	0.76	0.03
				2008/05/07	32.2/12.1	362	0.76	0.13
				2008/05/25	28.5/4.8	353	0.76	0.05
				2008/05/30	28.6/12.8	351	1.28	0.17
				2008/06/07	26.2/10.1	347	3.29	0.17
				2008/12/03	67.1/5.2	303	0.64	0.04
Walker Branch, TN	35.958/−84.287	365	Broadleaf forest	2001/08/14	31.3/2.3	308	2.46	0.27
Maricopa, AZ	33.069/−111.972	360	Broadleaf cropland/open shrubland	2001/05/24	23.3/5.5	317	1.11	0.06
				2001/07/27	26.4/5.4	295	2.77	0.07
				2001/08/28	32.3/5.0	287	2.76	0.09
				2001/12/02	58.6/4.6	275	0.98	0.05
				2001/12/18	60.5/4.7	280	0.41	0.04
Konza Prairie, KS	39.102/−96.610	341	Prairie grassland/cereal crop	2002/10/19	52.3/2.6	285	0.98	0.02
				2009/05/08	29.4/1.3	340	2.12	0.10
Sevilleta, NM	34.355/−106.885	1477	Semi-arid grassland/open shrubland/cereal crop	2001/10/19	48.8/5.0	275	0.46	0.02
				2009/01/16	61.5/16.9	290	0.50	0.02
				2009/09/25	40.8/4.3	278	0.91	0.03
				2009/10/05	45.2/18.6	277	1.11	0.02
				2009/11/05	54.2/12.1	271	0.69	0.02
				2009/12/06	60.8/5.7	276	0.47	0.04

[a] Elevation.
[b] θ_s: solar zenith angle, θ_v: view zenith angle.
[c] The values in these columns were obtained from the AERONET website (http://aeronet.gsfc.nasa.gov/) [21].
[d] Atmospheric water vapor.
[e] Aerosol optical thickness.

correction for molecular scattering, and ozone and water vapor absorptions, and (3) total correction including aerosol scattering and absorption. The continental aerosol model was assumed for all the aerosol corrections, based on the aerosol model selection criteria described in Ref. [23].

Three VIs were computed and evaluated in this study. The normalized difference vegetation index (NDVI) was computed from the TOA and atmospherically corrected reflectances as Ref. [24]:

$$NDVI = \frac{\rho_{NIR} - \rho_{red}}{\rho_{NIR} + \rho_{red}}, \tag{26.1}$$

where ρ_{red} and ρ_{NIR} are the red and near-infrared (NIR) reflectances, respectively. The enhanced vegetation index (EVI), developed as a standard satellite vegetation product for Terra and Aqua MODIS, was computed from the atmospherically corrected reflectances [25]:

$$EVI = 2.5 \frac{\rho_{NIR} - \rho_{red}}{\rho_{NIR} + 6\rho_{red} - 7.5\rho_{blue} + 1}, \tag{26.2}$$

where ρ_{blue} is the blue reflectance to correct for aerosol influences. Jiang et al. [26] have recently developed a two-band EVI which achieves the best similarity with the EVI without a blue band (EVI2) and, thus, is applicable to sensors without a blue band such as AVHRR:

$$EVI2 = 2.5 \frac{\rho_{NIR} - \rho_{red}}{\rho_{NIR} + 2.4\rho_{red} + 1}. \tag{26.3}$$

EVI2 was computed from the atmospherically corrected reflectances.

26.3 SPECTRAL COMPATIBILITY ANALYSES

One key sensor characteristic that varies widely among sensors is the spectral bandpass filters, and, thus, many previous studies have focused on this "spectral" issue (e.g., Refs. [27–31]). Figure 26.1 shows the normalized spectral response curves of red, NIR, and blue (when available) bands for moderate–coarse resolution satellite sensors designed or used for monitoring and biophysical characterization of global vegetation. The bandwidths of AVHRR/2 channels (Figure 26.1a) are the widest, followed by those of the AVHRR/3 sensors (Figure 26.1b), by SPOT-4 and -5 VEGETATION and ADEOS-II GLI 250 m bands (Figure 26.1c), and by Terra- and Aqua-MODIS (Figure 26.1c). The narrowest are the spectral bands of those sensors designed for both oceanic and terrestrial measurements, that is, SeaWiFS, ADEOS-II GLI (1 km), GOSAT CAI, and GOCM-C SGLI (Figure 26.1d). The blue bands of these oceanic/terrestrial sensors, except for GLI, are located at slightly longer wavelengths than those of the terrestrial sensors (i.e., MODIS, VEGETATION, and GLI 250 m). Uniquely positioned are the VIIRS spectral bands. While the VIIRS NIR band is similar to that of MODIS, the VIIRS red band is more similar to the AVHRR/3 counterpart than the MODIS counterpart, and the VIIRS blue band is positioned at slightly longer wavelengths similar to those of the oceanic/terrestrial sensors (Figure 26.1d).

Spectral convolution of hyperspectral data has been one of standard methodologies used for assessing and evaluating the effects of these spectral bandpass differences on VI compatibility and continuity [17,32–36]. This approach is advantageous because it allows examining continuity/compatibility of pairs of sensors which do not have actual overlapping periods of observations.

Previous studies that used this methodology can be divided into two major categories: (1) empirical studies and (2) theoretical studies. Empirical studies have focused on predicting target sensor's reflectance or VI values from those of a source sensor by regression. Polynomials have been assumed as an analytical form that relates reflectance or VI values from two different sensors. Some studies used first-order polynomials and concluded that simple linear relationships would hold for relating the NDVI from two different sensors [27,32,37], whereas other studies used second-order polynomials

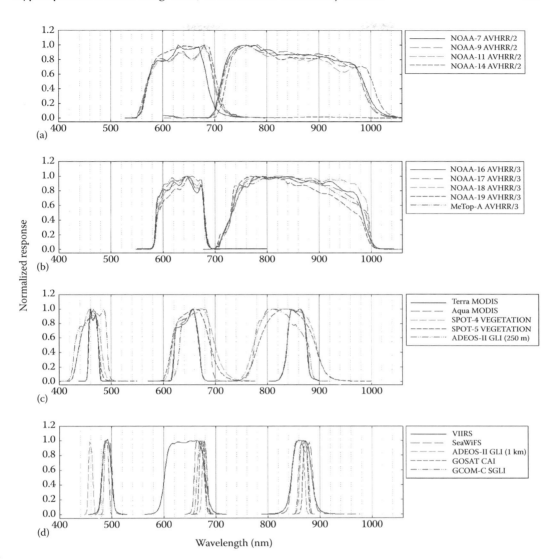

FIGURE 26.1 **(See color insert.)** Normalized spectral response curves of red, NIR, and blue bands for select moderate resolution sensors. They are plotted in four groups for clarity.

as they found nonlinearity in inter-sensor NDVI and reflectance relationships [17,33,35]. Recently, Kim et al. [34] showed that the EVI and EVI2 cross-sensor relationships were also modeled satisfactorily well with the first-order polynomial model. Whereas these studies used the ordinary least squares approach to fit polynomial models, Ji and Gallo [38] considered measurement errors in independent variables to more accurately characterize inter-sensor NDVI relationships (i.e., unbiased) and proposed a set of new statistics named "agreement coefficients."

Theoretical studies were motivated to take into account ecosystem parameters (e.g., LAI and soil brightness) in developing a spectral transformation algorithm that theoretically guarantees "exact" translations [16,17,39]. Based on the physics of atmosphere–vegetation–photon interactions, Yoshioka et al. [39] have theoretically justified the existence of and derived the functional form of interrelating VIs from two sensors. They have also shown that this "vegetation isoline" approach to interrelate VIs across sensors, resulted in a ~50% reduction in variability about the trend in cross-sensor VI relationships. The "exactness" of the translation results with this technique was also demonstrated using a simulated hyperspectral data set [40]. Noting that the isoline-based translation equation is a ratio of two polynomials, Yoshioka et al. [36] reduced the isoline-based translation equation into a quadratic polynomial. Although the coefficients of the polynomial could vary with surface and atmosphere conditions, this work theoretically justified the use of a polynomial form for multi-sensor translations of VIs.

An issue with these previous studies, however, is that each of them was limited in their spatial extent, seasonal coverage, and land cover type. Therefore, an extension of the results to different land cover types, geographic areas, and/or seasons is questionable. Satellite hyperspectral remote sensing has the great potential to be an excellent data source for expanding continuity analyses to global, full season analyses.

Using the Hyperion dataset described in Section 26.2, NDVI, EVI2, and EVI relationships of the sensors in Figure 26.1 to the Terra MODIS sensor are examined (MODIS VI minus source sensor VI plotted against source sensor values) in Figures 26.2 through 26.4. These comparisons assumed a total atmospheric correction scenario at 1 km spatial resolution with the MODIS point spread function (PSF) (see the next section for the MODIS PSF). The figures show how cross-sensor VI relationships (including magnitude and linearity) vary as a function of spectral bandpass. For example, Terra and Aqua MODIS are spectrally perfectly compatible for all the three VIs examined in this chapter (Figures 26.2j, 26.3j, and 26.4a); and it can also be seen that cross-sensor relationships to MODIS are generally more linear for EVI2 and EVI than for NDVI (Figures 26.2 through 26.4).

The simulated dataset if expanded to include more scenes can be divided into subsets based on geographic areas, seasons, and/or land cover types to examine geographic, seasonal, or land cover dependencies of cross-sensor VI relationships.

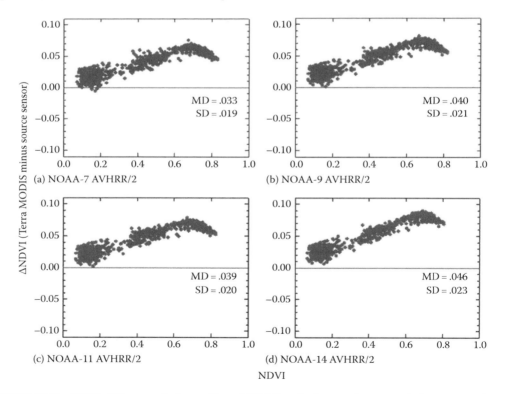

FIGURE 26.2 NDVI differences between Terra MODIS and other moderate resolution sensors (source sensors) plotted against source sensors. MD and SD are mean differences and standard deviations of the differences, respectively.

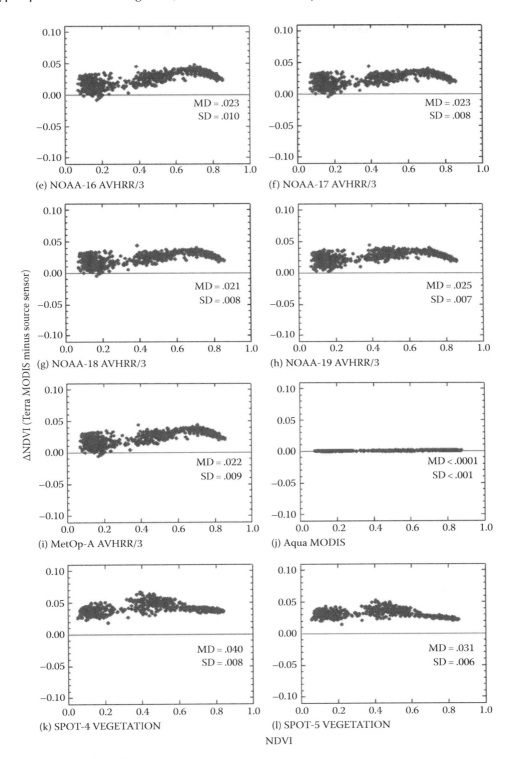

FIGURE 26.2 (continued)

(*continued*)

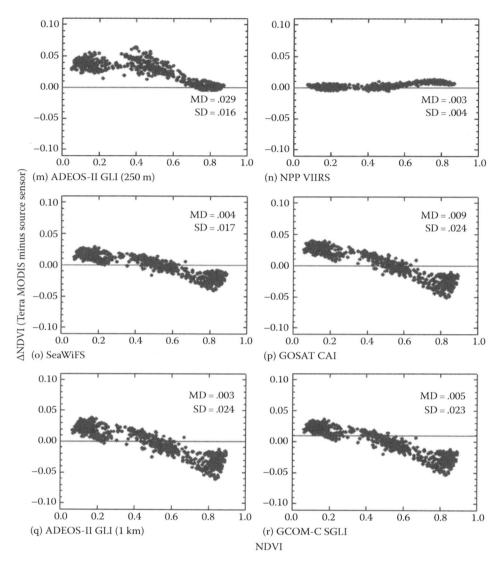

(m) ADEOS-II GLI (250 m)

(n) NPP VIIRS

(o) SeaWiFS

(p) GOSAT CAI

(q) ADEOS-II GLI (1 km)

(r) GCOM-C SGLI

NDVI

FIGURE 26.2 (continued)

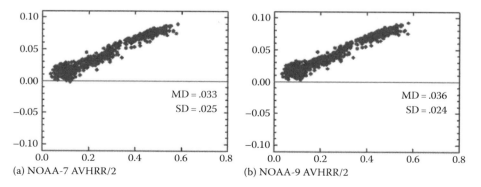

(a) NOAA-7 AVHRR/2

(b) NOAA-9 AVHRR/2

FIGURE 26.3 Same as Figure 26.2, but for the EVI2.

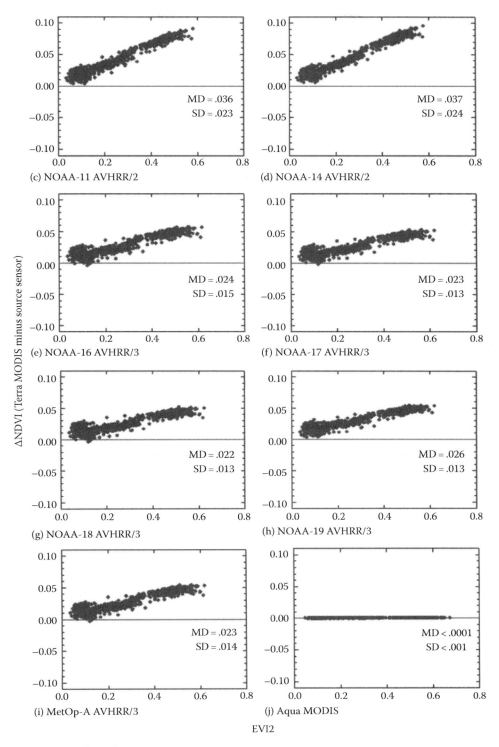

FIGURE 26.3 (continued)

(*continued*)

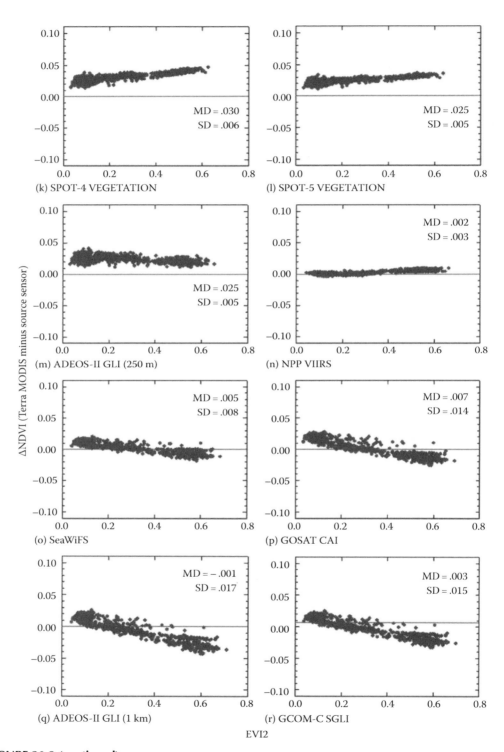

FIGURE 26.3 (continued)

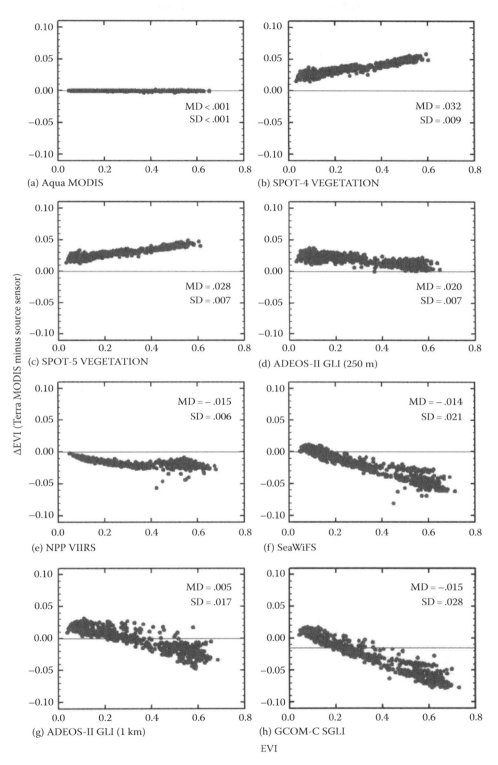

FIGURE 26.4 Same as Figure 26.2, but for the EVI.

26.4 SPATIAL COMPATIBILITY ANALYSES

Another key sensor characteristic that varies across sensors is spatial resolution (PSF). Although it is critical, the spatial issue of continuity has not been paid as much attention as the spectral issue. Central to VI spatial compatibility is the scale-invariance properties or scaling uncertainties of VIs with the influence of land surface heterogeneity [41–44]. The VI scaling uncertainties arise when a VI involves a non-linear transformation of input reflectance data. In other words, an average of fine-resolution VI values is not equal to a VI value computed from coarser-resolution reflectance [44] and the degree of this difference is expected to vary with surface heterogeneity and VI formula [18,42]. In the following, we compare the fine grain VIs and the coarse scale VIs using the NDVI as an example.

The fine grain NDVI can be aggregated to a coarser resolution pixel by [18,44]

$$\mathrm{NDVI}_{\mathrm{fine}} = f_1 \cdot \mathrm{NDVI}_1 + f_2 \cdot \mathrm{NDVI}_2$$

$$= \frac{f_1 \cdot (\rho_{\mathrm{NIR},1} - \rho_{\mathrm{red},1})}{\rho_{\mathrm{NIR},1} + \rho_{\mathrm{red},1}} + \frac{f_2 \cdot (\rho_{\mathrm{NIR},2} - \rho_{\mathrm{red},2})}{\rho_{\mathrm{NIR},2} + \rho_{\mathrm{red},2}}, \tag{26.4}$$

where two surface types with the fractional amounts of f_1 and f_2 ($f_1 + f_2 = 1$) are assumed. This quantity is not generally equal to the coarser resolution NDVI computed from the reflectances at the resolution analyzed, which can be expressed using the fine grain reflectances as

$$\mathrm{NDVI}_{\mathrm{coarse}} = \frac{(f_1 \cdot \rho_{\mathrm{NIR},1} + f_2 \cdot \rho_{\mathrm{NIR},2}) - (f_1 \cdot \rho_{\mathrm{red},1} + f_2 \cdot \rho_{\mathrm{red},2})}{(f_1 \cdot \rho_{\mathrm{NIR},1} + f_2 \cdot \rho_{\mathrm{NIR},2}) + (f_1 \cdot \rho_{\mathrm{red},1} + f_2 \cdot \rho_{\mathrm{red},2})} = \frac{f_1 \cdot (\rho_{\mathrm{NIR},1} - \rho_{\mathrm{red},1}) + f_2 \cdot (\rho_{\mathrm{NIR},2} - \rho_{\mathrm{red},2})}{f_1 \cdot (\rho_{\mathrm{NIR},1} + \rho_{\mathrm{red},1}) + f_2 \cdot (\rho_{\mathrm{NIR},2} + \rho_{\mathrm{red},2})} \tag{26.5}$$

and, thus,

$$\mathrm{NDVI}_{\mathrm{coarse}} \neq \mathrm{NDVI}_{\mathrm{fine}} \quad \text{or} \quad D \equiv \mathrm{NDVI}_{\mathrm{coarse}} - \mathrm{NDVI}_{\mathrm{fine}} \neq 0. \tag{26.6}$$

These two quantities are equal, or the quantity D is equal to zero (1) when either f_1 or f_2 is equal to zero (i.e., homogeneous case) [18] or (2) when the 1-norms of the two endmember spectra are equal (i.e., $\rho_{\mathrm{NIR},1} + \rho_{\mathrm{red},1} = \rho_{\mathrm{NIR},2} + \rho_{\mathrm{red},2}$) [45]. Theoretically, at least, the former applies to the EVI and EVI2 formula [18].

In practice, this implies that VIs derived at higher resolution could not be used simply as an enhanced resolution of VIs at a lower resolution, that is, their sensitivities to actual surface changes are different. This, in turn, requires an investigation on whether VIs from multiple sensors with various spatial resolutions and PSFs show compatible or different sensitivities to actual vegetation changes for VI-based change detections. The existing and planned hyperspectral sensors provide images at medium resolutions (30–60 m), which can spatially be aggregated to simulate various resolutions as low as their swath widths. Therefore, with hyperspectral imagery, scaling uncertainties can be analyzed separately (single factor analysis) and also simultaneously with the effect of spectral bandpass differences (two factor analysis). In the following, two examples of such hyperspectral data analyses are provided for the demonstration purpose.

In the first example, the Hyperion scenes in Table 26.1 were spatially aggregated to 60, 120, 240, 480, and 960 m spatial resolutions at the VI (fine-grain data) and reflectance (coarse-grain data) levels assuming a square PSF and a total atmospheric correction scenario. The fine-grain VIs were subtracted from the coarser-grain counterparts to assess scale-induced deviations (differences) at the different

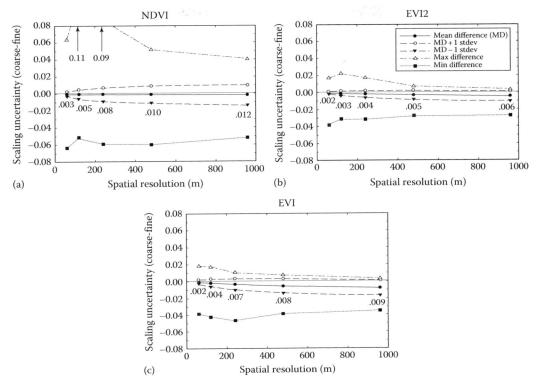

FIGURE 26.5 Univariate statistics of NDVI, EVI2, and EVI differences between their coarse- (MODIS band-passes) and fine-grain (MODIS bandpasses) averaged values computed from the May 24, 2001 Hyperion image over Maricopa, AZ. The numbers accompanying the filled triangles are standard deviations of the differences.

resolutions (D in Equation 26.6). In Figure 26.5, the derived differences where the MODIS spectral responses were assumed for both the fine- and coarse-grain data (single factor case) are plotted for the NDVI, EVI2, and EVI for the Maricopa scene of May 24, 2001. The plotted differences were the largest for the NDVI and the smallest for the EVI2, suggesting that the NDVI is subject to larger scaling uncertainties than the EVI2. For all the three VIs, in general, mean, maximum, and minimum differences decreased with increasing resolutions, while standard deviations of the difference increased (Figure 26.5). This indicates that the scale-induced deviation is generally larger for larger resolution differences (i.e., 30 m vs. 960 m) although extremely large deviations are more likely to be encountered for smaller resolutions (i.e., 30 m vs. 60–130 m). The nature of this monotonic change of NDVI scaling errors is discussed in detail for a two endmember linear mixture model in Ref. [45].

Plotted in Figure 26.6 are the derived differences for the same Maricopa scene in which the MODIS spectral responses were assumed for the coarse-grain data, but the Landsat-5 TM spectral responses for the fine-grain data, simulating synergistic applications of multi-resolution remote sensing (two factor case). Two differences from Figure 26.5 can be observed. First, large systematic differences were introduced due to spectral bandpass differences, which were ~.035 for the NDVI (Figure 26.6a), ~.03 for the EVI2 (Figure 26.6b), and ~.02 for the EVI (Figure 26.6c). Second, standard deviations of the differences for this two factor case (Figure 26.6) were larger and more uniform across the resolutions than those for the single factor case (Figure 26.5). These results simply suggest that scale-induced deviations in VIs can be larger when comparing multi-resolution sensor data with different spectral bandpasses than with the same bandpasses. A more thorough analysis is required to understand the mechanism by which scaling uncertainties are affected by spectral bandpass differences.

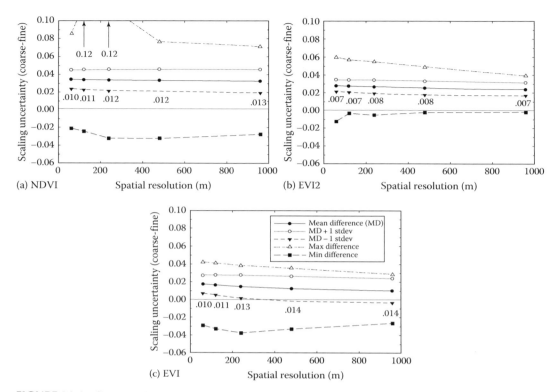

FIGURE 26.6   Same as Figure 26.5, but differences between the coarse-(MODIS bandpasses) and fine-grain (Landsat-5 TM bandpasses) averaged values.

In the second example, the Hyperion scenes in Table 26.1 were used to simulate the AVHRR Global Area Coverage (GAC) sampling scheme and the MODIS Climate Modeling Grid (CMG) aggregation for assessing the scaling uncertainties between these two products. A GAC pixel value represents the mean of four out of each five consecutive samples along the scan line and only data from each third scan line are processed and stored, which are performed onboard the sensor in real time [46]. As a result, the spatial resolution of GAC data near nadir is about 1.1 km × 4 km with a 3 km gap between pixels along track. In contrast, a MODIS CMG pixel is created by aggregating all of pixels inside the 0.05° CMG grid (~5 km × 5 km) (Didan, K., *personal communication*).

Spectrally convolved Hyperion scenes were first aggregated to AVHRR 1.1 km and MODIS 500 m resolutions, which were then aggregated to GAC and CMG pixels. We assumed a bell-shaped PSF and a triangular PSF in the scan direction for AVHRR and MODIS, respectively, and a rectangular PSF in the track direction for both AVHRR and MODIS [47,48]. VIs were computed from the simulated GAC pixels and five of these pixels were averaged to generate fine-grain VI values approximately equal to the CMG pixel size. These spatial aggregations and averaging of Hyperion pixels were performed carefully and systematically so that the derived fine-grain GAC and coarse-grain CMG pixels were co-located without mis-registration.

In Figure 26.7a and b, only the effect of the spectral bandpass difference between Terra MODIS and NOAA-14 AVHRR on the NDVI and EVI2, respectively, was assessed for the two spatial resolutions. MODIS-AVHRR cross-sensor NDVI and EVI2 relationships for the GAC resolution were basically the same as those for the CMG resolution, that is, the trends in the relationships were the same for the two resolutions. In fact, these trends in cross-sensor relationships were also very similar to the ones observed for 1 km resolution (see Figures 26.2d and 26.3d for the NDVI and EVI2, respectively). In Figure 26.7c and d, only the effects of the spatial resolution difference between the CMG and GAC resolutions (scaling uncertainties) was assessed for the

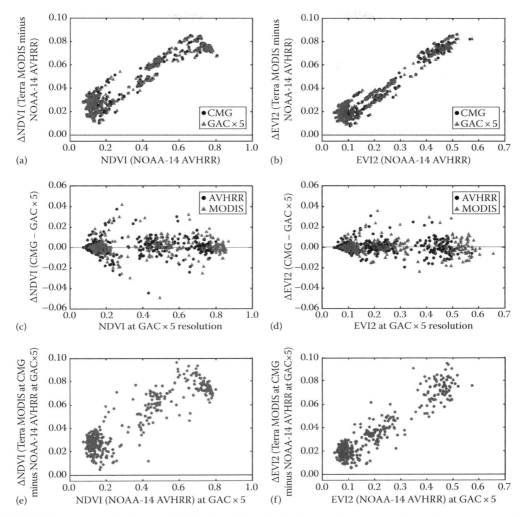

FIGURE 26.7 **(See color insert.)** (a) NDVI and (b) EVI2 differences due to spectral bandpass differences for CMG and GAC resolutions; (c) NDVI and (d) EVI2 differences due to resolution differences for AVHRR and MODIS spectral bandpasses; (e) NDVI and (f) EVI2 differences due to both spectral bandpass (MODIS vs. AVHRR) and spatial resolution (CMG vs. GAC) differences. Here, a GAC pixel is an average of five GAC pixels and, thus, written as "GAC × 5."

NDVI and EVI2, respectively, by fixing the spectral bandpasses. There were large variations in scale-induced differences for all the four cases (MODIS NDVI, AVHRR NDVI, MODIS EVI2, and AVHRR EVI2), ranging from −.05 to.05 at most for MODIS NDVI (Figure 26.7c); however, these scaling uncertainties did not appear to introduce any systematic differences (i.e., mean differences ≈ 0). In Figure 26.7e and f, the combined effects of the spectral bandpass and spatial resolution differences between MODIS CMG and AVHRR GAC VIs were assessed for the NDVI and EVI2, respectively. For both the NDVI and EVI2, the trends in cross-sensor relationships remained similar to those due only to the spectral bandpass difference; however, the secondary scattering about the mean trends became larger due to the scale-induced variations. These results suggest that MODIS CMG and AVHRR GAC VIs can be combined to generate a long-term data record, but would be accompanied by added uncertainties due to scaling differences.

The above two examples mentioned earlier can be expanded to a larger dataset to obtain more reliable estimates of scaling uncertainties. The demonstrated capability of hyperspectral data to analyze multiple factors one at a time and all at once will also allow to derive error budgets for those multiple factors. Although an example is not provided here, hyperspectral data can also be used in the same way as mentioned earlier to assess the impact of mis-registration on cross-sensor VI continuity.

26.5   ALGORITHM DIFFERENCES

The impacts of algorithm differences posed by sensor characteristic differences on multi-sensor VI continuity/compatibility are another area that requires careful and thorough investigations. Atmospheric correction is one key algorithm difference that exists among sensor products. Hyperspectral remote sensing can be an effective means to address the impacts of various atmospheric correction schemes on cross-sensor VI continuity/compatibility. In the following, we demonstrate it using the Hyperion scenes in Table 26.1.

Various atmospheric correction schemes for satellite remote sensing have been developed, but the work by Kaufman et al. [49,50] can be considered to have laid the foundation for operational atmospheric corrections of multi-spectral data in the solar reflective region over global land surface. The Global Inventory Modeling and Mapping Studies AVHRR NDVI product is corrected only for stratospheric aerosol effects [9], whereas other AVHRR products, including the Long-Term Data Records NDVI product [10] and the Conterminous USA and Alaska 1 km AVHRR product [12], are corrected for molecular scattering and ozone and water vapor absorptions. A total atmospheric correction scheme has been implemented for the MODIS VI and VEGETATION NDVI products; however, their algorithms and atmospheric data sources are different [51,52]. For NPP and JPSS VIIRS, two VI products are planned, that is, a TOA NDVI without any atmospheric correction and an atmospherically corrected top-of-canopy (TOC) EVI [53].

In Figure 26.8, cross-sensor NDVI relationships were examined for two atmospheric correction schemes: no correction (TOA NDVI) and total correction (TOC NDVI). The GAC sampling scheme was assumed for the simulated AVHRR data, whereas the CMG aggregation scheme was

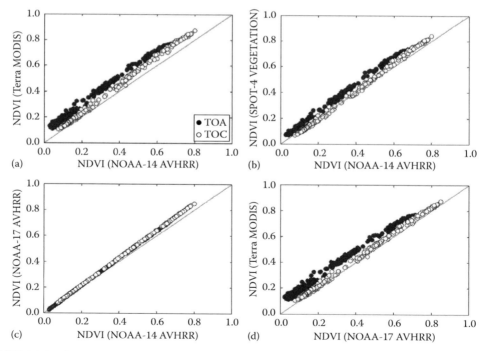

FIGURE 26.8   Cross-sensor NDVI plots for two atmospheric correction schemes among Terra MODIS, SPOT-4 VEGETATION, NOAA-14 AVHRR/2, NOAA-17 AVHRR/3, and NPP VIIRS.

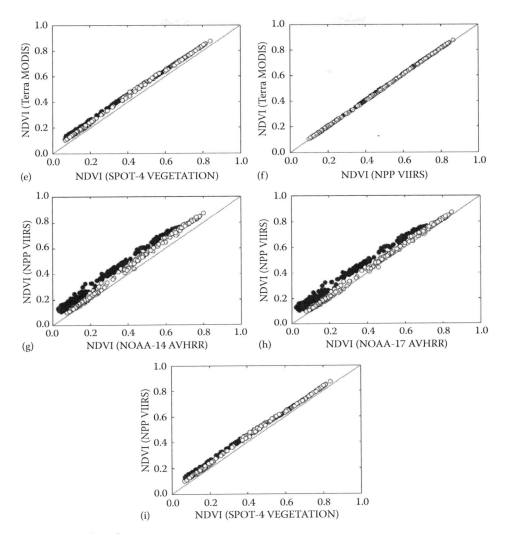

FIGURE 26.8 (continued)

used for all the other simulated sensor data (see Section 26.4). Cross-sensor NDVI relationships of both NOAA-14 AVHRR/2 and NOAA-17 AVHRR/3 with other sensors changed with atmospheric correction schemes (Figure 26.8a, b, d, g, and h), and, thus, separate cross-calibrations are required for establishing continuity for the TOA-NDVI and TOC-NDVI. In contrast, NOAA-14 AVHRR and NOAA-17 AVHRR, and Terra MODIS and NPP VIIRS had excellent continuity/compatibility; their cross-sensor NDVI relationships were the same regardless of the atmospheric correction schemes (Figure 26.8c and f). Once a cross-sensor NDVI relationship is established, this can be used to relate the NDVI at either the TOA or TOC level for these sensor pairs. The SPOT-4 VEGETATION sensor also showed relatively robust cross-sensor NDVI relationships with Terra MODIS and NPP VIIRS (Figure 26.8e and i).

In Figures 26.9 and 26.10, we examined a scenario where total atmosphere-corrected VIs from Terra MODIS and SPOT-4 VEGETATION were compared with partial atmosphere-corrected VIs from NOAA-14 AVHRR and NPP VIIRS. For all the sensor pairs examined, cross-sensor NDVI and EVI2 relationships varied with atmospheric corrections. For both the NDVI and EVI2, larger changes were observed for the relationships involving NOAA-14 AVHRR/2 (Figure 26.9a and b,

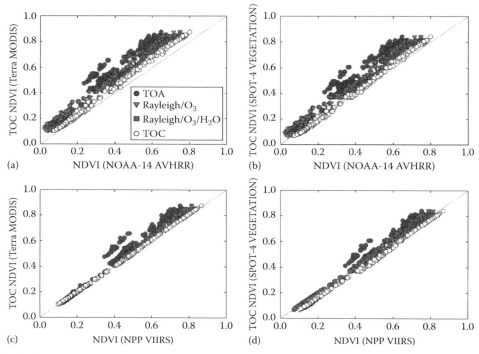

FIGURE 26.9 Cross-sensor NDVI plots for four different atmospheric correction schemes: (a) Terra MODIS vs. NOAA-14 AVHRR/2, (b) SPOT-4 VEGETATION vs. NOAA-14 AVHRR/2, (c) Terra MODIS vs. NPP VIIRS, and (d) SPOT-4 VEGETATION vs. NPP VIIRS. Only NOAA-14 AVHRR and NPP VIIRS values were subject to various atmospheric correction schemes.

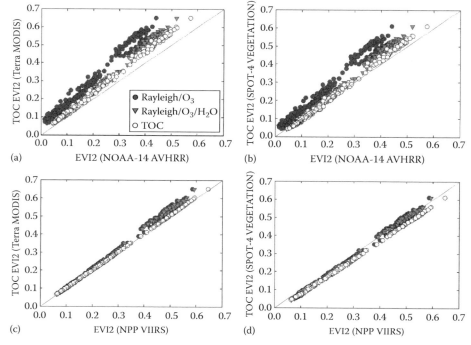

FIGURE 26.10 **(See color insert.)** Same as Figure 26.9, but for EVI2 with three different atmospheric correction schemes.

and Figure 26.10a and b). The smallest change was observed for cross-sensor EVI2 relationships involving NPP VIIRS (Figure 26.10c and d).

Overall, these example analyses have shown important implications for cross-sensor VI continuity/compatibility. It is worth while to note that, based on the Hyperion simulation analyses presented here, AVHRR/2 and AVHRR/3 as well as MODIS and VIIRS maintain excellent continuity for the NDVI regardless of atmospheric corrections.

26.6 ANGULAR EFFECTS

Other sensor and platform characteristics that require consideration are the sun-target-sensor geometry and temporal resolution. While polar-orbiting wide field of view sensors, such as AVHRR, MODIS, VEGETATION, and NPP discussed earlier, provide near daily global coverage, they acquire and build up sequential angular views over a period of hours to days [54]. For time series applications of VIs, this varying observation geometry is considered a source of noise, and, thus, various attempts have been made to normalize this noise, including multi-date temporal compositing [55,56] or BRDF-based adjustments to nadir-viewing geometry [57,58]. Since the BRDF changes as a function of wavelength and since orbital and scanning characteristics differ from sensor to sensor (including overpass time), the BRDF effect is another issue of multi-sensor VI continuity to be addressed. Satellite hyperspectral remote sensing has great potential to contribute to this issue.

Table 26.2 lists the current and planned satellite hyperspectral missions. As can be seen in Table 26.2, most of the hyperspectral sensors have the cross-track pointing capability to allow for imaging of the same target area at different viewing angles (e.g., [59]). The Hyperion sensor onboard the EO-1 satellite, for example, is capable of imaging within one adjacent World Reference System-2 (WRS-2) path in both the east and west direction, giving the sensor three imaging opportunities with three different view angles per its 16-day orbital repeat cycle (http://edcsns17.cr.usgs.gov/eo1/lookAngles.php). It is only the PROBA CHRIS which has the along-track pointing capability. Verrelst [60] used CHRIS/PROBA data to analyze angular sensitivities of the NDVI, and other narrowband greenness and light use efficiency indices. It is, however, unlikely that one can obtain a large enough number of multi-angular hyperspectral images over the same target for a short period of time from a single sensor, considering the attainable imaging frequency together with the influence of cloud cover [61]. Such observations may only be realized after several planned hyperspectral sensors (Table 26.2) are successfully launched.

26.7 DISCUSSIONS

In this chapter, we discussed the potential utilities of satellite hyperspectral remote sensing in multi-sensor VI continuity/compatibility studies for global long-term monitoring. Some aspects of such utilities have been demonstrated using the regional Hyperion dataset over the conterminous United States. They included spectral compatibility and spatial compatibility analyses, scaling uncertainty analysis, and the effects of atmospheric correction algorithm differences.

It is important to note that upcoming hyperspectral missions can serve not only as a means of detailed and highly precise characterization of terrestrial vegetation, but also as a spaceborne reference for establishing multi-sensor continuity and compatibility among current, past, and future multi-spectral sensors. The future hyperspectral missions will provide wider imaging swaths to cover larger land surface areas than the currently achieved coverage (Table 26.2). The three near-future missions of PRISMA, EnMAP, and HISUI will all provide 30 m spatial resolution hyperspectral images with a 30 km swath width, which may enable a provision of high temporal resolution, multi-angular hyperspectral observations over the same targets for the hyperspectral BRDF characterization of surface. The multi-angular hyperspectral observation capability may be one of next important steps in the field of hyperspectral remote sensing.

TABLE 26.2
Current and Future Spaceborne Hyperspectral Missions

Name	Launch Date	Swath (km)	Spatial Resolution (m)	Spectral Coverage (nm)	Nominal Bandwidth (nm)	No. of Bands	Pointing	Altitude, Overpass Time
Future missions								
MSMISat [59] (South Africa and Belgium)	2010 (TBD)	15	15	400–990	10	>200	±30° cross-track	~600 km
				940–2350	10			
PRISMA (Italy)	2012	30	30	400–1010	<10	92	±14.7° cross-track	620 km 10:30 am
				920–2505	<10	157		
EnMAP [62] (Germany)	2014	30	30	420–1000	6.5	94	±30° cross-track	643 km
				900–1390	10	139		
				1489–1760				
				1950–2450				
HISUI (Japan)	2015	30	30	400–970	10	57	±4.2° (±45 km) cross-track	618.2 km
				900–2500	12.5	128		
HyspIRI (USA)	>2016	150	60	380–2500	10	212	Cross-track	626 km 10:30 am
Resourcesat-3[a] (India)	2014	25	25			~200		817 km
Current missions								
EO-1 Hyperion (USA)	November 21, 2000	7.5	30	400–2400	10	196	±17.433° cross-track	705 km, drifting
PROBA CHRIS (ESA)	October 22, 2001	14×14	17	410–1050	5–12	18	Nadir ±36°	570–670 km
			34			63	±55° along track — ±30° cross-track	
HySI (India)	April 28, 2008	129.5	505.6	450–950	8	64	Nadir	635 km
HJ-1A (China)	September 6, 2008	50 (60)	100	450–950	5	128 (115)	Nadir	649 km

[a] Puckorius, T.J., personal communication.

ACKNOWLEDGMENTS

The authors would like to thank Alfredo Huete for his continuing discussions about this research topic and Akira Iwasaki for his comments on hyperspectral sensors. The authors would also like to thank Joshua Turner for his assistance in assembling the dataset used in this study. We thank the principal investigators of the five AERONET sites used in this study for their effort in establishing and maintaining these sites.

REFERENCES

1. Huete, A. R.; Huemmrich, K. F.; Miura, T.; Xiao, X.; Didan, K.; van Leeuwen, W.; Hall, F.; Tucker, C. J. Vegetation Index greenness global data set. *White Paper for NASA ESDR/CDR* 2006, http://landportal.gsfc.nasa.gov/Documents/ESDR/VI_Huete_whitepaper.pdf, last accessed on March 8, 2010.
2. Myneni, R. B.; Nemani, R. R.; Running, S. W. Estimation of global leaf area index and absorbed par using radiative transfer models. *IEEE Trans. Geosci. Remote Sens.* 1997, *35*, 1380–1393.
3. Sims, D. A.; Rahman, A. F.; Cordova, V. D.; El-Masri, B. Z.; Baldocchi, D. D.; Bolstad, P. V.; Flanagan, L. B. et al. A new model of gross primary productivity for North American ecosystems based solely on the enhanced vegetation index and land surface temperature from MODIS. *Remote Sens. Environ.* 2008, *112*, 1633–1646.
4. Jepson, W.; Brannstrom, C.; Filippi, A. Access regimes and regional land change in the Brazilian Cerrado, 1972–2002. *Ann. Assoc. Am. Geogr.* 2010, *100*, 87–111.
5. Bhattarai, K.; Conway, D.; Yousef, M. Determinants of deforestation in Nepal's central development region. *J. Environ. Manage.* 2009, *91*, 471–488.
6. Paudel, K. P.; Andersen, P. Assessing rangeland degradation using multi temporal satellite images and grazing pressure surface model in Upper Mustang, Trans Himalaya, Nepal. *Remote Sens. Environ.* 2010, *114*, 1845–1855.
7. Miettinen, J.; Liew, S. C. Estimation of biomass distribution in Peninsular Malaysia and in the islands of Sumatra, Java and Borneo based on multi-resolution remote sensing land cover analysis. *Mitig. Adapt. Strateg. Glob. Change* 2009, *14*, 357–373.
8. DeFries, R.; Achard, F.; Brown, S.; Herold, M.; Murdiyarso, D.; Schlamadinger, B.; de Souza, C. Jr. Earth observations for estimating greenhouse gas emissions from deforestation in developing countries. *Environ. Sci. Policy* 2007, *10*, 385–394.
9. Tucker, C. J.; Pinzon, J. E.; Brown, M. E.; Slayback, D. A.; Pak, E. W.; Mahoney, R.; Vermote, E. F.; El Saleous, N. An extended AVHRR 8-km NDVI dataset compatible with MODIS and SPOT vegetation NDVI data. *Int. J. Remote Sens.* 2005, *26*, 4485–4498.
10. Pedelty, J.; Devadiga, S.; Masuoka, E.; Brown, M.; Pinzon, J.; Tucker, C.; Roy, D. et al. Generating a long-term land data record from the AVHRR and MODIS Instruments. *IGARSS 2007*, 2007, 1021–1025, doi:10.1109/IGARSS.2007.4422974.
11. Yu, Y. Y.; Privette, J. L.; Pinheiro, A. C. Analysis of the NPOESS VIIRS land surface temperature algorithm using MODIS data. *IEEE Trans. Geosci. Remote Sens.* 2005, *43*, 2340–2350.
12. Eidenshink, J. A 16-year time series of 1 km AVHRR satellite data of the conterminous United States and Alaska. *Photogramm. Eng. Remote Sens.* 2006, *72*, 1027–1035.
13. Teillet, P. M.; Staenz, K.; Williams, D. J. Effects of spectral, spatial, and radiometric characteristics on remote sensing vegetation indices of forested regions. *Remote Sens. Environ.* 1997, *61*, 139–149.
14. Batra, N.; Islam, S.; Venturini, V.; Bisht, G.; Jiang, L. Estimation and comparison of evapotranspiration from MODIS and AVHRR sensors for clear sky days over the Southern Great Plains. *Remote Sens. Environ.* 2006, *103*, 1–15.
15. Swinnen, E.; Veroustraete, F. Extending the SPOT-VEGETATION NDVI time series (1998–2006) back in time with NOAA-AVHRR data (1985–1998) for Southern Africa. *IEEE Trans. Geosci. Remote Sens.* 2008, *46*, 558–572.
16. Yoshioka, H.; Miura, T.; Huete, A. R. An isoline-based translation technique of spectral vegetation index using EO-1 Hyperion data. *IEEE Trans. Geosci. Remote Sens.* 2003, *41*, 1363–1372.
17. Miura, T.; Huete, A.; Yoshioka, H. An empirical investigation of cross-sensor relationships of NDVI and red/near-infrared reflectance using EO-1 Hyperion data. *Remote Sens. Environ.* 2006, *100*, 223–236.
18. Huete, A.; Ho-Jin, K.; Miura, T. Scaling dependencies and uncertainties in vegetation index—Biophysical retrievals in heterogeneous environments. *Proceedings IGARSS '05*, 2005, *7*, 5029–5032.

19. Pearlman, J. S.; Barry, P. S.; Segal, C. C.; Shepanski, J.; Beiso, D.; Carman, S. L. Hyperion, a space-based imaging spectrometer. *IEEE Trans. Geosci. Remote Sens.* 2003, *41*, 1160–1173.

20. Ungar, S. G.; Pearlman, J. S.; Mendenhall, J. A.; Reuter, D. Overview of the Earth Observing One (EO-1) mission. *IEEE Trans. Geosci. Remote Sens.* 2003, *41*, 1149–1159.

21. Holben, B. N.; Tanré, D.; Smirnov, A.; Eck, T. F.; Slutsker, I.; Abuhassan, N.; Newcomb, W. W. et al. An emerging ground-based aerosol climatology: Aerosol optical depth from AERONET. *J. Geophys. Res.* 2001, *106*, 12067–12097.

22. Vermote, E.; Tanré, D.; Deuzé, J. L.; Herman, M.; Morcrette, J. J.; Kotchenova, S. Y. *Second Simulation of a Satellite Signal in the Solar Spectrum—Vector (6SV) User Guide Version 3*; 2006 (http://6s.ltdri.org/).

23. Kaufman, Y. J.; Tanré, D.; Remer, L. A.; Vermote, E. F.; Chu, A.; Holben, B. N. Operational remote sensing of tropospheric aerosol over land from EOS moderate resolution imaging spectroradiometer. *J. Geophys. Res.* 1997, *102*, 17051–17067.

24. Tucker, C. J. Red and photographic infrared linear combinations for monitoring vegetation. *Remote Sens. Environ.* 1979, *8*, 127–150.

25. Huete, A.; Didan, K.; Miura, T.; Rodriguez, E. P.; Gao, X.; Ferreira, L. G. Overview of the radiometric and biophysical performance of the MODIS vegetation indices. *Remote Sens. Environ.* 2002, *83*, 195–213.

26. Jiang, Z.; Huete, A. R.; Didan, K.; Miura, T. Development of a two-band enhanced vegetation index without a blue band. *Remote Sens. Environ.* 2008, *112*, 3833–3845.

27. Gallo, K.; Ji, L.; Reed, B.; Eidenshink, J.; Dwyer, J. Multi-platform comparisons of MODIS and AVHRR normalized difference vegetation index data. *Remote Sens. Environ.* 2005, *99*, 221–231.

28. Gao, B. C. A practical method for simulating AVHRR-consistent NDVI data series using narrow MODIS channels in the 0.5–1.0 μm spectral range. *Remote Sens. Environ.* 2000, *38*, 1969–1975.

29. Gitelson, A. A.; Kaufman, Y. J. MODIS NDVI optimization to fit the AVHRR data series—Spectral considerations. *Remote Sens. Environ.* 1998, *66*, 343–350.

30. Gunther, K. P.; Maier, S. W. AVHRR compatible vegetation index derived from MERIS data. *Int. J. Remote Sens.* 2007, *28*, 693–708.

31. Ji, L.; Gallo, K.; Eidenshink, J. C.; Dwyer, J. Agreement evaluation of AVHRR and MODIS 16-day composite NDVI data sets. *Int. J. Remote Sens.* 2008, *29*, 4839–4861.

32. Steven, M. D.; Malthus, T. J.; Baret, F.; Xu, H.; Chopping, M. J. Intercalibration of vegetation indices from different sensor systems. *Remote Sens. Environ.* 2003, *88*, 412–422.

33. Trishchenko, A. P.; Cihlar, J.; Li, Z. Effects of spectral response function on surface reflectance and NDVI measured with moderate resolution satellite sensors. *Remote Sens. Environ.* 2002, *81*, 1–18.

34. Kim, Y.; Huete, A. R.; Miura, T.; Jiang, Z. Spectral compatibility of vegetation indices across sensors: A band decomposition analysis with Hyperion data. *J. Appl. Remote Sens.* 2010, *4*, 043520, doi: 10.1117/1.3400635

35. Trishchenko, A. P. Effects of spectral response function on surface reflectance and NDVI measured with moderate resolution satellite sensors: Extension to AVHRR NOAA-17, 18 and METOP-A. *Remote Sens. Environ.* 2009, *113*, 335–341.

36. Yoshioka, H.; Miura, T.; Yamamoto, H. Investigation on functional form in cross-calibration of spectral vegetation index. *Proceedings of SPIE Remote Sensing and Modeling of Ecosystems for Sustainability III* 2006, *6298*, 629813–629819.

37. van Leeuwen, W. J. D.; Orr, B. J.; Marsh, S. E.; Herrmann, S. M. Multi-sensor NDVI data continuity: Uncertainties and implications for vegetation monitoring applications. *Remote Sens. Environ.* 2006, *100*, 67–81.

38. Ji, L.; Gallo, K. An agreement coefficient for image comparison. *Photogramm. Eng. Remote Sens.* 2006, *72*, 823–833.

39. Yoshioka, H.; Miura, T.; Yamamoto, H. Relationships of spectral vegetation indices for continuity and compatibility of satellite data products. In: *Multispectral and Hyperspectral Remote Sensing Instruments and Applications II, Proceedings of SPIE*, vol. 5655; Larar, A. M.; Suzuki, M.; Tong, Q. (Eds.), 2005, pp. 233–240, doi: 10.1117/12.578630

40. Miura, T.; Yoshioka, H.; Suzuki, T. Evaluation of spectral vegetation index translation equations for the development of long-term data records. *IGARSS 2008*, 2008, *3*, III-712–III-715.

41. Chen, J. M. Spatial scaling of a remotely sensed surface parameter by contexture. *Remote Sens. Environ.* 1999, *69*, 30–42.

42. Friedl, M. A.; Davis, F. W.; Michaelsen, J.; Moritz, M. A. Scaling and uncertainty in the relationship between the NDVI and land surface biophysical variables: An analysis using a scene simulation model and data from FIFE. *Remote Sens. Environ.* 1995, *54*, 233–246.

43. Hall, F. G.; Huemmrich, K. F.; Goetz, S. J.; Sellers, P. J.; Nickeson, J. E. Satellite remote sensing of surface energy balance: Success, failures, and unresolved issues in FIFE. *J. Geophys. Res.* 1992, *97*, 19061–19089.

44. Hu, Z.; Islam, S. A framework for analyzing and designing scale invariant remote sensing algorithms. *IEEE Trans. Geosci. Remote Sens.* 1997, *35*, 747–755.

45. Yoshioka, H.; Wada, T.; Obata, K.; Miura, T. Monotonicity of area averaged NDVI as a function of spatial resolution based on a variable endmember linear mixture model. *IGARSS 2008*, 2008, *3*, III-415–III-418.

46. Pinheiro, A. C. T.; Mahoney, R.; Privette, J. L.; Tucker, C. J. Development of a daily long term record of NOAA-14 AVHRR land surface temperature over Africa. *Remote Sens. Environ.* 2006, *103*, 153–164.

47. Schowengerdt, R. A. *Remote Sensing: Models and Methods for Image Processing*, Academic Press: New York, 2006.

48. Wolfe, R. E.; Nishihama, M.; Fleig, A. J.; Kuyper, J. A.; Roy, D. P.; Storey, J. C.; Patt, F. S. Achieving sub-pixel geolocation accuracy in support of MODIS land science. *Remote Sens. Environ.* 2002, *83*, 31–49.

49. Kaufman, Y. J.; Sendra, C. Algorithm for automatic atmospheric corrections to visible and near-IR satellite imagery. *Int. J. Remote Sens.* 1988, *9*, 1357–1381.

50. Tanré, D.; Holben, B. N.; Kaufman, Y. J. Atmospheric correction algorithm for NOAA-AVHRR products: Theory and application. *IEEE Trans. Geosci. Remote Sens.* 1992, *30*, 231–248.

51. Vermote, E. F.; Saleous, N. Z. Operational atmospheric correction of MODIS visible to middle infrared land surface data in the case of an infinite Lambertian target. In: *Earth Science Satellite Remote Sensing, Science and Instruments*, vol. 1; Qu, J. J. (Ed.), Tsinghua University Press: Beijing, 2006, pp. 123–153.

52. Maisongrande, P.; Duchemin, B.; Dedieu, G. VEGETATION/SPOT: An operational mission for the Earth monitoring; presentation of new standard products. *Int. J. Remote Sens.* 2004, *25*, 9–14.

53. Murphy, R. E.; Reed, B. *NPP VIIRS Update: Presentation to the Global Vegetation Workshop 2009*, 16–19 June 2009, Missoula, Montana, 2009.

54. Diner, D. J.; Asner, G. P.; Davies, R.; Knyazikhin, Y.; Muller, J.-P.; Nolin, A. W.; Pinty, B.; Schaaf, C. B.; Stroeve, J. New directions in earth observing: Scientific applications of multiangle remote sensing. *Bull. Am. Meteorol. Soc.* 1999, *80*, 2209–2228.

55. Holben, B. N. Characteristics of maximum-value composite images from temporal AVHRR data. *Int. J. Remote Sens.* 1986, *7*, 1417–1434.

56. van Leeuwen, W. J. D.; Huete, A. R.; Laing, T. W. MODIS vegetation index compositing approach: A prototype with AVHRR data. *Remote Sens. Environ.* 1999, *69*, 264–280.

57. Schaaf, C. B.; Gao, F.; Strahler, A. H.; Lucht, W.; Li, X.; Tsang, T.; Strugnell, N. C. et al. First operational BRDF, albedo nadir reflectance products from MODIS. *Remote Sens. Environ.* 2002, *83*, 135–148.

58. Vermote, E.; Justice, C. O.; Breon, F. M. Towards a generalized approach for correction of the BRDF effect in MODIS directional reflectances. *IEEE Trans. Geosci. Remote Sens.* 2009, *47*, 898–908.

59. Mutanga, O.; van Aardt, J.; Kumar, L. Imaging spectroscopy (hyperspectral remote sensing) in southern Africa: An overview. *S. Afr. J. Sci.* 2009, *105*, 193–198.

60. Verrelst, J.; Schaepman, M. E.; Koetz, B.; Kneubuhler, M. Angular sensitivity analysis of vegetation indices derived from CHRIS/PROBA data. *Remote Sens. Environ.* 2008, *112*, 2341–2353.

61. Galvão, L. S.; Roberts, D. A.; Formaggio, A. R.; Numata, I.; Breunig, F. M. View angle effects on the discrimination of soybean varieties and on the relationships between vegetation indices and yield using off-nadir Hyperion data. *Remote Sens. Environ.* 2009, *113*, 846–856.

62. Stuffler, T.; Kaufmann, C.; Hofer, S.; Förster, K. P.; Schreier, G.; Mueller, A.; Eckardt, A. et al. The EnMAP hyperspectral imager—An advanced optical payload for future applications in Earth observation programmes. *Acta Astronaut.* 2007, *61*, 115–120.

63. Korwan, D. R.; Lucke, R. L.; Corson, M.; Bowles, J. H.; Gao, B. G.; Li, R. R.; Montes, M. J. et al. The Hyperspectral Imager for the Coastal Ocean (HICO)—Design and early results. *2nd Workshop on Hyperspectral Image and Signal Processing: Evolution in Remote Sensing (WHISPERS)* 2010, Reykavyk, Iceland, pp. 1–4.

Part XI

Hyperspectral Remote Sensing
of Outer Planets

27 Hyperspectral Analysis of Rocky Surfaces on the Earth and Other Planetary Bodies

R. Greg Vaughan, Timothy N. Titus,
Jeffery R. Johnson, Justin J. Hagerty, Lisa R. Gaddis,
Laurence A. Soderblom, and Paul E. Geissler

CONTENTS

27.1 INTRODUCTION

This book is focused on studies of vegetation on Earth using hyperspectral remote sensing methods. However, it is appropriate to extend the application of these methods out to other rocky bodies in our Solar System for a variety of reasons. First, minerals, soils, and rocks form the substrate on which vegetation grows on Earth. Compositional analyses of these components with hyperspectral data provide essential background information for distinguishing, identifying, and removing their effects on vegetation spectra. Second, variation in distribution, chemical and physical properties among soil and rock, has been demonstrated to have a significant effect on factors such as moisture retention, dust production, and the presence and distribution of biological species ranging from bacteria, fungi, grasses, shrubs, trees, and small mammals to humans [1]. These factors in turn can have profound influences on human health.

Recent advances in the development and use of hyperspectral data for rocks, soils, and minerals have led to improvements in our understanding of how such data are collected, calibrated and otherwise processed, and applied to understand the geology and biology of materials on Earth and other planetary surfaces. Geologic remote sensing research over the last 40 years has developed and/or improved upon methods for image processing and analysis, data set validation, remote and in situ data collection, and field calibration, and has established nominal wavelengths for detection of common minerals [2–9]. Early studies used multispectral measurements in the visible, near-infrared, and shortwave infrared (VNIR/SWIR) range (400–2500 nm) to map weathering and alteration minerals [4,10–12]. With the advent of hyperspectral VNIR/SWIR measurements from airborne and spaceborne platforms and better calibration, imaging spectroscopy became possible, and images with high-resolution spectral data for each pixel could be directly compared to the reference spectra of pure minerals and used for remote mineral identification [6,13]. For the past 20 years, much research has been focused on the development of image analysis and processing techniques for hyperspectral VNIR/SWIR data, with particular attention to atmospheric correction and unique mineralogical identification [14–17].

Primary rock-forming minerals as well as many secondary weathering and alteration minerals exhibit wavelength-dependent, or spectral, absorption features throughout the visible and infrared wavelengths [18–22]. These features result from the selective absorption of photons with discrete energy levels and are dependent on the elemental composition, crystal structure, and chemical bonding characteristics of a mineral, and are therefore diagnostic of mineralogy [7,20,23,24]. The identification of a material based on its infrared spectrum requires access to the spectra of well-characterized specimens. Thus, the infrared spectral properties for many natural and man-made materials have been measured in the laboratory, and constitute the empirical basis for the surface mapping applications of infrared remote sensing applied toward Earth and other planets [18–22,25,26]. Archives of infrared spectral data for rocks, minerals, and vegetation are available for reference from the United States Geological Survey (USGS) spectral library (http://speclab.cr.usgs.gov/spectral.lib06), and the NASA Jet Propulsion Laboratory (JPL) Advanced Spaceborne Thermal Emission and Reflection Radiometer (ASTER) spectral library (http://speclib.jpl.nasa.gov) [26].

As described further in the following, geologic analyses of remotely acquired multi- and hyper-spectral data on planetary bodies have resulted in major scientific discoveries that help to put Earth science into a broader context. For example, correlating orbital remote sensing data for Mars from the Mars Reconnaissance Orbiter (MRO) CRISM (Compact Reconnaissance Imaging Spectrometer for Mars) instrument with those of high-resolution views of boulders, craters, and sediment layers by the High-Resolution Imaging Science Experiment (HiRISE) has helped to identify the likely presence of water at the surface of Mars through discovery of minerals such as gray hematite, clays, and sulfur-rich soils [27,28]. Similar nodules have been observed in southern Utah [29,30], where they appear to have formed by precipitation from fluid that leached out iron-rich minerals and precipitated them at more chemically favorable locations in the host sandstone. This discovery continues to be explored, in part because on Earth it is known that bacteria can make such concretions form more quickly [30,31]. The recent discovery of the presence of water on the surface of the Moon using hyperspectral Moon Mineralogy Mapper data [32] indicates that water has been much more common on that body and elsewhere in our Solar System than previously thought. These and other discoveries tell us that we still have much to learn about the geology and biology of Earth and other planets in our Solar System. Fortunately, remote sensing analyses such as those described in this book provide the tools needed for such future discoveries.

27.1.1 Planetary Bodies and Hyperspectral Instruments

27.1.1.1 Earth

There is currently a suite of Earth observing instruments that acquire hyperspectral images in the 400–2500 nm wavelength range [6,33–35]. As summarized in Table 27.1, these instruments vary in spatial resolution from 0.5 to 10 m, and spectral range from 380 to 2500 nm.

27.1.1.2 Mercury

The MErcury Surface, Space ENvironment, Geochemistry, and Ranging (MESSENGER) spacecraft performed two flybys of Mercury in 2008 and a third in September 2009. During both the first and second flyby, the Mercury Atmospheric and Surface Composition Spectrometer (MASCS) instrument obtained spectra of the surface. The purpose of the MASCS was to help determine surface mineral-ogy of Mercury, and to help characterize the exosphere (i.e., the outermost layer of Mercury's tenuous atmosphere). The MASCS consisted of two instruments: an UltraViolet-Visible Spectrometer (UVVS) designed primarily for observations of the exosphere, and a Visible-InfraRed Spectrograph (VIRS) intended to provide observations to determine surface mineralogy. Unfortunately, the VIRS instrument did not obtain surface data during the third flyby in 2009, although additional opportunities will be available when the MESSENGER spacecraft enters into an elongated orbit around Mercury in March, 2011. http://www.messenger-education.org/instruments/mascs.htm.

27.1.1.3 Moon

The NASA Moon Mineralogy Mapper (M^3) was a hyperspectral, pushbroom imaging spectrometer that was a guest instrument on Chandrayaan-1, the first mission to the Moon from the Indian Space Research Organization (ISRO). Intended for both global mapping and targeted operation, the main objective of M^3 was to allow scientists to examine lunar mineralogy at high spatial and spectral resolution [36]. M^3 targets included such features as outcrops exposed at the walls and central peaks of large craters, complex volcanic terrain, boundaries where different kinds of rocks converge, unusual or rare compositions, and lunar polar regions [36].

27.1.1.4 Mars

Mars has intrigued mankind since before the beginning of recorded history. Until the 1960s, Mars was believed to be a lush planet with canals and agricultural irrigation [37]. This belief became the basis for such science fiction classics as The War of the Worlds by H.G. Wells. The first flyby of Mars

TABLE 27.1

Summary of Earth and Planetary Hyperspectral Remote Sensing Instruments

	Hyperspectral Instrument	Spectral Range (nm)	Number of Channels	Spectral Bandpass (nm)	Spatial Resolution	Operational Dates
Earth						
Airborne	AVIRIS[a]	380–2500	224	10	4–20 m	1989–present
	ProSpecTIR-VS[b]	400–2450	256	2.3–20	1–10 m	~2000–present
	HyMap[c]	400–2500	128	15	2–10 m	~1997–present
	CASI[d]	400–1000	288	2–12	0.5–10 m	~1990–present
	SFSI[e]	1230–2380	230	10	0.5–10 m	1990–present
Spaceborne	EO-1 Hyperion[f]	400–2500	220	10	30 m	2001–present
Mercury	MESSENGER MASCS[g]	220–1450	768	0.2–0.5	1–650 km	2004–present
Moon	Chandrayaan-1 Moon Mineralogy Mapper[h]	400–2900	260	10	70–140 m	2008–2009
Mars	Mars Express OMEGA[i]	350–5100	352	7–20	300 m–4.8 km	2003–present
	Mars Reconnaissance Orbiter CRISM[j]	362–3920	545	6.55	15.7–200 m	2005–present
Jupiter	Galileo NIMS[k]	700–5200	1–408	12.5 and 25	50–500 km	1989–2003
Saturn	Cassini VIMS[l]	300–5100	352	7 and 14	10–20 km	1997–present

[a]   Airborne visible infrared imaging spectrometer (http://aviris.jpl.nasa.gov).

[b]   Spectral technology and innovative research corporation hyperspectral imaging spectrometer (http://www.spectir.com/assets/Images/Capabilities/ProspecTIR%20specs.pdf).

[c]   HyVista corporation hyperspectral mapper, developed by integrated spectronics (http://www.hyvista.com/main.html and http://www.intspec.com).

[d]   Compact airborne spectrographic imager (http://www.geomatics-group.co.uk/GeoCMS/Products/CASI.aspx).

[e]   SWIR full spectrum imager (http://www.borstad.com/sfsi.html).

[f]   Hyperion (http://eo1.gsfc.nasa.gov/Technology/Hyperion.html).

[g]   Mercury atmospheric and surface composition spectrometer (http://www.messenger-education.org/instruments/mascs.htm).

[h]   M³ (http://moonmineralogymapper.jpl.nasa.gov/INSTRUMENT/).

[i]   Observatoire pour la Minéralogie, l'Eau, les Glaces et l'Activité (http://sci.esa.int/science-e/www/object/index.cfm?fobjectid = 34826&fbodylongid = 1598).

[j]   Compact reconnaissance imaging spectrometer for Mars (http://crism.jhuapl.edu/).

[k]   Near-infrared mapping spectrometer (http://www2.jpl.nasa.gov/galileo/instruments/nims.html).

[l]   Visual and infrared mapping spectrometer (http://wwwvims.lpl.arizona.edu/).

by Mariner IV in 1965 put an end to such fantasies, revealing the red planet as a cold arid desert. While the use of remote sensing observations to track vegetation is meaningless for Mars, there are many other applications—such as the monitoring of seasonal changes in surface volatiles or the identification of minerals. The understanding of near-infrared observations of a planetary surface devoid of vegetation may provide insights into remote sensing of terrestrial surfaces with sparse or non-existent vegetation, for example, extremely arid deserts and ice-covered regions. Several spacecraft have visited Mars over the last five decades. Since 1997, three rovers, one lander, and four orbiters have successfully arrived at Mars and sent back valuable data to Earth. Most of these spacecraft far exceeded their nominal mission lifetimes (Table 27.1). Two of these spacecraft, Mars Express and MRO, carried VNIR/SWIR imaging spectrometers which are still functional at the time of this writing. The OMEGA (Observatoire pour la Minéralogie, l'Eau, les Glaces et l'Activité) instrument was a visible and infrared mineral mapper onboard the Mars Express orbiter designed to globally map minerals, water (hydrated minerals or ice), and ices. The CRISM is a high spatial resolution (20–200 m/pixel) imaging spectrometer designed to target regions to detect and map minerals and ices.

27.1.1.5 Jupiter

The Near-Infrared Mapping Spectrometer (NIMS) instrument on the Galileo spacecraft was the first imaging spectrometer to be sent to the outer Solar System. Galileo was launched in 1989 on a grand tour of the Solar System that crossed paths with Venus, Earth and the Moon (twice), and two small asteroids before finally arriving at Jupiter in late 1996, just in time to witness the kamikaze comet Shoemaker-Levy 9 crash into the giant planet. For more than 6 years, Galileo orbited Jupiter and made 29 successful close flybys of its major moons and many distant observations of Jupiter's atmosphere and rings and entourage of satellites. Each of the objects visited by Galileo became targets for scientific observations and calibration exercises for the novel NIMS instrument; here we focus on results from the Jupiter system, the primary science objective.

27.1.1.6 Saturn

The goals of the Visual and Infrared Mapping Spectrometer (VIMS) aboard the Cassini spacecraft include study of the composition, dynamics, clouds, and thermophysics of the atmospheres of Saturn and Titan and the identification and mapping of surface compositions of Titan, the other icy Saturnian satellites, and Saturn's rings. VIMS was an imaging spectrometer that covers the 350–5170 nm spectral region [38]. The VIMS visible-light channel is a multispectral imager that covers the spectral range from 350 to 1050 nm and uses a frame transfer charge-coupled device (CCD) detector on which spatial and spectral information is simultaneously stored. The VIMS infrared channel is a spatial-scanning spot spectrometer that covers the wavelength range from 850 to 5170 nm.

27.1.1.7 Future/Upcoming Instruments

One future Earth observing hyperspectral sensor mission recommended by the National Research Council, Decadal Survey (*Earth Science and Applications from Space: National Imperatives for the Next Decade and Beyond*) is the Hyperspectral Infrared Imager (HyspIRI) mission [39]. HyspIRI will have (1) a hyperspectral imaging spectrometer measuring radiance in the 380–2500 nm range with 10 nm spectral resolution, 60 m pixels, and a 19 day equatorial revisit time, and (2) a multispectral imager measuring radiance in seven channels in the 7,500–12,000 nm (thermal infrared region) plus one channel at 4000 nm (mid-infrared region) for measuring hot spots due to fires or volcanic activity, with 60 m pixels, and a 5 day revisit time (due to a wider swath than the VNIR/SWIR hyperspectral imager). The goals of the HyspIRI mission as related to VNIR/SWIR hyperspectral measurements are to detect responses of ecosystems to human land management and climate change variability, including studying the patterns and spatial distribution of ecosystems and their components; ecosystem function, physiology, seasonal activity, and relation to human health; biogeochemical cycles; and land surface composition [39]. See the HyspIRI web site for more information: http://hyspiri.jpl.nasa.gov/.

The Europa Jupiter System Mission (EJSM) planned for launch in 2020 will include the Jupiter Europa Orbiter (JEO) and the Jupiter Ganymede Orbiter (JGO). The mission will explore the Jupiter system, focusing on potential habitable environments among its icy satellites, with special emphasis on Europa and Ganymede because of the possibility they support internally active oceans. As part of the planned science payloads, hyperspectral imaging instruments are planned (via competitive selection) that will nominally cover the 800–2500 nm spectral region, with a targeted spectral resolution of 4 nm and instantaneous field of view smaller than 100 m (http://opfm.jpl.nasa.gov/files/JEO-Rpt_Public-Release_090203.pdf).

The Dawn visible and infrared (VIR) imaging spectrometer is an ion-drive spacecraft with instrumentation that includes VIR, an Italian hyperspectral imager that is a rebuild of the Visible and Infrared Thermal Imaging Spectrometer (VIRTIS) instrument—a mapping spectrometer on board the European spacecraft Rosetta. VIR has a spectral range from 250 to 5000 nm and will be used to map the mineral surface composition of the asteroids four Vesta and one Ceres. Neither asteroid has an atmosphere which often complicates remote sensing of planetary and terrestrial surfaces. On the other hand, while the lack of an atmosphere simplifies the identification of surface spectral features, the lack of an atmosphere results in wild temperature swings which can complicate the use of spectral features for mineral identification as many of these features are temperature sensitive. Additional laboratory work needs to be done to fully understand these issues.

27.1.1.8 Synergy with Remote Measurements outside the VNIR/SWIR Spectral Range

27.1.1.8.1 Thermal Infrared (TIR) Imaging and Spectroscopy

The VNIR/SWIR and TIR spectral ranges have traditionally been treated separately for a variety of reasons. Differences in the source of radiance (solar reflection vs. thermal emission) require different approaches to acquisition, calibration, and processing of the data. Also, infrared detector technology has played a role in the spectral ranges that are measured by different instruments. Recently the complementary nature of these two wavelength regions has been explored in detail by using instruments capable of making multichannel spectral measurements in both wavelength regions [40,41]. Some minerals exhibit diagnostic spectral features in one wavelength region, but not the other [42]. Also, some minerals that often occur together in nature have overlapping spectral features making unique mineral identification ambiguous using one wavelength region alone. Therefore, the synthesis of VNIR/SWIR and TIR provides a way to identify minerals more uniquely and remotely map more minerals and mineral assemblages [42].

27.1.1.8.2 MOLA and LOLA

Laser altimeters placed in orbit around Mars and the Moon provide not only high spatial resolution topographic models but are important in determining accurate placement of images and geometric registration among multiple imaging and hyperspectral planetary data sets. Sophisticated cartographic techniques allow spatial projection of images onto regional digital elevation models that improve overall accuracy of feature positions and minimize distortions or seams in large image mosaics. This is often essential to enable understanding particular surface features observed in multiple data sets (often acquired at different spatial resolutions). Further, knowledge of local topographic slopes is key in radiometrically calibrating data to compensate for photometric effects induced by mm-scale surface properties, light reflected from nearby features (e.g., crater rims), or light reflected by atmospheres (e.g., Mars, Titan).

27.1.1.8.3 Neutron/Gamma Ray

The use of gamma ray and neutron detectors to map surface and near-surface (typically, less than a meter deep) elemental abundances is gaining wide application on Mars [43,44], the Moon [45,46], and asteroids [47]. The detector footprint of the surface is altitude dependent but typically varies from 600 km diameter for Mars to 5 km for the Moon.

27.1.2 OVERVIEW OF HYPERSPECTRAL ANALYSIS TECHNIQUES

27.1.2.1 Radiometric Calibration

Radiometric calibration of hyperspectral imagers varies among instruments, but most employ some type of onboard calibration using integrating spheres illuminated either by sunlight or onboard illumination. Shutter systems are often used for acquisition of dark measurements to track real-time noise variations with sensor temperature and exposure times. Stray-light corrections can be necessary, often using a combination of pre-flight measurements or models and in-flight observations of stars or planetary surfaces with minimal spectral or textural variations to serve as working "flat field" observations as supplements to any similar onboard observations. Calibrated scene radiances are often converted to relative reflectance by convolving the solar flux at the planet's solar distance with specific bandpasses.

27.1.2.2 Atmospheric Correction/Compensation

In the VNIR/SWIR wavelength range, radiance measured at the sensor contains information about surface reflected radiance as well as radiance scattered by molecules and aerosols in the atmosphere, and absorbed by atmospheric gases. Of all the gaseous constituents of Earth's atmosphere, there are seven gasses that produce strong absorption of radiance in the VNIR/SWIR region: H_2O, CO_2, O_3, CH_4, N_2O, CO, and O_2 [48]. Although H_2O is not the most abundant gas, it is one of the most important in terms of how much it decreases the transmissivity of the atmosphere due to strong absorption. There is also a significant amount of variation in the distribution of these gasses, especially H_2O, both spatially and temporally. In addition to the absorbing effects of atmospheric gasses, molecules and aerosols in the atmosphere scatter solar radiance. The modeling of atmospheric absorption and scattering processes is known as radiative transfer theory and is reviewed in more detail by Hapke [49,50].

For the atmospheric correction of terrestrial VNIR/SWIR hyperspectral data a radiative transfer modeling approach is commonly used to correct for the effects of atmospheric conditions, backscattered radiance, albedo, and viewing geometry [51].

27.1.2.3 Spectral Indices

Similar to many terrestrial studies, first-order analyses of planetary data sets often involve the use of ratios or slopes between bands, often in combination with band depths computed at wavelengths diagnostic of particular minerals [52]. Such techniques provide computationally efficient means of identifying overall trends in spectral signatures, covering potentially large regions of planetary surfaces. Such products often provide necessary overviews of regions, from which anomalous spectral features subsequently can be investigated using higher spectral and/or spatial resolution observations.

In early lunar studies, ratios of narrowband telescopic images acquired at 400 and 560 nm were used to estimate weight percents of TiO_2 in the mare [53,54]. Early Martian spectroscopy made use of ratios and slopes to estimate dust contamination and movement on local and regional surfaces [55], whereas more recent analyses using CRISM hyperspectral data use a combination of slopes, ratio, and band depths to routinely provide maps of a suite of mineral and surface types [56]. In the outer Solar System, relatively lower albedo contaminants on Europa's trailing hemisphere exhibit steep positive spectral slopes (red color) up to 1000 nm. Slight differences in this slope are likely owing to variations in the water abundance and grain size of ice/contaminant mixtures [57].

27.1.2.4 Spectral Mixture Modeling

Natural geologic surfaces are often partially covered with non-geologic materials (e.g., vegetation) or composed of mixtures of minerals with varying grain sizes and differing degrees of compaction/solidification. These factors influence remote spectral measurements and can limit the number of pixels that can be classified and mapped accurately.

Mixing can exist at various scales and also affects the measured infrared spectral properties of an area. When minerals in a field of view are physically separated such that there is no scattering between components, the spectral signature of the area represents the sum of the fractions of each component and is thus a *linear mixture. Intimate mixing* occurs at smaller scales when different minerals are in close contact on a single scattering surface. The presence of rock coatings causes another type of mixing that varies depending on the thickness of the coating and the wavelength of the scattered radiation. And at the smallest scales, *molecular mixing* occurs when a liquid such as water is adsorbed onto a mineral surface or vegetation [24].

Even high spatial resolution (<2 m pixels) remote sensing measurements can have contributions from multiple sub-pixel-scale components. A simple semiquantitative approach can be used by calculating linear mixtures of reference spectra from spectral libraries and comparing them to remote and field spectra. Reference spectra for pure minerals can be chosen based on initial spectral analyses and/or a priori knowledge of the geology and surface minerals expected in a study area. Linear combinations of these pure mineral spectra can be compared to image spectra and spectral mixtures can be matched to image spectra to identify the dominant mineral phases present and estimate percentages of different mineral constituents. This semiquantitative treatment of linear mixtures for mineral mapping relies primarily on spectral feature shapes and locations, which are sufficient to detect the presence or absence of the dominant mineralogy of mixed pixels and produce mineral maps used to interpret the geology and geochemical environments. Using TIR data [58], and using both VNIR/SWIR and TIR data [42], showed that a linear unmixing technique that models the percentage of each end member composition can be used to identify individual surface minerals within a single pixel. More quantitative modeling requires well-calibrated spectral measurements and rigorous solutions to the equations radiative transfer theory [50].

27.2 HYPERSPECTRAL MISSIONS AND CASE STUDIES

27.2.1 EARTH

27.2.1.1 AVIRIS and Hyperion

The Airborne Visible Infrared Imaging Spectrometer (AVIRIS) is a hyperspectral imaging spectrometer developed by the NASA JPL. It measures radiance in 224 continuous channels between 380 and 2500 nm with a 10 nm spectral bandpass (Table 27.1). It has an instantaneous field of view (IFOV) of 1 mrad (0.057°) and a total field of view (TFOV) of 33° [14]. The signal-to-noise ratio (SNR) for channel 18 (550 nm) calculated using the mean/standard deviation method and normalized to 50% reflectance is ~500:1.

Hyperion is a spaceborne hyperspectral imager that was launched on the NASA EO-1 satellite in 2000 (Table 27.1). It measures radiance in 220 spectral channels from 400 to 2500 nm with 30 m pixels across a 7.5 km swath [35].

27.2.1.2 Calibration and Analysis Techniques

For AVIRIS data, the fast line-of-sight atmospheric analysis of spectral hypercubes (FLAASH) method can be used [51]. The FLAASH method uses a version of the moderate resolution atmospheric radiance and transmission (MODTRAN) model [59,60] to calculate the atmospheric parameters and at-surface reflectance. For the MODTRAN calculations the average atmospheric profile for a typical continental location at mid-latitudes during the summer, a visibility of 23 km, and a typical aerosol profile model for a rural area are commonly assumed. In addition, a spectral "polishing" routine [51,61] can be used to eliminate spectral artifacts that remain after atmospheric correction and to account for random channel to channel noise. The FLAASH method uses a running average across nine adjacent spectral channels to effect this polishing.

Calibrated surface reflectance spectra can be directly compared to reference spectra for pure minerals that are archived in spectral libraries. Both the USGS and JPL have spectral libraries available via the Internet and contain VNIR/SWIR spectra for over 2000 minerals and rocks combined. The most recent version of the USGS digital spectral library can be found at: http://speclab. cr.usgs.gov/spectral.lib06, and is described by Clark et al. [62]. The JPL ASTER spectral library is a compilation of spectra of rocks and minerals measured at JPL, the USGS, and John Hopkins University (JHU), and can be found at http://speclib.jpl.nasa.gov, and is described in Ref. [26].

A comprehensive analysis of multichannel image data incorporates classification techniques based on spectral variability within the scene, the physical laws governing radiative transfer and scattering, relation to field site measurements, and reliance on reference spectral libraries for matching band position, depth, and shape. The information contained in the spatial and spectral data can be displayed in three different ways: (1) in *image* space, where the spatial relationships between pixels are shown, (2) in *spectral* space, where spectral variations within a single pixel are shown, or (3) in *feature* space, where the spectral variations of each pixel are plotted as points (or vectors) in n-dimensional space, where n is the number of wavelength channels [63]. All three of these display methods can be utilized to maximize the amount of information extracted from multichannel image data.

There is a series of processing steps that has become standard in hyperspectral data analysis that yields reproducible results, although there are some subjective decisions that are required of the user [64,65]. These methods, from Refs. [65,66], can be implemented in the Environment for Visualizing Images (ENVI) software package [67]. The methods can be summarized by: (1) atmospheric correction and calibration to reflectance, (2) transformation to minimize noise and reduce data dimensionality, (3) location of spectrally "pure" pixels, (4) selection of spectral end members, and (5) pixel classification and mapping of spectral end members. The purpose of this methodology is to focus only on the information that is relevant to characteristic mineralogical features within the image.

27.2.1.3 Case Study

Hyperspectral VNIR/SWIR data have been used to identify and map a wide range of surface weathering and hydrothermal alteration minerals associated with acid mine drainage [68,69], mineral exploration targets and the surface expression of active and fossil geothermal systems [42,66,70].

In the example shown in Figure 27.1, AVIRIS data were used to map surface minerals associated with the active geothermal system at Steamboat Springs, Nevada. Steamboat Springs is an active geothermal system about 16 km south of Reno. It is characterized by exposures of recent siliceous sinter deposits, hydrothermally altered country rock, and structurally-controlled open fissures venting H_2S-rich steam. The Steamboat Springs system has been described as a modern analog to ancient hydrothermal systems associated with epithermal precious metal deposits throughout the Great Basin of the western United States [71–73]. Recent sinter deposits are composed of opaline silica, which transforms to β-cristobalite and chalcedony with increasing depth and age [73]. Also, as a result of the remote mineral mapping from Ref. [42], hydrous Na-Al sulfate crusts (tamarugite and alunogen) have been discovered forming around some active fumaroles. In the subsurface, sulfuric acid (H_2SO_4) solutions produced by H_2S reaction with atmospheric O_2 above the water table leaches the surrounding rock leaving opal, residual quartz, or quartz + alunite, adularia or kaolinite/ montmorillonite alteration assemblages. These minerals are characteristic of a steam-heated acid sulfate type alteration system [74] and are exposed at the surface locally as outcrops of argillized and acid-leached granodiorite and basaltic andesite.

In the Steamboat Springs region AVIRIS data mapped the distribution of opal, kaolinite and alunite in generally unvegetated areas (Figure 27.1). The mapped regions are displayed over a grayscale image of AVIRIS reflectance channel 28 (650 nm). AVIRIS spectrum in orange, from the main sinter terrace matches the reference spectrum from the USGS spectral library for opal. The broad spectral feature around 2250 nm and the strong features at 1400 and 1900 nm are indicative of opal. The Analytical Spectral Devices (ASD) field spectrometer data from both sites also match the opal spectrum. X-ray diffraction (XRD) analyses indicate that opal is the dominant mineral phase present at this site. The AVIRIS spectrum in blue, from the acid-sulfate alteration area, matches the USGS library reference spectrum for alunite. The broad, asymmetric doublet feature around 2200 nm and the secondary feature at 1760 nm are indicative of alunite. The ASD field spectrum from this site also matches the AVIRIS spectrum, and XRD analyses of samples from this site indicate the presence of alunite, quartz, kaolinite and minor opal. The AVIRIS spectrum in magenta, also from the acid sulfate alteration area, matches the reference spectrum for kaolinite. The sharp doublet feature around 2200 nm is indicative of kaolinite. The ASD spectrum from this site matches the AVIRIS spectrum and XRD analyses indicate the presence of kaolinite [42].

27.2.2 Mercury

27.2.2.1 MASCS

The MASCS (Table 27.1) instrument aboard the MESSENGER spacecraft uses a compact Cassegrain telescope to focus reflected light from Mercury or its atmosphere for analysis by both the UVVS and the VIRS. The UVVS employs a scanning grating monochromator with three spectral channels (115–190, 160–320, and 250–600 nm) that provide a spectral resolution varying from

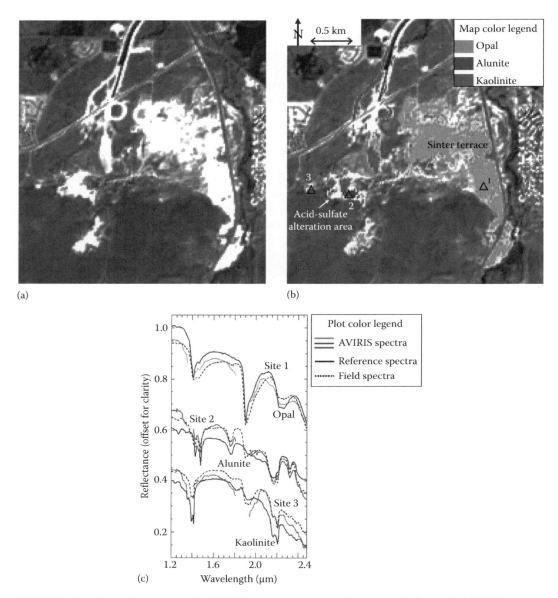

FIGURE 27.1   (See color insert.) (a) AVIRIS visible image over Steamboat Springs, (b) AVIRIS mineral map over same area, and (c) AVIRIS and field spectra of minerals mapped.

0.2 nm at UV wavelengths to 0.5 nm at visible wavelengths [75]. VIRS is a point spectrometer with a 0.023° field of view covering the wavelength range 320–1450 nm at 5 nm spectral resolution. It uses two linear diode arrays in the visible (320–950 nm) and near infrared (900–1450 nm) [76,77]. The first two channels of the UVVS instrument were used in combination with the VIRS instrument for surface studies (although the latter unfortunately did not return data from the third flyby in 2009). Nonetheless, results from the first two flybys included observations of the full disk of Mercury as well as hundreds of individual spectra covering ~1 × 4 km footprints [76,77].

27.2.2.2   Calibration and Analysis Techniques

Calibration of the VIRS data to radiance involves the use of dark current measurements and corrections for scattered light within the spectrograph. In-flight observations of standard stars validated laboratory data acquired pre-launch to within 10%. Reflectance was calculated as radiance factor (I/F), that is, the ratio of measured surface radiance to that from a perfect Lambertian surface normally illuminated by the Sun. Comparison of MASCS data of the Moon (acquired during its cruise to Mercury) to ground-based lunar observations [77] provided confidence in the calibration of the data. However, Domingue et al. [78] pointed out some differences between MASCS and ground-based observations of Mercury in the near infrared.

27.2.2.3 Case Study

The UVVS instrument's observations of the exosphere of Mercury demonstrated detailed spatial variations in sodium, calcium, and magnesium atoms and ions liberated by solar wind and micrometeorite interactions with the surface [79,80]. Examining the unusual differences between the temporal and spatial distributions of these elements will be ongoing while MESSENGER is in orbit around Mercury.

The MASCS visible/near-infrared spectra of Mercury acquired thus far do not exhibit any diagnostic spectral absorption features that can be attributed to specific minerals. D'Amore et al. [81] used principal component analyses to determine that the visible wavelengths in MASCS data contain more information than the near infrared. Holsclaw et al. [77] used analyses of oxygen–metal charge transfer absorptions and subtle spectral slopes in the visible wavelengths (and the lack of identifiable 1000 nm absorptions) to suggest a surface composition with low amounts of Fe^{2+}-bearing silicates, and relatively abundant, spectrally neutral opaque minerals. Such low-Fe^{2+} silicates could span the range from plagioclase feldspars to Mg-rich olivine and pyroxenes. Space weathering effects on Mercury are expected to be similar (if not enhanced) to those on the Moon, where vapor-deposited iron-rich nanophase particles coat many grains, resulting in few spectral features. However, early work with MASCS has demonstrated an unexpected, consistent absorption in the mid-ultraviolet region that should not be present owing to the effects of space weathering. As such, data acquired during the orbital phase of the MESSENGER mission will seek additional explanations for this observation.

27.2.3 Moon

27.2.3.1 M³

The M³ imaging spectrometer (Table 27.1) used a HgCdTe detector array for measuring electromagnetic radiation with wavelengths from 430 to 3000 nm (0.43 to 3 μm) [36]. This wavelength range covers the spectral range where diagnostic absorption features occur for all common lunar rock-forming minerals and hydrous phases. For global mapping, M³ obtained data in 86 spectral channels. For targeted operation, M³ divided the approximately 2600 nm range to which it is sensitive into 260 discrete bands, each of which is only 10 nm wide. This is considered very high spectral resolution, and was designed to enable M³ to detect the fine detail required for mineral identification. Unlike previous lunar spectrometers, M³ included sensitivity at the longer wavelength range from 2500 to 3000 nm, which is sensitive to small amounts of OH and H_2O. This spectral region is dominated by solar reflection, although a small component of emitted thermal radiation was also noted at these longer wavelengths when the lunar surface was warmer than ~250–300 K. M³ was intended to map the entire lunar surface from an altitude of 100 km at 140 m spatial sampling and 40 nm spectral sampling, with selected targets mapped at 70 m spatial and 10 nm spectral resolution. Although more than 80% coverage of the Moon was obtained by M³ at low-sun and a spatial resolution of ~140 m/pixel and significant scientific discoveries were made, the Chandrayaan-1 spacecraft suffered from technical difficulties throughout the mission; these difficulties precluded accomplishment of many of the goals of M³.

Designed for simplicity, reliability, and accuracy, M³ used a compact system of optics known as an "Offner" design, which produces little or no distortion, either spatially or spectrally [36]. M³ used the "pushbroom" method of image acquisition in which the instrument passively sweeps the scene below as it flies, recording 600 pixels of data simultaneously. Each of those 600 pixels simultaneously record images in each of 260 spectral channels. The primary M^3 product is an "image cube" that was 600 pixels wide, infinitely long over time as the instrument flies, and 260 spectral channels deep. The field of view (FOV) was 24° (or 40 km on the ground at 100 km altitude), allowing contiguous orbit-to-orbit measurements at the equator that minimized variations in lighting conditions.

27.2.3.2   Calibration and Analysis Techniques

Prior to launch, laboratory calibration measurements were made to determine the spectral, radiometric, spatial, and uniformity characteristics of M^3 [36]. The spectral range for the 260 channels was determined to span from 404 to 2993 nm with 9.96 nm sampling. The absolute radiometric calibration was determined with respect to a US National Institute of Science and Technology (NIST) traceable standard at the 5% uncertainty level. The FOV of the M^3 was measured to be 24.3°, and the cross-track sampling was measured as 0.698 mrad. Spectral cross-track uniformity and spectral IFOV uniformity of the M^3 are critical calibration characteristics.

The M^3 ground calibration data files allowed M^3 data to be calibrated to radiance-at-sensor and were further calibrated to the equivalent of reflectance data for scientific analysis. This calibration method involved dividing the radiance data by a solar spectrum and photometrically correcting all data to the same viewing geometry to eliminate variations due to lighting conditions [36]. For warm surfaces, a small thermal component at wavelengths beyond 2000 nm was removed [82]. Lunar samples returned to Earth by the Apollo missions were used as "ground truth" for calibrating the M^3 data. Assuming that lunar samples are representative of specific portions of the lunar surface, the sample properties were used to calibrate the remote sensing data [36]. The Apollo 16 site was well suited for this because it is largely dominated by one type of material (feldspathic breccias) unlike most other landing sites that contain diverse lithologies [36]. The Apollo 16 region is one of the prime Lunar International Science Calibration/Coordination Targets (LISCT) proposed for cross calibration of lunar data obtained by various missions [36].

27.2.3.3   Case Studies

27.2.3.3.1   Water

The search for water on the surface of the anhydrous Moon remained an unfulfilled quest for 40 years [32]. However, M^3 on Chandrayaan-1 detected absorption features near 2.8–3.0 μm on the surface of the Moon during some portions of the day. The 3.0 μm absorption feature was identified and the measurement was extended to longer wavelengths by two independent spacecraft: the NASA Cassini mission VIMS and the High-Resolution Instrument Infrared (HRI-IR) spectrometer on the NASA Deep Impact EPOXI mission [83,84]. For silicate bodies, such features are typically attributed to hydroxyl- and/or water-bearing materials. On the Moon, the feature is seen as a widely distributed absorption that appears strongest at cooler high latitudes and at several fresh feldspathic craters [32]. The general lack of correlation of this feature in sunlit M^3 data with neutron spectrometer hydrogen abundance data suggests that the formation and retention of hydroxyl and water are ongoing surficial processes [32]. The hydration signatures were observed by HRI-IR to be dynamic, with diurnal changes that differed for mare and highland units and returned to a steady state entirely between local morning and evening [84]. The observed hydration variation requires a ready daytime source of water group ions and is considered consistent with a solar wind origin [83,84]. In this scenario, hydrogen ions from the Sun are carried by the solar wind to the Moon where they interact with oxygen-rich minerals at the top millimeters of lunar soil to produce the observed H_2O and OH molecules [83,84]. Hydroxyl/water production processes may feed polar cold traps and make the lunar regolith a candidate source of volatiles for human exploration [32]. Such hydration by solar wind particles may occur throughout the Solar System on all airless bodies with oxygen-bearing minerals on their surfaces. Although abundances are not definitively known, as much as 1000 water molecules parts-per-million (0.1%) could be present [83].

27.2.3.3.2 Spinels

M[3] data were also used to search for unusual and in some cases new rock types. Recent research has identified two spinel-rich rock types at widely spaced locations on the lunar surface [85]. The first is a pink spinel rich in magnesium and iron and the second is a black or very dark chromite-rich spinel [85]. Both types of spinel have distinctive spectral absorption features in the M[3] data near 2 μm [85]. As observed in 2 μm band depth maps, the pink spinels occur near Mare Moscoviense on the lunar far side as several small, diffuse deposits that are not obviously dark and/or associated with any crater or steep slope that has exposed fresh material [85].

Elsewhere in the Solar System such spinel-rich surfaces have been observed in a few main-belt asteroids, where the spinel absorption feature near 2 μm is thought to indicate the presence of abundant calcium- and aluminum-rich inclusions (CAIs) such as those found in carbonaceous chondrite meteorites [86]. Current ideas about the origin of these deposits include the preservation of unusual deep-crustal or plutonic materials exposed on the surface or the presence of primitive material that has been deposited onto the lunar surface [87].

The dark, chromite-rich spinels are observed in the Sinus Aestuum region on the lunar near side [88]. These deposits are distinctly different from those on the far side in that they have lower albedo and additional spectral features at visible and 1000 nm wavelengths [88]. They are observed in a region with prominent pyroclastic volcanic deposits; these are believed to have been derived from several hundred km depths within the Moon and thus provide a link to mantle compositions [88].

27.2.4 MARS

27.2.4.1 Hyperspectral Instruments

27.2.4.1.1 OMEGA

The OMEGA instrument (Table 27.1) is a visible and infrared mineral mapper onboard the Mars Express orbiter designed to globally map minerals, water (hydrated minerals or ice). The OMEGA imaging spectrometer uses two bore-sighted telescopes for a spectral range from 500 to 5200 nm. The first telescope uses a silicon CCD (whiskbroom) to image the spectral range from 0.5 to ~1 μm. The second telescope uses two InSb detector arrays (pushbroom) to image the spectral range from ~1000 to 5200 nm [90]. OMEGA has an IFOV of 1.2 mrad, resulting in a spatial resolution of ~300 m close to periapsis (250 km) and ~5 km at an altitude of 4000 km.

27.2.4.1.2 CRISM

The CRISM (Table 27.1) is a high spatial resolution (20–200 m/pixel) imaging spectrometer designed to target regions to detect and map minerals and ices. CRISM uses a 10 cm diameter Ritchey-Critien telescope that feeds a pair of Offner convex-grating spectrometers. One spectrometer uses a Silicon detector array (VNIR) and the second spectrometer (IR) uses an array of HgCdTe diodes. A fully gimbaled optical system allows for motion compensation that enables a spatial resolution of 15.7–19.7 m. CRISM can also operate in mapping mode (no motion compensation) at 100 or 200 m/pix. CRISM's spectral range is from 362 to 3920 nm [9].

27.2.4.2 Calibration and Analysis Techniques

OMEGA Calibration. OMEGA data are processed using standard processes. DN levels are converted to radiances $[W \cdot sr^{-1} \cdot m^{-2}]$ and I/F is calculated by dividing by the solar flux $[W \cdot sr^{-1} \cdot m^{-2}]$ corrected for the appropriate Sun–Mars distance [89,90]. Unlike most hyperspectral imagers, where the calibrated data are provided, OMEGA provides the raw data and the calibration software to construct both calibrated radiances and geometric coordinates for each spectrum. The advantage of this methodology is that as calibration software is updated, users do not need to download new image cubes, but instead they download the new software and reprocess their raw cubes.

CRISM Calibration. A complete description of CRISM calibration can be found in Ref. [9]. CRISM uses a shutter and an integrating sphere as part of the internal calibration systems. The shutter in its closed position allows for the measurement of bias, dark current, and thermal background. A partially opened shutter acts as a mirror, allowing the full view of the integrating sphere.

Spectral Smile. Imaging spectrometers are ideally designed such that one dimension of the CCD is spatial and the other dimension is spectral. Ideally, this means that any row of pixels will correspond to common wavelength while any column corresponds to a common surface location. Due to minor flaws in optics and alignment, this is almost never the case, causing wavelength positions to shift. If one were to create a contour of a single wavelength across the CCD, it would appear as a smile (or frown).

Atmospheric Corrections. The atmosphere of Mars is composed of 95% CO_2 [91], which produces several absorption features at or near surface spectral features. The presence of these overlapping atmospheric features can introduce error into surface spectral indices by either creating a deeper absorption feature (atmospheric line directly overlaps the surface spectral feature) or introducing a slope into the continuum estimate (atmospheric line is adjacent to the surface feature). In addition, atmospheric scattering can introduce a color slope, causing the surface to appear bluer.

One simple technique to remove most of the atmospheric lines, but not spectral slopes due to scattering, is the volcano scan [92]. The volcano scan technique ratios the absorption line strengths observed at the top of a volcano (e.g., Olympus Mons) and the base of the volcano. If the surface on top of the volcano has the same photometric properties (e.g., albedo) as the base (which is the case for many volcanos on Mars since these regions are typically covered by thick layers of dust), then the ratio of the line strengths is due to the atmosphere. A few of the absorption lines are at wavelengths that do not overlap surface spectral features. These line strengths are then used to ratio the volcano scan line strengths, which can then be used to correct the atmospheric line strengths that do overlap with surface spectral features. The volcano scan method is useful for quick analysis of spectral features, but is inadequate to remove scattered light in the continuum. One approach is to use a Monte Carlo model designed to simulate multiple scattering of light from dust and ice aerosols in the atmosphere [93,94]. This approach is often combined with a volcano scan method to correct spectra for both scattering and gas absorption. Another method is to use a full-blown radiative transfer model that incorporates the optical properties of the gases and the aerosols (e.g., DISORT) [95]. This method requires a priori knowledge of atmospheric optical depths. This approach, while the most accurate (similar to MODTRAN), is also the most computer intensive.

One of the most common vegetation indices is the normalized difference vegetation index (NDVI), where one compares the reduction of reflectivity in the near IR when compared to red. This reduction in reflectivity is due to the absorption of water. Surface volatiles and exposed minerals can be identified in a similar manner, using absorption features unique to either a specific mineral or a suite of minerals. Pelkey et al. [56] outline several of the most common spectral indices used by OMEGA and CRISM to identify ices and minerals. Many of the spectral indices measure the depth of absorption bands, while others measure spectral slope. Many of these indices are less sensitive to atmospheric corrections, especially if the spectral feature is not near an atmospheric spectral feature.

27.2.4.3 Case Studies

Carr and Head [96] compiled an excellent review of the geological history of Mars based on both spacecraft and in situ measurements. Much of this story was a result of near-infrared and shortwave infrared spectroscopy identifying and mapping the distributions of phyllosilicates and sulfates.

27.2.4.3.1 *Phyllosilicates, Sulfates, and Alteration Minerals*

Prior to the arrival of imaging infrared spectrometers (e.g., OMEGA), Martian history was defined by epochs based on visible imaging—morphology, crater counts, etc. With the arrival of the French lead instrument OMEGA onboard the European spacecraft Mars Express, a new way of viewing Martian history immerged. This new view was based on mineralogical composition derived from spectral absorption features. Not only were broad classes of minerals identified and mapped (e.g., olivines and pyroxenes) but specific minerals (e.g., nontronite, Fe-rich chlorites, saponite, and montmorillonite) were identified and mapped [97,98]. The new view of composition loosely correlates with the timeline determined from visible imaging, but tells a story that surface materials changed from neutral-pH (or slightly alkaline) conditions early in Mars history to acidic alteration at a later epoch [99,100].

27.2.4.3.2 Ices

The polar caps are the most active regions on Mars, and the most dynamic processes visible from Earth. The annual cycling of atmospheric CO_2 into the seasonal CO_2 ice caps is a driving force of the Martian climate. The polar layered deposits (PLDs), with thousands of layers whose thickness is only resolvable with sub-meter spatial resolution from orbit, may contain a record of past climates. The polar regions contain the majority of known H_2O ice deposits, distributed between the residual caps and near-surface ice in the regolith. CO_2 ice has several spectral features throughout the infrared (e.g., 1435, 2340 nm) Unfortunately, CO_2 ice has many spectral features located at or near the same wavelengths as the gas bands (e.g., 1435 nm). One of the few exceptions is the doublet at 2345 nm. The presence of H_2O ice is easily detected by the presence of broad spectral features at 1500 and 2000 nm.

While the seasonal monitoring and mapping of the seasonal caps date back to Ref. [101], the ability to correctly identify the seasonal cap composition as CO_2 ice did not occur until the 1960s. See Refs. [102,103] for historical overview. In 2001, Kieffer and Titus [104] recognized that the northern seasonal cap was surrounded by a H_2O-ice annulus during the springtime retreat. This conclusion was based on comparing thermal observations (CO_2 ice is typically at ~145 K) with visible observations. The bright edge of the seasonal cap was too warm to be CO_2 ice and was hypothesized to be H_2O. This hypothesis was later spectrally confirmed by OMEGA in 2005. In addition to spectrally confirming the presence of the H_2O ice annulus, OMEGA also observed the presence and distribution of H_2O ice intimately mixed with the dominant CO_2 ice through much of the seasonal cap.

27.2.5 JUPITER

27.2.5.1 NIMS

A detailed description of the NIMS instrument (Table 27.1) design, calibration, and operation was presented by Carlson et al. [105]. NIMS's spectral range, from 700 to 5200 nm, allowed measurements of both reflected sunlight and emitted thermal radiation. The instrument used a 228 mm aperture, f/3.5 Ritchey-Chretien telescope with an equivalent focal length of 800 mm. The incident beam was reflected onto a wobbling secondary mirror that could scan through 20 fixed positions to provide whisk-broom spatial coverage in one direction, while the motion of a scan platform provided spatial coverage in either direction. The beam was directed through a chopper to a wobbling diffraction grating that was rotated in steps to generate spectral coverage at the commanded spectral resolution, up to 408 wavelengths. The chopper ensured that no light reached the detectors while the diffraction grating was in motion. The beam was focused onto an array of 17 detectors: 2 silicon detectors for near-infrared wavelengths and 15 indium antimonide (InSb) detectors sensitive to longer wavelengths. The focal plane assembly, including the detectors and their pre-amplifiers, was radiatively cooled to 64 K. The system had an IFOV of 0.5×0.5 mrad (leading to spatial resolutions of tens to hundreds of kilometers per "nimsel" for many of the Jovian satellite encounters) and an angular FOV of 0.5×10 mrad.

27.2.5.2 Calibration and Analysis Techniques

NIMS faced many challenges in its mission, including the harsh radiation environment of Jupiter. The instrument was shielded within a 3 mm thick tantalum enclosure, but even so, the signal-to-noise ratio degraded sharply during encounters with satellites in the inner reaches of Jupiter's radiation belts. The spacecraft suffered a downlink bottleneck caused by the failure of its High Gain Antenna, severely limiting the data volume returned by each instrument during the mission. Contamination was also a concern, and NIMS was equipped with covers, shields, and shades to protect it from thruster firings and thermal influences from the spacecraft. Spectral calibration (subject to change, due to thermal effects or vibration of the diffraction grating during launch) was provided by an onboard InGaAs LED lamp that emitted at a known wavelength. A radiometric calibration target, consisting of an extended, near-field blackbody source heated to a known temperature, was mounted on the spacecraft within the NIMS FOV. A photometric calibration target that diffusely reflected sunlight was also placed in view of the scan platform. Independent verification of many of the spectral detections made by NIMS took place during the millennium passage of the Cassini spacecraft en route to Saturn. Cassini's VIMS confirmed several weak absorption bands in the spectra of the Galilean satellites Io, Europa, Ganymede, and Callisto [106]. The NIMS instrument operated successfully throughout Galileo's 14 year mission despite failures of two of the InSb detectors and the temperature sensor of the radiometric calibration target.

27.2.5.3 Case Study

NIMS is credited with the first detection of ammonia ice clouds in the atmosphere of Jupiter [107] and made important observations of water vapor that demonstrated moist convection in Jovian lightning storms [108,109]. NIMS thermal observations were critical to understanding the nature and distribution of active volcanoes on Io [110] and the energetics of their eruptions [111]. NIMS did not identify the mineralogy of the lavas but provided an indirect indication that they must be made up of silicates, since their eruption temperatures were too high for other candidate materials. NIMS produced global maps of Io's surface SO_2 ice abundance and grain size at spatial resolutions from 100 to 350 km/nimsel, using the deep absorption band centered at 4100 nm together with the relative strengths of weaker bands [112,113]. Water ice grain sizes were measured on the icy satellites Europa, Ganymede and Callisto using a similar approach [114]. On Europa, NIMS found that the water ice bands were distorted, leading to suggestions of the presence of hydrated phases of salts [115] or sulfuric acid [116]. NIMS did not specifically identify the visibly red endogenic material that is associated with tectonic ridges and other young geologic features on Europa, but found associations between these features and the degree of distortion of the water ice bands. Several compounds attributed to radiolysis were identified on the surfaces of the icy moons, including H_2O_2 (through a band at 3.5 μm) [117], SO_2 (4050 nm), and CO_2 (4260 nm). CO_2 signatures were particularly strong on Callisto and the older terrain on Ganymede [118,119], perhaps due to radiolysis of meteoritic carbonaceous material. A weak feature attributed to CN (4570 nm) was also seen on Callisto.

NIMS's mission ended in September, 2003, when the Galileo spacecraft plunged deliberately into Jupiter in order to ensure that the risk of biological contamination of Europa was eliminated.

27.2.6 SATURN

27.2.6.1 VIMS

Cassini's VIMS (Table 27.1) is a primary orbital instrument to study the composition, dynamics, clouds, and thermophysics of the atmospheres of Saturn and Titan and the identification and mapping of surface compositions of Titan, the other icy Saturnian satellites, and the rings. VIMS consisted of two instrumental subsystems covering different spectral ranges [38]. The VIMS-VIS channel is a multispectral imager that covers the spectral range from 350 to 1050 nm with 96 spectral channels with a spectral resolution of 7.3 nm and uses a frame transfer CCD detector on which spatial and spectral information is simultaneously stored. The VIMS-IR channel covers a wavelength range from 850 to 5170 nm using 256 spectral channels with a spectral resolution of 16.6 nm. The two spectrometers have many modes but they typically employ effective IFOVs of 0.5×0.5 mrad. The largest spectral cubes made are 64×64 (lines and samples), although much smaller spatial views are often used to save on data volume. During Cassini's approach to Titan, VIMS observational sequences commence several hours before closest approach providing hemispheric mapping with full spectral cubes with spatial resolution ≤50 km/pixel. About 4 h before closest approach, 2×2 mosaics provide coverage of the disk. Resolutions ~10 km/pixel are achieved ~0.5 h before closest approach when regional mosaics can be acquired. Isolated cubes, rarely acquired during close flybys, can yield resolutions as high as 250 m/pixel.

27.2.6.2   Calibration and Analysis Techniques

The two imaging spectrometers that comprise VIMS were calibrated separately [38]. The VIMS-VIS or visible channel was calibrated in Italy prior to integration at JPL with the VIMS-IR channel that was also calibrated by JPL. Final tests of the integrated instrument included only filter transmission and mineral target measurements. Spectral calibration characterized the bandpass for each spectral channel over the FOV and as a function of temperature. Central wavelengths of the VIMS-IR channels varied by <1 nm; as the sampling interval is about 16 nm this is a small shift. Likewise the VIMS-VIS wavelength variations were <0.3 nm compared to the 7 nm sampling interval. Preflight calibration included radiometric/flat field response, geometric, polarimetric, spectral, and solar port response. Cooling the thermal vacuum chamber walls with liquid N_2 simulated the flight-like thermal environment. The simulated thermal environment was quite accurate as the in-flight temperatures of optics and focal planes were within a few K of those in the test chamber. In the case of analysis of VIMS data by authors of this chapter ISIS-3 (the USGS publically available Integrated Software for Imagers and Spectrometers) has been used for radiometric and geometric corrections and for subsequent spectral analysis and correlations with other Cassini data sets. For VIMS cubes, ISIS-3 provides a set of end-to-end tools (https://isis.astrogeology.usgs.gov/Installation/index.html) that start with raw data publicly available from the NASA PDS (Planetary Data System) and generate high-level products that have been radiometrically corrected and transformed to a wide variety of map projections. ISIS-3 also provides a suite of interactive tools for spectral analysis, generation of control nets, and mosaicing to cartographic standards.

27.2.6.3   Case Study: Ethane in a Titan Polar Lake

Titan affords a rich Solar System laboratory to study active organic synthesis on a global scale—it may exhibit chemical pathways holding clues to the primordial prebiotic organic chemistry that led to the emergence of life on Earth. Methane makes up a few percent of Titan's thick cold nitrogen atmosphere. Moving in a global cycle, the methane forms clouds, rain, rivers, lakes, and seas akin to Earth's hydrological cycle. In the upper atmosphere, methane and nitrogen are energetically broken down and recombine to form a vast spectrum of organics ranging from simple gases to large complex molecules. These form mixtures of organic liquids and aerosol solids that rain onto the surface. Spectral evidence for the composition of the lakes derived from VIMS observations is the subject here.

Able to penetrate the thickly absorbing, hazy atmosphere through a series of atmospheric windows, VIMS is the primary Cassini instrument used to study Titan's surface composition. Sunlight penetrates to the surface through narrow transmission windows that are separated by deep methane absorption bands. Within atmospheric windows with wavelengths <1000 nm, surface signals are swamped by multiple scattering from the aerosols. Aerosol scattering becomes decreasingly effective at longer wavelengths. As a result the surface is visible only through atmospheric windows centered at 940, 1080, 1280, 1600, 2000, 2700, 2800, and 5000 nm; all of these are in the spectral range of VIMS-IR channel.

Long before Cassini-Huygens arrived at Saturn in 2004 there was the expectation of finding hydrocarbon lakes. In mid-2005 the Cassini ISS (Imaging Science Subsystem) photographed a large dark feature suggestive of a lake near the south pole (later named Ontario Lacus) [120]. In 2006 Cassini SAR (synthetic aperture radar) images of the north polar region revealed a vast array of lake-like features north of 75° [121]. But the evidence for their being liquid was based morphological pattern and extremely low radar cross section. VIMS provided additional evidence for the composition of the lakes.

It had been predicted that the lakes would consist largely of methane and nitrogen with several tens-of-percent ethane [122]. Methane would be hard to detect owing to the abundant methane in the atmosphere and nitrogen exhibits no features in the atmospheric windows, but ethane was a possibility. During a close flyby of Titan in late 2007 VIMS observed Ontario Lacus collected a sequence of four spectral cubes [123]. Spectra, from inside and outside the lake, are quite similar dominated by the absorption bands in the atmosphere. Spectral differences do exist, however, in particular in the 2000 and 5000 nm windows. In ratios of spectra from the interior of the lake and from a nearby area outside cancel effects of the strong atmospheric absorptions and reveal spectral features in these two bands. Spectra with nearly identical path lengths were used to minimize residuals from strong atmospheric absorptions in the ratios. Ratios are shown for the dark lake interior and for a narrow annulus resembling a beach just inside the bright shoreline. The ratio spectra are nearly flat across the rest of the windows showing that the strong atmospheric absorptions were mostly canceled out.

Brown et al. [123] identified two well-developed spectral features in the ratios: a narrow absorption at 2018 nm and a broad absorption in the 5000 nm window that shows a steep drop at ~4800 nm continuing downward to the end of the VIMS spectral range. Clark et al. [124] provided derived optical constants for liquid ethane measured in the laboratory. Brown et al. [123] derived model ethane spectra from these optical constants (red line in Figure 7.3)—most of ethane's narrow absorption bands fall within the deep atmospheric window, with the narrow exception of a feature near the 2000 nm window that matches that is seen in the ratio spectra of the lake. The steep drop in reflectance beyond 4800 nm is a strong indicator of lake composition. This is characteristic of alkanes including ethane, propane, and butane [124,125]. The presence of propane, butane, and higher-order alkanes could explain the continued drop beyond the ethane absorption feature modeled in the 5000 nm window by Brown et al. [123]. Although the VIMS spectra cannot detect liquid methane theoretical work by Mitri et al. [126] on Titan's hydrocarbon lakes shows that if ethane is present, methane is most probably a major component as well.

27.3 CONCLUSIONS AND FUTURE CHALLENGES

Hypserspectral analyses of planetary surfaces in the visible/near-infrared wavelengths are used dominantly to constrain their geology and mineralogy and a means of understanding their history and evolution. The methods used to acquire and calibrate planetary data sets share many similarities with those used for terrestrial observations of vegetated and rocky surfaces on Earth. Simple methods such as ratios among individual bands provide efficient overviews of spectral variability over wide areas of interest. Spectral mapping of specific areas can be further explored using methods such as the principal component analyses in combination with spectral mixture models to detect specific bands associated with mineral classes and/or compositional variations.

Many of the techniques developed for the study of planetary hyperspectral data sets germinated from studies using Earth-based observations. As shown in this chapter, planetary bodies can range from those without atmospheres to those with atmospheres much thicker than on Earth. Despite the challenges associated with specific data sets, valuable information regarding composition and mineralogy of planetary surfaces has been derived. More importantly, many of the data sets acquired provide well-calibrated observations that will enable future researchers to explore these surfaces in more detail using analytical techniques yet to be developed. Such future work will include improved methods of eliminating interference from atmospheric contamination, particularly on bodies that experience atmospheric variability from seasonal changes. Ongoing work regarding unmixing of complicated rock and mineral mixtures using laboratory analyses and field studies will continue to improve the precision and accuracy of determining compositional variability and mineral abundances using remotely sensed hyperspectral data sets.

The instruments and case studies described here together document a highly active field of research that has important implications for remote sensing studies of Earth. Not only do the results of such planetary remote sensing studies help us to understand the nature and distribution of materials such as rocks and soils, they have been invaluable in detecting evidence of water in places in our Solar System we had long believed to be dry. There is a clear link between water and life in our Solar System [127]; see also http://mepag.jpl.nasa.gov/reports/MEPAG_Goals_Document_2010_ v17.pdf. Scientists are now in the process of unraveling and extending links between observed geological and biological materials such that we will soon be able to characterize and preserve habitable environments both on and off the Earth.

REFERENCES

1. Nielsen, U.N., Graham, H.R.O., Colin, D.C., David, F.R.P.B., and van der Wal, R. (2010). The influence of vegetation type, soil properties and precipitation on the composition of soil mite and microbial communities at the landscape scale. *Journal of Biogeography*, 37, 1317–1328.
2. Abrams, M.J., Ashley, R.P., Rowan, L.C., Goetz, A.F.H., and Kahle, A.B. (1977). Mapping of hydrothermal alteration minerals in the cuprite mining district, Nevada, using aircraft scanner images for the spectral region 0.46–2.36 μm. *Geology*, 5, 713–718.
3. Goetz, A.F.H., Rock, B.N., and Rowan, L.C. (1983). Remote sensing for exploration: An overview. *Economic Geology*, 78(4), 573–590.
4. Marsh, S.E. and McKeon, J.B. (1983). Integrated analysis of high-resolution field and airborne spectroradiometer data for alteration mapping. *Economic Geology*, 78(4), 618–632.
5. Taranik, J.V. (1988). Application of aerospace remote sensing technology to exploration for precious metal deposits in the Western United States. In: *Bulk Mineable Precious Metal Deposits of the Western United States*, R.W. Schafer, editor, *GSN Symposium Proceedings*, Geological Society of Nevada, Reno, NV.
6. Kruse, F.A. (1999). Visible-infrared sensors and case studies. Ch 11. In: *Remote Sensing for the Earth Sciences: Manual of Remote Sensing*, A.N. Rencz, editor, 3rd edn., vol. 3, John Wiley & Sons, New York.
7. Burns, R.G. (1993). *Mineralogical Applications of Crystal Field Theory*. 2nd edn., Cambridge University Press, New York.
8. Bibring, J.P., Langevin, Y., Mustard, J.F., Poulet, F., Arvidson, R., Gendrin, A., Gondet, B., Mangold, N., Pinet, P., and Forget, F. (2006). Global mineralogical and aqueous mars history derived from OMEGA/ Mars express data. *Science*, 312(5772), 400–404.
9. Murchie, S., Arvidson, R., Bedini, P., Beisser, K., Bibring, J.-P., Bishop, J., Boldt, J. et al. (2007). Compact reconnaissance imaging spectrometer for mars (CRISM) on mars reconnaissance orbiter (MRO). *Journal of Geophysical Research*, 112(E5), CiteID E05S03.
10. Abrams, M.J., Brown, D., Lepley, L., and Sadowski, R. (1983). Remote sensing for porphyry copper deposits in Southern Arizona. *Economic Geology*, 78(4), 591–604.
11. Podwysocki, M.H., Segal, D.B., and Abrams, M.J. (1983). Use of multispectral scanner images for assessment of hydrothermal alteration in the Marysvale, Utah, mining area. *Economic Geology*, 78(4), 573–590.
12. Hutsinpiller, A. and Taranik, J.V. (1988). Spectral signatures of hydrothermal alteration at Virginia City, Nevada. In: *Bulk Mineable Precious Metal Deposits of the Western United States*, R.W. Schafer, editor, Geological Society of Nevada Symposium Proceedings, pp. 505–530.
13. Swayze, G.A., Clark, R.N., Smith, K.S., Hageman, P.L., Sutley, S.J., Pearson, R.M., Rust, R.S. et al. (1998). Using imaging spectroscopy to cost-effectively locate acid-generating minerals at mine sites: An example from the California Gulch Superfund Site in Leadville, Colorado. *Proceedings of the 7th Airborne Earth Science Workshop*, JPL publication 97–21(1), pp. 385–389.
14. Vane, G., Green, R.O., Chrien, T.G., Enmark, H.T., Hansen, E.G., and Porter, W.M. (1993). The airborne visible/infrared imaging spectrometer (AVIRIS). *Remote Sensing of Environment*, 44, 127–143.
15. Gao, B., Heidebrecht, K.B., and Goetz, A.F.H. (1993). Derivation of scaled surface reflectances from AVIRIS data. *Remote Sensing of Environment*, 44, 165–178.
16. King, T.V.V. and Clark, R.N. (2000). Verification of remotely sensed data. Ch 5. In: *Remote Sensing for Site Characterization*, F. Kuehn, T. King, B. Hoerig, and D. Peters, editors, Springer-Verlag, Berlin, Germany.
17. Jacobsen, A., Heidebrecht, K.B., and Goetz, A.F.H. (2000). Assessing the quality of the radiometric and spectral calibration of *casi* data and retrieval of surface reflectance factors. *Photogrammetric Engineering and Remote Sensing*, 66(9), 1083–1091.

18. Lyon, R.J.P. (1965). Analysis of rocks by spectral infrared emission (8–25 μ). *Economic Geology*, 60, 715–736.
19. Farmer, V.C. (editor) (1974). *The Infrared Spectra of Minerals*. Mineralogical Society Monograph 4, Mineralogical Society, London.
20. Hunt, G.R. (1980). Electromagnetic radiation: The communication link in remote sensing. Ch 2. In: *Remote Sensing in Geology*, B.S. Siegal and A.R Gillespie, editors, John Wiley, New York, pp. 5–45.
21. Clark, R.N., King, T.V.V., Klejwa, M., and Swayze, G.A. (1990). High spectral resolution reflectance spectroscopy of minerals. *Journal of Geophysical Research*, 95(B8), 12653–12680.
22. Salisbury, J.W., Walter, L.S., Vergo, N., and D'Aria, D.M. (1991). *Infrared (2.1–25 μm) Spectra of Minerals*. Johns Hopkins University Press, Baltimore, MD.
23. Burns, R.G. (1993). Origin of electronic spectra of minerals in the visible to near-infrared region. Ch 1. In: *Topics in Remote Sensing 4—Remote Geochemical Analysis: Elemental and Mineralogical Composition*, C.M. Pieters and P.A.J. Englert, editors, Cambridge University Press, Cambridge, U.K.
24. Clark, R.N. (1999). Spectroscopy of rocks and minerals and principles of spectroscopy. Ch 1. In: *Remote Sensing for the Earth Sciences: Manual of Remote Sensing*, A.N. Rencz, editor, 3rd edn., vol. 3, John Wiley & Sons, New York.
25. Clark, R.N., Swayze, G.A., Wise, R., Livo, K.E., Hoefen, T.M., Kokaly, R.F., and Sutley S.J. (2003). USGS Digital Spectral Library splib05a. *USGS Open File Report*, 03–395.
26. Baldridge, A.M., Hook, S.J., Grove, C.I., and Rivera, G. (2009). The ASTER spectral library version 2.0. *Remote Sensing of Environment*, 113(4), 711–715.
27. Roach, L.H., Mustard, J.F., Swayze, G., Milliken, R.E., Bishop, J.L., Murchie, S.L., and Lichtenberg, K. (2010). Hydrated mineral stratigraphy of Ius Chasma, Valles Marineris. *Icarus*, 206, 253–268.
28. Milliken, R.E., Grotzinger, J.P., and Thomson, B.J. (2010). Paleoclimate of Mars as captured by the stratigraphic record in Gale Crater. *Geophysical Research Letters*, 37, L04201, doi:10.1029/2009GL041870
29. Chan, M.A. and Parry, W.T. (2002). *Rainbow of the Rocks: Mysteries of Sandstone Colors and Concretions in Colorado Plateau Canyon Country*, Public Information Series 77, Utah Geological Survey, 17 pp.
30. Chan, M.A., Johnson, C.M., Beard, B.L., Bowman, J.R., and Parry, W.T. (2006). Iron isotopes constrain the pathways and formation mechanisms of terrestrial oxide concretions: A tool for tracing iron cycling on Mars? *Geosphere*, 2(7), 324–332, doi: 10.1130/GES00051.1
31. Douka, C.E. (1977). Study of bacteria from manganese concretions. Precipitation of manganese by whole cells and cell-free extracts of isolated bacteria. *Soil Biology and Biochemisty*, 9(2), 89–97.
32. Pieters, C.M. et al. (2009). Character and spatial distribution of OH/H_2O on the surface of the Moon seen by M^3 on Chandrayaan-1. *Science*, 326(5952), 568–572, doi: 10.1126/science.1178658
33. Cocks, T., Jenssen, R., Stewart, A., Wilson, I., and Shields, T. (1998). The HyMap airborne hyperspectral sensor: The system, calibration and performance. In: *Proceedings of 1st EARSeL Workshop on Imaging Spectroscopy*, M. Schaepman, D. Schläpfer, and K.I. Itten, editors, pp. 37–43.
34. Watts, L.A., Davis, R.O., Granneman, R.D., LaVeigne, J.D., Chandos, R.A., Russell, E.E., and Cairns, B. (2001). Unique VISNIR-SWIR hyperspectral and polarimeter measurements. In: *Proceedings of the 5th Airborne Remote Sensing Conference*, San Francisco, CA.
35. Pearlman, J.S., Barry, P.S., Segal, C.C. et al. (2003). Hyperion, a space-based imaging spectrometer. *IEEE Transactions on Geoscience and Remote Sensing*, 41(6), 1160–1172.
36. Pieters, C.M. et al. (2009). The moon mineralogy mapper (M^3) on Chandrayaan-1. *Current Science*, 96(4), 500–505.
37. Antoniadi, E.M. (1930). *The Planet Mars*, Trans. Patrick Moore, Devon, UK: Keith Reid Ltd., 1975.
38. Brown, R.H., Baines, K.H., Bellucci, G., Bibring, J.-P., Buratti, B.J., Capaccioni, F., Cerroni, P. et al. (2004). The cassini visual and infrared mapping spectrometer (VIMS) investigation. *Space Science Reviews*, 115, 111–168.
39. National Research Council (NRC) Decadal Survey, (2007). Earth Science and Applications from Space: National Imperatives for the Next Decade and Beyond, Committee on Earth Science and Applications from Space: A Community Assessment and Strategy for the Future, National Research Council, ISBN: 978-0-309-10387-9, 456 pp.
40. Kruse, F.A. (2002). Combined SWIR and LWIR Mineral Mapping Using MASTER/ASTER. In: *Proceedings, IGARSS 2002, CD ROM*, Toronto, Canada, pp. 2267–2269.
41. Rowan, L.C. and Mars, J.C. (2003). Lithologic mapping in the mountain pass, California area using advanced spaceborne thermal emission and reflection radiometer (ASTER) data. *Remote Sensing of Environment*, 84, 350–366.
42. Vaughan, R.G. (2004). Surface Mineral Mapping at Virginia City and Steamboat Springs, Nevada with Multi-Wavelength Infrared Remote Sensing Image Data. PhD dissertation, University of Nevada, Reno, 273 pp.

43. Boynton, W.V., Feldman, W.C., Squyres, S.W., Prettyman, T.H., Brückner, J., Evans, L.G., Reedy, R.C. et al. (2002). Distribution of hydrogen in the near surface of Mars: Evidence for subsurface ice deposits. *Science*, 297(5578), 81–85.

44. Boynton, W.V., Taylor, G.J., Evans, L.G., Reedy, R.C., Starr, R., Janes, D.M., Kerry, K.E. et al. (2007). Concentration of H, Si, Cl, K, Fe, and Th in the low- and mid-latitude regions of Mars. *Journal of Geophysical Research*, 112(E12), CiteID E12S99.

45. Hasebe, N., Shibamura, E., Miyachi, T., Takashima, T., Kobayashi, M., Okudaira, O., Yamashita, N., et al. (2008). Gamma-ray spectrometer (GRS) for lunar polar orbiter SELENE. *Earth, Planets and Space*, 60, 299–312.

46. Mitrofanov, I.G., Bartels, A., Bobrovnitsky, Y.I., Boynton, W., Chin, G., Enos, H., Evans, L. et al. (2010). Lunar exploration neutron detector for the NASA lunar reconnaissance orbiter. *Space Science Reviews*, 150(1–4), 183–207.

47. Prettyman, T.H., Feldman, W.C., Barraclough, B.L., Capria, M.T., Coradini, A., Enemark, D.C., Fuller, K.R. et al. (2004). Mapping the elemental composition of Ceres and Vesta: Dawn's gamma ray and neutron detector, Instruments, Science, and Methods for Geospace and Planetary Remote Sensing, CA. Nardell, P.G. Lucey, J.-H. Yee, J. B Garvin, editors, *Proceedings of the SPIE*, 5660, 107–116.

48. Schott, J.R. (1997). *Remote Sensing: The Image Chain Approach*. Oxford University Press, New York.

49. Hapke, B. (1993). Combined theory of reflectance and emittance spectroscopy, Ch 2. In: *Topics in Remote Sensing 4—Remote Geochemical Analysis: Elemental and Mineralogical Composition*, C.M. Pieters and P.A.J. Englert, editors, Cambridge University Press, Cambridge, U.K.

50. Hapke, B. (1993). *Theory of Reflectance and Emittance Spectroscopy. Topics in Remote Sensing 3*, R.E. Arvidson and M.J. Rycroft, editors, Cambridge University Press, Cambridge, U.K.

51. Adler-Golden, S., Berk, A., Bernstein, L.S., Richtsmeier, S.C., Acharya, P.K., and Matthew, M.W. (1998). FLAASH, A MODTRAN 4 atmospheric correction package for hyperspectral data retrievals and simulation. *Proceedings of the 7th JPL Airborne Earth Science Workshop*, JPL, Pasadena, CA.

52. Clark, R.N. and Roush, T.L. (1984). Reflectance spectroscopy: Quantitative analysis techniques for remote sensing applications. *Journal of Geophysical Research*, 89(B7), 6329–6340, doi: 10.1029/JB089iB07p06329

53. Charette, M.P., McCord, T.B., Pieters, C., and Adams, J. (1974). Application of remote spectral reflectance measurements to lunar geology classification and determination of titanium content of lunar soils. *Journal of Geophysical Research*, 79, 1605–1613.

54. Johnson, J.R., Larson, S.M., and Singer, R.B. (1991). Remote sensing of potential lunar resources: 1. Near-side compositional properties. *Journal of Geophysical Research*, 96, 18861–18882.

55. Geissler, P.E., Singer, R.B., Komatsu, G., Murchie, S., and Mustard, J. (1993). An unusual spectral unit in West Candor Chasma—Evidence for aqueous or hydrothermal alteration in the Martian Canyons. *Icarus*, 106(2), 380–391.

56. Pelkey, S.M., Mustard, J.F., Murchie, S., Clancy, R.T., Wolff, M., Smith, M., Milliken, R. et al. (2007). CRISM multispectral summary products: Parameterizing mineral diversity on Mars from reflectance. *Journal of Geophysical Research*, 112, E08S14, 18, doi: 10.1029/2006JE002831

57. Geissler, P.E., Greenberg, R., Hoppa, G., McEwen, A., Tufts, R., Phillips, C., Clark, B. et al. (1998). Evolution of lineaments on Europa: Clues from Galileo multispectral imaging observations. *Icarus*, 135(1), 107–126, ISSN 0019-1035, doi: 10.1006/icar.

58. Ramsey, M.S. and Christensen, P.R. (1998). Mineral abundance determination: Quantitative deconvolution of thermal emission spectra. *Journal of Geophysical Research*, 103(B1), 577–596.

59. Berk, A., Bernstein, L.S., and Robertson, D.C. (1989). MODTRAN: A Moderate Resolution Model for LOWTRAN7. Rep. GL-TR-89-0122, Air Force Geophysics Lab, Bedford, MA.

60. Berk, A., Anderson, G.P., Bernstein, L.S., Acharya, P.K., Dothe, H., Matthew, M.W., Adler-Golden, S. et al. (1999). MODTRAN4 radiative transfer modeling for atmospheric correction. *Proceedings of the 8th JPL Airborne Earth Science Workshop*, JPL Publication 99–17, Pasadena, CA.

61. Boardman, J.W. (1998). Post-ATREM polishing of AVIRIS apparent reflectance data using EFFORT: A lesson in accuracy versus precision. *Summaries of the 7th JPL Airborne Earth Science Workshop*, JPL Publication 97–21, Pasadena, CA.

62. Clark, R.N., Swayze, G.A., Wise, R., Livo, E., Hoefen, T., Kokaly, R., and Sutley, S.J. (2007). USGS digital spectral library splib06a: U.S. Geological Survey, Digital Data Series 231, http://speclab.cr.usgs.gov/spectral.lib06

63. Landgrebe, D. (2000). Information extraction principles and methods for multispectral and hyperspectral image data. In: *Information Processing for Remote Sensing,* C.H. Chen, editor, World Scientific Publishing Co., Inc., River Edge, NJ.

64. Kruse, F.A. and Huntington, J.F. (1996). The 1995 AVIRIS Geology Group Shoot. *Proceedings of the Sixth JPL Airborne Earth Science Workshop*, JPL Publication 96–4 (1), Pasadena, CA, pp. 155–164.

65. Kruse, F.A., Boardman, J.W., and Huntington, J.F. (2003). Comparison of airborne hyperspectral data and EO-1 hyperion for mineral mapping. In: *Special Issue, IEEE Transactions on Geoscience and Remote Sensing*, 41(6), 1388–1400.

66. Kruse, F.A., Boardman, J.W., and Huntington, J.F. (1999). Fifteen years of hyperspectral data: Northern Grapevine Mountains, Nevada. In: *Proceedings of the 8th JPL Airborne Earth Science Workshop*, JPL publication 99–17, Pasadena, CA, pp. 247–258.

67. Research Systems Inc. (RSI), 2003, *ENVI User's Guide*, ENVI Version 4.0, September 2003: Research Systems Inc, 1084 p.

68. Swayze, G.A., Smith, K.S., Clark, R.N., Sutley, S.J., Pearson, R.M., Vance, J.S., Hugeman, P.L., Briggs, P.H., Meier, A.L., Singleton, M.J., and Roth, S. (2000). Using imaging spectroscopy to map acidic mine waste. *Environmental Science and Technology*, 34, 47–54.

69. Montero, I.C., Brimhall, G.H., Alpers, C.N., and Swayze, G.A. (2004). Characterization of waste rock associated with acid drainage at the Penn Mine, California, by ground-based visible to short-wave infrared reflectance spectroscopy assisted by digital mapping. *Chemical Geology*, 215(1–4), 453–472.

70. Vaughan, R.G. and Calvin, W.M. (2005). *Mapping Weathering and Alteration Minerals in the Comstock and Geiger Grade Areas Using Visible to Thermal Infrared Airborne Remote Sensing Data*, H.N. Rhoden, R.C. Steininger, and P.G. Vikre, editors, *Geological Society of Nevada Symposium 2005: Window to the World*, Reno, NV, May 2005.

71. White, D.E. (1981). Active Geothermal Systems and Hydrothermal Ore Deposits. In: *Economic Geology: 75th Anniversary Volume*, B.J. Skinner, editor, Economic Geology Publishing Co, New Haven, CT, pp. 392–423.

72. White, D.E., Thompson, G.A., and Sandberg, C.H. (1964). Rocks, Structure, and Geologic history of steamboat springs thermal area, Washoe County, Nevada. *USGS Professional Paper*, 458-B, 61.

73. Hudson, D.M. (1987). Steamboat Springs Geothermal Area, Washoe County, Nevada. In: *Bulk Mineable Precious Metal Deposits of the Western United States*, J.L. Johnson, editor, Geological Society of Nevada Field Trip Guidebook, pp. 408–412.

74. Rye, R.O., Bethke, P.M., and Wasserman, M.D. (1992). The stable isotope geochemistry of acid sulfate alteration. *Economic Geology*, 87(2), 225–262.

75. Solomon, S.C., McNutt, R.L., Gold, R.E., and Domingue, D.L. (2007). MESSNEGER Mission Overview, *Space Science Reviews*, 131, 3–39.

76. McClintock, W.E., Izenberg, N.R., Holsclaw, G.M., Blewett, D.T., Domingue, D.L., Head III, J.W., Helbert, J. et al. (2008). Flyby reflectance during MESSENGER's first Mercury spectroscopic observations of Mercury's surface, *Science*, 321, 62–65, doi: 10.1126/science.1159933.

77. Holsclaw, G.M., McClintock, W.M., Domingue, D.L., Izenberg, N.R., Blewett, D.T., and Sprague, A.L. (2010). A comparison of the ultraviolet to near-infrared spectral properties of Mercury and the Moon as observed by MESSENGER. Icarus, Volume 209, Issue 1, p. 179–194, doi: 10.1016/j.icarus.2010.05.001.

78. Domingue, D.L., Vilas, F., Holsclaw, G.M., Warell, J., Izenberg, N.R., Murchie, S.L., Denevi, B.W. et al. (2010). Whole-disk spectrophotometric properties of Mercury: Synthesis of MESSENGER and ground-based observations. Icarus, Volume 209, Issue 1, p. 101–124, doi: 10.1016/j.icarus.2010.02.022.

79. McClintock, W.E., Bradley, E.T., Vervack, R.J., Jr., Killen, R.M., Sprague, A.L., Izenberg, N.R., and Solomon, S.C. (2008). Mercury's exosphere: Observations during MESSENGER's first mercury Flyby. *Science*, 321, 92–94 [doi: 10.1126/science.1159467].

80. Vervack, R.J., Jr., McClintock, W.E., Killen, R.M., Sprague, A.L., Anderson, B.J., Burger, M.H., Bradley, E.T., Mouawad, N., Solomon, S.C., and Izenberg, N.R. (2010). Mercury's complex exosphere: Results from MESSENGER's third Flyby. *Science* [doi: 10.1126/science.1188572].

81. D'Amore, M., Helbert, J., Maturilli, A., Izenberg, N.R., Sprague, A.L., Holsclaw, G.M., Head, J.W., McClintock, W.E., and Solomon, S.C. (2010). Compositional units on Mercury from principal component analysis of Messenger reflectance spectra. In: *Lunar and Planetary Science Conference*, 41, abstract #2016.

82. Clark, R. et al. (2009). Thermal removal from Moon Mineralogy Mapper (M^3) data. In: *40th Lunar and Planetary Science Conference*, Abstract # 2136.

83. Clark, R. (2009). Detection of adsorbed water and hydroxyl on the Moon. *Science*, 326, #5952, 562–564, doi: 10.1126/science.1178105

84. Sunshine, J.M. et al. (2009). Temporal and spatial variability of lunar hydration as observed by the deep impact spacecraft. *Science*, 326, #5952, 565–568, doi: 10.1126/science.1179788

85. Taylor, L.A. et al. (2009). The role of spinel minerals in lunar magma evolution. In: *AGU Fall Meeting 2009*, Abstract #P34A-07.

86. Pieters, C.M. et al. (2009). New Mg-spinel rock-type on the lunar farside and implications for lunar crustal evolution, AGU Fall Meeting 2009, Abstract #P34A-05.

87. Pieters, C.M. et al. (2009). Lunar magma ocean bedrock anorthosites detected at Orientale basin by M3, AGU Fall Meeting 2009, Abstract #P34A-08.

88. Sunshine, J.M. et al. (2009). Hidden in plain sight: Spinel-rich deposits on the central nearside of the Moon, AGU Fall Meeting 2009, Abstract #P34A-06.

89. Bonello, G., Pierre Bibring, J., Soufflot, A., Langevin, Y., Gondet, B., Berthé, M., and Carabetian, C. (2005). The ground calibration setup of OMEGA and VIRTIS experiments: Description and performances. *Planetary and Space Science*, 53(7), 711–728.

90. Bibring, J.-P. et al. (2005). Mars surface diversity as revealed by the OMEGA/Mars Express observations. *Science*, 307, 1576–1581.

91. Owen, T., Biemann, K., Biller, J.E., Lafleur, A.L., Rushneck, D.R., and Howarth, D.W. (1977). The composition of the atmosphere at the surface of Mars. *Journal of Geophysical Research*, 82, Sept. 30, 4635–4639.

92. McGuire, P.C., Bishop, J.L., Brown, A.J., Fraeman, A.A., Marzo, G.A., Frank, M.M., Murchie, S.L. et al. (2009). An improvement to the volcano-scan algorithm for atmospheric correction of CRISM and OMEGA spectral data. *Planetary and Space Science*, 57(7), 809–815.

93. Vincendon, M. and Langevin, Y. (2010). A spherical Monte-Carlo model of aerosols: Validation and first applications to Mars and Titan. *Icarus*, 207(2), 923–931.

94. Vincendon, M., Langevin, Y., Poulet, F., Bibring, J.-P., and Gondet, B. (2007). Recovery of surface reflectance spectra and evaluation of the optical depth of aerosols in the near-IR using a Monte Carlo approach: Application to the OMEGA observations of high-latitude regions of Mars. *Journal of Geophysical Research*, 112(E8), CiteID E08S13.

95. McGuire, P.C., Wolff, M.J., Smith, M.D., Arvidson, R.E., Murchie, S.L., Clancy, R.T., Roush, T.L. et al. (2008). MRO/CRISM retrieval of surface Lambert Albedos for multispectral mapping of Mars with DISORT-Based radiative transfer modeling: Phase 1—using historical climatology for temperatures, aerosol optical depths, and atmospheric pressures. *IEEE Transactions on Geoscience and Remote Sensing*, 46(12), 4020–4040.

96. Carr, M.H. and Head, J.W. (2010). Geologic history of Mars. *Earth and Planetary Science Letters*, 294(3–4) 185–203.

97. Mustard, J.F., Murchie, S.L., Pelkey, S.M., Ehlmann, B.L., Milliken, R.E., Grant, J.A., Bibring, J.P. et al. (2008). Hydrated silicate minerals on Mars observed by the Mars Reconnaissance Orbiter CRISM instrument. *Nature*, 454, 305–309 (17 July 2008) doi: 10.1038/nature07097

98. McKeown, N.K., Bishop, J.L., Noe Dobrea, E.Z., Ehlmann, B.L., Parente, M., Mustard, J.F., Murchie, S.L., Swayze, G.A., Bibring, J., and Silver, E.A. (2009). Characterization of phyllosilicates observed in the central Mawrth Vallis region, Mars, their potential formational processes and implications for past climate. *Journal of Geophysical Research*, 114, E00D10, doi: 10.1029/2008JE003301

99. Soderblom, L.A. and Bell, J.F. III (2008). *Exploration of the Martian Surface: 1992–2007, The Martian Surface—Composition, Mineralogy, and Physical Properties*, Jim Bell, III, editor, Cambridge University Press, New York. 340 line figures, 40 halftones, 76 plates, 652 pages. 9780521866989, p. 3.

100. Bibring, J.-P. and Langevin, Y. (2008). *Mineralogy of the Martian Surface from Mars Express OMEGA Observations, The Martian Surface—Composition, Mineralogy, and Physical Properties*, Jim Bell, III. Cambridge University Press, New York. 340 line figures, 40 halftones, 76 plates, 652 pages. 9780521866989, p.153.

101. Herschel, W. (1784). On the remarkable appearances at the polar regions of the planet Mars, the inclination of its axis, the position of its poles, and its spheroidical figure; with a few hints relating to its real diameter and atmosphere. *Philosophical Transactions of the Royal Society of London*, 74, 233–273.

102. James, P.B., Kieffer, H.H., and Paige, D.A. (1992). The seasonal cycle of carbon dioxide on Mars, In: *Mars* (A93-27852 09–91), p. 934–968.

103. Titus, T.N., Calvin, W.M., Kieffer, H.H., Langevin, Y., and Prettyman, T.H. (2008). *Martian Polar Processes, The Martian Surface—Composition, Mineralogy, and Physical Properties*. Jim Bell, III, editor, Cambridge University Press, New York. 340 line figures, 40 halftones, 76 plates, 652 pages. 9780521866989, p. 578.

104. Kieffer, H.H. and Titus, T.N. (2001). TES mapping of Mars' North seasonal cap. *Icarus*, 154(1), 162–180.

105. Carlson, R.W., Weissman, P.R., Smythe, W.D., and Mahoney, J.C. (1992). Near-infrared mapping spectrometer experiment on Galileo. *Space Science Reviews*, 60(1–4), May 1992, 457–502.

106. McCord, T.B., Brown, R.H., Baines, K., Bellucci, G., Bibring, J., Buratti, B., Capaccioni, F. et al. (2001). Galilean Satellite Surface Composition: New Cassini VIMS Observations and Comparison with Galileo NIMS Measurements. American Geophysical Union, Fall Meeting 2001, abstract #P12B-0504.

107. Baines, K.H., Carlson, R.W., and Kamp, L.W. (2002). Fresh ammonia ice clouds in Jupiter I. Spectroscopic identification, spatial distribution, and dynamical implications. *Icarus*, 159(1), 74–94.

108. Gierasch, P.J., Ingersoll, A.P., Banfield, D., Ewald, S.P., Helfenstein, P., Simon-Miller, A., Vasavada, A., Breneman, H.H., and Senske, D.A., Galileo Imaging Team, (2000). Observation of moist convection in Jupiter's atmosphere. *Nature*, 403(6770), 628–630.

109. Ingersoll, A.P., Gierasch, P.J., Banfield, D., and Vasavada, A.R., Galileo Imaging Team (2000). Moist convection as an energy source for the large-scale motions in Jupiter's atmosphere. *Nature*, 403(6770), 630–632.

110. Lopes-Gautier, R., Douté, S., Smythe, W.D., Kamp, L.W., Carlson, R.W., Davies, A.G., Leader, F.E. et al. (2000). A close-up look at Io from Galileo's near-infrared mapping spectrometer. *Science*, 288(5469), 1201–1204.

111. Davies, A.G. (2003). Volcanism on Io: Estimation of eruption parameters from Galileo NIMS data. *Journal of Geophysical Research*, 108(E9), 10–1, doi: 10.1029/2001JE001509

112. Carlson, R.W., Smythe, W.D., Lopes-Gautier, R.M.C., Davies, A.G., Kamp, L.W., Mosher, J.A., Soderblom, L.A. et al. (1997). Distribution of sulfur dioxide and other infrared absorbers on the surface of Io. *Geophysical Research Letters*, 24, 2479.

113. Douté, S., Schmitt, B., Lopes-Gautier, R., Carlson, R., Soderblom, L., and Shirley, J. (2001). Mapping SO_2 frost on Io by the modeling of NIMS hyperspectral images. *Icarus*, 149(1), 107–132.

114. Stephan, K., Jaumann, R., Hibbitts, C.A., and Hansen, G.B. (2005). Band depths ratios of water ice absorptions as an indicator of variations in particle size of water ice on the surface of Ganymede. American Astronomical Society, DPS meeting #37, #58.16. *Bulletin of the American Astronomical Society*, 37, 754.

115. McCord, T.B., Hansen, G.B., Fanale, F.P., Carlson, R.W., Matson, D.L., Johnson, T.V., Smythe, W.D. et al. (1998). Salts on Europa's surface detected by Galileo's near infrared mapping spectrometer. *Science*, 280(5367), 1242.

116. Carlson, R.W., Johnson, R.E., and Anderson, M.S. (1999). Sulfuric acid on Europa and the radiolytic sulfur cycle. *Science*, 286(5437), 97–99.

117. Carlson, R.W., Anderson, M.S., Johnson, R.E., Smythe, W.D., Hendrix, A.R., Barth, C.A. et al. (1999). Hydrogen peroxide on the surface of Europa. *Science*, 283(5410), 2062.

118. McCord, T.B., Carlson, R., Smythe, W., Hansen, G., Clark, R., Hibbitts, C., Fanale, F. et al. (1997). Organics and other molecules in the surfaces of Callisto and Ganymede. *Science*, 278, 271–275.

119. Hibbitts, C.A., Pappalardo, R.T., Hansen, G.B., and McCord, T.B. (2003). Carbon dioxide on Ganymede. *Journal of Geophysical Research* (*Planets*), 108(E5), 2–1, doi: 10.1029/2002JE001956

120. Turtle, E.P., Perry, J.E., McEwen, A.S., DelGenio, A.D., Barbara, J., West, R.A., Dawson, D.D., and Porco, C.C. (2009). Cassini imaging of Titan's high-latitude lakes, clouds, and south-polar surface changes. *Journal of Geophysical Research Letters,* 36, L02204, 6.

121. Stofan, E.R., Elachi, C., Lunine, J.I., Lorenz, R.D., Stiles, B., Mitchell, K.L., Ostro, S. et al. (2007). The lakes of Titan. *Nature*, 445, 61–64.

122. Lunine, J.I., Stevenson, D.J., and Yung, Y. L. (1983). Ethane Ocean on Titan. *Science*, 222, 1229–1230.

123. Brown, R.H., Soderblom, L.A., Soderblom, J.M., Clark, R.N., Jaumann, R., Barnes, J.W., Sotin, C., Buratti, B., Baines, K.H., and Nicholson, P.D. (2008). The identification of liquid ethane in Titan's Ontario Lacus. *Nature*, 434, 607–610.

124. Clark, R.N., Curchin, J.M., Hoefen, T.M., and Swayze, G.A. (2009). Reflectance spectroscopy of organic compounds I: Alkanes. *Journal of Geophysical Research*, 114, E03001.

125. Grundy, W.M., Schmitt, B., and Quirico, E. (2002). The temperature-dependent spectrum of methane ice I between 0.7 and 5 mm and opportunities for near-infrared remote thermometry. *Icarus*, 155, 486–496.

126. Mitri, G., Showman, A.P., Lunine, J.I., and Lorenz, R.D. (2007). Hydrocarbon lakes on Titan. *Icarus*, 186, 385–394.

127. Hoehler, T.M. and Westall, F. (2010). Mars exploration program analysis group goal one: Determine if life ever arose on Mars. *Astrobiology*, 10(9), 859–867, doi: 10.1089/ast.2010.0527

Part XII

Conclusions and Way Forward

28 Hyperspectral Remote Sensing of Vegetation and Agricultural Crops: Knowledge Gain and Knowledge Gap After 40 Years of Research

Prasad S. Thenkabail, John G. Lyon, and Alfredo Huete

CONTENTS

28.1 CRITICAL NEEDS IN ADVANCING UNDERSTANDING, MODELING, AND MAPPING OF VEGETATION USING HYPERSPECTRAL DATA

Hyperspectral remote sensing has been initially used for detecting and mapping minerals. In recent years, numerous researchers have used hyperspectral remote sensing in studies pertaining to natural vegetation and agricultural crops—specifically to characterize, model, and map (a) species composition, (b) biophysical and biochemical properties, (c) health, (d) disease and stress, (d) nutrients, (e) moisture, and so on. The chapters of this volume examine state of the art of hyperspectral remote sensing of terrestrial vegetation. These advances can be seen as inevitable, given that between 1995 and 2005, the global demand for plant matter increased by approximately 5%. That is, in 1995, humans consumed 20.3% of the plant material that planet Earth produced (the photosynthetic capacity of the land) but by 2005, that number increased to 25.6% both because of increased consumption of plant material per person and an increase in population [1]. Imhoff and his team reached these conclusions by comparing the rate at which people require plant products, in terms of carbon, to the rate that the Earth can produce plant carbon [1]. Increase of carbon in the atmosphere over the years and the resulting global warming due to greenhouse effects are expected to increase global net primary productivity (NPP) or carbon stored in biomass, thus potentially cooling the planet and mitigating about 10%–15% of the 2.5°–4° rise in global temperature predicted by various climate models to occur over the next century [2]. However, uncertainties in biomass carbon modeling and mapping are highly significant and contribute to "missing carbon" in global carbon budgets. As we see in various chapters, the advances made in the study of vegetation using hyperspectral remote sensing offer one of the best opportunities to reduce these uncertainties in terrestrial carbon modeling and mapping of vegetation through in-depth study of biophysical and biochemical properties, and classification and mapping of vegetation types and species composition.

Hyperspectral imaging, or imaging spectroscopy (IS), gathers data in near contiguous spectra consisting of hundreds or thousands of narrow wavebands along the electromagnetic spectrum. It is fast emerging as a technology to provide practical solutions in characterizing, quantifying, modeling, and mapping natural vegetation and agricultural crops. But as reviewed by Qi et al. in Chapter 3, and from definitions of Goetz and Shippert, it is important to have narrowbands that are contiguous for strict definition of hyperspectral data and not so much the number of bands alone.

Recent research has demonstrated the advances made and great value of using hyperspectral data in a wide range of applications pertaining to vegetation. Even though these accomplishments and capabilities have been reported in various places, the need for a collective "knowledge bank" that links these various advances in one place is missing and is one of the most significant goals of this book. Further, most scientific papers address only specific aspects of research, rather than a comprehensive assessment of advances that have been made, or discussion of how the professional can incorporate these technologies into their work.

For example, in-depth scientific journals report practical applications of hyperspectral narrowbands, yet one has to canvass the literature broadly to obtain the pertinent facts. Since several chapters report this, these findings need to be synthesized so that the reader can determine the best wavebands for their particular study (e.g., Table 28.1) for their practical applications. Also, studies differ in methodology most suited for detecting parameters such as crop moisture variability, chlorophyll content, and stress levels. Professionals need the sort of synthesis and the details provided here to adopt best practices for their own work.

Therefore, the overarching goal of this effort is to critically look at the contents and discuss the advances made by hyperspectral narrowband data in modeling and mapping terrestrial vegetation including agricultural crop characteristics. Here, we highlight the current state of knowledge and opportunities for future further advancement in order to understand, model, and map vegetation characteristics such as species type, biophysical and biochemical quantities, phenology, and productivity. This is accomplished looking at local to global levels using data from a wide array of ground-based, truck-mounted, and airborne and space-borne sensors.

TABLE 28.1

Optimal Hyperspectral Narrowbands in 400–2500 nm to Study Vegetation and Agricultural Crop Biophysical and Biochemical Properties and for Classification of Their Types and Species[a,b,c,d,e]

Serial Number (#)	Wavebands Centers (nm)	Plant Variable Name	Authors
A. Blue bands			
1	405	*Nitrogen, senescing*: sensitivity to changes in leaf nitrogen. Significant absorption due to chlorophyll and carotenoids; reflectance changes due to pigments is moderate to low. Sensitive to senescing (yellow and yellow green leaves)	Stroppiana et al. (Chapter 11)
2	450	*Chlorophyll, carotenoids, senescing*: sensitive to chlorophyll a and b. Significant absorption due to chlorophyll and carotenoids; reflectance changes due to pigments is moderate to low. Sensitive to senescing (yellow and yellow green leaves)	Alchanatis and Cohen (Chapter 13), Roberts et al. (Chapter 14), Galvão et al. (Chapter 17), Ramsey and Ramgoolama (Chapter 21)
3	490	*Carotenoid, LUE, stress in vegetation*: Sensitive to senescing and loss of chlorophyll\browning, ripening, crop yield, and soil background effects	Thenkabail et al. (Chapter 1), Gitelson (Chapter 6), Middleton et al. (Chapter 12)
B. Green bands			
4	515	*Pigments (carotenoid, chlorophyll, anthocyanins), nitrogen, vigor*: positive change in reflectance per unit change in wavelength of this visible spectrum is maximum around this green waveband	Thenkabail et al. (Chapter 1), Gitelson (Chapter 6), Galvão et al. (Chapter 17)
5	531	*LUE, xanophyll cycle, stress in vegetation, pest and disease*: Senescing and loss of chlorophyll\browning, ripening, crop yield, and soil background effects	Gitelson (Chapter 6), Middleton et al. (Chapter 12), Slonecker (Chapter 23), and Yao et al. (Chapter 25)
6	550	*Anthocyanins, chlorophyll, LAI, nitrogen, LUE*: sensitive to numerous vegetation variables	Thenkabail et al. (Chapter 1), Gitelson (Chapters 6 and 15), Middleton et al. (Chapter 12), Alchanatis and Cohen (Chapter 13), Roberts et al. (Chapter 14)
7	570	*Pigments (anthrocyanins, chlorophyll), nitrogen*: negative change in reflectance per unit change in wavelength is maximum as a result of sensitivity to vegetation vigor, pigment, and N	Thenkabail et al. (Chapter 1), Gitelson (Chapters 6 and 15), Stroppiana et al. (Chapter 11), Middleton et al. (Chapter 12), Slonecker (Chapter 23)
C. Red bands			
8	650	*Pigment, nitrogen*: moderate to high sensitivity to changes in pigments (chlorophyll, anthocyanins) and nitrogen	Stroppiana et al. (Chapter 11), Roberts et al. (Chapter 14)
9	687	*Biophysical quantities, chlorophyll, solar induced chlorophyll fluorescence*: LAI, biomass, yield, crop type\discrimination. Greatest soil-crop contrast. Actively induced emission peaks in red\far-red 687 and 740 nm	Thenkabail et al. (Chapter 1) and numerous others

(continued)

TABLE 28.1 (continued)
Optimal Hyperspectral Narrowbands in 400–2500 nm to Study Vegetation and Agricultural Crop Biophysical and Biochemical Properties and for Classification of Their Types and Species[a,b,c,d,e]

Serial Number (#)	Wavebands Centers (nm)	Plant Variable Name	References
D. Red edge bands			
10	705	*Stress in vegetation detected in red edge, stress, drought*: Nitrogen stress, crop stress, crop growth stage studies. Blueshift in case of stress. Shift toward NIR for healthy vegetation	Zhang (Chapter 7), Gitelson (Chapters 6 and 15), Stroppiana (Chapter 11), Thomas (Chapter 20), Yao et al. (Chapter 25)
11	720	*Stress in vegetation detected in red edge, stress, drought*: Nitrogen stress, crop stress, crop growth stage studies. Blueshift in case of stress. Shift toward NIR for healthy vegetation	Thenkabail et al. (Chapter 1), Gitelson (Chapters 6 and 15), Zhang (Chapter 7), Stroppiana et al. (Chapter 11), Roberts et al. (Chapter 14), Thomas (Chapter 20)
12	700–740	*Chlorophyll, senescing, stress, drought*: first-order derivative index over 700–740 nm has applications in vegetation studies (e.g., blueshift during stress and redshift during healthy growth)	Thenkabail et al. (Chapter 1), Gitelson et al. (Chapters 6 and 15), Yao et al. (Chapter 25)
E. NIR bands			
13	760	*Biomass, LAI, Solar-induced passive emissions*: NIR reference band for many indices. Solar-induced passive emissions with retrievals made in O_2 atmospheric features at 687 and 760 nm	Gitelson (Chapter 15), most other chapters in the book
14	855	*Biophysical/biochemical quantities, heavy metal stress*: LAI, biomass, yield, crop\discrimination, chlorophyll, anthocyanin, carotenoids. Sensitive to heavy metal stress due to reduction in chlorophyll. High stability in NIR band for developing indices	Most chapters in the book
15	970	*Water absorption band*: most prominent water absorption trough. Also useful in quantifying most biophysical and biochemical properties	Thenkabail et al. (Chapter 1), Colombo et al. (Chapter 10), Alchanatis and Cohen (Chapter 13), Roberts et al. (Chapter 14), Pu (Chapter 19)
16	1045	*Biophysical and biochemical quantities*: leaf area index, wet and dry biomass, plant height, grain yield, crop type, crop discrimination, total chlorophyll, anthocyanin, carotenoids	Zhang (Chapter 7) Stroppiana et al. (Chapter 11)
E. Far near infrared (FNIR) bands			
17	1100	*Biophysical quantities*: sensitive to biomass and leaf area index. A point of most rapid rise in spectra with unit change in wavelength in far near infrared (FNIR)	Thenkabail et al. (Chapter 1), Colombo et al. (Chapter 10)
18	1180	*Water absorption band*	Colombo et al. (Chapter 10), Roberts et al. (Chapter 14)
19	1245	*Water sensitivity*: water band index, leaf water, biomass. Reflectance peak in 1050–1300 nm	Thenkabail et al. (Chapter 1), Colombo et al. (Chapter 10)

TABLE 28.1 (continued)

Optimal Hyperspectral Narrowbands in 400–2500 nm to Study Vegetation and Agricultural Crop Biophysical and Biochemical Properties and for Classification of Their Types and Species[a,b,c,d,e]

Serial Number (#)	Wavebands Centers (nm)	Plant Variable Name	References
F. Early short-wave infrared (ESWIR) bands			
20	1450	*Water absorption band*: very high moisture absorption trough in early short wave infrared (ESWIR). Use as an index with 1548 or 1620 or 1690 nm	Thenkabail et al. (Chapter 1), Colombo et al. (Chapter 10), Alchanatis and Cohen (Chapter 13), Ben-Dor (Chapter 22)
21	1548	*Lignin, cellulose*: plant biochemical properties	Thenkabail et al. (Chapter 1), Pu et al. (Chapter 19)
22	1620	*Lignin, cellulose*: plant biochemical properties. Peak reflectance in SWIR 1 for vegetation	Numata (Chapter 9), Colombo et al. (Chapter 10), Roberts et al. (Chapter 14), Galvão et al. (Chapter 17)
23	1650	*Heavy metal stress, Moisture sensitivity*: Heavy metal stress due to reduction in chlorophyll. Sensitivity to plant moisture fluctuations in ESWIR. Use as an index with 1548 or 1620 or 1690 nm	Numata (Chapter 9), Colombo et al. (Chapter 10), Roberts et al. (Chapter 14), Galvão et al. (Chapter 17)
24	1690	*Lignin, cellulose, sugar, starch, protein*: plant biochemical properties	Numata (Chapter 9), Pu (Chapter 19)
25	1760	*Water absorption band, senescence, lignin, cellulose*: high to moderate moisture absorption in ESWIR for moisture in plant leaves. Use as an index with 1548 or 1620 or 1690 nm	Roberts et al. (Chapter 14), Galvão et al. (Chapter 17), Pu (Chapter 19)
G. Far short-wave infrared (FSWIR) bands			
26	1950	*Water absorption band*: highest moisture absorption trough in FSWIR. Use as an index with any one of 2025, 2133, and 2213 nm. Affected by noise at times	Alchanatis and Cohen (Chapter 13), Colombo et al. (Chapter 10), Pu (Chapter 19), Ben-Dor (Chapter 22)
27	2025	*Litter (plant litter), lignin, cellulose*: litter-soil differentiation	Nagler et al. (Chapter 16)
28	2050	*Water absorption band*: high moisture absorption trough in FSWIR. Use as an index with any one of 2025, 2133, and 2213 nm. Not affected by noise	Alchanatis and Cohen (Chapter 13), Colombo et al. (Chapter 10), Pu (Chapter 19), Ben-Dor (Chapter 22)
29	2133	*Litter (plant litter), lignin, cellulose*: typically highest reflectivity in FSWIR for vegetation. Litter-soil differentiation	Roberts et al. (Chapter 14), Nagler et al. (Chapter 16), Galvão et al. (Chapter 17)
30	2145	*Water absorption band*: moderate moisture absorption trough in FSWIR. Use as an index with any one of 2025, 2133, and 2213 nm. Not affected by noise	Ben-Dor (Chapter 22)
31	2173	*Water absorption band*: moderate to low moisture absorption trough in FSWIR. Use as an index with any one of 2025, 2133, and 2213 nm. Not affected by noise	Thenkabail et al. (Chapter 1)

(continued)

TABLE 28.1 (continued)

Optimal Hyperspectral Narrowbands in 400–2500 nm to Study Vegetation and Agricultural Crop Biophysical and Biochemical Properties and for Classification of Their Types and Species[a,b,c,d,e]

Serial Number (#)	Wavebands Centers (nm)	Plant Variable Name	References
32	2205	*Litter, lignin, cellulose, sugar, startch, protein; Heavy metal stress*: typically, second highest reflectivity in FSWIR for vegetation. Heavy metal stress due to reduction in chlorophyll	Numata (Chapter 9), Nagler et al. (Chapter 16), Slonecker (Chapter 23)
33	2295	*Stress and soil iron content*: sensitive to soil background and plant stress	Thenkabail et al. (Chapter 1), Galvão et al. (Chapter 17), Pu (Chapter 19), Ben-Dor (Chapter 22)

[a] Wavebands were selected based on research and discussions in the chapters.

[b] When there were close wavebands (e.g., 960 and 970 nm), only one waveband (e.g., 970 nm) was selected based on overwhelming evidence as reported in various chapters. This would avoid redundancy.

[c] A nominal 5 nm waveband width can be considered optimal for obtaining best results with aforementioned wavebands as band centers. So, for 970 nm waveband center, we can have a band of range of 968–972 nm. Ideal bandwidth is about 3 nm. But, noise levels in lower bandwidths can be a significant problem.

[d] The aforementioned wavebands can be considered as optimal for studying vegetation. Adding more waveband will only add to redundancy. Vegetation indices can be computed using aforementioned wavebands. Some of the typical biophysical and biochemical vegetation indices for hyperspectral narrowband data are defined, for example, in Table numbers: 14.1, 17.2, 19.2, 23.2, 23.3, 10.1, and 8.7 of this book. These indices can be computed using wavebands provided in this Table.

[e] Thirty-three wavebands lead to a matrix of $33 \times 33 = 1089$ two band vegetation indices (TBVIs). Given that the indices above the diagonal and below diagonal replicate and indices along diagonal are redundant, there are 528 unique TBVIs.

The book itself is organized to provide the comprehensive and systematic advances made in hyperspectral remote sensing of vegetation and agricultural crops. Chapter 1 provides an overview of hyperspectral technology. Chapters 1 through 3 also discuss all the hyperspectral sensors of past, present, and the immediate future. Chapters 3 through 5 as well as Chapter 1 deal with methods and approaches in overcoming the Hughes effect, whereby redundant wavelengths combined with system noise reduce accuracy when too many wavelengths are included in the analysis. Chapters 6 through 15 focus on characterizing biophysical and biochemical properties of leaves and plants. This includes characterizing variables such as leaf area index (LAI), biomass as well as nitrogen, water, chlorophyll, and lignin content using hyperspectral data from various platforms. These chapters also discuss in-depth critical hyperspectral narrowbands and indices that have been most influential in characterizing and modeling these leaf and plant biophysical and biochemical properties.

Chapters 16 and 17 examine evapotranspiration studies, actual water use by plants, phenology, light use efficiency (LUE), and gross primary productivity (GPP). Chapters 18 through 20 report on species identification and Chapters 21 through 24 on broader land cover classifications. From these, we can also understand the higher accuracies (and low errors and uncertainties) in species and land cover classification using selected sets of hyperspectral narrowbands when compared to multispectral broadbands. Chapters 25 through 27 describe crop management issues involving insect and disease infestation studies, herbicide-pesticide-nitrogen applications, precision farming, and crop stress. Chapter 26 explains how the understanding we have developed can be applied to larger areas and global level studies through cross-sensor calibration. Hyperspectral data and methods applied in terrestrial studies are almost the same as those applied in the study of other planets. Further, the study of soils and rocks (background material for vegetation) will be approached along similar lines. These linkages are discussed in Chapter 27. The study of sensor characteristics, preprocessing, and methods and approaches adopted by planetary scientists and terrestrial scientists can complement\supplement our knowledge base, and, we can thereby learn from each other and advance. Finally, the discussions are organized based on thematic subject areas (e.g., instruments, data handling, methods and approaches, applications) of hyperspectral remote sensing of vegetation. The following sections are organized in a thematic framework. In each of these sections, we highlight unique contributions of each chapter and point out the chapters that complement/supplement each other.

28.2 HUGHES PHENOMENON, HYPERSPECTRAL DATA PROCESSING METHODS AND ALGORITHMS, AND OVERCOMING DATA REDUNDANCY

An exhaustive review of space-borne, airborne, unmanned aerial vehicles, helicopter-mounted, truck-mounted, and ground-based hyperspectral sensors is provided in Table 1.1; and a comprehensive review follows in Chapter 2 by Ortenberg and Chapter 3 by Qi et al. Ground-based and truck-mounted measurements are made using spectroradiometers or spectrometers that typically gather data approximately every nanometer between 400 and 2500 nm. Airborne sensors include AVIRIS, HYDICE, AISA, HyMAP, ARES, CASI 1500, and AisaEAGLET. These sensors gather data in tens or hundreds of narrow wavebands (<10 nm bandwidth), typically over 400–2500 nm. The spatial resolution of airborne hyperspectral images are generally in the meters range and can change depending on the flight characteristics and the sensor equipment used. Hyperspectral data gathered from airborne sensors, at low heights, reaches submeter spatial resolutions.

The advantages of airborne, ground-based, and truck-mounted sensors are that they enable relatively cloud-free acquisitions that can be gathered on demand anywhere. Over the years, they have also allowed careful study of spectra in controlled environments to advance the genre. There are 18 (see Table 2.1) space-borne hyperspectral sensors, either already in operation or planned over next few years, acquiring data in 20–400 narrowbands. Most of these sensors acquire data at a 20–30 m spatial resolution in 1 to 10 nm bandwidth and have 10–30 km swath widths. These sensors include Hyperion onboard EO-1, CHRIS onboard PROBA, ARTEMIS onboard TacSat-3, and others including commercial, defense, or dual use systems (see Table 2.1). The advantages of space-borne systems are their capability to acquire data (a) continuously, (b) consistently, and (c) over the entire globe. A number of system design challenges of hyperspectral data are discussed in Chapter 3 by Qi et al. Challenges include cloud cover and large data volumes. However, as we see in various chapters of this book, the need for hyperspectral data to characterize, model, and map vegetation and agricultural crops of diverse ecosystems is critical and provides great advances over conventional multispectral remote sensing. We are only in the beginning of an era where data from commercial hyperspectral space-borne sensors will become routine and will be widely used in many terrestrial applications. Chapters 1 through 3 provide many perspectives of hyperspectral data, their characteristics, and their value for vegetation studies with some overlapping, but mostly unique perspectives.

Hyperspectral narrowbands provide near continuous spectra leading to spectral signatures of every target. For any given target, hyperspectral narrowband data also provide a significant number of redundant bands. Neighboring bands, in particular, have a high probability of redundancy (both bands providing similar information). This is widely known as the "curse of high dimensionality" or the Hughes phenomenon. This is because, if the number of bands remained high, the number of observations required to train a classifier increases exponentially to maintain classification accuracies. Chapters 3 through 5, as well as some other chapters like Chapter 1, provide methods and approaches to overcome data redundancy.

You will find many unique approaches in these methods, enriching the reader with several different perspectives. This is an important first step in handling hyperspectral data. Chapters 3 and 4 discuss in detail feature selection and information extraction methodologies. Feature selection is necessary in any data mining effort. Feature selection reduces the dimensionality of data by selecting only a subset of measured features (predictor variables). Feature selection methods recommended in Chapter 4 are based on (a) information content (e.g., selection based on theoretical knowledge, band variance, information entropy); (b) projection-based methods (e.g., principal component analysis [PCA], independent component analysis [ICA]); (c) divergence measures (e.g., distance-based measures); (d) similarity measures (e.g., correlation coefficient, spectral derivative analysis); and (e) other methods (e.g., wavelet decomposition method). Information extraction methodologies include (a) multivariate and partial least square regression; (b) discriminant analysis; (c) unsupervised classification; (d) supervised classification. All these approaches have merit; it remains for the user to apply them to the situation of interest.

Chapter 5 clearly demonstrates the Hughes phenomenon as a result of limited training samples per class in supervised full-pixel classifications. This chapter also demonstrates ways and means of overcoming the Hughes phenomenon with kernel approaches. Firstly, supervised support vector machines (SVM) classification was used with only 1% training pixels per class to attain accuracies as high as 90%. Second, unsupervised linear and nonlinear unmixing algorithms were applied to determine fractional abundance of classes within a pixel with high accuracies. These algorithms take advantage of both spectral and spatial characteristics as well as machine learning techniques. The key contributions of Chapters 4 and 5 are in (a) highlighting data redundancy, or Hughes phenomenon, in hyperspectral narrowband data, (b) presenting methods and approaches of feature selection and information extraction, (c) demonstrating means of overcoming the Hughes phenomenon, and (d) developing or presenting algorithms to attain high accuracies in hyperspectral narrowband data classification.

Data mining algorithms are also discussed in Chapter 1 by Thenkabail et al. and Chapter 3 by Qi et al. Apart from the methods and approaches described in Chapters 2, 4, and 5, Thenkabail et al. introduce a unique method of Lambda 1 (λ_1: 400–2500 nm) vs. Lambda 2 (λ_2: 400–2500 nm) plots to determine redundant bands vs. most useful bands to study various crop characteristics such as biomass, LAI, yield, chlorophyll, carotenoid, and nitrogen. These Lambda vs. Lambda plots are contour plots of R-squared values plotted along the 400–2500 nm spectral range (λ_1: 400–2500 nm vs. λ_2: 400–2500 nm) with an R-squared value for every narrowband combination (e.g., 1 nm band width in many spectroradiometers and 10 nm in Hyperion). The "bull's eye" regions of the Lambda vs. Lambda plots are the most useful bands to study a particular vegetation characteristic. Such Lambda by Lambda plots are also presented in Chapters 8, 11, and 18, reflecting its wide applicability to identify redundant bands and to select the best bands to study various vegetation characteristics.

Hyperspectral data composition needs to take into account the concept of mega-file data cubes (MFDCs) that compress hundreds or thousands of bands into a single file as discussed in Chapter 1. It is important to master these concepts since most of the further processing of hyperspectral data to gather useful information for various applications will make use of MFDCs.

28.3 BIOPHYSICAL AND BIOCHEMICAL MODELING USING HYPERSPECTRAL VEGETATION INDICES AND NARROWBANDS

Chapters 1 and 6 through 15 provide us with up-to-date knowledge of hyperspectral narrowband application and advances with regard to characterizing, modeling, and mapping of biophysical properties (e.g., LAI, biomass, crop height, canopy volume, yield), biochemical properties (e.g., nitrogen, chlorophyll, lignin, salinity, leaf water content, phosphorous, potassium), species composition, and land cover classifications of vegetation and agricultural crops. Hyperspectral narrowbands also provide significant advances in quantifying these properties relative to point measurements in the field. Other chapters provide discussions of this method.

Chapter 1 by Thenkabail et al. comprehensively reviews the current state of the art in the study of biophysical and biochemical properties of vegetation and agricultural crops using hyperspectral-narrowband data. The focus is on a detailed assessment in the study of biophysical properties of natural vegetation as well as agricultural crops using hyperspectral narrowband data gathered from ground-based spectroradiometers as well as the space-borne Hyperion sensor. Vegetation parameters such as LAI, wet and dry biomass, plant height, and crop yield are studied. A variety of hyperspectral vegetation indices (HVIs) are described: (a) hyperspectral two band vegetation index (HTBVI) and hyperspectral multiple-band models (HMBMs), (b) hyperspectral derivative greenness vegetation indices (HDGVIs), (c) hyperspectral hybrid vegetation indices (HHVIs), (d) soil-adjusted hyperspectral two band vegetation indices (SA HTBVIs), (e) atmospherically resistant hyperspectral two band vegetation indices (AR HTBVIs), (f) HVI of short wave infrared (SWIR) and thermal infrared (TIR) bands (HVIST), and (g) other methods of hyperspectral data analysis (e.g., ICA, wavelet transforms, radiative transfer (RT) models, minimum noise fraction (MNF) transformation, and spectral unmixing analysis). The results are compared with multispectral broadband vegetation indices (BBVIs; e.g., NIR- and red-based indices; soil-adjusted indices; atmospheric-resistant indices; and midinfrared-based indices). Typically, HVIs explain 10%–20% greater variability in data when compared with BBVIs. Based on this analysis mostly focused on modeling biophysical quantities, 28 optimal hyperspectral narrowbands (see Table 1.2), each of nominal 10 nm wide in the 400–2500 nm spectral range, are recommended for use.

Chapter 15 by Gitelson discusses the performance of vegetation indices (VIs) in estimating six biophysical quantities derived from three crops (wheat, maize, and soybeans). The biophysical quantities studied were (a) vegetation fraction, (b) fraction of absorbed photosynthetically active radiation (fAPAR) absorbed by photosynthetically active ("green") vegetation, (c) total canopy chlorophyll content, (d) green LAI, and (e) gross primary production. The performance was studied using measurements from a spectroradiometer hoisted 6 m above the top of canopy (TOC). VIs were derived and scaled up to correspond with wavebands of 250 m moderate resolution imaging spectrometer (MODIS) and the 300 m medium resolution imaging spectrometer (MERIS): NIR (841–876 nm), red edge (707–717 nm), red (620–670 nm), and green (545–565 nm). Nine VIs, selected from 30 recommended indices in the literature, were as follows:

1. Red edge normalized difference vegetation index ($NDVI_{red\ edge}$) involving an NIR band and a red edge band in a normalized difference
2. Green NDVI ($NDVI_{green}$) involving a green band and a red band in a normalized difference
3. Wide dynamic range vegetation index (WDRVI), which is the same as the NDVI and includes an NIR and a red band in the numerator and denominator but also includes a weighting coefficient to attenuate the contribution of the NIR and reduce NDVI saturation at moderate-to-high biomass levels
4. Visible atmospherically resistant vegetation index ($VARI_{green}$), a normalized difference of green and red bands
5. $VARI_{red\ edge}$ involving red edge, red, and a blue bands with weighing factors (see Equation 15.6)
6. EVI2 involving an NIR and red bands with weighting factors (see Equation 15.3)
7. Two chlorophyll indices, CI_{green} involving green and NIR bands and $CI_{red\ edge}$ involving red edge and NIR bands, and (h) the MERIS terrestrial chlorophyll index (MTCI) involving NIR, red, and red edge bands.

The study determined that indices that best estimated biophysical variables were (a) $NDVI_{red\ edge}$ and WDRVI for vegetation fraction and $fAPAR_{green}$; (b) $CI_{red\ edge}$, MTCI, and $NDVI_{red\ edge}$ for total chlorophyll content (product of leaf chlorophyll and LAI); (c) NDVI with $LAI_{green} < 1.5\ m^2/m^2$ and MTCI, CI_{green} and $CI_{red\ edge}$ with $LAI_{green} > 1.5\ m^2/m^2$; and (d) NDVI with GPP $< 7\ gC/m^2/d$ and CI_{green}, $CI_{red\ edge}$, MTCI, EVI, and WDRVI with GPP $> 7\ gC/m^2/d$. It was shown that the VIs that were optimal at a close range were also accurate in estimating biophysical characteristics from MODIS and TM sensors.

In Chapter 13, Alchanatis and Cohen not only investigate the value of hyperspectral narrowband data in crop biophysical and biochemical properties, but also evaluate the advantages obtained if hyperspectral narrowband data are fused with hyperspatial data. They highlight the fact that by studying biophysical and biochemical properties, one can also evaluate other factors such as pest damage, plant diseases, and weed infestations. They describe four main methods of spectral analysis:

1. Band selection (e.g., unsupervised: PCA, band-band correlation; supervised: stepwise discriminant analysis)
2. Use of spectral indices (e.g., two band ratio or normalized difference indices)
3. Linear multivariate statistics and models (e.g., multivariate regression, partial least squares regression, principal component regression)
4. Nonlinear methods (e.g., spectral angle mapper, wavelets). Their review and studies showed the following wavebands were important in the study of crop biophysical and biochemical components

These were (a) strong chlorophyll absorption at 450 and 680 nm, (b) water absorption bands (760, 970, 1450, 1940–2350 nm), (c) LAI and biomass at 550 and 740 nm, and (d) nitrogen content through a first order derivative between 400 and 900 nm. Generally, they recommended use of wider spectral ranges rather than single bands; a spectral range of 400–900 nm for biochemical properties such as nitrogen and chlorophyll, and for biophysical properties such as biomass and LAI, and 1300–2500 nm for vegetation water content studies.

There were several wavebands useful in the study of biochemical properties that also had capabilities for study of biophysical properties and vice versa (see recommendation in this chapter and compare with Thenkabail et al., Chapter 1). The biggest, potential future advance in agricultural applications is likely through a fusion of spectral, spatial, and temporal data. Alchanatis and Cohen describe spectral and spatial data integration accomplished through algorithms such as SVMs and adjusted modes of the Beamlet approach. These can segment a farm field into homogeneous zones for management (e.g., division into different N levels that may be used to determine the appropriate fertilization levels required in different parts of a farm).

In Chapter 14, Roberts et al. provide a comprehensive overview of the performance of narrowband HVIs in the study of vegetation biophysical, biochemical, and water absorption properties. Biophysical properties are best studied using the narrowband HVI versions of the broadband indices by taking an NIR and a red band. The best results are obtained by taking narrowbands ($\leq$10 nm) centered at 682 and 845 nm (see Chapter 1 by Thenkabail et al.). In addition, biophysical properties are also studied using derivative indices such as the first-order derivative greenness vegetation index (DGVI) that integrates contiguous or near-contiguous spectra from 626 to 796 nm wavelengths. However, as Roberts et al. point out, the spectral range used to derive DGVI can vary and needs to be optimized for each vegetation variable. One can also consider a narrowband centered at the NIR peak at 900 nm. Planophile leaf angle (<30° leaf angle) of vegetation leaves, typically, has a near flat NIR shoulder (760–910 nm) when compared to sharp increases in reflectance from 760 to 910 nm for the case of erectophile (>60° leaf angle) leaf angular distributions (see Chapter 1). Vegetation water is studied using (a) water band index involving the ratio of 900 nm (NIR peak) to 970 nm (water absorption trough), (b) normalized difference water index that involves narrow wavebands at 857 nm (NIR shoulder) and 1240 nm (water absorption trough), (c) moisture index involving a ratio of 830 and 1650 nm, (d) another normalized difference water index that involves narrow wavebands at 1650 or 2200 nm (one of the SWIR water absorption bands) and 1240 nm (water absorption trough), and (e) thin tissue water index involving 960 and 1180 nm.

Lignin and cellulose cause many prominent absorption features in the SWIR (Roberts et al., Chapter 14). The cellulose absorption bands are centered at 2101 and 1754 nm. In deriving indices, a nonabsorbing band centered at 1680 nm (for 1754 nm), 2031, and 2211 nm (for 2101 nm) is typically used in conjunction with the absorption band of interest.

Vegetation biochemical properties include three important pigments: (a) chlorophyll that absorbs in the blue and red, (b) anthocyanins that absorb in green, and (c) carotenoids that absorb in the blue. Chapter 6 by Gitelson recommends two relatively narrowbands (green: 540–560 nm, red edge: 700–720 nm), one narrowband (centered at 510 nm with a band width of 10 nm), and one relatively broad band (NIR: 760–900 nm), a total of just four bands, for accurate retrieval of three foliar pigments (chlorophylls, carotenoids, and anthocyanins). In order to select these optimal bands, they proposed and implemented a three-band model. Roberts et al. (Chapter 14) studied pigment indices that model all of these pigments or combination of these pigments, such as carotenoids-to-chlorophyll ratio. Roberts et al. found that the key wavebands for studying these pigment indices were 445, 650, 675, or 680, and 800 nm. As per their review, the best chlorophyll indices consisted of wavebands centered at 550, 670, 720, and 845 nm and the best anthocyanin indices consisted of wavebands centered at 550, 720, and 845 nm.

In Chapter 13, Alchanatis and Cohen not only investigate the value of hyperspectral narrowband data in crop biophysical and biochemical properties, but also evaluate the advantages one would have when hyperspectral narrowband data are fused with hyperspatial data. They highlight the fact that by studying biophysical and biochemical properties one also evaluates other factors such as pests damage, plant diseases, and weeds infestation. There are three main methods for spectral analysis: (a) band selection (e.g., unsupervised: PCA, band-band correlation; supervised: stepwise discriminant analysis), (b) use of spectral indices (e.g., two band ratio or normalized difference indices), (c) linear multivariate statistics and models (e.g., multivariate regression, partial least squares regression, principal component regression), and (d) nonlinear methods (e.g., spectral angle mapper, wavelets) (Alchanatis and Cohen, Chapter 13). Their review and studies showed the following wavebands are important in study of crop biophysical and biochemical components. These were (a) strong chlorophyll absorption in 450 and 680 nm; (b) water absorption bands (760, 970, 1450, 1940, and 2350 nm); (c) LAI and biomass in 550 and 740 nm; and (d) nitrogen content through first order derivative over 400–900 nm. Generally, they recommend to use wider spectral range rather than single bands: spectral range of 400–900 nm for biochemical properties such as nitrogen and chlorophyll, and for biophysical properties such as biomass and LAI, and 1300–2500 nm for vegetation water content studies. However, there are several wavebands useful in study of biochemical properties also useful for biophysical properties and vice versa (see recommendation in this chapter and compare with Thenkabail et al., Chapter 1). However, the biggest advance in coming years in agricultural application is likely to come through a fusion of spectral, spatial, and temporal data. The spectral and spatial data integration is performed through algorithms such as SVMs and adjusted modes of the Beamlet approach, which have advances in segmenting a farm field into homogeneous zones for management (e.g., division into different N levels that may be used to determine fertilization levels required in different parts of a farm).

28.4 CROPLAND TYPE AND SPECIES DISCRIMINATION AND CLASSIFICATION

Chapter 1 by Thenkabail et al. and Chapter 17 by Galvão et al. investigated the capability of hyperspectral narrowbands to differentiate vegetation types and plant species and enumerated on the classification accuracies relative to broadband data. Galvão et al. provide a comprehensive insight into the capability of hyperspectral narrowband Hyperion satellite sensor data in discriminating crop types and their varieties keeping in mind various management practices. They selected six distinct crops including (a) annual crops like flooded rice (*Oryza sativa* L.), corn (*Zea mays* L.), soybean (*Glycine max* (L.), bean (*Phaseolus vulgaris* L.), and pasture, (b) semiperennial crops like sugarcane (*Saccharum* spp.), and (c) perennial crops like coffee (*Coffea arabica* L.). Each has their own unique characteristics, which influence the reflectance properties in different wavebands.

For example, soybeans and beans have planophile leaf structure, are leguminous, and have high N content leading to high reflectance in the NIR and strong absorption in red. Corn, sugarcane, and pasture have erectophile leaf structure and have relatively lower NIR reflectance. However, reflectance from crops is not only dependent on their intrinsic biophysical and biochemical properties, but also on management practices. For example, the cropping calendars (planting, growth period, and harvest) of the crops vary. This will lead to certain time periods when different crops are more distinguishable than at other times. Other management properties can significantly influence reflectance behavior. These factors include: tillage or no-tillage (e.g., soybean, corn in selected crops), burning prior to harvest (sugarcane), type of pruning (coffee, orchards), and harvest timing (e.g., sugarcane can be harvested almost all year long). Galvão et al. used all of the above crop characteristics and as well as numerous other characteristics such as cultivar type and canopy cover (see Chapter 17) to establish the capability of hyperspectral narrowbands to discriminate crop types and their cultivars within a crop type. Crop discrimination can be visualized by plotting a visible band centered at 660 nm vs. an NIR band centered at 864 nm.

In an example from Chapter 17, six crops cluster in a similar way to a conventional green vegetation, shadow, and soil plot from spectral mixture analysis. Soybeans and beans have high reflectance in NIR and high absorption in red as a result of planophile leaf structure and high N content of leguminous plants or high chlorophyll content in a diaxial leaf surface. Coffee and flooded rice cluster with low reflectivity (or high absorption) in both NIR and red bands due to water background and shadow effects within and between canopies. Finally, corn, sugarcane, and pasture cluster on the bright side with high reflectivity in both NIR and red bands due to their erectophile structure, lower N content, and spectral influence of the nonphotosynthetic vegetation within their canopies. Galvão et al. also show equally good separability between crops by plotting waveband ratios of 864/660 vs. 1649/569; these graphic devices, while seemingly simple, are useful to develop hypotheses for further testing. Other bands they identified were (a) 671 nm (chlorophyll absorption), (b) 983 and 1205 nm (leaf water), and (c) 2103 and 2304 nm (lignin-cellulose). They used 36 narrowbands to classify seven distinct soybean cultivars and attained an accuracy of 84.1%, which was about 20% higher than any broadband classification accuracy. The 35 important narrowbands found useful were identified in all portions of the electromagnetic spectrum, including visible, NIR, and SWIR.

In Chapter 1, Thenkabail et al. demonstrate the ability of specific hyperspectral narrowbands to separate vegetation types when such differentiation is infeasible using any of the broadbands. For example, they are able to use 10 nm wide narrowbands centered at 675 nm (red) and 910 nm (NIR) to separate wheat from barley whereas this was not possible using the red and NIR broadbands of Landsat ETM+. They recommend stepwise discriminant analysis (e.g., Wilks' lambda and Pillai trace) to evaluate class separability. Typically, some combination of 10–15 narrowbands (each of 10 nm wide) in the 400–2500 nm spectral range will provide optimal separation between various vegetation categories. They showed that accuracies in classifying 5–6 crop types using 15–22 narrowbands were about 30% higher than the classification accuracies obtained using six Landsat ETM+ broadbands. They recommended discriminant model error matrices to achieve this goal.

28.5   FOREST TYPE AND SPECIES DISCRIMINATION AND CLASSIFICATION USING HYPERSPECTRAL DATA

Chapter 18 by Clark deals with tropical forest species characterization using hyperspectral and hyperspatial data. He reviews the fundamental biochemical, structural (biophysical), and phenological factors that can help discriminate species based on their reflectance spectra. Tropical forest canopies show strong chlorophyll-*a* absorption between 410 and 430 and 600 and 690 nm, chlorophyll-*b* absorption between 450 and 470 nm, and leaf water absorptions at 970 and 1200 nm. Clark reviews recent findings from laboratory spectrometers, which show the potential of hyperspectral sensors to distinguish between species of trees and vines (lianas). In one early study by Clark and colleagues, linear discriminant analysis (LDA) with 10 narrowbands attained an accuracy of 90% in separating seven tree species in a tropical moist forest. This accuracy reached 100% when 40 narrowbands were used. However, when the number of species increased, the accuracies decreased. For example, other recent studies have shown that an accuracy of 85% is possible for 20 species, but accuracies decrease to 80% when 40 species were involved. The chapter also provides guidelines to extrapolate leaf level knowledge gained in the laboratory out to the full canopy.

RT models are the best route to establish proper understanding of how leaf-level biochemical/structural properties are expressed in canopy-scale reflectance spectra, especially when considering the effects of structural and phenological variability (e.g., changes in LAI). Clark reports that one recent study found high correlation between foliar chlorophyll, carotenoids, and specific leaf area (SLA) with leaf reflectance measured in the laboratory ($r = 0.89$–0.91). There was also moderately high correlation between foliar N, P, and water with leaf level measurements ($r = 0.79$–0.85). These correlations generally increased in RT-modeled canopy-scale spectra due to amplification of leaf biochemical and structural properties by 3D scattering of photons among tissues in the crown. Actual airborne hyperspectral data, which includes confounding factors such as atmospheric contamination and variable sun and view angles, have been used to distinguish native, introduced, and invasive, N-fixing and nonfixing tree species based on spectral differences related to biochemical and structural properties within groups, and advances have been made in linking foliar biochemical/structural diversity to overall canopy species richness.

Clark reviews existing analyses that have focused on separating individual species with airborne and space-borne hyperspectral sensors. Major themes that emerge from the literature are that full-spectrum (visible to shortwave infrared) data provide optimal species discrimination, data reduction, and classification techniques can be used to optimize separation among species and discard redundant or noisy data and finally changes in leaf phenology (e.g., leaf area) are important to consider in operational mapping of tree species. This chapter also discusses the high value of fusing hyperspatial data with hyperspectral data. For example, in one study, hyperspatial data were used to delineate tree crowns. Then five species were separated using 25 narrowbands to obtain an accuracy of 100%. However, tropical forest canopy reflectance is quite complex and only the initial stages have been achieved. Less well-understood factors include (Clark, Chapter 18): tropical forest tree structure (e.g., leaf thickness, air spaces), nonphotosynthetic tissues (e.g., bark, flowers, seeds), and other photosynthetic canopy organisms (e.g., vines, epiphytes, epiphylls). This is a work in progress.

Forests cover nearly 30% of terrestrial Earth and play important roles in global carbon cycling. As pointed out in Chapter 20 by Thomas, multispectral imagery at best monitors forest location, type, and changes over time. In contrast, hyperspectral data provide the opportunity to expand remote characterizations of forests to examine species, forest health and condition, stand structure, and possibly forest ecosystem function. In Chapter 20, Thomas highlights advances made using hyperspectral data, when compared to multispectral data, in studying various forest characteristics such as (a) forest inventories (e.g., species, height, and diameter at breast height, stand basal area, biomass, volume, age, health, and carbon), (b) forest species mapping using spectral matching algorithms (for pure pixels) and spectral mixture analysis (for mixed pixels), (c) forest biophysical variables (LAI, biomass, or volume) wherein hyperspectral indices are known to explain up to 82% variability in models with the best relationships obtained using hyperspectral images acquired at high signal-to-noise ratios such as from AVIRIS sensor, (d) carbon flux assessments at national/global-scale by calculating NPP from the fraction of photosynthetically active radiation absorbed by the canopy (fPAR) and LUE, and (e) wildfire fuel types, amounts, their moisture content, greenness, dry carbon (nonphotosynthetic vegetation), and their spatial distribution leading to improved fire risk maps. The future is bright.

However, as previously noted, the complexities of forest canopies are many, and they can include (a) horizontal and vertical species mixtures (causing layering in the canopy), (b) complex canopy architecture, (c) variable height and biomass, (d) within- and across-species variability in leaf area and foliar biochemistry, and (e) variability in leaf morphology and foliar biochemistry within a single canopy, particularly between sunlit leaves near the top of the canopy and shaded leaves below (Thomas, Chapter 20). A review of the literature has shown great promise for studying these forest complexities by fusing Lidar data (providing better information on canopy structure) with hyperspectral data (providing better information on canopy biochemistry) through a geometric-optical RT models.

Chapter 7 by Zhang provides an extensive review of forest leaf and canopy level studies to estimate chlorophyll content. Methods for estimating leaf and canopy chlorophyll content include (a) empirical methods and (b) physically based model inversion methods (e.g., modeling method for broadleaf and needle leaf chlorophyll content estimation). Wavebands considered crucial in empirical statistical models of chlorophyll estimation were centered at 434, 502, 672, 699, 701, 705, 720, 722, 728, 749, and 750 nm. We note a number of very close or near-overlapping bands. If these bands are discarded, the most suitable bands for chlorophyll estimation are 434, 502, 672, 699, 720, 728, and 750 nm.

Some examples illustrate the utility of these bands. Further, the chlorophyll SPAD meter is based on measuring leaf transmittance in two wavebands centered at 650 and 940 nm, which correlate well with the leaf chlorophyll content (Stroppiana et al., Chapter 11). PROSPECT, LIBERTY, and LEAFMOD are three physically based models found to be very useful. However, these models require accurate data on characteristics such as leaf thickness, leaf reflectance, leaf transmittance, LAI, and leaf structure. PROSPECT has been the most useful and widely used model to date. However, it tends to overestimate leaf chlorophyll content and is most accurate when plants have high chlorophyll accumulation periods (e.g., summer months in Canadian forests).

Spatiotemporal mapping of forest characteristics over large geographic areas requires consistent image acquisitions from space-borne sensors. Currently, NASA's Hyperion and ESA's CHRIS PROBA are the only two commercial hyperspectral sensors with global repeat coverage. In coming years, these will be joined by three other systems: (1) Italian space agency's (ASIs) PRISMA (planned launch: 2012), (2) Germany's EnMAP (2014), and (3) NASA's HyspIRI (date of launch still uncertain, but probably in year 2015 or soon after).

28.6　NPP, CARBON FLUX, AND LIGHT USE EFFICIENCY MODELS USING HYPERSPECTRAL DATA AND INDICES

One of the increasingly sought after applications of hyperspectral data are in understanding, modeling, and mapping carbon flux (i.e., NPP). First, because of the specific narrowbands available for quantifying specific ecosystem variables (e.g., biophysical and biochemical variables, 3-D geometric properties of vegetation, and soils) and second, because of the ability of remote sensing to spatially extrapolate models and to map data over large areas consistently and repeatedly. Further this extrapolation is based on physical sampling of phenomena rather than on a purely mathematical basis such as Kriging.

Carbon is removed from the atmosphere via photosynthesis by plants. Upon entering the terrestrial ecosystem, it is termed GPP, with the difference between carbon gain via GPP and carbon loss through plant respiration defined as NPP [3]. In Chapter 12, Middleton et al. introduce an approach for modeling canopy or ecosystem photosynthesis based on the photosynthetic LUE concept. An LUE model, rather than describing complex biochemical, canopy structure, and meteorological information, attempts to describe the overall ability of plants/canopies/ecosystems to convert solar energy into useable carbohydrates and biomass. NPP is difficult to measure (in situ) over large areas owing to spatial variability of environmental conditions and limitations in the accuracy of allometric equations [3]. Currently, carbon fluxes are mainly measured by a global network of about 400 flux towers, carbon accounting, global vegetation models, atmospheric measurements, and satellite-based techniques (McCallum et al., 2009, Middleton et al., Chapter 12).

Satellite remote sensing provides the only data that can consistently map carbon fluxes. Over space and time, and over large areas including the entire globe, researchers often use production efficiency models (PEMs). PEMs are based on the theory of LUE, which states that a relatively constant relationship exists between photosynthetic carbon uptake and radiation receipt at the canopy level. In addition to LUE, PEMs typically require input of meteorological data (i.e., radiation, temperature and others) and the satellite-derived fraction of absorbed photosynthetically available radiation (fAPAR) [3].

In Chapter 12, Middleton et al. present a bioindicator approach and LUE model that combines remote sensing and in situ measures and provides an advanced method to estimate carbon flux. Spectral bioindicators have the capacity to supply unique information about ecosystem health and physiology that can be incorporated into process models and/or coupled climate carbon models to better assess the feedbacks between ecosystems and the climate system.

Narrow or continuous high spectral resolution bands, including a narrowband centered at 531 nm along with an appropriate reference band, are essential to obtaining viable instantaneous estimates of ecosystem stress responses, such as LUE, for use in carbon cycle models. The proposed bioindicators (in Chapter 12) may be best determined using a photochemical reflectance index (PRI), which typically uses two narrowbands in the visible spectrum. These bands include 531 nm and another reference band (also see Section 14.2.3). The reference bands can be centered at 488, 551, 570, or at 678 nm (usually referred to as PRI_{488}, PRI_{551}, PRI_{570}, and PRI_{678}) with band width ranging from 3 to 10 nm. PRI is well correlated with LUE with R^2 values as high as 0.72 to determine LUE based on PRI_{678} (i.e., a PRI computed using wavebands 531 and 678 nm). Hence, PRI is used to quantify LUE. Nevertheless, studies have also shown how viewing geometry, soil background and satellite data preprocessing can significantly affect the retrieved measurements, for which generalized approaches for mitigating these effects have not yet been developed (Middleton et al., Chapter 12).

28.7 PRECISION FARMING APPLICATIONS

Crop nitrogen (N) status provides key information for the application of variable rate technology in precision agriculture, which aims to maximize productivity and, at the same time, limit the environmental impact of excessive fertilization (Stroppiana et al., Chapter 11). Chapter 8 by Zhu et al. analyzes the ability of hyperspectral data to estimate nitrogen content in cereal crops (wheat and rice). They found two critical wavebands (and waveband widths): $\lambda_1 = 722$ nm ($\Delta\lambda_1 = 5$ nm) and $\lambda_2 = 812$ nm ($\Delta\lambda_2 = 10$ nm) used in indices like the soil-adjusted vegetation index (SAVI) and ratio vegetation index (RVI) that best model N. They show that increases in N will result in linear decreases in SAVI and RVI. The best models of rice and wheat N for SAVI and RVI explained 75%–87% variability. The results were based on different N rates, water regimes, cultivars, and growing seasons.

Stroppiana et al. (Chapter 11) present an exhaustive review of various studies, as well as their own research on the determination of N content from VIs. The methods generally used to determine N is based on (i) individual or multiple waveband reflectance, (b) simple or normalized difference VIs, (c) chlorophyll-sensitive indices, and (d) red edge indices. The most widely used approaches to estimate N content are based on regressive models relating in situ measurements and VIs. The wavebands that are most sensitive to N content are often found in the visible and red edge regions of the spectrum where the relation between photosynthetic pigments and nitrogen content is strongest.

Stroppiana et al. (Chapter 11) also found another interesting result: the performance of the VI-N relation can be significantly influenced by other variables such as LAI or biomass. The authors observed from field data that indices that are based on NIR wavebands can model N content only when it is correlated to LAI/biomass; if these structural parameters are not correlated to N, the same indices are not able to explain its variability. In summary, VIs suitable for estimating and monitoring N content should rely on the combination of those wavebands where the influence of the optical properties of photosynthetic pigments is more significant and the influence of the structural parameters is minimized.

Detecting and mapping N content in agricultural crops is important to assess crop productivity in terms of biophysical characteristics (e.g., LAI, biomass) and grain yield. It is clearly established that higher N applications to crops will lead to greater accumulation of chlorophyll in plant leaves, resulting in greater absorption in the visible bands and greater reflectance in the NIR. However, Perry and Roberts [4] showed that adding too much N can actually reduce productivity. Li et al. (Chapter 24) showed that sensitivity of hyperspectral reflectance between cropland fields with varying N application was clearly discernable during early vegetative and senescing growth stages when fields with lower N application have higher red band reflectivity. However, with full canopy cover, the statistical differences in reflectivity across crops with varying N application were not significant since red band absorption was the same and NIR reflectance was similar. Overall, Li et al. (Chapter 24) identified three key wavebands for distinguishing varying N applications in cotton crops. These wavebands were centered at (a) 550 nm, (b) 680 nm, and (c) continuous spectra in 550–750 nm range.

In Chapter 25, Yao et al. reported the advances that can be made in using hyperspectral narrowband data for precision farming applications involving (a) soil management zoning, (b) weed sensing and control, (c) crop nitrogen stress detection, (d) crop yield estimation, and (e) pest and disease detection. Different sets of wavebands and/or indices derived from them model soil properties (e.g., organic matter, moisture, salinity) and/or their mineral contents (e.g., Ca, Mg, P, and Zn), and/or their other properties (e.g., texture, color). Typically, spectral regions from 400 to 800 and from 950 to 1500 nm are sensitive to soil nutrient composition. These sensitivities can help us use specific narrowbands to map soil characteristics such as in-field pH map in order to target location-specific application of nutrients, pesticides, and herbicides.

In order to classify crop types and their growth stages, optimal narrowbands (e.g., Thenkabail et al., Chapter 1) must be used. The study and review of the literature by Yao et al. (Chapter 25) also highlights the following hyperspectral narrowbands as key to detecting various crop and farm parameters. These included wavebands centered at (a) 430, 550, 670 (or 680), and 780 (or 801) nm for nitrogen status evaluation, (b) 674 and 755 nm for the LAI and percent green cover, (c) 680 and 900 nm for biomass and yield, (d) 550, 670, 700, and 800 nm for crop chlorophyll content prediction, and (e) 745 nm or a combination of 531 and 570 nm for certain pest and disease infestation. Overall, they found it advantageous to use a generic approach such as stepwise discriminant analysis, artificial neural networks, or other multivariate statistical procedures to select the best wavebands for specific applications.

28.8   HEAVY METAL EFFECTS ON VEGETATION

Slonecker (Chapter 23) highlights the value of hyperspectral data and indices in the studies of heavy metals (or toxic metals) and their impact on vegetation. Indeed, early studies by Horler, Collins, and others (as reported in Chapter 23) all focused on heavy metals. Some metals such as selenium, copper, and zinc are micronutrients that are required by most plant and animal life in very small doses while others, such as mercury, arsenic, and lead, are toxic and with no known benefit to living organisms (Slonecker, Chapter 23). Typically, the presence of heavy metals leads to plant stress and a "blueshift" (shift in red edge, 680–760 nm, position toward shorter wavelengths) is observed. The degree of blue shift is proportional to amount of heavy metals in soils and vegetation.

In addition, the presence of heavy metals leads to greater reflectance (and less absorption), relative to healthy vegetation that was unaffected by stress from heavy metals throughout the 400–2500 nm spectral range, but could be specifically be tracked at (a) 480 nm, (b) 680 nm, and (c) throughout the SWIR bands. When the plant is healthy, there is a shift of the red edge toward longer wavelengths, increased absorption at 680 nm, and lower reflectivity throughout SWIR.

Slonecker (Chapter 23) also shows the capability of hyperspectral data for detecting and mapping arsenic distributions on lawn grasses based on vegetation stress. Key VIs that provided high R^2-values when detecting and modeling heavy metal stress were (a) the normalized pigment chlorophyll index (NCPI) involving wavebands centered at 430 and 680 nm, (b) the PRI centered at 531 and 570 nm, (c) the red edge position (REP) calculated by looking at position of maximum reflectivity in 690–740 nm, and (d) value of the continuum-removed spectra at 1730 nm.

28.9 WETLAND MAPPING USING HYPERSPECTRAL DATA

Chapter 21 by Ramsey and Rangoonwala provides a comprehensive overview of wetland mapping using hyperspectral narrowband data and highlights its relative advantages over broadband data. First, hyperspectral data were unable to spectrally distinguish wetland mangrove vegetation types: black, red, and white mangroves (*Rhizophora mangles*). However, when background litter was considered, the spectral differences were pronounced. Second, other background factors such as background salt water mangrove flooding and seasonal leaf-off conditions played a key role in mangrove forest species separation and classification studies. This highlights the need to consider background reflectance (and not just the TOC or TOC reflectance) in the studies of mangrove vegetation. Third, hyperspectral narrowband data were also considered ideal in early detection and monitoring of mangrove forest stress and dieback. Fourth, carotene (CAR) and chlorophyll a and b (CHL) pigments of wetland vegetation exhibited high absorptions in (a) the blue band with a maximum between 440 and 460 nm and (b) a secondary CHL absorption with peak near 670 nm. Green and red edge bands were located in the tails of these absorption bands. Fifth, spectrally varying strengths of pigment absorptions indicated that small changes in CHL content could be represented as dramatic changes in the green and red edge bands but relatively small changes in the blue and red bands (see also Chapters 6 and 15). Further, Chapter 21 highlights the challenges involved in mapping wetland invasive species like Chinese tallow (*Sapium sebiferum*). Often these invasive species are (a) below mangrove canopies occupying only a fraction of a pixel area, (b) inconsistent in spatial distribution over space and time, and (c) in wide variety of leaf colors, including green, yellow, red, and brown and every gradation and mixture of those colors, even on the same tree. The factors that can overcome these complexities include (a) acquisition of hyperspectral data at sufficiently high spatial resolution, (b) adopting unmixing analysis, (c) timing or seasonality of image acquisition, and (d) fusing hyperspectral data with hyperspatial data, and/or radar data, and/or Lidar data.

28.10 PASTURE CHARACTERISTICS

In many ways, pastures are ideal for characterization using hyperspectral data. Chapter 9 by Numata clearly outlines these factors. First, typically, pasture vegetation is less than 1 m in height and is ideal for field spectral measurements (unlike forest vegetation). Second, this vegetation is usually contiguous over large areas with a single or a few species with distinctive characteristics (see Chapter 9). The physical characteristics important for remote sensing of pasture include (a) height and variation in height, (b) proportion of bare soil, (c) leaf area, (d) leaf orientation or leaf angle distribution, (e) density of reflective or absorptive structures (i.e., canopy cover percentage), (f) proportion of live and dead materials, and (g) spatial arrangement of structures. Controlling factors in studying biophysical and biochemical properties of pasture are (a) structure (e.g., homogeneous or heterogeneous cover type, erectophile or planophile leaf structure, broad leaves or needle leaves), (b) foliar chemical composition (e.g., chlorophyll a and b, protein, lignin, cellulose, water), (c) nonphotosynthetic vegetation (NPV; nongreen material)), and (d) background effects (e.g., litter, wood and bark). Hyperspectral approaches for pasture characterization include (a) VIs, (b) red edge indices, (c) derivative indices, (d) spectral mixture analysis, and (e) statistical modeling. These indices or methods are used to determine pasture biophysical quantities (e.g., biomass, LAI), biochemical properties (e.g., nutrients such as N, P), degradation analysis, and species identification or classification. Pasture biomass, for example, can be estimated with an R^2 of >0.80 using stepwise linear regression of red edge narrowbands relative to best R^2 values of 0.32 using broadbands. Narrowband ratios involving 895/730 nm provided an R^2 value of 0.76 for LAI, a significantly better result than any from broadband indices.

28.11 INVASIVE SPECIES SEPARATION FROM NATIVE PLANTS

The detrimental effects of invasive species on ecosystems and their management cost over 137 billion dollars per year in the United States alone (Pu, Chapter 19). Hyperspectral narrowband data present an opportunity in data and methods for species-level detection and mapping of invasive species as reviewed, studied, and presented for aquatic species, grasses and weeds, scrubs and shrubs, and trees in this chapter. Invasive species may have distinct biophysical and biochemical properties when compared to native species. Further, the phenological and physiological characteristics of the invasive species may vary from native species. These factors are used to help distinguish invasive species from native species.

Pu suggests the following methods for detecting invasive species: (a) derivative analysis, (b) spectral matching, (c) VIs, (d) knowledge of absorption features, (e) hyperspectral transformation, (f) spectral mixture analysis, and (g) hyperspectral image classification. Pu found that lower order derivatives (e.g., the first order derivative) were less sensitive to noise and hence more effective in operational remote sensing, especially contiguous spectra in 650–800 nm range. Quantitative spectral matching techniques involve methods such as spectral correlation similarity, spectral similarity value, and Euclidean distance. This also involved matching the shapes as well as magnitudes of class spectra with ideal spectra of a species.

A number of two-band NDVI-like indices (see Chapter 1) were very useful in detecting, modeling, and mapping invasive species. For example, Pu et al. reported that photochemical and carotenoid reflectance indices (PRI, CRI) explained differences in LUE between invasive and native tree species and thus helped separate the invasive tree species (*Myrica faya*) from a native tree species (*Metrosideros polymorpha*). A number of absorption wavebands are used in these indices and explain specific vegetation characteristics. The characteristic absorption features reviewed and reported were (a) chlorophyll-*a* absorption centered at 450 and 670 nm; (b) water absorption centered at 970, 1200, 1400, and 1940 nm; (c) N absorption centered at 1510, 2060, 2180, 2300, and 2350 nm; (d) lignin absorption centered at 1120, 1420, 1690, and 1940 nm; and (e) cellulose absorption centered at 1200, 1490, 1780, 1820, 2270, 2340, and 2350 nm.

Invasive species classification was performed by eliminating redundant bands and using methods such as PCA and its modified version: maximum noise fraction (MNF). Hyperspectral narrowband data distinguished and provided significantly higher accuracies in separating vegetation species when compared to broadband data as was apparent in many studies reported in several chapters. However, in practice, for hyperspectral image classification, it can be difficult to obtain adequate training samples for supervised methods and for labeling unsupervised spectral clusters because this requires extensive and costly fieldwork (Pu, Chapter 19). Pu points out that the high accuracy of mapping invasive species is possible when hyperspectral data are also hyperspatial (e.g., 0.5–5 m data from airborne hyperspectral sensors such as AVIRIS, HyMap, and CASI) and hypertemporal (frequent visits). The airborne hyperspectral systems have infrequent acquisition and there are only a few operational commercial hyperspectral sensors: Hyperion (30 m resolution) onboard EO-1 satellite, CHRIS (Compact High Resolution Imaging Spectrometer) (18 m resolution) onboard the ESA's PROBA satellite with a planned mission of HyspIRI (60 m resolution).

The great economic and ecological loss caused by invasive species is well known (see Chapter 19 by Pu) and is further enumerated with examples of riparian shrubs and fire-promoting grasses in Chapter 16 by Nagler et al. The strong cellulose/lignin absorption features centered at 2050 nm were most prominent when studying invasive species of the Sonoran Desert compared with relatively milder absorption at this wavelength for native species. In addition to hyperspectral data, two key factors that often help in accurate mapping of invasive species are (a) high spatial resolution and/or (b) seasonality of image acquisition. These factors distinctly improve the accuracy of invasive species mapping. For example, the well-known water absorption bands centered at 975, 1150, 1450, and 1950 nm have prominent absorption features for invasive species when compared with relatively weaker absorption troughs in native species at specific growth periods. At other times in the season, such separation may not be possible. The 1450 and 1950 nm water absorption bands are excellent for discrimination but are adversely impacted by atmospheric water vapor bands centered at 1400 and 1900 nm (see Chapter 3 by Qi et al.) and hence need to be used after careful observation of spectra.

28.12 PLANT STRESS AND RED EDGE BANDS

Plant stress was best studied using red edge bands. As discussed in Chapter 14, a blueshift (shift of the REP toward shorter wavelengths) occurs when there was stress due to many causes such as drought and heavy metals (also see Chapter 23) and a redshift (shift of the red edge position toward longer wavelengths) occurred during chlorophyll accumulation (Chapters 6 and 15). The bands used in plant stress studies were (a) integration of spectra in a derivative index along the red edge from 650 to 750 nm, (b) a red edge vegetation stress index that involves taking average reflectance of two bands centered at 714 nm (red edge) and 752 nm (NIR) and deducting their reflectance using another red edge band centered at 733 nm (Chapter 14), and (c) calculating a red edge index involving one red edge band centered at 720 nm and another band in NIR (845 or 900 nm) (Chapter 1).

28.13 PLANT LITTER MAPPING

Chapter 16 by Nagler et al. focused on detecting, modeling, and mapping vegetation litter and invasive species using hyperspectral remote sensing. The importance of litter mapping is highlighted by its contribution in enriching soil nutrients, stabilizing energy cycles, controlling soil erosion, and reducing soil CO_2 release to the atmosphere especially from tilled agricultural croplands. The spectral reflectance of litter throughout 400–2500 nm was, generally, higher than healthy green vegetation. However, the greatest difficulty in quantifying and mapping plant litter using remote sensing lies in its spectral similarity with many background soils throughout the 400–2500 nm spectral range. Nagler et al. (Chapter 16) recommended a cellulose absorption index (CAI) that involves narrowband centers at 2023, 2100, and 2215 nm, taking advantage of the spectral "absorption trough" centered at 2100 nm as a result of the cellulose/lignin absorption peak that clearly distinguishes different soils from various plant litter. Positive values of CAI represent the presence of cellulose and hence plant litter whereas negative values of CAI indicate the absence of cellulose and plant litter (Nagler et al., Chapter 16).

28.14 VEGETATION WATER CONTENT

Chapter 10 by Colombo et al. deals with determining vegetation water content at leaf, canopy, and landscape level using ground-based, airborne, and space-borne sensors. They point out that vegetation water content can be expressed in many ways such as equivalent water thickness of a leaf (EWT_L; gm/cm^2), canopy EWT ($EWT_C = EWT_L * LAI$ expressed in kg/m^2), and so on. The SWIR absorption bands around 1450, 1940, and 2480 nm are more sensitive to leaf water content variations than the weaker water absorption bands of the NIR region. However, Colombo et al. point out that atmospheric water vapor absorption greatly reduces the energy reaching the ground surface around 1450 and 1950 nm and thus water absorption bands are generally not exploited for landscape level studies (further supported by Galvão et al., Chapter 17). Thereby, water absorption bands typically used are 983 and 1205 nm (Galvão et al., Chapter 17).

Colombo et al. suggested that at the landscape level, spectral indices based on reflectance around 1175 nm and other techniques that exploit contiguous spectral bands within this water absorption peak are better suited for the prediction of water content than those based on 975 nm and on longer wavelengths in the SWIR. The hyperspectral narrowband methods used for vegetation water content determination that are recommended by Colombo et al. are (a) NIR- and SWIR-based spectral indices (with $\Delta\lambda$ or band width of about 10 nm), (b) derivative indices, (c) absorption-band-depth analysis and stepwise linear models, (d) curve fitting techniques, and (e) inversion of RT models.

28.15   SOIL CHARACTERISTICS MODELING AND MAPPING

Understanding soil background reflectance is a key to discerning characteristic spectra of vegetation, soils, and combined effects of these. Chapter 22 by Ben-Dor reviews the advances made by IS and near infrared spectroscopy (NIRS) in the study of soils over the last three decades. Studying soil spectra requires in-depth understanding of soil chemical process (e.g., clay minerals, carbonates, organic matter, water, iron, and soil salinity), soil physical processes (e.g., particle size and sample geometry), imaging geometry (view angle, radiation intensity, incidence angle, and azimuth angle of the source), and the factors affecting soil reflectance (e.g., higher vegetation, lower vegetation, soil cover, moisture). For example, iron content in a series of smectite minerals were best identified at wavelengths centered at 2294, 2259, 2291, and 1266 nm. The spectral response of soils can be significantly impacted by water in soils and how it is held (e.g., hydration water which is incorporated into the lattice of the mineral, such as limonite ($Fe_2O_3 \cdot 3H_2O$) and gypsum ($CaSO_4 \cdot 4H_2O$), hygroscopic water that is adsorbed on the soil surface as a thin layer, and free water that occupies the soil pores). Hydration water has strong OH absorption features at around 1400 and 1900 nm. In comparison, OH features of free water (H_2O) are found at 1456 nm. Others have suggested that the 900–1220 nm region is well suited for mapping organic matter in soils.

Recent studies have shown that the best interpretation of water content emerges when both dry and wet soil samples are spectrally measured. For example, the correlation coefficient was greater than 0.92, when the 1926, 1954, and 2150 nm wavelengths were used in NIRS analysis for soil moisture. In the highly sensitive 1900 nm region, a water OH combination band showed excellent nonlinear fit to the increase in water content (Ben-Dor, Chapter 22). They also highlight the features—directly associated with the OH group in the water molecule at 1400 and 1900 nm, and some that are indirectly associated with the strong OH group in the TIR region (around 2750–3000 nm), that affect the lattice OH in clay (at 2200 nm) and CO_3 in carbonates (at 2330 nm).

Chapter 22 also points out that whereas soil taxonomy and soil mapping require knowledge of the entire soil profile, remote sensing measures only the thin (about 50,000 nm) upper surface layer. This limitation becomes more prominent when the natural surface becomes altered by activities such as agriculture or masked by natural vegetation, desert varnish, and/or snow cover. Other factors like view angle, changes in illumination, and atmospheric conditions affect IS and NIRS spectra. For example, in the VIS region atmospheric attenuation affects all spectral regions as a result of aerosol scattering (monotonous decay from 400 to 800 nm), absorption of ozone (around 600 nm), water vapor (730 and 820 nm), and oxygen (760 nm). The NIR-SWIR regions are influenced by absorption of water vapor (940, 1140, 1380, and 1880 nm), oxygen (around 1300 nm), carbon dioxide (around 1560, 2010, and 2080 nm) and methane (2350 nm) (Ben-Dor, Chapter 22). Apart from these limitations, the other difficulties in applying soil-IS-NIRS applications are in extrapolating and mapping point understanding to the spatial domain, low signal to noise ratios, mixed pixels, and bidirectional reflectance distribution function (BRDF) effects.

28.16   CROSS-SENSOR CALIBRATION

The consistent study of any terrestrial process requires a clear understanding of intersensor relationships. Chapter 26 by Miura uses space-borne hyperspectral data from sensors such as Hyperion and PROBA/CHRIS to investigate cross-sensor calibration with a multitude of sensors such as AVHRR, MODIS, VEGETATION, and VIIRS. One of the great strengths of hyperspectral narrowband data is its ability to be used to simulate spectral data from another sensor and help develop intersensor calibrations to enable, for example, continuous long-term study of vegetation characteristics over space and time irrespective of the sensors used. In Chapter 26, Miura and Yoshioka demonstrate cross correlation of multiple sensors considering various spectral, spatial, algorithmic, and angular characteristics of various sensors that are derived or simulated using hyperspectral narrowband data. First, intersensor relationships were used to study spectral characteristics including (a) empirical methods that provide linear or nonlinear regressions between VIs derived from two sensors (both can be simulated using hyperspectral data) and (b) theoretical studies taking into account ecosystem parameters such as LAI and biomass in developing a spectral transformation algorithm that theoretically guarantees 'exact' translations (Miura and Yoshioka, Chapter 26). Second, intersensor relationships were used to study spatial characteristics taking, for example, hyperspectral data with 30–100 m and aggregating them to match other sensors (e.g., 500 m of MODIS) and then studying their cross correlations. Generally, it was shown that uncertainty in VIs was higher at finer spatial resolutions than at coarser resolutions. Third, a surprising result was that the vegetation index cross correlations between sensors were similar for top of atmosphere (TOA) and TOC, indicating that atmospheric correction had little influence. However, this may be due to the fact that the hyperspectral data used in the study were acquired on clear, cloud-free days. Fourth, the off-nadir viewing effects on data can also be studied for cross sensors using hyperspectral narrowband data, but are not reported in the book due to inadequate data. Overall, these cross-sensor relationships are most reliable and robust when acquisitions from hyperspectral and other sensors cover multidate images over same geographic areas, and multisite acquisitions representing different land use/land cover.

28.17 STUDY OF OUTER PLANETS AND CROSS LINKAGES WITH PLANET EARTH: HYPERSPECTRAL APPROACHES

Chapter 27 by Vaughan et al. provides an overview of hyperspectral remote sensing sensors and their applications to other planetary bodies and compares them with hyperspectral observations of planet Earth. Even though other planets in our solar system almost certainly do not have vegetation like Earth (the focus of this book), there are a number of reasons for scientific interactions between those studying imaging spectroscopy (or hyperspectral remote sensing) applications for Earth and those studying other planets using similar sensors. First, most of the design of hyperspectral sensors that gather data in 400–2500 nm is the same whether they are deployed to study other planets or the Earth. Second, the methods and techniques of hyperspectral data interpretation and analysis are similar irrespective of which planets are being studied. For example, hyperspectral data analysis methods discussed in various chapters of this book such as subpixel analysis, spectral matching techniques, linear and nonlinear unmixing, and various classification techniques can be applied across planetary hyperspectral data sets. Third, preprocessing algorithms such as the radiometric, geometric, and atmospheric corrections are similar and/or can be used with slight modifications for various planetary bodies. Fourth, as we have learned from previous chapters, the background influence from factors such as soils on spectral reflectivity is significant. So, the lessons learned in soil spectral reflectance studies, considering their chemical-physical-moisture properties, on one planet should be applicable to other planets as well. For example, there are certain narrowbands or their combinations that best predict soil organic matter or other soil properties. Fifth, hyperspectral narrowbands provide an opportunity to compute hundreds or thousands of material identification indices, such as indices that provide the best results in modeling and mapping soil characteristics. This understanding can also be used across planets. Sixth, the spectral libraries available for soils, rocks, vegetation, and soils with varying moisture content measured from terrestrial materials are also useful in spectral matching techniques for other planetary surfaces to identify and label targets by matching ideal spectra with target spectra. For example, an absorption feature found around 2800–2900 nm in data gathered by the M3 hyperspectral sensor onboard Chandrayaan-1 led to the detection of hydroxyl/water (about 0.1% of total volume of material) on the moon.

Finally, as scientists continue to discover numerous exo-planets, some of which may be habitable, the knowledge of hyperspectral signatures of various organic and inorganic materials described in this book will help shed light on possible future discoveries of planetary bodies that may exhibit spectral evidence for life or the conditions for life. Vaughan et al. (Chapter 27) provide an excellent overview of the spectral–spatial–radiometric characteristics of hyperspectral sensors that have been used in studies of the surface composition of other planets (e.g., Mercury, Earth's Moon, Mars, Jupiter, Saturn, and the moons of the outer planets) as well as on planet Earth.

28.18   CONCLUDING THOUGHTS

This volume summarizes advances made over forty years of understanding, modeling, and mapping terrestrial vegetation using hyperspectral remote sensing (or imaging spectroscopy). These efforts have used sensors that are ground-based, truck-mounted, airborne, and space-borne. Advances have included (a) significantly improved characterization and modeling of a wide array of biophysical and biochemical properties of vegetation, (b) ability to discriminate plant species and vegetation types with high degree of accuracy, (c) reduced uncertainties in determining net primary productivity (NPP) or carbon assessments from terrestrial vegetation, (d) improved crop productivity and water productivity models, (e) ability to assess plant stress resulting from causes such as management practices, pests and disease, water deficit or water excess, and (f) establishing more sensitive wavebands and indices to detect plant water/moisture content. The advent of space-borne hyperspectral sensors (e.g., NASA's Hyperion, ESA's PROBA, and upcoming Italy's ASI's PRISMA, Germany's DLR's EnMAP, Japanese HIUSI, NASA's HyspIRI) and numerous methods and techniques espoused in various chapters of this book to overcome the Hughes phenomenon or data redundancy when handling large volumes of hyperspectral data have generated tremendous interest in advancing the knowledge base of hyperspectral applications to larger spatial extents such as at regional, national, continental, and global scales. There is also increasing interest in using hyperspectral narrowband data to simulate broadband data (e.g., Landsat, Resourcesat, SPOT). Thereby, hyperspectral sensors not only help advance remote sensing science through imaging spectroscopy, but also facilitate data continuity of broadband sensors such as Landsat, SPOT, and IRS.

Studies reported in various chapters of this book have found that the large number of hyperspectral narrowbands and the hyperspectral vegetation indices (HVIs) derived from these wavebands offer far greater opportunities in finding an appropriate index for studying a given vegetation variable and performed significantly better when compared with broadband data. However, as established in a number of chapters a large number of narrowbands and their indices are redundant in studying vegetation. For example, three narrowbands centered at as 540 nm, 550 nm, and 560 nm are almost perfectly correlated to one another when studying vegetation characteristics. So, using one of these bands (one that is slightly better than the other two) will suffice. Indeed, when we have 100 s or 1000 s of narrowbands many of them often provide similar information leading to redundancy on numerous wavebands. So, selection of nonredundant optimal wavebands to study a wide array of vegetation biophysical and biochemical properties and use these non-redundant bands to classify, model, and map vegetation characteristics was considered important as reported in several chapters of this book.

A significant and unique strength of hyperspectral narrowband data is in the availability of specific targeted wavebands that are most suitable to study particular biophysical and/or biochemical properties. As examples, plant moisture sensitivity is best studied using a narrowband (5 nm wide or less) centered at 970 nm, while plant stress assessments are best made using a red edge band centered at 720 nm (or a first-order derivative index derived by integrating spectra over the 700–740 nm range), and biophysical variables are best retrieved using a red band centered at 687 nm. These bands are often used along with a reference band to produce an effective index such as a two-band normalized difference vegetation index involving a near infrared (NIR) reference band centered at 890 nm and a red band centered at 687 nm. Chapters in the book report hyperspectral narrowband study of wide array of biochemical and biophysical parameters from around the world for a large number of vegetation/crop type and species. These include biochemical parameters such as chlorophyll a, chlorophyll b, total chlorophyll, carotenoids, anthocyanins, nitrogen, water, and plant structure (e.g., lignin, cellulose). Biophysical parameters include biomass, leaf area index, plant height, canopy cover, fraction of photosynthetically absorbed radiation absorbed by photosynthetically active vegetation (fAPAR), net primary productivity (NPP), and grain yield.

There are always alternative bands that one can select for many of the wavebands in Table 28.1. For example, a water absorption band centered at 960, 970, or 983 nm all provide nearly similar results. Researchers often recommend one of these bands over the others depending on conditions such as crop types, crop conditions, and vegetation species. However, final selection is based on factors such as frequency of occurrence of bands (e.g., 970 nm occurs more frequently than 960 or 983 nm) as reported in existing literature and/or importance of a specific targeted band to study a specific feature (e.g., 531 nm for photochemical reflectance index of PRI), and/or a delicate compromise of selecting a band that equally meets the need of studying more than one variable (e.g., 2205 nm for litter, lignin, cellulose, and heavy metals). Selecting 2200 nm to study heavy metal impact on vegetation is often recommended whereas 2213 nm is commonly recommended for plant litter studies. However, a delicate compromise of 2205 nm performs equally well in studying heavy metals as well as plant litter. Similarly, chlorophyll absorption for healthy vegetation peaks at 682 nm for most vegetation. But if you center this band at 670, 675, or 687 nm, the results are almost the same as a band centered at 682 nm. Here we selected 687 nm in Table 28.1 because it is (a) as good as 670, 675, or 687 nm bands in modeling biophysical and biochemical variables and (b) centered at actively induced emission peaks in red/far-red. Such compromise is not recommended in certain cases such as to study PRI and/or light use efficiency (LUE). For PRI and LUE a waveband centered at 531 nm is considered critical and a slight shift even by 1 or 2 nm is considered detrimental for optimal results.

Discrimination of subtle biochemicals such as the starches, proteins, lignin, and cellulose requires finer (5 nm or even less) spectral bandwidths. For example, even a 0.5 nm shift in the position of a wavelength can complicate separating water vapor from liquid water in an image. This is specially so for shortwave infrared (SWIR) bands where sensitivity to biochemical and biophysical properties change swiftly over a very narrow wavelength range. Based on extensive studies conducted by various researchers and reported in this book, it can be easily concluded that specific biophysical and biochemical properties are best studied using specific narrowbands (≤5 nm) at targeted waveband centers (see Table 28.1). Even though, it may be even preferable to have as narrow a band as 3 nm wide for vegetation studies in visible, near infrared (NIR), and SWIR, signal-to-noise ratio of such narrowbands can cause problems and hence a 5 nm wide narrowband is suggested as optimal band width. Within this book, individual chapters have defined narrowband widths slightly differently anywhere from 3 nm to 10 nm (in overwhelming number of cases). However, we recommend a consistent use of 5 nm (or as less as 3 nm if one can address the issue of signal to noise ratios) bandwidths to define narrowbands.

Methods of classifying vegetation classes or crop types or vegetation species using hyperspectral narrowbands are discussed extensively in this book and include multivariate or partial least square regressions (PLS), discriminant analysis, unsupervised classification, supervised approaches, spectral angle mapper (SAM), artificial neural networks, and support vector machines (SVM). Generally, about 20 narrowbands provided optimal classification accuracies which, typically, provided 10%–30% higher accuracies than broadband data from sensors such as Landsat. Studies have shown that through only 1% of training pixels per class, almost 90% overall classification accuracies are obtained using SVM methods. Nevertheless, some of the vegetation classification and discrimination is best achieved by integrating hyperspectral data with hyperspatial data. When adequate spatial resolution does not exist, spectral mixture analysis (SMA) was successfully implemented to unmix the spectra and highlight spectral abundances in individual pixels.

A thorough review revealed a set of hyperspectral narrowbands (Table 28.1) that are most useful in study of various vegetation characteristics. It must be noted, however, that there is not universal agreement on the most suitable bands. This is expected, given the wide differences in types of vegetation studied (e.g., forest vs. croplands), crop types (e.g., cotton, soybeans, corn), methodologies used, sensors from which data are acquired (e.g., space-borne or ground-based), and calibration issues. For example, the optimal bands defined in Chapter 1 are mainly applicable to agricultural crops. Other chapters (e.g., Chapters 6, 10, 14, 15, 17, 19, 23) have recommended wavebands for specific applications such as N, anthocyanins, and LAI. However, a synthesis of wide range of studies from last 40+ years of research lead us to carefully choose the optimal wavebands to study wide range of terrestrial vegetation (Table 28.1). These wavebands have following unique characteristics:

A. *Overcoming the Hughes phenomenon (or the curse of high dimensionality of hyperspectral data)*

   Reduce data volumes significantly by eliminating redundant bands and focusing on the most valuable hyperspectral narrowbands (Table 28.1) to study vegetation. If the number of bands remained high, the number of observations required to train a classifier increases exponentially to maintain classification accuracies. Data volumes are reduced through data mining methods such as feature selection (e.g., principal component analysis, derivative analysis, wavelets), lambda by lambda correlation plots, and vegetation indices. Data mining methods lead to: (a) reduction in data dimensionality, (b) reduction in data redundancy, and (c) extraction of unique information.

B. *Narrowbands targeted to study specific vegetation biophysical and biochemical variable*:

   Each waveband in Table 28.1 is uniquely targeted to study specific vegetation biophysical, and biochemical properties and/or captures specific events such as plant stress. For example, a waveband centered at 550 nm provided excellent sensitivity to plant nitrogen, a waveband centered at 515 nm is best for pigments (carotenoids, anthocyanins), and a waveband centered at 970 or 1245 nm was ideal to study plant moisture fluctuations. Lignin, cellulose, protein, and nitrogen have relatively low reflectance and strong absorption in SWIR bands by water that masks other absorption features.

C. *Improved models of vegetation biophysical and biochemical variables*:

   The combination of wavebands in Table 28.1 or HVIs derived from them provide us with significantly improved models of vegetation variables such as biomass, LAI, NPP, leaf nitrogen, chlorophyll, carotenoids, and anthocyanins. For example, stepwise linear regression with a dependent plant variable (e.g., LAI, Biomass, nitrogen) and a combination of "N" independent variables (e.g., chosen by the model from wavebands in Table 28.1) establish a combination of wavebands that best model a plant variable. A wide array of empirical (e.g., simple linear and non linear regression, multiple linear regression (MLR), partial least square regression (PLSR), and artificial neural networks (ANN)), and physically based models (e.g., PROSPECT or LEAFMOD for leaf scale, SAIL for canopy scale, and geometric optical and radiative transfer model or GORT for canopy scale) are extensively discussed in different chapters.

D. *Improved accuracies in vegetation type or species classification*

   Combination of narrow wavebands in Table 28.1 help provide significantly improved accuracies (10%–30%) in classifying vegetation types or species types compared to broadband data.

E. *Distinct separation of vegetation types or species*

   Separating vegetation specific narrowbands, often, help discriminate two crop types or their variables distinctly when compared with broadbands.

F. *Strength of HVIs*

The selected 33 hyperspectral narrowbands (Table 28.1) still offer 528 unique two band HVIs when compared to just 21 unique indices from a seven-band Landsat TM image. Some vegetation biophysical and biochemical HVIs that can be computed using the wavebands listed in Table 28.1 are defined in tables in various chapters (e.g., Tables 14.1, 17.2, 19.2, 23.2, 23.3, 10.1, and 8.7). These indices are often targeted to study specific biophysical and biochemical quantities such as photochemical reflectance index (PRI), normalized difference water index (NDWI), leaf chlorophyll index (LCI), red-edge vegetation stress index (RVSI), anthocyanin reflectance index (ARI), derivative chlorophyll index (DCI), and numerous other indices with unique names. The wavebands listed in Table 28.1 provide options to compute other unique two-band or multiband indices not defined/named in the aforementioned tables. Thus, there is sufficient scope to expand on research to come up with additional hyperspectral two-band vegetation indices (HTBVIs) and hyperspectral multiband vegetation indices (HMBVIs). This would often mean opportunities to investigate and study a specific crop variable with improved variability explained in models leading to decreased uncertainties in derived models and maps.

The selection of optimal wavebands to study terrestrial vegetation is a consensus view based on broad range of literature reported in this book. This will provide an educated guidance for future generation of scientists. However, uncertainties still remain, especially in studying complex tropical forest canopies that have diverse overstory, understory, tree crowns, shadowing, species composition, phenology, various levels of photosynthetic and nonphotosynthetic vegetation, and background influences.

A number of chapters discuss the usefulness and utility of using whole spectra (e.g., continuous and entire spectra over 400–2500 nm) for analysis using such methods as partial least squares regression (PLSR), wavelet analysis, continuum removal, and spectral angle mapper (SAM). Whole spectra is ideal in applying spectral matching techniques (SMTs). However, advances in applications of whole spectral analysis through methods such as SMTs will require well-characterized and understood spectral libraries of features of interest (e.g., spectral libraries of plant species, crop type, and soil type). Whole spectral analysis also helps facilitate applications that use multiple narrowbands by using methods such as integrating area under spectra, multiple linear/nonlinear regression, and principal component analysis. Another advantage of whole spectral analysis is observing specific features of the spectrum at specific wavelengths. For example, studying the structure of plant canopy (e.g., erectophile vs. planophile) through slope of the spectra in the NIR shoulder (760–900 nm). Another example is a blueshift in the red-edge (700–740 nm) portion of the spectrum indicates stress due to many causes such as drought and heavy metals and a redshift (shift of the red-edge position toward longer wavelengths) indicates chlorophyll accumulation.

Finally, strengths of hyperspectral data in biophysical and biochemical characterization of vegetation are well known. However, better characterization and modeling of the vegetation height/depth, crown sizes, basal area, biomass, and structure will require LIDAR. Further plant water properties are better understood using thermal data. Given these facts, simultaneous acquisition and integration of hyperspectral data along with LIDAR and thermal data are considered the future of remote sensing.

ACKNOWLEDGMENTS

The authors of this chapter are grateful to comments provided by all lead authors of the individual chapters on content specific to their work and to Table 28.1. The authors are grateful to Prof. Anatoly Gitelson and Prof. Dar Roberts, leading authorities in hyperspectral remote sensing of vegetation, for their critical review of this material. Authors also thank Dr. Susan Benjamin of U.S. Geological Survey for valuable editing and useful comments.

REFERENCES

1. Imhoff, M., Bounoua, L., and Zhang, P. (2010). Satellite supported estimates of human rate of NPP carbon use on land: Challenges ahead (pdf). Presented at the Fall Meeting of the American Geophysical Union, abstract #B31H-05.
2. National Academy of Sciences. (2006). Testimony to U.S. House of representatives—Climate change: Evidence and future projections.
3. McCallum, I., Wagner, W., Schmullius, C., Shvidenko, A., Obersteiner, M., Fritz, S., and Nilsson, S. (2009). Satellite-based terrestrial production efficiency modeling. *Carbon Balance and Management*, 4: 8, doi: 10.1186/1750-0680-4-8
4. Perry, E.M. and Roberts, D.A. (2008). Sensitivity of narrowband and broadband indices for assessing nitrogen availability and water stress in annual crop. *Agronomy Journal*, 100(4): 1211–1219.

Index

Y

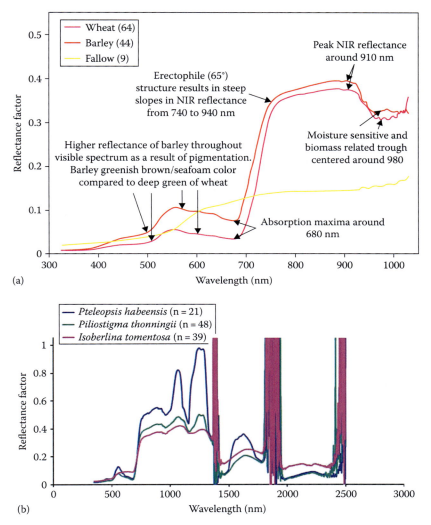

FIGURE 1.1 Hyperspectral characteristics illustrated for few vegetation and agricultural crops. Hyperspectral narrowband data obtained from ASD spectroradiometer illustrated for (a) agricultural crops, (b) shrub species, and

(continued)

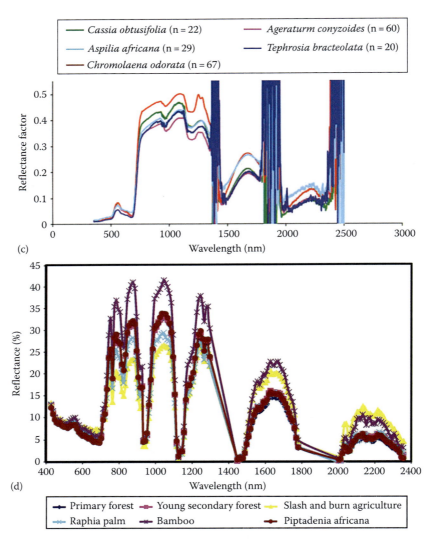

FIGURE 1.1 (continued) (c) weed species. Hyperspectral narrowband data obtained from the Hyperion sensor onboard Earth Observing-1 (EO-1) satellite illustrated for (d) tropical rainforest vegetation. (From Thenkabail, P.S. et al., *Remote Sens. Environ.*, 91, 354, 2004; Thenkabail, P.S. et al., *Remote Sens. Environ.*, 90, 23, 2004; Thenkabail, P.S. et al., *Remote Sens. Environ.*, 71, 158, 2000. With permission.)

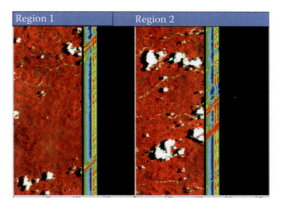

FIGURE 1.2 Hyperspectral data cube (HDC). The 242 band Hyperion data composed as HDC for two areas of African rainforests. Spectral signatures derived for few classes from this figure are illustrated in Figure 1.1d.

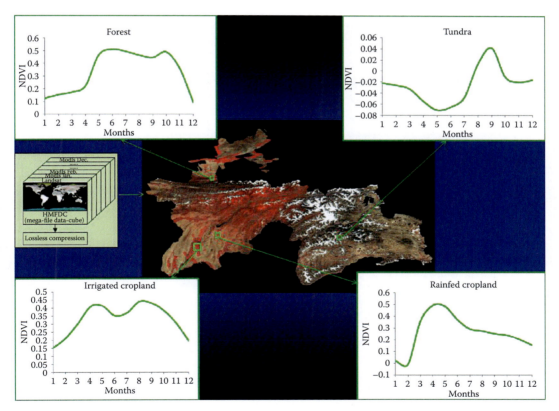

FIGURE 1.3 Mega-file data cube (MFDC). Spectral signatures extracted for few classes from the 12 band MODIS monthly NDVI maximum value composite time-series MFDC. Akin to hyperspectral data (see Figures 1.1 and 1.2), this multispectral time-series is composed as hyperspectral data cube. All hyperspectral data analysis techniques can be applied here. Note: Illustrated for Tajikistan.

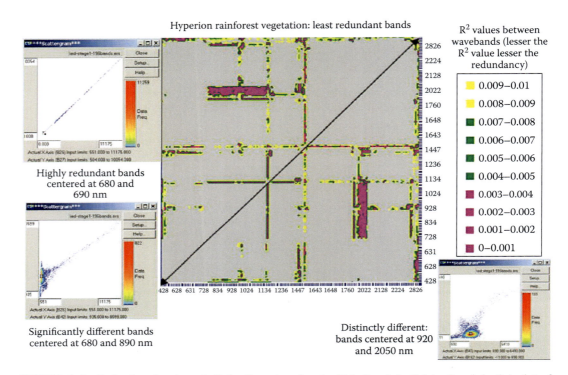

FIGURE 1.4 Redundant bands and distinctly unique bands. This Lambda (λ_1) by Lambda (λ_2) plot of Hyperion bands show redundant bands (higher the correlation higher the redundancy) and distinctly unique bands (lower the correlation between the bands greater the uniqueness). For example, an R-square value of 1 indicates that the two wavebands are perfectly correlated and provide the same information. Thus, only one of the two bands should be used. An R-square value of 0 indicates that the two wavebands provide unique information, and thus it is relevant to use both wavebands.

Contour plot of coefficient of determination (R^2) between vegetation indices at various wavebands versus WBM of: (a) cotton crop (bottom of 45° line) and (b) soybeans crop (top of 45° line)

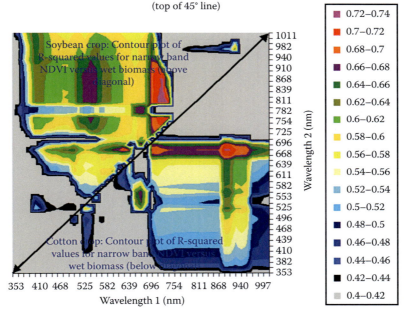

FIGURE 1.5 The hyperspectral two-band vegetation index (HTBVI) versus the crop biophysical variable depicting areas of rich information content. The HTBVIs are correlated with crop wet biomass (WBM) and contour plots of R^2 values depicted for the soybean crop (above diagonal) and the cotton crop (below diagonal). The "bulls-eye" features help us determine waveband center and waveband width with highest R^2 values. These are the best bands to model the biophysical and biochemical quantities of crops. (From Thenkabail, P.S. et al., *Remote Sens. Environ.*, 71, 158, 2000. With permission.)

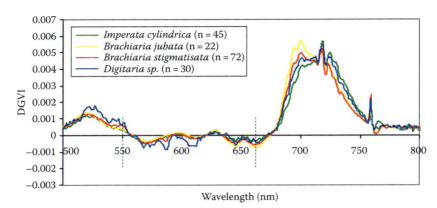

FIGURE 1.6 First-order hyperspectral derivative greenness vegetation index (HDGVI1) computed along the 500–800 nm range for certain weed species.

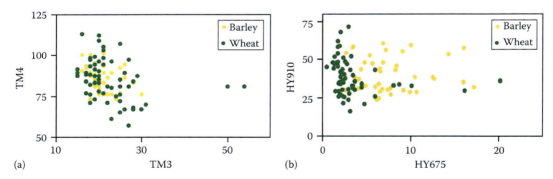

FIGURE 1.7 Separating two agricultural crops using broadbands versus narrowbands. The two broadbands fail to separate wheat from barley (Figure 1.7a) whereas two distinct narrowbands separate wheat from barley (Figure 1.7b).

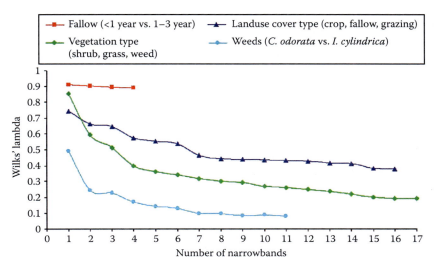

FIGURE 1.8 Class separability using hyperspectral narrowbands determined based on Wilks' lambda. Lower the Wilks' lambda, the greater is the separability. So, with the addition of wavebands, the separability increases, reaching an optimal point beyond which the addition of wavebands does not make a difference. (From Thenkabail, P.S. et al., *Remote Sens. Environ.*, 91, 354, 2004. With permission.)

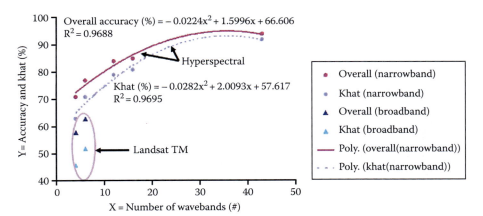

FIGURE 1.9 Classification accuracies using hyperspectral narrowbands versus Landsat broadbands. The accuracies increased by 25%–30% when about 25 hyperspectral narrowbands were used to classify five agricultural crops when compared to six broad-Landsat bands. Classification accuracies reach about 95% with 30 bands, beyond which accuracies do not increase. (From Thenkabail, P.S. et al., *Remote Sens. Environ.*, 91, 354, 2004. With permission.)

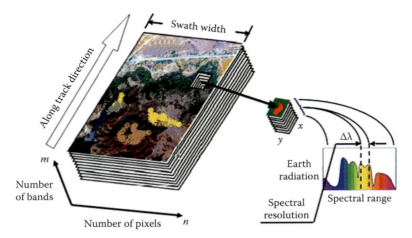

FIGURE 2.2 Imaging spectroradiometer concept—data cube composed of individual images recorded over "m" spectral bands.

FIGURE 2.6 Landscape fantastic HS control.

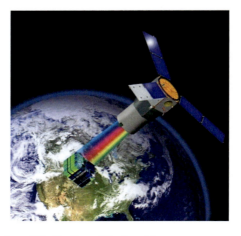

FIGURE 2.9 Artist's impression of TacSat-3 satellite in orbit.

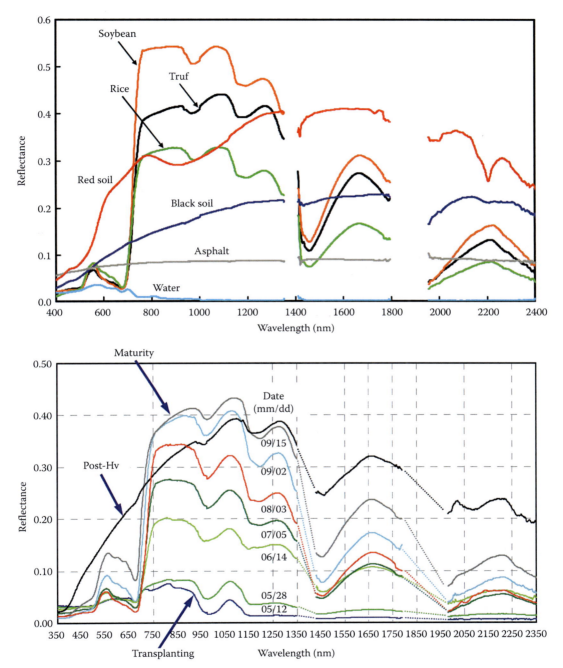

FIGURE 3.1 Typical reflectance spectra in agro-ecosystem surfaces (upper) and seasonal change of spectra in a paddy rice field (lower).

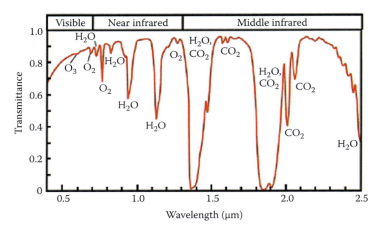

FIGURE 3.2 Plot of atmospheric transmittance versus wavelength for typical atmospheric conditions.

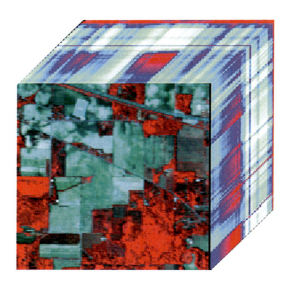

FIGURE 4.1 A hyperspectral image cube of AVIRIS scene from June 12, 1992 (https://engineering. purdue.edu/~biehl/MultiSpec/hyperspectral.html) displayed with bands 40, 25, and 15 as R, G, and B for the top layer.

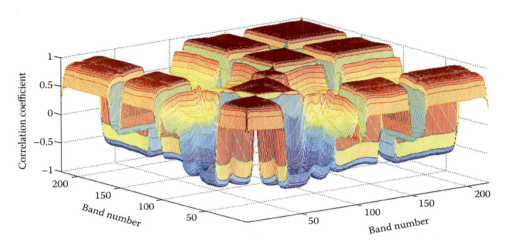

FIGURE 4.2 Correlation between the bands of hyperspectral image shown in Figure 4.1. Only the alternate bands are used to compute the correlation.

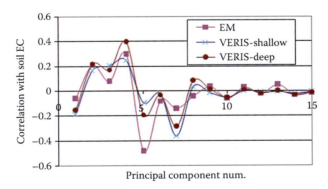

FIGURE 4.6 Correlation between principal components of a hyperspectral image of an agricultural field and apparent soil electrical conductivity at two different depths. (From Bajwa, S.G. et al., *Trans. ASABE*, 47, 895, 2004. With permission.)

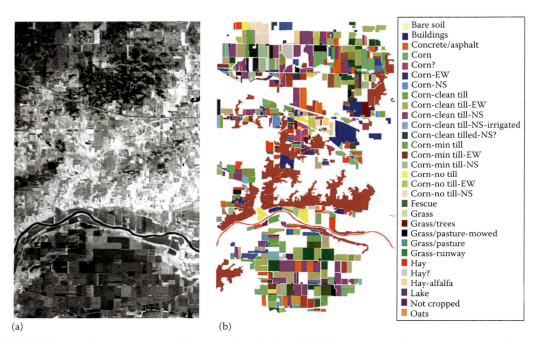

FIGURE 5.4 (a) Spectral band at 587 nm wavelength of an AVIRIS scene comprising agricultural and forest features at Indian Pines region. (b) Ground-truth map with 30 mutually exclusive land-cover classes.

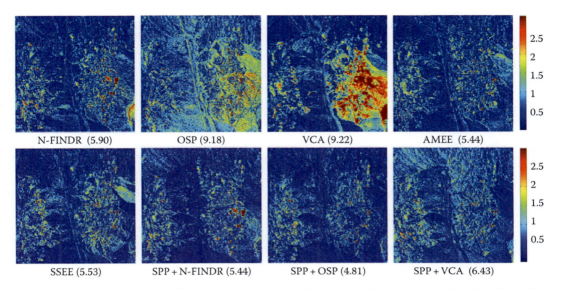

FIGURE 5.6 RMSE reconstruction errors (in percentage) for various endmember extraction algorithms after reconstructing the AVIRIS Cuprite scene.

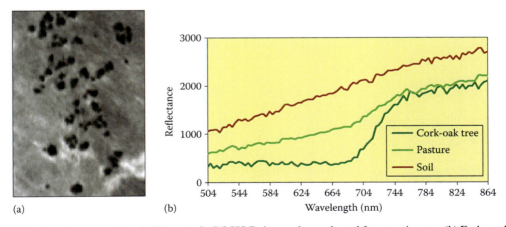

(a)   (b)

FIGURE 5.8 (a) Spectral band (584 nm) of a ROSIS Dehesa subset selected for experiments. (b) Endmember signatures of soil, pasture, and cork-oak tree extracted by the AMEE algorithm, where scaled reflectance values are multiplied by a constant factor.

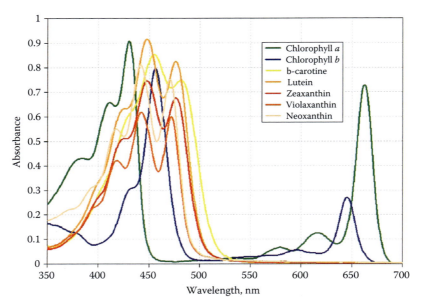

FIGURE 6.1   Absorbance spectra of pigments extracted in 90% acetone. (Merzlyak, M.N., unpublished.)

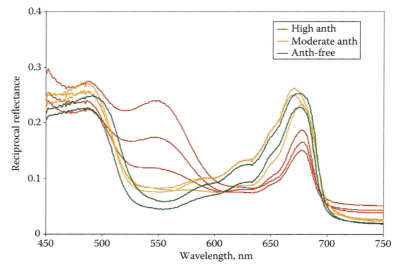

FIGURE 6.8   Reciprocal reflectance spectra of Virginia creeper leaves.

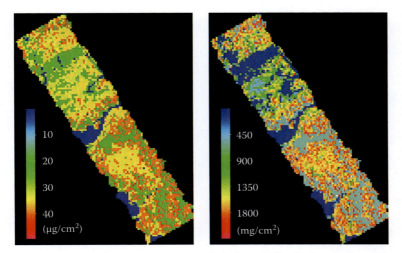

FIGURE 7.5 CASI leaf chlorophyll$_{a+b}$ content (per unit leaf area) distribution (left) and forest chlorophyll content (per unit ground surface area) distribution (right). The images were produced based on the retrieved chlorophyll$_{a+b}$ content for the three vegetated cover types. The spatial resolution of the image is 20×20 m.

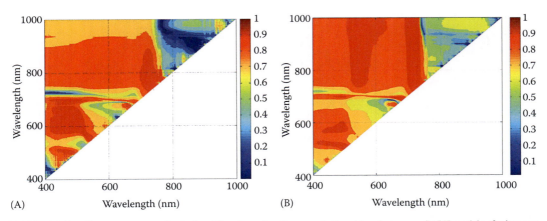

FIGURE 8.2 Contour maps of relative R^2-values for linear relationships between SAVI and leaf nitrogen concentration for rice (A) and wheat (B).

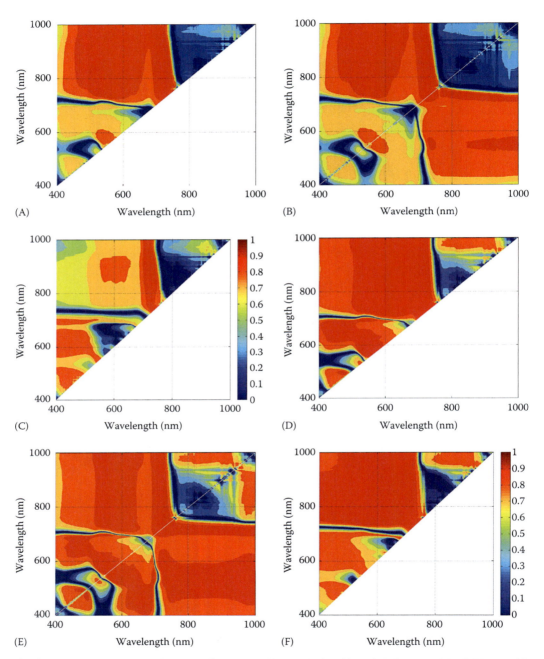

FIGURE 8.3 Contour maps of relative R^2-values for linear relationships of NDVI (A and D), RVI (B and E), and DVI (C and F) against canopy leaf nitrogen concentration during mid-late growing periods of rice (A–C) and wheat (D–F).

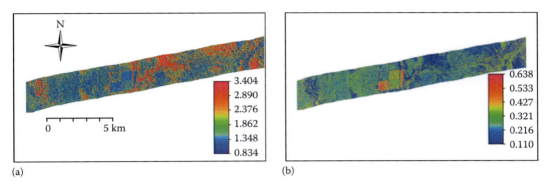

(a)   (b)

FIGURE 9.6 Maps showing spatial distribution of concentration (%) of (a) nitrogen and (b) phosphorus and scatterplots obtained from the best-trained neural network used for mapping. (From Mutanga, O. and Skidmore, A.K., *Remote Sens. Environ.*, 90, 104, 2004; Mutanga, O. and Kumar, L., *Int. J. Remote Sensing.*, 28, 21, 2007.)

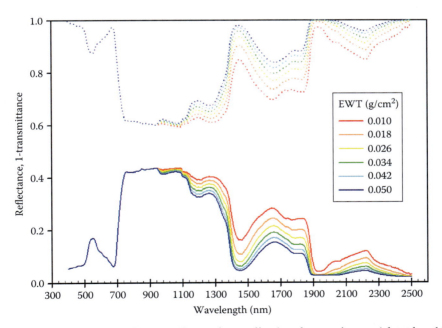

FIGURE 10.2 Example of leaf reflectance (R, continuous lines) and transmittance (plotted as 1-T, dotted lines) spectra, simulated with the PROSPECT model for different levels of EWT.

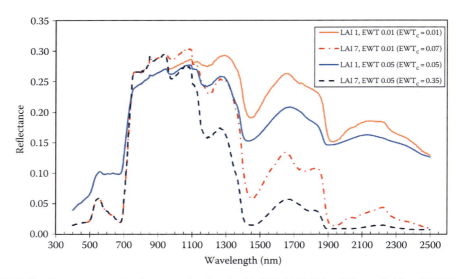

FIGURE 10.3 Canopy spectral reflectance simulated with PROSAILH for different LAI and EWT values.

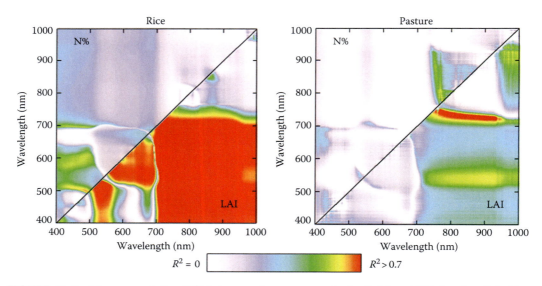

FIGURE 11.2 Linear correlation (R^2) between nitrogen concentration/LAI and ND for rice (left; From Stroppiana, D. et al., *Field Crops Res.*, 111, 119, 2009) and SR for pasture (right; From Fava, F. et al., *Int. J. Appl. Earth Obs. Geoinf.*, 11, 233, 2009. With permission.) for field canopy spectra acquired with a FieldSpec FR PRO spectroradiometer.

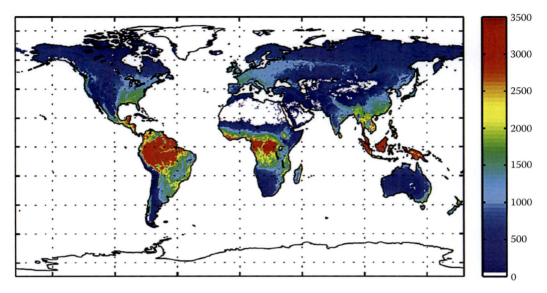

FIGURE 12.1 Spatial variations of the global median annual GPP (gC/m²/a) from various spatially explicit approaches. (From Beer, C. et al., Terrestrial gross carbon dioxide uptake: Global distribution and covariation with climate, *Science*, 329 (5993), 834–838, 2010. Reprinted with permission of AAAS.)

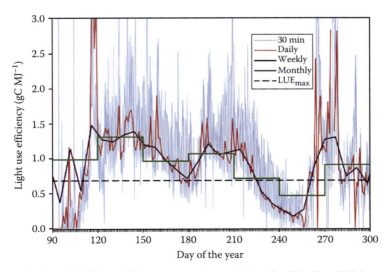

FIGURE 12.3 Variations in LUE at different temporal resolutions for Shindler, Oklahoma, C4 grassland from green-up in April through senescence in October in 1998. LUE calculated using GEP and incident PAR from flux tower, with green *f* PAR estimated from broadband NDVI using PAR and shortwave radiation sensors. (Flux data from Verma, S., University of Nebraska, Lincoln, NE.)

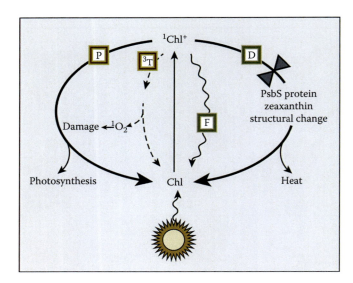

FIGURE 12.4 A schematic showing the interrelationship of photosynthesis and two photoprotective mechanisms—the xanthophylls cycle and chlorophyll fluorescence. Sunlight is absorbed by chlorophyll (Chl) in the light harvesting complexes, producing an excited singlet chlorophyll (^{1}Chl*). Energy is used either for: (i) photochemistry (P) via electron transport to yield photosynthesis (left circuit); (ii) safe dissipation of excess excitation energy as heat (D) via the xanthophyll cycle (right circuit); or (iii) fluorescence emission (F), wavy line. (T3) is the triplet pathway leading to the formation of singlet oxygen (^{1}O$_2$*) and photooxidative damage. (Reprinted by permission from Macmillan Publishers Ltd. [*Nature*] (Demmig-Adams, B. and Adams, W.W., Photosynthesis: Harvesting sunlight safely, 403, 371), Copyright (2000).)

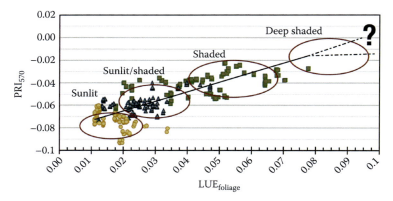

FIGURE 12.10 This figure describes the relationship between the Photochemical Reflectance Index (PRI) and photosynthetic light use efficiency (LUE$_{foliage}$, µmol C µmol^{-1} APAR) for foliage exposed to a range of illumination conditions in a Douglas-fir forest in Canada. The lowest PRI and LUE$_{foliage}$ values are associated with sunlit foliage throughout the 2006 growing season. The highest PRI and LUE$_{foliage}$ values measured were associated with shaded foliage, but high values are also expected for foliage residing in the deeply shaded canopy sectors that could not be measured. (From Middleton, E.M. et al., *Can. J. Remote Sens.*, 35(2), 166, 2009.)

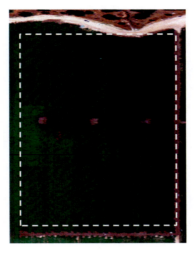

FIGURE 13.1   RGB (670, 550, and 420 nm) image of the experimental plot.

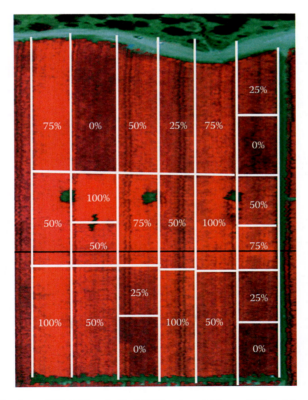

FIGURE 13.2   Combination of IR (750 nm), Red (670 nm), and Green (550 nm) bands of the experimental plot overlaid by the borders of the N treatments.

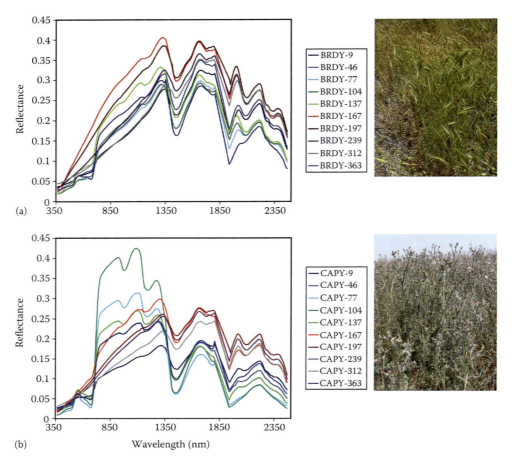

FIGURE 14.5 Reflectance spectra of *Brachypodium distachyon* (BRDI) and *Carduus pycnocephalus* (CAPY) for 2009. The number to the right on the legend reports Julian day.

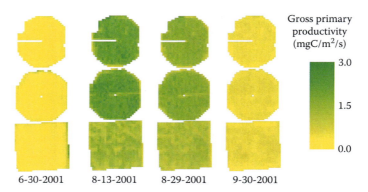

FIGURE 15.20 Midday gross primary production in maize (first and second rows) and soybean (bottom row) retrieved from atmospherically corrected Landsat-7 ETM+ imagery taken over Nebraska in 2001.

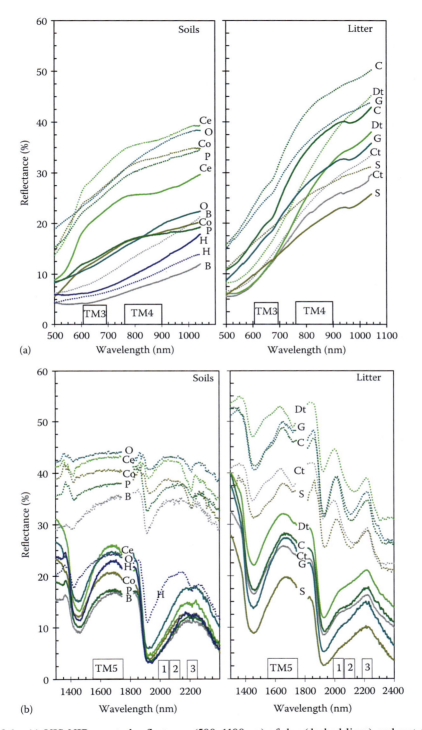

FIGURE 16.1 (a) VIS-NIR spectral reflectance (500–1100 nm) of dry (dashed lines) and wet (solid lines) soils and litter types. (b) SWIR spectral reflectance (1300–2400 nm) of dry (dashed lines) and wet (solid lines) soils and litters. The soils are Othello (O, o), Cecil (E, e), Codorus (C, c), Portneuf (P, p), Barnes (B, b), and Houston Black Clay (H, h). The plant litters are corn (M, m), soybean (S, s), deciduous tree (D, d), coniferous tree (C, c), and grass (G, g). (From *Remote Sensing of Environment*, 71, Nagler, P.L., Daughtry, C.S.T., Goward, S.N., Plant litter and soil reflectance, 207–215, Copyright (2000), with permission from Elsevier.)

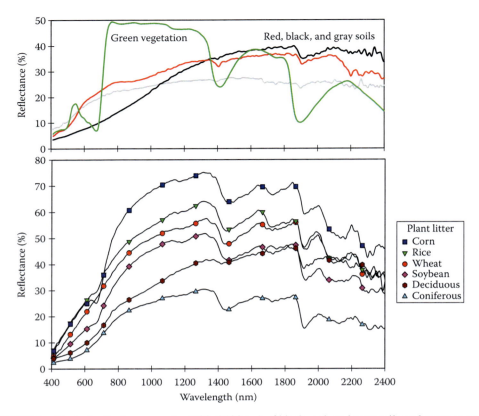

FIGURE 16.3 Typical reflectance spectra (400–2400 nm) of black, red, and gray soils and green vegetation (top) and spectra of four crop residues and two tree litters (bottom). The symbols in the bottom figure do not represent data sampling points, but rather are placed on the spectral lines for clarity. (From *Remote Sensing of Environment*, 87, Nagler, P.L., Inoue, Y., Glenn, E.P., Russ, A., Daughtry, C.S.T., Cellulose absorption index (CAI) to quantify mixed soil-plant litter scenes, 310–325, Copyright (2003), with permission from Elsevier.)

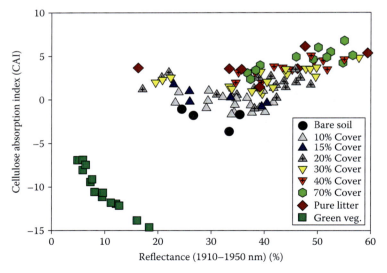

FIGURE 16.5 CAI of all the pure soils, mixed scenes of plant litter with different levels of cover, and green vegetation as a function of percent reflectance (%) in the water absorption band (1910–1950 nm). (From *Remote Sensing of Environment*, 87, Nagler, P.L., Inoue, Y., Glenn, E.P., Russ, A., Daughtry, C.S.T., Cellulose absorption index (CAI) to quantify mixed soil-plant litter scenes, 310–325, Copyright (2003), with permission from Elsevier.)

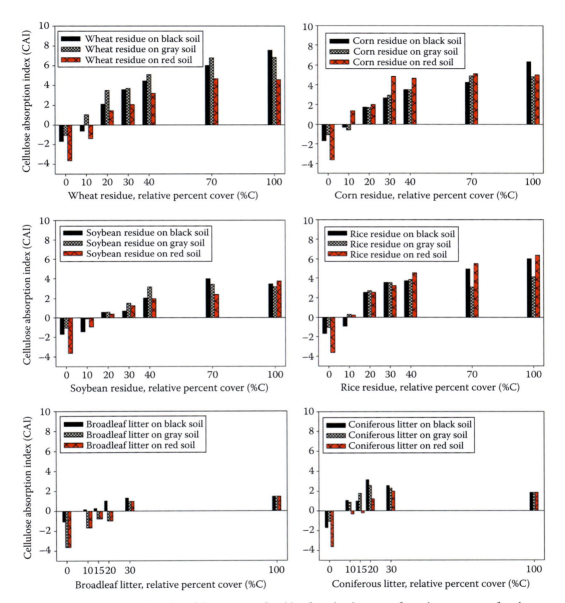

FIGURE 16.6 CAI as a function of the amount of residue for mixed scenes of varying amounts of each crop residue and tree litter, shown for each of the three soils. (From *Remote Sensing of Environment*, 87, Nagler, P.L., Inoue, Y., Glenn, E.P., Russ, A., Daughtry, C.S.T., Cellulose absorption index (CAI) to quantify mixed soil-plant litter scenes, 310–325, Copyright (2003), with permission from Elsevier.)

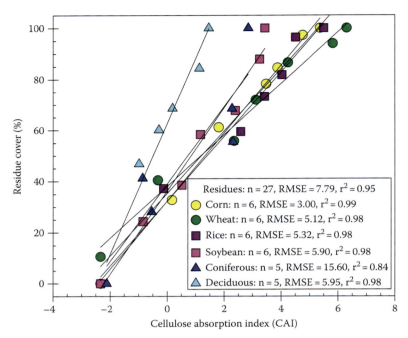

FIGURE 16.7 Crop and forest litter levels estimated by their weight (Rel.% cover) averaged over three soils shown as a function of CAI. (From *Remote Sensing of Environment*, 87, Nagler, P.L., Inoue, Y., Glenn, E.P., Russ, A., Daughtry, C.S.T., Cellulose absorption index (CAI) to quantify mixed soil-plant litter scenes, 310–325, Copyright (2003), with permission from Elsevier.)

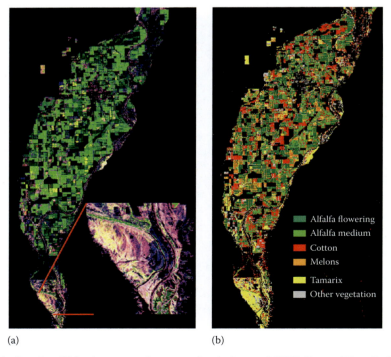

FIGURE 16.15 Landsat TM color composite spectral ratio image (NDVI, $R_{1,5}$ and $R_{1,7}$ displayed as BGR, respectively) of Lower Colorado River Region (a). The Cibola National Wildlife Refuge (CWR) was shown in the insert image. Results of unsupervised classification where NDVI, $R_{1,5}$ and $R_{1,7}$ were used as ratio inputs (b).

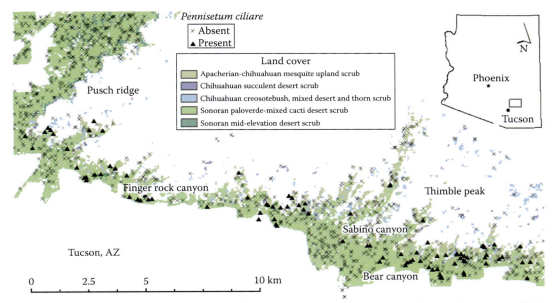

FIGURE 16.16 Buffelgrass study area.

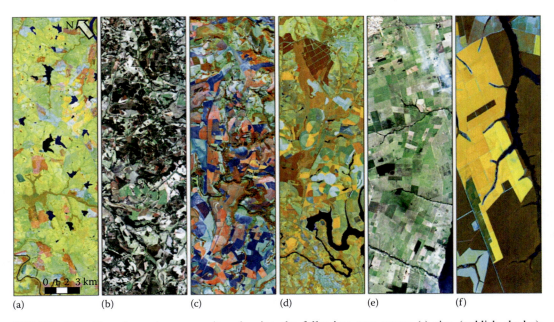

FIGURE 17.1 Hyperion color composites showing the following crop types: (a) rice (reddish shades); (b) coffee (dark green color); (c) sugarcane (reddish shades); (d) corn and bean (pivots with dark and bright reddish shades, respectively); (e) pasture; and (f) soybean. In (a), (c), (d), and (f), bands centered at 864 nm (red), 1649 nm (green), and 671 nm (blue) were used in the false color composites. In (b) and (e), the true color composites refer to bands at 671 nm (red), 569 nm (green), and 487 nm (blue).

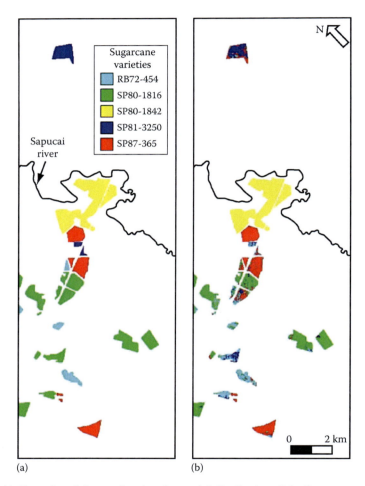

FIGURE 17.17 (a) Ground truth image showing the spatial distribution of the five sugarcane varieties under study in the central portion of the study area. (b) Classification image derived from MDA. (Adapted from Galvão, L.S. et al., *Remote Sens. Environ.*, 94, 523, 2005.)

FIGURE 18.1 Image of old-growth forest at the La Selva Biological Station, Costa Rica.

FIGURE 18.2 (a) Liana (darker green) on a tree crown, (b) flowering *D. panamensis,* and (c) deciduous tree with epiphytes.

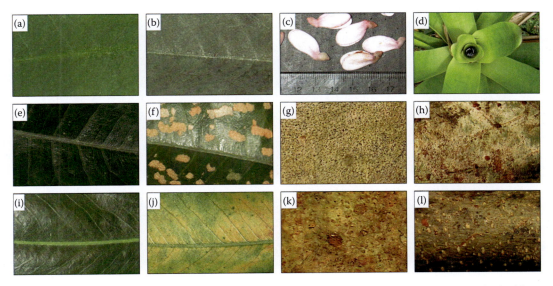

FIGURE 18.3 (a) *H. mesoamericanum* leaf without surface lichen, (b) *H. mesoamericanum* leaf without surface lichen, (c) *D. panamensis* flowers, (d) canopy epiphytic bromeliad, (e) *L. ampla* without herbivory, (f) *L. ampla* with herbivory, (g) *B. elegans* trunk bark, (h) *B. elegans* branch bark, (i) *D. panamensis* mature leaf, (j) *D. panamensis* senescing leaf, (k) *D. panamensis* trunk bark, and (l) *D. panamensis* branch bark.

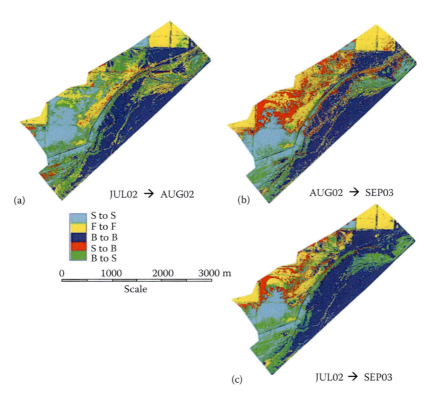

FIGURE 19.3 Change detection resultant maps produced by ANN with principal components extracted from the CASI data (a) from JUL02 to AUG02, (b) from AUG02 to SEP03, and (c) from JUL02 to SEP03. In legend, "S to S," "F to F," and "B to B" represent no-changes of saltcedar, farmland, and bare/wildland, respectively; "S to B" means saltcedar changed to bare/wildland while "B to S" means bare/wildland changed to saltcedar. (Reprinted with kind permission from Springer Science + Business Media: *Environmental Monitoring and Assessment*, Invasive species change detection using artificial neural networks and CASI hyperspectral imagery, 140, 2008, 15–32, Pu, R., Gong, P., Tian, Y., Miao X., Carruthers, R., and Anderson, G.L.)

FIGURE 20.1 Ecosystem, economic, and cultural services of forests.

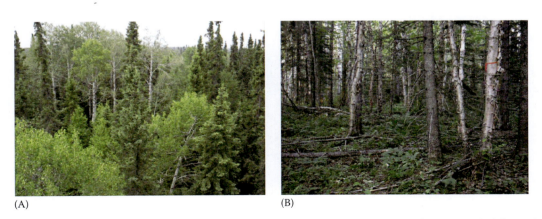

(A)   (B)

FIGURE 20.2 Complex forest canopy architecture. (A) Evident of variable species, canopy height, and layering in a boreal mixedwood forest. (B) Below-canopy variability of a boreal mixedwood forest.

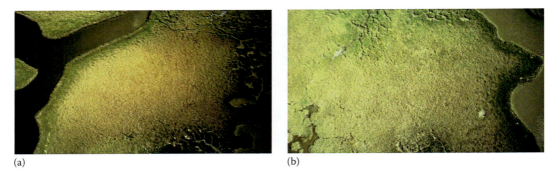

(a)   (b)

FIGURE 21.12 Two marsh sites that could be visually classified as (a) brown-green and (b) green-brown dieback marshes. Although the descriptive visual aerial interpretation of dieback progression is helpful, to provide comparability and repeatability needed for regional mapping and monitoring, additional quantitative measures of the marsh condition are needed. Hyperspectral canopy reflectance provided more quantitative assessments of dieback progression and determination of dieback onset in the highly temporally and spatially complex coastal marshes. The shown marsh sites were located in coastal Louisiana (Figure 21.2). (Adapted from Ramsey, E., III and Rangoonwala, A., Mapping the onset and progression of marsh dieback, in *Remote Sensing of Coastal Environments*, Wang, Y. (Ed.), CRC Press, Remote Sensing Applications Series, Boca Raton, FL, 2009.)

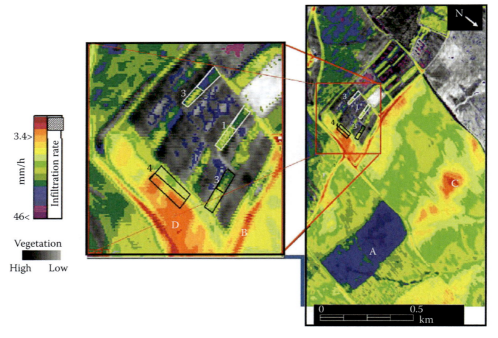

FIGURE 22.16 The infiltration image of a loess soil as generated on the basis of soil reflectance information and rain simulator measurements. (After Ben-Dor, E. et al., *Soil Science Society of American Journal*, 72, 1, 2008.)

FIGURE 22.17 The extended isothermal maps of the burned soil based on spectral measurements and prediction models to reconstruct the surface temperature. (After Lugassi, R., *Remote Sensing of Environment*, 114, 322, 2010.)

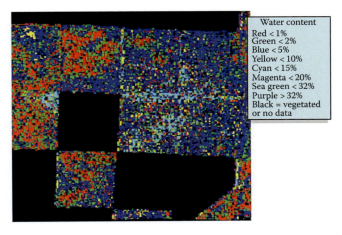

FIGURE 22.22 Surface water content (gravimetric) from AVIRIS data (May 3, 2003, near Lemoore, California) as estimated with the SMGM. (After Whiting, M.L. et al., *Remote Sensing of Environment*, 89, 535, 2004.)

(a)

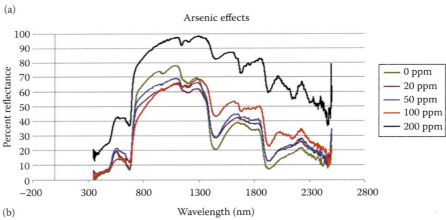

(b)

FIGURE 23.2 (a) Visual effects of arsenic stress on *Nephrolepis exaltata* (Boston fern). Ferns are planted in clean sand amended with, from left to right, 0, 20, 50, 100, and 200 ppm sodium arsenate. (From Slonecker, E., *Remote Sensing Investigations of Fugitive Soil Arsenic and Its Effects on Vegetation Reflectance*, George Mason University, Fairfax, VA, 2007.) (b) Laboratory reflectance spectra of arsenic-affected ferns in (a). Spectra were collected with an ASD full range spectrometer from 6 in. above the canopy of each plant. Note the loss of photosynthetic absorption at 680 nm, causing higher reflectance, the blueshift, and the general increase in reflectance in SWIR (due to loss of water) with increasing soil arsenic. (From Barcelo, J. and Poschenrieder, C., *J. Plant Nutr.*, 13, 1, 1990.)

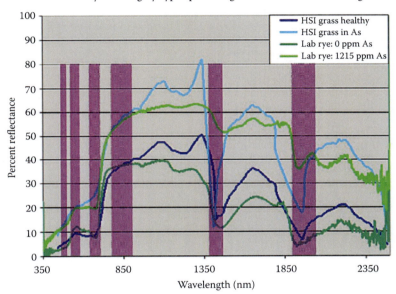

FIGURE 23.5 Healthy and stressed grass signatures from both the laboratory and hyperspectral imagery. The same critical areas in the green, red, near-infrared, and SWIR show the patterns of spectral separation between the healthy and stressed grass that enable the image processing algorithm to separate, identify, and map arsenic stressed grasses. (From Slonecker, E., *Remote Sensing Investigations of Fugitive Soil Arsenic and Its Effects on Vegetation Reflectance*, George Mason University, Fairfax, VA, 2007.)

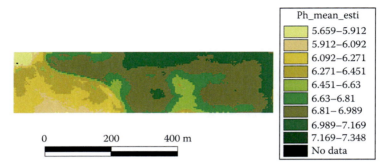

FIGURE 25.2 In-field variability of soil pH as indicated by the sequential Gaussian cosimulation method using aerial hyperspectral data. (From Yao, H., Hyperspectral imagery for precision agriculture, PhD dissertation, University of Illinois at Urbana-Champaign, 2004.)

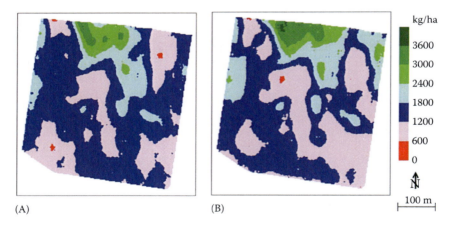

FIGURE 25.5 In-field yield variability maps generated from (A) an airborne hyperspectral image using a nine-band regression model and (B) yield monitor data for a 16 ha cotton field [70]. The nine bands are 499, 546, 601, 702, 717, 738, 771, 778, and 826 nm. (From Yao, H., Hyperspectral imagery for precision agriculture, PhD dissertation, University of Illinois at Urbana-Champaign, 2004.)

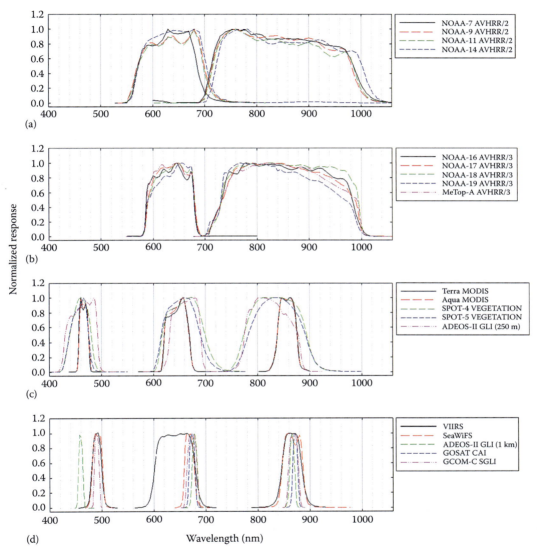

FIGURE 26.1 Normalized spectral response curves of red, NIR, and blue bands for select moderate resolution sensors. They are plotted in four groups for clarity.

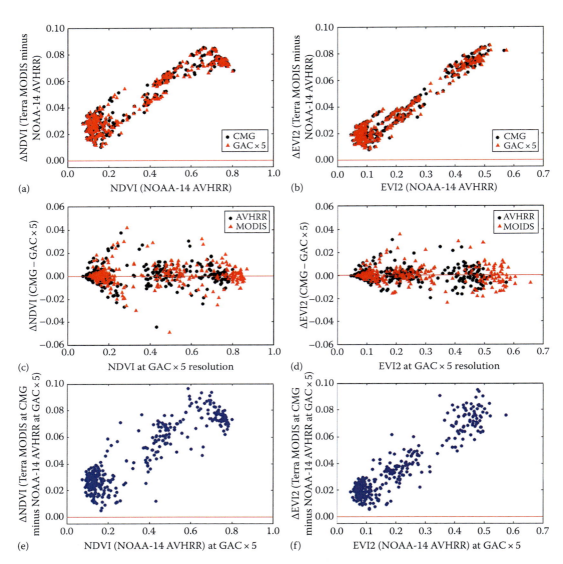

FIGURE 26.7 (a) NDVI and (b) EVI2 differences due to spectral bandpass differences for CMG and GAC resolutions; (c) NDVI and (d) EVI2 differences due to resolution differences for AVHRR and MODIS spectral bandpasses; (e) NDVI and (f) EVI2 differences due to both spectral bandpass (MODIS vs. AVHRR) and spatial resolution (CMG vs. GAC) differences. Here, a GAC pixel is an average of five GAC pixels and, thus, written as "GAC × 5."

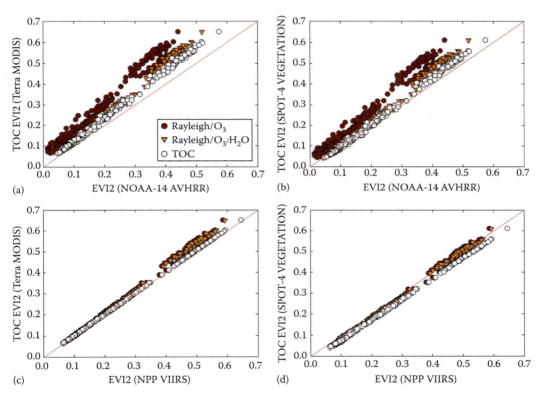

FIGURE 26.10 Same as Figure 26.9, but for EVI2 with three different atmospheric correction schemes.

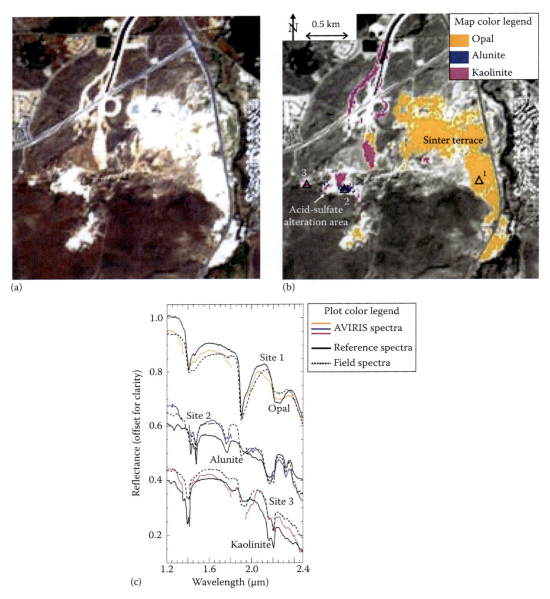

FIGURE 27.1 (a) AVIRIS visible image over Steamboat Springs, (b) AVIRIS mineral map over same area, and (c) AVIRIS and field spectra of minerals mapped.

An environmentally friendly book printed and bound in England by www.printondemand-worldwide.com

PEFC Certified

This product is
from sustainably
managed forests
and controlled
sources

www.pefc.org

PEFC/16-33-415

This book is made entirely of sustainable materials; FSC paper for the cover and PEFC paper for the text pages.

#0530 - 041114 - C40 - 254/178/41 [43] - CB